Topley and Wilson's

Principles of bacteriology, virology and immunity

Seventh edition in four volumes

Volume 4

Edward Arnold

General Editors

Sir Graham Wilson
MD, LLD, FRCP, FRCPath, DPH, FRS

Formerly Professor of Bacteriology as Applied to Hygiene, University of London, and Director of Public Health Laboratory Service, England and Wales.

Sir Ashley Miles CBE
MD, FRCP, FRCPath, FRS

Deputy Director, Department of Medical Microbiology, London Hospital Medical College, London.
Emeritus Professor of Experimental Pathology, University of London, and formerly Director of the Lister Institute of Preventive Medicine, London.

M. T. Parker
MD, FRCPath, Dip Bact

Formerly Director, Cross-Infection Reference Laboratory, Central Public Health Laboratory, Colindale, London.

Principles of bacteriology, virology and immunity

W. W. C. Topley, 1886–1944

Volume 4

Virology

Edited by

F. Brown
and
Sir Graham Wilson

First published 1929
by Edward Arnold (Publishers) Ltd
41 Bedford Square, London WC1B 3DQ.

Reprinted 1931, 1932, 1934
Second edition 1936
Reprinted 1937, 1938, 1941 (twice), 1943, 1944
Third edition 1946
Reprinted 1948, 1949
Fourth edition 1955
Reprinted 1957, 1961
Fifth edition 1964
Reprinted 1966
Sixth edition 1975
Seventh edition in four volumes 1983 and 1984

Volume 1 ISBN 0 7131 4424 6
Volume 2 ISBN 0 7131 4425 4
Volume 3 ISBN 0 7131 4426 2
Volume 4 ISBN 0 7131 4427 0

British Library Cataloguing in Publication Data

Topley, William Whiteman Carlton
Topley and Wilson's principles of bacteriology, virology and immunity. – 7th ed.
Vol. 4. Virology
1. Medical microbiology
I. Title II. Wilson, *Sir* Graham.
III. Miles, *Sir* Ashley IV. Parker, M. T.
V. Brown, Fred
616′.01 QR46
ISBN 0-7131-4427-0

To EM, BRP and the memory of JW

Filmset in 9/10 Times New Roman
and printed and bound in Great Britain by
Butler & Tanner Ltd
Frome and London

Volume Editors

F. Brown MSc, PhD, FRS
Head of Virology, Wellcome Research Laboratories, Beckenham, Kent; Formerly Deputy Director and Head of Biochemistry Department, Animal Virus Research Institute, Pirbright, Surrey.

Sir Graham Wilson MD, LLD, FRCP, FRCPath, DPH, FRS
Formerly Professor of Bacteriology as Applied to Hygiene, University of London, and Director of Public Health Laboratory Service, England and Wales.

Contributors

June D. Almeida DSc
Principal Scientist, Wellcome Research Laboratories, Beckenham, Kent.

J. E. Banatvala MA, MD, FRCPath
Professor of Clinical Virology, St. Thomas's Hospital Medical School, London.

Derrick Baxby BSc, PhD
Senior Lecturer, Department of Medical Microbiology, University of Liverpool.

A. J. Beale MD, FRCPath, Dip Bact (London)
Director of Biological Products, Wellcome Research Laboratories, Beckenham, Kent.

Jennifer M. Best BSc, PhD, MRCPath
Senior Lecturer, Department of Virology, St. Thomas's Hospital Medical School, London.

F. Brown MSc, PhD, FRS
Head of Virology, Wellcome Research Laboratories, Beckenham, Kent; formerly Deputy Director and Head of Biochemistry Department, Animal Virus Research Institute, Pirbright, Surrey.

D. R. Gamble MB, BS, FRCPath, Dip Bact
Director, Public Health Laboratory, Epsom, Surrey.

David J. Garwes MSc, PhD
Principal Scientific Officer, Microbiology Department, AFRC Institute for Research on Animal Diseases, Compton, Berkshire.

Colin R. Howard PhD, MSc
Senior Lecturer in Virology, London School of Hygiene and Tropical Medicine.

Richard H. Kimberlin BSc, PhD
Head of Pathogenesis Section, ARC and MRC Neuropathogenesis Unit, Edinburgh.

C. R. Madeley MD, MRCPath
Professor of Clinical Virology, Head of Department of Virology, University of Newcastle upon Tyne.

Walter Plowright CMG, DVSc, FRCVS, FRS
Head, Department of Microbiology, Institute for Research on Animal Diseases, Compton, Berkshire.

James S. Porterfield MD
Reader in Bacteriology, Sir William Dunn School of Pathology, University of Oxford.

C. R. Pringle BSc, PhD
Professor of Virology, Department of Biological Sciences, University of Warwick, Coventry.

Geoffrey G. Schild PhD, FIBiol
Head of Division of Viral Products, National Institute of Biological Standards and Control, Holly Hill, London.

Robert F. Sellers ScD, PhD, MRCVS
Director, Animal Virus Research Institute, Pirbright, Woking, Surrey.

D. I. H. Simpson MD, MRCPath, FIBiol
Director, Special Pathology Reference Laboratory, Public Health Laboratory Service Centre for Applied Microbiology and Research, Porton, Salisbury, Wiltshire

J. J. Skehel BSc, PhD, FRS
Co-Director, World Influenza Centre, and Member of Staff, Division of Virology, National Institute for Medical Research, London.

Harry Smith, PhD, DSc, FRCPath, FIBiol, FRS
Professor and Head of Department of Microbiology, University of Birmingham, Birmingham.

E. J. Stott BA, PhD
Principal Scientific Officer, AFRC Institute for Research on Animal Diseases, Compton, Newbury, Berkshire.

Sir Charles H. Stuart-Harris CBE, MD, DSc, FRCP
Emeritus Professor of Medicine, University of Sheffield, Formerly Professor of Medicine and Postgraduate Dean, Medical School, University of Sheffield.

Clive Sweet BSc, PhD
Research Fellow, Department of Microbiology, University of Birmingham, Birmingham.

G. S. Turner BSc, PhD
Formerly Virologist at The Lister Institute and Blood Products Laboratory, Elstree, Hertfordshire.

D. H. Watson BSc, PhD
Professor of General Microbiology, University of Leeds.

J. A. Wyke Vet MB, PhD, MRCVS
Head, Laboratory of Tumour Virology, Imperial Cancer Research Fund, London.

General Editors' Preface to 7th edition

After the publication of the 6th edition in 1975 we had to decide whether it would be desirable to embark on a further edition and, if so, what form it should take. Except for the single-volume edition of 1936, the book had always appeared in two volumes. We hesitated to alter this arrangement but reflection made us realize that a change would be necessary.

If due attention was to be paid to the increase in knowledge that had occurred during the previous ten years two volumes would no longer be sufficient. Not only had the whole subject of microbiology expanded greatly, but some portions of it had assumed a disciplinary status of their own. Remembering always that our primary concern was with the causation and prevention of microbial disease, we had to select that part of the newer knowledge that was of sufficient relevance to be incorporated in the next edition without substantial enlargement of the book as a whole.

One of the subjects that demanded consideration was virology, which would have to be dealt with more fully than in the 6th edition. Another was immunology. Important as this subject is, much of it is not directly concerned with immunity to infectious disease. Moreover, numerous books, reviews and reports were readily available for the student to consult. What was required by the microbiologist and allied workers was a knowledge of serology, and by the medical and veterinary student a knowledge of the mechanisms by which the body defends itself against attack by bacteria and viruses. We resolved, therefore, to provide a plain straightforward account of these two aspects of immunity similar to but less detailed than that in the 6th edition.

The book we now present consists of four volumes. The first serves as a general introduction to bacteriology including an account of the morphology, physiology, and variability of bacteria, disinfection, antibiotic agents, bacterial genetics and bacteriophages, together with immunity to infections, ecology, the bacteriology of air, water, and milk, and the normal flora of the body. Volume 2 deals entirely with systematic bacteriology, volume 3 with bacterial disease, and volume 4 with virology.

To this last volume we would draw special attention. It contains 27 chapters describing the viruses in detail and the diseases in man and animals to which they give rise, and is a compendium of information suitable alike for the general reader and the specialist virologist.

The first two editions of this book were written by Topley and Wilson, and the third and fourth by Wilson and Miles. For the next two editions a few outside contributors were brought in to bridge the gap that neither of us could fill. For the present edition we enlisted a total of over fifty contributors. With their help every chapter in the book has been either rewritten or extensively revised. This has led to certain innovations. The author's name is given at the head of each chapter; and each chapter is prefaced by a detailed contents list so as to afford the reader a conspectus of the subject matter. This, in turn, has led to a shortening of the index, which is now used principally to show where subjects not obviously related to any particular chapter may be found. A separate but consequently shorter index is provided for each of the first three volumes and a cumulative index for all four volumes at the end of volume 4. Each volume will be on sale separately. As a result of these changes we shall no longer be able to ensure the uniformity of style and presentation for which we have always striven, or to take responsibility for the truth of every factual statement.

We are fortunate in having Dr Parker, who has been associated with the 5th and 6th editions of the book, as the third general editor of all four parts of this edition and as editor of volume 2. Dr Geoffrey Smith with his extensive knowledge of animal disease has greatly assisted us both as a contributor and as editor of volume 3. Dr Fred Brown, formerly of the Animal Virus Research Institute, Pirbright, has organized the production of volume 4, and Professor Heather Dick the immunity section of volume 1.

Two small technical matters may be mentioned. Firstly, in volume 2 we have retained many of the original photomicrographs and added others at similar magnifications because they portray what the student sees when he looks down an ordinary light microscope in the course of identifying bacteria. Electronmicrographs have been used mainly to illustrate general statements about the structure of the organisms under consideration. Secondly, all temperatures are given in degrees Celsius unless otherwise stated.

Apart from those to whom we have just expressed our thanks, and the authors and revisers of individual chapters, we are grateful to the numerous workers who have generously supplied us with illustrations; to Dr N. S. Galbraith and Mrs Hepner at Colindale for furnishing us with recent epidemiological information; to Dr Dorothy Jones at Leicester for advice on the *Corynebacterium* chapter and Dr Elizabeth Sharpe at Reading for information about *Lactobacillus*; to Dr R. Redfern at Tolworth for his opinion on the value of different rodent baits; to Mr C. J. Webb of the Visual Aids Department of the London School of Hygiene and Tropical Medicine for the reproduction of various photographs and diagrams; and finally to the Library staff at the London School and Miss Betty Whyte, until recently chief librarian of the Central Public Health Laboratory at Colindale, for the continuous and unstinted help they have given us in putting their bibliographical experience at our disposal.

GSW
AAM

Volume Editors' Preface

Our knowledge of virology has grown tremendously during the last 30 years. This has been particularly true for the chemical nature of the agents and is the result of the explosive increase in the knowledge of nucleic acids and proteins and the general acceptance that viruses are essentially nucleoprotein molecules. The single most important event in virology in the thirty year period was the discovery by the groups at the Virus Laboratory in California and the Max Planck Institute in Tübingen that the RNA extracted from tobacco mosaic virus was itself infectious. Following closely on the heels of the discovery by Hershey and Chase in 1952 that only the DNA of the infecting bacteriophage entered the cell and the epoch-making paper by Watson and Crick on the structure of DNA, the two papers on tobacco mosaic virus put virology on the crest of a wave on which it is still riding. Classification is now firmly based on the chemical properties of the viruses and our understanding of the way in which the agents multiply, although far from complete, rests on this fundamental chemical knowledge. Alas, the processes of virus disease are much less well defined but already molecular virologists are turning their attention away from model systems to the study of disease.

After an introduction on the history of virology, the criteria on which viral taxonomy is based, and a list of the seventeen families of viruses affecting vertebrates that are now recognized internationally, there comes a chapter on morphology abundantly illustrated with tables and figures and containing a critical examination of the methods by which the shape and structure of viruses may be determined. This is followed by five more general chapters describing the ways in which different viruses multiply, the genetics of viruses, their pathogenic properties, the epidemiology of viral diseases, and the means by which they may be combated and controlled by vaccines and drugs. The greater part of the volume consists of a detailed account of the individual viruses met with in temperate and tropical regions affecting not only man but also a variety of animals and birds.

The volume aims at giving a balanced view of the subject ranging from molecular aspects of the viruses to their aetiological role in a miscellany of diseases. It is furnished with a plenitude of tables, figures and electronmicrographs with numerous references to papers, reviews, monographs and books in which more detailed information is available. Altogether it is a compendium of knowledge suitable for both the general reader and the expert virologist.

London — FB
1984 — GSW

Contents of volume 4

Virology

Contents of volumes 1, 2 and 3

Contents of volume 1
General microbiology and immunity

Contents of volume 2
Systematic bacteriology

Contents of volume 3
Bacterial diseases

79

The nature of viruses

Fred Brown

Introductory

Although virus diseases have been known for many centuries, the science of virology only started to emerge during the last decade of the nineteenth century. In 1892 Ivanovski discovered that tobacco mosaic disease was caused by an agent which could pass through a filter that retained the smallest bacteria. However, it was Beijerinck (1898) who introduced the concept of an agent which differed fundamentally from a bacterium. He showed that the agent causing tobacco mosaic would diffuse through agar and concluded that it was liquid or soluble and not corpuscular. Beijerinck introduced the term 'contagium vivum fluidum' for this agent. He also showed that only those organs of the plant that are growing and whose cells are dividing are capable of being infected. Beijerinck postulated that the agent must be incorporated into the living protoplasm of the cell in order to propagate and that it cannot multiply outside the cell.

Independently Loeffler and Frosch (1898) demonstrated that foot-and-mouth disease of cattle could also be transferred by material which could pass through a filter; this was, indeed, the first animal disease shown to be caused by a virus. Many years later Twort (1915) and d'Hérelle (1917) recognized that bacteria also could be infected by filter-passing agents, namely the bacteriophages.

While it was recognized, therefore, before the turn of the century that viruses are different from bacteria, it is only since the mid 1930s that the real nature of viruses has been elucidated. In 1935 Stanley crystallized tobacco mosaic virus (Stanley 1935). This was a major step in reaching our present concept that a virus is not living. However, it was the observations of Schlesinger (1936) with bacteriophage and Bawden *et al.* (1936) with tobacco mosaic virus which established the fact that viruses contain nucleic acid and are in fact nucleoproteins.

Further advances in our concept of viruses had to await the discovery by Hershey and Chase (1952) that only the DNA of bacteriophage T2 entered the cell when it infected its bacterial host and therefore that only the DNA was necessary for infection. Shortly afterwards the experiments of Fraenkel-Conrat (1956) in the USA and Gierer and Schramm (1956) in Germany proved without doubt that all the information necessary for the growth of tobacco mosaic virus is carried in its RNA. From that time there has been an enormous upsurge in our knowledge of the nature of viruses and this has advanced alongside the spectacular advances in molecular biology. Indeed viruses have provided extremely useful model systems for exploring the problems of replication, control mechanisms etc.

Viruses can be distinguished from other living things by five characters (Lwoff and Tournier 1966):

1. Possession of only one type of nucleic acid, either DNA or RNA but not both.
2. Reproduction solely from nucleic acid, whereas other agents grow from the sum of their constituents and reproduce by division.
3. Inability to undergo binary fission.

4. Lack of genetic information for the synthesis of essential cellular systems.
5. Use of ribosomes of their host cells.

These criteria clearly distinguish viruses from other micro-organisms, the most important being that viruses contain only one type of nucleic acid and are completely dependent on the host cell for their reproduction.

Physical structure

Morphology

Viruses occur in many shapes and sizes as can be seen from the diagrams in Chapter 80 taken from Matthews' review (1982) on their classification and nomenclature. Electron microscopy of negatively stained particles (Chapter 81) played a vital role in the characterization of viruses in the 1950s and 1960s and has been indispensable in the study of the details of virus structure. The method has also enabled us to obtain an overall picture of how viruses infect and replicate in the cell.

The size of the vertebrate viruses varies between *ca* 20 nm in diameter for the smallest DNA viruses (the parvoviruses) and *ca* 25 nm for the smallest RNA viruses (the picornaviruses) to 300 nm for the poxviruses. Members of the latter group are thus larger than the smallest micro-organisms, the mycoplasmas. Electron micrographs of members of all the virus groups are shown in Chapter 81. It is interesting to note that the volume of viruses can differ by a factor of 1000 but that despite of this they can be arranged into a relatively small number of groups (see Chapter 80).

Construction of viruses

The principles involved in the construction of viruses were laid down in the early 1960s by Caspar and Klug (Klug and Caspar 1960; Caspar and Klug 1962, Caspar 1965); their papers have remained the basis for our concept of the architecture of viruses. These aspects of virus structure are considered in detail in Chapter 81.

The simple viruses consist of nucleic acid enclosed within or built into a protein coat, the *capsid*. The capsid and its enclosed nucleic acid constitute the *nucleocapsid*. In its simplest form the capsid consists of a single layer of similar protein molecules arranged in an icosahedral shell or a helical tube. Clusters of similar or different structure units form the morphological units or *capsomeres* which are seen in the electron microscope.

Some viruses have an envelope of lipoprotein surrounding the nucleocapsid. The envelope is acquired as the virus passes through or buds from one of the cellular membranes and thus contains host cell components.

Chemical structure

Purification

The chemical composition of viruses could not be determined until they had been obtained in a purified form. Although tobacco mosaic virus and some other plant viruses were obtained in a highly purified state in the 1930s, e.g. Stanley crystallized tobacco mosaic virus in 1935, it was not until the 1950s, with the introduction of more refined methods of purification, that animal viruses were purified sufficiently for their analysis to become meaningful. Another reason for the delay in purifying animal viruses was the small amount of material available, but the advent of tissue culture methods for their cultivation and the large scale production of a poliovaccine from virus grown in tissue culture demonstrated the possibilities of growing sufficient virus for chemical analysis. Schwerdt and Schaffer crystallized poliomyelitis virus in 1955 and showed that it was a nucleoprotein containing 30 per cent RNA and 70 per cent protein. This work by Schwerdt and Schaffer gave great impetus to the purification of other animal viruses, and since that time many of them have been purified and their chemical composition determined.

In the purification of all viruses advantage is taken of the fact that they are much larger than even the largest components of the cell. The application of differential ultracentrifugation leads to considerable purification but probably the most important technique introduced in the last twenty years makes use of the different rates of sedimentation in gradients of sucrose of viruses and cell components. This technique was introduced by Brakke in the 1950s (Brakke 1953) and was used in the purification of poliomyelitis virus referred to above. Since a device for producing continuous gradients was described in 1961, the method has had widespread application.

Another method which has proved of great value makes use of the buoyant density of virus particles in salts such as caesium chloride and potassium tartrate. Density gradients of these salts are prepared and the mixture of virus and host cell components is centrifuged in a high speed centrifuge. The different particles take positions in the gradient corresponding to their buoyant density.

Criteria of purity

Though these methods remove most host cell components, traces of cell material often adhere to virus particles. Such traces can be removed from non-enveloped viruses by the use of mild detergents because *these* agents do not damage the virus particles. However, enveloped viruses are disrupted by detergents, which cannot therefore be used for the purification of this large group of viruses. Consequently there is often a lingering doubt about the purity of enveloped viruses.

The extent of contamination of viruses with host nucleic acid and protein can be measured by growing the virus in host cells that have been pre-labelled with radio-active precursors of these substances. Provided the purification procedure is satisfactory the 'pure' virus should not contain any radioactivity.

All viruses contain protein and some contain carbohydrate and lipid. However, the most important feature is that they contain DNA or RNA, but not both. These similarities between viruses exist irrespective of whether the host is animal, plant or bacterial.

Viruses can be conveniently divided into those that contain a lipid envelope (enveloped viruses) and those that do not (naked viruses). The envelope consists of lipid and carbohydrate, both of which are derived from the cell in which the virus is grown. The enveloped viruses are readily disrupted by lipid solvents or mild detergents into sub-units, some of which have biological properties distinct from those of the intact viruses. The naked viruses are stable in lipid solvents and even in strong detergents such as sodium dodecyl sulphate. Disruption into their constituents usually requires severe treatment such as heating above 50° or treatment with protein denaturants such as urea or phenol.

Nucleic acid

The virus nucleic acid, which, as mentioned above, may be DNA or RNA, may also be single- or double-stranded and the genome may consist of one or several molecules. In most of the DNA viruses described so far the genomes or genetic information consists of a single molecule of nucleic acid but the genomes of many RNA viruses consist of several different molecules. If the genome consists of a single molecule, this may be linear or have a circular configuration. The way in which the genome of those viruses containing several different segments of nucleic acid is organized in the particles is not known. The molecular weight of the DNA of different viruses varies from 1 to over 200×10^6. However, the range of molecular weights of virus RNA is much less, ranging from *ca* 2 to 15×10^6. The variety of forms in which the genetic material may occur is summarized in Table 79.1 together with examples of the viruses in which they are found.

Table 79.1 Nature of the genetic material of viruses

Nucleic acid	*Examples*
Single-stranded DNA	feline panleucopenia, adeno-associated
Double-stranded DNA—linear	pox, herpes
—circular	Shope papilloma, polyoma
Single-stranded RNA—unsegmented	polio, rubella, measles, rabies
—segmented	influenza, Rift Valley fever, lymphocytic choriomeningitis
Double-stranded RNA—segmented	reo, rota, bluetongue

In viruses whose nucleic acid consists of single-stranded RNA molecules, the virus nucleic acid is either a positive strand, e.g. poliovirus RNA, or a negative strand, e.g. rabies virus RNA. With some RNA viruses, the nucleic acid can be extracted in an infectious form. In others the isolated nucleic acid is not infectious, even though it contains all the necessary genetic information. This is because its transcription depends on a virion-associated transcriptase which is separated from the nucleic acid by the extraction procedure. In viruses whose nucleic acid consists of single-stranded RNA molecules, the virus nucleic acid is either positive-stranded, i.e. one that can function as its own messenger, or negative-stranded, i.e. one that produces mRNA as a first step in its replication. Some viruses with single-stranded DNA contain either the positive or the complementary strand.

Most viruses contain only virus DNA or RNA. However, some viruses contain host cell nucleic acids. For example, some papovaviruses contain host cell DNA and the arenaviruses contain cellular ribosomes.

Proteins

All viruses contain proteins and indeed proteins form the major part of most viruses. Although there are viruses which contain only one species of protein, e.g. the caliciviruses, most contain several. Thus the smallest RNA viruses (the picornaviruses such as poliovirus and foot-and-mouth disease virus) contain four distinct species of protein. The more complex viruses such as herpes virus contain many distinct protein species, the molecular weight of the different species ranging from *ca* 5×10^3 to 150×10^3. Whereas there is normally only one copy of the virus nucleic acid in a virus particle, there are usually many copies of each protein. Thus there are 60 copies of each of the four proteins of poliovirus in each particle. Although the proteins provide a protective shell for the genome, they have other properties as well. For example, the surface proteins have an affinity for the specific receptors on the surface of susceptible cells and they also contain the antigenic determinants which are responsible for the production of protective antibody in the infected or vaccinated animal. Some virus proteins have enzymic activity. For example, there is a protein in the negative-stranded RNA viruses, e.g. rabies virus, which acts as a transcriptase.

In addition to the proteins of the virus particle, many virus-induced non-structural proteins are found in cells infected with viruses. Indeed, most of the virus genome codes for these non-structural proteins. The functions of these proteins are known in only a few instances although it seems clear that they are required for virus multiplication. This brief outline of the nature of viruses is expanded in subsequent chapters on their classification, morphology, replication and genetics (Chapters 80–83).

Further reading

For further information on general virology the following books may be consulted.

Andrewes, C. and Pereira, H. G. (1972) *Viruses of Vertebrates*, 3rd Edn. Baillière Tindall, London.

Baltimore, D., Huang A. S. and Fox, C. F. (1976) *Animal Virology*, 4th Edn. Academic Press, New York and London.

Fenner, F. *et al.* (1976) *The Biology of Animal Viruses*, 2nd Edn.

Howard, C. R. (Ed.) (1982) *New Developments in Practical Virology*. Alan R. Liss Inc., New York.

Lauffer, M. A., Bang, F. B., Maramorosch, K. and Smith, K. M. (1982) *Advances in Virus Research*. Academic Press, New York.

Lennette, E. H. and Schmidt, N. J. (1979) *Diagnostic Procedures for Viral, Rickettsial and Chlamydial Infections*, 5th Edn. American Public Health Associaton, Washington, DC.

Luria, S. E., Darnell, J. E. Baltimore, D. and Campbell, A. (1978) *General Virology*, 3rd Edn. John Wiley & Sons, London and New York.

Primrose, S. B. and Dimmock, N. J. (1980) *Introduction to Modern Virology*, 2nd Edn. Blackwell Scientific Publications Ltd, Oxford.

Report (1980) Interactions between Virus and Host Molecules *Proc. R. Soc. Lond. B.* **210,** 317–476.

Rowson, K. E. K., Rees, T. A. L. and Mahy, B. W. J. (1981) *A Dictionary of Virology*. Blackwell Scientific Publications Ltd, Oxford.

Waterson, A. P. and Wilkinson, L. (1979) *An Introduction to the History of Virology*. Cambridge University Press, London.

References

Bawden, F. C., Pirie, N. W., Bernal, J. D. and Fankuchen, I. (1936) *Nature*, **138,** 1051.

Beijerinck, M. W. (1898) *Zentr. Bakteriol. Parasitenk. Abt. ii*, **5,** 27.

Brakke, M. L. (1953) *Arch. Biochem. Biophys.*, **45,** 275.

Caspar, D. L. D. (1965) In: *Viral and Rickettsial Infections of Man*, 4th Edn, p. 51. Ed. by F. L. Horsfall and I. Tamm. Lippincott, Philadelphia.

Caspar, D. L. D. and Klug, A. (1962) *Cold Spring Harbour Symposium Quant. Biol.*, **27,** 1.

Fraenkel-Conrat, H. (1956) *J. Amer. chem. Soc.*, **78,** 882.

Gierer, A. and Schramm, G. (1956) *Nature*, **177,** 702.

d'Herelle, F. (1917) *C.R. Acad. Sci.*, **165,** 373.

Hershey, A. D. and Chase, M. (1952) *J. gen. Physiol.*, **36,** 39.

Ivanowski, D. (1892) *Bull Acad. Sci.* St Petersbourg.

Klug, A. and Caspar, D. L. D. (1960) *Advanc Virus Res.* **7,** 25.

Loeffler, O. and Frosch, P. (1898) *Zbl. Bakt. I. Abt. Orig.*, **23,** 371.

Lwoff, A. and Tournier, P. (1966) *Annu. Rev. Microbiol.*, **20,** 45.

Matthews, R. E. F. (1982) *Intervirology*, **17,** 7.

Schlesinger, M. (1936) *Nature*, **138,** 508.

Schwerdt, C. E. and Schaffer, F. L. (1955) *Virology*, **2,** 665.

Stanley, W. M. (1935) *Science* **81,** 644.

Twort, F. W. (1915) *Lancet* **ii,** 1241, 165, 373.

80

Classification of viruses

Fred Brown

Classification of viruses

The aim of virus classification is to make an ordered arrangement of viruses that will indicate their similarities and differences. Viruses can be classified on the basis of any of their properties but it is logical to use a system that can be applied to all viruses. Such a scheme has considerable value in identification so that the unknown properties of a virus can be predicted by analogy with the known properties of similar viruses. This enables related viruses to be united into the same category. It also has value in virus taxonomy.

The value of a classification scheme is indisputable but the best method to achieve it led to much debate, some of it acrimonious. A system has emerged which is now fairly well accepted universally and it is interesting to summarize the major steps in its development because it has a firm place in the history of virology.

Historical introduction

Infectious diseases were known before the agents which cause them so it was natural that the agents were named according to the disease. Early efforts to classify viruses arranged them according to host symptoms or type of disease and tissue affinities. For example in 1939 Bennett, on behalf of the Committee for Virus Nomenclature of the Council of the American Phytopathological Society, proposed the following criteria for classification:

Criteria for classification

1. Type of symptoms produced on different species and varieties of susceptible plants.
2. Morphological and cytological disturbances produced.
3. Relation of insect vectors to virus transmission.
4. Antigenic reactions in animals and plants.
5. Chemical and physical properties of the viruses themselves.

Such a system has obvious deficiencies. For example, (1) the same virus can produce a different syndrome in different hosts; (2) different strains of the same virus can produce different syndromes in the same host; (3) different viruses can produce the same clinical picture (e.g. foot-and-mouth disease, vesicular stomatitis, vesicular exanthema and swine vesicular disease).

Some years later Holmes (1948) devised a Latin binomial system based on the biological properties of viruses, primarily their symptomatology and host range, but also on their mode of transmission and vector specificity. Very little attempt was made to consider the properties of the viruses themselves. With hindsight, it is obvious that Holmes used poor taxonomic criteria.

As early as 1939, Bawden had suggested that virus nomenclature should be based on morphological, chemical and serological information. In other words, the properties of the virus particle should be used. However there was little response to his suggestion, partly because of the lack of knowledge at that time of the basic properties of virus particles. With the explosion of information on viruses which occurred from about 1950, it became increasingly apparent that a scheme of classification which separated viruses from bacteria and higher organisms was necessary. A Subcommittee on Virus Nomenclature was established under the chairmanship of C. H. Andrewes in 1950 at the International Congress of Microbiology held in Rio de Janeiro. This subcommittee decided that the Bacteriological Code could not be applied to virus nomenclature although they realized that the use of any comprehensive system would be unwise until more information on their properties became available. Nevertheless they recommended that the naming of a few better known virus groups should be attempted on the basis of characters such as morphology, chemical composition and physical and antigenic properties rather than host range, tissue tropisms, pathology and symptomatology; and such names as Poxvirus, Herpesvirus, Adenovirus, Picornavirus, Reovirus, Myxovirus and Paramyxovirus were proposed.

A considerable step forward was made by Cooper in 1961 when he proposed a hierarchy of structure as the basis for grouping animal viruses. By far the most important proposal was the division into RNA or DNA containing viruses. However a secondary subdivision using sensitivity to ether, as suggested by Andrewes and Horstmann (1949), allowed the recognition of four groups of viruses and resulted in an ordering which has proved to be an extremely valuable basis for their subsequent classification.

A year later, Lwoff, Horne and Tournier (1962) proposed that a system (the LHT system) should be established for the classification of all viruses. While it is not the intention to discuss in detail in this chapter viruses other than those infecting vertebrates, it is worth remembering that these represent only a fraction of the viruses infecting different hosts. The development of classification has taken account of viruses infecting all known hosts and the unification of the subject has been stimulating and worthwhile in showing resemblances which might otherwise have been overlooked.

The LHT system was based on the properties of the virus. Four characters were selected:

(1) the nature of the nucleic acid, DNA or RNA, as in Cooper's system;
(2) Symmetry of the virus, helical or cubical, or binary as in the phages where there are different symmetries in the head and tail;
(3) the presence or absence of an envelope;
(4) the diameter of the nucleocapsid for helical viruses or the number of capsomeres for cubic viruses.

These proposals were largely adopted by a Provisional Committee on Nomenclature of Viruses, set up in 1965, which replaced the Subcommittee of Virus Nomenclature mentioned earlier. The suggestions of the Provisional Committee were contested by Gibbs *et al.* (1966) who proposed a system on an Adansonian classification in which the nomenclature to be adopted would consist of vernacular names together with a cryptogram consisting of four pairs of characters:

(1) type of nucleic acid/strandedness of nucleic acid;
(2) molecular weight in millions of nucleic acid/percentage of nucleic acid in particle;
(3) outline of virus particle/outline of nucleocapsid;
(4) kinds of host/kinds of vector.

Thus the cryptogram of poliovirus would be R/1:2.5/30:S/S:V/O where R = RNA, S = spherical, V = vertebrate and O = no vector.

In 1966 the International Committee on Nomenclature of Viruses (ICNV) was formally established to find a taxonomic system for all viruses, and subcommittees for vertebrate, invertebrate, plant and bacterial viruses were established with an additional subcommittee devoted to exploring the use of the cryptogram. The gradual emergence of a useful system of classification has taken place from these beginnings and has been reviewed by Wildy (1971), Fenner (1976) and Matthews (1979, 1982), the three successive Presidents of the International Committee. The first of these reviews by Wildy is masterly in its description of the first five years of existence of the International Committee, with all its difficulties in formulating a common policy.

As he pointed out, two principal philosophies prevail in virus nomenclature. In the first, viruses are classified on the basis of a few important properties and named on a hierarchic basis. In the second, viruses are grouped on the basis of many criteria. Although the second approach should lead eventually to a definitive classification, it will take many years to complete. On the other hand the first approach, which is certain to produce mistakes, has an immediate use. The ICNV adopted a middle course by comparing individual viruses by the use of many criteria, grouped them and selected names for them. The Committee drew up a set of rules which are given in Wildy's report (1971) on the Classification and Nomenclature of Viruses, but as experience was gained in naming and grouping it became clear that certain of the rules required alteration. The rules in use at present are:

Rule 1—The code of bacterial nomenclature shall not be applied to viruses.
Rule 2—Nomenclature shall be international.
Rule 3—Nomenclature shall be universally applied to all viruses.
Rule 4—An effort will be made towards a latinized nomenclature.
Rule 5—Existing latinized names shall be retained whenever feasible.
Rule 6—The law of priority shall not be observed.
Rule 7—Sigla may be accepted as names of viruses or virus groups, provided that they are meaningful to workers in the field and are recommended by international virus study groups.
Rule 8—No person's name shall be used.
Rule 9—Names should have international meaning.
Rule 10—The rules of orthography of names and epithets are listed in Chapter 3, Section 6 of the proposed international code of nomenclature of viruses (Appendix D; Minutes of 1966 (Moscow) meeting).
Rule 11—A virus species is a concept that will normally be represented by a cluster of strains from a variety of sources, or a population of strains from a particular source, which have in common a set or pattern of correlating stable properties that separates the cluster from other clusters of strains.
Rule 12—The genus name and species epithet, together with the strain designation, must give an unambiguous identification of the virus.
Rule 13—The species epithet must follow the genus name and be placed before the designation of strain, variant or serotype.
Rule 14—A species epithet should consist of a single word or, if essential, a hyphenated word. The word may be followed by numbers or letters.
Rule 15—Numbers, letters, or combinations thereof may be used as an official species epithet where such numbers or letters already have wide usage for a particular virus.
Rule 16—Newly designated serial numbers, letters or combinations thereof are not acceptable alone as species epithets.
Rule 17—Artificially created laboratory hybrids between different viruses will not be given taxonomic consideration.
Rule 18—Approval by ICTV of newly proposed species, species names and type species will proceed in two stages. In the first stage, provisional approval may be given. Provisionally approved proposals will be published in an ICTV report. In the second stage, after a 3-year waiting period, the proposals may receive the definitive approval of ICTV.
Rule 19—The genus is a group of species sharing certain common characters.
Rule 20—The ending of the name of a viral genus is '... virus'.
Rule 21—A family is a group of genera with common characters, and the ending of the name of a viral family is '... viridae'.
Rule 22—Approval of a new family must be linked to approval of a type genus; approval of a new genus must be linked to approval of a type species.

Four of the subcommittees mentioned above deal with viruses that affect primarily hosts of one or other of the four major groups. However, many viruses that infect vertebrate animals and several viruses that infect plants also multiply in invertebrates, notably insects, ticks and mites, which may act as vectors. The virus families are dealt with by a Coordination Subcommittee so constituted that it includes virologists interested in the viruses in each class of host affected.

The name of the committee was changed in 1973 to International Committee for the Taxonomy of Viruses (ICTV) and an additional subcommittee, to consider the viruses affecting fungi, was set up in 1975. This covers a highly specialized area of virology which did not fit naturally into any of the other main groups. The work of the subcommittees on bacterial, fungal, invertebrate and plant viruses need not concern us here. The viruses of most interest to the medical student are included in the work of the subcommittee on vertebrate viruses and of the coordinating subcommittee. However, the rate at which information on viruses is being accumulated means that methods for storing this knowledge are becoming increasingly important. The old Cryptogram Subcommittee is now known as the Code and Data Subcommittee and its work will be considered below. First, however, the present position in the classification of the viruses infecting vertebrates will be considered.

Viruses infecting vertebrates

The taxonomy of viruses that primarily infect vertebrates is considerably more advanced than that of any of the other groups. There are 12 Study Groups operating under the chairman of the Vertebrate Virus

Subcommittee. In addition there are five Study Groups on viruses which infect other kinds of host as well as vertebrates. Those viruses affecting hosts distributed in different phyla are dealt with by the Coordination Subcommittee. All these groups are listed in Table 80.1 and represented diagrammatically in the descriptions of the individual families.

Table 80.1 Study group of viruses infecting vertebrates and other hosts*

	Study Group
DNA viruses	Adeno
	Herpes
	Irido*
	Papova
	Parvo*
	Pox*
RNA viruses	Arena
	Bunya
	Calici
	Corona
	Orthomyxo
	Paramyxo
	Picorna
	Reo*
	Retro
	Rhabdo*
	Toga

A vast amount of work has been done on the wide range of viruses that infect man and his domestic animals, but wild vertebrates have not been examined to anything like the same extent. Although there are some viruses affecting man and his domestic animals which are still unclassified (e.g. the Marburg agent) the majority of the several hundred viruses so far identified can be assigned to one or another of the 17 families.

Most of the families consist of several related but readily distinguishable genera. Because of lack of agreement, some families have not been subdivided but it seems to be only a matter of time before this subdivision takes place.

Storage of data on viruses

It is now generally accepted that a system for storing information is essential if we are to derive maximum benefit from all the information that has been accumulated. When Gibbs *et al.* (1966) proposed the use of a system based on an Adansonian classification in which the nomenclature would consist of vernacular names together with a cryptogram, they pointed out that computers would allow such a system to be used easily and accurately. The cryptogram is a useful shorthand description of the properties of viruses and a list of cryptograms of all the better known viruses was provided in Wildy's 1971 report on the Classification and Nomenclature of Viruses. It seemed at that time that it would soon become universally accepted. Although there was the criticism, particularly from those working with well characterized viruses, that the cryptogram was unnecessary, it was clear that virologists working with a wide range of viruses would be the group to gain most benefit from it. The imprecision of the cryptogram was also criticized but its enlargement to include more terms would have rendered it cumbersome. More precise definition was obtained by indicating in the third term whether the virus is enveloped and in the fourth term the mode of transmission.

However in 1976, the role of the cryptogram was reconsidered. It was recognized that the cryptogram had played a valuable role in drawing attention to the need for basic information on viruses but its inflexibility, particularly in the light of the important newly discovered characteristics of viruses such as their mode of replication, persuaded the Executive Committee of ICTV to discontinue its use. The subcommittee turned its attention to the development of a virus data sheet, suitable for all viruses, that can be used to collect data for computer storage and retrieval. Agreement has been reached on the extent and nature of the data to be collected and this will be standardized with the data collection form developed by the WHO Centre for the Collection and Evaluation of Data on Comparative Virology, directed by P. A. Bachmann in Munich.

Description of viruses

The main characteristics of the 17 virus families listed in Table 80.2 are described by Matthews (1979, 1982). These descriptions contain the information that was accepted by the ICTV at its meetings in the Hague in 1978 and Strasburg in 1981. The amount of information available for most virus families is now considerable (Matthews 1979, 1982,1983).

As described in the Introduction, viruses exhibit a wide diversity of morphology and nucleic acid structure. There are vertebrate viruses containing single-stranded RNA, double-stranded RNA, single-stranded DNA or double-stranded DNA. The single-stranded RNA can be present as one piece or segmented whereas the double-stranded RNA is segmented. No segmented DNA viruses have been described. Quite clearly the mode of replication in these groups is different. This feature, which is often termed the *strategy*, promises to have increasing importance in the classification of viruses. Cooper (1974) has pointed out that members of the same virus family have closely similar strategies whereas the strategies of different families are so unlike that phylogenetic relationships between them would appear to be unlikely. In fact he has suggested that, in those instances where the strategy of an unclassified virus is not evident, caution should be exercised in allocating it to new or existing families, whatever its particle structure suggests. The different strategies are described in detail in chapter 82. A summary of the main properties of the virus families is given below. The wide range given for some of the physico-chemical properties arises because of the variation between members.

Table 80.2 Families of virus infecting vertebrates and other hosts

Adenoviridae

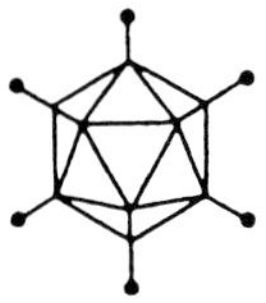

Morphology	Non-enveloped isometric particle with icosahedral symmetry, 70–90 nm in diameter.
Nucleic acid	Single molecule of linear ds DNA, mol.wt 20–30 × 10^6.
Protein	At least 10 polypeptides in virion.
Lipid	None
Carbohydrate	Fibres are glycoprotein.

Arenaviridae

Morphology	Enveloped spherical or pleomorphic particle, 50–300 nm in diameter with surface projections. Ribosome-like particles, 20–25 nm in diameter, present in envelope.
Nucleic acid	Two virus-specific linear or circular negative ss RNA molecules, mol.wt 2.1–3.2 × 10^6 and 3 of host origin, probably the two ribosomal RNAs and 5S RNA.
Protein	3 major, 2 surface and one associated with the RNA; in addition several enzymes are present.
Lipid	Present.
Carbohydrate	Present.

Bunyaviridae

Morphology	Enveloped spherical or oval particle, 90–100 nm in diameter with surface projections and three circular nucleocapsids with helical symmetry.
Nucleic acid	3 segments of circular negative ss RNA, mol.wt 3, 2 and 0.5 × 10^6; 1–2 per cent of particle.
Protein	3 major, 1 minor of which two are glycosylated.
Lipid	33 per cent of particle.
Carbohydrate	7 per cent of particle, associated with glycoproteins and glycolipid.

Caliciviridae

Morphology	Unenveloped, spherical nucleocapsids 35–39 nm in diameter with 32 cup-shaped depressions arranged in icosahedral symmetry.
Nucleic acid	Single molecule of linear infectious positive ss RNA, mol.wt 2.6–2.8 × 10^6, 20 per cent of particle. Polyadenylated at 3′ end, covalently linked protein at 5′ end.
Protein	One major polypeptide, mol.wt c 70 × 10^3 (180 copies); trace of polypeptide mol.wt c 15 × 10^3. Also one small polypeptide attached covalently to 5′ end of genome.
Lipid	None.
Carbohydrate	None.

Table 80.2—*continued*

Coronaviridae

Morphology	Pleomorphic enveloped particle, 75–160 nm in diameter with club-shaped surface projections.
Nucleic acid	Single molecule of positive ss RNA, mol.wt $5.5–6.1 \times 10^6$.
Protein	4–6 polypeptides; surface polypeptides are glycosylated.
Lipid	Present in envelope.
Carbohydrate	Present in surface glycoproteins.

Herpesviridae

Morphology	Enveloped virus particle, 120–150 nm in diameter, consisting of 4 structural components: (a) the core in which the DNA is wrapped; (b) the icosahedral nucleocapsid; (c) a tegument surrounding the capsid and (d) the envelope.
Nucleid acid	Single molecule of linear ds DNA, mol.wt $80–150 \times 10^6$.
Protein	More than 20 polypeptides.
Lipid	Present in envelope.
Carbohydrate	Present in surface glycoprotein.

Iridoviridae

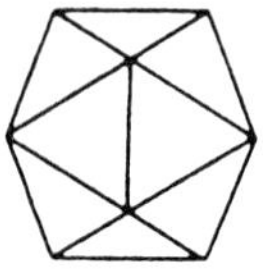

Morphology	Enveloped, icosahedral particle, 125–300 nm in diameter with spherical nucleoprotein core.
Nucleic acid	Single molecule of linear ds DNA, mol.wt $100–250 \times 10^6$.
Protein	13–25 structural proteins and several enzymes.
Lipid	5–9 per cent of particle.
Carbohydrate	Present on some surface proteins.

Orthomyxoviridae

Morphology	Enveloped pleomorphic particle, 80–120 nm in diameter with surface projections. Nucleocapsid has helical symmetry.
Nucleic acid	8 pieces of linear, negative ss RNA, total mol.wt c 5×10^6.
Protein	7–9 polypeptides and enzymes.
Lipid	18–37 per cent of particle, mainly in virus envelope.
Carbohydrate	5–9 per cent of particle, present on surface glycoproteins.

Table 80.2—*continued*

Papovaviridae

Morphology	Non-enveloped icosahedral particle, 45–55 nm in diameter.
Nucleic acid	Single circular molecule of ds DNA, mol.wt 3–5 × 10^6. 10–12 per cent of particle weight.
Protein	5–7 polypeptides, some cellular.
Lipid	None.
Carbohydrate	None.

Paramyxoviridae

Morphology	Enveloped pleomorphic but roughly spherical particle, 150 nm in diameter with surface projections. Nucleocapsid has helical symmetry.
Nucleic acid	Single molecule of negative ss RNA mol.wt 5–7 × 10^6; about 0.5 per cent of particle weight.
Protein	5–7 polypeptides and enzymes.
Lipid	20–25 per cent of particle, derived from host.
Carbohydrate	6 per cent of particle, on glycoprotein.

Parvoviridae

Morphology	Non-enveloped isometric particle with icosahedral symmetry, 18–26 nm in diameter.
Nucleic acid	Single molecule of ss DNA, mol.wt 1.5–2.2 × 10^6.
Protein	3 polypeptides.
Lipid	None.
Carbohydrate	None.

Picornaviridae

Morphology	Unenveloped spherical nucleocapsids, 22–30 nm in diameter with icosahedral symmetry. Surface is featureless.
Nucleic acid	Single molecule of linear infectious positive ss RNA, mol.wt c 2.5 × 10^6; 30 per cent of particle. Polyadenylated at 3′ end, covalently linked protein at 5′ end. Some members have polycytidylic acid tract near 5′ end.
Protein	4 major polypeptides (60 copies of each) and one small polypeptide attached covalently to the 5′ end of the genome.
Lipid	None.
Carbohydrate	None.

Table 80.2—*continued*

Poxviridae

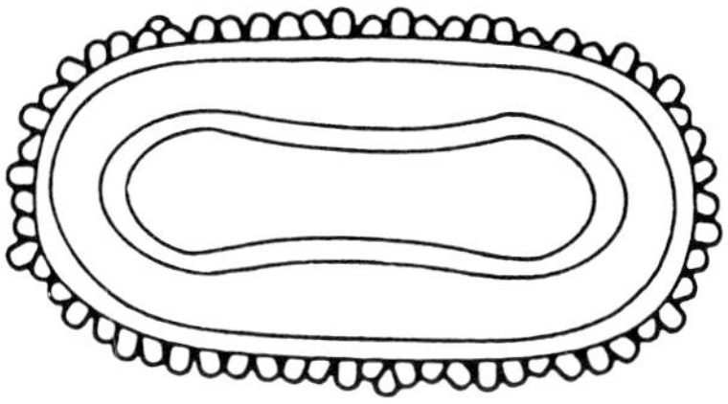

Morphology	Large, brick-shaped or ovoid particle, 300–450 × 170–260 nm with external coat containing lipid and tubular or globular protein structures. The coat encloses a core which contains the genome.
Nucleic acid	Single molecule of ds DNA, mol.wt 80–240 × 10^6.
Protein	More than 30 structural proteins and several enzymes.
Lipid	About 4 per cent of particle.
Carbohydrate	About 3 per cent of particle.

Reoviridae

Morphology	Unenveloped icosahedral particle, diameter 60–80 nm with two protein coats.
Nucleic acid	10–12 pieces of linear ds RNA; about 14–22 per cent of particle weight.
Protein	6–10 polypeptides, including enzymes.
Lipid	None.
Carbohydrate	Some may be associated with polypeptides.

Retroviridae

Morphology	Enveloped, spherical particle, 80–100 nm in diameter with surface projections and an internal icosahedral capsid containing a helical ribonucleoprotein.
Nucleic acid	Inverted dimer of linear positive ss RNA, mol.wt 3 × 10^6 (for monomer). Polyadenylated at the 3′ end and with a cap structure at the 5′ end; 2 per cent of particle. Virus RNA is not infectious. A tRNA, serving as primer for reverse transcription, is bound to the virus RNA by base pairing.
Protein	4 internal non-glycosylated proteins, 2 surface glycoproteins and reverse transcriptase; 60 per cent of particle.
Lipid	35 per cent of particle, derived from plasma membrane.
Carbohydrate	3.5 per cent of particle, present as glycoprotein and glycolipid.

Rhabdoviridae

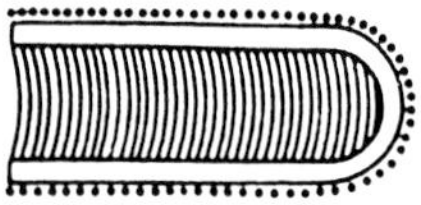

Morphology	Enveloped particle, usually bullet-shaped or bacilliform, 130–380 × 50–95 nm with surface projections and a nucleocapsid with helical symmetry.
Nucleic acid	Single molecule of negative linear ss RNA, mol.wt 3.5–4.5 × 10^6; 1–2 per cent of particle.
Protein	5–6 structural polypeptides and several enzymes; 65–75 per cent of particle.
Lipid	15–25 per cent of particle, derived from host.
Carbohydrate	3 per cent of particle, associated with spike glycoprotein and glycolipid.

Table 80.2—*continued*

Togaviridae		
	Morphology	Enveloped spherical particle, 40–70 nm in diameter, with envelope in close association with nucleocapsid. Surface projections protrude from envelope.
Nucleic acid	Single molecule of positive ss RNA, mol.wt c 4×10^6; 5–8 per cent of particle weight.	
Protein	3–4 polypeptides, of which one or more are glycosylated.	
Lipid	Cell-derived lipid envelope.	
Carbohydrate	Glycoproteins.	

Conclusions

The upsurge in interest in the classification of viruses which took place from the mid 1960s has now abated.

Clearly a great deal of the work has been done but it would be wrong to think that the task is complete. New groups continue to be recognized and the recent work on the caliciviruses, which has resulted in the creation of a new virus family because its properties are so different from those of existing families, provides a good example of the continuing work on classification. Information is updated frequently by Melnick in editions of *Progress in Medical Virology*.

Those interested in a much fuller classification should turn to the reports of the ICTV referred to earlier, to Wilner's (1964) book *A Classification of the Major Groups of Human and Other Animal Viruses*, and to a recently published book *A Critical Appraisal of Viral Taxonomy* edited by Matthews (1983), which surveys all aspects of the subject.

References

Andrewes, C. H. and Horstmann, D. M. (1949) *J. gen. Microbiol.* **3**, 290.

Bawden, F. C. (1939) *Plant Viruses and Virus Diseases*. Chronica Botan Co., Leiden.

Bennett, C. W. (1939) *Phytopathology* **29**, 422.

Cooper, P. D. (1961) *Nature* **190**, 302.

Cooper, P. D. (1974) *Intervirology* **4**, 317.

Fenner, F. (1976) Second Report of the International Committee on Taxonomy of Viruses: Classification and Nomenclature of Viruses. *Intervirology* **7**, 1–115.

Gibbs, A. J., Harrison, B. D., Watson, D. H. and Wildy, P. (1966) *Nature* **209**, 450.

Holmes, F. O. (1948). In *Bergey's Manual of Determinative Bacteriology*, 6th Edn., pp. 1127–1286. Ed. by R. E. Breed, E. G. D. Murray and A. P. Hitchens. Williams and Wilkins, Baltimore.

Lwoff, A., Horne, R. and Tournier, P. (1962) *Cold Spring Harbor Symposia quant. Biol.* **27**, 51.

Matthews, R. E. F. (1979) Third Report of the International Committee on Taxonomy of Viruses: Classification and Nomenclature of Viruses. *Intervirology* **12**, 150–280; (1982) Fourth Report of the International Committee on Taxonomy of Viruses: Classification and Nomenclature of Viruses. *Intervirology* **17**, 7–199; (1983) *A Critical Appraisal of Viral Taxonomy*. CRC Press.

Wildy, P. W. (1971) First Report of the International Committee on Taxonomy of Viruses: Classification and Nomenclature of Viruses. *Monogr. Virol.* **5**. Karger, Basel.

Wilner, B. I. (1964) *A Classification of the Major Groups of Human and other Animal Viruses*. Burgess Publishing Co.

81

Morphology: virus structure

June D. Almeida

Introduction and morphology: virus structure

Although reputedly the simplest possible form of life, the virus is nevertheless a structure of such complexity that even the term 'virus structure' is open to interpretation. To the biochemist it presumes nucleotide sequences or structural polypeptides. To the immunologist it is the arrangement of individual antigens that build up to give the whole immunogenic virion. To the simple morphologist it presents the far from simple problem of understanding the arrangement of protein, lipoprotein, or glycoprotein components with which the virus covers its vulnerable nucleic acid during transit from one host or cell to another. It is this last aspect, that of basic morphological structure, with which this chapter will deal.

Methods of examination

Viruses exist in the size range 18–300 nm. This means that light microscopy, with a resolution limit around 200 nm can never be employed in the study of virus structure. Only two techniques are available which allow the investigation of virus substructure to be carried out at a satisfactory level of resolution. These are x-ray diffraction, which was the first technique used in this field and which remains as offering the highest level of resolution (Klug and Caspar 1960) and direct visualization in the electron microscope (EM) (Horne and Wildy 1961). Each technique has its strengths and its limitations. X-ray diffraction yields far greater detail, resolving down to the 1.5 Å level, and is able to elucidate structures within the particle but can be applied only to highly purified virus. The transmission EM has a theoretical resolution around 2 Å but with biological material a practical limit in the 5–10 Å range, considerably less than that of x-ray diffraction. On the other hand the EM, depending on the method of preparation used, allows direct visualization of subunit construction and can be carried out on individual virus particles. Complementary information from these two methods, together with the information from biochemical studies, allows a reasonably coherent understanding of the morphological characteristics of virus particles.

Although the results of x-ray diffraction studies of plant viruses have been basic to understanding the concepts of virus construction, with very few exceptions it has not been possible to apply this method to animal viruses, as only a few have been purified to the level required for this technique. It is therefore from the EM that we have obtained the greatest amount of information/on animal virus morphology.

When an untreated virus is examined in the EM it shows up as a poorly defined 'shadowgraph' on the screen. The reason for this is that biological material in general is almost transparent to the electron beam. In order to increase electron density, staining methods are required, but to understand the kind of staining that is needed, it is first necessary to consider the properties of the transmission electron microscope itself (Chapman 1980).

Accelerated electrons pass down the column of the microscope, are formed into a proper beam by the condenser lens system and then further pass through a series of magnifying lenses to be projected eventually on to a phosphorescent screen. During this transit down the column the electrons additionally pass through the specimen itself, thus yielding a transmission image of the object under study. The exact mechanism of image production is one of electron scatter with dense material producing a high level of scatter and lucent substances leaving the beam unaltered. This means that both upper and lower surfaces of a particle contribute to the final image; and when it is remembered that the focal depth of the electron microscope is such that the whole of a virus particle is always in focus, then it can be appreciated that there is a certain art in interpreting the transmission image provided by the EM. This will be discussed in more detail later. In order to obtain clearer images of viruses it is necessary to increase the contrast between the particles and the background.

In light microscopy the image is improved by the use of pigments which specifically stain the structures of interest. In the EM exactly the same approach is used but the only 'stains' possible are those that alter the electron density of a specimen. Hence EM 'stains' invariably contain heavy metal. If we confine our considerations to the transmission EM, there are three basic techniques by which heavy metal staining can be accomplished.

Shadow casting

The EM had become technically practical by the end of the 1930s but it was not until 1946 that Williams and Wyckoff described the method of shadow casting that made viruses a more realistic subject for electron microscopy. Virus particles are adsorbed to a suitable coated grid and then placed under vacuum. Pure metal is evaporated off a heated filament at a known angle to the specimen. Since this is carried out *in vacuo* the metal atoms travel in a straight line to attach themselves only to that part of the specimen which projects from the surface of the grid. An area behind such a projection will receive no metal at all, hence a shadow will be cast. From the discussion on electron density, this means that a raised part of the specimen with attached metal will have increased ability to scatter electrons and will appear dark on the screen, or the final print, while the metal-free shadow will appear

Fig. 81.1 Shadow-cast preparation of adenovirus particles. The particles have been coated with platinum-carbon. Both the geometric form of the particles and the arrangement of subunits are clearly illustrated. ×250 000. (Micrograph kindly supplied by Dr. M. V. Nermut, Nat. Inst. Med. Research, Mill Hill, London).

white. However, the human eye does not like to see black three-dimensional objects casting light shadows, so the very simple step of a photographic reversal was introduced; the observer should always be aware of this hidden step in understanding the mechanism of this method. The limiting factor of this technique is the granularity of the metal used to coat the particles. Platinum-carbon is the most widely used coating at present, and with this reagent it is possible to resolve down to the level of the morphological subunit (Fig. 81.1) of cubic viruses (Roseto *et al* 1979, Nermut 1980).

The greatest advantage of the shadow casting technique is that it allows the transmission EM to be used to investigate the third dimension of a virus (Hayat 1972). Scanning electron microscopy also does this, but the resolution of this machine at present does not allow it to be used at the level of virus substructure. In the study of virus morphology, shadow casting has yielded two variant techniques that have contributed significantly to the general understanding of the subject. The first of these was described in 1958 by Williams and Smith, who shadow-cast a large insect virus, Tipula iridescent virus, not from one but from two different point sources. This showed that the shape of shadow cast by a particle was dependent on the angle of the particle to the source of the metal. Examination of the different shapes of shadow showed clearly that, although in outline Tipula iridescent virus appeared hexagonal, in three dimensions it had the morphology of an icosahedron. This is the first occasion that EM findings implicated the icosahedron in virus morphology.

The second development of the shadow casting technique is the one that was introduced by Kleinschmidt and his associates in 1962 and which is known as rotary shadow casting. For this method the metal is evaporated from a single stationary point but the specimen grid is slowly rotated on a turntable. As every aspect of the specimen is equally coated with metal there is no pronounced shadow formation but raised structures, on account of their angle to the source, receive more metal coating than the background and hence become more electron dense. Because of photographic reversal these objects appear light against a darker background. The major use of this method is the visualization of nucleic acid; there is no other EM technique which enables this to be done satisfactorily. It is this technique alone that has yielded the information now available on the morphology of viral nucleic acids.

(For Cryo-electronmicroscopy, see Adrian *et al.* 1984.)

Thin sectioning

The second means by which viruses can be examined in the EM is that of thin sectioning. This is perhaps an obvious extension of the established paraffin-wax embedded microtome-cut section used for histological and histopathological studies. However, before the same concept could be applied to electron microscopy, several difficult technical problems had to be solved. Basically these were concerned with the preservation of the cells; (a) light-microscope fixatives are of little value at the EM level, and (b) the difficulty of cutting sections thin enough to yield reasonable resolution in the EM. The first problem has been solved by finding that glutaraldehyde is a good basic EM fixative; and that osmium tetroxide is an excellent secondary fixative which, in addition, adds the heavy metal osmium to tissue components and hence not only fixes but also stains cellular components. In order to understand the second group of problems, i.e. the need to produce very thin sections, it is necessary to consider again the mechanism by which the EM image is formed. The beam passes through the specimen forming an image by means of electron scatter; the thicker the specimen the greater the degree of scatter, both primary and secondary, the level of resolution dropping with increasing scatter. There is a simple axiom stating that the resolution of the EM can never be better than one-tenth the thickness of the object being studied (Chapman 1980). The thinnest sections that can be produced for the light microscope are in the 2 μm range. One-tenth of this would yield a resolving power of around 200 nm, a very long way from the 20 nm resolving power available. The use of epoxyresins, ultramicrotomes, and diamond knives has now made it possible to produce thin sections in the range of 2000 nm with resulting resolution in the 200 nm range. However, even with these refinements the problem remains that the full resolving power of the electron microscope cannot be used when thin sectioning is used as a preparatory technique. This method is thus limited to the study of relations between virus and host cell (Fig. 81.2), although within this limitation a wealth of information has been obtained (Dalton and Haguenau 1973). The method has also been extended through the use of ferritin or enzyme-tagged antibody to the study of antigens both on the virus and in the infected cell.

Negative staining

The techniques discussed so far have altered the electron density of the virus particle by incorporating the

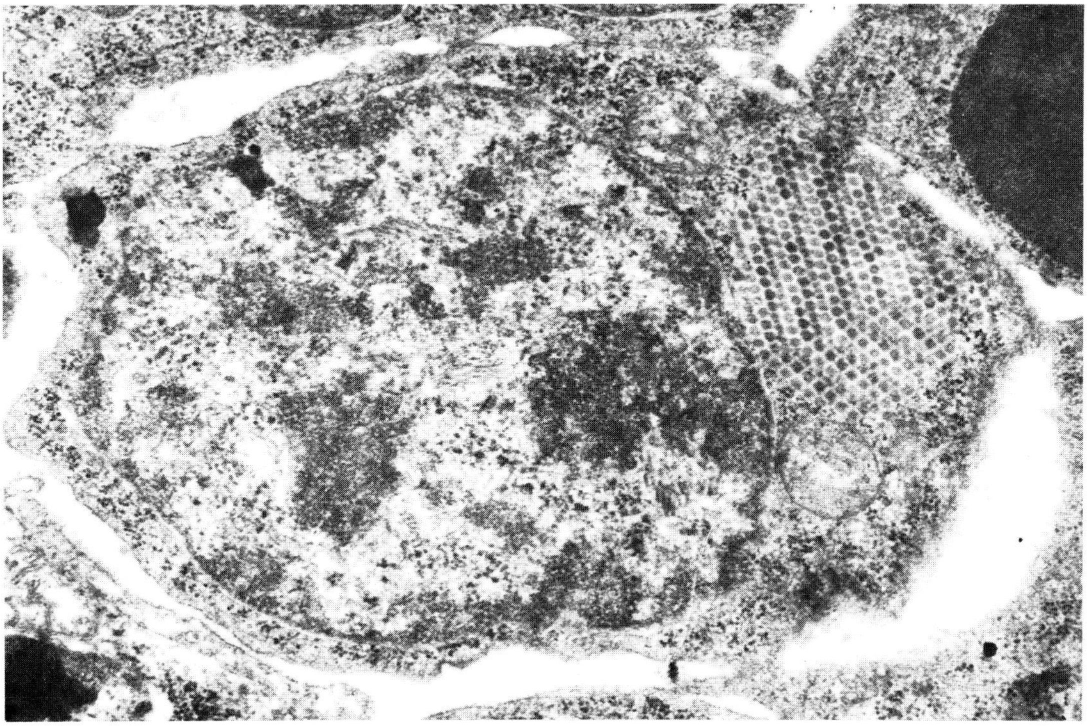

Fig. 81.2 Thin sectioning allows virus to be visualized in relation to the host cell. The cell shown here has a crystalline arrangement, virus particles lying to the right of the nucleus. The tissue is chicken bursa and the virus is that of infectious bursal disease. × 30 000.

heavy metal stain either on or in the virus. Negative staining, as its name suggests, reverses this by leaving the virus alone and staining the background space around the particles. This is achieved by mixing the virus suspension with an electron-dense stain, which is nearly always a compound of tungsten or uranium. When placed on a microscope grid the stain dries to an amorphous film with electron-transparent 'holes' that are occupied by the virus. Expressed differently, it is a displacement technique with organic material forming areas of lucency in a film that would otherwise be uniformly electron dense. The advantages of the method are numerous. Because the virus needs no other preparatory step than mixing it with the stain, it is rapid. Similarly, since it is examined in an unaltered state, only the size of the virus itself is implicated, and at least with the smaller viruses resolution approaching that of the microscope can be obtained. By extension, when virus components rather than the whole particle are used, even better resolution can be achieved. The nature of the negative stain is such that it is able to flow into the surface structure of the particles allowing the morphology to be studied in detail. In addition, because of the distinctive virus morphology revealed by negative staining, there is no need for the particles to be purified from contaminating cell material.

If a shadow casting technique provided the first visual proof that the icosahedron was implicated in virus structure, then negative staining offered the first visual evidence that the icosahedron was formed from subunits (Brenner and Horne 1959). From this original work it was realized that by vizualising the substructure of a virus it was possible to place it within a small number of morphological groups. These groupings supplied information not only on the symmetrical properties of the viruses but, by analogy with other perhaps better characterized members of the group, additional information on their biochemical nature and behaviour. However, it should be stressed that negative staining delineates only the outer covering of an intact virus; it may or may not yield information on the internal components if and when they are exposed to the stain.

Having considered the techniques available for the study of viruses in the EM we must now return to the era before these methods, particularly negative staining, became available. Plant viruses have always been more amenable to study than animal viruses, simply because of the greater ease with which they can be obtained. By the mid 1950s several plant viruses had been obtained in sufficient quantity and purity for them to be crystallized (Fig. 81.3). X-ray diffraction studies of these crystals showed the presence of two types of symmetry, cubic and helical (Klug and Caspar 1960). Comparison of these results with early electron micrographs showed that the individual particles forming the crystals with helical symmetry were rod shaped, whereas the particles forming cubic crystals were, at least superficially, round or isometric.

At almost the same time Crick and Watson (1956) published their fundamental article on the principles of virus structure. They pointed out that a virus would have insufficient nucleic acid to code for large structural proteins, and the only means by which economy and efficiency could be maintained would be by the

Fig. 81.3 Hexagonal bipyramidal crystals of cowpea mosaic virus. The symmetry shown by the crystals extends to the individual virus particle. (Photograph kindly supplied by Dr. J. E. Johnson, Purdue University, West Lafayette, Indiana, USA).

use of small identical protein subunits assembled in such a way that the environment of one subunit would be identical with any other. A corollary of this was that, just as the subunits would be identical and identically arranged, so also the virus particles would appear as identical structures. They next suggested that the symmetry obtained by x-ray diffraction studies would be extended from the level of the crystal to the individual particle. Not only would a crystal of tobacco mosaic virus (TMV) have helical symmetry, but the individual rod-shaped particles of which it was formed would likewise have subunits displaying helical symmetry. Similarly, the isometric particles forming crystals of cubic symmetry would be found, not to be truly round, but to be formed of subunits arranged according to the dictates of cubic symmetry. A further prediction was that, again through lack of nucleic acid, the only types of virus construction that could exist would be either helical or cubic in type. The first corroboration that round viruses were not really spherical came from the previously described experiments of Williams and Smith (1958), who showed that double shadow casting of Tipula iridescent virus revealed an icosahedral form.

One year later in 1959, Brenner and Horne published their classic paper on the structure of adenovirus as revealed by negative staining. While shadow casting had shown a three dimensional icosahedron, negative staining showed both the facets of the same solid and also a geometric arrangement of subunits (see Fig. 81.7). The Crick and Watson model of a particle formed from small repeating subunits had been rendered apparent.

Once all this information had become available in the early 1960s, Caspar and Klug, in a series of articles which have remained as classics in the field, laid down the general rules by which viruses are constructed (Klug and Caspar 1960, Caspar and Klug 1962, Caspar 1965). The years since then have seen the description of many additional virus types, but the basic principles of construction have remained unaltered.

Symmetry

The word symmetry was first associated with virus structure through x-ray diffraction studies. The symmetry detected was produced by the mass action of virus either in a crystal or in an oriented gel. An extrapolation was made that, if the crystal had symmetrical properties, then the same symmetry must be present in the individual virus of which the crystal was formed. Negative staining not only confirmed but showed that helical and cubical were the only types of symmetry found in relation to viruses. Helical symmetry is well demonstrated by the everyday spiral staircase. If one admits that the building brick of the staircase is the individual step, then it is seen that there is a series of identical units identically arranged in relation to each other apart from those on the bottom and top turn of the staircase. If one hangs over the balustrade at the top of the staircase until the head is exactly in the middle of the central space (without too much personal risk), then one would be looking down the central axis of the helix. If it were now possible to rotate the staircase about this central axis it would be found that, as soon as the space occupied by one step was replaced by the next step, then the staircase would look exactly as it did before; in other words there is rotational axial symmetry, although were the turning to be continued through the number of steps contained in one revolution of the staircase, then the whole structure would have moved up or down by the distance between the levels of the staircase. This can be presented more mathematically by saying that the properties of a helix can be stated if one knows the number of identical subunits in one turn of the helix and the distance (pitch) between turns. As for tobacco mosaic virus it is not necessary to have an integral number of subunits in one turn of the helix. The helix displays two types of symmetry: the rotational one present in the subunits when they are viewed along the line of the central axis; and a translational one when the structure is viewed at right angles to this axis. Translational symmetry is the type that occurs in a step ladder. There are identical rungs to the ladder and these occupy identical positions on the ladder as one's eye progresses from top to bottom of the structure. Similarly, along the length of the viral helix it is possible to see the regular spacing associated with the pitch of the helix.

The term cubic symmetry applies to a group of solids all of which possess at least one set of four three-fold axes which have the same relation to each other as have the body diagonals of a cube. The regular solids that belong in this grouping are the tetrahedron, the cube itself, the octahedron, and the icosahedron. The symmetrical properties of these solids are described as the tetrahedron 2:3, the cube 4:3:2, the octahedron 4:3:2, and the icosahedron 5:3:2. Each of these numbers describes the axes that the solid possesses and refers to rotational symmetry. That is, possession of twofold symmetry means that one of the axes of the polyhedron allows the solid to be rotated through 180° and still look the same. Similarly, the threefold axis allows the solid to be rotated three times through 120° and still look the same. The 5:3:2 axes of the icosahedron are demonstrated in the accompanying diagram with the axes accentuated by symbols to show the number of identical points they possess (Fig. 81.4). As shown in this figure, the icosahedron is a regular polyhedron having 20 sides and 12 apices. X-ray diffraction studies of virus crystals had shown isometric viruses to have cubic symmetry; and early shadow cast and negative staining electron microscopy showed that isometric viruses had icosahedral form. Since these early results no animal virus with cubic symmetry has been found that does not also have icosahedral symmetry. It can then be asked what

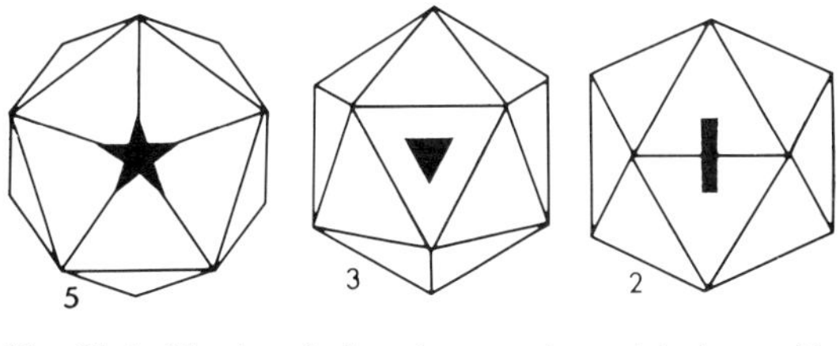

Fig. 81.4 The icosahedron is a regular polyhedron with twenty sides and twelve apices. It has rotational axes of fivefold, threefold and twofold symmetry. Here these three axes are marked by symbols which highlight the number of identical positions obtained during rotation.

special properties of the icosahedron make it the preferred form for this virus group. The answer appears to be a mathematical one, namely that the icosahedron allows the use of the greatest number of symmetrically arranged subunits in forming a shell.

Basic virus construction

Even although further divisions are possible, a virus particle is basically an inert two component system. The two components are the nucleic acid, which may be either deoxyribonucleic acid (DNA) or ribonucleic acid (RNA), and the protective covering. It is inert because it contains no energy-rich system which would allow a replicative cycle to be carried out. As far as the simple viruses are concerned the protective covering is protein alone; more complex animal viruses contain lipoprotein and glycoprotein. The protective protein of the simple viruses must satisfy three basic aspects of virus requirements. Firstly, it must be formed from small repeating subunits that will display the properties of either helical or cubic symmetry, depending on whether the particle is rod-shaped or isometric. Secondly, this protective covering must offer adequate protection to the nucleic acid when the virus is released into the environment. Thirdly, the structure must be so formed that it can be taken apart by a host cell in order to release the nucleic acid for the next replicative cycle. The protective protein layer of the virus is known as the *capsid*, and the subunits visible to the EM are known as *capsomeres* or morphological subunits. In the more complex animal viruses this helical or cubic capsid may have an additional outer membrane or envelope that contains cell-derived but virally altered lipoprotein and glycoprotein. The word *virion* refers to the intact biologically active virus.

The virus capsid, although produced by organisms that we consider to be at the bottom of the evolutionary scale, is probably the most effective and economical protective covering yet developed. It would be wrong in a chapter on virus structure not to mention the similarity of cubic virus construction to the geodesic domes of R. Buckminster Fuller, an illustrious contemporary architect (Marks 1960). These domes have been accepted architecturally as one of the most effective and economical ways of producing an enclosed space, yet their strategy is precisely that of the *Reoviridae* which have been using this approach for the last few million years.

However the helical plant viruses are probably the most straightforward of the morphological groups; these will now be considered in greater detail.

Helical viruses

It might be argued that the tobacco plant has done more good for mankind through the fact that it becomes infected with tobacco mosaic virus than harm through the habit of smoking. Tobacco mosaic virus (TMV), generic name tobamovirus, is the single virus that has contributed most to the understanding of virus structure and function. It has the distinction of being the first virus to be recognized, purified, yield infectious nucleic acid, and to be fully characterized morphologically and biochemically (Waterson and Wilkinson 1978; Jonathan *et al.* 1978). Much of this is due to the fact that the virus can be obtained in large amounts and be purified simply.

By negative staining, TMV appears as a rigid rod 300 nm long and 15 nm in diameter. There is a central hole, shown dark by negative staining, that is approximately 4 nm (40Å) in diameter; and a regular pitch suggesting an underlying helical arrangement (Finch 1964; Fig. 81.5(**a**)). However, the ordinary micrograph does not provide further information, and it is neces sary to turn to x-ray diffraction studies to gain further detail of the construction of the virus. By using both intact and RNA-free virus particles, the nucleic acid of the particle is seen to be in the form of a single strand that lies at a radius of 4 nm from the long axis and is arranged in a helix. The particle visualized by negative staining is indeed a helix that x-ray diffraction shows to have a pitch of 2.3 nm. It is built up of identical protein subunits which are approximately 7 nm long and 2.3 nm wide with an internal groove at a radius of 4 nm. This groove corresponds to the radius of the RNA strand and is the protected environment within which it lies (Fig. 81.5(**b**)). In one turn of the helix there are $16\frac{1}{3}$ subunits, so that an integral number, 49, is reached in every third turn. These subunits, which are in themselves asymmetric, are arranged identically in relation to each other, each being surrounded by six others except for those on the terminal turns of the helix which will have only five adjacent neighbours. Because the rod is rigid, it appears that the bonding between neighbouring subunits is identical in every direction; this identical bonding can also be described as equivalence. The simplicity of the bonding is illustrated by the fact that the basic helix can be polymerized *in vitro* either from the subunits present in infected cells or from degraded virus particles. Polymerization is not dependent on the presence

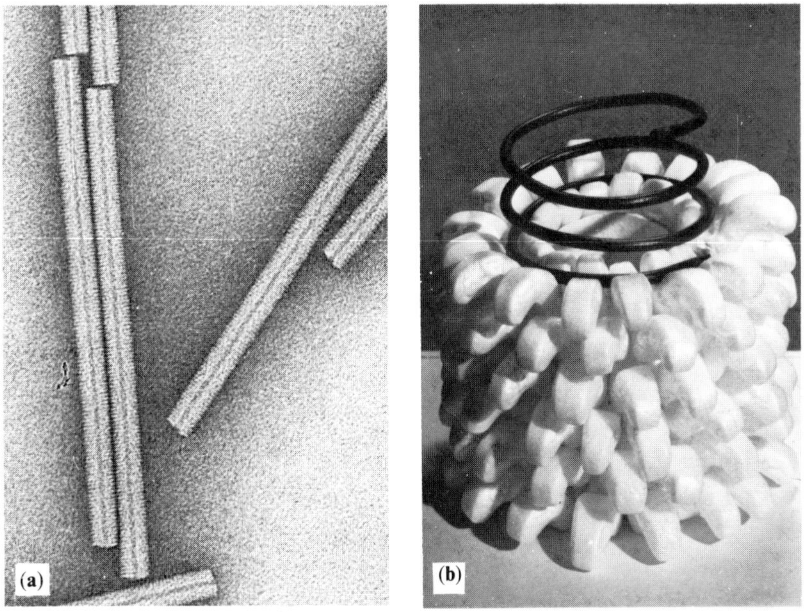

Fig. 81.5(a) Two complete and some partial rods of tobacco mosaic virus as seen by negative staining. The central axis of the particles and the 23A pitch are clearly resolved. ×216 000. (Micrograph kindly supplied by Dr. J. T. Finch, Lab. Mol. Biol., Cambridge, England). (**b**) A three-dimensional model of tobacco mosaic virus based on the information supplied by x-ray diffraction and electron microscopy. Asymmetric subunits protect the single helical strand of RNA. The environment of each subunit is identical with that of the others.

of the nucleic acid, although the TMV rod that contains nucleic acid is physically more stable than the one that does not. This shows that though the protein subunits both protect and stabilize the RNA, there is an effect in the opposite direction inasmuch as the presence of the nucleic acid produces a more stable arrangement of protein subunits. Tobacco mosaic virus is the classical example of the helical virus as predicted by Crick and Watson, namely with asymmetric but identical protein subunits, equivalently bonded, protecting a nucleic acid strand from the environment. However, consideration of other rod-shaped plant viruses shows that there are more viruses with variations on this classical theme than those possessing the original form. The genera represented by the closterovirus, carlavirus, potyvirus, and potexvirus groups all have rod shapes with diameters in the range 10–13 nm, but they all differ from TMV inasmuch as the rods, instead of being rigid structures, are flexuous. This presumes that the condition of equivalence can no longer be maintainted, since the bonding between subunits on different turns of the helix cannot be identical with that between subunits lying adjacent to each other on the horizontal plane. An indication of the rather looser bonding between turns of the helix in these viruses is shown by the fact that the pitch in these groups is 3.7 nm for closterovirus and 3.4 nm for all the others. This concept of strict equivalence being replaced by quasi-equivalence is a recurrent one and in animal viruses having helical symmetry is the predominant form.

However, more important than the substitution of strict equivalence by quasi-equivalence is the fact that the simple arrangement of nucleic acid and protein found in the helical plant viruses does not exist in animal viruses. Animal viruses with helical symmetry always have an additional lipoprotein envelope. The internal nucleic acid protein helix is referred to as ribonucleoprotein or RNP. It may be that the additional envelope component is a result of the evolution of the immune system in the animal kingdom. The simple helical virus can survive in the presence of the primitive defence mechanisms of the plant but not in that of the immunologically sophisticated animal. The major group of these animal viruses is represented by the *Paramyxoviridae*. Morphologically, the RNP of animal viruses with helical symmetry closely resembles the flexuous plant viruses, although the diameter of the helix is somewhat wider at 18 nm (Fig. 81.6(**a**) and (**b**)). In a well resolved micrograph of a disrupted helix it is possible to visualize a cog wheel appearance in single turns of the helix. This is a direct visualization of the structural subunit. It can be presumed that the theoretical structure postulated for viruses such as TMV will be closely similar to that present in this virus group. However, even greater variation exists, and although the *Orthomyxoviridae* can be seen to have an RNP with an obvious helical conformation the helix

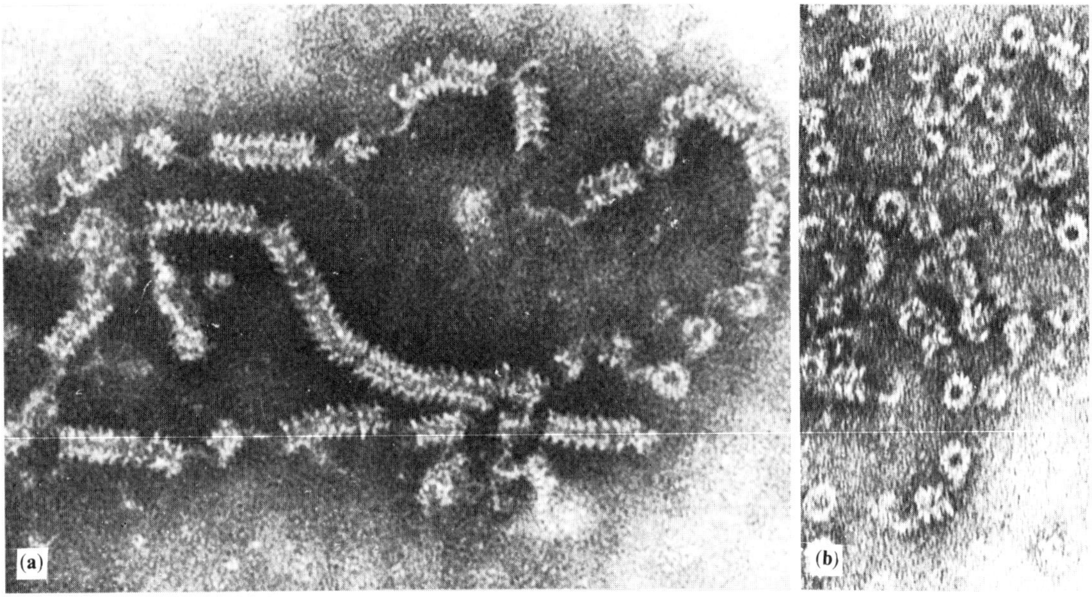

Fig. 81.6(a) Ribonucleoprotein (RNP) of parainfluenza virus. The single helical structure has a diameter of 18 nm, and displays a distinctive herring-bone appearance. At various points the strand can be seen unwinding. ×275 000. **(b)** A fortuitous micrograph of an RNP helix showing several single turns that are lying flat on the grid. It is just possible to see the cog-wheel pattern which is the visualization of individual structural subunits. ×295 000.

is dissimilar to that of the plant viruses and the exact arrangement of RNA and protein subunits has not yet been established. One important point about animal helical viruses is that, invariably, their nucleic acid is RNA.

Cubic viruses

As stated above the shell of a cubic virus is known as the capsid and the visible subunits forming this capsid are the capsomeres or morphological subunits. However, although negative staining allows direct visualization of the basic structural subunit forming helical viruses, this is not usually true for cubic viruses. The subunits which are visualized by the EM on cubic viruses are not identical subunits, nor are they identically arranged. If one considers the location of subunits forming the icosahedral shell, it is obvious that those at the apices cannot be identical with those at other sites on the surface. Each apical subunit, because it is on an axis of five-fold symmetry, must, of necessity, have five other subunits around it. The remaining subunits are surrounded by six others. Since it is important at the EM level to recognize the difference between these two types of subunit, they have been designated as *pentons* on the apices and *hexons* elsewhere. Figure 81.7 shows both the icosahedral form of an adenovirus and the clearly distinguishable capsomeres, both hexons and pentons. Consideration of the cubic viral capsid makes it obvious that if there were only hexons there would be a planar arrangement with no ability to 'turn the corners'; to make the three dimensional structure, the second type of subunit, the penton, must be introduced. However, if we now assume that there is a level of construction below that of the *morphological subunit*, we come to the concept of

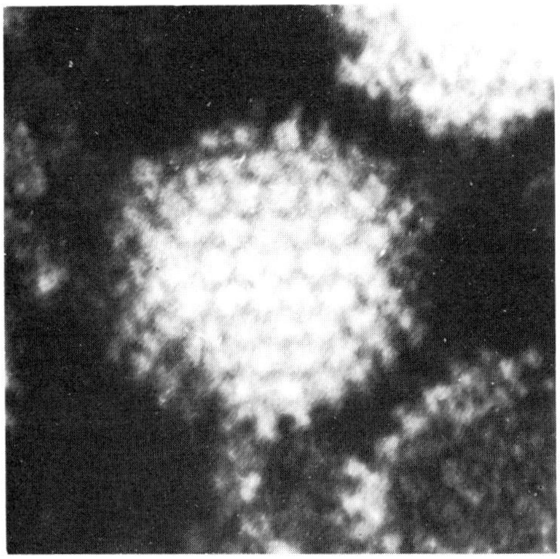

Fig. 81.7 One individual adenovirus particle showing the icosahedral form and subunit arrangement of this virus type. The particle is viewed along the two-fold axis and the six subunits between the two adjacent apices are clearly resolved. ×500 000 (see Fig. 81.4).

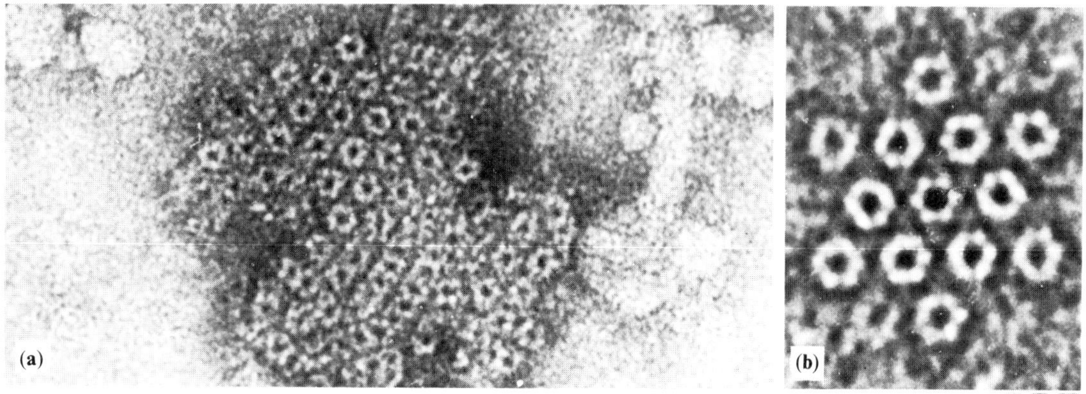

Fig. 81.8(a) An array of capsomeres from a disrupted herpes virus capsid. The upper part of the micrograph contains only a single layer of capsomeres and this has allowed high resolution to be obtained. Each capsomere in this area shows a hexagonal outline, and it is possible to see that the capsomeres are built up from structural subunits. ×400 000. (**b**) By use of a photographic rotational technique, it is possible to reinforce the structural subunit pattern shown in Fig. 81.8(**a**). This pattern should now be related to Fig. 81.9(**a**). ×800 000.

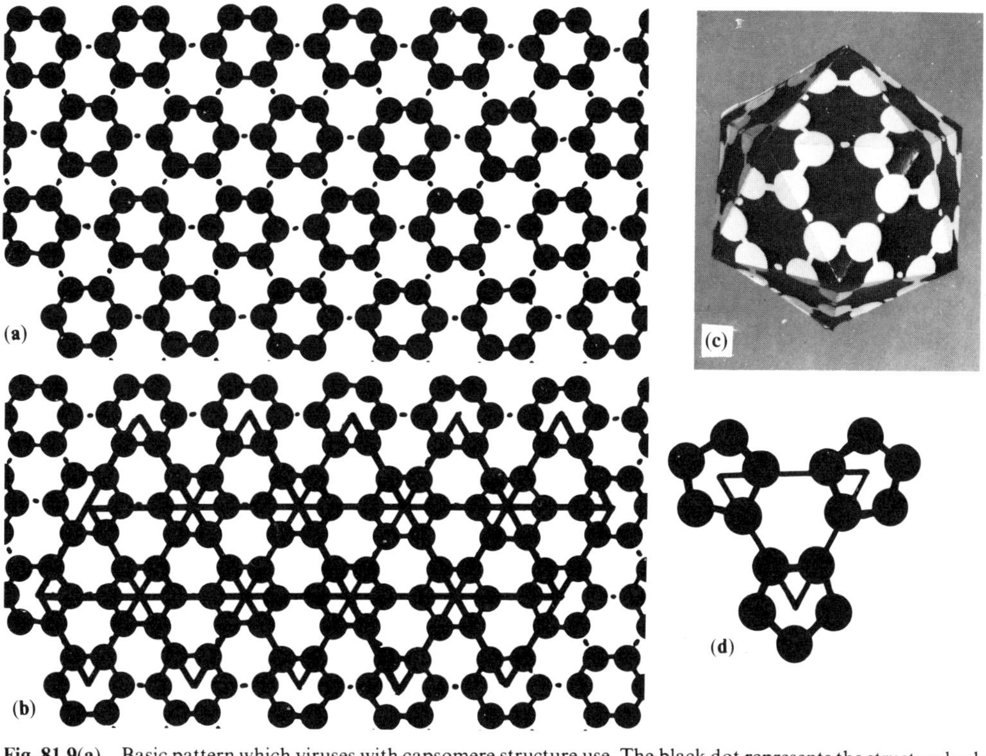

Fig. 81.9(a) Basic pattern which viruses with capsomere structure use. The black dot represents the structural subunit which, in groups of six, form the morphological subunit. Each structural subunit is shown as having two strong bonds within the morphological unit and one weak bond between units. (**b**) When the net for an icosahedron is imposed on the pattern shown in Fig. 81.9(**a**) various arrangements of subunits can be obtained. The simplest is that shown here where the side of the icosahedron occupies the distance between two adjacent morphological subunits. When this pattern is cut out, it can be seen that all of the original six-fold subunits are converted to five-fold units and the resulting model consists of twelve five-fold capsomeres or pentons. (**c**) The model built from the two-dimensional pattern shown in Fig. 81.9(**b**). In order to simulate negatively stained virus appearance the pattern has been photographically reversed. Twelve morphological subunits built from 60 structural subunits form the icosahedral structure. (**d**) The pattern on one face of this icosahedron shows that there are six half units located within the triangle. Since this is the simplest possible arrangement and cannot be subdivided, it is described as having a triangulation number of one. (T = 1).

a *structural subunit* that approaches the ideal of a single entity arranged in an equivalent fashion. If every penton is built up from five, and every hexon from six, of these structural subunits, then we will have a structure that satisfies a theoretical construction plan for cubic viruses. Although the concept of a structural subunit was put forward on theoretical grounds, the occasional fortuitous micrograph extends further than the morphological subunit and allows direct visualization of this entity. Figure 81.8(**a**) shows a disrupted herpes virus capsid so spread out that only one layer of subunits is seen. A photographic reinforcement technique (Fig. 81.8(**b**)) allows the detail to be made even clearer and it is seen that the geometrically arranged rows of morphological units are made up of six underlying structural subunits. In this micrograph only hexons are present. From micrographs such as that it is possible to draw up a representation of the basic arrangement of structural and morphological subunits (Fig. 81.9(**a**)). If the net for an icosahedron is now imposed on this basic pattern, it is seen how a three-dimensional solid can be built from an arrangement of standard subunits identically arranged in relation to each other. The accompanying figure shows this for the simplest possible cubic virus, that with 12 subunits (Fig. 81.9(**b**,**c**)).

However, though the majority of cubic viruses display these two levels of morphological subunits, or capsomeres, with underlying structural subunits, recent findings suggest that at least one cubic virus group does not have morphological subunits at all but uses only structural subunits, which are all identical and have an almost equivalent arrangement. This group is the *Reoviridae* and Fig. 81.10(**a**) shows a disrupted rotavirus capsid which allows the arrangement of subunits to be determined. Unlike the herpes micrograph, there is no suggestion of capsomeres but, instead, there is a continuous lattice arrangement of subunits. Examination of fragments of this lattice shows linked groupings of subunits clustered into arrangements of five or six. These fragments further reveal the relationship of the subunits within these clusters, so that it is possible to draw the basic hexagonal tessellation (Fig. 81.10(**b**)) and once again impose on it the net for an icosahedron (Fig. 81.11). The two patterns shown in Figs 81.9(**a**) and 81.10(**b**) are

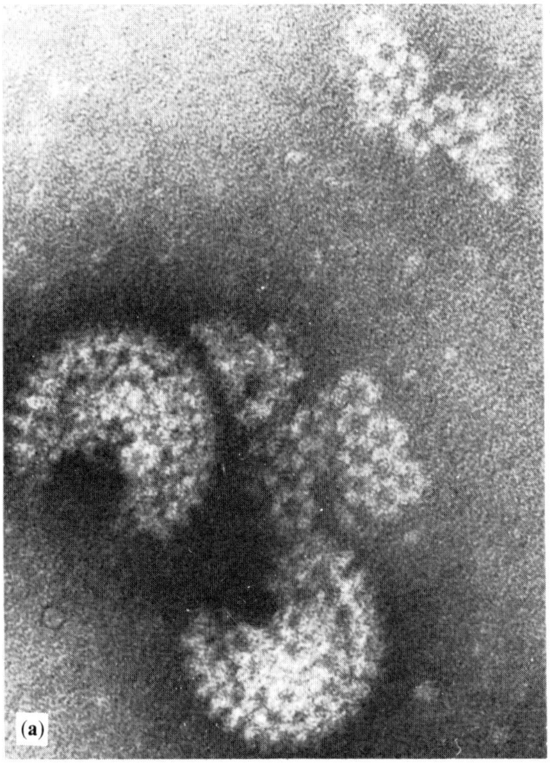

Fig. 81.10(a) Rotavirus is based on a lattice of structural subunits rather than on capsomere construction. Here a disrupted rotavirus particle allows the arrangement of this lattice to be seen. Light structural subunits cluster in groups of five or six round a central dark space. ×400 000.

Fig. 81.10(b) The basic arrangement of structural subunits as seen in Fig. 81.10(**a**). Each subunit takes part in three adjacent clusters of six and, since this illustration is planar, only six-fold clusters are present.

Fig. 81.11 Imposition of an icosahedral net on the basic lattice shows how a three-dimensional structure can be formed. In this diagram there would be four structural subunits on each triangular facet and three holes or cluster centres between apices.

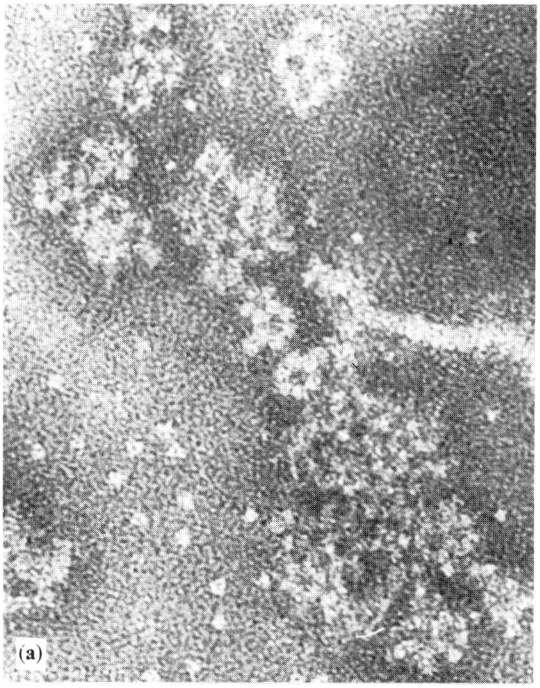

Fig. 81.12(a) Rotavirus lattice, on breakdown, gives rise to individual structural subunits. ×400 000.

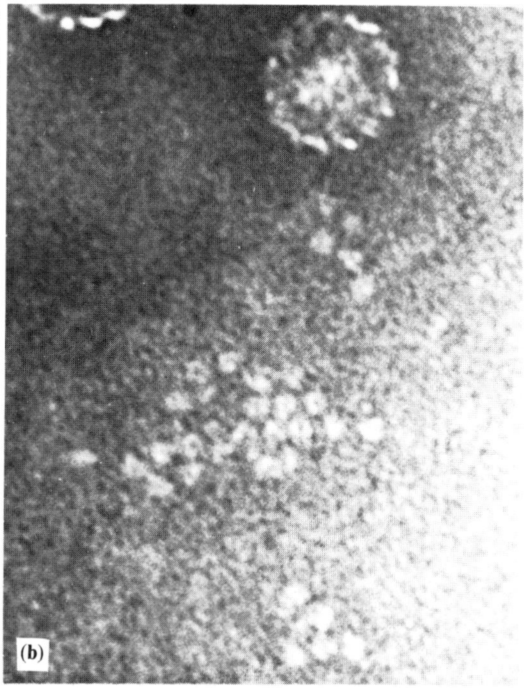

Fig. 81.12(b) Virus with capsomere construction, represented here by wart virus, on breakdown, gives rise to individual morphological subunits. ×400 000.

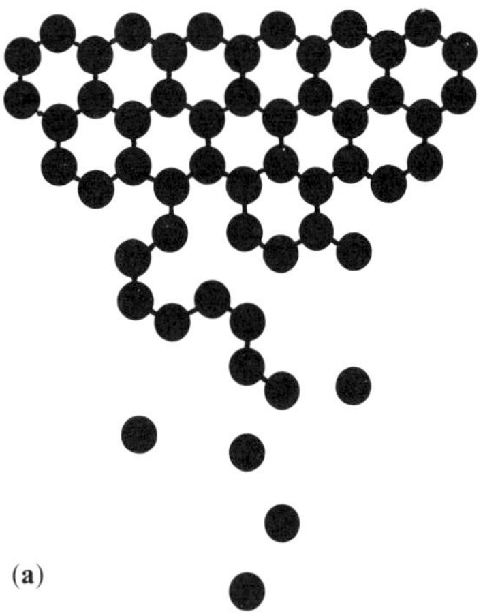

(a)

Fig. 81.13(a) A diagrammatic representation of Fig. 81.12(a).

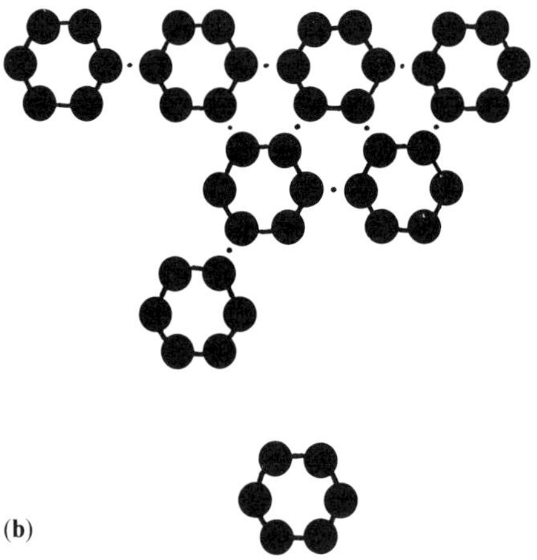

(b)

Fig. 81.13(b) A similar diagram for Fig. 81.12(b).

the only arrangements of subunits that have been found for cubic viruses. An interesting difference between them is that, when disrupted, the construction typified by rotaviruses yields individual structural units, whereas the other type of construction breaks down to give capsomeres (Figs 81.12 and 81.13).

If we now consider these two types of cubic virus construction from the viewpoint of the spatial relation between subunits and underlying icosahedral symmetry we find that the *Reoviridae* constitute a single example of a straightforward arrangement adhering to the rules of cubic symmetry. Cubic viruses using capsomere construction, on the other hand, have several different means of aligning their subunits in relation to the underlying icosahedral structure. As the *Reoviridae*, typified by rotavirus, furnish a fairly simple introduction to this problem of subunit orientation, they will be discussed first.

Examination of the disrupted rotavirus and the corresponding lattice in Fig. 81.10 shows that the pattern is based on the clustering of five or six subunits around a central dark space. Each subunit is identically bonded to three others, no matter what location it occupies on the capsid. However, because of the angular nature of the icosahedron it is obvious that there must be a change in the angle of bonding of those subunits that take part in a grouping of five as well as adjacent sixes. In spite of this, the concept of identical subunits identically arranged is approached. Examination of micrographs also shows that the subunits are aligned along the edges of the icosahedral facets. Figure 81.14 shows both a rotavirus and a model icosahedron that corresponds most closely to the substructure present. This model has five holes or centres

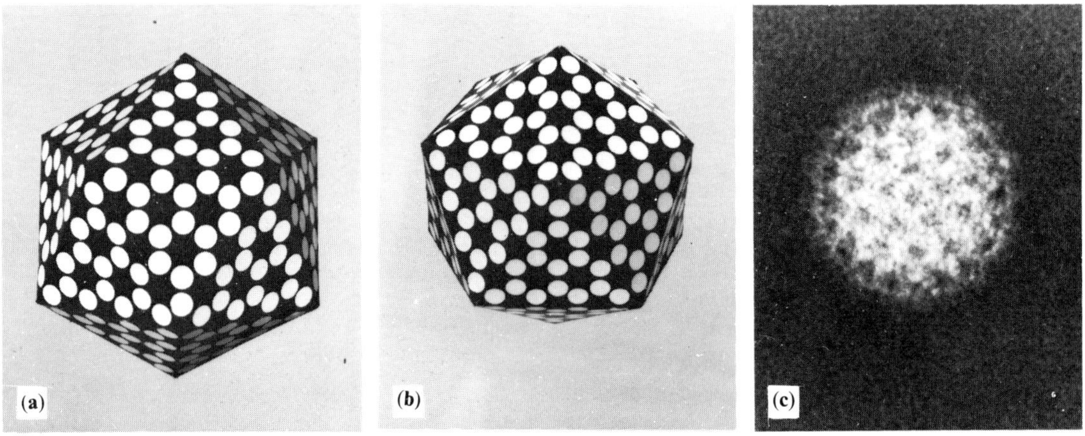

(a) (b) (c)

Fig. 81.14(a) A model showing the number of structural subunits that agrees best with micrographs of rotavirus. There are five centres or holes between apices and the subunit number is 320. The model is viewed along the three-fold axis. (b) The same model viewed along the axis of five-fold symmetry for comparison with the micrograph shown in (c). (c) A single rotavirus particle viewed along the five-fold axis. Although the individual structural subunits are not resolved, the arrangement of dark centres approximates well to the model in (b). ×400 000.

between adjacent apices of the icosahedron. Knowing this, it is possible to employ a simple formula giving the total number of subunits forming the capsid. This formula is $20(n-1)^2$ where n represents the number of centres between two adjacent apices; with the 5-hole model the number of structural subunits is 320 (Esparza and Gil 1978). It has been suggested by these authors that the subunits concerned with *Reoviridae* structure are themselves trimers that can be further broken down, and this may indeed be so. The rotavirus model described is the simplest form of cubic virus so far characterized. The requirements of equivalence are almost met and the relation of the subunits to the underlying icosahedron is straightforward. Although only rotavirus with its proposed 320 subunits has actually been visualized, it is possible to see how a series with the same basic arrangement could be built up. The formula $20(n-1)^2$ gives a series of capsids of 20, 80, 180 and 320 subunits. Figure 81.15 shows one triangular facet of each of this series imposed on the basic hexagonal net.

If we now turn to the other type of cubic virus construction, that with subunits at both the morphological and structural level, the situation is more complex. Pictures of disrupted virions show that the hexons and pentons are stable structures that retain their integrity. Bonding is clearly quasi-equivalent rather than equivalent, as the intracapsomere bonding appears more stable than the intercapsomere one. So far, three different arrangements of capsomeres with their underlying structural units relative to the underlying icosahedral structure have been described. These are: first, a series of increasing number as the number of capsomeres along the edge of the triangular facet is increased; second, a series that varies as the number of hexons placed centrally in the triangular facet is increased; and third a skew arrangement where there is a more complex relationship between the triangular facets and the subunits. Each of these arrangements belongs to a different icosahedral class; these are given P numbers which are as follows: edge-oriented arrangement, P = 1; face centred, P = 3; skew, P = 7. It should be pointed out here that the models illustrated are true icosahedra; that is, they have both icosahedral symmetry and form. However, viruses with a smaller subunit number, while bearing icosahedral symmetry, do not have definite icosahedral form and appear approximately spherical in the EM. Only when there are larger subunit numbers as in adenovirus does a distinctly icosahedral form appear.

Fig. 81.15 The basic lattice associated with rotavirus construction can give rise to a series of icosahedra having increasing subunit number. Here the arrangement on one triangular facet of a series that would have 20,80,180 and 320 subunits on the whole icosahedron is shown.

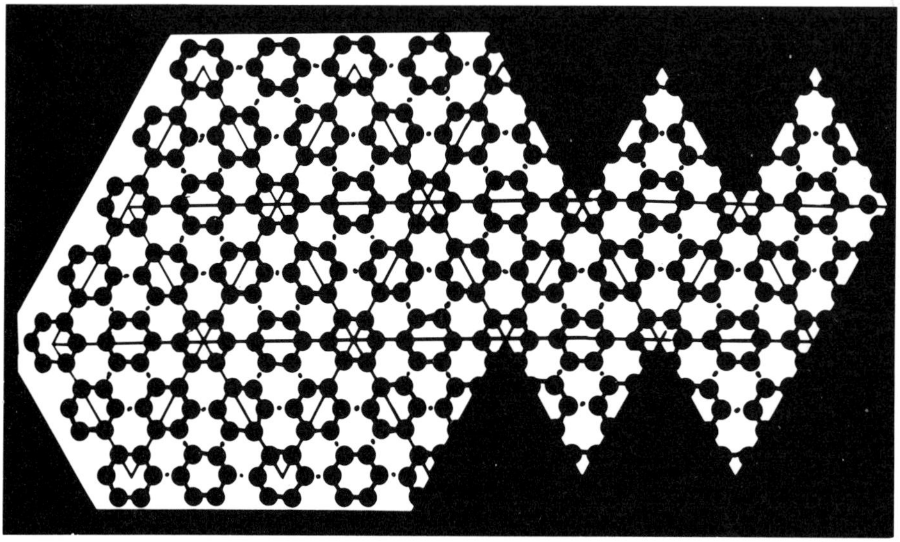

Fig. 81.16 For the P = 1 icosahedral class, subunit number increases by the addition of morphological subunits along the edges of the solid. Here the pattern for a 42 subunit structure is shown. In order to build the model the net is cut out as shown here and folded at the lines.

Starting with icosahedral class P = 1, the simplest possible arrangement for this type of construction is one in which there are twelve apical pentons and no hexons at all. This structure is the one found in the bacteriophage $\phi \times 174$. A model of this structure is shown in Fig. 81.9(**c**). One hexon inserted between the pentons on the edge of the triangular facet gives rise to a structure with 42 morphological subunits or capsomeres, 12 pentons and 30 hexons (Fig. 81.16). Additional members of this series can be calculated from the formula $10(n-1)^2+2$, where n is the number of subunits between two adjacent five-fold axes. This brings us to the next important concept in the understanding of cubic virus structure, that of triangulation number. When a series such as the one just described is built up, the pattern can be seen either as an increase in subunit number or as a multiple of the pattern displayed by the simplest member of the series. For the P = 1 series the simplest member is the twelve subunit bacteriophage $\phi \times 174$. Examination of the model of this virus illustrated in Fig. 81.9(**c**) and the drawing in Fig. 81.9(**d**) shows that one triangular facet of this structure contains six half subunits. Since this is the simplest possible arrangement for the P = 1 series, it is said to have a triangulation number of one, T = 1. If one now considers the next possible member of this series, a capsid having 42 subunits, it is possible to subdivide the face into four triangles, each of which contains the arrangement present in the simplest member of the series. Hence the 42 subunit structure is said to have a triangulation number of 4, or T = 4 (Fig. 81.17). By use of this T number it is possible to calculate the number of morphological subunits (M) by the formula $M = 10T + 2$, and the number of structural subunits (S) by the formula $S = 60T$. The triangulation number itself is calculated from the formula $T = Pf^2$, where f is any number in the series 1,2,3,4---------, and P is the icosahedral class. A corollary of this is that with every icosahedral class, for the first member of the series the triangulation number will be the same as the class number because f = 1. Returning to the P = 1 class, the T numbers of the series will be 1,4,9, 16,25. From the formula $M = 10T + 2$, the number of morphological subunits of this series will be 12,42, 92,162,252. Of these, 12,162 and 252 are known to be associated with virus structure. The group of viruses in the P = 1 series is fairly simple, because the triangulation pattern coincides with the triangular facets of the icosahedron itself. Figure 81.18 shows the basic arrangement, and Fig. 81.19 the triangulation pattern for the first member of the P = 3 series. From a visual viewpoint, the distinctive feature of this series is that the subunits are added to the centre of the icosahedral facet rather than along the edge. From the triangulation pattern viewpoint, each of the subdividing triangles is dissected by the icosahedral edge and, as can be seen from Fig. 81.19, T = 3. If the formula, $M = 10T + 2$ is applied the number of subunits is given as 32. The

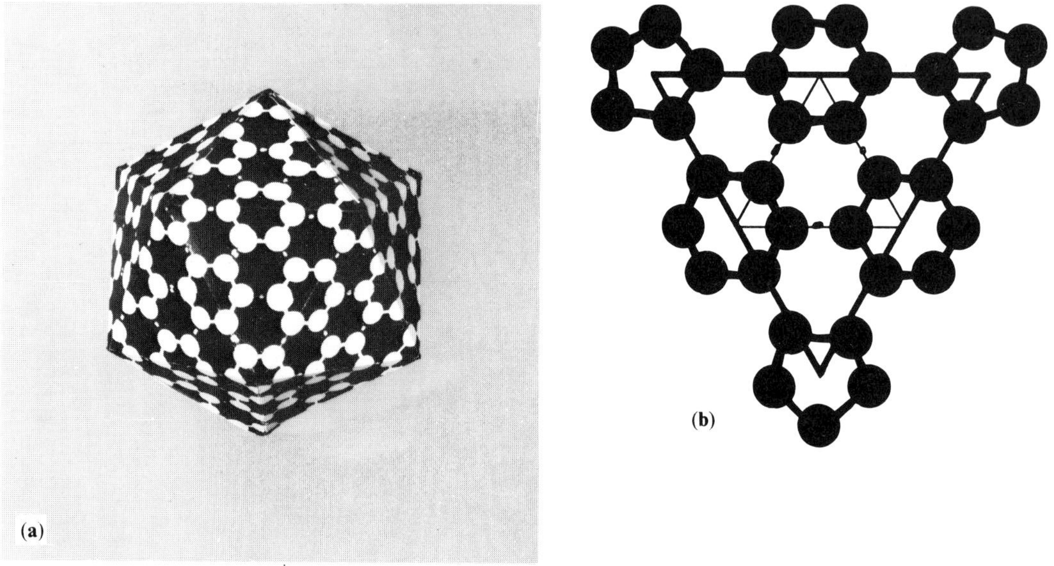

Fig. 81.17(a) The completed model from the arrangement shown in Fig. 81.16. There are twelve apical pentons and 30 hexons. (**b**) The pattern on one face of the 42 subunit arrangement. It is possible to subdivide the major triangle into four minor triangles each of which displays the basic arrangement shown in Fig. 81.9(**d**); therefore, for the 42 subunit construction T = 4.

Fig. 81.18 For the P = 3 series the number of subunits is increased by the addition of face centred hexons. The two-dimensional arrangement for the first member of this series M = 32 is shown here.

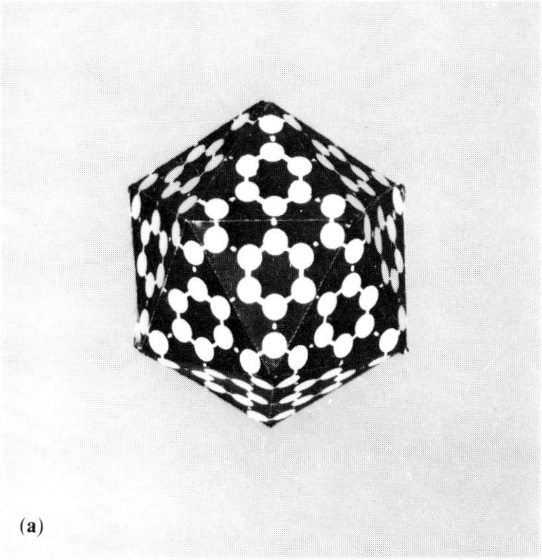

Fig. 81.19(a) The three-dimensional construction from the arrangement shown in Fig. 81.18. The model is viewed along the three-fold axis and shows a distinctive, one surrounded by six, pattern (see Fig. 81.30).

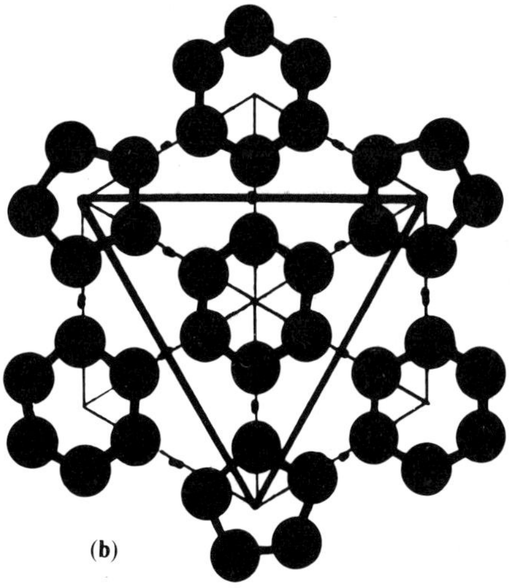

Fig. 81.19(b) One facet of the 32 subunit model showing the triangulation pattern. The minor triangles are subdivided by the edges of the major triangle so that one face contains six half triangles, making T = 3.

type virus displaying this morphology is turnip yellow mosaic virus. The next member of this series would be M = 122, but no virus with this arrangement has been found. The last class of icosahedron implicated in virus structure is P = 7. This class has the distinction of existing both in the laevo and dextro forms (Fig. 81.20). Examination of Fig. 81.21 shows that the relation of the triangulation pattern to the icosahedral face is complex, and that the progression from one penton to the next is mediated by two hexons forward and one to the side. The illustration shows the left-handed form and, since it is the first member of this series, T = 7 and M = 72. Human wart virus is based on the 72d arrangement; it is proposed that the rabbit papilloma virus is 72l.

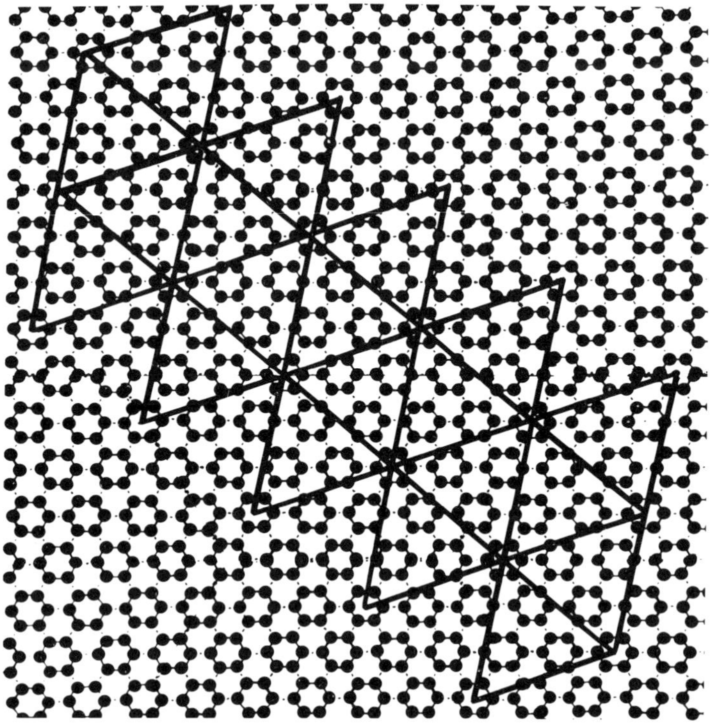

Fig. 81.20 This figure shows the arrangement for the P = 7, M = 72 icosahedron. Since this is an asymmetric, skew arrangement it exists both in the dextro and the laevo form. The illustration here is the laevo arrangement. A photocopy of this and similar figures would enable the reader to construct his own three dimensional models.

Fig. 81.21(a) The three dimensional model constructed from the arrangement shown in Fig. 81.20, viewed along the three-fold axis. There is no direct line of capsomeres from one apex to the next; the relationship between pentons is two hexons forward and then one to the left.

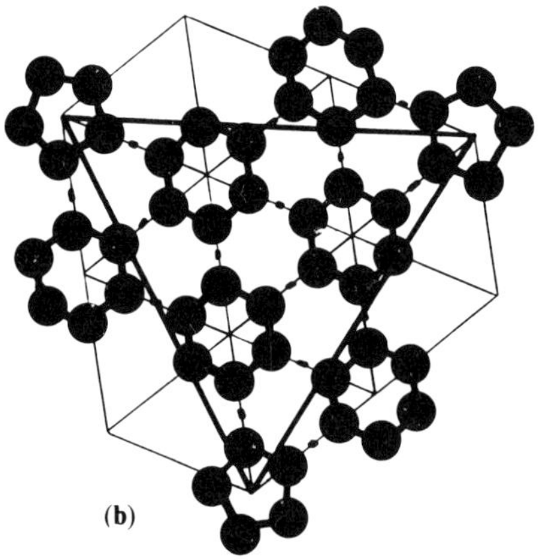

Fig. 81.21(b) For the 72 capsomere model, T = 7, but, because the subunits are not directly aligned with the icosahedral facet, there is a complex relation between these and the minor triangles.

If one turns from the theoretical to the practical side of virus substructure, in order to determine the geometry of any virus it is necessary to visualize the subunit arrangement on one facet of the icosahedral capsid or at least between two adjacent five-fold axes. This is not always a simple matter. Figure 81.22 shows two adenovirus particles. The lower particle has capsomeres that are clearly delineated, while the other, because of superimposition effects, shows very poor capsomere resolution, rendering it impossible to establish the nature of the subunit construction. The magnitude of this superimposition problem seems to vary with the particular virus being studied. Adenovirus preparations always contain numerous particles displaying good subunit arrangement, but smaller viruses such as the human common wart very rarely do, the majority of patterns being superimposition ones. Another problem is concerned with deciding which icosahedral class is implicated in a particular virus structure. Fortunately, the majority of viruses belong to the P = 1 category, which gives the number of morphological subunits by the simple $10(n-1)^2+2$ formula. This series is also easy to recognize, as it is the only one with capsomeres arranged along the edges of the icosahedral facets. In turn, this means that it is the only series where the structures actually look like an icosahedron, as the subunits outline the underlying form. The P = 3 class can be recognized by the presence of a subunit in the centre of the icosahedral face, and only the first member of this series, that with 32 subunits, has so far been recognized. The skew class P = 7 can be recognized in a negative fashion by the lack of direct linking capsomeres between adjacent five-fold axes.

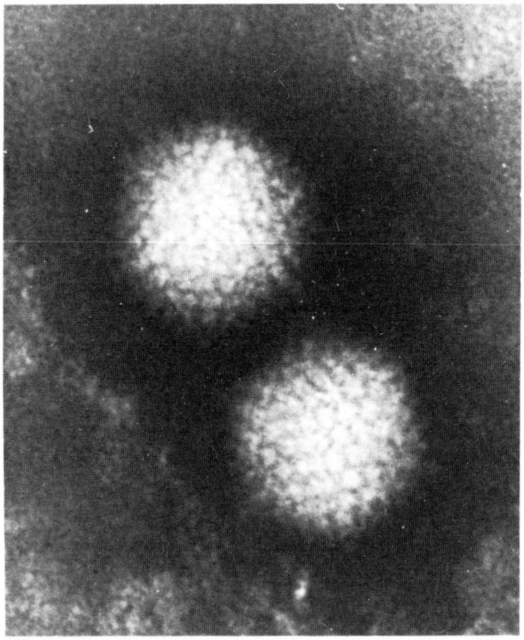

Fig. 81.22 Since the EM is a transmission instrument, superimposition is a constant problem in elucidating substructure. Both upper and lower surfaces of a virus are projected on to the screen and the ability to interpret structure is greatly dependent on the orientation of the particle. This micrograph shows two adjacent adenovirus particles, the lower one shows good subunit resolution, but the upper one is so arranged that it would be impossible to define the dimensions or arrangement of the subunits, although icosahedral facets are distinguishable × 350 000.

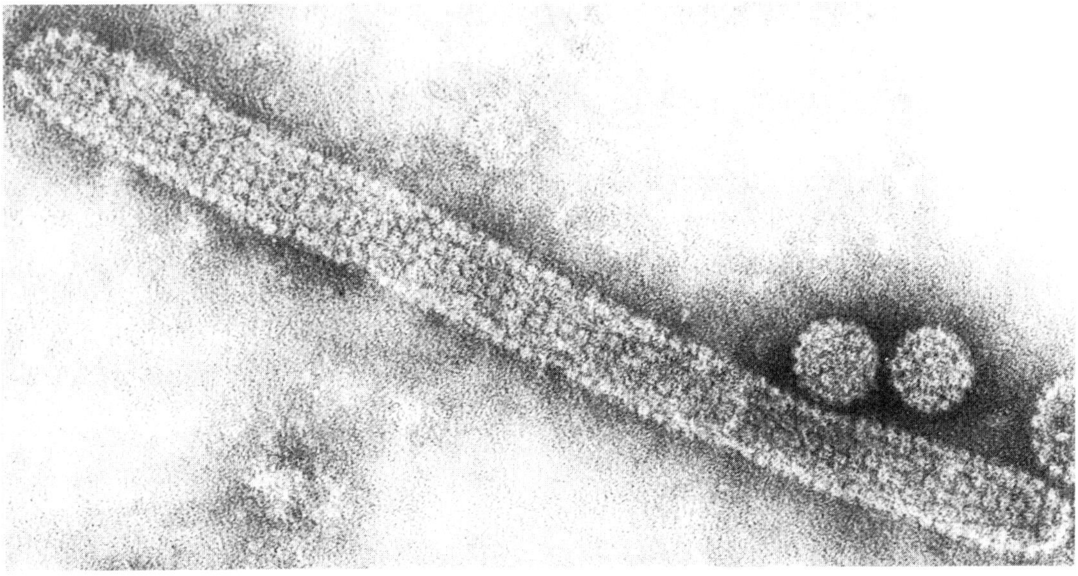

Fig. 81.23 Most cubic viruses form tubular, or aberrant, particles. The tubular form shown here is of polyoma virus and although closed at both ends the elongated portion consists only of hexons. × 275 000.

A feature of almost all cubic viruses is that, in addition to the icosahedral forms, tubular forms occur (Fig. 81.23). It has been thought that these tubules are aberrant forms of virus construction with extended incorporation of six-fold units, to turn the corners. However, it has also been suggested that a tubular form could be an intermediate stage of synthesis for cubic viruses (Diosi and Georgescu 1979). Future studies may well show that both these views have validity, as sometimes the virus has lost the ability to insert pentons at the correct location, while at others tubules might be a convenient way of storing hexons prior to final virus formation.

Much of the EM evidence on which the theoretical aspects of virus structure is based takes this instrument to the limit of its performance and it is therefore not surprising that controversies have arisen. Although rotavirus has been described here as a lattice construction of 320 structural units, it has also been described as having a triangulation number of 13 and 132 morphological units (Roseto *et al.* 1979). Similarly, the papovaviruses have been put in the P = 1 class with capsomere numbers of either 42 or 92 (Almeida *et al.* 1966), as well as the P = 7 class with 72 subunits (Klug and Finch 1965). Fortunately, although there may be some inconsistencies to clear up, it is a simple matter to recognize any of these viruses from their appearance in the EM, and for most workers this is the factor of major importance.

Yet another problem is presented by a large group of viruses usually described as small cubic viruses which do not display any substructure at all when examined by negative staining. The reason that they can be described as cubic viruses arises from X-ray diffraction studies of poliovirus, one of the few animal viruses which has been crystallized. These studies have shown that poliovirus has cubic symmetry and a subunit number that is either 60 or a multiple of 60. Unfortunately the negative stains at present available show this virus group, which contains the enteroviruses, the rhinoviruses and the aphthoviruses as approximately spherical particles with no distinctive surface structure. The particles show so little distinctive morphology that it is frequently difficult to distinguish them from background proteinaceous material, which frequently appears in the form of small round particles. The only distinguishing feature that these particles may display is that of rim, or ring, staining. This refers to the appearance that is obtained when the capsid is in some way damaged and the negative stain is allowed to penetrate into the interior of the particle. This means that the capsid itself appears as a lightly staining ring or rim, with a dark, stain filled interior (Fig. 81.24). Protein particles, which are solid, do not show this effect. However, in practice when a specimen is suspected of containing a small cubic virus it is advisable to use immune electron microscopy which will allow the particles to be seen grouped in the form of immune complexes and hence make identification simpler (Fig. 81.25).

With simple negative staining, the majority of viruses display some aspect of the symmetries that have been discussed. A few, for example influenza virus, need special treatment before the symmetry-bearing component can be visualized. Yet others such as the pox viruses have not yet been shown to possess symmetry properties. In fact, the pox group of viruses displays an appearance similar to certain bacteria when these are examined by negative staining. It might be suggested that viruses with a larger complement of nucleic acid are no longer limited to the strict confines of geometric construction and from the morphological view point, are no longer classical viruses. In this context the word 'classic' refers to the possession of either cubic or helical symmetry as described in the 1956 Crick and Watson paper.

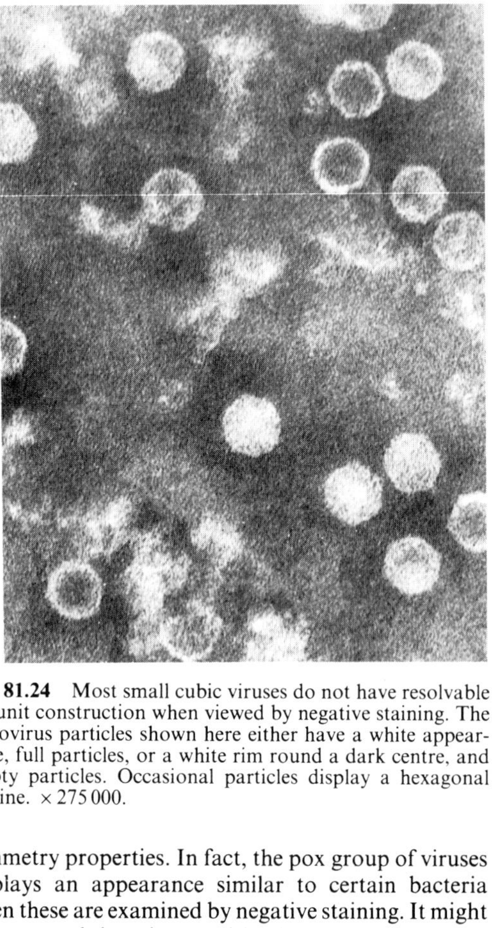

Fig. 81.24 Most small cubic viruses do not have resolvable subunit construction when viewed by negative staining. The poliovirus particles shown here either have a white appearance, full particles, or a white rim round a dark centre, and empty particles. Occasional particles display a hexagonal outline. × 275 000.

When we now turn from the consideration of the molecular construction of viruses to the use of their distinctive morphology as a means of identification, it appears that the EM offers a method for grouping viruses into meaningful categories. Viruses with identical appearance have the same basic properties, such as type of nucleic acid and location of synthesis within the cells. Serological relationships occur only between viruses that appear alike, although viruses that look alike need not be serologically related. According to the groupings put forward in the Third Report of the International Committee on Taxonomy of Viruses (Matthews 1982), the morphological characteristics of animal viruses will be described and the more common members of each group given. In accordance with this system viruses are designated into families, then genera.

As has already been done for basic symmetry properties, first helical, and then cubic viruses will be considered. Finally, those viruses with less certain or no symmetry properties will be discussed.

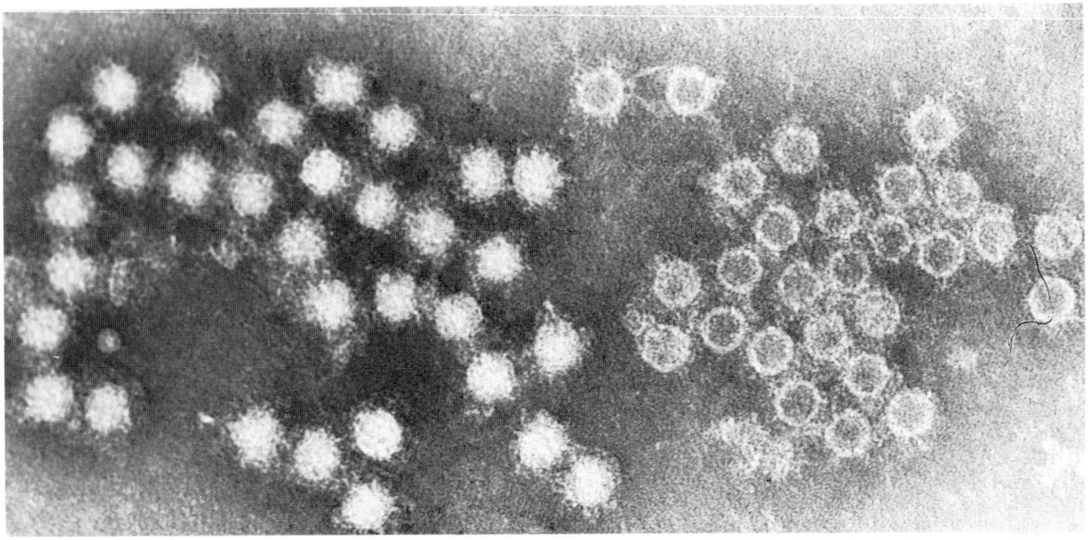

Fig. 81.25 The use of immune electron microscopy makes the recognition of small cubic viruses easier. Here the specific antibody has formed the poliovirus particles into two discrete complexes, one containing 'fulls' and the other 'empties', as these two forms of poliovirus are antigenically distinct. × 230 000.

Helical viruses

Family: *Paramyxoviridae*

This is probably the largest group of the animal helical viruses, and also the one that displays symmetry closest to that described for the classical helical viruses. The family contains three genera, paramyxovirus, morbillivirus, and pneumovirus. Of these, the first two are morphologically identical but differ inasmuch as paramyxoviruses have neuraminidase whereas morbilliviruses do not.

Genera: Paramyxovirus and Morbillivirus
Common members: Paramyxovirus; Parainfluenza viruses 1–5; mumps; Newcastle disease virus, morbillivirus; measles; rinderpest; distemper viruses; peste des petits ruminants, Sendai virus.

These are two-component viruses comprising a helical RNP, which contains single stranded (ss) RNA, and a cell-derived but virally altered lipoprotein envelope.

Although the intact particle has a diameter of approximately 100 nm and is roughly spherical, most particles examined by negative staining show some degree of breakdown and frequently appear flattened on the grid with diameters of up to 500 nm (Fig. 81.26(**a**)).

The helix is approximately 18 nm in diameter with a length of up to 1 μm (Fig. 81.26(**b**)); however, most micrographs show the helix broken into numerous short lengths. The helix can be seen at various stages of degradation, varying from a tightly arranged rigid structure similar to a TMV particle to a completely opened-up arrangement revealing the underlying single RNP strand of which the helix is formed. It is also possible to visualize a single turn of the helix that has broken off and see the radiating subunits and the central canal of the helix. The majority of micrographs show the helix slightly extended and a herring-bone pattern is seen. The lipoprotein enveloped carries short glycoprotein projections, approximately 10 nm long on the surface of the particle.

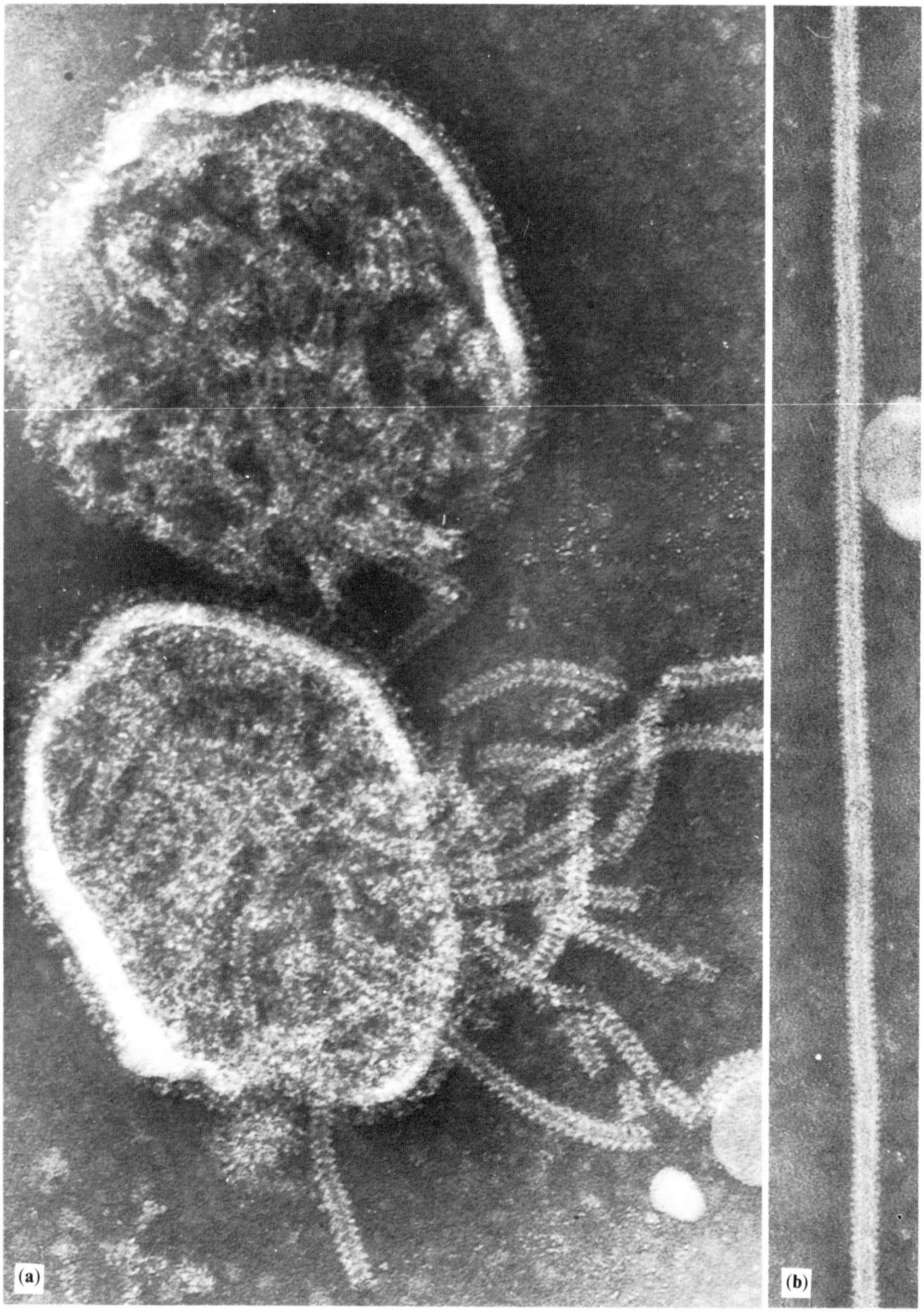

Fig. 81.26(a) Two partly disrupted parainfluenza virus particles. The cell-derived lipoprotein envelope has a fringe of virus-specific glycoprotein projections on the surface. In the lower particle the RNP component has spilled out, and it is possible to see the distinctive herring-bone pattern of the single helix. × 221 000. (**b**) Occasionally, it is possible to find very long lengths of RNP. The one shown here, for lack of space, is only part of the original structure, which measured 2μm in length.

Genus: Pneumovirus
Common member: Respiratory syncytial virus
Similar in general construction to paramyxovirus but with much more prominent projections, approximately 15 nm long, on the surface of the viral envelope. The helix is a more fragile structure than that associated with the paramyxovirus and is more difficult to visualize. The diameter of this helix is in the 12–15 nm range.

Family: *Orthomyxoviridae*
Genus: Influenzavirus
Common members: Influenzaviruses A, B and C
Highly pleomorphic lipoprotein-bound particles, with very distinct surface projections 10 nm in length. The most frequently encountered form is a roughly spherical particle of 70–80 nm diameter, although curved forms are also common (Fig. 81.27(**a**)). However, considerable variation both in shape and size occur and filamentous forms of the virus are common. These filaments are approximately 70 nm in diameter and can be as much as a micrometre in length. Unlike the *Paramyxoviridae*, influenza particles are physically stable and the internal component is not visible in the majority of particles. The few particles that are disrupted reveal an internal helix of 50 nm diameter (Fig. 81.27(**b**)) and a variable number of turns ranging from as few as five in spherical particles to as many as 40 in elongated forms. These helices are formed from a strand of 6 nm diameter, and it is this structure that must contain the arrangement of ss-RNA and protein subunits in a helical conformation that has not yet been established. Degradation studies show that the second order helix, i.e. the one of 50 nm diameter, is itself a double helix (Almeida and Brand 1975, Murti *et al.* 1980). Also through degradation studies (Fig. 81.27(**c**)), it can be shown that the surface projections display two distinct types of morphology. One which is rod shaped corresponds to the haemagglutinin while the other, which has a mushroom shape represents the neuraminidase of the virus (Wrigley 1979).

Overleaf
Fig. 81.27(a) A field of Influenza A particles showing the heterogeneous nature of this virus. The characteristic feature of influenza is the prominent 10 nm projections on the surface. These projections represent the haemagglutinin and neuraminidase of the virus. In this micrograph no internal component is visible. × 230 000. (**b**) One influenza virus particle which has been penetrated by stain to reveal the internal helix. This helix is formed from a strand 6 nm wide and the diameter of the helix is approximately 50 nm. × 275 00. (**c**) Detergent treatment degrades the influenza virus and allows the internal helix to be visualized on its own. As can be seen, it is a double helix formed from the 6 nm diameter RNP strand. During the detergent treatment the surface haemagglutinin re-forms to give the stars and rosettes seen in the background of this micrograph. × 275 000.

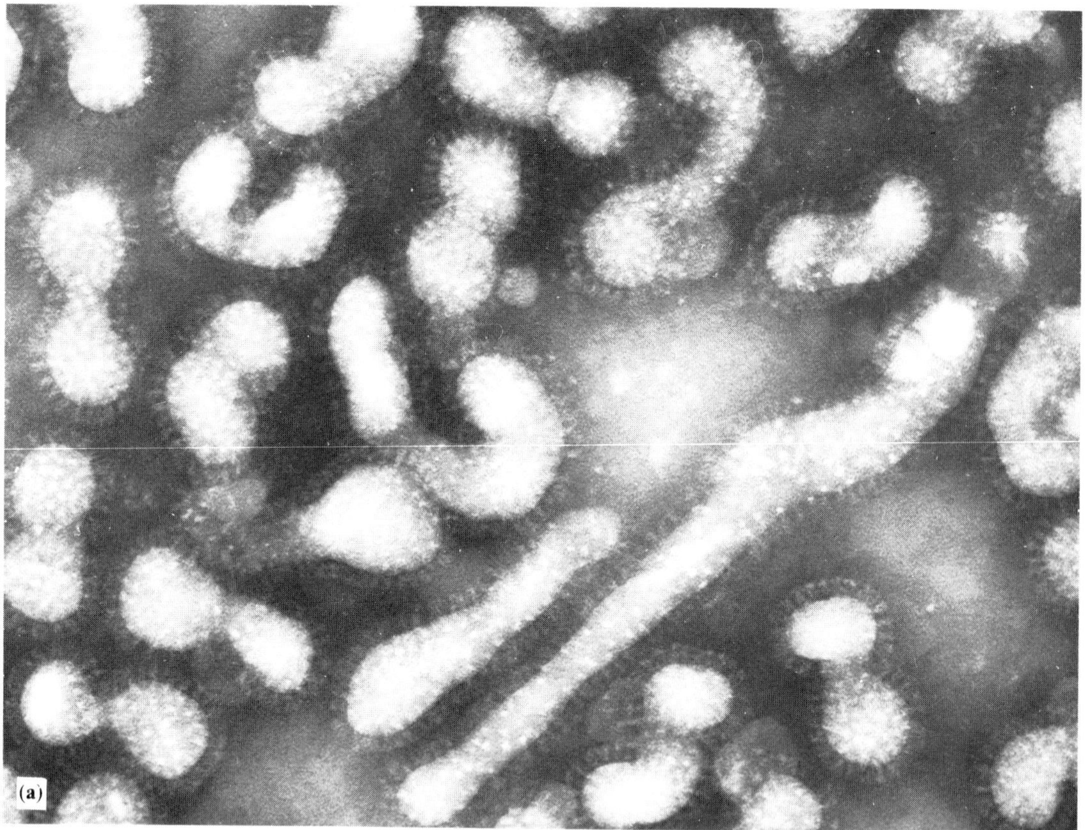

(**a**)

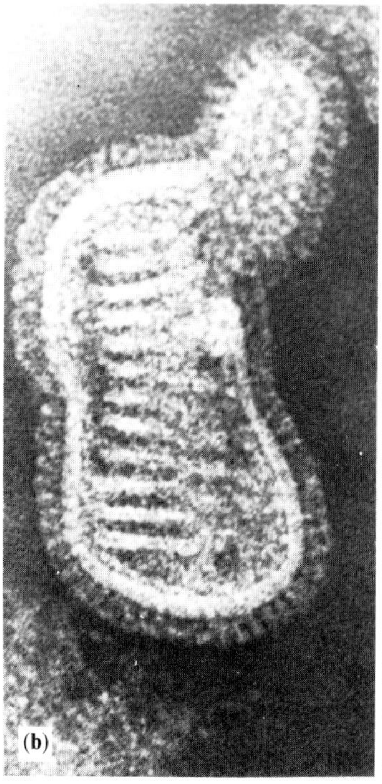

(**b**)

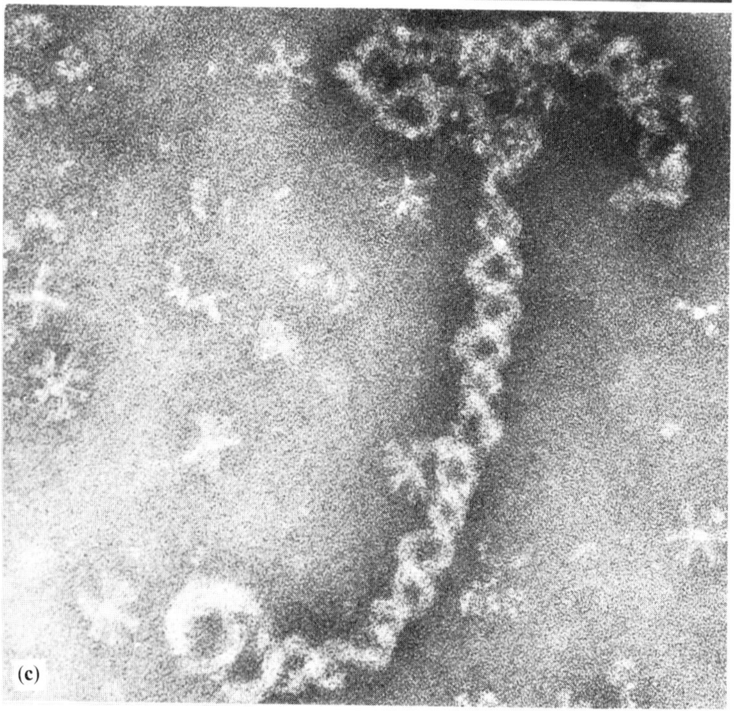

(**c**)

Family: ***Rhabdoviridae***
Genus: Vesiculovirus
Common members: Vesicular stomatitis virus, Cocal; Piny
Bullet-shaped particles 60–80 × 130–220 nm in dimension rounded at one end and flat at the other. The flat end frequently has a tail of cell-derived but virally altered membrane. The outer surface of the particle bears fine projections 8–10 nm long, which are glycoprotein and are set in the cell-derived lipoprotein membrane (Fig. 81.28). Within this membrane is a regular helical structure with a periodicity of 4.5 nm. This structure has a diameter of 50 nm; and when particles are either spontaneously or deliberately degraded an RNP strand, with a morphology not unlike that of the paramyxoviruses, is revealed. Nucleic acid is ss-RNA.

Genus: Lyssavirus
Common member: Rabies virus
Similar to vesiculovirus but with less prominent surface projections that, when suitably oriented, can be seen to have a honeycomb arrangement (Murphy and Harrison 1979).

Unclassified virus with helical symmetry
Marburg-Ebola virus. A large RNA virus consisting of a fringed filament of 100 nm diameter and a length of up to several micrometres. This filament is frequently curved to give either a 'comma' appearance or even a closed ring structure. The ring structure is found more frequently with the Marburg agent and the elongated form with Ebola. A partial serological cross exists between the two viruses. Degraded particles show that there is a helix, presumably built up of an RNP, inside the lipoprotein envelope (Ellis *et al.* 1978).

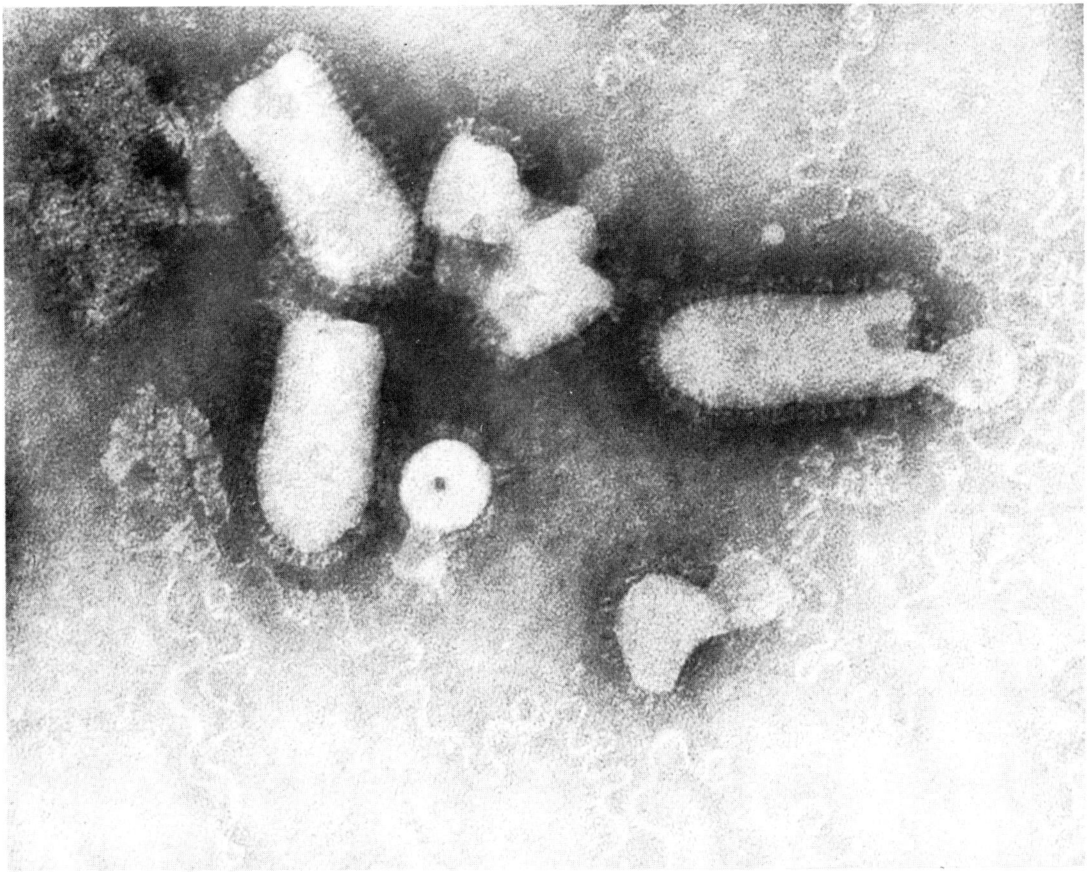

Fig. 81.28. Rhabdoviruses, represented here by vesicular stomatitis virus, have bullet-shaped particles with one round and one flat end. Dimensions are approximately 80 × 180 nm. The lipoprotein surface of the particles bears a 10 nm virus-specific glycoprotein fringe. Within this envelope there is a protein skeleton that has a regular periodicity of 4.5 nm. This structure can just be seen in the particle fragment at the top left of the micrograph. Internal to this there is an RNP helix that here can be seen unwinding. × 230 000.

Cubic viruses

Unlike animal helical viruses, cubic viruses may contain either RNA or DNA as their nucleic acid. Also unlike helical viruses, animal cubic viruses may or may not have an additional lipoprotein membrane enclosing the symmetry-bearing component. It is therefore convenient to divide cubic viruses, first into the nucleic acid type, and then according to whether or not they possess a lipoprotein outer membrane. Those which do not possess an outer membrane are described as simple, or naked; those with a membrane are either compound or enveloped.

Simple cubic RNA viruses

Family: ***Picornaviridae***
Genus: Enterovirus
Common members: polioviruses; Coxsackieviruses; ECHO-viruses; various human and animal enteroviruses.
Possible member: human hepatitis A virus.
By current methods of negative staining none of these viruses displays detectable subunit structure. All appear as smooth round structures of 27 nm diameter, except for the individual particles that show ring staining (Fig. 81.29(**a**)). Occasionally, some particles show an hexagonal outline – additional proof of the underlying but unresolved cubic symmetry. For practical purposes it is advisable to use immune electron microscopy to visualize these viruses (Almeida 1980). All members of the *Picornaviridae* have virions that consist of a protein capsid and ss-RNA. The enteroviruses are pH stable and are negatively stained by the routine phosphotungstic acid solution which is adjusted to pH6. Rhinoviruses on the other hand are labile below pH6 and must be visualized with a negative stain adjusted to above this value; normally pH8 is used.

Genus: Cardiovirus
Common member: Encephalomyocarditis (EMC) virus.
Similar in size and appearance to enterovirus.

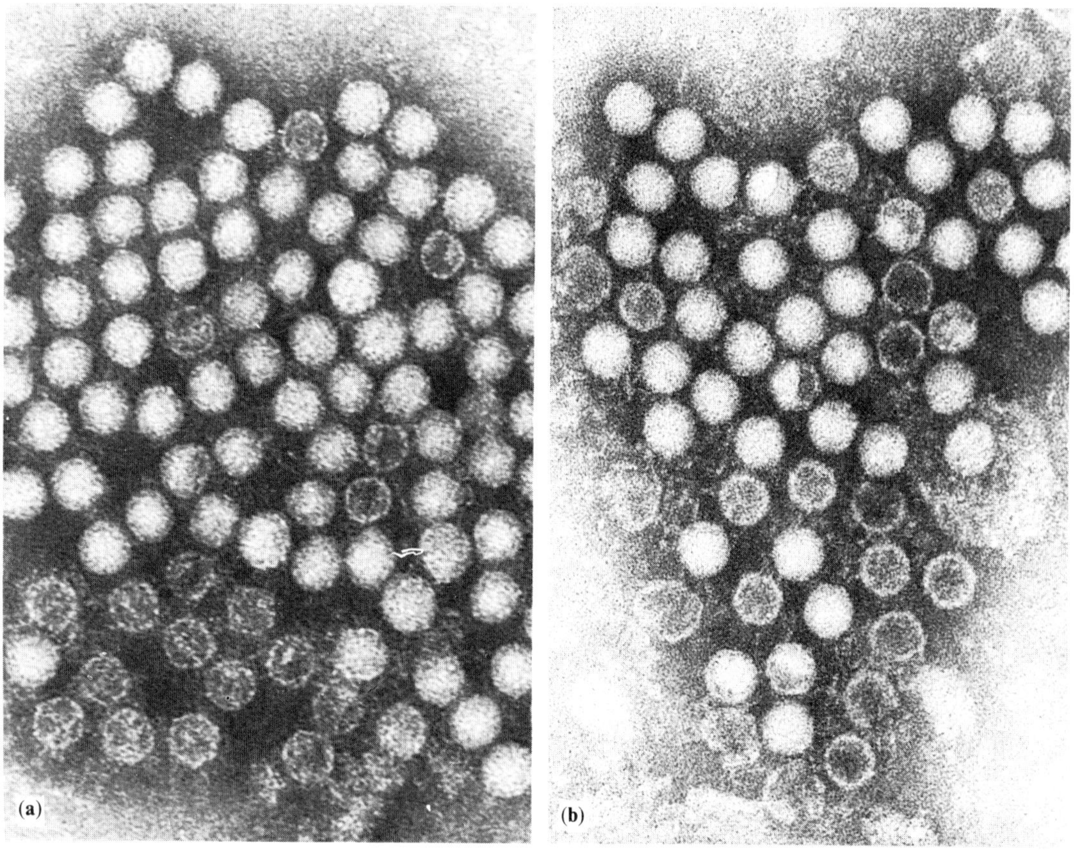

Fig. 81.29(a) An immune complex of poliovirus particles showing the structureless appearance of the enterovirus group. The virus measures 27 nm in diameter and occurs both as full and empty particles. × 275 000. (**b**) An immune complex of rhinovirus particles. Like the enteroviruses, these particles are 27 nm in diameter and show little or no substructure. However, this virus is pH labile and the negative stain has to be adjusted accordingly. Rhinoviruses are structurally fragile; round the outside of the aggregate distintegrating particles can be seen. × 275 000.

Genus: Rhinovirus
Common members: human and bovine common cold rhinoviruses.
Similar to enteroviruses and still with 27 nm diameter, but the particles appear more delicate than the enteroviruses and because of their pH lability can easily be lost (Fig. 81.29(**b**)).

Genus: Aphthovirus
Member: foot-and-mouth disease virus of cattle.
Smooth round particles of 24 nm diameter with no suggestion of substructure. This virus group is very pH sensitive and staining solutions must be carefully adjusted to values between pH 7–8 before satisfactory preparations can be obtained.

Family: *Caliciviridae*
Genus: Calicivirus
Common members: vesicular exanthema of swine; feline calicivirus.
Distinctive particles of 35–39 nm diameter with a subunit structure that appears as a pattern of dark units rather than the usual light capsomeres present on most cubic viruses (Fig. 81.30). The type of substructure, although not completely elucidated, may well be similar to that described for the *Reoviridae* with the holes in a lattice being the most prominent feature of the capsid. The pattern of holes or depressions is compatible with a 32 subunit structure. The 'diagnostic' patterns seen are one subunit surrounded by six, and a diamond. This establishes caliciviruses as being based on a class P = 3 icosahedron, with T = 3, and M = 32. The nucleic acid of this group is ss-RNA.

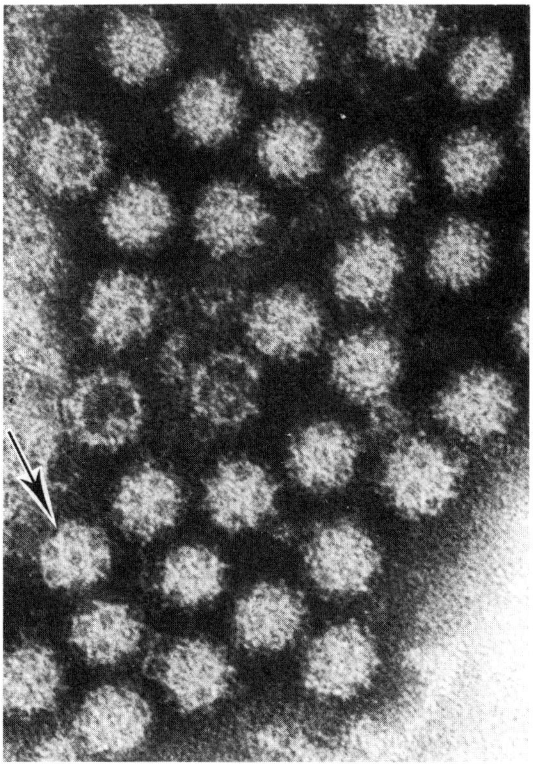

Fig. 81.30 A group of feline calicivirus particles showing the distinctive pattern of dark centres. The arrow points to a particle showing the 'one surrounded by six' arrangement, and the particle immediately below this shows a diamond arrangement. × 275 000.

Family: *Reoviridae*
Within the *Reoviridae* family there are subtle, but distinct, morphological differences between the different genera. Reovirus is larger than rotavirus and has separate radiating external subunits. Orbiviruses display a less distinct structure than the other members

Genus: Reovirus
Common members: human and animal reoviruses, serotypes 1, 2 and 3.
Reoviruses contain a genome of segmented ds-RNA. The capsid consists of a double-shelled construction of protein subunits on a smooth-walled internal core. The diameter of the core is 40–45 nm, the internal capsid 55 nm and the complete particle 70–80 nm. Both the internal capsid and the complete virion have subunits that project separately round the edge of the particle. As discussed in the section on basic morphology, all of the *Reoviridae* members appear to have a substructure based on a lattice formation of structural subunits with the dark dots on the surface formed by the centres of the clusters. In reovirus the exact number of these subunits has not yet been established with certainty. The double-shelled construction based on a lattice gives the reovirus a distinctive appearance, best described as resembling a lace doily (Fig. 81.31(**a**)).

Genus: Rotavirus
Common members: Rotaviruses of man and a wide range of animals, causing diarrhoea particularly in neonates.
The particle consists of a double capsid containing 11 pieces of ds-RNA. There is an inner core of 45 nm, an internal capsid of 50 nm, described as rough because of the projecting subunits, and an external smooth capsid measuring 65 nm in diameter. The external capsid is described as smooth, because there appears to be a continuous line round the periphery of the particle (Fig. 81.31(**b**)). As discussed, the two layer capsid is based on a lattice of structural subunits and there would appear to be 320 of these. The rough internal capsid bears the group-specific antigens, while the type-specific are present on the smooth external capsid.

Genus: Orbivirus
Common members: bluetongue virus; African horse sickness
Basic construction similar to other members of the *Reoviridae* with an internal capsid showing distinct subunit construction. The outer capsid has a diameter of 65–80 nm and a slightly 'fuzzy' appearance which distinguishes this group from the other two.

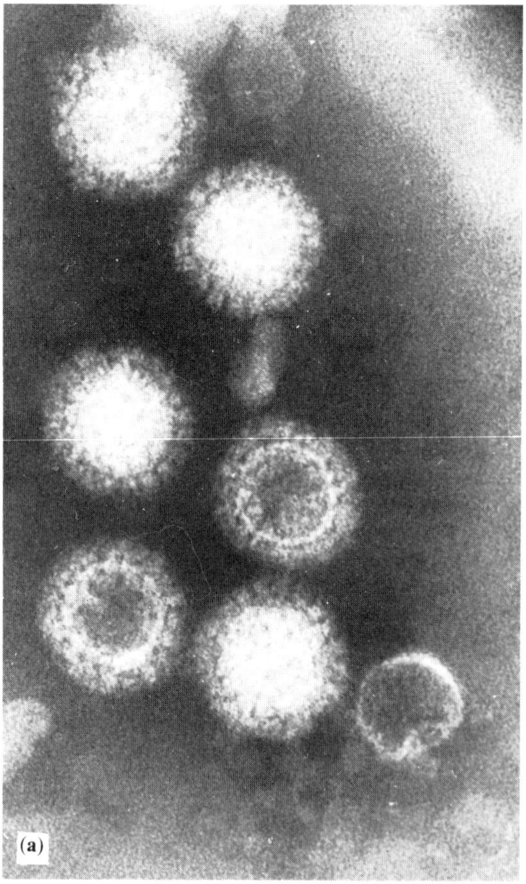

Fig. 81.31(a) A small group of reovirus particles which have a diameter of 75 nm. It is possible to see the internal capsid in two of the particles; and a smooth, 45 nm, inner core can be seen at the bottom right of the micrograph. Note that the outer subunits of the complete particle are clearly separated. × 275 000.

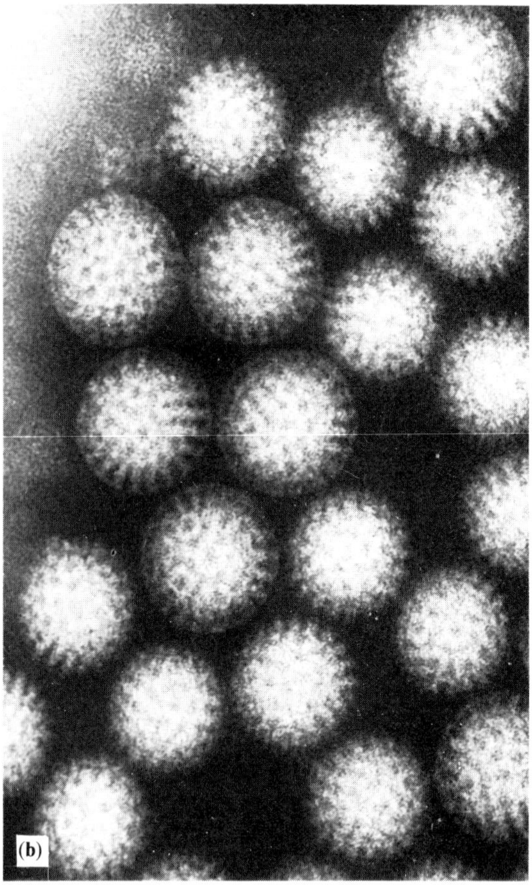

Fig. 81.31(b) A group of bovine rotavirus particles showing the 65 nm diameter complete particles and the 50 nm diameter internal capsid. The subunits on the complete virus are joined to give a smooth appearance, while the inner capsid appears rough. The pattern of dark spots is produced by the holes in the lattice forming the particles. × 275 000.

Simple cubic DNA viruses

Family: ***Parvoviridae***
Genus: Parvovirus
Common members: Kilham rat virus; adeno-associated virus; minute virus of mice; bovine parvovirus.
A small cubic virus 18–22 nm in diameter containing ss-DNA. The virus has 12 distinct large subunits which frequently display a diamond arrangement similar to that seen on the bacteriophage $\phi \times 174$ (Williams and Fisher 1974).

Family: ***Papovaviridae***
Genus: Polyomavirus
Common members: Polyomavirus; mouse K virus; SV40; various human viruses (e.g. BK and JC) isolated from cases of multifocal leucoencephalopathy and also from immunosuppressed patients

Distinctive rough-surfaced particles of 45 nm diameter. (Fig. 81.32(**a**)). Capsomeres are readily resolved but, because of the comparatively small size of the virus and the fact that it maintains a three-dimensional structure rather than flattening on the grid, it is not easy to determine the arrangement of these subunits. The polyoma virus group is based on the right-handed form of a P = 7 icosahedron with T = 7 and 72 morphological subunits. Aberrant tubular forms of the virus are common (Fig. 81.23).

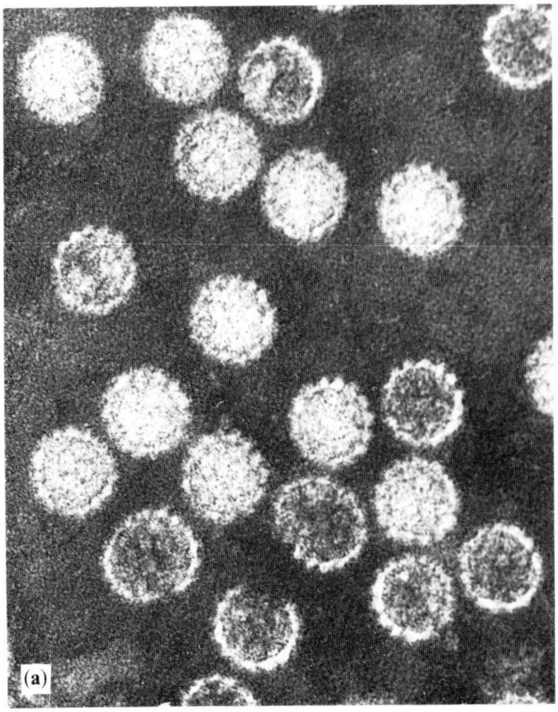

Fig. 81.32(a) Polyoma virus displays both full and empty forms. The pronounced capsomeres give a rough 'blackberry' appearance to the particles. The virus is approximately 45 nm in diameter, and it is not easy to find particles that allow the capsomere arrangement to be determined. Polyoma virus has 72 morphological subunits based on the right-handed form of a P = 7 icosahedron. × 275 000.

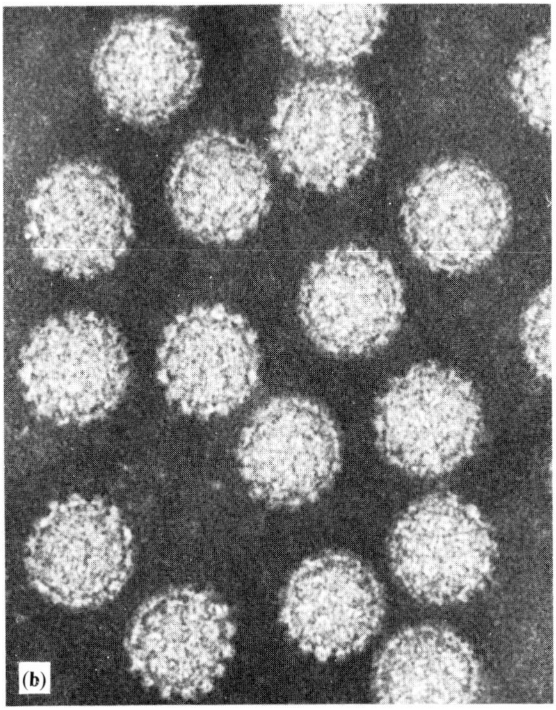

Fig. 81.32(b) Similar to poloyma virus but, with a diameter of 55 nm, a much more robust looking particle. Superimposition effects make it difficult to determine subunit number, but wart virus is given as M = 72 on the left-handed form of a P = 7 icosahedron. × 275 000.

Genus: Papillomavirus
Common members: Shope papilloma virus; human common and genital wart virus; viruses associated with various animal papillomata.

The diameter of the papilloma viruses is 55 nm compared with 45 nm for polyoma virus, although this difference does not seem great, the wart viruses appear as considerably larger, more robust particles in the electron microscope (Fig. 81.32(**b**)). All of the papilloma viruses are designated as having 72 subunits on a P = 7, T = 7 icosahedron. Since this is a skew structure, it is possible to distinguish between human wart virus which displays the right-handed structure and the rabbit papilloma virus which is left handed. Like polyoma, wart virus produces many aberrant forms which may be tubular, drumstick, or dumb-bell in shape.

Family: ***Adenoviridae***
Genus: Mastadenovirus
Common members: Numerous serotypes of human and animal adenoviruses all having a group-specific antigen.
Genus: Aviadenovirus
Common member: CELO virus of chickens
Of all viruses examined by negative staining, the adenoviruses provide the best example of icosahedral structure. The capsid has 252 morphological subunits, and P = 1, T = 25 (Figs. 81.1, 81.7 and 81.22). The nucleic acid is ds-DNA and the diameter is 80 nm. The hexons and pentons are serologically distinct and have separate functions. By special staining it is possible to see that each penton has a fibre attached to it and that this fibre terminates in a spherical knob (Valentine and Pereira 1965, Nermut 1980).

Cubic DNA viruses with lipoprotein envelope

Family: ***Herpesviridae***
Genera are unnamed although alphavirinae has been suggested for herpes virus hominis and betavirinae for human cytomegalovirus.
Common members of herpesvirus groups: herpesvirus hominis, types I & II; varicella-zoster; cytomegalovirus; Epstein-Barr virus, Marek's disease virus.

All the herpes viruses contain ds-DNA and are composed of a geometric capsid with surrounding lipoprotein membrane. The capsid is a P = 1; T = 16 structure with 162 morphological subunits each of which displays a distinct central hole. In intact particles, the surrounding envelope is closely applied and itself reproduces the hexagonal or pentagonal outline of the underlying icosahedral structure (Fig. 81.33). However, more frequently the envelope is disrupted and the particle resembles a fried egg with the capsid as yolk and the envelope as the white (Fig. 81.34). The outer envelope

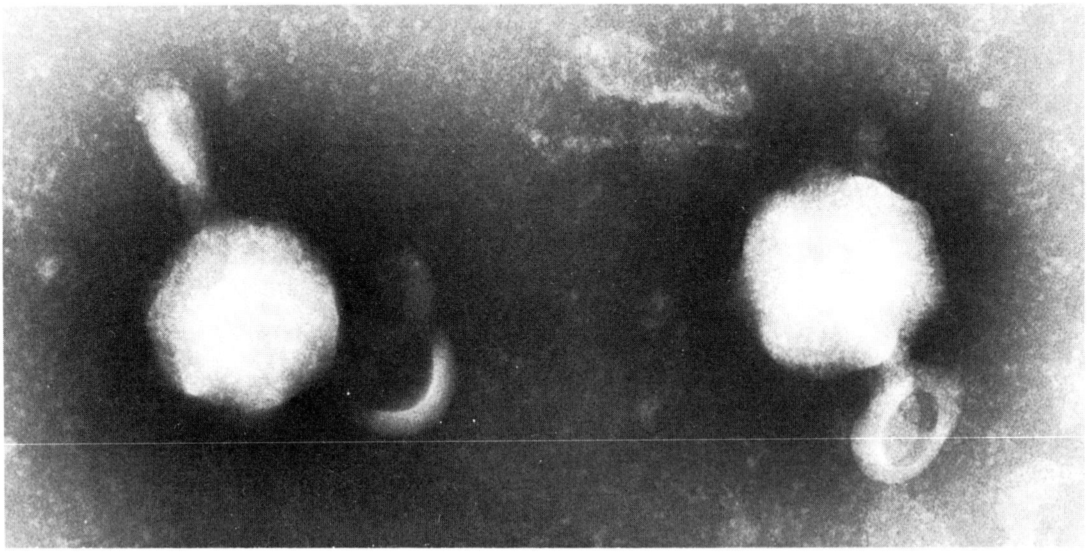

Fig. 81.33 Intact herpes virus particles have a tightly applied lipoprotein membrane around the internal geometric capsid. In this form they measure just over 100 nm in diameter, and it is possible to see the hexagonal or pentagonal outline that is imposed by the internal icosahedron. × 230 000.

carries fine projections of 12 nm in length. The capsid diameter is 100 nm and the intact particle 120 nm; however, the majority of enveloped particles show disruption and are in the size range 150–200 nm.

Family: *Iridoviridae*

Only one genus, iridovirus, has been established for Tipula iridescent virus, an insect virus, but African swine fever has a large complex particle approximately 200 nm in diameter. The complete particle has a plasma membrane-derived envelope and a pronounced hexagonal outline. Within this outer covering there is an inner capsid displaying geometrically arranged subunits but the exact number of these subunits has not yet been satisfactorily established.

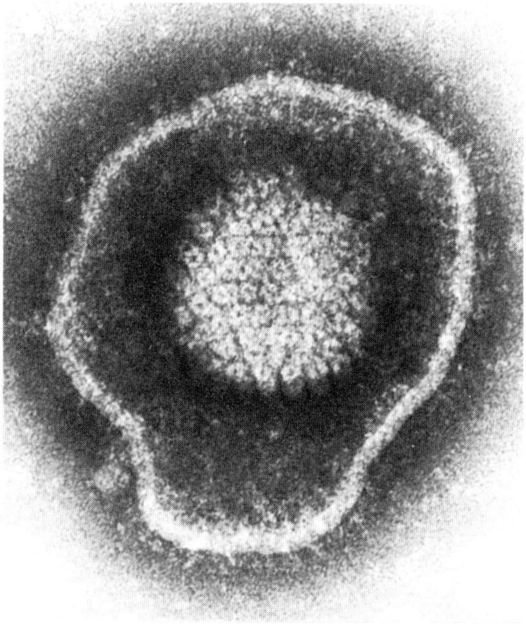

Fig. 81.34 A single herpes virus particle that has degraded sufficiently to allow visualization of the internal capsid. The complete particle now measures over 200 nm in diameter, and the capsid is 100 nm. The icosahedral form of the capsid is apparent and there are 162 capsomeres on a P = 1 icosahedron. × 275 000.

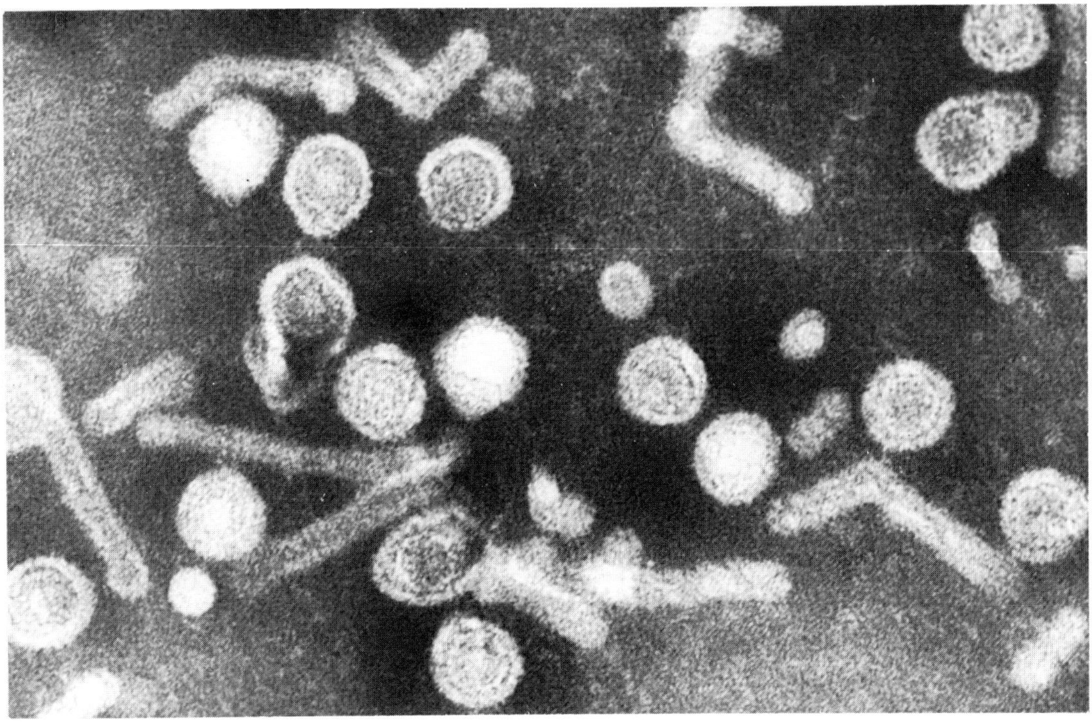

Fig. 81.35 This micrograph shows the various morphological forms associated with the virus of human viral hepatitis type B. Collectively the forms are known as hepatitis B antigen (HBAg), and this comprises small spheres and tubules of 20 nm diameter as well as double-shelled particles of 42 nm diameter. The double-shelled particles, also known as Dane particles, contain a core (HBcAg) that has both DNA and DNA polymerase. × 275 000.

Unclassified compound cubic DNA virus

Hepatitis B virus of man. Although not yet propagated in the laboratory, the double-shelled Dane particle of hepatitis B antigen (HBAg) possesses the characteristics of viruses in this group (Fig. 81.35). The core (HBcAg) of the Dane particle contains both ds-DNA and DNA polymerase. The capsid of the core is protein; the outer covering (HBsAg) is lipoprotein. The core has a diameter of 27 nm and, when viewed without the outer envelope, is closely similar in appearance to the eneteroviruses, although it is a DNA virus. The intact Dane particle measures 42 nm in diameter. Electron micrographs of HBAg in addition to the Dane particles show numerous pleomorphic spheres and tubules with an approximate diameter of 20 nm. These particles do not contain nucleic acid and are formed from excess hepatitis B surface antigen (HBsAg).

Viruses not yet accorded symmetry properties

While all of the preceding viruses have either definite or presumptive symmetry, there is a residual group with less definite or no evidence of symmetry. Many of these viruses can still be identified in the EM by means of their general appearance or distinctive surface structure. As before, these viruses will be divided into those containing RNA and those containing DNA. All of the viruses in this section contain lipid.

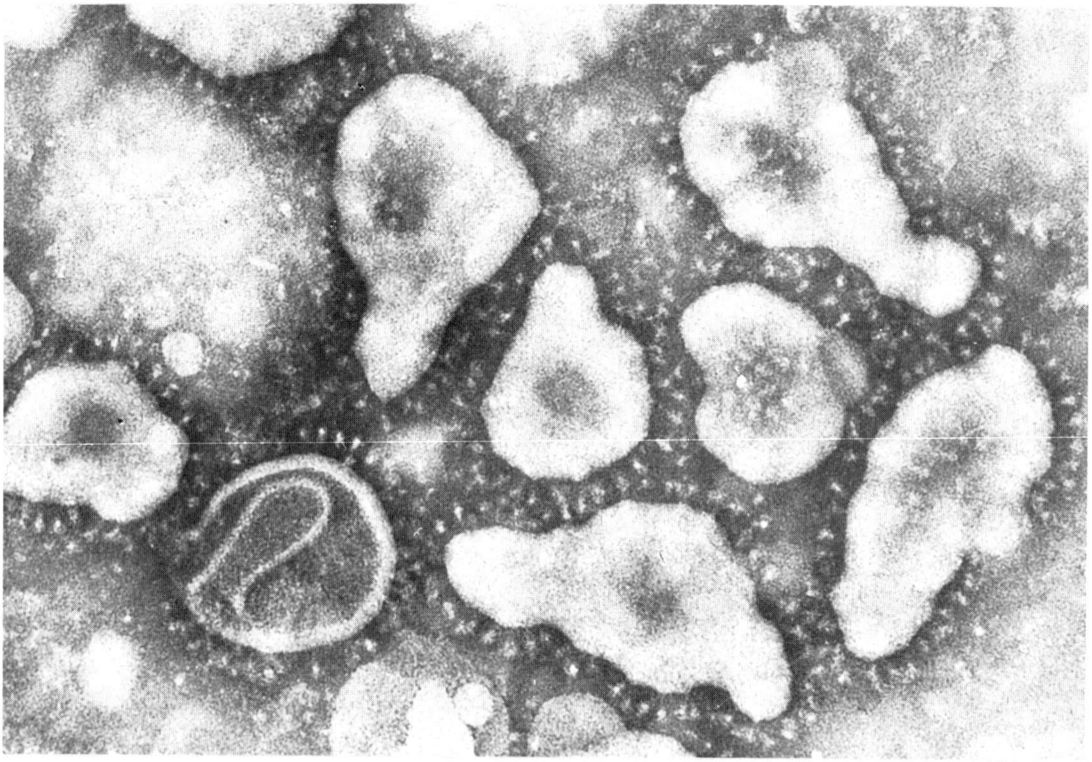

Fig. 81.36 The name coronavirus was used for this virus group because of the corona of prominent spikes that surround the pleomorphic lipoprotein outer membrane. One of the avian infectious bronchitis particles shown here has been penetrated by stain to reveal an inner tongue or bottle-shaped structure. ×230 000.

Unestablished RNA viruses

Family: ***Coronaviridae***
Genus: Coronavirus
Common members: avian infectious bronchitis virus; human coronaviruses associated with respiratory infections; mouse hepatitis virus; transmissible gastro-enteritis of pigs.

Predominantly spherical, but nevertheless pleomorphic lipoprotein-covered particles frequently showing an umbilicate appearance. Particle diameters range from 75–160 nm. The outstanding physical feature of this virus group is the distinctive club, or leaf-shaped, projections on the surface (Fig. 81.36). These projections range from 12–24 nm in length and in many instances readily detach themselves from the virus surface. Coronaviruses without surface projections are impossible to identify, as their lipoprotein envelopes look like any cell fragment. Reports have described a bottle-shaped internal structure for avian infectious bronchitis virus (Bingham and Almeida 1977); and a helix has been described for some of the human respiratory viruses (MacNaughton *et al.* 1977). Morphologically, this is one of the most difficult virus types to identify, many fringed particles have been put forward as coronaviruses, although they do not possess the true characteristics of this virus group.

Family: ***Togaviridae***
Genus: Alphavirus
Common members: Sindbis; eastern equine encephalomyelitis; western equine encephalomyelitis; Venezuelan equine encephalomyelitis; Semliki Forest virus. These and other members are all group A arboviruses.

More or less spherical particles of 60–70 nm diameter. The surface bears fine projections approximately 10 nm long which are seen as a halo around the particle (Fig. 81.37(**a**)). Although the particles have no particular physical features apart from this fringe, they can still be recognized even when examined in the presence of cellular debris. This is in contradistinction to the next three genera.

Genera:
Flavivirus
Common members: yellow fever virus; tick-borne encephalitis; louping ill. All of these are group B arboviruses, many other viruses belong in this group.

Rubivirus
Member: rubella

Pestivirus
Member: bovine viral diarrhoea virus

These three genera will be considered together, as they appear similar by negative staining.

All of the viruses are seen as pleomorphic but approximately spherical nondescript membranous sacs 50–70 nm diameter. Their surface bears small insignificant projections that are usually not well resolved. Unless very pure preparations are available, it is advisable to use immune electron microscopy for their identification, because only when they are present in the form of complexes is it possible to be certain of their nature (Fig. 81.37(**b**)).

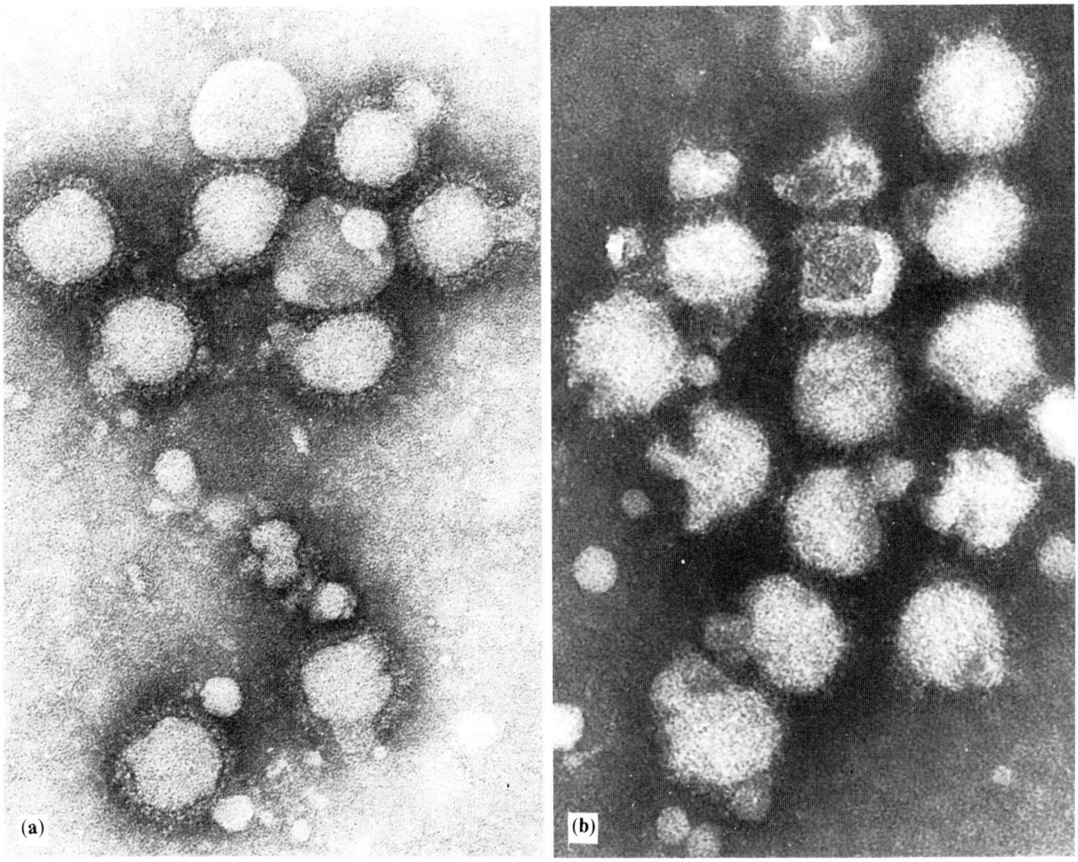

Fig. 81.37(a) Group A arboviruses, illustrated here by Sindbis virus, are 60–70 nm in diameter. They consist of an approximately spherical lipoprotein sac with a surface halo of fine projections. ×230 000. (**b**) Flavivirus; rubivirus; and pestivirus all appear similar by negative staining. Here these genera are illustrated by rubella virus. Approximately spherical lipoprotein sacs of 60 nm diameter are seen with so few distinguishing features that they are best visualized when they have been aggregated by the use of immune serum. The surface of the particles has a fringe of very fine projections that are only a few nanometres in length. ×230 000.

Family: ***Arenaviridae***
Genus: arenavirus
Common members: lymphocytic choriomeningitis virus; Lassa fever virus; Junin; Pichinde

A ss-RNA, lipid-containing virus group given the name arenavirus, because on thin sectioning the particles contain granular material that gives a 'sandy' appearance. By negative staining the particles appear as pleomorphic, but predominantly spherical enveloped structures with short club-shaped projections 10 nm long. Particles fall in the size range 110–130 nm.

Family: ***Retroviridae***
Sub-family Oncovirinae
Genus: Type B oncovirus group
Common member: Bittner mouse mammary tumour virus

Pleomorphic lipoprotein-bound particles roughly spherical in shape but frequently having a tail of outer membrane. Particles penetrated by stain show an internal spherical component that appears to have an underlying subunit construction (Fig. 81.38). The diameter of the particles is around 100 nm; the internal component is 75 nm. The external surface of the particle bears 10 nm long projections that closely resemble those on the influenza virus particle.

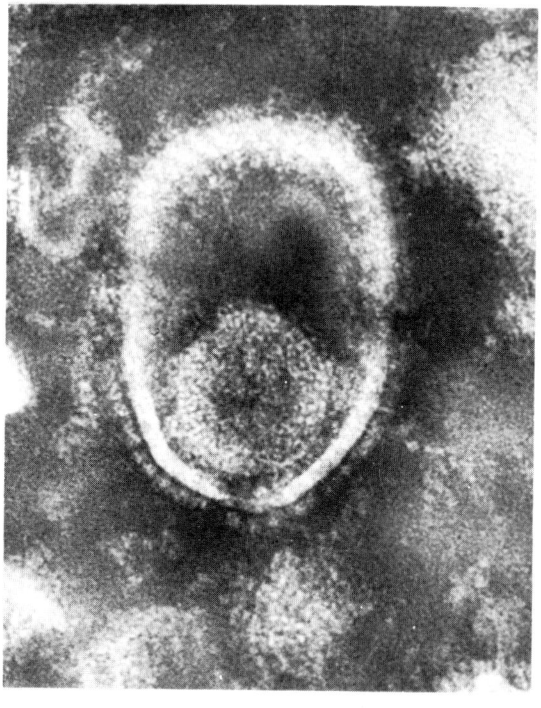

Fig. 81.38 The Bittner mouse mammary tumour virus is one of the few oncoviruses that have a distinctive appearance when examined by negative staining. The particles, approx. 100 nm in diameter, consist of an outer cell-derived membrane with distinctive 10 nm projections. The internal component is a spherical structure of 75 nm that shows the presence of substructure. This substructure has not yet been defined. × 275 000.

By negative staining the oncovirinal agents, apart from the mouse mammary tumour agent, show little distinctive morphology and are much better characterized by the thin-sectioning technique (Dalton and Haguenau 1973). By this method they can be shown to have an intracytoplasmic stage known as the type A particle, which is doughnut shaped and approximately 70 nm in diameter, and a type C particle which is usually seen budding from the plasma membrane. The mature C type particle has an inner nucleoid placed symmetrically in the type C oncovirus and eccentrically in the type B oncovirus.

Genus: Type C oncovirus group
Common members: Bovine leukosis; murine sarcoma and leukaemia; feline sarcoma and leukaemia; woolly monkey sarcoma.

Roughly spherical lipoprotein-bound particles of 80–100 nm diameter bearing non-distinctive projections of 8 nm on the surface. The internal component has not been well characterized, although there is a suggestion that an internal capsid contains a helical RNP.

Yet other compound RNA viruses include the *Bunyaviridae* and the *Spumavirinae*, but their morphology is so little characterized that they have not been included.

Unestablished DNA viruses

Family: ***Poxviridae***
Genus: Orthopoxvirus
Common members: vaccinia; cowpox; monkeypox, variola.

Brick-shaped particles of 200 × 250 nm. The mature virion displays a distinctive 'rough' surface with ridges that appear to lie just under the surface membrane of the particle (Fig. 81.39). The centre of the particle appears raised in many instances; this is the direct effect of the internal core. The core is only occasionally visualized by negative staining, but can be shown readily in thin section preparations of pox-infected cells. By this technique the core consists of three components, a bi-concave lozenge-shaped structure with the long axis corresponding to the long axis of the particle; and two roughly spherical lateral bodies lying in the concavities of the larger structure (Dalton and Haguenau 1973). The genetic material of the virus, which is ds-DNA, lies within these core structures.

These mature virions are frequently described as showing a mulberry appearance, because of the surface corrugation of the particles, but specimens also contain immature particles which display a different appearance. Immature viruses are larger than the mature particles and do not display a ridged surface structure. An inner and outer membrane can be visualized. The edge of the outer layer is crinkled giving an almost lacy appearance; the inner membrane is relatively smooth, and the whole particle has a generally grey appearance by negative staining. When the outer membrane of these immature or smooth particles is stripped off, distinct subunits can be seen on the surface of the inner membrane (Fig. 81.39).

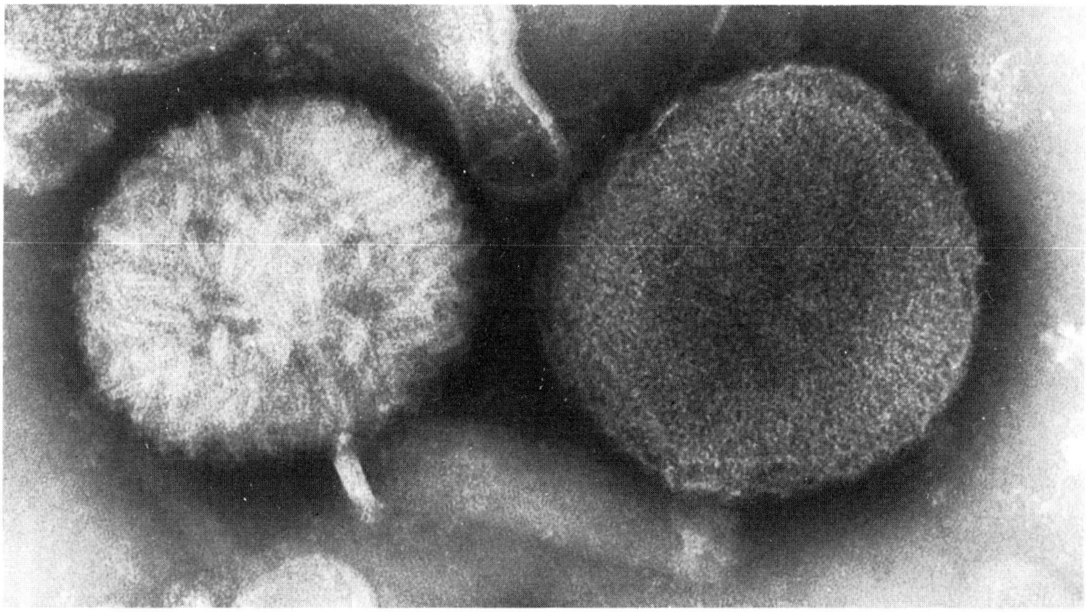

Fig. 81.39 Two vaccinia virus particles; that on the left shows the mature form of the virus. It is approximately 200 × 250 nm in size, and the surface is covered by fine strands that give the particle the appearance of a ball of string. This has also been described as a 'mulberry' appearance. The particle on the right is the immature form of the virus. It is slightly larger than the mature form and, because it is penetrated by stain, has a grey appearance. It is possible to see an inner membrane; the outer surface displays a frilled edge. × 230 000.

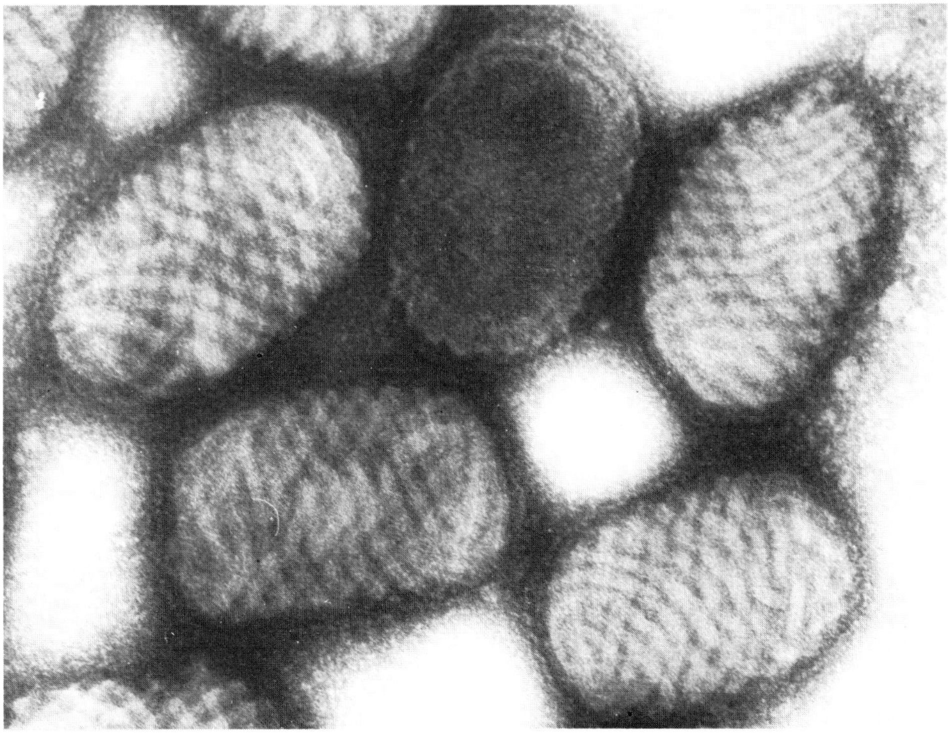

Fig. 81.40 A group of Orf particles, also known as the virus of sheep pustular contagious dermatitis. Like other pox viruses, Orf has both a mature and an immature form. The mature particle is approximately 300 × 200 nm and the surface ridges form a distinctive helical pattern. × 191 000.

Genus: parapoxvirus
Common members: Orf (Sheep pustular contagious dermatitis); the virus of milkers' nodules.

Like the orthopoxviruses, this group shows both mature and immature particle morphology. However, the particles are longer than the other group, having dimensions of 300 × 200 nm. In addition, the surface striations of the mature particles are arranged in a helical pattern giving the particles a distinctive and aesthetically pleasing appearance (Fig. 81.40).

Although containing DNA, the pox group has the distinction of being the only viruses with this nucleic acid which are synthesized in the cytoplasm of the cell.

Conclusions

The ability to visualize the fine structure of viruses by the negative staining technique has now been available for over twenty years. During this time the electron microscope and viruses have complemented each other. The electron microscope is limited to the study of small structures if its potential resolution is to be realized. Not only did viruses fall into the right size range, but they displayed a geometric construction that made full use of the potential of the instrument.

Two divergent uses of this interaction between virus and microscope have developed. First, direct visualization of the symmetrical properties of viruses has led to a better understanding of the principles governing physical construction in many biological systems.

Secondly, the method of negative staining, coupled with distinctive virus morphology, has opened up a whole new field of practical virus diagnosis. To quote only one example, the totally different visual appearance of the smallpox and the chicken pox virus has enabled many problem cases to be diagnosed quickly and authoritatively because, in this context, one picture really is worth a thousand words.

References

Adrian, M, Dubochet, J., Lepault, J. and MeDowell, A. W. (1984) *Nature*, **308,** 32.
Almeida, J. D. (1980) *Yale J. Biol. Med.* **53**, 5.
Almeida, J. D. and Brand C. M. (1975) *J. gen. Virol.* **27**, 313.
Almeida, J. D., Waterson A. P., Fletcher, E. W. L. (1966) *Progr. exp. Tumor Res.* **8,** 95.
Bingham, R. W., and Almeida, J. D., (1977) *J. gen. Virol.* **36,** 495.
Brenner, S. and Horne, R. W. (1959) *Biochim. biophys. Acta* **34,** 103.
Caspar, D. L. D. (1965) In: *Viral and Ricketisial Infections of Man.* p. 51. Ed. by F. L. Horsfall and I. Tamm, Lippincott, Philadelphia.
Caspar, D. L. D. and Klug, A. (1962) *Cold Spr. Harb. Symp. quant. Biol.* **27,** 1.
Chapman, S. K. (1980) *Understanding and Optimising Electron Microscope Performance. 1.* Transmission Microscopy. 2nd Edn. Science Reviews Ltd., London and Cambridge.
Crick, F. H. C. and Watson, J. D. (1956) *Nature (London)* **177,** 473.
Dalton, A. J. and Haguenau, F., (Eds) (1973) *Ultrastructure of Animal Viruses and Bacteriophages*, Vol. 5. Academic Press, New York and London.
Diosi, P. and Georgescu, L. (1979) *Arch. Roum. Path. exp. Microbiol.* T. **39,** 141.
Ellis, D. S., Simpson, D. I. H., Francis, D. P., Knobloch, J. Bowen, E. T. W., Lolik, P. and Deng, I. M. (1978) *J. clin. Path.* **31,** 201.
Esparza, J. and Gil, F. (1978) *Virology* **91,** 141.
Finch, J. T. (1964) *J. molec. Biol.* **8,** 872.
Hayat, M. A. (Ed) (1972) *Principles and Techniques of Electron Microscopy. Biological Applications*, Vol. 2 Van Nostrand Rheinhold, London.
Horne, R. W., and Wildy, P. (1961) *Virology* **15,** 348.
Jonathan, P., Bulter, G., and Klug, A. (1978) *Scient. Amer.* Nov. p. 52.
Kleinschmidt, A. K., Lang, D., Jackerts, D., and Zahn, R. K. (1962) *Biochim. biophys. Acta* **61,** 857.
Klug, A., and Caspar, D. L. D. (1960) *Advances in Virus Research*, Vol. 7, p. 225. Academic Press, New York.
Klug, A. and Finch, J. T. (1965) *J. molec. Biol.* **11,** 403.
MacNaughton, M. R., Madge, M. H., Davies, H. A., and Dourmashkin, R. R. (1977) *J. Virol.* **24,** 821.
Marks, R. W. (1960) *The Dymaxion World of Buckminster Fuller.* Reinhold, New York.
Matthews, R. E. F. (1982) *Classification and Nomenclature of Viruses.* Karger, Basel, New York.
Murphy, F. A. and Harrison, A. K. (1979) *Rhabdoviruses*, Vol. 1. CRC Press Inc. Florida.
Murti, K. G., Bean, Jr. W. J., Webster, R. G. (1980) *Virology* **104,** 224.
Nermut, M. V. (1980) *Arch. Virol.* **64,** 175.
Roseto, Escaig J., Delain E., Cohen J. and Scherrer R. (1979) *Virology* **98,** 471.
Valentine, R. C. and Pereira, H. G. (1965) *J. molec. Biol.* **13,** 13.
Waterson, A. P., and Wilkinson, L., (1978) *An Introduction to the History of Virology.* Cambridge University Press, Cambridge.
Williams, R. C., and Fisher, H. W. (1974) *An Electron Micrographic Atlas of Viruses.* Charles C Thomas, Illinois.
Williams, R. C. and Smith, K. M. (1958) *Biochim. biophys. Acta.* **28,** 464
Williams, R. C. and Wyckoff, R. W. G. (1946) *J. appl. Phys.* **17,** 23.
Wrigley, N. G. (1979) *Brit. med. Bull.* **35,** 35.

82

Virus replication

J. J. Skehel

Introductory

Viruses are intracellular parasites. They contain either RNA or DNA which may be single-stranded or double-stranded and they express this genetic information during replication using the protein synthetic apparatus of the host cells. The complete cycle of their replication, although differing in detail from virus to virus, may be divided into five stages which involve binding of the infecting virus particle to the cell and transfer of the virus genome to the cellular site of replication; transcription of virus nucleic acid to produce messenger RNAs; translation of messenger RNA into virus specific proteins; genome replication; and encapsidation and release of progeny virus particles. In this chapter each of these stages is considered in turn using a number of viruses as examples for each stage chosen to illustrate the variety of the mechanisms involved in the replication of different viruses.

Stage 1, virus infection

The initial interactions between viruses and cells occur between components of their surfaces. In the case of viruses containing membranes this is best understood in infections by myxoviruses in which defined glycoproteins of the virus surface bind specifically to sialic acid residues of cell membrane glycoconjugates. Each influenza virus membrane for example contains about 400 molecules of a rod-shaped glycoprotein, the haemagglutinin, which binds sialic acid in a pocket at the tip of the molecule. Since the three dimensional structure of the haemagglutinin has been determined, the structure of this binding site is known in detail. This situation contrasts with that for many other viruses in which neither the cellular nor the viral components involved in binding are known. However, a number of observations indicate that different cell membrane glycoproteins are recognized specifically by different groups of viruses. For example among the picornaviruses, polioviruses bind only to cells of primates and analyses of mouse-human hybrid cells indicate that only those containing the human chromosome 19 express the poliovirus receptor. Coxsackie B viruses on the other hand do not recognize these receptors and appear to bind to a much wider range of cells (Lonberg-Holm and Philipson 1980).

After such virus-cell associations the genetic material of the virus is transferred across the cellular membrane to initiate infection. For viruses containing membranes this process involves membrane fusion. Again the most detailed information on the virus components concerned in mediating transfer is available for myxoviruses. The membranes of the parainfluenza virus, Sendai virus, for example, contain two sorts of glycoprotein, one of which is specifically implicated in the fusion of the virus membrane with the plasma membrane of the cell to be infected. Similarly, the haemagglutinin glycoproteins of influenza viruses, in addition to binding the viruses to cell membrane sialic acid-containing receptors, are also directly involved in fusion. In these cases, however, as for many other viruses, fusion occurs between virus membranes and the membranes of endocytic vesicles in which the adsorbed viruses are taken inside by endocytosis, and is specifically triggered by changes in endosomal pH (Helenius *et al.* 1980).

Introduction of the genetic components of non-enveloped viruses into cells is poorly understood, as indeed is the transmembrane movement of any macromolecule. From work with bacteriophage it may be imagined that specific membrane proteins such as those normally involved in carrier functions may facilitate entry of either virus or virus nucleic acid into the cell but supportive evidence for these possibilities is not available.

The final targeting of the virus genetic components requires their delivery to the nuclear or cytoplasmic sites at which replication begins. The signals for differential distribution presumably are located on the proteins of the infecting virus complex but the processes concerned in selection or intracellular movement of the complexes are not understood.

Stage 2, messenger RNA production

For some RNA viruses the infecting nucleic acid is a messenger RNA and translation of the virus genome into the proteins required for virus replication proceeds directly. For others, the so-called negative-stranded RNA viruses, and the viruses which contain DNA or double-stranded RNA, messenger RNAs are produced by transcription of the infecting virus genome.

In general, transcription of DNA into messenger RNA in eukaryotic cells involves a number of processes including recognition of a specific DNA sequence by RNA polymerase II; initiation of transcription and capping of the transcript at or near the initiating 5′-terminal nucleotide; elongation of the complementary polynucleotide chain; nucleolytic processing of the primary transcript; polyadenylation at the 3′-terminus of the processed RNA; and splicing of the capped and polyadenylated RNA to remove intervening sequences and ligate the exons of the mature messenger RNA. This scheme, which is summarized in Fig. 82.1, appears to be accurately undertaken through the recognition by the appropriate enzymes of certain sequences in the DNA and the transcripts. Many genes have regions of homologous structure within fifty residues of the messenger RNA synthesis initiation site containing the sequence TATAAA which may be concerned in initiation. Similarly many messenger RNAs contain the sequence AAUAAA at their 3′-termini which may specify the site for polyadenylation; splicing also appears to occur at regions of the transcript containing short homologous

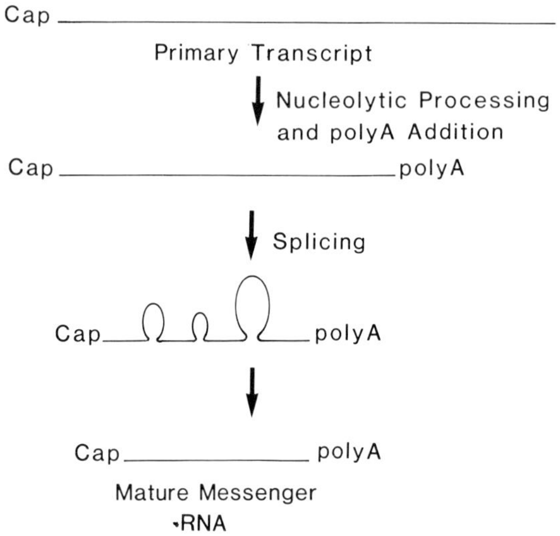

Fig. 82.1(a) Messenger RNA synthesis is divided into three stages in this scheme involving (1) The initiation of DNA transcription about 25 nucleotides downstream from a TATA sequence occurs with 'Cap' addition at the 5′-terminus of the primary transcript and is followed by transcription which often terminates beyond sites at which poly A is subsequently added; (2) The primary transcript is nucleolytically processed and poly A is added at sites near the AAUAAA sequences; (3) Intervening sequences are removed during splicing to produce mature messenger RNA. GU sequences at the 5′-termini of intervening regions and AG sequences at their 3′ termini are the most usual features of 'consensus sequences' at which splicing occurs.

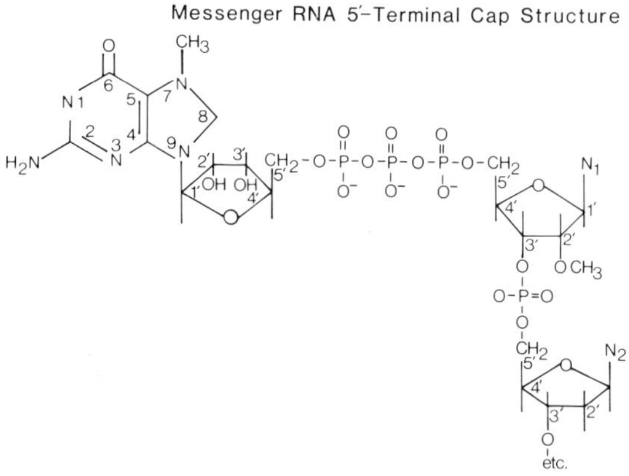

Fig. 82.1(b) The terminal 7-methylguanosine residues of messenger RNAs and the penultimate nucleotides are joined by their 5′-hydroxyl groups through a triphosphate bridge. The nature of the penultimate residue varies with the messenger RNA with regard to the nucleotide base and its degree of methylation. In some messenger RNAs the penultimate nucleotide is not methylated; in others the 2′ hydroxyl is methylated, and in some cases the 2′ hydroxyl of the third nucleotide is also methylated.

sequences: —GU at the 5′— and AG— at the 3′ ends of the intron sequences (Breathnach and Chambon 1981).

The transcription of DNA viruses other than poxviruses occurs in the nucleus of the infected cell and similar processes appear to be responsible for the generation of virus messenger RNAs. Indeed the discovery of transcript splicing was made in the course of analyses of adenovirus transcription. This virus is considered as an example of how DNA-containing viruses express their genetic information (Ziff 1980).

The transcription of DNA viruses is divided into two phases, early and late, by the time of onset of virus DNA replication. In the early phase of adenovirus replication messenger RNAs are produced from five distinct regions of virus DNA; in three of these regions one strand of virus DNA is used as template, in the other two regions the opposite strand is transcribed. Transcription of each region leads to the production of several different species of messenger RNA which have common capped 5′-termini differing in length and sequence as a result of splicing and polyadenylation of the primary transcript at different positions.

Most messenger RNAs produced late in infection also fall into five groups. However, in these cases the groupings are based on the sharing of common sites of polyadenylation. All the messenger RNAs appear to share a common 5′-terminal sequence which is formed by the splicing of three short regions of the primary transcript. This situation is shown diagrammatically in Figure 82.2. The primary transcripts are initiated at position 16, and extend beyond the polyadenylation sites on occasions to the end of the genome at position 100. They are then nucleolytically processed, polyadenylated at one of the five specified sites, each of which is near an AAUAAA sequence, and finally spliced to generate the multiple messenger RNAs all of which contain the same capped 5′ terminal sequence. Description of the controls concerned in adenovirus messenger RNA production is far from complete. The available evidence is based on studies of transcription in cells incubated in the presence of inhibitors of protein or DNA synthesis or in cells infected with temperature-sensitive mutants. These provide evidence indicating that early messenger RNA synthesis can occur in the absence of protein synthesis and that the normal shut-off of its production is prevented under these conditions. Thus, newly synthesized protein presumably specified by the virus is required to regulate early gene expression. This conclusion is supported by studies of temperature-sensitive mutants which indicate particularly that regulation of adenovirus early messenger RNA production is influenced by a virus-specific DNA binding protein which is a product of an early virus gene. How this regulation is accomplished is, however, not yet known. Inhibitors of virus DNA synthesis on the other hand do not affect the controlled transcription of the early genes. They prevent the change in transcription pattern from early to late by blocking the synthesis of late messenger RNAs; this conclusion is also supported by studies of temperature-sensitive mutants blocked in DNA synthesis.

Transcription of poxvirus DNA, which differs from other DNA viruses in occurring in the cytoplasm, is also divided into early and late phases before or after the onset of DNA replication. However, in this case, at least early in infection, there is no evidence that cell enzymes play a part in either messenger RNA synthesis, capping or polyadenylation. Instead, viral enzymes which fulfil all these functions are packaged in the infecting virus particle. Discovery of the pox-

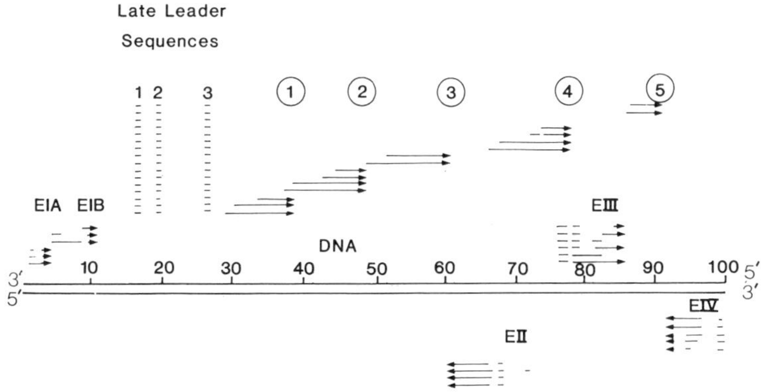

Fig. 82.2 Adenovirus messenger RNA synthesis. Different messenger RNAs are produced during the 'early' and 'late' stages of virus replication, before and after the onset of virus DNA synthesis respectively. The 'early' messenger RNAs are in five transcription units, designated E1A, E1B, E11, E111 and E1V, from each of which several different messenger RNAs are produced by splicing at different sites in the primary transcripts. 'Late' messenger RNAs form five groups (①-⑤) with the members of each group sharing a common poly A addition site. In addition the messenger RNAs in all five groups have a common 5′ capped sequence. The adenovirus genome can be divided into 100 units as shown and the common 5′ capped sequence is derived from sequences at positions 16, 19 and 27, which together, by splicing, form the tripartite 5′ leader sequences.

virus transcriptase in the late 1960s marked the initial awareness of the importance in virus replication generally of virion-associated enzymes. In further distinction, as far as is known at present, poxvirus messenger RNA production does not demand RNA splicing which, therefore, appears to be a specifically nuclear event.

Those RNA viruses whose infecting genomes do not function as messenger RNAs all have transcriptase activity. The most famous is without doubt the reverse transcriptase of retroviruses which converts virus RNA into double-stranded DNA. As far as messenger RNA synthesis is concerned, the virus DNA, after its integration into cellular chromosomes, appears to be transcribed by similar processes to those normally concerned in nuclear transcription. In addition, since the RNA in virus particles has the same polarity as messenger RNA, complete transcripts of the integrated double-stranded DNA are also the final products of genome replication.

The genomes of viruses found to contain double-stranded RNA all appear to be segmented (Joklik 1974). In the case of reovirus ten double-stranded RNAs which function as ten distinct genes are present in each virus particle. All of them are transcribed into messenger RNA by a virion polymerase in a process similar to DNA transcription which does not require displacement of either strand of the double-stranded RNA segment. Only one strand of each segment is used as a template for messenger RNA. The messengers produced are complete transcripts of that strand; they are capped by virion enzymes (discovery of which led to the finding of cap structures on all cell messenger RNAs) but unusually they appear not to be polyadenylated.

Negative-stranded RNA viruses contain either a single RNA molecule or, as with influenza viruses, a number of distinct RNA segments. They can also be divided on the basis of the cellular sites at which transcription occurs; only influenza virus transcription occurs in nuclei. Vesicular stomatitis virus, a rhabdovirus, is perhaps the most studied of those viruses containing a single RNA; it was the first single-stranded RNA virus to be shown to contain a transcriptase. Its RNA molecule has a molecular weight of about 3.6×10^6 and is transcribed into five messenger RNAs, each of which is capped and polyadenylated by enzymes contained in the virus particle. The precise method of production of the five distinct messenger RNAs is however not known. The two possibilities are that either the genome is completely transcribed into a single RNA which is then nucleolytically processed to produce the five messenger RNAs, or that each messenger RNA is independently initiated and terminated. RNA sequence analyses indicate that each messenger RNA contains a common 5′-terminal sequence AACAG and that the virion RNA contains five copies of the sequence $AUACU_7G/CA$ which mark the 3′-ends of the messenger RNAs. Whether these sequences are used as signals for initiation and termination or are recognition sites for processing nucleases is unknown.

Transcription of the eight RNAs of influenza viruses occurs in the nucleus; the messenger RNAs produced are capped and polyadenylated. However, although virus particles contain RNA transcriptase activity, they do not contain enzymes responsible for capping. Rather they contain a nuclease which appears to generate capped primers about 12 nucleotides in length from nuclear DNA transcripts synthesized by RNA polymerase II. These primers are used to initiate the virus transcripts and, as a consequence, each virus messenger RNA contains a capped 5′-terminal extension, heterogeneous in sequence and derived from nuclear DNA transcripts. As in VSV transcription, polyadenylation at the 3′-terminus of messenger RNAs is signalled by U rich sequences which are present 16 nucleotides before the 5′-terminus of the template RNAs. Whether termination of messenger RNA synthesis occurs at these sequences or whether complete transcripts are processed nucleolytically has not been determined. In addition to eight messenger RNAs which are co-linear with their eight RNA templates, three more capped and polyadenylated virus-specific messenger RNAs are produced from transcripts of the two smallest virus RNAs by splicing. Probably as a consequence of the nuclear location of transcription, therefore, the variety of influenza messenger RNAs appears to be increased.

The control of transcription in RNA virus replication is less well substantiated than for DNA viruses. In negative-stranded RNA viruses production of messenger RNA by transcription of the infecting genome is often called primary transcription, and subsequent transcription of newly synthesized virus RNA is termed secondary. Possibly as the result of an increased number of virus RNA templates, once RNA replication is underway, secondary transcription usually occurs at an enhanced rate compared with primary transcription. In addition, in a number of virus infections, transcription of different virus genes occurs at different rates at different times after infection. The mechanism of how such control is brought about, and its significance for the efficacy of virus replication are not known.

Finally in this section, although messenger RNAs are the only products resulting from virus DNA transcription, RNA virus replication requires the synthesis of transcripts which serve as templates for virion RNA synthesis. Unlike messenger RNAs, by definition such transcripts are complete copies of virus RNAs, and the transcriptase complex from which they are produced, although containing the same template virus RNA, must differ from those taking part in messenger RNA synthesis. The nature of this difference is not known but newly synthesized virus proteins appear to be continuously required during complete transcript RNA synthesis which may modify transcriptase activity.

Stage 3, virus protein synthesis

In eukaryotic cells, protein synthesis is initiated by the formation of an 80S ribosome-messenger RNA complex which requires a number of initiation factors, GTP, and the initiating methionyl-transfer RNA, as well as specific recognition of messenger RNA 5′-terminal cap structures. Polypeptide chain elongation proceeds with the binding of amino acyl-transfer RNAs and peptide bond formation directed by different elongation factors until a chain termination codon in the messenger RNA. At this point the peptidyl-transfer RNA bond is hydrolysed in a reaction involving termination factors, and the completed polypeptide is released from the ribosome complex (Kerr 1975). Virus messenger RNAs, especially those synthesized by transcription, are completely compatible with this synthetic system and in some cases appear to be particularly effective in directing polypeptide translation. In other instances, however, they differ from cellular messenger RNAs in either lacking a cap structure at their 5′-termini, in lacking poly A sequences at their 3′ termini, and in containing the information for

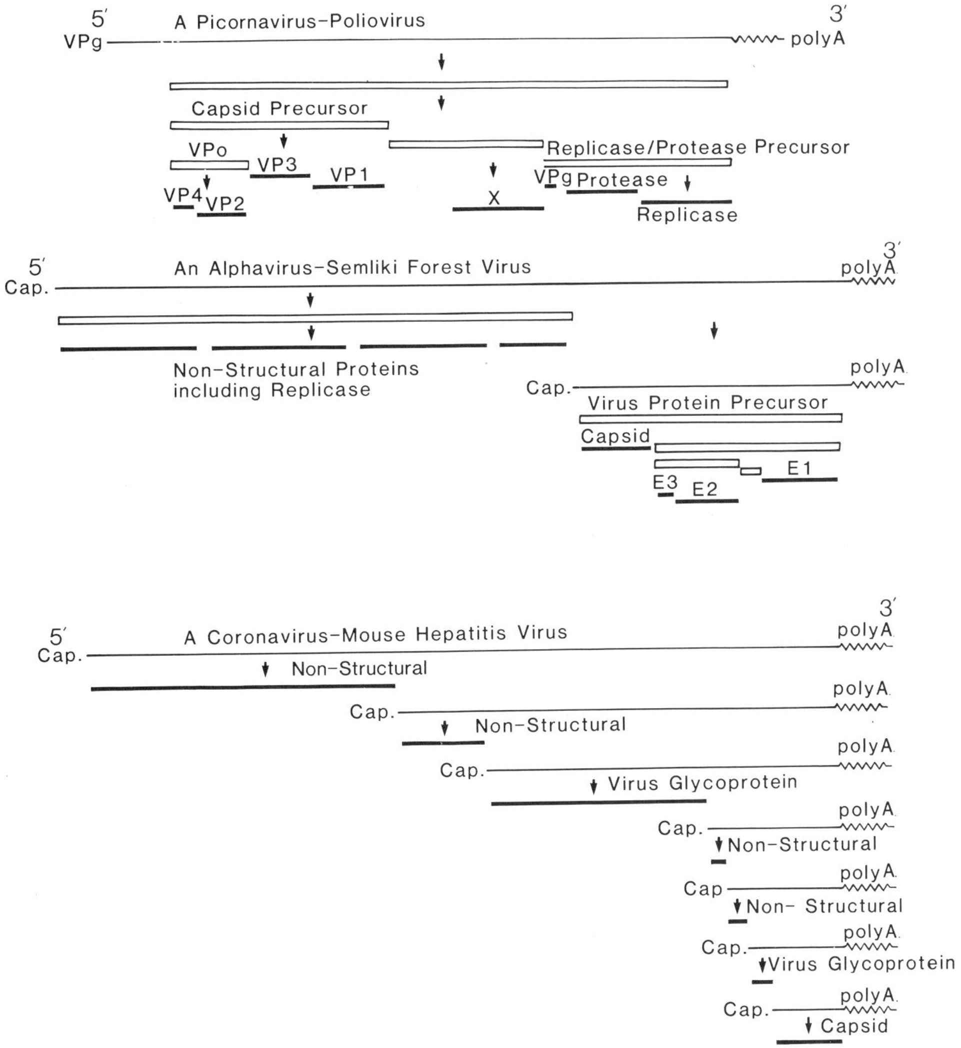

Fig. 82.3 Translation of plus-stranded RNA virus messenger RNAs. 1) The genome RNAs of picornaviruses are translated into polypeptide precursors which are proteolytically processed to generate the enzymes and capsid proteins required for virus replication. The proteases involved appear to be virus-coded. 2) Alphavirus protein synthesis requires two messenger RNAs; virus RNA and a 3′-terminal short RNA which contains the information for virion protein synthesis. The enzymes involved in virus RNA synthesis are produced by direct translation of virus RNA. 3) In coronavirus replication distinct messenger RNAs are produced for each virus-specific protein.

the synthesis of more than one protein. Picornavirus RNAs, for example, which are infectious and therefore can function as messenger RNAs, contain in one messenger RNA molecule all the information required for the replication of the virus. Furthermore, instead of being capped they have a small protein covalently linked to the 5′-terminal nucleotide. The consequences of these properties are firstly that individual picornavirus proteins are formed by proteolytic cleavage of a precursor polypeptide which is the product of translation of the entire virus genome and secondly that the translation machinery of the cell is modified to accept the uncapped virus messenger RNA. A scheme for translation of poliovirus RNA is shown in Fig. 82.3.

Other infectious virus RNAs appear to utilize similar proteolytic processes in expressing their genetic information (Fig. 82.3). Alphavirus genomes for example are capped and function directly as messenger RNAs, although it appears that in these cases only the enzymes concerned in RNA synthesis and precursor polypeptide processing are translated from the complete virus RNA. The proteins found in virus particles are translated from smaller messenger RNA, which are the more abundant virus messenger RNAs in polysomes, and are derived from the 3′ terminal one-third of the virus RNA. A sufficient supply of proteins for virus assembly may be ensured in this way. Regulation of the amounts of different virus proteins produced is achieved in a similar manner by viruses which preferentially transcribe different genes at specific times after infection, as described for DNA viruses in the previous section.

The protein products of virus messenger RNA translation, at least of the smaller viruses, are mainly implicated as enzymes in virus nucleic acid synthesis and as components of virus particles. They are transferred to the cellular sites at which they function presumably by normal cellular processes. For example, the glycoproteins of virus membranes are associated with membranes during their synthesis by hydrophobic amino-terminal signal sequences just as their cellular counterparts, and they are glycosylated during transfer across the membranes by cellular enzymes which recognize sites for glycosylation equivalent to those present on cellular glycoproteins. Other virus proteins may be modified by phosphorylation or activated by proteolytic cleavage, and it can be expected that when the proteins of larger viruses, such as those of the herpes and poxvirus groups are characterized, virus proteins of a considerably greater variety of structure and functions will be recorded.

Stage 4, genome replication

Virus DNAs are either single-stranded or double-stranded, linear or circular, and the replication processes are somewhat different for each type (Challberg and Kelly 1982). The small linear single-stranded DNAs of parvoviruses are replicated by cellular polymerases most effectively in rapidly dividing cells. They are initially converted into double-stranded molecules, replicative forms, in a reaction primed by the 3′-terminus of the infecting DNA, which is presented in a loop structure formed by base pairing with the neighbouring sequence (Fig. 82.4). It is proposed that the hairpin-like double-stranded intermediate is then nicked at a position opposite the initiation site to generate a 3′-hydroxyl from which the chain is elongated to complete the replicative form. Single-stranded DNA replication subsequently necessitates copying of the template strand of replicative form with concomitant displacement of virus DNA.

Replication of the linear double-stranded adenovirus DNA also requires specialized terminal structures. Replication can be initiated at either end of the double strand and is primed by an early adenovirus protein, part of which remains covalently attached to the 5′ terminus of the new DNA strand. Synthesis from 5′ to 3′ then proceeds continuously with displacement of the parental strand of the same polarity. In turn the displaced strands serve as templates in the formation of double-stranded molecules by synthesis of complementary DNA strands. Cellular DNA polymerases again appear to be concerned in chain elongation but, in addition to the virus protein used in initiation, another virus specified protein with affinity for single-stranded DNA may also be required in this process.

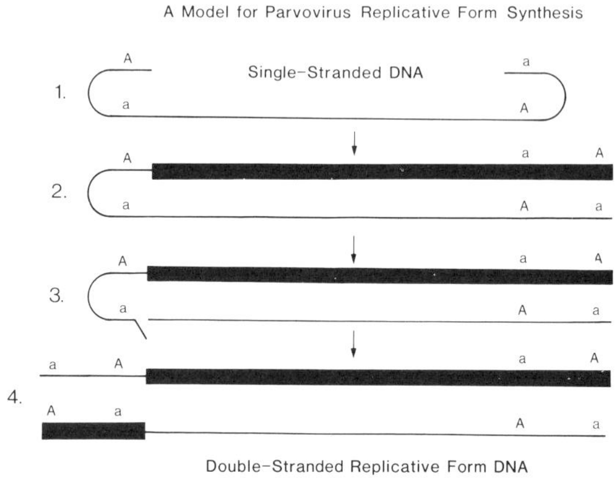

Fig. 82.4 The first stage in the replication of single-stranded DNA virus genomes consists of the synthesis of double-stranded replicative form (RF) DNA molecules. Synthesis of the DNA strand complementary to virus DNA is initiated by self priming at the 3′-terminus, 1. A and a represent regions of complementary sequence in virus DNA at both 5′ and 3′ termini which allow terminal loop structures to form. DNA synthesis proceeds to the 5′ terminus of virus DNA to produce the hairpin structure, 2. The virus DNA strand of this duplex is nucleolytically cleaved opposite the site of initiation, 3, and DNA synthesis re-initiates at the 3′-OH group so produced to complete the synthesis of the replicative form DNA, 4.

For viruses containing circular double-stranded DNAs, the papovavirus SV40 may be taken as an example. In this case replication begins at a unique site at which an early virus protein, the T antigen, appears to promote unwinding of the DNA helix. DNA synthesis by host polymerases proceeds in both directions from this origin and both strands are copied simultaneously. Since all nucleic acid biosynthesis proceeds in a 5′ to 3′-terminal direction, the simultaneous synthesis of both strands requires that one must grow in the direction opposite to the movement of the other (Fig. 82.5). This is accomplished by repeated initiation

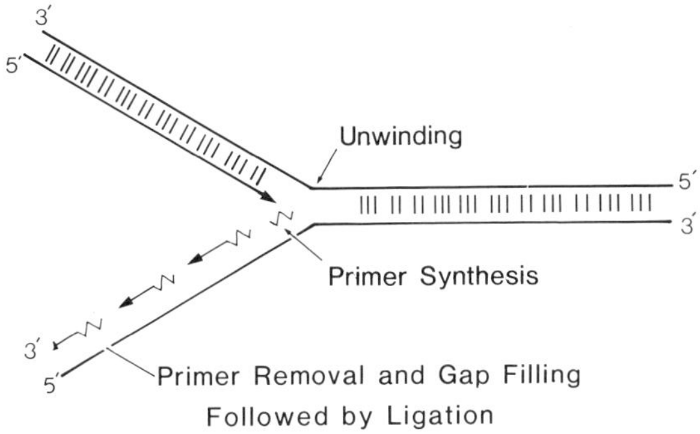

Fig. 82.5 A summary of the processes involved in discontinuous DNA replication. In this scheme one strand of the double-stranded DNA template is supposed to be copied continuously, the other discontinuously. Discontinuous synthesis is primed by small RNA molecules nine to ten nucleotides long which are removed when the length of the nascent DNA fragments is between 100 and 200 nucleotides. The gaps produced are then filled in before ligation to complete the new DNA strand.

and synthesis of short DNA segments complementary to one strand of the template, followed by ligation of the segments to produce a complete strand. Replication is therefore semi-discontinuous; in each direction one strand is produced as a continuous polymer, the other in a discontinuous fashion in segments up to 300 deoxynucleotides long. As in cellular DNA replication, discontinuous synthesis is primed by short RNA molecules about ten nucleotides in length and ligation of the DNA products is preceded by primer removal and DNA segment extension to fill the resulting discontinuity.

Retrovirus replication also needs the synthesis of double-stranded DNA, in this case by reverse transcription of virus RNA, but unlike other virus DNAs this replicative intermediate requires to be integrated into cellular chromosomes to function in replication (Weiss *et al.* 1982). Initiation of DNA synthesis by reverse transcriptase is primed by host transfer RNA molecules packaged in virus particles which contain sequences complementary to and associated with a short region of virus RNA near the 5′-terminus (Fig. 82.6). Reverse transcription proceeds from this site of initiation to the 5′-end of virion RNA. The diagram of virus RNA structure in Fig. 82.6 shows that at the 5′ terminus there is a sequence about 80 nucleotides long which is exactly repeated at the 3′-terminus; the next stage in reverse transcription appears to involve displacement of the DNA-enzyme complex from this 5′-terminal repeat and its transfer to form a similar association with the 3′-terminal repeat. Synthesis of the complementary DNA strand then proceeds by reverse transcription towards the 5′-terminus of the template virion RNA. Synthesis of the second DNA strand is initiated once reverse transcription has proceeded beyond a specific site near the 3′-terminus of the RNA template. In this case the template for double-stranded DNA synthesis is the nascent reverse transcript, and extension of the second strand beyond the site of initiation of this template occurs after completion of reverse transcription. Synthesis of the first strand does not however terminate at this point but is continued by copying the second DNA strand to the position equivalent to the 5′-terminus of virus RNA. In this way the completed double-stranded DNA product contains copies of the 5′-terminal RNA sequence, the terminal repeat and the 3′ terminal RNA sequence at both ends (Figure 82.6). Minor modifications occasioning the removal of 2 base pairs from each end complete the structure of the provirus which is found integrated in many different regions of the cell genome, and transcription of provirus DNA leads to the production of virus RNA.

By comparison the replication of other virus RNAs appears much less tortuous. Reoviruses replicate their double-stranded RNA by encapsidating messenger RNAs and copying each of them once to form double-stranded molecules within the subviral particle. All other RNA viruses replicate their genomes by forming complete transcripts of virion RNA which are then used as templates for RNA replication. In many cases details of the enzymes responsible are not known but, for example in poliovirus RNA synthesis, it appears that initiation of replication involves the covalent linkage of the 5′-terminal nucleotide to a virus polypeptide. The polypeptide is subsequently proteolytically cleared, possibly to release the replicase required for RNA elongation, leaving a short peptide attached to the 5′-terminus of the newly synthesized RNA chain. Consequently, in this process each replicase would produce a single virus RNA. For negative-stranded RNA viruses the mechanism of distinction between complete transcript production and the synthesis of the subgenomic or incomplete transcripts which function as messenger RNA is not known. It has been suggested that certain virus proteins interact with specific sequences in virus RNAs to influence the course of transcription and that the choice between complete or incomplete transcript formation may depend on the abundance of such proteins in infected cells at different times after infection. With regard to selection of the correct transcript as a

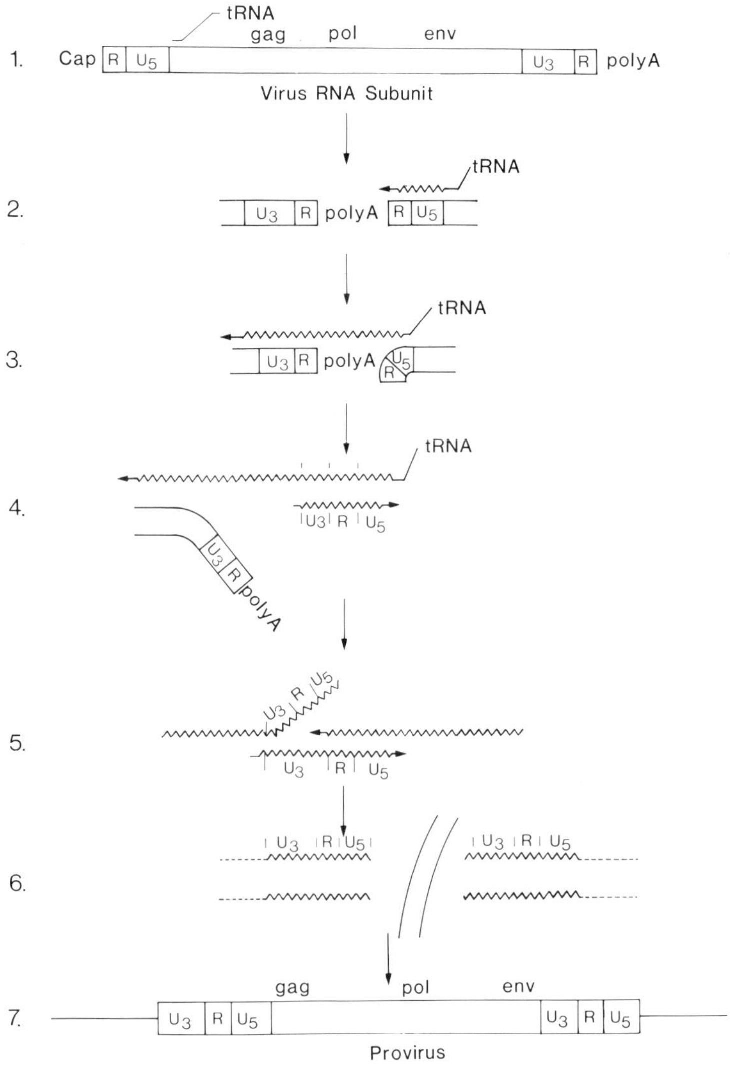

Fig. 82.6 Reverse transcription of retrovirus RNA to produce double-stranded DNA. 1) One of the two identical subunits of a retrovirus genome to show the 5′-terminal cap and the 3′-terminal poly A; R, a short sequence about 80 nucleotides long present at both ends of the subunit; U5 and U3, sequences unique to the 5′ and 3′ ends of virus RNA, respectively, which vary in length from about 80 to 1000 nucleotides in different viruses; transfer RNA bound to the site near U5 at which DNA synthesis begins; and the location of the genes for the virus specific proteins, gag, pol, and env, which encode the virus core proteins, the reverse transcriptase and the virus membrane proteins respectively. 2) Complementary DNA, cDNA, synthesis is initiated from the host tRNA primer bound at a site adjacent to U5, and reverse transcription proceeds to the 5′ terminus of the virus RNA subunit. 3) The nascent cDNA strand, indicated by ←∿∿∿, associates with the repeat sequence R at the 3′-terminus of either the same virus RNA subunit or possibly the other subunit of the diploid genome. This requires displacement of the 5′-terminal R sequence, and is followed by extension of the reverse transcript using the newly associated region of RNA as template. 4) Synthesis of the second DNA strand begins when reverse transcription has passed the U3 region. The primer for initiation of this strand is not known. Using the newly synthesized cDNA as template, the second strand is extended to the tRNA primed initiation site of cDNA. 5) cDNA synthesis is completed but is extended using the second DNA strand as template so that it contains U3 R U5 sequences at both ends. Second strand synthesis continues using the completed cDNA as template. The primer tRNA is removed. 6) Double-stranded DNA synthesis is complete with the linear molecule having identical sequences at both ends. 7) Double-stranded DNA is integrated into cell chromosomes at many sites as provirus DNA, which lacks only two base pairs from each end of linear double-stranded DNA.

template for replication, it is certainly a common observation that terminal RNA sequences are conserved within groups of viruses, which suggests that they have a functional importance. They might serve as recognition signals for the replicase molecules.

Stage 5, virion assembly

The final stages of virus replication necessitate interactions of virus nucleic acids with proteins to form structures with either helical or icosahedral symmetry which are released from the cell during cell lysis by budding through cell membranes, or in some cases by unknown methods of transmembrane movement. The nucleic acids concerned in forming the symmetrical structures may be naked as, for example, the RNA of alpha viruses which combines with a single type of virus protein to form the icosahedral virus nucleocapsid; or more rarely they may be complexed with proteins, as is SV40 DNA, which appears to associate with cellular histones to form a minichromosome structure. The molecular basis of the specificity of encapsidation is not known. It seems reasonable to assume that the initial reaction between capsid protein and virus RNA in the formation of helical nucleocapsids occurs at a specific sequence on the virus RNA and, from studies of rhabdovirus nucleocapsid formation using RNA fragments, it seems that in this case a site near the 5′-terminus of virus RNA is required for assembly.

Sometimes subviral protein structures may be assembled before nucleic acid association. Such a situation appears to occur in poliovirus replication where a procapsid is formed which subsequently binds and encapsidates virus RNA. Proteolytic cleavage of specific procapsid polypeptides which is required for the formation of infectious virus particles follows RNA encapsidation. Similarly, DNA-free protein shells serve as precursors to virions in the nuclei of adenovirus-infected cells and a number of proteins are processed by proteolysis during virus maturation.

In addition to the nucleic acid-protein and protein-protein interactions concerned in nucleocapsid formation, the assembly of viruses which contain membranes requires modification of cell membranes by the insertion of virus-specific transmembrane glycoproteins (Simons and Garoff 1980). In the budding of viruses from cell membranes it is surmised that nucleic acid-containing subviral particles interact with the cytoplasmic domains of the glycoproteins and in so doing induce the modified membrane to envelope them. The specificity of these reactions appears to vary between groups of viruses. The icosahedral nucleocapsids of alphaviruses, for example, interact specifically with their membrane-associated glycoproteins to yield virus particles containing equal numbers of virus capsid proteins and each membrane glycoprotein. The interaction between the helical nucleocapsids of myxo-, paramyxo- and rhabdoviruses on the other hand is mediated by additional virus proteins, the M proteins, and, although M protein-nucleocapsid associations are virus-specific, in mixed infections viruses can be obtained containing the glycoprotein components of unrelated viruses. This implies that M protein-glycoprotein interactions are less specific. Nevertheless it is striking that cellular membrane glycoproteins are excluded from the virus assembly process. In the final maturation of a number of membrane-containing viruses proteolytic cleavage of specific virus glycoproteins is also required to generate infectious virus particles. Examples of glycoproteins processed in this way are the fusion glycoprotein of parainfluenza viruses and the haemagglutinin of influenza viruses.

Division of viruses into replication groups

This overview of the processes taking part in the replication of viruses and particularly the considerations of virus messenger RNA transcription and translation provide a basis for dividing viruses into six groups (Fig. 82.7). Each group contains viruses with a range of genetic complexity but all members of a group employ similar mechanisms of gene expression.

1 The first group is made up of viruses containing double-stranded DNA. These have the widest range of genome size from about 3×10^6 in molecular weight for papovaviruses to 160×10^6 for poxviruses and the DNAs may be linear or circular. They mainly replicate in the nucleus, only poxviruses being cytoplasmic; some such as herpes viruses and poxviruses have membranes; others such as adenoviruses are assembled as naked icosahedra. All produce messenger RNAs by DNA transcription as in cellular gene expression. The enzymes used for transcription may be packaged in virus particles as in the poxviruses, or the nuclear polymerases of the cell may be used with

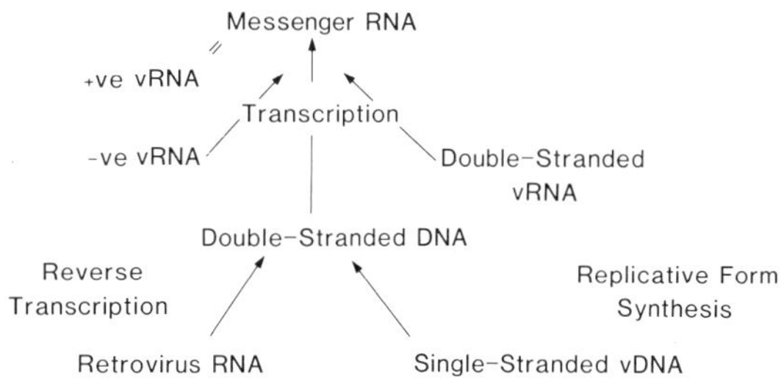

Fig. 82.7 Strategies for the expression of virus genomes.

or without modification, as in papova-, adeno- and herpes virus replication.

2 Members of the second group also contain DNA, in this case single-stranded between 1 and 2×10^6 in molecular weight. These are the parvoviruses. Their replication involves the formation of double-stranded DNA replicative forms; messenger RNA appears to be transcribed from one strand of this template. Cell polymerases are required in transcription, which occurs in the nucleus.

3 Group three contains the retroviruses which also produce messenger RNA by transcription of a double-stranded DNA replicative intermediate. In these cases, however, the DNA is produced by reverse transcription of genome RNA and is subsequently integrated into cell chromosomes. Integrated DNA then appears to be transcribed by cell enzymes like other cellular genes.

Viruses in the remaining three groups contain only RNA and their replication is independent of DNA synthesis.

4 The double-stranded RNA viruses of the fourth group, which includes reoviruses and rotaviruses, produce messenger RNA by conservative transcription of double-stranded virus RNA using virus-specified enzymes which form part of the virus particles. All double-stranded RNA viruses reported contain segmented genomes; their replication appears to be completely cytoplasmic and host enzymes are not involved in RNA synthesis.

5 The negative-stranded RNA viruses are the fifth group. Their replication can either be exclusively cytoplasmic as for the rhabdoviruses or may involve the nucleus as in influenza replication. The genomes may consist of a single RNA molecule as in the paramyxoviruses and rhabdoviruses or may contain a number of RNAs, for example eight in the case of influenza viruses, three for the bunyaviruses. Messenger RNA is produced by transcription of genome RNA; for the cytoplasmic viruses, only virus-specified enzymes are implicated. The primers used to initiate myxovirus transcription in nuclei, however, are formed by cellular polymerases and both virus and cell polymerases are, therefore, concerned in influenza messenger RNA synthesis.

6 Finally those viruses which contain infectious RNA, the plus stranded viruses, form the sixth group. Initial expression of the genome in these cases requires direct translation of the infecting nucleic acid to produce the proteins concerned in RNA replication. Certain members of the group such as the picornaviruses use only complete virus RNA as messenger RNA throughout replication; others produce subgenomic messenger RNAs, possibly to amplify the synthesis of particular gene products. Thus, for example, alphaviruses form the proteins of their virus particles by translation of a 1.4×10^6 molecular weight messenger RNA equivalent to the 3′-terminal end of the 4×10^6 molecular weight virus RNA, and coronavirus translation involves the formation of five or six 3′-co-terminal messenger RNAs which appear to contain the information for distinct virus products. Apart from Nodamura virus, which contains two positive-stranded RNAs, all members of group six contain a single RNA molecule.

This grouping system provides a simple framework for comparing the properties of different viruses and, together with an appreciation of the differences in size and morphology of virus particles, is an introduction to the variety in virus replication strategies.

References

Breathnach, R. and Chambon, P. (1981) *Annu. Rev. Biochem,* **50,** 349.

Challberg, M.D. and Kelly, T.J. (1982) *Annu. Rev. Biochem.* **51,** 901.

Helenius, A., Kartenbeck, J., Simons, K. and Fries, E. (1980) *J. cell Biol.* **84,** 404.

Joklik, W.K. (1974) Reproduction of Reoviridae. In: *Comprehensive Virology*, Vol. 2, pp. 231–4. Ed. by H. Fraenkel-Conrat and R.R. Wagner. Plenum Press, New York.

Kerr, I.M. (1975) Translation Control of Protein Synthesis. In: *Society for General Microbiology Symposium.* **25,** pp. 29–50. Ed. by D.C. Burke and W.C. Russell. Cambridge University Press, Cambridge.

Lonberg-Holm, K. and Philipson, L. (1980) Molecular aspects of virus receptors and cell surfaces. In: *Cell Membranes and Viral Envelopes*. Ed. by H.A. Blough and J.M. Tiffany. Academic Press, London.

Simons, K. and Garoff, H. (1980) *J. gen. Virol.* **50,** 1.

Weiss, R., Teich, N., Varmus, H. and Coffin, J. (Eds.) (1982) *RNA Tumor Viruses.* Cold Spring Harbor Laboratory, New York.

Ziff, E.B. (1980) *Nature, Lond.* 287, 491.

83

The genetics of viruses

C. R. Pringle

1 Introductory

Viruses display a range of genetic diversity unparalleled in any other group of organisms. In this chapter the viruses discussed are limited to those which infect man and higher animals. The genetic systems considered here do not represent all the known categories, and only indirect reference will be made to the wide range of viruses infecting invertebrates, higher and lower plants, fungi, protozoa, algae, mycoplasmae and bacteria. It is worth remembering, however, that these wide taxonomic gaps are bridged by some viruses; for instance, many of the insect-transmitted viruses which infect vertebrates are able to multiply in their invertebrate hosts and a few plant viruses such as lettuce necrotic yellows virus—a rhabdovirus—are able to multiply in their invertebrate vector.

Virus genetics is concerned with elucidation of the precise structure of the genomes of the various groups of viruses and the extent to which this determines the biological properties and disease-producing potential of viruses. 'The strategy of the viral genome' is a concept which has been invoked to embrace the organization and function of genetic information in viruses (Subak-Sharpe 1971, Baltimore 1971, Subak-Sharpe and Pringle 1975). The other major concern of virus genetics is delineation of the pattern and origin of virus variation, both in terms of virus evolution and the temporal changes of the antigenicity and pathogenicity of viruses which contribute to the prevalence of human and animal disease.

The methodology of virus genetics has relied predominantly on the isolation of induced or spontaneous mutations usually as conditional lethal mutants, followed by their functional classification by complementation test or by the construction of genetic recombination maps (see Section II). Recently genetic analysis has undergone a profound technological revolution with the use of restriction endonucleases in the analysis of viral DNA, the molecular cloning of fragments of viral nucleic acids and their amplification in plasmid vectors, and the introduction of rapid methods of sequencing nucleic acids. As a consequence it has become possible to determine the precise structure of viral genes and even entire genomes, in terms of their nucleotide sequences, and to construct mutants to order (Weissmann *et al.* 1979, Shortle *et al.* 1979). Virus genetics has entered a phase of rapid expansion and this chapter can provide no more than an appreciation of the principles and potential of these new approaches; space does not permit a detailed account of recent advances.

Sufficient sequencing of viral nucleic acids has been completed to show that the universality of the genetic code remains inviolate, although there may be preferences in the use of particular codons (see Section 4.2). Irregular coding assignments, such as the reading of the termination codon UGA as tryptophan in mitochondrial DNA (Barrell *et al.* 1979), have not been found so far in human or animal viruses, whether DNA-containing or RNA-containing.

In recent years research in animal virology has promoted a more flexible concept of the basic unit of heredity—*the gene*. In the 1950s analysis of the fine structure of bacteriophage genes replaced the original concept of Beadle and Tatum (1941) of one gene—one enzyme, by the more precise concept of one gene—one polypeptide chain (protein monomer). Recent research on animal viruses has necessitated further modification to provide a less rigid concept of the gene. This was a consequence of the discovery that the same DNA (or RNA) sequence may specify more than one gene product. The transcription of messenger RNA (mRNA) in two or even three reading frames is possible so that two or three polypeptides may be derived from the same nucleic acid sequence. In other words genes may overlap adjacent genes and are blurred, and the genome can no longer be equated to a linear array of discrete and independent elements like beads on a string. Additional complexity is introduced by the splicing together of non-adjacent segments of genetic material, either during the synthesis of mRNA or, at least with eukaryotic genes, as a result of rearrangement of the genome itself during differentiation. Overlapping genes and discontinuous genes, where a common leader sequence is spliced to different unique regions, are devices which contribute to the remarkable genetic economy of animal viruses. The most obvious consequence of overlapping and discontinuous genes is that the coding capacity of small genomes is increased, although both these features may be an attribute of all genomes irrespective of their size (Dulbecco 1979). Perhaps the greatest single contribution of animal virology to biology, however, has been the discovery and characterization of the reverse transcriptase enzyme of RNA tumour viruses (Temin and Mizutani 1970, Baltimore 1970). For the first time it became apparent that information transfer in biological systems was not exclusively in the direction DNA→RNA→PROTEIN, but that in certain situations information could flow in the reverse direction RNA→DNA→RNA→PROTEIN (see also Fig. 83.2).

One of the most important roles of genetics in virology has been its contribution to the clear discrimination of virus-specific processes from those which are part of the biosynthetic apparatus of the host cell and essential for the viability of both the virus and its host. Successful chemotherapy of virus infection depends on the development of anti-viral drugs which affect only virus-specific processes and not those common to virus and cell. The isolation of virus mutants is the best and often the only means of identifying virus-specific processes and defining appropriate targets for pharmacological attack.

2 Methods of genetic analysis

2.1 Mutation and mutants

Genetic variants or mutants are the raw material of genetic analysis and the source of variation upon which evolution depends. Mutation can be defined as a discontinuous event which results in a change in information content. Virus mutants occur spontaneously or can be induced by chemical or physical agents (mutagens). Every nucleotide along the genome is capable of alteration and each alteration is a potential change of phenotype (the discernible properties of the virus). The degeneracy and wobble in the genetic code, however, mean that not all nucleotide changes will bring about the introduction of a different amino acid during translation into protein. Perhaps half of all single base changes do not result in a detectable mutation and exist only as *silent mutations*.

Point mutations are substitutions of one base (or base pair in double-stranded molecules) by another, resulting in change of nucleotide sequence without any change in molecular size. Reversal of this change restores the original sequence and results in true *reversion*. *Frameshift mutations* arise by insertion or deletion of single base pairs. This has the effect of altering the entire gene from the site of mutation to the end of the molecule. Because the triplet reading frame is displaced by one base the effect of frameshift mutation is thus more extensive than point mutation. Reversion of frameshift mutations occurs by restoration of the original reading frame by a compensating addition or deletion of a base distal to the original mutation. Thus reversal of phenotype occurs by *suppression* rather than reversion.

Extensive changes in polynucleotide sequence can

occur by rearrangement, duplication and deletion. Reversion of *deletion mutants* rarely, if ever, occurs and non-revertibility is a diagnostic feature. Spontaneous deletion mutants of viruses are common and occur in all virus groups. Particles with sequences deleted frequently interfere specifically with the replication of the intact normal genome and are known as DI (defective interfering) particles. It has been suggested that DI particles play an important natural role in moderating the course of infection until the immune response becomes effective (see section 7.1).

Mutants may have special phenotypes such as resistance to heat inactivation or an inhibitory drug. For instance guanidine-resistant mutants of poliovirus and foot-and-mouth disease virus have played a key role in mapping the genome of these viruses (see section 6.1). Plaque morphology mutants are also common and mutants of herpes simplex virus which induced formation of a syncytial plaque instead of the normal clear lytic plaque have contributed to the genetic mapping of this virus (see section 4.4). In general, however, mutants with specific phenotypes have proved less useful than *conditional lethal mutants*. The vast majority of conditional lethal mutants of animal and human viruses are *temperature-sensitive* (*ts*) mutants. These mutants are unable to multiply in susceptible cells at high temperature, but are unrestricted at lower temperatures. The restrictive temperature depends on the particular virus type and even strain of virus, but it is usually in the range 38–40°. The permissive temperature is arbitrary and usually in the range 31–33°. The value of *ts* mutants is that a single phenotype is sufficient to yield a wide range of mutants. In principle, conditional lethal mutation can be obtained in any gene whose function is essential for normal replication. This is because a change in an amino acid may produce a conformational change which affects the stability of the whole molecule at high temperature.

Host-restricted mutants, analogous to the amber mutants of prokaryotes, are rare, probably because mammalian cells with the characteristics of the suppressor strains of prokaryotes have not yet been identified. Mutants of herpes simplex virus have been described by Cremer *et al.* (1979) which have characteristics suggesting they may in fact be *nonsense* (*chain-terminating*) *mutants*. These workers found that the phenotype of thymidine kinase-deficient mutants of HSV-1 was suppressed in an *in vitro* translation system supplemented with tRNA from suppressor strains of yeast. Normal enzyme activity was restored by serine-inserting UAG suppressor tRNA in the case of one mutant and an intact but inactive thymidine kinase protein was synthesized by addition of serine- and leucine-inserting UGA suppressor yeast tRNAs. It was concluded that the former mutant (HSV TK4$^-$) was an amber (UAG) mutant and the latter (HSV TKJ$^-$) an opal (UGA) mutant. In general, however, *host restriction* of animal viruses is usually due to lack of an appropriate receptor at the cell surface. Conditional host range mutants of negative-strand RNA virus have been isolated capable of replicating in some types of cells at the restrictive temperature but not in others. The *td*CE mutants of VSV and Chandipura virus are mutants of this type, which are unable to multiply in chick cells at 39°. There is some evidence that this phenotype is determined by mutation affecting the viral RNA polymerase (Pringle 1978).

Spontaneous mutants occur at low frequency, but the frequency of mutation can be increased by the use of mutagenic agents. *Mutagens* which have proved effective with human and animal viruses include base analogues (e.g. 5-fluorouracil and 5-azacytidine for RNA viruses and 5-bromodeoxyuridine for DNA viruses), alkylating agents (e.g. ethyl methane sulphate, diethyl sulphate), intercalating agents (proflavine and N-nitro-N-nitrosoguanidine), deaminating agents (nitrous acid), hydroxylamine and ultraviolet light irradiation. The base analogues are incorporated into the viral genome and mutations are produced by miscoding during replication. The other mutagens induce mutation by direct chemical change. Frameshift and nonsense (polypeptide chain-terminating) mutants of human or animal viruses have rarely been unequivocally identified and most induced mutants appear to be *mis-sense mutants* where substitution of an amino acid modifies the functional activity of a gene product.

2.2 Genetic interactions

Analysis of the genetic interactions which occur between mutant viruses provides information about the structure of viral genomes, the control of expression of genetic information and the mechanism by which genes are arranged into new combinations. The ultimate aim of genetic analysis is to identify all the genes of a virus and to assign each gene to a specific synthetic or structural function. This goal can be approached in two stages. Initially mutants can be allocated to specific genes by *complementation analysis*, which is a test of non-identity of function and does not require any prior knowledge of gene functions. Once different genes are identified, the properties of individual mutants representing these genes can be characterized biochemically to identify the precise nature of the genetic lesions.

Complementation is observed when cells are co-infected by mutants with lesions in different genes. The defective gene product of one virus is compensated by the normal gene product of the other and vice versa, with the result that viable progeny is produced. The progeny virus remains genotypically mutant, since the interaction between the two viruses occurs only at the level of gene products. *Ts* mutants can be classified into complementation groups by infecting cells with pair-wise combinations of mutants and then incubating them at the restrictive temperature. If the *ts* mutants affect the same function no progeny will be produced, but if they are located in different genes infectious virus will be produced. The extent of complementation can be gauged by calculation of a *complementation index* (CI) i.e.

$$\mathrm{CI} = \frac{\text{yield } (\mathrm{A}+\mathrm{B})^{\mathrm{RT}}}{\text{yield } \mathrm{A}^{\mathrm{RT}} + \text{yield } \mathrm{B}^{\mathrm{RT}}}$$

where A and B are any pair of *ts* mutants, and the superscript RT is the restrictive temperature of incubation. Since the progeny by complementation will be mutant, the assay of the yields must be carried out at the permissive temperature. A positive complementation index indicates non-identity of function. No complementation is merely a negative result, rather than proof of identity.

The various genes can then be orientated by *recombination analysis*. Mutation is the source of genetic variation, and recombination the principal mechanism by which mutations are spread through a population of viruses. Recombination rearranges the genetic constitution of different viruses to produce new combinations and the number of combinations will be $2n$ for any two parental viruses differing at n genetic sites. Recombination can take two forms. Viruses with segmented genomes (see sections 7 and 8) undergo recombination by exchange of complete genome subunits. This form of recombination is known as *reassortment* and is formally equivalent to the random segregation of different chromosomes in eukaryotes. No DNA virus has a segmented genome; therefore the phenomenon of reassortment is confined to RNA viruses. The other form of recombination is *molecular recombination* involving the exchange of genetic material between homologous nucleic acid molecules. A *cross-over* event produces a recombinant molecule with sequences derived from different parents. The frequency of cross-over events is a function of the position of two mutations on the genome. The closer the two mutations the lower the probability of a cross-over event and the separation of the two mutations by recombination. This is the basis of genetic mapping. Mutations in close proximity show *linkage*, whereas distant mutants behave as if they are unlinked and randomly reassort. Genetic maps are constructed by measuring the frequency of recombination between pairwise combinations of mutants and ranking them in order. Theoretically, recombination frequencies should be additive over the entire genome if the possibility of cross-over is constant per nucleotide. However, it is usually observed that recombination frequencies are additive only over short distances. *Three-factor crosses* are necessary to orientate these additive regions and obtain an unambiguous linkage map (see section 4).

Recombination is a phenomenon common to all DNA viruses and genetic maps have been constructed for several mammalian DNA viruses (see section 4). Recombination has also been detected with two different viruses belonging to the picornavirus group (see section 6), but no other RNA virus exhibits recombination apart from the reassortment associated with segmented RNA viruses. The mechanism of recombination between non-segmented RNA genomes remains to be elucidated. High-frequency recombination is also observed with RNA tumour viruses, and recombination is a normal consequence of replication of the genome (Hunter 1978).

Recombination between *ts* mutants is measured as a frequency (percentage):

$$\text{Recombination frequency (RF)} = 2 \times 100 \times \frac{\text{yield } (\mathrm{A}+\mathrm{B})^{\mathrm{RT}} - (\text{yield } \mathrm{A}^{\mathrm{RT}} + \text{yield } \mathrm{B}^{\mathrm{RT}})}{\text{yield } (\mathrm{A}+\mathrm{B})^{\mathrm{PT}}}$$

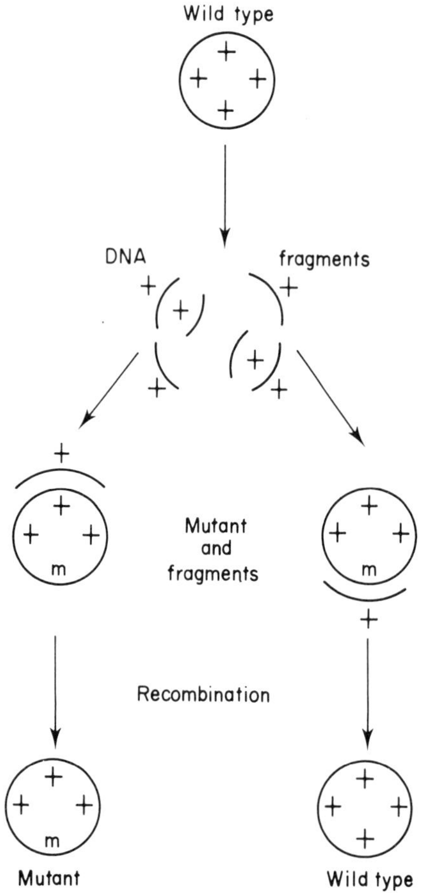

Fig. 83.1 Marker rescue by DNA fragments. Cells are infected with specific restriction endonuclease-generated fragments of wild type DNA and a *ts* mutant. Only the fragment carrying the wild type allele of the mutated gene (m) can give wild type recombinants.

where A and B are any pair of *ts* mutants and RT is the restrictive temperature, and PT the permissive temperature. The single and mixed infections are carried out at permissive temperatures and yields assayed at the temperatures indicated by the superscript. The factor 2 is included because only the wild type recombinant is measured by this assay. It is assumed that the other recombinant (the double *ts* type) is present at equal frequency.

Restriction endonucleases are enzymes which cleave DNA molecules at specific sites. Each restriction endonuclease recognizes a specific sequence of 4–6 nucleotide base pairs. Since these sequences will occur only rarely, discrete numbers of fragments are generated by enzymes with different specificities and physical maps of the genome can be constructed from the patterns of fragments.

The genetic and physical maps of DNA viruses can be correlated by inter-type recombination (see section 4) or by using the technique of marker rescue. The principle of *marker rescue* is illustrated in Fig. 83.1. Cells are infected with intact DNA from a *ts* mutant, and specific fragments of wild type DNA. Wild type recombinants will be obtained when the DNA fragment contains sequences homologous to the region of the genome carrying the *ts* mutation. Thus the *ts* lesion can be located at a specific region of the genome. *Reverse marker rescue* has been used to introduce mutations into specific genes of several DNA viruses. In this instance the cells are infected with wild type virus and specific fragments of DNA which have been modified by treatment with a chemical mutagen (e.g. nitrous acid). Recombination incorporates the mutated sequence into the wild type genome, and the recombinants are identified and cloned by screening for a convenient phenotype such as temperature sensitivity.

3 The animal virus genome

The viruses of man and animals fall into three distinct classes; namely, the DNA-containing viruses, the RNA-containing viruses and the retroviruses which alternate between RNA in the virion and proviral DNA in the host cell. The coding of genetic information in RNA, rather than DNA, is unique to viruses (animal, plant and bacterial) and RNA is the predominant molecular form of the genome in both animal and plant viruses. Indeed it was thought at one time that all plant viruses were RNA viruses.

The *molecular weights* (*mol.wt*) of the genomes of DNA viruses extend over a 100-fold range from the 1.5×10^6 of the parvoviruses to the 160×10^6 of the poxviruses. By contrast the RNA viruses are more uniform, since the mol. wt of the genome vary within a three-fold range only (or four-fold if the retroviruses are included). The DNA viruses, with the exception of the iridoviruses and the poxviruses, multiply in the nucleus of the host cell and RNA is transcribed from the viral DNA by the host-specified transcription apparatus. As a consequence the naked viral DNA stripped of its protective protein coat is infectious when introduced into an appropriate host cell. This is true both for the minute single-stranded parvoviruses and the large double-stranded herpesviruses which have genomes of considerable structural complexity (see section 4). The hepatitis B virus genome with a chain length of 3200 nucleotides ($\sim 1 \times 10^6$ mol.wt) is the smallest animal virus genome.

Viral nucleic acid molecules may be linear or circular, single-stranded or double-stranded, covalently bonded to protein, or present as unique or diploid subunits, according to the type of virus. The structural characteristics of the genomes of the different virus groups are given in Tables 83.1A and B together with the relative coding capacity of the various viral genomes. The *relative coding capacity* provides a measure of the minimum amount of genetic information present in any viral genome and is derived in the following manner. The ratio of the average nucleotide mol.wt to the average amino acid mol.wt is taken to be 321:110 or 2.9:1 approximately. The triplet genetic code assigns 3 nucleotides to each amino acid. Therefore the relation between nucleic acid mol.wt and polypeptide mol.wt is $2.9 \times 3{:}1$, or 10:1 as a convenient approximation in view of the uncertainties inherent in the calculation. As a rule of thumb calculation the relation between genome mol.wt and gene product mol.wt is 10:1 for single-strand viruses and 20:1 for double-strand viruses. Thus the 4×10^6 mol.wt genomic RNA of vesicular stomatitis virus (*Rhabdoviridae*) is sufficient to code for polypeptides with a total mol.wt of 400 000, which is in fact close to the 396 000 total mol.wt of the known viral proteins. This is probably exceptional, however, since in other viruses the coding capacity of the genome is greatly expanded by the existence of overlapping and discontinuous genes. Nevertheless, although the relative coding capacity assumes that genes are unique and non-overlapping, it remains a useful index for comparison of the minimum genetic information content of the various virus groups.

The diverse nature of the viral genome can be rationalized by regarding the genome as an intermediate stage in the cycle of replication of viral nucleic acid that has been sequestered in an extracellular particle (Fig. 83.2). Figure 83.2 excludes consideration of the segmentation of the viral genome which probably represents a modification towards greater control of transcription. The only stage in the replication cycle

Table 83.1 The physical and genetic characteristics of the genomes of animal viruses

A The DNA viruses

			Nucleic Acid					Genetic properties			
Group	Example	Site of replication	Type	Strand in virion	Form	Infect-tivity	Mol. wt $\times 10^6$	Relative* coding capacity	Comple-mentation groups	Recom-bination frequency	Special features
Parvo	Murine Minute	Nucleus	SS-DNA	–	Linear	ND	1.5	1.5×10^5	ND	ND	Inverted terminal repeats. Mitosis-dependent. Helper stimulated in some cells. Genes overlap.
	Adeno-associ-ated	Nucleus	SS-DNA	– and + in separate particles	Linear	Yes	1.5	1.5×10^5	ND	ND	Inverted terminal repeats. Helper and mitosis-dependent in all cells. Genes overlap.
Unassigned	Hepatitis B	?	Partially DS-DNA	(+)/–	Circular	Yes	1.6 (2.1)	10^5	ND	ND	Smallest viral genome. Nicked complete –ve strand, incomplete (55–85%) +ve strand. Particle DNA polymerase converts partial DS-DNA (1.6×10^6) to complete DS-DNA (2.1×10^6). Circular but not supercoiled. Overlapping coding sequences.
Papova†	SV 40	Nucleus	DS-DNA	+/–	Super-coiled helix	Yes	3.0	1.5×10^5	5	Low	Genes overlap. Genome read in more than one reading frame.
Adeno	Adeno 5	Nucleus	DS-DNA	+/–	Linear	Yes	23.0	1.2×10^6	17	Variable, Linear map	Inverted terminal repeats. Protein linked to both termini.
Her-peto	HSV-2	Nucleus	DS-DNA	+/–	Linear	Yes	100.0	5×10^6	23	Variable, Linear map	Inverted terminal and internal repeats.
Pox	Vaccinia	Cyto-plasm	DS-DNA	+/–	Linear	No	160.0	8×10^6	ND	Variable	Inverted terminal repeats. Cross-linked at ends. DNA dep. RNA polymerase and other enzyme activities.
Irido	Insect irridescent	Cyto-plasm	DS-DNA	+/–	Linear	ND	160.0	8×10^6	ND	ND	ND

* The total polypeptide specifying capacity of the genome read in one frame with no overlaps or splicing (See Text).

† The papillomavirus sub-group of the papovavirus group have a larger genome (5×10^6 mol. wt), but these viruses have not been propagated *in vitro* and their genetic characteristics are not well defined.

ND = no data.

B The RNA viruses

			Nucleic acid					Genetic properties			
Group	Example	Site of replication	Type	Strand in virion	Form	Infect-tivity	Mol. wt $\times 10^6$	Relative* coding capacity	Comple-mentation groups	Recom-bination frequency	Special features
Pic-orna	Polio 1	Cytoplasm	SS-RNA	+	Linear	Yes	2.6	2.6×10^5	Weak	Low Linear map	One polycistronic-message; post-translational cleavage. Protein at 5′ terminus.
Toga (Alpha)	Sindbis	Cytoplasm	SS-RNA	+	Linear	Yes	4.0	4×10^5	6	None	Two polycistronic messages; post-translational cleavage.
Toga (Flavi)	Dengue	Cytoplasm	SS-RNA	+	Linear	ND	4.0	4×10^5	ND	None	(Polycistronic with internal initiation of translational?)
Cor-ona	IBR	Cytoplasm (Nucleus?)	SS-RNA	+	ND	Yes	6–8.0	$6–8 \times 10^5$	ND	ND	ND
Bunya	Snowshoe hare	Cytoplasm	SS-RNA	–	3 unique sub-units	No	6.4	6.4×10^5	(3)	High 3 Groups	Virion RNA dep. RNA Polymerase. Circular nucleocapsids.
Arena	Pichinde	Cytoplasm	SS-RNA	–	2 unique sub-units	No	3.2	3.2×10^5	ND	High 2 Groups	Virion RNA dep. RNA Polymerase
Ortho-myxo	Influ-enza A	Cytoplasm (and Nucleus?)	SS-RNA	–	8 unique sub-units	No	5.0	5×10^5	(8)	High 8 Groups	Virion RNA dep. RNA Polymerase
Para-myxo	Sendai	Cytoplasm	SS-RNA	–	Linear	No	7.0	7×10^5	7	None	Virion RNA dep. RNA Polymerase
Rhab-do	Vesicular Stomatitis	Cytoplasm	SS-RNA	–	Linear	No	4.0	4×10^5	6	None	Virion RNA dep. RNA Polymerase.
Dip-lorna	Reo	Cytoplasm	DS-RNA	+/–	10 unique sub-units	No	15.0	7.5×10^5	(10)	High 10 Groups.	Virion DS-RNA dep. RNA Polymerase.
Retro	Rous Sarcoma	Nucleus and cytoplasm	SS-RNA	+	2 identical sub-units	RNA No DNA Yes	6.0	3×10^5	ND	High	Virion RNA dep. DNA Polymerase. Post-translational cleavage.

* The total polypeptide specifying capacity of the genome read in one frame with no overlaps or splicing (See Text).
ND—No data.

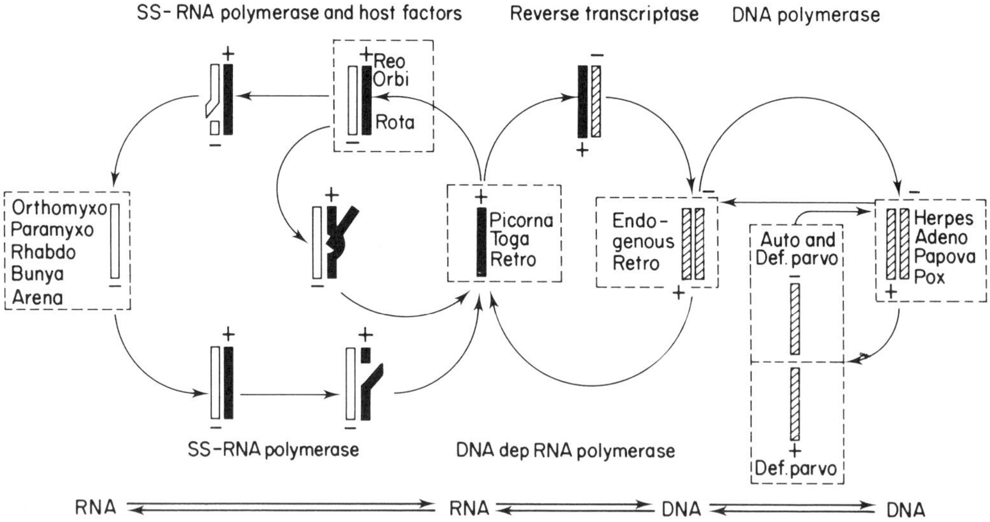

Fig. 83.2 The stages of the replication cycle of viral nucleic acid sequestered in virions.

RNA replication follows the five-intermediate model with the plus strands in black and the negative strands in white. DNA strands are cross-hatched. The stages sequestered in virions are indicated by the dashed boxes.

The segmentation of the genome of some negative and all double-stranded RNA viruses is not indicated, because subdivision of the genome is probably a device to enhance control of transcription. The only form not represented among known viruses is the RNA/DNA intermediate in the reverse transcription cycle.

of viral nucleic acids, excluding the replicative intermediates, not represented among known viruses is the RNA:DNA intermediate in the reverse transcription pathway.

The RNA viruses are predominantly cytoplasmic viruses and some but not all can multiply in enucleated cells (Pringle 1977*a*). It is now an accepted convention to designate a single-strand RNA molecule as the *plus-strand* if it functions as mRNA in protein synthesis, and its complement as the *negative-strand* or anti-message (Baltimore 1971). Figure 83.2 shows that the genome of RNA viruses may be represented by either a plus- or a negative-strand, or both. The genome of the plus-strand non-segmented RNA viruses is infectious because it can function directly as mRNA for synthesis of the entire complement of viral proteins including the polymerase necessary for replication of the viral genome. The genome of negative-strand RNA viruses, on the other hand, is not infectious because it cannot function as mRNA or be replicated. Since mammalian cells contain no enzymes which can replicate or transcribe RNA templates, RNA viruses must introduce these enzymes (or the information for their synthesis in the case of plus-strand viruses) into the host cell with the infecting virion. The nucleocapsid of negative-strand viruses includes an RNA-dependent RNA polymerase as an integral part of its structure so that mRNA synthesis can be initiated. The genome of retroviruses also functions as mRNA and is thus the plus-strand. Retrovirus RNA is not infectious, probably because it is dependent on preformed virion-associated reverse transcriptase for its replication. The proviral DNA form of the genome is infectious, however, since it can be transcribed by the host cell DNA-dependent RNA polymerase and includes the genetic information for synthesis of the viral reverse transcriptase. The virus-specified origin of the retrovirus and negative-strand virus polymerases has been verified by the isolation of conditional lethal virus mutants with defects in polymerase activity (Szilagyi and Pringle 1972, Verma *et al.* 1974).

The DNA of the cytoplasmic poxviruses is not infectious and several enzyme activities including a DNA-dependent RNA polymerase are associated with the virion. The presence of the virion RNA polymerase appears to be essential for infectivity. Indeed the discovery of the first virion-associated enzyme, the DNA-dependent RNA polymerase of vaccinia virus, by Kates and McAuslan (1967) was a crucial step in the understanding of the organization and function of viral genomes.

4 The DNA viruses

4.1 The parvovirus genome

The parvoviruses have single-stranded DNA genomes in the range 1.4–1.8 × 10^6 mol.wt (Table 83.1A). These viruses have the lowest information content of all mammalian viruses excluding the unassigned hepatitis B virus, with a coding capacity equivalent to polypeptide sequences of 1.5 × 10^5 total mol.wt (read in the one reading frame). This is more than adequate to accommodate the 9 × 10^4 mol.wt protein which is the only known gene product. The virions yield three distinct structural polypeptides A, B and C but tryptic peptide mapping indicates that their sequences overlap. In the autonomous parvoviruses the A protein, which is the largest, is a minor species which appears to be metabolically stable during maturation. Protein C is derived by proteolytic cleavage of B molecules in the virion. In the defective parvovirus, on the other hand, both proteins A and B are minor species and the main constituent of the virion is protein C (Tattersall 1978). The mode of synthesis of these sequence-related polypeptides is still the subject of research, but there appears to be only one gene product, a 9 × 10^5 mol.wt RNA, which is probably the messenger for the 9 × 10^4 mol.wt polypeptide. The role of the remainder of the genome is unknown, since replicative and transcriptive functions are provided by the host cell or by a helper adenovirus.

Unlike the oncogenic DNA viruses, parvoviruses are unable to induce cellular DNA synthesis. The *non-defective or autonomous* parvoviruses (e.g. Kilham rat virus, feline panleukopenia virus, minute virus of mice) require a cellular function expressed transiently during the late S phase of the cell cycle. Therefore non-cycling cells cannot support parvovirus replication. The *defective parvoviruses* (the adeno-associated viruses (AAV)) are totally dependent on functions supplied by a helper virus. In the absence of helper adenovirus, adsorption and penetration occur, but there is no AAV-specific synthesis of DNA, RNA or protein. By studying the helper activity of adenovirus *ts* mutants it has been established that the helper activity is a late function of the adenovirus genome and that adenovirus DNA synthesis is not required. The adeno-associated virus is probably defective in more than one function, since an incomplete helper function can be provided by herpes simplex virus. AAV DNA synthesis, transcription and protein synthesis occur in cells co-infected with herpesvirus, but no infectious AAV is formed and the block appears to be in the packaging of progeny DNA.

The stimulation of AAV synthesis in helper co-infected cells is accompanied by inhibition of the helper virus. This is probably the basis of the inhibition of the *in vivo* oncogenicity of adenovirus 12 in hamsters by AAV (or H − 1, an autonomous parvovirus isolated from human tumour cells).

The DNA found in the virions of defective AAV consists of nearly equimolar amounts of complementary single-stranded molecules, which anneal to form double-stranded molecules on extraction. The individual plus and minus strands are packaged in separate virions. The *non-defective (or autonomous) parvoviruses* only encapsidate the minus strand (i.e. the strand which is transcribed into mRNA). The structure of the 3′-end of the genome of the defective and non-defective parvoviruses differs (Fig. 83.3). The ends of both are inverted terminal repeat sequences so that the molecule can fold back on itself to form a hairpin structure at each terminus. In non-defective parvoviruses the 5′- and 3′-terminal repeats are dissimilar, whereas they are identical in the defective parvoviruses. The palindromic region at each end of the molecule of Kilham rat virus DNA, for instance, is estimated to be 90–150 nucleotides long. Therefore these structures constitute about 5 per cent of the genome and may be required for replication, since DNA polymerase could initiate polymerization from the hydrogen-bonded 3′-end. Indeed duplex molecules in the form of dimers and circles are intermediates in the replication of parvoviruses (Rose 1974, Tattersall and Ward 1978). The terminal redundancy of the ends of the defective parvoviruses may be responsible for the non-selective inclusion of complementary DNA strands in virions, in contrast to the selectivity of the autonomous parvoviruses with their unique 5′-and 3′-termini (Berns and Hauswirth 1978).

The parvoviruses provide two clear examples of *conservation of genetic information* by small genomes. On the one hand information capacity is conserved by the substitution of specific replicative functions by host- or helper virus-specified functions, and on the

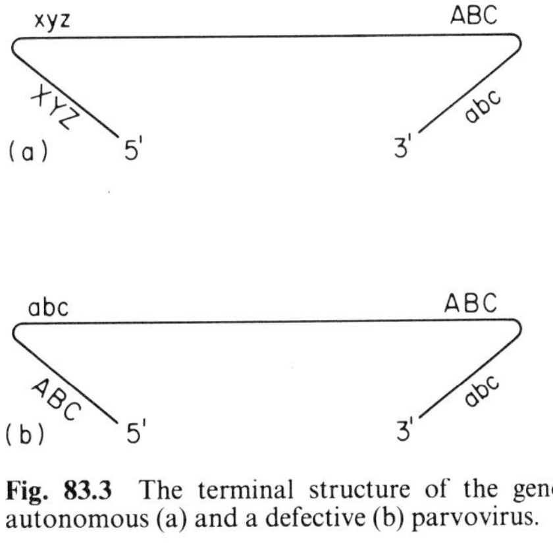

Fig. 83.3 The terminal structure of the genomes of an autonomous (a) and a defective (b) parvovirus.

other hand additional structural components are provided by proteolytic cleavage of a primary gene product.

The formal genetics of parvoviruses has not been investigated.

4.2 The papovavirus genome

The papovaviruses are the smallest of the double-stranded DNA viruses if hepatitis B virus is excluded. (Table 83.1A). The DNA is a single covalently closed circular DNA molecule of $3\text{–}3.5 \times 10^6$ mol.wt.* The sequence of the entire genome of three papovaviruses is known and consists of some 5290 nucleotide pairs (Fiers *et al.* 1978, Soeda *et al.* 1980, Yang and Wu 1979). Since the molecule is double-stranded the relative coding capacity is no greater than that of the parvovirus genome. However, the actual coding capacity is greatly expanded by the overlapping of genes. The identified gene products consist of 3 virion structural proteins and 3 early non-structural proteins (the T antigens), which have an affinity for DNA and play an important role in the interaction of the virus with the host cell.

Papovavirus circular DNA has approximately 20 histone particles associated with it when released from the virion. Complete deproteinization of papovavirus DNA produces *super-coiling* of the molecule. Nicking of either one of the DNA strands allows the super-coiled molecule to relax to a circular form. The super-coiled helix of deproteinized DNA is an indication that the binding of histone produces a three-dimensional alteration in DNA structure, which is compensated for by supercoiling of the molecule when the histone is removed (Germond *et al.* 1975). A double-strand cut produces a linear molecule. The super-coiled, circular and linear DNA molecules are all infectious.

Physical maps of the papovavirus genome have been obtained by using restriction endonucleases (Griffin 1980). The enzyme EcoRl produces a single cut which provides a convenient zero point for orientation of the genome, and a physical map of the genome can be constructed by using a variety of other restriction enzymes which produce multiple cuts in DNA. Conventionally the genome is marked out in decimal units in clockwise fashion from the EcoRl site (Figs. 83.5 and 83.6). The location of individual virus mutants can be determined by a variation of the *marker rescue* technique (see section 2). A fragment of wild-type DNA generated by restriction endonuclease cleavage is annealed to a single-stranded circular DNA obtained from a mutant. Cells infected with such a partial *heteroduplex* molecule will yield wild-type progeny if the fragment overlaps the mutated gene; if not, only mutant progeny will result.

Alternatively complete heteroduplex molecules can be produced by annealing single-stranded circular molecules obtained from wild-type and mutant viruses. The mismatched sequence at the site of mutation is susceptible to cleavage by the enzyme S1 nuclease which degrades single-stranded regions of DNA. Both DNA strands will be cut and the pattern of fragments produced after restriction endonuclease digestion will be altered. The map position of the mutated sequence can be established by comparison of the restriction endonuclease patterns produced with and without S1 nuclease treatment.

Figure 83.4 indicates the location of mutants representing four classes of *ts* mutants of SV40 virus. When a map of mRNA transcription is superimposed on the physical and genetic map it is apparent that the genome is divided into early (0.16–0.65 map co-ordinates) and late (0.65–0.15 map co-ordinates) regions. The early and late mRNA are transcribed from different DNA strands, and early reading frames do not correspond to late reading frames. During *the late lytic cycle* of polyoma virus, three viral RNAs appear in the cytoplasm after commencement of DNA replication, which code for the coat proteins VP1, VP2 and VP3.

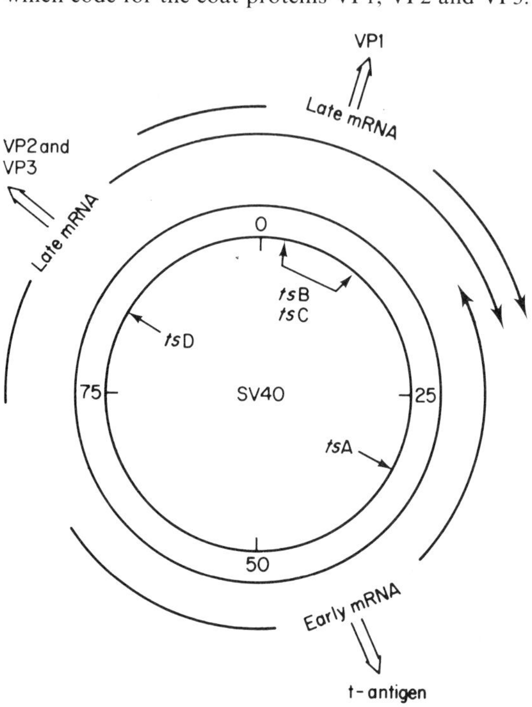

Fig. 83.4 A simplified functional and genetic map of the SV40 genome.

The locations of the four groups of *ts* mutants are indicated internally, and the regions coding early and late mRNA externally.

*The papilloma viruses which are responsible for common and genital warts have larger genomes (4.9×10^6 mol.wt). None of these viruses can be propagated *in vitro*; therefore nothing is known about their genetic organization, although physical maps have been prepared (zur Hausen 1977).

VP3 lies wholly within VP2 in the proximal part of the late region, and VP1 is located in the distal part (Fig. 83.5). The N terminus of VP1 and the C terminus of VP2/VP3 overlap and are read in different reading frames. In polyoma virus the overlap is 32 nucleotides long, and 122 nucleotides long in SV40. The VP coding regions of polyoma virus and SV40 show considerable homology, but contrary to expectation not the overlap region. *Overlapping genes* or sequences in different reading frames should be highly conserved, since any change in sequence would affect both genes simultaneously. Presumably the region of overlapping sequence does not code for functionally significant regions of the protein molecules (Soeda *et al.* 1980).

Figure 83.5 shows the structures of the mRNA for these three proteins. VP2 and VP1 have a 5′-terminal leader sequence spliced to the main body of the mRNA, and all three mRNAs terminate at a common site, i.e. they represent an overlapping set of 3′-coterminating molecules. They have variable non-coding 5′-terminal regions and the mRNAs for VP2 and VP3 have extensive untranslated 3′-regions. This shows that all three mRNA molecules are derived from the same segment of the genome. The 3′ half of the mRNA for VP2 is untranslated and there is a large deletion of sequences at the 5′ end of the VP1 mRNA. The mRNA for VP3 has both an untranslated 3′-region and deletion of sequences at the 5′-end. It has not been determined yet whether the signals for mRNA processing operate at the DNA level by control of transcription, or at the RNA level with a precursor RNA transcript controlling its own splicing. The late region of the genome cannot be read in a single reading frame; two are necessary.

The *early region of the genome* is transcribed before DNA replication begins and seems to be responsible for transformation of cells *in vitro* and induction of tumours *in vivo*. Three proteins are coded by this region and are designated as the *small*, *middle* and *large T antigens*. The structure and processing of the mRNA for these proteins is still a matter for research, but the genetic information for these three proteins is encoded

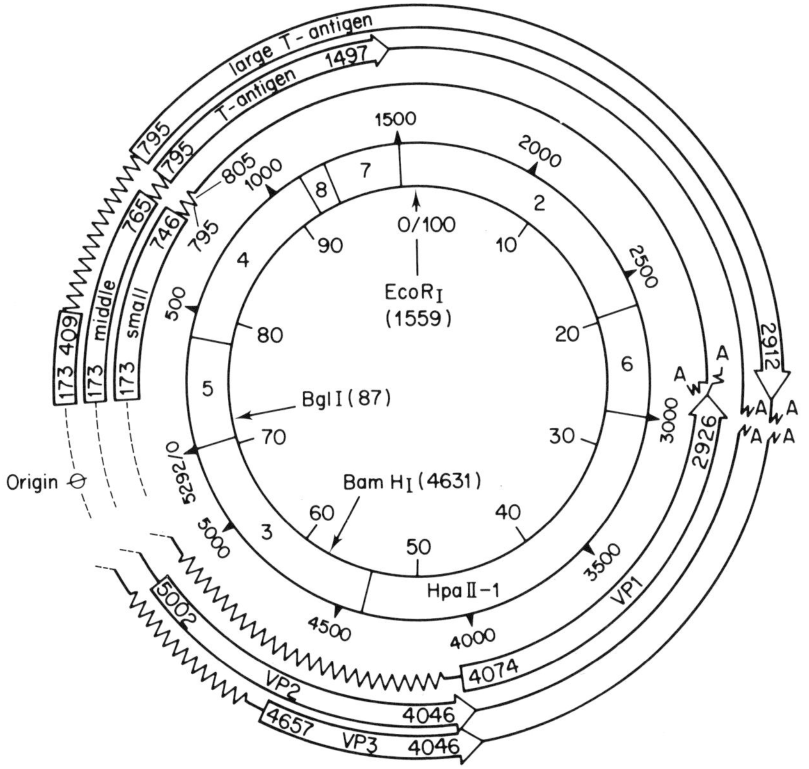

Fig. 83.5 Landmarks in the polyoma virus genome. The probable location of the three early virus-coded proteins (the small t, middle T and large T antigens) and the three late proteins (VP1, VP2, and VP3) are indicated as curved bars. The numbers in the curved bars are nucleotide indices (numbered in a clockwise sequence from an arbitrary site). Parts of mRNA not coding for proteins are indicated as single curved lines. DNA sequences absent from mRNA as a result of splicing are indicated by jagged lines. VP2/VP3 are encoded in different reading frames as are the central regions of middle T and large T. The coding frames are indicated in Fig. 83.6. (Reproduced with permission from Soeda *et al.* (1980), *Nature* **283,** 449).

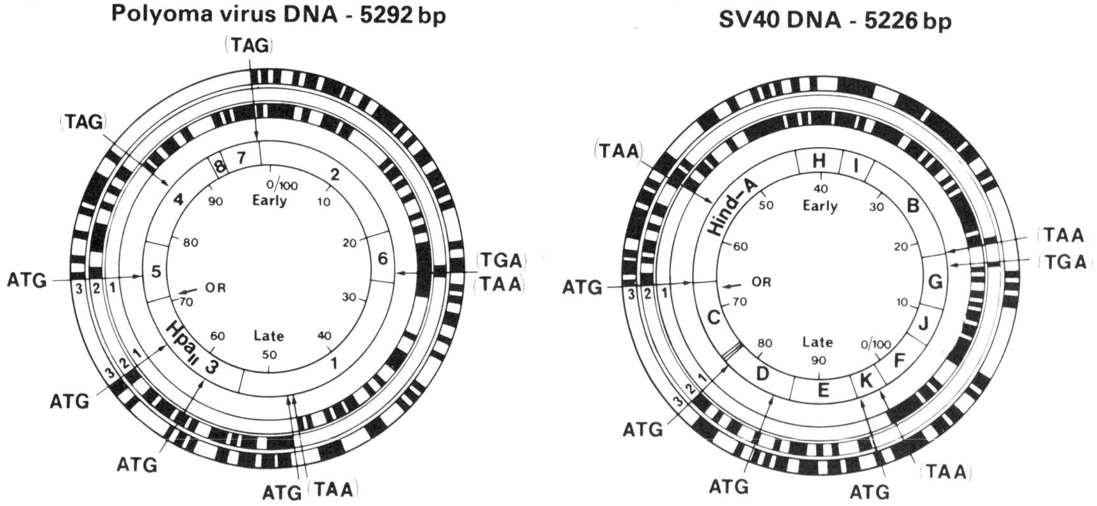

Fig. 83.6 A diagram indicating the three coding frames of the genome of polyoma virus (left) and SV40 (right) relative to the physical map.

The black boxes indicate termination codons (either TAA, TAG or TGA) in the sequence. The diagram for SV40 corresponds to that of polyoma virus except that there are no signals for a middle T antigen. For polyoma virus, starting at the origin of replication (OR), the single ATG initiation codon and the TAG, TGA and TGA termination codons for small, middle and large T are indicated in a clockwise direction on the outside. The ATG initiation codons for VP2, VP3 and VP1 and their respective termination codons (TAA) are indicated in a counter-clockwise direction (read from the opposite strand). See text for additional explanation. (Reproduced with permission from Soeda *et al.* (1980) *Nature*, **283,** 449).

in overlapping genes, with more than one reading frame in use. It is believed that all three proteins begin in the same reading frame (frame 1) and share the same N-terminus. The main portion of the middle and large T antigen are encoded in frames 2 and 3 respectively and the tail of the small T antigen may be in frame 1. The coding potential of the early regions of polyoma virus and SV40 is illustrated schematically in Fig. 83.6, where the entire sequence of the DNA genome is represented in its three reading frames with black blocks indicating the termination codons TAA, TAG and TGA (Soeda *et al.* 1980). For instance, frame 1 is the only reading frame open between map co-ordinates 74 and 85. This frame is then closed and frames 2 and 3 are open from map co-ordinates 85 to 99. Frame 2 is the only reading frame open from position 99 to the end of the early region.

As well as these coding regions there are also two distinct *non-coding regions*. One region lies around the origin of replication and the other is the early region between nucleotides 746 and 795 (Fig. 83.6). Presumably the latter contains signals regulating splicing. The other non-coding region is presumed to contain signals for viral replication and initiation of transcription.

The complete sequence of the genome of *BK virus* (Yang and Wu 1979)—a papovavirus frequently isolated from the urine of immunosuppressed patients which can transform hamster cells and induce tumours *in vivo*—is also now available in addition to that of *SV40* and polyoma virus. The sequences of the DNAs of the human BK virus and the monkey SV40 virus are very similar. The late coding regions of polyoma virus also resemble the late regions of SV40 and BK virus, and about 60 per cent of the nucleotide base pairs are homologous. The common sequences in the overlapping late genes show the same homology as the unique sequences, which is surprising as greater conservation of sequence would be expected in the overlap region because of the constraints on mutation. The early region of polyoma virus also shows considerable homology in terms of preferred amino acids. In both SV40 and BK virus, however, there is no sequence comparable to the unique portion of the middle T antigen (and consequently part of large T). This is the region of the polyoma genome which is necessary for transformation of cells and tumour induction. Therefore it is probable that transformation of cells is achieved by different mechanisms, and the presumptive transformation proteins may be quite different.

The *codons* used in the polyoma virus genome differ from random expectation. There is a deficiency of codons containing the CG doublet, although the CG combination occurs with average frequency between codons (i.e. NNC–GNN). By contrast analysis of the codon usage in the haemagglutinin gene of influenza virus revealed that there was a deficit of CG dinucleotides between codons as well as within codons, suggesting that the last base of a codon influenced the choice of the first base of the next codon in this gene at least (Threlfall *et al.* 1980). The low CG frequency is in agreement with the deficiency of CpG dinucleotides observed in the nearest neighbour analysis of both polyoma and mammalian cell DNA (Subak-Sharpe 1967).

Ts mutants of polyoma virus and SV40 have been isolated after exposure to a variety of mutagens (Tegtmeyer 1980) and the SV40 mutants have been classified into five groups (A, B, C, BC and D) by *complementation analysis.* Some of the properties of these mutants are given in Table 83.2. Mutants in groups B, C and BC synthesize a defective VP1 (see Fig. 83.4) and are distinguished by their ability to complement one another. This indicates that these mutations lie in the same gene and that the complementation is *intragenic complementation.* Mutants of group D are defective in VP3 and do not complement unless the infected cells are incubated for a period at permissive temperature. This is because uncoating of the input genome is blocked at the restrictive temperature. The group A mutants are early mutants which produce an unstable A protein (large T antigen) and are defective for *induction of transformation.* The corresponding *ts* mutant of polyoma virus is also defective for initiation of transformation, but not for the maintenance of the transformed state once it has been induced. Thus the A gene product is necessary for initiation of the transformed state in both viruses, but only for the maintenance of transformation in SV40.

Table 83.2 Functional defects of *ts* mutants of SV40

	Ts Mutant Group		
Function	A	BC	D
DNA infectivity	−	−	+
Virion infectivity	−	−	−
Productive infection:			
early RNA	+	+	−
A-gene protein	+	+	−
viral DNA	−	+	−
late RNA	−	ND	ND
late protein	−	+	−
cell DNA	±	+	−
Transforming infection:			
A-gene protein	+	+	−
initiation	−	+	−
maintenance	±	+	+

ND = no data.

Another type of mutant known as *hr-t* (host range transformation), was located in the early region of the genome (Benjamin 1972). These mutants were selected for their ability to grow in polyoma virus-transformed cells and some other types of cells, but not in untransformed mouse 3T3 cells. Consequently they were unable to transform cells, although they were able to complement the *ts* A mutant. In SV40 an F gene which specifies the small T antigen has been identified among viable deletion mutants constructed *in vitro*, and may correspond to the *hr-t* gene of polyoma virus. Probably therefore there are two transforming functions. Recent research suggests that the *hr-t* mutation affects the middle and small T antigen, but not large T, i.e. it is located in sequences which are removed by splicing in the formation of the large T mRNA. The *ts* A mutation has now in fact been mapped in the unique large T region. Therefore in polyoma virus middle T antigen is sufficient for transformation, although at least part of the large T antigen plays a role in maintenance of transformation in some cells. In SV40 large T is responsible because there is no middle T.

Recombination has been detected between pairs of mutants of both polyoma virus and SV40. However, the frequency of recombination was low and no genetic maps have been constructed. This is due in part to the nature of the mutants, many of which exhibit *multiplicity-dependent leakage*, i.e. a cytopathic effect is observed at high multiplicity of infection (Eckhart 1977).

Defective particles of papovaviruses which have a lower buoyant density than competent virions and contain less DNA arise during continuous passage at high multiplicity. The DNA of these particles is circular with between 10 and 50 per cent of the genome deleted. Although unable to complete a productive infection, defective virus can induce T antigen synthesis and tumours *in vivo.*

Segments of the genome which have been deleted may be replaced by DNA sequences derived from the host cell genome with on average 80 per cent of the genome substituted by host DNA. The precise origin of the *substituted DNA* has not been defined, but recombination must be involved.

Pseudovirions are papovavirus particles which contain linear DNA fragments derived from the host cell in place of the circular viral genome. The proportion of pseudovirions depends on cell type, and is usually about 20–40 per cent of the particle yield in polyoma virus stocks.

Exceptionally, papovavirus may recombine with non-homologous DNA to produce a *hybrid virus.* Passage of adenovirus type 7 in SV40-contaminated African green monkey kidney cells, which were not permissive for adenovirus, resulted in the isolation of 'adapted' adenoviruses that were later shown to be true hybrid viruses in which part of the adenovirus genome was covalently linked to part of the SV40 genome. A few *adeno-SV40 hybrids* have been described in which either the adenovirus or the SV40 genome is intact. The adeno 7-SV40 hybrid virus described initially by Huebner *et al.* (1964), however, was incomplete and dependent on an intact adenovirus type 7 as a helper virus. This adeno 7-SV40 hybrid was highly oncogenic and the tumour cells induced antibodies to the T antigens of adenovirus 7 and SV40.

Subsequently, propagation of adenovirus type 2 in monkey kidney cells yielded hybrid viruses which contained the complete SV40 genome and incomplete adenovirus genomes linked in tandem. These hybrid viruses were dependent on a competent helper adenovirus for their replication, although they yielded intact infectious SV40 virus derived wholly from the hybrid genome (Lewis and Rowe 1970).

In addition *helper-independent* adeno 2-SV40 hybrids have been obtained, e.g. Ad2ND1 which contained a non-defective, though incomplete, adenovirus genome covalently linked to a SV40 sequence (Levine *et al.* 1973). The inserted SV40 sequence provided the information allowing adenovirus type 2 to multiply in monkey cells. The amount of the SV40 genome inserted varied from 7 to 43 per cent, and the dispensable amount of the adenovirus genome deleted varied between 4.5 and 7.6 per cent. In all these hybrids the SV40 sequences began at the same point in the adenovirus genome, namely at a site 14 per cent from one end (Grodzicker 1980).

These phenomena show that the genome of papovaviruses, and perhaps other DNA viruses, can be modified by substitution of host DNA sequences and in rare circumstances can contribute DNA sequences to other DNA viruses.

The genome of papovavirus becomes stably incorporated into the host cell and, as described previously, transcription of a viral gene is necessary to maintain the transformed phenotype. Recent research on the sequences flanking integrated virus has established that there are no specific *integration sites* in the host DNA (Topp *et al.* 1980). Therefore papovaviruses cannot be regarded as analogues of the *transposable elements* of prokaryotes, since the latter are inserted at specific sites and have duplicated host sequences at their insertion points. As will be described later, the retroviruses are more akin to transposable elements (see section 5).

4.3 The adenovirus genome

The adenovirus genome is a linear DNA duplex of 20–25 × 10^6 mol.wt, with inverted repeat sequences of approximately 100–140 nucleotides at each terminus (Table 83.1A). Adenovirus DNA is unique in possessing a protein covalently linked to both ends of the genome (Robinson and Bellett 1975).

The adenovirus genome was the first large mammalian DNA virus to be mapped comprehensively by both physical and genetical methods (Flint 1980*a*). *Ts*, host range, and plaque morphology mutants of several adenoviruses have been isolated, and details of the mutagens used and their efficiency have been reviewed by Ginsberg and Young (1977). Seventeen non-overlapping complementation groups of adenovirus type 5 have been defined; their phenotypes are given in Table 83.3. A conventional linear *genetic map* of the genome of adenovirus type 5 was constructed by pairwise crosses of *ts* mutants obtained mainly by chemical

Table 83.3 Phenotypic characteristics of *ts* mutants of adenovirus type 5

Class	Number of complementation Groups	Phenotype	
		DNA	Protein
I	4	+	All capsid antigens present. *ts* 18 and *ts* 19 defective in interferon induction in CE cells.
II	2	−	No capsid antigen present. *ts* 36 and *ts* 125 defective in initiation of transformation.
III	3	+	Fiber antigen defective.
IV	4	+	Hexon antigen defective. *ts* 17 and *ts* 20 hexon assembly defective.
V	2	+	Abnormal hexon transport. Decreased synthesis of hexon and penton base.
VI	2	+	Unknown.

mutagenesis. A heat-resistant mutant was used in three-factor crosses to increase the precision of the map distances and confirm the orientation of the mutants (Williams *et al.* 1975). Subsequently the *linearity* of the map and the orientation of the mutants was confirmed by direct correlation of the genetic and physical maps (Sambrook *et al.* 1975). This was effected initially by restriction endonuclease analysis of the physical structure of recombinant genomes isolated from mixed infections of *ts* mutants of adenovirus type 5 and *ts* mutants of a strain of adenovirus type 2 which contained covalently linked SV40 sequences (Ad2$^+$ND1). The approximate site of a recombinational event could be pinpointed by comparison of the patterns of fragments obtained by restriction endonuclease digestion of DNA from the recombinant clone with the two parental viruses. Since the recombinants had been produced by crossing two *ts* mutants, the position of the cross-over event enabled the *ts* mutants to be located in a specific region of the genome. The location of individual mutants was defined more precisely by additional crosses, until it was established clearly that the genetic and physical maps were co-linear.

The technique of *marker rescue* (see section 2) was introduced subsequently to obtain finer mapping of the adenovirus genome without recourse to intertypic crosses, or the necessity to have *ts* mutations in more than one adenovirus (Frost and Williams 1978). Figure 83.7 illustrates the concordance of the genetic map and the physical map of the adenovirus type 5 genome. *Transcription maps* have also been produced defining the regions from which individual mRNAs are transcribed (Fig. 83.8), and the proteins encoded by these mRNAs have been identified by hybridization-arrested translation (Paterson *et al.* 1977) to give an accurate *translation map* and ultimately a *functional map* of the adenovirus genome (Flint 1977, 1980*a*).

A unique feature of the adenovirus map is the location at the left-hand end of sequences essential for transformation of cells *in vitro*. No more than 7 per cent of the genome of adenovirus type 5 is required to induce transformation of the growth properties of cells in culture (Flint 1980*b*). The function of this region of the genome is largely unknown; and the only mutants which have been unambiguously mapped in this region are *host range mutants* (Fig. 83.7). These mutants are able to multiply in transformed cells, but cannot transform or multiply in normal cells. However, study of these mutants has not yet identified the transformation function. A *cold-sensitive* host-range mutant which multiplied and transformed normal cells at 38°, but not at 32°, has been isolated by Williams and colleagues (1975) which maps in the left-hand end. This suggests that the left-hand end of the adenovirus genome could code for DNA binding proteins, since *cold-sensitivity* has been associated with conformational change affecting the binding of protein to DNA.

Another striking feature which has had great influence in molecular biology is the pattern of transcription and processing of mRNA (Flint and Broker 1980). The late mRNA molecules are coded predominantly from the same strand and have a common non-coding *leader* sequence at the 5′ end (Fig. 83.8). The non-coding leader molecule is itself a tripartite structure linked by two internal splices. The leader does not code for protein products, but it contains the ribosome binding site, signals for initiation of transcription and messenger capping. *Splicing* is a mechanism for generating multiple messengers from the same sequence of DNA. In adenovirus at least 17

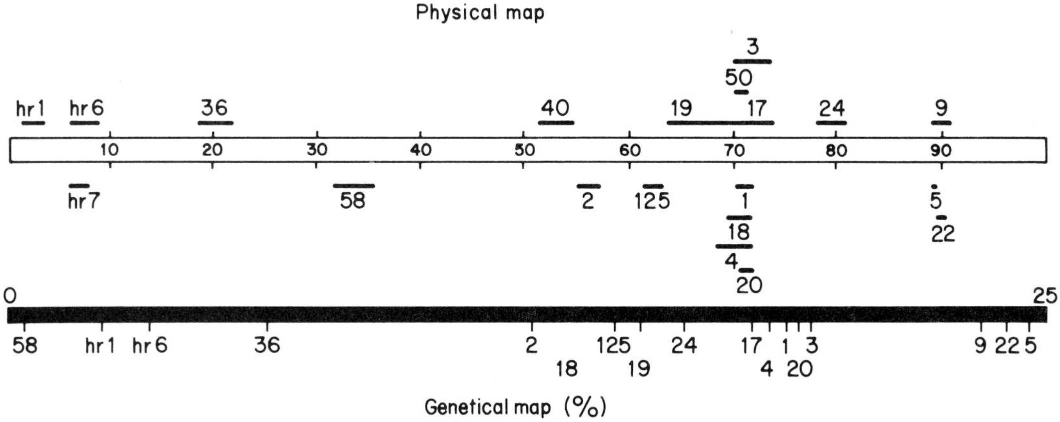

Fig. 83.7 Comparison of the genetical and physical maps of the genome of adenovirus type 5. The loci of *ts* and host-range (*hr*) mutants deduced from recombination analysis are indicated in the lower line. The physical location of these mutants is indicated above. In general there is a close correspondence between the two maps, apart from the discordant location of *ts* 58 at the extreme left end of the genetical map in the transformation region. (Compiled from data supplied by Richard Galos and Jim Williams).

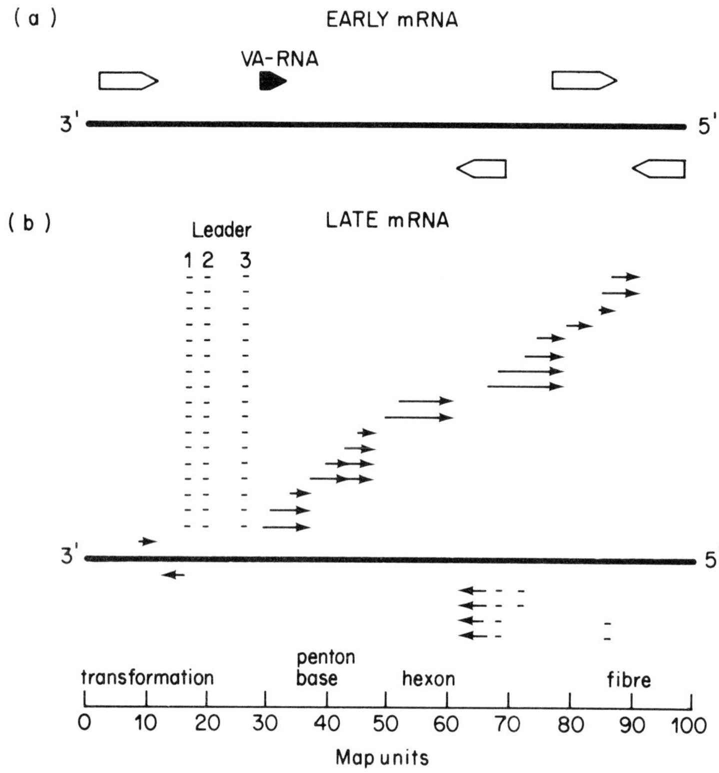

Fig. 83.8 A simplified transcription map of the adenovirus type 2 genome. (a) Early mRNA synthesized primarily 1–8 hours after infection; (b) late mRNA. The region of the genome coding for the major structural proteins and the transforming functions are indicated at the bottom.

The arrows show the 5′–3′ direction of transcription. The dots indicate the short leader sequences. (Adapted from Flint 1977 and Broker *et al* 1980).

different mRNAs falling into 5 families are produced from the late region (Fig. 83.8). During transcription of the late region the growing chain is cloven at one of five termination points, a poly A sequence is added at the 3′-end and a 5′-methyl-G cap at the 5′-end. Subsequently further splicing occurs in each class of mRNA, in such a way that a common 5′-end leader sequence is produced with variable deletion of sequences between the leader splice point and other splice points towards the 3′-end. The whole of the late region of the adenovirus genome contains coding sequences, but a sequence that is read in one message (the formal equivalent of the exon sequences in eukaryotic DNA) may be an intervening sequence (the formal equivalent of the intron of eukaryotic DNA) in another. Splicing is probably a mechanism for providing control of expression of different regions of the genome. Splicing might also influence the frequency of formation of particular mRNAs (i.e. their abundance). It also provides a simple means of changing the reading frame and amplifying the information content of the genome, as has already been discussed in relation to the early and late genes of polyoma virus.

Discontinuous genes and splicing were discovered first in adenovirus and later found to be a common feature of eukaryotic genetic systems. With viruses, however, splicing links a common leader sequence to sequences transcribed from different down-stream regions of the genome. In the host cell, splicing also links together non-contiguous segments of a gene as well as providing a common leader. Different coding regions of a single gene (exons) may be separated by intervening non-coding sequences (introns), e.g. the avian ovalbumin gene is distributed over seven non-contiguous regions of the genome.

Until recently it was considered that early mRNA originated from four discrete regions at the extremities of the genome. However, the mapping of one of the two early group mutants (*ts*36) centrally (Fig. 83.7) revealed additional early genes, not previously detected by transcription mapping owing to the low concentration of the complementary mRNAs (Galos *et al.* 1979). The other group of early mutants also map centrally, one of which (*ts*125) has the unique ability to induce transformation of cells at high frequency, possibly owing to its reduced cytotoxicity.

A number of small *non-coding RNAs* (known as VA-RNAs) betwen 130 and 200 nucleotides in length are synthesized during productive infection in addition to mRNA and its precursors. VA-RNA I and VA-RNA II are encoded in adjacent regions of the genome (at position 29, Fig. 83.8) and are transcribed by ribonuclease III. VA-RNAs appear early during infection and VA-RNA I accummulates in very large amounts. The function of all VA-RNA is unknown, although it has been speculated that they may be intermediates in mRNA splicing.

Examination of the location, number and organization of the sections of adenovirus type 2 genome which are present in transformed cells has established that *integration* can occur at different sites in the host cell genome. No common feature has been discerned in the host DNA sequences which flanked the inserted viral DNA. In some cells transformed by adenovirus type 2 the viral sequences appeared to have undergone amplification after integration, and suffered deletion or more rarely inversion. The flanking cellular DNA sequences may undergo rearrangement also, but there is no clear correspondence with the transposable elements of prokaryotes (Sambrook *et al.* 1980).

The apparent inability of adenovirus to cause tumours in their natural hosts, together with the lack of site-specific integration in transformed rodent cells, leaves the question of the potential oncogenicity of adenoviruses in man unresolved. The possibility cannot be entirely excluded at present that the integration of viral sequences represents a cellular defence mechanism in which foreign DNA (in this case viral DNA) is swept up and sequestered in the host DNA. However, the ubiquitous expression of early genes (T antigen etc.) in transformed cells weighs against this interpretation.

4.4 The herpesvirus genome

The genomes of herpesviruses are double-stranded DNA molecules of 80–140 × 10⁶ mol.wt (Table 83.1A). This DNA is infectious, therefore no viral enzymes are necessary for transcription of the genome in the host cell. In terms of its structure the genome of herpes simplex virus (HSV), despite its name, is the most complex viral genome so far examined. The genome of HSV consists of two stretches of *unique sequences*—the long unique sequence and the short unique sequence—each of which possesses repeated terminal sequences which are inverted with respect to one another (Sheldrick and Berthelot 1974). Thus, when HSV DNA is denatured and reannealed, the ends of the molecule can hybridize to the *internal repeats* which link the two unique sequences, thus forming a structure with two loops connected by a double-stranded stem (Fig. 83.9**a**). The genomes of other members of the herpesvirus group have similar features, but are less complex. The structure of six

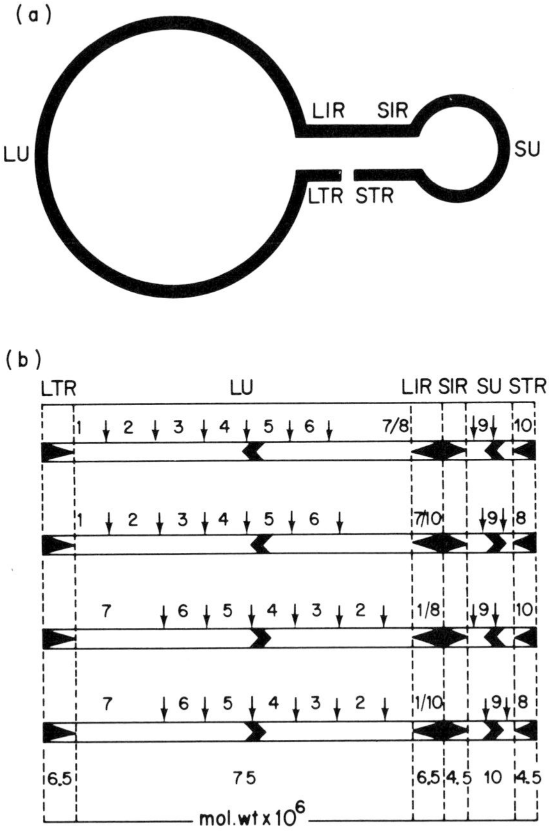

Fig. 83.9 The physical structure of the genome of herpes simplex virus. **(a)** The homology of the internal and terminal repeat regions produces a double-ring structure when single-stranded DNA is self-annealed. Inversion of the long and short region could occur by internal recombination in the redundant sequences. LTR—long terminal repeat; LU—long unique region; LIR—long internal repeat; SIR—short internal repeat; SU—short unique region; STR—short terminal repeat. **(b)** The four arrangements of the genome which produce different molar proportions of unique fragments (indicated by numbers 1–10) after endonuclease cleavage. The orientation of the unique and repeat regions are indicated by the arrows.

herpesvirus DNAs is shown in Fig. 83.10. Channel catfish virus has the simplest structure with one unique region flanked by terminal repeats. *Herpesvirus saimiri* also has a unique region, but with extensively reiterated repeat sequences. The other four viruses (EBV, PRV, CMV, HSV) have a second short unique sequence flanked by different terminal structures. In HSV the inverted sequences immediately flanking the joint (the a′/a′ region) and at the termini (the a region) are identical repeats, unlike the remainder of the short and long repeat segments. The total mol.wt range from 85×10^6 for CCV to 142×10^6 for CMV, and the long unique segments from 65×10^6 to 100×10^6 respec-

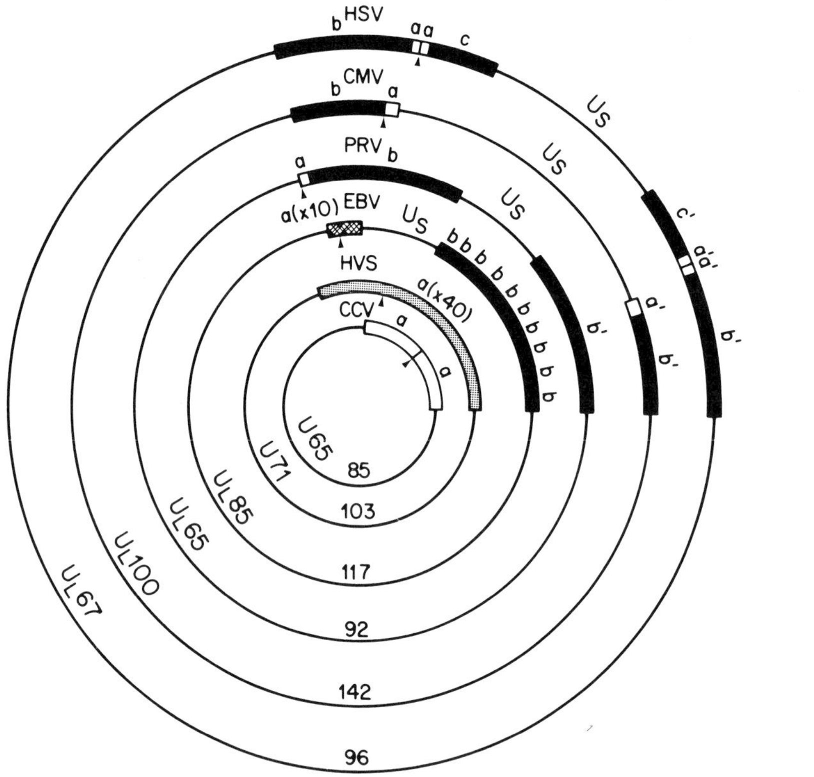

Fig. 83.10 Comparison of the physical structures of genomes of six herpesviruses. The DNA molecules are represented diagrammatically as circles (not to scale). The viruses are herpes simplex virus (HSV), cytomegalovirus (CMV), pseudorabies virus (PRV), Epstein-Barr virus (EBV), *Herpesvirus saimiri* (HVSL) and Channel Catfish virus (CCV). Light bars (a) stippled bars (b) and shaded bars (c) represent repeated, terminal and inverted sequences respectively. U, U_L and U_S denote unique, long unique and short unique sequences respectively. Arrows indicate the postions of the termini of linear molecules. Numbers on the vertical axis in the lower half of the figure indicate the approximate mol. wt of the DNAs in millions; numbers to the left of vertical indicate the approximate mol. wt of the U or U_L sequences of the DNAs. (Reproduced with the permission of the Cold Spring Harbor Laboratory).

tively. The biological relevance of these differences in genome structure remains to be elucidated.

A striking feature of the HSV genome is the random orientation of the two unique sequences, in such a way that *four forms of the genome* exist simultaneously. Figure 83.9(**b**) illustrates these four orientations and the types and relative proportions of the fragments (1.0M; 0.5M; 0.25M) generated by restriction endonuclease cleavage. The four orientations are a consequence of the structures of the molecule when reciprocal recombination occurs at high frequency between the terminal sequences and their internal repeat. Another consequence of the genome structure is that all mutants in the long unique sequence will map equidistantly from cell mutants in the short unique region. The mutants within each region will form a *linear map*, and, if recombination is unrestricted in the internal repeat regions, the two linear maps will appear unlinked, i.e. the HSV genome will appear to consist of two chromosomes (Subak-Sharpe and Timbury 1977).

Linear genetic maps of the genomes of HSV-1 (Brown *et al.* 1973, Schaffer *et al.* 1974, Schaffer 1975*a*, *b*), HSV-2 (Benyesh-Melnick *et al.* 1974, Timbury and Calder 1976) and pseudorabies virus (Pringle *et al.* 1973) have been derived from two-factor crosses of *ts* mutants and three-factor crosses with syncytial plaque mutants of HSV-1 and HSV-2. These mutants and drug resistant mutants have been located on the physical map of the herpes simplex virus genome by the technique of *inter-typic marker rescue* (Stow and Wilkie 1978). There was good agreement between the genetic and physical maps of HSV-2. The physical mapping of gene products has also been achieved by *inter-typic recombination*. The genome structure of recombinants can be deduced by restriction endonuclease analysis, and the map locations of specific viral polypeptides can be located, since many of the corresponding viral polypeptides of HSV-1 and HSV-2 can be distinguished by slight differences in their electrophoretic mobility in polyacrylamide gradient gels (Preston *et al.* 1978, Marsden *et al.* 1978, Morse *et al.* 1978, Chartrand *et al.* 1980, Halliburton 1980). Figure 83.11 is a *composite map* of the herpesvirus genome constructed by N. M. Wilkie and colleagues from information available up to mid 1980. The genome is represented as a 150 Kilobase circle. Various mutants and the determinants of some biological properties are located on the outer ring; type 1 above and type 2 below. The inner rings indicate the location of late, early and immediate early polypeptides with respect to the unique and repeat sequences. The significant features are that most of the *ts* mutants map in the long unique region, but *ts*D, *ts*K, *ts*T and *ts*C75, which are defective in *immediate early functions*, are all located in the short repeat regions (TR_S and IR_S). The immediate early (IE) polypeptides, that is those which appear in infected cells before viral DNA synthesis, also map in this region. This region of the genome is also responsible for transformation of cells. The viral DNA polymerase activity maps together with several *ts* mutants with lesions in DNA synthesis and mutants

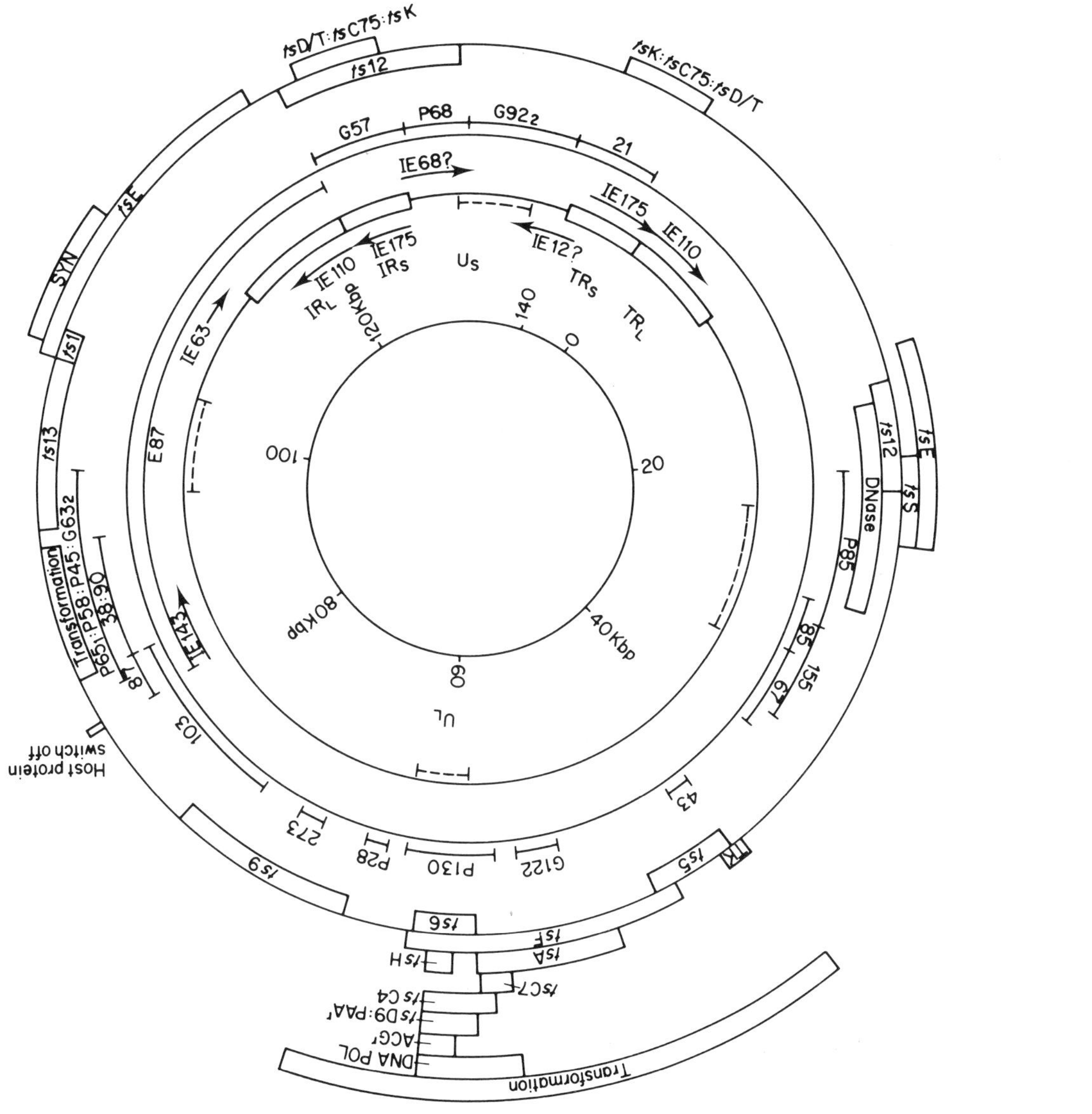

Fig. 83.11 A composite map of the genome of herpes simplex virus. The DNA of herpes simplex virus is represented diagrammatically as a circle; further explanation is given in the text. (Reproduced with the permission of N. M. Wilkie and colleagues, Institute of Virology, Glasgow University).

resistant to acycloguanosine (ACG^r) and phosphonoacetic acid (PAA^r), drugs which inhibit the viral polymerase. The deoxypyrimidine (usually described as thymidine) kinase gene and polypeptide 43 the thymidine kinase protein are coincident.

A larger proportion of the *ts* mutants of herpesviruses than of adenoviruses have DNA-negative phenotypes, i.e. they do not induce viral DNA synthesis at the restrictive temperature. Consequently some of the extra information content of herpesviruses is devoted to ensuring that viral DNA synthesis is more independent of the host than with adenovirus. An insufficient number of genes have been located to ascertain whether there is any clustering of functionally related genes.

Thymidine-kinase deficient cells can be 'transformed' by exposure to HSV DNA or restriction fragments. DNA from these transformed cells can in turn transform other thymidine-kinase deficient cells and it has been shown that cell genetic information can be co-transferred simultaneously with the viral thymidine-kinase gene which thus behaves as a vector. The frequency of *transformation* is low and it is thought that only a minority of cells in the population are competent (Wigler *et al.* 1979). Co-transformation of cells using the herpesvirus thymidine kinase gene as a vector could become a general method for transforming genes between heterologous somatic cells.

The unique feature of herpesviruses is their propensity to become latent. The prolonged sojourn of HSV-1 in the trigeminal ganglia, with only rare recrudescence in cold sore lesions, is well established. The factors which control *latency* are largely unknown, but Lofgren *et al.* (1977) found that five *ts* mutants of HSV-1 differed strikingly in their ability to establish latent infection in the nervous system of mice. There was no relation between ability to establish latent infection and DNA phenotype. It was also concluded from studies of a latency-negative mutant (*ts*K) (in neuroblastoma cells) that synthesis of immediate-early polypeptides was insufficient for establishment of latency (Gerdes *et al.* 1979).

Mutants of herpesvirus have been used also as *genetic probes* for detection of the presence of herpesvirus-specific genetic information in a non-reactivable or defective form in neurological and other tissues. Reactivable virus was detected in explanted human trigeminal ganglia, which had failed to release virus spontaneously, by the complementation of superinfecting *ts* mutants of HSV-1 at non-permissive temperature (Brown *et al* 1979). Likewise the presence of herpesvirus genetic information in HSV-2-transformed cells has been confirmed by the complementation of HSV-2 *ts* mutants in these cells (Macnab and Timbury 1976) and the retrieval of HSV-2 sequences by intertypic recombination following superinfection with *ts* mutants of HSV-1 (Park *et al.* 1980).

4.5 The poxvirus genome

The poxviruses and the lesser studied iridoviruses are unique among DNA viruses, because their multiplication cycles take place entirely within the cytoplasm. The genome is a large double-stranded DNA of 160×10^6 mol.wt approximately (Table 83.1A), which endows these viruses with greater coding potential than any other virus. In vaccinia virus there are long ($7-8 \times 10^6$ mol.wt) inverted repeats at the termini which encode early functions just as in herpes simplex virus (Garon *et al.* 1978, Clements, Watson and Wilkie 1977). Restriction endonuclease mapping of different orthopoxvirus DNAs indicates that the *physical maps* of their genomes differ predominantly in the terminal regions with the unique intervening sequences showing little variation (Wittek *et al.* 1977, Garon *et al.* 1978, Mackett and Archard 1979).

The DNA strands of vaccinia virus are covalently linked at their ends, so that the molecule becomes a single-stranded circle on denaturation (Geshelin and Berns 1974). The DNA is not infectious, because mRNA synthesis must be initiated by a virus-associated DNA-dependent RNA polymerase.

Complementation and genetic recombination have been observed between *ts* mutants of rabbit pox virus (Padgett and Tompkins 1968). A class of mutants (*u* mutants) of rabbit pox virus, which produced white nodular pocks on chorioallantoic membranes instead of the usual haemorrhagic ulcerated pocks, also underwent recombination (Gemmel and Fenner 1960, Fenner and Sambrook 1966), although recently Lake and Cooper (1980) reported failure to reproduce these findings. A category of white pock mutants (*p* mutants) failed to grow on pig kidney cells (Fenner and Sambrook 1966). These mutants mapped at a single locus and paradoxically exhibited a wide range of phenotypes (plaque morphology, DNA synthesis, antigen production). Analysis of restriction endonuclease generated maps of *u* and *p* mutants has revealed that all *p* mutants are deletion mutants with extensive deletion ($10-20 \times 10^6$ mol.wt) at either end of the genome (Lake and Cooper 1980). Delections at the right-hand end produced the *u* phenotype and deletion at the left-hand end produced the *p* phenotype. There was no relation between the extent of the left-hand deletion and specific phenotype. The terminal deletions explain the inability of Lake and Cooper to obtain recombination between *u* mutants, and of Fenner and Sambrook (1966) to obtain recombination between *p* mutants. Recombination between *u* and *p* mutants was readily obtained by both groups because the deletions are at opposite ends of the genome. These results indicate that orthopoxvirus can tolerate extensive deletions ($\sim 10\%$) at or near the termini. Analysis of gene function has only just begun and no information is available. Members of the vaccinia sub-group of poxviruses can recombine with each other and, similarly, recombination has been observed within the myxoma sub-group. Recombination has not been observed, however, between viruses belonging to the different sub-groups of the *Poxviridae* (Bedson and Dumbell 1964).

The history of the introduction of myxoma virus into Australia provides one of the clearest examples of the selective forces which modify the disease potential of viruses in the natural environment (Fenner and Ratcliffe 1965). Myxoma virus was extremely lethal for European rabbits when first introduced into Australia in 1950 as a pest control measure. Mortality rates from *myxomatosis* were in excess of 99 per cent. However, more attenuated variants of the virus soon appeared and within a few years became the predominant field virus. The reduced virulence of the virus contributed to its own survival by allowing some rabbits to survive from one season to the next to re-establish a susceptible population each year. A corresponding increase in genetic resistance of the host was observed shortly after the appearance of attenuated virus variants, so that the mortality induced by a standard virulent strain declined from 90 to 25 per cent over a five-year period. Thus a catastrophic epidemic disease was modified to a milder endemic form within a few years by changes both in the disease-producing potential of the virus and the genetic resistance of the host. This example emphasizes the delicacy of the relationship between virus and host in the absence of outside intervention.

5 The retrovirus genome

The retroviruses bridge the gap between the cytoplasmic RNA viruses and the DNA viruses. The retroviruses alternate between RNA in the extracellular particle and proviral DNA in the infected cell. In the virion the genome is in the form of two identical sub-units and is therefore diploid (Table 83.1B). The two sub-units are single-stranded RNA molecules of 3×10^6 mol.wt linked by hydrogen bonding between short repeats at their 5′ ends. The virion RNA has messenger activity in an *in vitro* protein synthesizing system (i.e. it is the plus strand), but it is not infectious, unlike plus strand RNA from conventional cytoplasmic RNA viruses because the enzyme required to initiate reverse transcription of RNA into DNA is part of the protein complement of the virion. Once formed, however, the double-stranded DNA form of the genome is infectious. Proviral DNA can become stably integrated and transmitted from cell to cell indistinguishable from any other cellular gene. *Endogenous retroviruses* are transmitted from one generation to the next via the germ line, and can evolve with their host to establish multigenic families. It has been estimated that 0.01–0.1 per cent of genome of wild and laboratory mice comprises retroviral sequences and it is likely that primates have as much maternally transmitted retroviral information as rodents (Todaro 1980*a*). Many of the endogenous viruses appear to have become established as cross-infections by viruses from distantly or unrelated hosts which are able to integrate into the germ line. For instance, Benveniste and Todaro (1974*a*, *b*) deduced that RD114, an endogenous virus of cats, represented horizontal transfer of an endogenous retrovirus from primates to felines at some time in the past, possibly in the Pliocene. Indeed the presence of specific endogenous retroviral sequences has been used to explore phylogenetic relationships (Todaro 1980*b*), and Benveniste and Todaro (1976) have concluded from viral data that man evolved from the apes in Asia and has only penetrated to the African continent relatively recently.

The major distinction between the retroviruses and the nuclear DNA viruses is that the former do not kill their host cell while maintaining productive infections, whereas the latter always kill their host cell and can become stably integrated only in a non-productive form. Further study of the phenomenon of latency of herpesviruses, however, may blur this distinction.

The *provirus* is composed of viral genes arranged in the order found in the RNA genome of the virion with direct terminal repeats of several hundred nucleotide pairs derived from sequences unique to both ends of the viral RNA. At the cell-virus junction there is a five-base direct repeat of cellular DNA next to a three-base pair inverted repeat of viral DNA. This is a structure similar to that at the termini of several bacterial transposons. Therefore retroviruses, unlike the oncogenic DNA viruses, have properties in common with the transposable elements of prokaryotes (Shimotohno *et al.* 1980). This is consistent with the theory that retroviruses themselves may have evolved from regions of the cell genome concerned in normal transfer of genetic information between cells (Temin 1976, Strayer and Gillespie 1980).

Many of the retroviruses induce tumours in animals, and these viruses are known collectively as the RNA tumour viruses to distinguish them from other RNA viruses. These viruses have host-derived genetic material inserted into their genome which confers *oncogenic potential*. The sub-unit of the genome of Rous sarcoma virus (RSV), the first discovered and probably least representative retrovirus, consists of a single-stranded polyadenylated molecule of 9–10 kilobases. The genome can be considered as four genes (see Table 83.4 and Fig. 83.14) which in 5′–3′ order determine the major internal structural proteins by post-translational cleavage of a precursor polypro-

Table 83.4 Structure of the avian sarcoma virus genome

Genome region	Size in nucleotides	Gene Product Primary	Gene Product Final	Function
5′-cap				
R	17–21	Non-coding		Terminal repeat: DNA polymerase transfer
U	80			Unknown
PB	16			Primer binding site
gag gene	2200	Pr 76	p19 p12 p27 p15	Virion internal proteins
pol gene	3000	Pr 180	p100 p70	Proviral DNA synthesis
env gene	2400	Pr 95	gp85 gp35	Virion envelope proteins
src gene	1800	p60	p60	Cell transformation
C	200	Non-coding		Control of RNA synthesis
R	17–21			Terminal repeat: DNA polymerase transfer
3′poly A tail				

tein (the *gag* gene), the RNA dependent DNA polymerase (the *pol* gene), the viral envelope protein (the *env* gene) and the sarcomagenic protein (the *src* gene) (Coffin 1980). The *src* gene is followed by a constant region (the *c* region) which appears to be conserved in all avian sarcoma and leucosis viruses. The avian leucosis viruses lack the *src* sequences and transformation defective (*td*) mutants of RSV have been isolated which have more than 70 per cent of the *src* gene deleted. Viruses known as rescued (*r*) RSV have been recovered from tumours arising in chickens or quails infected with *td* mutants and these viruses appear to have reacquired the full *src* gene from cellular sequences (known as the sarc sequences) by genetic recombination. The *src* product, a 6×10^5 mol.wt protein kinase was fully expressed in transformed cells. *Td* mutants with complete deletion of the *src* gene were not capable of generating rRSV, however, and it is probable that sequences at the 3′-end of the *src* gene are essential for recombination with the endogenous sarc gene.

The *src* gene product of Rous sarcoma virus is an autophosphorylated protein kinase, which has the novel property of phosphorylating tyrosine residues rather than the usual serine or threonine residues (Collett and Erikson 1978, Hunter and Sefton 1980). Protein kinase activity of this type has been found associated with other transforming viruses (adenovirus, polyoma virus and SV40) and may be responsible for production of malignant transformation.

Most but not all strains of RSV are defective owing to deletion of *env* sequences and are dependent on the presence of an avian leucosis virus helper which provides the envelope protein by phenotypic mixing. Avian leucosis viruses (ALV) of this type produce non-acute disease and differ in this respect from endogenous ALV which do not cause disease. They also differ in their envelope protein and their *c* region. A study of recombinants between endogenous and exogenous ALV indicated that the *c* region was more important in conferring (non-acute) disease potential than the envelope gene (Robinson *et al.*, 1979).

A virion with the genotype of one virus and the coat of another is known as a *pseudotype*. Pseudotypes can be obtained experimentally with the genome of a lytic virus like VSV and the coat of a retrovirus (Zavada 1972). Conversely the detection of pseudotypes (as neutralization-resistant virus) in VSV preparations indicates the presence of a cryptic virus. VSV/retrovirus pseudotypes have been used successfully for a study of host range and assay of non-cytopathic agents, etc. (Boettiger 1979, Weiss 1980). Host range in avian retroviruses is a property of the virion envelope; however, in murine viruses the restriction occurs after entry into the cell.

The agents which cause acute avian leucosis, and transform fibroblasts and haematopoietic cells *in vitro*, e.g. avian erythroblastosis virus (AEV), avian myeloblastosis virus (AMV), and avian myelocytomatosis virus strain 29 (MC 29), differ from RSV in that oncogenic information is inserted into the interior of the viral genome at the expense of the *pol* gene and portions of the *gag* and *env* genes (Fig. 83.12). The transformation specific sequences of AEV, AMV and MC 29 are denoted by the symbols *erb*, *mac* and *myo* respectively in Fig. 83.12, because these viruses transform haematopoietic cells to erythroblast-like cells, macrophage-like cells or myeloblast-like cells. One of the gene products of these viruses is a fusion polyprotein representing an uninterrupted transcript of the

RSV	5^{1}-GAG-POL-ENV-SRC-C-3^{1}
RSV endogenous	5^{1}-GAG-POL-ENV^{E}-SRC-C^{E}-3^{1}
RSV defective	5^{1}-GAG-POL-SRC-C-3^{1}
RAV	5^{1}-GAG-POL-ENV-C-3^{1}
AEV	5^{1}-(GAG)-ERB-(ENV)-C-3^{1}
MC29	5^{1}-(GAG)-MAC-(ENV)-C-3^{1}
AMV	5^{1}-(GAG)-MYO-(ENV)-C-3^{1}
MLV	5^{1}-GAG-POL-ENV-3^{1}
MSV	5^{1}-(GAG)-ONC-(ENV)-3^{1}

Fig. 83.12 The arrangement of genetic information in the genomes of several retroviruses.

gag and transforming sequence (e.g. (*gag*) *erb* for AEV, etc.), which is believed to be the transforming protein (Graf and Beug 1980, Klein 1982).

The murine (MSV) and feline (FSV) sarcoma viruses resemble the avian acute leucosis viruses in that the transforming sequences are inserted into the viral genome internally with the loss of viral genes, rather than terminally as in RSV. Consequently MSV and FSV are defective and require the presence of a helper virus (MLV or FLV) for their replication. The murine sarcoma-producing agents were all initially derived from stocks of competent virus (MLV) often continuously passaged in a foreign host. Scolnick *et al.* (1973) established that the Kirsten and Harvey strains of MSV had acquired their sarcomagenic potential during passage in rats and that the newly acquired sequences (ras) corresponded to endogenous rat retrovirus sequences. Other sarcoma-producing agents have been isolated directly which appear to be genetic recombinants between MLV (helper) virus and host cell sequences (Rapp and Todaro 1978).

Hybridization studies with cDNA have shown that all avian sarcoma isolates have *sarc sequences* which are highly conserved, whereas the sarc sequences of murine viruses are diverse. The avian sarc cDNA also hybridized with the DNA of normal cells of several avian species including Japanese quail which has neither inducible endogenous virus nor sequences homologous to the rest of the ASV genome. The thermal stability of these hybrids decreased with the evolutionary distance of these species from domestic fowl. The prevalence of the sarc-related sequences suggested that the *src* gene of avian sarcoma viruses was derived ultimately from the host cell. Normal cells also contain a 6×10^5 mol.wt protein that is structurally related to the *src* gene product and likewise appears to have protein kinase activity specific for tyrosine residues (Collett *et al.* 1978). The amount of this protein is about 50-fold less than the amount of *src* gene product in transformed cells. Thus transformation might be a consequence of a change of a cell phosphorylation reaction involved in some normal cell function like control of cell division.

Other retroviruses do not transform cells or induce tumours, and are not associated with pathological states. Nevertheless formation of proviral DNA and integration into the host cell genome is probably a normal feature of their multiplication cycle (Chiswell and Pringle 1978). Two sheep retroviruses, Visna and Maedi, do induce progressive degenerative diseases, however. The progressive nature of the disease induced by Visna virus may be related to the propensity of the virus to undergo *sequential antigenic mutation*, whereby the virus continuously overcomes the immune response of the host by antigenic change. During the course of the disease a whole series of antigenic variants can be isolated from a single infected sheep (Clements *et al.* 1980*a*). Three classes of *ts* mutants of Visna virus have been distinguished by their time of expression during the growth cycle. Complementation was observed between mutants in these three classes (Clements *et al.* 1980*b*).

Conditional lethal (*ts*) and transformation defective (*td*) mutants of retroviruses can be isolated in the laboratory and have contributed to functional analysis of the genome and the nature of oncogenesis (Vogt 1977, Wyke 1975, Friis 1978). The isolation of a *ts* mutant with a thermo-sensitive reverse transcriptase established that the α sub-unit of this enzyme was a virus specified protein, and that it was necessary for transformation and production of progeny virus (Verma *et al.* 1974).

High frequency *recombination* is a feature of the retrovirus genome. The diploid nature of the retroviruses means that heterozygous genomes can exist and indeed may normally be intermediates in the generation of stable recombinants. In crosses of exogenous retroviruses recombination has been detected both within and between genes, and between exogenous and endogenous viruses. The precise mechanism of recombination, however, has not been elucidated, but it does not appear to be analogous to the reassortment of genome sub-units associated with the segmented genome RNA viruses (see sections 7 and 8). Explanations of recombination in retroviruses have to accommodate three phenomena, (1) the existence of heterozygous particles, (2) the formation of circular proviral DNA during reverse transcription of single-stranded RNA into double-stranded DNA and (3) the location of the tRNA primer for DNA synthesis near the 5′-end of the tandemly associated RNA molecules of the genome. The location of the primer means that the reverse transcriptase must jump from the 5′-end of the molecule to the 3′-end of the same or the adjacent molecule soon after initiation of synthesis. Thus every round of synthesis becomes an act of recombination (Hunter 1978).

Whatever the mechanism the significance of recombination is that it provides an opportunity for the circulation of variation and incorporation of host sequences. It has been suggested that retroviruses may play a role in transducing host genes between cells and that this could be exploited in somatic cell genetics (Stephenson *et al.* 1979). However, there appears to be a strict limitation on the amount of material incorporated, presumably by packaging constraints, since viral information is invariably lost in the acquisition of transforming information.

6 The plus-strand RNA viruses

Three groups of RNA viruses, the picornaviruses, the togaviruses (alphaviruses and flaviviruses) and the coronaviruses, have single-stranded RNA genomes which are infectious and have messenger activity and are therefore defined as plus-strand genomes. The entire coding capacity is used to specify the structural components of the virion and the RNA dependent RNA polymerase activity which is indispensable for replication. Mammalian cells have no enzymic apparatus which can replicate RNA molecules; therefore the genome of RNA viruses, whether single or double stranded, or plus or negative sense, must carry the information for synthesis of a polymerase with this specificity (Table 83.1B).

6.1 The picornavirus genome

The genetics of two picornaviruses—poliovirus and foot-and-mouth disease virus (FMDV)—have been investigated in detail. The genome of poliovirus is a single-stranded RNA molecule of 2.6×10^6 mol.wt with a 12000 mol.wt protein (VP) covalently bonded at the 5′ end in place of the 7-methyl-GppN cap of normal mRNA. The VP_g protein is thought to be a virus-specified protein and to function as an initiator of RNA replication. The negative-strand template, which is an intermediate in replication, also has VP_g covalently linked at the 5′ end. The genome RNA likewise has a poly-A tail and a corresponding poly-U tract at the 5′ end of the negative-strand intermediate. Picornaviruses are unique among RNA viruses in that the poly-A sequences are encoded in the viral genome and not added terminally, as is normal with the mRNA transcripts of negative-strand RNA virus.

The primary gene product of the poliovirus genome is a poly-protein which is processed by nascent and subsequent *post-translational cleavages* into the structural proteins of the virion and the replicative enzymes. The poliovirus genome behaves as a single polycistronic message. The possible existence of a second small cistron has been inferred from *in vitro* translation experiments (Celma and Ehrenfeld 1974). Abraham and Cooper (1975) also estimated that 15 per cent of the poliovirus genome was either untranslated or translated from a second initiation site as a second cistron at the 3′-end. Complementation tests, however, have failed to detect a second cistron.

Recombination between pairwise combination of *ts* mutants of both poliovirus and FMDV occurs at low but reproducible frequency and genetic maps of the genomes of both viruses have been constructed (Cooper 1977, Lake *et al.* 1975). The mechanism of recombination and the enzymes involved are still unknown, but recombination is an early event in the multiplication cycle (Ledinko 1963, Cooper 1968, Pringle 1968). Both maps are linear and approximately additive, the frequency of recombination being equivalent to 1 per cent per 1250 nucleotide pairs for poliovirus compared to 1 per cent per 1750 nucleotide pairs for *Escherichia coli* (Cooper 1977). *Three-factor crosses* by means of two *ts* mutants and a third marker (resistant to the drug guanidine HCl) were used to establish the orientation of adjacent mutants. In both maps there was a relation between mutant phenotype and map position. Mutants with properties suggesting defects in structural components were located in one half of the maps, and mutants affecting synthetic properties in the other half (see Fig. 83.14). An unexpected feature was the location of the *guanidine resistance* marker among structural protein mutants, since guanidine HCl was considered to inhibit polymerase activity. This is illustrated in Fig. 83.13 which is an alignment of the genetic map of FMDV and a biochemical map derived from a study of the peptide profiles of individual mutants (King *et al.* 1980). Subsequent analysis indicated that guanidine HCl combined with a structural protein preventing synthesis of the poliovirus replicase, RNA synthesis being thus dependent on protein conformation. Simultaneous infection of cells with guanidine resistant and sensitive mutants showed that sensitivity was dominant. (Dominance tests have been applied widely in virus genetics and serve to distinguish mutants affecting catalytic functions from those affecting synthetic functions. For instance, if cells are co-infected with wild type (ts^+) virus and increasing amounts of

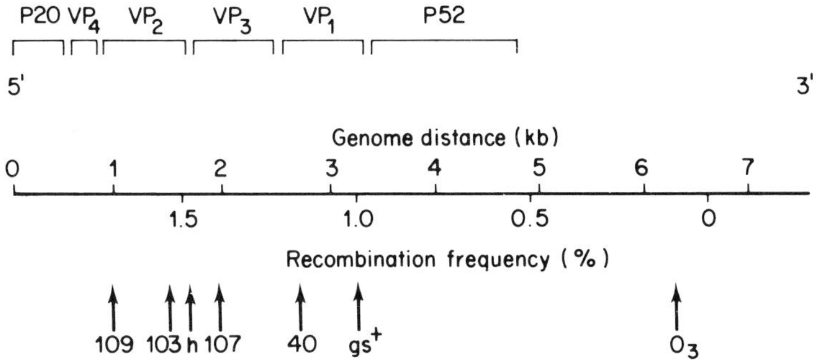

Fig. 83.13 A genetical and biochemical map of foot-and-mouth disease virus. The loci of individual *ts* mutants are indicated by arrows. h is a constellation of closely linked mutants. The putative gene products are indicated above. VP1, 2, 3 and 4 are virus proteins. (After King *et al.* 1980).

a *ts* mutant affecting a non-catalytic function, the yield of ts^+ virus in the progeny will be depressed, because defective components will be assembled into particles irrespective of their genotype. Conversely if the *ts* mutant affects a catalytic function the yield of ts^+ progeny will not be related to the titre of the mutant.)

Two types of *phenotypes* were recognized among *ts* mutants mapped in the other half of the poliovirus genome, which corresponded with the presumed functions of the viral replicase. One class induced synthesis of double-stranded RNA at restrictive temperature, whereas the other class was unable to induce synthesis, indicating that the viral replicase activity determined both synthesis of the negative strands from the plus-strand genome and plus strands on the negative-strand template.

The genome of these viruses has also been mapped by an independent procedure involving the antibiotic pactamycin (Taber *et al.* 1971). *Pactamycin* inhibits initiation of protein synthesis, but not the extension of initiated chains. Consequently polypeptides radio-labelled in pulse-chase experiments in the presence of pactamycin will be increasingly deficient in label in relation to their proximity to the 5′-end initiation site. Thus the orientation and order of synthesis of the cleavage products could be established, and it was shown that the coat protein sequences were located in the 5′-end half of the genome.

Recombination in picornaviruses was thought to be type-specific, but recently intertypic recombinants have been obtained (Agol, personal communication) and recombinants between serologically distinct types

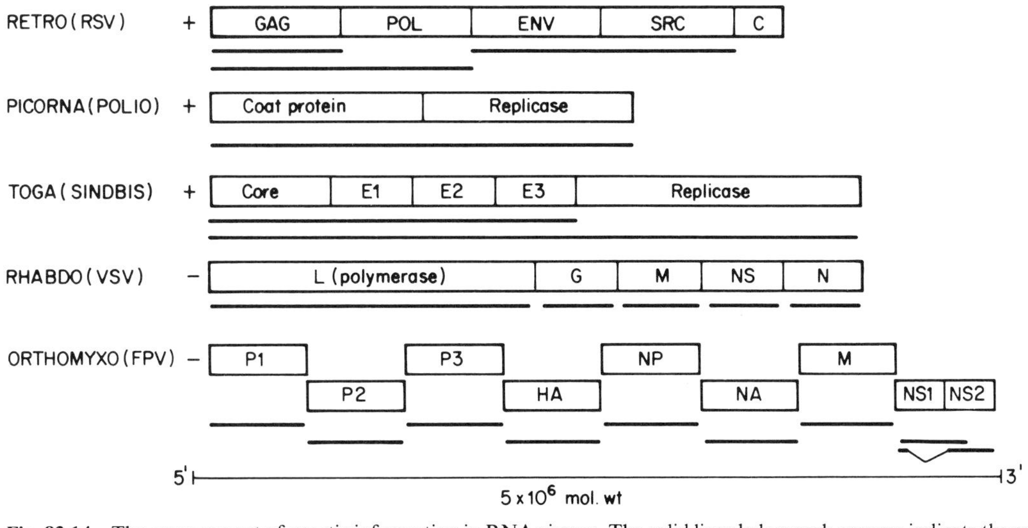

Fig. 83.14 The arrangement of genetic information in RNA viruses. The solid lines below each genome indicate the primary transcription product (mRNA). In the case of polycistronic mRNA functional gene products can be derived either by post-translated cleavage of polyproteins (polio) or internal initiation of translation (Sindbis), or both (RSV and Sindbis).

N.B. Transcription of the sub-genomic Sindbis mRNA is initiated internally on the −ve strand transcript of the genome. The 5′ end of the genome is at the left in this diagram for all viruses except Sindbis where it is at the right.

of FMDV have been identified unequivocally (MacKenzie and Slade, 1975; King *et al.* in preparation).

6.2 The alphavirus genome

The genome of alphaviruses with an approximate mol. wt of 4×10^6 can accommodate 50 per cent more information than the picornavirus genome. The pattern of transcription differs, however. The parental genome being the plus strand has messenger activity and can be translated into a large precursor molecule from which the viral polymerase is derived by post-translational cleavage. The viral polymerase initiates synthesis of both complete genome transcripts and sub-genomic plus strands, by way of negative-strand templates. The *sub-genomic RNA* is a molecule with a sedimentation coefficient of 26s, which *in vitro* codes for a *polyprotein* subsequently cloven to form the structural components of the virion (a nucleocapsid protein and three envelope proteins) (Fig. 83.14). It is thought that the sub-genomic 26S RNA arises by internal initiation of transcription, providing a control mechanism for amplification of the portion of the genome coding for structural proteins (Burke and Russell 1975).

Whatever the mechanism of replication and transcription of the alphavirus genome, there are important genetic consequences. Genetic recombination has not been detected in mixed infections with *ts* mutants of Sindbis or Semliki forest viruses (frequency >0.01 per cent), whereas *ts* mutants of Sindbis virus could be classified into five, or possibly six, groups by genetic *complementation*. Mutants with an RNA-negative phenotype (unable to synthesize RNA at permissive temperature) were classified into groups A and B. Most of the RNA-negative mutants were placed in group A, although sporadic complementation interpreted as *intragenic complementation* was observed between certain pairs of mutants. The status of one mutant was unresolved; this was segregated into group A^1. Complementation was ineffective in general, except for mixed infections involving mutant *ts* 11 (the sole representative of group B) which produced yields 10–50 per cent of normal. Mutants with RNA-positive phenotypes were classified into groups C, D and E which were concordant with their physiological properties. The three mutants in group C all exhibited the same defect, namely failure of cleavage *in vivo* or *in vitro* of the structural protein precursor at restrictive temperature. The group C mutation is evidence for the existence of a virus-coded protease, although other explanations are possible. Group D appeared to represent mutation of the viral haemagglutinin, and group E a non-haemagglutinating envelope protein. The functions of the RNA-negative mutants have not been clearly defined, but the efficient complementation between groups A and B suggested two viral proteins were involved in viral RNA synthesis. This analysis stemmed from the classical studies of Burge and Pfefferkorn who pioneered the genetic analysis of non-recombining genomes (Pfefferkorn 1977). Other workers using different *ts* mutants of both Sindbis virus and Semliki forest virus have failed to obtain complementation. It remains to be decided whether host or virus strain differences are responsible for this anomaly.

Experiments with RNA-negative *ts* mutants of Semliki forest virus (SFV) revealed that inhibition of host macro-molecular synthesis was dependent on prior viral RNA synthesis. In this respect togaviruses (SFV) differ from picornaviruses (polio) where parental RNA can induce inhibition of host RNA synthesis.

The dichotomy in the transcription of mRNA from the alphavirus genome represents a degree of organization above that of the picornavirus pattern. A separation of synthetic functions from structural functions has been accomplished by transcription of 42S (genomic) and 26S mRNA, providing separate control of RNA replicase and capsid protein synthesis. This introduces an inherent flexibility lacking in the host-restricted picornaviruses, which may account for the greater host range of togaviruses and their ability to alternate between vertebrate and invertebrate hosts (Luria *et al.* 1978).

6.3 The flavivirus genome

The flaviviruses (formerly group B arboviruses) are less amenable in the laboratory and have been less well studied. The information content of the genome is similar to that of alphaviruses but the method of processing the information may differ. There is some evidence, though not yet conclusive, that discrete functional polypeptides are produced directly from the polycistronic genome by *independent initiation of translation* at multiple sites, in contrast to the single initiation site of all other non-segmented RNA viruses.

6.4 The coronavirus genome

The single-stranded genome of coronaviruses has a mol.wt of $6\text{–}8 \times 10^6$, giving it considerably greater genetic potential than other plus-strand viruses (Robb and Bond 1979). Little is known about these viruses, but it appears that the expression of coronavirus information is mediated through multiple sub-genomic mRNAs, which form a 3′ co-terminal nested set (Siddell *et al.* 1983). *Ts* mutants of a murine coronavirus have been isolated and partly characterized but no complementation was observed between mutants of different phenotype (Robb *et al.* 1979). One *ts* mutant of this virus induced acute and persistent demyelinization after intracranial inoculation (Haspel *et al.* 1978).

7 The negative-strand viruses

The negative-strand RNA viruses represent a higher level of genome organization. These viruses form a large and diverse group and include viruses with segmented genomes (the orthomyxoviruses, the bunyaviruses and the arenaviruses) and with unsegmented genomes (the paramyxoviruses and the rhabdoviruses). High-frequency recombination occurs in all negative-strand virus groups with multipartite genomes by the reassortment of genome sub-units. Recombination occurs between mutants located in different sub-units, but not between mutants in the same sub-unit, the number of groups corresponding to the number of sub-units, ranging from 2 in the arenaviruses to 8 in the orthomyxoviruses (Table 83.1B).

The mechanism of assembly of the correct complement of genome sub-units remains an intriguing problem. Recombination has not been observed in negative-strand viruses with unsegmented genomes, but complementation analysis has contributed to elucidation of genome structure.

All negative-strand RNA viruses include a virus-specified polymerase in the virion which initiates plus-strand (messenger) RNA synthesis. The genome is not therefore infectious. The control of RNA synthesis and the switch in function of the viral polymerase from synthesis of discrete sub-genomic mRNA to synthesis of complete plus- and negative-strand copies in replication of the genome is still being investigated, and *ts* mutants are playing a role in unravelling this control.

7.1 Unsegmented genome viruses

7.1.1 The rhabdovirus genome

The genome of vesicular stomatitis virus (VSV), the prototype of the group, is a single-stranded RNA molecule of approximately 10 000 nucleotides, all but 70 of which, as well as an untranslated leader sequence, are transcribed into mRNA (Rose 1980). The genome consists of 5 genes in the sequence 3′-N-NS-M-G-L-5′ (Ball 1980) preceded by a *leader sequence* of 47/48 nucleotides at the 3′ end (Fig. 83.14 and 83.15). The virion-associated polymerase transcribes these 5 genes into 5 capped, methylated and polyadenylated monocistronic messages, which are translated into the N (nucleocapsid), NS (a core protein, initially thought to be a non-structural protein), M (matrix), G (glycosylated), and L (large) polypeptides (Figure 83.14). The N, NS and L proteins are components of the helical core and are all required for normal polymerase activity. The G and M are envelope proteins and neutralizing specificity is a property of the virion spikes (G protein). The discrete mRNAs and the leader sequence might be generated either by cleavage of a complete plus-strand transcript of the whole genome, or by separate initiation of each. The situation is at present unresolved. *Replication* occurs by synthesis of complete plus strands as templates for synthesis of progeny negative strands. Genetic experiments with various *ts* mutants and inhibitors of protein synthesis have indicated that the same viral proteins are involved in both transcription and replication with the possible participation of host-specified functions (Pringle 1975, 1977*b*, 1978). Perleman and Huang (1973) separated transcription and replication experimentally by using appropriate *ts* mutants and cycloheximide treatment, demonstrating a more complex control of viral synthesis than that observed in the plus-strand viruses. This may be the attribute which endows the rhabdoviruses with their wide host range, enabling them to multiply in plants, vertebrates and invertebrates, and is responsible for the tendency of other negative-strand viruses (particularly the parainfluenza viruses) to establish stable non-cytocidal persistent infections.

The *inter-genic sequences* between all 5 genes of VSV have been determined; 4 of the 5 junctions show extensive homology and a common structure. There is a

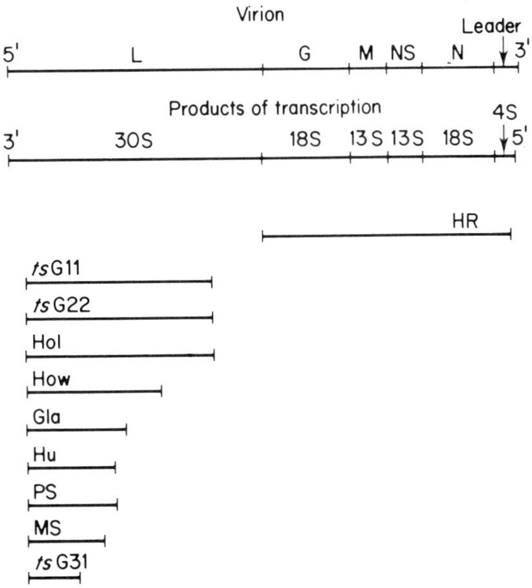

Fig. 83.15 Physical maps of the nucleic acid of interfering particles of VSV in relation to the viral genome.

The letters on the complete genome indicate the protein coded by each viral cistron. The DI particles are arranged in order of increasing genome size. (After Reichmann and Schnitzlein 1979).

sequence of 7 U at the mRNA-poly A tail junction, which suggests that synthesis of the poly A tail is initiated in this region by reiterative copying (or 'slippage' polymerization) (McGeoch 1979). This poly U sequence is part of a unadecamer common to all four junctions (UAUGUUUUUUU) which is followed by a dinucleotide (CA or GA) not represented in the mRNA transcript. This dinucleotide is followed by a sequence common to the 5′ end of the mRNAs. The 70 nucleotides of the VSV (Indiana) genome which are not represented in mRNA transcripts include these 4 dinucleotide spacers, 3 nucleotides at the leader-N gene junction and 59 nucleotides (transcribed but not translated) at the 5′ end of the genome (Rose 1980). Thus the VSV genome is constructed with great economy. Oligonucleotide finger-printing of the VSV mRNAs indicated that there was no splicing of mRNA (Freeman *et al.* 1979).

Ts mutants of VSV and Chandipura virus, a human rhabdovirus serologically unrelated to VSV, have been obtained as induced and spontaneous mutations and classified into a maximum of 6 groups by *complementation* tests. These groups correspond to the 5 genes determining structural proteins and assignments have been made accordingly. For instance, groups I, II, III, IV and V of VSV Indiana corresponding to the genes coding for the L, NS, M, N and G polypeptides respectively (Pringle 1977*b*). In the absence of recombination, unambiguous assignment is difficult to effect without recourse to sequencing (see Discussion by Pringle and Szilagyi 1980); the evidence in favour of these assignments is drawn from many experiments.

The possibility of a sixth as yet undiscovered virion protein was suggested by the existence of 6 complementation groups. However, reinvestigation of complementation indicated that *intra-genic complementation* (within the L gene) could account for the supernumerary group (Gadkari and Pringle 1980*a*). Intragenic complementation indicates either that the gene product exists as a multimer in its functional form, or that the gene product is a multifunctional protein, with mutation affecting different active sites in the same molecule.

Rose (1980) observed that there was an out-of-phase AUG initiation codon within the N mRNA sequence of VSV Indiana with a UAA termination codon downstream in the same reading frame, which could code for a 30 residue polypeptide. However, it is unlikely that any single VSV gene codes for more than one gene product, since McGeoch *et al.* (1980) found that there was only one open reading frame in the serologically related VSV New Jersey strain.

Complementation group E of VSV New Jersey contains 3 mutants all of which exhibit abnormal mobility of the NS polypeptide in polyacrylamide gel (Evans *et al.* 1979). The NS polypeptide is an essential component of the polymerase complex; comprehensive determination of the phenotype of these mutants indicated that the NS polypeptide was *multifunctional* with a role in transcription, replication and maturation (Szilagyi and Pringle 1979).

Certain mutants of VSV known as *td* CE mutants are *ts* only in avian and some other cells, but not in hamster cells. This mutation was associated also with a defect in virion polymerase activity *in vitro*, implicating a host-derived factor in the normal functioning of the viral polymerase (Szilagyi *et al.* 1977). One of these mutants was also restricted in undifferentiated embryonal carcinoma (EC) cells, but not in differentiated cells, indicating that the putative *host factor(s)* which interact with the polymerase were elaborated during differentiation (Pringle 1978). Embryonal carcinoma cells have been observed to restrict the growth of several other viruses including murine leukaemia virus (MLV) (Teich *et al.* 1977). Gautsch (1980) has shown by cell-fusion experiments that restriction of MLV is a recessive trait, indicating that a host function essential for MLV growth is absent in these cells.

In broader terms these mutants of VSV are useful in delineating the truly virus-specific aspects of virus multiplication; this methodology is being applied to the rabies-complex viruses which are more directly relevant to human medicine (Flamand 1980).

Many *ts* mutants of VSV induced enhanced *neuro-virulence* in new-born hamsters but there was no clear association with any specific complementation group (i.e. gene product). The change of disease pattern was a consequence of the increased duration of survival of animals after infection with *ts* mutants (Stanners and Goldberg 1975).

Infection of mice with *ts* G31 (group III), a late mutant with an M protein defect, specifically induced a unique CNS disease characterized by status spongiosus and hind-limb paralysis. However, it is now believed that this disease was induced by a variant which appeared in the CNS after infection and was defective in polymerase function and RNA synthesis (Hughes *et al.* 1979, Hughes and Johnson 1980). Presumably the initial *ts* mutation prevented normal maturation, but allowed virus replication to proceed without causing the death of the animal, until the neurological variant appeared after spread of infection to the CNS.

A *ts* mutant of Chandipura virus, possibly also defective in maturation (Gadkari and Pringle 1980*b*), specifically induced primary demyelination in mice (Dal Canto *et al.* 1979). It is not known whether this was a direct or secondary consequence of infection, but it emphasizes the fact that disease potential can be affected drastically by mutation in a single gene.

The *defective interfering (DI) particles* of rhabdoviruses are more easily characterized than those of other RNA viruses, because the dimensions of the particle are determined by the size of nucleocapsid. DI particles have large portions of the genome deleted and are unable to replicate in the absence of complete virus. They are generated by replication at high multiplicities of infection and represent mainly deletions of the 5′ end of the genome with the termini preserved. They are considered to originate by loop-back synthesis rather than by a recombinational mechanism (Huang 1977). A feature of DI particles is their ability to interfere specifically with replication of their homologous helper virus, but not with heterologous virus. Another type of DI particle (HR in Fig. 83.15) with deletion of the 3′-end of the genome (again with the extreme termini preserved) is unique in retaining some gene functions since complete cistrons remain intact (see Fig. 83.15). This DI is able to interfere with replication of heterologous VSV as well as homologous helper virus. Virus infection is typically self-limiting and it is likely that DI particles play a role in moderating the disease process under natural conditions (Huang and Baltimore 1970). DI particles have also been implicated in the initiation and maintenance of *persistent infection in vivo* and *in vitro* by many RNA viruses including VSV, rabies virus, reovirus, measles virus, LCM virus, JE virus, and some DNA viruses, although other factors such as *ts* mutations, interference, and host resistance are also involved. For example Holland and Villareal (1974) observed that a specific DI particle associated with mutant *ts* G31 of VSV Indiana, which contained only 10 per cent of the genome, was a good inducer of persistent infection although it was not required for maintenance of the persistent state. The association of DI particles with DNA viruses (e.g. SV40, cytomegalovirus, and herpes simplex virus) is significant, because these viruses have the option of persisting by integration into cell DNA. This suggests that expression of the genome of a DNA virus in a persistent infection may be possible only if the genome exists as an episome with a defective genome acting as an auto-regulator.

In VSV at least persistent infection is associated with extensive and *progressive mutation* of the genome. Oligonucleotide mapping of VSV from a persistent infection of BHK-21 cells indicated that about 200 base changes had accumulated during continuous propagation for 5 years. Comparison of the proteins of the initiating *ts* G31 mutant and the 5-year virus by peptide mapping detected changes in all proteins confirming that these mutations were by no means all 'silent' third base changes (Rowlands *et al.* 1980). Sequencing of the 3′ end of the genome also revealed one base substitution in the leader RNA and one (UUU→GUU) in the 3-base link between the leader RNA and the N message, as well as 7 base changes in the non-coding region at the 5′-end. Thus the non-coding leader and inter-cistronic regions were no more conserved than the rest of the genome during persistent infection. Similar changes were not observed during repeated cycles of lytic infection with either the initiating mutant or the derivative 5-year virus (Holland *et al.* 1979). Thus the persistent state was associated with *mutability* over the whole VSV genome and persistence may play an important role in the generation of variation particularly in viruses like VSV which cannot acquire variation by recombination.

7.1.2 The paramyxovirus genome

The paramyxovirus genome is functionally similar to that of the rhabdoviruses, but contains approximately one-third more coding capacity. The number of complementation groups identified, which are indicative of the number of gene products, is consequently greater (e.g. nine in the Newcastle disease virus (NDV), and seven in both Sendai virus and respiratory syncytial virus). Recombination does not occur and new variation can be generated only by mutation. A slow growth cycle and a tendency to induce persistent non-lytic infection have impeded genetic analysis; nevertheless mutants of Sendai virus have contributed to an understanding of *pathogenesis*. The envelope of para-influenza viruses contains three virus-specificed proteins—the haemagglutinin (HN), the matrix (M), and the fusion (F) proteins. The F protein is involved in penetration, cell fusion and haemolysis and is activated by cleavage into two disulphide-bonded subunits F1 and F2. The ability of the virus to infect cells, spread and cause disease is dependent on cleavage of the F protein by proteases in the cell membrane. Cleavage is highly specific, because mutants have been isolated the F protein of which is susceptible to cleavage by a single protease, e.g. certain mutants are susceptible to cleavage by elastase, but not by chymotrypsin and vice versa, etc. (Choppin *et al.* 1975). Thus the ability of the virus to infect particular cells and spread is dependent on the presence of an appropriate protease in the cell membrane. Conversely, the establishment of a persistent infection may depend on selection of host cells lacking a specific protease, or on mutation of the F protein to resistance to the host cell proteases. Thus the host range of para-influenza viruses is also determined by the susceptibility of their F proteins to cleavage by specific proteases. Likewise, viruses isolated from nature may become adapted to *in vitro* culture by selection of a mutant with an appropriate susceptible F protein. It follows that the properties of the virus may have been irrevocably altered by the act of isolation. This exemplifies one of the phenomena which complicate attempts to relate properties of virus strains maintained *in vitro* to disease potential and virulence *in vivo*.

A unique form of host restriction has been described by Choppin *et al.* (1980) which affects disease potential. Measles virus can be isolated from the brain of some SSPE patients only after co-cultivation with susceptible cells. These patients have high serum antibody titres against all measles virus proteins except the matrix (M) protein. Radio-labelling experiments with brain explants indicated that M protein synthesis was also defective. Thus normal maturation did not occur in neural cells and chronic non-lytic infection ensued. It is likely that mutation affecting the M protein is responsible for modification of a virus producing an acute exanthematous disease into an agent producing a chronic neurological disorder.

In measles virus infection of hamsters, hydrocephalus was associated with a particular *ts* mutant. In this instance a reduction in lethality induced symptoms normally masked by the dominant neurological effects of wild type virus.

7.2 Segmented genome viruses

Three groups of negative-strand RNA viruses have segmented genomes. The orthomyxoviruses have a genome with 8 sub-units, which represent 8 distinct viral genes. Segmentation of the genome allows independent transcription of each gene and opportunity for independent control of synthesis of each gene product. However, segment 8 of influenza A virus which codes for the NS1 polypeptide also contains coding sequences for a second smaller polypeptide NS2, partly co-linear but out-of-phase with NS1. The functions of the NS1 and NS2 proteins are unknown, but the genome is more than an assemblage of eight completely independent genes. Likewise, one of the three segments of the bunyavirus genome codes for both viral envelope glycoproteins, and the other two for the N and L proteins. These glycoproteins are unique molecules, but it is not known yet whether they are co-linear or in series. Genetic evidence of *intragenic complementation* suggested and nucleotide sequencing confirmed that the N gene segment could code also for an additional viral protein (The arenaviruses have two segments and, although little is known about their coding functions, again more than one polypeptide may be encoded in the same segment).

The most striking consequence of segmentation is the occurrence of genetic recombination at high frequency by *reassortment* of genome sub-units. *Ts* mutants of all these viruses fall into freely recombining groups which correspond to the number of genome sub-units. Mutants in these groups do not recombine or complement one another, although they can complement mutants in other groups. (Intra-genic complementation has been described only with mutants of the N gene of Maguari virus (*Bunyaviridae*)). Recombination in these viruses should properly be termed reassortment to distinguish it from the low-frequency recombination of plus-strand RNA-viruses, but the term recombination has survived in common usage.

7.2.1 The orthomyxovirus genome

Influenza virus types A and B have 8 genome sub-units and influenza virus type C seven sub-units. Each genomic segment synthesizes a monocistronic mRNA, apart from the NS segment which codes for two polypeptides (Fig. 83.14). The mRNAs are distinguishable from replicative template RNA by the absence of sequences complementary to the 5′-end of genomic RNA and cannot therefore function as templates. The orthomyxoviruses have an early dependence on the cell nucleus for their multiplication (Pennington and Pringle 1978).

The functions of the 8 sub-units of the influenza A virus genome have been established by a combination of genetic and biochemical analyses (Figs. 83.14 and 83.16). The RNA sub-units of different strains of influenza A virus have different electrophoretic mobilities in polyacrylamide gel, as do the viral polypeptides. Figure 83.16 illustrates how the coding assignments can be deduced (Palese 1977). The accuracy of *gene assignment* by this procedure has been confirmed by other approaches, e.g. molecular hybridization (Scholtissek *et al.* 1976, Scholtissek 1978), and specific inhibition (arrest) of *in vitro* translation by individual RNA sub-units (Barry *et al.* 1978).

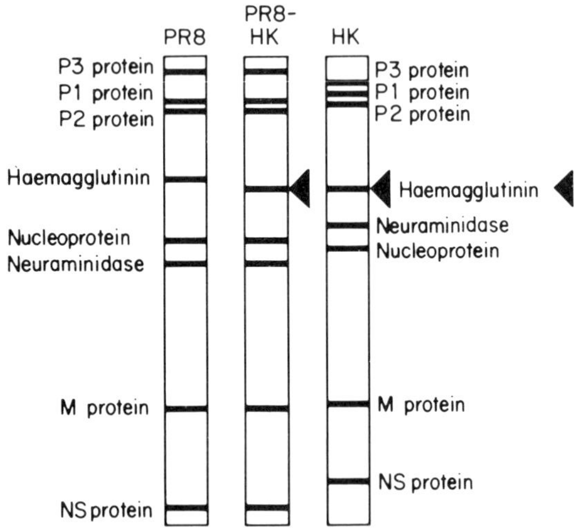

Fig. 83.16 Assignment of influenza virus genes. The eight sub-units of the influenza virus genome can be separated by polyacrylamide gel electrophoresis. The relative mobilities of the RNA sub-units of the PR8 (A/PR/8/34) and HK (A/HK/8/68) strains of influenza virus type A are illustrated diagrammatically in the left and right hand columns. The assignment of gene function for each strain is illustrated to the right and left. The centre track is a recombinant clone which derives the haemagglutinin gene (arrowed) from the HK strain and all others from PR8 strain. Analyses of the RNAs and proteins of recombinants have enabled all the influenza gene functions to be identified unequivocally.

One outcome of this is that it is possible now to construct strains of known genetic constitution in the laboratory. Considerable effort has been directed towards identifying an avirulence gene which could be introduced into epidemic strains to provide *vaccine* virus (Murphy *et al.* 1975). These efforts have been frustrated by the polygenic nature of virulence, whereby disease potential is influenced by constellations of genes rather than by individual genes (Rott *et al.* 1977). Furthermore, although a live virus vaccine has logistic advantages, there is always the hazard that a virus tailored for good growth properties could enhance by recombination the disease potential of naturally existing strains of low virulence.

The epidemiology of influenza virus has been interpreted in terms of antigenic drift—the gradual accumulation of small antigenic changes in response to developing immunity—responsible for the epidemic behaviour of influenza virus, and antigenic shift—the abrupt appearance of a new antigenic type—heralding the onset of a pandemic. Oligopeptide and amino acid sequence analysis have confirmed that minor changes in amino acid composition occur in *antigenic drift* and major changes in *antigenic shift* (Laver *et al.* 1980). Analysis of oligonucleotide fingerprints gives a more sensitive measure of mutational change. Young *et al.*, 1979) observed that nucleotide sequences changes in strains isolated subsequent to the appearance of the H1N1 pandemic virus in 1977 were consistent with sequential mutation. Mutation was not restricted to the genes coding for the HA and NA surface antigens alone, however, but occurred throughout the genome. Similarly the rate of mutational change in pandemic strains was not very different for genes under selective pressure (e.g. the HA gene) and other genes (e.g. the M gene). Therefore selective antigenic pressure cannot be solely responsible for the emergence of genetic variants (see also section 7.1).

The antigenic shifts responsible for human pandemics have two divergent explanations. On the one hand, massive mutational changes might occur over a short period generating new virus strains. Alternatively, new human strains might arise by transmission of an animal strain to man, or of a recombinant between a human and an animal virus. The evidence favours the latter explanation. For instance, molecular studies have shown that the HA of the Hong Kong influenza virus introducing the pandemic of 1969 was similar to the HA of viruses isolated from ducks in the Ukraine and horses in Miami in 1963. Reconstruction experiments demonstrated that recombinants between the Hong Kong virus (A/Hong Kong/68 (H3N2)) and a swine influenza virus (A/Seine Iowa/30 (Hsw1N1)) were produced during simultaneous infection of a pig by these two viruses (Webster and Campbell 1972).

Migratory birds are probably responsible for the spread of avian influenza, since species of birds—terns, shearwaters and ducks—which migrate between the northern and southern hemispheres carry influenza viruses. A virus isolated from domestic chickens in Scotland in 1959 was closely related antigenically to a virus isolated from terns in South Africa in 1961. Antibody to this tern virus was subsequently detected in wild birds in the USSR and Alaska in 1971. Although there is no clear evidence of infection of man by avian viruses, there is opportunity for transmission, as avian influenza viruses multiply in the alimentary tract and are excreted in large amount, remaining viable in cold fresh water for several weeks. The large number of antigenic types provides a reservoir of variation potentially available for transfer to human viruses by recombination. Indeed, it is likely that influenza virus may have originated in the bird kingdom and has evolved a stable relationship with its host in contrast to its behaviour in human populations.

The evidence for antigenic shift by recombination as the origin of pandemics is still circumstantial, however, and a mutational origin cannot be discounted entirely. It is possible that under some circumstances massive mutational change could occur in the interval between pandemics, perhaps during sequestration of the virus in foci of persistent infection (Holland *et al.* 1979). Likewise genetic variation in influenza virus strains of the same serotype is not restricted to mutation alone, but can also involve recombination (Young and Palese 1979).

There are however, limitations to the recombination of influenza viruses. Reassociation of sub-units appears to be possible only between strains of the same type and no inter-type recombinants have been isolated.

7.2.2 The bunyavirus genome

The total mol.wt of the single-stranded RNA genome is $\sim 4\text{–}6 \times 10^6$, giving these viruses a coding capacity equivalent to the orthomyxoviruses. However, the coding potential is distributed over three segments (known as L, M and S), and the known viral proteins account for only 72 per cent of the coding capacity (Obijeski 1976*a*). The M RNA segment encodes both surface glycoproteins, but 14 per cent of the coding capacity remains unaccounted for. The N protein has been assigned to the S RNA segment and the L protein to the L RNA sub-unit, leaving 43 per cent and 38 per cent respectively of each segment unaccounted for (Bishop 1979, Bishop and Shope 1979).

The nucleocapsids of bunyaviruses are circular; and circular as well as linear RNA molecules are present in extracts of nucleic acids. The circular molecules are non-covalently linked and inverted terminal repeats are responsible for the circularization (Obijeski *et al.* 1976*b*).

High-frequency *recombination* occurred in homologous crosses of *ts* mutants of individual viruses belonging to the California encephalitis group (Gentsch and Bishop 1976, Ozden and Hannoun 1978) and the Bunyamwera group (Iroegbu and Pringle 1981; Ozden and Hannoun 1980). Recombination also occurred between certain combinations of California encephalitis group viruses, and analysis of recombinants by oligonucleotide fingerprinting confirmed that recombination occurred by reassortment of genome sub-units. In the Bunyamwera antigenic group heterologous recombination was less restricted and there appeared to be no barrier to exchange of two at least of three genome segments between three viruses originating from different continents and with no known overlap in their range (Iroegbu and Pringle 1981).

Nothing is known about the generation of variation in this large and heterogeneous group of viruses, but it is apparent that there is scope for antigenic shifts by recombination.

7.2.3 The arenavirus genome

Recombination occurred at high frequency between *ts* mutants of Pichinde virus, and two recombination groups have been defined presumably corresponding to the genome segments (Vezza and Bishop 1977). The genome of arenaviruses awaits detailed study, but it is evident that recombination is mediated by sub-unit reassortment.

8 Double-stranded RNA viruses

Both families of double-stranded RNA viruses, the *Reoviridae* and the *Orbiviridae*, have segmented genomes (Table 83.1B). Each of the ten segments of the reovirus codes for a unique polypeptide (Table 83.5). Recombination occurs at high frequency by reassortment of genome sub-units and there are 10 recombination groups corresponding to the 10 RNA sub-units.

Table 83.5 The reovirus type 3 genome: gene assignment.*

Genome sub-unit	Gene product	Function
L1	$\lambda3$	Minor core protein
L2	$\lambda2$	Major core protein
L3	$\lambda1$	Major core protein
M1	$\mu2$	Minor outer capsid protein
M2	$\mu1$ cloven to $\mu1c$	Minor – Major core proteins
M3	$\mu4$	Non-structural protein
S1	$\delta1$	Minor-outer capsid protein
S2	$\delta2$	Major core protein
S3	$\delta2A$	Non-structural protein
S4	$\delta3$	Major outer capsid protein

* From McCrae and Joklik 1978; Mustoe *et al.* 1978.

The *replication* of the reovirus genome is conservative, because the parental double-stranded RNA remains sequestered in a sub-viral particle. Plus-strand RNA is synthesized by a virion-associated double-strand RNA dependent RNA polymerase, functioning either as mRNA or template RNA for synthesis of progeny double-stranded RNA when sequestered in nascent sub-viral structures. *Reassortment* must occur at the time of synthesis of the plus strands, therefore, and not between parental genomes.

Seven (groups A to G) of the ten groups of *ts* mutants were obtained by screening of clones after exposure to several chemical mutagens (Cross and Fields 1977). The missing three groups were obtained as a by-product of analysis of reversion. Ramig and Fields (1977, 1979) established that reversion of *ts* mutants did not usually occur by direct back mutation, but that the wild type phenotype was restored by a suppressor mutation at another site. These *pseudorevertants* were the result of a second mutation in the same gene (intra-genic suppression) or another gene (extra-genic suppression) and suppressed pseudorevertants were identified in all seven groups of *ts* mutants. Some of the suppressor mutations were themselves temperature-sensitive. Eight of these *ts* mutants were characterized and six belonged to two previously undetected recombination groups (H and I) (Ramig and Fields 1979) and later the tenth recombination group (J) was discovered in the same way.

Suppressed pseudorevertants have also been identified in persistently infected cultures of L cells initiated by a *ts* mutant belonging to group C. The suppressor *ts* mutants all belonged to group H. Thus in reovirus pseudoreversion is a prolific source of new variation.

The phenomenon of pseudoreversion high-lights the difficulty of distinguishing phenotype and genotype in animal virus systems. The high frequency of suppression of *ts* mutants by *extra-genic suppression* in reovirus (ten-fold greater than intra-genic suppression) suggests that this may be a general phenomenon. Recombination in double- and negative-stranded RNA viruses is limited to reassortment of sub-units. Therefore the suppression of mutations by pseudoreversion may be an alternative method of genetic repair. The presence of suppressor mutations in all ten reovirus genome sub-units indicates that all the components of the virion are inter-dependent.

Fields *et al* (1980) have used inter-typic recombinants of reovirus types 1 and 3 to identify factors involved in the *pathogenesis* of reovirus infection. The haemagglutinin protein ($\sigma1$) reacts with cell membrane receptors and, in recombinants between type 3 reovirus (neuronal-tropic) and type 1 reovirus (ependymal-tropic), neuronal-tropism in the mouse brain was associated solely with the $\sigma 1$-determining sub-unit of the genome.

The M2 gene also affects the disease potential of reovirus. Type 1 reovirus is non-lethal by the intranasal route in sucking mice, whereas type 3 is lethal. Neither virus is lethal if introduced into the stomach. However a recombinant with the $\sigma 1$ protein of type 3 and the $\mu 1$ protein of type 1 was virulent when introduced by this route. *In vitro* reconstruction experiments confirmed that the type 1 M2 gene product was chymotrypsin-resistant and the type 3 M2 gene product chymotrypsin-sensitive. Apparently the type 1 M2 gene product enables the recombinant to survive in the stomach and the neuronal-tropism of the S1 gene product of type 3 endows it with full virulence.

A change in disease pattern was also associated with a specific *ts* mutant (Fields 1972). Normally reovirus type 3 is lethal for new-born rats, whereas *ts* mutants generally exhibit reduced virulence. In mutant *ts* B1 the majority of animals survived even high doses of virus, but after apparent recovery developed progressive neurological disease later in life which culminated in premature death. The assembly of the outer capsid is defective in *ts* B1-infected cells and it was presumed that accumulation of intracellular nucleocapsids was responsible for the altered disease pattern.

This exemplifies again one of the themes of this chapter, namely that virulence and avirulence are complex genetic phenomena. In favourable cases, however, as in the examples given above the pathogenic potential of virus strains can be related to individual viral genes.

References

Abraham, G. and Cooper, P. D. (1975) *J. gen. Virol.* **29,** 199.

Ball, L. A. (1980) In: *Rhabdoviruses*, Vol. 2, p. 61. CRC Press, Boca Raton.

Baltimore, D. (1970) *Nature* **226,** 1209; (1971) *Bact. Rev.* **35,** 235.

Barrell, B. G., Bankier, A. T. and Drouin, J. (1979) *Nature* **282,** 189.

Barry, R. D., Almond, J. W., McGeoch, D. J., Inglis, S. C. and Mahy, B. W. I. (1978) In: *Negative Strand Viruses and the Host Cell*, p. 1. Academic Press, London.

Beadle, G. W. and Tatum, E. L. (1941) *Proc. nat. Acad. Sci., Wash.* **27,** 499.

Bedson, H. S. and Dumbell, K. R. (1964) *J. Hyg. Camb.* **62,** 147.

Benjamin, T. (1972) *Curr. Top. Microbiol. Immunol.* **59,** 107.

Benveniste, R. E. and Todaro, G. J. (1974*a*) *Nature* **252,** 170; (1974*b*) *Proc. Nat. Acad. Sci., Wash.* **71,** 4513; (1976) *Nature* **261,** 101.

Benyesh-Melnick, M., Schaffer, P. A., Courtney, R. J., Esparza, J. and Kimura, S. (1974) *Cold Spring Harbor Symp. quant. Biol.* **39,** 731.

Berns, K. I. and Hauswirth, W. W. (1978) In: *Replication of Mammalian Parvoviruses*, p. 13. Cold Spring Harbor Laboratory, New York.

Bishop, D. H. L. (1979) *Curr. Top. Microbiol. Immunol.* **86,** 1.

Bishop, D. H. L. and Shope, R. E. (1979) In: *Comprehensive Virology*, Vol. 14, p. 1. Plenum Press, New York.

Boettiger, D. (1979) *Prog. med. Virol.* **25,** 37.

Broker, T. R., Grodzicker, T., Sambrook, J. and Solnick, D. W. (1980) *Genetic Maps*. National Cancer Institute.

Brown, S. M., Ritchie, D. A. and Subak-Sharpe, J. H. (1973) *J. gen. Virol.* **18,** 329.

Brown, S. M., Subak-Sharpe, J. H., Warren, K. G., Wroblewska, Z. and Koprowski, H. (1979) *Proc. nat. Acad. Sci., Wash.* **76,** 2364.

Burke, D. C. and Russell, W. C. (1975) In: *Control Processes in Virus Multiplication*, p. 253. Cambridge University Press, Cambridge.

Celma, M. L. and Ehrenfeld, E. (1974) *Proc. Nat. Acad. Sci., Wash.* **71,** 2440.

Chartrand, P., Crumpacker, C. S., Schaffer, P. A. and Wilkie, N. M. (1980) *Virology* **103,** 311.

Chiswell, D. J. and Pringle, C. R. (1978) *Virology* **90,** 344.

Choppin, P. W., Richardson, C. D., Merz, D. C., Hall, W. W. and Scheid, A. (1980) *J. Supramol. Structure* Suppl. 4, 232.

Choppin, P. W., Scheid, A. and Mountcastle, W. (1975) *Neurology* **25,** 494.

Clements, J. B., Watson, R. J. and Wilkie, N. M. (1977) *Cell* **12,** 275.

Clements, J. E., D'Antonio, N. and Narayan, O. (1980*b*) *Virology* **102,** 46.

Clements, J. E., D'Antonio, N., Narayan, O., Pederson, F. S. and Haseltine, W. S. (1980*a*) *J. supramol. Structure* Suppl. 4, p. 231.

Coffin, J. M. (1980) In: *Molecular Biology of RNA Tumor Viruses*, p. 199. Academic Press, New York.

Collett, M. S., Brugge, J. S. and Erikson, R. L. (1978) *Cell* **15,** 1363.

Collett, M. S. and Erikson, R. L. (1978) *Proc. nat. Acad. Sci., Wash.* **75,** 2021.

Cooper, P. D. (1968) *Virology* **35,** 584; (1977) In: *Comprehensive Virology*, Vol. 9, p. 133. Plenum Press, New York.

Cremer, K. J., Bodemer, M., Summers, W. P., Summers, W. C. and Gesteland, R. F. (1979) *Proc. nat. Acad. Sci., Wash.* **76,** 430.

Cross, R. K. and Fields, B. N. (1977) In: *Comprehensive Virology*, Vol. 9, p. 291. Plenum Press, New York.

Dal Canto, M. C., Rabinowitz, S. G. and Johnson, T. C. (1979) *J. neurol. Sci.* **42,** 155.

Dulbecco, R. (1979) *Microbiol. Rev.* **43,** 443.

Eckhart, W. (1977) In: *Comprehensive Virology*, Vol. 9, p. 1. Plenum Press, New York.

Evans, D., Pringle, C. R. and Szilagyi, J. F. (1979) *J. Virol.* **31,** 325.

Fenner, F. and Ratcliffe, F. N. (1965) *Myxomatosis*. Cambridge University Press, London and New York.

Fenner, F. and Sambrook, J. F. (1966) *Virology* **28,** 600.

Fields, B. N. (1972) *New Engl. J. Med.* **287,** 1026.

Fields, B. N., Weiner, H. L., Drayna, D. A. and Sharpe, A. H. (1980) *J. Supramol. Structure* Suppl. 4, p. 230.

Fiers, W. *et al.* (1978) *Nature* **273,** 113.

Flamand, A. (1980) In: *Rhabdoviruses*, Vol. 2, p. 115, CRC Press, Boca Raton.

Flint, J. (1977) *Cell* **10,** 153.

Flint, S. J. (1980*a*) In: *DNA Tumor Viruses*, p. 383. Cold Spring Harbor Laboratory, New York; (1980*b*) In: *The*

DNA Tumour Viruses, p. 547. Cold Spring Harbor Laboratory, New York.
Flint, S. J. and Broker, T. R. (1980) In: *The DNA Tumor Viruses*, p. 443. Cold Spring Harbor Laboratory, New York.
Freeman, G. J., Rao, D. D. and Huang, A. S. (1979) *Gene* **5,** 141.
Friis, R. R. (1978) *Curr. Top. Microbiol. Immunol.* **79,** 261.
Frost, E. and Williams, J. F. (1978) *Virology* **91,** 39.
Gadkari, D. A. and Pringle, C. R. (1980*a*) *J. Virol.* **33,** 100; (1980*b*) *Ibid.* **33,** 107.
Galos, R., Williams, J., Binger, M. H. and Flint, S. J. (1979) *Cell* **17,** 945.
Garon, C. R., Barbora, E. and Moss, B. (1978) *Proc. nat. Acad. Sci., Wash.* **72,** 4810.
Gautsch, J. W. (1980) *Nature* **285,** 110.
Gemmell, A. and Fenner, F. (1960) *Virology* **11,** 219.
Gentsch, J. and Bishop, D. H. L. (1976) *J. Virol.* **20,** 351.
Gerdes, J. C., Marsden, H. S., Cook, M. L. and Stevens, J. G. (1979) *Virology* **94,** 430.
Germond, J. E., Hirt, J. E., Oudet, P., Gross-Bellard, M. and Chambon, P. (1975) *Proc. nat. Acad. Sci., Wash.* **72,** 1843.
Geshelin, P. and Berns, K. I. (1974) *J. med. Virology* **88,** 785.
Ginsberg, H. S. and Young, C. S. H. (1977) In: *Comprehensive Virology*, Vol. 9, p. 27. Plenum Press, New York.
Graf, T. and Beug, H. (1980) *J. supramol. Structure.* Suppl. 4, p. 227.
Griffin, B. E. (1980) In: *The DNA Tumor Viruses*, p. 61. Cold Spring Harbor Laboratory, New York.
Grodzicker, T. (1980) In: *The DNA Tumor Viruses*, p. 577. Cold Spring Harbor Laboratory, New York.
Halliburton, I. W. (1980) *J. gen. Virol.* **48,** 1.
Haspel, M. W., Lampert, P. W. and Oldstone, M. B. A. (1978) *Proc. nat. Acad. Sci., Wash.* **75,** 4033.
Holland, J. J. and Villareal, L. P. (1974) *Proc. nat. Acad. Sci., Wash.* **71,** 2956.
Holland, J. J., Graban, E. A., Jones, C. L. and Semler, B. L. (1979) *Cell* **16,** 495.
Huang, A. S. (1977) *Bact. Rev.* **41,** 811.
Huang, A. S. and Baltimore, D. (1970) *Nature* **226,** 325.
Huebner, R. J., Chanock, R. M. Rubin, B. A., and Casey, M. J. (1964) *Proc. nat. Acad. Sci., Wash.* **52,** 1333.
Hughes, J. V. and Johnson, T. C. (1980) *J. supamol. Structure.* Suppl. 4, p. 280.
Hughes, J. V., Johnston, T. C., Rabinowitz, S. G. and Dal Canto, M. C. (1979) *J. Virol.* **29,** 312.
Hunter, E. (1978) *Curr Top. Microbiol. Immunol.* **79,** 295.
Hunter, T. and Sefton, B. M. (1980) *Proc. nat. Acad. Sci., Wash.* **77,** 1311.
Iroegbu, C. U. and Pringle, C. R. (1981) *J. Virol.* **37,** 383.
Kates, J. R. and McAuslan, B. R. (1967) *Proc, nat. Acad. Sci., Wash.* **58,** 134.
King, A. M. Q., Slade, W. R., Newman, J. W. I. and McCahon, D. (1980) *J. Virol.* **34,** 67.
Klein, G. (1982) *Advances in Viral Oncology*, Vol. 1. Oncogene Studies. Raven Press, New York.
Lake, J. R. and Cooper, P. D. (1980) *J. gen. Virol.* **48,** 135.
Lake, J. R., Priston, R. A. J. and Slade, W. R. (1975) *J. gen. Virol.* **27,** 355.
Laver, W. G., Air, G. M., Dopheide, T. A. and Ward, C. W. (1980) *Nature* **283,** 454.
Ledinko, N. (1963) *Virology* **20,** 107.
Levine, A. S., Levin, M. J., Oxman, M. N. and Lewis, A. M. (1973) *J. Virol.* **11,** 672.
Lewis, A. M., Jr. and Rowe, W. P. (1970) *J. Virol.* **5,** 413.
Lofgren, K. W., Stevens, J. G., Marsden, H. S. and Subak-Sharpe, J. H. (1977) *Virology* **76,** 440.
Luria, S. E., Darnell, J. E. Jr., Baltimore, D. and Campbell, A. (1978) *General Virology*, 3rd edn., John Wiley and Sons, New York.
McCrae, M. A., and Joklik, W. K. (1978) *Virology* **89,** 578.
McGeoch, D. J. (1979) *Cell* **17,** 673.
McGeoch, D., Dolan, A., and Pringle, C. R. (1980) *J. Virol.* **33,** 69.
MacKenzie, J. S. and Slade, W. R. (1975) *Aust. J. exp. Biol. med. Sci.* **53,** 251.
Mackett, M. and Archard, L. C. (1979) *J. gen. Virol.* **45,** 683.
Macnab, J. C. M. and Timbury, M. C. (1976) *Nature* **261,** 233.
Marsden, H. S., Stow, N.D., Preston, V. G., Timbury, M. C. and Wilkie, N. M. (1978) *J. Virol.* **28,** 624.
Morse, L. S., Pereira, L., Roizman, B. and Schaffer, P. A. (1978) *J. Virol.* **26,** 389.
Murphy, B. R., Spring, S. B., Richman, D. D., Tierney, E. L., Kasel, J. and Chanock, R. M. (1975) *Virology* **66,** 533.
Mustoe, T. A., Ramig, R. F., Sharpe, A. H. and Fields, B. N. (1978) *Virology* **89,** 594.
Obijeski, J. F., Bishop, D. H. L., Murphy, F. A. and Palmer, E. L. (1976*a*) *J. Virol.* **19,** 985.
Obijeski, J. F., Bishop, D. H. L., Palmer, E. L. and Murphy, F. A. (1976*b*) *J. Virol.* **20,** 664.
Ozden, S. and Hannoun, C. (1978) *Virology* **84,** 210; (1980) *Ibid.* **103,** 232.
Padgett, B. L. and Tomkins, J. K. N. (1968) *Virology* **36,** 161.
Palese, P. (1977) *Cell* **10,** 1.
Park, M., Lonsdale, D. M., Timbury, M. C., Subak-Sharpe, J. H. and Macnab, J. C. M. (1980) *Nature* **285,** 412.
Paterson, B. M., Roberts, B. and Kuff, E. L. (1977) *Proc. nat. Acad. Sci. Wash.* **74,** 4370.
Pennington, T. H. and Pringle, C. R. (1978) In: *Negative Strand Viruses and the Host Cell*, p. 457. Academic Press New York.
Perleman, S. M. and Huang, A. S. (1973) *J. Virol.* **12,** 1395.
Pfefferkorn, E. R. (1977) In: *Comprehensive Virology*, Vol. 9, p. 209. Plenum Press, New York.
Preston, V. G., Davison, A. J., Marsden, H. S., Timbury, M. C., Subak-Sharpe, J. H. and Wilkie, N. M. (1978) *J. Virol.* **28,** 499.
Pringle, C. R. (1968) *J. gen. Virol.* **2,** 199; (1975) *Curr. Top. Microbiol. Immunol.* **69,** 85; (1977*a*) *Ibid.* **76,** 50; (1977*b*) In: *Comprehensive Virology*, Vol. 9, p. 239. Plenum Press, New York; (1978) *Cell* **15,** 597.
Pringle, C. R., Howard, D. K. and Hay, J. (1973) *Virology* **55,** 495.
Pringle, C. R. and Szilagyi, J. F. (1980) In: *The Rhabdoviruses*, Vol. 2. p. 141, CRC Press, Boca Raton.
Ramig, R. F. and Fields, B. N. (1977) *Virology* **81,** 170; (1979) *Ibid.* **92,** 155.
Rapp, U. R. and Todaro, G. J. (1978) *Proc. nat. Acad. Sci., Wash.* **75,** 2468.
Reichmann, M. E. and Schnitzlein, W. M. (1979) *Curr. Top. Microbiol. Immunol.* **86,** 123
Robb, J. A. and Bond, C. W. (1979) In: *Comprehensive Virology*, Vol. 14, p. 193, Plenum Press, New York.
Robb, J. A., Bond, C. W. and Leibowitz, J. L. (1979) *Virology* **94,** 385.

Robinson, A. J. and Bellett, A. J. D. (1975) *Cold Spring Harbor Symp. quant. Biol.* **39,** 523.

Robinson, H. L., Pearson, M. N., Disimone, D. W., Tsichlis, P. N. and Coffin, J. M. (1979) *Cold Spring Harbor Symp. quant. Biol.* **44,**

Rose, J. A. (1974) In: *Comprehensive Virology*, Vol. 3, p. 1. Plenum Press, New York.

Rose, J. K. (1980) *Cell* **19,** 415.

Rott, R., Scholtissek, C., Klenk, H-D. and Orlich, M. (1977) In: *Negative Strand Viruses and the Host Cell*, p. 653. Academic Press, London

Rowlands, D., Grabau, E., Spindler, K., Jones, C., Semler, B and Holland, J. (1980) *Cell* **19,** 871.

Sambrook, J., Williams, J. F., Sharp, P. A. and Grodzicker, T. (1975) *J. molec. Biol.* **97,** 369.

Sambrook, J. F., Botcham, M., Hu, S-L. and Stringer, J. (1980) *J. supramol. Structure* Suppl. 4, p. 223.

Schaffer, P. A. (1975*a*) In: *Oncogenesis and Herpesviruses II*, p. 195. IARC, Lyon; (1975*b*) *Curr. Top. Microbiol. Immunol.* **70,** 51.

Schaffer, P. A., Tevethia, M. J. and Benyesh-Melnick, M. (1974) *Virology* **58,** 219.

Scholtissek, C. (1978) *Curr. Top. Microbiol. Immunol.* **80,** 139.

Scholtissek, C., Harms, E., Rohde, W., Orlich, M. and Rott, R. (1976) *Virology* **74,** 332.

Scolnick, E., Rands, E., Williams, D. and Parks, W. (1973) *J. Virol.* **12,** 458.

Sheldrick, P. and Berthelot, N. (1974) *Cold Spring Harbor Symp. quant. Biol.* **39,** 667.

Shimotohno, K., Mizutani, S. and Temin, H. M. (1980) *Nature* **285,** 550.

Shortle, D., Pipas, J., Lazarowitz, S., Di Maio, D. and Nathans, D. (1979) In: *Genetic Engineering*, Vol. 1, p. 73. Plenum Press, New York.

Siddell, S., Weye, H. and Ter Meulen, V. (1983) *J. gen. Virol.* **64,** 761.

Soeda, E., Arrand, J. R., Smolar, N., Walsh, J. E. and Griffin, B. E. (1980) *Nature* **283,** 445.

Stanners, C. P. and Goldberg, V. J. (1975) *J. gen. Virol.* **29,** 281.

Stephenson, J. R., Khan, A. S., van de Ven, W. J. M. and Reynolds, F. H. Jnr. (1979) *J. nat. Cancer Inst.* **63,** 1111.

Stow, N. D. and Wilkie, N. M. (1978) *Virology* **90,** 1.

Strayer, D. R. and Gillespie, D. H. (1980) *The Nature and Organisation of Retroviral Genes in Animal Cells.* Springer-Verlag, Wien and New York.

Subak-Sharpe, J. H. (1967) *Brit. med. Bull.* **23,** 161; (1971) In: *The Strategy of the Viral Genome*, p. 1. Churchill-Livingstone, Edinburgh.

Subak-Sharpe, J. H. and Pringle, C. R. (1975) In: *Control Processes in Virus Multiplication*, p. 363. Cambridge University Press, Cambridge.

Subak-Sharpe, J. H. and Timbury, M. C. (1977) In: *Comprehensive Virology*, Vol. 9, p. 89. Plenum Press, New York.

Szilagyi, J. F. and Pringle, C. R. (1972) *J. molec. Biol.* **71,** 281; (1979) *J. Virol.* **30,** 692.

Szilagyi, J. F., Pringle, C. R. and McPherson, T. M. (1977) *J. Virol.* **22,** 381.

Taber, R., Rekosh, D. and Baltimore, D. (1971) *J. Virol.* **8,** 395.

Tattersall, P. (1978) In: *Replication of Mammalian Parvoviruses*, p. 53. Cold Spring Harbor Laboratory, New York.

Tattersall, P. and Ward, D. (1978) In: *The Replication of Mammalian Parvoviruses*, p. 1. Cold Spring Harbor Laboratory, New York.

Tegtmeyer, P. (1980) In: *The DNA Tumor Viruses*, p. 297. Cold Spring Harbor Laboratory, New York.

Teich, N. M., Weiss, R., Martin, G. R. and Lowy, D. R. (1977) *Cell* **12,** 973.

Temin, H. M. (1976) *Science* **192,** 1075.

Temin, H. M. and Mizutani, S. (1970) *Nature* **226,** 1211.

Threlfall, G., Barber, C., Carey, N and Spencer, E. (1980) *Cell* **19,** 683.

Timbury, M. C. and Calder, L. (1976) *J. gen. Virol.* **30,** 179.

Todaro, G. J. (1980*a*) *J. supramol. Structure* Suppl. 4, p. 233; (1980*b*) In: *Molecular Biology of RNA Tumor Viruses*, p. 47. Academic Press, New York.

Timbury, M. C. and Calder, L. (1976) *J. gen. Virol.* **30,** 179.

Todaro, G. J. (1980*a*) *J. supramol. Structure* Suppl. 4, p. 233; (1980*b*) In: *Molecular Biology of RMA Tumor Viruses*, p. 47. Academic Press, New York.

Topp. W. C., Lane, D. and Pollack, R. (1980) In: *The DNA Tumor Viruses*, p. 297. Cold Spring Harbor Laboratory New York.

Verma, I. M., Mason, W. S., Drost, S. D. and Baltimore, D. (1974) *Nature* **251,** 27.

Vezza, A. C. and Bishop, D. H. L. (1977) *J. Virol.* **24,** 712.

Vogt, P. K. (1977). In: *Comprehensive Virology*, Vol. 9, p. 341. Plenum Press, New York.

Webster, R. G. and Campbell, C. H. (1972) *Virology* **48,** 528.

Weiss, R. A. (1980) In: *Rhabdoviruses*, Vol. 3 p. 51, CRC Press.

Weissmann, C., Nagata, S., Taniguchi, T., Weber, H. and Meyer, F. (1979) In: *Genetic Engineering*, Vol. 1, p. 133. Plenum Press, New York.

Wigler, M., *et al.* (1979) *Cell* **16,** 777.

Williams, J. F., Young, C. S. H. and Austin, P. (1975) *Cold Spring Harbor Symp. quant. Biol.* **39,** 427.

Wittek, R., Menna, A., Schümperli, D., Stoffel, S., Müller, H. K., Wyler, R. (1977) *Virology* **23,** 669.

Wyke, J. A. (1975) *Biochem. biophys. Acta* **417,** 91.

Yang, R. C. A. and Wu, R. (1979) *Science* **206,** 456.

Young, J. F., Desselberger, U. and Palese, P. (1979) *Cell* **18,** 73.

Young, J. F. and Palese, P. (1979) *Proc. nat. Acad. Sci., Wash.* **76,** 6547.

Závada, J. (1972) *J. gen. Virol.* **15,** 183.

Zur Hausen, H. (1977) *Curr. Top. Microbiol. Immunol.* **78,** 1.

84

The pathogenicity of viruses

Harry Smith and Clive Sweet

Introductory

Pathogenicity or virulence is the capacity of some micro-organisms to produce disease which not only entails entry into a host and multiplication within its tissues (infectivity), but also the production of harmful, sometimes lethal, effects. At first sight it might appear that all viruses are pathogenic because they are obligate parasites. However, many viruses are either non-pathogenic (avirulent) or poorly pathogenic because they lack either the ability to infect animals or they do so without causing harm. Also, viruses show a great diversity in the manifestations and severity of the diseases that they cause. This chapter describes our current knowledge of the factors that determine pathogenicity. Oncogenesis is excluded since it is best discussed in relation to specific viruses (Chapter 104). In accordance with editorial policy, many of the references are review articles which: (a) will allow the reader to explore the subject further and (b) provide the numerous references to original work which could not be cited here because of lack of space.

The methods of studying virus pathogenicity

There are two methods. Natural infections can be observed either in single animals or within populations. The latter is called epidemiology (Chapter 85); it involves pathogenicity, i.e. the capacity to produce disease in one host, but is predominantly concerned with communicability, the ability of viruses to spread from one host to another. More discerning studies of pathogenicity can be made by deliberately introducing pathogenic viruses into animals. This Chapter deals mainly with such experimental studies but the pattern of natural disease in individuals is summarized first as background.

General pattern of virus disease in individuals

Entry to the host occurs occasionally through the skin by trauma or vector bite. Most infections, however, begin over mucous surfaces, e.g. those of the respiratory and alimentary tracts, with virus-host interactions rather than external agencies providing the impetus for entry. Sometimes, infection is largely confined to the mucous surfaces with minimal penetration to other tissues, e.g. in influenza and in the common cold. Many viruses, however, spread to other tissues either after initial replication in epithelial cells, e.g. Newcastle disease virus (NDV), or by penetrating the mucosal surface without causing infection, e.g. rinderpest virus (Burrows 1972).

In the tissues, viruses are faced with anti-viral substances in the body fluids and in the cells they infect. They must also contend with phagocytes, either polymorphonuclear (PMN) or mononuclear (MN). These cells are derived from bone marrow and can ingest and kill some viruses and virus-infected cells. Some of these anti-viral mechanisms are present at entry and others are mobilized within a few hours as a result of inflammation. At this primary lodgement phase of infection the viruses are most vulnerable to host defences and this is where the success or failure of most infections is decided. Successful viruses may localize at the initial site, but mostly they spread, first through the lymph channels which run close to the mucous surfaces and then in the blood stream. The spreading virus may be free or on or within white or red blood cells. Spread is helped by the circulatory systems but it may be hindered by MN phagocytes, the so-called macrophages, fixed in filtration systems set across lymph and blood vessels in the lymph nodes, spleen and liver. These organs, together with the bone marrow and the tissue and blood-borne phagocytes, are called the reticulo-endothelial system, which has a protective function. After spreading from the initial site some infections become generalized as in Lassa fever, but others tend to localize in particular tissues, such as the salivary glands in mumps.

Pathogenic viruses can damage host tissues on mucous surfaces, at primary lodgement, during spread and after localization. This damage can cause unpleasant clinical effects and may be fatal. Eventually, the host eliminates the virus in most cases and recovers from the disease. Recovery is often, but not always, accompanied by immunity to further attacks. There are two exceptions to this pattern of damage and recovery. Sometimes damage is so slight that the host is unaware of the disease in subclinical infections; and occasionally the host recovers from overt disease but virus persists for months or years. In chronic infections significant quantities of virus are continually produced and may be shed—the so called 'carrier state' as occurs in hepatitis B virus infections. In the latent state smaller quantities of virus are retained in a form from which the disease can be reactivated, e.g. in herpes infections.

This summary of the infectious process indicates the properties that pathogenic viruses must possess: ability to survive on and penetrate mucous surfaces; ability to replicate in host tissues; ability to interfere with the humoral and cellular defence mechanisms on mucous surfaces, at primary lodgement and during spread; and a capacity to damage tissue.

Experimental study of virus pathogenicity

The causative virus must be identified and then the mechanisms of pathogenicity recognized and explained in biochemical terms. The latter step can be studied with single virus isolates but the task is made easier by comparing the properties of strains of differing virulence for one host species.

Identifying the causative virus: Koch's postulates

The criteria below were used by Koch in 1891 when he identified *Bacillus anthracis* as the cause of anthrax. Later they were applied to virus diseases, and even now fulfilment of Koch's Postulates is sought for any newly appearing disease, e.g. Lassa fever.

There are three postulates though these often cannot be completely fulfilled.

(1) The organism should be found in all cases of the disease in question and its distribution in the body should be in accordance with the lesions observed. Ideally, as for influenza and smallpox, the virus can be recognized in pathological samples by electron microscopy or fluorescent-antibody staining and its presence confirmed by isolation in tissue culture and testing *in vitro*. However, in some diseases, such as hepatitis B, the causative virus cannot be easily recognized or cultured *in vitro*. Fortunately, in such cases, immunological methods, not available to Koch, usually indicate that the virus is or has been present.

(2) The virus should be cultivated outside the body of the host in pure culture for several generations. This has been accomplished for most viruses but not all, such as those of hepatitis B and some diarrhoeas and the putative viruses of some human cancers.

(3) The organism so isolated should reproduce the disease when introduced into other susceptible animals. The major difficulties occur here. It is easily accomplished for veterinary diseases where the natural host can be inoculated and the typical effects observed, e.g. with foot-and-mouth disease in cattle. Regarding human diseases, ethics restrict experimentation. Volunteers have been inoculated for diseases like the common cold and influenza but this clearly cannot be done with lethal viruses such as those of Lassa and Ebola fevers. Experimental animals must be used. Unfortunately, only in a few cases, such as influenza in the ferret, can a disease typical of that seen in man be reproduced in small laboratory animals. Although the Lassa fever virus kills mice we cannot be certain the disease syndrome sufficiently resembles that of human infection. In some cases primates suffer human-like illnesses, e.g. poliomyelitis in monkeys. Nevertheless, the inability to test in the natural host a virus putatively responsible for human sickness sometimes prevents the fulfilment of Koch's Postulates, notably for those viruses that possibly cause human cancers.

Comparisons of the pathogenicity or virulence of different viruses

Strictly the terms pathogenicity and virulence should be used only in a comparative sense: a virus population is more or less pathogenic or virulent than another population. Although the two terms are synonymous, Miles (1955) has suggested that pathogenicity should be used with respect to species of microbes, e.g. the rabies virus is more pathogenic than influenza virus and virulence with respect to strains within the species, e.g. the A/Finland strain of influenza virus is more virulent than the A/Okuda strain.

The first requirement for comparisons of virulence is a suitable host. Ideally, this should be the natural host but often experimental animals must be used for viruses causing human disease. The choice of experimental animal is governed by the similarity of its disease syndrome to the human disease, the availability of sufficient numbers for statistical analysis, and reliability as regards lack of indigenous diseases. The virus strains to be compared must be grown in a standard manner with respect to host (tissue culture cells, eggs, sucking mice), temperature and time of incubation. After harvesting, the concentration of live virus in the suspension to be inoculated must be assessed e.g. by focal or quantal assays (Chapter 19). Unfortunately, these methods detect only a few—1 in 10 to 1 in 1000 particles according to the virus species—of the total virus particles present (see later). The route of inoculation affects virulence and thus it must be standardized. The most natural route should be used but often it is not, because of the difficulties of infecting the respiratory, alimentary and urogenital tracts quantitatively. Known quantities of aerosol-borne virus can be delivered to the upper and lower respiratory tract (Burrows 1972) but the apparatus is expensive and requires expert use. Some respiratory viruses are instilled intranasally but, as a rule, virus strains are compared by injecting them intravenously, subcutaneously, intraperitoneally or even intracerebrally. The advantage of such routes is that standard amounts of virus are received by the inoculated animal. However, in most cases these routes are completely unnatural except perhaps for vector-transmitted viruses, and they by-pass one important step in pathogenesis, namely survival on and penetration of mucous surfaces.

Quantitative methods of assessing disease effects are also required. With a fatal disease assessment is simple: the death rate in a batch of animals or the average time to death if all are killed. When the disease is not fatal, e.g. influenza, less definite assessments must be used. Examples are the extent of skin lesions and rashes, clinical signs and symptoms (Beare and Reed 1977), duration and height of pyrexia (Sweet and Smith 1980), and the extent of lung infection and damage (Potter and Oxford 1977). A manifestation of cell damage must be assessed; demonstration that a virus is infective for animals is neither proof nor a measure of pathogenicity.

With these requirements fulfilled, the virulence of strains can be compared either qualitatively or quantitatively. For the former, the same dose of each strain is inoculated and the disease effect assessed, e.g. strain A kills 16 of 20 animals and strain B 4 of 20 animals, or strain A produces a more severe fever than strain

Table 84.1 The multifactorial nature of virulence

Virulence determinants	Possession of virulence determinants by: Virulent strain U	Attenuated or avirulent strains W	X	Y	Z
A	A	—	A	A	A
B	B	B	—	B	B
C	C	C	C	—	C
D	D	D	D	D	—

B. This indicates only that strain A is more virulent than strain B. To make quantitative comparisons, graded doses of the two strains are inoculated and the doses of each which produce the same disease effect are determined. For lethal viruses, the doses that kill 50 per cent of the animals—the 50 per cent lethal dose (LD_{50})—are obtained from the sigmoid curve which relates the response of batches of animals to increasing inocula of the virus. Comparisons of the LD_{50}s for different strains provide the quantitative relation between the virulence of the two strains, e.g. strain A, 100 plaque-forming units (pfu); strain B, 10 000 pfu; thus strain A is $10\,000 \div 100 = 100 \times$ more virulent than strain B. A similar procedure can be applied to non-lethal viruses, e.g. by comparing the doses which produce a lesion of a certain size or a fever above a chosen level. These comparisons of virulence suffer from the usual variation of biological experiments and any quantitative difference must be qualified by its statistical variation. Also, the methods for estimating infectious virions usually detect only a small proportion of the total virus particles as revealed by electron microscopy and may not therefore measure all of those, which could vary for different strains, capable of multiplying in experimental animals. For example, a suspension of Semliki Forest virus (SFV) could be diluted to a stage where plaques were not formed on chick embryo fibroblasts, yet the suspension killed sucking mice (Smith 1972). Thus, an LD_{50} based on a plaque count may be considerably different from that based on the number of particles capable of replicating in the infected animals. For this reason even quantitative differences between virus strains may be misleading and only virus strains which show large differences in virulence should be used in comparative experiments designed to identify virulence mechanisms. Clearly, assessing quantitative differences in virulence between virus strains is a time-consuming and expensive business. Unfortunately, therefore, most comparisons of virus virulence are qualitative rather than quantitative.

Virulence determinants: the multifactorial nature of virulence

As mentioned previously, pathogenic viruses must be able to: (1) enter the host, which often entails surviving on and penetrating mucous surfaces; (2) replicate in some cells of the host; (3) interfere with the action of humoral and cellular defence mechanisms on mucous surfaces, at primary lodgement and during spread in the lymph and blood; and (4) damage tissues, thereby producing unpleasant and possibly lethal effects.

The virus components or host cell factors responsible for these steps in disease production are *virulence determinants*. Virulent strains can accomplish all four steps; avirulent or attenuated—used here for strains of low virulence both naturally occurring and laboratory-derived—strains fail at one or more of them. Each step is complex involving many facets of virus-host interaction. Hence, virulence is multifactorial, i.e. viruses must possess several different determinants in order to accomplish the whole disease process. This cardinal point means that: (a) a strain is fully virulent only when it possesses a full complement of virulence determinants; (b) strains attenuated by loss of one or more determinants, i.e. a gene product is modified so that it does not contribute to virulence, will still possess the remaining members of the complement; and (c) an avirulent strain can be converted into a virulent strain if the gene(s) coding for its missing virulence determinant(s) can be transferred from another strain which may also itself be avirulent. For example, if the virulence of strain U is dependent upon 4 determinants A, B, C and D (Table 84.1), attenuated or avirulent strains may be formed by loss of any one of these determinants, but each of the strains (W, X, Y, Z; Table 84.1) would possess 3 of the 4 determinants. Virulent strains would result from transference of the gene coding for determinant A from attenuated strain X to attenuated strain W or that coding for determinant B from attenuated strain Y to attenuated strain X.

The virulence determinants of viruses are more complex than those of bacteria, which are often clearly defined products such as the capsular polysaccharides which prevent ingestion of pneumococci by phagocytes (Smith 1978). In some cases, virion components are virulence determinants like the surface glycoproteins of the para- and ortho-myxoviruses which, if cleaved* proteolytically during the final stages of virus maturation, ensure the infectivity of progeny virus

*Editorial note.
Though the correct spelling of the past participle of the verb to 'cleave' is 'cloven' (e.g. cloven hoof), the authors of this chapter prefer to use the word 'cleaved', as it is so widely used in the literature on virology.

(Rott 1979; Klenk 1980). However, other determinants of virulence may require intimate interaction between the virus and the host cell. Such processes may inactivate phagocytes (Smith 1972, 1980) or result in the insertion of virus components or products into host cell membranes which may cause cell fusion (Poste 1972), or evoke harmful immunological reactions (Webb and Hall 1972; Burns and Allison 1977). Despite the complexity of these processes, the essential character of a virulence determinant remains, i.e. a compound or process which is produced or induced by a fully virulent strain and is wholly or partly absent from some attentuated strains.

Recognition and identification of virulence determinants by comparing strains of differing virulence

Some knowledge of virulence determinants, particularly the general mechanisms of replication, may be gained from studying one virus strain. However, the classical method of investigating the determinants of virulence of any microbe is to compare the properties of virulent and attenuated strains. Studies are conducted at two levels. Firstly, strains are compared in biological tests related to the four essential steps in disease production to see if virulent strains have a greater capacity than attenuated strains to accomplish one or more of these steps. Examples are: (1) the greater ability of virulent influenza viruses to attack the epithelium and damage the cilia of respiratory tissues in organ culture (Hoke *et al.* 1980; Sweet and Smith 1980); (2) the more rapid replication of some virulent strains of NDV in chick cells (Waterson *et al.* 1967); (3) the greater capacity of virulent strains of ectromelia virus to replicate in mouse phagocytes (Mims 1964); and (4) the greater capacity of virulent strains of NDV to induce cytopathological changes and cell fusion in chick cells (Reeve and Poste 1971). The second level is to identify the biochemical determinants of the relevant biological differences detected between the strains of high and low virulence. Unfortunately, replication is the only facet of virus pathogenicity to receive much attention at the biochemical level. Almost all knowledge concerning the other three facets of pathogenicity is at the observational level of study and not at the determinant stage. This will become evident as this chapter continues.

The possession of strains of high and low virulence allows the search for *virulence markers*, i.e. properties associated with virulent rather than attenuated strains. These markers are not necessarily virulence determinants (Smith 1978). The aim is to have convenient tests, preferably *in vitro* but sometimes in experimental animals, for recognizing virulent and attenuated strains in epidemiological and vaccine studies so that virulence tests, particularly those in man, need not be carried out so frequently. Cultural and serological properties can suffice as markers. Examples are: (1) the large plaques formed in cell cultures by virulent strains of poliovirus, Coxsackie A9 virus and Venezuelan equine encephalitis (VEE) virus (Holland 1964); and (2) the resistance of virulent strains of pseudorabies virus to temperature and urea (Golais and Sabo 1975). Sometimes convenient biological tests relating to virulence have been used as markers, e.g. the replication of influenza virus strains in tracheal organ cultures (Sweet and Smith 1980, Hoke *et al.* 1980). Although most strains may comply with a particular marker system, usually there are one or two that do not.

Recombination, genetic constitution and virulence determinants

One technique for deriving strains of differing virulence is recombination, i.e. genetic exchange between two parent viruses (Fenner *et al.* 1974). The consequences of the multifactorial nature of virulence outlined above have been underlined by experiments with recombinants of influenza virus and of reoviruses where genetic analysis has been related to comparisons of virulence or facets of it. These RNA viruses have segmented genomes and, under appropriate conditions, not only can the different gene segments be distinguished from one another by polyacrylamide gel electrophoresis, but also the parental origin of the gene segment can be identified in a recombinant virus (Palese 1977; Fields *et al.* 1978). Furthermore, the polypeptide products of the gene segments—potential virulence determinants—have been identified (Table 84.2). Thus, by comparing the virulence of recombinants of known genetic constitution, the influence of different gene segments and their polypeptide products can be examined. In this examination the following points are relevant as well as those in the two previous sections. Although each designated gene segment in any particular strain has features in common with the corresponding segment of other strains of the same virus, their different molecular weights as determined on polyacrylamide gels indicate subtle differences between them, the products of which may influence virulence. Thus, in exchange of gene segments during recombination, the parental origin as well as the specific gene segment might influence pathogenicity. Also, non-pathogenic recombinants may be derived from two pathogenic parents for the following reasons. Although all gene segments may be concerned to some extent in pathogenicity, some will probably have a greater influence than others and cooperation between gene products may be required for full expression of the virulence determinant. Thus, during recombination where any gene segment may be exchanged, a recombinant derived from two pathogenic parents may receive gene segments which are not particularly concerned in pathogenicity, or a segment may be acquired without its cooperating segment.

Table 84.2 RNA segment-polypeptide assignments for influenza A viruses and reoviruses

Influenza virus Gene segment	Virus-specified polypeptide		**Reovirus** Gene segment	Virus-specified polypeptide	
1	P1 }		L1	λ3,	core component
2	P2 }	polymerases	L2	λ2,	core component
3	P3 }		L3	λ1,	core component
4	HA,	haemagglutinin			
5	NP,	nucleoprotein*	M1	μ2,	core component
6	NA,	neuraminidase	M2	μ1†,	capsid component
7	M,	matrix	M3	μNS,	non-structural
8	NS1 and NS2, non-structural				
			S1	σ1,	capsid component
			S2	σ2,	core component
			S3	σNS,	non-structural
			S4	σ3,	capsid component

* nucleoprotein, together with the gene products P1, P2 and P3, forms the polymerase complex.
† μ1 is proteolytically cleaved to the secondary gene product μ1c.

The single-stranded RNA (ss RNA) genome of influenza A virus has 8 RNA segments coding for 9 gene products (Table 84.2) (Palese 1977; Sweet and Smith 1980). Experiments with recombinants indicate that no single gene product determines either full virulence or particular facets of it; virulence seems dependent upon an optimal combination of gene segments, some of which may be present in avirulent and attenuated strains.

Recombinants of avian influenza viruses—particularly fowl plague virus (FPV)—made between themselves or with human and swine influenza viruses have provided the main information. Exchange of any gene segment could modify virulence (Rott *et al.* 1979), which was influenced both by the parental source as well as the particular segment exchanged (Scholtissek *et al.* 1977). Thus, FPV recombinants in which segment 1 alone was derived from strain A/PR/8/34 were avirulent for chickens, whereas recombinants with this segment replaced by that from strain A/Swine/1976/31 were fully virulent. When segment 2 alone originated from strain A/Swine/1976/31 then the recombinant was avirulent, but when segment 2 was from a human A_2 or avian influenza virus a virulent recombinant resulted. In gene exchange between a virulent and an attenuated strain, recombinants with larger numbers of gene segments from the latter parent tended to be attenuated for chickens; but, as expected, virulent recombinants were obtained from two attenuated parents and attenuated recombinants from two virulent parents (Rott *et al.* 1979; Sweet and Smith 1980). An optimal combination of gene segments was required for a virus to be virulent: FPV recombinants which derived some gene segments for the RNA polymerase complex (segments 1, 2, 3 and 5) from one virulent parent and some from another were avirulent for chickens, while, with one exception, recombinants which derived all these segments from one or other parent were virulent. In addition, all of the virulent recombinants derived gene segments 4 and 6 from the same parent, whereas all but two of many avirulent recombinants did not (Rott *et al.* 1979).

Similar results to those above have been obtained with small numbers of recombinants derived from human influenza viruses, the virulence tests being performed with ferrets and to a lesser extent with man (Sweet and Smith 1980). In one recombinant series (A/PR/8/34-A/England/939/69) the attenuated parent (A/PR/8/34) had some virulence attributes such as the ability to replicate in the upper respiratory tract; although substitution of gene segments from the virulent parent (A/England/939/69) produced recombinants of intermediate virulence, one recombinant containing 4 gene segments from each parent was more attenuated than A/PR/8/34 and another recombinant was more virulent than A/England/939/69. In contrast, with another recombinant series (A/Okuda/57-A/Finland/4/74) there was no evidence that the attenuated parent (A/Okuda/57) provided gene segments which substantially increased pathogenicity. Recombinants tended to become less virulent as the number of gene segments from A/Okuda/57 increased. When gene segments 4 and 6 were derived from this strain, the recombinants were attenuated, whereas those having these gene segments from A/Finland/4/74 were virulent or of intermediate virulence.

Turning to particular facets of virulence, replication *in vivo* appears to depend upon gene segments 4 and 6 of the correct parental origin. However, with recombinants of the A/PR/8/34-A/England/939/69 series, no correlations with gene segments could be seen for other facets of virulence, such as the persistence of nasal infection against host defences and the height and duration of fever (Sweet and Smith 1980).

The ten double-stranded RNA (ds RNA) segments of reovirus types 1, 2 and 3 are: 3 large segments (L1, L2, L3), 3 medium segments (M1, M2, M3) and 4 small segments (S1, S2, S3, S4). Their corresponding polypeptides are shown in Table 84.2. The function of many of the virion proteins has not yet been determined. However, reovirus type 1 agglutinates human but not bovine erythrocytes, whereas type 3 agglutinates bovine but not human erythrocytes, and the haemagglutinating properties of recombinants with single gene exchanges showed the σ1 polypeptide from gene segment S1 to be the haemagglutinin (Fields *et al.* 1978). Furthermore, intracerebral inoculation of newborn mice with such recombinants showed that it determined cell tropism in the central nervous system. Type 3 reovirus produced an acute lethal encephalitis associated with destruction of neuronal cells, whereas type 1 produced a non-fatal infection involving ependymal cells but not neurones. Recombinants with their S1 segment from type 3 reovirus produced neuronal destruction and acute encephalitis, whereas those with their S1 gene segment from type 1 virus destroyed ependymal cells and produced hydrocephalus (Fields *et al.* 1978; Weiner *et al.* 1980). Thus, with reoviruses, the product of a single gene segment determines a particular facet of virulence-target cell attack and damage. However, intracerebral inoculation by-passed many steps in the infectious process. If a more natural route of infection was used with older mice having more mature host defence mechanisms, virulence might require the products of other gene segments.

Difficulties of investigating the determinants of viral pathogenicity

Most difficulties stem from two facts. First, for any one virus there are several determinants. To recognize all the determinants, comparisons between several different pairs of strains may be needed. For virus U (Table 84.1) at least four comparisons between it and virus strains W, X, Y and Z would be necessary to reveal determinants A, B, C and D. This assumes that strains W, X, Y and Z are available or can be produced by genetic manipulation. Even when the biological activity of the determinant has been recognized, identifying the virus component or process involved is a difficult biochemical task, especially in separating one determinant from another or possibly relating their activities.

Other difficulties arise from the fact that virus virulence is influenced by changes in cultural condition, e.g. the infectivity of ortho- and para-myxoviruses is dependent on the nature of the host cells in which they have been grown (Rott 1979; Klenk 1980). Viruses are most conveniently studied in tissue culture systems but, unfortunately, when viruses are removed from infected animals and grown in the de-differentiated cells of such systems, the change in environment may select genotypes and induce phenotypes which are different from those found *in vivo*. They may lack one or more virulence determinants and indeed there are many examples of viruses being attenuated by repeated growth in tissue culture (Smith 1972). Thus, for studies of virulence determinants, viruses grown in established cell lines may be incomplete, even when the original source of these lines was the appropriate host. Also, with regard to host factors which influence pathogenicity, such as those involved in host and tissue specificity of infection (see later), cell susceptibility to virus infection can also change with growth in tissue culture. Although studies with tissue culture systems have added to knowledge of pathogenicity, notably with regard to replication, their possible deficiencies for investigating all facets of the subject must be recognized. Often, it may be advisable to avoid such systems and combine experiments in animals with studies using either primary cell cultures or, better, organ cultures where small pieces of tissue are held *in vitro* for short periods (Hoorn and Tyrrell 1969). These culture systems are most likely to preserve the *in vivo* character of viruses and host cells which is needed for successful studies of pathogenicity.

Entry of the host through the skin and mucous membranes

Most viruses enter the host through either the skin or the mucous membranes of the respiratory and alimentary tracts; a few infections begin in the urogenital tract and on the conjunctiva (Burrows 1972; Mims 1966, 1977).

Entry through the skin

The skin is a strong barrier to infection. Viruses that alight on it are reduced in number or destroyed by desiccation and by acids and other inhibitors formed by indigenous (commensal) micro-organisms (Mims 1966). The continuously shed, cornified epidermis is impermeable unless broken by cuts, abrasions or punctures such as occur in vaccination against smallpox and in hepatitis B. Insect bite is perhaps the main method of penetration. During feeding, vectors such as mosquitoes, mites, ticks, fleas and sand flies inoculate arboviruses directly into the blood stream. Some, e.g. myxoma virus, are transmitted mechanically by the contaminated mouth-parts of the arthropod. Others are transmitted biologically, i.e. they replicate in the vector before transference to a fresh host, e.g. yellow fever virus in the saliva of mosquitoes. Occasionally, bites from larger animals such as dogs, wolves and bats deliver saliva containing virus, e.g. rabies virus, directly into the tissues. Entry through the skin is purely mechanical; virus components or properties play little part.

Entry through mucous membranes

Here virus properties are involved, although usually the determinants of entry have not been recognized because of lack of techniques for studying the early stages of infection, i.e. the behaviour of a few highly dispersed virus particles on mucous surfaces which are often covered with commensal micro-organisms. What information is available relates largely to the respiratory tract.

Respiratory tract

Descriptions can be found elsewhere (Burrows 1972) of the respiratory tract defences, particularly the scavenging role of alveolar macrophages in the lower tract and the mucociliary blanket which removes particles from all areas including extruded alveolar macrophages with their contents. These mechanisms are summarized here only as a background to a discussion of how virulent viruses might counteract them.

The initial lodgement site of airborne virus depends on the anatomy of the respiratory tract and the size of the droplet to which the virus is attached (Burrows 1972). The latter is of prime importance; large particles are trapped by nasal hairs, those of about 10 μm in diameter are deposited on the nasal epithelium, and those less than 5 μm diameter reach lung alveoli (Druett 1967). Further infection of the lower tract may follow from aerosols formed by passage of air over an infected upper tract.

No matter where they land the virus particles must make contact with the surfaces of epithelial cells to initiate infection. This will occur only by chance, because the particles are entrapped in mucus and passed upwards to the pharynx or backwards from the nose by the mucociliary escalator. Host and environmental factors that may affect the chance of contact include variations in thickness, flow rate and viscosity of mucus and gaps in the mucociliary blanket (Kilburn 1968; Platt 1970). The mucus is thinner and more slowly moving in the lower parts of the tract. Obstructions such as bronchial junctions produce local changes in flow rate and there is a velocity gradient from small to large airways. Mucus secretion and ciliary action may be inhibited by changes in temperature, ion concentration and humidity of the air (Platt, 1970; Webster 1970). Experimentally, reduction of nasomucociliary activity of chicks by exposure to low temperature, or by injections of cocaine or pilocarpine, increased infection rates with a standard dose of NDV (Bang and Bang 1969; Bang and Foard 1969). However, in epidemics, respiratory infection occurs in too many persons to implicate simultaneous impairment of their mucociliary action. Virus properties must be involved in the increased infection. The neuraminidases of influenza viruses might facilitate access to the surface cells by degrading mucus, and neuraminidases of virulent and attenuated strains might vary in this respect. Influenza viruses also adhere strongly to cilia of epithelial cells and sometimes clump them; these mechanisms could inhibit removal by the mucus blanket allowing time for virus penetration into the cells (Sweet and Smith 1980).

While in the mucus, virulent viruses must resist antiviral materials and phagocytic cells found within it. Non-specific virus inhibitors have been found in mucus and mucosal homogenates from other animals (Burrows 1972). Mucus also contains PMN and MN phagocytes and sometimes antibodies arising from previous infections or immunization. Alveolar macrophages ingest virus particles and migrate either to the regional lymph node or upwards in the mucus to the outside (Burrows 1972). We discuss later how virulent viruses might overcome these humoral and cellular defences.

Commensal micro-organisms, present in the upper respiratory tract but not usually below the larynx (Mims 1977), could influence the primary attack of viruses. Fungi, bacteria and their products could induce interferon and thus inhibit virus replication (Finter 1973; Burrows 1972). Mycoplasmas might inhibit the replication of some viruses, e.g. measles virus and adenoviruses, by competing for arginine (Burrows 1972). On the other hand, mycoplasmas might enhance virus infection by damaging ciliated epithelium and by preventing interferon production (Cole *et al.* 1978, Taylor-Robinson 1973). Similarly, *Haemophilus influenzae*, a persistent colonizer of the bronchus of chronic bronchitics, destroys ciliated epithelium (Denny 1973). There is no proof that such mechanisms operate in natural infection but evidence suggests that commensals may be protective. Germ-free mice were more susceptible to influenza than ordinary mice (Dolowy and Muldoon 1964) and antibiotic treatment of pigeons enhanced their susceptibility to airborne VEE virus, perhaps because of a reduction in commensal bacteria (Miller 1966). If commensals have a protective function, we do not know how it operates or how virulent viruses overcome it.

As regards penetration of the epithelium, some viruses (rhinoviruses, influenza viruses and adenoviruses) attach to, penetrate and infect the epithelial cells, whereas others (rinderpest, African swine fever) appear to pass through without causing infection (Burrows 1972). Attachment to and penetration of epithelial cells depend on cooperation between virus surface components, such as the haemagglutinin of influenza virus or a capsid protein of adenovirus (Mims 1977) and receptors on susceptible cells. Such cells may not be evenly distributed through the tract. Differential susceptibility of the upper and lower respiratory tract to various viruses has been demonstrated by supplying the same dose of virus as nasal drops or as a fine aerosol. The upper respiratory tract of man was more susceptible to infection with a Coxsackie virus and a rhinovirus, the lower to an adenovirus, and both tracts were equally infected with influenza virus (Burrows 1972). The mechanisms of penetration without infection are unknown; one possibility is ingestion by macrophages and drainage to the local lymph node (Burrows 1972).

Do virulent strains survive on and penetrate the respiratory mucosa *in vivo* better than attenuated strains? Not necessarily; some attenuated strains of influenza virus infected the mucosa of ferrets as well as virulent strains, the differences in virulence being expressed later in the results of infection (Sweet and Smith 1980). However, in organ cultures of respiratory tissues some virulent strains of influenza and other viruses replicated better and produced more damage to the cilia than attenuated strains (Burrows 1972). These experiments with organ cultures indicate an approach which may reveal some determinants of mucosal attack.

Alimentary tract

Viruses enter the alimentary tract in food, drink and in mucus from the respiratory tract. Host defences are similar to those in the respiratory tract. Moving lumen contents will remove viruses not adhering to the epithelium although perhaps not so effectively as the mucociliary action. The surface mucus is acid in the stomach, alkaline in the intestine and contains phagocytes and virus inhibitors such as proteolytic enzymes, bile and antibodies. The abundant commensals of the lower alimentary tract may have a protective function.

In keeping with their survival during passage through the stomach and duodenum, viruses that infect the alimentary tract, e.g. enteroviruses, reoviruses, coronaviruses and rotaviruses, are more resistant to acid and bile than are rhinoviruses, influenza viruses and other enveloped viruses that do not usually produce intestinal infections (Mims 1977). The viruses that adhere to, penetrate and infect the cells of the lower alimentary tract must also possess surface components which specifically interact with receptors on these cells. Little is known of the chemical basis for the stability of intestinal viruses to acid and bile or of the virus components and host cell receptors that determine adherence and penetration. The acid stability of poliovirus may be due to stronger binding between virus RNA and protein compared to that of the acid-labile rhinoviruses and foot-and-mouth disease virus (FMDV) (Rowlands *et al.* 1971). Also, the findings described later for the interaction of poliovirus with primate nervous tissue may have some relevance to the primary attack of this virus on intestinal epithelium.

Regarding the possible protective influence of commensals, intestinal virus infections do not appear to be a significant problem with patients undergoing antibiotic therapy.

Urogenital tract A few viruses infect this tract e.g. in man **Herpes simplex** virus types 1 and 2 (HSV 1 and 2) and the agent of genital warts. Most viruses seem unable to establish themselves there presumably because of lack of interaction with host cell receptors which is required for adherence and cell penetration. In addition, the urethra is frequently flushed with sterile urine and, although the vagina has no particular cleansing mechanisms, from puberty to menopause commensal lactobacilli produce lactic acid which would inhibit acid-labile viruses.

Conjunctiva The conjunctiva is protected by secretions from the lachrymal and other glands and the continuous wiping effect of the eyelids; microbes are swept away via the tear ducts to the nasal cavity and thence to the alimentary tract. Occasionally viruses infect the eye, e.g. conjunctivitis caused by adeno, herpes and vaccinia viruses, but the processes involved are unknown except that trauma is implicated in 'shipyard eye'.

To sum up, little is known of the mechanisms and determinants that enable viruses to survive on and penetrate mucous surfaces.

Replication *in vivo*

An essential but not the sole attribute for virus pathogenicity is an ability to replicate in the environment of the host tissues. Two aspects of replication are of prime importance. Firstly, the reasons for the ability of viruses to replicate in some, susceptible or permissive, but not other, insusceptible or non-permissive, cells are important since they determine whether or not a virus can attack a particular host or tissue. Secondly, the factors controlling the extent and rate of virus production *in vivo* need to be understood; the greater and more rapid the production, the more likely some virions may escape destruction by defence mechanisms that are present or quickly induced within the tissues. The amount and rate of production will depend not only on replication kinetics within the cell but also on the number of infectious virions produced and released per cell together with the total number of susceptible cells in the mixed cell population of host tissues. A connection between a high rate of production and virulence has been shown by observations *in vivo* and *in vitro* on virulent and attenuated strains of some viruses (Smith 1972). Often successful pathogenic viruses are those that multiply rapidly in surface epithelial cells; virus is produced, causes damage and is shed to the exterior before the immune response (see later) can influence events.

All stages of virus replication, namely attachment and penetration, uncoating, synthesis of virus nucleic acid and proteins, assembly and release, require cooperation between virion components and factors in the host cell. The degree to which cooperation occurs determines whether replication ensues, its rate and the infective quality of the progeny. It may cause an 'all or nothing' response but often lack of full cooperation reduces the yield of infectious virus per unit time but does not eliminate production. Cooperation, and therefore replication and virulence, is affected by variations in either the virus or the host cell and by the environmental conditions encountered by the latter. The host-cell characters which determine the various stages of virus replication have been called 'replication factors' (Smith 1972). They are the virological counterparts of nutrients required for bacterial multiplication. The term was coined to distinguish between them and intracellular defence mechanisms (see later) which may also determine the outcome of virus replication. Distinguishing between replication factors and host defence mechanisms is particularly difficult with obligate intracellular parasites. For example, when a virus infects macrophages but does not produce infectious particles, this may be due to a lack of a particular replication factor or to inhibitory or virucidal mechanisms produced within these cells. In some cases the distinction has been made; the lack of an uncoating mechanism—a replication factor—prevents mouse hepatitis virus (MHV) from replicating in the macrophages of resistant mice, while interferon—a host defence mechanism—prevents replication of many viruses in many different cells. More distinctions of this type would provide further understanding of the ability of virulent viruses to cooperate with replication factors and to interfere with host defence.

The intricacies of virus replication are described in Chapter 82. The aim here is to demonstrate how virus and host factors might affect production of infectious virus *in vivo* and therefore pathogenicity. The best examples to illustrate particular points are taken from the numerous experiments with established cell lines. A few examples have been quoted from the less frequent use of primary cell cultures and macrophages. However, most of the latter experiments and those using organ cultures are discussed later in the section on host and tissue specificity where replication *in vitro* has been related directly to the presence and pattern of infection in animals.

Influence of host replication factors and virion components on virus production

Adsorption Here, the replication factors are cell surface receptors for virion surface components. Poliovirus infects primates but not usually non-primates and, within the former, intestinal and central nervous tissues appear to be the only susceptible tissues. In 'classical' experiments, Holland (1964) showed that these differences in susceptibility occurred in primary cell cultures *in vitro* and were determined by cell receptors. Whole cells and homogenates, only from those tissues susceptible *in vivo*, adsorbed virus. However, insusceptible cells, e.g. mouse cells, could transcribe and translate poliovirus RNA when this was introduced directly into them either *in vitro* or *in vivo* in mouse brain. The nature of the receptor for poliovirus is still unclear but it may be a lipo- or glycoprotein coded for in man by a structural gene on chromosome 19 (Smith 1977*a*). Not all virus attachment is as specific as that of poliovirus. The pox viruses attach themselves indiscriminately to many cell types and even to charged surfaces (Dales 1973). Similarly, for both ortho- and paramyxoviruses the cell receptors are N-acetyl neuraminic acid residues of glycoproteins or glycolipids (Schulze 1975) that are found on the membranes of most cells including those insusceptible to infection.

The role of the virion component in such virus-receptor interactions is also well illustrated by the work on poliovirus. When poliovirus RNA was transcribed and translated in insusceptible cells (see above), only one cycle of replication took place both *in vitro* and in mouse brain, because the released virions possessed a capsid which did not permit further binding to and penetration of mouse cells. However, when poliovirus RNA was enclosed within the capsids of a mouse enterovirus, Coxsackie virus B1, the hybrid virus produced one cycle of replication in mouse cell culture and in mouse brain. Here the enterovirus capsids could interact with specific receptors on mouse cells, an interaction that was prevented by anti-serum to Coxsackie B1 virus. Virulent strains of poliovirus adhered more strongly than attenuated strains to the receptors of primate nerve tissue (Smith 1972), indicating that subtle differences in surface components contribute to virulence. Structural differences in the capsids of virulent and attenuated strains have been observed but whether they relate to differences in adsorption is unknown. Surface components of other viruses that are important in infectivity include the capsid glycoproteins of vesicular stomatitis virus (VSV) (Smith 1977*a*) and the haemagglutinin of ortho- and paramyxoviruses. With the latter viruses attachment is mediated by the carbohydrate moiety (Scholtissek and Klenk 1975), which may vary with the virus strain and the cell type in which it is grown (Nakamura and Compans 1979).

Penetration After attachment of virus, penetration occurs by endocytosis of intact virions (viropexis) or by fusion between the virus envelope and the host cell membrane (Bachi *et al.* 1977, Rott and Klenk 1977). The membrane components which promote penetration are largely unknown. In some cases they may differ from those which mediate adsorption. Thus, some retroviruses adsorb to both susceptible and resistant cells but penetrate only susceptible cells (Smith 1977*a*); a different binding seems necessary for penetration. Similarly, the lack of infectivity of Epstein-Barr virus for T lymphocytes appeared to be due to a block at the penetration rather than the adsorption stage (Menezes *et al.* 1977). In other cases, the membrane components mediating attachment appear to be involved in both penetration and uncoating (see below).

Penetration by NDV, Sendai virus, Simian virus 5 and other paramyxoviruses is mediated by the surface glycoprotein F, which induces fusion with the cell membrane (Klenk *et al.* 1977) and must be cleaved proteolytically to function effectively. An envelope glycoprotein of HSV-1 is required for penetration and also possibly mediates virus-cell fusion (Sarmiento *et al.* 1979). Finally, the haemagglutinin of influenza virus may be concerned in penetration as well as in adsorption (Klenk 1980).

Uncoating A strain of MHV-2 which infected PRI but not C_3H mice also infected and destroyed liver and peritoneal macrophages of the susceptible but not the resistant mice. Virus adsorbed to and penetrated both susceptible and resistant macrophages, but eclipse and replication ensued only in the susceptible cells; a block in uncoating occurred in resistant cells (Mogensen 1979).

The receptor for adsorption of poliovirus also appears to promote penetration and uncoating since membrane fractions from susceptible but not insusceptible cells adsorbed virus and induced uncoating (Smith 1977*a*).

Macromolecular synthesis Different cell lines may use the same genetic message—virus or complementary RNA—with varying degrees of efficiency. Thus, yields of mengovirus and bovine enterovirus 1 from a range of cell types varied by more than a thousand fold, even though adsorption, penetration and uncoating proceeded normally (Buck *et al.* 1967). A similar result occurred with NDV in 3 cell types (Gresland *et al.* 1979). The reasons for the differences are unknown. Replication factors concerned with transcription and translation may have been deficient in some cells, but, equally, host cell nucleases—a host defence mechanism—may have degraded virus RNA differentially (Buck *et al.* 1967). This is a good example of the dilemma outlined above. The failure of human adenoviruses to replicate productively in monkey kidney cells was due to lack of transcription or translation of virus mRNA (Farber and Baum 1978). A similar block occurred in some rat brain cells infected with HSV-1 (Adler *et al.* 1978*b*). Two neuronal cell lines produced much less virus than two permissive glioma lines. In the neuronal lines, thymidine kinase, DNA polymerase and structural proteins were produced but synthesis of virus DNA did not proceed efficiently.

The virus products most likely to affect macromolecular synthesis are the enzymes involved. Their absence or inhibition at the normal temperatures of host tissues can lead to attenuation, as in thymidine kinase deficient mutants of HSV-1 and -2 (Field and Wildy 1978), and temperature-sensitive (*ts*) mutants of some but not all viruses (Fenner *et al.* 1974; Preston and Garland 1979).

Maturation and assembly The two surface glycoproteins HN and F of the paramyxoviruses, which determine adsorption and penetration respectively, are incorporated into the virus envelope at maturation and budding (Klenk *et al.* 1977; Scheid and Choppin 1974). Proteolytic cleavage of the glycoproteins before incorporation into the envelope is essential for infectivity of the progeny, and cleavage depends on the virus strain and the host cell. Thus, the virulence of NDV strains for chickens or chick embryos is associated with the ubiquity of cleavage of their glycoproteins. For virulent strains, cleavage occurred not only in primary cultures of chick embryo chorioallantoic membrane (CAM) cells but also in cell lines (MDBK and BHK) not derived from the chicken. In contrast, with attenuated strains, cleavage did not occur in the latter cells or even in chick embryo cells but only in CAM cells (Klenk *et al.* 1977; Klenk 1980). Furthermore, in chick embryos the glycoproteins of a virulent strain were cleaved in cells of the endoderm, mesoderm and ectoderm thus allowing spread of a fatal infection, whereas those of an attenuated strain were cleaved only in endodermal cells so that infection was restricted to this tissue (Nagai *et al.* 1979).

The infectivity of influenza virus also depends on post-translational proteolytic cleavage of the surface haemagglutinin (Klenk 1980; Scholtissek and Klenk 1975). As with the paramyxoviruses, the virulence of avian influenza viruses is associated with cleavage of the haemagglutinin in a broad spectrum of cells including fibroblasts from various avian genera and MDCK cells (Bosch *et al.* 1979; Rott 1979). Attenuated strains were produced in a non-infectious form in most of these cells owing to non-cleavage of the haemagglutinin and *in vivo*, unlike virulent strains, they were restricted to the endodermal layers of chick embryo membranes and the respiratory and intestinal tracts of chickens with no fatal consequences (Bosch *et al.* 1979; Rott 1979). However, cleavage of the haemagglutinin is not the only factor responsible for virulence, since some attenuated strains of influenza virus possess haemagglutinins that are cleaved satisfactorily (Sweet and Smith 1980).

In the macrophages of adult mice HSV-1 underwent an abortive cycle of replication; both virus DNA and proteins were produced, yet electron microscopy showed few naked or enveloped virions (Stevens and Cook 1971*a*). There appeared to be an error in the final assembly of virus. This did not occur in macrophages of infant animals which, in contrast to adult animals, can be infected by HSV-1 (Mims 1972).

Release Simian virus 5, a paramyxovirus, budded freely from monkey kidney cells, whereas much smaller amounts of infective virus were released from BHK cells despite intracellular accumulation of virus subunits (Klenk and Choppin 1969). Studies on the related orthomyxoviruses suggest that release may be related to the synthesis of a controlling factor—virus M protein—the production of which can vary with different cell types (Compans and Choppin 1975; Rott and Klenk 1977). Moreover virus neuraminidase may be concerned in release of orthomyxoviruses (Sweet and Smith 1980).

Defective interfering particles (DIPs) Lack of replication factors may also result in the formation of incomplete virus particles. Some may be incapable of replication but others, called defective interfering particles (DIPs), replicate in the presence of standard infectious virus and interfere specifically with its replication. DIPs are formed in tissue culture and their quantity and degree of interference with infectious virus are dependent upon the host cell (Huang 1973). Thus, SFV produced DIPs when grown in mouse L cells but not in chick embryo fibroblasts (Levin *et al.* 1973). Do DIPs operate *in vivo* to reduce virulence by specifically interfering with the replication of fully infectious virus? Animals have been protected against infection with various viruses, e.g. VSV, by prior or simultaneous infection of the corresponding DIPs (Doyle and Holland 1973; Rabinowitz *et al.* 1977). However, specific interference with production of infectious virus may not be the explanation, at least not always; inactivated DIPs protected mice against VSV infection by a short-term non-specific effect followed by a long-term one, probably antibody-mediated (Crick and Brown 1977). Thus, DIPs may exert an effect *in vivo* by stimulating non-specific and immunospecific defences rather than by specific interference with virus replication. In this they would be akin to other incomplete and antigenic but non-replicating particles resulting from a lack of replication factors.

Influence of host cell environment on virus production

Temperature Small variations in temperature affect virus replication *in vitro* and virus virulence *in vivo* (Lwoff 1959; Mims 1972; Fenner *et al.* 1974). The temperatures of the upper respiratory tract and the extremities are significantly lower than those of internal organs and these differences influence virus behaviour. Thus, as the optimal temperature for growth of rhinoviruses is 33°, this explains, to some extent, their localization in the upper respiratory tract (Mims 1977). Temperature-sensitive mutants of influenza virus have a defect in polymerase function at lung temperature which restricts replication to the upper respiratory tract with consequent loss of virulence (Sweet and Smith 1980). Similarly, alveolar macrophages taken from calves infected with para-influenza virus contained virus antigens and nucleocapsids but little infectious virus until incubated *in vitro* at 32° (Tsai 1977). The influence of fever on virus virulence is described later.

pH Variations of pH also affect virus stability and replication (Fenner *et al.* 1974); their effects on mucosal membrane interactions have already been discussed. The pH of an inflammatory exudate tends to be low because of glycolysis within phagocytes. Ability to replicate in these conditions would be an asset to an invading virus and the lower optimum pH for attachment of virulent strains of some alphaviruses is interesting in this respect (Marker and Jahrling 1979).

Materials of low molecular weight Such materials stimulated the replication of certain viruses in tissue culture, e.g. arginine for HSV, leucine for vaccinia virus, proline and serine for VEE virus, fatty acids for poliovirus (Smith 1977*a*), and a material in egg fluids for several viruses (Eylan and Gazit 1978). Similar but fewer observations have been made *in vivo*, e.g. leucine enhanced vaccinia virus infection in mice (Smith 1977*a*), but how far such materials influence normal infections is not known.

In summary, much is known of the mechanisms and determinants of virus replication. Most of the information has been obtained from experiments *in vitro*, but often it appears to apply to infection *in vivo*.

Interference with non-specific host defences

There has been much work on host defence against virus infections, with an increasing awareness that it is directed against virus-infected cells as well as virions themselves (Smith 1980; Blanden *et al.* 1977). In contrast, the mechanisms by which virulent viruses counteract these defences and how cells infected with them escape destruction have received little experimental attention (Mims 1972; 1977). This section deals with possible mechanisms of resistance to non-specific defences largely in relation to viruses *per se*. Possible methods of survival of virus-infected cells are discussed later because, although these cells may be destroyed non-specifically by natural killer (NK) cells (Anderson 1978; Santoli *et al.* 1978; Fujimiya *et al.* 1978; Morahan *et al.* 1980) or by PMN phagocytes and complement (Grewal *et al.* 1980), they appear particularly prone to immunospecific defences (Blanden *et al.* 1977; Rawls and Tompkins 1975).

Non-specific defence mechanisms against viruses

These mechanisms are more effective in adults than in young animals and involve humoral, cellular and environmental factors that are present before infection or are induced by it (Mims 1972; 1977). At primary lodgement and during cell-to-cell transfer viruses encounter inhibitors or virucidal materials in the tissue fluids. When viruses attack cells, host nucleases may destroy their nucleic acid (Newton 1970), also antiviral interferons are induced. Furthermore, virus replication evokes an inflammatory response (Mims 1977). Mediators released from damaged tissue cause contraction of vascular endothelial cells and plasma contents—including virus inhibitors and phagocytes—leak through the resulting gaps. Later, chemotactic factors released from the virus-infected tissue cause active passage of phagocytes through the endothelium to the site of infection. Initially, many PMN phagocytes arrive with some MN phagocytes and lymphocytes. Later, the response consists predominantly of MN phagocytes and lymphocytes. Virus-phagocyte interaction can produce more interferon (Finter, 1973) and endogenous pyrogen, the mediator of fever which can adversely affect virus replication (Bernheim *et al.* 1979). The acidity of the inflammatory exudate has been mentioned previously. Virus escaping from the primary lodgement site encounters MN phagocytes in the blood and lymph and also in the tissues where they have differentiated into more effective phagocytes, the macrophages. These macrophages are present in or on the surface cells that line the sinuses of the lymph nodes, liver (Kupffer cells), spleen, bone marrow, adrenal glands and alveoli. Macrophages and PMN phagocytes ingest and destroy viruses and virus-infected cells. They form the main non-specific defence against virus attack. Occasionally lymphocytes play a non-specific role but their main function is in the immune response.

Overcoming the non-specific defences depends not only on their effectiveness in relation to the viruses' capacities to counteract them but also on the number of invaders. Thus, Rift Valley fever virus killed mice in 6 hr when a high dose was given intravenously, because the Kupffer cells were unable to prevent infection of hepatic cells which resulted in rapid death (Mims 1972). However, in natural disease, a small infecting dose must increase to a population sufficiently large to cause host damage despite the non-specific defence mechanisms that are, at first, heavily weighted against the few invading virions. Obviously virulent viruses counteract the defences, but how they do it and the determinants concerned are largely unknown. Apart from the relative lack of interest in this important topic, there are two reasons for our ignorance. Firstly, there have been few attempts to distinguish between the influence of replication factors alone and virus resistance to cellular host defence mechanisms in the successful infection of phagocytes and other host cells. Secondly, there have been few comparisons of the interaction of virulent and attenuated strains with non-specific defences despite their recognized importance in bacterial pathogenicity.

Resistance to humoral inhibitors

Non-specific virus inhibitors including complement have been described in sera, urine and milk of animals and in homogenates of lung and intestinal mucosae (Burrows 1972; Hirsch *et al.* 1980). Some, such as the lipid and non-lipid inhibitors of normal sera, inhibit many viruses, while others are fairly specific, e.g. the α, β and γ inhibitors of myxoviruses and the mucoid inhibitors of mumps virus and pneumonia virus of mice (Burrows 1972). The most studied are the α, β and γ inhibitors of influenza virus which prevent virus attachment to cell receptors *in vitro*. Their role in host defence is equivocal and, in any case, little is known of how the influenza virus counteracts them. In man, resistance to β inhibitors may have allowed Asian strains of influenza virus to infect the lung more readily (Mulder and Hers 1972). If such resistance is important, virus neuraminidase may be the virulence determinant since it digests some non-specific inhibitors (Sweet and Smith 1980).

Resistance to host cell nucleases

Host cell nucleases may be free in the cytoplasm or released into vacuoles by lysosomes (Newton 1970). Clearly, the nucleic acids of virulent viruses are protected from these nucleases but the processes involved are not known. Possible mechanisms of resistance to nucleases include association of the nucleic acid with virus protein—vaccinia virus—or intracellular membranes—myxoviruses, picornaviruses and reoviruses—and the possession of a double-stranded nuclease-resistant structure (reoviruses).

Inhibition of interferon

Tests *in vitro* show that antiviral glycoproteins, the interferons, are induced in virus-infected cells and that they prevent virus replication in further cells (Finter 1973). Originally, interferon was regarded as a single substance but this is not so (Borden 1979). Three human interferons have been characterized: two type I interferons produced by leucocytes and fibroblasts respectively, and type II 'immune' interferon produced by T-lymphocytes (Borden 1979). Leucocyte and fibroblast interferons can be distinguished from each other and from type II interferon by physicochemical and antigenic criteria. For example, antisera to leucocyte and fibroblast interferons do not neutralize the biological activity of type II interferon and the latter, unlike leucocyte and fibroblast interferons, is labile at pH 2. All three interferons have similar biological properties but type I interferons are more potent antiviral agents, whereas type II interferon is more active as an immunomodulator (see later). Here, we consider interferon as one substance, because all interferons are antiviral and most information was obtained before separate interferons were characterized.

Interferon is produced in virus-infected animals. It has been demonstrated in rabbit skin after infection with vaccinia virus, in the throat washings of patients with influenza and in mouse brain after infection with togavirus (Fenner *et al.* 1974). Often there is a temporal association between the presence of virus and interferon in different tissues but this is not proof of protection *in vivo* (Gresser *et al.* 1976*a*). In influenza of man the quantities of interferon are usually related directly to the degree of replication and the severity of illness (Douglas 1975). Nevertheless, in some cases, interferon is protective. Its administration has protected animals against virus infection (Finter 1973), and anti-interferon immunoglobulin enhanced infections of mice with encephalomyocarditis (EMC) virus (Gresser *et al.* 1976*a*), HSV, Maloney sarcoma virus, VSV and NDV (Gresser *et al.* 1976*b*).

Virus species and strains within species differ both in their induction of and susceptibility to interferon. In some cases, virulent strains induce less interferon or are more resistant to it than attenuated strains, suggesting that ability to counteract interferon is important in virulence. Unfortunately, in other cases, this is not so (Finter 1973; Holland 1964). However, a strict correlation would not be expected from the multifactorial nature of virulence. Interferon almost certainly contributes to host defence against some diseases, and a capacity to reduce its production or to resist its action would be advantageous to an invading virus. How could this be achieved?

Early inhibition of host cell mRNA and protein syntheses may depress interferon production, e.g. for enteroviruses. In experiments *in vitro*, 'persistent fractions' of the virus population are often detected, which replicate despite the presence of interferon. The reason for their refractory nature is unknown, but they may be significant *in vivo* (Finter 1973). Also, cells infected with vaccinia virus, HSV, ECHOvirus or VSV at a level greater than one infectious particle per cell were not protected by interferon (Stitz and Schellekens 1980). The mechanism of resistance is unknown but this type of resistance may occur *in vivo* if a local virus population becomes high. Virus-induced antagonists of interferon—termed stimulons, enhancers and anti-interferons—have been reported in tissue culture and in infected tissue (Smith 1972; Mims 1972). For example, substances of this nature have been detected in the serum of mice infected with EMC virus, SFV, influenza virus, HSV-2 and murine cytomegalovirus (CMV) (Stringfellow *et al.* 1977). The nature of these antagonists is unknown. Whether they are virus virulence determinants is doubtful, because similar substances have been found in uninfected tissue (Mims 1972).

Inhibition of PMN phagocytes

Although they appear first in many inflammatory exudates and might limit infection at an early stage, little attention has been paid to the role of PMN phagocytes in defence against virus attack, or to possible virus mechanisms for countering them. Experiments *in vitro* show that PMN phagocytes adsorb and ingest many viruses but virus replication in these cells has never been observed (Smith 1972). Thus, they could be powerful factors in host defence since viruses are either inhibited or destroyed by them. If they have this role, inhibition of their function by virulent viruses would enhance invasion. That PMN phagocytes are affected by interaction with various viruses is shown by a reduced capacity for chemotaxis and for ingestion and killing of bacteria (Smith 1980). Such a reduced capacity would probably apply to phagocytosis of viruses as well. The unknown determinants of this detrimental effect may be revealed by more research on virus cytotoxicity (see later). Phagocytosis experiments with virulent and attenuated strains of viruses might add to knowledge in this field, although in one investigation with ferret PMN phagocytes and a virulent and an attenuated strain of influenza virus both strains were adsorbed and inactivated equally (Sweet and Smith 1980).

Inhibition of MN phagocytes

Blood MN phagocytes and tissue macrophages ingest and destroy some viruses but not others. Some viruses are not killed because they resist ingestion (Mims 1972). Many others are ingested but resist killing, and persist or replicate intracellularly. Ability to replicate in macrophages has been related to the virulence of strains of pox viruses, HSV, MHV and lymphocytic choriomeningitis (LCM) virus (Mims 1972; Mogensen 1979). The Hampstead mouse strain of ectromelia virus infected and replicated in mouse macrophages and Kupffer cells better than the Hampstead egg strain and was correspondingly more virulent (Mims 1972). In contrast to HSV-1, HSV-2 produced progressive focal necrotizing hepatitis in mice and, in macrophages *in vitro*, replication of HSV-1 was restricted more than that of HSV-2. Change of macrophage function sometimes relates to increase in host resistance with age. Thus, HSV-1 was ingested by peritoneal and tissue macrophages of sucking and adult mice but productive infection occurred only in macrophages from the former, allowing virus spread to the brain and death from encephalitis (Johnson 1964). Selective blockade of macrophage function in adult mice by silica and anti-macrophage serum caused death from HSV-1 infection (Zisman *et al.* 1970). Anti-macrophage serum also enhanced infections with yellow fever virus and VSV (Smith 1972). Survival within MN phagocytes provides protection from humoral inhibitors and antibody, a means of spread through the blood and lymph systems and, for fixed macrophages, an initial focus of infection in organs such as the liver (Mims 1972).

Although viruses survive and replicate within macrophages, the intracellular environment is not propitious for replication (Mims 1972). Thus, when a population is inoculated many macrophages do not become infected and, when replication occurs, yields of infectious virus are low compared with those from other cell types. When mouse macrophages were infected with HSV, Sendai virus, ectromelia virus or LCM virus, large amounts of virus antigens were produced but virus yields were low (Mims 1972).

At present, we are ignorant of the virus mechanisms or products which determine resistance to ingestion by MN phagocytes or survival or replication within them. A virion component may promote intracellular survival and replication of its own virion nucleic acid by directly interfering with cellular inhibitors. However, survival of part of the invading population of virions may be due to inhibition of macrophage function by preliminary infection with virions which may or may not replicate fully. Some viruses such as myxoviruses, mumps virus, Coxsackie virus, vaccinia virus and measles virus are cytotoxic to MN phagocytes inhibiting chemotaxis and their phagocytic activity towards bacteria (Gresser and Lang 1966; Notkins *et al.* 1970; Sawyer 1969; Sweet and Smith 1980). Other viruses such as MHV can kill phagocytes (Bang 1972). We must learn more about the biochemical basis of virus cytotoxicity to recognize the determinants of survival and replication of some viruses in MN phagocytes.

Inhibition of lymphocyte function

Some viruses such as influenza and Coxsackie virus are inactivated by lymphocytes non-specifically. However, others, such as HSV, VSV, yellow fever virus, ECHOvirus, pox viruses, LCM virus and adenoviruses replicate in lymphocytes and this aids virus spread (Mims 1972). As for macrophages, the reasons for the ability of some viruses to survive within lymphocytes while others are destroyed are unknown and similar speculations apply.

Resistance of virus-infected cells to natural killer (NK) cells Virus-infected cells are killed *in vitro* by lymphocytes which act without the need for previous sensitization or antibody (Anderson 1978; Santoli *et al.* 1978; Fujimiya *et al.* 1978; Morahan *et al.* 1980). These NK cells probably operate *in vivo* as another non-specific mechanism against virus infection, but whether host cells infected with virulent strains of viruses survive better against them than cells infected with attenuated strains is unknown.

Resistance to pyrexial temperatures As mentioned previously, the replication of some viruses is affected by small changes in temperature. Based on work with strains of poliovirus of differing neurovirulence for primates, Lwoff (1959, 1969) suggested that fever was a host defence mechanism against virus infections and that virulent strains replicated at pyrexial temperatures better than attenuated strains. When animals infected with Sindbis, vaccinia, ectromelia, Coxsackie B and other viruses were kept at elevated environmental temperatures, virus infection was reduced compared with normal temperatures, especially for attenuated strains (Fenner *et al.* 1974; Roberts 1979; Drillien *et al.* 1978). Thus, fever could be a host defence mechanism but there is little experimental evidence causally relating fever to cessation of virus growth *in vivo*. However, reduction of fever in vaccinia pneumonia of rabbits by drugs enhanced virus production in the lungs and increased the percentage mortality (Drillien *et al.* 1978). Furthermore, investigations into differences in virulence for ferrets of two clones of a recombinant influenza virus showed, for both clones, significant correlations between rise in rectal temperature and subsequent reduction of virus titres in nasal washings (Sweet and Smith 1980). Furthermore, these correlations occurred earlier for the attenuated clone than for the virulent clone and replication of the former in organ cultures of nasal turbinates was more restricted at pyrexial temperatures than that of the latter. Similar results were obtained with four other strains of differing virulence from the same recombinant system.

If fever is a host defence mechanism how do virulent viruses resist it? Lwoff (1969) suggested that the raised temperature inhibited the RNA polymerase activity of attenuated strains of poliovirus and also, with other factors, may have promoted the release of lysosomal nucleases which destroyed the virus nucleic acid. Virulent strains had replicases that worked rapidly at high temperature, so that virus replication occurred before nucleases were released. Whether this is so for influenza virus and other viruses remains to be determined.

The remarks made at the beginning of this section regarding the lack of knowledge of how viruses counteract non-specific host defence should now be clear.

Virus spread within the host

Virus spread can be considered under three headings: local extension of virus replication to neighbouring cells or tissues; spread through the lymph and blood systems; and transport along nerves.

Localized extension of infection

Within tissues, cell-to-cell infection occurs with or without a distinct extracellular phase of the transferred virus (Fenner *et al.* 1974). As mentioned previously, virus can be spread over mucous surfaces by secondary aerosols and moving mucus and lumen contents. Such translocation of infected fluid is not possible on the dry skin but some spread may occur by scratching, rubbing and transfer by fingers (Mims 1977).

Spread through the lymph and blood systems

Potentially these systems are excellent vehicles for spreading viruses. Failure or success in producing generalized infections depends upon virus interaction with three barriers: between the primary lodgement site and the lymph/blood system; in the lymph/blood system and its associated reticulo-endothelial system; and at blood-tissue junctions.

Breaching the barrier between the primary lodgement site and the lymph/blood system

Most viruses that have survived the initial inflammatory response enter the local lymphatics, either directly through damaged vessels, or in association with phagocytes which actively transport them to these lymphatics (Johnson and Mims 1968; see previous section). Viruses enter the blood stream directly through damaged capillaries at the initial lodgement site or later with the lymph.

Surviving the action of defence mechanisms in the lymph, blood and the reticulo-endothelial system

With few exceptions, such as rhinoviruses in the nasal tract and warts virus in the skin, all viruses, including those described as causing localized infections, breach the first barrier to some extent. However, the second barrier is formidable and spread to other tissues occurs only if the virus can resist humoral and cellular defence mechanisms in and associated with the lymph and blood systems. Of particular importance is a capacity to counteract the MN phagocytes and lymphocytes in the lymph and blood and the fixed macrophages in the lymph nodes, spleen and liver (Mims 1964, 1972, 1977). The methods used by viruses for counteracting these have already been described. Viruses lacking this capacity remain largely confined to the initial lodgement site although there may be transient spread or even viraemia, e.g. with influenza in man and ferrets (Sweet and Smith 1980). The effectiveness of the second barrier was illustrated by the experiments of Mims (1964). After intravenous inoculation of vaccinia, influenza or myxoma viruses into mice and ectromelia virus into rats, all the viruses disappeared rapidly owing to phagocytosis and destruction by liver macrophages. The liver parenchyma, shown to be susceptible to virus attack by bile duct inoculation, was unharmed.

The size of the virus particle appears to influence the efficacy of ingestion by the reticulo-endothelial macrophages. The circulating half-life of large viruses, e.g. pox viruses, is less than that of small viruses like arboviruses (Mims 1964). Another factor is the speed of lymph or blood flow; the slower the rate the greater the chance of virus uptake (Mims 1977).

Viruses such as smallpox, measles, polio- and adeno-viruses can replicate in phagocytes in lymph nodes, and when virions are discharged into the efferent lymph they enter the blood to produce a generalized infection (Mims 1977). Blood-borne virus can be protected from humoral defence mechanisms by associating with circulating MN phagocytes, lymphocytes (see previously), erythrocytes, e.g. Colorado tick fever virus, and other cells, e.g. leukaemia viruses infect megakaryocytes and platelets derived from them (Mims 1964, 1977). Uptake by and replication in Kupffer cells appeared to be a prerequisite for hepatic cell infection with ectromelia virus in mice, distemper virus in dogs and yellow fever virus in monkeys (Mims 1964). Similarly, replication in macrophages seemed to be the first phase of infection of mouse liver and spleen by the WE_3 strain of LCM virus. In contrast, the Armstrong strain failed to infect the liver and spleen and this was related to non-ingestion by their macrophages (Smith 1972). Here, resistance to ingestion appeared to be a mechanism of attenuation. On the other hand, a virulent strain of mouse CMV infected hepatic cells, apparently because, in contrast to an attenuated strain, it resisted ingestion by Kupffer cells (Mims and Gould 1978). Also, in mice congenitally infected with LCM virus, lack of virus ingestion by liver macrophages appeared to explain the persistent viraemia (Mims 1977). Although the presence or absence of infection of Kupffer cells is an important factor influencing subsequent infection of hepatocytes, ultrastructural studies (Wisse 1970) have indicated that liver sinusoids are lined by endothelial cells with the macrophages distributed in or on the surface; hence the presence or absence of ability to infect those endothelial cells may be another factor involved in hepatic cell infection.

Breaching the blood-tissue junctions

Any mechanism which increases virus concentration in the lymph or blood increases the chance of breaching these barriers, e.g. escape in large amounts from the primary lodgement site and replication within circulating and fixed MN phagocytes and lymphocytes. Some viruses, such as ectromelia, West Nile and Sindbis viruses in mice and distemper virus in dogs, appear to maintain their circulating concentration by replicating in the endothelial cells lining the lymph and blood vessels (Johnson and Mims 1968; Mims 1964). Often, initial escape from the primary lodgement site results in a low or undetectable viraemia, but this primary viraemia is followed by a much higher secondary viraemia, which is due to replication of virus in a primary target organ, e.g. the liver. This latter viraemia allows infection of secondary target organs such as the brain, placenta or skin (Mims 1977).

The sites of virus penetration into the tissues are the capillaries where vessel walls are thinnest and circulation slowest. The mechanisms of penetration are obscure and the determinants unknown. Presumably, adherence to the endothelial surface is important and susceptibility of endothelial cells to virus attack would help penetration. The virus might then grow through the capillary wall; already this has been mentioned as a possible factor in invasion of the liver. When the endothelium is insusceptible, leakage through pores or passive ferrying of virions within cells of the wall may occur (Johnson and Mims 1968). Some viruses may be transported across cells in pinocytotic vesicles (Johnson and Mims 1968). After passage through the capillary walls the susceptibility of the tissue cells determines whether infection will proceed. All extravascular tissues can be infected, but which ones depends upon the virus. Infections of the brain, skin and fetus will be considered as examples.

Infection of the central nervous system might follow from blood-borne virus entering the cerebrospinal fluid by either passing or growing through the choroid plexus, the capillaries of which have a porous endothelium (Johnson and Mims 1968). Alternatively, virus might enter the cerebrospinal fluid directly from the blood (Johnson and Mims 1968). HSV, yellow fever virus and measles virus may grow through the walls of cerebral capillaries while ECHOvirus, Coxsackie virus, LCM virus and mumps virus may enter the cerebrospinal fluid by leakage or by being ferried across (Mims 1977). Viruses could be transported to all parts of the central nervous system in the fluid, but spread might also occur by neural pathways as appears to happen for polioviruses. Invasion of the brain and spinal cord can take place across the ependymal lining of the ventricles and spinal canal, or across the pia mater in the subarachnoid spaces (Mims 1977). Viruses seem to traverse the blood-brain barrier more easily in the immature host, perhaps because of a thinner basement membrane (Johnson and Mims 1968).

One method of breaching the blood-skin barrier may occur through local inflammation, which is commonplace in the skin; blood-borne viruses could localize in small vessels at sites of inflammation and pass across the capillary endothelium (Mims 1966).

The strength of the maternal blood-fetal junction in the placenta varies with the type of placenta and the stage of pregnancy, since the barrier usually becomes thinner in late pregnancy. Few viruses can breach this barrier. The placenta may be infected but the fetus may still be protected, e.g. in some cases of rubella (Rawls *et al.* 1968), bluetongue infection of sheep (Blattner *et al.* 1973) and influenza in ferrets (Sweet *et al.* 1977). Nevertheless, some viruses such as rubella virus, CMV and HSV cross this junction to infect the fetus (Lowrie *et al.* 1977; Mims 1968). Virus particles may leak across the barrier, be ferried across in barrier cells or be carried across in or on mobile maternal cells (Mims 1968). However, the most likely mechanism is replication in placental cells with release of infectious progeny virus into the fetal circulation (Lowrie *et al.* 1977). This holds for rubella virus, CMV and HSV (Lowrie *et al.* 1977). The factors that localize viruses to the placenta and govern transmission to the fetus are not understood.

Virus transport along nerves

Spread from peripheral sites to the central nervous system can occur along nerve fibres. Thus, spread of rabies virus, HSV and varicella-zoster virus was delayed or prevented by cutting appropriate peripheral nerves (Field and Hill 1974); and, with nerves intact, the cells of the appropriate dorsal root-ganglion were the first to be infected (Johnson and Mims 1968). The exact pathway of virus transport along the nerve is a matter of doubt and debate and may vary for different viruses. Possible pathways include sequential infection of Schwann cells, transit along the tissue spaces between nerve fibres and carriage up the axon. Sequential infection of Schwann cells seems unlikely for rabies virus, because it does not seem to infect these cells (Hill 1979). Moreover, although pseudorabies virus does infect Schwann cells, its rate of travel between cells (1–2 mm/day) is too slow compared with the rate of spread of virus along the nerve in infected animals (1.7 mm/hr) (Field and Hill 1975). HSV has been observed in Schwann cells (Lascano and Berria 1980) but its rate of travel between cells was not established. This virus has also been observed in intercellular spaces (Lascano and Berria 1980) but it is uncertain whether they play any role in virus transport. Carriage up the axon is possible for rabies virus and HSV. Thus hormones and transmitters move up axons from the periphery to the central nervous system at about 3 mm/hr and rabies virus and HSV were carried at a similar rate: 3 mm/hr for rabies (Dean *et al.* 1963) and 1.7 mm/hr for pseudorabies (Field and Hill 1975). In addition, HSV has been seen in axons by electron microscopy (Mims 1977).

In summary, the main processes that influence virus spread are a capacity to resist host defence mechanisms associated with the lymph and blood vascular systems, an ability to replicate in or pass through capillary endothelium and, for some viruses, a capacity to travel along nerve fibres. The mechanisms and determinants concerned are largely unknown.

Counteracting immunospecific defence mechanisms

A few days after primary infection or immediately, if the host has been immunized artificially or by previous natural infection, viruses and virus-infected cells must contend with the immune response. This response is described in Chapter 10; only the aspects relevant to virus infection are summarized here as background to discussion of how viruses and virus-infected cells might counteract it.

Viruses stimulate both antibody- and cell-mediated immunity (CMI). Observations on immunosuppressed patients, or those with B- or T-cell defects, and experiments using animals with impaired B- or T-cell functions, indicate that for some diseases, such as entero- and arbovirus infections and influenza, antibody appears to be the main immunospecific defence; for others, such as poxvirus, HSV and CMV infections, CMI seems more important (Allison 1972; Bloom and Rager-Zisman 1975; Fenner *et al.* 1974).

When elicited by previous infection or immunization, secretory IgA may limit virus attack on mucous surfaces. In a host not previously immunized, the role of circulating IgM or IgG in terminating infection at the primary lodgement stage is uncertain because these antibodies may not be induced early enough. However, they limit spread of some viruses, such as poliovirus, to target organs and contribute to recovery from disease (Ogra *et al.* 1975). Experiments *in vitro* indicate that antibody might act in various ways *in vivo*: neutralization of infectious virus by preventing virus adsorption and penetration; opsonization of virions for phagocytosis and killing by PMN and MN phagocytes; and, by interaction with virus-induced antigens on their membranes, rendering virus-infected cells prone to lysis by complement or destruction by phagocytes or K lymphocytes—cytotoxic cells requiring antibody for their activity (Mims 1972; Smith 1972; 1980; Rawls and Tompkins 1975; Burns and Allison 1977).

Unless infection occurs in a previously primed host, CMI, like antibody, operates at a later stage of disease than primary lodgement: it prevents infection of target organs and promotes recovery by destroying virus and virus-infected cells. Lymphocyte activation by antigen releases lymphokines which attract activated phagocytes to, and inhibits their migration from, the site of infection. These activated macrophages destroy some but not all viruses better than macrophages from uninfected animals (Smith 1972). Also, they are very effective against antibody coated virus-infected cells. In addition to macrophages, cytotoxic or killer T cells destroy infected target cells bearing virus antigens to which the lymphocytes have been stimulated. Their destructive effect requires an identity between their histocompatibility antigens and those of infected target cells, a connection which operates for both mouse (H-2) and human (HLA) antigens (Doherty 1980). Thus, killer T lymphocytes from mice immunized with LCM or ectromelia virus will lyse the respective infected target cells only if they share with them either the D or K ends of the H-2 complex (Blanden *et al.* 1975). T cells may recognize either histocompatibility antigens that have been modified by virus-induced membrane antigens or two separate groups of receptors, i.e. the histocompatibility antigens and the virus induced antigens (Doherty 1980). T-cell generation may be augmented by T-helper cells (Ashman and Mullbacher 1979) while immune interferon, released from T cells, may be immunosuppressive by stimulating suppressor T-cell function (Johnson and Baron 1977).

Usually viruses have only limited success in counteracting the immunospecific defences since hosts recover from most virus diseases. When there is recovery, virulent strains merely elude or delay the specific defences for sufficient time to produce their disease effects. However, in some cases, the immune response is present but ineffectual, and virus persists to form chronic or latent infection (see previously). Infection with HSV-1 and HSV-2 provides the classical example of latency in man and experimental animals (Stevens and Cook 1971*b*; Stevens *et al.* 1972; Docherty and Chopan 1974; Fenner *et al.* 1974). After subsidence of the initial infection, some virus, unaffected by the immune response, is maintained in a quiescent state from which it can be reactivated to produce overt disease. Experiments with mice and rabbits (Stevens and Cook 1971*b*; Stevens *et al.* 1972) showed that virus persisted in sensory ganglia of the nervous system and this also occurs in man. Reactivation occurs under many conditions which tend to stress the host (Docherty and Chopan 1974) and it may be enhanced if immunity is reduced by immunosuppression (Price and Schmitz 1978). The methods by which viruses either do or might achieve limited or total success in persisting against the immune response are summarized below.

A low capacity for evoking immune responses

Virus strains, or host cells infected with different strains, may vary in their ability to induce immune responses. Variation in antibody responses to virus strains has been noted (Waterson *et al.* 1967). Some viruses, such as the scrapie agent, appear not to evoke an antibody response (Worthington and Clark 1971). However, there is no evidence that the immune response to virulent strains, or cells infected with them, is less effective than that evoked by attenuated strains; in fact the reverse is often true (Smith 1972). Why some virion components or products induced by them in cell membranes are less antigenic than others is unknown.

Suppression of the immune responses

Some virus infections, e.g. measles, rinderpest and LCM, depress but do not eliminate antibody synthesis (Mims 1977). Occasionally synthesis may be increased, e.g. in VEE of mice and guinea-pigs (Notkins *et al.* 1970). Suppression of CMI, as indicated by depression of hypersensitivity reactions and delayed allograft rejection, occurs in man and experimental animals infected with many viruses including enteroviruses, mumps virus, NDV, CMV, influenza virus, measles virus and varicella-zoster virus (Mims 1972, 1977; Woodruff and Woodruff 1975; Wheelock and Toy 1973). The mechanisms of suppression may vary with the particular infection and include one or more of the following. In all cases the determinants of suppression are unknown.

Lymphocyte cytotoxicity Several viruses are cytotoxic for lymphocytes *in vitro* and *in vivo*. In lactic dehydrogenase (LDH) virus infection of mice the lymph nodes and spleen showed necrotic changes with T-cell depletion, and necrosis of both B- and T-lymphocyte areas occurred with rinderpest and VEE (Woodruff and Woodruff 1975). B and T cells may be affected independently, e.g. in mice infected with EMC virus, B cells rather than T cells were affected (Faden *et al.* 1978). An effect on T cells can suppress thymus-dependent antibody production, e.g. antibody against HSV (Burns 1975). Destruction of lymphocytes may result from direct virus infection by mechanisms that are discussed later and also from host components arising as a response to infection. An immunoglobulin-like factor, toxic for lymphocytes, was present in sera from patients with virus infections (Woodruff and Woodruff 1975). Thus in measles, DNA synthesis in lymphocytes was depressed and cells infected with measles virus *in vitro* released a factor which depressed DNA synthesis in other cells (Woodruff and Woodruff 1975).

Alteration of lymphocyte traffic Some virus infections affect lymphocyte distribution. There is a temporary abrogation of their ability to migrate into lymphoid tissues and this may reduce interaction between immunologically competent cells and antigen. Affected lymphocytes may travel to the liver and be phagocytosed with resultant lymphocytopenia. Killer T cells circulating in blood and lymph may be directed away from infected cells (Woodruff and Woodruff 1975). The cause of the disruption of traffic is unknown although the effects induced by influenza virus and NDV were attributed to changes in lymphocyte membranes induced by virus neuraminidases (Woodruff and Woodruff 1975).

Effects on cellular modulators In man, mumps reduces delayed type hypersensitivity responses, apparently by stimulating suppressor T-cell activity (Woodruff and Woodruff 1975); and HSV infection and measles depressed antibody production by suppressing helper T cells (Pelton *et al.* 1980; McFarland 1974). Murine CMV infection of macrophages leads to immunosuppression of T-cell responses possibly because of impairment of their capacity for processing antigen (Loh and Hudson 1980).

Interferon induction Type I interferon can augment CMI and depress antibody responses (Borden 1979). It also delays monocyte maturation to macrophages (Lee and Epstein 1980). Type II interferon reduces delayed type hypersensitivity responses by stimulating suppressor T-cell activity (Johnson and Baron 1977). As mentioned previously, virus strains differ in their capacity to induce interferons, and the degree to which this occurs might influence the absence or occurrence of immunological defects.

Depression of complement activity Depressed complement concentrations were found in dengue fever and acute viral infectious hepatitis (Minta and Movat 1979). However, we do not know whether depression of complement activity plays a significant role in persistence of virus disease, although mice decomplemented by cobra venom factor suffered a more prolonged infection with influenza virus (Hicks *et al.* 1978).

In almost all the experiments reported above immunosuppression was detected by tests using antigens other than the infecting virus. Hence, any assumption that the immunosuppression would result in persistence of the inducing virus should be made with caution.

Resistance to the immune response

Some viruses form complexes with antibody which are still infectious; this occurs in equine infectious anaemia, Aleutian disease of mink and in mouse infections with LDH virus, all of which are persistent (Porter 1975). Whether retention of infectivity is due to inadequate amounts of antibody or to its failure to react with critical sites on the virion is unknown. In some cases, this non-neutralizing antibody can interfere with the action of neutralizing antibody (Fenner *et al.* 1974).

HSV, CMV and varicella-zoster viruses induce a glycoprotein receptor for the Fc portion of human or rabbit IgG in the membranes of cells from mice, hamsters, monkeys and man (Burns 1975; Baucke and Spear 1979). This receptor may block immune destruction of the cell by binding non-immune globulin (Adler *et al.* 1978*a*).

The binding of anitbody to virus antigens in the surface of infected cells can depress virus DNA and antigen synthesis to a point where the immune defences may no longer recognize the infected cell.

This process seems to occur in Burkitt-lymphoma cells *in vitro* and may explain the persistence of HSV infections in ganglion cells (Rawls and Tompkins 1975). Similarly, when HeLa cells infected with measles virus *in vitro* were treated with specific IgG, in the absence of complement, which would result in cell lysis, the virus antigens were redistributed as a cap at one end of the cell membrane. This cap was then shed leaving a persistently infected cell with no detectable virus antigens on its membrane. If complement activity was depressed or rendered less effective in measles infection, then such a mechanism might lead to virus persistence *in vivo* in the presence of the immune response (Crumpacker 1980). A similar mechanism may be involved in the persistence of canine distemper (Ho and Babiuk 1980).

Avoidance of recognition by the immune response

Viruses may effect this in several ways.

Avoidance of antibody by remaining intracellular HSV, CMV, measles and simian foamy viruses spread to contiguous uninfected cells by cell fusion or virus budding (Notkins 1974). The viruses are rarely extracellular and are thus protected from antibody. Viruses that replicate in macrophages are similarly protected (Mims 1972).

Replication in immunologically privileged sites Parts of the brain and some glands such as the salivary and mammary glands are relatively inaccessible to antibody and CMI. Virus replication here is protected from the immune response (Mims 1977).

Antigenic variation Periodically some viruses change their surface antigens, and antibody and CMI directed against the previous antigens are unable to recognize the new antigens. This occurs with influenza virus between epidemics and contributes towards the ability of the virus to attack new hosts. It also happens during the course of infection in equine infectious anaemia and visna virus infection of sheep and contributes to persistence of infection within the individual (Smith 1980). Measles virus may also undergo antigenic variation during persistence in man which culminates in subacute sclerosing panencephalitis (SSPE) (ter Meulen *et al.* 1978). The triggers for such antigenic changes in persistent infections are unknown, but antibody probably contributes to the demise of the old variant and selection of the new.

Reduced numbers of antigenic sites on infected cells Virus antigens on the membranes of infected cells vary in number and nature according to the strain (Burns and Allison 1977), e.g. some variants of vaccinia virus failed to produce surface antigens on infected cells that are normally found early during infection (Burns 1975). If immune mechanisms directed against such antigens were important in protection then lack of antigen production would be a virulence mechanism. Thus, cells infected with simian foamy virus were not lysed by antiviral antibody and complement because of a low number of individual antigenic sites, possibly of unfavourable distribution (Hooks *et al.* 1976). Unfortunately, comparisons between virulent and attenuated strains with regard to the number and distribution of antigenic sites on infected cells have not yet been made for any virus.

Reduction in histocompatibility antigens A reduction in the number of histocompatibility antigens on the cell surface might impair the effectiveness of cytotoxic T cells for such cells. VSV infected mouse cells had a 70 per cent reduction in their ability to adsorb cytolytic antibodies directed against H-2 antigens and NDV infected L cells lost 50 per cent of their H-2 antigens (Burns 1975). Whether these infected cells were less susceptible to cytotoxic T cells was not tested. Clearly, appropriate tests with cells infected with virulent and attenuated strains of the same virus should be carried out.

The nature of the virus in persistent infections

The ways by which viruses may avoid or counteract the immune defences to become persistent have been considered. However, many viruses are cytolytic in tissue culture, so how do they maintain themselves *in vivo* without destroying the cells they infect? The form of the persisting virus probably varies with the system under consideration (Rosenthal 1979). There may be a dynamic equilibrium between continuing replication of small amounts of virus and removal of it or virus-infected cells by the immunospecific defences; and in some cases evidence exists for this equilibrium (Docherty and Chopan 1974). Alternatively, experiments *in vitro* suggest that virus may perhaps persist in forms other than replicating infectious virus, i.e. a static rather than a dynamic state may occur (Docherty and Chopan 1974; Rosenthal 1979; Friedman and Ramseur 1979). Possible mechanisms for such persistence are described below.

Integration into the host cell genome In cell cultures the genomes of many oncogenic viruses become incorporated into the host cell DNA similar to lysogenic bacteriophages (Docherty and Chopan 1974). Similar integration of proviral DNA may perhaps occur in latent infections of cells produced by HSV-1 and 2 and human CMV but the evidence is inconclusive (Docherty and Chopan 1974; Rosenthal 1979; Friedman and Ramseur 1979).

Production of defective interfering particles Selection of DIPs seems to be the mechanism of persistence of some viruses *in vitro* (Rima and Martin 1976; Holland *et al.* 1979). Co-infection of BHK-21 cells with DIPs and infectious virus of VSV produced persistent infection which underwent continuous evolutionary changes involving many mutations usually expressed as poorly replicating, temperature-sensitive, small plaque mutants of increased interferon sensitivity (Holland *et al.* 1979). Similarly, persistence could be initiated and maintained by DIPs of measles virus in VERO cells; this was probably due to a reduction of virus-specific proteins in the plasma membrane to a level below that required for virus-induced cell fusion and budding (Rima *et al.* 1977).

The role of DIPs in persistence *in vivo* is unclear. They may initiate persistent infection as shown in mice with influenza virus (Frolov *et al.* 1978) and Rift Valley fever virus (Mims 1956). However, DIPs can augment antibody responses which could clear virus or virus-infected cells (Rabinowitz and Hupriker 1979; Mims 1956; Crick and Brown 1977; Jacobson *et al.* 1979). Also, persistent LCM infections in mice seem to be maintained by the generation of virus particles that are uninhibited by DIPs (Jacobson and Pfau 1980).

Production of mutants Production of *ts* and interferon-sensitive mutants have been associated with virus persistence in some cell cultures (Friedman and Ramseur 1979). It is unknown whether they explain persistence *in vivo*.

In summary, the mechanisms by which viruses and virus-infected cells may counteract the immunospecific defences are varied and numerous. However, many have been suggested by experiments *in vitro* and their role *in vivo* is still unclear. Whether virulent strains are better able than attenuated strains to counteract these defences is largely unknown.

Host and tissue specificity

Host specificity is the ability of viruses to attack under normal conditions certain animal species in preference to others, e.g. variola virus infects man and FMDV domestic animals. Similarly, tissue specificity is the predilection of viruses for certain tissues, e.g. rhinoviruses for the upper respiratory tract. There are many examples of the phenomena (Bang 1972; Fenner *et al.* 1974). They are not necessarily 'all-or-nothing'; they describe situations where differences in virus replication in different hosts or tissues are substantial and significant in relation to production of disease. Limited replication may occur in a host or tissue normally considered resistant when virus challenge is abnormally high (Fenner *et al.* 1974). Also, in the same host species, susceptibility often varies with age, e.g. sucking mice are more susceptible than adult animals to arboviruses (Smith 1977*a*; Mogensen 1979).

The susceptibility of a host or tissue depends initially on the ability of individual cells to support virus replication, i.e. on their complement of replication factors (see previously). If this replication cannot occur, then the host or tissue is insusceptible; even if the virus can replicate, infection of the host or tissue may not occur because of two other factors. Firstly, incoming virus may be removed or inactivated by defence mechanisms before reaching potentially susceptible cells. Secondly, incoming virus may be denied access to these cells by anatomical and other barriers. Variations between hosts or tissues in any of the defence mechanisms or barrier phenomena described previously could lead to susceptibility or resistance. The difficulty is to devise experimental systems to investigate these variations in a manner relevant to natural infection. Animal experiments are necessary, first to establish differences in host and tissue specificity and then to identify the influence on susceptibility of route of entry, barrier phenomena, and the immune response. Parallel studies *in vitro* using cell and organ culture permit analysis of the abilities of cell types to support virus replication provided they retain the cellular susceptibilities observed in the animal. Established cell lines are, in general, unsuitable because of changes in susceptibility that occur during prolonged cultivation *in vitro*. Primary, and occasionally secondary, cell cultures derived from susceptible target tissues should be used, or, better, organ cultures where the cell-to-cell relation of the original tissue is maintained (Hoorn and Tyrrell 1969). In such cultures, other than those with phagocytes and lymphocytes, the relation of virus replication to susceptibility can be studied in the absence of the other factors that influence susceptibility *in vivo*. When the specificities found in the intact host are not manifest in primary cell or organ cultures, they are probably determined by variation of host defences and/or barrier phenomena. Infection *in vitro* of phagocytes (particularly macrophages), lymphocytes and organ cultures of reticulo-endothelial tissues taken from normal and immunized animals can provide valuable information on the relation of defence mechanisms to susceptibility. However, experiments with intact animals are also needed to investigate these mechanisms as well as the role of barrier phenomena in tissue specificity. Thus, the experimental approach to host and tissue specificity involves: (a) demonstration of the phenomena in animals; (b) primary cell and organ culture studies to establish whether similar specificities are manifest *in vitro*; and (c) more detailed studies *in vitro* or in the intact host when these are indicated. Sometimes infection of the chick embryo which, for some viruses shows the tissue specificity found in fowls and man, has proved useful (Bang 1972). Also, investigations have been helped by the development of inbred strains of mice in which the genetics of resistance and susceptibility to certain viruses are known (Mogensen 1979).

The influence of replication factors on host and tissue specificity

The section on replication described how the presence or absence of replication factors on or in host cells could allow or inhibit virus production at any stage. Now we consider this subject in relation to host and tissue specificity, excluding replication in specialized host defence cells such as macrophages which is discussed later.

Receptors for adsorption and penetration Experiments similar to those already described for poliovirus have shown that the presence or absence of the correct receptors on individual cells largely determines the specificity of infection of five antigenic subgroups of avian RNA tumour viruses for different lines of chickens and of feline RNA tumour virus type A for cats but not for dogs (Smith 1977*a*). In each case the nature of the receptor is unknown.

Host factors involved in virus macromolecular syntheses Recombinants of FPV and three strains of human influenza virus differed in their virulence for chickens and mice. Virulence for mice and chickens differed and only mouse virulent strains replicated in mouse embryo brain cells, possibly because a specific host factor was necessary for the function of the virus RNA polymerase (Scholtissek *et al.* 1979; Rott 1979).

After intranasal inoculation of ferrets with influenza viruses tissue specificity is similar to that in human influenza, namely a severe infection of the upper respiratory tract, some lung involvement, but little virus in extra-respiratory tissues. In organ cultures, nasal turbinate tissue produced more virus than lung tissue, thus simulating the situation *in vivo*. In organ cultures, individual susceptible cells of nasal mucosa produced about 10-fold more virus than the alveolar type I and II cells which became infected in lung tissue. The differences seemed to be determined, not by adsorption and penetration, but by cell factors influencing either the amount of virus components synthesized or their assembly (Sweet and Smith 1980). Alveolar Type I cells are relatively devoid of ribosomes and mitochondria and this may have limited virus production. The lungs of neonatal ferrets are more prone to infection with respiratory syncytial virus (RSV) than the lungs of adult animals; this difference in susceptibility was reflected in organ cultures or cell monolayers from lung tissue (Porter *et al.* 1980). Similarly, the susceptibility and resistance of two strains of mice to mouse CMV was reflected in infection of tracheal organ cultures (Nedrud *et al.* 1979). In neither instance have the reasons for the differences been investigated but they may be similar to those discussed for ferret influenza.

Differential assembly and maturation In the example quoted above, assembly of influenza virus may have been inefficient in alveolar type II cells; these have little smooth endoplasmic reticulum, which appears to be important in transfer of envelope proteins of influenza virus to the plasma membrane (Sweet and Smith 1980). The restriction of attenuated strains of NDV and influenza virus to the endoderm of the chorio-allantoic membrane of chick embryos appears to depend on the lack of ability of mesodermal and ectodermal cells to cleave the haemagglutinin during maturation and assembly (Nagai *et al.* 1979; Rott 1979).

Differential release This may be as important in the tissue specificity of influenza virus in the ferret as the differential production of virus in nasal turbinate and lung cells. In the organ culture studies described above with one strain of virus, lung tissue released only about 6 per cent of the total virus produced, whereas nasal turbinates released approximately 30 per cent. Also, ferret neonatal lung, which is more prone to influenza virus infection than adult lung *in vivo*, released more of the virus produced in organ culture. The mechanisms preventing release from adult lung tissue are at present unknown but poor production of neuraminidase is one possibility (Sweet and Smith 1980).

Differential production of defective interfering particles The resistance of C3H/RV mice and the susceptibility of C3H/He mice to West Nile virus, can be reproduced in cell culture *in vitro* (Smith 1977*a*). Brain explants, spleen cells, peritoneal exudate cells and embryo fibroblasts from resistant mice produced far less virus than those from susceptible mice. Thus, cellular factors were responsible for the differences in specificity. Antibody and interferon production, complement, amount of and sensitivity of infected cells to immune cytolysis did not differ significantly in the two types of mice. Darnell and Koprowski (1974) showed that virus uptake and the initial stages of replication occurred similarly in embryo fibroblasts from both strains but that more DIPs were produced in the cells from the resistant mice. Such DIPs might be responsible for the reduced infection *in vivo*. However, similar experiments with brain or spleen cells from adult mice should be done before finally concluding that differential production of DIPs is responsible for the host specificity.

Different numbers of susceptible cells in particular tissues The production of virus by a tissue will be determined by the number of susceptible cells as well as by their individual ability to support replication. A tissue with 10^8 susceptible cells per gram will appear more susceptible than one with 10^3 susceptible cells per gram even if the amount of virus produced in each susceptible cell of the latter is 100-fold higher than for those of the former. Though immunofluorescent studies of tissues from natural infections show that some cells produce virus while others do not, there have been few comparisons of the numbers of susceptible cells in various tissues or in the same tissues of different hosts.

Differential role of environmental factors Rhinoviruses infect only the upper respiratory tract because they replicate well at the lower temperature of this tract and poorly, if at all, at the normal body temperature of other tissues (Fenner *et al.* 1974). Substantial differences in body temperature exist between species and these may influence host specificity of viruses (Fenner *et al.* 1974). Variations of pH also influence tissue specificity (see later).

Influence of host defence mechanisms on host and tissue specificity

Defences on mucosal surfaces It has been mentioned previously that the acid pH of the stomach may determine the type of viruses that infect the intestine. However, despite their potential in this respect, there are no examples of other mucosal defences influencing host and tissue specificity.

Non-specific defences: interferon Some cases of host and tissue specificity may be determined by differential induction or activity of interferon (Smith 1977*a*). Susceptible infant mice infected with para-influenza virus type I or Coxsackie B virus produced less interferon than the more resistant adult mice. The increased resistance of older chick embryos to VSV and to myxoviruses was related to the increased activity of interferon in their cells. Similarly, the resistance of A_2G mice to influenza virus was related to a greater activity of interferon against this virus, but not against VSV, in their macrophages compared to macrophages from susceptible mice (Haller *et al.* 1980). In contrast, no role could be found for interferon in the differential susceptibilities of mice to cowpox, West Nile or Sindbis viruses (Smith 1977*a*).

Non-specific defences: macrophages The presence or absence of virus replication in macrophages has been associated with both genetic and age-determined host specificity (Mogensen 1979). The mechanisms of susceptibility or resistance are specific to one virus/host system. Thus, PRI mice and their macrophages are susceptible to MHV-2 but resistant to West Nile virus while for C3H mice and their macrophages the position is reversed.

As mentioned previously for MHV-2 in PRI (susceptible) and C3H (resistant) mice, the virus fails to uncoat in the macrophages of the resistant mice. The factor(s) which determines the difference in macrophages is unknown but resistant C3H macrophages were converted to susceptibility by lymphokines excreted from splenic lymphocytes from allogeneic susceptible mice. Furthermore, susceptible PRI macrophages were rendered resistant by products from concanavalin A-stimulated lymphocytes and treatment with this reagent rendered PRI mice resistant to infection (Mogensen 1979).

The host specificities of MHV-3 for different mouse strains and HSV-2 for GR/A (resistant) and BALB/c (susceptible) mice were reflected in the relative susceptibilities of their macrophages to infection but the reasons for the differences are unknown. In contrast, no relation was found between infection of macrophages and differences in susceptibility of mice to infection with HSV-1 and ectromelia virus (Mogensen 1979).

The increase in resistance of mice with age to virus-induced encephalitis after peripheral inoculation with HSV-1 was related to restricted release of virus from infected macrophages (Mogensen 1979). For HSV-2, age-related resistance was determined by restricted replication in liver and peritoneal macrophages; inactivation of macrophages with silica conferred susceptibility on adult mice and sucking mice were made resistant by infusion of adult macrophages (Mogensen 1979). Similar findings have been obtained for young and old mice with MHV-2 and yellow fever virus but not for murine CMV. The maturation process which allows macrophages from adult mice to confine virus infections is not yet understood. However, macrophages from newborn mice are lacking in rough endoplasmic reticulum—an area where some virus components are produced or processed.

Non-specific defences: natural killer cells The resistance or susceptibility of some mouse strains to HSV-1 infection was related to the numbers of NK cells; and in two patients with severe herpes infection antibody was not depleted but the number of NK cells was significantly lower than in normal subjects or persons infected with other viruses (Ching and Lopez 1979).

Immunospecific defences The ability of the murine leukaemia viruses to induce tumour formation in some but not other mouse strains is controlled initially by replication factors which determine whether or not virus infection occurs. However, the infected cells produce large tumours only in those strains of mice which are incapable of developing an adequate immune response because they lack the necessary immune response (Ir) gene. The relative resistance of different mouse strains to ectromelia virus was not related to replication in macrophages but to the speed with which they mounted both an antibody and cell-mediated immune response (Mogensen 1979). Differences in immune responsiveness have also been implicated in the resistance and susceptibility of different mouse strains to murine CMV infection (Smith 1977*a*).

The influence of barriers on tissue specificity

Some viruses show a tissue specificity for tissues at their primary lodgement site because spread is prevented to tissues which are potentially susceptible, as shown by infection of primary cell or organ cultures and local inoculation of animals (Bang 1972; Fenner *et al.* 1974; Smith 1977*a*). As mentioned previously, the barriers preventing spread include layers of insusceptible cells and virus destruction by the reticulo-endothelial system. Already we have noted, as an example of the former, that attenuated myxoviruses are confined to the endoderm of chick embryos and possibly the respiratory and alimentary tracts of chickens; and, as an example of the latter, the lack of ability of blood-borne vaccinia and ectromelia virus to infect the liver parenchyma of mice and rats respectively because of the protective effect of Kupffer cells. An example of the simultaneous influence of both types of barriers is summarized below.

Prevention of spread of influenza virus to urogenital and fetal tissues

Influenza in man and ferrets is primarily an upper respiratory tract infection; viraemia and infection of extra-respiratory tissues is rare. In particular, urogenital and fetal tissues do not usually become infected despite the fact that, in organ culture, they are highly susceptible. Barriers must therefore prevent the spread of virus from the respiratory tract to susceptible tissues in all but a few individuals on rare occasions. These barriers were investigated by using influenza in the ferret as a model for human infection (Sweet and Smith 1980).

After intranasal inoculation, influenza virus was found in the cervical or mediastinal lymph nodes, but only small amounts of virus escaped, as virus antigen could not be detected by fluorescent antibody in the spleen of animals with severe respiratory infection, although it could after blood stream inoculation of moderate quantities of virus. With regard to the blood and reticulo-endothelial barrier, amounts of virus smaller than the maximum capable of escaping from the respiratory tract were inactivated rapidly *in vitro* by virus inhibitors in blood, and large amounts of virus inoculated into the bloodstream were undetectable after 30 min. Thus, any viraemia which may have developed from the respiratory infection was quickly reduced. The barrier between the blood and urogenital tissues seemed impenetrable in non-pregnant ferrets. These tissues could be infected *in vivo* by local but not by blood stream inoculation. One factor in this barrier may have been the insusceptibility of the endothelial cells lining blood vessels to infection with influenza virus. The barrier is modified in pregnancy and, in ferrets, was breached when a high viraemia was artificially induced by blood stream inoculation of virus. Virus was isolated from the uterus, bladder, placenta, umbilical cord, amnion and chorion and, when inoculation was at late gestation, fetal infection ensued.

Thus, in ferrets and probably man, the confinement of the virus to the respiratory tract is due to limited escape of virus from the respiratory tract, the rapid removal of blood-borne virus by non-specific inhibitors and reticulo-endothelial tissues and the strength of the barrier at the blood/urogenital junction, which appears to be weakened in pregnancy, at least in ferrets. Faults in any of these barriers in individuals may explain the rare occurrence of viraemia and extra-respiratory infection.

Damage to the host

Virus damage to the host can result in mild, severe, or fatal illness, teratogenic effects, behavioural changes and, as described in Chapter 104, oncogenesis. The aggregate disease produced is the culmination of damage to individual cells. There are now indications of how this damage occurs and the determinants responsible. They are summarized first as a prelude to descriptions of the various manifestations of cell damage seen in adults, fetuses and neonates.

Mechanisms of cell damage

Cells can be damaged by a more or less passive role of the virus, as a result of direct cytotoxicity and immunopathological processes.

Passive virus damage The requirements of virus replication may deplete cells of components essential for life. This may take time; diploid cells infected with rubella virus died slowly after many weeks of continuous release of newly synthesized virus (Koch *et al.* 1980). Cells may also be harmed by excessive accumulation of virus components within them as occurs for some virulent strains of NDV (Smith 1980).

Virus cytotoxicity There is no convincing evidence that viruses produce circulating toxins like bacteria, despite occasional reports that damage may occur without virus replication (Smith 1980). However, there is increasing awareness that infected cells can be damaged by the purely local action of virion components and possibly virus-specified products, both of which might be called virus cytotoxins by virtue of their deleterious effects. Cell damage may be expressed as a visible cytotoxic (cytopathic) effect when cells round up, become detached from glass surfaces, fuse into syncytia (polykaryocytes, giant cells) or die with or without lysis (Bablanian 1975). However, macromolecular synthesis may be depressed or altered without apparent morphological effect but nevertheless cause major pathological effects, e.g. in nerve cells.

The occurrence of cytopathic effects and/or depression of host-cell macromolecular synthesis in inoculated cells in the absence of active virus replication (Bablanian 1972) indicates that both effects could be due to either preformed virion components or virus-induced products. Thus, influenza virus, NDV, a murine picornavirus and mengovirus were cytopathic for cells which were either incapable or poorly able to support virus replication; vaccinia, rabbit pox, and reoviruses were cytopathic after inactivation by ultraviolet light; and chemical inhibitors of virus replication in permissive cells did not prevent poliovirus, vaccinia virus and rabbit pox virus from evoking cytopathic effects or poliovirus from inhibiting macromolecular synthesis (Bablanian 1975).

Some virion components responsible for cytotoxicity have been identified. The capsid penton of adenovirus causes cells to round up and detach from glass (Bablanian 1972), although its role in cell death is equivocal, since lethal infection with adenovirus can occur without penton synthesis (Koch *et al.* 1980). The fibre antigen of adenovirus is not cytotoxic but large quantities inhibit host-cell macromolecular synthesis (Bablanian 1975). Late fusion and degeneration of cells infected with vaccinia virus appear to be connected with virus release and to be induced by a surface tubular protein of the virion. This protein, extracted from intact virions and infected cells, was toxic when introduced into uninfected cells (Burgoyne and Stephen 1979). It appears not to be responsible for the early cell rounding seen in vaccinia virus infection or for the inhibition of host-cell protein synthesis. Capsid proteins of reovirus may be concerned in cytotoxicity (Smith 1977*b*), and a structural protein of frog virus 3 may inhibit host-cell macromolecular syntheses (Drillien *et al.* 1980). Virion components may also be responsible for the cytotoxicity and depression of host-cell macromolecular synthesis produced by VSV (Smith 1980). Inactivated Sendai virus, NDV and HSV, like the infectious viruses, caused cells to fuse into syncytia but only when large amounts were used, implicating preformed virion components. Fragments of the virus envelopes also induced fusion of cells. The fusion factors for both HSV and NDV appear to be liproproteins but, at present, all envelope fragments possess less fusing activity than intact viruses (Smith 1980).

The evidence, obtained from cell culture experiments suggesting that preformed virion components cause damage, is supported by observations *in vivo*. Large quantities of influenza and pox viruses produced toxic effects in animals too rapidly for replication to have had much influence; also influenza virus was toxic for mouse lung tissue; this has been attributed to virus RNA which may code for cytotoxic proteins (Sweet and Smith 1980). Similarly, frog virus 3 failed to replicate in mice but produced damage to hepatic cells within 3 hr and death within 18–36 hr (Kirn *et al.* 1972), and soluble proteins from this virus killed mice (Aubertin *et al.* 1977).

Evidence that non-structural virus-induced products cause cell damage is not so strong as that for virion components; it rests largely on observations of the effects on cell damage of selectively inhibiting with drugs virus replication, nucleic acid production and protein synthesis (Bablanian 1975). Early cell rounding caused by vaccinia virus seems to be produced by virus-induced polypeptides (Bablanian 1975; Smith 1980) and inhibition of host protein synthesis is associated with early virus-induced RNA synthesis (Schrom and Bablanian 1979*a*). The cytotoxic effects caused by poliovirus appear to be due to virus-induced polypeptides, but it is not clear whether host macromolecular synthesis is restricted by a capsid protein or an induced inhibitor (Smith 1980). WEE virus may damage cells by inducing an inhibitor of host DNA synthesis (Koizumi *et al.* 1979). However, none of these putative, non-structural, virus cytotoxins has yet been isolated.

How do virion components and virus-induced non-virion products damage cells? A close connection between inhibition of macromolecular synthesis and cytopathic effects does not always occur (Bablanian 1975); sometimes they may be independent events originating from different virus determinants. Either event might be due to the direct action of the virion component or induced product on a vital host cell process, e.g. poliovirus components or products appear to interfere with or inactivate the initiation factor eIF-4B, which is required for host but not poliovirus protein synthesis (Rose *et al.* 1978). A similar mechanism has been proposed for the interference with protein synthesis by vaccinia virus, although the role of the early, virus-induced RNA (see above) is not clear (Schrom and Bablanian 1979*a*, *b*). An alternative suggestion for the inhibition of host protein synthesis by this virus is that the virus-induced RNA competes with cellular mRNA (Rosemond-Hornbeak and Moss 1975; Cooper and Moss 1979). Cytopathological effects or inhibition of macromolecular synthesis may also follow from indirect effects of the cytotoxins on internal and external cellular membranes, e.g. changes in lysosomal membranes may release autolytic enzymes (Mallucci and Allison 1965). Poliovirus and vaccinia virus may also induce changes in external membranes which result in elevated amounts of intracytoplasmic Na^+ which, in turn, may favour translation of virus in preference to host cell mRNA (Smith 1980).

Finally, an important but unexplained point regarding virus cytotoxicity should be mentioned. Virus infection can cause harmful effects in some animals and tissues but not in others and, *in vitro*, viruses may damage some but not other cells (Bablanian 1975; Fenner *et al.* 1974; Mims 1977). Thus, arboviruses usually replicate in insect cells without noticeable perturbation, whereas infection of mammalian cells is cytopathic (Koch *et al.* 1980). Similarly, poliovirus replicates in but does not harm primate intestinal epithelial cells, in contrast to its effect on nerve cells.

Immunopathological damage We have seen that the immune response can destroy virus-infected cells. Usually the numbers are small and the cell damage makes little impact. Sometimes, however, the numbers of cells destroyed may be sufficiently large to cause distress or even death to the host. Virus-infected cells may be damaged by either type II (antibody mediated) or type IV (immune cell-mediated) cytotoxic reactions (Coombs and Gell 1968). Also, destruction of *normal* host cells may add to the total damage because antibody or CMI directed against them may be evoked by host antigens in virus envelopes or in the membranes of infected cells (Webb and Hall 1972). Uninfected host cells may also be damaged by immunopathological reactions of types I and III (Coombs and Gell 1968). In type I (anaphylactic) reactions virus interacts with reaginic (IgE) antibody located on mast cells, and vaso-active compounds are released from the damaged cells. In type III reactions immune complexes, derived from the interaction of antibody with viruses or their antigens on tissue debris, circulate in the blood stream and are deposited in the walls of small blood vessels such as those in the glomeruli. Complement is activated continuously by these complexes and chronic inflammation results. Tissue cell destruction may follow from the action of lysosomal enzymes liberated from phagocytes of the inflammatory response.

Primary and secondary manifestations in adult animals of virus-induced damage to cells

Now we consider the impact of cell damage on the host. The subject is considered broadly because details of the signs, symptoms and pathology of specific virus diseases are given in appropriate chapters. In attempting to understand the consequences of cell damage two questions should be asked. Firstly, which cells are damaged by virus replication or immunopathological reactions and secondly does this damage explain the pathological effects of the disease?

Location of damaged cells Histopathological techniques have been used for many years to detect cell damage, e.g. anterior horn cells damaged by poliovirus, respiratory epithelium destroyed by influenza virus, and giant cell formation in lesions caused by HSV. Electron microscopy and fluorescent antibody techniques allow more detailed investigations of lesions including those that might result from immunopathological reactions (Oldstone and Dixon 1975). These techniques have shown that virus replication can occur *in vivo* without producing morphological damage, e.g. poliovirus in intestinal epithelium. However, biochemical changes may have occurred in such cells with important pathological effects. Hence, although overtly damaged cells should receive attention first in attempts to explain the pathology of the disease, it would be unwise to disregard any cell type showing evidence of virus replication.

Primary and secondary manifestation of cell damage Some manifestations of virus disease are primary effects of cell damage, e.g. the paralysis caused by destruction of anterior horn cells by poliovirus, the neurological effects following damage of nerve cells by rabies virus, the nasal tract discomfiture resulting from denudation of epithelial cells by rhinoviruses and influenza virus and the diarrhoea caused by rotavirus and coronavirus infection of intestinal epithelium. On the other hand, many manifestations are secondary effects of the original cell damage. These effects are the general reactions of the body to injury which occur in other infectious diseases (Stoner 1972) and sickness from diverse causes (Stoner and Threlfall 1960). Inflammation is one such general reaction resulting from complement activation or liberation of endogenous permeability factors from damaged cells. Another is the homeostatic response to a lowered circulating blood volume and also the fatal secondary shock syndrome which occurs when loss of blood volume is excessive. A lowered circulating blood volume can result either from an enlargement of the capillary bed by the action of vasodilators liberated from damaged cells or from leakage of blood fluid forming oedema and haemorrhage (Stoner and Threlfall 1960). This leakage can be brought about by direct damage to endothelial cells or liberation of endogenous permeability factors from other cells. Oedema and haemorrhage may have serious mechanical consequences in the brain and lung. Virus infections of the intestinal tract can also bring about fluid disturbances. Fever is another general reaction resulting from the liberation of endogenous pyrogen from PMN and MN phagocytes by many agents including viruses (Bernheim *et al.* 1979). The liberation of endogenous pyrogen may also account for the headache, myalgia, shivering, nausea and malaise that occur in some virus diseases such as influenza, for they were induced, along with fever, in human volunteers by the inoculation of human endogenous pyrogen (Rawlins and Cranston 1973). Experiments with influenza of ferrets as a model for the human disease indicate that all the usual constitutional symptoms could follow from the liberation of endogenous pyrogen after interaction of virus with phagocytes in the upper respiratory tract (Sweet and Smith 1980).

The relative importance of direct cytotoxicity and immunopathology This is a difficult question because primary and secondary effects can follow from cell damage derived from either cause and the precise mechanisms of cell damage *in vivo* are not easily investigated. The primary manifestations seen in acute disease such as paralysis in poliomyelitis probably result from direct cytotoxicity as do some secondary manifestations, such as fever and other constitutional effects in influenza and secondary shock in pox virus infections. However, inflammatory and vascular disturbances occurring later in infection may have an immunopathological origin. Evidence that immunopathology is responsible for the main disease effect is not easily obtained from human diseases. Demonstration that a patient is immunologically sensitive to virus products is not proof of the implication of immunopathology in the pathological effects. Usually the required experiments can be set up only in animals: the pathological effects must be simulated by immunological reactions invoked either in a sensitized host by virus, or virus-infected cells, or in an infected host by antibody or lymphocytes and macrophages from an immune host. For LCM in mice and Aleutian disease of mink there is little doubt that immunopathology plays a major role in the disease effects (Webb and Hall 1972, Nathanson *et al.* 1975). For example, in acute LCM of mice cytotoxic type II or IV reactions predominate, while in the chronic form of the disease type III reactions occur leading to glomerulonephritis (Webb and Hall 1972; Oldstone and Dixon 1975; Nathanson *et al.* 1975).

In human disease the following immunopathological conditions have been suggested: type I reactions in the pathogenesis of upper respiratory tract infections and some skin rashes; type II reactions in some of the liver necrosis occurring in serum hepatitis and yellow fever; type III reactions in the prodromal rashes seen in exanthematous virus infections and serum hepatitis and the intravascular coagulation caused by yellow fever virus and smallpox; and type IV reactions in lesions caused by HSV and pox viruses (Webb and Hall 1972; Oldstone and Dixon 1975; Nathanson *et al.* 1975). In some cases the evidence is reasonably good but investigations of dengue fever show that caution is needed. Fatal haemorrhagic shock in children infected with dengue virus was thought to occur only during secondary attacks of the disease and to be due to immunopathological causes. However, haemorrhagic shock occurs in primary dengue and appears to follow from a direct action of the virus, possibly by replicating in and damaging endothelial cells (Smith 1980). Antibody present from a previous infection may enhance the disease by aiding infection of cells by the virus (Halstead 1979), not by inducing immunopathological reactions.

In general, it appears that direct cytotoxicity is more important than immunopathology in producing the pathology of virus disease in adults.

Damage to the fetus

Fetal damage may result directly from virus infection or indirectly from the effects of virus disease on the mother. Depending upon the virus and the host, the infection may result in death *in utero* with subsequent abortion, resorption, or mummification; stillbirth; early neonatal death; malformations or retarded development of surviving fetuses. Rubella virus and CMV definitely cause fetal damage in man; and influenza, mumps, Coxsackie, polio, varicella, and hepatitis viruses are suspect (Tobin *et al.* 1977).

Fetal infection

Cell damage probably results from the direct action of virus and the primary and secondary manifestations are basically similar to those for adult animals. However, they may be amplified by the rapid metabolic and developmental changes occurring in the fetus, e.g. mitotic inhibition and chromosomal damage may assume more importance in the fetus than in the adult. Rubella virus produced a lowered mitotic rate in human embryonic diploid cells and curtailed the number of passages for which the culture could be maintained (Lowrie *et al.* 1977). *In vivo*, a depressed mitotic rate would reduce cell production during the gestation period, which could lead to smaller organs and runted offspring (Mims 1968, Lowrie *et al.* 1977). Chromosomal abnormalities that could cause malformations have been reported in human fetal cells infected with rubella virus *in vitro* and also in a naturally infected fetus (Mims 1968).

The timing of infection during fetal development is of paramount importance in determining the nature and extent of damage (Lowrie *et al.* 1977). Early in pregnancy, major organogenesis is occurring in the fetus and infection may well produce malformations to the chief organ systems. In man, infections at different periods of gestation are associated with a prevalence of defects in organs undergoing organogenesis at these periods (Lowrie *et al.* 1977). For rubella infections in the first trimester, the malformations followed directly from virus destruction of certain cells within the affected organ (Lowrie *et al.* 1977). When organs develop later in pregnancy malformations may be induced later in fetal life. Thus, cerebellar hypoplasia produced by Kilham's rat virus and feline panleucopenia virus appears to result from the cytolytic action of the viruses on the external germinal layer of the cerebellum late in fetal life (Mims 1968; Lowrie *et al.* 1977).

The interdependence of different tissues during fetal development raises the possibility of virus replication in one organ resulting in malformations in another. Influenza of chick embryos produced neural abnormalities but replication was limited to extra-neural sites (Johnson *et al.* 1971). Vasculitis, with concomitant thrombosis leading to organ hypoxia and hypoplasia, may be the major cause of fetal damage in rubella, CMV, and HSV-1 infections (Catalano and Sever 1971). The vasculitis may be present in a fetal tissue which does not become malformed or may even occur in the placenta.

Maternal illness Virus disease of the mother can affect fetal development (Coid *et al.* 1977). Fever can produce abortions, stillbirths and fetal malformations including anencephaly, microencephaly, and hydroencephaly (Edwards 1972). Changes in placental circulation such as vasoconstriction, congestion, or haemorrhage would have rapid and severe effects on the fetus (Mims 1968). Fetal wastage and growth retardation seen with Coxsackie B3 virus infections of pregnant mice was principally due to malnutrition. The mother could not digest and metabolize protein because of virus-induced pancreatic acinar atrophy and hepatitis (Coid *et al.* 1977).

Damage to the infant

Maternal antibody acquired transplacentally or in colostrum protects infants against most virus infections and reduces their severity if they occur. Nevertheless, some viruses do adversely affect the young child. Rotaviruses, relatively uncommon in adults, can produce enteritis in the first two years of life. Respiratory viruses such as RSV, parainfluenza virus type 3 and influenza virus have been implicated in bronchiolitis, croup, pneumonia and the sudden infant death syndrome (SIDS) in infants less than 1 year old (Fenner and White 1976, Collie *et al.* 1980). There is no evidence that these viruses cause more damage to infant than to adult cells. However, their diseases appear more hazardous in babies than adults. This is probably because of anatomical and physiological differences between the immature and mature respiratory tree. In infants, the mucus and inflammatory response resulting from relatively minor damage to the epithelium of the upper and middle respiratory tract may cause obstruction because of their disproportionately narrow air passages. This will be compounded by shallow and irregular breathing and inability to take breath orally (Swift and Emery 1973). Thus, death may occur by occlusion of the airways, and this may happen in SIDS (Collie *et al.* 1980). With regard to the lung, collateral channels for ventilation such as the pores of Kohn, are deficient in both number and size in the infant lung and patchy atelectasis is more likely to develop (Wohl and Chernick 1978). In addition, the lungs of neonates have a low proportion of alveolar cells in comparison with those of adults (Doershuk *et al.* 1975). Damage to a few alveoli may affect gaseous exchange and thus pneumonia may have a more severe effect.

Conclusion

Knowledge of the mechanisms and determinants of virus pathogenicity is still incomplete. This applies particularly to attack on mucous surfaces, interference with host defence, and the cause of damage. In future research there should be a change of emphasis away from replication towards these other equally important aspects of pathogenicity.

References

Adler, R., Glorioso, J. C., Cossman, J. and Levine, M. (1978*a*) *Infect. Immun.* **21**, 442.
Adler, R., Glorioso, J. C. and Levine, M. (1978*b*) *J. gen. Virol.* **39**, 9.
Allison, A. C. (1972) *Ann. Inst. Pasteur* **123**, 585.
Anderson, M. J. (1978) *Infect. Immun.* **20**, 608.
Ashman, R. B. and Mullbacher, A. (1979) *J. exp. Med.* **150**, 1277.
Aubertin, A. M., Anton, M., Bingen, A., Elharrar, M. and Kirn, A. (1977) *Nature, Lond.* **265**, 456.
Bablanian, R. (1972) *Symp. Soc. gen. Microbiol.* **22**, 359; (1975) *Progr. med. Virol.* **19**, 40.
Bachi, T., Deas, J. E. and Howe, C. (1977) *Virus Infection and the Cell Surface.* North Holland Publishing Co., Amsterdam.
Bang, B. G. and Bang, F. B. (1969) *J. exp. Med.* **130**, 105.
Bang, F. B. (1972) *Symp. Soc. gen. Microbiol.* **22**, 415.
Bang, F. B. and Foard, M. A. (1969) *J. exp. Med.* **130**, 121.
Baucke, R. B. and Spear, P. G. (1979) *J. Virol.* **32**, 779.
Beare, A. S. and Reed, S. E. (1977) *Chemoprophylaxis and Virus Infections of the Respiratory Tract*, Vol. 2. CRC Press Inc., Ohio.
Bernheim, H. A., Block, L. H. and Atkins, E. (1979) *Ann. intern. Med.* **91**, 261.
Blanden, R. V., Doherty, P. C., Dunlop, M. B. C., Gardner, I. D., Zinkernagel, R. M. and David, C. S. (1975) *Nature, Lond.* **254**, 269.
Blanden, R. V., Pang, T. E. and Dunlop, M. B. C. (1977) *Virus Infection and the Cell Surface.* North Holland Publishing Co., Amsterdam.
Blattner, R. J., Williamson, A. P. and Heys, F. M. (1973) *Progr. med. Virol.* **15**, 1.
Bloom, B. R. and Rager-Zisman, B. (1975) *Viral Immunology and Immunopathology.* Academic Press, London.
Borden, E. C. (1979) *Ann. intern. Med.* **91**, 472.
Bosch, F. X., Orlich, M., Klenk, H.-D. and Rott, R. (1979) *Virology* **95**, 197.
Buck, C. A., Granger, G. A., Taylor, M. W. and Holland, J. J. (1967) *Virology* **33**, 36.
Burgoyne, R. D. and Stephen, J. (1979) *Arch Virol.* **59**, 107.
Burns, W. H. (1975) *Viral Immunology and Immunopathology.* Academic Press, London.
Burns, W. H. and Allison, A. C. (1977) *Virus Infection and the Cell Surface.* North Holland Publishing Co., Amsterdam.
Burrows, R. (1972) *Symp. Soc. gen. Microbiol.* **22**, 303.
Catalano, L. W. and Sever, J. L. (1971) *Annu. Rev. Microbiol.* **25**, 255.
Ching, C. and Lopez, C. (1979) *Infect. Immun.* **26**, 49.
Coid, C. R., Lansdown, A. B. G. and McFadyen, I. R. (1977) *Infections and Pregnancy.* Academic Press, London.
Cole, B. C., Lombardi, P. S., Overall, J. C. and Glasgow, L. A. (1978) *Proc. Soc. exp. Biol. Med., N.Y.* **157**, 83.
Collie, M. H., Rushton, D. I., Sweet, C. and Smith, H. (1980) *J. med. Microbiol.* **13**, 561.
Compans, R. W. and Choppin, P. W. (1975) *Comprehensive Virology*, Vol. 4. Plenum Publishing Corpn., New York.
Coombs, R. R. A. and Gell, P. G. H. (1968) *Clinical Aspects of Immunology.* Blackwell Scientific Publications, Oxford.

Cooper, J. A. and Moss, B. (1979) *Virology* **96**, 368.
Crick, J. and Brown, F. (1977) *Infect. Immun.* **15**, 354.
Crumpacker, C. S. (1980) *Rev. infect. Dis.* **2**, 78.
Dales, S. (1973) *Bact. Rev.* **37**, 103.
Darnell, M. B. and Koprowski, H. (1974) *Mechanisms of Virus Disease*. Benjamin, Menlo Park, California.
Dean, D. J., Evans, W. M. and McClure, R. C. (1963) *Bull. World Hlth. Org.* **29**, 803.
Denny, F. W. (1973) *Airborne Transmission and Airborne Infection*. Oosthoek Publishing Co., Utrecht.
Docherty, J. J. and Chopan, M. (1974) *Bact. Rev.* **38**, 337.
Doershuk, C. F., Fisher, B. J. and Matthews, K. W. (1975) *Pulmonary Physiology of the Fetus, Newborn and Child*. Lea and Febiger, Philadelphia.
Doherty, P. C. (1980) *The Molecular Basis of Microbial Pathogenicity*. Verlag Chemie, Weinheim.
Dolowy, W. C. and Muldoon, R. L. (1964) *Proc. Soc. exp. Biol. Med., N.Y.* **116**, 365.
Douglas, R. G. (1975) *The Influenza Viruses and Influenza*. Academic Press, London.
Doyle, M. and Holland, J. J. (1973) *Proc. nat. Acad. Sci., Wash.* **70**, 2105.
Drillien, R., Keller, F., Koehren, F. and Kirn, A. (1978) *Viruses and Environment*. Academic Press, London.
Drillien, R., Spehner, D. and Kirn, A. (1980) *FEMS Microbiol. Lett.* **7**, 87.
Druett, H. A. (1967) *Symp. Soc. gen. Microbiol.* **17**, 165.
Edwards, M. J. (1972) *Lancet* **i**, 320.
Eylan, E. and Gazit, A. (1978) *Arch. Virol.* **56**, 47.
Faden, H., Glasgow, L. A. and Overall, J. C. (1978) *Infect. Immun.* **19**, 94.
Farber, M. S. and Baum, S. G. (1978) *J. Virol.* **27**, 136.
Fenner, F., McAuslan, B. R., Mims, C. A., Sambrook, J. and White, D. O. (1974) *The Biology of Animal Viruses*. Academic Press, London.
Fenner, F. and White, D. O. (1976) *Medical Virology*. Academic Press, London.
Field, H. J. and Hill, T. J. (1974) *J. gen. Virol.* **23**, 145; (1975) *Ibid.* **26**, 145.
Field, H. J. and Wildy, P. (1978) *J. Hyg., Camb.* **81**, 267.
Fields, B. N., Weiner, H. L., Ramig, R. G. and Ahmed, R. (1978) *Persistent Viruses*. Academic Press, London.
Finter, N. B. (1973) *Interferon and Interferon Inducers*. North Holland Publishing Co., Amsterdam.
Friedman, R. M. and Ramseur, J. M. (1979) *Arch. Virol.* **60**, 83.
Frolov, A. F., Scherbinskaya, A. M., Maksimovich, N. A. and Kuz'menkova, L. V. (1978) *Bull. exp. Biol. Med.* **85**, 475.
Fujimiya, Y., Babiuk, L. A. and Rouse, B. T. (1978) *Canad. J. Microbiol.* **24**, 1076.
Golais, F. and Sabo, A. (1975) *Acta Virol.* **19**, 387.
Gresland, L., Niveleau, A. and Huppert, J. (1979) *J. gen. Virol.* **45**, 569.
Gresser, I. and Lang, D. J. (1966) *Progr. med. Virol.* **8**, 62.
Gresser, I., Tovey, M. G., Bandu, M.-T., Maury, C. and Brouty-Boye, D. (1976*a*) *J. exp. Med.* **144**, 1305.
Gresser, I., Tovey, M. G., Maury, C. and Bandu, M.-T. (1976*b*) *J. exp. Med.* **144**, 1316.
Grewal, A. S., Rouse, B. T. and Babiuk, L. A. (1980) *J. Immunol.* **124**, 312.
Haller, O., Arnheiter, H., Lindemann, J. and Gresser, I. (1980) *Nature, Lond.* **283**, 660.
Halstead, S. B. (1979) *J. infect. Dis.* **140**, 527.
Hicks, J. T., Ennis, F. A., Kim, E. and Verbonitz, M. (1978) *J. Immunol.* **121**, 1437.
Hill, T. J. (1979) *Soc. gen. Microbiol.: Quarterly* **6**, 56.
Hirsch, R. L., Griffin, D. E. and Winkelstein, J. A. (1980) *J. infect. Dis.* **141**, 212.
Ho, C. K. and Babiuk, L. A. (1980) *Immunol.* **39**, 231.
Hoke, C. H., Hopkins, J. A., Meiklejohn, G. and Mostow, S. R. (1980) *Rev. infect. Dis.* **1**, 946.
Holland, J. J. (1964) *Symp. Soc. gen. Microbiol.* **14**, 257.
Holland, J. J., Grabau, E. A., Jones, C. L. and Semler, B. L. (1979) *Cell* **16**, 495.
Hooks, J. J., Burns, W., Hayashi, K., Geis, S. and Notkins, A. L. (1976) *Infect. Immun.* **14**, 1172.
Hoorn, B. and Tyrrell, D. A. J. (1969) *Progr. med. Virol.* **11**, 408.
Huang, A. S. (1973) *Annu. Rev. Microbiol.* **27**, 101.
Jacobson, S., Dutko, F. J. and Pfau, C. J. (1979) *J. gen. Virol.* **44**, 113.
Jacobson, S. and Pfau, C. J. (1980) *Nature, Lond.* **283**, 311.
Johnson H. M. and Baron, S. (1977) *Pharmacol. Ther.* **1**, 349.
Johnson, K. P., Klasnja, R. and Johnson, R. T. (1971) *J. Neuropath. exp. Neurol.* **30**, 68.
Johnson, R. T. (1964) *J. exp. Med.* **120**, 359.
Johnson, R. T. and Mims, C. A. (1968) *New Engl. J. Med.* **278**, 23, 84.
Kilburn, K. H. (1968) *Amer. Rev. resp. Dis.* **98**, 449.
Kirn, A., Gutt, J. P., Bingen, A. and Hirth, C. (1972) *Arch. gesamt. Virusforsch.* **36**, 394.
Klenk, H.-D. (1980) *The Molecular Basis of Microbial Pathogenicity*. Verlag Chemie, Weinheim.
Klenk, H.-D. and Choppin, P. W. (1969) *Virology* **38**, 255.
Klenk, H.-D., Nagai, Y., Rott, R. and Nicolau, C. (1977) *Med. Microbiol. Immunol.* **164**, 35.
Koch, M. A. *et al.* (1980) *The Molecular Basis of Microbial Pathogenicity*. Verlag Chemie, Weinheim.
Koch, R. (1890) **I.** Hirschwald, Berlin; (1891) Ueber bakteriologische Forschung *Verhandlungen X int. med. Kongr.*
Koizumi, S., Simizu, B., Hashimoto, K., Oya, A. and Yamada, M. (1979) *Virology* **94**, 314.
Lascano, E. F. and Berria, M. I. (1980) *Arch. Virol.* **64**, 67.
Lee, S. H. S. and Epstein, L. B. (1980) *Cell. Immunol.* **50**, 177.
Levin, J. G., Ramseur, J. M. and Grimley, P. M. (1973) *J. Virol.* **12**, 1401.
Loh, L. and Hudson, J. B. (1980) *Infect. Immun.* **27**, 54.
Lowrie, D. B., Toms, G. L. and Pearce, J. H. (1977) *Infections and Pregnancy*. Academic Press, London.
Lwoff, A. (1959) *Bact. Rev.* **23**, 109; (1969) *Ibid.* **33**, 390.
McFarland, H. F. (1974) *J. Immunol.* **113**, 1978.
Mallucci, L. and Allison, A. C. (1965) *J. exp. Med.* **121**, 477.
Marker, S. C. and Jahrling, P. B. (1979) *Arch. Virol.* **62**, 53.
Menezes, J., Seigneurin, J. M., Patel, P., Bourkas, A. and Lenoir, G. (1977) *J. Virol.* **22**, 816.
Miles, A. A. (1955) *Symp. Soc. gen. Microbiol.* **5**, 1.
Miller, W. S. (1966) *Bact. Rev.* **30**, 589.
Mims, C. A. (1956) *Brit. J. exp. Path.* **37**, 129; (1964) *Bact. Rev.* **28**, 30; (1966) *Ibid.* **30**, 739; (1968) *Progr. med. Virol.* **10**, 194; (1972) *Symp. Soc. gen. Microbiol.* **22**, 373; (1977) *The Pathogenesis of infectious Disease*. Academic Press, London.
Mims, C. A. and Gould, J. (1978) *J. gen. Virol.* **41**, 143.
Minta, J. O. and Movat, H. Z. (1979) *Curr. Top. Path.* **68**, 135.
Mogensen, S. C. (1979) *Microbiol. Rev.* **43**, 1.

Morahan, P. S., Morse, S. S. and McGeorge, M. B. (1980) *J. gen. Virol.* **46**, 291.

Mulder, J. and Hers, J. F. Ph. (1972) *Influenza.* Wolters-Noordhoff, Groningen.

Nagai, Y. *et al.* (1979) *J. gen. Virol.* **45**, 263.

Nakamura, N. and Compans, R. W. (1979) *Virology* **95**, 8.

Nathanson, N., Monjan, A. A., Panitch, H. S., Johnson, E. D., Petursson, G. and Cole, G. A. (1975) *Viral Immunology and Immunopathology.* Academic Press, London.

Nedrud, J. G., Collier, A. M. and Pagano, J. S. (1979) *J. gen. Virol.* **45**, 737.

Newton, A. A. (1970) *Symp. Soc. gen. Microbiol.* **20**, 323.

Notkins, A. L. (1974) *Cell. Immunol.* **11**, 478.

Notkins, A. L., Mergenhagen, S. E. and Howard, R. J. (1970) *Annu. Rev. Microbiol.* **24**, 525.

Ogra, P. L., Morag, A. and Tiku, M. L. (1975) *Viral Immunology and Immunopathology.* Academic Press, London.

Oldstone, M. B. A. and Dixon, F. J. (1975) *Viral Immunology and Immunopathology.* Academic Press, London.

Palese, P. (1977) *Cell* **10**, 1.

Pelton, B. K., Duncan, I. B. and Denman, A. M. (1980) *Nature, Lond.* **284**, 176.

Platt, H. (1970) *Resistance to Infectious Diseases.* Modern Press, Saskatoon.

Porter, D. D. (1975) *Viral Immunology and Immunopathology.* Academic Press, London.

Porter, D. D., Muck, K. B. and Prince, G. A. (1980) *Clin. Immunol. Immunopathol.* **15**, 415.

Poste, G. (1972) *Int. Rev. Cytol.* **33**, 157.

Potter, C. W. and Oxford, J. S. (1977) *Chemoprophylaxis and Virus Infections of the Respiratory Tract.* CRC Press Inc., Ohio.

Preston, K. J. and Garland, A. J. M. (1979) *J. Hyg., Camb.* **83**, 319.

Price, R. W. and Schmitz, J. (1978) *Infect. Immun.* **19**, 523.

Rabinowitz, S. G., Dal Canto, M. C. and Johnson, T. C. (1977) *J. infect. Dis.* **136**, 59.

Rabinowitz, S. G. and Hupriker, J. (1979) *J. infect. Dis.* **140**, 305.

Rawlins, M. D. and Cranston, W. I. (1973) *The Pharmacology of Thermoregulation.* S. Karger, Basel.

Rawls, W. E., Desmyter, J. and Melnick, J. L. (1968) *J. Amer. med. Ass.* **203**, 627.

Rawls, W. E. and Tompkins, W. A. F. (1975) *Viral Immunology and Immunopathology.* Academic Press, London.

Reeve, P. and Poste, G. (1971) *J. gen. Virol.* **11**, 17.

Rima, B. K., Davidson, W. B. and Martin, S. J. (1977) *J. gen. Virol.* **35**, 89.

Rima, B. K. and Martin, S. J. (1976) *Med. Microbiol. Immunol.* **162**, 89.

Roberts, N. J. (1979) *Microbiol. Rev.* **43**, 241.

Rose, J. K., Trachsel, H., Leong, K. and Baltimore, D. (1978) *Proc. nat. Acad. Sci., Wash.* **75**, 2732.

Rosemond-Hornbeak, H. and Moss, B. (1975) *J. Virol.* **16**, 34.

Rosenthal, L. J. (1979) *Canad. J. Microbiol.* **25**, 239.

Rott, R. (1979) *Arch. Virol.* **59**, 285.

Rott, R. and Klenk, H.-D. (1977) *Virus Infection and the Cell Surface.* North Holland Publishing Co., Amsterdam.

Rott, R., Orlich, M. and Scholtissek, C. (1979) *J. gen. Virol.* **44**, 471.

Rowlands, D. J., Sangar, D. V. and Brown, F. (1971) *J. gen. Virol.* **13**, 141.

Santoli, D., Trinchieri, G. and Koprowski, H. (1978) *J. Immunol.* **121**, 532.

Sarmiento, M., Haffey, M. and Spear, P. G. (1979) *J. Virol.* **29**, 1149.

Sawyer, W. D. (1969) *J. infect. Dis.* **119**, 541.

Scheid, A. and Choppin, P. W. (1974) *Virology* **57**, 475.

Scholtissek, C. and Klenk, H.-D. (1975) *The Influenza Viruses and Influenza.* Academic Press, London.

Scholtissek, C., Rott, R., Orlich, M., Harms, E. and Rohde, W. (1977) *Virology* **81**, 74.

Scholtissek, C., Vallbracht, A., Flehmig, B. and Rott, R. (1979) *Virology* **95**, 492.

Schrom, M. and Bablanian, R. (1979*a*) *J. gen. Virol.* **44**, 625; (1979*b*) *Virology* **99**, 319.

Schulze, I. T. (1975) *The Influenza Viruses and Influenza.* Academic Press, London.

Smith, H. (1972) *Bact. Rev.* **36**, 291; (1977*a*) *Virus Infection and the Cell Surface.* North Holland Publishing Co., Amsterdam; (1977*b*) *Bact. Rev.* **41**, 475; (1978) *Essays in Microbiology.* John Wiley and Son, Chichester; (1980) *Scand. J. infect. Dis.* **24**, 119.

Stevens, J. G. and Cook, M. L. (1971*a*) *J. exp. Med.* **133**, 19; (1971*b*) *Science* **173**, 843.

Stevens, J. G., Nesburn, A. B. and Cook, M. L. (1972) *Nature, New Biology* **235**, 216.

Stitz, L. and Schellekens, H. (1980) *J. gen. Virol.* **46**, 205.

Stoner, H. B. (1972) *Symp. Soc. gen. Microbiol.* **22**, 113.

Stoner, H. B. and Threlfall, C. J. (1960) *The Biochemical Response to Injury.* Blackwell Scientific Publications, Oxford.

Stringfellow, D. A., Kern, E. R., Kelsey, D. K. and Glasgow, L. A. (1977) *J. infect. Dis.* **135**, 540.

Sweet, C. and Smith, H. (1980) *Microbiol. Rev.* **44**, 303.

Sweet, C., Toms, G. L. and Smith, H. (1977) *Brit. J. exp. Path.* **58**, 113.

Swift, P. G. F. and Emery, J. L. (1973) *Arch. Dis. Child.* **48**, 947.

Taylor-Robinson, D. (1973) *Airborne Transmission and Airborne Infection.* Oosthoek Publishing Co., Utrecht.

ter Meulen, V., Hall, W. W. and Kreth, H. W. (1978) *Persistent Viruses.* Academic Press, London.

Tobin, J. O'H., Jones, D. M. and Fleck, D. G. (1977) *Infections and Pregnancy.* Academic Press, London.

Tsai, K.-S. (1977) *Infect. Immun.* **18**, 780.

Waterson, A. P., Pennington, T. H. and Allan, W. H. (1967) *Brit. med. Bull.* **23**, 138.

Webb, H. E. and Hall, J. G. (1972) *Symp. Soc. gen. Microbiol.* **22**, 383.

Webster, A. J. F. (1970) *Resistance to Infectious Disease.* Modern Press, Saskatoon.

Weiner, H. L., Powers, M. L. and Fields, B. N. (1980) *J. infect. Dis.* **141**, 609.

Wheelock, E. F. and Toy, S. T. (1973) *Advanc. Immunol.* **16**, 123.

Wisse, E. (1970) *J. ultrastruct. Res.* **31**, 125.

Wohl, M. E. B. and Chernick, V. (1978) *Amer. Rev. resp. Dis.* **118**, 759.

Woodruff, J. F. and Woodruff, J. J. (1975) *Viral Immunology and Immunopathology.* Academic Press, London.

Worthington, M. and Clark, R. (1971) *J. gen. Virol.* **13**, 349.

Zisman, G., Hirsch, M. S. and Allison, A. C. (1970) *J. Immunol.* **104**, 1155.

85

Epidemiology of viral infections

Charles Stuart-Harris

Introductory

The science of epidemiology is based on the study of health and disease as it occurs in the community either in groups of persons or the entire population. Its aim is to define the circumstances under which disease occurs and the factors influencing the latter's rate of development in the community. Applied to infectious disease, the study by epidemiological methods has provided knowledge of the transmission and spread of infection through the community and of the relative susceptibility or resistance of individuals to infection. Though its study may properly be said to have begun with the collection of statistics relating to the influence of age, sex and locality on the rate of mortality in the population, the observation of patients with illness and the use of indirect methods of determining morbidity have been of great significance. Even before bacteriological methods were available, it was a London physician, John Snow, and a country doctor in the west of England, William Budd, who provided the key observations from which the transmission by contaminated water of the diseases respectively of cholera and typhoid fever were deduced. In more modern times Walter Reed showed the insect transmission of yellow fever and the filtrability of the infectious agent at a time when very little was known about viruses as agents of human disease.

The contribution of the laboratory worker to the accuracy of epidemiological data has, however, become immense. The value of laboratory results in differentiating human diseases bearing a clinical resemblance yet possessing different causal agents is now accepted. Of all the many different agents causing acute respiratory illnesses in man, the first virus to be transmitted to laboratory animals, the influenza virus A, opened the door to the separation of epidemic influenza from the diseases caused by other agents. Recognition successively of the adenoviruses, the paramyxoviruses, rhinoviruses and coronaviruses has enabled this differentiation to continue and has permitted the construction of a picture of their different epidemiological patterns. Again, the laboratory has uncovered the frequent occurrence of subclinical virus infection so that the chain of transmission of the virus can be established more certainly and the total amount of infection in the community can be measured. Some of the work of the epidemiologist is, therefore, the detection of infection and in this the laboratory is an essential aid to the clinician and statistician. In this chapter the epidemiology of virus infections as aided by the laboratory will be described in general terms, specific conditions being described individually, as required, because of particular aspects in the transmission of infection or evolution of their community infections. (See also Chapter 48 on general epidemiology).

Methods of study

Epidemiological studies require methods which are applicable to the populations of entire countries, or to small community groups, since their aim is to acquire information from which a comprehensive account of disease in the community under observation can be constructed. No matter how large or small the scope of the study may be, epidemiological inquiries are based first upon the accurate ascertainment of illness. A single cross-sectional study of a community conducted at a particular time should provide the number of those who are infected and ill, those who are infected without illness (subclinical infection) and the total population from which the cases of infection were derived. The ratio of the numerator, the number ill or infected, compared with the denominator, the total population at risk, gives the prevalence of disease or infection in the community at the time of the inquiry. When surveillance of the same community is continued, it is possible to count the number of new cases of infection over a selected period of time. This number, compared with the total community, gives the incidence or 'attack rate' of new cases over this time. Usually incidence is expressed as an attack rate per hundred or thousand or more over a specified period of weeks, months or years. By including personal records of age, sex, location and other characters of both infected persons and the total population, attack rates by age and sex and so on can be constructed. In the case of illnesses in which the clinical diagnosis alone is unreliable, surveillance requires additional laboratory findings.

Thus, the clinical diagnosis of influenza virus infection is often unreliable because of diagnostic confusion with other respiratory virus infections. Ascertainment of infection in the community can be pursued by determining the frequency of influenza viruses in throat swabs from persons with acute respiratory illnesses collected at random from a defined population. These will give an assessment of the number of illnesses yielding viruses but inevitably will fail to reveal subclinical infections. To estimate the latter, serum samples are obtained from healthy persons and also those with respiratory infections to provide evidence of recent infection. The method of surveillance of influenza used by the Public Health Laboratory Service in England and Wales is based upon a register of all new acute respiratory illnesses reported by each practitioner from each of 60 different general practices. On the estimated attack rate on the total practice population, which is approximately 150 000, the rate of ascertainment of influenza virus infection from a parallel laboratory survey is superimposed so that a virologically estimated rate of influenza virus infection can be deduced (Public Health Laboratory Service 1977). In Houston, Texas, the rate of community infection with influenza viruses has been ascertained by the alternative method of testing all persons with acute febrile respiratory tract illnesses attending selected primary care clinics or emergency treatment centres of hospitals (Glezen and Couch 1978). When such methods are used, the total prevalence of influenza virus A infection in the community is found to exceed the total number estimated on a clinical diagnostic basis.

The study of localized outbreaks

Epidemiological studies of an outbreak of illness in a particular locality or a population group such as persons resident in an institution, a camp or barracks or a school, are perhaps the commonest of all inquiries at the present time. They are conducted when the infection is one of a dangerous illness with a high risk of death. Control measures may then be put into force by quarantining contacts in order to limit the transmission of infection or by raising the level of immunity by immunization.

When the outbreak is not one of a dangerous infection, the method of inquiry to be pursued is essentially a field study of the whole or of a large sample of the community. Studies to identify the infectious agent may then be supplemented by serological tests to determine how widespread the infection has been in the past and the proportion of susceptible persons remaining uninfected.

Serological epidemiology

The use of population surveys by the examination of sera collected at random and tested by various methods for the presence of antiviral antibodies received a great impetus from two groups of American authors. The first, John Paul, used the technique for the neutralization of poliovirus with sera collected in Alaska in 1949. He showed (Paul and Riordan 1950) that sera from persons of different ages were sharply different in regard to the presence of antibody to poliovirus type II. Those less than 20 years of age did not possess serum neutralizing antibody to the virus whereas those aged 20 years or more usually possessed antibody. Paul deduced from this that an epidemic of poliomyelitis occurring 20 years earlier had been caused by poliovirus type II which did not persist further in the community. Others have used similar tests with all three polioviruses to determine the age of acquisition of antibody in different communities (Melnick *et al.* 1955).

The second group of workers who used serological screening on persons of different ages was the Ann Arbor influenza group led by Francis and his associates (1953). Their pioneer studies showed that antibodies detected by the haemagglutination-inhibition test with each of the major serotypes of influenza A virus differed in their age-distribution. A study of the latter showed that antibody acquired in infancy by exposure to the particular haemagglutinin antigen of the influenza A virus then circulating persisted throughout life. Moreover antibody was not usually present in sera from those born after the virus A had undergone antigenic variation and had been replaced by a virus of a different subtype bearing a different haemagglutinin. These authors named this orientation of the antibody response to the infecting virus of infancy, the 'original antigenic sin' of the human species towards influenza virus. This phenomenon can be explained on the basis of the property of B together with T lymphocytes to exhibit an immunological memory (Virelizier *et al.* 1974). Reference to the use of this technique for discerning the viruses of past influenza pandemics by serological surveys is made below. Serological surveys of influenza virus antibodies using current viruses have also been made to enable a prediction of the probable occurrence of future epidemics of the strains of virus A and B used in the test. This method of epidemic prediction was first developed in the USSR (Zhdanov 1967).

Serological surveys with arboviruses have been used extensively in the past to detect the occurrence of infection in areas either known to be infected or apparently free from the viruses used. Such screening was of great value in yellow fever research and was used to indicate the desirability of introducing measures to control the mosquito population or to offer immunization with yellow fever vaccine.

Screening of sera from various community groups for the hepatitis B virus antigen or of antibody against it has been used to compare the prevalence of infection in different countries as well as to detect sources of possible transmission of the virus. It was by this means that the virus was found to be much more prevalent in African countries than in Europe or the USA. Many more examples exist of the value of sero-epidemiological studies, as, for instance, the antibody distributions to hepatitis A and B viruses in South Pacific communities that exhibit considerable differences in prevalence (Wong *et al.* 1979).

The characters of epidemics

An infection which is endemic in a community is present throughout the year or a major portion of the year at a more or less unvarying rate of prevalence. If the rate increases above its previous or expected level, the infection is said to have become epidemic. Yet this usage of the word is more readily justified when either the size of the increase or the rate at which it has occurred makes the change obvious. In fact, the word 'epidemic' covers a variety of experiences and it is difficult to formulate a definition to embrace occurrences ranging from a dozen or so cases to one of many thousands of illnesses. The localized epidemic, often called an outbreak, may in any case conceal a much larger amount of community infection occurring subclinically in the families or neighbourhood of the manifestly ill persons. Thus a single case of paralytic poliomyelitis has become recognized as the signal of a threatened outbreak; and measures to control the spread of virus by giving oral poliovirus vaccine to contacts and to those in the neighbourhood of the patient may be initiated.

When the daily number of new cases of illness rises rapidly to a peak and then falls to its pre-epidemic level the epidemic may be termed explosive. Such outbreaks may follow the simultaneous infection of a large number of susceptible persons from a single source spread by food or water. But explosive epidemics can also develop when the infection exhibits a short incubation period of 24 to 48 hours, is highly infectious and threatens many susceptible persons. Thus in semi-closed residential communities airborne infections of the respiratory tract, such an influenza, often produce an explosive epidemic. So many persons fall ill, for instance, from influenza in a residential school during only a few days that it is difficult to postulate that the virus has passed from person to person. This difficulty arising from explosive outbreaks also occurs with swine influenza, and the possibility of an inapparent pre-seeding of virus before the actual appearance of illness was originally suggested by Shope (1944). Shope's own work (1941) on the provocation of virus multiplication after latent infection of swine by virus carried on lungworms, though unconfirmed, was a possible basis for spread of virus before an outbreak. Hope-Simpson (1979, 1981) has argued for the existence of latent infection by human influenza virus as an explanation of the influence of season upon prevalence but such latency has never been demonstrated directly.

The kinetics of epidemics

(a) Fluctuations in prevalence: Seasonal and periodical changes

Changes in the prevalence of an infection at regular or irregular intervals are of two forms. The first is a variation in prevalence according to the season of the year. Some infections increase during the summer and others during the winter, and the peak season is related to the mode of transfer and source of virus. In vector-borne infection it is the season that encourages the breeding of the vectors which exhibits increased epidemicity. In faecally-spread outbreaks the summer and autumn are favoured. The majority of airborne virus infections increase during the winter and are at their lowest level in the summer in temperate climates. But in the tropics, where the temperature does not vary greatly in the different seasons, it is the rainy season which favours respiratory infections.

Accordingly, where infections are spread by droplets or droplet nuclei, it is the effect of the temperature and the humidity in favouring the survival of virus which appears to be the chief environmental reason for the seasonal increase in prevalence. The respiratory myxoviruses, influenza and the paramyxoviruses, are thermolabile at room temperature and the decay of virus in droplets held under conditions of different relative humidity with constant temperature has been found to be more rapid at high than at low relative humidity levels. Loosli *et al.* (1943) and Schulman and Kilbourne (1963) have shown this to be the case also in studies on experimental influenza virus infection in mice. Hemmes *et al.* (1960) suggested that indoor heating caused a reduction in the relative humidity indoors which would improve the survival of virus. Tyrrell (1965) favours the view that the seasonal rise in incidence of common colds is due to an increase in the efficiency of transfer of virus. In the winter people spend longer hours indoors perhaps in poorly ventilated rooms. They run a greater risk of exposure to airborne infection because of social contacts; and when heating is inadequate they may huddle together. The degree of success in infecting volunteers with secretions containing rhinoviruses at the Common Cold Research Unit of the Medical Research Council varied within only narrow limits, being slightly lower in winter than in summer (Roden 1963). It is important also to realize that the secondary attack rate of colds is only doubled in winter compared with the rate in summer (Brimblecombe *et al.* 1958); this could readily be the result of increased virus transfer.

Although these are valid reasons for regarding the seasonal prevalence of respiratory infections as being due to the environment, it must be admitted that a successfully spreading virus, such as the Influenza A/Asian virus of 1957, appears to be able to ignore the season. Thus the H2N2 virus, which originated in China in March 1957, reached Singapore in April and Indonesia, the Pacific Islands and Japan in May, and was able to cause outbreaks without interference from weather or climate. Also the first wave of the 1918 epidemic caused epidemics of influenza in western Europe in June, though its more virulent outbreak developed in the late autumn and following spring of 1919.

From time to time it has been suggested that individual host factors are concerned in the seasonal prevalence of infection. This may be interpreted in relation to the spread of faecally-transmitted viruses which are prevalent in the summer or autumn. The host is not so much adversely affected by the higher temperature at such times as by the change in eating habits with a preference for uncooked cold food, with more frequent use of swimming-pools and greater exposure of food to fly-borne contamination. All these changes favour the transfer of virus. But the opposite season of the winter involves chilling of the host and this may reduce the ciliary activity concerned in the disposal of airborne organisms or particulate matter deposited on the surface of the respiratory epithelium. However chilling of volunteers has not been shown to increase their susceptibility to infection by common cold viruses. (Jackson *et al.* 1960).

The flora of the respiratory tract alters seasonally and *Haemophilus influenzae* can be more readily isolated from the human trachea at post mortem in the winter than during the summer (Rosher and Cole 1939). Such apparent colonization of the lower respiratory tract is not, however, a factor in the transmission of the primary virus infection, though it may alter the outcome in the individual. It must be clear from the above that there is still doubt concerning the basis for the seasonal alteration in the prevalence of respiratory illnesses, which is such a constant and striking characteristic.

The second major variation in the prevalence of virus infections is that referred to as periodicity. By this is meant that, excluding seasonal changes, the size of an epidemic varies from one year to the next so that major epidemics occur at regular intervals, as is the case with measles, or less regularly with influenza. The biennial character of epidemics of measles in large industrial populations has been widely recognized. Alternate years between major prevalences are occupied by sporadic cases and small outbreaks, but the 18 months interval between the end of the major epidemic and the start of the next one is fairly constant. Such regular periodicity in England and Wales was interrupted in the early years of the second world war, when epidemics occurred in two successive years, 1940 and 1941 perhaps because children were evacuated from cities to the countryside. Another interruption occurred in 1968–69 two years after the previous peak in 1966–67 (Figure 85.1). This change was attributed to the introduction of immunization in infants in 1968 and its continuation thereafter. One and a quarter million doses of live measles vaccine had been used by March 1969 when a temporary interruption occurred in immunization. Over the next few years, fewer children received vaccine in proportion to the number of births and biennial periodicity with a series of lesser epidemics returned. There is general agreement that biennial periodicity results from the exhaustion of susceptible hosts during a major wave of measles and the disease does not reappear until there are enough newly susceptible infants. The solid immunity resulting from measles in childhood is another requisite factor in the equation.

Much speculation and mathematical theory have been pressed into use to explain the genesis of epidemics and their decline; those interested should consult Bailey's *Mathematical Theory of Infectious Disease* (1975). Basically, the shape of an epidemic wave turns on a relatively simple mathematical expression $R = \lambda SI$, where R is the number of new cases expressed as an incidence rate and S and I represent the concentrations of susceptible and immune persons in the community. The constant (λ) is the average number of new cases produced by a single infectious person (transmission parameter).

Mathematical models have also been developed to predict the course of influenza epidemics and success has been obtained in fitting a calculated set of weekly data to actual epidemic experiences in the past. In the USSR, Baroyan *et al.* (1977) base their predictions on the weekly notifications of influenza in persons attending polyclinics. A surprising finding is that the product of the proportion of susceptible persons and virus transmissibility on which the spread of an epidemic within a city depends is the same for every city in the USSR thus studied. Spicer (1979) has used the weekly deaths from influenza and influenzal pneumonia in England and Wales to calculate the expected deaths by a formula predicting the number of deaths likely week-by-week. He was able to fit expected deaths to the observed deaths during epidemics experienced between 1958 and 1972 even though the indicator of influenza virus activity in the community was the indirect one of mortality.

The periodicity of influenza epidemics is now known to be related in part to the occurrence of anti-

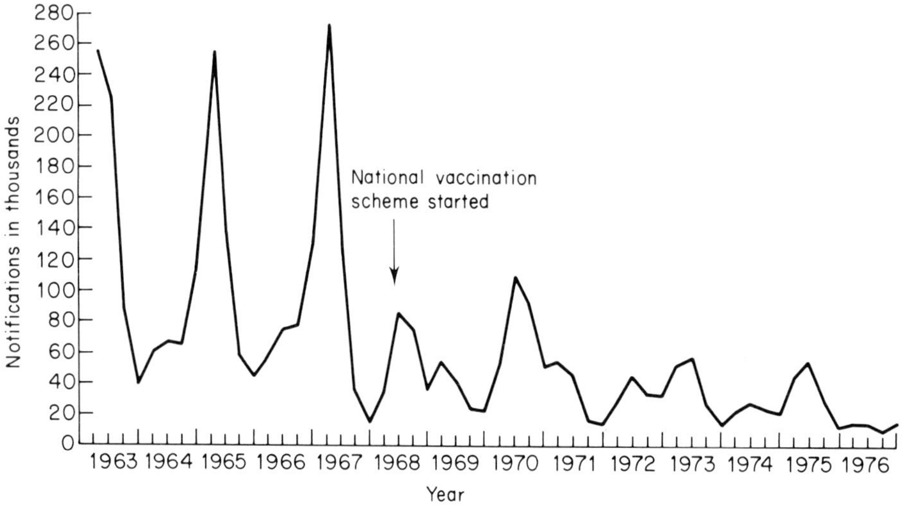

Fig. 85.1 Quarterly notifications of measles in England and Wales from 1963 to 1976. (Courtesy of the Department of Health and Social Security, London.)

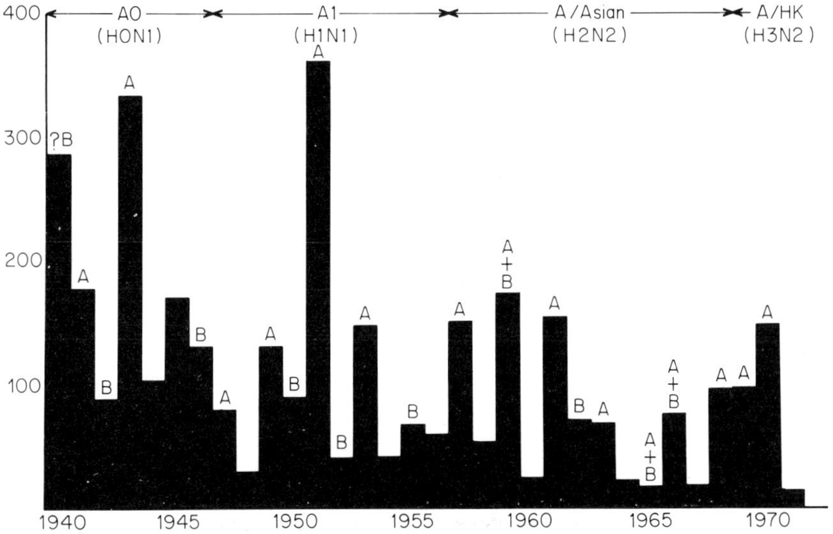

Fig. 85.2 Influenza death rates per million 1940 to 1970, England and Wales. Prevalence of subtypes of influenza virus A shown thus [←—→]. Occurrence of influenza A or B shown in relation to annual deaths. (After Stuart-Harris and Schild 1976.)

genic variation in the influenza viruses prevalent in the community. This variation is of two varieties, a step-by-step change termed *antigenic drift* whereby the surface antigens of haemagglutinin (H) and neuraminidase (N) change slightly or more considerably every one to three years in the case of influenza A and at longer intervals in influenza B. The second antigenic variation is a sudden appearance, usually in one geographical area, of an influenza A virus with a haemagglutinin or a neuraminidase antigen widely divergent from those of the previous virus. In the past, these large changes known as *antigenic shift* have occurred at not less than ten year intervals, the new virus antigens being unfamiliar (Fig. 85.2). It is because of the latter that the serological immunity of the population is readily breached, the virus spreads widely and causes both explosive outbreaks and more protracted epidemics. These may develop as spreading epidemics throughout the world so that pandemics are produced either soon after the new virus has appeared or within a year or so. There is thus no doubt that pandemics are related to antigenic variation but the larger epidemics in between pandemics can also be interpreted as the result of major divergence in antigens resulting from drift. The largest pandemic of influenza ever experienced, which caused 10 to 20 million deaths, occurred in 1918–19, long before the first isolation of an influenza virus. In consequence its aetiology remains unproved though serological evidence in man suggests that the Hsw1.N1 virus that Shope recognized in swine in 1931 is a persisting reservoir of the 1918 virus (see below).

(b) The variation in observed attack rates during epidemics

Although it is clear from observing epidemics and from experimental studies that the ratio of susceptible to immune persons in a community is the essential element in determining the form of an epidemic, other factors connected with the host and the causative virus also play a part. This is obvious when a comparison is made of the attack rate in similarly constituted communities caused by the same agent. Often there are considerable differences in these rates which are not readily explained. Examples are numerous but those of influenza outbreaks already mentioned in relation to explosive epidemics may be quoted.

The semi-closed residential communities, such as those of servicemen, institutions or schools, experience attack rates during explosive epidemics which are much higher than those found in the same age group of persons living at home. This has often been described as the effect of crowding together susceptible persons which may facilitate the transfer of airborne infection. Yet in these very communities explosive epidemics of influenza A or B exact a different toll in different places which cannot be explained on the basis of susceptibility or of crowding. Thus in 1957, the antigenically novel A/Asian virus, with its previously unknown antigens haemagglutinin H2 and neuraminidase N2, caused large outbreaks in many of the residential 'public schools' in England and Wales where boys and girls aged 13 to 18 were housed. The rates varied from 40 to 80 per cent according to the Committee on Influenza Virus Vaccine of the Medical Research Council (1958). In the analogous first year's experience of the new subtype A/Hong Kong/68

(H3N2) virus in 1968–69, schools similar to those just mentioned experienced outbreaks with attack rates ranging from 27 to 82 per cent (Tyrrell *et al.* 1970). At this particular time the community incidence of A/Hong Kong infection was comparatively low though protracted in duration and the contrasting behaviour of the infection in the different schools and between the schools and the community was remarkable. Influenza virus B causes a similarly varied incidence in schools with rates in any one epidemic ranging from 14 to 60 per cent (Stuart-Harris and Schild 1976). As far as the influenza A outbreaks quoted are concerned, it can be assumed that there was no humoral immunity specific to the 'new' haemagglutinins before the epidemics developed and the scholars should have had approximately the same degree of susceptibility to the virus. It therefore seems likely from these experiences that the transmission of virus even during such explosive outbreaks in susceptible hosts may somehow be interrupted.

It is a fact that the general population hardly ever experiences attack rates as high as those quoted in residential groups. The Asian epidemic in the autumn of 1957 caused the highest rates for the previous several years, but Fry (1958) found that only 17 per cent of his suburban practice population then experienced clinical illness. Woodall and colleagues (1958) made a special study in another suburban area near London and recorded an average rate of clinical illness of 30 per cent during the same epidemic, the highest rate, (49 per cent), being in children 5 to 14 and the lowest, 12 per cent, being in those aged 60 and over. Subclinical infection requires serological surveys for its detection and Clarke *et al.* (1958) found that 44 per cent of sera from persons of different age after the same epidemic had haemagglutination-inhibiting antibodies to the A/Asian virus, 4 per cent only being positive in a pre-epidemic collection from the same general population. Accordingly, unless infection, which fails to register by means of a serum antibody response detected by the tests used, is prevalent, at least half the community escapes infection during epidemics even when the assault of a 'new' virus is apparently fierce. This again suggests that the transmission of infection due to influenza virus ceases before the susceptible portion of a community is fully saturated. The corollary can be deduced that the multiplicity of contacts within a residential community favours the spread of the virus and that infection at night-time may be a significant factor in this enhancement.

(c) The age incidence of infections

Two major variations occur in the observed age-incidence of different virus infections. In the one, infection is confined to infancy and early childhood, and in the other the disease occurs in persons of all ages, though its actual frequency may vary in the different decades. These variations are largely concerned with host factors and particularly with immunity. Infections which produce durable, perhaps life-long immunity, the viruses of which are always present in the community, characteristically occur in childhood. Thus *varicella* and *measles* are found maximally in those aged 1 to 4 and only occasionally cause infection in adult life. The introduction of measles vaccine has altered the age incidence of the residue of measles infections and in the USA outbreaks in the 1970s have affected older children. Thus among more than 55 000 notified cases of measles in 1978, 60 per cent of cases occurred in children and adults 10 years of age and over (Report 1979). This shift in age incidence may reflect the fact that immunization of children in their first or second year of life had hindered the transmission of virus, so that those who for some reason or other were not vaccinated in infancy acquired the infection later in childhood or even in adult life.

Measles is also noteworthy in regard to the effectiveness of maternal immunity so that childhood infection rarely occurs before 6 months of age. The duration of the immunity which follows an attack of measles was well shown by the experience of an epidemic of the disease in 1846 in the Faröe Islands recorded by Panum (1846). The epidemic affected all age groups, 6000 of the 7782 islanders being attacked in the first six months of the outbreak. Among the population who did not suffer from the disease were a number over the age of 65 who had contracted measles during the last previous outbreak in 1781. Quarantine had been practised from 1781 onwards and this apparently kept the islands free from infection over the intervening years. The uniformity of the attack rate at all ages is a striking instance of the ability of measles to attack adults in the absence of immunity resulting from childhood infection.

The occurrence of infection at all ages is a feature of many of the respiratory viruses and notably influenza and the rhinoviruses. *Influenza* outbreaks are characteristically accompanied by a higher attack rate in childhood than in adult life, the highest incidence being from 5 to 15 years of age. However, because infants under 5 years of age may be only mildly ill, influenza in this age group is often missed, or diagnosed as croup, and it is only when investigations are made on throat swabs or upon sera that the rate of infection may be found to be as high or higher in this age group than in older children. Adult influenza is often regarded as evidence that the viruses produce only temporary immunity. Yet it is now apparent that adult infection should often be regarded as a consequence of antigenic variation either by drift or shift. The H1N1 subtype of influenza A viruses, which was prevalent from 1946 to 1957, disappeared but reappeared 20 years later. This recurrent virus of 1977 was found to be antigenically similar to that found in 1949 and 1950 (Kendal *et al.* 1978). It caused epidemics in China, the USSR, South-East Asia, western Europe and the Americas. Everywhere that the 1977 H1N1 virus appeared there were infections and epidemics in schools and colleges. In spite of a high attack rate in those less than 20 years of age, there was an almost complete lack of H1N1 influenza in persons greater than 22 years of age born before 1956. This resistance to illness was also exhibited in 1978 and indicates that immunity to influenza can persist under such conditions for many years.

Rhinoviruses also infect persons of all ages but the multiplicity of rhinovirus serotypes (Tyrrell 1968, Gwaltney 1975) circulating in the community probably accounts for the apparent impotence of acquired immunity.

The age incidence of *poliomyelitis* was formerly maximal in those less than 5 years of age, hence the old name – infantile paralysis. Large epidemics of the disease occurred in Scandinavia and the USA from 1910 onwards and their extent and clinical severity steadily increased. From 1940 onwards when poliomyelitis began to affect the military forces in the Middle East engaged in the second world war, cases of poliomyelitis during epidemics were increasingly found in older children and adults. From 1944 to 1950 almost equal proportions of cases occurred in England and Wales in those less than 5 years of age, those aged 5 to 14 and those of 15 years of age and over. As the likelihood of severe residual paralysis or death appears to increase in adult life, this postponement of the disease from infancy was one of its most alarming epidemiological features. The explanation for this change in age incidence appears from serological surveys to be the lack of specific neutralizing antibody particularly against type I virus in children and adults. The reason for this development is considered below.

The age incidence of infection by the gastro-enteritis viruses is not yet well enough defined for detailed comment. *Rotavirus* infections have been detected among babies in the perinatal period in hospital after delivery and infection in families exhibits a peak attack rate in infants 6 months to 3 years of age. However older children and adults are susceptible and may become infected during family or institutional outbreaks (Kapikian *et al.* 1976). The duration of immunity after infection is not known and the fact that the virus possesses more than one serotype might be a factor in permitting re-infection. The age incidence of *hepatitis A virus*, exhibited in the general population during epidemic prevalence in autumn and early winter, is highest in children aged 5 to 15. The seasonal prevalence led to the assumption that the virus was spread via the respiratory route. However wartime epidemics in military groups in the Middle East and elsewhere occurred at times when faecal contamination of food was most likely to be the source of infection. Also epidemics spread by eating shell-fish contaminated by sewage confirm the view that infection can be derived from faeces.

(d) Long-term trends in the prevalence of virus infections

The existence of secular changes in the incidence of infections, observed over an interval of half-a-century or so, has been more readily apparent in bacterial infections such as tuberculosis or pneumonia than in virus diseases. Nevertheless long-term trends in mortality have occurred and are particularly evident in measles and influenza. Measles was formerly a serious cause of infant mortality. The death-rate per million living was as high as 700 to 900 in major epidemics of the second half of the nineteenth century and was still from 300 to 400 per million in the alternate years. Even at the turn of the century the rate was as high as 400 per million; but thereafter a decline occurred. From a case-fatality rate based on notifications of measles of 20 to 30 per 10 000 cases, the numbers of deaths continued to fall. In 1961 a study of measles was conducted by the Public Health Laboratory Service (1963) on 132 deaths occurring in the first six months of that year. Half of these occurred in children with genetic disorders such as mongolism and mental retardation, and by 1968 when measles vaccine was introduced in England and Wales on a large scale, the previous epidemic of 1967 caused only 100 deaths.

The reasons for the fall in mortality from measles are not obvious. The decline occurred in the prechemotherapeutic era and, though chemotherapy may have accelerated the reduction after 1940, measles pneumonia is not necessarily responsive to antibacterial therapy. In African countries measles continues to exact a heavy toll in infants, particularly in those suffering from malnutrition of the type known as kwashiorkor. On these grounds it has been deduced that the improvement in the nutrition and general health in children during the twentieth century in western countries has been a major factor in reducing measles mortality. It is necessary to point out that good nutrition is not a guarantee of resistance to other virus infections. Foot-and-mouth disease characteristically attacks well nourished cattle, and the child affected with paralytic poliomyelitis when the latter was epidemic was often well grown and healthy.

The long-term trends in mortality from influenza are shown for England and Wales in the accompanying Fig. 85.3. The curiously low level of deaths from influenza in the period from 1850 onwards has never been explained, particularly as some epidemics were recorded during that period (Beveridge 1977). The low mortality was abruptly altered with the large 1890–91 epidemics believed to have been due to influenza A viruses of the Asian (H2N2) or Hong Kong (H3N2) antigenic subtypes. The enormous epidemics of the 1918–19 period dominate the entire scale of Fig. 85.3.

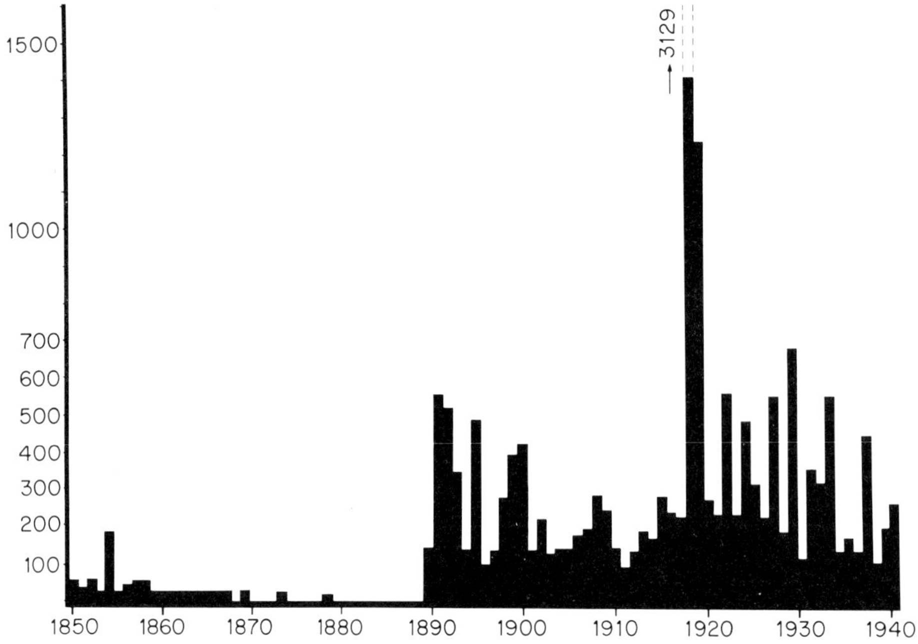

Fig. 85.3 Influenza death rates per million 1850 to 1940, England and Wales. (After Stuart-Harris and Schild 1976.)

The death rate formerly maximal in the period before 1918 in those over 60 altered dramatically in October 1918 in that large numbers of deaths occurred in young and middle-aged adults 20 to 40 years of age. The next largest epidemic in 1929 was accompanied by a death-rate approximating to 700 per million living, but the elderly section of the population bore the brunt of mortality. In the years from 1930 to 1947 there were several large epidemics, those of 1933, 1937 and 1943 being associated with the viruses of the WS/PR8 subtype (HON1) which were replaced by the H1N1 family in 1947. This subtype caused no apparent immediate impact upon mortality, the large outbreak of H1N1 infection in 1951 being four years after the virus was first detected in Britain. The first three years after the Asian (H2N2) and the Hong Kong (H3N2) subtypes of influenza A spread in the community were accompanied by increases in the mortality rate but in any one year this did not reach 200 per million living. There can be no doubt, however, that a fall has occurred in the mortality rates from influenza in the epidemics since 1929, even though the trend has been interrupted by periodic outbreaks probably related to the circulation of new antigenic subtypes.

Death from influenza is, of course, related in part to bacterial secondary infection of the lung; the pneumococcus is numerically the commonest of the respiratory organisms concerned. Chemotherapy against the pneumococcus first became available in 1939 and some effect would again be expected in the mortality from influenza with the introduction of penicillin after 1945 and of other antibiotics from 1951 onwards. In consequence it is difficult to attribute the changes in mortality since the second world war to trends in either host susceptibility or the virulence of the causative virus. One very significant factor which has become apparent, however, is that viruses with little impact on middle-aged or elderly adults such as influenza B or the H1N1 strains of 1977 and 1978 produce much less enhancement of the mortality rate.

Long-term trends in other virus infections are less apparent. The variations of vector-borne disease turn largely on changes in the size of the vector population, and of the infections which are faecally transmitted upon the presence or absence of good hygiene. Awareness of the importance of variation in the serotypes of the enteroviruses and respiratory viruses helps in the interpretation of sudden changes in the prevalence of illness such as aseptic meningitis or outbreaks of febrile respiratory disease. The appearance in the 1920s, and later disappearance of the world-wide epidemic of encephalitis lethargica has, at present, no explanation, for no causative virus was identified by the investigations then possible upon patients with this illness.

(e) Epidemics and human ecology

The influence of epidemic disease upon population dynamics is a subject which can be appreciated only with a historical perspective. The effect of the periodical incursions of bubonic plague into Europe was, of course, obvious at the times when they occurred and was a major cause of mortality. Similarly the impact of smallpox on the size of the population of Great Britain could have been appreciated in the eighteenth century by the contemporary opinion that sooner or later everyone caught it. But the fact that smallpox restricted the size of the population by the mortality which it occasioned, particularly among infants and children, was apparent only when the disease itself began to be controlled. Razzell (1977) thus considers that it was the deliberate inoculation with variola virus which preceded Jennerian prophylaxis, and not the Industrial Revolution that led to the burgeoning of the population in England late in the eighteenth century. The effect of the global eradication of smallpox by WHO is unlikely to be of similar dimensions. What new successes or perils may await the human race in the future cannot be foretold. The need for constant endeavour to maintain both epidemiological studies and a search for means of specific prevention and treatment of virus infections of man and of animals is obvious.

The transmission and epidemiology of human virus infections

It is proposed to describe here some of the features of these infections, including the transmission and character of infections by viruses which are exclusively concerned in human disease. A later section will describe the epidemiology of the viruses of vertebrates specifically related to human infection (Table 85.1).

Human and other animals infected by viruses contaminate their environment in ways which depend on the particular mode of shedding of the virus from the body surface. Thus there are three major routes for the transmission of viruses from infected to non-infected persons. These are transmission by physical contact with contaminated skin or mucous membranes; transmission by aerial droplets or dust bearing virus particles; and faecal-oral transmission of viruses excreted in the faeces. Infection can also be acquired from infected blood or by a foetus from the mother. Table 85.1 lists these routes and names the viruses concerned. There are instances where viruses infect the body in such a manner that they may subsequently spread by more than one route. Transmission by vectors is described in the section dealing with animal viruses.

(a) Infection derived by contact with the skin or mucosae

Variola and vaccinia

The simplest mode of virus transmission occurs when a susceptible person makes physical contact with a person from whose skin virus is being shed on dried skin scales or in crusts from actual lesions. The variola virus of *smallpox* spreads in this manner and close contact between an infected person and unimmunized contacts enables it to spread through the community as a narrow stream of infection. This occurs in spite of the possession of immunity by most of the population. Smallpox patients contaminate their clothes and bed-linen with dried saliva and infected skin particles but, though such dust might convey infection over short distances, more remote transmission has never been satisfactorily proved.

Smallpox in Britain ceased to be endemic in 1935 after ten years of outbreaks of alastrim (variola minor). Thereafter importations from Asia and elsewhere caused a series of outbreaks (Report Ministry of Health 1963). The eradication campaign was effected by intensive search for cases of variola and immunization of all possible contacts. Successively the continents and countries with endemic variola were thus freed from the virus until it was confined to Ethiopia and Somaliland by 1976. (Arita 1979).

The disappearance from the world of smallpox since October, 1977 was confirmed by WHO in October 1979. So far as present knowledge extends, the only sources of variola virus which remain are the four WHO - designated laboratories where the virus is stored or cultivated. The hazard of future outbreaks from such sources becomes more serious as vaccination with vaccinia virus is discontinued so that the community becomes more susceptible. The sad events of 1978 when a photographer in Birmingham contracted fatal smallpox as a result of laboratory infection underlined the problem of laboratory containment of this dangerous virus. The fact that this person had had no direct contact with the laboratory, which was located one floor below the one is which she worked, indicates the difficulty in deciding upon the vehicle for such laboratory infections. Meanwhile in Africa the disease monkeypox might be capable of being transferred from person to person as a variola-like disease should the virus undergo adaptation to man but this is very unlikely. (See Section on zoonoses, p. 142).

Vaccination with vaccinia virus is still being used in a few countries for the protection of health service workers and travellers. This practice should diminish with the passage of time. Meanwhile the hazard to persons, particularly children with eczema, of acquiring vaccinia virus by contact with a recently vaccinated person remains. Generalized vaccinia or the more

Table 85.1 Source, material and route of transfer of principal human viruses (other than vector-borne)

Viruses	Source	Material	Route
Poxviridae:			
Variola, Vaccinia	Skin	Dried crusts	Intimate contact
Molluscum contagiosum		scales	? air at close range
Herpesviridae:			
Herpesvirus types 1 and II	Skin	Vesicle fluid (mucosae)	Intimate contact Skin, genitalia
Varicella-zoster		Vesicle fluid (crusts)	
Papovaviridae:			
Warts	Skin	Dried scales	Immediate contact fomites
Respiratory viruses:			
Influenza A, B and C, Parainfluenza 1 to 4 Resp. syncytial (RS) Rhinovirus. Coronavirus. Adenovirus.	Nasopharynx	Nasopharyngeal secretions or sputum or saliva	Airborne droplets Droplet nuclei Fingers Handkerchiefs
Systemic viruses:			
Measles. Rubella. Mumps.	Nasopharynx	,,	,,
Herpesviridae:			
Herpesvirus type I Cytomegalovirus, EB virus Varicella-zoster	Nasopharynx	,,	,,
Picornaviridae. Enterovirus:			
Poliovirus I, II, III Echovirus Coxsackie A and B Adenovirus Hepatitis A	Faeces	Faeces Sewage	Faecal-oral to mouth via fingers Water, Food } contaminated Flies
Gastroenteritis viruses:			
Rotavirus, Norwalk agent, Astrovirus. Intestinal Coronavirus	Faeces	Faeces	Faecal-oral
Inoculation			
Hepatitis B	Blood	Blood serum or blood products	Direct inoculation (syringes) contaminated cuts, etc.
Cytomegalovirus	Blood		
Congenital			
Rubella Cytomegalovirus Variola, Vaccinia, Varicella	Maternal	Maternal blood	Transplacental

serious gangrenous form can occur also in persons receiving treatment with immunosuppressive drugs.

Varicella-zoster

Varicella-zoster virus is ubiquitous, particularly in children during the first six years of life, and those who escape infection may later contract chicken-pox as adults, when varicella pneumonia may be a complication. Persons under treatment by immunosuppression for malignant disease may develop zoster which is often more extensive than that which occurs in previously healthy persons as they grow old. The relation between chickenpox and zoster is discussed below in the section on latent virus infections (p. 139). Nosocomial infection with varicella occurs in childrens' wards in hospital and is particularly severe in children with maligant disease such as leukaemia. This is the principal reason for the prophylactic use of zoster convalescent immunoglobulin in such children; the use is also being explored of an attenuated live varicella virus vaccine for a similar purpose (Takahashi *et al.* 1974).

The other diseases mentioned in Table 85.1 as being spread by actual contact include some such as the poxvirus infection of molluscum contagiosum, herpesvirus types 1 and 2, and papillomatous warts in which skin lesions are the source of virus. The *herpesvirus type 1* is acquired early in childhood from skin vesicular lesions or from facial skin of adults contaminated by infected saliva. Recurrent infections occur from virus latent in nerve ganglia throughout life. *Herpesvirus type 11* is acquired by contact with genitalia contaminated by skin lesions in the male or by cervical secretions in the female.

It must be pointed out that only fairly stable viruses such as the enterovirus group of the *Picornaviridae* are able to resist environmental conditions outside the body sufficiently long for transmission to be derived from dust contaminated by skin or body secretions and carried by fomites or through the air.

(b) Airborne infection

Many viruses enter the body through inhaled droplets previously expelled by infected persons from the mouth or nose in the acts of talking, sneezing or coughing. The viruses concerned are those which infect the nasopharyngeal epithelium or the epithelium of the air passages from which they are wafted upwards to reach the nasopharynx. That they are expelled into the air on droplets is shown clearly by photographs taken during a brief exposure. The larger droplets have a short trajectory downwards to the floor or to the clothing. They may be heavily contaminated with virus and can convey infection over short distances, whereas the smaller droplets evaporate rapidly but persist as droplet-nuclei floating in the atmosphere for variable periods of time. Viruses that lose infectivity by drying are not thus transferred. A sneeze or a cough probably conveys more infection than does talking, even though the distance between speaker and listener may be short enough for direct inhalation of expired air. There are few experimental data concerning the aerial transmission of infection under controlled conditions. Andrewes and Glover (1941) found that ferrets housed so as to exclude the transmission of large droplets between them spread influenza through the air from one ferret previously infected with influenza virus to normal ferrets. Mice can also be infected by air contaminated by small droplets containing influenza virus.

The winter season clearly favours the transmission of all the respiratory viruses listed in Table 85.1. The most likely explanation for this is that in some way viral transfer is facilitated (Tyrrell 1965). Factors concerning the survival of virus particles in the air in an infective state were mentioned above in connection with the influence of season upon epidemics. Either the quantity of virus in respiratory secretions is greater in the winter season, or the winter condition of the air slows down the decay of the infectivity of the virus particles. It appears unlikely that the susceptibility of the host is enhanced by the season of the year sufficiently to account for the observed phenomenon.

Respiratory tract viruses

The many different virus families which cause acute illnesses of the respiratory tract and which are transmitted by airborne droplets are represented by seasonal prevalences experienced in all parts of the world. The incidence rises from September onwards in temperate climates with a peak between January and March and then it subsides to the nadir in August. Common colds, febrile sore throats (tonsillitis and pharyngitis), febrile illnesses including or resembling influenza, croup, laryngo-tracheitis, acute bronchitis, bronchiolitis and pneumonia are the principal syndromes, and any of the respiratory viruses may be concerned in particular patients (Stuart-Harris 1977). However in each syndrome one or more of the families predominates, as, for instance, respiratory syncytial virus as a cause of bronchiolitis in infants under 3 years old. Correlation of the result of cultivation of throat swabs or nasopharyngeal secretions in appropriate tissue cultures with the prevalence of acute respiratory disease in the community reveals a different picture from that seen by the clinician. This is because the smoothed-out curve of prevalence of individual syndromes appears as a series of discrete outbreaks by different viruses or different serotypes of the same virus (Hope-Simpson and Higgins 1969, Higgins 1974, Horn *et al.* 1975). Similar correlated studies in children with acute lower respiratory tract illness tell the same story, parainfluenza, and the influenza viruses A and B being chiefly found (Glezen and Denny 1973, Foy *et al.* 1973, Martin *et al.* 1978). In those patients with signs of pneumonia the bacterium *Mycoplasma pneumoniae* and organisms such as pneumococci and staphylococci causing secondary bacterial infection are commonly found in the sputum; a pure virus pneumonia is uncommon.

Nearly all these viruses cause their greatest impact on the children in the community. It is accepted that these, particularly when infants, are not only the most susceptible but also the members most concerned in the spread of viruses. The age-specific incidence of *RS virus* is maximal in infants less than 3 years of age, and babies with bronchiolitis are mostly less than one year of age. However there is clear evidence of recurrent infections by this virus in childhood (Henderson *et al.* 1979) and even in adult life, and it is now known that adults caring for babies with respiratory syncytial infection may themselves become infected and may suffer illnesses with a temporary impairment of the pulmonary function (Hall, W. J. *et al.* 1978). Similar recurrences probably occur with parainfluenza viruses during adult life.

Viruses such as *rhinoviruses*, which cause copious nasal secretions, contaminate the external nares, the face and the fingers. It is then possible for the fingers of another person in direct contact with the hands of an infected person to acquire the virus. Gwaltney *et al.* (1978) cultivated rhinovirus from the self-contaminated hands of a volunteer inoculated intranasally with rhinovirus. They also found virus on the hands of a person whose contact with the volunteer's hands lasted for only ten seconds. The rhinovirus is sufficiently stable to endure for some time outside the body, thus contrasting with other more labile respiratory viruses such as respiratory syncytial virus. Both rhinoviruses and coronaviruses cause common colds at all seasons of the year though their infections are more prevalent from September to May. It is hardly possible to define the period of maximum age incidence; children suffer more frequently than adults mainly because of their limited experience of the many rhinovirus serotypes. Adenoviruses, some of the serotypes of which are responsible for outbreaks in military groups, are less commonly cultured routinely from children but may be found as sporadic infections in both winter and summer. The respiratory illnesses associated with certain ECHO viruses or Coxsackie type A viruses are often indistinguishable from those due to other respiratory viruses which tend to cause family episodes of febrile illnesses. Infection by adenoviruses is considered again under faecally-transmitted viruses.

Death from the respiratory viruses occurs primarily in babies less than one year and in adults over 65 years of age. Persons with chronic illnesses of the lung and heart are particularly at risk from influenza, but an excess mortality, particularly during influenza A epidemics, is present from all causes as well as from respiratory causes of death. About two-thirds of the total mortality attributable to influenza occurred in those over 65 in all of the epidemics of the past 40 years. Persons living in retirement in homes or institutions may die during actual outbreaks of influenza. This does not mean that there is no risk to the aged from other viruses, for even respiratory syncytial virus has been isolated from respiratory illnesses in institutions housing geriatric patients. At the other end of life the perinatal period is one where babies are particularly susceptible to respiratory syncytial virus and also to adenoviruses. Both of these cause fatal illnesses with pneumonia in babies, though respiratory syncytial virus infection is not as deadly as adenovirus pneumonia in the first few days of life when maternal immunity is usually most effective. The view is gaining ground that all of the respiratory viruses cause recurrent infections throughout life and this contrasts greatly with the infections caused by viruses which spread throughout the body and are not confined to the respiratory tract.

Measles, rubella and mumps

The systemic virus infections of measles and rubella, though spread by the air as with viruses confined to the respiratory tract, are probably transmitted during the pre-eruptive phase of illness when children appear to be most infectious. These viruses are readily cultivable from nasopharyngeal secretions, and also from the saliva when mouth lesions are present. Measles, rubella and mumps are rarely encountered as a clinical illness more often than once in a life-time, thus indicating the effectiveness of acquired immunity. The age incidence and periodicity of measles and the effect upon these of immunization with measles vaccine has already been discussed.

Rubella occurs in epidemics at irregular intervals, the largest outbreaks being separated by an interval of perhaps nine years in countries such as the USA, where rubella has been notifiable for a long time. However, in the intervening years sporadic cases and smaller outbreaks are experienced in all countries. The age incidence of rubella covers a wider span than measles, namely from school entry with a peak between 5 and 10 years to adolescence and early adulthood. Serological studies before the introduction of rubella vaccine in 1970 in the UK showed that 20 per cent of girls aged 18 to 20 had no antiviral antibodies and were susceptible to infection. Selective immunization of up to 80 per cent of girls at 11 to 13 years of age has reduced this percentage to 10 or less but the problem remains of congenital rubella in infants born to women exposed to rubella during pregnancy and acquiring the infection with or without symptoms. Immunization of girls which leaves boys unprotected, has not led to a reduction in the epidemicity of rubella in the UK, and the risk of congenital rubella persists in spite of screening programmes to detect those who are sero-negative. In the USA, however, immunization of both male and female infants has greatly reduced epidemicity. In some countries such as the West Indies and in South-East Asia a larger proportion of women of child-bearing age are sero-negative than in the USA or UK.

Mumps is an epidemic infection primarily of school-children, but adults are affected during localized outbreaks and are most likely then to develop complications such as meningitis, orchitis or ovariitis.

Herpesviruses Viruses which appear regularly in the saliva, such as mumps and the herpesviruses of *herpes febrilis* (type 1), *cytomegalovirus* and *EB virus* of infectious mononucleosis, may be acquired from droplets, by direct contact as in kissing with exchange of saliva, or by contamination of the fingers. A good instance of the latter is the acquisition of herpesvirus type 1 infection by nurses as a result of contact with patients' mouths during toilet or surgical procedures. The resulting lesion which develops on their fingers resembles a whitlow (Stern *et al.* 1959). There is evidence that EB virus which seems frequently to cause infectious mononucleosis in children and adolescents is present in the saliva of infected persons in a cell-free state (Morgan *et al.* 1979).

(c) Faecally-transmitted infection

Viruses known to be excreted in the faeces are listed in Table 85.1. They can be transferred to the mouth of susceptible persons by contact of their fingers with the self-contaminated fingers of the infected person. They are found in sewage and thus may contaminate water supplies or food. Domestic flies may also convey virus from faeces to food. Faecal-oral infection is most often acquired during the summer or autumn as with hepatitis virus A and the enteroviruses.

Poliomyelitis

The *polioviruses*, like the adenoviruses, are found in the nasopharynx during the earliest phase of clinical illness but persist in the faeces for two or more weeks after the onset of illness. The early work of Kling and his colleagues in Sweden in 1912 indicated that virus was present at autopsy in both the nasopharynx and the contents and wall of the small intestine. Faecal excretion is prolonged, and sewage has many times been found to contain polioviruses. Indeed its examination can be used to monitor the presence of viruses in the community. This is particularly important because of the frequency of inapparent poliovirus infection. Viruses thus cultured can be examined also for their biological properties such as neurovirulence.

Modern sanitation has altered the epidemiology of faecally-transmitted virus infections. Instead of eliminating the clinical disease, it appears sometimes to have led to recurrent epidemics. This was certainly true of poliovirus infections which became epidemic in the earlier years of the twentieth century, as already described. This epidemic pattern experienced in the UK from 1946 onwards subsided when immunization, first with inactivated virus and then with live attenuated oral vaccine, was adopted routinely. Epidemics have begun even more recently in tropical countries including Asia where improvement in sanitation and hygiene of food and water has lagged behind that in western countries. Ubiquity of the viruses apparently prevented the occurrence of epidemics of poliomyelitis by promoting serological immunity which in turn prevented invasion of the blood and central nervous system. Thus, the paradoxical effect of improved sanitation by hindering the faecal-oral acquisition of poliovirus infection by infants has rendered persons more susceptible to infection later in life. Certainly without immunization recurrent epidemics of poliomyelitis have occurred in all countries with good hygiene, the virus spreading through the community in those susceptible to its invasion.

Other enteroviruses

The enteroviruses of the ECHOvirus and *Coxsackievirus* groups are spread by both the respiratory and the faecal route. It is not yet clear whether they have similarly become more serious as a cause of childrens' disease since the improvement in sanitation. Epidemics of aseptic meningitis associated with ECHOviruses and Coxsackie B viruses which occur seasonally and at intervals are fortunately benign and rarely cause death from encephalitis. The viruses are found at least as frequently in healthy as in ill children and unusual clinical manifestations such as myocarditis in the newborn, exanthemata with rubelliform or vesicular rashes and pericarditis are now known to be usually associated with particular serotypes of the ECHOviruses and Coxsackie A or B viruses. As with polioviruses, there is usually subclinical infection in healthy members of families where one member has been taken ill. Outbreaks occur primarily in the summer or autumn at seasons favouring faecal transmission, yet virus can be cultivated as readily from throat swabs in the early days of illness as from the faeces. Coxsackie A viruses mostly behave like respiratory viruses and cause family illnesses and outbreaks of febrile pharyngitis.

Adenoviruses

These cause respiratory infections in the winter and seem to be spread by swimming-baths in the summer. It is indeed difficult to be certain whether airborne or faecal infection is the more important for the spread of these viruses. They cause sporadic infections throughout the year with outbreaks of respiratory illnesses in military communities or residential schools during the winter season. In children they are responsible for about 7 per cent of respiratory illnesses (Brandt *et al.* 1969). The serotypes causing outbreaks, chiefly types 3, 4, 7, 14 and 21, differ from those found in tonsils and adenoids removed surgically (types 1, 2 and 5). The latter probably represent residual persistent virus following acute infection in infancy; the same types may be found in the faeces of infants in residential homes. An association exists between intestinal adenovirus infection and intussusception (Potter 1964).

Hepatitis A virus

Hepatitis type A virus is one of the three viruses associated with hepatitis. The disease occurs primarily in children as a community outbreak, though adults are susceptible and may suffer more serious illnesses. The frequency of subclinical infection in the community is believed to be high, for laboratory investigations have disclosed anicteric infections which are unaccompanied by jaundice. Though chronic active infection of the liver is an occasional sequel, fatalities rarely occur.

Hepatitis type A virus occurs epidemically in military groups under wartime conditions where faecal contamination is likely. Infection has also been traced to contaminated water and shell-fish. Volunteers have been infected by swallowing faecal extracts from jaundiced patients, but nasopharyngeal secretions have not been similarly infective. Now that the virus has been found to replicate in tissue cultures (Provost and Hilleman 1979) knowledge will doubtless be gained on the usual mode of transfer in normal communities. The fact that contact with an infected person may need to be only brief for hepatitis virus to be transferred was emphasized by Pickles (1939) in his study of rural communities in England. The incubation period is approximately three weeks.

Gastro-enteritis viruses

The groups of unrelated viruses now known to produce intestinal infection and symptoms such as vomiting, diarrhoea and abdominal pain are difficult to study because cultivation in tissue cultures is generally unsuccessful. In all probability these agents cause recurrent infections in childhood beginning in the neonatal period. Recurrence is sometimes due to a different serotype of the same virus but this is not invariable. Immunity after a particular alimentary virus infection may not be durable and, as in the respiratory tract, the turn-over of cells of the intestinal epithelium may lead to a return of susceptibility to viruses causing only surface infections. The intestinal coronaviruses, whose particles also occur in the faeces, may be excreted for weeks or months after infection (Clarke *et al.* 1979). (These viruses are not identical with the respiratory coronaviruses, (Tyrrell *et al.* 1975).

Presence of the viruses associated with gastro-enteritis in faeces seems good evidence that faecal spread is the usual route of transfer. Yet the commonest virus of the group—rotavirus—though present in very large amounts in faeces—usually causes winter outbreaks of diarrhoea (Brandt *et al.* 1979). The unrelated Norwalk agent (Kapikian *et al.* 1976. Wyatt *et al.* 1978) has been transferred to volunteers by oral administration of faecal filtrates. It appears to infect somewhat older children and adults than rotaviruses (Greenberg *et al.* 1979). An epidemic of gastro-enteritis in Australia after eating oysters was associated with the Norwalk agent (Murphy *et al.* 1979).

Reoviruses are found in both respiratory secretions and the faeces. They appear to be ubiquitous even in healthy children, but an association with disease has been found only in children with diarrhoea. In addition to the human intestinal coronaviruses, rotaviruses and reoviruses found in the faeces, particles of a calicivirus (Flewett 1976) and of an astrovirus (Madeley and Cosgrove 1975) have been reported in the faeces of patients with gastro-enteritis. Their significance and source is unknown.

(d) Infection acquired from blood

Hepatitis B virus

Outbreaks of jaundice traced to injections of measles convalescent serum, yellow fever vaccine containing human serum, or blood transfusions were historically found to be due to the transfer of infection by blood or blood products before the particles of virus and of hepatitis B virus antigen were demonstrated therein. Any mode by which blood even in minute quantities can be transferred beneath the skin such as by self-injection of drugs or blood collection in contaminated imperfectly sterilized syringes can also transfer the virus. The incubation period is from 6 weeks to 3 months.

Hepatitis B virus antigen is found in varying proportions of persons in all countries where serological investigations have been made. High rates of carriage of antigen or of antibodies are found in some areas of Africa or the Pacific. Because of the possibility of transfer of blood via bites from insects, mosquitoes have been trapped and both those from East and West Africa have been found to contain hepatitis B antigen. Among persons tested in the UK and the USA, those attending clinics for sexually-transmitted infections have a higher rate of infection than healthy persons. Practising homosexuals are particularly likely to become infected, but though the antigen can be found in semen as well as the saliva and faeces, blood is the probable source of infection. Drug addicts, particularly those using self-injection, are especially liable to infection with hepatitis B virus. Infections with this virus are in general more serious than those due to hepatitis A virus; in the outbreaks which have occurred after blood transfusion in renal dialysis centres there have been deaths among the patients or attendants concerned.

The virus now termed non-A non-B hepatitis agent (Prince *et al.* 1974), like hepatitis B, is also transmitted by blood, chiefly as a result of blood transfusion. It has a shorter incubation period than hepatitis B but a more prolonged one than hepatitis A. Little is yet known about other possible modes of transfer (Purcell *et al.* 1976).

Cytomegalovirus

Cytomegalovirus is transferable by blood transfusion, though this is not the usual mode of infection by the virus. Though it is possible to screen blood collected from healthy donors by testing for hepatitis B virus antigen (HBsAg) and thus prevent its transfer to patients by blood transfusion, it is not yet possible to identify similarly the healthy carrier of cytomegalovirus. Cytomegalovirus infection causes childhood Inclusion disease, usually acquired from a mother with infected saliva or cervical secretions. The more serious congenital disease is accompanied by abnormalities of the liver, brain and eye. The study of antibodies to this virus shows that infection is ubiquitous. Antibodies develop in gradually increasing proportions during adult life, so that there is a risk of congenital infection from latent virus in apparently healthy women as well as from those becoming infected for the first time during pregnancy. Viruria has been found in from 2.5 to 6.3 per cent of pregnant women in studies in the USA (Hanshaw and Dudgeon 1978), but in Japan it has been demonstrated in a much higher proportion of urine specimens from infants. (Numazaki *et al.* 1970).

There are still many gaps in our knowledge of the precise epidemiology of diseases caused by this virus. Antibodies are reported in from 7 to over 60 per cent of children in different communities. After renal transplantation immunosuppressive therapy increases the risk of death from cytomegalovirus pneumonia. Presumably the virus had been acquired from infected blood or had been present in the patient himself before transplantation was undertaken.

(e) Congenital infections

The viruses which give rise to congenital disease in the offspring are transmitted vertically from one generation to the next as opposed to the more usual horizontal spread. They may cause no illness in the fetus, or they may cause lesions which can be either mild or severe. They are transmitted transplacentally from the mother's blood. In congenital rubella organic lesions of the fetus are most likely during the first three months of pregnancy. The risk is greatest in the first four-five weeks after conception when it ranges from 20 to 50 per cent. Thereafter it declines until the fourth month onwards when transmission of virus becomes increasingly improbable. After birth the congenitally infected baby may excrete virus in the nasopharyngeal secretions, the urine or saliva, and thus can infect those in direct contact who are susceptible to infection. Contact of pregnant mothers with a source of infection may lead to inapparent infection, which is as dangerous to the fetus as a clinical rubella illness.

Congenital cytomegalic disease of the fetus is at least as important as that of congenital rubella. It is difficult to rule out the additional possibility of contamination of an unaffected baby during birth by cervical secretions, or of aquisition from breast milk (Stagno *et al.* 1980). It is said, however, that cytomegalovirus infection is most probable during the middle or later stages of pregnancy. The congenitally infected baby excretes virus in both the saliva and the urine.

Congenital varicella is uncommon, and the former instances of congenital variola and vaccinia are not likely to be seen again. The vertical transmission of virus in the absence of symptoms has been mentioned for cytomegalovirus and hepatitis B but the great potential importance of this mode is instanced by certain animal tumour viruses.

The role of persistent, chronic and latent virus infections

The majority of the human viruses considered above cause acute infections in which the host is temporarily infectious for a few days or weeks and then becomes free from virus. But reference has already been made to the *Herpesviridae* and other viruses which may persist in the infected host; the role of such infections in the transfer of virus will now be considered.

Herpesvirus types I and II cause life-long infections with recurrent skin lesions, but in between these episodes the viruses remain latent in cells of sensory ganglia of the cranial or spinal nerves. A shock such as fever, menstruation, an injection of a vaccine such as TAB, or a respiratory infection may reactivate the virus, which then recommences replication and spreads peripherally along the respective sensory nerve to the skin of the corresponding dermatome. There vesicles develop from which virus can be transferred to susceptible persons by contact. The Gasserian ganglion, for example, has been shown to shelter type I virus (Bastian *et al.* 1972). The sacral nerve ganglion similarly conceals latent type II virus (Baringer 1974).

Cytomegalovirus may similarly persist latently, but its whereabouts is known only by its presence in blood, urine and saliva. The infected person may exhibit no symptoms or signs and thus remains a hidden source of virus for transmission of infection. It is not yet certain whether latent virus persists after an initial illness perhaps resembling infectious mononucleosis, or is acquired by transplacental infection without clinical reactions being apparent.

EB virus is carried by apparently healthy persons and is excreted by the saliva. Its presence in the tumour cells of Burkitt lymphoma or nasopharyngeal carcinoma is probably derived from saliva in the first instance. Though the lymphoblasts or carcinoma cells contain EB antigen, vegetative virus can only rarely be demonstrated. It is thus possible that the virus genome is integrated within the DNA of the host cells as in the case of some animal tumour viruses.

Latent infection by *varicella-zoster virus* persists after childhood infection manifested as chickenpox. The evidence for this was marshalled by Hope-Simpson (1965), who showed that, even when zoster rash occurs in children, an attack of chickenpox had already occurred previously. The regional distribution of zoster rash is limited to dermatomes and is analogous to the centripetal distribution of the rash of chickenpox. Hope-Simpson believes that the virus lies latent in sensory ganglia, but this has not been supported by direct demonstration. However, zoster is a manifestation which is increasingly common with age;

it is also more severe in immunosuppressed persons. Both these facts support the concept of latency. It must be admitted that zoster lesions are infective for children with resultant varicella.

Hepatitis B virus persists not only in the blood of those without overt signs of past or present illness, but as a chronic infection after an acute illness with jaundice. The hidden reservoir of infection is particularly dangerous in persons who become donors of blood, for they then transmit the virus to those thus transfused. Infection is also believed to be latent after maternal infection of the fetus.

Certain chronic infections may follow acute illnesses as, for instance, the tonsillar infection of adenovirus types 1, 2 or 5, but then the virus may remain entirely latent. *Measles virus* is now known to remain as a persisting infection in the brain after an ordinary attack of measles in childhood. The neurological disease known as subacute sclerosing panencephalitis (SSPE) ensues some years later and is ultimately fatal. The brain virus can be recognized in the affected cells but is difficult to cultivate except by co-culture techniques. The failure of virus to escape from infected cells is believed to be due to absence of the gene in the nucleic acid responsible for the M protein. The virus is thus defective (Wechsler and Field 1978, Hall, W. W. *et al.* 1978). SSPE is thus non-infectious for other persons.

A group of agents belonging to the *Papovaviridae* has been found in patients suffering from chronic neurological disease or being treated by immunosuppression (Narayan *et al.* 1973, Coleman *et al.* 1973). Though related to the SV40 viruses from Rhesus monkey kidney cultures, their source and mode of transfer is unknown. The viruses in brain tissue are associated with the lesions of progressive multifocal leucoencephalopathy, which occurs secondarily to malignant disease. Others, however, appear in the urine of patients who have received renal transplants.

Obscurity also exists over the source and transfer of the agent causing the chronic neurological disease known as Creutzfeldt-Jakob disease. The spongiform encephalopathy found in this disease resembles that associated with scrapie in sheep or kuru in New Guinea Islanders. The Creutzfeldt-Jakob agent has been transmitted to other persons by a corneal graft and by the use of imperfectly sterilized stereotactic electrodes in neurosurgical procedures (Bernoulli *et al.* 1977). The epidemiology of these examples of slow viruses in man is not known, though kuru is believed to have been transmitted by cannibalism among the Fore tribe of New Guinea. (See Chapter 103.)

The epidemiology of vector-borne infections

Table 85.2 lists certain diseases and causative viruses,

Table 85.2 Vector-Borne infections

Family and virus	Reservoir	Vector
Togaviridae		
(i) **Alphavirus**		
Eastern, Western, Venezuelan equine encephalitis	Avian-Mosquito Cycles	*Aëdes* } *Culex* } mosquitoes
African—Chikungunya etc.	Monkeys, birds? rodents	Various mosquito species
Australian—Ross River	? Avian	*Culicine* mosquitoes
(ii) **Flavivirus**		
Jungle Yellow fever	Mosquito-monkey cycle	*Aëdes* mosquitoes
Urban Yellow fever	Mosquito-man cycle	*Aëdes* mosquitoes
Dengue	Mosquito-man	*Aëdes aegypti*
St. Louis, Japanese B, Murray Valley encephalitis	Avian-mosquito	Various *Culex* mosquitoes
West Nile	Avian (mosquitoes; ticks)	*Culicine* mosquitoes
Russian Spring-Summer, European Encephalitis	Ixodid ticks	} *Ixodid* ticks
Louping-ill	Sheep, Ixodid ticks	} *Ixodid* ticks
Powassan	Rodents	} *Ixodid* ticks
Kyasanur Forest Disease	Ticks	*Haemaphysalis* ticks
Omsk Haemorrhagic fever	Ticks	*Dermacentor* ticks
Bunyaviridae		
Bunyavirus	Mosquito	*Aëdes* mosquitoes
California encephalitis group	Rodents	*Aëdes* mosquito
Crimean Haemorrhagic fever (Congo virus)	Small mammals	*Hyalomma* ticks
Rift Valley fever	Sheep, goats	*Culex* and *Aëdes* mosquitoes
Sandfly fever	Man	*Phlebotomus papatasi*
Reoviridae (Orbivirus) Colorado tick fever	Rodents, ground squirrel	*Dermacentor* ticks

the transmission of which from one host to another occurs by means of a biting insect or arachnid tick. Vector-borne virus is acquired by a mosquito, midge, sandfly or tick by contamination from blood or superficial tissues during feeding upon an infected host. When later feeding upon a previously uninfected host the same vector introduces virus into or beneath the skin and transmits the infection. Vector-transmission of virus may be accidental and mechanical but in other instances the vector itself undergoes infection and serves as a host. The virus then replicates in the vector and transmission to the primary host is facilitated. Moreover, in three instances the virus in the vector passes vertically to the next generation by transovarial transmission. According to Downs (1976) the ticks infected by the virus of Russian Spring-Summer encephalitis, mosquitoes concerned in Californian encephalitis and, doubtfully, sandflies concerned in transmitting sandfly fever, perpetuate their respective viruses outside their normal breeding season in this manner. This mode of survival of virus in the vector is known as 'over-wintering'. It may well apply to other infections listed in the Table.

Table 85.2 also shows under the heading 'reservoir' the probable source of virus and the vectors concerned. In certain of the best known *Togaviridae* a primate or avian host maintains the virus in a continual vector-animal-vector cycle but in other instances the virus survives in vector cycles with domesticated mammals or rodents. In *Urban yellow fever*, *Dengue* and *Sandfly fever* the only known vertebrate host is man himself and the vectors, which are *Aëdes* mosquitoes or biting flies of the *Phlebotomus* variety, transmit the viruses from man to man. A clue to the possible manner of evolution of a virus from infection in the animal kingdom exists in the case of yellow fever.

Jungle yellow fever occurs when a susceptible person enters a tropical forest in Africa, Central or South America and comes into contact with a mosquito-monkey-mosquito cycle of yellow fever virus. In South America, *Haemagogus* mosquitoes feeding in the tree-tops on monkeys maintain the cycle of infection. When a forester thus infected returns to his village or urban area he develops yellow fever and if the locality contains *Aëdes egypti* mosquitoes, the latter may act as a second vector to spread the virus to other persons. In Ethiopia during yellow fever epidemics in the 1960s according to Sérié *et al.* (1968), the African monkeys in the jungle were bitten by *Aëdes africans* mosquitoes but *Aëdes simpsoni* was responsible for the human epidemics. Similar cycles of jungle yellow fever with particular mosquitoes of the locality keep the virus active in the tropical forests of Africa and the Americas and periodic epidemics continue to occur outside the forests.

A similar jungle cycle involving primates and *Aëdes* mosquitoes is suspected in the case of dengue in Malaysia and other Asian countries but certainly does not occur in America. Nor has an animal reservoir ever been demonstrated in sandfly fever (Taylor *et al.* 1956).

Geographical occurrence and epidemicity

Vector-borne viruses exist widely in Africa, Asia, the United States of America and other countries. They occur chiefly in tropical or sub-tropical zones where the respective mosquito or other vectors are resident. Tick-borne viruses occur in temperate zones particularly in Europe and throughout the USSR and in Canada (Powassan virus), Scotland (Louping-ill) and India (Kyasanur Forest Disease). Knowledge concerning the prevalence of particular infections has been gained chiefly by virus isolation from animals, vectors or patients themselves. Serological tests have proved of less value among the *Togaviridae* than with other groups largely because of cross-neutralization. But knowledge has slowly been gained on the relative frequency of subclinical infections or simple pyrexial illnesses and of severe cases including encephalitis. The relative proportion of these differs in each infection.

The influence of vector transmission on the epidemiology of the particular infections is seen in the fact that seasonality depends upon the available number of vectors and thus upon the breeding season. So far as the host is concerned, resistance following previous infection by the same or an antigenically related virus influences the age of those taken ill during epidemics and the size of the latter. Children thus suffer particularly in epidemics of encephalitic viruses. Animal hosts are said to exert a multiplying effect on the amount of virus carried by the vector particularly when the vector itself becomes infected. *Kyasanur Forest disease* appeared in Mysore State in India as an epizootic with many fatalities among the monkey population. Foresters acquired the disease even without handling dead monkeys and ticks of the *Haemaphysalis* species were found to carry the virus; small forest mammals were thought to act as a reservoir (Work 1958). The various *Equine encephalitis* viruses of North America and Venezuela first attracted attention because of *sickness* in horses. The mortality in horses varies in the different infections being highest in Eastern equine encephalitis of the United States. The severity of illness also varies in the different infections in man, and though encephalitis due to the Eastern equine encephalitis is frequently fatal, the actual occurrence is fortunately rare (Downs 1976; Andrewes *et al.* 1978).

The development of epidemics in areas previously unaffected by vector-borne viruses has in two instances been due to a geographical spread of virus perhaps associated with a change in the animal-vector reservoir. Thus *Murray Valley encephalitis* appeared in South-Eastern Australia in the early 1950s following a long interval after its previous visitation in 1917–18 when it was termed Australian-X disease. The vector in the 1950s epidemics was identified as *Culex annulirostris* and the reservoir was believed to be in water birds such as herons and cormorants whose unusual southward migration had followed excessive rainfall in the tropical areas of the Northern Territory. According to Miles (1960) the virus is endemic in the latter area and in New Guinea.

Rift Valley fever virus has been known to cause epizootics in sheep and goats with fatalities in lambs and kids in the Rift Valley of East Africa. In the past few years the virus has spread to South Africa, the Sudan and elsewhere. Though not previously identified as a frequent cause of infections in man, there occurred in Egypt a large epidemic in 1977 which has since recurred. Hundreds of cases of dengue-like febrile illnesses with haemorrhagic rashes and ocular and neurological lesions have occurred and there have been many deaths (Meegan 1979*a*, Meegan *et al.* 1979*b*, Hoogstraal *et al.* 1979). According to Brès (1978) the virus is resident in rats, and *Culex* mosquitoes are the vectors but its main hosts are still sheep and goats. The considerable increase in irrigation in Egypt of recent years may have been a factor in the increase of the mosquito population which was observed during the epidemic.

A third example of geographical variation of particular viruses is the spread of *dengue haemorrhagic fever* from the Philippines to Thailand, Malaya and India in 1960 (Hammon *et al.* 1960). The vector is *Aëdes egypti* and it is suggested that the different clinical picture of this from that of ordinary dengue is caused by the fact that it represents a second infection by one dengue serotype of virus in persons previously infected by another type. Finally, Ross River virus, first identified in Australia by Doherty and co-workers (1966), causes an epidemic exanthem and polyarthritis (Shope and Anderson 1960). It has recently spread to various areas of the Pacific. The vectors are Culicine mosquitoes. The reservoir may be avian. (For the part played by wind in the carriage of infected vectors, see Sellers 1980).

Zoonotic infections not transmitted by vectors

Many other virus infections are acquired by man from contact with infected animals without the transfer of virus through an intermediate vector. Table 85.3 lists the viruses, the animal reservoirs and the mode of transfer. It is obvious that infection must be acquired by contact with an infected animal and that there is an occupational risk; this may be slight in the poxvirus infections or exceedingly hazardous in those handling specimens in the laboratory from Marburg and Ebola infections. In the case of zoonoses causing widespread infections in the field, such as the various haemorrhagic fevers, the mode of contact may be obscure and then spread from man to man may be suspected or even proved. Indeed the likelihood of transmission of a zoonosis to contacts of patients varies with each virus and may even vary with different strains of the same virus. The mode of contact and method of infection must be considered separately therefore with each of the virus groups.

The *animal pox viruses* spread when persons handle animals possessing actual skin lesions. *Cowpox* pustules on cows' udders have in the past enabled virus to spread to the fingers of milkers. However, Baxby (1977) has recorded recent instances of cowpox pustular eruptions in persons who had had no direct contact with cows and this suggests that the virus may at times spread from person to person. *Camelpox* occurs in drivers of camels in the Middle East and contact with lesions on the nose and lips may be the mode of infection. *Contagious pustular dermatitis* (*orf*) spreads in a similar way from infected lambs or kids to shepherds who develop nodular lesions on the hands. *Monkeypox* has become a matter of concern since the successful eradication of smallpox. Arita (1979) has described the surveillance of its occurrence in man in Zaire and elsewhere. Though resembling smallpox clinically its occurrence among hunters and its low secondary attack rate indicate a probable source in the animal kingdom. Curiously enough the virus has been isolated only from monkeys in captivity so that the precise source of infection in the field is not known. Studies in the laboratory of monkeypox virus indicate that there are distinguishing features from variola virus. The virus of *Tanapox* isolated from monkeys and persons in Kenya with a febrile illness and a single pock-like lesion is not related to the *Poxviridae* (Downie *et al.* 1971). Though a zoonosis, its occurrence as an epidemic along the Tana river suggested to Manson-Bahr and Downie (1973) that it might be transmitted by mosquitoes.

Rabies is contracted usually from bites by rabid dogs, wolves, cats, jackals or skunks. Though bites from rabid foxes are theoretically possible the fact that the virus has spread widely in Western Europe in the fox population without thereby causing a similar increase in human cases of rabies suggests that it rarely spreads to man in this way. The transmission of rabies virus through saliva has been appreciated since early in the nineteenth century and, though experimental infection in the laboratory has been obtained by inhalation, natural transmission to man from dogs involves bites or contamination by saliva of a skin wound (Kaplan 1969). In Mexico, Central and South America and Trinidad, rabies virus is transmitted by bites from vampire bats. When caves lodging bats are entered infection has occurred through inhalation (Frederickson and Thomas 1965; Kaplan 1971).

Though this formidable virus is still uncontrolled

Table 85.3 Zoonotic virus infections

Viruses	Animal reservoir	Mode of transfer to man
(a) **Poxviridae**		
Cowpox	Cows	Animal skin lesion to human skin (abraded)
Contagious pustular dermatitis (Orf)	Sheep	
Camelpox	Camels	
Monkeypox	Monkeys (Zaire, Nigeria)	Contact and? other mode
(b) **Rhabdoviridae**		
Rabies	Dogs, cats, bats, wolves, foxes, skunks, raccoons	Bite contaminated with saliva
Marburg	*Cercopithecus* monkeys (Uganda)	Contact infected blood
Ebola	Source not discovered (Zaire, Sudan)	Person-to-person contact blood and ? airborne
Herpesviridae		
Herpesvirus B (Simiae)	*Rhesus* monkeys	Saliva by bite or contact
(c) **Arenaviridae**		
Haemorrhagic fever—Junin	Field mice (Argentina)	Contact urine/faeces with abraded skin
Haemorrhagic fever—Machupo	Field mice (Bolivia)	
Lassa fever	Multimammate rat (*Mastomys*)	Unknown ? urine, saliva or faeces. Close contact or airborne
Lymphocytic choriomeningitis	Wild mice	Contact urine/faeces
(d) **Picornaviridae**		
Encephalomyocarditis	Rodents and monkeys	Unknown ? contamination by urine or faeces
Footh and Mouth Disease	Cattle, pigs, sheep	Rare human transfer
Swine vesicular disease	Human ? pigs	Reverse transfer of Coxsackie B5 from man to pig
(e) **Orthomyxoviridae**		
Swine influenza virus	Pigs	Sporadic human infections. Fort Dix epidemic, 1976
(f) **Paramyxoviridae**		
Newcastle disease of fowls (NDV)	Birds, poultry	Accidental laboratory infection of conjunctiva

throughout large areas of the world, the history of rabies in Britain shows the effectiveness of quarantine, providing that illegal importation of animals from areas where rabies is endemic is avoided. After years when numbers of cases of rabies occurred in the nineteenth century, the imposition of the Muzzling Order of 1887 for London and in 1889 its extension to the whole country, combined with the destruction of stray dogs and the import Regulations, led to the complete eradication of rabies. The country was then free from the disease until 1918 when it was re-introduced, probably by an illegally imported dog. Eradication was achieved by 1922. The duration of quarantine imposed on all imported dogs and cats in 1922 was six months but in 1969 it was extended to 12 months after a single case of rabies occurred in a dog nine months after importation (Hill 1971). Quarantine has since been reduced to six months in duration. A case has been recorded of person-to-person transmission of virus following a corneal transplant from a person later shown to have died from rabies (Houff *et al.* 1979).

Marburg virus and the related virus of *Ebola fever* indicate the severity of some zoonotic infections. The original outbreak in 1967 was of a series of 25 illnesses in West Germany and Yugoslavia with seven fatalities in persons exposed to monkeys or monkey tissues from a batch of vervet (*Cercopithecus*) monkeys imported from Uganda. There were subsequently six contact cases and another one occurred in the wife of one patient eighty-three days after the onset of his illness and this was believed to have been contracted through semen; this was later demonstrated to contain virus (Simpson 1977). Search for the source of the monkey infections was unsuccessful (Kissling *et al.* 1970). Marburg virus was again found in 1975 in the fatal case of a traveller through Rhodesia and South Africa but the source of virus was unknown. A nurse in contact with the patient also developed fever but recovered.

In 1976 an extensive epidemic of *haemorrhagic fever* occurred in the Sudan and Zaire and this was shown to be due to a Marburg-like virus termed Ebola virus.

Factory and agricultural workers were affected and also patients in hospital (Bowen *et al.* 1977; Simpson 1977). The origin of this new virus is still unknown. The mortality of the disease was high and in one hospital 41 of the 76 infected members of the staff died. As with Marburg disease the Ebola virus is present in blood, urine, semen and possibly other secretions. It appears that close physical contact is usually necessary for the transmission of the virus. In a laboratory infection acquired in England from working with the virus, treatment with interferon and convalescent serum obtained from the Sudan appeared to reduce the titre of virus in the blood and may have aided recovery (Emond *et al.* 1977).

Herpesvirus B (Simiae) is a relatively common herpesvirus carried by Rhesus monkeys. Inoculated into man by bites or acquired from saliva contaminating cuts, the virus produces a severe meningoencephalitis. This virus has given rise to much anxiety in handling monkeys imported for work on poliovirus and the preparation of poliovirus vaccines.

Among the zoonoses due to *Arenaviridae* are exotic infections acquired in Africa (*Lassa fever*) and the *haemorrhagic fevers of Bolivia* (Machupo virus), Brazil and the Argentine (Junin and Tacaribe viruses). The source of the virus of Lassa fever, which occurs endemically in Nigeria, Liberia and Sierra Leone, appears to be the multimammate rat (*Mastomys natalensis*). Rodent urine or faeces appear to be the source of infection, for this animal enters homes. Transmission of infection from person to person is reported to have occurred in hospital. The virus is present in the pharynx and also in blood and urine (Monath 1974) and airborne spread is therefore possible. In Sierra Leone, Keane and Gilles (1977) recorded a mortality of about 16 per cent in patients admitted to hospital. Convalescent serum is said to be therapeutically valuable.

The *haemorrhagic fevers of the Americas* are acquired from mouse-like rodents (*Calomys*) house and field mice and also bats, possibly through the medium of mites. The infections are severe and often fatal. Convalescent plasma may assist recovery (Maiztegui *et al.* 1979). *Lymphocytic choriomeningitis virus*, which is one cause of the aseptic meningitis syndrome, occurs in many countries and is acquired from wild mice in the urine or faeces. These is no evidence of transmission from man to man.

There are few examples of zoonoses among the viruses of the *Picornaviridae*. However, *encephalomyocarditis virus*, which occurs in Africa in mice and other rodents, has been isolated from the blood and stools of sporadic cases of pyrexial illnesses with meningitis occurring in laboratory workers.

Foot-and-mouth disease virus has been reported to cause minor infections with vesicular lesions in persons in contact with infected animals.

Though there are no recorded zoonoses among the numerous enterovirus infections, the example must be quoted of probable transfer of a human strain of virus to a domestic animal. The virus isolated during epidemics of *swine vesicular examthem* has a close relation but is not identical with Coxsackie B5 virus (Brown *et al.* 1973). In view of the frequent feeding of pigs on uncooked garbage it seems probable that pigs may have acquired the virus originally from a human source.

The relation between swine and human influenza

Swine influenza virus first isolated in the laboratory by Shope (1931*a*) was shown to be the cause, in association with *Haemophilus influenzae suis*, of epidemic influenza in pigs (Shope 1931*b*). Its close serological relation to the WS strain of human influenza A virus led to partial cross-immunity in convalescent ferrets (Smith *et al.* 1933). The surface antigens H and N of the swine virus have shown little change from 1931 among virus strains isolated at intervals from swine in the USA (Nath *et al.* 1975). The isolation in 1976 of several strains of virus (A/New Jersey 1976) with the antigenic formula Hsw1. N1 during an epidemic of influenza at the Army Camp of Fort Dix in New Jersey, USA, raised the possibility of a large epidemic being caused by this virus. There had been a few sporadic instances of the isolation of swine influenza virus from illnesses or at autopsy in the USA before 1976 (Smith *et al.* 1976, Thompson *et al.* 1976); these supported the view that the swine virus was capable of causing an epidemic in the human population. Such infections did not spread to contacts; but within Fort Dix, however, but not outside its walls, serological evidence existed of swine virus infection in 500 persons. A simultaneous outbreak of influenza due to H3N2 subtype of human influenza virus was also in progress.

Fear that the New Jersey virus might cause a pandemic similar to that of 1918 arose partly from the serological evidence of previous infection by a virus resembling antigenically the swine virus of Shope in most persons alive during 1918–19 but not in those born after 1922 (Andrewes *et al.* 1935). Stuart-Harris and Schild (1976) discuss the evidence that human influenza virus of the A/Hong Kong/68 subtype passed rapidly into the domestic pig population in Great Britain after the major epidemics of 1968 and 1969. This is analogous to the hypothetical transfer of the 1918–19 virus from man to pigs. The zoonotic return of the Shope virus to man in 1976 represented for the most part a 'dead end' non-transmissible infection except in Fort Dix. At some future date when the population has lost its content of anti-Hsw1 antibody, a similar return of virus from pigs might give rise to a spreading pandemic.

References

Andrewes, C., Pereira, H. G. and Wildy, P. (1978) *Viruses of Vertebrates*, 4th edn. Baillière Tindall, London.

Andrewes, C. H. and Glover, R. E. (1941) *Brit. J. exp. Path.* **22,** 91.

Andrewes, C. H. Laidlaw, P. P. and Smith, W. (1935) *Brit. J. exp. Path.* **16,** 566.

Arita, I. (1979) *Nature (Lond.)* **279,** 293.

Bailey, N. T. J. (1975) *Mathematical Theory of Infectious Diseases and its Applications*, 2nd edn., p. 413. Charles Griffin & Co. Ltd., London.

Baringer, J. R. (1974) *New Engl. J. Med.* **291,** 828.

Baroyan, O. V., Rvachev, L. A. and Ivannikov, Yu. G. (1977) *Publ. N. F. Gamaleya Institute Epidemiol. and Virol.* Moscow.

Bastian, F. O., Rabson, A. S., Yee, C. L. and Tralka, T. S. (1972) *Science* **178,** 306.

Baxby, D. (1977) *Brit. med. J.* **i,** 1379.

Bernoulli, C., Siegfried, J., Baumgartner, G., Regli, F., Rabinowicz, T. Gajdusek, D. C. and Gibbs, C. J. J. (1977) *Lancet* **i,** 478.

Beveridge, W. I. B. (1977) *Influenza—the Last Great Plague.* Heinemann, London.

Bowen, E. T. W., Platt, G. S., Lloyd, G., Baskerville, A., Harris, W. J. and Vella, E. E. (1977) *Lancet* **i,** 571.

Brandt, C. D., Kim, H. W., Vargosko, A. J., Jeffries, B. C., Arrobio, J. O., Rindge, B., Parrott, R. H. and Chanock, R. M. (1969) *Amer. J. Epidem.* **90,** 484.

Brandt, C. D. *et al.* (1979) *Amer. J. Epidem.* **110,** 243.

Brès, P. (1978) *J. Egypt. publ. Hlth. Ass.* **53,** 147.

Brimblecombe, F. S. W., Cruickshank, R., Masters, P. L., Reid, D. D. and Stewart, G. T. (1958) *Brit. med., J.* **i,** 119.

Brown, F., Talbot, P. and Burrows, R. (1973) *Nature (Lond.)* **245,** 315.

Clarke, S. K. R., Caul, E. O. and Egglestone, S. I. (1979) *Postgrad. med. J.* **55,** 135.

Clarke, S. K. R., Heath, R. B., Sutton, R. N. P. and Stuart-Harris, C. H. (1958) *Lancet* **i,** 814.

Coleman, D. V., Gardner, S. D. and Field, A. M. (1973) *Brit. med. J.* **iii,** 371.

Committee on Influenza and other Respiratory Virus Vaccines of the MRC (1958) *Brit. med. J.* **i,** 415.

Doherty, R. L., Gorman, B. M., Whitehead, R. H. and Carley, J. G. (1966) *Aust. J. exp. Biol. med. Sci.* **44,** 365.

Downie, A. W., Taylor-Robinson, C. H., Caunt, A. E., Nelson, G. S., Manson-Bahr, P. E. C. and Matthews, T. C. H. (1971) *Brit. med. J.* **i,** 363.

Downs, W. G. (1976) In: *Viral Infections of Humans: Epidemiology and Control*, p. 71. Ed. by A. S. Evans. John Wiley and Sons, London and New York.

Emond, R. T. D., Evans, B., Bowen, E. T. W. and Lloyd, G. (1977) *Brit. med. J.* **ii,** 541.

Flewett, T. H. (1976) *Proc. R. Soc. Med.* **69.** 693.

Foy, H. M., Cooney, M. K., McMahan, R. and Grayston, J. T. (1973) *Amer. J. Epidem.* **97,** 93.

Francis, T., Jr., Davenport, F. M. and Hennessy, A. V. (1953) *Trans. Ass. Amer. Phycns.* **66,** 231.

Frederickson, L. E. and Thomas, L. (1965) *Publ. Hlth. Rep., USA.* **80,** 495.

Fry, J. (1958) *Brit. med. J.* **i,** 259.

Glezen, W. P. and Couch, R. B. (1978) *New Engl. J. Med.* **298,** 587.

Glezen, W. P. and Denny, F. W. (1973) *New Engl. J. Med.* **288,** 498.

Greenberg, H. B. *et al.* (1979) *J. infect. Dis.* **139,** 564.

Gwaltney, J. M., Jr. (1975) *Yale J. Biol. Med.* **48,** 17.

Gwaltney, J. M., Jr., Moskalski, P. B. and Hendley, J. O. (1978) *Ann. intern. Med.* **88,** 463.

Hall, W. J., Hall, C. B., and Speers, D. M. (1978) *Ann. intern. Med.* **88,** 203.

Hall, W. W., Kiessling, W. and Ter Meulen, V. (1978) *Nature (Lond.)* **272,** 460.

Hammon, W. M., Rudnick, A. and Sather, G. E. (1960) *Science* **131,** 1102.

Hanshaw, J. B. and Dudgeon, J. A. (1978) *Viral Diseases of the Fetus and Newborn.* W. B. Saunders, Philadelphia.

Hemmes, J. H., Winkler, K. C. and Kool, S. M. (1960) *Nature (Lond.)* **188,** 430.

Henderson, F. W., Collier, A. M., Clyde, W. A. Jr. and Denny, F. W. (1979) *New Engl. J. Med.* **300,** 530.

Higgins, P. G. (1974) *J. Hyg. Camb.* **72,** 255.

Hill, F. J. (1971) *Proc. R. Soc. Med.* **64,** 231.

Hoogstraal. H., Meegan, J. M., Khalil, G. M. and Adham, F. K. (1979) *Trans. roy. Soc. trop. Med. Hyg.* **73,** 624.

Hope-Simpson, R. E. (1965) *Proc. R. Soc. Med.* **58,** 9; (1979) *J. Hyg. Camb.* **83,** 11; (1981) *Ibid.* **86,** 35.

Hope-Simpson, R. E. and Higgins, P. G. (1969) *Progr. med. Virol.* **11,** 354.

Horn, M. E. C., Brain, E., Gregg, I., Yealland, S. J. and Inglis, J. M. (1975) *J. Hyg. Camb.* **74,** 157.

Houff, S. A. *et al.* (1979) *New Engl. J. Med.* **300,** 603.

Jackson, G. G., Dowling, H. F., Anderson, T. O., Riff, L., Saporta, J. and Turck, M. (1960) *Ann. intern. Med.* **53,** 719.

Kapikian, A. Z. *et al.* (1976) *Acute Diarrhoea in Childhood.* In: *Ciba Found. Symp.* **42,** New Series, p. 273, Elsevier/Excerpta Medica, North Holland.

Kaplan, M. M. (1969) *Nature (Lond.)* **221,** 421; (1971) *Proc. R. Soc. Med.* **64,** 225.

Keane, E. and Gilles, H. M. (1977) *Brit. med. J.* **i,** 1399.

Kendal, A. P., Noble, G. R., Skehel, J. J. and Dowdle, W. R. (1978) *Virology* **89,** 632.

Kissling, R. E., Murphy, F. A. and Henderson, B. E. (1970) *Ann. N.Y. Acad. Sci.* **174,** 932.

Kling, C., Pettersson, A. and Wernstedt, W. (1912) *Commun. Inst. méd. État., Stockholm* **3,** 5.

Loosli, C. G., Lemon, H. M., Robertson, O. H. and Appel, E. (1943) *Proc. Soc. exp. Biol. Med.* **53,** 205.

Madeley, C. R. and Cosgrove, B. P. (1975) *Lancet* **ii,** 451.

Maiztegui, J. I., Fernandez, N. J. and de Damilano, A. J. (1979) *Lancet* **ii,** 1216.

Manson-Bahr, P. E. C. and Downie, A. W. (1973) *Brit. med. J.* **ii,** 151.

Martin, A. J., Gardner, P. S. and McQuillin, J. (1978) *Lancet* **ii,** 1035.

Meegan, J. M. (1979*a*) *Trans. roy. Soc. trop. Med. Hyg.* **73,** 618.

Meegan, J. M., Niklasson, B. and Bengtson, E. (1979*b*) *Lancet* **ii,** 1184.

Melnick, J. L., Paul, J. R. and Walton, M. (1955) *Amer. J. Pub. Hlth.* **45,** 429.

Miles, J. A. R. (1960) *Bull. Wld Hlth Org.* **22,** 239.

Monath, T. P. (1974) *WHO Chronicle* **28,** 212.

Morgan, D. G., Niederman, J. C., Miller, G., Smith, H. W. and Dowaliby, J. M. (1979) *Lancet* **ii,** 1154.

Murphy, A. M., Grohmann, G. S., Christopher, P. J., Lopez,

W. A., Davey, G. R. and Millsom, R. H. (1979) *Med. J. Australia* **ii,** 329.
Narayan, O., Penney, J. B., Jr., Johnson, R. T., Herndon, R. M. and Weiner, L. P. (1973) *New Engl. J. Med.* **289,** 1278.
Nath, D. M., Rodkey, L. S. and Minocha, H. C. (1975) *Arch. Virol.* **48,** 245.
Numazaki, Y., Yano, N., Morizuka, T., Takai, S. and Ishida, N. (1970) *Amer. J. Epidem.* **91,** 410.
Panum, P. L. (1846) Reprinted 1940, Delta Omega Society, publ. American Public Health Assoc., printed by F. H. Newton, New York.
Paul, J. R. and Riordan, J. R. (1950) *Amer. J. Hyg.* **52,** 202.
Pickles, W. N. (1939) *Epidemiology in Country Practice*, p. 69, John Wright & Sons Ltd., Bristol.
Potter, C. W. (1964) *J. Path. Bact.* **88,** 263.
Prince, A. M., Brotman, B., Grady, G. F., Kuhns, W. J., Hazzi, C., Levine, R. W. and Millian, S. J. (1974) *Lancet* **ii,** 241.
Provost, P. J. and Hilleman, M. R. (1979) *Proc. Soc. exp. Biol. Med.* **160,** 213.
Public Health Lab. Service (1963) *Monthly Bull. Ministry Health Lab. Service* **22,** 167.
Public Health Laboratory Service. Advisory Committee on Influenza (1977) *J. Hyg. Camb.* **78,** 223.
Purcell, R. H., Alter, H. J. and Dienstag, J. L. (1976) *Yale J. Biol. Med.* **49,** 243.
Razzell, P. (1977) *The Conquest of Smallpox*. Caliban Books, Sussex.
Report Ministry of Health (1963) Smallpox 1961–62, *Report on Public Health and Medical Subjects* No. 109, HMSO, London.
Report (August 31, 1979) *Morbid. Mortal.* **28,** 410. US Dept. HEW Publ. CDC.
Roden, A. T. (1963) *J. Hyg. Camb.* **61,** 231.
Rosher, A. B. and Cole, W. T. (1939) Observations on the occurrence of *H. influenzae* in the Trachea, p. 107. *Minist. Hlth. Rep.* No. 90, HMSO, London.
Schulman, J. L. and Kilbourne, E. D. (1963) *J. exp. Med.* **118,** 267.
Sellers, R. F. (1980) *J. Hyg. Camb.* **85,** 65–102.
Sérié, C., Lindrec, A., Poirier, A., Andral, L. and Neri, P. (1968) *Bull, Wld Hlth Org.* **38,** 835.
Shope, R. E. (1931*a*) *J. exp. Med.* **54,** 373; (1931*b*) *Ibid.* **54,** 349; (1941) *Ibid.* **74,** 41, 49; (1944) *Medicine* **23,** 415.
Shope, R. E. and Anderson, S. G. (1960) *Med. J. Aust.* **i,** 156.
Simpson, D. I. H. (1977) Marburg and Ebola Virus Infections. *World Hlth. Org. Publ.* No. 36.
Smith, T. F., Burgert, E. O., Jr., Dowdle, W. R., Noble, G. R., Campbell, R. J. and Van Scoy, R. E. (1976) *New Engl. J. Med.* **294,** 708.
Smith, W., Andrewes, C. H. and Laidlaw, P. P. (1933) *Lancet* **ii,** 66.
Spicer, C. C. (1979) *Brit. med. Bull.* **35,** No. 1, 23.
Stagno, S., Reynolds, D. W., Pass, R. F. and Alford, C. A. (1980) *New Engl. J. Med.* **302,** 1073.
Stern, H., Elek, S. D., Millar, D. M. and Anderson, H. F. (1959) *Lancet* **ii,** 871.
Stuart-Harris, C. H. (1977) *Chemoprophylaxis and Virus Infections of the Respiratory Tract*, Vol. 1, p. 1. Ed. by J. Oxford, CRC Press, Inc., Cleveland, Ohio.
Stuart-Harris, C. H. and Schild, G. C. (1976) *Influenza, the Viruses and the Disease*, Edward Arnold, London.
Takahashi, M., Otsuka, T., Okuno, Y., Asano, Y., Yazaki, T. and Isomura, S. (1974) *Lancet* **ii,** 1288.
Taylor, R. M., Work, T. H., Hurlbut, H. S. and Risk, F. (1956) *Amer. J. trop. Med. Hyg.* **5,** 579.
Thompson, R. L., Sande, M. A., Wenzel, R. P., Hoke, C. H., Jr. and Gwaltney, J. M., Jr. (1976) *New Engl. J. Med.* **295,** 714.
Tyrrell, D. A. J. (1965) *Common Colds and Related Diseases*, 1st edn., p. 41. Edward Arnold, London; (1968) Rhinoviruses *Monographs in Virology* **2,** 67 Springer-Verlag, Vienna/New York.
Tyrrell, D. A. J., Buckland R., Rubenstein, D. and Sharpe, D. M. (1970) *J. Hyg. Camb.* **68,** 359.
Tyrrell, D. A. J. *et al.* (1975) *Intervirology* **5,** 76.
Virelizier, J-L., Allison, A. C. and Schild, G. C. (1974) *J. exp. Med.* **140,** 1571.
Webster, R. G., Yakhno, M., Hinshaw, V. S., Bean, W. J. and Murit, K. G. (1978) *Virology* **84,** 268.
Wechsler, S. L. and Field, B. N. (1978) *Nature (Lond.)* **272,** 458.
Wong, D. C., Purcell, R. H. and Rosen, L. (1979) *Amer. J. Epidem.* **110,** 227.
Woodall, J., Rowson, K. E. K., and McDonald, J. C. (1958) *Brit. med. J.* **ii,** 1316.
Work, T. H. (1958) *Progr. med. Virol.* **1,** 248, Karger, Basel.
Wyatt, R. G. *et al.* (1978) *Perspect. Virol.* **10,** 121. Ed. by M. Pollard. Academic Press, New York.
Zhdanov, V. M. (1967) 1st int Conf. Vaccines against Viral and Rickettsial Diseases of Man. Pan-American Health Org. Sci. Publ. No. 147, p. 34.

86

Vaccines and antiviral drugs

A. J. Beale

Introductory

During this century there has been a notable reduction in the impact of virus diseases of man and animals. Much of this has come about as a result of sanitary methods of control. The provision of clean water and modern methods of sewage disposal led to a reduction of virus infections as well as bacterial disease. Similarly, the changes associated with a higher standard of living, a reduction in overcrowding and better nutrition led to a delay in the acquisition of infection from most viruses.

Nevertheless these measures may lead to the emergence of substantial numbers of non-immune persons and thus to epidemics of disease when the virus is introduced. The delay in the age of infection with polioviruses led to recognition of poliomyelitis epidemics in the early part of this century in Sweden. Similarly, the delay in the onset of infection with Epstein-Barr (EB) virus led to outbreaks of infectious mononucleosis in students, and the delay in infection with hepatitis viruses to considerable outbreaks of hepatitis among the military personnel in World War Two and the later wars in Korea and Vietnam.

In veterinary medicine, sanitary methods of control of virus infections have often been applied with great success. The control of rinderpest, rabies, foot-and-mouth disease, swine vesicular disease and swine fever has been effectively accomplished in the United Kingdom by slaughter of affected animals and their contacts, supplemented by standstill orders and by banning imports or permitting them only under strict control and quarantine regulations. These methods can be successful in man also; for example, the control of hepatitis B virus infections in renal dialysis units has been effected by identifying carriers, screening of blood donors and tight aseptic control measures to prevent cross-contamination, as shown by Polakoff (1981).

The main method of control of virus diseases and one that has been triumphantly successful is by means of vaccines; more recently some measure of success has been achieved with drugs. This chapter will consider these methods of controlling virus diseases and virus infections.

Virus vaccines

Before embarking on vaccine development the question must be asked, is the disease sufficiently important to be worth the effort required to prevent it? As we shall see, there is always some risk associated with a vaccination programme and the risk and expense must be worth it. In human medicine the decision is often clear cut because of the severity of the disease in terms of mortality or morbidity, as in rabies, yellow fever or poliomyelitis or the later complications of infection such as the effect of rubella virus on the fetus. The effect on special population groups may make a case for selective immunization; such as for travellers abroad or for those at special risk from hepatitis B virus infection. The economic effects of the disease are also important, as for respiratory diseases in man where, in addition to the morbidity and mortality from influenza, mild disease has severe economic effects. The economic factors are usually decisive for immunization in animals. The decision is not always easy; the need for a vaccine against mumps has been accepted in the USA where it is a component of the childhood measles, mumps, rubella attenuated living vaccine (Hayden *et al.* 1978). The reason for using it is the universal nature of the infection and existence of serious complications, such as meningo-encephalitis and orchitis. The frequency and severity of these complications are not generally considered sufficient to warrant the expense of a vaccine campaign in the United Kingdom, although if the age of mumps infection rises this decision may change. In the USA the trend of the incidence of the disease after vaccination is likely to follow that established for measles; that is, a fall in the number of cases (already a fall of 90 per cent has been reached), a rise in age of those attacked and a break in the transmission of the disease. Other examples of the difficulty in deciding whether to vaccinate are chickenpox and cytomegalovirus infections. Vaccines against both these diseases could be developed and are undergoing development. Natural infections with these viruses lead to life-long latent infection and, since the vaccines being developed are living ones, they also must be assumed to cause latent infection; it is the unknown consequences of this latent infection that give cause for delay and further consideration. The frequency of recurrence with vaccine strains is unknown and so are the long-term consequences, especially because other members of the herpesvirus group can cause tumours. Considerations of this sort make the decision difficult and cause some workers to prefer killed vaccines for this group of viruses. There is evidence that such killed herpes virus vaccines could be effective for the prevention or limitation of the primary infection (Hilleman *et al.* 1980; Kapoor *et al* 1982). The possibility that EB virus causes Burkitt's lymphoma, nasopharyngeal carcinoma or Hodgkin's disease has encouraged some workers to try and develop a vaccine but the cost of preventing a case by this means may be high.

Given that there is a cause for immunization, what sorts of vaccines are available? There are two main types, living attenuated and killed, with quite different characteristics which are described below.

Living vaccines

The major triumphs in the control of virus diseases have come from the use of living attenuated vaccines against diseases like smallpox, yellow fever and poliomyelitis. The eradication of smallpox from the world by means of vaccination must rank as one of the greatest achievements in medicine. Its origin goes back to the work of Edward Jenner who, in 1798, published 'An inquiry into the causes and effects of variolae vaccinae a disease discovered in some of the western counties of England particularly Gloucestershire and known by the name of cowpox'. Jenner observed that cowpox conferred protection against smallpox and deliberately inoculated cowpox matter as a means of protection against the more severe disease, thus introducing the procedure of vaccination. Vaccine and vaccination have since acquired a much wider meaning, namely the use of living or dead micro-organisms to produce active immunization against disease. The principle that Jenner introduced was the production of a mild form of the disease before exposure to the virulent disease. He did it by using a related virus, cowpox, which provides protection from the more severe disease. This is not the only approach possible; for example, the frequency of subclinical infections showed that milder attenuated non-pathogenic variants of a virus existed in nature and might be adapted for use as vaccines.

Thus, the aim with a living virus vaccine is to produce a mild virus infection and to reproduce, as far as is possible, the complete range of immune responses to natural infection without incurring the risk of disease. The intention, therefore, is to confer a similar degree of immunity both in terms of quality and duration. This ideal of solid clinical immunity and freedom from production of disease is closely approached by many living viral vaccines; the main ones for use in man are listed in Table 86.1.

Table 86.1 Living attenuated virus vaccines for use in man

Vaccinia	Measles
Yellow fever	Rubella
Poliomyelitis	Mumps

Methods for attenuating viruses

Several methods have been used to attenuate viruses for vaccine production.

1. Use of a related virus from another animal The classic example is the use of cowpox to prevent smallpox. The origin of the vaccinia viruses that are now used for vaccine production is obscure; they seem by modern analysis to be more like smallpox than cowpox virus. Vaccinia may be a host range mutant like the majority of living attenuated virus vaccines now in use.

Another example of using a related virus in an unnatural host is the use of measles viruses to prevent distemper in young puppies (Baker 1963). These two viruses are related morphologically and serologically; measles virus antibodies neutralize distemper virus effectively, but measles virus is not neutralized by distemper virus antibodies. Measles virus can therefore replicate in the presence of antibodies to distemper virus in the puppy derived from the bitch and confer active immunity, whereas living distemper virus vaccine would be inhibited, and not confer immunity. The inhibition of vaccine virus replication by passively transmitted maternal antibody is a serious problem when using living vaccines in settings where early infection is common. Examples are measles infection in the third world, and distemper in boarding kennels for stray dogs. The use of a related vaccine virus is one way of solving the problem; another is to use repeated doses of living vaccine during the period of waning maternal immunity and when the risk of natural infection is highest. A third possibility is the use of a potent killed vaccine.

Another example of the use of a related virus as a vaccine is the use of turkey herpes virus to protect against Marek's disease of chickens (Okasaki, Purchase and Burmeister 1970). Marek's disease is a malignant lymphomatous disorder in chickens; the related turkey herpes virus produces a persistent non-pathogenic infection in chickens and confers protection against the more serious Marek's disease; although it does not prevent superinfection it does prevent disease.

2. Administration of pathogenic or partly attenuated virus by an unnatural route The virulence of a virus is often reduced when administered by an unusual route. This method is not used for human vaccines because the margin of safety is low, but it has been employed for veterinary vaccines, such as for virulent infectious laryngo-tracheitis virus which has been given intra-ocularly as a vaccine.

3. Administration of virus with specific antiserum or immunoglobulin: combined active and passive immunization This method was much employed in the early days of vaccine development. Two examples will suffice: the use by Dalling (1931) of a distemper virus and antiserum to protect against distemper, and the combined use of the imperfectly attenuated Edmonston B strain of measles virus with human immunoglobulin for measles vaccination (Enders *et al.* 1960). These methods have been superseded by the development of satisfactory attenuated strains of distemper and measles viruses which can be used alone for vaccination. The combined method of measles vaccine and immunoglobulin for immunization proved unsatisfactory because the balance of doses required to allow the virus to replicate and stimulate immunity without causing disease was too fine. During investigation of outbreaks of measles in immunized children in the USA it was found that there were two common causes of failure. First, using the vaccine at too early an age, when maternal antibody interferes with vaccine virus replication; thus, the vaccine is now recommended to be given at 15 months of age or older. Second, where vaccine had been given simultaneously with immunoglobulin. Another cause of failure more common in the third world is instability of measles vaccine; this has been solved by the use of freeze-dried vaccine with improved stabilizers so that a vaccine dose is still present after holding at 37° for seven days.

4. Passage of the virus in an 'unnatural' host or host cell The major vaccines used in man and animals have all been derived in this empirical way. There are numerous examples of this approach; yellow fever 17D strain developed by passage in mice and then in chick embryos (Theiler and Smith 1937); polioviruses passaged in monkey kidney cells (Sabin 1956); measles virus in chick embryo fibroblasts, and rubella virus in WI38 cells or other foreign cells, like rabbit kidney cells or duck embryo fibroblasts. After a number of such passages virus is given to the natural host and a passage level chosen that gives good immunity without causing disease. Despite the intellectually unsatisfactory nature of this procedure, because of its lack of a known scientific basis, it has proved outstandingly successful. The production of vaccine strains of measles virus reviewed by Dudgeon (1969) is a good example of the method. The molecular basis for host range mutants is not known. One possible mechanism is that the enzymes concerned with processing the viral gene products in different host cells are slightly different. It has been found that protease cleavage of the precursor fusion peptide (F_0) of paramyxoviruses into F_1 and F_2 is required for infectivity and a similar cleavage is required to yield haemagglutinins HA_1 and HA_2 of influenza viruses. The enzymes in cells may be lost on passage of primary cells (Silver, Scheid and Choppin 1978) but can be replaced by exogenous protease. Mutants of Sendai virus have been isolated which are cloven by different proteases and have a different host range. Thus, the lack of enzymes involved in processing may greatly reduce the capacity of a virus to replicate.

5. Cold adaptation Efforts have been made to produce attenuated strains by means of passage of the virus at a lower temperature, sometimes as an adjunct to the selection of host range mutants. Examples are the use of low temperatures (a maximum of 34°) for the propagation of attenuated polioviruses. Deliberate attempts to produce cold-adapted strains of poliovirus type 3 in order to reduce the tendency of this strain to revert to virulence have not been successful. Maassab (1967) has used cold-adapted strains of influenza virus as the means of deriving experimental living influenza vaccines; these strains show promise. In particular, they have not been found to revert to virulence, either in ferrets or in infants.

6. Temperature sensitive (ts) mutants Chanock and Murphy (1980) have attempted to exploit the temperature differential in the respiratory tract to derive mutant virus strains incapable of growing in the lower respiratory tract whilst still growing in the upper respiratory tract and producing immunity as a result. They used fluorouracil to increase mutation rates and produced a series of ts mutants for experimental influenza and respiratory syncytial (RS) virus vaccines. These have not gained acceptance, mainly because they have shown a tendency to revert to virulence, probably on account of their tendency to produce point mutations. In an effort to prevent reversion the two most defective viruses containing ts lesions in different RNA segments of influenza virus were combined by genetic reassortment to yield a double ts mutant in gene segments coding for the P1 and P3 proteins of influenza virus. Despite a high degree of attenuation enabling the virus to be given safely to babies lacking pre-existing antibodies to both the haemagglutinin and neuraminidase of influenza viruses, revertants occurred probably owing to suppressor mutations in other genes that led to loss of ts phenotype.

7. Genetic reassortment The use of genetic reassortment to obtain attenuated vaccine strains for viruses that have segmented RNA genes has been proposed. Thus, master attenuated strains of influenza virus derived by conventional passage experiments in eggs, by cold adaptation, or by selection of ts mutants can be characterized and then used as the basis for producing new vaccine strains by genetic reassortment, so that only the RNA segment coding for the haemagglutinin and neuraminidase surface antigen are derived from the new wild virus, all the rest being from the master strain. Unfortunately, so far no really acceptable strain has been produced. Thus, with a host range master strain A/Okuda/57, a variety of strains were tested in ferrets and some were found to be too virulent despite having all the appropriate RNA segments. The problems encountered with ts mutants have been described above. The use of host range mutants which have also been cold adapted have likewise been used. Because they probably have more genetic changes, there is a presumption that the attenuation will be more stable.

Genetic reassortment is also a feasible means for developing attenuated rotavirus strains as candidate vaccine strains, in the same way as for influenza, especially as human and animal rotaviruses can be propagated in cell culture. Rotaviruses seem to share a high genetic lability similar to that of influenza viruses, and developing vaccines by well established but empirical approaches will not be easy.

All the techniques described above are empirical but many of them are very successful. Since the cardinal principle in immunization is 'Does it work in practice?', they are none the worse for their empirical origin. It is, however, unsatisfactory that the scientific basis for attenuation is so poorly understood. The ability to clone viral genes provides a powerful new tool to attempt to understand the basis for attenuation and to engineer strains with a lesser chance of reversion to virulence. The discovery of the primary structure and gene organization of poliovirus, for example, opens the way to the deliberate construction of poliovirus strains which are more precisely attenuated (Kitamura *et al.* 1981). It may be possible to delete segments of the RNA responsible for virulence, thus greatly reducing the probability of emergence of virulent revertants. It can be expected that a new generation of living vaccines will emerge from a fuller understanding of the genetic and molecular basis for virulence and attenuation.

Production and control of living vaccines

The most important aspects of the production and quality control for a living attenuated vaccine are to preserve the correct level of attenuation and to exclude contamination with other micro-organisms, especially viruses, that are more pathogenic.

An important means of maintaining the correct level of attenuation is the use of a seed lot system. Thus, the seed for making vaccine is stored in a manner and at a temperature that preserves the virus unaltered. If possible, master seeds are preserved freeze-dried or in the gaseous phase over liquid nitrogen at approximately −150°. In manufacturing establishments these seed viruses are fully documented and stored in locked secure refrigerators. The master seeds are used to make working seed cultures for actual vaccine production.

The seed virus or vaccine made from it will ideally have been used in initial clinical trials to establish the effectiveness and safety of the vaccine. The seed virus is a uniform preparation of virus which has been dispensed into many small portions for storage, so that a single ampoule can be removed in order to make vaccine. In this way, the variations from batch to batch of vaccine are minimized. Another measure to reduce the possibility of variation is to limit the number of passages of the seed virus permitted before vaccine production. It is known that variation can occur in virulence during such passage; indeed, this is the means for deriving most attenuated living vaccines. Passage generally reduces virulence and can lead to over-attenuation. It can also result in increased virulence; for example, yellow fever vaccine used in Brazil was found to be more neurovirulent and caused many cases of encephalitis (see Wilson 1967); also some attempts to make a leucosis-free yellow fever 17D strain failed because of an increase in neurovirulence which resulted from the passaging of the virus required to remove the contaminating virus. Similarly, the Sabin strains of poliovirus, especially of type 3, can increase in neurovirulence on passage in cell culture in the laboratory, and regularly do so after passage in man. In addition to using a seed lot system for manufacture, all the other conditions of growth are maintained as constant as possible, in particular the multiplicity of infection and during virus growth as well as the medium composition, and the pH and temperature used for the propagation of the substrate.

Another hazard resulting from passage of viruses in laboratories is contamination with more virulent strains or with extraneous unrelated agents. The means to safegard against these hazards have evolved over the years and are now enshrined in the requirements for the manufacture of vaccines published by the World Health Organization as a guide for National Control Authorities. In the United Kingdom the powers of control are vested in the Medicines Act and the laboratory control is carried out by the National Institute for Biological Standards and Control.

There are several possible sources of contamination. First, the laboratory area for production of vaccines has to be kept separate and can handle only the vaccine virus strains: thus, virulent strains, research and diagnostic work are excluded from production areas and only identified staff concerned with production are permitted to enter it. A second level of control of the environment is the provision of sterile, filtered air to the working areas, and monitoring by air sampling and settle plates.

Another important source of contamination is the cell substrate used to grow the viruses. This problem came into prominence with the large-scale production of polioviruses from monkey kidney cells. It was soon established that such cells were frequently contaminated, not only with already known simian pathogens, like virus B, but with a wide spectrum of newly revealed simian viruses representing most of the major virus families, as first described by Hull and his colleagues (1958). Virus contamination of other cell sources used for vaccine production has also been revealed by more sensitive techniques: thus, for example, fertile hens' eggs were originally thought to be free from viral contamination. It is now clear that they can be contaminated with a whole range of avian viruses.

There have been two main approaches to solve this problem. First, the use of animals bred in a controlled environment to provide embryos or tissues for vaccine production. These animals are kept as a closed colony and are monitored regularly, usually by a range of serological tests. In this way, flocks of chickens are maintained for the manufacture of yellow fever, measles and mumps vaccine, and ducks and rabbits have been maintained for rubella vaccine. More recently monkeys have been bred and kept under controlled conditions for similar reasons. A second approach has been the use of cells that can be maintained frozen as a cell bank and then, after clearance by extensive testing to exclude contamination with infective agents, they can be used for vaccine production. The establishment of a cell bank brings to the cell substrate the advantages of consistency and freedom from contamination attained for the virus seed by the seed lot system. The cells so far accepted for the production of living vaccines for man are the human diploid cell strains WI38 and MRC5. These cells were both derived from human embryo lung and are normal in karyology and will multiply only in culture for a limited number of generations, approximately 50 cell doublings. These cells are also unable to multiply in immunosuppressed experimental animals. For living veterinary vaccines, notably in the USA, cell lines that can be indefinitely passaged in the laboratory and no longer have a normal karyotype, have been permitted provided that the cells are free from microbial contaminants and do not produce tumours in the species for which the vaccine is used. Cell substrates for killed vaccine production raise different problems which are considered below.

Other components used for vaccine production can usually be effectively sterilized to remove possible sources of contamination. Two components frequently employed for vaccine production which are sterilized by filtration are trypsin used to disperse cells, and animal serum used in the cell growth medium. Both have been identified as possible sources of contamination of vaccines, especially with mycoplasma, bacteriophage and other viruses, in particular reoviruses and parvoviruses in serum. Pre-testing for freedom from microbial contaminants is always carried out but only relatively small samples can be tested. Recently gamma irradiation with 2.5 mega rads has been introduced for sterilization of trypsin and serum as an extra measure for reducing the chance of contamination.

When a batch of vaccine has been produced, it is desirable to carry out tests to establish that it has the same characters as the seed virus from which it is derived. It is rarely possible to do this directly, except to show that they react specifically with the appropriate antiserum. Attenuated poliovaccines are tested by comparison with a standard preparation for neurovirulence in monkeys. Other tests carried out for attenuated polioviruses are the rct 40 test, which compares the titre of virus grown at 35° and at 40.5°, and the d test, which compares the titres in low and high bicarbonate concentrations. Similarly, for other viruses, the plaque morphology and growth characters can be used to identify the vaccine and seed virus. In future, these relatively crude tests will be replaced or supplemented by more exact biochemical or immunological tests such, as by analysis by monoclonal antibodies as has been done for polioviruses by Minor *et al.* (1982), isoelectrofocusing for foot-and-mouth disease viruses and characterization of nucleic acids. Already the mapping by two-dimensional electrophoretic separation of enzymically derived oligonucleotides has been applied to the analysis of poliovirus (Nottay *et al.* 1981) and foot-and-mouth disease strains (King *et al.* 1981) isolated in the field to establish whether they are of vaccine origin. These techniques will be increasingly used in future to ensure that vaccine strains have the correct characters. Similarly, restriction endonuclease analysis has been used by Plotkin *et al.* (1983) to partly characterize cytomegaloviruses from patients given experimental vaccine in an effort to establish whether they are of vaccine or other origin. For viruses with segmented RNA genomes, like influenza viruses, the presence of the appropriate vaccine virus segments can be established analytically. The extension of these techniques to include complete sequencing of a crucial part of the genome may well be the means of vaccine strain characterization in future. So far these techniques have usually been applied to characterize strains from the community in epidemiological studies, but they are clearly applicable to vaccine virus identification in the course of manufacture.

Potency of living attenuated vaccines

The efficacy of a living virus vaccine will often have been shown in a controlled clinical trial, such as the Medical Research Council's studies on measles vaccines (1965, 1966 and 1968). They compared the incidence of measles in three groups of children; one group of about 10 000 children received killed measles vaccine followed by living vaccine (Schwartz strain); another of similar size had the living vaccine only; and the third group of 16 000 were controls. These studies showed the living vaccine to be about 90 per cent effective in preventing disease. The productuion of neutralizing antibody and its persistence can be used as a good guide to efficacy for living virus vaccines and also as a means for assessing the duration of immunity. In the evaluation of an attenuated virus vaccine a dose response study should be made to establish the infective dose for man; it is usual to provide at least a hundred-fold more virus in the vaccine as issued for sale. For example, the amount of measles, mumps, rubella and yellow fever virus in a dose is 1000 TCID50 or PFU. Since an attenuated vaccine contains living virus, the maintenance of full viability under field conditions is of the utmost importance. Problems have occurred in the past with the stability of living vaccines, especially in hot countries. Special methods have been employed by manufacturers, for example, the addition of Molar $MgCl_2$ to poliovaccine (Wallis and Melnick 1961) or the use of freeze drying medium for yellow fever, rubella and measles vaccines (Burfoot, Young and Finter 1977). WHO established a stability requirement for measles vaccine in 1982 to be used in conjunction with their expanded programme for immunization. Vaccine is exposed at 37° for one week and at the end must have lost no more than 1 log unit of virus infectivity and to have a residual titre of over 10^3 TCID50; similar stability requirements should be instituted for other living vaccines wherever possible.

It is rare for living attenuated viruses to need purification. The virus can be harvested in medium that lacks added serum and is usually capable of dilution to make up vaccine. In this respect living vaccines have considerable advantages over dead ones which often need concentration and purification to provide a sufficient mass of antigen to be effective.

Safety of living attenuated vaccines

Provided a living attenuated vaccine is prepared from a master seed of established safety and efficacy under controlled conditions of manufacture, there are few risks associated with their use. The manufacture must be carried out in adequate premises by competent staff and for this reason all control authorities have a regular system of inspection of the staff and premises of vaccine manufacturers. The main hazards that can occur have been reviewed by Wilson (1967). As far as the manufacturing process for living vaccines is concerned, these are:

1. Use of the wrong seed. In order to prevent this, master seeds are prepared and kept by authorized staff in special locked refrigerators. As already mentioned, the manufacturing premises are totally committed to vaccine production and only the handling of vaccine strains is permitted. All experimental, diagnostic or field trial work is excluded from vaccine production premises.
2. Use of altered seed virus. Before it was possible to store viruses readily at low temperatures, or freeze-dried, they were periodically passaged and thus could alter in virulence. Storage of master seeds, preferably freeze-dried, solves this problem.
3. Contamination by extraneous agents. This problem is tackled by the control of the sources of contam-

ination, by providing an environment where the handling of cultures can take place with a minimum risk of contamination. The actual working environment is arranged so that there is controlled access for staff and equipment. Moreover, the handling of cultures is carried out in an atmosphere supplied with filtered air and shown by settle plates to contain few bacterial colonies.

The cell substrate is controlled by being derived from an enclosed group of animals that are monitored regularly for microbial infection. Alternatively, the cells are stored as master cell seed. Wherever possible apparatus and medium components are heat-sterilized. Particular attention has to be paid to the conditions of sterilization to ensure that the particular cycles of heat exposure are actually effective for sterilizing large containers and volumes of liquid. Reliance on 'rule of thumb' formulae for sterilization is to court certain disaster.

The use of animal serum and trypsin for production poses problems because neither can be heat-sterilized. Usually serum is pre-tested for contamination with extraneous viruses and other microbial contamination, but this is not totally satisfactory because of the limitations on the sample size that can be tested. For other than the most fastidious cells, pre-tested serum can be sterilized by 2.5 mega rads gamma irradiation. The use of human serum as a stabilizing agent for living vaccines has caused serious outbreaks of hepatitis owing to the presence of hepatitis B virus in some serum batches. Stabilizers for vaccines that can be sterilized are now employed.

4. Inherent virulence of the virus. In selecting a strain of virus for vaccine production cautious trials have to be carried out in progressively increasing numbers of individuals. During this process of evaluation, strains that have too much residual virulence are progressively eliminated. Thus, for example, the Dakar strain of yellow fever has been replaced by the 17D strain developed by Theiler and Smith (1937). Similarly, the Sabin (1956) strains of attenuated poliovirus are used instead of those developed by Cox (1954). Modification of the Sabin type 3 would be desirable.

Many attenuated virus vaccines are associated with little, if any, untoward reactions. However, continuous surveillance of immunization programmes is required. The incidence of paralytic poliomyelitis temporally associated with the administration of Sabin strains of attenuated poliovaccine is commonly expressed in terms of numbers of cases per millions of doses distributed, rather than as doses administered to susceptible individuals. As the disease comes progressively under contriol, the importance of the rare case associated with vaccine increases, so that by the late 1970s approximately half the cases of poliomyelitis in the USA were vaccine associated (Salk, D. 1980). Moreover, as time goes on much more of the vaccine is given to individuals without prior exposure to polioviruses, so that new assessments of risk are required.

Living veterinary vaccines can usually be established as effective by direct challenge studies in the host species and as a result a more accurate assessment of the dose required to prevent disease and death can be established.

Killed vaccines

Methods of preparation

Killed virus vaccines work on a different principle from living attenuated ones. A sufficient amount of viral antigen is administered, usually in a series of doses by intramuscular or subcutaneous injection to produce an immune response sufficient to prevent disease. Although the term killed vaccine is commonly used to describe such vaccines containing non-replicating viral immunogens it is incorrect, since as the technology for production improves they may be chemically extracted, or biologically or chemically synthesized molecules not involving the growth of virulent virus at all. Prevention of disease rather than infection is often the primary goal; indeed this may be the aim for living vaccines also. Thus, Marek's disease vaccines prevent disease but not infection. Originally, killed vaccines were derived from crude preparations of virus from animal tissues. Gradually, as means to grow viruses to high titre in the laboratory were developed, first in fertile hens' eggs and later in cell culture, the virus content got higher and it became possible to purify virus and viral antigens. Indeed, it is now possible with some vaccines to control the amount of immunogen in a vaccine by weight as well as by biological activity. The antigen may be the intact virus, or extracted from the virions (a subunit vaccine). In future such vaccines will increasingly comprise fully characterized protective immunogens derived either from virus grown conventionally in cell culture, by expression of the appropriate gene in bacteria, yeasts or mammalian cells, or by synthesis of peptides representing the immunogenic determinant.

The killed vaccines that have proved most successful in man are shown in Table 86.2. A whole range of effective killed vaccines for animals exists. A general feature of such vaccines is that circulating antibodies are sufficient to confer immunity. They have so far proved less successful where other components of the immune response are essential for protection, such as local secretory IgA in the respiratory or intestinal tract, or cell-mediated immunity.

Table 86.2 Killed virus vaccines for use in man

Poliomyelitis	Hepatitis B
Rabies	Japanese B encephalitis
Influenza	Tick-borne encephalitis

The main requirements for developing a vaccine containing a non-replicating immunogen are considered below.

1. Identification and measurement of protective antigens

Since relatively large amounts of antigen are required, it is desirable to know which antigens are essential to provide protection and to have accurate *in vitro* methods to measure them. The identification of antigens enables them to be measured and their role in immunity assessed. Clearly, preparations that lack the essential antigens for protection will be useless and may indeed be harmful, as has been found to our cost. Killed vaccinia virus vaccines have not proved successful in preventing pox virus infections in animals even when large quantities of virus became available and could be purified. Rabbit pox is a good model for the study of prophylactic immunization to this group of viruses. Boulter and Appleyard and their colleagues in a series of investigations (summarized in 1973) established that vaccinia virus infection and convalescent serum conferred respectively active and passive immunity against rabbit pox. Antiserum against killed vaccinia was ineffective for prevention of rabbit pox, despite its having a much higher titre of neutralizing antibodies when tested in a conventional tissue culture assay. They later showed that the qualitative difference between the antisera was explained by the presence of antibodies to extracellular virus in convalescent sera. Thus, the crucial protective antigen was found only on extracellular virus. The presence of the new antigen could be detected by electron microscopy, by immunofluorescence and by neutralization, provided this was a plaque assay with a liquid overlay. In these conditions antibody to 'extracellular' virus produces tight plaques, whereas antibody to 'intracellular' virus allows virus to spread and produce satellite plaques because virus released from the cell is not neutralized.

Other examples of the incomplete protection afforded by a killed vaccine lacking an important protective antigen are killed measles and respiratory syncytial virus vaccines. Both these vaccines caused clinical hypersensitivity rather than clinical immunity when used in field trials in man. For measles, the vaccine was found in early trials to produce haemagglutinin-inhibiting antibodies and protection against the disease (Winkelstein *et al.* 1965). Later, on exposure to measles virus, a new syndrome characterized by pneumonia and a petechial haemorrhagic rash occurred (Fulginiti *et al.* 1967). Norrby, Enders-Ruckle and ter Meulen (1975) showed that after killed measles vaccine, antibodies to the haemagglutinin, but not the haemolysin or fusion protein of measles, were produced, whereas living virus, either natural measles or attenuated vaccine, stimulated antibodies to both components. In studies with the paramyxovirus, SV5, Merz, Scheid and Choppin (1980) showed that antibodies to the fusion protein, but not the haemagglutinin alone, were capable of preventing the spread of the virus in a susceptible cell sheet. Probably an effective killed vaccine against these viruses could be prepared if a sufficient quantity of the protective antigens were provided in it. Similarly, the protective antigens of a number of other viruses have been identified and can be used to measure antigenic content, for example, the haemagglutinin of influenza or the surface glycoprotein of rabies virus.

Progressive purification of viral antigens as the basis of vaccines has considerable attractions. The dose of antigen can be measured and adjusted to be sufficient to produce the appropriate immune response for long lasting clinical immunity. Unwanted, and perhaps toxic, components can be removed, lessening the risk of untoward reactions, either clinical hypersensitivity or immunosuppression resulting from vaccination. Crystalline hexons of adenovirus type 5 have been successfully used in experimental vaccines.

The purification of antigens has now progressed to the next stage where peptide sequences encompassing major antigenic sites of foot-and-mouth disease have been identified and shown to give rise to neutralizing antibodies to the virus and protection against the disease (Bittle *et al.* 1982). It is not known yet whether this will be applicable to other viruses, but it seems certain that new principles will emerge from such studies. A frequent consequence of purification and isolation of the antigen from intact virions or separation into smaller and smaller subviral components is a loss of immunogenicity. This may necessitate reaggregation of subunits, or, for the peptides, coupling to a larger immunogenic carrier protein. Pure antigens can also be enhanced in immunogenicity by the use of adjuvants, like aluminium salts for use in man, and saponin and mineral oils for use in animals. An understanding of the molecular basis of immunogenicity and of the best means of presenting immunogens to the immune system to secure a protective immune response still requires more investigation.

A variety of methods is available to measure the antigen content of vaccines. At present these rely mainly on serological tests with specific antibodies to the protective antigen. It is necessary to standardize these tests against reference preparations. The test methods may use immunodiffusion in agar-based solidified media which can be enhanced in sensitivity by using radio-labelled reagents, or radio isotope or enzyme-labelled reagents in competitive immunological assays. Such accurate tests for antigen content need to be supplemented by tests for the immunological potency of the vaccine in animals. These tests must also be used with a reference preparation. It is usual to assess potency in terms of neutralizing antibody, and a dose schedule that produces a linear increase in response over the dose range to be tested is required. This usually, but not always, implies the use of a single dose.

2. Host immune response for protection

The host responses to infection serve two distinct roles, first, to limit the infection and cure the disease and second, to prevent second attacks. It is usually the second of these roles we aim to play with a vaccine. Knowledge of the aspects of the immune response preventing disease is therefore of the greatest interest for the development of a killed vaccine. Whereas an attenuated vaccine can be expected to reproduce the whole range of immune responses, this cannot be assumed for a killed vaccine. Circulating antibody and immunological memory are sufficient to prevent many diseases, especially when spread of virus from the portal of entry to the target organs is via the blood stream. It is characteristic of these diseases that passively administered serum is protective. Under these circumstances it is to be expected that a killed vaccine will be effective in preventing disease, and that a killed measles vaccine will be successful, because passive immunization is successful. Its failure to do so was therefore surprising until it was found that the vaccine used failed to contain the protective antigen. Killed polio vaccines were shown by the trial in 1954 (Francis *et al.* 1955) to be effective in preventing paralysis, and the dose of antigen in the vaccine was sufficient to produce immunological memory. Killed poliovaccine, unlike the attenuated vaccine, does not produce local IgA immunity in the intestinal tract, and as a result produces an inferior gut immunity. Attenuated poliovaccine stimulates local immunity by local multiplication in the gut (Ogra and Karzon 1971). Despite this clear difference, killed poliovaccine has been found to have a profound effect on the spread of polioviruses. This is because the killed vaccine virtually stops excretion of virus from the nasopharynx and greatly reduces the duration and titre of virus excretion in the faeces (Salk, D. 1980). When killed poliovaccine was introduced a number of manufacturing problems were encountered. First there was a failure to inactivate the virus properly. The steps taken to resolve this problem resulted in vaccine of less than optimal potency. Technical improvements have enabled vaccine of predetermined antigen content to be made (van Wezel, van Herwaarden and van de Heuvel-de Rijk 1979). This potent killed poliovaccine containing sufficient antigen, for example 40, 8 and 16 D antigenic units for types 1, 2 and 3 respectively, will give rise to antibodies after a single dose and stimulate immunological memory (Salk, J.E. *et. al.* 1980). It can be reasonably assumed that such vaccines will prevent the disease.

Similarly, killed rabies vaccine grown on human diploid cell cultures will produce high levels of neutralizing antibody (Plotkin 1980). For years rabies vaccines were unsatisfactory. The first rabies vaccine developed by Pasteur in the 1880s (Pasteur 1885) employed virus derived from CNS tissue dried to lose virulence. Later, killing agents were added and the most satisfactory was the Semple vaccine employing phenol treatment (Semple 1911). The vaccine contained a rather small quantity of virus antigen and many doses were required to produce antibodies. Moreover, it was associated with serious side effects owing to reactions to the nervous tissue present in the vaccine (Wilson 1967). Later the virus was produced in duck embryos. This was still a crude vaccine consisting of homogenized embryo tissue. It was improved in purity and produced fewer severe reactions but was of low and uncertain potency as judged by its capacity to stimulate antibody production. The later development of tissue culture vaccines have overcome these problems. For animal use, killed rabies vaccines can be made in a continuous cell line, like BHK clone 21 cells (MacPherson and Stoker 1962), which can be grown in suspension. As a result, highly potent killed vaccines can be made which give good protection in animals. This type of vaccine is not acceptable for use in man at present because of the employment of a continuous cell line as the substrate. For human use the virus has been adapted to human diploid cells. The result has been the development of a highly immunogenic vaccine for routine prophylaxis or for treatment of individuals bitten by an animal which might be rabid (Plotkin 1980). This vaccine produces circulating antibody and prevents disease in experimental animals. It also stimulates the production of interferon which may help in the prevention of rabies when the vaccine has to be given to unimmunized individuals after exposure to infection.

Killed influenza vaccines are less completely successful in controlling the disease than the others listed in Table 86.2. The most important reason is the great antigenic variability of the viruses that allows them to circumvent the natural post-infection immunity. The antigenic drift under the selective pressure of antibody and the larger antigenic shifts are the main challenge to those working to develop truly effective influenza vaccines. The problem is compounded by the fact that the first experience with influenza determines the subsequent immune response to later antigenic variants, so that there is a predominant response to earlier antigenic determinants and a lesser one to the novel antigens of strains encountered subsequently. Natural infection does appear to produce a more durable immunity and cope better with minor drifts than killed vaccines. Even after receiving living attenuated vaccine, serum neutralization titres seem to offer a better indication of immunity to influenza, (Freestone *et al.* 1972) but the exact fit of antibody to infecting virus must be of crucial importance since relatively small antigenic drifts result in infection and disease.

Destruction of viral infectivity

The virus infectivity in a killed vaccine has to be reliably destroyed whilst the antigenicity is retained. The traditional agent for this purpose is formalin. It was introduced for toxoiding diphtheria toxin and subsequently became established as the inactivating agent of first choice. It has limitations, especially for the killing of relatively crude viral vaccines. Soon after the introduction of killed poliovaccine there was a serious incident in the USA due to the distribution of vaccine containing residual living poliovirus. In all there were 260 cases of paralytic poliomyelitis (Nathanson and Langmuir 1963). This incident led to a review of the formalin inactivation procedure and of the principles underlying the safety testing. In reality it was a rediscovery of the limitations of formalin as a disinfectant in the presence of organic material (Wilson 1967). In his initial development of killed poliovaccine Salk (Salk *et al.* 1954) had experimented with different concentrations, temperatures and times of inactivation before deciding empirically that 1/4000 formalin at 37° gave a consistent, apparently first-order, inactivation reaction. Subsequently, using purified virus, van Wezel and his colleagues (1979) clearly demonstrated the satisfactory performance of 1/4000 formalin for the inactivation of poliovirus. In the interim, the limitations of formalin as a means of inactivation of crude microbial preparations such as poliovirus in the presence of viral clumps and protein had been rediscovered. These deficiencies apply not only to poliovirus but also to other viruses. Indeed, some outbreaks of foot-and-mouth disease have been traced to failure to inactivate vaccine virus (King *et al.* 1981). Formalin also probably destroys the protective antigen in some vaccines, as it does with the fusion protein of measles virus.

There are other possible inactivating agents. Acetylethyleneimine (AEI) has been widely used for the inactivation of foot-and-mouth disease virus. Since its introduction by Brown *et al.* (1963) AEI has also been successfully used for the inactivation of rabies virus grown on BHK cells. This agent has the advantage of inactivating viruses with first-order kinetics even in the presence of contaminating proteins. A third widely used agent is β-propiolactone. This agent is highly effective and has been used for rabies and for influenza vaccines. During inactivation, especially with AEI and β-propiolactone, it is essential to maintain the pH of the virus suspension at around neutrality. If not, the pH will be lowered and will destroy those virus antigens that are sensitive to acid.

The use of a viral inactivating agent is often considered to be a means for controlling the danger from any extraneous virus that may contaminate a virus preparation particularly from the host cells. This safety net depends upon the sensitivity of the contaminating virus to the inactivating agent and should never be relied upon. The most striking illustration of this was the discovery that SV40 virus, now known to be a frequent contaminant of monkey kidney cells, was not inactivated by formalin used for the processing of killed polio vaccines and that antibodies were produced by vaccines. The problems of destroying viral infectivity will in future be solved by using defined viral antigens isolated from intact virion and free from viral nucleic acid.

Manufacture of killed virus vaccines

The requirements for the manufacture of a killed vaccine are similar to those for a living attenuated vaccine, although differing in important respects. The use of dedicated laboratories with control of the environment to avoid microbial contamination is required for both vaccines. The scale of manufacture of antigen for killed vaccines makes for some differences. Thus, for killed vaccines it is desirable to use large fixed stainless steel tanks for cell culture rather than laboratory glassware. For veterinary vaccines continuous cell lines that will grow in suspension culture can be used; these can be propagated in tanks of 1000 litres capacity or more, thus providing large quantities of viral antigen for processing (Capstick *et al.* 1962). So far such cell lines have not been generally licensed for use in the manufacture of human vaccines, although WHO have drawn up minimum requirements for their use. The problem is that cell lines that will grow in suspension culture are tumorigenic in animals. This means that the cells from such an immortal cell line, when implanted into animals that are unable to mount an immune response, will produce tumours. Some of these cells will also metastasize to distant organs, whereas others, even in this environment, are able to undergo only limited replication without tumour development. Examples of cells that produce metastases under these conditions are the baby hamster kidney cell line, BHK 21, and the human cell line, HeLa. Some other continuous cell lines, such as the monkey kidney cell line, Vero, which produce local tumours, do not metastasize. Other cells, like the human diploid lung fibroblast cell strains WI31, and MRC 5, do not produce tumours at all.

Primary cells, like monkey kidney cells, can be grown on microcarriers (van Wezel, van Herwaarden and van de Heuvel-De Rijk 1979) and thus be cultivated on a big scale in homogeneous systems in culture tanks holding 1000 litres or so. Vero cells are superior to human diploid cells for use in this system for the production of some viruses—rabies and poliovirus—because they produce higher yields of virus. There is as yet no consensus as to whether such cells would prove acceptable for the preparation of killed vaccines. There would seem to be no objection to their use provided the product of such cells could be purified by a process shown to exclude cellular proteins or nucleic acid. The problem is that it is impossible to prove a negative, but techniques exist to show that for each gram of antigen there is less than a nanogram of such impurities.

The cell substrate must be derived from enclosed animal colonies if primary cells are used, or preferably, from a thoroughly tested and characterized master cell seed bank. The choice of cells and virus should depend upon the highest yield of protective antigen. For human vaccines this choice is limited because only human diploid cells are generally accepted, although consideration of the purification and treatment steps that may allow the use of continuous cell lines is under consideration.

The virus seed needs to be standardized in the same way as for a living vaccine, and a seed lot system is one of the measures taken to ensure that there is consistency of manufacture. Consistency and reproducibility of each step in the chain of manufacture is vital for the production of safe and effective vaccines. In particular, the purification and inactivation procedures have to be controlled to ensure that the usual course is being followed. This requires tests for the purity and quality of viral antigen at each step and a check on the kinetics of viral inactivation to ensure that there are no deviations. After inactivation, the virus is passed through a physical barrier so that it is in an area where no living virus is present. In this area only staff that do not handle living virus are permitted, so that contamination of the inactivated vaccine by living virus is prevented. Regulations always require that if there is any break in consistency of manufacture, then this has to be re-established by successful processing of a series of batches before vaccine can be released.

The two most important factors in assessing a non-replicating immunogen as a vaccine are safety and efficacy. The safety depends on adequate killing of the living virus when this is the means of obtaining the viral antigen. The principles of inactivation and testing of vaccines for residual virus are well understood and, provided the lessons are not forgotten, as they have unfortunately been in the past over the use of formalin, then killed vaccines can be safely and consistently made. The second problem is ensuring that the correct antigen is present to confer protection. A test to measure antigen content throughout the process of preparation is required; there are now adequate techniques to do this. Finally, an animal potency test is needed using as a control a preparation that has proved effective in a controlled field trial.

Serotherapy and prophylaxis

The control of virus infections by antiserum has been largely displaced by vaccines because of the short-lived protection it afforded. The half-life of immunoglobulin in the blood is about three weeks. Preformed antibodies, however, can be effective both for the treatment and prevention of virus diseases and they still have a limited role to play. Antisera prepared in animals and partly purified are rarely used for this purpose. In earlier times before viruses could be grown easily to high titre in cell culture, potent antisera were difficult to prepare. Moreover, foreign serum gives rise to hypersensitivity and, in addition to the dangerous manifestation of hypersensitivity like anaphylactic and serum sickness; the antibodies are more rapidly cleared from the recipient's blood stream on a second administration. Fractionation of immunoglobulin to produce F(ab) fragments has not been employed for viral antisera. The major use of antisera from animals in virus disease has been for the treatment of rabies. Unfortunately, the highly potent and effective tissue culture vaccine is at present too expensive for widespread use in the pre-exposure immunization of subjects at risk in the third world. Rabies virus can be propagated in cells, for example, primary dog kidney cells or the continuous cell line derived from African green monkeys, Vero, which will grow on microcarriers for production of killed poliovaccine. Specific immunoglobulin is therefore still required for the post-exposure treatment. Although it is technically feasible to immunize volunteer donors and produce specific human antirabies immunoglobulin, in most parts of the world animal antisera are used.

Specific immunoglobulin is also used for prevention of hepatitis A virus infections. Infection with this virus is very prevalent and usually subclinical in developing areas of the world, but it represents a significant hazard for visitors from areas where infection is not acquired in childhood. Normal immunoglobulin is used and will confer protection for six months, but as tests for virus neutralizing antibodies become available for hepatitis A virus so the specific antibody content of the immunoglobulin must be standardized.

Specific immunoglobulin against varicella-zoster virus obtained from patients convalescent from an attack of zoster is used to prevent chicken-pox, especially among immunodeficient patients. Similarly, immunoglobulin with high levels of antibody to cytomegalovirus has been employed to prevent cytomegalovirus infections. These immunoglobulin preparations are of value in prophylaxis but are of unproven value for treatment.

Specific immunoglobulin for measles is rarely used nowadays for the prevention of the disease because vaccine is so effective. It is still of value in childrens' wards when immunization campaigns are inadequate and outbreaks of disease can occur.

Antiviral chemotherapy

The development of effective antiviral drugs has lagged behind the success of antibacterial drugs. There are many reasons for this; most important is the close association of viruses and their host cells in the biochemical processes of replication. Consequently, the targets for selective action of drugs are more limited. A second problem for antiviral chemotherapy is the short-lived nature of many viral diseases. When the disease is diagnosed the virus multiplication is largely completed. This problem is well illustrated by the short time course of recurrent herpes simplex infections, either herpes labialis or genitalis, which last for a few days compared with primary infections. Sometimes, especially for primary infections, virus replication continues for many days and there is time for successful intervention with an effective antiviral drug.

Very few drugs have become established in clinical use for the control of virus diseases. A list of those that have been approved for clinical use in a country with exacting requirements is shown in Table 86.3.

Table 86.3 Antiviral drugs

Methisazone	Idoxuridine
Amantadine	Vidarabine
(and Rimantadine)	Trifluorothymidine
Cytarabine	Acyclovir

Testing for antiviral activity

The screening of compounds for antiviral activity is carried out in a manner similar to the search for antibacterial drugs. A nearly confluent infection of a monolayer of susceptible cells is established in dishes under a solidified medium overlay. Discs soaked in the compounds under test are added and the drug will diffuse, and if it is antiviral will produce a zone of protected cells. The diameter of protection is roughly proportional to the potency of the compound. Such plates can be stained and kept as a permanent record. More precise estimation of an antiviral activity can be obtained by determining the dose that reduces the plaque count by 50 per cent. The effective inhibitory dose (EID50) expressed in micromoles is an essential tool for the chemist in guiding the synthesis of more active compounds. The appearance of such a plaque reduction assay for an active compound 4′,6-dichloroflavan against rhinoviruses is shown in Figs. 86.1 and 86.2.

The use of cell cultures to test for antiviral activity is convenient because it is fast and accurate and many samples can be tested. For a number of reasons it does not necessarily reflect antiviral activity *in vivo*. An inactive drug may be converted to an active one *in vivo*; for example, acyclovir is active *in vitro* and *in vivo*, but there are a number of inactive compounds *in vitro* that are enzymically converted to acyclovir *in vivo*. Other drugs may be converted to inactive metabolites *in vivo* or they may not reach the site of viral replication. Interferon, for example, when given parenterally, does not prevent virus infections of the respiratory tract, yet when given by the respiratory route it prevents common cold infections (Scott *et al.* 1982*a*, *b*). It is, therefore, essential to supplement cell culture screening of antiviral compounds by tests in a suitable model of the virus infection.

The antiviral drugs that have been developed so far have been derived more or less empirically. The first successful antiviral drug was methisazone which was described by Bauer (1977) for the prevention of smallpox and treatment of complicated vaccinia infections. It was discovered as a result of following up an observation by Brownlie and Hamre (1951) that *p*-aminobenzaldehyde thiosemicarbazone inhibited the replication of vaccinia virus. It is now only of historical importance since smallpox has been eradicated.

The next antiviral compounds were products of intensive efforts in the USA to find antitumour drugs by synthesizing analogues of compounds involved in the cellular synthetic pathways for nucleic acids. These compounds, idoxuridine, trifluorothymidine and vidarabine, are not truly selective antiviral compounds but depend for their differential toxicity on the greater viral as opposed to cellular DNA synthesis. They have been found to be of value for the treatment of local infections with the herpes group of viruses. Their greatest value has been for eye infections with herpes simplex. Vidarabine has also been licensed for use systemically for the treatment of herpes encephalitis and neonatal herpes, but it is far from perfect and needs to be given early if it is to be effective in curing the infection without serious sequelae (Whitley *et al.* 1980, 1981).

A considerable advance in antiviral chemotherapy occurred with the discovery of acyclovir. This drug is truly selective for the treatment of herpes simplex since it was shown by Elion and her colleagues (1977) that it is preferentially phosphorylated by herpesvirus-specified thymidine kinase and that the triphosphate of acyclovir acts to inhibit viral DNA polymerase and

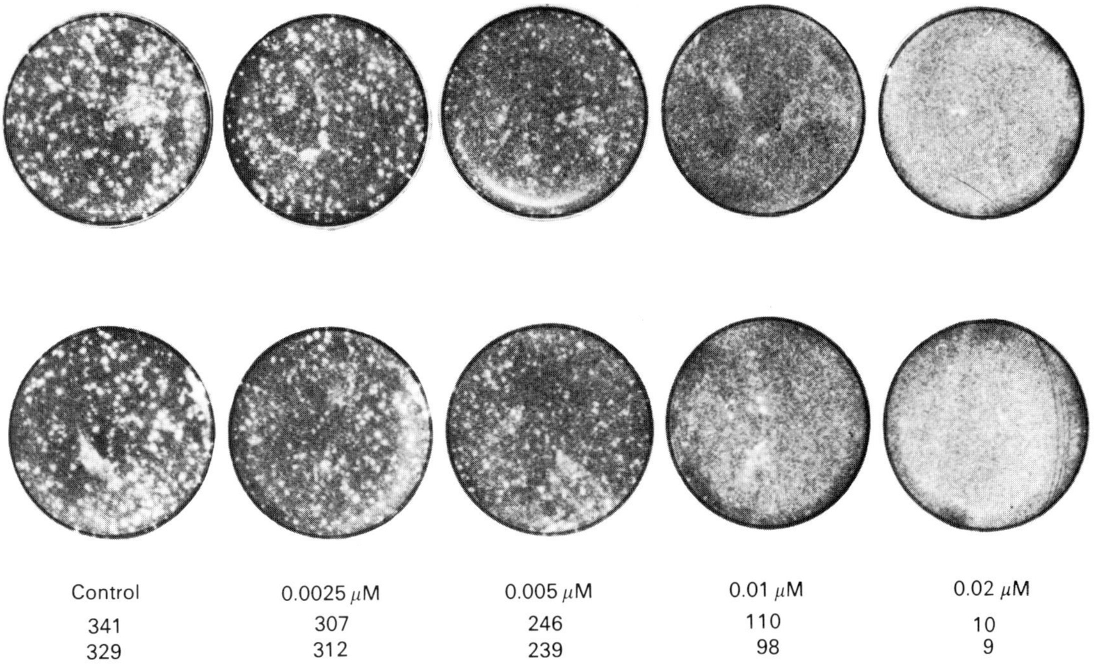

Fig. 86.1 Plaque reduction assay with 4′,6-dichloroflavan against rhinovirus type 1B

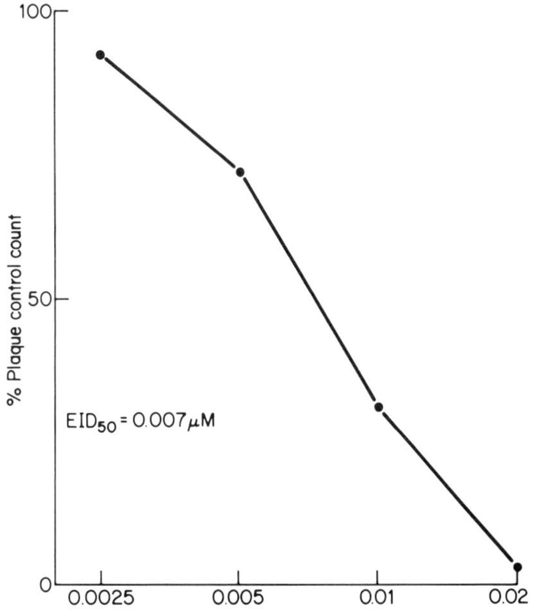

Fig. 86.2 Dose response line for 4′,6-dichloroflavan against rhinovirus type 1B in tissue culture

as a chain terminator in DNA synthesis. This compound has been shown in controlled studies to be effective in both primary and recurrent infections with herpes simplex virus as well as varicella zoster when administered topically or more successfully systemically by mouth or intravenously. It is also effective in rabbits against herpes simiae. It does not appear to be effective against cytomegalovirus and, although it may be effective against EB virus, this is not yet proven by controlled field trials.

The final proven antiviral drug is amantadine which prevents infections with influenza A viruses. It is effective at a very early stage of viral replication.

Clinical assessment of antiviral drugs

There has been an unfortunate tendency for antiviral drugs to be studied in a few acutely ill patients and, on the basis of a clinical impression, for them to be advocated as effective in a particular disease. Later controlled trials showed that the early claims could not be sustained. For example, Stevens *et al.* (1973) showed that cytarabine at a dose level of 100 mg/m^2 intravenously was actually harmful in the treatment of varicella zoster in immunodeficient patients, but not till after its use had been widely advocated.

Carefully controlled studies on the other hand have shown the effectiveness of acyclovir in the prevention of herpes simplex and varicella zoster infections in the immunodeficient patient. It is also effective in treatment. Effectiveness is greatest in controlled studies in the more severe disease and when assessed by objective criteria like the excretion of virus. It is less effective for recurrent disease especially when it is seen late on. Thus, in studies of herpes zoster it is more effective in elderly patients who have more severe disease than in younger ones. Similarly, in the treatment of recurrent herpes simplex the earlier the drug is given the more effective it is; for example, if it is given during a prodromal phase recurrence is usually aborted. Similarly, the drug is more effective given systemically, intravenously or orally rather than locally. The results of studies with acyclovir are summarized by Field and Phillips (1983).

Amantadine has been shown to be effective in controlled studies for the prevention of influenza (Stuart-Harris and Schild 1976) but for several reasons has not become established for prevention of the disease. First, it is not effective against influenza B infections and the degree of effectiveness is not good enough in relation to the side effects to encourage its use. Moreover, the drug has not been consistently effective in all trials. More recent work, especially in the USA, shows that the related drug, rimantadine, is as effective with no more side effects than a placebo.

Resistance to antiviral drugs

Viruses resistant to antiviral drugs are easy to derive *in vitro*, especially for amantidine and acyclovir. Resistant mutants to acyclovir which are easy to generate in the laboratory are usually mutant viruses that lack viral-induced thymidine kinase; there is evidence that these viruses are less pathogenic and less neurotropic, at least in animals. Resistant mutants to acyclovir of four different classes have been described.

1. TK minus strains.
2. Strains producing less viral TK.
3. Strains with modified TK with less capacity to phosphorylate acyclovir.
4. Strains with modified viral DNA polymerase with a lesser ability to recognize acyclovir.

Field and Phillips (1983) have reviewed these mechanisms, which reveal a complex series of possible interactions, some mutants being resistant to other drugs that interfere with DNA synthesis and some not. Studies to date reviewed by Dekker *et al.* (1983) show that the emergence of resistant mutants has not proved to be an important clinical problem. Animal experiments show that such resistant mutants can be selected by continuous treatment of herpes infection in the immunodeficient. The emergence of resistant viruses in recurrent disease is unlikely, since sensitive viruses are latent in the central nervous system. The future is less clear and experience with antibacterial chemotherapy certainly behoves us to monitor isolates for the possible emergence of resistance. The study of resistant mutants may also lead to the development of new antiviral drugs.

Interferon as an antiviral agent

At present interferon, despite all the work that has been done, remains a promise for the future rather than part of the armamentarium of physicians combatting infections.

The types of interferon available are shown in Table 86.4. Interferons can now be produced in large quantities either as a mixture of alpha interferons from lymphoblastoid cells, or as monocomponent individual molecular species of alpha interferons, by recombinant DNA technology in *Esch. coli.* Beta and gamma interferons are also now readily available as a result of the application of recombinant DNA techniques. The complexity of the biological activities of interferons, both antiviral and on the immune system, is steadily emerging (Stewart 1979, Meager 1983); it is clear that they play a major role in the regulation of host defence mechanisms. It is not yet certain what role, if any, interferon will play in the control of virus infections. Although it is clearly active in a number of settings, for example, in rhinovirus infections of the respiratory tract, herpes simplex and varicella infections, especially herpes keratitis, its usefulness in re-

Table 86.4 Classification of interferons

Name	Alternative	No. of species	Inducers	Major producing cells
Alpha (α)	Leucocyte Lymphoblastoid	At least 16	Viruses Bacteria Virus infected cells Mitogens	B lymphocytes Macrophages
Beta (β)	Fibroblast	1	Viruses Polynucleotides	Fibroblasts Epithelial cells
Gamma (γ)	Immune	1	Antigen Virus infected cells	T lymphocytes

lation to toxicity is not established. There are indications that interferon may be beneficial especially in papilloma virus infections and probably in hepatitis B virus infections. The need for carefully controlled trials is an overriding requirement. The assumption that interferon as a natural product would be free from toxicity is clearly incorrect; the toxicity indeed so closely resembles the systemic effects of influenza virus infection as to encourage the speculation that they are due to the stimulation of endogenous interferon by the virus. Nevertheless, studies of the biology and mode of action of interferon promises to open up pathways for new methods in the control of virus infections.

The future

The advances in molecular biology of viruses, their structure in relation to their function and mode of replication, are revealing new targets for antiviral chemotherapy. The existence of selective agents like amantadine and acyclovir and the immense research effort they have spawned suggests that we may well be on the brink of a new era of antiviral chemotherapy.

References

Baker, J.A. (1963) The proceedings of the 12th Gaines Veterinary Symposium, Kankakee, Illinois. *Vet. Rec.* 30 November 1963.

Bauer, D.J. (1977) *Specific treatment of virus diseases.* MTP, Lancaster.

Bittle, J.L. *et al.* (1982) *Nature Lond.* **298,** 30.

Boulter, E.A. and Appleyard, G. (1973) *Progr. med. Virol.* **16,** 86.

Brown, F., Hyslop, N. St. G., Crick, J. and Morrow, A.W. (1963) *J. Hyg., Camb.* **61,** 337.

Brownlie, K.A. and Hamre, D. (1951) *J. Bact.* **61,** 127.

Burfoot, C., Young, P.A. and Finter, N.B. (1977) *J. biol. Standard.* **5,** 173.

Capstick, P.B., Telling, R.C., Chapman, W.G. and Stewart, D.L. (1962) *Nature Lond.* **195,** 1163.

Chanock, R.M. and Murphy, B.R. (1980) *Rev. infect. Dis.* **2,** 421.

Cox, H.R. (1954) *Brit. med. J.* **ii,** 259.

Dalling, T. (1931) *Vet. Rec.* **18,** 617.

Dekker, C., Ellis, M.N., McLaren, C., Hunter, G., Rogers, J. and Barry, D.W. (1983) Virus resistance in clinical practice. *J. antimicrob. Chemother.* **12,** Suppl. B, 137

Dudgeon, J.A. (1969) *Brit. med. Bull.* **25,** 153.

Elion, G.B., Furman, P.A., Fyfe, J.A., di Miranda, P., Beauchamp, L. and Schaeffer, H.J. (1977) *Proc. nat. Acad. Sci. Wash.* **74,** 5716.

Enders, J.F., Katz, S.L., Milovanovic, N.V. and Holloway, A. (1960) *New Engl. J. Med.* **263,** 153.

Field, H.J. (1983) *J. antimicrob. Chemother.* **12,** Suppl. B, 129.

Field, H.J. and Phillips, I. (1983) *J. antimicrob. Chemother.* **12,** Suppl. B, 1–199.

Francis, T. Jr. *et al.* (1955) *Amer. J. publ. Hlth.* **45,** No. 5 Suppl.

Freestone, D.S., Hamilton Smith, S., Schild, G.C., Buckland, R., Chinn, S. and Tyrrell, D.A.J. (1972) *J. Hyg. Camb.* **70,** 531.

Fulginiti, V.A., Eller, J.J., Downie, A.W. and Kempe, C.H. (1967) *J. Amer. med. Ass.* **202,** 1075.

Hayden, G.F., Preblud, S.R., Orenstein, W.A. and Conrad, J.L. (1978) *Pediatrics* **62,** 965.

Hilleman, M.R., Larson, V.M., Lehman, E.D., Salerno, R.A., Coward, P.G. and McLean, A.A. (1980) Proc. Internat. Conf. on Herpes Viruses. Emery University School of Medicine, Atlanta, Georgia.

Hull, R.N., Minner, J.R. and Mascoli, G.C. (1958) *Amer. J. Hyg.* **68,** 31.

Jenner, E. (1798). *An enquiry into the causes and effects of the variolae vacciniae, a disease discovered in some of the Western counties of England, particularly Gloucestershire, and known by the name of the Cowpox.* London.

Kapoor, A.K., Nash, A.A., Wildy, P., Phelan, J., McLean, C.S. and Field, H.J. (1982) *J. gen. Virol.* **60,** 225.

King, A.M.Q., Underwood, B.O., McCahon, D., Newman, J.W.I. and Brown, F. (1981) *Nature Lond.* **293,** 479.

Kitamura, N., *et al.* (1981) *Nature Lond.* **291,** 547.

Maassab, H.F. (1967) *Nature Lond.* **213,** 612.

MacPherson, I. and Stoker, M. (1962) *Virology* **16,** 147.

Meager, A. (1983) Interferons: their role in natural resistance to virus infections. In: *Human Immunity to Viruses.* Ed. by F. Ennis. Academic Press, New York and London.

Merz, D.C., Scheid, A. and Choppin, P.W. (1980) *J. exp. Med.* **151,** 275.

Minor, P.D. *et al.* (1982) *J. gen. Virol.* **61,** 167.

MRC Trials of Measles. MRC Measles Vaccine Committee, 1965, 1966 and 1968. *Brit. med. J.*, 1965, **1,** 817; 1966, **1,** 441; 1968, **1,** 441.

Nathanson, N. and Langmuir, A.D. (1963) *Amer. J. Hyg.* **78,** 16–81.

Norrby, E., Enders-Ruckle, G. and ter Meulen, V. (1975) *J. infect. Dis.* **132,** 262.

Nottay, B.K., Kew, O.M., Hatch, M.H., Heyward, J.T. and Obijeski, J.F. (1981) *Virology* **108,** 405.

Ogra, P.L. and Karzon, D.T. (1971) Formation and function of poliovirus antibody in different tissues. *Progr. med. Virol.* **13,** 156.

Okasaki, W., Purchase, H.G. and Burmeister, B.R. (1970) *Avian Dis.* **14,** 413.

Pasteur, L. (1885) *C. R. Acad. Sci.* **101,** 765.

Plotkin, S.A. (1980) *Rev. infect. Dis.* **2,** 433.

Plotkin, S.A., Friedman, H.M., Starr, S.E., Smiley, M.L., Grossman, R.G. and Barker, C. (1983) The prevention of cytomegalovirus disease. In: *Human Immunity to Viruses.* Ed. by F.A. Ennis. Academic Press, London and New York.

Polakoff, S. (1981) *J. Hyg., Camb.* **87,** 443.

Sabin, A.B. (1956) *J. Amer. med. Ass.* **162,** 1589.

Salk, D. (1980) Parts I, II and III. *Rev. infect. Dis.* **2,** 228.

Salk, J.E., Krech, U., Youngner, J.S., Bennett, B.L., Lewis, L.J. and Bazeley, P.L. (1954) *Amer. J. publ. Hlth* **44,** 563.

Salk, J.E., *et al.* (1980) *Develop. biol. Standard.* **47,** 181.

Scott, G.M., Phillpotts, R.J., Wallace, J., Ganci, C.L., Tyrrell, D.H.J. and Gremier, J. (1982*a*) *Lancet* **ii,** 186.

Scott, G.M., Phillpotts, R.J., Wallace, J., Secher, D.S., Cantell, K. and Tyrrell, D.H.J. (1982*b*) *Brit. med. J.* **284,** 1822.

Semple, D. (1911) The preparation of a safe and effective antirabies vaccine. *Sci. Mem. Medical Sanitation Department India*, No. 44.

Silver, S.M., Scheid, A. and Choppin, P.W. (1978) *Infect. Immun.* **20,** 235.

Stevens, D.A., Jordan, G.W., Waddell, T.F. and Merigan, T.C. (1973) *New Engl. J. Med.* **289,** 873.

Stewart, W.E. (1979) *The Interferon System.* Springer-Verlag, New York.

Stuart-Harris, C.H. and Schild, G.C. (1976) *Influenza, the Viruses and the Disease.* Edward Arnold, London.

Theiler, M. and Smith, H.H. (1937) *J. exp. Med.* **65,** 767.

van Wezel, A.L., van Herwaarden, J.A.M. and van de Heuvel-de Rijk, E.W. (1979) *Develop. biol. Standard* **42,** 65.

Wallis, C. and Melnick, J.L. (1961) Texas Reports. *Biology and Medicine* **19,** 683.

Whitley, R.J., Nahmias, A.S., Soong, S-J., Gallaso, G.G., Fleming, C.L. and Alford, C.A. (1980) *Pediatrics* **66,** 495.

Whitley, R.J. *et al.* and the Niaid Collaborative Antiviral Study Group (1981) *New Engl. J. Med.* **304,** 313.

Wilson, G.S. (1967) *The Hazards of Immunizations.* The Athlone Press.

Winkelstein, W. Jr., Karzon, D.T., Rush, D. and Mosher, W.E. (1965) *J. Amer. med. Ass.* **194,** 106.

87

Poxviruses

Derrick Baxby

Introductory

The last endemic case of human smallpox occurred in 1977, and in 1980 the Assembly of the World Health Organization declared that smallpox had been eradicated. Consequently this chapter deals with a virus group whose most important member no longer exists naturally. Previous editions of this and other books have had to take account of smallpox as an important disease, endemic in some countries, which might be introduced into non-endemic areas at any time.

Of the diseases produced by poxviruses, smallpox and molluscum contagiosum are specifically human. Infection with vaccinia virus is artificial, and the morbidity and mortality caused by complications of smallpox vaccination will be eliminated as vaccination is discontinued. Human diseases caused by other poxviruses are relatively trivial infections transmitted from animals. Although goatpox, sheeppox and camelpox are of veterinary importance, greatest attention has always been paid to smallpox. Consequently a rather more detailed account is given of it than might be expected of an extinct disease. Although smallpox has gone, hopefully forever, its historical importance should not be overlooked. Quite apart from being the first virus disease of human beings to be eradicated, it was the first disease to be controlled by immunization and the first virus disease for which effective chemoprophylaxis was made available.

Recognizing a poxvirus

Poxviruses are the largest and most complex agents accepted as viruses. Different members infect a wide range of animal species. Although not all poxviruses are morphologically identical (see below), a newly isolated virus can quickly be assigned to the poxvirus group on the basis of size and brick-shaped morphology. Further information can be obtained by using inhibitors of DNA synthesis and histological techniques. The only DNA viruses, other than poxviruses, that replicate in the cytoplasm are those icosahedral viruses which include the virus of African swine fever (see Chapter 105).

Woodroofe and Fenner (1962) showed that four genera of poxviruses shared a cross-reactive internal antigen and suggested that possession of the antigen could be used to recognize a poxvirus. Another character which can be used is that an infectious suspension of one poxvirus will 'reactivate' a heat-inactivated suspension of another poxvirus (Fenner and Woodroofe 1960) by the mechanism known as 'non-genetic reactivation' (see below). Poxviruses contain many enzymes (see later) and it has been proposed that their possession could be used as a means of assigning an unknown virus to the poxvirus group (Pogo *et al.* 1971).

Taxonomy and nomenclature

The latest proposals were accepted by ICTV in 1978 (Matthews 1979). Poxviruses are placed in one family, *Poxviridae*, which is divided into two sub-families, *Chordopoxvirinae*, i.e. poxviruses of vertebrates, and *Entomopoxvirinae* i.e. poxviruses of insects. There is no evidence that the poxviruses from insects can infect vertebrates and they will not be discussed here. They have been the subject of comprehensive reviews (Granados 1973, 1981).

The poxviruses of vertebrates are placed in six genera or sub-groups. These and the species or members within them are listed in Table 87.1. Table 87.1 also lists other poxviruses which have yet to be assigned to genera, and others which can be assigned to genera but whose relation to existing species is uncertain.

Poxviruses are assigned to genera on the basis of close serological and genetic relationship. In addition the morphology of the members of the genus Parapoxvirus is conspicuously different from that of other genera, although they are still recognizable as poxviruses by electron-microscopy. The general properties of poxviruses of vertebrates have been summarized in a number of useful reviews (Mayr *et al.* 1972, Nakano 1977, Andrewes *et al.* 1978). Those poxviruses which are known to be pathogenic for man are listed in Table 87.2 on page 165.

Structure

Most information is available for Orthopoxvirus, and vaccinia in particular. This section describes features which are probably common to all vertebrate poxviruses or which are representative of genera. Features important for distinguishing species within genera are discussed in later sections. Excellent reviews have appeared recently which deal with poxvirus biochem-

Table 87.1 Family Poxviridae

A. Sub-family *Chordopoxvirinae* (Poxviruses of vertebrates)

Genus	Species[1]/members
Orthopoxvirus	*Vaccinia*,[2] Smallpox,[3] Monkeypox, Cowpox, Buffalopox, Camelpox, Mousepox viruses.
Parapoxvirus	*Orf* (contagious pustular dermatitis) Paravaccinia (pseudocowpox, Milkers' nodes, Bovine Papular Stomatitis) viruses.
Capripoxvirus	*Sheeppox*, Goatpox, Lumpy Skin Disease viruses.
Leporipoxvirus	*Myxoma*, Rabbit (Shope) Fibroma, Hare Fibroma, Squirrel Fibroma viruses.
Avipoxvirus	*Fowlpox*, Turkeypox, Pigeonpox, Canarypox viruses, (also isolates from starling, sparrow, etc.).
Suipoxvirus	*Swinepox* virus.

B. Poxviruses not yet officially assigned to genera

i 'Elephant virus', 'Carnivore virus' (both related to cowpox), 'Whitepox' viruses (variants of smallpox or monkeypox?), Racoonpox virus, Gerbil-poxvirus—All Orthopoxviruses.

ii 'Ausdyk' (Contagious ecthyma of camels), Seal poxvirus—Probably Parapoxviruses.

iii Molluscum contagiosum.

iv Tanapox and Yaba monkey tumour virus—serologically related.

v Cotia virus.

C. Sub-family *Entomopoxvirinae* (Poxviruses of insects).
(3 genera based on morphology, DNA content and host-range—*see* Granados 1973, 1981).

1. Species italicized are the type species, and give the common subgroup names to the genera e.g. *Vaccinia* subgroup of poxviruses.
2. Vaccinia includes rabbitpox.
3. Smallpox includes alastrim.

Table 87.2 Poxviruses pathogenic for man

Virus	Comment
Smallpox	Generalized infection with vesicular rash. Mortality 1–50% depending on virus strain and immunity. Last endemic case, Somalia October 1977. *Total eradication announced October 1979.*
Monkeypox	101 cases of clinical human smallpox detected in W. Africa 1970–83. Mortality 14%. Little evidence of case-to-case spread.
Vaccinia	Rare complications of vaccination. Encephalitis and Eczema vaccinatum (mortality in untreated 40%). Vaccinia gangrenosa (mortality in untreated 80–100%).
Cowpox	Restricted to Europe. Localized haemorrhagic ulcer with pyrexia. Reservoir host unknown.
Orf, Paravaccinia	Skin lesions similar to cowpox but painless. Occupational infections.
Tanapox	Localized skin nodules with pyrexia. Incidence affected by environmental factors? Spread by insects?
Molluscum	Multiple skin nodules. Long lasting. Buffalopox, camelpox, 'elephant virus' may occasionally cause infections resembling cowpox or vaccination. Sheeppox, goatpox may occasionally cause infections resembling Orf—many reports anecdotal.

istry, replication and biogenesis (Moss 1978; Dales and Pogo 1981).

Infective virions accumulate in infected cells and can be obtained in great numbers for analysis by disrupting the cells. Only a small amount of virus is released naturally into the extracellular medium. At one time this was assumed to be identical with the artificially-released virus, but important differences were later detected between artificially-released ('intracellular') and naturally-released ('extracellular') virions (Appleyard *et al.* 1971, Turner and Squires 1971). Unless otherwise stated, the work described below was done on artificially-extracted virus.

Size and morphology

Virions are somewhat pleomorphic and the size and morphology may be affected during specimen preparation. Nagington and Horne (1962) gave the size of vaccinia virus (Orthopoxvirus) as 303 × 240 nm, whereas Westwood and his colleagues (1964) found it slightly smaller (254 × 201 nm). Virions of orf (Parapoxvirus) are about 250 nm long, and narrower (158 nm) than those of vaccinia (Fig. 87.1**a, b**) (Nagington and Horne 1962). Sheeppox virus is shorter and narrower (194 × 114 nm) than vaccinia (Abdussalam 1957). In all three viruses there was considerable variation in the size of individual virions; size alone cannot be regarded as a diagnostic character.

Two types of particle can usually be seen in negatively-stained preparations. They are referred to as the M ('Mulberry') and C ('Capsule') forms and are interconvertible. The differences are caused by variation in the rate of drying and the degree of penetration of the negative stain (Harris and Westwood 1964). The M form of viruses other than Parapoxvirus has a 'beaded' surface covered with short rods or tubules about 10 nm wide (Fig. 87.1**a**), and usually composes 70–90 per cent of the virus population. The C form is an electron-dense body somewhat larger than the M form, and is bounded by a capsule or membrane about 20–25 nm thick (Fig. 87.1**c**).

The surface of Parapoxvirus species is covered by a single long thread which coils round the virion in a spiral; the criss-cross pattern (Fig. 87.1**b**) is caused by superimposition of the top and bottom surfaces of the virion in an electronmicrograph (Nagington *et al.* 1964).

The internal structure of the virion can be seen in Fig. 87.1**d, e**. In the centre is the biconcave 'nucleoid' or core. This contains the genome and is bounded by a membrane. Two lens-shaped 'lateral bodies' fit into the concavities, and the virion is bounded by an outer membrane. Inside the cell it is often enclosed within a double cellular membrane.

In 1966 Easterbrook described a simple method for degrading vaccinia virions *in vitro*. By using this technique pure preparations of outer coat components, subsurface components and cores (Fig. 87.1**f**) can be produced and analysed. For example, the cores produced *in vitro* can be compared with the subviral elements that appear *in vivo* during uncoating and morphogenesis (see e.g. Sarov and Joklik 1973).

Chemistry

The chemical analysis of vaccinia virus (Zwartouw 1964) shows that, apart from large amounts of protein and 3–6 per cent DNA, the virions contain about 6 per cent lipid and trace amounts of carbohydrate and RNA.

The poxvirus genome is one long molecule of double-stranded DNA. It has covalent cross-links at the ends, so that on denaturation a closed circle of single-stranded DNA is produced (Geshelin and Berns 1974). There are significant differences in the properties of DNA from different genera. For example Orthopoxvirus DNA has a molecular weight of about 120×10^6, a G+C ratio of 36 per cent and is about 80 μm long (Joklik 1962, Geshelin and Berns 1974). The DNA of most orthopoxviruses, but not that of variola, has inverted terminal repeat segments which cross-hybridize (Mackett and Archard 1979). That of Parapoxvirus has a molecular weight of 85×10^6 and a G+C ratio of 63 per cent (Wittek *et al.* 1979). Pox-

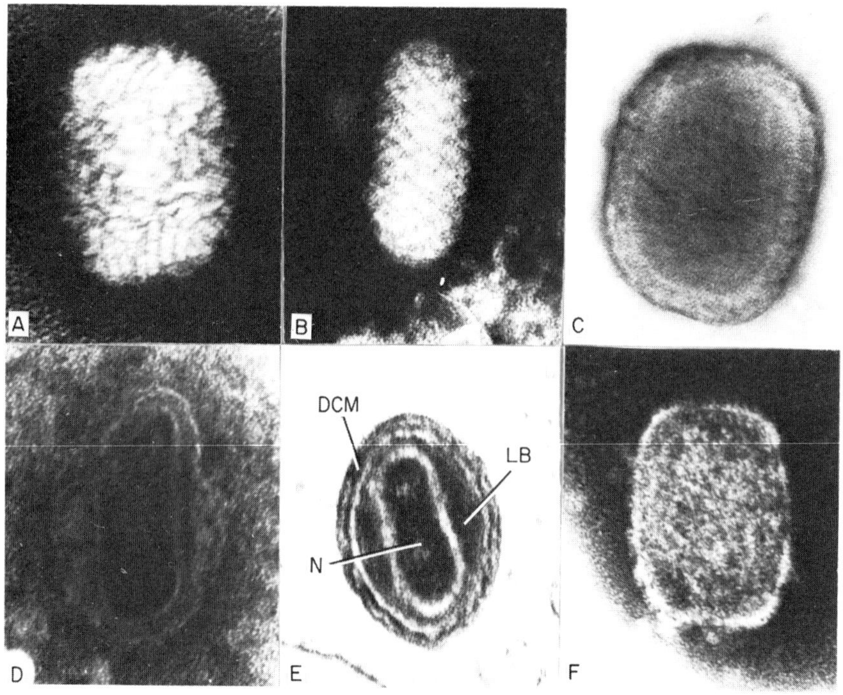

Fig. 87.1 (**a**) M form of vaccinia virus. (**b**) M form of Orf virus. (**c**) C form of vaccinia virus. (**d**) Whole virion of vaccinia, negatively-stained and showing internal structure. (**e**) Thin section of vaccinia virus invested by double cell membrane (DCM). LB = lateral body, N = Nucleoid or core. (**f**) Isolated core, negatively-stained (× 100 000).

virus DNA has been comprehensively reviewed by Holowczak (1982).

Early reports that poxviruses contained a small amount of RNA have been confirmed (Roening and Holowczak 1974). Its significance is unknown, but the small amount present (0.1 per cent) is greater than the total genome of the smallest RNA virus.

Considerable progress has been made in the characterization of *poxvirus polypeptides*. Some workers have studied the total viral polypeptides present in infected cells and determined whether they are produced before or after DNA replication (e.g. Pennington 1974). Others have determined the location of polypeptides extracted from purified virions by the Easterbrook technique (e.g. Sarov and Joklik 1972). Such an approach has shown that many of the polypeptides which distinguish Orthopoxvirus species are surface components (Turner and Baxby 1979). The workers mentioned above detected polypeptides by one-dimensional polyacrylamide electrophoresis. Recently the 40 or so structural polypeptides detected in this way have been separated into over 100 by two-dimensional techniques (Essani and Dales 1979). Some polypeptides are subject to post-translational modification such as glycosylation, phosphorylation or cleavage (see review by Moss 1978). Naturally-released virions have glycopeptides not present in virions released artificially (Payne 1978).

Information is available on the biological behaviour of some components. For example Pennington (1973) detected virus-specific serological activity in two polypeptides, and Stern and Dales (1976) isolated a polypeptide from the surface of intracellular virus that could elicit neutralizing antibody for that virus. Payne has recently suggested that a glycopeptide, of molecular weight 89 000, is responsible for Orthopoxvirus haemagglutination (Payne 1979).

Investigations of the polypeptides of species of Parapoxvirus, Leporipoxvirus and Avipoxvirus (Arita and Tagaya 1980) indicate their potential value in determining the intra- and inter-generic relationships of poxviruses.

Enzymes located in the core of poxviruses include a DNA-dependent RNA polymerase, deoxyribonucleases, a poly-A polymerase, phosphohydrolases, and guanyl- and methyl- transferases, which together are capable of producing fully functional mRNA. Also present is a protein kinase. Many of these enzymes have been isolated and their substrate specificities and viral origins confirmed in cell-free systems. Their discovery and properties have been described by Moss (1978) and Dales and Pogo (1981), who provide references.

Orthopoxvirus species are 'ether-resistant' when tested by the method of Andrewes and Horstmann (1949). However infectivity is less when more vigorous methods of lipid extraction are used (Zwartouw 1964), thus showing the structural importance of lipid.

Antigenic properties

Early work on vaccinia was reviewed by Smadel and Shedlovsky (1942). Briefly it concerned three classes of antigen; the 'nucleoprotein' antigen prepared by alkaline extraction of the virions, the antigen that elicited neutralizing antibody, and the 'LS' antigen. Most attention was focused on the LS antigen, which was believed to be a molecule with two antigenic sites, one heat-labile (L) and one heat-stable (S). This was present on the virus surface and in the virus-free soluble antigens. Because antisera absorbed with LS still agglutinated virions, the existence of other 'X'-agglutinogens was postulated.

The actual picture is more complex. Immunodiffusion techniques showed that up to twenty soluble antigens could be detected, seven of which could also be extracted from virions (Westwood *et al.* 1965). Wilcox and Cohen (1969) reviewed and extended this work and determined the time at which structural and soluble antigens were produced. Some 'early' and 'late' antigens were incorporated into virions, but not all of the late antigens elicited neutralizing antibody. Soluble antigens were examined by Rondle and Williamson (1968), who concluded that a molecule corresponding to the classical LS antigen of vaccinia did exist. It was detectable in fresh material, but was often degraded during fractionation. The LS antigen is a surface component of vaccinia but not of cowpox viruses, and antisera to it have no neutralizing activity (Baxby 1972*a*). Neutralization of intracellular virus is a complex process involving more than one antigen (see e.g. Baxby 1982).

Naturally-released virus possesses an extra surface antigen not present in intracellular virus; antibody to this antigen will neutralize extracellular virus (Appleyard *et al.* 1971, Turner and Squires 1971). The importance of these antigens will be discussed later.

Orthopoxvirus species produce a haemagglutinating antigen (HA) which acts on the erythrocytes of those fowls which are also agglutinated non-specifically by tissue lipids. Burnet (1946) suggested that the HA, easily separable from the virion without loss of infectivity, was a lipoprotein. This was confirmed by the reconstitution of HA from protein and lipid components that had been separated by solvent extraction and chromatography (Smith *et al.* 1973). More recent studies have shown that the HA is a glycopeptide, mol. wt 89 000, which will attach to, but not agglutinate erythrocytes (Payne 1979). This suggests that the glycopeptide is a 'monovalent' molecule which needs to be incorporated into a lipoprotein to provide multiple sites which will agglutinate. The HA glycopeptide is a major component of the extra envelope of extracellular virus.

Woodroofe and Fenner (1962) showed that antisera to the nucleoprotein (NP) antigens of vaccinia and myxoma viruses would neutralize the homologous virus, but precipitate the NP antigen of both viruses. They also showed that alkaline extracts of vaccinia and myxoma viruses were precipitated by antisera to Orthopoxvirus, Parapoxvirus, Leporipoxvirus, and Avipoxvirus, and suggested that the NP antigen was common to all poxviruses.

Replication

Replication takes place in virus 'factories' referred to as B-type inclusions. In addition Avipoxvirus and cowpox and mousepox viruses produce extra inclusions referred to as A-type inclusions (Fig. 87.2). These latter homogeneous eosinophilic inclusions correspond to the classical Bollinger and Marchal bodies (Kato 1964). The general features of replication revealed by electron microscopy were described by Dales and Siminovitch (1961), who thought that the whole virion entered the cell by pinocytosis. Recent work (Chang and Metz 1976), however, indicates that the outer coat of some virions fuses with the cell membrane, so that partly uncoated virions enter the cell.

Uncoating is a two-stage process. The first stage, release of the core, is brought about by pre-existing cellular enzymes; the second stage, release of DNAase-sensitive DNA, requires RNA and protein synthesis. It was originally thought that the coated genome could not be transcribed. Joklik suggested that second-stage uncoating was effected by a new cellular enzyme 'induced' by a viral protein released during first-stage uncoating. However the genome in the core produced by first-stage uncoating can be transcribed, and it is believed that the virus codes for second-stage uncoating (Joklik 1968). The existence of core enzymes explains why poxvirus DNA is not infectious *per se*, and also explains the mechanism of non-genetic reactivation—the Berry-Dedrick phenomenon. If, for example, heated non-infectious vaccinia virus is inoculated into a cell with another infectious poxvirus the vaccinia will replicate, i.e. it is reactivated. That the mechanism is 'non-genetic' is shown by using as the reactivating agent a poxvirus with damaged DNA but intact protein, (see e.g. Fenner 1962). The core enzymes provide a number of important components whose inactivation would render the virion non-infectious, and whose functions could be supplied by concurrent infection with another poxvirus. Sam and Dumbell (1981) have shown that purified poxvirus DNA can produce infectious progeny virus in cells co-infected with another closely related poxvirus.

Early transcription leads to the production of DNA polymerase, thymidine kinase, a deoxyribonuclease, the poly-A polymerase and some structural and non-structural antigens. After DNA replication late mRNA is transcribed, which leads to production of the remaining virion enzymes and antigens (see Wilcox and Cohen 1969, Moss 1978). Because different proteins are produced before and after DNA replication, some control mechanisms must be concerned. Such control is complex, and Cooper and Moss (1979)

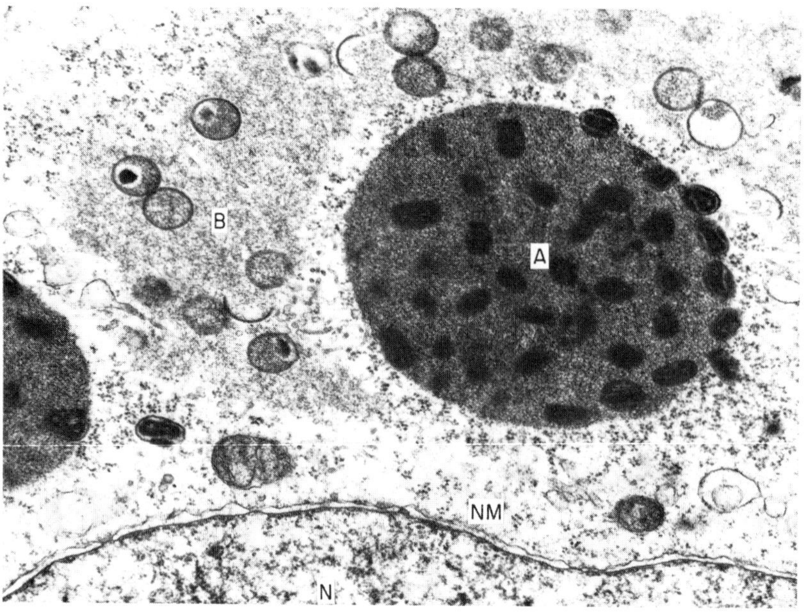

Fig. 87.2 Thin section of portion of a chick cell infected with cowpox virus. A = A-type inclusion, containing mature virions. B = B-type inclusion containing virions in various stages of maturation. N = nucleus. NM = nuclear membrane. **Note:** the A-type inclusions of some strains of cowpox and ectromelia viruses do not contain virions (Baxby 1975). (× 20 000), (from Baxby 1981).

have stressed that the mechanisms may be multiple. A model for the replication of poxvirus DNA has been proposed (Moyer and Graves 1981). This accounts for the production of mutants with terminal deletions and of genomes with inverted terminal repeat segments.

Virus in the extracellular fluid is seldom wholly 'extracellular' as defined above. The fraction released from cells which behaves as intracellular virus may lose its labile membrane after release or may be released without it. Virions may become invested with a double membrane (Fig. 87.1**e**) which they lose as they are released. Extracellular virus retains a membrane as it leaves the cell, and it is this membrane that carries the extra antigens (Payne and Kristensson 1979).

Genetics

Studies in the 1950s and 1960s showed that hybrids and recombinants could be obtained easily (see e.g. Fenner 1962). Pock mutants of cowpox, monkeypox, and rabbitpox occur with high frequency, and a linkage map for recombinants of rabbitpox has been produced (Gemmell and Cairns 1959). Recently it has been shown that the genome of pock mutants of cowpox lack a terminal segment which represents 11 per cent of the genome (Archard and Mackett 1979).

Some work has also been done on conditional lethal mutants. Production by them, under restrictive conditions, of DNA, antigens, and polypeptides is being studied (Lake and Cooper 1980). Correlation between this work and analysis of genome fragments by endonuclease digestion will enable the genes coding for various functions to be recognized (McFadden, Essani and Dales 1980).

Orthopoxvirus

The property which unites all species of Orthopoxvirus is that antiserum to one will neutralize all the others. This property can be used to assign a newly isolated virus to the genus. More convenient techniques such as gel precipitation are not so certain because minor cross-reactions occur between Orthopoxvirus and Suipoxvirus and Cotia poxvirus (see below). The methods used to identify Orthopoxvirus species (Baxby 1975, Nakano 1977, 1979) are summarized in Table 87.3 and need not be discussed in detail here. All species grow well in cell cultures. However the results obtained with some viruses in certain types of cell have not always been consistent. These anomalies, and the types of cytopathic effect produced by different species have been discussed in detail (Baxby 1975).

One major problem was that of developing a simple laboratory test which would distinguish between the viruses responsible for the two clinical varieties of smallpox, variola major with a mortality rate up to 50 per cent and alastrim (variola minor) with a mortality rate of less than 1 per cent. This was solved most conveniently when Nizamuddin and Dumbell (1961) introduced the ceiling temperature test. This determines the highest temperature at which pocks will form on the chick chorioallantois (CAM). Variola major will do so at 38.5°, whereas alastrim will not do so above 37.5°. However, further work was to show that the separation of smallpox viruses into two distinct varieties was an oversimplification (see below).

Until about 1970, apart from the problem of distinguishing between varieties of smallpox, the then known orthopoxviruses (i.e smallpox, monkeypox, cowpox, mousepox, and vaccinia) could be identified simply by examining their behaviour on the CAM, particularly at restrictive temperatures (Bedson and Dumbell 1961), and on the rabbit skin (Table 87.3). Since then a number of new orthopoxviruses have been isolated, some of which have properties previously thought to be characteristic of existing species. These include viruses from buffaloes which are similar in some respects to both smallpox and vaccinia, viruses from camels which superficially resemble smallpox virus (Table 87.3), and viruses from captive animals in European zoos which are related to cowpox (see below). Of particular significance is the isolation from non-human primates and rodents of viruses (the so-called 'Whitepox' viruses) which are so far indistinguishable from smallpox virus. This caused those concerned with the Smallpox Eradication Campaign to re-examine one of the most important criteria upon which the Campaign was based, namely that smallpox virus has no animal reservoir. At the same time the occurrence, in smallpox-free areas of Africa, of clinical human smallpox caused by monkeypox virus, stimulated studies on this virus.

In the 1970s use was made of serological and biochemical techniques for the identification of the increasing number of orthopoxviruses in the hope that the results might be more certain than those obtained by a rather arbitrary combination of biological tests. Although the results may be more certain, on the whole they confirm rather than extend the results of biological tests.

Although antigenic relations were very close, the work of Downie and his colleagues (see e.g. Downie and McCarthy 1950) indicated that Orthopoxviruses were not antigenically identical. The value of serological methods in orthopoxvirus identification was therefore re-examined. From the results of gel-diffusion tests on cross-absorbed sera Gispen and Brand-Saathof (1974) detected three particular antigens (Table 87.3). One, *Mo*, was produced only by monkeypox, which failed to produce the other two. Smallpox produced *Va*, and vaccinia, buffalopox, and camelpox produced both *Va* and *Vc*; cowpox and mousepox did not react in these tests.

By cross-neutralization tests on cross-absorbed antisera Baxby (1982) detected five surface antigens, the distribution of which distinguished between vaccinia, monkeypox, mousepox, and camelpox, and grouped buffalopox with cowpox (Table 87.3).

Harper and her colleagues (Harper *et al.* 1979), studying the polypeptides produced in infected cells, divided orthopoxviruses into four groups, namely smallpox, vaccinia (which included buffalopox),

Table 87.3 Laboratory characters of orthopoxvirus species[1]

	Small-pox	Vaccinia	Monkey-pox	Cow-pox	Mouse-pox	Buffalo-pox	Camel-pox
Pock on CAM[2]	SOW	PHU[2]-LOW	SHU[2]	H[2]	ISOW	PHU[2]	SOW
Ceiling temperature[3]	37.5–38.5	41	39	40	39	38.5	38.5
Chick mortality	+	++++	++	+++	++	++	+
Rabbit skin[4]	TrN	V	IPC	IPC	TrN	ERN	TrN
A-type inclusion	–	–	–	+	+	–	–
Diffusible LS antigen	+	+	+	–	–	+	+
Specific antigens[5]	Va	Va, Vc	Mo	–	–	Va, Vc	Va, Vc
Surface antigens[6]	?	Vac	Mp	Cp	Msp	Cp	Cmp
Polypeptides[7]	Spx	Vac	Mp	Cp	Msp	Vac	Cmp

1. Data from Baxby (1975) unless stated.
2. SOW = small opaque white, PHU = pale haemorrhagic ulcer, LOW = large opaque white, SHU = small haemorrhagic ulcer, H = haemorrhagic, ISOW = irregular small opaque white. Those indicated [2] produce white pock mutants.
3. Highest temperature at which pocks are produced on CAM.
4. TrN = transient nodule, V = variable, IPC = indurated lesion with purple centre, ERN = erythematous red nodule.
5. Distribution of 3 antigens detected by Gispen and Brand-Saathof (1974).
6. Grouping suggested by neutralization tests on absorbed sera (Baxby 1982).
7. Grouping suggested by analysis of total (Harper *et al.* 1979) and structural (Turner and Baxby 1979) polypeptides.

monkeypox, and cowpox (which included mousepox). Turner and Baxby (1979) examined structural polypeptides and confirmed the specificity of monkeypox and the close relations of vaccinia and buffalopox. In addition they established the specificity of camelpox and separated cowpox and mousepox.

Although these studies tend to confirm the biological groupings some anomalies remain to be resolved. For example, the polypeptides of buffalopox are indistinguishable from those of vaccinia, but it behaves immunologically as cowpox, yet all three viruses can be separated by biological tests (Table 87.3). Similar anomalies have been found with some of the viruses not yet assigned to species (see below).

In the final analysis it is the nucleotide sequence of the genome which will define the basic relations between the viruses, and it will be a long time before these sequences are known. However a start has been made with the construction of physical maps of genomes based on data obtained from the hybridization of genome fragments isolated after digestion with restriction endonucleases (see e.g. Mackett and Archard 1979). Such work has shown that a large central portion of the genome is conserved, and that this presumably codes for the common functions. Differences between the species were located mainly in the terminal regions. Smallpox, monkeypox, vaccinia, cowpox, and mousepox were all quite distinct; information on camelpox, buffalopox and unassigned viruses is being sought.

Smallpox

Smallpox has been greatly feared for at least four centuries. Its origins are unknown; the further back one looks the less is it certain that the disease in question really was smallpox. It was during the seventeenth century that smallpox began to have a definite recorded impact on Europe. It is sometimes difficult to imagine how great this was. Only brief examples can be given, but further information can be found in *Smallpox* (Dixon 1962) and *Jenner's Smallpox Vaccine* (Baxby 1981). If the smallpox mortality in England in the eighteenth century is corrected for the increase in population, it is equivalent to over 100 000 smallpox deaths *per annum* in the 1970s. Similar mortality is caused by heart disease or cancer. However, whereas these kill mainly adults, smallpox, before the introduction of vaccination, killed mainly children. Deaths from smallpox in England accounted for 1/14–1/8 of all deaths, and 1/8–1/2 of smallpox patients died. In addition the survivors were often disfigured, and smallpox was a major cause of blindness.

Attempts were made to prevent and control smallpox by the ancient practice of inoculating smallpox virus (variolation) in the hope that a mild, protective and non-transmissible disease would result. An analysis of this topic and of the equally controversial introduction of vaccination has recently appeared and been reviewed by Baxby (1981).

In this century, variola major persisted in Asia and parts of Africa. A less virulent form, variola minor (alastrim), became established in Europe and America but was gradually eliminated.

There have been occasional importations of variola major into Britain and America since the second world war, causing small outbreaks which were quickly controlled. For example, there were outbreaks on Merseyside in 1946 which killed 14/56 and in Brighton in 1951 which killed 10/37. During 1961–2 there were six separate importations into England and Wales with 99 cases and at least 24 deaths. In 1973 and 1978, three deaths occurred from smallpox in very small outbreaks which originated in English smallpox laboratories.

Clinical features

Though smallpox has now been eradicated, we shall describe the clinical features of the disease in the present tense. Because smallpox virus has a limited host range, information about the pathogenesis and pathology of the disease is in part deduced from experimental studies on mousepox (Fenner 1949), rabbitpox (Bedson and Duckworth 1963), and monkeypox (Cho and Wenner 1973).

Smallpox is an acute systemic disease with a characteristic pustular eruption. Infection is via the upper respiratory tract, and during the incubation period the virus spreads, presumably by a transient viraemia, and multiplies in the internal organs. At the end of this period, typically twelve days, virus can be isolated from the blood and internal organs (Downie *et al.* 1953); the first signs of illness, a prodrome with fever headache and prostration, appear, and the patient becomes infectious. The prodrome lasts three to five days, after which the characteristic rash develops, and the temperature returns to normal. The rash appears first on the face, forearms and hands, and then the lower limbs and trunk. Its distribution is characteristically centrifugal—an important diagnostic feature. The lesions are more numerous on the face, hands and forearms than on the upper arms, and more numerous on the feet and calves than on the thighs and trunk. They start as macules and progress through papules to vesicles, pustules and crusts, which become detached in about three weeks.

The clinical picture in individual cases may vary considerably. Although virus strains differ in virulence, host factors, particularly vaccination status, modify the disease. Attempts have been made to classify clinical types based on their severity. Christie (1980), for example, separates clinically obvious smallpox into four categories. *Haemorrhagic smallpox* is invariably fatal before focal lesions develop. *Flat* or *malignant smallpox* also has a very high mortality but focal lesions, usually described as being 'velvety', do develop. In *ordinary* or *classical smallpox* the distribution and development of the eruption corresponds to the 'typical' account. *Modified smallpox* is usually found in previously vaccinated patients. In some cases there may be very few lesions. Data on the incidence of these types, and the effect of previous vaccination on them are given by Downie (1970).

In patients whose immunity is high but who nevertheless become infected the prodromal illness only may occur (*variola sine eruptione*) with or without a conjunctivitis which has occasionally been the source of infection. Sero-conversion without clinical illness is probably more common than usually thought (Kempe *et al.* 1969). Such cases, although not infectious, contribute to herd immunity.

Variola major and alastrim Alastrim is usually milder than variola major. However, severe cases of alastrim may occur, and it is not possible to distinguish the two types in an individual case.

Variola major was endemic in Asia, and alastrim in South Africa and South America. In other areas of Africa the mortality was intermediate between the two classical types. Virus strains from Tanzania produce fewer pocks at 38.3° on the CAM than variola major strains, but alastrim strains produce none (Bedson *et al.* 1963). This suggested that there was a third 'intermediate' variety of smallpox. Since then, results with other biological markers and strains of virus from other areas (Dumbell and Huq 1975, Shelukhina *et al.* 1979) have shown that there is wide variation in the virulence of human smallpox associated with differences in some biological characters of the viruses. Consequently it is reasonable to regard variola major and alastrim as opposite ends of a spectrum rather than distinct diseases.

Diagnosis

Now that smallpox has been eradicated, increased attention will be paid to vesicular rashes occurring in recently endemic areas. Specimens from suspicious cases will be examined in WHO Reference Laboratories at Moscow and Atlanta. The methods available for laboratory diagnosis have been discussed by Downie and Kempe (1969) and Nakano (1979).

Chickenpox is the commonest source of confusion; examination of vesicle fluid or crust extracts by electronmicroscopy enables a poxvirus to be recognized. Orthopoxvirus antigens may be demonstrated in such extracts by gel-precipitation and complement-fixation tests. However, for unequivocal diagnosis the virus must be isolated. At present confirmation is provided by biological tests, particularly the type of pock produced on the CAM (Fig. 87.3) at restrictive temperatures (Table 87.3). Precise immunochemical methods for recognition of the specific antigens in smallpox, vaccinia or monkeypox-infected tissues are now avail-

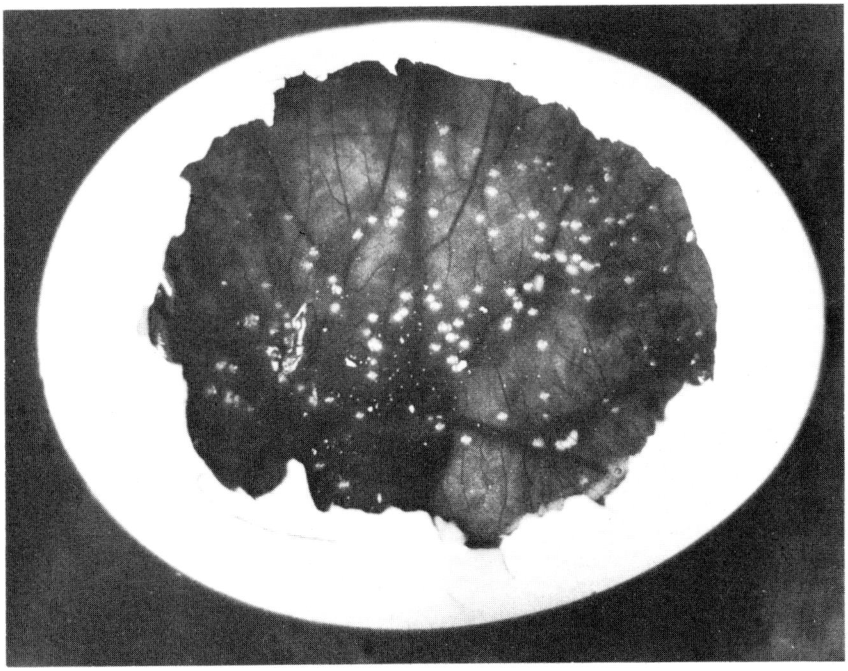

Fig. 87.3 Chorioallantoic membrane *in situ* three days after inoculation with smallpox virus and incubation at 35°.

able (Gispen and Brand-Saathof 1974, Esposito *et al.* 1977, Hutchinson *et al.* 1977), and may in future be used for precise identification.

When virus is not isolated, tests for antibody are a poor and late alternative. The results may be affected by antibody induced by previous vaccination. However, precipitating antibodies, not present after vaccination, are usually found in smallpox patients. Recently, tests have been developed which give specific evidence of monkeypox infection in cases where virus is not isolated (Walls *et al.* 1981).

Treatment No specific therapy is effective once the clinical disease develops. Vaccination and γ-globulin, when given early in the incubation period, may prevent the disease altogether in contacts or convert it into a mild form (see e.g. Hanna 1913, Kempe 1960). Methisazone (n-methyl isatin β-thiosemicarbazone) is effective in post-exposure prophylaxis. There were 3 cases of smallpox with no deaths among 1101 treated contacts, whereas 12 deaths and 78 cases occurred in 1126 untreated control contacts (Bauer *et al.* 1963).

Immunity and vaccination

The importance of circulating antibody was shown by the effectiveness of γ-globulin in preventing smallpox in treated contacts, and in aiding the recovery of patients with vaccination complications. The importance of cell-mediated immunity was demonstrated by the poor response of patients with thymic aplasia who, when vaccinated, produced antibody but not delayed hypersensitivity, and who developed progressive vaccinia (Fulginiti *et al.* 1966).

Animals immunized with inactivated intracellular virus do not completely resist challenge; their sera will not passively protect, although they have strong neutralizing activity against intracellular virus (Boulter *et al.* 1971). However, animals immunized with the extra antigen from extracellular virus are immune to challenge, and their sera will passively protect. They neutralize extracellular but not intracellular virus (Payne 1980). The important circulating antibody appears to be that elicited by the extra antigen on naturally-released virus. Cell-mediated immunity is stimulated by a virus-specific antigen which is located on the surface of infected cells. Animals immunized with this antigen resist challenge, although their sera neutralize neither intra- nor extracellular virus and do not passively protect (Kozinowski and Ertl 1976).

After infection, neutralizing, complement-fixing and haemagglutinin-inhibiting antibodies are produced. The last two do not persist for long, so that their presence can be used as an indicator of recent infection.

The effectiveness of vaccination need not be stressed. It is as evident from the safety with which adequately vaccinated persons could handle smallpox patients and virus as from the tragic deaths of those similarly occupied who were not vaccinated. However, vaccination must be done properly and subsequently inspected to ensure that a satisfactory 'take' has been obtained. A serious disadvantage of the WHO International Vaccination Certificate was that it had no provision for recording the results of a re-vaccination. A vaccination which produces a vesicle may be regarded as a 'major' reaction and any other response as 'equivocal' (see WHO, 1964). The vesicle indicates that virus replication has occurred and that immunity has been induced. The major response on re-vaccination usually develops quicker (an 'accelerated take') than on primary vaccination and constitutional disturbance is less common. Equivocal reactions on revaccination are usually hypersensitivity reactions, and may be obtained with heated antigen.

The duration of immunity varies, and even when it has waned it may prevent death or serious illness. The initial effectiveness of primary vaccination and the need for subsequent re-vaccination is seen by considering the age incidence of smallpox (see e.g. Downie 1970). The highest morbidity and mortality in the unvaccinated occurs in young children. In those vaccinated in childhood but not re-vaccinated, the highest morbidity and mortality shifts to the adults.

In countries such as the UK and USA the morbidity and mortality from vaccination in recent years was higher than that from smallpox itself. For example at least eighteen people died from vaccination complications in the rush for vaccination during the 1961–2 outbreaks in England and Wales (Dick 1971). Most countries have now discontinued routine vaccination and this will eliminate the serious complications of vaccination which have such a high mortality in untreated patients (Table 87.2). Complications such as eczema vaccinatum and vaccinia gangrenosa occur in those with deficient host responses and the mortality can be reduced significantly by treatment with γ-globulin and methisazone (Kempe 1960, Bauer 1977). Neurological complications are less amenable to treatment and there is evidence that they may be caused by mutants of increased virulence which are present in the vaccine (Ehrengut and Sarateanu 1975). Further information on vaccination complications is provided by Dixon (1962), Wilson (1967), Ferry (1977), Christie (1980), and Dittmann (1981).

Vaccination should be offered to those working with Orthopoxviruses. It will reduce the effect of, but not necessarily prevent, accidental infection with more virulent viruses such as cowpox and monkeypox.

Epidemiology and eradication

A number of features combine to make smallpox an ideal candidate for eradication. There is no animal reservoir (but see below) and recovery from smallpox results in complete elimination of the virus from the patient. Consequently the source of infection is the smallpox patient. Patients are not infectious during the relatively long incubation period and epidemics develop slowly. Consequently control measures can be organized to deal with a potential epidemic. Finally smallpox is caused by an immunologically homogeneous and stable virus and an effective vaccine is available.

Therefore the key to successful smallpox control in non-endemic areas is the early diagnosis and effective isolation of cases, as well as the prompt tracing, vaccination and isolation of their contacts. Extensive vaccination should not be necessary. Although mass vaccination was practised in endemic areas and did reduce the incidence of smallpox, experience in Nigeria indicated that endemic smallpox could be controlled by aggressive search, vaccination and containment techniques (see Foege *et al.* 1975). Such an approach is possible because of the way in which smallpox spreads. Although infection via contaminated fomites such as bedding can occur, the main mode of transmission is by respiratory virus to very close contacts. Although virus will survive for long periods under artificial conditions, virus in crusts is inactivated within about three weeks in realistic conditions of temperature and humidity (Huq 1976). Smallpox also spreads slowly because, contrary to popular belief, it is not very communicable. Henderson (1976) estimated that one patient generally infects only two to five other people, and the observations of Foege and his colleagues in Nigeria showed that it could take as long as seven weeks for all the susceptible contacts in family groups to become infected (Foege *et al.* 1975).

Airborne transmission of smallpox has always been a controversial issue and missing cases probably account for distant spread. One outbreak in which airborne spread might have occurred was investigated in Germany in 1970 (Wehrle *et al.* 1970).

Epidemiological evidence indicated that a once-endemic area could be declared free of smallpox if no cases occurred in the subsequent two years. The last case in Asia, the reservoir of variola major, occurred in Bangladesh in 1975, and the last natural case anywhere occurred in Somalia in October 1977. Consequently the world was declared free of smallpox in October 1979, a conclusion which the WHO General Assembly endorsed in May 1980.

Certification was issued only after the most extensive surveys. For example in the two years following the last case in Bangladesh extensive house-to-house searches were made, three in India and eight in Bangladesh. Each search covered 98 per cent of the 726 811 villages and towns in the two countries (Arita 1979). Brief accounts of the Eradication Campaign have been published (e.g. Fenner 1977, WHO 1980), and a final exhaustive report and analysis will be published by WHO.

An animal reservoir for smallpox? The Eradication Campaign was mounted on the assumption that smallpox had no animal reservoir. The virus has an extremely limited host range (see Herrlich *et al.* 1963); suckling mice can be infected intracerebrally, and monkeys can be infected by a variety of routes. However experimental transmission in monkeys by aerosol cannot be maintained for more than a few passages (Noble and Rich 1969). The strongest evidence against an animal reservoir is provided by the failure of smallpox to recur in countries from which it has been eradicated.

However six viruses have been isolated which all tests, including endonuclease analysis, have failed to distinguish from variola major. These are usually referred to as the 'whitepox' viruses. Two strains, originally thought to be monkeypox virus, were isolated in 1964 from the kidneys of healthy monkeys (Gispen and Brand-Saathof 1972). The remainder were isolated from primates and rodents trapped in Zaire during 1970–3 (Marennikova *et al.* 1972, 1976). The implication of these findings has been discussed elsewhere (Baxby 1977*a*, Fenner 1977). Briefly, if whitepox viruses are *not* strains of smallpox virus then increasingly sophisticated techniques are failing to separate them. Should whitepox viruses prove to be naturally pathogenic for man this will present enormous diagnostic problems. If they *are* strains of smallpox virus then total eradication may be impossible and occasional outbreaks of human smallpox may occur in the future. These viruses have been isolated from areas which have remained free from smallpox and under intensive surveillance for a number of years. The WHO Smallpox Eradication Unit did not consider them to be sufficient of a problem to delay the certification of eradication. The whitepox viruses were isolated in smallpox laboratories so that the possibility of cross-contamination is difficult to exclude. Re-evaluation of the two strains of whitepox isolated in Holland in 1964 led Dumbell and Kapsenberg (1982) to conclude that they arose from contamination of monkey tissue culture with a particular strain of smallpox virus. It is also suggested that whitepox strains are mutants of monkeypox virus (see below). For further information on smallpox see Dixon (1962, Downie (1970), and Christie (1980).

Monkeypox

Monkeypox virus was first isolated from an outbreak in captive monkeys in 1958 (von Magnus *et al.* 1959); since then another nine outbreaks have been reported in captive Asiatic monkeys but no virus isolations have been made from free-living animals (Arita *et al.* 1972). Monkeypox virus has a wider host range than smallpox virus and will grow readily in the rabbit skin. Its higher ceiling temperature is also a useful diagnostic character (Table 87.3). Particular attention has been paid to monkeypox virus since 1970 when it was found to be the cause of a disease clinically resembling human smallpox. Up to 1980 forty-eight cases of human monkeypox had been detected in West and Central Africa, seventy-six of them from Zaire (Breman *et al.* 1980; Arita *et al.* unpublished). Clinically the disease resembles smallpox although the capacity for case-to-case spread is much lower. Most of the patients were children and only four of the cases had vaccination scars. Eleven died during the acute illness. Some cases may have gone unnoticed and other factors such as malnutrition have contributed to give an unnaturally high mortality.

Human monkeypox seems to be a relatively localized problem and it is not thought to pose a serious threat to the smallpox eradication programme. Despite its ability to cause human infection, monkeypox virus is otherwise quite distinct from smallpox virus (Table 87.3). Nevertheless the existence of a virus which can cause clinical human smallpox is disturbing. In the short term the occurrence of such cases complicates post-eradication monitoring. Because of this, and the remote possibility that a patient incubating the disease may be imported into a heavily populated area, it is important that rapid and specific diagnostic methods are developed. Attempts are being made to produce monoclonal antibodies specific for the monkeypox-specific *Mo* antigen. In the long term one must wonder whether there may be a shift in the properties of monkeypox virus which may result in it becoming a more widespread human pathogen. The differences seen now between smallpox and monkeypox viruses make this eventuality unlikely. However the position is being carefully monitored in both the field and laboratory.

Mutants of monkeypox virus which produce white pocks on the CAM have been isolated (Gispen and Brand-Saathof 1972). However in 1978 it was reported that some white pock mutants, isolated in Moscow, were indistinguishable from variola major virus (Marennikova and Shelukhina 1978). It was suggested that these results explained the origin of whitepox and, more important, smallpox virus. If true this would have enormous implications. However smallpox and monkeypox viruses differ in many respects and wholesale transformation of one virus into the other would have to occur if the above explanation is correct. Analysis of further white pock mutants isolated in London indicated that some resembled smallpox virus in some superficial characteristics but that their genomes were clearly identifiable as typical of monkeypox virus (Dumbell and Archard 1980). On the other hand, analysis of the smallpox-like mutants of monkeypox isolated in Moscow has shown that their genomes are indistinguishable from those of smallpox virus (Fenner *et al.* 1980). Clearly there is a dilemma here which carefully-controlled experiments with cloned virus should resolve. Again it is difficult to exclude the possibility that cross-contamination had occurred in some experiments.

It is not known how human monkeypox infections occur and intensive efforts are being made to find the reservoir of the virus. Asiatic monkeys are evidently very susceptible indicator species and the reservoir host or hosts must be sought in Africa. Early surveys provided no evidence that monkeypox was enzootic in monkeys (see e.g. Arita *et al.* 1972). Since then surveys carried out in the vicinity of human cases have detected antibody to Orthopoxvirus in monkeys and rodents (Breman *et al.* 1977, Marennikova 1979). However it is possible that the animals had been infected with some other Orthopoxvirus. Antibody to the monkey-specific *Mo* antigen has been detected in three wild monkeys and this represents the first unequivocal evidence that monkeypox has occurred in free-living monkeys (Gispen *et al.* 1976). It is possible that monkeys are not the sole or even the major reservoir host of monkeypox virus and more information is required on the natural history of this virus.

Vaccinia

Vaccinia virus is an artificially-derived virus propagated in research and vaccine laboratories. It has no natural animal reservoir but is responsible for sporadic infections in animals such as cows, camels, pigs, and buffaloes. In these cases the virus has been transmitted to the animals from recently-vaccinated attendants. Confusion is often caused by the tendency to refer to vaccinia infections in animals as 'cowpox', 'buffalopox' etc., and this hinders studies on the poxviruses normally associated with the animals.

Vaccine strains are easily grown either on the CAM or in cell cultures and are extensively used as 'representative' DNA viruses. They can usually be distinguished from other accepted Orthopoxvirus species by suitable tests (Table 87.3), although problems sometimes arise. For example in the survey that led to the isolation of some of the whitepox viruses, a virus designated MK-10 was isolated from a non-human primate (Shelukhina *et al.* 1975). Biologically this behaves as vaccinia and has the polypeptide composition of vaccinia (Turner and Baxby 1979) but behaves immunologically as cowpox (Baxby 1982).

Strains of vaccinia virus differ both in laboratory characteristics, such as mouse and rabbit virulence, pock character and heat resistance (Fenner 1958), and also in human pathogenicity. Certain strains such as the Copenhagen vaccine which produced relatively severe lesions with a high risk of complications have been discontinued (see e.g. Polak *et al.* 1963). Since the early 1960s all routine vaccines were produced from the British (Lister/Elstree), American (Wyeth), or Russian (EM63) strains. Vaccines have traditionally been prepared in animals (calves, sheep, buffaloes) but experimental vaccines have been made in cell cultures or CAM. Attenuated vaccines have been used in patients at risk for whom normal vaccines were contra-indicated (Kempe 1968).

Very occasional outbreaks of Orthopoxvirus infection have occurred in laboratory rabbits and the virus isolated has been referred to as rabbitpox virus. No such outbreaks have been recognized in free-living rabbits. There are minor differences among the strains (Fenner 1958, Baxby 1975) but they are all very closely related to vaccinia and are currently classified as strains of vaccinia virus. The term rabbitpox is sometimes still used to maintain a continuity of identity where particular strains have been used in long term studies.

The origins of vaccinia are unknown and have been the subject of much speculation. The alternatives considered are that it has been derived from smallpox or cowpox, either separately or by hybridization, or from a virus such as horsepox which is now extinct. The various alternatives have recently been discussed in detail in the light of current knowledge (Baxby 1981).

Cowpox

Cowpox virus has been isolated only from Great Britain and western Europe and was so named because it was isolated from cattle and farmworkers in contact with them. Human infection is usually more severe than primary vaccination and of ten cases recently reviewed, four patients were admitted to hospital and two more were unable to work. Lesions occur on the hand and/or face and are often haemorrhagic; a clinical diagnosis of anthrax is not uncommon (Baxby 1977*b*). Bovine cowpox is usually restricted to the teats and the infection needs to be distinguished from paravaccinia and bovine herpes mammillitis (Gibbs *et al.* 1970). In Holland, but not Britain, bovine vaccinia has also been reported (Baxby 1977*b*).

Although it has been assumed that cowpox is enzootic in cattle recent observations indicate that this is not so. Serological and clinical surveys indicate that bovine cowpox is rare. Human cowpox is also very rare; only one or two cases are reported each year in England and Wales and human cases occur without contact with cattle. The situation contrasts strongly with diseases such as paravaccinia which are known to be enzootic in cattle. Consequently it has been suggested that the reservoir of cowpox is not the cow but some as yet unidentified small wild mammal, with both human and bovine infections being accidental (Baxby 1977*a*, *b*). The reservoir has not yet been identified. It is likely that infection in the natural host will be inapparent and may be detected only by serological surveys.

Cowpox infection in okapis was reported from Rotterdam Zoo (Zwart *et al.* 1970). More recently outbreaks of cowpox have killed valuable cheetahs in two English zoos (Baxby *et al.* 1982) but investigations failed to trace the source of infection. This susceptibility of captive animals to infection with cowpox is part of a more general pattern. Two outbreaks of pox have occurred in Moscow Zoo and killed carnivores (Marennikova *et al.* 1977) and a number of outbreaks in Germany have infected and killed elephants (Baxby and Ghaboosi 1977). Natural cowpoxvirus infections have also been recognized in domestic cats in England (Thomsett, Baxby and Denham 1978).

Cowpox virus, the Russian carnivore virus and the German elephant virus are all closely related but not identical (Baxby *et al.* 1979*b*). Ceiling temperature tests separate cowpox (40°) from the other two (39.5°) and tests on the rabbit skin and chick embryo cells separate all three. The reservoir of the Russian virus is a wild gerbil (Marennikova and Shelukhina 1976) but those of the elephant virus and cowpox virus are not yet known. These episodes provide an interesting example of how animals which have been introduced into an unnatural habitat because they were in danger in their natural one are exposed to new and unexpected pressures. The laboratory characteristics and natural histories of the viruses involved require further study.

Camelpox

Outbreaks of camelpox occur in the Middle East, and information is available on virus strains isolated in Iran (Baxby 1972*a*), Russia (Marennikova *et al.* 1974), Kenya (Davies *et al.* 1975), and Somalia (Kriz 1982). Although some infections are caused by parapoxvirus (see below) and occasional cases may be caused by vaccinia many of the outbreaks are caused by a particular orthopoxvirus—camelpox virus.

The disease in camels resembles smallpox in man (Baxby *et al.* 1975, Kriz 1982). Young animals are usually affected. The mortality is usually 4–10 per cent but can reach 28 per cent (Kriz 1982). There is a superficial resemblance between camelpox and smallpox viruses (Baxby 1972*b*) particularly in those characters, such as growth on the CAM and rabbit skin, which are usually considered of diagnostic importance (Table 87.3). However camelpox virus, unlike smallpox virus, produces multinucleate giant cells in some infected cell cultures (Baxby 1975). The differences in antigenic and polypeptide composition indicate that camelpox is probably no more related to smallpox than are other Orthopoxviruses. Nevertheless if human camelpox was less rare, diagnostic problems might arise. Although there are anecdotal accounts of human infection with camelpox virus, recent surveys in Iran (H. Ramyar, unpublished data) and Somalia (Kriz 1982) indicate that it is probably rare. Despite this lack of human pathogenicity, camelpox is the most important Orthopoxvirus infection now that smallpox has been eradicated.

Buffalopox

Not a lot is known about the natural history of buffalopox. Outbreaks have been recorded in India, and the work of Mathew (1967) and Singh and Singh (1967) showed that Orthopoxviruses were responsible. Viruses isolated by Singh and Singh were examined further by Baxby and Hill (1971) who concluded that, while some outbreaks were caused by vaccinias a new orthopoxvirus, buffalopox virus, was also involved. The relative importance of the two viruses is not known.

Buffalopox virus shares a number of properties with vaccinia including a common polypeptide pattern. It might be regarded as a temperature-sensitive variant of vaccinia except that immunological differences between vaccinia and buffalopox have been found (Baxby 1982, Table 87.3).

A virus, termed 'Lenny' virus, was isolated from a case of clinical human smallpox in an undernourished patient in Nigeria (Bourke and Dumbell 1972). The virus is very similar to vaccinia but like buffalopox has the ceiling temperature of smallpox virus, and Bourke and Dumbell suggested that it might be a hybrid of vaccinia and smallpox. Such an origin might also be possible for buffalopox virus.

Mousepox (infectious ectromelia)

The term infectious ectromelia was given to a disease of laboratory mice by Marchal (1930) after whom the conspicuous A-type inclusions were named. The disease is caused by an Orthopoxvirus easily distinguished from the other species (Table 87.3).

The early work on mousepox has been reviewed by Fenner (1949) who did considerable work on the pathogenesis of the disease and on its epidemiology in laboratory mice. Early workers described two forms of the disease and Fenner's work showed the similarities between them. One form is an acute infection which kills mice within a day of becoming ill. The second form, which gave mousepox its alternative name, is more prolonged and is characterized by ulcerating lesions on the feet and tail which often become detached. Virus multiplication in the liver and spleen is common to both types and mice which recover from the acute phase develop the second type. The primary route of infection is through small abrasions in the skin and the incubation period is about eight days. In other respects mousepox can be used as a model for smallpox.

However, recovery from mousepox does not lead to elimination of the virus. Virus can be isolated from the faeces and tail skin of mice 100 days or so after infection, including some which have never been ill (Gledhill 1962). This explains how 'spontaneous' outbreaks of mousepox occur in isolated mouse colonies, or appear after the introduction of new mice. The infection is often triggered off by trauma. This occasionally led to the isolation of ectromelia virus from animals inoculated with other agents (see e.g. Palmer *et al.* 1968). Ectromelia virus is not pathogenic for man. It was the active agent in a vaccine for murine typhus which was used in man without effect (Packalan 1947).

Ectromelia virus has a narrow host range and it is not known whether wild mice or other rodents are involved. McGaughey and Whitehead (1933) found sick wild mice in an animal house when an outbreak occurred, and antibody to Orthopoxviruses was detected in small numbers of randomly caught woodmice and bank-voles in England and Wales (Kaplan *et al.* 1980). However the tests used would not distinguish between different Orthopoxviruses and it is possible that these animals were infected with cowpox.

Other orthopox viruses

A poxvirus similar in many respects to alastrim virus has been isolated from gerbils in Dahomey; it requires further study (Lourie *et al.* 1975).

Horsepox occurred in Europe during the nineteenth century and was often confused with 'grease', a disease of unknown aetiology. Successful smallpox vaccines were derived from horsepox but no virological studies have been made on the European disease this century. However a disease of horses in Kenya known as 'Uasin Gishu' has been shown to be caused by an orthopoxvirus related to cowpox and vaccinia (Kaminjolo *et al.* 1974).

Not a lot is known about Racoonpox, or the virus which causes it. It is clearly an orthopoxvirus although its relation to established members of the genus is uncertain (Thomas *et al.* 1975). The virus is enzootic in the American racoon (Alexander *et al.* 1972), and is the only orthopoxvirus indigenous to N. America.

Conclusions

Now that smallpox has been eradicated, Orthopoxviruses other than camelpox are of relatively little importance for human or veterinary medicine. However human monkeypox, and infections in man and valuable captive animals by cowpox and closely related viruses, although uncommon, are still of interest. Their occurrence, the isolation of various incompletely characterized viruses, and in particular the isolation of the 'whitepox' viruses means that problems concerning the laboratory characteristics and natural history of Orthopoxviruses still require investigation.

Parapoxvirus

Poxviruses with the characteristic Parapoxvirus morphology (Fig. 87.1**b**) cause infections of cattle, sheep and goats, and have a world-wide distribution.

Human cases occur in those in contact with infected animals.

The viruses cannot be isolated on the CAM and do not grow as well as Orthopoxviruses in cell cultures. Diagnosis is often made by electron microscopy and specimens are not usually routinely cultured.

Clinically the basic division has been made between ovine infections (orf = contagious pustular dermatitis = contagious ecthyma), bovine teat infections (paravaccinia = pseudocowpox = milkers' nodes) and bovine mouth infections (bovine papular stomatitis = ulcerative stomatitis).

The viruses

On primary isolation ovine strains grow in sheep but not bovine testis cultures or in human amnion cells, whereas strains isolated from persons in contact with infected sheep and also bovine strains show no such specificity. The specificity of ovine strains is lost on subculture but a permanent but slight difference was noted in the nature of the inclusions produced in cell cultures (Nagington 1968). Although the viruses are related antigenically, cross-immunity tests in animals showed that slight differences existed which did not correlate with the source of virus (Horgan and Haseeb 1947). This is confirmed by kinetic neutralization tests which show extensive cross-reactivity between all the strains and do not allow their classification (Wittek *et al.* 1980).

Comparisons have recently been made of Parapoxvirus DNA by restriction endonuclease analysis (Wittek *et al.* 1980). Although these preliminary results indicated that bovine and ovine strains might be separated, later experiments with more strains showed such heterogenicity that classification of the strains was not possible (R. Wyler, personal communication). Nevertheless the suspicion persists that bovine and ovine strains may be different although the rapid changes that take place on subculture *in vitro* may lead to the loss of such characters. Cross-infection experiments suggest that there may be differences in the infectivity of ovine and bovine isolates for sheep and cattle (Huck 1966) and it is possible that any differences may be lost on subculture.

Disease in animals

Orf in sheep and goats occurs throughout the world. Lesions are commonly found around the lips and mouth but in serious cases may spread to the eyes, genitals and legs. It is in these seriously affected animals that a diagnosis of sheeppox or goatpox could be made which might lead to confusion when the viruses isolated are characterized. Young lambs are usually affected but the infection may be transmitted to the teats and udders of the ewe. Orf is an economic problem in many countries, including the USA where it has been given high priority by the Sheep Industry Development Programme (Renshaw and Dodd 1978).

Knowledge of paravaccinia started with Edward Jenner who described different forms of 'spurious cowpox' which did not protect against smallpox (Baxby 1981). Infection is usually restricted to the teats and it is not always easy to distinguish clinically between cowpox and paravaccinia. Guidance and illustrations have been provided by Gibbs and his colleagues (1970).

Bovine papular stomatitis has been studied in Africa where virus infections causing stomatitis may confuse the diagnosis of foot-and-mouth disease (Plowright and Ferris 1959). A similar clinical picture has been seen in Britain (Nagington *et al.* 1967) and the same virus is believed to cause this disease and paravaccinia; stomatitis in calves occurs when the animals suckle a cow with paravaccinia. Teat infections can be reduced by attention to hygiene during milking (Gibbs and Osborne 1974).

Human infection Orf and paravaccinia viruses produce the same type of lesion in man, and a decision as to which virus is responsible is usually made on epidemiological grounds (see e.g. Nagington *et al.* 1966). Contact with infected animals or their products can usually be proved and infections occur in farmworkers, abattoir workers, veterinary students etc. The typical 'milkers' node' begins as an inflammatory papule, enlarges to form a granulomatous blister and usually crusts and regresses without vesicle formation. A detailed account of their development is given by Leavell and his colleagues (1968). Although the lesions may take several weeks to heal they are surprisingly painless; this is a useful diagnostic feature.

Other Parapoxviruses Contagious ecthyma or 'Ausdyk' in camels is caused by a Parapoxvirus (Roslyakov 1972). In a recent survey in Somalia thirty-five isolations were made of 'true' camelpox and five of Ausdyk (Kriz 1982). The relative importance of the two viruses is at present under investigation.

Parapoxviruses have been seen in lesions in free-living and captive sea lions (Wilson and Sweeney 1970). The relationship of these viruses to other Parapoxviruses is not known.

Capripoxvirus

The viruses of sheeppox, goatpox, and lumpy skin disease are placed in the genus Capripoxvirus. Differences have been detected in antigenic properties and host specificity and there is some geographical variation.

In Africa outbreaks of 'sheeppox' and 'goatpox' occur naturally together and are caused by the same virus (Davies 1976). Up to 60 per cent of infected animals show clinical illness and almost all develop antibody. Mortality is below 2 per cent. Lesions are most common on the hairless areas. The ether-sensitive virus can be isolated in lamb testis cell cultures and only occasional strains can be adapted to the CAM. The virus is not related to orf and the relationships between the African isolates and those from elsewhere is not known. A number of viruses have been isolated from lumpy skin disease of cattle, one of which is a poxvirus. Immunity to this variety of lumpy skin disease has been produced by infection with the 'Kedong' strain of African sheep/goatpox (Coakley and Capstick 1961).

In the Middle East and India 'sheeppox' and 'goatpox' are distinct infections and under natural conditions are species-specific (Rafyi and Ramyar 1959, Sharma and Dhanda 1972). The diseases are of considerable economic importance and vaccination is common. In the past, vaccines containing live virus adsorbed to aluminium hydroxide gel have been used. They produce little local reaction and act essentially as inactivated vaccines, giving short-term immunity. Attenuated vaccines are now available which give a longer-lasting immunity (Martin *et al.* 1973).

The viruses can be isolated in cell cultures, particularly lamb and kid testis, although better attenuation is obtained in calf kidney cells (Martin *et al.* 1973). Infected cells show cytoplasmic inclusions and the characteristic condensation of chromatin at the nuclear membrane first noted by Borrel; this has been examined by electron microscopy (Murray *et al.* 1973).

Differences have been reported in the immunological relationships between sheep and goatpox viruses. In India strains of sheeppox and goatpox cross-immunize (Sharma and Dhanda 1971) but in Iran, although goatpox confers immunity to sheeppox, reciprocal protection is lacking (Rafyi and Ramyar 1959).

It is possible that there are antigenically distinct strains of virus in different countries which produce a similar clinical picture. However it would be unwise to speculate on the reasons for these anomalies until isolates from various sources have been compared by means of up-to-date serological, biochemical and biological methods under controlled conditions.

Leporipoxvirus

The different members of this genus produce fibromas in American rabbits and squirrels and in European hares, and myxomatosis in rabbits. The viruses are all closely related and their interrelationships and laboratory characteristics have been discussed by Woodroofe and Fenner (1965) and Fenner (1965). The viruses are related antigenically but can be separated by the degree of cross-neutralization and the specificity of certain precipitating antigens. Myxoma virus grows well in chick embryo fibroblasts whereas the other viruses do not. All grow, but to different extents, in rabbit cells. By using these methods Woodroofe and Fenner have separated the fibroma viruses from myxoma and also distinguished Californian and South American myxoma viruses.

There is no evidence that these viruses are pathogenic for man. The main interest concerns the way in which virus-host interrelationships have evolved following attempts to control rabbits in Australia by the deliberate introduction of myxomatosis. These developments have been described in detail by Fenner and Ratcliffe (1965).

Myxoma virus originated in South America where it produces a relatively trivial infection in its natural host *Sylvilagus*, the Brazilian wild rabbit. Infection is spread by mosquitoes which pick up virus on their mouth parts and transmit it mechanically to susceptible hosts. Myxoma virus produces a serious infection with very high mortality in *Oryctolagus*, the European rabbit. The virus was introduced into Australia in 1950 and, because the mosquito vector could transmit the disease over long distances, spread rapidly. At first the mortality approached 100 per cent but within a few years virus strains were isolated which killed 90 per cent or less of the control susceptible rabbits. Rabbits infected with these strains would survive longer and would be available to the next season's mosquitoes; these would then spread the attenuated strains. Epidemics occurred in summer when the vectors were most active. It was also found that rabbits developed an innate genetic resistance to myxomatosis. Within seven years survival rates of 70 per cent were recorded with a standard virus strain which had killed 90 per cent of the original rabbits.

Myxomatosis was introduced into Britain and Europe in 1953. In Britain particularly, the vector is the rabbit flea, and its year-round activity explains the lack of seasonal epidemics. Again there has been natural emergence of attenuated virus strains and resistant rabbits (Fenner and Chapple 1965, Ross and Sanders 1977). Unlike the position in Australia, where the spread of myxomatosis is official policy, there was considerable opposition to the practice in Europe and it was made illegal in Britain.

It is generally agreed that myxomatosis can never eradicate the rabbit and a reasonably stable equilibrium has emerged. The numbers are being controlled, and the exercise has proved a valuable experiment in evolution which still continues (see e.g. Shepherd and Edmonds 1979, Fenner 1983).

Avipoxvirus

Antigenically-related poxviruses which produce A-type inclusions (Bollinger bodies) have been isolated from many avian species, but it is not known how many distinct viruses are responsible. Differentiation by biological tests such as growth in chick fibroblasts, pock production, host range etc. have not always given consistent results and it has been suggested that cross-neutralization tests give the most satisfactory results (Tripathy *et al.* 1973). Using this technique fowlpox, turkeypox, pigeonpox and canarypox viruses can be separated. However it is not known how well the many poxviruses isolated from wild birds fit into this scheme (Kirmse 1967).

Fowl, pigeon, and turkeypox viruses produce similar infections in their natural hosts. Warty skin nodules are found on the head and comb, and lesions are common in the mouth and throat. A pseudo-membrane may be formed ('Avian diphtheria'). The mortality is not usually high (Cunningham 1978). Canarypox is a more serious infection with more generalized lesions and the mortality is high (Giddens *et al.* 1971).

Although cross-neutralization tests show apparent specificity, experiments *in vivo* show that fowlpox, turkeypox and to a lesser extent pigeonpox will cross-protect (Mayr *et al.* 1972). Consequently heterologous as well as attenuated homologous vaccines can be used. For example, an attenuated turkeypox vaccine will give good protection against fowlpox (Baxendale 1971). Although vaccines are traditionally inoculated into feather follicles, a successful vaccine has been developed which can be added to drinking water (Mayr and Danner 1975).

Insects are important in the spread of Avipoxvirus infections. More information on avian poxviruses is provided by Cunningham (1978).

Suipoxvirus

The virus of swinepox has a limited host range both *in vivo* and *in vitro*. It does not grow on the CAM but can be isolated in pig cell cultures (Kasza *et al.* 1960, Datt 1964*a*). Vacuolation of the nuclei of infected cells is common (see e.g. Cheville 1966). It is immunologically distinct. A slight cross-reaction with vaccinia detected by gel-diffusion techniques is probably due to the common poxvirus antigen (Datt 1964*b*, Fenner 1965). Swinepox affects young animals, and the lesions commonly occur on areas covered by thin skin. It is a mild disease with low mortality; vaccination is not recommended. The pig louse transmits the disease mechanically (Shope 1940). Vaccinia infects pigs producing a disease very similar to swinepox, but porcine vaccinia is probably rare (Cheville 1966). The importance of distinguishing foot-and-mouth disease from swinepox has been stressed by Kasza (1970).

Unclassified poxviruses

Molluscum contagiosum

Molluscum contagiosum is an infectious human disease characterized by the production of small, pearly, flesh-coloured skin nodules. Their number varies in individual patients; they may occur in most skin areas, except the palms and soles. They can persist for many months and often regress spontaneously. A caseous white material can be expressed from the central core of the lesion; examination of this and of infected cells shows that a poxvirus is the causative agent. The disease is of world-wide but variable distribution (see Postlethwaite 1970). Transmission is by direct contact or by fomites and there is increasing evidence that the disease may be transmitted sexually (Brown, Nalley and Kraus 1981). Attempts to transmit infection to animals have failed. However Douglas and his colleagues (1967) reported the 'spontaneous' occurrence of a molluscum-like disease in chimpanzees.

The disease is of interest particularly because of the many repeated failures to obtain evidence of its replication in cell culture. Early attempts have been thoroughly reviewed by Postlethwaite (1970). Although a 'cytopathic effect' has been reported to occur in various types of cell, this is not transmissible, and is a toxic effect induced by the inoculum. Nevertheless molluscum virus will adsorb, penetrate, and express some activity in certain types of cells. For example, interferon is induced in mouse embryo cell cultures (Postlethwaite and Lee, 1970) and new antigen is synthesized in a number of primate cell lines (McFadden *et al.* 1979). However the virus will not reactivate, nor be reactivated by, another poxvirus (LaPlaca *et al.* 1967, McFadden *et al.* 1979). Electronmicroscopic studies indicate that the second stage of virus uncoating does not take place and virus cores remain in the cells until gradually broken down. There is no evidence of virus spread to uninfected cells in explants of infected tissues. Postlethwaite has likened molluscum virus to a conditional lethal mutant, limited by conditions other than the unique circumstances in the lesion *in vivo*. There has been one recent report that molluscum virus will undergo a complete replicative cycle and produce plaques in human amnion cells after initial isolation in WI-38 cells (Francis and Bradford 1976). This has not been confirmed (McFadden *et al.* 1979), and it is perhaps of interest that the virus isolated was resistant to inhibitors of DNA synthesis and sensitive to inhibitors of picornaviruses.

Tanapox

The name Tanapox was given to a mild infectious human disease caused by a poxvirus which occurs in the Tana River Valley in Kenya. The disease is characterized by a short prodromal illness and the production of usually one, and never more than two, nodular lesions. The poxvirus responsible will not grow on the CAM or in chick cells but can be isolated in a variety of human and simian cells (Downie *et al.* 1971).

Small epidemics were first noticed in 1957 and 1962. Neutralizing antibodies are not long-lasting. Consequently detection of positive sera in 1971 and 1977 and in children born since 1962 indicates that human cases continue to occur (Axford and Downie 1979).

Serological surveys show that Tanapox is enzootic in various species of African and Malaysian, but not Indian or New World, monkeys (Downie 1974). Epidemiological evidence suggests that human infection may be transmitted by mosquitoes. The small outbreaks probably occurred because of close association between monkey and human populations during floods.

Poxvirus infections occurred in captive monkeys during 1965–6 in Oregon, Texas, and California. The clinical picture in monkeys and human contacts was the same as with Tanapox. A direct comparison of the viruses with Tanapox virus showed their identity (Downie and España 1972).

Yaba poxvirus

A poxvirus was shown to be the cause of benign subcutaneous growths found in monkeys at Yaba, Nigeria (Niven *et al.* 1961). As with Tanapox, antibody to Yaba virus has been found in African and Malaysian but not Indian or New World monkeys (Downie 1974). There is a superficial resemblance between the Yaba tumour and the infections reported in America with Tanapox, which were referred to as 'Yaba-like' or 'Yaba-related' but the two diseases are histologically distinct. Yaba and Tanapox viruses have been compared directly by Downie and España (1973). They are antigenically related, but serum adsorption experiments show that each virus has specific, in addition to shared, antigens. It is probable that, because of this close relation, they will eventually be placed in a separate poxvirus genus. They are not immunologically related to other poxviruses, such as monkeypox and whitepox viruses, which have also been isolated from monkeys.

Cotia virus

Cotia poxvirus was isolated from sentinel mice used in arbovirus surveys in Cotia County in Brazil. It grows well in a number of human, monkey and rabbit cell lines. Although light microscopy suggested that A-type inclusions were produced, they did not look like typical A-type inclusions when examined by electronmicroscopy. A slight antigenic relation to vaccinia was noticed (Ueda *et al.* 1978). As mentioned above (Suipoxvirus), such slight cross-reactions are regarded by Fenner (1965) as being caused by the common poxvirus antigen.

References to further occasional isolations of poxviruses from various species of animals have been listed by Andrewes and his colleagues (1978).

References

Abdussalam, M. (1957) *Amer. J. vet. Res.* **18,** 614.

Alexander, A. D., Flyger, V., Herman, Y. F., McConnell, S. J., Rothstein, N. and Yager, R. H. (1972) *J. wildl. Dis.* **8,** 119.

Andrewes, C. H. and Horstmann, D. M. (1949) *J. gen. Microbiol.* **3,** 290.

Andrewes, C. H., Pereira, H. G. and Wildy, P. (1978) *Viruses of Vertebrates*, 4th edn., p. 356. Ballière, London.

Appleyard, G., Hapel, A. J. and Boulter, E. A. (1971) *J. gen. Virol.* **13,** 9.

Archard, L. C. and Mackett, M. (1979) *J. gen. Virol.* **45,** 51.

Arita, I. (1979) *Nature, Lond.* **279,** 293.

Arita, I. *et al.* (1972) *Bull. World Hlth Org.* **46,** 625.

Arita, M. and Tagaya, I. (1980) *Arch. Virol.* **63,** 209.

Axford, J. S. and Downie, A. W. (1979) *J. Hyg. Camb.* **83,** 273.

Bauer, D. J. (1977) *The Specific Treatment of Virus Diseases*, p. 135. MTP Press, Lancaster.

Bauer, D. J., St Vincent, L., Kempe, C. H. and Downie, A. W. (1963) *Lancet* **ii,** 494.

Baxby, D. (1972*a*) *J. Hyg., Camb.* **70,** 353; (1972*b*) *Lancet* **ii,** 1063; (1975) *Progr. med. Virol.* **19,** 215; (1977*a*) *Arch. Virol.* **55,** 169; (1977*b*) *Brit. med. J.* **i,** 1379; (1981) *Jenner's Smallpox Vaccine*, Heinemann Educational, London; (1982) *J. gen. Virol.* **58,** 251.

Baxby, D., Ashton, D. G., Jones, D., Thomset, L. R. and Denham, E. M. (1982) *J. Hyg. Camb.* **89,** 365.

Baxby, D. and Ghaboosi, B. (1977) *J. gen. Viol.* **37,** 407.

Baxby, D. and Hill, B. J. (1971) *Arch. ges. Virusforsch.* **35,** 70.

Baxby, D., Ramyar, H., Hessami, M. and Ghaboosi, B. (1975) *Infect. and Immun.* **11,** 617.

Baxby, D., Shackleton, W. B., Wheeler, J. and Turner, A. (1979) *Arch. Virol.* **61,** 337.

Baxendale, W. (1971) *Vet. Rec.* **88,** 5.

Bedson, H. S. and Duckworth, M. (1963) *J. Path. Bact.* **85,** 1.

Bedson, H. S. and Dumbell, K. R. (1961) *J. Hyg. Camb.* **59,** 457.

Bedson, H. S., Dumbell, K. R. and Thomas, W. R. G. (1963) *Lancet* **ii,** 1085.

Boulter, E. A., Zwartouw, H. T., Titmuss, D. H. J. and Maber, H. B. (1971) *Amer. J. Epidem.* **94,** 612.

Bourke, A. T. C. and Dumbell, K. R. (1972) *Bull. World Hlth Org.* **46,** 621.

Breman, J. G., Kalisa-Ruti, Steniowski, M. V., Zanotto, E., Gromyko, A. I. and Arita, I. (1980) *Bull. World Hlth Org.* **58,** 165.

Breman, J. G., Nakano, J. H., Coffi, E., Godfrey, H. and Gautin, J. C. (1977) *Amer. J. trop. Med. Hyg.* **26,** 273.

Brown, S. T., Nalley, J. F. and Kraus, S. J. (1981) *Sex. trans. Dis.* **8,** 227.

Burnet, F. M. (1946) *Nature, Lond.* **158,** 119.

Chang, A. and Metz, D. H. (1976) *J. gen. Virol.* **32,** 275.

Cheville, N. F. (1966) *Path. vet.* **3,** 556.

Cho, C. T. and Wenner, H. A. (1973) *Bact. Rev.* **37,** 1.

Christie, A. B. (1980) *Infectious Diseases*, 3rd edn., p. 222. Churchill Livingstone, Edinburgh.

Coakley, W. and Capstick, P. B. (1961) *Res. vet. Sci.* **2,** 369.

Cooper, J. A. and Moss, B. (1979) *Virology*, **96,** 368.

Cunningham, C. H. (1978) In: *Diseases of Poultry*, 7th edn., p. 597. Ed. by M. S. Hofstad, Iowa State University Press, Ames.

Dales, S. and Pogo, B. T. (1981) *The Poxviruses*, Springer-Verlag, Vienna.

Dales, S. and Siminovitch, L. (1961) *J. biophys. biochem, Cytol.* **10,** 475.

Datt, N. S. (1964*a*) *J. comp. Path.* **74,** 62; (1964*b*) *Ibid.* **74,** 70.

Davies, F. G. (1976) *J. Hyg. Camb.* **76,** 163.

Davies, F. G., Mungai, J. N. and Shaw, T. (1975) *J. Hyg. Camb.* **75,** 381.

Dick, G. (1971) *Brit. med. J.* **ii,** 163.

Dittmann, S. (1981) *Beitr. Hyg. Epidem.* **25,** 26.

Dixon, C. W. (1962) *Smallpox.* Churchill Livingstone, London.

Douglas, J. D. *et al.* (1967) *J. Amer. vet. med. Assoc.* **151,** 901.

Downie, A. W. (1970) In: *Infectious Agents and Host Reactions*, p. 487. Ed. by S. Mudd, Saunders, Philadelphia; (1974) *J. Hyg. Camb.* **72,** 245.

Downie, A. W. and España, C. (1972) *J. Hyg. Camb.* **70,** 23; (1973) *J. gen. Virol.* **19,** 37.

Downie, A. W. and Kempe, C. H. (1969) In: *Diagnostic Procedures for Viral and Rickettsial Infections*, p. 281. Amer. publ. Health Ass., New York.

Downie, A. W. and McCarthy, K. (1950) *Brit. J. exp. Path.* **31,** 789.

Downie, A. W., McCarthy, K., MacDonald, A., MacCallum, F. O. and Macrae, A. D. (1953) *Lancet* **ii,** 164.

Downie, A. W., Taylor-Robinson, C. H., Caunt, A. E., Nelson, G. S., Manson-Bahr, P. E. C. and Matthews, T. C. H. (1971) *Brit. med. J.* **i,** 363.

Dumbell, K. R. and Archard, L. C. (1980) *Nature, Lond.* **286,** 29.

Dumbell, K. R. and Huq, F. (1975) *Trans. R. Soc. trop. Med. Hyg.* **69,** 303.

Dumbell, K. R. and Kapsenberg, J. G. (1982) *Bull. World Hlth Org.* **60,** 381.

Easterbrook, K. B. (1966) *J. ultrastruct. Res.* **14,** 484.

Ehrengut, W. and Sarateanu, D. E. (1975) *Dtsch. med. Wschr.* **100,** 1457.

Esposito, J. J., Obijeski, J. F. and Nakano, J. H. (1977) *J. med. Virol.* **1,** 47.

Essani, K. and Dales, S. (1979) *Virology* **95,** 385.

Fenner, F. (1949) *J. Immunol.* **63,** 341; (1958) *Virology* **5,** 502; (1962) *Proc. roy. Soc. B.* **156,** 388; (1965) *Aust. J. exp. Biol. med. Sci.* **43,** 143; (1977) *Progr. med. Virol.* **23,** 1; (1983) *Proc. roy. Soc., B.*, **218,** 259

Fenner, F. and Chapple, P. J. (1965) *J. Hyg. Camb.* **63,** 175.

Fenner, F., Dumbell, K. R., Marennikova, S. S. and Nakano, J. H. (1980) *WHO Smallpox Eradication Document* WHO/SE/80.154.

Fenner, F. and Ratcliffe, F. N. (1965) *Myxomatosis.* Cambridge University Press, London.

Fenner, F. and Woodroofe, G. M. (1960) *Virology* **11,** 185.

Ferry, A. J. (1977) *Med. J. Aust.* **ii,** 180.

Foege, W. H., Millar, J. D. and Henderson, D. A. (1975) *Bull. World Hlth Org.* **52,** 209.

Francis, R. D. and Bradford, H. B. (1976) *J. Virol.* **19,** 382.

Fulginiti, V. A. *et al.* (1966) *Lancet* **ii,** 5.

Gemmell, A. and Cairns, J. (1959) *Virology* **8,** 381.

Geshelin, P. and Berns, K. I. (1974) *J. molec. Biol.* **88,** 785.

Gibbs, E. P. J., Johnson, R. H. and Osborne, A. D. (1970) *Vet. Rec.* **87,** 602.

Gibbs, E. P. J. and Osborne, A. D. (1974) *Brit. vet. J.* **130,** 150.

Giddens, W. E. *et al.* (1971) *Vet. Path.* **8,** 260.

Gispen, R. and Brand-Saathof, B. (1972) *Bull. World Hlth Org.* **46,** 585; (1974) *J. infect. Dis.* **129,** 289.

Gispen, R., Brand-Saathof, B. and Hekker, A. C. (1976) *Bull. World Hlth Org.* **53,** 355.

Gledhill, A. W. (1962) *Nature, Lond.* **196,** 298.

Granados, R. R. (1973) *Misc. Publ. ent. Soc. Amer.* **9,** 73; (1981) In: *Pathogenesis of Invertebrate Microbial Diseases*, p. 101. Allanheld Osmun, Totowa, New Jersey.

Hanna, W. (1913) *Studies in Smallpox and Vaccination* Wright, Bristol.

Harper, L., Bedson, H. S. and Buchan, A. (1979) *Virology* **93,** 435.

Harris, W. J. and Westwood, J. C. N. (1964) *J. gen. Microbiol.* **34,** 491.

Henderson, D. A. (1976) *Int. J. Epidemiol.* **5,** 19.

Herrlich, A., Mayr, A., Mahnel, H. and Munz, E. (1963) *Arch. ges. Virusforsch.* **12,** 579.

Holowczak, J. A. (1982) *Curr. Top. Microbiol. Immunol.* **97,** 28.

Horgan, E. S. and Haseeb, M. A. (1947) *J. comp. Path.* **57,** 1.

Huck, R. A. (1966) *Vet. Rec.* **78,** 503.

Huq, F. (1976) *Bull. World Hlth Org.* **54,** 710.

Hutchinson, H. D., Ziegler, D. W., Wells, D. E. and Nakano, J. H. (1977) *Bull. World Hlth Org.* **55,** 613.

Joklik, W. K. (1962) *Virology* **18,** 9; (1968) *Annu. Rev. Microbiol.* **22,** 359.

Kaminjolo, J. S., Nyaga, P. N. and Gicho, J. N. (1974) *Zbl. vet. Med. B.* **21,** 592.

Kaplan, C., Healing, T. D., Evans, N., Healing, L. and Prior, A. (1980) *J. Hyg. Camb.* **84,** 285.

Kasza, L. (1970) In: *Diseases of Swine*, p. 257. Ed. by H. W. Dunne, Iowa State University Press, Ames.

Kasza, L., Bohl, E. H. and Jones, D. O. (1960) *Amer. J. vet. Res.* **21,** 269.

Kato, S. (1964) *Acta path. jap.* **14,** 189.

Kempe, C. H. (1960) *Pediatrics* **26,** 176; (1968) *Yale J. Biol. Med.* **41,** 1.

Kempe, C. H., Dekking, F., St Vincent, L., Rao, A. R. and Downie, A. W. (1969) *J. Hyg. Camb.* **67,** 631.

Kirmse, P. (1967) *Wildl. Dis.* **49,** Suppl. 1.

Kozinowski, U. and Ertl, H. (1976) *Europ. J. Immunol,* **6,** 679.

Kriz, B. (1982) *J. comp. Path.* **92,** 1.

Lake, J. R. and Cooper, P. D. (1980) *J. gen. Virol.* **47,** 243.

LaPlaca, M., Portolani, M. and Rosa, A. (1967) *Arch. ges. Virusforsch.* **18,** 251.

Leavell, U. W. Jr., McNamara, M. J., Muelling, R., Talbert, W. M., Rucker, R. C. and Dalton, A. J. (1968) *J. Amer. med. Ass.* **204,** 657.

Lourie, B., Nakano, J. H., Kemp, G. E. and Satzer, H. W. (1975) *J. infect. Dis.* **132,** 677.

McFadden, G., Essani, K. and Dales, D. (1980) *Virology* **101,** 277.

McFadden, G., Pace, W. E., Purres, J. and Dales, S. (1979) *Virology* **94,** 297.

McGaughey, C. A. and Whitehead, R. (1933) *J. Path. Bact.* **37,** 253.

Mackett, M. and Archard, L. C. (1979) *J. gen. Virol.* **45,** 683.

Marchal, J. (1930) *J. Path. Bact.* **33,** 713.

Marennikova, S. S. (1979) *Bull. World Hlth Org.* **57,** 461.

Marennikova, S. S., Maltseva, N. N., Korneeva, V. I. and Garanina, N. M. (1977) *J. infect. Dis.* **135,** 358.

Marennikova, S. S., and Shelukhina, E. M. (1976) *Acta virol.* **20,** 442; (1978) *Nature, Lond.* **276,** 291.

Marennikova, S. S., Shelukhina, E. M., Maltseva, N. N. and Ladnyi, I. D. (1972) *Bull. World Hlth Org.* **46,** 613.
Marennikova, S. S., Shelukhina, E. M. and Shenkman, L. S. (1976) *Acta virol.* **20,** 80
Marennikova, S. S., Shenkman, L. S., Shelukhina, E. M. and Maltseva, N. N. (1974) *Acta virol.* **18,** 423.
Martin, W. B., Ergin, H. and Koylu, A. (1973) *Res. vet. Sci.* **14,** 53.
Mathew, T. (1967) *Acta virol.* **14,** 513.
Matthews, R. E. F. (1979) *Intervirology* **12,** 132.
Mayr, A. and Danner, K. (1975) *Develop. biol. Standards* **33,** 249.
Mayr, A., Mahnel, H. and Munz, E. (1972) *Zbl. vet. Med. B.* **19,** 69.
Moss, B. (1978) In: *Molecular Biology of Animal Viruses*, p. 849. Ed. by D. P. Nayak, Dekker, New York.
Moyer, R. W. and Graves, R. L. (1981) *Cell,* **27,** 391.
Murray, M., Martin, W. B. and Koylu, A. (1973) *Res. vet. Sci.* **82,** 477.
Nagington, J. (1968) *Vet. Rec.* **82,** 477.
Nagington, J. and Horne, R. W. (1962) *Virology* **16,** 248.
Nagington, J., Lauder, I. M. and Smith, J. S. (1967) *Vet. Rec.* **81,** 306.
Nagington, J., Newton, A. A. and Horne, R. W. (1964) *Virology* **23,** 461.
Nagington, J., Tee, G. and Smith, J. S. (1966) *Vet. Rec.* **78,** 305.
Nakano, J. H. (1977) In: *Comparative Diagnosis of Viral Diseases*, Vol. 1, p. 287. Ed. by E. Kurstak and C. Kurstak, Academic Press, New York; (1979) In: *Diagnostic Procedures for Viral, Rickettsial and Chlamydial Diseases*, p. 257. Amer. Publ. Health Ass., New York.
Niven, J. S. F., Armstrong, J. A., Andrewes, C. H., Pereira, H. G. and Valentine, R. C. (1961) *J. Path. Bact.* **81,** 1.
Nizamuddin, M. and Dumbell, K. R. (1961) *Lancet* **i,** 68.
Noble, J. and Rich, J. A. (1969) *Bull. World Hlth Org.* **40,** 279.
Packalan, T. (1947) *Acta path. microbiol. scand.* **24,** 375.
Palmer, E. L., Ziegler, D. W., Kissing, R. E., Hutchinson, H. D. and Murphy, F. A. (1968) *J. infect. Dis.* **118,** 500.
Payne, L. G. (1978) *J. Virol.* **27,** 28; (1979) *Ibid.* **31,** 147; (1980) *J. gen. Virol.* **50,** 89.
Payne, L. G. and Kristensson, K. (1979) *J. Virol.* **32,** 614.
Pennington, T. H. (1973) *J. gen. Virol.* **19,** 65; (1974) *Ibid.* **25,** 433.
Plowright, W. and Ferris, R. D. (1959) *Vet. Rec.* **71,** 718.
Pogo, B. T., Dales, S., Bergoin, M. and Roberts, D. W. (1971) *Virology* **43,** 306.
Polak, M. F., Beunders, B. J. W., Werf, A. R., Saunders, E. W., Klaveren, J. N. and Brans, L. M. (1963) *Bull. World Hlth Org.* **29,** 311.
Postlethwaite, R. (1970) *Arch. environm. Hlth* **21,** 432.
Postelthwaite, R. and Lee, Y. S. (1970) *J. gen. Virol.* **6,** 117.
Rafyi, A. and Ramyar, R. (1959) *J. comp. Path.* **69,** 141.
Renshaw, H. W. and Dodd, A. G. (1978) *Arch. Virol.* **56,** 201.
Roening, G. and Holowczak, J. A. (1974) *J. Virol.* **14,** 704.
Rondle, C. J. M. and Williamson, J. D. (1968) *J. Hyg. Camb.* **66,** 415.
Roslyakov, A. A. (1972) *Vopr. Virus* No. 17, 26.
Ross, J. and Sanders, M. F. (1977) *J. Hyg. Camb.* **79,** 411.
Sam, C. K., and Dumbell, K. R. (1981) *Ann. Virol.* **132E,** 135.
Sarov, I. and Joklik, W. K. (1972) *Virology* **50,** 579; (1973) *Ibid.*, **52,** 223.
Sharma, S. N. and Dhanda, M. R. (1971) *Indian J. anim. Sci.* **41,** 267; (1972) *Ibid.* **11,** 39.
Shelukhina, E. M., Maltseva, N. N., Shenkman, L. S. and Marennikova, S. S. (1975) *Brit. vet. J.* **131,** 746.
Shelukhina, E. M., Marennikova, S. S., Shenkman, L. S. and Froltsova, A. E. (1979) *Acta virol.* **23,** 360.
Shepherd, R. C. H. and Edmonds, J. W. (1979) *J. Hyg. Camb.* **83,** 285.
Shope, R. E. (1940) *Arch. ges. Virusforsch.* **1,** 457.
Singh, I. P. and Singh, S. B. (1967) *J. Res. Ludh.* **4,** 440.
Smadel, J. E. and Shedlovsky, T. (1942) *Ann. N.Y. Acad. Sci.* **43,** 35.
Smith, E. C., Pratt, B. P. and Baxby, D. (1973) *J. gen. Virol.* **18,** 111.
Stern, W. and Dales, S. (1976) *Virology* **75,** 232.
Thomas, E. L., Palmer, E. L., Obijeski, J. F. and Nakano, J. H. (1975) *Arch. virol.* **49,** 217.
Thomsett, L. R., Baxby, D. and Denham, E. M. (1978) *Vet. Rec.* **103,** 567.
Tripathy, D. N., Hanson, L. E. and Killinger, A. H. (1973) *Avian Dis.* **17,** 325.
Turner, A. and Baxby, D. (1979) *J. gen. Virol.* **45,** 537.
Turner, G. S. and Squires, E. J. (1971) *J. gen. Virol.* **13,** 19.
Ueda, Y., Dumbell, K. R., Tsuruhara, T. and Tagaya, I. (1978) *J. gen. Virol,* **40,** 263.
Von Magnus, P., Anderson, E. K., Peterson, K. B. and Birch-Anderson, A. (1959) *Acta path. microbiol. scand.* **46,** 156.
Walls, H. H., Ziegler, D. W. and Nakano, J. H. (1981) *Bull. World Hlth Org.* **59,** 253.
Wehrle, P. F., Posch, J., Richter, K. H. and Henderson, D. A. (1970) *Bull. World Hlth Org.* **43,** 669.
Westwood, J. C. N., Harris, W. J., Zwartouw, H. T., Titmus, D. H. J. and Appleyard, G. (1964) *J. gen. Microbiol.* **34,** 67.
Westwood, J. C. N., Zwartouw, H. T., Appleyard, G. and Titmuss, D. H. J. (1965) *J. gen. Microbiol.* **38,** 47.
WHO (1964) *World Hlth Org. techn. Rep. Ser.* No. 283; (1968) *Ibid.* No. 393; (1980) *The Global Eradication of Smallpox*, WHO, Geneva.
Wilcox, W. C. and Cohen, E. H. (1969) *Curr. Top. Microbiol. Immunol.* **47,** 1.
Wilson, G. S. (1967) *The Hazards of Immunization*, pp. 157, 224. Athlone Press, London.
Wilson, T. M. and Sweeney, P. R. (1970) *J. wildl. Dis.* **6,** 94.
Wittek, R., Herlyn, M., Schumperli, D., Bachmann, P., Mayr, A. and Wyler, R. (1980) *Intervirology* **13,** 33.
Wittek, R., Kuenzle, C. C. and Wyler, R. (1979) *J. gen. Virol.* **43,** 231.
Woodroofe, G. M. and Fenner, F. (1962) *Virology* **16,** 334; (1965) *Aust. J. exp. Biol. med. Sci.* **43,** 123.
Zwart, P., Gispen, R. and Peters, J. C. (1970) *Brit. vet. J.* **127,** 20.
Zwartouw, H. T. (1964) *J. gen. Microbiol.* **34,** 115.

88

The herpesviruses

D. H. Watson

Introductory

The herpesviruses (formally internationally approved family name *Herpesviridae*) are a group of over 70 viruses most of which have vertebrates as their natural host. They are grouped together on the basis of common features of morphology, genome structure and cytopathology (Matthews 1979, 1983). Although no common group specific antigen has been consistently demonstrated, there are some grounds for believing that one may exist.

The family has been recently divided into three sub-families (Matthews 1979), termed alphaherpesvirinae, betaherpesvirinae and gammaherpesvirinae. The alphaherpesvirinae are characterized by a relatively short replicative cycle (12 to 18 h), frequently cause latent virus infection of ganglia and show variable host range, from very wide to very narrow. Betaherpesvirinae have a slower reproduction cycle (at least 24 hours), produce enlargement of the infected cell *in vivo* and often *in vitro* (cytomegalia), often cause latent infection of the salivary gland and/or other organs and have a narrow host range, usually growing best in fibroblasts. Their DNA is usually of higher mol. wt (130 to 150×10^6) than that of alpha- or gammaherpesvirinae (85 to 110×10^6). The gammaherpesvirinae will all replicate in lymphoblastoid cells, being specific for either B or T lymphocytes, often cause latent infection in lymphoid tissue and have a narrow host range, being usually restricted to the same order as the natural host.

There are five herpesviruses whose natural host is man; three, namely herpes simplex type 1, herpes simplex type 2 and possibly varicella-zoster virus are included in the alphaherpesvirinae; one, human cytomegalovirus, is grouped with the betaherpesvirinae; and the fifth, Epstein-Barr virus (Epstein and Achong 1979) belongs to the gammaherpesvirinae. The viruses are almost invariably known by their vernacular titles. Any attempt to adopt formal 'approved' names on the supposedly ideal latinized binomial basis is almost doomed to failure for a group of viruses already possessing names well established by long usage, even if it may be considered unsatisfactory, that three are derived from descriptions of the related disease, one from cytopathological effects caused by the virus and the fifth from the name of its discoverers. The neatest approach to formal names comes from the provisional labels (Table 88.1) suggested by a study group (Roizman *et al.* 1973), after they had recognized the hopelessness of their task of producing formal names on more conventional lines.

Table 88.1 Human herpesviruses: provisional labels

Virus	Label
Herpes simplex type 1	Human herpesvirus 1
Herpes simplex type 2	Human herpesvirus 2
Varicella-zoster virus	Human herpesvirus 3
EB virus	Human herpesvirus 4
Cytomegalovirus	Human herpesvirus 5

Other herpesviruses can occasionally infect man, the most notorious (although fortunately rare) example being the B virus of monkeys, which causes almost invariably fatal infections in man. The pseudorabies virus of pigs has been reported to infect man with less severe results.

This chapter begins with a summary of current knowledge of the molecular biology of the virus, partly because this fundamental knowledge must become increasingly relevant to attempts at understanding the unusual features of the pathogenesis of the viruses, and partly because many of the techniques used in fundamental studies of the viruses are becoming increasingly important in modern laboratory studies of human infections with these viruses.

Molecular biology of herpesviruses

This outline attempts to give an impression of current knowledge without necessarily giving a critical and fully documented view of conflicting reports in areas of incomplete understanding.

Morphology

All herpesviruses exhibit a similar distinctive morphology (Fig. 88.1), sufficiently distinctive for it to be the main criterion for including many viruses in the family. The mature enveloped particle is seen in negatively stained preparations—whether of vesicle fluid from a cold sore or of a homogenate of experimentally infected tissue culture cells—as a pleomorphic membrane-bound envelope, containing an inner particle, diameter 100 nm, termed the nucleocapsid (Fig. 88.1(a)). This can be seen more clearly in the free or naked form (Fig. 88.1(b)). The nucleocapsid, which contains the virus DNA, shows hollow, i.e. stain-penetrated capsomeres approximately 10 nm in diameter disposed on the outer surface of the capsid. Detailed examination of the capsomere array on

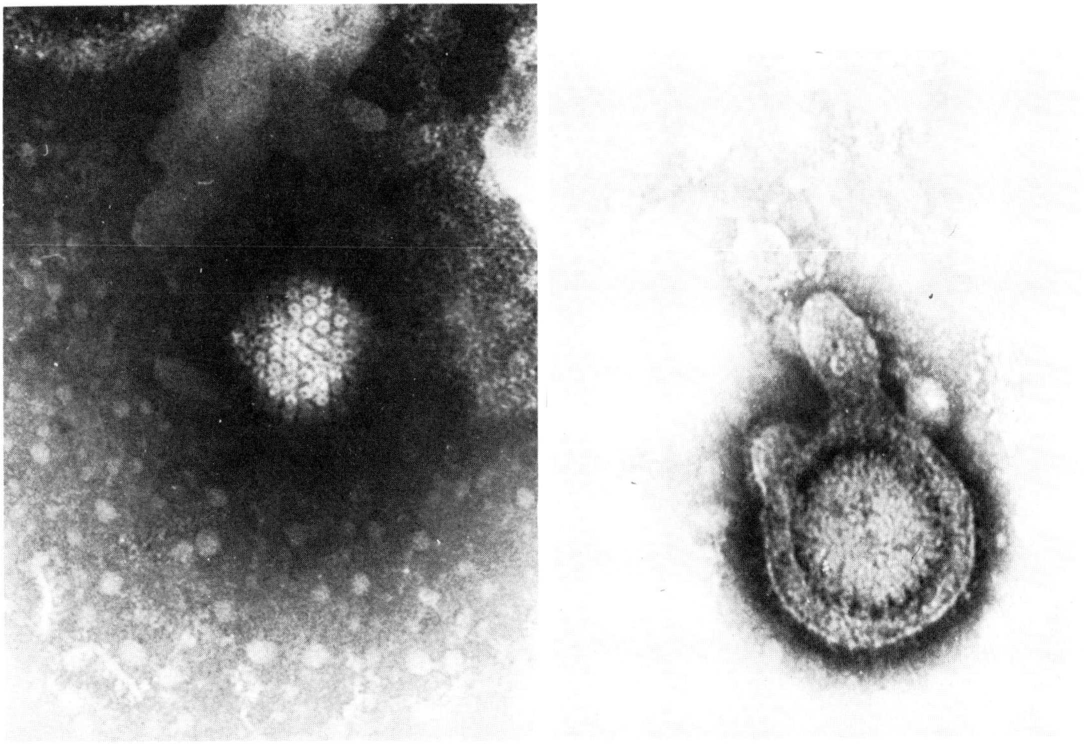

(a) (b)

Fig. 88.1 Electron micrographs of (a) enveloped and (b) naked particles of herpes simplex virus type 1. (Reproduced with permission of Academic Press.)

nucleocapsids of herpes simplex type 1 (Wildy *et al.* 1960) led to the conclusion that the shell was composed of 162 capsomeres. It is presumed that all herpes viruses are similarly composed, although this has been proved definitively for only a very small number of the members of the family.

Examination of particles in thin section shows other components not readily seen in negative contrast pictures. Thus there is clearly electron-dense material between the envelope and the capsid which has been given the general name of the tegument. A densely stained region or core can be discerned within the capsid. This can be removed by deoxyribonuclease digestion showing that it contains the virus DNA. It has been suggested that the core is in the form of a hollow toroid of DNA spooled round an inner core of material of lower electron density than the outer layer of DNA. The structure and composition of this inner plug of the core is unknown.

Chemical composition

Nucleic acid

The human herpesviruses share with other members of the family the possession of linear double-stranded DNA molecules. The molecular weight of herpes simplex DNA is approximately 1×10^8; of the other human viruses it is likely that only cytomegalovirus has DNA of significantly larger molecular weight. The members of the *Herpesviridae* family show great diversity in mean percentage of guanine + cytosine (under 40 per cent to over 70 per cent). Of the human herpesviruses, the herpes simplex viruses have 67 per cent and 69 per cent G + C for type 1 and type 2 respectively, EB virus and cytomegalovirus have 58 per cent and varicella-zoster virus 46 per cent G + C.

Several herpesvirus DNAs have been shown to fragment after denaturation with alkali. This has been ascribed by some to gaps in the DNA strands and by others to occasional ribonucleotides in deoxyribonucleotide chains.

Although herpes simplex virus types 1 and 2 DNAs show extensive homology (up to 50 per cent) by DNA hybridization, none can be observed between any other pair of human herpesviruses. The DNA of each herpesvirus can be distinguished from that of other viruses of the family by the characteristic patterns of cleavage with restriction enzymes (Fig. 88.2). Smaller variations in cleavage patterns are demonstrable be-

Fig. 88.2 Separation by agar gel electrophoresis of fragments of herpes simplex viruses types 1 (left) and 2 (right) DNAs produced by cleavage with E co RI. (Reproduced by kind permission of Dr I. W. Halliburton.)

tween different isolates of a virus. Thus it appears that epidemiologically unrelated isolates of herpes simplex type 1 virus from different patients can be distinguished by restriction enzyme cleavage patterns, although epidemiologically related isolates from different patients have similar patterns. Similar variations are observed in different isolates of other herpes viruses and, since restriction enzyme cleavage patterns can now be developed from small amounts of material, it is likely that this powerful method will find increasing diagnostic application. Detailed analysis of restriction enzyme cleavage patterns has given valuable information on sequence arrangement in herpesvirus DNAs. When intact single strands of herpes simplex type 1 DNA are self annealed, two single-stranded loops of different size are formed, linked by a double stranded region (Fig. 88.3). This led Sheldrick and Berthelot (1974) to suggest that the two loops represented segments of unique sequences flanked by inverted repeats. These workers had also shown that the ends of the DNA were terminally repetitive, since exonuclease digestion of the nucleic acid allowed circularization of the molecules. They suggested that the existence of inverted repeats could allow an intramolecular recombinational event which might lead, on subsequent breakage, to independent inversion of each unique region with respect to the other and hence to the existence of four isomeric forms of the molecules. Analysis of restriction enzyme fragments has confirmed this conclusion. Not all the fragments resulting from restriction enzyme cleavage are present in equimolar proportions, some being present in half and others in quarter molar proportions. This can be explained on the basis of physical maps which depict the different fragments in order on the genome. Derivation of these maps is a complicated procedure requiring cleavage of the nucleic acid with a series of enzymes. The terminal fragments can always be identified from their susceptibility to exonuclease digestion. Fragments from enzyme cleavage with one enzyme are then related to fragments from other cleavages containing similar sequences by hybridization analysis.

Figure 88.4 depicts a physical map for cleavage of the DNA of the herpes simplex type 1 strain HFEM by the enzyme. The larger of the two unique sequences is labelled U_L and the smaller U_S. The terminal repetitive regions are represented by a; b′a′ and a′c′ represent internal inverted repetitions of the two terminal regions ab and ca respectively. The diagram shows the four isomeric arrangements with the relative inversions between each form designated by the arrows above U_L and U_S respectively. The fragments are designated by letters with A representing the largest and B, C...... successively smaller fragments. In the top isomeric form it will be seen that fragments D and M are juxtaposed to the internal inverted repeat regions and, since there is no cleavage site between them, digestion of this isomeric form gives a compound fragment DM which arises only from this form and thus is present in quarter molar amounts if the four forms occur in equimolar proportions. In forms I_L and I_{SL}; D occurs at the terminus and, since it is present in two of the four forms, it occurs in half molar amounts.

Herpes simplex virus type 2 has similar sequence arrangements and once again four forms are possible. It now seems likely that a similar arrangement occurs in cytomegalovirus DNA. By contrast varicella-zoster virus DNA is based on a different sequence arrangement in which only the sequences from the terminus

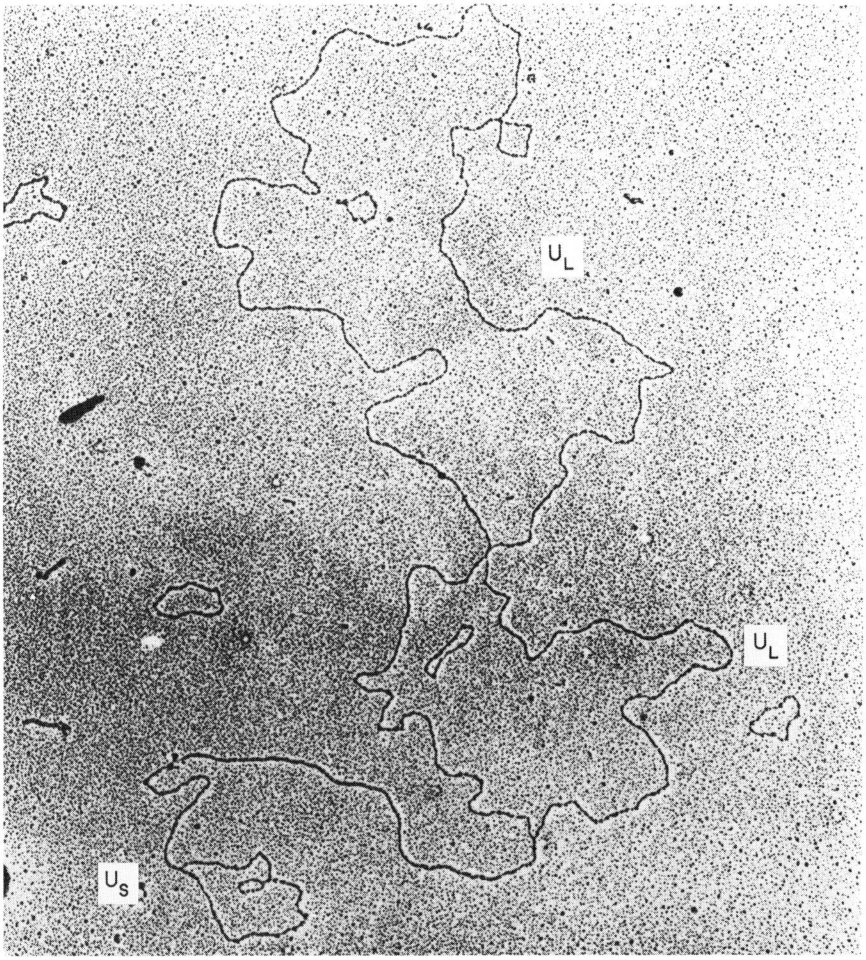

Fig. 88.3 Electron micrograph of single strand of DNA of herpes simplex virus type 1 (Strain F) at magnification of 10 000. The smaller circles are marker ϕX174 DNA molecules. The larger herpes DNA molecule consists of two loops, the larger representing single stranded U_L (= large unique) and the smaller single stranded U_S (= small unique). They are linked by a double stranded region resulting from bonding of terminal regions with internal sequences which are inverted with respect to the terminal regions. (Reproduced by kind permission of Dr P. Sheldrick.)

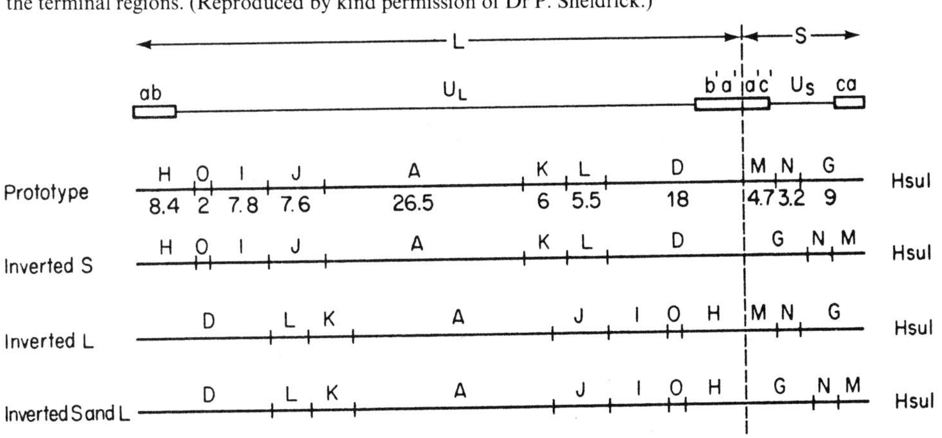

Fig. 88.4 Physical map of restriction enzyme (Hsu I) cleavage fragments of herpes simplex virus type 1. The maps show from top to bottom: Prototype (P) Inverted S (I_S) Inverted L (I_L) and Inverted S and L (I_{SL}) forms. (Reproduced by kind permission of Dr I. W. Halliburton and National Academy of Sciences, USA.)

of the U_S component are inverted internally. In consequence re-arrangement allows only the U_S component to invert relative to U_L and there are as a result only two isomeric forms, the existence of which is deduced from the presence of 0.5 M but not 0.25 M fragments in restriction enzyme digests. Epstein-Barr virus DNA exists in only one form; the termini are marked by a series of reiterated sequences. A further series of reiterated sequences, differing from those at the termini, divides long and short unique sequences. Some strains of the virus possess a third shorter series of reiterated sequences within the long unique region.

Proteins

Structural polypeptides of purified enveloped particles of herpes simplex virus type 1 have been analysed by disruption of the virus particles with sodium dodecyl sulphate (SDS) and subsequent separation of the released polypeptides by electrophoresis on SDS polyacrylamide gels (Fig. 88.5). Up to 33 polypeptides have been reported by separation in conventional single dimension gels but additional separation in a second dimension based on isoelectric focusing has revealed a greater number. One difficulty in interpretation of the data is that the apparently large number of polypeptides makes it likely that some will be present in such small proportion (less than 1 per cent of the total mass) as to make them indistinguishable from minor contaminants. The number of components may be exaggerated by the possibility that precursors to fully processed products co-exist in the virus particles.

Electrophoretic analysis of naked non-enveloped particles isolated from the nuclei of infected cells, particularly in the presence of detergent to remove adherent tegument proteins, has allowed clearer identification of nucleocapsid polypeptides. At least one of these, labelled VP 22 in Fig. 88.5, is absent in DNA free ('empty') nucleocapsids and, is therefore deduced to be a core protein associated with the DNA. Other polypeptides are glycosylated and are located in the envelope surface, because they can be selectively iodinated in the presence of a lactoperoxidase catalyst incapable of penetrating membranes. It is presumed, with varying degrees of certainty, that most of the residual polypeptides are present in the tegument. The glycoproteins of the virus particle have been labelled gA (VP8 in Fig. 88.5), gB (VP7), gC (VP8.5), gD (VP17) and gE (migrating in region marked VP 12 in Fig. 88.5). There is some evidence to suggest that gA and gB are related and truly independent components. There are clearly other glycoproteins in the virus particle which remain to be fully documented.

Our knowledge of the function of the different structural polypeptides is rudimentary. Most information is available for the external glycoproteins. Many of these, as will be seen later, react with homologous antibodies with loss of infectivity. The glycoprotein gE seems to be responsible for the Fc binding reactivity of virus particles and infected cell membranes. The significance, if any, of this reactivity, which causes binding of immunoglobulin by the Fc terminus, in the biology of the virus is still a matter for speculation. Perhaps gC has some role in promoting virus directed membrane fusion and, because mutants lacking gC seem to be viable, at least in experimental conditions, it cannot be essential for virus infectivity. By contrast, mutants lacking gB, although able to attach to the cell surface, are unable to penetrate it and are thus non-

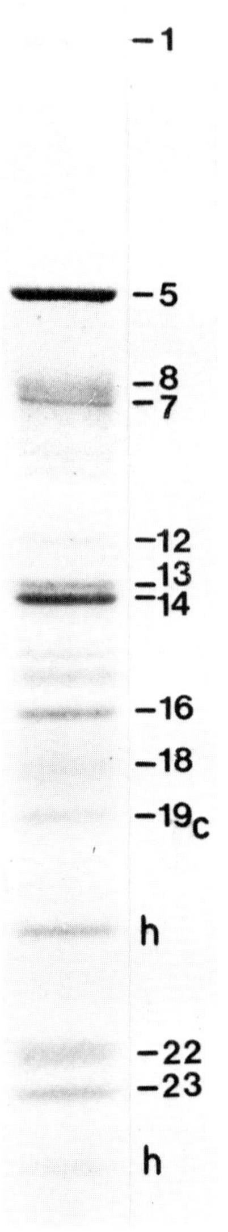

Fig. 88.5 Structural polypeptides of herpes simplex virus type 1 separated by SDS polyacrylamide gel electrophoresis.

infective. Possibly gB promotes the fusion between the cell membrane and virus particle necessary for penetration. The glycoprotein exists in the virus particle in an oligomeric form which is dissociated in heated but not cold SDS.

Comparison of the structural polypeptide patterns of different herpesviruses is complicated by two factors. First, individual isolates of herpes simplex type 1 and, to a lesser extent of type 2, show characteristic differences in the mobility of certain structural polypeptides. Strain variation in the structural polypeptide pattern of other human herpesviruses has been less well investigated but certainly exists. Secondly, the structural glycoproteins of a given virus strain may differ after growth in different host cells. Nevertheless, differences in mobility between analogous structural proteins of different herpesviruses are clearly greater than that between individual isolates of any one virus. Figure 88.6 depicts diagrammatically the structural polypeptides of four known herpesviruses. For the two types of herpes simplex virus, although the molecular weights of apparently analogous polypeptides differ, there is an obvious one-for-one correspondence. By contrast, the polypeptides of cytomegalovirus and of EB virus differ extensively from each other and from the two types of herpes simplex virus.

Non-structural polypeptides SDS polyacrylamide gel electrophoresis of radioactively labelled polypeptides from cells infected with herpes simplex type 1 reveals about 50 virus specific polypeptides in single-dimensional gels; there are substantially more in two-dimensional gels, of which approximately half correspond to structural polypeptides. Enumeration of the different polypeptides is complicated by the existence of precursor-product relationships between differently glycosylated forms of the same glycoprotein. Some polypeptides are phosphorylated or sulphated and once again precursor-product relationships can be discerned.

As will be seen later, the non-structural polypeptides can be divided into three temporal groups, α, β, and γ from the time of their synthesis in the growth cycle.

Our knowledge of the functions of individual non-structural polypeptides is even more perfunctory than for structural polypeptides. As will be described in a later section, some can be correlated with virus specific enzymes or proteins with DNA binding properties.

Other components *Lipids:* Ether sensitivity of virus infectivity was one of the earliest recognized characteristic properties of herpesviruses. It is therefore not surprising that lipid should have been identified by chemical analysis of virus particles. It has been suggested that lipids present in the cell before infection are incorporated into the virus envelopes. *Polyamines:* As for several other viruses, polyamines have been detected in herpesviruses. Spermine occurs in the nucleocapsid, or is presumed to do so from failure to remove it by detergent action on virus particles. Since the amounts of spermine present would neutralize at least half the phosphate of the virus DNA, it probably plays a role in condensation of the DNA. Spermidine is removed from particles by detergent and may therefore be present in the envelope or tegument.

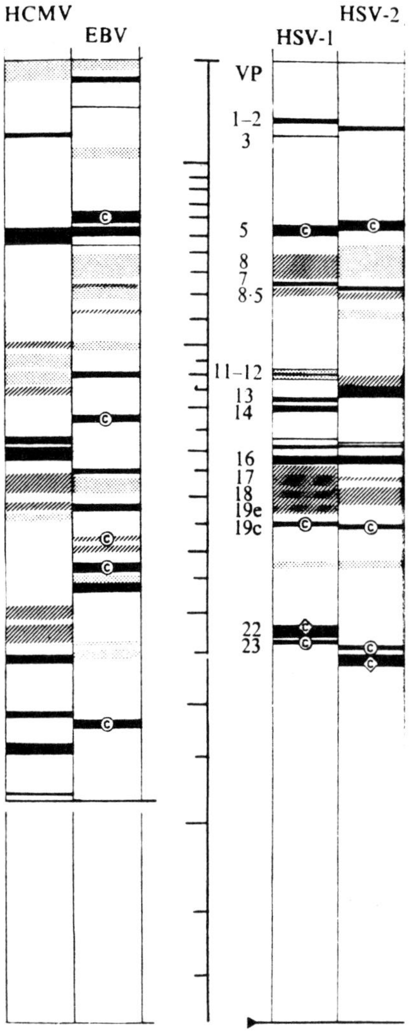

Fig. 88.6 Diagrammatic representation of structural polypeptides of herpes simplex virus types 1 and 2, human cytomegalovirus and EB virus. (Reproduced by permission of the Society for General Microbiology.)

Virus-specific enzymes

Complex viruses, such as herpesviruses, would be expected to require a large number of enzymes for the varied biosynthetic processes in the infected cell. The genome is plainly large enough to specify many such enzymes and the first of these virus-specific enzymes was identified nearly 20 years ago. In the intervening years comparatively few enzymes have been conclusively identified as being virus-specific, although there is fragmentary information on a number of enzymes which might be expected to be concerned in virus biosynthesis.

Prima facie evidence for the specificity of an enzyme comes from observations of raised levels of activity following infection. This might be the result of, for example, stabilization of an unstable enzyme following infection, or even induction of a new but host-coded activity in the infected cell. While virus specificity has often been claimed from observations of mutant viruses lacking ability to produce the particular enzyme, or of inhibition of enzyme activity by an antiserum reacting uniquely with virus-specified components, such evidence cannot be regarded as absolutely conclusive. Modern technology should allow in-vitro translation of individual gene transcripts and thus provide definitive evidence if such products could be demonstrated to possess similar enzyme activities to in-vivo products; evidence of this kind is now available.

Thymidine kinase Increased amounts of this enzyme were first observed after infection with herpesvirus type 1 by Kit and Dubbs (1963). They further showed that the enzyme, which catalyses phosphorylation of thymidine to its monophosphate, is not essential for the growth of the virus in cultured cells, since bromodeoxyuridine-resistant virus mutants, lacking the ability to manufacture kinase were able to grow in mutant cells that lacked the normal host thymidine kinase. It was first pointed out by Kaplan that the natural habitat of herpes simplex virus was in resting cells where phosphatase levels are high and, indeed, augmented after virus infection. He reasoned that the kinase might allow the virus to grow in this environment and therefore be an essential requirement in the natural rather than under laboratory conditions. Subsequently kinaseless mutants of the virus were shown incapable of growing in stationary cells in culture, while more recently several workers have demonstrated diminished pathogenicity of kinaseless mutants in mice.

Supporting evidence for the virus specificity, assumed from the existence of kinaseless mutants, has come from the observation that kinases specified by different herpesviruses can be distinguished from each other and from the host cell enzyme on the basis of serological specificity. Thus an antiserum, prepared by immunization of rabbits with RK13 cells infected with herpes simplex virus type 1, inhibits kinase in infected but not uninfected BHK 21 cells. Similar antisera were used to show a difference in the antigenic specificity of herpes simplex type 1 and 2 kinases. The thymidine kinase activity of herpes simplex type 1 is associated with a non-structural polypeptide whose molecular weight is 43 000, whereas the analogous type 2 polypeptide differs slightly in molecular weight. In-vitro translation studies on transcripts of the thymidine kinase gene (see below) have confirmed many of the less precise results described above. Additionally the nucleotide sequence of the gene has now been reported.

The biochemical characters—Michaelis constant, sensitivity to feed-back inhibition and thermostability—of herpes simplex type 1 thymidine kinase differ from the host enzyme. Types 1 and 2 virus-specific kinases differ significantly in thermostability; this has been used as a means of distinguishing the two types. A further unusual feature of this enigmatic enzyme is that the polypeptide carrying kinase activity apparently possesses in addition deoxycytidine kinase activity. This was concluded from the demonstration that BUDR-resistant mutants which, as noted above, lacked the ability to specify thymidine kinase, failed to produce deoxycytidine kinase; similar properties are demonstrable for mutants resistant to cytosine arabinoside.

Though this section has reflected the concentration in published work on the thymidine kinases of the two herpes simplex viruses, it should be noted that varicella-zoster virus also appears to specify such an enzyme. By contrast there is no evidence for virus specific kinases of cytomegalovirus or EB virus.

DNA polymerase All the human herpesviruses produce a virus-specific DNA polymerase in infected cells. As for thymidine kinase, the virus specific polymerases can be distinguished biochemically and antigenically from the host cell polymerases. The enzymes from cells infected with herpes simplex virus of both types have been extensively purified and the activities in both cases are associated with a polypeptide whose molecular weight is 150 000. The enzyme is active, at least *in vitro*, in its monomeric form, although there is evidence for the enzyme being present in infected cells in a multimeric form. The polypeptides of the two types show considerable cross-reaction in serological tests but it seems likely that they are not identical. As might be expected, the DNA polymerase, unlike the thymidine kinase, is essential for virus growth. Of considerable significance is the differential (i.e. greater than that of cell polymerase), sensitivity of the herpes simplex polymerase to potential chemotherapeutic agents such as phosphonoacetate and acycloguanosine (acyclovir).

Other enzymes There are a number of reports on enhanced deoxyribonuclease activity in cells after infection with herpes simplex virus of both types. At least one component of alkaline exonuclease is thought to be virus-specified from antiserum inhibition tests. This is supported by the thermosensitivity of the exonuclease present in cells infected with the temperature-sensitive mutant. An acid deoxyribonuclease activity may represent a different enzyme. There are reports of nucleolytic activity being associated with purified DNA polymerase. Although at times the activity seems likely to be due to contaminating proteins, at least some deoxyribonuclease activity seems to be closely associated with purified polymerase and, significantly, this activity, like that of the polymerase itself, is inhibited by phosphonacetate.

Ribonucleotide reductase activity has been observed to increase after infection with herpes simplex virus. The infected cell enzyme is more resistant to inhibition by dTTP than the uninfected cell enzyme, whereas temperature-sensitive mutations in two different genes give rise to altered expression of the enzyme.

There are a number of reports on protein kinase in herpes-infected cells and it has been stated that an enzyme is closely associated with a core polypeptide of herpes simplex virus type 1, possibly as VP23. An alkaline phosphatase has been associated with the envelope surface of the virus.

Among other enzymes reported in infected cells are thymidylate synthetase, deoxcytidine deaminase, uridine kinase and choline kinase.

Despite a number of reports on increased DNA-dependent RNA polymerase in infected cells there is no clear evidence for a virus-specified protein with such activity. The observation of protein-free infective DNA shows that the virus can use host polymerase in this situation at least, while the sensitivity of herpes simplex growth to α-amanitine strongly suggests that the host cell DNA polymerase II is necessary for virus infection. The particle-associated polymerase of vaccinia virus is insensitive to this drug.

Herpesvirus antigens

In view of the extensive range of polypeptides with molecular weights extending to over 100 000 which are present in cells infected with herpesviruses, the number of potential antigenic determinants is formidably vast. Reports on herpesvirus antigens present a picture of bewildering nomenclatural complexities. Introduction of monoclonal antibody methods, while providing extensive and interesting information, is in the early stages, perhaps adding to the confusion.

An attempt is made in the succeeding paragraphs to summarize the essential information relevant to the differences and similarities between human herpes viruses and the clinical picture, including possible oncogenicity of the viruses. The experimental basis for such a discussion may be stated briefly. By means of a rabbit antiserum to infected RK 13 cells, i.e. a general or poly-specific antiserum, it is possible to obtain multiple precipitin bands in classical agar gel immunodiffusion tests with lysates of infected cells. Such antisera precipitate most of the virus-specific polypeptides in infected cells. Antisera of more restricted specificity may be derived after separation of infected cell antigens by electrophoretic or chromatographic methods, and also by immunizing rabbits with separated structural or sometimes non-structural polypeptides derived from SDS polyacrylamide gel electrophoresis. Though this technique clearly denatures some, if not most, of the antigenic determinants of the native molecule, this is of some advantage in deriving antisera reactive with a restricted range of determinants.

Turning first of all to antigens of the two herpes simplexviruses, there is now a considerable body of evidence relating these to homologous antigens of other herpesviruses; little of this, however, relates to the other human herpes viruses. Of these, there is little available information on antigens of varicella-zoster virus and of cytomegalovirus, while the extensive work on EB-virus antigens has developed, to a certain extent, independently of work on herpes simplex viruses, although recent introduction of biochemical methods for characterizing antigens and relating them to polypeptides is leading to easier recognition of unifying principles. It is accordingly convenient to describe work on EB-virus antigens separately.

Polypeptide composition of antigens: type-specific and type-common (cross-reacting) determinants

Antiserum to both herpes simplex virus types precipitates virtually all the available polypeptides present in cells infected with either type (Fig. 88.7). Nevertheless, as noted in a previous section, virtually all analogous polypeptides of the two types have slightly different mobilities in SDS polyacrylamide gel electrophoresis and it is therefore not surprising to find type-specific antigenic determinants on polypeptides. These can be demonstrated by immune precipitation of polypeptides by antiserum of the homologous type after absorption with the heterologous type to remove cross-reacting antibodies. These experiments show that virtually all the polypeptides carry type-specific as well as type-common (or cross-reacting) determinants.

We know that a very restricted range of the common determinants shared by the two herpes simplex virus types are also shared with other non-human herpes viruses. In particular, it appears that the major DNA binding protein, termed ICP8 for herpes simplex type 1 and ICSP 11–12 for type 2, and the major capsid protein (VP5) of a number of herpesviruses may be precipitated by heterologous antisera. Correspondingly, an antiserum to the purified major DNA-binding protein of herpes simplex virus type 2 will precipitate the analogous polypeptide from cells infected with other herpesviruses. There is some suggestive evidence that this antiserum cross-reacts with an antigen in cells infected with human cytomegalovirus. Gel immunodiffusion tests suggest a cross-reaction between varicella-zoster virus and herpes simplex virus but it is not certain if this relates to the major DNA-binding protein.

Two other antisera of more restricted specificity,

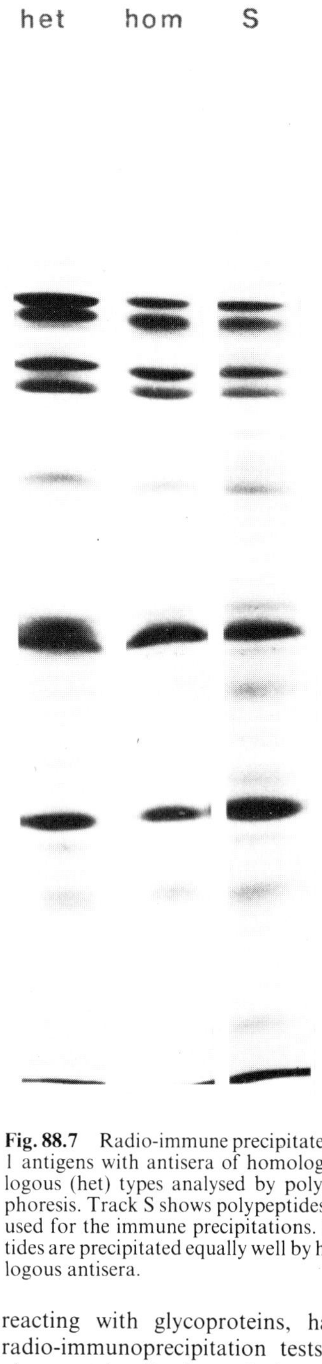

Fig. 88.7 Radio-immune precipitates of herpes simplex type 1 antigens with antisera of homologous (hom) and heterologous (het) types analysed by polyacrylamide gel electrophoresis. Track S shows polypeptides in antigen preparation used for the immune precipitations. Note that the polypeptides are precipitated equally well by homologous and heterologous antisera.

reacting with glycoproteins, have been tested in radio-immunoprecipitation tests. Antiserum to the glycoprotein gD of type 1 virus precipitates both the homologous and heterologous type gD i.e. cross-reacts, while an antiserum to denatured gA + gB + gC of type 1 virus reacts with the homologous type only. This restricted reactivity must be a reflection of the SDS denaturation of the antigen, since, as noted above, the corresponding native glycoproteins carry type-common as well as type-specific determinants. In accord with this, an antiserum prepared against native or less denatured gA + gB does cross-react. As will emerge later, both this antiserum and the antiserum to gD can be shown to contain antibodies reacting with type-specific as well as type-common components. It is appropriate to describe here three other antigens identified in herpes infected cells that we shall see later have been claimed to be of relevance in studies on carcinomas possibly associated with herpes simplex viruses. These antigens have been fully described by Pearson and Aurelian (1981).

The first antigen, variously termed VP143 or ICSP 11/12, corresponds to the major DNA-binding protein found in cells infected with herpes simplex virus type 2. It has been observed in biopsies of cervical carcinoma, and it is claimed that there are higher concentrations of antibody to this antigen in sera from cervical carcinoma patients than in those from patients with other or no cancers.

The second antigen has been designated AG-4 and corresponds to a polypeptide, ICP6, with a slightly greater mobility in SDS polyacrylamide gel electrophoresis than the major capsid protein of herpes simplex virus. It is unusual in being described as an immediately early (α) protein (see below under replication), although also structural. It has been detected in cervical tumour biopsies and IgM antibody to it has been claimed to be prognostic of cervical carcinoma, whereas patients with primary type 2 infections show only transient antibodies to Ag-4.

The third antigen of this kind has been labelled Ag-e. It is stated to consist of two polypeptides of mol. wt 130 000 and 140 000, possessing type-common reactivity. Once again the antigen is present in cervical carcinomas but, unlike Ag-4, antibodies to it are not prognostic.

A further antigen of mol. wt between 40 000 and 60 000 is found in cells infected with either herpes simplex virus type. This has been claimed to cross-react with an antigen HSV-TAA (tumour associated antigen) isolated from squamous cell carcinomas. Complement-fixation tests show higher antibody activity against the antigen in sera from patients with squamous cell carcinomas of the cervix, head and neck than in sera from patients with no malignancies or with other carcinomas.

Type-specific and type-common components in neutralization reactions

Studies on neutralization of virus infectivity allow more precise delineation of some of the type-specific and type-common determinants detected in the immune precipitation as described above. General antisera to both herpes simplex types cross-neutralize. Correspondingly, antisera to gD and to undenatured or partly denatured gA + gB cross-neutralize, whereas an antiserum to highly denatured gA + gB + gC of type 1 neutralizes the homologous type only. While the type-specific activity of the latter antiserum could reside in the gC element, it can be shown that gA + gB and gD contain type specific epitopes which lead to neutralization when a reaction with homologous antibody occurs. Thus if antiserum to gD or to gA + gB (undenatured or only partly denatured) from type 1 virus is absorbed with excess heterologous (type 2) antigen, the resulting absorbed antiserum will neutralize the homologous virus type only.

One important result of these neutralization reactions is that it is not possible to determine uniquely prior infection with type 2 virus by means of a neutralization test on a patient's serum. This becomes of some importance in sero-epidemiological studies which have been used to claim an association between cervical carcinoma and prior infection with herpes simplex virus type 2. It is desirable that such studies should test for the type 2 specific neutralizing activity which alone is diagnostic of prior type 2 infections. Obviously this can be done by absorbing the serum with excess type 1 antigen, thus removing cross-neutralizing activity which obscures the type-specific reactions.

EB-virus antigens

A number of antigens have been described in cells infected with EB virus. The reagents originally used in their identification were of necessity human sera and most of the antigens described are of significance in studies of Burkitt's lymphoma and nasopharyngeal carcinoma. The original observations generally used immunofluorescence techniques but more recently radio-immune precipitation methods have been used to relate the antigens to individual polypeptides.

Virus capsid antigen (VCA) This antigen was described as a brilliant intranuclear antigen in immunofluorescence tests on cell lines derived from Burkitt's lymphoma in the presence of human sera from lymphoma or normal patients. The proportion of immunofluorescent cells corresponded closely to the proportion shown by electron microscopy to contain intranuclear nucleocapsids and it was shown later that, when individual VCA positive cells were picked individually and examined in the electron microscope after sectioning, they always contained large numbers of intranuclear nucleocapsids, whereas VCA negative cells showed none. Sera containing VCA antibody also agglutinate naked nucleocapsids.

Membrane antigen (MA) Membrane antigen was first detected on the membranes of fresh biopsy cells from Burkitt's patients in indirect immunofluorescence tests with human sera. Since other herpesviruses express glycoproteins on the surface of the infected cell and since antibodies to such glycoproteins have been shown to neutralize virus infectivity of these viruses, it is not surprising that there is good correlation between EBV neutralizing activity of a serum and anti-MA but not anti-VCA activities. Polypeptides in the MA complex have been identified by immunoprecipitation, with antisera containing anti-MA activity, of polypeptides from surface radio-iodinated polypeptides from cells possessing MA. Four polypeptides were observed. One showed little variation in its observed mol. wt of 85 000, irrespective of the source of the cell line used. Two, with mol. wt quoted as 240 000–270 000 and 320 000–340 000 showed minor mobility differences in SDS polyacrylamide gel electrophoresis, dependent on the origin of the cell line used, while a fourth, mol. wt 160 000, was found only in two cell lines. These antigens are of significance with respect to current interest in producing an experimental vaccine to EB virus.

Early antigen (EA) This antigen was first detected in 'non-producer' Raji cells which were superinfected with the P_3HR-1 strain of EB virus. At early stages after superinfection the cells contained no VCA but reacted with sera from patients with active EB virus-associated diseases but not with sera from normal patients, who nevertheless have antibodies to other ES virus antigens.

The EA complex shows two different antigenic specificities. One, termed restricted (R), is confined to the cytoplasm whereas the other, diffuse (D), is described as more finely dispersed and is present in both nuclei and cytoplasm. Acute infectious mononucleosis patients contain predominantly anti-D whereas sera from Burkitt's lymphoma patients rarely show anti-D activity on its own but may show either anti-R activity alone or antibodies to both determinants.

EB nuclear antigen (EBNA) This antigen was detected using an anti-complement immunofluorescence test in which a fluorescent conjugate of anti-C3 is used to detect complement bound to the appropriate antigen–antibody complex. It is present in the nuclei of all known EB virus infected or transformed cells and has DNA binding properties. The antigen has been purified to a single component of mol. wt 180 000 which when electrophoresed on SDS polyacrylamide-gels, is shown to consist of one polypeptide of mol. wt 48 000.

Lymphocyte detected membrane antigen (LYDMA) This antigen is another membrane antigen on cells carrying EB virus, distinct from MA. It is detected by EB virus specific killer T cells derived from active infectious mononucleosis patients after removal of non-specific killer cells by Fc receptor positive natural killer cells. It is thought that the antigen appears at an early stage of B cell transformation by EB virus since this is prevented by an addition of the anti-LYDMA effector cells.

Monoclonal antibodies Monoclonal antibody methods are now being widely deployed in studies of herpesvirus antigens. A bewildering array of reagents is now available for each of the five human herpes viruses. It is difficult to give an account of results obtained with these reagents within the limits of this chapter, particularly at this expanding stage of their use. It may be noted briefly that it has in many cases been possible to separate a number of the complex reactivities of polyspecific antisera. Thus, several different monoclonal antisera to gD of herpes simplex virus type 1 have virus neutralizing activity. Some of these will neutralize type-specifically, while others also cross-neutralize type 2 virus. One of the main applications of these reagents will be in the purification of components appropriate for experimental vaccines or for specific diagnostic reagents.

Antigenic variation In concluding this section it is perhaps necessary to emphasize that there is now abundant evidence for antigenic variation between different isolates or strains of at least four of the human herpesviruses.

For herpes simplex virus type 1 it has already been noted that restriction enzyme cleavage patterns and electrophoretic mobility of analogous virus polypeptides vary between individual isolates. Corresponding antigenic differences can also be demonstrated. Thus absorption of an antiserum to one strain or isolate with antigen of a second will frequently leave residual neutralizing activity against the first and in some cases also other strains. By a series of such experiments it has been shown that epitopes reacting with neutralizing antibody but which are found only on type 1 strains may be either (a) present on particles of many or all type 1 strains, (b) present in particles of only a few strains, or (c) present on particles possibly of only one strain. Similar patterns are emerging from studies with monoclonal antibodies.

The same antigenic diversity with respect to sites involved in virus neutralization has not been shown in type 2 strains and it appears that polypeptide mobility differences are less marked in these than for type 1 strains.

Cytomegalovirus strains have been claimed to differ antigenically from a comparison of neutralization kinetics of homologous and heterologous strains with a given antiserum. There are also reports of EB virus strain heterogeneity detected by monoclonal antibodies to glycoprotein determinants. So far with none of the viruses studied have inter-strain antigenic differences been related to any clinical attributes such as the site of isolation of herpes simplex viruses, although some distinction may perhaps be made between different patterns of polypeptide mobility of different isolates in terms of their site of isolation.

Genetics

Genetic and physical maps Classical recombination analysis using temperature-sensitive mutants of herpes simplex virus types 1 and 2, either with or without an additional factor such as plaque morphology, has produced a number of genetic maps for different strains. However, recently, most interest has centred on physical mapping, either of *ts* mutations or of gene functions. Correlation between map locations of transcripts or of polypeptides with their temporal class will be deferred to a later section, as will consideration of the value of genetic techniques in addressing the problems of the functional equivalence or otherwise for the four isomers of the herpes simplex gene. This section will be confined to a brief outline of the methods used.

Marker rescue In this method the physical map position of restriction enzyme fragments which will 'rescue' a *ts* mutant, i.e. allow infective virus to be made under restrictive conditions in cells infected with a *ts* mutant together with the wild type restriction enzyme fragment, is used to position the mutation of the map. Obviously, when a number of different overlapping fragments derived by cleavage with different enzymes are used, it is possible to map the mutation with some precision. If a mutation can be linked with failure to produce one unique polypeptide or enzyme then obviously the gene for this product is automatically mapped.

In-vitro translation of mapped transcripts If mRNA can be mapped by hybridization tests with different fragments, and the translation product of the messenger can be made *in vitro*, this product is automatically mapped. This method will probably prove to be the most powerful of those considered here.

Use of intertypic recombinants This method makes use of the fact that, since restriction enzyme cleavage sites are different for the two types of herpes simplex virus, it is possible by restriction enzyme analysis of intertypic recombinants to detect the region(s) of cross-over with a precision limited only by the number of available restriction enzymes. Since additionally polypeptides of the two types have mobilities in SDS polyacrylamide gel electrophoresis characteristic of the type, it is possible for each intertypic recombinant to assign each polypeptide to one or the other parent. Comparison for a number of recombinants of cross-over sites with parental type of polypeptides allows location of the particular polypeptides on the map.

Replication of herpesviruses

The replication of herpesviruses has been most fully investigated for herpes simplex virus type 1. As might be expected, the main features of the type 2 replicative cycle are identical with type 1. Of the other three human herpesviruses experimental difficulties have left knowledge of the varicella-zoster virus cycle in a very rudimentary form. Molecular events in cytomegalovirus-infected cells seem likely to follow similar principles to those established for the herpes simplex virus. Finally, for EB virus, while the outcome of the host-cell virus interaction is very different, with the addition of a transforming function and the absence of any truly permissive cell system making analogous studies impossible, it seems at least possible that the separation into early and late gene functions will follow similar principles to those to be discussed here for the herpes simplex viruses.

Attachment and penetration Growth of the virus in its normal habitat must be initiated by the enveloped particle and, although nucleocapsids and naked DNA can infect cells in experimental conditions, their specific infectivities are much lower. It is therefore reasonable to base a discussion of the early events of infection on the fate of enveloped particles in experimental studies. Our knowledge of the virus attachment site and of cellular receptors for the virus is minimal. It seems likely that one or more of the surface glycoproteins must be involved in attachment. Since particles without gC can infect cells, at least one of the other glycoproteins must be able to mediate infection; a similar argument could be applied to gB in the light of evidence that particles lacking it can attach to cells even when unable to penetrate. However the incomplete understanding of the relation between gA and gB makes it difficult to interpret this evidence in a conclusive manner.

In relation to cellular receptors, since there is evidence that cells saturated with one herpes simplex virus type allow attachment of the other type, there would seem to be some grounds for believing that specific receptors exist. It may be noted here that the lack of a suitable fully permissive experimental cell system for EB virus stems from the fact that few cells which can be grown in culture possess appropriate receptors such as are found on B lymphocytes.

The next stage, that of penetration, has been the subject of controversy. It is clear that in some cell systems virus can enter cells in pinocytotic vesicles. However, since these studies inevitably use high ratios of particles to cells, it cannot be regarded as certain that virus particles identified in pinocytotic vesicles are not moving along an irrelevant cul-de-sac. While there are some problems in interpretation of electron microscopic images of thin sections apparently depicting particles entering by fusion of the envelope and cell membranes, this mechanism may well be the correct one, particularly since it is observed to occur rapidly and is therefore consistent with the times at which virus-specific protein synthesis is detected. Nevertheless, while the fusion process successfully delivers an uncoated nucleocapsid to the cytoplasm, there is still a veil over events leading to the presumed transport of the virus DNA to the nucleus.

Protein synthesis It is convenient to consider this element of the growth cycle before the more logical transcription stage. Rates of incorporation of radioactively labelled amino acids into acid insoluble proteins show a rapid decrease immediately after the infection, owing to the inhibition of host cell protein synthesis by the infecting virus. Thereafter, the rate increases as virus-directed synthesis begins. Finally the rate declines slowly as the production of progeny virus reaches its peak. Identification of new virus-specific polypeptides in infected cells was once a major stumbling block but higher resolution methods of gel electrophoresis have made the distinction easier. Differentiation of these polypeptides from uninfected cell polypeptides now rests mainly on the recognition of an increasing rate of synthesis after infection, together with specific precipitation of the polypeptides by appropriate antisera.

Virus-specific polypeptides can be arranged in a minimum of three sequentially ordered groups, α, β, γ, immediate early, early and late. Production of each later group requires not only prior synthesis of the previous group but also its inhibition.

The first temporal group, α polypeptides, are defined operationally as those polypeptides made immediately after removal of cycloheximide or puromycin present during infection and for several hours thereafter. These polypeptides are still made if actinomycin D is added to cells when the protein synthesis inhibitors are removed. It is concluded that immediate early mRNA is made in the presence of protein synthesis inhibitors; α polypeptides are likely to exert regulatory functions in the growth cycle but their exact role needs to be determined.

***β* polypeptide synthesis** This begins soon after that of α polypeptides. The rate of production of the latter then declines, apparently through the action of β polypeptides. Production of β polypeptides requires fresh transcription of appropriate regions of the genome in the presence of α polypeptides; when actinomycin D is added to cells on removal of protein synthesis inhibitors present from the time of infection and for several hours thereafter, the production of β polypeptides is inhibited. This group of polypeptides includes the virus specific DNA polymerase and thymidine kinase.

The transition from β to γ polypeptide production follows a similar pattern to the α to β transition. The γ polypeptides form by far the largest group and include most of the structural proteins. These proteins, unlike α and β polypeptides, are made at the onset of DNA replication. Gamma polypeptides can be synthesized in the presence of inhibitors of DNA synthesis but only in reduced amounts. This is probably because the smaller size of the virus DNA pool limits transcription and the rate at which protein can be manufactured; mRNAs for these late proteins have a much shorter half life than α and β mRNAs and this may explain why maximum rates of synthesis of these polypeptides require the optimum size of the DNA pool.

Genes specified in the three temporal groups of proteins are not characteristically clustered on the physical map of the genome. Thus, although many of the late γ polypeptides are mapped on the long unique sequence (U_L), one of the glycoproteins, gD, is located in the short unique (U_S) region. Other glycoproteins are located in U_L, with gA and gB mapped in the same region, consistent with their close relationship. Distribution of β polypeptide genes is similar.

The α functions are widely distributed on the map but are noteworthy for location, at least partly, in reiterated sequences flanking the unique regions. For one of these, a phosphorylated polypeptide, molecular weight 175 000, designated ICP4, there are two copies of the gene per DNA template and both are expressed.

Transcription Herpes simplex virus type 1 mRNAs share with host cell mRNA the basic properties of nuclear synthesis, polyadenylation and cap structures. Transcripts have been located on the physical genome by a number of elegant techniques, particularly 'Southern blot' hybridization, in which restriction enzyme cleavage fragments of DNA are transferred to nitrocellulose sheets after electrophoretic separation and then transcripts mapped by hybridization to the 'blots' of ^{32}P labelled mRNAs. It is clear that splicing of messengers, while not extensive in the herpesviruses, does occur.

Immediate early messengers are transcribed from regions at or near repetitive sequences. At least some are spliced and not all the mRNAs are transcribed in the same direction.

Early mRNAs are transcribed from regions scattered over the whole genome. One of those which has been unambiguously located is the thymidine kinase at map co-ordinate 0.3 in U_L corresponding to that determined from polypeptide mapping (0.300 to 0.309). The thymidine kinase mRNA is unspliced.

As expected there are a large number of late mRNAs transcribed in either direction from regions throughout the genome. There is at least some representation of spliced messengers.

Replication of DNA A fundamental question on replication of herpesvirus DNA is not completely resolved. As explained in a previous section, four of the human herpesvirus DNAs exist in isomeric forms and it is obviously important to know for a proper understanding of DNA replication if they are functionally equivalent i.e., whether all participate equally in replicative processes. At the moment the consensus is shifting from non-equivalence to equivalence. The earlier view was based on a detailed analysis of cross-overs in DNAs of a number of intertypic (herpes simplex type 1 × type 2) recombinants by Morse *et al.* (1977). These workers labelled the isomeric forms with reference to a P (prototype) form with the other three arrangements being termed I_S(U_S segment inverted with respect to P), I_L (U_L segment inverted with respect to P) and I_{SL} (both segments inverted with respect to P). They argued that a single cross-over in U_L of the P arrangement would be unaffected by inversion of U_S and so would still appear as a single cross-over in I_S. However in the other two forms, I_L, and I_{SL}, the inversion of U_L would inevitably require a further cross-over at the U_S, U_L junction to conform with the correct parental gene type in U_S (Fig. 88.8). Such an apparent increase in the number of cross-over sites would be an inevitable consequence of an odd number of cross-overs. Morse *et al.* (1977) displayed all cross-overs in the arrangement which minimized their number and found that this arrangement was always P, which was in fact so defined as the result of this analysis. The weak link in the argument is whether such recombinants do really possess odd numbers of cross-overs or do so only apparently, through one cross-over, perhaps near the end of the molecule, being undetected. Since the collection of available restriction enzyme cleavage sites can never completely saturate the genome, it is always possible to miss a cross-over and, in the last analysis, only sequencing can completely overcome this problem. In fact more detailed analysis of at least some recombinants thought to possess an odd number of cross-over sites has revealed additional sites, thus undermining the basic premise of the case for non-equivalence.

Attempts have been made to resolve the problem by analysis of genetic linkage between markers in U_S and U_L. If only one arrangement initiates replication, then fully consistent linear maps should be obtained whether the markers are in different or the same segments. If other arrangements participate, then all mar-

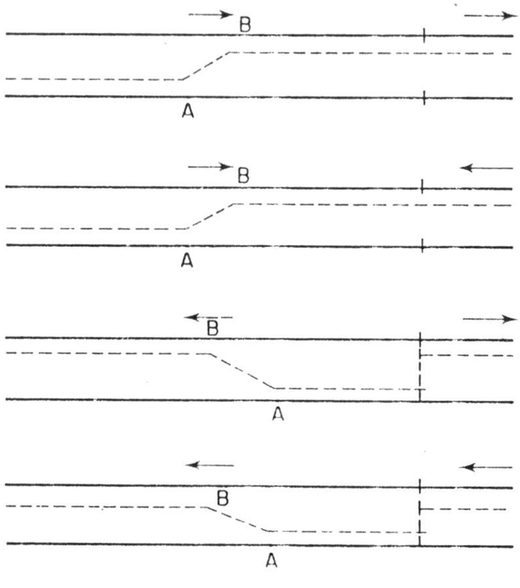

Fig. 88.8 Diagrammatic representation of the effect in an intertypic recombinant of herpes simplex virus of a cross-over between restriction enzyme cleavage site A on the type 2 genome and a similar site B on the type 1 genome. For each of the four isomeric forms P, I_S, I_L, and I_{SL} the upper solid line depicts the type 1 genome and the lower the type 2 genome with the dotted line showing the recombinant genome. Note that a single cross-over in the P, I_S forms of necessity appears as a double cross-over in I_L and I_{SL}.

kers of U_L should form one linkage group, those in U_S another, and all markers in one group should give identical recombination frequencies with all markers in the other. In the event the results cannot be claimed to give a clear-cut answer one way or the other.

A very careful analysis of DNA of a non-human herpesvirus, pseudorabies virus of pigs, whose DNA like that of varicella-zoster virus exists in two isomeric forms, showed that each form participated equally, not only in infection of the cell, but also in subsequent replication (Ben Porat *et al.* 1980). This would at the very least suggest the possibility that the isomeric forms of herpes simplex DNA are functionally equivalent.

The origin of herpes simplex DNA replication has been studied by electron microscopic examination of DNA extracted at the time of DNA synthesis. The positioning of loop structures on the molecules is interpreted as indicating initiation sites in the U_S segment and also at or near the middle of U_L. The loops near the end of the molecule may not reflect true initiation sites, as they could result from exonuclease digestion, which is known to produce tiny loops near the ends of DNA molecules. However it cannot be finally established whether the site in U_L is a unique site of initiation or whether two sites occur. Newly synthesized DNA appears in a rapidly sedimenting form. DNA synthesized during the first round of replication sediments more slowly than that synthesized later which is presumed to be concatemeric. Restriction enzyme analysis of these concatemers shows them to link head to tail. It is thus assumed that the DNA replicates by circularization with head to tail concatemers being generated in a rolling circle mode which would allow participation of all four isomeric forms and generate progeny, again with the four forms represented in equimolar proportions.

Assembly and release Our knowledge of the assembly processes involved in production of progeny particles in cells infected with herpesviruses is scanty. Electron microscopic studies of thin sections of infected cells show aggregates of capsids in the nucleus. Coreless particles and two forms of cored particles are seen. Genetic evidence supports the notion that the 'empty' particles are precursors to the 'full' particles containing DNA. Mutants which are unable to synthesize virus DNA frequently produce empty particles, as do infected cells treated with DNA inhibitors. However, some mutants able to synthesize DNA still do not produce full particles; it is assumed that they have a defect in a presumed mechanism for excision of 'head full' segments of DNA from concatemers and their insertion into empty capsids by analogy with processes well characterized for T-even bacteriophages. There is however little understanding of the processes whereby all herpes viruses assemble the icosahedral capsids of 162 capsomeres so characteristic of the family, although the observed antigenic cross-reaction between major capsid proteins of a number of herpes viruses may be of significance.

Though electron microscopic examination of thin sections makes it relatively easy to propose a mechanism of envelopment of nucleocapsids by cellular membranes and most workers have favoured the inner nuclear membrane as fulfilling this role, it is clear that this process has varying levels of efficiency in different cell lines. Thus, in infected BHK 21 cells large numbers of non-enveloped capsids of herpes simplex virus type 1 can be discerned in the cytoplasm. This may be the result of the observed rapid destruction of nuclear membranes and therefore subsequent spillage of nuclear content into the cytoplasm of these cells. Nevertheless, evidence has been adduced for envelopment of nucleocapsids by cytoplasmic or even plasma membranes. Obviously the problems of interpretation of static images of a dynamic process do not allow us to exclude the possibility that some progeny particles proceed into a pseudo-infectious cycle and become de-enveloped by fusion with cytoplasmic membrane. The role of the tegument components in virus assembly is difficult to perceive from the thin section studies, while mechanisms for insertion of virus specific glycoprotein in membranes destined for enlistment as virus envelopes is in no way so clear as, for example, with influenza virus which assembles at the plasma membrane.

Processes of release of virus from infected cells also seem to vary widely between different cell types. Thus, BHK 21 cells infected with herpes simplex virus type 1 frequently lyse and the released virus in the supernatant medium contains a high proportion of naked nucleocapsids. HEp2 cells infected with the same virus strain seem less prone to lysis and the supernatant medium contains a high proportion of enveloped particles. It is therefore easy to envisage virus as being pumped out of the infected cell by a reverse pinocytotic process. It has also been suggested that in some cases the virus may proceed after envelopment at the inner nuclear membrane through an inter-connecting labyrinth of intra-cisternal channels to the cell exterior.

Cell transformation by herpesviruses

Earlier sections have implied a contrast between EB virus, for which no fully permissive cell system exists, the virus always having a transforming function, and the other four human herpesviruses, undergoing a lytic cycle in permissive cells. Over the past decade it has become clear that all of these viruses, with the exception so far of varicella-zoster virus, can in appropriate cases initiate a cellular transformation response. To date, it has not been possible to produce with these viruses fully 'immortalized' cell lines retaining the whole virus genome without expressing all the functions of the lytic cell leading to production of progeny viruses and cell death. Generally, cell lines containing only a proportion of the virus DNA have been reported. These cells have been produced, for example, by treating cells with UV-irradiated virus, temperature-sensitive mutants at restrictive temperature, or by transfection of cells with restriction enzyme cleavage fragments of the virus genome.

The overworked term 'transformation' has been used to describe cells obtained by a variety of methods and exhibiting equally varied properties. Thus, while some possess the classical transformed cell properties such as altered morphology and transplantability in experimental animals to form tumours, others can only be described as transformed in the sense of 'altered in character, disposition' given as alternatives in the Concise Oxford Dictionary to the more familiar 'altered in appearance, outward form'.

Thus it will be recollected that it was possible to produce cell lines, by selection with BUDR, containing no native thymidine kinase and that these cells have been used to obtain transformants which have been induced to manufacture herpes simplex virus specific thymidine kinase. The enzyme made in these cells can be characterized as having all the physical, biochemical and antigenic properties of the virus-specific thymidine kinase found in virus-infected cells. These transformation reactions have been useful in developing methods for investigating the molecular biology of cells which have been transformed in morphological and transplantability properties. Thus it is possible to develop methods to show that herpes simplex virus genes in these cells map on the virus genome at precisely the map co-ordinates previously established for the thymidine kinase gene. With the methods thus validated it has been possible to map genes causing morphological transformation to the 0.311–0.415 region of the type 1 genome and the 0.582–0.682 region of the type 2 genome.

These methods have also been used to search for virus genes in cells from tumours where an aetiological role for herpesviruses has been suspected. One problem in these experiments has been raised by the controversy as to whether the continued presence of virus genes is necessary for maintenance of the transformed phenotype. Thus the herpes simplex type 2 transformed cell line 333-8-9, which displays altered morphology and transplantability, showed substantial elements of virus genetic information in early passages, with decreasing or undetectable amounts in later passages, although the transformed phenotype was maintained. While the sensitivity of the hybridization probe methods used to detect virus genes may have been too insensitive to detect small but nevertheless significant methods of virus DNA, this does not fully explain the progressive loss of such sequences. However it may be noted that the original cell lines were established from foci and were not cloned and it cannot be excluded that with continued passage a cell population with lower amounts of integrated virus genes may have been selected.

Studies on the expression of integrated virus genes in transformed cells are also of relevance in considering analogies with similar investigations on tumour cells. Expression of virus thymidine kinase and glycoproteins has been noted in some morphologically transformed cells. Generally the gene products expressed in such cells are known from mapping experiments to derive from gene sequences detected in cells.

Characteristics of experimentally produced cells containing EB virus Despite the ubiquitous properties of EB virus, most cells which can be grown in culture cannot be infected with the virus. Receptors for the virus particle have been found only on B lymphocytes of man and some primates—gibbon, squirrel monkey and marmoset—and on epithelial cells of nasopharyngeal carcinomas. There is a noteworthy difference between events in transformation experiments with lymphocytes from umbilical cord blood as opposed to those from adults, in that the former rarely contain the EA and VCA found in the latter. Transformation of cord blood lymphocytes can be used as an assay for EB virus since they can be considered unlikely to harbour the virus before infection. The assay can be used to detect neutralizing activity which inhibits transformation.

As with the other herpesviruses, studies on detection of virus genomes in experimentally transformed cells are useful in establishing and assessing methods used to detect the virus DNA in tumour cells. While certain human lymphoid cells apparently contain small numbers of EB virus genome copies (from 1–5), in other experiments clones of transformed cells isolated after exposure to less than 0.1 virus particle/cell contain up to 100 copies.

Though evidence has been presented for association of these virus DNA molecules in transformed cells with the host cell chromatin, a number of workers have shown that at least a majority of persisting EB-virus genomes in non-producer transformed cells can be separated from host DNA by centrifuging in alkaline sucrose gradients. It is concluded that the virus DNA exists in an independent episomal form. Supercoiled and relaxed circular duplex molecules have been found. These experiments do not of course exclude the possibility that small amounts of virus DNA, below the detection limits of the techniques used, are integrated into cellular DNA. (For review, see Sugden 1982.)

Disease patterns with human herpesviruses

Herpes simplex virus type 1

Historical Although the family name of 'herpes' is nowadays used only in connection with infections with the two herpes simplex virus types, this is a comparatively recent development. The word herpes (from ερπε, to creep—although some classical scholars prefer a translation 'to move slowly') has been used in clinical descriptions for at least 25 centuries, although Hippocrates, who used the term, would apparently not have used it for what we know as herpes simplex infections, even although he was apparently aware of them. Paradoxically probably in his time and for centuries thereafter, 'herpes' was generally used in connection with shingles.

For many years the orthodox belief that the pathogenesis of herpes infections was mediated by excretion of acrid bile from the skin was scarcely challenged—save perhaps by Shakespeare:

'O'er ladies lips who straight on kisses dream
Which oft the angry Mab with blisters plagues
Because their breaths with sweetmeats stained are'
—Romeo and Juliet, Act I Scene 4.

We may perhaps note in passing that those fascinated by cryptographical statements in Shakespeare's works, perhaps through unfamiliarity with virological concepts, seem to have missed the almost prescient connection between kissing and herpes infection which was to be substantiated in our own time with the recognition of EB virus as 'kissing virus' or indeed Nahmias' description of herpesviruses as the 'viruses of love'.

In any event the restrictions in use of the term 'herpes' are of recent development; illustrated reference manuals as late as 1938 included ringworm and erythema multiforme as herpes infections—perhaps reflecting Galen's definition of herpes as 'an inflammatory skin disease characterized by formation of small vesicles in clusters'.

What we now recognize as herpes simplex was shown to be infectious by human inoculation in the last quarter of the nineteenth century. The infection was transferred to rabbits early in the present century and the agent shown to be filtrable by Luger and Lauda (1921). In 1930 Andrewes and Carmichael showed that large proportions of adults possessed neutralizing antibodies in their sera and that only these persons were liable to recurrent infections. These remained a complete enigma in that they were apparently unrelated to contact infection, until Burnet and Williams (1939) explained the nature of primary and recurrent infections, although, as we shall see, the mechanisms of recurrence continue to exercise the skill and imagination of virologists deploying the full panoply of modern molecular biology in its investigation.

Oral infections The primary oral infection with herpes simplex virus type 1 may manifest itself purely subclinically. Repeated surveys show that large proportions of those possessing neutralizing antibodies deny any knowledge of primary or indeed of any recurrent infection. Such surveys have supported the original observation of Andrewes and Carmichael (1930) that the proportion of serologically positive persons increases with declining socio-economic group and suggests that with rising standards of living the incidence of infection is likely to fall.

The primary infection usually presents as a gingivostomatitis (Fig. 88.9) which in children has a peak incidence in the second and third years of life. In such young children the infection is usually mild or even trivial. In older children it may be slightly less mild and in adults can be severe. This tendency for severity of primary infection to increase with age is, as we shall see, a common characteristic of herpes viruses. Primary infections in young adults are often accompanied by the discomfort of general 'influenza-type' symptoms.

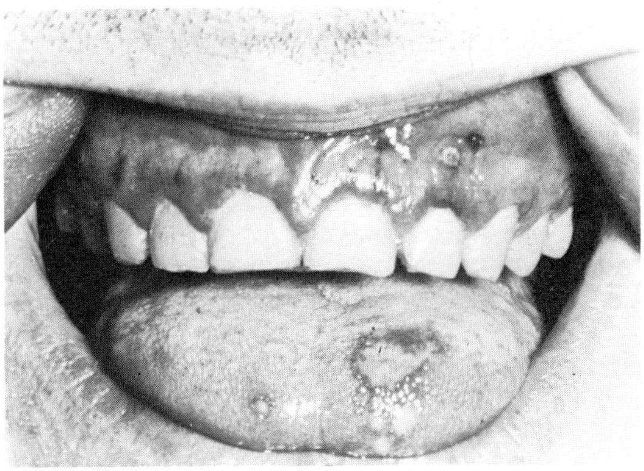

Fig. 88.9 Primary gingivostomatitis resulting from herpes simplex type 1 infection of 25-year-old male patient. Note lesions on tongue and gums. (Reproduced by kind permission of Dr B. E. Juel-Jensen and Heinemann Medical Books.)

Recurrence and latency The recurrent infection occurs usually at the site of primary infection although a recurrent lip infection is quite common after a primary infection in infancy in the oral cavity. During the prodromal phase, a tingling or itching at the site of subsequent eruption is commonly reported. Thereafter, small clusters of papules appear which become vesicular within approximately one day. Vesicles subsequently burst and a sore forms. Duration of recurrent attacks varies from two days to three weeks but is most commonly a little over a week. There is a tendency for the duration to be relatively constant for a given individual as does the interval between recurrences, which may be anything from days to years.

Recurrences are often associated with identifiable physical or emotional trauma. Extreme heat or cold, hormone treatments, allergic reactions or fever associated with other infections—pneumococcal pneumonia, malaria, meningococcal meningitis, and influenza, seem to provoke recurrence more often than other fevers—are examples of physical stimuli. Stress associated with menstruation or approaching examinations for students is often associated with recurrence.*

Persistence of the virus (latency) between recurrent infections, often in the presence of high levels of neutralizing antibody, obviously suggests that latent virus is in sites physically protected from the action of antibodies. Observations on the effects of surgical resection of the trigeminal ganglia of herpetic carriers for treatment of associated neuralgia have provided some clues. Such patients often suffered herpes infections in the area supplied by the maxillary and mandibular branches of the nerves on the operated side. Observations of the effects of section on either side of the ganglia suggested that virus resided in the sensory cells of the ganglia. Nevertheless, animal experiments showed that, when latent infections were established, no infective virus could be isolated from the ganglia. This problem was overcome when it was discovered that virus could be cultured if such ganglia were co-cultivated with cells sensitive to herpes simplex infection. That these findings were likely to be relevant in man was shown by recovery of virus from sensory ganglia taken from cadavers at post-mortem by similar co-cultivation methods.

This evidence suggests that the virus genome, but not fully infective virus, is present in the sensory ganglion. A number of ingenious experiments have led to the conclusion that the neuronal body harbours the latent virus. Thus reactivation of virus has been studied in ganglia implanted in Millipore chambers, and virus antigen, particles and virus DNA detected in neurones before satellite cells. Others reactivated virus latency induced by a temperature-sensitive virus mutant. Reactivation at a non-permissive temperature led only to development of virus antigen in single neurones.

The proposition that latency is mediated by virus genomes present in ganglion tissue left the way open for use of molecular biological methods to detect virus DNA. Sensitivity problems arose but by the use of ganglia from a large number of animals it was shown that herpes-specific nucleotides were present. *In situ* hybridization methods have detected virus-specific RNA. The transcripts present in ganglia harbouring latent virus have recently been demonstrated to derive from the left hand 30 per cent of the genome although transcripts from other U_L regions were observed occasionally. No transcripts from U_S regions were detected.

Controversy over the nature of transport between the peripheral site and the neurones now seems to have been resolved and it seems clear that an axonal route is followed. There is less agreement on mechanisms of reactivation. This has been provoked in experimental animals by UV treatment and 'Sellotape' stripping of peripheral skin. It has also been suggested that virus-specific antibody may hold virus in the latent state. What is then difficult to explain is why the well-established trauma can provoke recrudescence and, by implication, reactivation. The difficulties of explaining the mechanisms of such 'ganglion trigger' mechanisms have been circumvented by the ingenious suggestion that virus is circulating in small quantities all the time between ganglia and peripheral sites. This would lead to microfoci of infection on the skin not provoking a full-blown recrudescent lesion. Support for the notion comes from frequent reports of isolation of virus from saliva of herpetics during apparent latent periods of the infection. It has been suggested that local physiological factors e.g. prostaglandin levels in the skin may convert a microfocal infection to a full recrudescent lesion, and a link has been suggested between known effects of UV radiation from sunlight which increases local concentration of prostaglandins.

While it has been suggested that antibody to the virus is of limited importance in the control of latency, it is important to consider other immunological mechanisms, particularly important in view of the increasing use of clinically immuno-suppressive regimens which have often been associated with the provocations of recrudescence, frequently of a severe kind, leading sometimes to generalized infections. Cell-mediated immunity is important for protection of the host against herpes infections and defects in this arm of the immune system are thought to be responsible for recurrences. Although T lymphocytes in those undergoing recrudescence are competent in respect of lymphoproliferative response, they exhibit lymphokine deficiency in terms of both macrophage migration and leucocyte inhibition tests. Lymphokine production is normally seen as a product of delayed-type hypersensitivity reactions. Here some experiments on mice may be of considerable relevance. It has been noted that delayed-type hypersensitivity, developed after herpes infection of experimental mice, can be transferred only for the first ten days after infection to syngeneic recipients which have never encountered virus. This has been explained by the development of cells which suppress expression of delayed-type hypersensitivity. This suppressor activity is restricted to the draining lymph node and is not found in the contralateral lymph nodes; this asymmetry has been linked with latency.

Herpetic whitlow Primary herpetic infection of the finger (Fig. 88.10) has been regarded as a particular hazard for doctors, nurses and especially dentists. Other cases have been reported in children—sucking finger with a mouth infected with primary gingivo-

*I have followed general practice in referring to recurrent infections or recurrence. Wildy *et al.* (1982) have correctly distinguished between successive stages in the process of awakening of latent infections as follows: (1) Reactivation: stimulation of the latent state giving rise to infective virus, and (2) Recrudescence—initiation of the recrudescent lesion. They would reserve recurrence for the passage of virus to peripheral sites with no noticeable lesion being formed.

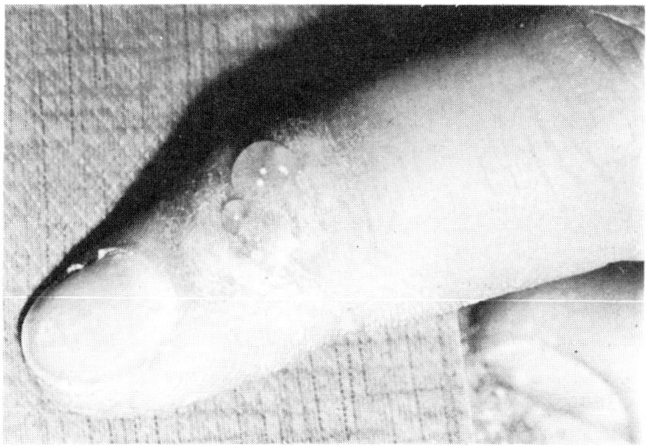

Fig. 88.10 Herpetic whitlow produced by herpes simplex type 1 infection of 21-year-old female patient. (Reproduced by kind permission of Dr B. E. Juel-Jensen and Heinemann Medical Books.)

stomatitis—and in adults who have cut or grazed their finger, often in the course of manual work. Similar infections of the toe have been described. As with oral cold sores, intense itching is often reported before development of deep vesicles. Whitlows are often extremely painful and enlarged lymph nodes and pain extending up the arm are common occurrences; in others rigors, headaches, and general 'influenza type' symptoms are reported. Recurrences are common, often at closely spaced intervals.

Herpetic whitlows are by no means rare among those handling the virus, either directly in laboratories, or by contact with patients e.g. tracheotomy patients frequently excrete virus in the stoma and dental patients may have virus in the saliva even in the absence of overt infections. Since many of those with these occupational hazards are likely to have no herpes antibody, which could well protect, this painful infection should be very much in their minds.

Ocular keratitis (keratoconjunctivitis) This manifestation of herpes simplex virus type 1 may be a serious hazard to the patient's sight. Primary infection usually takes the form of a conjunctivitis with oedema of the lids where herpetic vesicles sometimes occur. When the infection is limited to this stage it will last a few days only. If the cornea is affected it will last much longer with development going on to coarse corneal opacities. Ocular infections may recur at other sites, usually in the form of keratitis, often with dendritic ulceration of the cornea. Amoeboid ulcers occupying an extensive proportion of the corneal surface have usually been ascribed to topical application of steroids. These are frequently used to alleviate nonspecific inflammations of the eye (iritis) and have been claimed to be helpful where these result from ophthalmic zoster (see below). However, their use in infections resulting from herpes simplex is clearly contra-indicated. Fortunately primary ocular herpetic infections and secondary epithelial keratitis and ulceration associated with secondary infections can be resolved by speedy chemotherapy. These were among the first virus infections to yield to such therapy. Nevertheless inflammatory responses associated with the stromal keratitis in recurrent infections do not respond to anti-viral compounds, presumably because these responses involve complex immunopathological mechanisms rather than cytopathic effects mediated directly by growth. Steroid therapy has been recommended in such cases but, as indicated, this needs very careful assessment. The dilemma indicates the fine balance of immunological factors in herpes infections. Repression of immune responses is clearly helpful in the inflammatory response characteristic of stromal keratitis but in ulceration from epithelial keratitis this may also suppress responses which help to limit infection.

Herpes encephalitis Herpes encephalitis has occasionally been reported as occurring in a mild and diffuse form but the difficulty in reviewing cases of this kind is the absence of definite evidence for infection, the mildness of the affliction precluding biopsy techniques.

The more serious form with poor prognosis, often referred to as focal encephalitis, has sometimes been described as the commonest form of sporadic encephalitis in temperate climates, although accurate information on case occurrence is scanty. The British Herpes Encephalitis Working Party suggest that about 20 deaths per annum in England and Wales in the period 1967–71 may possibly be attributable to herpes encephalitis. Retrospective assessment of case histories indicated 99 cases of the disease in England and Wales in the period 1967–72. Nevertheless a further 150 deaths per annum were returned from encephalitis generally and there is little information on the nature of these cases. Areas of the brain affected are usually the frontal temporal lobes of the cerebrum but the brain stem can also be involved. High fever nearly always occurs and severe headaches are a frequent early symptom. Personality changes occur in some patients. Neck rigidity and fits and coma are observed later. Local neurological symptoms, e.g. hemiplegia, have often been reported. It has been estimated that the fatality rate is 50 per cent with half the survivors showing considerable disability.

Early unambiguous diagnosis requires examination of brain biopsy material. Infective virus is rarely isolated from the cerebrospinal fluid (see herpes simplex type 2 meningitis below) and the value of examination of such fluid is in differential diagnosis. However, it has recently been claimed that virus-specific glycoproteins are found in the cerebrospinal fluid and another report claims the presence of herpes-specific thymidine kinase. Retrospective serological examination of cerebrospinal fluid is useful since, in the herpes encephalitis, the ratio of antibody in the fluid to that in the serum increases. Unfortunately, such tests at the moment are too insensitive until at least the tenth day of illness, underlining the need for more sensitive methods and reagents. Immunofluorescence examination of brain tissue is regarded as the optimum method, electron microscopy of section material being too time-consuming, insensitive and incapable of differential diagnosis between different herpesviruses. Virus culture is regarded as the final reference method, although difficulties in isolation have been reported. Co-cultivation methods may be helpful.

Once again the difficulties underline the need for sensitive and rapid diagnostic methods by means of, for example, appropriate specific antibodies in specific and simple tests.

The disease causes a disastrous progressive necrosis of the brain. Treatment (see below under chemotherapy) is very effective when applied as early as possible. Mortality in herpes encephalitis is reported at between 50 and 70 per cent, even with successful anti-viral chemotherapy; a high degree of specialized care is also necessary to avoid respiratory, cardiovascular and endocrine complications. Decompression to avoid death from oedema is mandatory. This can be done by surgical intervention but steroids (e.g. dexamethasone) have been suggested, although not without controversy. Once again, as with ocular herpes infections, the concern is over repression of helpful immune responses.

Generalized infections with herpes simplex Infections in patients with impaired immunity have already been mentioned; neonatal infections are generally more related to herpes simplex virus type 2, although very occasionally type 1 virus has been implicated. Severe generalized infections in adults are uncommon.

Other possible manifestations of herpes simplex type 1

Chronic psychiatric illness A number of reports over the years have claimed that, in cases of psychotic depression or aggressive psychopathy, there are raised levels of herpes simplex-specific antibodies as compared with control populations. Similar elevations of antibodies to other viruses are not found. Of interest here is an observation of herpes simplex nucleic acid sequences in smears of brain tissue of chronic psychiatric patients. Plainly this finding might be of importance in relation to whether herpes encephalitis represents a bizarre recrudescence of an earlier conventional infection with virus or arises from establishment of a latent infection in the brain itself.

There have also been reports of isolation of herpes simplex virus type 1 from the cerebrospinal fluid of patients with schizophrenia or senile dementia.

Recurrent duodenal ulcer Analogies have been drawn between the clinical patterns of recurrent cold sores and recurrent duodenal ulcers. One survey has reported that those suffering from recurrent duodenal ulcers exhibit increased type 1 specific antibody activity in the same way as those suffering from recurrent cold sores, without corresponding increases in type 2 specific antibody responses. Control populations showed no type-specific bias of this kind.

Herpes simplex virus type 2 infections

Genital infections Genital infections with herpes simplex virus were first described at least as early as the first part of the nineteenth century. Such infections have been described as the most rapidly increasing venereal disease.

In the male the lesions may be on the shaft of the penis, or the inside of the urethra, giving rise to urethritis. In the female, lesions may occur in the cervix, where primary infection may lead to severe necrosis of the vagina, urethra or labia. Lesions on the perineum and inner aspects of buttocks and thighs are common (Fig. 88.11). There is extensive discomfort and these infections have been described as the 'most common and psychosocially distressing of the sexually transmitted diseases' (Nahmias *et al* 1981). While genital

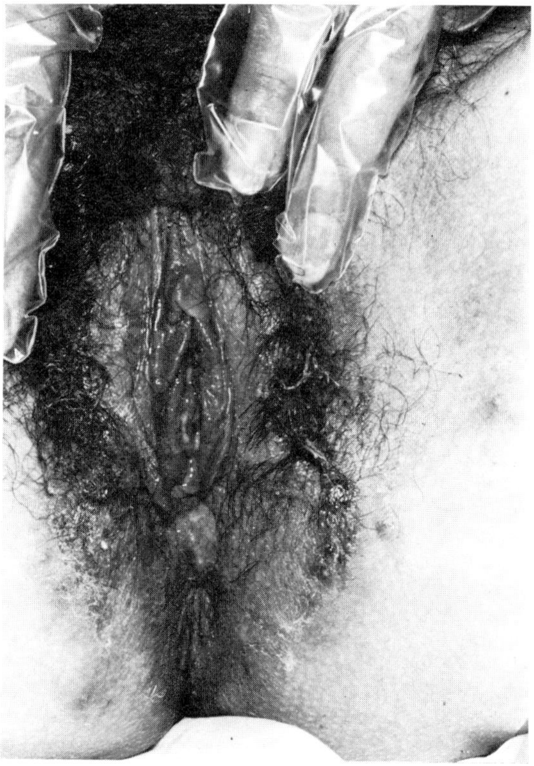

Fig. 88.11 Genital lesions produced by herpes simplex type 2 infection of female patient aged 19 years. Note lesions on labia, perianal region, perineum, inner aspects of buttocks and thighs. (Reproduced by kind permission of Dr B. E. Juel-Jensen and Heinemann Medical Books.)

infections are considered to be predominantly due to type 2 virus there are frequent reports of isolation of type 1 virus. The differential proportions vary extensively depending on the nature and practices of the populations surveyed. A laboratory complication is undoubtedly the differential thermolability of the two viruses; it has been claimed that this, together with the predilection of type 1 virus infections for more readily accessible sites than type 2, has distorted the differential proportion in some surveys. A further variable is the method of typing used; in general, the more precise methods require specialized facilities and/or reagents, leading to the use of the more convenient and accessible but less rigorous methods (see below). A recently published survey by Nahmias *et al.* (1981) of 557 female and 342 male genital isolates has shown 91 per cent of the female and 97 per cent of the male isolates to be type 2, although frequencies of up to 30 to 40 per cent and even 60 per cent, type 1 have been reported by others.

Neonatal infections Neonatal infections with herpes simplex virus type 2 are associated with a high mortality. When the infection is localized in the central nervous system, skin, mouth or eye or, as occurs rarely, is asymptomatic, the mortality rate is lower, just over 20 per cent. Such localized or asymptomatic infections represent about one-third of the case histories with the remaining two-thirds resulting in more generalized infections and mortality approaching 100 per cent. Many of those surviving neonatal infections show impaired development.

The major source of infection in neonatal cases is thought to be the genital tract of the mother. Caesarian section is advised when maternal herpes infection is diagnosed. Chemotherapy is inadvisable in view of the mutagenic potential of many chemotherapeutic agents.

It has been claimed that neonatal type 2 infections are less common in the United Kingdom than in the United States which is the source of most of the well documented surveys. Once again, a great deal may depend on the exact status of the population surveyed and the intensity of laboratory investigation.

Herpes meningitis Infection of the meninges, rather than encephalitis, is now usually more associated with type 2 than with type 1 infections. Isolation of virus from the cerebrospinal fluid is commonly reported in association with this infection. Clinical features of this usually benign infection are febrile illness with meningeal irritation and headaches. Once again this syndrome has only been reported rarely in the United Kingdom.

Herpes simplex type 2 and cervical carcinoma Cervical neoplasia is generally stated to arise in the squamocolumnar junction. Colposcopy suggests that there is a normal replacement of columnar epithelial by metaplastic squamous cells, often noted at the time of the first pregnancy. It has been suggested that conversion from such 'normal' to atypical metaplastic cells is caused by an oncogenic agent and that progression proceeds from these, either via carcinoma *in situ* or dysplasia to invasive carcinoma. It should be emphasized that whatever the route there is not an inevitable progression between these stages, and regression at any intermediate stage can occur.

In considering the possible link between herpes simplex type 2 infection and cervical carcinoma it is important to note that both are strongly related to sexual activity e.g. age at first intercourse, number of sexual partners, etc., and the problem of unlinked co-variation confounds all sero-epidemiological surveys. A number of workers have attempted to enunciate modified versions of Koch's postulates to cover criteria to be satisfied in attempting to forge an aetiological link between a virus and a tumour. One version proposed by Rawls *et al.* (1977) suggests that

1. There should be an epidemiological or biological association between the malignancy and the suspected virus.
2. Purified preparations of the virus should induce tumours in a natural host or in a similar species and the resulting tumours should be similar to the naturally occurring ones.
3. Naturally occurring malignancies in animals should be caused by similar viruses.
4. The virus should cause *in vitro* transformation of cultured cells and these should induce tumours similar to naturally occurring ones.
5. It should be possible to prevent the malignancy by preventing infection with the suspected virus.

While these postulates are eminently reasonable, the use of 'similar species' of host or 'similar viruses' raises particular problems in view of the exaggerated differential response of even closely related hosts to heterologous viruses. Thus herpes B virus of monkeys, as benign in its natural host as herpes simplex in man, causes an invariably fatal infection of man. Similar comments can be made on the severity of herpes simplex infection of chimpanzees. Of particular relevance to the oncogenic role of herpesviruses is the oncogenic effect of herpesvirus saimiri, whose natural host is the squirrel monkey, in heterologous but not homologous species of monkey.

Sero-epidemiological studies

Sero-epidemiological surveys purporting to link prior herpes simplex virus type 2 infection with cervical carcinoma are notable for their variations in differential sero-positivity between cases and controls. Rawls and Campione-Piccardo (1981) summarized the results of 28 surveys where the sero-positivity in cases ranged from 15 to 97 per cent, and in controls from 10 to 77 per cent although the ratio of case to control sero-positivity ranged from 1.14 to 5.55 in the different populations. Some of the variation is undoubtedly due to differences in the method used; while many of the studies were based on tests for neutralizing antibodies not all were. More significantly, many of the studies did not eliminate the confusion of cross-reacting antibody to type 1 virus. In such surveys the diagnostically significant type 2 specific activity should be sought. However many of the differences between the different reports relate to real population differences, since these have been confirmed in surveys of the different populations in one laboratory using the same method. Selection of the matched population may have caused difficulties, because these may vary depending on sexual behaviour of the subjects and their sexual partners. Nevertheless Rawls and Campione-Piccardo (1981) have analysed the data from these surveys on the hypothesis that a constant proportion of cervical carcinoma cases arises from other causes than prior type 2 infections, and have shown significant correlation between the incidence of cervical carcinoma and the fraction of cases with type 2 antibody, with the data showing good fit at the 0.1 per cent level of significance to a theoretical curve based on the hypothesis.

Other studies have surveyed antibodies to specific antigens or virus components. Thus Aurelian's group have reported that antibody to AG-4 is present in sera of more than 75 per cent untreated cervical carcinoma patients and only 10 to 15 per cent of controls. Similar differences between cancer and control populations have been reported with the use of ICSP 11/12, the major DNA binding protein from cells infected with type 2 virus, or its type 1 analogue ICSP 8. (This protein is known to be strongly conserved between different herpes viruses. While this allows the substitution of a type 1 for a type 2 reagent, it does give rise to specificity problems, i.e. the results would be equally consistent with the aetiology based on another herpesvirus). Similar success has been claimed with the transformation-associated antigen of herpes simplex type 1 and type 2 viruses and with Ag-e.

Detection of virus antigens in tumour cells Early immunofluorescence studies with general (polyspecific) antisera to type 2 antigen showed increases in the percentage of positives in cells exfoliated from carcinoma than in those from control cases. More recent studies with the more specific antisera to Ag-e and to ISCP 11/12 showed higher percentages of positives in cases than in controls. There have also been reports of an antiserum made to tumour tissues showing a line of identity in immunodiffusion tests with a precipitin produced by type 2 antiserum, although the antisera to the tumour antigen surprisingly did not react with type 2 infected cells. The previously described AG-4 and herpes simplex tumour-associated antigens have also been reported in cervical carcinoma cells.

Detection of virus nucleic acids in cervical carcinoma cells Virus DNA and RNA were first reported as being present in cervical carcinoma cells ten years ago in one isolated case by means of kinetic analysis of liquid hybridization data. Many other groups reported negative results in later experiments using similar methods. More recently *in situ* hybridization methods have been used to detect herpes simplex type 2 specific RNA transcripts. The methods can be made very sensitive by using gene-cloning methods to obtain increased quantities of restriction enzyme fragments of virus DNA. After labelling, these can be used to search for small regions of the genome. Transcripts from the left-hand end of the genome (map position 0.07–0.3) are expressed as in latently-infected ganglia, but additional transcripts mapping at 0.3–0.4, 0.58–0.63 and 0.82–0.85 are found which are not expressed in ganglion cells. The 0.82–0.85 segment is in the repeat sequences flanking the unique short region and is therefore also represented by sequences mapping at 0.94–0.96. The 0.58–0.62 region would seem to overlap type 2 genes capable of causing *in vitro* transformation, whereas the 0.3–0.4 region is thought to include the sequences specifying the major DNA-binding protein which has been found in cervical carcinoma but not in latently infected ganglion cells.

Conclusion Despite the ever-increasing number of reports on possible involvement of herpes simplex virus type 2 in cervical carcinoma, these would appear to be of significance only in relation to points (1) and (4) of the criteria quoted above from Rawls *et al.* (1977).

Diagnosis and chemotherapy of herpes simplex infections

Diagnosis The specialized problems of diagnosis of herpes encephalitis have already been mentioned. For other herpes simplex infections the diagnosis is more often than not made on clinical grounds, with later virus isolation and in some cases typing.

Typing is frequently done by means of differential size of pocks formed on the chorioallantoic membrane of fertile hens' eggs (see Figs. 85.4 and 85.5 in 6th edn., pp 2417 and 2418). Type 2 virus causes larger lesions. Unfortunately aberrant strains are known which would classify them with the wrong type on the basis of this test; the same may be true also for the thermosensitivity tests which rest on the differential inhibition of growth in tissue culture cells of type 2 but not type 1 viruses at 38°. A number of experimental antisera are now available which will neutralize one type only; indeed such reagents can be easily made by heterologous absorption of general, or polyspecific, antisera. With developments in serological methods, associated particularly with work on monoclonal antibodies, there seems no reason in principle why it should not be possible to manufacture, reasonably cheaply, reagents for some form of sensitive and rapid agglutination test on inert particles coated with type-specific antibody. Such tests have the advantage over more precise and sensitive radio-immunoassays or enzyme-linked immuno-adsorbent assays in not requiring special apparatus.

Chemotherapy A number of compounds are now available which are potential chemotherapeutic agents for herpes simplex viruses. All of them inhibit one of the stages in biosynthesis of virus DNA but vary in the extent to which they inhibit cellular DNA synthesis. The earliest agent, idoxyuridine (iodo-deoxyuridine), is fairly unselective and thus too cytotoxic for general use, but was initially used as the first agent to treat ocular herpes infections successfully. In this situation where the host cells are not metabolizing actively, the cytotoxic effects are minimal and the inhibition of biosynthesis of virus DNA produces the appropriate therapeutic effects. The compound inhibits activity of virus-specific thymidine kinase and DNA polymerase but perhaps, more importantly, is incorporated into the virus DNA as a base analogue, rendering it non-functional. Bromovinyldeoxyuridine has been stated to be more selectively inhibitory to virus than to host-cell biosynthesis, apparently because it is phosphorylated selectively by the virus-specific thymidine kinase, being relatively unaffected by the host cell kinase. Cytarabine (cytosine arabinoside), vidarabine (adenine arabinoside), and trifluorothymidine also all inhibit virus-specific enzymes but again are principally active through incorporation as base analogues. They are moderately cytotoxic, whereas acyclovir (acycloguanosine) has relatively little effect on host cells and hence is recognized as the most promising of the potential chemotherapeutic agents for herpes viruses. It has a high activity after its phosphorylation to the reactive form by the virus-specific thymidine kinase, through selective inhibition of the virus-specific DNA polymerase.

Phosphonoacetate and its homologue phosphonoformate also exert a selective effect on the virus-specific polymerase. Phosphonoacetate is likely to be less promising owing to its toxic effects.,

Of the agents listed, bromovinyldeoxyuridine, phosphonoformate and acyclovir have been used on a trial basis, but there is yet no fully documented survey on their usefulness. Such information as is available seems to be very promising.

Chemotherapeutic treatment of herpes simplex infections

Ocular infections Idoxyuridine has been the principal agent used for ocular infections, although it is now being replaced by more recently developed agents. Drops 0.1 per cent hourly to two-hourly have proved effective in early dendritic ulcers. Prolonged application may cause contact dermatitis and other undesirable toxic effects. It does not generally diminish the frequency of recurrences and furthermore is unstable with a short half-life when applied as drops. Ointment (0.5 per cent) has been claimed to be more effective. Failure of idoxyuridine therapy has been ascribed to resistant strains, inadequate treatment or development of toxicity and allergy.

Vidarabine has been shown to be equally effective despite the fact that its relative potency is considerably lower than that of idoxyuridine. It is metabolized quickly to hypoxanthine arabinoside, which is less effective and like idoxyuridine is relatively insoluble. However toxic effects in the eye are rare and continuous treatment with 3.3 per cent vidarabine can be tolerated for up to two months.

Trifluorothymidine is much more soluble than idoxyuridine or vidarabine and can therefore be used in higher concentrations. Corneal ulcers have been reported to be resolved more quickly by trifluourothymidine than by idoxyuridine.

There have been a number of sporadic reports on the use of acyclovir in ocular infections. Dendritic ulcers were claimed to be cleared by minimal wiping débridement within periods ranging from two to seven days. Effective levels are found in tears after oral administration. Bromovinyldeoxyuridine has in preliminary reports been claimed to be effective as a 0.1 per cent solution.

Skin infections Cold sores have been treated with 5 per cent idoxyuridine in dimethyl sulphoxide; the duration of lesions is said to be reduced. Herpetic whitlows are also said to respond well to 40 per cent idoxyuridine in dimethylsulphoxide.

Primary infection in patients with atopic eczema who may suffer generalized eruptions have been treated successfully by intravenous treatment with cytarabine. Such infections would be suitable for acyclovir treatment.

Treatment of genital infections has generally proved less successful. *In-vitro* studies show that the type 2 strains generally have a higher resistance to antiviral compounds than type 1 strains, although trifluorothymidine is unusual in showing the opposite effect, while acyclovir is only slightly less effective *in vitro* against type 2 than against type 1 strains.

Herpes encephalitis Cytarabine was found in a British trial to be relatively ineffective in a survey of 11 patients compared to 12 placebo patients.

Vidarabine, like acyclovir, is known to cross the blood brain barrier, although this may be of little relevance in cases of encephalitis where the barrier is unlikely to be intact.

An American trial of vidarabine showed seven out of ten fatal cases in a placebo group and only five out of eighteen in drug recipients. However when the results were assessed at 30 days and at 60 days the drug mortality had increased to 40 per cent. The original trial was abandoned on ethical grounds, i.e. the preliminary results showed such apparently favourable effects of the drug that a continuation of the placebo group seemed unethical. A point of some interest arose when both the original trial and a later extended group of patients were examined on the basis of state at start of trial. Patients who had a biopsy and were given vidarabine when only lethargic, i.e. before coma, fared best—30 per cent mortality with 70 per cent of the survivors able to lead a useful life. By contrast, of patients treated when in coma, 60 per cent died and approximately half of the survivors were unable to lead a normal life.

It is important to comment on the comparison between the British cytarabine and the American vidarabine trial. Biopsy was mandatory before treatment in the British trial and this had the effect of allowing only desperate cases to be admitted to the trial group. By contrast, US patients were often treated when the disease was in an earlier stage. This would lead to recruitment to the trial of patients with only mild diffuse encephalitis who would have had an improved chance of survival anyway.

Conclusion The future direction of chemotherapy of herpes simplex infections will remain very much in the balance until the results of surveys on newer agents such as acyclovir have been reported. Other promising developments may come from bromovinyldeoxyuridine and phosphonoformate.

Varicella-zoster virus infections

Varicella or chicken pox has been recognized for centuries as a mild affliction of children, although it was considered to be a less severe form of smallpox until the late eighteenth century. The derivation of the 'chicken' of its vernacular name is unclear—suggestions made include from old Erse *gican*, to itch, which would accord with a familiar symptom; from *cicer*, a chick pea, possibly in accord with the smaller size of vesicles compared with smallpox, the word being closely related to the Anglo-Saxon *cicen*, a young fowl, from which the English word chicken clearly stems.

Zoster (Greek, girdle) or shingles (from *cingulus*) the Latin equivalent of zoster) was known again for centuries as a creeping infection which girdled the body, although the connection with varicella was not documented until the turn of the century, when it was observed that chicken pox could occur after exposure to shingles. Zoster vesicle fluid was later observed to transmit varicella and the similar histopathology of the two diseases was noted. Weller and his colleagues in the 1950s (Weller and Coons 1954; Weller and Witton 1958; and Weller, Witton and Bell 1958) showed vesicle fluid of each disease to contain cross-reactive antigens and demonstrated the identity of isolates from each by immunofluorescence. Identity of the DNA molecular weight and restriction enzyme-cleavage patterns has been demonstrated within the last few years.

Primary varicella-zoster infection

Until recently no good animal model has existed for varicella zoster infection, although the virus of Macaque monkey and the Delta virus of Patas monkeys which have recently been shown to cause an exanthematous infection, together with a varicella zoster virus of guinea-pigs, have recently been proposed as model systems. General epidemiological observations—rapid spread of varicella infections in schools, nurseries etc.—suggest that the primary infection is spread by the respiratory route. There is no evidence, however, for the presence of the virus in upper respiratory passages during the incubation period. A recent report described isolation of the virus from the throat in one of a group of 12 exposed children who subsequently contracted the disease but this was found only on the first day after exposure. Viraemia has been observed during both the incubation period and the active infection of immuno-suppressed but not normal children. Nevertheless, the general absence of virus in respiratory passages after exposure suggests rapid transport to a secondary site where it replicates before spreading to the skin through the blood stream. Virus is then demonstrable in the endothelium with epidermal involvement and accumulation of fluid to form a vesicle between the prickle cell layer and the outer epidermis. The vesicle is infiltrated with inflammatory cells, usually polymorphonuclear leucocytes, although mononuclear cells migrate later.

The early stages of the infection are characterized by a mild fever, following a median incubation period of approximately 14 days (range 10 to 23 days). The rash appears first at the back of the head, spreading later to face, neck, trunk and extremities. Repeated crops of lesions appear on the same part of the body over two to four days so that the vesicles, crust, and pustules may be observed simultaneously at one site. The lesions heal rapidly and recovery is complete within two weeks of onset of the rash. Mild cases, with very little rash, are known, which may often be unrecognized. Primary varicella in adults is more likely to be severe, as are infections of neonates or other persons with defective immune reactions. Progressive varicella in immuno-deficient patients produces severe skin eruptions, haemorrhage, high fever, invasion of visceral organs, leading to hepatitis, pneumonitis, obstruction of the small bowel and encephalitis.

Pneumonia is the most likely complication in infected adults but is produced less often in the infection of children. The extent of varicella pneumonia in adults is not known but it has been reported that radio-diagnostically visible changes in the lung were observed in 16 per cent of US army personnel having varicella, although most of these were asymptomatic.

Neurological complications—meningo-encephalitis, optic neuritis, paralysis of cranial nerves, myelitis—are known but only in about 1 in 1000 cases.

Zoster

Zoster exemplifies classical herpes latency. Hope-Simpson (1965) has perhaps made the greatest contribution to understanding the nature of this latency on the basis of extensive experience as a country practitioner. He has in particular pointed out the enormous survival advantage to the virus of its particular life style; whereas a virus such as measles apparently requires a sizeable population (up to 3×10^5) to survive, varicella zoster can manage with only about 10^3.

The association of the skin rash of zoster with histopathological lesions of the dorsal root ganglia was noted towards the end of the last century, and later extensive anatomical investigations substantiated the conclusions drawn that zoster represented a reactivation of latent varicella, years before the confirmation of the identity of the virus responsible for the two diseases.

Clinical onset of zoster is marked by localized burning pain in the dorsal spinal or cranial sensory nerves. This suggests that virus reactivation occurs in the ganglia and intranuclear inclusions; characteristic herpesvirus particles have been observed in the ganglia. Reactivation can sometimes be associated with trauma such as fractures, knocks, or intramuscular injections, and there is an association with immunosuppressive therapy. Pain can often occur in the absence of subsequent skin rash but in the normal development a generalized vesicular rash appears in a week (Fig. 88.12). In contrast to varicella there is usually only a single crop of vesicles. Zoster is most often unilateral but bilateral infections can occur without the dire consequences associated with a completely encircling girdle in popular folk lore. Approximately half the cases affect the thoracic region, while infections from the three divisions of the trigeminal branch more commonly infect the first ophthalmic branch. Distribution from lumbar, cervical or sacral nerves occurs less commonly.

Apart from persisting pain, paradoxically anaesthesia of the affected area commonly occurs. Neuralgia following infection may persist for years.

While zoster can affect all ages, the incidence undoubtedly increases with age. Hope-Simpson (1965) reported a rate of over 10 per 1000 in octogenarians—more than 14 times greater than for children under 10.

Complications of zoster are facial palsy, paralysis of motor nerves and myelitis. Meningitis and encephalitis occur most frequently in patients with zoster infections of the upper nerve regions.

All immuno-suppressed patients are at higher risk and those with Hodgkin's disease have a particularly high risk—up to 50 per cent. Extensive skin disease with gangrenous necrosis may be sufficiently severe to require skin grafting. Visceral involvement occurs in up to 10 per cent of immuno-suppressed patients. The morbidity among immuno-suppressed patients from disseminated zoster is quoted at 3–5 per cent.

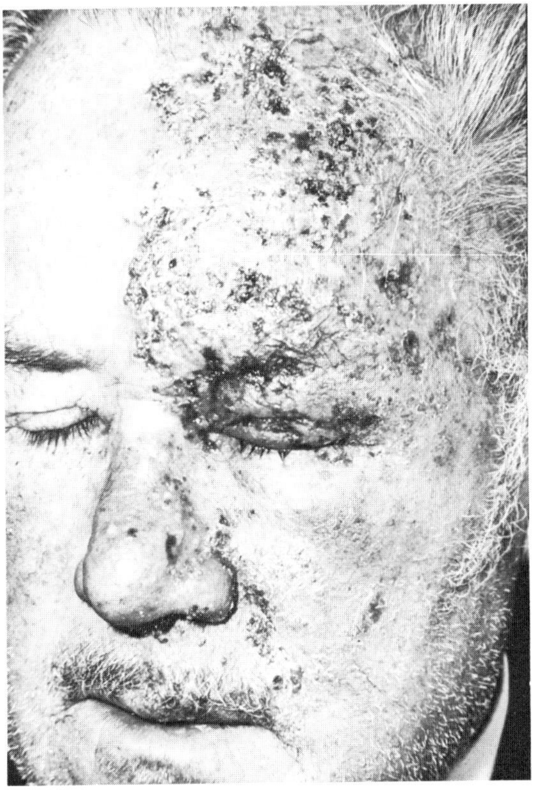

Fig. 88.12 Dense eruption due to zoster in male patient aged 68 years. The infection involved the 1st and 2nd divisions of the trigeminal nerve. (Reproduced by kind permission of Dr B. E. Juel-Jensen and Heinemann Medical Books.)

Diagnosis and treatment of varicella-zoster infections Differential diagnosis from smallpox is now less critical and can most speedily be accomplished by electron microscopy.

Inoculation of vesicle fluid on human fibroblast cultures gives foci of irregular rounded and occasionally multinucleated cells. The virus remains cell-associated in tissue culture and is usually rapidly inactivated on cell disruption.

Specific complement-fixing or neutralizing antibodies develop within a week of occurrence of rash. Neutralization tests are more demanding; most serological investigations use the complement-fixation test, although the antibody reactive in this test declines rapidly after recovery. There is also the possible problem of cross-reaction with other human herpesvirus antigens although the effects of this have never been fully documented. Experimental studies have certainly shown such cross-reaction with herpes simplex virus, and there is always a possibility, with more sensitive radioimmuno-assay and ELISA tests currently being used, of problems arising.

While chemotherapy is not considered necessary for varicella infections of the normal child, it may become more necessary with adult infections. Parenteral administration of vidarabine at a daily dose of 11 mg/kg has been reported to resolve successfully severe varicella in adult patients within three days. Vidarabine treatment has been used in a number of severe zoster infections. Immuno-suppressed children with severe varicella have been treated with cytarabine but, apart from its toxicity, this is undesirable because of its immuno-suppressive properties. Passive immunization with zoster-immune globulin has been used to limit infections in populations regarded as particularly at risk, for example leukaemic children.

Recent reports on topical application of interferon in zoster have given a most encouraging picture of relief from pain.

Cytomegalovirus infections

In contrast to the 'old' herpesviruses associated with long recognized viruses, cytomegalovirus was first isolated just over a quarter of a century ago, being first associated with congenital cytomegalic inclusion disease, a severe but rather rare infection first described at the turn of the century. The disease was characterized by large cells (cytomegalia) in the affected organs, (kidney, liver, lungs) of infants examined post mortem. The cells contained inclusions which were noted as resembling those caused by herpes simplex virus. Inclusion-bearing cells were detected in the urine of infected children and allowed the first diagnoses before death. What was not initially realized was that infants suffering from this disease represented only about 1 per cent of those born each year with systemic cytomegalovirus infections.

Congenital, neonatal and perinatal infections are now recognized to be very common with a significant proportion, possibly up to 20 per cent, being initiated during birth. The frequent shedding of virus displayed by infected adults, e.g. the mother, gives ample opportunity for infection of the infant either during birth, via an infected genital tract, or after birth through infected breast milk into which virus is frequently shed. Infection of infants may give rise, on the one hand, to generalized damage to many organs and, on the other, to asymptomatic infections which may later result in extensive abnormality, hearing defects and impaired development. Important clinical features may include jaundice, microcephaly, chorioretinitis, pneumonia, seizures, diarrhoea and premature birth. Laboratory investigations reveal anaemia, inclusion-bearing cells in the urine and isolation of virus from the throat and urine.

After infancy a primary infection may often be manifested as cytomegalovirus infectious mononucleosis. At one time this was frequently characterized as being negative in a Paul–Bunnell test but improvements in the quality of this test suggest the view that cytomegalovirus-associated infectious mononucleosis is rarely negative. Fever and malaise in the absence of cervical adenopathy and pharyngitis are features of cytomegalovirus as opposed to EB virus infectious mononucleosis. Development of the infection is frequently associated with involvement of the liver and other organs. Virus can be isolated from leucocytes. Blood transfusion may often be associated with development of the mononucleosis syndrome.

As with other herpesviruses the immuno-suppressed state is often associated with infection. A particular problem is seen with renal transplants where there is the double risk of transfusing cytomegalovirus from the donor and of reactivating the latent virus in the recipient through immuno-suppressive regimens. Evidence of cytomegalovirus has been found in up to 91 per cent of groups of patients receiving renal transplants although most of these were asymptomatic. Evidence of cytomegalovirus infection is to be found on autopsy of renal transplant patients.

The virus has also been frequently isolated from or found at autopsy in patients with leukaemia and Hodgkin's disease where it has been a frequent cause of terminal complications.

Cytomegalovirus has been shown to transform cells in culture and has been associated, at least loosely, with some human tumours, for example of the colon and the prostrate; and it has been isolated from Kaposi's sarcoma.

In summary, cytomegalovirus has emerged in the last decade as the most common infectious cause of neonatal and perinatal brain damage, the main cause of morbidity and death in organ transplantation and a cause of terminal complications in several malignancies. The virus is shed frequently and effectively in the absence of apparent clinical symptoms, so that ample opportunities exist for its spread. Inevitably the question must arise as to whether we have only scratched the surface in recognizing the diseases associated with this pervasive virus. Treatment of cytomegalovirus infections with chemotherapeutic agents has generally been of little effect; at best a temporary suppression of virus excretion has been noted. Interferon has been used prophylactically in renal transplant patients and has resulted in decreased viraemia. This was accomplished by delayed excretion of virus, so that the advantages of interferon treatment were somewhat limited. Treatment of established congenital infections usually gave only a temporary response.

Epstein-Barr virus associated disease

EB virus was discovered after descriptions by Burkitt (1958) in African children of lymphoid tumours of a kind which had not been previously described but which he recognized as a manifestation of one lymphoma syndrome. Regional variations in the incidence led Burkitt (1962) to suggest that it might be caused by a climate-dependent arthropod vector transmitted virus. This prompted Epstein and his colleagues to search for a virus in biopsy samples of the lymphoma. Later, a herpesvirus was identified in cells cultured *in vitro* from lymphoma biopsies by electron microscopic examination of thin sections (reviewed by Epstein and Achong 1979). Immunofluorescent studies of these cells showed this virus to be unique, differing antigenically from other known herpesviruses. Further immunofluorescent studies with a variety of human sera showed that antibodies to the virus were distributed widely throughout the world, although the lymphoma itself was geographically restricted. A series of investigations initiated by the Henles (reviewed by Henle and Henle 1979) showed that seroconversion to EB positive reactions in human sera accompanied infectious mononucleosis.

Infectious mononucleosis

Congenital infection seems to be rare. Although infants at birth are often seropositive for EB-virus antigens, these are presumed from their rapid decline to be maternally transferred; in accord with this, 'immortalized' lymphoblastoid lines can be established from fetal cord lymphocytes only after exposure to EB virus *in vitro*. Seroconversion in children to EB positive reactions is common but is not usually associated with any apparent infection and it is assumed, as happens with some other common virus infections, that severity of primary infections increases with age.

The association with infectious mononucleosis is confirmed on other grounds besides seroconversion. Thus it occurs only in those who are previously seronegative; seropositive persons are immune. It may occur in seronegative patients after blood transfusion. Oropharyngeal secretion of EB virus occurs more frequently in infectious mononucleosis patients (up to 90 per cent) than in healthy control patients (less than 30 per cent).

Infectious mononucleosis is presumed to be acquired through intimate contact, particularly kissing, with an excreter of the virus, who may or may not have symptoms of the disease. It has been suggested that the incubation period is approximately 90 days, but epidemiological investigation of the actual contact transmission is clearly difficult to establish. More reliable is an incubation period of 3–5 weeks for infectious mononucleosis acquired through transfusion. The onset of illness is insidious and is usually marked by fever, which may last for up to six weeks, and sore throat, together with enlargement of the cervical lymph nodes and hyperplasisa of pharyngeal lymphoid tissues. Haematological examination reveals lymphocytosis with the presence of atypical lymphocytes. Heterophile antibody agglutination of sheep or horse erythrocytes in the Paul–Bunnell test is absorbable by ox erythrocytes but not by guinea-pig kidney powders. Antibodies which neutralize EB virus, and to the virus capsid and membrane antigens of the virus, appear during acute illness and persist for years thereafter. Antibodies to the diffuse early antigen component are often found but the titres fall within several months. Both the Paul–Bunnell and the EB antibody tests are positive in over 90 per cent of the cases, but the heterophile antigen is quite different from the EB antigen (see Crumpacker 1982). (For pathogenesis, see Epstein Achong 1979, and for a fuller description of the disease, see 6th edition p. 2285).

Burkitt's lymphoma Although Burkitt's lymphoma has been associated particularly with certain climatic regions of Africa, e.g. West Nile district of Uganda, it is observed also in New Guinea, but sporadic Burkitt's lymphoma has rarely been found in other parts of the world e.g. North America. In such sporadic Burkitt lymphoma cases the organ distribution differs, the age peak is later and there seem to be better prospects for long-term therapy.

All African Burkitt lymphoma patients have antibodies to virus capsid antigens. Tumour cells do not tend to contain virus capsid antigens, although the majority contain EBNA and can be shown by nucleic acid hybridization to contain multiple copies of EB virus DNA. The universal presence of anti-VCA activity in lymphoma patients is to be compared with the similar activity in 80 per cent of African control populations. It would appear that EB virus is a necessary but not a sufficient aetiological factor for lymphoma. It has been suggested that malaria might be a co-factor since the climatic restrictions on both malaria and Burkitt's lymphoma are the same. Hyperendemic malaria occurs in the very regions of Africa and New Guinea where the lymphoma is most frequent. Tumour incidence has fallen in areas where malaria has been reduced or eradicated, while the sickle-cell trait which has been associated with resistance to malaria is generally associated with a lower incidence of lymphoma.

An IARC prospective sero-epidemiological study was conducted in the West Nile region of Uganda, where there is a high incidence of Burkitt lymphoma, in the period 1972–4. Sera from 42 000 children up to the age of eight were collected. By the end of 1977, 14 of the bled children had contracted lymphoma. The geometric mean titre of antibody to virus capsid antigen at the time of bleeding, which ranged from seven to 54 months before diagnosis of lymphoma, was approximately three times the geometric mean titre of the same antibody in control cases. The VCA titre remained fairly constant between the survey and the time of clinical diagnosis but antibodies to the early antigen (EA) rose in 8 of the 14 cases.

Nasopharyngeal carcinoma Nasopharyngeal carcinoma, like Burkitt's lymphoma, shows a restricted geographical distribution. It is most frequently seen in South-East Asia, particularly among the Cantonese Chinese, East and North-East Africa. In all these endemic areas it is twice as common in males as in females. Patients show raised antibody activity to the membrane and early antigens as well as to the virus capsid antigen of EB virus.

EB virus DNA has been found in virtually every biopsy and *in situ* hybridization methods have been used to confirm the observation that DNA in the tumour is located in epithelial tumour cells rather than in lymphocytes infiltrating the tumour tissues.

At the present stage it is difficult to establish whether the association of EB virus with nasopharyngeal carcinoma is purely casual or indicative of an aetiological relationship.

Conclusion The linking of EB virus with Burkitt's lymphoma and nasopharyngeal carcinoma leaves a number of problems to be faced in ascribing an aetiological role to the virus. In the first place, it is difficult to reconcile an almost ubiquitous virus causing rare infections with restricted geographical distribution. Even the invocation of malaria as a possible co-factor raises problems, since in many of the high incidence Burkitt's lymphoma areas virtually every child will have had malaria. Strain variation in the virus and variations in genetic susceptibility in the host have all been suggested as a basis for the paradox. In general, all of these explanations merely move the problem one stage back; thus, to take one example, if Burkitt's lymphoma or indeed nasopharyngeal carcinoma is caused by a strain variant of EB virus, it is then necessary to explain why this variant is restricted to geographical areas of high incidence. Once again, with a herpesvirus which is a candidate for a role in a malignant tumour, we are left with the proposition that if the virus has a role in the malignancy, elimination of the virus should eliminate the malignancy. This raises the whole question of herpesvirus vaccines which is considered in the following section.

Prospects

Herpesvirus vaccines

The prevention of Marek's disease of chickens by vaccination inevitably drew attention to the possibility of human herpesvirus vaccines, although the resemblances between the diseases caused by these two viruses are somewhat imprecise. Thus, to immunize chickens against an infection causing a mortality of up to 90 per cent in affected flocks, where the possible long-term effects of the vaccine are barely significant in terms of expected span of life, can scarcely be considered to create an adequate precedent for immunization against infections whose effect is considerably less absolute and where, in some cases at any rate, it is longevity of life itself which exacerbates the problem. Thus Nahmias has commented that cervical carcinoma can scarcely have been a problem for ancient women; indeed the aphorism might be expressed better by saying that, like most other cancers detectable from the age of 40, it could only be regarded as a problem for comparatively modern woman.

Vaccines for all human herpesviruses have been canvassed in the past decade. The first to be seriously proposed was against cytomegalovirus. Vaccination has been justified in terms of the risks associated with congenital infection with this virus. Thus, up to 3000 infants in the United Kingdom might be expected to be born each year congenitally infected with cytomegalovirus and about 500 of these might suffer severe brain damage. This figure compares with an expected 200 cases of congenital rubella and 1300 cases of mongolism each year.

The experimental vaccine of Elek and Stern (1974) was of the attenuated type. Proof of attenuation is plainly difficult for a virus with a propensity for asymptomatic infections. Immunization of volunteers without pre-existing antibody with a virus strain passaged 55 times in human embryo lung fibroblasts gave minimal reactions, and no virus could be detected in throat washings or urine. No clinical or haematological abnormalities were noted. Seroconversion was successful.

The main criticisms of the vaccine related to the possible establishment of latent infections with later reactivation or possible oncogenic effects. Doubts have also been raised on the likely effectiveness of the vaccine.

In relation to possible latency, it has been noted that reactivation of cytomegalovirus is more frequent where primary infection is acquired early; and it has been argued that reactivation in pregnancy is less likely with a vaccine to be administered at 12–14 years of age. Studies on patients receiving a U.S. attenuated vaccine of the same type, who were later immunosuppressed in connection with renal transplants, showed no reactivation; it was argued that this was evidence against latent infection following vaccination. Further, although vaccinated patients who later received transplants from seropositive donors excreted cytomegalovirus, restriction enzyme analysis showed that the excreted virus differed from the vaccine strain. It must nevertheless be arguable that latent infections can scarcely be ruled out without considerably more case studies. The problem of oncogenicity must inevitably be even more intractable and the possibilities, however remote, of oncogenic effects whenever DNA of attenuated or inactivated viruses is introduced must inevitably require investigations of the potential of sub-unit vaccines. Earlier doubts on effectiveness of immune responses to these have been dispelled by the demonstration of both humoral and cell-mediated responses to trial vaccines containing envelope proteins free of virus DNA.

Vaccines against varicella-zoster virus were produced almost simultaneously with the first experimental cytomegalovirus vaccine. The motivation here was the severe effects of varicella in leukaemic children. There are now a number of reports of use of such trial vaccines. The live vaccine has been reported to produce no untoward side effects and to confer protection against varicella infection in comparison with control populations. No enhanced relapse of leukaemia was reported in the vaccinated groups. A more recently developed live vaccine is claimed to be effective in immunosuppressed children without serious side effects.

In relation to latent infection following vaccination, it may be significant that a small number of cases of zoster were recorded in vaccinated children, although there was no clear evidence that this occurred in any higher proportion than in the control population.

Numerous experimental herpes simplex virus vaccines have been formulated and tested in animal models. While an early version was a killed vaccine, most recent preparations have been derived by solubilizing envelope glycoproteins. Thus Skinner *et al.* (1980) extracted virus glycoproteins from virus grown in human diploid cells, in the absence of serum. The glycoprotein extract was treated with formalin and ultracentrifuged to remove virus DNA. Glycoproteins were precipitated from the supernatant fluid and freeze-dried. Exhaustive tests showed that the vaccine had no infective or transforming activity on cultured cells, and proved effective in animal experiments. Similar success has been reported over a limited time period in a small number of human subjects considered to be at risk of type 2 infection, through consorting with partners with recurrent genital infection. Seroconversion was detected in most of the subjects and none contracted genital infection in the six months following vaccination, whereas during this period there was a significant number of infections in unvaccinated subjects consorting with partners with genital infection.

Inevitably EB virus has been suggested as a prime candidate for a vaccine to test for elimination of an associated malignancy. Here there is the advantage that the time scale of initiation of associated malignancy is short. Once again the candidate vaccines consist of solubilized envelope glycoproteins.

It is clear from the development of several vaccines against different viruses that the accent is now on sub-unit type vaccines. These must alleviate earlier fears on reactivation and oncogenic effects. Nevertheless recent studies have shown the extreme complexity of immune reactions in experimental herpes infections of animals (reviewed by Wildy *et al.* 1982). As outlined above, in the section on mechanisms of latency of herpes simplex infections, there is a complex system of checks and balances. The unravelling of these complex equilibria will shed light on latency, but it also must be of import in consideration of vaccines. Bearing in mind the recurring picture of inapparent primary infections, the possible effects of vaccination of a subject in whom some immune reactions may already have been provoked must be open to question. Perhaps the evolution from attenuated to sub-unit vaccines must go further with the aim of formulating selective immunizing reagents which will provoke the optimum protective effect in the absence of any other reaction.

Future role of molecular biological studies The increasing technological power of molecular biology has already produced intriguing information on the latent state, on virus-specified products in malignant cells, and allowed nosocomial infections to be traced from the characteristic restriction enzyme cleavage patterns of the DNA of virus isolates. Hopefully, continuation of these studies will provide the key to the classical problem of latency. In the case of the malignant cell, the question of driver or passenger may still confuse the situation.

Molecular biological methods may also assist in providing sensitive and precise diagnostic reagents through the exploitation of monoclonal antibody techniques to delineate appropriate test reagents. Use of gene cloning and in-vitro protein synthesis techniques to produce them in quantity will be a desirable adjunct in such investigations. Similar methods might identify and then manufacture the ideal immunogen, which would promote the desired protection without any undesired side reaction. Such reagents may then, as has so often been suggested recently, provide the final answer to the role of herpesviruses not only in relation to malignancy, in which they are already suspect, but also in the many other, apparently unrelated, diseases in which they have unexpectedly but increasingly been incriminated.

Acknowledgements

I am grateful to Dr I. W. Halliburton, Dr B. E. Juel-Jensen, Dr R. A. Killington, Dr K. L. Powell and Dr P. Sheldrick for kindly providing illustrations used in this chapter. My colleagues Ian Halliburton, Dick Killington, Ken Powell and Dorothy Purifoy very kindly read the draft manuscript and made many helpful comments. I am grateful to Mrs C. I. Moorhouse, Mrs J. Killington, Mrs B. Boyle and Mrs W. M. Webster for typing the manuscript so efficiently at a time when they were under very great pressure. Finally I am indebted to the Editors for their forebearance.

Text references

Andrewes, C. H. and Carmichael, E. A. (1930) *Lancet* **i**, 857.
Ben Porat, T., Ladin, B. and Veach, R. A. (1980) In: *Microbiology 1980*, p. 270. Amer. Soc. for Microbiol. Washington D.C.

Burkitt, D. P. (1958) *Br. J. Surg.* **46,** 218. (1962) *Nature* **194,** 232.
Burnet, F. M. and Williams, S. W., (1939) *Med. J. Aust.* **i,** 637.
Crumpacker, C. S. (1982) *Rev. infect. Dis.* **4,** 1071.
Elek, S. D. and Stern, H. (1974) *Lancet* **i,** 1.
Epstein, M. A. and Achong, B. G. (1979) In: *The Epstein-Barr Virus*, p. 1. Ed. by M. A. Epstein and B. G. Achong, Springer-Verlag, Berlin.
Henle, W. and Henle, G. (1979) In: *The Epstein-Barr Virus*, p. 61. Ed. by M. A. Epstein and B. G. Achong, Springer-Verlag, Berlin.
Hope-Simpson, R. E. (1965) *Proc. R. Soc. Med.* **58,** 9.
Kit, S. and Dubbs, D. R. (1963) *Biochem. biophys. Res. Commun.* **13,** 500.
Luger, A. and Lauda, E. (1921) *Z. gesamte exp. Med.* **24,** 289.
Matthews, R. E. F. (1979) *Intervirology* **12,** 132; (1983) *Ibid.* **17,** 1.
Morse, C. S., Buchman, T. G., Roizman, B. and Schaffer, P. A. (1977) *J. Virol.* **24,** 231.
Nahmias, A. J., Dannenburger, J., Wickliffe, C., and Muther, J. (1981) In: *The Human Herpesviruses*, p. 3. Ed. by A. J. Nahmias, W. R. Dowdle and R. F. Schinazi. Elsevier, New York.
Pearson, G. R., and Aurelian, L. (1981) In: *The Human Herpesviruses*, p. 297. Ed. by A. J. Nahmias, W. R. Dowdle and R. F. Schinazi. Elsevier, New York.
Rawls, W. E., Bacchetti, S. and Graham, F. L. (1977) *Curr. Top. Microbiol. Immunol.* **77,** 71.
Rawls, W. E., Campione-Piccardo, J. (1981) In: *The Human Herpesviruses*, p. 137. Ed. by A. J. Nahmias, W. R. Dowdle and R. F. Schinazi. Elsevier, New York.
Roizman, B. *et al.* (1973) *J. gen. Virol.* **20,** 417.
Sheldrick, P. and Berthelot, N. (1974) *Cold Spr. Harb. Symp quant. Biol.* **39,** 667.
Skinner, G. R. B., Buchan, A., Hartley, C. E., Turner, S. P. and Williams, D. R. (1980) *Med. Microbiol. Immunol.* **169,** 39.
Sugden, B. (1982) *Rev. infect. Dis.* **4,** 1048.
Weller, T. H. and Coons, A. H. (1954) *Proc. Soc. exp. Biol. Med.* **86,** 789.
Weller, T. H. and Witton, H. M. (1958) *J. exp. Med.* **108,** 869.
Weller, T. H., Witton, H. M. and Bell, E. J. (1958) *J. exp. Med.* **108,** 843.
Wildy, P., Field, H. J. and Nash, A. A. (1982) *Soc. gen. Microbiol. Symp.* **33,** 133.
Wildy, P., Horne, R. W. and Russell, W. C. (1960) *Virology* **12,** 204.

General review references

Molecular biology

Halliburton, I. W. (1980) *J. gen. Virol.* **48,** 1 (Genetics, mapping)
Hampar, B. (1981) *Advanc. Cancer Res.* **35,** 27 (Transformation by herpes simplex virus)
Honess, R. W. and Watson, D. H. (1977) *J. gen. Virol.* **37,** 15 (Comparisons between herpesviruses)
Schaffer, P. A. (1981) In: *The Human Herpesviruses*, p. 53. Ed. by A. J. Nahmias, W. R. Dowdle and R. F. Schinazi. Elsevier, New York. (Molecular Genetics)
Spear, P. G. and Roizman, B. (1980) In: *DNA Tumour Viruses*, p. 615. Cold Spring Harbor Laboratory, New York. (Comprehensive overall review of molecular biology of herpes simplex viruses)
Zur Hausen, H. (1980) In: *DNA Tumour Viruses*, p. 747. Cold Spring Harbor Laboratory, New York (Oncogenic Herpesviruses)

Diseases

General overviews of clinical aspects, epidemiology, immunology, diagnosis, prevention and treatment of all human herpesviruses given in *The Human Herpesviruses*. Ed. by A. J. Nahmias, W. R. Dowdle and R. F. Schinazi, Elsevier, New York (1981).

Other references apart from those quoted above under text references.

Bauer, D. J. (1980) *Recent Advances in Clinical Virology* **2,** 111 (Chemotherapy of Herpesvirus infections)
Epstein, M. A. and Achong, B. (Eds) (1979) *The Epstein Barr Virus*. Springer Verlag. (Comprehensive survey of all aspects of Epstein Barr virus by major contributors to the field)
Longson, M., Bailey, A. S. and Klapper, P. (1980) *Recent Advances in Clinical Virology* **2,** 159. (Herpes encephalitis)
Nahmias, A. J. and Roizman, B. (1973) *New Engl. J. Med.* **289,** 667, 719 and 781 (Herpes simplex infections)
Stern, H. (1977) *Recent Advances in Clinical Virology* **1,** 117 (Cytomegalovirus vaccine)

89

Vesicular viruses

Robert F. Sellers

Introductory

The vesicular viruses include foot-and-mouth disease, swine vesicular disease, vesicular stomatitis and vesicular exanthema viruses. Others such as variola, cowpox, bovine papular stomatitis, members of the herpes group, and Coxsackie viruses also give rise to vesicles at some time after infection. However, the term is used here for those vesicular viruses that give rise to major epidemics in cattle, sheep, goats and pigs.

The virus of foot-and-mouth disease enjoys the distinction of having been the first animal virus to be discovered. Loeffler and Frosch (1898) showed that the filtration of preparations from disease material caused foot-and-mouth disease. Vesicular stomatitis virus was recognized in 1926 (Cotton 1926, Olitsky, Schoening and Traum 1927), and vesicular exanthema virus in 1932 (Madin and Traum 1955). The virus responsible for swine vesicular disease was first isolated in 1966 (Nardelli *et al.* 1968), and in 1973 San Miguel sea-lion virus, related to vesicular exanthema, was discovered in sea-lions off the coast of California (Smith *et al.* 1973).

Despite the clinical similarity of the diseases caused, especially in pigs, the viruses belong to different genera and families. Foot-and-mouth disease is an aphthovirus and swine vesicular disease an enterovirus, both belonging to the family *Picornaviridae*. Vesicular stomatitis is classified as vesiculovirus belonging to the family *Rhabdoviridae*. Vesicular exanthema and San Miguel sea-lion viruses are caliciviruses belonging to the family *Caliciviridae*.

Although primarily diseases of animals, infection of man in the field and in the laboratory has been found with foot-and-mouth disease and vesicular stomatitis viruses, and in the laboratory with swine vesicular disease and San Miguel sea-lion viruses.

Various reviews of the viruses have been published. Among these may be cited: for foot-and-mouth disease, Brooksby (1958), Bachrach (1968); for vesicular stomatitis, Hanson (1952) and Mason (1978); for vesicular exanthema, Madin and Traum (1955) and Bankowski (1965). In addition, many aspects of the viruses are covered in the series of books on Comprehensive Virology.

Foot-and-mouth disease

Foot-and-mouth disease is a highly infectious disease of ungulates, primarily cattle, sheep, goats and pigs. It also affects wild animals such as buffalo and deer and causes disease in hedgehogs. Man is infrequently infected, with production of vesicles, but may act as a mechanical carrier transferring infection to animals. Seven types of virus have been identified: O, A, C, SAT 1, SAT 2, SAT 3 and Asia 1.

Physico-chemical properties of the virus (Fig. 89.1)

Foot-and-mouth disease virus has a diameter of 24 nm and icosahedral symmetry. There are 32 capsomeres. The buoyant density is 1.43 g/ml in caesium chloride and the molecular weight is 8.3–8.4 × 10^6. The virus is acid-sensitive, being inactivated at pH 5.0,

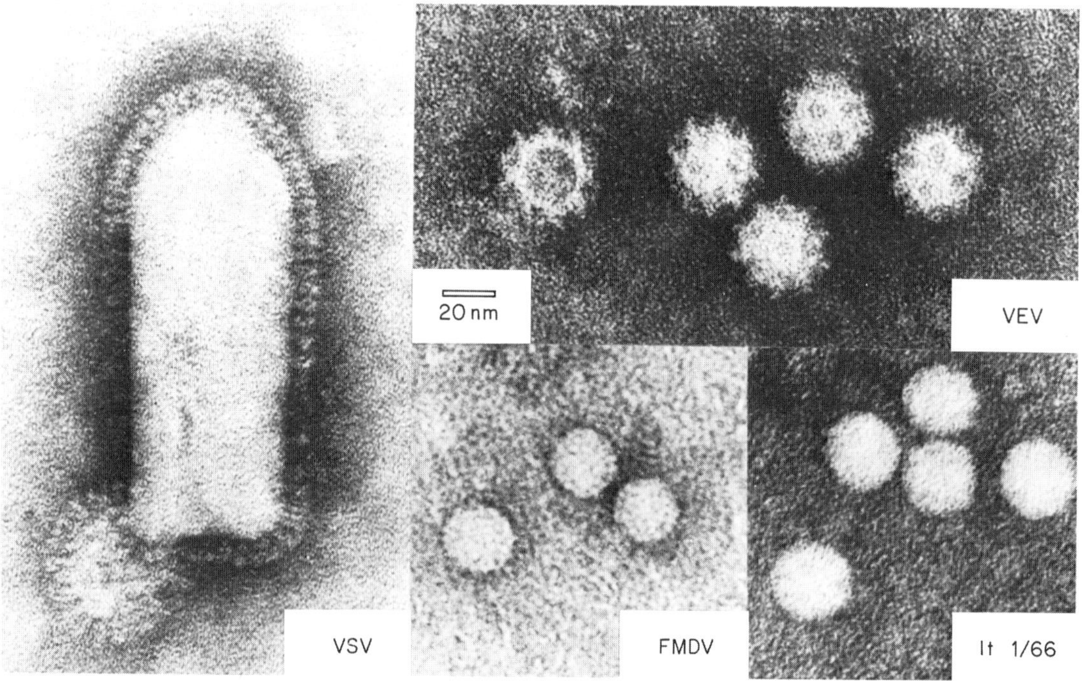

Fig. 89.1 Electronmicrograph of viruses causing vesicular diseases: VSV—vesicular stomatitis virus, FMDV—foot-and-mouth disease virus, VEV—vesicular exanthema virus, It 1/66—swine vesicular disease virus (strain Italy 1/66). (Photograph by C. J. Smale. Reproduced by courtesy of *Nature*.)

Table 89.1 Physico-chemical properties of vesicular viruses

	Foot-and-mouth disease	Vesicular stomatitis	Swine vesicular disease	Vesicular examthema
Stability at pH 5	Labile	Stable	Stable	Stable
Stabilization, 1M $MgCl^2$, 50°C	Not stabilized	—	Stabilized	Not stabilized
Ether resistance	Resistance	Labile	Resistant	Resistant
Buoyant density	1.43	1.20	1.32	1.37
Nucleic acid	RNA (Infectious)	RNA (Non-infectious)	RNA (infectious)	RNA (infectious)
Size	24 mm	175 × 65	30 nm	35–40
Sedimentation coefficient	146	620–625	150	160–170

but is stable in the presence of magnesium chloride. Three sizes of particle have been described, one with a sedimentation coefficient of 146S, an empty particle with a sedimentation coefficient of 75S, and a capsid subunit 12S (Sangar 1979; Table 89.1).

The virion consists of 30 per cent RNA, the remainder being protein. No lipid is present. The RNA, which is infectious, has positive polarity, being a single-stranded molecule of about 8000 bases with a molecular weight of 2.8×10^6. At the 3′ end of the molecule is found a poly A tract and close to the 5′ end is a poly C tract (Sangar 1979). There is no cap at the 5′ end of the RNA but instead a protein called VPg. The protein coat consists of four polypeptides—Vp_1, VP_2, VP_3 and VP_4, although traces of VP_0 and the polypeptide for RNA polymerase may be found. There are 60 copies of the four major structural polypeptides in each particle.

Foot-and-mouth disease virus multiplies readily in cells. After attachment it enters the cell as a result of phagocytosis and uncoating takes place. Shortly after entry into the cell, protein and RNA synthesis is initiated. The virion RNA also acts as m-RNA and fresh RNA (either virion RNA or m-RNA) is replicated via a structure termed a replicative intermediate. Virion RNA contains VP_g but m-RNA may or may not contain VP_g.

Protein synthesis involves the translation of the RNA into a polypeptide the size of the coding capacity of the genome and called the polyprotein. It is cloven into primary polypeptides or primary products. The primary products are cloven by virus-specified proteases into polypeptides. The primary products are from the 5′ end—p20, p88, p52 and p100. P20 is further cloven into p16, p88 into p38 (VP_0) which is subsequently cloven into VP_4 and VP_2, and p56b which becomes VP_3 and VP_1. P52 becomes p34 and p20 and p100 is cloven into p72 and p20 and eventually into p56a. RNA polymerase contains p56a together with trace amounts of p72 (Lowe and Brown 1981).

Assembly of polypeptides takes place and the RNA is inserted to form the whole particle, which is then released from the cell on lysis. Replication takes place in the cytoplasm independently of cellular DNA functions. Intermediate products are also released when the cells lyse. VIA antigen is now recognized as being RNA polymerase (Cowan and Graves 1966, Polatnick *et al.* 1967, Newman *et al.* 1979).

Biological properties of the virus

Foot-and-mouth disease virus can be grown in laboratory animals. It can be adapted to guinea-pigs, the virus giving rise to vesicles when inoculated into the foot pad. Generalization occurs and vesicles are subsequently found on the other pads and on the tongue. Unweaned mice are readily infected by intraperitoneal or intramuscular inoculation (Skinner 1951), after which they show paralysis and die. The virus multiplies in the heart and skeletal muscle and high titres of virus are produced. The virus may also be adapted to grow in older mice, giving rise to paralysis, especially of the legs. Lesions are also found in the pancreas (Platt, 1956).

Of other laboratory animals, young hamsters, newborn rabbits and newborn rats can be readily infected, with the production of paralysis and death (Mowat *et al.* 1978).

The virus does not grow in eggs unless it has first been adapted. However, adaptation occurs when the virus is inoculated by the intravenous route and the eggs incubated at 35°. The virus also infects day-old chicks when given by the intravenous and intramuscular routes. In older chicks, vesicles are produced on the tongue as a result of intradermolingual inoculation.

Early work with tissue cultures indicated that foot-and-mouth disease virus grew in surviving tissues of embryo guinea-pigs. Subsequently, Frenkel (1950) demonstrated the multiplication of virus in surviving fragments of bovine tongue epithelium, and the presence of virus was demonstrated by titration in cattle, detection of complement-fixing antigen or by inoculation of mice. Multiplication of virus, with production of cytopathic effects, has been demonstrated in pig, calf or bovine, lamb, guinea-pig and other tissues.

Titrations of infectivity have been carried out by observation of the cytopathic effect, the metabolic inhibition test, microplate tests and plaque tests. Cell lines which have been found most susceptible include BHK 21 and IB-RS-2. The most sensitive tissues to cattle strains are calf thyroid monolayer cultures and these are used extensively for isolating virus from the field (Mowat and Chapman 1962, Snowdon 1966).

Various serological tests are used in investigations on foot-and-mouth disease. In complement-fixation tests, antigen or antibody can be examined qualitatively or quantitatively. Serum neutralization tests can be carried out in large or laboratory animals or in tissue cultures. The latter tests include plaque reduction, inhibition of cytopathic effect, metabolic inhibition and microplate tests. Antigen and antibody can also be detected or measured in single radial immunodiffusion, single radial haemolysis, RIA, ELISA, immunofluorescence and DID tests. The advantages and disadvantages of some of these tests have been discussed by Crowther (1978).

Such tests are used in conjunction with whole virus, with preparations derived from virus-infected cells, e.g. VIA antigen (Cowan and Graves 1966), 146S and 12S antigen, and with antibodies to such antigens or with IgG or IgM antibodies.

Variation in foot-and-mouth disease viruses

Foot-and-mouth disease is regarded as one of the most variable viruses. Variation may occur in the nature of the disease in the host, the hosts affected, the behaviour in laboratory animals and eggs, the growth in tissue culture under varying conditions, the antigenicity in animals and the structure of the RNA and protein.

Biological variation Historically, it was the behaviour of strains in different species that was first noticed. Certain strains were found that were naturally adapted to pigs, others that were cattle strains, and intermediate strains occurred between (Brooksby 1958). In Africa, variations in resistance to the development of foot-and-mouth disease have been found between local and imported animals; while this may have been partly due to previously acquired immunity, most of it is due to innate resistance developed by selection over the years among the local animals.

Strains vary in the readiness with which they may be adapted to laboratory animals; for example, some require many passages before causing generalization in guinea-pigs or paralysis in adult mice, others only a few passages. After adaptation, the strains may lose their pathogenicity for their original host. Similar findings have emerged from work with passage of virus in chick embryos.

In tissue cultures numerous variations have been found. Some echo the variation found in host susceptibility. For example, a type O strain isolated from pigs in Hong Kong infected cattle with difficulty; it is also difficult to grow in calf thyroid tissue culture but can be propagated with cytopathic effect in IB-RS-2 cells. Plaque size variation can be found extensively, this being due to a number of circumstances: the degree of adaptation of a particular strain of virus to the tissue concerned, differences between growth patterns in monolayer and suspension culture, the type of overlay used and the temperature at which the culture is incubated.

A molecular basis for variation in behaviour or virulence has not yet been found. However, a strain less virulent for mice was associated with a mutation in that part of the genome coding for VP_3 (McCahon 1981). Increased ability to grow in BHK 21 cells and lower ability to grow in calf kidney cells and produce lesions in cattle was associated with the poly-C tract (Harris and Brown 1977).

Antigenic variation Seven types of foot-and-mouth disease virus have been identified: O, A, C, SAT 1, SAT 2, SAT 3, and Asia 1. This is based on the finding that animals recovered from infection with one type are still susceptible to infection with a new type. Subtypes, over sixty in number, have been identified mainly as a result of developments in vaccines, since animals vaccinated with a preparation based on one subtype will not give complete protection to challenge with another subtype (Table 89.2).

The main method used for classifying the strains of virus has been the complement-fixation test, and the basis for typing and subtyping has been the degree of relatedness. However, the difficulties of dealing with

Table 89.2 Family, genus, types and subtypes of vesicular viruses

Virus:	Foot-and-mouth disease	Vesicular stomatitis	Swine vesicular disease	Vesicular exanthema San Miguel sea-lion
Family	*Picornaviridae*	*Rhabdoviridae*	*Picornaviridae*	*Caliciviridae*
Genus	Aphthovirus	Vesiculovirus	Enterovirus	Calicivirus
	7	2	1	VE 13+
Types	O. A. C. SAT 1, SAT 2, SAT 3, Asia 1	Indiana New Jersey		SMSL 14+
Subtypes	60+	Indiana, 3	0	—
Groups	—	—	Several	—

large numbers of subtypes have led to a reassessment (Pereira 1977).

Other tests used include the serum neutralization test, which measures the reaction between antiserum and the trypsin-sensitive site on the outside of the virion associated with the polypeptide VP_1. Competition RIA and ELISA tests can also be used but, unless specific sera and antigens are employed, will tend to measure reaction of all the surface polypeptides (Crowther 1978).

Advances in molecular biology have made possible analysis of the molecular basis for antigenic variation (Brown 1980). The methods are based on analysis of the components of the RNA genome, the primary protein products and the polypeptides produced which make up the virus. Such tests include RNA hybridization, RNA ribonuclease T_1 mapping, nucleotide sequencing, tryptic peptide analysis of the primary products (proteins) and the capsid proteins, electrofocusing of the polypeptides and analysis of the RNA polymerase (p56a—VIA antigen). The results indicate that, although mutations may be found affecting the genome coding for the p56a polymerase, most variation occurs in the structural region of the genome; of this part the VP_4 is most conserved.

Hybridization results from the RNA show 60–70 per cent homology between types O, A, C and Asia 1 and between types SAT 1, SAT 2 and SAT 3. Between O, A, C and Asia 1 and the SAT types homology is 26–40 per cent. Between subtypes homology is 70–80 per cent and within subtypes is 87 per cent. Antigenic variation occurs mainly in VP_1 and VP_3 for types O and A, but for SAT 2 and A_{22} variation has been found in VP_2.

As mentioned before, VP_1 is responsible for eliciting neutralizing antibodies and thus for protection to animals (Laporte *et al.* 1973, Bachrach *et al.* 1975). The cloning and expression of such antigens in *Esch. coli* have led to the production of clones which can be analysed by sequencing of the nucleotides and eventually of the amino acids, so that the basis of antigenic variation can be established more securely at the molecular level (Küpper *et al.* 1981, Strohmaier *et al.* 1978, Boothroyd *et al.* 1981).

From recent studies of field outbreaks and analyses of strains kept over the years (King *et al.* 1981), it would appear that mutations are occurring all the time at a rate not different from that of other viruses (McCahon 1981). Presumably, therefore, there must be epidemiological factors which enable such strains to be selected for and to survive in the field. One may be the evolution of heterotypic strains in the presence of antibody (Hyslop and Fagg 1965), although variation also occurs in the absence of an immune or semi-immune population.

Clinical signs of foot-and-mouth disease in animals

In foot-and-mouth disease, clinical signs may vary according to the species of animal, the breed, the physical and nutritional status, the degree of previous experience of infection and immunity, and the strain of virus concerned.

Classically, the first signs of disease in cattle are dullness, anorexia and rise in temperature. This is followed shortly by nasal discharge and the appearance of vesicles in the mouth and/or on the feet. In the mouth the vesicles are present on the tongue, the hard palate and dental pad, the lips and muzzle. Excessive salivation develops. On the feet, vesicles may be found in the interdigital space or along the coronary band and give rise to lameness. Vesicles may also be found on the teats. Initially the vesicles appear as raised white areas containing serous fluid; they may coalesce but later rupture, the epithelium becoming detached. Fresh epithelium grows on healing but, if secondary infection occurs or if the lesions have involved a wide area, scar tissue may be formed. In some cases, as a result of foot lesions lameness may persist permanently; mastitis in the udder may lead to loss of quarters. Loss of condition may occur and in some animals in tropical countries 'panting' or 'dyspnoea' due to heart lesions is found. In pregnant animals abortion sometimes occurs.

With some strains of foot-and-mouth disease in cattle of low productive capacity and nutritional status the lesions are often milder and the animals may not lose condition. Such cattle may recover quickly.

The incubation period may vary from 1 to 14 days, depending on the route of infection and the dose. In experimental cattle inoculated on the tongue, lesions develop within 24 hours. Within a farm the normal incubation period is 3–6 days and between farms 3–14 days.

In sheep the lesions are less apparent than in cattle and their extent depends on the strain. Lesions seen in laboratory-infected sheep are less severe than those seen in the field. Inapparent infections also often occur. The first sign is usually a nasal discharge. Vesicles are found on the feet in the interdigital space and the coronary band, and in the mouth on the tongue, the dental pad and the lips. Lesions on the tongue resolve quickly. Loss of condition occurs and abortion may be seen in pregnant sheep.

In goats nasal discharge is also found and lesions in the mouth and on the feet are similar to those in sheep. However, even more than in sheep, lesions are difficult to detect. In both sheep and goats the incubation period is 2–14 days.

In pigs the first signs are loss of appetite and fever and the pig often lies down. Lameness may develop and the animal is reluctant to move. Vesicles are found on the tongue, the lips, the snout and the feet. In very severe cases vesicles occur on the hocks and on the belly and teats. On healing, the claw of the feet may be shed. Abortions may occur in infected sows. The incubation period varies from 2 to 14 days. In experimental animals lesions may appear at 24 hours but on farms a 2–6 day interval between animal infections is likely. Some pigs may develop inapparent infection.

In lambs, calves, kids and piglets some strains may give rise to sudden death as a result of heart failure. In these cases no vesicles may be visible.

Wild animals are also affected. Buffalo, yaks, roe, muntjac and sika deer, occasionally fallow deer, impala and kudu among wild ruminants show clinical lesions of foot-and-mouth disease; inapparent infection may occur in other ruminant species (Hedger, 1976, Gibbs *et al.* 1975). Warthogs and bush pigs have suffered from disease, as have elephants and camels. Hedgehogs also develop lesions of foot-and-mouth disease, and other small mammals are susceptible (Capel-Edwards, 1971).

Pathology The vesicular lesions show swelling and ballooning degeneration of the cells of the epithelium just above the Malpighian layer. Subsequently, these cells necrose, leading to reaction from the cells of the lymphoid series.

Where death is due to heart failure, myocarditis can be found, with degeneration and necrosis of the heart muscle, leading to the formation of white streaks in the muscle.

Pathogenesis Up to the 1960s it was generally believed that initial infection with foot-and-mouth disease was through an abrasion in the skin, leading to local multiplication of virus and lesion formation, with subsequent generalization through the body. However, critical work by Korn (1957) and later by British and American workers has caused a revision of these views (Burrows 1972, McVicar 1977).

The pathogenesis of the disease may be considered in six stages:

(i) route of infection
(ii) early incubation period
(iii) late incubation period
(iv) early period of lesions
(v) late period of lesions
(vi) recovery and persistence of virus.

(i) Route of infection The three main routes of infection are inhalation, ingestion or penetration of the epithelium, or a combination of all three. The likelihood of infection occurring by one or more of these routes depends on the source of infection and the dose of virus required (Sellers 1971). Cattle require 10–100 infective particles of virus for infection by the respiratory route, whereas more than 10^5 particles are required to set up infection by ingestion. In a susceptible site such as the tongue, one infectious unit is sufficient, but the susceptible epithelium is protected by a keratinous layer and the placing of virus on it does not result in infection (McVicar 1977). Sheep and goats have similar requirements to cattle, being more readily infected by the respiratory route. Pigs require more than 10^5 infectious units (particles $>3\,\mu m$) to become infected by inhalation and ingestion. Animals can also be infected by the intramammary and conjunctival routes.

(ii) Early incubation period In naturally infected animals the virus enters the body, and probably the cells, where it starts to multiply. The most common site is in the pharyngeal area and soft palate (cattle, sheep and pigs), the tonsils (sheep) and the bronchi and bronchioles (pigs). In experimentally infected animals the virus multiplies at the site of inoculation—on the tongue (cattle) or the bulb of the foot (pigs). It then travels to the local lymph nodes and from there to the rest of the body.

(iii) Late incubation period This is the period just before the appearance of lesions. In natural infections the virus has spread from the initial site of multiplication to infect other sites. At this time virus to high titre is found in the pharynx, in the milk and at a lower titre in the blood, and the excretions and secretions. It is also found in the air exhaled from infected animals, especially sheep, and in the skin, especially in those sites where lesions are about to develop.

(iv) Early period of lesions Virus is found to high titre in the animal. The greatest amount is in the epithelium of the lesion and in the vesicular fluid but high titres are found in the blood, milk, semen, urine, faeces and exhaled air.

(v) Late period of lesions During this period the animal responds to infection by the production of antibodies. At the same time, the amounts of virus in the tissues, secretions and excretions diminish. However, virus is still present in the epithelium and in the faeces.

The antibodies found are neutralizing antibodies. Of the classes of antibody, IgA and IgM appear first and reach a peak about three days later. IgG is found two days after the appearance of IgA and IgM and peaks about two weeks afterwards. IgG persists at a high level.

(vi) Recovery and persistence of virus carriers With the recovery of the animal, foot-and-mouth disease virus usually disappears from the body. However, in cattle, sheep and goats virus may persist in the pharyngeal area (van Bekkum *et al.* 1959, Sutmöller and Gaggero 1965, Burrows, 1966, 1968). In 50 per cent of cattle and of sheep and goats it may persist for nine months and nine weeks, respectively. In individual cattle, virus has been isolated 2 years 3 months after infection, in sheep 9 months and in goats 11 months. In pigs virus does not persist beyond one month. Antibodies are found in the animal for several years.

Foot-and-mouth disease in man

Man can become exposed to foot-and-mouth disease virus in a number of ways. During outbreaks of the disease he may handle the affected animals or breathe in infected air. As a result, virus attaches to the exposed surface of the skin, to the conjunctiva and to his clothes, and virus is present in the nostrils and the upper respiratory tract (Sellers *et al.* 1970). He may also drink unpasteurized milk or consume other dairy products from infected cows. The amounts of virus may be considerable; for example, pigs excrete as an aerosol up to 6×10^4 ID_{50} of foot-and-mouth disease virus per minute and infected milk may contain over 10^6 ID_{50} per ml (Sellers 1971). Workers in laboratories handling the virus, in vaccine production plants especially those in open systems, and slaughterers handling infected animals and carcases may also be exposed to large amounts of virus.

Despite this, the number of subjects who develop antibodies or show lesions is surprisingly small. Most of the virus is carried away by the cilia in the nasal passage, the small particles may be carried into the lungs, the larger particles are trapped. Virus drunk in milk and dairy products will be inactivated by the acid in the stomach.

In many of the cases reported in the literature no attempts at virus isolation or antibody studies have been carried out. These older records have been reviewed by Hyslop (1973). However, 37 cases have been described in which, apart from clinical signs, diagnosis was based on virus isolation alone (6 cases), virus isolation together with rise of antibody (5 cases), virus isolation and demonstration of antibody in convalescent serum (17 cases), demonstration of antibody in convalescent serum (9 cases). In another case (1921) in a laboratory worker, diagnosis was based on clinical signs alone. Type O virus was isolated in 23 cases, type A in 1 and type C in 6; in the remainder the type was not given (Armstrong *et al.* 1967, Bojlen 1941, Eissner *et al.* 1967, Garbe *et al.* 1959, Heinig and Neumerkel 1964, Melendez 1961, Michelsen and Suhr-Rasmussen 1959, Pilz and Garbe 1966, Pilz *et al.* 1962, Platt 1958, Suhr-Rasmussen 1969, Vetterlein 1954*a*).

Of the patients involved, 13 had close contact with infected cattle or with pigs, i.e. examining or milking infected animals, and a further 4 had broken skin while in close contact. Three were some distance from infected animals. In 6, the clinical disease was attributed to drinking raw milk or cream. Six were concerned with vaccine manufacture or laboratory procedures and in 4 the cause of infection was not given.

The ages of the infected persons ranged from 1 year 2 months to 60; 23 were males and 13 females (no sex given for 2). Three had a history of concurrent dermatitis and 2 of herpes.

Clinical signs in man vary. The incubation period may be from 1 to 10 days, but most commonly between 2 and 7 days. The initial symptoms may be fever, sore throat, headache and general malaise; later, vesicles are found on the hands, the feet and in the mouth, often preceded by heat and itching. In some instances vesicles appear first on one site, to be followed by fever and generalization of vesicles. These vary in size, depending on the site, and contain clear serous fluid which later becomes turbid. The blisters burst and heal and recovery is usually complete. There does not appear to be cross-infection between patients and contacts.

In cases where contact with animals is concerned, the first symptoms are usually seen in the throat, suggesting that the virus entered through the nasal passages. In those who drink raw milk, the first blisters usually appear in the mouth. In laboratory workers who had pricked themselves or in veterinarians or attendants who had handled infected animals, the vesicles were first seen on the hands.

There is one instance recorded of a worker in a Waldmann type vaccine laboratory being infected twice, the first time with type C virus and the second with type O virus. The second time, in addition to the development of antibodies to type O, there was also an anamnestic response to type C (Pilz and Garbe 1966). Clinical signs of disease were seen on both occasions.

Studies of antibody levels in laboratory workers or in workers handling infected animals have been carried out. In Polish workers in areas of high exposure to virus, antibodies were found to types O, A and C (maximum titre 1/40) (Wisniewski and Jankowska 1968). Low levels have also been found by Vetterlein (1954*b*) and Suhr-Rasmussen (1969) in laboratory workers. Low levels of antibodies have been reported in other laboratories handling foot-and-mouth disease but the number and levels have been very low.

The symptoms of foot-and-mouth disease infection in man have to be distinguished from herpes, hand-foot-and-mouth disease, and other syndromes. In these cases a history of contact with animals infected with foot-and-mouth disease or with large doses of virus should be sought.

Epidemiology of foot-and-mouth disease

Geographical distribution At present there is no foot-and-mouth disease in North and Central America, Chile, Scandinavia, Northern Russia, Japan, Australia or New Zealand. Strains of types O, A and C occur in South America; O, A and C in Europe, O, A, C, SAT 1, SAT 2 and SAT 3 in Africa; O, A, C and Asia 1 in the Middle East, India and other parts of Asia. SAT 1 was found in the Middle East in 1962 and for a few years afterwards. Various subtypes occur, but the main ones found at the moment are the O_1 strain in South America, Europe and the Middle East with related O_1 in India and South-East Asia. Two main subtypes of type C are found: C_1 (Europe) and C_3 (South America). SAT strains vary according to the area in which they are found in Africa, and there appear to be Southern, Central and West African variants. With Asia 1 western and south-eastern variants can be distinguished. Type A varies considerably, with A_{24} variants in South America, A_5 in Europe, A_{22} in the Middle East including Sudan and the Indian subcontinent; other A variants are found in Western and Central Africa.

Evolution of variants continues with the development of new strains in response to pressure exerted by low levels of immunization and to the natural changes that are continually occurring (McCahon 1981).

Spread of disease

Foot-and-mouth disease spreads in a number of ways—by direct contact, through animal products such as meat and milk, by the airborne route, by mechanical means such as on people, wild animals and birds, vehicles, fomites and possibly by carriers (Hyslop 1970, Sellers 1971).

Direct contact Direct contact is possibly the most important and this may occur on farms, in markets, through movement within a country and from one country to another. Spread may occur at the time of disease or during the incubation period, when virus is in the excretions and secretions.

Animal products When susceptible species are slaughtered during the later stages of the incubation period or at the time of development of lesions, virus is present in the tissues. In the muscle, virus is inactivated because of the acid formed but virus survives in lymph nodes, bone marrow and offal (Henderson and Brooksby 1948). Susceptible animals coming into contact with such infected material may develop disease.

Virus is present in milk in the later stages of the incubation period and in the early stages of lesions. If such milk is fed to calves and piglets, foot-and-mouth disease may result (Hedger and Dawson 1970). Similarly, semen collected from bulls just before the appearance of lesions contains virus which may infect cows and heifers at normal service or artificial insemination (Cottral *et al.* 1968, Sellers *et al.* 1968).

Airborne spread The most sensitive method of infection for cattle, sheep and goats appears to be by inhalation. Cattle are the most susceptible; the infective dose is small and the respiratory rate (100 litres per minute) means that they breathe in greater volumes of air than the other species (sheep and goats, 10 litres per minute). Pigs require a large dose to become infected by inhalation or by ingestion (Sellers and Parker 1969).

Pigs excrete the most virus as an aerosol, amounts ranging from 4×10^5 to 10^8 infectious units per day; cattle, sheep and goats excrete from 5×10^3 to 2×10^5 infectious units per day. Amounts vary according to the strain of the virus; most virus has been found with O and C strains of Western European origin (Sellers *et al.* 1973).

Most of the airborne particles are 6μm or more in diameter and, once emitted, are subject to normal disposition. However, turbulence in the atmosphere together with wind will spread the virus as a plume horizontally and vertically (Gloster *et al.* 1981).

Survival of the virus particles depends on the relative humidity and, when this remains above 60 per cent (Barlow 1972; Donaldson 1972), the particles survive in the air for many hours. Atmospheric conditions are more stable and relative humidities above 60 per cent are common at night; thus, airborne spread is more likely to take place at that time. High temperatures and sunlight do not appear to have any effect on the virus except to the extent that they contribute to lowered relative humidity.

Some investigators (Smith and Hugh-Jones 1969) have associated rain with airborne spread; however, such spread often occurs in the absence of rain. The apparent relationship may be due to the maintenance of high relative humidity and the stability of the atmosphere in warm fronts associated with rain and winds at the time of outbreaks (Gloster *et al.* 1981).

Most spread over land is within 10 kilometres and is affected by topography, the nature of which may also play a part in spread over longer distances—60 km or more—over land. However, spread over sea may occur for distances of 250 km (Hurst 1968). The criteria for spread are a high virus output, low virus dispersion, high virus survival and large numbers of susceptible livestock exposed to virus for many hours. Such criteria have been satisfied in spread from Denmark and Sweden and from France to the south coast of England. Turbulence is less over sea than over land and stability of atmosphere above the sea may occur for periods sufficient to allow the virus to be carried over the sea (Gloster *et al.* 1982, Donaldson *et al.* 1982; see also Sellers and Gloster 1980).

Airborne spread has been found mainly in temperate climates, where relative humidities are commonly above 60 per cent and livestock concentrations high. In tropical and sub-tropical countries, the concentrations of livestock may be less but airborne spread over shorter distances may occur, especially at night and at waterholes where animals congregate.

Mechanical spread Virus which is shed from animals with foot-and-mouth disease through their secretions and excretions results in the contamination of surrounding areas. Vehicles in which animals have been carried are a source of infection to animals subsequently conveyed. The clothing of persons who have looked after infected animals becomes contaminated with virus and virus may persist in the nose (Sellers *et al.* 1970). Virus may survive on hay and straw and in slurry and dung.

Spread has also been attributed to carriage of virus

by birds (Wilson and Matheson 1952), but their role has been disputed (Murton 1964, Hurst 1968). As with airborne spread, successful transmission by mechanical means, whether on people, birds, wild animals, vehicles or fomites, demands contamination by high levels of virus and survival in sufficient concentration to initiate infection in susceptible animals with which the various agents come into contact. This may be satisfied in some cases but the chances are probably less than with other methods of spread.

Animal carriers As described previously, virus may survive in the pharyngeal area of cattle for long periods. Experimentally no-one has demonstrated the transmission of such virus to animals held in contact. However, since such transmission may occur only occasionally, this type of spread is difficult to establish experimentally but it may be one of the methods by which virus is maintained over long periods in the field.

Role of domestic animals and wildlife In South America foot-and-mouth disease appears to be maintained by continual spread of the virus among the susceptible population of domestic animals. However, in Africa the wild game population plays a part, especially in southern Africa, where, for example, on the borders of the Kruger National Park there is extensive contact between game and cattle. In England, however, deer and hedgehogs do not appear to be important, although they may become infected and pass on the disease on occasions (Gibbs *et al.* 1975).

The role of sheep varies in different parts of the world. In England outbreaks in sheep are less common than in cattle and pigs and in Kenya also they do not appear to transmit infection. However, sheep, together with goats, form one of the major pathways of infection in the Middle East, where they are the dominant animals.

In effect, therefore, the extent of disease and the methods of spread reflect the type of husbandry, movements of trade, livestock and wild life and the climate of the areas concrened.

Diagnosis

Diagnosis of foot-and-mouth disease is based on recognition of clinical disease, with laboratory tests to establish the type and subtype. Vesicular stomatitis would present a problem in differential diagnosis in the Americas and, on farms where only pigs are affected, swine vesicular disease and vesicular exanthema should also be considered (Table 89.3).

Laboratory diagnosis is based on complement-fixation tests on vesicular fluid or vesicular epithelium. When this is unsuccessful, the virus is passaged in tissue cultures and the harvests retested by complement fixation. On occasions, passage in unweaned mice and in cattle and pigs has been carried out; harvests can be examined under the electron microscope. Relations with other strains can be investigated by one-way complement fixation, neutralization, RIA or ELISA tests. More recently, electrofocusing and finger-printing of ribonuclease T_1 oligonucleotides have been used (Brown 1981, King *et al.* 1981).

Control

Spread of disease may be lessened by banning or controlling the movement of animals. The effect of this is not usually seen until at least one incubation period (on average 10 days) has elapsed. Especially important in this type of control is the ban on markets and abattoirs. Movements of animals should be traced to determine the areas at risk.

Spread through animal products may be controlled by ensuring that meat is deboned and offal is heated to destroy virus. In addition, meat should be imported

Table 89.3 Differential laboratory diagnosis of vesicular viruses (Source of virus: tissue, vesicular epithelium, vesicular fluid)

Complement fixation	Foot-and-mouth disease, FMD
	Vesicular stomatitis, VS
	Swine vesicular disease, SVD
	Vesicular exanthema, VE
Growth of tissue culture	
Calf thyroid	FMD, VS
BHK	FMD, VS
IB-RS-2	FMD, VS, SVD, VE
Chick embryo cells	VS
Serum neutralization test	FMD
	VS
	SVD
	VE
Morphology	FMD – spherical, 24 nm
(electron microscope)	VS – bullet shaped, 175 × 65 nm
	SVD – spherical, 30 nm
	VE – spherical, dark staining areas, 35–40 nm

from countries where foot-and-mouth disease does not exist or where an adequate vaccination policy is carried out. Scraps of meat or swill should be heated before being fed to pigs. Milk should be heat-treated before being fed to calves or piglets. Semen should be held for a period to ensure that no foot-and-mouth disease occurs in the donor animal or in others on the same farm.

For animals at risk from airborne spread of virus it may be an advantage to put them indoors. Indications of where such spread may occur can be obtained by estimating output of virus, wind speed and other meteorological factors and topography (Gloster *et al.* 1981). Measures may then be taken to prevent movement of such animals.

Vehicles, fomites and other contaminated material may be sprayed or washed with disinfectants. Suitable disinfectants contain acid, alkali, formalin or combinations of these with detergents (Sellers 1968). People should be encouraged to avoid animals; if they do have to approach them, they should be required to wear protective clothing.

In many countries the aim is to break the virus cycle by action on the animal. This may be done by slaughter of affected animals and their contacts or by vaccination, or by a combination of both methods. When slaughter is decided upon, this should be done as soon as possible after diagnosis, to avoid further excretion of virus, especially into the air.

Vaccination

Vaccination, however, is practised in many parts of the world (Mowat *et al.* 1978). The majority of vaccines are inactivated; attenuated live vaccines are no longer used. The source of virus for vaccine is that grown in tissue cultures of surviving fragments of epithelium from cattle tongues, monolayer calf kidney cells, or monolayer or suspension cultures of BHK 21 cells (Frenkel, 1950, Ubertini *et al.* 1960, Mowat and Chapman 1962, Capstick *et al.* 1962). The virus is harvested 24–48 hours after seeding and is then inactivated with formalin, acetylethyleneimine, binary ethyleneimine, or both preparations (Wesslen and Dinter 1957 Brown *et al.* 1963, Bahnemann 1975). An adjuvant such as aluminium hydroxide, saponin or oil is added. The vaccine is tested for innocuity and potency in animals. Standards for vaccine now demand at least 3 PD_{50} but preferably more should be present in a vaccine. Vaccines may contain one (monovalent) or more (bivalent, polyvalent) types or subtypes of virus.

The choice of virus strain for the vaccine is important. It should be a good immunogen and give protection against the subtype present in the country and as many others as possible. New strains arising in a country should be tested against the vaccine by serological or cross-protection tests to see if the vaccine will protect against the new strains.

Vaccines given to cattle and sheep give protection for about three months after the first dose but a second dose gives an anamnestic response and protection will then be provided for a year in cattle, although possibly for a shorter time in sheep. Young animals usually have maternal antibody either from previous infection or vaccination and can be vaccinated at 4–6 months of age.

The vaccination programme adopted usually depends on the weight of infection in the country and the type of animal husbandry. Where the expected challenge is high, animals are vaccinated three times a year; when medium, twice a year; when low, young animals are vaccinated twice and then annually. Vaccination is best done at the time before maximum challenge may be expected. This may be just before and just after the rainy season in tropical countries and before animal movements to other parts in other countries. In the case of outbreaks on a farm, animals are revaccinated. In Europe, however, the animals on the farm are slaughtered and the surrounding farms vaccinated.

In some countries, barrier or ring vaccination is practised. This is usually carried out after disease has been reported; the areas to be vaccinated have to be carefully defined. Stocks of vaccine are kept in reserve. The use of vaccines in the control of the disease has been described by Boldrini (1978).

The inactivated vaccines as developed protect against generalization of the virus. They do not necessarily protect the primary site of multiplication, unless two or more inoculations have been carried out (Garland 1974). Consequently, animals exposed to infection, even though no disease is shown, may harbour virus in the pharyngeal area for some time (Sutmöller *et al.* 1968). However, this was considerably less than when no animals were vaccinated. For example, in Kenya the incidence of carriers in vaccinated areas was 0.49 per cent, compared with 3.34 per cent in a non-vaccinated area (Anderson *et al.* 1974). No carriers have been detected in France among the many animals that have been exported overseas since 1967.

Future developments in vaccine production include the possible use of genetically engineered clones for production of synthetic peptides, the polypeptide VP_1 or the appropriate part of VP_1 for conferring protection (Kupper *et al.* 1981, Boothroyd *et al.* 1981, Kleid *et al.* 1981, Bittle *et al.* 1982).

Vesicular stomatitis

Vesicular stomatitis is a vesicular disease of horses, cattle, sheep and pigs; its mode of transmission is not clear but spread by arthropods and by contact infection is surmised. The causal agent is a virus belonging to the family *Rhabdoviridae*; other members of that family include rabies and other lyssaviruses, bovine ephemeral fever, fish rhabdoviruses and other viruses of vertebrates, invertebrates and plants (Brown *et al.* 1979). Vesicular stomatitis virus belongs to the genus Vesiculovirus, to which Chandipura, Isfahan and Piry viruses belong.

Physico-chemical properties of the virus

The virion of vesicular stomatitis virus contains RNA, protein and lipid. Its buoyant density is 1.20 mg/ml in caesium chloride and it has a sedimentation coefficient of 620–625S and a molecular weight of 2.9 to 3.85 × 10^6. It is unstable at pH 3 but stable between pH 5 and pH 10 and is rapidly inactivated at 56°. Lipid solvents reduce the infectivity of the virus (see Table 89.1). Vesicular stomatitis virus is also inactivated by ultraviolet light and X-irradiation (Howatson 1970, Rowlands 1979, Wagner 1975).

The virion is bullet-shaped with a diameter of 65 nm and a length of 175 nm and consists of a nucleocapsid surrounded by an envelope containing surface projections. The nucleic acid is a single-stranded linear RNA with negative polarity of molecular weight 4 × 10^6; it is non-infectious. The ribonucleoprotein core consists of 40–45S RNA with a single polypeptide, N. There are five polypeptides: L, G, N, NS and M. The nucleocapsid contains the ribonucleoprotein (RNA–N) complexed with L and NS polypeptides surrounded by polypeptide M. The nucleocapsid has helical symmetry, the outer diameter being 49 nm and the inner 29 nm, with 35 sub-units per turn. It is infectious. The envelope contains lipid incorporating carbohydrate and protein in the form of a glycosylated polypeptide. The surface projections (G protein) are 5–10 nm in length. An axial channel runs through the centre of the core of the projections.

G protein forms the type-specific immunizing antigen. N is the complement-fixing antigen. M may also be antigenic. L and NS proteins function as transcriptase. Truncated forms are produced in the course of infection (defective interfering, DI, or T particles) and these are about one-third the length of the virion (B particles) with RNA of molecular weight 1.2–1.3 × 10^6.

Multiplication of the virus proceeds in a number of steps. Adsorption to the cell is accomplished through the glycoprotein spikes containing G protein. Phagocytosis of the whole particle then occurs in a phagocytic vacuole leading to uncoating of the particle. Fusion of the virion envelope with the cell wall may also take place. Inside the cell 40S RNA is transcribed by RNA transcriptase into two size classes of RNA—28S and 12–16S m-RNA. These viral messengers are responsible for translation into the structural polypeptides, L, G, N, NS and M. The polypeptides N, L and NS are probably synthesized in the cytoplasm and polypeptides M and G in the plasma membrane. The method of replication of the progeny virion RNA is not completely known but it may need a 40S RNA–N template. The nucleocapsid is assembled at the membrane, where G and M proteins are synthesized and the cellular membrane lipids assembled. Finally, the newly formed virion is rapidly released from the cell.

Harvests from cultures infected with vesicular stomatitis virus contain intact virion and defective interfering particles as well as two subviral constituents of 20S and 6S (Huang 1977). The 20S is related to N protein and the 6S reacts with viral envelope antibodies. The percentage of defective interfering virus depends on the multiplicity of initial infection, the passage level of the strain and the cells used.

Biological properties of the virus

Guinea-pigs inoculated on the foot pad with vesicular stomatitis virus develop vesicles at the site within 48 hours; Subsequently, generalization occurs, with lesions developing on other foot pads and on the tongue. Adult mice inoculated intracerebrally develop an encephalitis with spastic paralysis and ataxia in the hind legs. Death occurs in 3–5 days. Suckling mice can be infected by the intraperitoneal and intracerebral routes, the virus multiplying in the brain. Seven to 10 day old chick embryos inoculated on the chorioallantoic membrane or into the allantois die in 24–48 hours. The embryo is congested but the fluids remain clear. Chickens develop vesicles on the tongue after inoculation (Skinner 1959).

Wild animals of Central America have been studied extensively by Tesh *et al.* (1970) and many have been found susceptible. Bats have also been infected in the laboratory (Donaldson 1970).

Many kinds of tissue culture from different species of animal—man, monkey, calf, pig, sheep, guinea-pig and chick embryo—support the growth of vesicular stomatitis virus, with the production of large amounts of virus and the development of cytopathic effects. The virus may also be grown in BHK 21, HeLa and L cells and in cell lines derived from mosquitoes and moths. Vesicular stomatitis virus readily produces plaques on monolayers of a number of cells, the plaques often being detectable within 24 hours.

The presence of antibodies in serum can be detected in complement-fixation, neutralization and plaque reduction tests as well as in agar gel precipitin tests. Antibodies to whole virion, defective virus or to purified antigens of the virus are induced in guinea-pigs or rabbits. Haemagglutination with goose erythrocytes has been demonstrated but the test has not been used extensively because of the large amounts of virus required.

Because of its ability to grow in different cell cultures and because of its susceptibility, vesicular stomatitis virus has been used extensively for the titration of interferon.

Vesicular stomatitis virus was initially thought, on cross-serological tests, to consist of two types—VSV-Indiana and VSV-New Jersey. Subsequently, new isolations of virus have been made in various parts of America and, on the basis of cross-immunity tests in cattle and pigs, complement-fixation tests and cross-neutralization tests, three sub-types of VSV-Indiana have been determined. These are represented by the following strains of virus: (i) Indiana C, (ii) Cocal and an Argentinian isolate (Salto) and (iii) a Brazilian isolate, Alagoas (Federer *et al.* 1967) (see Table 89.2).

Clinical signs of the disease

In domestic animals the clinical disease caused by vesicular stomatitis virus resembles that caused by foot-and-mouth disease and other vesicular viruses. In horses and in cattle the principal lesions are found in the mouth. Vesicles develop on the tongue, the gums and lips and salivation is excessive. In some instances severe lesions may be seen, but often erosive necrotic lesions are found which may heal rapidly. Extension of lesions at a site is uncommon. In milking cows vesicles are seen on the teats. However, vesicles on the feet of cattle and horses are uncommon. There is a fever but death rarely occurs. Loss of condition is slight except where extensive vesiculation of the mouth or udder is seen.

In pigs, fever is the first sign of disease and is followed by vesicles on the tongue, snout and coronary band and interdigital space of the foot. The vesicles rupture quickly, leaving a raw edge. Sick pigs may be reluctant to move. Sometimes the vesicles are shallow and break easily, leaving necrotic epithelium. Death is unusual but outbreaks of vesicular stomatitis due to the New Jersey type have led to high mortality among pigs in Central America.

Vesicular stomatitis in sheep and goats is rare.

Pathology In the epithelium there is an accumulation of oedematous fluid in the intercellular spaces of the Malpighian layer. These accumulations coalesce to form papules and then vesicles. The epithelial cells undergo degenerative changes as a result of the oedema and pycnosis, and karyolysis occurs. With severe lesions the dermis may become oedematous and undergo inflammatory changes. No myocarditis in young animals is seen.

Pathogenesis The pathogenesis of vesicular stomatitis is obscure. The virus may enter the body through a break in the mucosal epithelium of the nasal cavity, by the nasopharyngeal route, by a break in the integument of the skin, or through the bite of an insect. Probably local multiplication takes place and in the mouth vesicles are formed after 2–4 days. The vesicular fluid and epithelium contain high titres of virus. Viraemia is rarely, if ever, present. However, antibodies are found, so presumably virus or viral antigen is carried to antibody-forming sites. Complement-fixing antibodies are detected a week after infection, reach a peak at 9–16 days and then decline, being no longer found after 50–110 days. Neutralizing antibodies form at the same time but remain high with fluctuations in level over a long time.

No persistence of virus in the animal has been found. Animals can be reinfected experimentally within 30–60 days of recovery with the same strain of virus (Mason 1978).

Vesicular stomatitis in man

Cases of vesicular stomatitis have been reported in man. The persons afflicted included, in the laboratory, workers handling virus on the bench in eggs and tissue cultures and animal attendants and investigators looking after and examining diseased animals and, in the field, farmers, farm workers and veterinarians associated with the disease in animals on farms (Fellowes *et al.* 1955, Fields and Hawkins 1967, Johnson *et al.* 1966, Hanson *et al.* 1950, Patterson *et al.* 1958, Shelokov *et al.* 1961).

The symptoms vary from those of a mild 'flu-like illness to severe forms. The onset is usually sudden with fever, chills and severe sweating. Aching of muscles of arms, legs and eyes is accompanied by headaches, weakness and vertigo. Anorexia is often seen, together with nausea in some cases. Sore throats may also occur. In some persons blisters and vesicles develop on the lips and in the pharyngeal area or occasionally at the site of infection. Conjunctivitis is also seen.

Symptoms last for 3–4 days but in some cases there is a diphasic response with symptoms occuring initially for 1–2 days, then an apparent recovery followed by a return of symptoms. The incubation period is from 28 hours to 6 days after exposure to virus in laboratory infections and up to 9 days in field infections.

Infection may occur by the nasopharyngeal route, through the conjunctiva or, in laboratory infections, by accidental inoculation with fluids containing virus. In tropical and sub-tropical areas infection may result from the bite of an infected arthropod vector. The sites of virus multiplication are not known. Attempts to isolate virus from blood, serum, nose and throat washings and faeces have been made in many cases but all were unsuccessful, apart from one isolation from the blood on the day after development of clinical signs. Complement-fixing antibodies appeared at 10 days after infection and reached a high titre at 14–22 days and then declined. Neutralizing antibodies appeared and reached a peak about the same time as complement-fixing antibodies, and persisted for several years. Diagnosis of the disease in man is thus based on isolation of the virus (rarely) or, more often, demonstration of a rise in antibody with paired serum samples.

Spread of virus from person to person does not occur, nor do people in contact develop antibodies. However, in laboratories and in the field infection with the virus from animals with the disease, from insect bites and from handling the virus in the laboratory often take place. Antibodies have been demonstrated in farm personnel in south-eastern Georgia (USA) where disease had occurred, and in other areas of that State. Antibodies have also been found in persons in New Mexico and Colorado. In Panama antibodies are found more often in older people and in those who work with cattle. In Central America antibodies to vesicular stomatitis are widespread (Brody *et al.* 1967, Mason 1978, Tesh *et al.* 1969).

Clinical cases occur after infection with either New Jersey or Indiana type; the extent depends on the type of virus handled in the laboratory at a particular time or that responsible for an epidemic or present in an endemic area.

Epidemiology

Geographical distribution Vesicular stomatitis is confined to North, Central and South America, although vesicular disease in horses was recorded in Europe (1915–1917), in South Africa (1881 and 1897) and possibly in Asia (Hanson 1952). Both New Jersey and Indiana types are found throughout the area.

A number of areas of disease occurrence can be distinguished. In the northern States of America the disease occurs at intervals of about 10 years or more. For example, there were outbreaks in Wisconsin in 1926, 1937 and 1949 affecting large numbers of cattle and horses. The disease occurred in the latter half of the year, from July to October. Between the years no other outbreaks were detected but antibodies persisted in affected animals.

In the southern States outbreaks may occur each year. In Georgia the disease is usually first seen in pigs and may then spread to cattle. The initial outbreaks take place in May and the disease disappears in November. On farms, about 28–80 per cent of animals have antibodies. In the USA there was an apparent movement westwards of Indiana and New Jersey types between 1963 and 1966 and between 1964 and 1966, respectively (Jenney 1967).

In Mexico there are areas in the south-east and across the isthmus of Tehuantepec where vesicular stomatitis is found regularly. In Panama infection is endemic, antibodies being found in animals in many rural districts.

In South America outbreaks occur regularly in all the countries.

Methods of spread

Contact The importance of this type of spread is disputed, since there are many instances of outbreaks on farms where disease did not spread to neighbouring animals on the farm. On the other hand, if large amounts of virus are present in the epithelium and vesicular fluid, virus could enter through breaks in the epithelium of in-contact animals. Virus can also be spread by milkers and milking machines used on infected teats. Animals fed on infected offal can be infected through abrasions on the tongue.

Persons associated with infected animals or working in the laboratory become infected by the nasopharyngeal route and, in the field, animals in contact with others may probably become infected by this route. Spread by contact between swine is more common than between cattle.

Arthropods In the field vesicular stomatitis virus, Indiana type, has been isolated from sandflies in Panama and from *Aedes* mosquitoes in New Mexico (Mason 1978). Cocal virus has been isolated from *Gigantolaeleps* mites in Trinidad and from *Culex* mosquitoes in Brazil and Trinidad. New Jersey virus was isolated from gnats (*Hippelates pusio*) in Colorado. Multiplication of Indiana type virus has been demonstrated in *Aedes aegypti* and *Culex pipiens*. Virus could also be transmitted mechanically on tabanids and mosquitoes (Ferris *et al.* 1955). Thus, insects might be biological or mechanical transmitters of the virus. However, viraemia is rarely found in domestic animals, but has been demonstrated in bats and marmosets.

Persistence of infection It is thought that no persistence of virus occurs; however, it is difficult to explain the transatlantic carriage of infection to France in 1915 or to South Africa at the end of the nineteenth century without some form of persistence. It may be that virus remains latent and is activated by stress, as with herpes infection (Mason 1978).

It would appear that several methods of spread of virus can occur to varying extents in the different areas (Sellers 1980). In the endemic area of Panama and in other parts of Central and northern South America, the virus is maintained in a phlebotomus-mammal cycle (Indiana), insect–mammal cycle (New Jersey and Indiana) in forest mammals and in local susceptible domestic animals. Spread also occurs through contact.

Transovarial transmission of the Indiana type in Phlebotomus may occur (Tesh *et al.* 1971).

In epidemic areas of Mexico and North America and in South America, spread of virus may occur by movement of animals or of infected insects on the wind (Hanson 1968). Virus probably does not persist between outbreaks where these occur every ten years or more, but in areas where they occur every year the virus may persist during the winter months.

A number of hypotheses have been put forward to explain the epidemiology of the disease. These include the suggestion that vesicular stomatitis is a plant virus already present in the pasture in the plants (Jonkers 1967) or transmission by phlebotomus (Johnson *et al.* 1969). The epidemiology is, however, far from being resolved and may embrace a number of factors.

Diagnosis Clinically, vesicular stomatitis affects horses as well as cattle and pigs. However, horses are kept on few farms and it is necessary to distinguish the virus by other means.

Complement-fixation of the sample of suspect vesicular fluid or vesicular epithelium is the first test and, where there is insufficient antigen, passage is carried out in tissue culture. Vesicular stomatitis grows in many tissue cultures with cytopathic effect. It will also grow in chick embryos and chick embryo tissue cultures (See Table 89.3).

Paired sera may be tested by serum neutralization tests.

Control Since the epidemiology of the disease is not fully understood, rational discussion of control is impossible. Clearly it is necessary to establish first that the vesicular lesions are not due to foot-and-mouth disease virus.

Measures such as the control of animal movement can be taken. Vaccines have been investigated and it has been shown that vesicular stomatitis virus given intramuscularly does not cause lesions but induces the production of neutralizing antibodies (Lauerman *et al.* 1963). In vaccine trials, administration of a vaccine reduced the number of clinical cases. Duration of immunity is unknown and may not be long since, although antibodies persist for months and years, natural immunity is of short duration.

Persons handling virus in the laboratory should work with the virus under containment, and people in the field such as veterinarians, cattlemen and pig handlers should take precautions when handling infected animals.

Swine vesicular disease

Swine vesicular disease is an infectious disease of pigs characterized by the appearance of vesicles indistinguishable from those caused by foot-and-mouth disease on the tongue, in the mouth and on the feet and hocks. The disease is caused by a porcine enterovirus belonging to the family *Picornaviridae* and was first recognized in 1966 in Italy (Nardelli *et al.* 1968). Since then, swine vesicular disease has been seen in many countries of Europe as well as in Hong Kong and Japan.

Physico-chemical characteristics of the virus

The virion is spherical with icosahedral symmetry and has a diameter of 30 nm. The buoyant density is 1.32 g per ml in caesium chloride. The virus is stable in 1 M magnesium chloride and at acid and alkaline pH values and is not inactivated by ether or by exposure to trypsin (0.2 per cent) for 2 hours at 37°. Two populations of complement-fixing particles have been separated, one with a sedimentation coefficient of 150S, the other 80S (Delagneau *et al.* 1975). The latter probably represents empty capsids.

The replication of swine vesicular disease virus is similar to that of other members of the enterovirus group and takes place in the cytoplasm. The RNA is infectious and acts as the messenger RNA for translation of protein (see Table 89.1).

Biological characters

Newborn mice can be infected by intraperitoneal or intracerebral inoculation of swine vesicular disease virus, but 7-day old and adult mice are not affected (Nardelli *et al.* 1968).

The virus multiplies in tissue cultures derived from pig tissues, for example, pig kidney monolayers and cell lines such as IB-RS-2 and PK 15. No multiplication is seen in the BHK 21 cell line but multiplication in HeLa cells has been observed.

Serological tests include complement fixation, serum neutralization, double immunodiffusion (DID) and radio immunodiffusion (RID). Recently, RIA, ELISA and counter-current electrophoresis tests have been developed.

Variation does not occur to the same extent in swine vesicular disease virus as in foot-and-mouth disease virus. However, differences in field isolates have been shown in neutralization, immunodiffusion, competition RIA, polyacrylamide gel electrophoresis and T_1 mapping of oligonucleotides (Harris *et al.* 1979). Swine vesicular disease isolates from the United Kingdom could be placed in four groups and isolates from Hong Kong in five groups.

Swine vesicular disease virus is neutralized by antiserum to Coxsackie B5 virus (Graves 1973). Pigs inoculated with Coxsackie B5 develop antibodies to both swine vesicular disease and Coxsackie B5 viruses but do not develop the disease. There is speculation that swine vesicular disease virus was originally derived from Coxsackie B5 virus, which also causes paralysis and death in newborn mice.

Swine vesicular disease virus resembles the porcine enteroviruses in its physico-chemical characteristics, but cross-complement fixation and neutralization tests indicate that it forms a serotype of its own within the eleven serotypes into which porcine enteroviruses have been divided (Knowles and Buckley 1980) (see Table 89.2).

Clinical signs of the disease

The lesions resemble those caused by foot-and-mouth disease and other vesicular diseases in pigs.

The incubation period varies from 2 to 7 days and perhaps longer if the infecting dose is small. At times affected pigs show a disinclination to take food and may have a temperature rise of 1–2°. However, in many cases no distinct change in the pig's behaviour is seen and signs of disease go unnoticed.

Vesicles develop on the bulbs of the feet, the interdigital space and coronary band. Often the supernumerary digits are involved together with the hocks and the skin of the lower limbs. Vesicles are found on the snout, lips and tongue and, on occasions in severe cases, on the abdomen and teats. The severity often depends on the housing of the pigs. On concrete floors pigs develop more vesicles which, on rupture, become sores. Usually, however, recovery is uncomplicated and—unlike foot-and-mouth disease—no loss of condition occurs. Subclinical infection sometimes takes place.

Abortion does not occur in pregnant sows but the death of newborn pigs with extensive lesions after contact with an affected mother has been seen. Myo carditis in piglets is not a feature as with piglets affected with foot-and-mouth disease. Nervous signs have been found in pigs in Italy and in England (Zoletto *et al.* 1974), Lenghaus *et al.* 1976).

Cattle do not become infected, although virus can be recovered from the pharynx of steers in contact with pigs. Sheep, however, undergo subclinical infection when in contact with large amounts of virus (Burrows *et al.* 1974b).

Pathology The epithelial vesicles show essentially the same changes as in foot-and-mouth disease, although slight differences were noted (Lenghaus and Mann 1976). In the central nervous system meningo-encephalitis with perivascular cuffing of the blood vessels and formation of neuroglia cell foci have been described.

Pathogenesis Pigs become infected by coming into contact with other infected pigs, with infected tissues, secretions and excretions, and by feeding on infected meat or meat products. The various routes of infection have been investigated and the results indicate that a dose of $10^{6.8}$ ID_{50} but not $10^{5.3}$ is sufficient to cause disease in pigs infected via the mouth, nose, tonsil or eye. Infection through the skin is obtained by a dose of $10^{3.6}$ ID_{50} (Mann and Hutchings 1980).

Towards the end of the incubation period in pigs infected by contact, virus is found in the tonsil and lymph nodes of the head and neck. Inoculation of the foot leads to the presence of virus in popliteal lymph nodes. Thus, virus would appear to spread via the lymph channels to the local lymph nodes. From there it is carried to other parts of the body by the lymph and blood vessels, with preferential sites of multiplication in the epithelial tissues, where vesicles and high titres of virus are found. The virus also multiplies in the epithelium of the upper digestive tract (Mann and Hutchings 1980).

Virus is present in the tissues, secretions and excretions before and during the appearance of lesions. Antibodies develop 5–7 days after the appearance of lesions; at the same time, the amounts of virus isolated from the pig diminish and after 7–12 days no virus is cultured from the nose and pharynx. Virus may persist in the epithelial lesions, and in the faeces virus is present for up to one month or possibly longer (Burrows *et al.* 1974*a*).

Infection of man

During the handling of infected pigs, aerosols are generated which may be breathed in by the handlers. The amounts of virus excreted in this way are less than those due to foot-and-mouth disease (Sellers and Herniman 1974). In addition, virus from the infected pigs may penetrate the human skin, especially where there are cuts and abrasions.

No vesicles have been seen among workers with pigs affected by swine vesicular disease. In laboratories, inapparent infection has taken place and antibodies have been detected. In addition, there are four human cases in which presumptive evidence of infection with swine vesicular disease could be adduced. In one person there was a severe illness with fever, generalized muscle pains, weakness and stomach pains; diarrhoea also occurred. In two others the disease was mild. In the fourth, aseptic meningitis was seen and the patient was put into hospital. In three, infection occurred after exposure to pigs; in one, exposure to virus in the laboratory took place.

In laboratory diagnosis, infection due to swine vesicular disease virus has to be distinguished from that due to Coxsackie B5 virus. Neutralization, immunodiffusion, RIA, PAGE and T_1 mapping of oligonucleotides are available for the testing of paired sera but it is better to attempt isolation and identification of the virus. Brown *et al.* (1976) analysed the cases of infection that occurred in workers handling swine vesicular disease and Coxsackie B5 viruses and showed that four were due to swine vesicular disease infection and two to Coxsackie B5 infection.

No cases have been seen in persons exposed to natural disease in pigs or to infected pig products.

Epidemiology

Geographical distribution Swine vesicular disease virus was first isolated from affected pigs in Italy in 1966. The disease recurred in Italy in 1972 and has been found every year since then. In 1971 the virus was isolated from pigs with vesicular lesions in Hong Kong; subsequent isolations have been made in 1974, 1975, 1977 and 1980. In Europe the disease is or has been present in the United Kingdom (1972–77, 1979–81), France (1973–75), Austria (1972–76, 1978), West Germany (1973, 1975–78, 1980), Poland (1972–73), Holland (1975), Belgium (1978), Switzerland (1973), Malta (1975) and Greece (1979). In Asia, disease was reported from Japan in 1973 and 1975.

Methods of spread

Swine vesicular disease is spread through contact with infected pigs. This may occur on the farm, in markets or in vehicles. Virus excreted in pens at markets or in vehicles may infect pigs subsequently introduced to those pens or carried in those vehicles.

Pigs killed in the latter part of the incubation period or with lesions have large amounts of virus in the skin, the epithelial tissues and the muscle (pH of the mucsle after rigor mortis does not inactivate the virus). If discarded meat or offal containing virus is fed to pigs without heat treatment, the pigs may become infected, most probably through the skin. Contamination of fomites or of persons coming into contact with infected pigs may also lead to transfer of infection. Although virus is found in the air around infected pigs, spread by the aerosol route over long distances does not occur.

On farms, spread occurs among pigs within a pen but not necessarily between pens. Some pigs exposed to infection from diseased pigs do not show lesions but antibodies are found. Recovered pigs with high levels of antibody do not pass on infection to fresh pigs placed in contact with them.

Spread between farms occurs through the movement of infected pigs, through movements by way of markets and in contaminated vehicles, and by contact with contaminated persons.

Pigs which have been fed low doses of virus often have an inapparent infection without virus or antibody being detected. However, if such pigs are mixed with other pigs or undergo some form of stress such as fighting, the virus multiplies in the animal and antibodies are found (Mann *et al.* 1975). In some circumstances such pigs may pass on virus to others in contact.

Virus has been found to be excreted in the faeces for up to 5 weeks and 3 months in some instances. It was thought that virus persisted in the pigs by this means; however, as previously mentioned, recovered pigs do not pass the virus on to pigs placed in contact and the amounts of virus are not enough to maintain infection. Rather, it is believed that the virus is maintained by a continuous cycle of infection in pigs and by the survival of virus in meat and meat products and in the environment. Apparent persistence can be explained by the fact that disease is not detected until several cycles of transmission have occurred. This is probably the result of the difficulty of observing lesions, especially in sows, and the intensive methods of pig husbandry where pigs are not seen daily or are kept under conditions where observation is difficult.

Diagnosis As the clinical disease is similar to foot-and-mouth disease, the latter has to be excluded and this depends on laboratory tests. The tests used are complement fixation on antigen prepared from the epithelium of lesions. Where there is insufficient antigen in the epithelium, the suspension of epithelium is inoculated into pig tissue cultures and, if a cytopathic effect develops, the supernatant fluid is tested by complement fixation (see Table 89.3).

Serological tests are also carried out; the ones of choice are double immunodiffusion, ELISA and the neutralization test. For large numbers of samples and for rapid results, the first two tests are preferable. An immuno-osmoelectrophoresis test has been developed for serological surveys.

Control To avoid spread of disease, the main methods of control include the banning or limitation of the movement of pigs to other farms, markets or abattoirs.

Treatment of waste food by heat to inactivate the virus can be successfully applied at a temperature of 69°. A range of disinfectants (Herniman *et al.* 1973) is available for disinfection of contaminated vehicles, markets, infected farms and workers handling pigs.

Slaughter of affected pigs is carried out in the United Kingdom and in other European countries. Vaccines against swine vesicular disease have been developed. The virus was grown in pig tissues, inactivated with formalin, betapropiolactone or acetylethyleneimine, and administered with an adjuvant. Pigs were protected against generalization (Gourreau *et al.* 1975, Mowat *et al.* 1974). Attempts to develop attenuated vaccines have not proved very successful (Preston and Garland 1979).

Vesicular exanthema

Vesicular exanthema is an acute contagious disease of pigs characterized by the appearance of vesicles mainly on the tongue, in the mouth and on the feet. It is caused by a virus which belongs to the family *Caliciviridae*. The disease was first seen in 1932, persisted for a number of years in the United States of America, mainly in California, and was eradicated in 1956 (Madin and Traum 1955, Bankowski 1965). However, in 1972 a virus closely related to vesicular exanthema virus, San Miguel sea-lion virus, was isolated from sea-lions off the southern coast of California (Smith *et al.* 1973, Smith and Akers 1976).

Physico-chemical characters of the virus

Vesicular exanthema virus and San Miguel sea-lion virus are caliciviruses and belong to the family *Caliciviridae*, which also includes feline caliciviruses (Studdert 1978, Schaffer 1979, Schaffer *et al.* 1980).

The diameter of the virus is 35–40 nm and the surface of the virus shows the characteristic hollows. The buoyant density is 1.36–1.39 mg/ml in caesium chloride and the sedimentation coefficient is 160–170S. The virus is stable at pH 5 and resistant to lipid solvents (see Table 89.1).

The virus contains RNA and protein. The RNA is linear and forms about 18 per cent of the virion; infectious RNA is obtained by phenol treatment. At the 5′ end of the RNA there is a polypeptide mol.wt 10 000—covalently linked VPg. There is a major polypeptide, mol.wt 60 000—80 000, and a minor polypeptide, mol wt 15 000. There are 180 major protein subunits in the virion (Wawrzkiewicz *et al.* 1968, Newman *et al.* 1973, Bachrach and Hess 1973, Burroughs and Brown 1974, Black *et al.* 1978, Burroughs *et al.* 1978*b*).

Multiplication of the virus takes place in the cytoplasm of the cell. After adsorption and penetration, a number of RNAs can be found—a major SRNA (3637) and a minor single-stranded RNA (22S), a major and a minor double-stranded RNA, and an 18S RNA. Polypeptides found during replication include p115 and p86. The p86 may be the precursor of the capsid polypeptide p60 (Schaffer 1979).

Biological characters of the virus

There are no laboratory hosts in which vesicular exanthema virus grows regularly. Some strains produce vesicles on the abdomen of hamsters. Limited multiplication of virus occurs in dogs and horses.

Tissue cultures of pig origin support the growth of virus, with cytopathic effect, and plaques are produced. Cytopathic effect is also produced in Vero cells; and one strain gave rise to a cytopathic effect in dog kidney cultures (Madin 1975).

Thirteen types of vesicular exanthema virus have been recognized on the basis of complement-fixation and cross-neutralization tests (Bankowski 1965). There may have been more since a number of strains isolated before 1948 are not available. Complement-fixation tests show cross-reaction between the types and with San Miguel sea-lion virus (see Table 89.2).

The San Miguel sea-lion virus grows in pig tissue cultures, in Vero cells and in cell lines from ruminants. It also infects monkeys.

At least 14 serotypes have been distinguished by complement-fixation and neutralization tests and there are relations with vesicular exanthema virus in immunoprecipitin tests (Burroughs *et al.* 1978*a*).

Clinical signs

Pigs may be infected by contact or by feeding of infected waste food, the virus gaining entry by the respiratory or alimentary tract or by the skin (Madin and Traum 1955).

After an incubation period of two or more days, the pig develops a fever and vesicles appear on the snout, tongue, lips and mouth, on the feet, the coronary band, the interdigital space, the sole and occasionally on the skin of the joints. Vesicles may also be found on the teats of sows. The vesicles rupture, with discharge of fluid. The extent of the vesicles varies; if the underlying tissues are affected, scars are formed on healing.

During the formation of vesicles the animal is listless and disinclined to eat, and lameness occurs when the feet are affected. Diarrhoea sometimes occurs. Pregnant sows may abort and in lactating sows the milk yield drops.

Recovery usually occurs but, if the feet have been badly affected, shedding of the hoof lasting 1–3 months may take place.

The main pathological signs are in the lesions. Changes take place in the Malpighian layer of the epithelium, where ballooning of epithelial cells is found. This leads to degeneration of the cell, with the development of pycnotic and karyorhexic nuclei. Strains of vesicular exanthema vary in virulence but there appears to be no correlation with the serological type responsible. Strains virulent for pigs have been shown to produce large plaques and avirulent strains small plaques on pig kidney tissue cultures (McClain *et al.* 1958).

San Miguel sea-lion virus produces vesicles on the flippers of fur seals and pups. It has also been isolated during a period of abortion in sea-lions. When inoculated into pigs, the virus gives rise to vesicular lesions on the feet and snout (Smith *et al.* 1973, Akers *et al.* 1974).

Infection of man Laboratory infection of man with San Miguel sea-lion virus was reported but no disease was seen (Smith *et al.* 1978, Soergel *et al.* 1978).

Pathogenesis

Most work has been done on animals infected experimentally on the snout or exposed to animals previously infected (Madin 1975).

In pigs experimentally infected, the virus multiplies locally and vesicles are produced within 24–48 hours. Virus is found in the blood 24 hours after infection and thereafter until about 36 hours after the last vesicle has appeared. Vesicles appear at secondary sites 48 hours after the primary lesion, and virus can be found in the epithelium of the feet and skin and in the spleen, lymph nodes, bone marrow, muscle and viscera before the development of lesions and for up to 7 days after their development.

Pigs are infectious for other pigs for 12–120 hours after experimental inoculation. Faeces and urine contain little virus and animals in contact with faeces may become infected and develop immunity without development of lesions.

The dose for infecting by feeding is about 10^4–10^5 TCD_{50} of vesicular exanthema virus.

Epidemiology

Geographical distribution Vesicular exanthema was first noted in California in 1932 among pigs fed garbage. Outbreaks were seen in the following three years until the middle of 1936, but never more than 31 in any one year, and were confined to farms where garbage was fed. Disease was seen again in 1939 and was present in various parts of California until 1955. The number of outbreaks in each year varied and affected over 40 per cent of the pigs in the State. At times when the number of outbreaks was high, vesicular exanthema was found on farms that fed grain and also in slaughterhouses. The disease was also found in pigs being exported to Honolulu in 1946 and 1947. In 1952 vesicular exanthema was recognized in Nebraska, Wyoming and Omaha and subsequently in 42 States of the USA. In 1955 it occurred in Iceland. The last cases in California occurred in 1955, and the last in the USA occurred in New Jersey in 1956 (Bankowski 1965, Madin 1975).

San Miguel sea-lion virus appears to maintain itself among the sea-lions and seals on the Western Pacific coast of the USA and, as these animals migrate as far north as Alaska and the Bering Sea, virus may be found there.

Method of spread

Vesicular exanthema virus is spread by the feeding of uncooked meat or offal containing virus, by direct or indirect contact between swine and by carriage in contaminated vehicles.

The number of infected pigs on a farm varied from 10 to 100 per cent, depending on the strain of virus concerned.

Pigs recovered from one of the 13 types of virus were susceptible to infection by another type.

Spread of San Miguel sea-lion virus occurs among animals by contact, probably on land where the animals are resting.

It is possible that vesicular exanthema originally came from contamination of garbage with virus from sea-lions and seals and that this was transmitted through the pig population (Prato *et al.* 1974).

Serological surveys carried out indicate that whales, foxes, wild donkeys and sheep have antibodies to San Miguel sea-lion virus. The virus has also been isolated from liver fluke of a sea-lion and from fish (Smith *et al.* 1981). It has been suggested that both vesicular exanthema and San Miguel sea-lion virus are caliciviruses of fish which have spread through the food chain or parasites to sea-lions, seals and pigs.

Diagnosis Diagnosis of vesicular exanthema in swine is based on clinical signs. These, however, can be confused with foot-and-mouth disease, swine vesicular disease and vesicular stomatitis. Some differentiation can be made by complement-fixation tests but in new cases virus isolation can be attempted in tissue cultures of pig origin, where a cytopathic effect is found. The virus-infected tissue culture harvests can be examined under the electron microscope for the characteristic cup-shaped hollows. The supernatant can be tested in complement-fixation and neutralization tests against the various types of vesicular exanthema and San Miguel sea-lion virus (see Table 89.3). Acute and convalescent serum samples can be examined for a rise in antibody.

Control The main methods of control include a ban on the movement of pigs and the prior treatment of garbage by heat before feeding to pigs. Vehicles carrying pigs should be disinfected.

Affected pigs can be slaughtered and farms disinfected before re-stocking with fresh pigs.

No vaccines have been developed and, if they were, the plurality of types would make adequate protection difficult.

References

Akers, T. G., Smith, A. W., Latham, A. B. and Watkins, H. M. S. (1974) *Arch. ges. Virusforsch.* **46,** 175.

Anderson, E. C., Doughty, W. J. and Anderson, J. (1974) *J. Hyg. Camb.* **73,** 229.

Armstrong, R., Davie, J. and Hedger, R. S. (1967) *Brit. med. J.* **iv,** 529.

Bachrach, H. L. (1968) *Annu. Rev. Microbiol.* **22,** 201.
Bachrach, H. L. and Hess, W. R. (1973) *Biochem. biophys. Res. Commun.* **55,** 141.
Bachrach, H. L., Moore, D. M., McKercher, P. D. and Polatnick, J. (1975) *J. Immunol.* **115,** 1635.
Bahnemann, H. G. (1975) *Arch. Virol.* **47,** 47.
Bankowski, R. A. (1965) *Advance vet. Sci.* **10,** 23.
Barlow, D. F. (1972) *J. gen. Virol.* **15,** 17.
Bittle, J. L. *et al.* (1982) *Nature* **298,** 30.
Black, D. N., Burroughs, J. N., Harris, T. J. R. and Brown, F. (1978) *Intervirology* **10,** 51.
Bojlen, K. (1941) *Ugeskr. Laegerb* **103,** 497.
Boldrini, G. M. (1978) *Vet. Rec.* **102,** 194.
Boothroyd, J. C., Highfield, P. E., Cross, G. A. M., Rowlands, D. J., Lowe, P. A., Brown, F. and Harris, T. J. R. (1981) *Nature, Lond.* **290,** 800.
Brody, J. A., Fischer, G. F. and Peralta, P. H. (1967) *Amer. J. Epidem.* **86,** 158.
Brooksby, J. B. (1958) *Advanc. Virus Res.* **5,** 1.
Brown, F. (1980) *Ann. N.Y. Acad. Sci.* **354,** 202; (1981) *Ann. Neurol.* **9,** *Suppl.* 39.
Brown, F., Goodridge, D. and Burrows, R. (1976) *J. comp. Path.* **86,** 409.
Brown, F., Hyslop, N. St G., Crick, J. and Morrow, A. W. (1963) *J. Hyg., Camb.* **61,** 337.
Brown, F. *et al.* (1979) *Intervirology* **12,** 1.
Burroughs, J. N. and Brown, F. (1974) *J. gen. Virol.* **22,** 281.
Burroughs, J. N., Doel, T. and Brown, F. (1978*a*) *Intervirology* **10,** 51.
Burroughs, J. N., Doel, T. R., Smale, C. J. and Brown, F. (1978*b*) *J. gen. Virol.* **40,** 161.
Burrows, R. (1966) *J. Hyg. Camb.* **64,** 81, (1968) *J. Hyg., Camb.* **66,** 633, (1972) *Symp. Soc. gen. Microbiol.* **22,** 303.
Burrows, R., Mann, J. A. and Goodridge, D. (1974*a*) *J. Hyg. Camb.* **72,** 135; (1974*b*) *J. Hyg. Camb.* **73,** 101.
Capel-Edwards, M. (1971) *Vet. Bull.* **41,** 815.
Capstick, P. B., Telling, R. C., Chapman, W. G. and Stewart, D. L. (1962) *Nature, Lond.* **195,** 1163.
Cotton, W. E. (1926) *J. Amer. vet. med. Ass.* **23,** 168.
Cottral, G. E., Gailiunas, P. and Cox, B. F. (1968) *Arch. ges. Virusforsch.* **23,** 362.
Cowan, K. M. and Graves, J. H. (1966) *Virology* **30,** 528.
Crowther, J. (1978) *Bull. Off. int. Epiz.* **89,** 831.
Delagneau, J. F., Bernard, S. and Lenoir, G. (1975) *Biochem. biophys. Res. Commun.* **66,** 226.
Donaldson, A. I. (1970) *Amer. J. Epidem.* **92,** 132; (1972) *J. gen. Virol.* **15,** 25.
Donaldson, A. I., Gloster, J., Harvey, L. D. J. and Deans, D. H. (1982) *Vet. Rec.* **110,** 53.
Eissner, G., Boehm, H. O. and Huelich, E. (1967) *Dtsch. med. Wschr.* **92,** 830.
Federer, K. E., Burrows, R. and Brooksby, J. B. (1967) *Res. vet. Sci.* **8,** 103.
Fellowes, O. N., Dimopoullos, G. T. and Callis, J. J. (1955) *Amer. J. vet. Res.* **16,** 623.
Ferris, D. F., Hanson, R. P., Dicke, R. J. and Roberts, R. H. (1955) *J. infect. Dis.* **96,** 184.
Fields, B. N. and Hawkins, K. (1967) *New Engl. J. Med.* **277,** 989.
Frenkel, H. S. (1950) *Amer. J. vet. Res.* **11,** 371.
Garbe, H. G., Hussong, H. J. and Pilz, W. (1959) *VetMed. Nachr.* **3,** 136.
Garland, A. J. M. (1974) *Ph.D. thesis.* University of London.
Gibbs, E. P. J., Herniman, K. A. J., Lawman, M. J. P. and Sellers, R. F. (1975) *Vet. Rec.* **96,** 558.
Gloster, J., Blackall, R. M., Sellers, R. F. and Donaldson, A. I. (1981) *Vet. Rec.* **108,** 370.
Gloster, J., Sellers, R. F. and Donaldson, A. I. (1982) *Vet. Rec.* **110,** 47.
Gourreau, J. M., Dhennin, L. and Labie, J. (1975) *Recueil Méd. vét.* **151,** 85.
Graves, J. H. (1973) *Nature, Lond.* **245,** 314.
Hanson, R. P. (1952) *Bact. Rev.* **16,** 179; (1968) *Amer. J. Epidem.* **87,** 264.
Hanson, R. P., Rasmussen, A. F., Brandly, C. A. and Brown, J. W. (1950) *J. Lab. clin. Med.* **36,** 754.
Harris, T. J., Underwood, B. O., Knowles, N. J., Crowther, J. R. and Brown, F. (1979) *Infect. Immun.* **24,** 593.
Harris, T. J. R. and Brown, F. (1977) *J. gen. Virol.* **34,** 75.
Hedger, R. S. (1976) In: *Wildlife Diseases*, p. 235. Plenum Publishing Company, New York.
Hedger, R. S. and Dawson, P. S. (1970) *Vet. Rec.* **87,** 186.
Heinig, A. and Neumerkel, H. (1964) *Dtsch. Gesundh. Wschr.* **19,** 485.
Henderson, W. M. and Brooksby, J. B. (1948) *J. Hyg. Camb.* **46,** 394.
Herniman, K. A. J., Medhurst, P. M., Wilson, J. N. and Sellers, R. F. (1973) *Vet. Rec.* **93,** 620.
Howatson, A. F. (1970) *Advanc. Virus Res.* **16,** 195.
Huang, A. S. (1977) *Bact. Rev.* **41,** 811.
Hurst, G. W. (1968) *Vet. Rec.* **82,** 610.
Hyslop, N. St G. (1970) *Advanc. vet. Sci. comp. Med.* **14,** 261; (1973) *Bull. World Hlth Org.* **49,** 577.
Hyslop, N. St G. and Fagg, R. H. (1965) *J. Hyg. Camb.* **63,** 357.
Jenney, E. W. (1967) *Proc. U.S. Livestk San. Assoc.* **71,** 371.
Johnson, K. M., Tesh, R. B. and Peralta, P. H. (1969) *J. Amer. vet. med. Ass.* **155,** 2133.
Johnson, K. M., Vogel, J. E. and Peralta, P. H. (1966) *Amer. J. trop. med. Hyg.* **15,** 244.
Jonkers, A. H. (1967) *Amer. J. Epidem.* **86,** 286.
King, A. M. Q., Underwood, B. O., McCahon, D., Newman, J. W. I. and Brown, F. (1981) *Nature, Lond.* **293,** 479.
Kleid, D. G. *et al.* (1981) *Science,* **214,** 1125.
Knowles, N. J. and Buckley, L. S. (1980) *Res. vet. Sci.* **29,** 113.
Korn, G. (1957) *Arch. exp. vet. Med.* **11,** 637.
Küpper, H. *et al.* (1981) *Nature, Lond.* **289,** 555.
Laporte, J., Grosclaude, J., Wantyghen, J., Bernard, S. and Rouze, P. (1973) *C. R. Acad. Sci., Paris, Serie D* **276,** 3399.
Lauerman, L. H., Kuns, M. L. and Hanson, R. P. (1963) *Proc. U.S. Livestk Sanit, Assoc.* **67,** 483.
Lenghaus, C. and Mann, J. A. (1976) *Vet. Path.* **13,** 186.
Lenghaus, C., Mann, J. A., Done, J. T. and Bradley, R. (1976) *Res. vet. Sci.* **21,** 19.
Loeffler, F. and Frosch, P. (1898) *Zbl. Bakt. Parasit. Abt. I. Orig.* **22,** 257.
Lowe, P. A. and Brown, F. (1981) *Virology* **111,** 23.
McCahon, D. (1981) *Arch. Virol.* **69,** 1.
McClain, M. E., Hackett, A. J. and Madin, S. H. (1958) *Science* **172,** 1391.
McVicar, J. W. (1977) *Bol. Cent. Panamer. Fieb. Aft.* **26,** 9.
Madin, S. H. (1975) In: *Diseases of Swine.* Ed. by Dunne and Lemon, Iowa, USA.
Madin, S. H. and Traum, J. (1955) *Bact. Rev.* **19,** 6.
Mann, J. A., Burrows, R. and Goodridge, D. (1975) *Bull. Off. int. Epiz.* **83,** 117.

Mann, J. A. and Hutchings, G. H. (1980) *J. Hyg. Camb.* **84,** 355.
Mason, J. (1978) *Bol. Cent. Panamer, Fieb. Aft.* **29-30,** 35.
Melendez, L. (1961) *Bol. Ofic. Sanit. Panam.* **50,** 135.
Michelsen, E. and Suhr-Rasmussen, E. (1959) *Ugeskr. Laegev,* **121,** 98.
Mowat, G. N. and Chapman, W. G. (1962) *Nature, Lond.* **194,** 253.
Mowat, G. N., Garland, A. J. M. and Spier, R. E. (1978) *Vet. Rec.* **102,** 190.
Mowat, G. N., Prince, M. J., Spier, R. E. and Staple, R. F. (1974) *Arch. ges. Virusforsch.* **44,** 350.
Murton, R. K. (1964) *Ibis* **106,** 289.
Nardelli, L. *et al.* (1968) *Nature, Lond.* **219,** 1275.
Newman, J. F. E., Cartwright, B., Doel, T. R. and Brown, F. (1979) *J. gen. Virol.* **45,** 497.
Newman, J. F. E., Rowlands, D. J. and Brown, F. (1973) *J. gen. Virol.* **18,** 171.
Olitsky, P. K., Schoening, H. W. and Traum, J. (1927) *N. Amer. Vet.* **8,** 42.
Patterson, W. C., Nott, L. O. and Jenney, E. W. (1958) *J. Amer. vet. med. Ass.* **133,** 57.
Pereira, H. S. (1977) *Develop. biol. Standard.* **35,** 167.
Pilz, W. and Garbe, H. G. (1966) *Zbl. Bakt. Orig.* **198,** 154.
Pilz, W., Garbe, H. G. and Beck, W. (1962) *Vet. Neder. Nachr.* **4,** 224.
Platt, H. (1956) *J. Path. Bact.* **72,** 299; (1958) *Med. Press* **240,** 1195.
Polatnick, J., Arlinghaus, R. B., Graves, J. H. and Cowan, K. M. (1967) *Virology* **31,** 609.
Prato, C. M., Akers, T. G. and Smith, A. W. (1974) *Nature, Lond.* **244,** 108.
Preston, K. J. and Garland, A. J. M. (1979) *J. Hyg. Camb.* **83,** 319.
Rowlands, D. J. (1979) *Proc. Nat. Acad. Sci. Washington* **76,** 4793.
Sangar, D. V. (1979) *J. gen. Virol.* **45,** 1.
Schaffer, F. L. (1979) *Comprehensive Virology,* Vol. 14, p. 249. Plenum Press, New York.
Schaffer, F. L. *et al.* (1980) *Intervirology* **14,** 1.
Sellers, R. F. (1968) *Vet. Rec.* **83,** 504; (1971) *Vet. Bull.* **41,** 431; (1980) *J. Hyg. Camb.* **85,** 65.
Sellers, R. F., Barlow, D. F., Donaldson, A. I., Herniman, K. A. J. and Parker, J. (1973) *Fourth Int. Symp. Aerobiol.* p. 405. Osthoek Publishing Company, Utrecht, Netherlands.
Sellers, R. F., Burrows, R., Mann, J. A. and Dawe, P. S. (1968) *Vet. Rec.* **83,** 303.
Sellers, R. F., Donaldson, A. I. and Herniman, K. A. J. (1970) *J. Hyg. Camb.* **68,** 565.
Sellers, R.F. and Gloster, J. (1980). *J Hyg. Camb.* **85,** 129.
Sellers, R. F. and Herniman, K. A. J. (1974) *J. Hyg. Camb.* **72,** 61.
Sellers, R. F. and Parker, J. (1969) *J. Hyg. Camb.* **67,** 671.
Shelokov, A. I., Peralta, P. H. and Galindo, P. (1961) *J. clin. Invest.* **40,** 1081.
Skinner, H. H. (1951) *Proc. R. Soc. Med.* **44,** 1041; (1959) *Arch. ges. Virusforsch.* **9,** 92.
Smith, A. W. and Akers, T. G. (1976) *J. Amer. vet. med. Ass.* **169,** 700.
Smith, A. W., Akers, T. G., Madin, S. H. and Vedson, N. A. (1973) *Nature, Lond.* **244,** 108.
Smith, A. W., Prato, L. and Skilling, D. E. (1978) *Amer. J. vet. Res.* **41,** 1846.
Smith, A. W., Skilling, D. E. and Latham, A. B. (1981) *Amer. J. vet. Res.* **42,** 693.
Smith, L. P. and Hugh-Jones, M. E. (1969) *Nature, Lond.* **223,** 712.
Snowdon, W. A. (1966) *Nature, Lond.* **210,** 1079.
Soergel, M. E., Schaffer, F. L., Sawyer, J. C. and Prato, C. M. (1978) *Arch. Virol.* **57,** 271.
Strohmaier, K., Wittmann-Liebold, B. and Geissler, A-W. (1978) *Biochem. biophys. Res. Commun.* **85,** 1640.
Studdert, M. J. (1978) *Arch. Virol.* **58,** 157.
Suhr-Rasmussen, E. (1969) *Medlemsbl. danske. Dyrlaegeforen* **52,** 498.
Sutmöller, P. and Gaggero, A. (1965) *Vet. Rec.* **77,** 968.
Sutmöller, P., McVicar, J. W. and Cottral, G. E. (1968) *Arch. ges. Virusforsch.* **23,** 327.
Tesh, R. B., Chaniotis, B. N. and Johnson, K. M. (1971) *Science* **175,** 1477.
Tesh, R. B., Peralta, P. H. and Johnson, K. M. (1969) *Amer. J. Epidem.* **90,** 255; (1970) *Amer. J. Epidem.* **91,** 216.
Ubertini, B., Nardelli, L., Santero, G. and Panina, G. (1960) *J. Biochem. microbiol. Technol. Eng.* **2,** 327.
Van Bekkum, J. G., Frenkel, H. S., Frederiks, H. H. J. and Frenkel, S. (1959) *Tijdschr. Diergeneesk.* **84,** 1159.
Vetterlein, W. (1954*a*) *Arch. exp. VetMed.* **8,** 541; (1954*b*) *Zbl. Bakt. Orig.* **162,** 8.
Wagner, R. R. (1975) In: *Comprehensive Virology,* Vol. 4, pp. 1–93. Plenum Press, New York.
Wawrzkiewicz, J., Smale, C. J. and Brown, F. (1968) *Arch. ges. Virusforsch.* **25,** 337.
Wesslen, T. and Dinter, Z. (1957) *Arch. ges. Virusforsch.* **7,** 394.
Wilson, W. W. and Matheson, R. C. (1952) *Agriculture* **59,** 213.
Wisniewski, J. and Jankowska, J. (1968) *Bull. vet. Inst. Pulaway* **12,** 1.
Zoletto, R., Kadoi, K., Carlotto, F., Turilli, C. and Corazzola, S. (1974) *Third Internat. Pig Vet. Congr. Lyons.*

90

Togaviridae

D. I. H. Simpson

Introductory

The viruses described in this and the next chapter belong to what are colloquially known as the Arboviruses. The following account of their properties applies to the viruses in both these chapters.

An arthropod-borne or arbovirus has been defined as a virus that is maintained in nature principally through biological transmission between susceptible vertebrate hosts by haematophagous arthropods; arboviruses multiply and produce viraemia in the vertebrate, multiply in the tissues of arthropods and are passed on to new vertebrates by the bites of arthropods after a period of extrinsic incubation (World Health Organization 1967). Most arboviruses fulfil the criteria laid down in this definition, but the group is very heterogeneous, containing viruses which, because they have not been fully classified on morphological or physicochemical grounds, are included among the arboviruses for convenience. There are currently known

Table 90.1 Relationship between antigenic groups and the taxonomic status of arboviruses

Family	Viruses
Togaviridae	
Alphaviruses	25
Flaviviruses	63
Ungrouped	2
Bunyaviridae	
Bunyavirus	121
Nairovirus	22
Phlebovirus	31
Uukunvirus	5
'Bunyavirus-like'	30
Reoviridae	
Orbivirus	49
Rhabdoviridae	
Vesiculovirus	11
Unassigned or possible members	27
Coronaviridae	2
Herpesviridae	1
Iridoviridae	1
Nodaviridae	1
Orthomyxoviridae	2
Paramyxoviridae	1
Poxviridae	3
Unclassified	50

to be 474 arboviruses of which just over 80 are capable of infecting man. This very large group contains representatives from several different viral families (Table 90.1 and Calisher *et al.* 1980), the most important of which are the families *Togaviridae*, *Bunyaviridae*, *Reoviridae* and *Rhabdoviridae* (see Porterfield 1980*b*).

By definition arboviruses have at least two different hosts, one a vertebrate species and the other an invertebrate arthropod species, but many arboviruses have complex life cycles involving several different vertebrates and some are capable of transmission by more than one vector. All arboviruses, with perhaps a very few exceptions such as dengue and O'nyong nyong, are actual or potential zoonoses, being maintained in nature by hosts other than man. Together with the hosts that maintain them, they have evolved to a state of mutual tolerance or symbiosis. As arboviruses rely only on the production of a viraemia in the vertebrate host, usually animals or birds, for successful transmission, the causation of disease in the host would be of no advantage. For this reason, arboviruses seldom cause recognizable disease in their maintenance hosts, including man, and such disease in man or his domestic animals is often the only overt sign of the presence of these viruses. The diseases caused by arboviruses range from mild febrile illnesses, which may or may not be accompanied by a skin rash and sometimes by polyarthritis, to severe and often fatal encephalitis or haemorrhagic fever. The same virus may produce different disease patterns in different subjects. The illnesses often follow a biphasic course. The mild fever, which is often unrecognized, occurs during the initial viraemic stage. This may be followed by a much more serious type of illness, at which stage viraemia may have ceased and immunological responses, including antibody formation, have taken place. Frequently, only a small proportion of persons infected with potentially encephalitogenic arboviruses in epidemics develop encephalitis in the second phase. The great majority develop only the first phase, or the infection may even be asymptomatic.

Mosquitoes are the most important arbovirus vectors, followed by ticks, with Phlebotomines, Culicoides and Cimicidae being involved in the transmission of some arboviruses. In certain instances the same virus may be transmitted by more than one type of vector. Although the great majority of arbovirus infections of man follow exposure to the bite of an arthropod, exceptionally infection may follow the ingestion of infected cows' or goats' milk, by inhalation of infected secretions or laboratory-produced aerosols, or by close contact with infected animals and, on rare occasions, through direct contact with infected patients' blood.

Laboratory diagnosis of arbovirus infections

Specific diagnosis of arbovirus infections is expensive and requires well developed virus laboratory facilities. Except for epidemiological research and for the investigation of major epidemics, facilities for routine diagnosis cannot at present be justified in many developing countries, especially as the diagnosis often cannot be established before the patient is dead or has recovered, and there is no specific treatment for these infections. In severe infections, particularly those complicated by bleeding, there may be a considerable risk to medical and nursing staff and to laboratory workers. Well designed and meticulously applied safety procedures are therefore essential in handling such infections. Diagnosis depends either on isolation and identification of the causative virus or on the demonstration of a specific antibody response coincident with the course of the illness.

Virus isolation and identification

Specimens

These are usually either blood taken during the first febrile phase or post-mortem tissue. Bone marrow has been successfully used in Venezuelan encephalitis. Cerebrospinal fluid has often been tried but seldom with success except in Kyasanur Forest disease, where there is an unusually intense viraemia and contaminant blood is the probable source of virus in the CSF. Because much of the viraemia may be over before the patient is seen, blood for virus isolation must be taken as early as possible in the course of the illness as viraemia generally lasts for only 2 to 3 days after the onset of the illness. Isolation from the blood is very rare in encephalitis patients because the viraemia, which occurs during the often unrecognized first phase, is over by the time of onset of CNS symptoms. By special arrangement with the laboratory a finger-prick sample can be used but a specimen of venous blood is preferable, especially as it gives an opportunity for the collection of a first serum sample, which is essential whether the diagnosis is to be established by virus isolation or serology. At least 10 ml of blood should therefore be withdrawn with sterile precautions, a small sample removed and either sent immediately to the virus laboratory or stored in a sealed container at low temperature on dry ice, in a -65° refrigerator, or in liquid nitrogen. Serum should be separated from the remainder of the blood and suitably stored until a further specimen is obtained 2–3 weeks later. There may be considerable risks in taking specimens in serious infections. Staff taking blood and post-mortem specimens should always wear gloves and gowns to avoid skin contact.

Post-mortem material should be obtained as soon as possible after death. In general, small representative pieces of as many tissues as possible should be placed in screw-cap containers and either sent immediately to the virus laboratory or stored as above. For short periods of travel the virus may survive in tissue if it is sent in pH 7.4 buffered 50 per cent glycerol saline in a screw-cap container packed in ice in a vacuum flask. The success rate in virus isolation from tissue depends very much on the duration of the illness. Buescher (1963), in a study of 27 fatal cases of Japanese encephalitis, isolated virus from the brain of seven of eight patients dying after 1–2 days' illness, from four of seven patients dying after 3–5 days' illness, but from only one of ten dying after more than 5 days of illness. This progressive fall is mainly because of the increase of circulating antibody which neutralizes the virus when the tissue is ground up.

As with many other infectious diseases, post-mortem examinations must be made with great care, the operator wearing adequate protective clothing, gloves and preferably face-shields, otherwise dangerous infections may be transmitted to the operator by contact, cuts on sharp bone edges, aerosols or splashing. Great care must also be taken in cleaning down the table, instruments, etc. with a suitable disinfectant (preferably 5 per cent hypochlorite) after the autopsy.

Isolation

The blood, usually defibrinated, or serum, or suspensions of the tissues in a buffered protein-containing saline diluent, usually with added antibiotics, are inoculated into baby mice, other laboratory animals, or into suitable cell cultures. Ninety-four per cent of the arboviruses were first isolated in baby mice, aged 1–4 days, which have been most commonly used. On primary isolation the incubation period in these animals is variable, depending both on the dose of virus they receive and on the type of virus. Usually, however, some mice become sick or die within 14 days and often much sooner. The tissues of sick or dead mice are usually passaged in further groups of baby mice in which a shorter and more uniform incubation period is likely; antigens made from their tissues may enable a preliminary serological identification of the infecting virus to be made. Sometimes, however, several passages are required for sufficient adaptation.

Cell cultures are being increasingly used but there is still no cell type as susceptible as baby mice to such a very wide range of arboviruses on primary isolation. In local situations, however, a cell culture may be suitable for isolation of most of the likely viruses: for example, chick embryo cell cultures are generally suitable for alphaviruses and a wide variety of monkey kidney cell lines are being increasingly used. One such line, Vero, is being used routinely in many laboratories and interesting developments with mosquito cell lines are in progress (Varma *et al.* 1974). However, a range of cell cultures may be necessary.

Identification

Once a sufficient titre of virus has been attained either in an experimental animal or in a cell culture, an antigen can be prepared and tested in one or more of the serological tests (below) against antisera to known viruses. If the laboratory possesses the appropriate antisera, an identification can sometimes be made within 2–3 weeks of the collection of the specimen. When, however, the virus is an unfamiliar one, the identification will take much longer. In any case in certain virus groups, e.g. flaviviruses, where a number of viruses are closely related antigenically, precise identification may be slow although a preliminary rough identification may be possible earlier. The virus isolated must also be confirmed as the cause of the disease. This can be done in two ways, the first of which is applicable particularly to post-mortem material; the second is much more rigorous:

1 Re-isolation of the same virus from the same material.
2 Testing of paired sera from the patient against the virus isolated so as to demonstrate a rise in antibody during the course of the illness.

Confirmation is essential lest the virus isolated is a contaminant from the laboratory or an adventitious infection of the patient.

Serology

Three main serological tests are used in diagnosis: the complement-fixation (CF), the haemagglutinin-inhibition (HI) and the neutralization tests. Fluorescent antibody methods are being used widely; Emmons and Lennette (1966) could diagnose Colorado tick fever within hours by testing for antigen in acute-phase blood clot. Other tests such as the agar gel precipitation (Clarke 1964) and immunoelectrophoresis tests (Cuadrado and Casals 1967) are used for special purposes and may come into more frequent use for rapid diagnosis. Standardization of all these tests is important in order to obtain comparable results, but in any case paired sera from a patient should always be titrated together in the same test. Whenever possible, a strain of virus locally isolated should be used in serological tests, as minor antigenic differences may affect the success rate in diagnosis.

The complement-fixation test This can be carried out by a variety of techniques, the only special feature being the production of suitable antigens from the tissues of infected animals or cell cultures. The CF antibody response is usually later than that of HI or neutralizing antibody. In a study of 84 proved cases of Japanese encephalitis, Buescher (1963) found that by two weeks after the onset of disease, CF antibody appeared in fewer than half the patients but HI antibody in over 90 per cent of them. The CF test can therefore sometimes demonstrate a significant increase in antibody—at least fourfold—when the initial serum sample has been taken so late that it already contains haemagglutinin-inhibition (HI) and neutralizing antibody. The general problem of specificity is discussed below.

The haemagglutinin-inhibition test (Clarke and Casals 1958). This depends on the ability to prepare a haemagglutinin (HA) for the appropriate virus from the tissues of infected animals or from infected cell cultures. This has been successful with many but by no means all arboviruses. Some viruses have proved very difficult, even with elaborate procedures, e.g. Ardoin and Clarke (1967). The antigens are generally stable at high pH (9.0), but each virus agglutinates goose erythrocytes at a specific acid pH. Sera are treated with kaolin or are acetone-extracted to remove non-specific HA inhibitors; they are heat-inactivated (56°) to prevent erythrocyte lysis, and the goose-cell agglutinins are absorbed out. The sera are then diluted twofold in pH 9.0 buffered saline and a measured amount of HA, usually 8 units, added in a similar buffer. After incubation at pH 9.0, a goose erythrocyte suspension is added in an adjusting buffer which brings the mixture to the optimum pH for haemagglutination by the particular virus. Haemagglutinin-inhibition tests can be carried out either by testing serum dilutions, usually twofold starting at 1:10, against 8 units of antigen, as above, or alternatively by testing serum, usually at 1:10, against a range of dilutions, usually twofold, of antigen. O'Reilly *et al.* (1968) compared the two methods in sheep after infection and reinfection with louping ill virus. Both were suitable but the second procedure was the more sensitive.

Neutralization tests There are two stages in any neutralization test:

1 A measured dose of virus is incubated with a measured amount of serum or serum dilution in the presence of any necessary accessory factors. For maximum neutralization of at least some arboviruses, a heat-labile factor present in normal serum is required. Sera are therefore usually heat-inactivated (56°) before testing and a measured amount of normal serum is added to provide the same amount of accessory factor in all the reaction mixtures.

2 After incubation, the serum-virus mixtures are tested for the presence of virus which has not been neutralized, by inoculation into susceptible animals, usually mice, or into cell cultures.

(a) Dilutions of virus, usually tenfold (or $\sqrt{10}$-fold), are incubated with standard amounts of undiluted serum. The residual log virus titre can then be calculated and compared with a control. The difference is the log neutralizing index of the serum. A log difference between the neutralizing indices of two sera of 1.7 (fiftyfold) is regarded as significant, although in well controlled tests differences of 1.1 or 1.2 (thirteen- or sixteen-fold) may be significant.

(b) Dilutions, usually twofold, of serum are incubated with a standard amount of virus, usually 100 LD50 or TCD50, and scored as either neutralizing or failing to neutralize this amount of virus. Differences of fourfold or greater are usually significant.

Tests done in tissue culture tubes and judged by cytopathic effects are of about the same accuracy as those done in similar numbers of animals, although generally cheaper. However, there must be available a cell culture which gives a clear cut cytopathic effect when infected by the relevant virus, and there are therefore still some viruses which have to be tested in mice. Cell culture methods in which the remaining virus is enumerated by plaque counts are considerably more accurate and therefore enable smaller differences in antibody titres to be established as significant.

In all these tests the timing and spacing of the paired sera in relation to the course of the illness are critical for successful diagnosis. In diseases which are seen mainly in the second phase, notably encephalitis, the CF test is often more successful in diagnosis because CF antibody increases later than HI or neutralizing antibody. No reliable serological diagnosis can be made on the basis of a single serum specimen because, as will become apparent, subclinical infections with arboviruses are common. Thus only a significant increase in antibody coincident with the course of the illness can be a basis for serological diagnosis.

In areas where a number of antigenically related viruses co-exist, the problem of diagnosis is greatly aggravated; this is particularly true of flaviviruses of which several are found in most tropical areas. When a person is first infected with a flavivirus, his antibody response is reasonably specific for the infecting virus, although some antibody cross-reacting with other flaviviruses is likely to be formed. When, however, he is later infected with a second flavivirus he will probably have a very broadly cross-reacting antibody response in which it may be difficult or even impossible by normally available methods to make a precise diagnosis. Sometimes it may be possible to say only that the infecting virus was a flavivirus.

Theiler and Casals (1958) attempted serological diagnosis of patients with yellow fever in Trinidad from which yellow fever virus had actually been isolated and the diagnosis was therefore not in doubt. In those not previously infected with a flavivirus, the CF test gave a clear cut diagnosis, and the HI, though less specific, generally gave higher titres against yellow fever than related viruses, especially in earlier convalescent sera. In those with previous flavivirus infections, however, very broad cross-reactions were found by both tests and the only diagnosis generally possible was of a flavivirus infection. The neutralization test is usually most specific, but it is difficult to predict the relative specificity of the CF and HI tests, as this varies with the reagents used, the history of previous infections and the timing and spacing of the paired sera available for testing.

Methods used to study interrelationships among arboviruses

The taxonomic status of an arbovirus can best be determined by the use of biochemical tests and electron microscopy; these, however, are generally very sophisticated methods. Biochemical studies will determine the nature and strandedness of the viral nucleic acid, while the electron microscope establishes the size, shape, symmetry and presence or absence of an envelope. Much more reliance in the early stages of identification is placed on serological tests. Neutralization, complement-fixation and haemagglutination-inhibition tests have already been mentioned, and there are various refinements of these already in use. Additionally gel precipitation tests, radio-immunoassay and immunoelectro-osmophoresis have been applied successfully.

Tissue culture methods are increasingly replacing *in vivo* assays. Plaquing methods cannot be bettered in their precision and sensitivity, and microtechniques have been used successfully in several arbovirus antibody systems (Madrid and Porterfield 1969, 1974, Karabatsos 1975, Chanas *et al.* 1976).

Togaviridae

Togaviridae are small to medium sized, 25–50 nm diameter, spherical, enveloped viruses having a genome of linear single-stranded RNA with a molecular weight of $2\text{–}4 \times 10^6$ Daltons. The name is derived from the Latin 'toga', a cloak or mantle, indicating the lipid viral envelope. The family contains four genera and some additional members; the genera are: Alphavirus, formerly known as the Group A arboviruses, the type species being Sindbis virus; Flavivirus, named after yellow fever virus, the type species, and replacing the earlier name of Group B arboviruses; Rubivirus, containing a single member, rubella virus, and the fourth genus, Pestivirus, which includes three viruses producing important diseases in domestic animals—Hog cholera (swine fever), Mucosal disease of cattle and Border disease of sheep. Rubella and the Pestiviruses are not arthropod-borne, so cannot be termed arboviruses; and there are several viruses within the Flavivirus genus, isolated from small rodents and bats, which appear not to have any arthropod vector, so are not arboviruses. (For properties of *Togaviridae*, see Porterfield 1980*a*, Schlesinger 1980.)

Alphaviruses

There are 25 members of the alphavirus group (Table 90.2) all of which are transmitted by mosquitoes. Eight members produce significant disease in man (Table 90.3).

Clinical features

Chikungunya Chikungunya virus was first isolated from patients and mosquitoes by Ross (1956) during an epidemic in the Newala district of Tanzania in 1952–3. The native name is derived from the main symptom, being 'doubled-up' as a result of excruciating joint pains. Since then Chikungunya virus has been frequently isolated from man and mosquitoes during epidemics in India and south-east Asia as well as in eastern, western, central and southern Africa. The largest epidemics in recent years have been in cities of the Indian sub-continent. Sharma *et al.* (1965) estimated that there were 300 000 cases of illness in a population of nearly two million in Madras.

Table 90.2 Alphaviruses

Name	Abbreviation
Aura	AURA
Bebaru	BEB
Cabassou	CAB
Chikungunya	CHIK
Eastern Equine Encephalitis	EEE
Everglades	EVE
Fort Morgan	FM
Getah	GET
Highlands J	HJ
Kyzylagach	KYZ
Mayaro	MAY
Middelburg	MID
Mucambo	MUC
Ndumu	NDU
O'nyong nyong	ONN
Pixuna	PIX
Ross River	RR
Sagiyama	SAG
Semliki Forest	SF
Sindbis	SIN
Tonate	TON
Una	UNA
Venezuelan Equine Encephalitis	VEE
Western Equine Encephalitis	WEE
Whataroa	WHA

After an infectious mosquito bite, there is an incubation period of 3–12 days followed by the sudden onset of fever and crippling joint pains which may incapacitate the patient within a few minutes to a few hours of onset. The pain in the limbs and spine in so severe as to cause patients to be doubled-up and immobile. Headache is usually mild, there is no retro-orbital or eye pain and patients have anorexia and constipation. The disease has a biphasic course; after 1 to 6 days of fever the temperature returns to normal for 1 to 3 days and then there is a second period of fever for a few days. In the second phase of illness 80 per cent of patients display a maculopapular, pruritic rash on the trunk and extensor surfaces of the limbs. After 6 to 10 days patients recover completely although, rarely, joint pains may persist. A leukopenia is the only unusual laboratory finding.

In India and south-east Asia, Chikungunya virus has been implicated in outbreaks of haemorrhagic fever, often in association with dengue viruses. Thiruvengadam *et al.* (1965) reviewed 242 cases during the Madras epidemic of Chikungunya fever. Although mostly mild infections, 11 per cent of patients had haemorrhagic manifestations, none of which was severe. Sarkar *et al.* (1965) reported 11 patients with haemorrhagic fever from whom Chikungunya virus was isolated during an outbreak in Calcutta in 1963–4; 9 had haematemesis and melaena, 4 had petechiae and 2 died of shock. Paired sera from 7 of these patients had rising antibody titres against Chikungunya and two patients also had dengue virus type 2 antibodies. Chikungunya has been repeatedly isolated as well as all four dengue serotypes from patients during haemorrhagic fever outbreaks in Thailand and Singapore but Nimmannitya *et al.* (1969) suggested that, if cases displaying 'shock' were the only ones accepted as true haemorrhagic fever, then Chikungunya would be excluded. No haemorrhagic complications in Chikungunya infections have ever been reported in Africa.

Chikungunya virus is transmitted in Africa by *Aedes africanus* and *Aedes aegypti*, while *A. aegypti* transmits the disease in the urban centres of India and south-east Asia. No vertebrate host other than man has been discovered although McIntosh *et al.* (1963) found evidence that monkeys might be a maintenance host in Africa.

Eastern encephalitis Eastern encephalitis occurs widely along the eastern seaboard states of the USA and South America. Small outbreaks have occurred in the United States, the Dominican Republic, Cuba and Jamaica. In the USA equine cases occur each summer in coastal regions bordering the Atlantic and the Gulf

Table 90.3 Alphaviruses known to cause human disease

Virus	Probable transmission to man	Geographical distribution of viruses	Other features
Chikungunya	Mosquito	Tropical Africa, S and SE Asia, Philippines	Epidemics E Africa, India
Eastern encephalitis	Mosquito	N and Central America, Trinidad, Guyana, Brazil, Argentina	?Present in SE Asia and Philippines. Epidemics only in N America
Mayaro	Mosquito	Trinidad, Brazil	
O'nyong nyong	Mosquito	E and W Africa, Zimbabwe	Epidemics E Africa and Zimbabwe only
Ross River	Mosquito	Australasia, Fiji	Epidemics of polyarthritis
Sindbis	Mosquito	Africa, E Mediterranean, S and SE Asia, Borneo, Philippines, Australia, Sicily	Disease recognized only in Africa
Venezuelan encephalitis	Mosquito	Venezuela, Colombia, South and Central America	
Western encephalitis	Mosquito	N America, Mexico, Guyana, Brazil, Argentina	Epidemics only in N America

of Mexico and in other eastern states. Cases in eastern Canada were first recorded in 1972. Outbreaks in man are generally sporadic with only a small number of cases but in horses and pheasants there is considerable morbidity and mortality. The virus is probably maintained by wild birds and mosquitoes, but the maintenance species in tropical areas have not been elucidated. *Culiseta melanura* seems to be the main mosquito infecting birds in North America; outbreaks are often associated with *Aedes sollicitans* and *Aedes vexans*. In the tropics, *Aedes taeniorhynchus Culex taeniopus* and *Culex nigripalpus* appear to be involved.

Clinical features In the United States encephalitis occurs mainly in young children. Inapparent cases are rare. Onset is abrupt with high fever (39–41°), headache and vomiting followed by drowsiness, coma and severe convulsions. On examination there is neck stiffness, spasticity and, in infants, bulging fontanelles. Oedema of the legs and face and cyanosis have been described. The CSF is under pressure and contains increased protein and up to 1000 cells/mm^3. Death can occur within 3–5 days of onset. Sequelae are common in non-fatal cases and include convulsions, paralyses and mental retardation. Older patients usually recover more completely.

No vaccine is yet available for human use although a formalin-inactivated vaccine prepared in chick embryos has been found effective in horses.

Mayaro virus This virus has been isolated from man and various mosquito species in Trinidad and Brazil. The virus causes a 2–6-day illness in man with fever, headache, conjunctivitis, prostration, joint and muscle pains and a rash.

O'nyong nyong virus This virus caused a major epidemic which began in Uganda during 1959 and quickly spread to Kenya, Tanzania and Malawi affecting an estimated 2 million people. This virus is closely related to Chikungunya and the disease it produces also resembles this infection. Transmission to man is by *Anopheles gambiae* and *Anopheles funestus*, which are predominantly malaria transmitting species.

Clinical features After an incubation period of up to 8 days illness begins abruptly with fever, rigors and sometimes epistaxis, followed by backache, severe joint pains, headache, pain in the eyes, generalized lymphadenopathy and an irritating rash beginning on the face and spreading to the trunk and limbs. High fever is uncommon. The rash generally lasts for 4–7 days but the joint pains and malaise are protracted. There are no sequelae and no deaths have been directly attributed to the infection.

O'nyong nyong virus is still active in western Kenya (Bowen *et al.* 1973), young children having evidence of recent infection and the virus again being isolated there in 1979 (B.K. Johnson, personal communication).

Ross River virus This infection has caused epidemics of febrile illnesses with a rash and polyarthritis (Doherty *et al.* 1971) in Australia, south-western Pacific Islands and more recently in Fiji (WHO 1979). Ross River virus has been isolated from *Aedes vigilax* in Australia and New Guinea and from *Aedes aegypti* in Fiji. Epidemics in Murray Valley of Australia have involved several thousand people. The patients present with fever, arthralgia of the small joints of the hands and feet, sore throat, skin rash and paraesthesiae of the palms and soles of the feet. The rash, which may cover the whole body, has sometimes begun as discrete macules, progressed to papules and occasionally to small vesicles. Petechiae and an enanthem have also been seen. Arthritis can last for 2–28 days and in one case lasted for 8 months.

Sindbis virus Although this virus is widely distributed in Africa, India, tropical Asia and Australia, it has only occasionally been associated with overt human disease. It has been isolated from several *Culex* mosquito species—*C. tritaeniorhynchus*, *C. pseudovishnui*, *C. univittatus* and from wild birds. Five cases in Uganda had fever, headache and myalgia with slight jaundice in two patients. McIntosh *et al.* (1964) described 12 cases of febrile illness with an associated rash in South Africa, and Malherbe *et al.* (1963) reported a more severe illness in one patient who developed vesicles on the feet. Virus was isolated from fluid taken from the vesicles.

Venezuelan encephalitis virus This virus causes large epizootics of encephalitis in horses, mules and donkeys. In man encephalitis occurs in a small proportion of cases but the majority have only a non-fatal 'influenza-like' febrile illness. At least four subtypes of Venezuelan encephalitis are separable on the basis of serological and physicochemical studies (Monath 1979). Four antigenic variants can be distinguished within subtype I and are designated IAB, IC, ID, and IE. Venezuelan IAB and IC viruses have been found in equine epizootics and are more pathogenic for man and horses than the enzootic strains ID and IE and subtypes II, III, and IV. Only about 4 per cent of human cases caused by epizootic viruses develop encephalitis; this occurs particularly in children below the age of 15 years, but the case-fatality rate is of the order of 20 per cent. A similar spectrum of disease from inapparent infection with fever to fatal encephalitis is seen in equines. The enzootic strains, ID, IE, II and III, may occasionally produce sporadic disease but the virulence for equines and man is much reduced.

Epizootic strains of Venezuelan encephalitis are distributed in the north of South America (Colombia, Ecuador, Guyana and Venezuela) occurring at intervals when sufficient susceptible equine populations have accumulated. In 1969 Venezuelan encephalitis appeared in Guatemala and spread in huge epizootic waves reaching north to southern Texas in 1971 and south to Costa Rica (Report 1972). Over 200 000 horses died in this outbreak and there were several thousand human infections.

A wide variety of mosquitoes including *Aedes*, *Mansonia* and *Psorophora* species can transmit the virus. Enzootic strains are continually active in subtropical and tropical areas of the Americas having small rodent and marsupial hosts and are transmitted by *Culex* (*melanoconion*) mosquitoes. Birds may also act as hosts.

Clinical features There is a brief febrile, occasionally influenza-like illness with a sudden onset characterized by malaise, nausea or vomiting, headache which may be severe, and myalgia (Sanmartin-Barberi *et al.* 1954). Fever lasts up to 4 days and convalescence may take 3 weeks with generalized asthenia. A small proportion of patients develop encephalitis, which can be severe or fatal.

Vaccines An effective and safe live attenuated vaccine has been used in laboratory workers. An attenuated vaccine was used extensively to protect horses and mules in the 1969–71 epizootic and was found to be very effective.

A variety of new strains of Venezuelan encephalitis have been reported from time to time, some of which cause disease in man. They include Everglades and Mucambo viruses.

Western encephalitis virus The virus has been isolated in the United States, Canada, Guyana, Brazil and Argentina but human disease has been recognized only in North America and Brazil. Epidemics have occurred in the United States where encephalitis in horses is common. Since 1955, 947 human cases have been reported in the USA, the largest recent outbreak occurring in 1975 in North and South Dakota, Minnesota and adjacent Manitoba. This outbreak was precipitated by extensive flooding in early summer which encouraged the breeding of *Culex tarsalis*. This mosquito is the principal vector in western USA, while *Culiseta melanura* is more common in the east. The virus has been isolated from a variety of wild birds which appear to be the maintenance hosts at least in North America. Equine epizootics generally precede the appearance of human cases but horses play no part in the transmission cycle (Monath 1979). Very little is known about the maintenance cycle of the virus in tropical regions.

Clinical features The disease varies with age but fever and drowsiness are common at all ages. Convulsions occur in 90 per cent of affected infants and in 40 per cent of those aged between 1 and 4 years. Convulsions are rare in adults. Typically there is fever, headache, vomiting, stiff neck and backache. Restlessness and irritability are commonly seen in children. Drowsiness, severe occipital headache, mental confusion and coma are seen in up to 40 per cent of adults. In milder cases recovery takes place in 3–5 days and in severe cases within 5–10 days. Convalescence may be protracted. In adults sequelae are rare but become more common with reducing age. In infants almost half are left with convulsions. Mortality varies from outbreak to outbreak from 2–15 per cent.

No vaccine is yet available for use in man. A formalin inactivated chick embryo vaccine has been used effectively in horses. In man this vaccine has produced only minor reactions in volunteers (Bartelloni *et al.* 1970).

Structure and antigenic relationships of alphaviruses

Serological grouping of alphaviruses is based upon their cross-reactivity by haemagglutination-inhibition (HI), complement-fixation (CF) and neutralization (N) tests. Alphaviruses possess at least three antigens. They share a group reactive nucleoprotein antigen which does not participate in HI or N reactions. The other two antigens associated with two envelope glycoproteins show complex and type-specific reactivity. Alphavirus particles are spherical, 60–65 nm in diameter, and consist of a 35- to 39-nm cubic nucleocapsid, a lipoprotein envelope derived from host-cell membrane, and a surface projection layer made up of glycoprotein units. Nucleocapsids are assembled in the cytoplasm, and maturation occurs by budding through host cell membranes—primarily plasma membranes.

The naked 42S RNA genome of alphaviruses is infectious and has been shown with certain members of the genus to be capped at its 5′ end and polyadenylated at its 3′ end. It presumably serves as messenger RNA in primary translation, and is thus plus in sense. One or more of the non-structural polypeptides may represent a virus-coded polymerase which transcribes the input RNA into a complementary strand. This

Table 90.4 Alphaviruses and their complexes and subgroups

Complex or sub-group	Virus	Geographic variants
EEE	EEE	At least two
WEE	WEE	At least three
	FM	
	SIN	At least three
	WHA	
	AURA	
	MID	
	NDU	
VEE	VEE	At least five
	EVE	
	MUC	
	PIX	
	SF	
	CHIK	
	ONN	
	BEB	
	GET	
	RR	
	SAG	
	MAY	
	UNA	

minor-sense RNA is the template for plus-strand progeny RNA and presumably for the 26S (+) RNA, the latter species encoding the structural polypeptides. Polyprotein formation and its subsequent post-translation cleavage have been demonstrated for both the structural and non-structural proteins induced in alphavirus infections.

By use of antigenic comparison of alphaviruses the group can be divided into at least three complexes or sub-groups (Table 90.4). The first contains only EEE which has at least two geographic variants. The second sub-group contains six viruses with NDU and MID being much less closely related than the others. The largest sub-group contains the remaining alphaviruses including VEE, SF, CHIK and ONN. (See also Porterfield 1980*b*.)

Flaviviruses

There are just over 60 members of the flavivirus group: 30 are mosquito-borne (Table 90.5), 15 are tick-borne (Table 90.6) while the remainder have no known arthropod vector (Table 90.7). Twenty-six flaviviruses can cause human disease (Table 90.8) but several of them have produced only laboratory-acquired infections or isolated cases of disease in man. The range of clinical manifestations produced by flaviviruses is similar to that of the alphaviruses—febrile illnesses with or without a rash or encephalitis. In addition, Yellow fever, Kyasanur Forest disease, Omsk haemorrhagic fever and the dengue viruses can cause haemorrhage. Only those viruses that have produced substantial numbers of human infections are discussed in detail.

Table 90.5 Mosquito-borne flaviviruses

Name	Abbreviation
Alfuy	ALF
Bagaza	BAG
Banzi	BAN
Bouboui	BOU
Bussuquara	BSQ
Dengue 1	DEN-1
Dengue 2	DEN-2
Dengue 3	DEN-3
Dengue 4	DEN-4
Edge Hill	EH
Ilheus	ILH
Japanese encephalitis	JE
Jugra	JUG
Kokobera	KOK
Kunjin	KUN
Murray Valley encephalitis (Australian encephalitis)	MVE
Naranjal	NJL
Ntaya	NTA
Rocio	ROC
Sepik	SEP
St Louis encephalitis	SLE
Spondweni	SPO
Stratford	STR
Tembusu	TMU
Uganda S	UGS
Usutu	USU
Wesselsbron	WSL
West Nile	WN
Yellow fever	YF
Zika	ZIKA

Table 90.6 Tick-borne flaviviruses

Name	Abbreviation
Absettarov	ABS
Hanzalova	HAN
Hypr	HYPR
Kadam	KAD
Karshi	KSI
Kumlinge	KUM
Kyasanur Forest disease	KFD
Langat	LGT
Louping ill	LI
Omsk haemorrhagic fever	OMSK
Powassan	POW
Royal Farm	RF
Russian spring-summer encephalitis	RSSE
Saumarez Reef	SRE
Tyuleniy	TYU

Table 90.7 Flaviviruses with no known arthropod vector

Name	Abbreviation
Apoi	APOI
Aroa	AROA
Carey Island	CI
Cowbone Ridge	CR
Dakar bat	DB
Entebbe bat	ENT
Israel Turkey meningo-encephalitis	IT
Jutiapa	JUT
Koutango	KOU
Modoc	MOD
Montana myotis leuk.	MML
Negishi	NEG
Phnom-Penh bat	PPB
Rio Bravo	RB
Saboya	SAB
Sal Vieja	SV
San Perlita	SP
Sokuluk	SOK

Table 90.8 Flaviviruses known to cause human disease

Virus	Probable transmission to man	Geographical distribution of viruses	Other features
Banzi	Mosquito	S and E Africa	One case only
Bussuquara	Mosquito	Brazil, Colombia, Panama	One case only
Dengue types 1–4	Mosquito	S, SE Asia, Pacific Is, New Guinea, Caribbean area, Venezuela, Colombia, W. Africa	Tropics and subtropics wherever the virus and a *Stegomyia* vector exist
Ilheus	Mosquito	Central America, Trinidad, Colombia, Brazil, Argentina	
Japanese encephalitis	Mosquito	E, SE and S Asia, W Pacific	
Kunjin	Mosquito	Australia, Sarawak	Laboratory case only
Kyasanur Forest	Ixodid tick	Mysore, India	
Langat	Ixodid tick	Malaysia	Only experimental cases proven
Louping ill	Ixodid tick	N and W British Isles	
Murray Valley encephalitis	Mosquito	Australia, New Guinea	
Omsk haemorrhagic fever	Ixodid tick	Central USSR	
Powassan	Ixodid tick	Canada and USA	
Rio Bravo	?Bat saliva	USA, Mexico	Laboratory cases more severe
Rocio	Poss. mosquito	Brazil	Epidemics of encephalitis
St Louis encephalitis	Mosquito	N America, Panama, Jamaica, Trinidad, Brazil, Argentina	
Sepik	Mosquito	New Guinea	One case only
Spondweni	Mosquito	E, W and S Africa	
Tick-borne encephalitis (Central European)	Ixodid tick	Central Europe from Scandinavia to Balkans & from Germany to W USSR	
(Far Eastern)	Ixodid tick	E USSR and sometimes in W Russia and Czechoslovakia	
Wesselsbron	Mosquito	E, W and S Africa, Thailand	
West Nile	Mosquito	E and W Africa, S and SE Asia, Mediterranean area	Disease recognized mainly in Israel and S France
Yellow fever	Mosquito	W and Central Africa, S and Central America	Periodical epidemics in neighbouring areas, e.g. Ethiopia
Zika	Mosquito	E and W Africa, Malaysia, Philippines	One case in Uganda

Mosquito-borne flaviviruses

Dengue

There are four dengue virus serotypes, all of which are endemic throughout the tropics, particularly in Asia, the Caribbean, the Pacific and in some areas of West Africa. The various types—dengue 1, dengue 2, dengue 3 and dengue 4—are closely related and no significant biological differences are known between them. Many of the largest epidemics have been caused by dengue 1 as in the recent outbreaks in the Caribbean. However, in many situations several types co-exist and successive epidemics may be due to different types. Many of the same patients may be affected in each outbreak, since Sabin (1952) showed that cross-protection between dengue types in man lasted only a short time.

Epidemics of dengue have been known since the late eighteenth century, and waves of urban epidemics occurred in tropical and sub-tropical regions during the nineteenth and early twentieth centuries. According to Smith (1956*a*) these epidemics seem to have followed the migration of *Aedes aegypti* along trade routes from Africa to India and around the coast of Asia to reach Hong Kong and across the Pacific to Hawaii. Originally, dengue was probably a mainly rural infection in tropical Asia transmitted by indigenous *Stegomyia.* As no virological diagnosis was possible, 'dengue' has been a symptom complex which includes a large number of tropical febrile illnesses.

Dengue is endemic in tropical areas where *Stegomyia* species are constantly active. The boundaries are the winter isotherms for 64°F (17.8°). Large epidemics occur outside these areas from time to time, e.g. Brisbane (1906), Durban (1927) and Athens (1928). *Aedes aegypti* is the most important vector, particularly in urban areas, but other *Stegomyia* species play a role in rural areas of Asia and the Pacific Islands. These include *Ae. Albopictus*, *Ae. polynesiensis* and *Ae. scutellaris.* There is some evidence that monkeys may be concerned in virus maintenance but there is no convincing evidence of a vertebrate maintenance host other than man.

The clinical picture of 'classical' dengue fever was described by Siler *et al.* (1926) and Simmons *et al.* (1931) following experimental infection of volunteers. The 'classical' form usually affects adults and older children. After an infective mosquito bite there is an incubation period of 5 to 8 days followed by the sudden onset of fever, which often becomes biphasic, severe headache, pain behind the eyes, backache, chilliness and generalized pains in the muscles and joints. A maculopapular rash generally appears on the trunk between the third and fifth day of illness and spreads later to the face and extremities. Lymphadenopathy, anorexia, and constipation and altered taste sensation are common. Occasionally, petechiae are seen on the dorsum of the feet, legs, hands, axillae, and palate late in the illness. In young children, upper respiratory tract symptoms predominate and dengue is rarely suspected. The illness generally lasts for about 10 days after which recovery is usually complete although convalescence may be protracted. Laboratory findings are leukopenia, a mild thrombocytopenia and a relative lymphocytosis.

In the past two decades there have been an increasing number of epidemics of a severe disease syndrome caused by dengue viruses throughout south-east Asia, India and the western Pacific. It occurs most frequently in young children aged between 2 and 13 years and is associated with numerous haemorrhagic manifestations and quite often terminates fatally. Since its recognition in the Philippines in 1953 (Quintos *et al.* 1954) dengue haemorrhagic fever has occurred in Thailand, Burma, Cambodia, Indonesia, Malaysia, Vietnam, Singapore, eastern India and several western Pacific Islands (WHO 1974). Outbreaks tend to occur most frequently in primarily affected areas. Over 500 cases a year are admitted to hospital in the Philippines. In the Bangkok-Thonburi area in Thailand, 10 000 cases were hospitalized in the period 1958–63 and all but 25 of these were younger than 14 years; 694 of these children died—an indication of the severity of the disease (WHO 1966). Epidemics continue to occur annually in Thailand, the highest number of cases in a single year being 8288 in 1973 with 310 deaths (WHO 1974). Only indigenous populations are involved in these epidemics, with neither ethnic origin nor socio-economic conditions apparently having any effect on the incidence of the disease. Outbreaks of classical dengue are uncommon during haemorrhagic disease epidemics, but immigrants from non-endemic areas often suffer from classical dengue while haemorrhagic disease occurs in the indigenous population. (For further references, see Schlesinger 1977, Hotta 1978.)

Dengue haemorrhagic fever

This syndrome is almost entirely confined to indigenous children, usually orientals, often as young as 6 months. In the initial phase the child may present with fever, upper respiratory symptoms, headache, vomiting and abdominal pain. Myalgia and arthralgia are uncommon. This minor illness, during which the child is often not confined to bed, lasts 2 to 4 days and many children recover without any further symptoms. In a proportion of children and initial phase is followed by an abrupt collapse with hypotension, peripheral vascular congestion, petechiae and sometimes a rash. There are varying degrees of shock. The child is often restless, sweating and has cold, clammy extremities and a hot feverish trunk. The fourth and fifth days are critical; purpura, ecchymoses, epistaxis, haematemesis, melaena, coma, convulsions and severe shock indicate a poor prognosis. Should the patient survive this period, recovery is generally complete. Laboratory studies show thrombocytopenia, a prolonged bleeding time, an elevated prothrombin time, a raised haematocrit, hypoproteinaemia and a positive tourniquet test. The liver is often enlarged, soft and tender. Nimmannitya (1978) suggested that the acute onset of shock, and the rapid and often dramatic clinical recovery when the shock is treated properly, together with the absence of inflammatory vascular lesions, suggest short-term vascular lesions. The central role of complement activation with the formation of immune complexes has been demonstrated. C3a and C5a anaphylatoxins, which are products of complement activation, are thought to be the cause of plasma leakage. Their rapid inactivation and elimination from the circulation are consistent with the short duration of shock. There is increased fibrinogen consumption, which indicates disseminated intravascular coagulation. Both leukopenia and leukocytosis have been reported. There may be maturation arrest of megakaryocytes in the bone marrow and phagocytic activity of reticulum cells. Immunoelectrophoretic studies of serum proteins have shown the disappearance of the βIC line at the onset of shock suggesting an immunological phenomenon—a massive antigen-antibody reaction. There is a very rapid rise in flavivirus antibody in the early stages of the disease, which suggests that patients may have been previously sensitized to the infecting virus by earlier infection with a closely related virus, probably another type of dengue virus.

Post-mortem studies show that vascular changes predominate with vasodilatation, congestion, oedema, and haemorrhages. Pleural and peritoneal effusions are seen with haemorrhages in the stomach and intestines; there are widespread petechial haemorrhages.

Treatment of dengue Nimmannitya (1978) maintains that the management of dengue haemorrhagic fever is entirely symptomatic, with the basic principle being directed towards correction of plasma leakage. In Bangkok satisfactory results have been obtained with the following regimen: immediate replacement of plasma loss with isotonic salt solution and plasma or plasma expanders in cases of profound shock; further plasma leakage is continually replaced to maintain the circulation volume for another 12–24 hours and allow extravasated plasma to be reabsorbed; correction of the electrolytic and acid-base disturbance; transfusions of fresh blood in cases of massive bleeding.

With this regimen the case-fatality rate in Bangkok fell from 9 per cent in 1964 to 2 per cent in 1974. Nimmannitya claims that the extensive use of the microhaematocrit has been invaluable as a guideline. Corticosteroids have not been found to be of great benefit and heparin is not generally indicated in the management of the dengue shock syndrome, even when there is evidence of disseminated intravascular coagulation.

Control of dengue The control of dengue depends principally on control of the vector, particularly *Ae. aegypti*. A great deal could be done by eliminating the peridomestic breeding places such as flower pots, old jars and tin cans around houses and by using insecticides carefully. Rural breeding sites are much more difficult to control. No vaccine is yet available although several live attenuated strains are under development.

Ilheus virus Ilheus virus is active over a wide area of Central and South America and in Trinidad. It is probably maintained in a forest complex involving birds and mosquitoes. Our knowledge of human disease is limited to five naturally occurring cases in Brazil and Trinidad and nine cancer patients infected experimentally by Southam and Moore (1951). Half of the infections were asymptomatic, three had mild febrile illnesses and four had a more severe illness with affection of the CNS. These patients displayed fever, headache, myalgia, photophobia and signs of encephalitis. No deaths or sequelae have been reported.

Japanese encephalitis virus This virus occurs over the whole of the eastern seaboard of Asia and the offshore islands from the maritime province of Russia to South India and Sri Lanka. Epidemics have been recognized for years, generally occurring in late summer in the more temperate regions; in tropical areas the virus remains enzootic throughout the year. Some of the largest outbreaks in recent years have taken place in India.

Rice-field breeding mosquitoes are the main vectors. They include the *Culex vishnui* complex, especially *C. tritaeniorhynchus*, *C. annulus* and *C. annulirostris*. In tropical areas, *C. gelidus*, which breeds in close association with pigs and cattle, is also involved. During extensive studies in Japan, Scherer *et al.* (1959) showed that there is intense virus activity in the spring among young herons and *C. tritaeniorhynchus*. These mosquitoes then infect pigs, which act as amplifying hosts infecting more *C. tritaeniorhynchus*; these, in turn, bite man causing outbreaks in late summer. Japanese workers can predict human cases in time and place in any year simply by monitoring abattoir pigs for evidence of infection (antibody conversion, especially IgM) which precedes outbreaks of human disease by 2–3 weeks. In the tropics the vertebrate maintenance hosts are not clearly known, but in Sarawak (Macdonald *et al.* 1965, 1967, Simpson *et al.* 1976) it appears that the principal maintenance hosts are pigs and *C. gelidus*. After the flooding of paddy fields prior to planting, there is intense breeding of *C. tritaeniorhynchus*, which produces human infections.

Clinical features Generally onset is sudden with fever, headache and vomiting. Fever is usually continuous but subsides after 2–4 days. Lethargy is a common feature, faces are expressionless and there are sensory and motor disturbances affecting speech, the eyes and limbs. There may be confusion and delirium progressing to coma. Convulsions may be the first sign in children. Weakness and paralysis can affect any part of the body, the lesions generally being upper motor neurone in character. Neck rigidity and a positive Kernig's sign are found and reflexes are abnormal. An initial leucocytosis is followed by a leucopenia; the CSF is clear and under pressure with increased cells and protein. The duration of illness is variable. Fatal cases usually die within 10 days. Convalescence is often protracted and sequelae are common especially in young children. They include incoordination, tremors, mental impairment and personality changes. Residual paralysis, aphasia and cerebellar ataxia can occur. Southam (1956) estimated that only one case of encephalitis occurred in every 500–1000 infections in Japan but the risk may be higher in immigrants.

Post mortem, oedema and congestion of the central nervous system are apparent. Histologically there is neuronal degeneration and necrosis and perivascular cuffing. The Purkinje cells are severely affected.

There is no specific treatment. No really satisfactory vaccine is yet available although formolized vaccines prepared in mouse brain and hamster kidney cultures show some promise. Vaccination of the amplifier hosts, i.e. pigs, is worth considering as a means of breaking the infection link to man. An attenuated vaccine for use in pigs is being evaluated in Japan (Fujisaki *et al.* 1975).

Murray Valley encephalitis This disease was originally called Australian X disease. The virus causes epidemics in south-eastern Australia and southern Queensland and sporadic cases in New Guinea. Outbreaks occur in late summer, February to April, and *Culex annulirostris* is the principal arthropod vector. *Aedes normanensis* and *C. bitaeniorhynchus* may be concerned in transmission in enzootic tropical and subtropical areas, with birds acting as the vertebrate hosts (Doherty *et al.* 1961). Clinically this disease closely resembles Japanese encephalitis. Inapparent infections are common. Several deaths have occurred mostly in children. It has a low morbidity but a high case-fatality rate.

Rocio virus In February 1975 an acute infective illness affecting the central nervous system appeared in coastal areas in São Paulo State, Brazil (Lopes *et al.* 1978). This illness was characterized by fever, headache, vomiting, encephalitis and meningitis. During 1975 and 1976 in 825 recorded cases there were 95 deaths. The average attack rate was 15 per 1000 population and the mortality rate was 2 per 1000; the case-fatality rate was 13 per cent (Monath 1979). There was a high incidence in adults.

A newly recognized flavivirus, named Rocio, was isolated from nine patients. It seems most likely that the virus is mosquito-borne and wild birds may be important vertebrate hosts. Only one isolate from mosquitoes has been made—from *Psorophora ferox*—and isolates have also been obtained from sentinel mice and from a wild-caught sparrow.

St Louis encephalitis virus St Louis encephalitis is the most important mosquito-borne disease in the United States and has caused major epidemics in recent years. It is widely distributed from Canada to Argentina. The virus occurs in endemic form west of the Mississippi, in the eastern USA it reappears in epidemic form especially in the Mississippi-Ohio basin, Texas and Florida. Outbreaks have also occurred in Canada and northern Mexico. In Central and South America human infections are frequent but epidemics are unknown (Monath 1979).

In western USA the important mosquito host is *Culex tarsalis* which breeds in irrigated or flooded dryland areas. It is widely distributed and causes many infections in rural areas. In eastern USA the principal vectors are *Culex pipiens pipiens* and *C.p. quinquefasciatus* which breed in polluted water and produce large populations in urban-suburban areas. In Florida, *C. nigripalpus* is the epidemic vector. Wild birds are the major vertebrate host.

There were large outbreaks in the early 1960s in Florida, Houston, Illinois and the Delaware Valley. After a quiet 8-year period the virus reappeared in epidemic form in 1974 and outbreaks have occurred in each year since then, particularly in the Ohio-Mississippi basin.

Clinical features Most patients have a febrile illness with severe headache which lasts a few days and is followed by complete recovery. A variable number of cases suffer from aseptic meningitis or encephalitis. There is sudden onset of fever, weakness and nausea; the headache becomes severe and is followed by confusion, drowsiness and vomiting. There may be convulsions, which have a poor prognosis. Fever may last for 3–10 days. A stiff neck is common, and the CSF is under increased pressure and has increased cells and protein. There may be muscular weakness, pains, tremors, spasticity, dysphasia, photophobia and visual disturbances. Occasionally dramatic recovery can occur in even severely ill patients. In the elderly there is a higher incidence of disease, greater severity and higher mortality; severe sequelae are more frequent.

At post-mortem examination a few haemorrhages are seen in the CNS with neuronal damage occurring particularly in the mid-brain and brain stem together with perivascular cuffing.

No vaccine is yet available.

Wesselsbron virus This virus is widely distributed in Africa and causes epizootics in sheep, producing abortion and death in new-born lambs and ewes. The mosquito vectors are *Aedes caballus* and *Ae. circumluteolus* (Kokernot *et al.* 1960). Man may also be infected, developing fever, headache, muscular pains and retro-orbital pain. Convalescence may be prolonged. Splenomegaly and hepatomegaly were found in one patient while another had visual disturbances. Several laboratory-acquired infections have occurred.

West Nile virus This virus has been isolated in many parts of Africa and in Israel, Cyprus, France, India and Borneo. Recognizable disease due to West Nile virus infection has been observed in Israel where epidemics have occurred between May and October (Marburg *et al.* 1956). In France it has caused febrile illness in man and encephalitis in horses. The disease in Egypt is generally a mild febrile illness, mainly of young children. The virus is probably maintained by mosquitoes, especially *Culex* species, and wild birds, although *Argas* ticks may also play a role in the maintenance cycle.

Clinical features After an incubation period of 3–6 days there is a sudden onset of fever, headache, myalgia, a maculopapular rash mainly on the trunk, and lymphadenopathy. Although most cases recover without untoward effects, the disease is more severe in elderly patients who may develop neurological signs or myocarditis. Death may occasionally result. No vaccine is yet available.

Yellow fever The first reported outbreak of yellow fever was in Barbados in 1647 and since then innumerable appalling epidemics occurred in the West Indies, Central and South America and the southern United States throughout the seventeenth, eighteenth and nineteenth centuries as well as in seaports in more temperate regions of the western hemisphere. This virus is believed to have originated in Africa and to have been carried to the Americas by trading and slaving ships, which may have also introduced one of its important vectors, *Aedes aegypti*. Epidemics generally took place in urban conurbations, being transmitted from man to man by *Ae. aegypti*. The elimination of this vector almost completely eradicated yellow fever from towns, but sporadic cases of the disease continued to occur in rural areas, particularly those bordering on forest zones. It was later discovered that yellow fever virus was maintained in a sylvan cycle involving monkeys and forest-dwelling mosquitoes such as *Haemagogus* and *Sabethes* species in South America and *Aedes africanus* in East Africa where *Ae. simpsoni* provides the link between monkey and man; in West Africa a variety of *Aedes* species appears to be involved. In recent outbreaks *Aedes furcifer/taylori* has been strongly implicated. (For the discovery of the mosquito carriage of yellow fever, see 6th edn, p. 2556.)

Yellow fever is still the most important cause of

viral haemorrhagic disease, being active in several South American and African countries. Two devastating epidemics took place in Africa in the last two decades. The largest of these was in Ethiopia between 1960 and 1962 when there were enormous numbers of cases and between 15 000 and 30 000 deaths (Sérié *et al.* 1964, 1968). *Ae. simpsoni* was the mosquito host involved in the man-mosquito-man cycle. The other large epidemic occurred in Senegal in 1965 (Chambon *et al.* 1967), with several thousand cases and several hundred deaths. *Aedes aegypti* was the main mosquito vector. Sporadic cases continue to occur in rural areas in West Africa and South America.

Clinical features

The disease in man varies from an inapparent infection in native Africans to a fulminating disease terminating in death. After an incubation period of 3 to 6 days the illness begins suddenly with fever, rigors, headache and backache. The patient is intensely ill and restless with flushed face, swollen lips, bright red tongue, congested conjunctivae and suffers from nausea and vomiting. A tendency to bleeding may be seen early in the course of the disease. This stage of active congestion is followed quickly by one of stasis. The facial oedema and flushing is replaced by a dusky pallor, the gums become swollen and bleed easily, and there is a pronounced bleeding tendency with black vomit ('vomito negro'), melaena and ecchymoses. The pulse rate is slow despite high fever, the blood pressure falls leading to albuminuria, oliguria and anuria. Death, when it occurs, is usually within 6 to 7 days of onset and is rarely seen after 10 days of illness. The jaundice, which gives the disease its name, is generally apparent only in convalescing patients. Mortality may be quite low, often of the order of 5 per cent. *Post mortem* the organs most particularly affected are the liver, spleen, kidneys and heart. Typically a mid-zone necrosis is apparent in the liver affecting cells around the periphery of the lobule and sparing the area around the central vein. Hyaline necrosis is evident and typical Councilman bodies have been described.

Treatment

Treatment is largely symptomatic with maintenance of fluid and electrolytic balances. Blood transfusion is sometimes required to correct blood loss through haemorrhage.

Vaccines

Two live attenuated virus vaccines are available; the 17D strain which was attenuated in tissue culture and is prepared in chick embryos; and the French neurotropic strain which was attenuated and is produced in mouse brain. The 17D strain is given by subcutaneous injection and the French neurotropic strain by scarification. During the Senegal epidemic of 1965 about 1.9 million people received the French neurotropic vaccine and almost 120 000 were given 17D vaccine. Of the former group at least 246 subsequently developed encephalitis with 23 deaths. About 90 per cent were children. Only two children who received 17D vaccine developed encephalitis and both recovered (Chambon *et al.* 1967). Largely as a result of these complications the French neurotropic vaccine is not now used to such a large extent. Immunity following yellow fever vaccination is long lasting.

Apart from vaccination, control of yellow fever depends on controlling the vectors. In urban areas where *Aedes aegypti* is the vector, peridomestic breeding sites can be controlled but in rural areas, where forest dwelling mosquitoes are the vectors, little can be done.

Flaviviruses contain a non-segmented single-stranded infectious 42S RNA and three virion polypeptides, one of which is a glycoprotein (Porterfield *et al.* 1978). Much less study has been carried out on flaviviruses in contrast to the alphaviruses, but the processes of maturation appear to be similar. Virion diameters range from 36–44 nm with the core measuring 25–30 nm.

Antigenic relationships among the flaviviruses are shown in Table 90.9.

Tick-borne flaviviruses

Kyasanur Forest disease This virus, like Omsk haemorrhagic fever virus, is a member of the tick-borne encephalitis complex but only rarely causes disease involving the central nervous system. The virus was first isolated in Mysore State, India, in 1957 (Work 1958); human infections, which still occur, are limited to villages surrounding Kyasanur Forest. The virus is now known to be widely distributed in India but human infections do not occur outside Mysore.

Clinical features

After an infectious tick bite there is an incubation period of 3 to 7 days before the sudden onset of fever, frontal headache, severe myalgia and prostration. This is quickly followed by nausea, vomiting, confusion and restlessness. The conjunctivae are injected and the palate is suffused and often covered with maculopapular haemorrhagic spots. A generalized lymphadenopathy has been noted. Many patients have bronchiolar involvement. The fever generally lasts for 5 to 12 days and sometimes follows a biphasic course with a mild meningo-encephalitis seen occasionally during the second phase (Webb and Rao 1961). Epistaxis, haematemesis, haemoptysis, melaena and bleeding gums are common and sometimes there may be uterine

Table 90.9 Flaviviruses and their complexes or sub-groups

Complex or sub-group	Virus
Tick-borne viruses	All members (Table 90.6)
Viruses with no known vector	All members (Table 90.7)
Mosquito borne	
WN sub-group	ALF
	BUS
	ILH
	JE
	KUN
	KOK
	MVE
	SLE
	SEP
	STR
	USU
	WSL
	WN
	YF
Spondweni sub-group	SPO
	ZIKA
Ntaya sub-group	NTA
	BAG
	IT
	TMU
Uganda S sub-group	BAN
	BOU
	EH
	ROC
	UGS
Dengue sub-group	DEN-1
	-2
	-3
	-4

bleeding. Albuminuria, leukopenia and thrombocytopenia are usual findings. A small proportion of patients may die, usually 8 to 12 days after the onset of illness, developing coma or bronchopneumonia before death. The majority of cases make an uneventful and complete recovery. (For a clinical study, see Webb and Rao 1961.)

In 1957 there were probably about 500 cases in a 70 square mile area. The death rate was about 10 per cent. Numerous laboratory infections have occurred but there have been no deaths.

Boshell (1969) reviewed the ecology of this disease extensively. The virus is transmitted by *Haemaphysalis* ticks—especially *H. spinigera*—and is maintained in small mammals. In Mysore State the symptomless enzootic situation was dramatically altered by man's need for more grazing land. Cattle were put to graze around the forest and provided the *Haemaphysalis* tick with a new and plentiful source of blood meals which produced a population explosion among the ticks. The abundant ticks fed on other mammalian species such as monkeys, and these became infected with Kyasanur Forest disease virus and developed viraemia and an illness from which they died. It was noted in 1957 that human infection was preceded by illness and death in forest-dwelling *Langur* and *Macacus* monkeys which acted as amplifiers of the virus.

Louping ill So called from the old Norse word meaning to leap, Louping ill is a member of the tick-borne encephalitis complex. It has been known in the British Isles for many years as a disease of sheep characterized by CNS manifestations, particularly cerebellar ataxia, including paralysis and death. It has never been a serious disease hazard to man, most reported cases being the result of laboratory infections. A few natural infections have occurred in persons closely associated with sheep; the illness is generally biphasic with encephalitic involvement in the second phase (Webb *et al.* 1968). The vector is *Ixodes ricinus* and the virus is maintained in rodents and ground-living birds.

Omsk haemorrhagic fever This virus is also antigenically related to the tick-borne encephalitis complex. An epidemic of severe haemorrhagic fever occurred in Omsk in Novosibirsk Oblast in Siberia between 1945 and 1948. The virus was transmitted by ticks, *Dermacentor pictus*, and by contact with infected muskrats (*Ondrata zibethica*). Most of the more recent cases of human disease appear to have been acquired through direct contact with muskrats. Infections originate generally in the northern forest-steppe-lake belt of western Siberia, which contains much wet grassland and swamp.

Clinical features After an incubation period of 3 to 7 days the illness begins abruptly with fever, which often follows a biphasic course, headache, vomiting and diarrhoea. An enanthem of the palate, sometimes haemorrhagic, generalized lymphadenopathy and meningism are common findings. Epistaxis, haematemesis, melaena and uterine bleeding may occur accompanied by a pronounced leukopenia and thrombocytopenia and albuminuria. The central nervous system is rarely affected. The case-fatality rate is low (0.5–3 per cent). Convalescence may be prolonged but there are no sequelae.

The precise epidemiology of Omsk haemorrhagic fever virus is still doubtful. A biological cycle of unknown complexity, which may involve rodents and ticks, exists. Muskrats which were introduced into the region some 60 years ago for hunting purposes are somehow infected and are capable of transmitting the virus by direct contact (Casals *et al.* 1970).

Powassan virus This virus is a rare cause of acute viral CNS disease in Canada and the USA. It was first isolated from the brain of a 5-year-old child who died in Ontario in 1958. Between 1970 and 1978 eight cases have been seen in the USA mostly in upper New York State. The clinical manifestation is usually encephalitis; some residual neurological problems have been described in recovered patients. The virus is maintained between ixodid ticks and wild animals—woodchucks and squirrels. Man is rarely infected; less than 1 per cent of residents of enzootic areas have demonstrable antibodies.

Tick-borne encephalitis This disease occurs

throughout Russia, eastern and central Europe and as far west as Alsace. The complex of viruses that make up the group causes illnesses which range from severe paralytic encephalitis, such as occurs in Siberia and is transmitted by *Ixodes persulcatus*, to a less paralytic encephalitis generally following a biphasic course, which occurs in central Europe. This form is transmitted by *Ixodes ricinus*. Agricultural and forestry workers are most frequently affected, as the foci of infection are in and around forest areas, the vertebrate maintenance hosts being rodents and ground-living birds.

Far East Russian encephalitis This is also known as Russian spring-summer encephalitis (Silber and Soloviev 1946), is mainly confined to eastern USSR but a few viruses occur around Leningrad and western USSR. In Czechoslovakia, related strains Absettarov and Hypr have been isolated.

Clinical features The incubation period ranges from 7 to 10 days and the illness often follows a biphasic course. The first phase of fever and headache lasts around 5 to 10 days and is followed by an afebrile period of 4 to 10 days. In a proportion of cases the second phase is marked by intense headache and high fever followed by severe CNS manifestations of varying degrees of severity ranging from meningitis to encephalitis and death. Flaccid paralysis followed by atrophy is common and there are frequently symptoms due to bulbar involvement. Mortality can be as high as 30 per cent. Nystagmus, vertigo, somnolence and visual disturbances indicate the development of encephalitis which may cause delirium and coma. In patients who recover, convalescence is prolonged. Headache and debility are common and residual paralysis may persist in 3–5 per cent of cases.

Central European encephalitis This is often called bi-undulant meningo-encephalitis or diphasic milk fever and occurs in central Europe from Scandinavia to the Balkans and from Alsace and West Germany to western Russia. The virus can be transmitted through drinking infected goats' milk. Goats are infected by tick-bite. Popov (1967) reported that cases occurring early in the season were due to the drinking of infected milk, whereas later cases were caused by tick bite. Several cases have occurred in campers and picnickers in forested areas of central Europe.

Clinical features This infection always follows a biphasic course. An afebrile period of 4 to 10 days intervenes between the first influenza-like symptoms and the second phase of meningitis or meningo-encephalitis. Mild or inapparent forms are common. In the more severe cases there may be transient or permanent paralysis and the bulbo-spinal form is often fatal. Generally the disease is not as severe as the Far Eastern form.

Langat virus This virus is a member of the group which was isolated from ricks in Malaysia (Smith 1956*b*). Its comparatively low pathogenicity for mice and monkeys suggested its possible use as an immunizing agent against the tick-borne viruses. Smorodintzev *et al.* (1968) gave a strain to over 1000 persons with satisfactory results. Thind and Price (1966) used a promising avirulent strain grown in chick embryos. Recently, low passage strains of Austrian isolates of tick-borne encephalitis have been successfully grown in chick embryo cells and, when inactivated with formalin, have produced excellent antibody responses in volunteers.

(For references to Arctic and Tropical Arboviruses, see E. Kwistak (1979), Academic Press, New York.)

References

Ardoin, P and Clarke, D.H. (1967) *Amer. J. trop. Med. Hyg.* **16,** 357.

Bartelloni, P.J., McKinney, R.W., Duffy, T.P. and Cole, F.E. (1970) *Amer. J. trop. Med. Hyg.* **19,** 123.

Boshell, M.J. (1969) *Amer. J. trop. Med. Hyg.* **18,** 67.

Bowen, E.T.W. *et al.* (1973) *Trans. R. Soc. trop. Med. Hyg.* **67,** 702.

Buescher, E.L. (1963) Paper presented at *7th int. Congr. trop. Med. Malaria,* Rio de Janeiro.

Calisher, C.H. *et al.* (1980) *Intervirology,* **14,** 229.

Casals, J., Henderson, B.E., Hoogstraal, H., Johnson, K.M. and Shelokov, A. (1970) *J. infect. Dis.* **122,** 437.

Chambon, L. *et al.* (1967) *Bull. Wld Hlth Org.* **36,** 113.

Chanas, A.C., Johnson, B.K. and Simpson, D.I.H. (1976) *J. gen. Virol.* **32,** 295.

Clarke, D.H. (1964) *Bull. Wld Hlth Org.* **31,** 45.

Clarke, D.H. and Casals, J. (1958) *Amer. J. trop. Med. Hyg.* **7,** 561.

Cuadrado, R.R. and Casals, J. (1967) *J. Immunol.* **98,** 314.

Doherty, R.L., Barrett, E.J., Gorman, B.M. and Whitehead, R.H. (1971) *Med. J. Aust.* **1,** 5.

Doherty, R.L., Carley, J.G., Mackerras, M.J., Trevethan, P. and Marks, E.N. (1961) *Aust. J. Sci.* **23,** 302.

Emmons, R.W. and Lennette, E.H. (1966) *J. Lab. clin. Med.* **68,** 923.

Fujisaki, Y., Sugimori, T., Morimoto, T., Miura, Y., Kawakami, Y. and Nakawa, K. (1975) *Nat. Inst. Anim. Hlth Quart. (Tokyo)* **15,** 55.

Hotta, S. (1978) *Dengue and Related Tropical Diseases.* Yukosha Printing House, Kobe, Japan.

Karabatsos, N. (1975) *Amer. J. trop. Med. Hyg.* **24,** 527.

Kokernot, R.H., Smithburn, K.C., Paterson, H.E. and De Meillon, B. (1960) *S. Afr. med. J.* **34,** 871.

Lopes, O.de S., Sacchetta, L.de A., Coimbra, T.L.M., Pinto, G.H. and Glasser, C.M. (1978) *Amer. J. Epidem.* **108,** 394.

Macdonald, W.W., Smith, C.E.G. Dawson, P.S., Ganapathipillai, A. and Mahadevan, S. (1967) *J. med. Entomol.* **4,** 146.

Macdonald, W.W., Smith, C.E.G. and Webb, H.E. (1965) *J. med. Entomol.* **1,** 335.

McIntosh, B.M., Harwin, R.M., Paterson, H.E. and Westwater, M.L. (1963) *Cent. Afr. J. Med.* **9,** 351.

McIntosh, B.M., McGillivray, G.M., Dickinson, D.B. and Malherbe, H. (1964) *S. Afr. med. J.* **38,** 291.

Madrid, A.T. de and Porterfield, J.S. (1969) *Bull. Wld Hlth Org.* **40,** 113; (1974) *J. gen. Virol.* **23,** 91.

Malherbe, H., Strickland-Cholmley, N. and Jackson, A.L. (1963) *S. Afr. med. J.* **37,** 547.

Marburg, K., Goldblum, M., Sterk, V.V., Jasinska-Klingberg, W. and Klingberg, M.A. (1956) *Amer. J. Hyg.* **64,** 259.

Monath, T.P. (1979) *Bull. Wld Hlth Org.* **57,** 513.

Nimmannitya, S, (1978) *Asian J. infect. Dis.* **2,** 67.

Nimmannitya, S., Halstead, S.B., Cohen, S.N. and Margiotta, M.R. (1969) *Amer. J. trop. Med. Hyg.* **18,** 954.

O'Reilly, K.J., Smith, C.E.G., McMahon, D.A., Bowen, E.T.W. and White, G. (1968) *J. Hyg. Camb.* **66,** 217.

Popov, V.F. (1967) *Medskaya Parazitologia,* **36,** 288.

Porterfield, J.S. (1980*a*) In: *The Togaviruses, Biology, Structure and Replication.* Ed. by R.W. Schlesinger, Academic Press, London, New York; (1980*b*) Arboviruses in perspective. In: *Arboviruses in the Mediterranean Countries.* Ed. by J. Vesenjak-Hirjan *et al.* Fischer Verlag, Stuttgart, New York.

Porterfield, J.S. *et al.* (1978) *Intervirology,* **9,** 129.

Quintos, F.N., Lim, L., Juliano, L., Reyes, A. and Lacson, P. (1954) *Philip. J. Paediat.* **3,** 1.

Report (1972) *Venezuelan Encephalitis.* Pan. Amer. Hlth Org. Scientific Publication No. 243. Wash. DC.

Ross, R.W. (1956) *J. Hyg. Camb.* **54,** 177.

Sabin, A.B. (1952) *Amer. J. trop. Med. Hyg.* **1,** 30.

Sanmartin-Barberi, C., Groot, H. and Osorno-Mesa, E. (1954) *Amer. J. trop. Med.* **3,** 283.

Sarkar, J.K., Chatterjee, S.N., Chakravarti, S.K. and Mijram, A.C. (1965) *Indian J. med. Res.* **53,** 921.

Scherer, W.F., Buescher, E.L. and McClure, H.E. (1959) *Amer. J. trop. Med. Hyg.* **8,** 644.

Schlesinger, R.W. (1977) *Dengue Viruses.* Springer Verlag, Wien.

Schlesinger, R.W. Ed. (1980) *The Togaviruses, Biology, Structure and Replication* Academic Press, London and New York.

Sérié, C., Andral, L., Lindrec, A. and Neri, P. (1964) *Bull. Wld Hlth Org.* **30,** 299.

Sérié, C., Lindrec, A., Poirier, A., Andral, L. and Neri, P. (1968) *Bull. Wld Hlth Org.* **38,** 835.

Sharma, H.M., Shanmugham, C.A.K., Iyer, S.P., Ramachandra Rao, A. and Kuppuswami, S.A. (1965) *Indian J. med. Red.* **53,** 720.

Silber, L.A. and Soloviev, V.D. (1946) *Amer. Rev. Soviet Med. Special Supplement,* pp. 6–80.

Siler, J.F., Hall, M.W. and Hitchens, A.P. (1926) *Philip. J. Science,* **29,** 1–302.

Simmons, J.S., St. John, J.H. and Reynolds, F.H.K. (1931) *Philip. J. Science.* **44,** 1–247.

Simpson, D.I.H. *et al.* (1976) *Trans R. Soc. trop. Med. Hyg.* **70,** 66.

Smith, C.E.G. (1956*a*) *J. trop. Med.* **59,** 243; (1956*b*) *Nature,* **178,** 581; (1967) Symposia Series in Immunological Standardisation, **4.** *Immunological Methods of Biological Standardisation,* p. 263. Karger, Basel/New York.

Smorodintsev, A.A., Prozorova, I.N. and Platonov, V.G. (1968) *Bull. Wld Hlth Org.* **39,** 425.

Southam, C.M. (1956) *J. infect. Dis.* **99,** 155.

Southam, C.M. and Moore, A.E. (1951) *Amer. J. trop. Med.* **31,** 724.

Thiruvengadam, K.V., Kalyanasundarum, V. and Rajopal, J. (1965) *Indian J. med. Res.* **53,** 720.

Theiler, M. and Casals, J. (1958) *Amer. J. trop. Med. Hyg.* **7,** 585.

Thind, I.S. and Price, W.H. (1966) *Amer. J. Epidem.* **84,** 193, 214, 225.

Varma, M.G.R., Pudney, M. and Leake, C.J. (1974) *Trans. R. Soc. trop. Med. Hyg.* **68,** 574.

Webb, H.E., Connolly, J.H., Kane, F.F., O'Reilly, K.J. and Simpson, D.I.H. (1968) *Lancet,* **ii,** 255.

Webb, H.E. and Rao, R.L. (1961) *Trans. R. Soc. trop. Med. Hyg.* **55,** 284.

Work, T.H. (1958) Russian spring-summer virus in India: Kyasanur Forest disease. *Progress in Medical Virology,* **1,** 248–279.

World Health Organization (1966) *Bull. Wld Hlth Org.* **35,** 1–103; (1967) *WHO tech. Rep. Ser.* No. 369; (1974) *Weekly epidem. Rec.* **49,** 277; (1979) *Ibid,* **24,** 191; (1980) *Ibid,* **55,** 52.

91

Bunyaviridae

James S. Porterfield

Introductory

The family *Bunyaviridae* contains more than 200 separate viruses and is by far the largest of the families which contribute to the biologically determined set of arthropod-borne animal viruses, or arboviruses, and one of the largest of all virus families. The name is derived from the type species, Bunyamwera virus, which was isolated in Uganda from a pool of *Aedes* mosquitoes collected during field studies on the epidemiology of yellow fever (Smithburn *et al.* 1946). Bunyamwera was clearly unrelated to yellow fever virus, nor to any other arbovirus known at that time. When relationships were found between Bunyamwera virus and newly isolated agents from other parts of the world, the term 'the Bunyamwera serogroup' was introduced, in contradistinction to the groups A, B and C arboviruses (Casals 1957, Casals and Whitman 1960). As initially conceived, serological groups were, by definition, composed of related viruses while viruses in different groups were unrelated. However, when low level cross-reactions were demonstrated between viruses which had been placed in different serogroups, it became necessary to construct a 'supergroup' composed of viruses which were distantly related to Bunyamwera virus (Whitman and Shope 1962, WHO 1967). Later, when the family *Bunyaviridae* was proposed (Murphy *et al.* 1973), the Bunyamwera supergroup became the genus Bunyavirus of some 87 viruses (Porterfield *et al.* 1975–76). Still later, three further genera were established: Phlebovirus, named after the Phlebotomus or Sandfly fever viruses, Nairovirus, named after Nairobi sheep disease virus, and Uukuvirus, named after the Finnish isolate, Uukuniemi virus (Bishop *et al.* 1980, Matthews, 1982).

Morphology

As defined at present, *Bunyaviridae* are 90–100 nm in diameter, have an envelope with surface projections, and contain a genome of single-stranded, negative sense RNA, divided into three segments with the total molecular weight of the genome being 4–6 × 10^6. Until recently, all known *Bunyaviridae* were arboviruses, members in the Bunyavirus genus being essentially all mosquito-borne, the viruses in the Phlebovirus genus being mostly transmitted by Phlebotomines, and the viruses in the Nairovirus and Uukuvirus genera being tick-borne. However, the recent placing of Hantaan virus and related viruses within the *Bunyaviridae*, although outside existing genera for the present, has broadened the family to include some non-arthropod-borne viruses (McCormick *et al.* 1982; White *et al.* 1982).

Structural, biochemical and genetic aspects

Recent knowledge of the structural, biochemical and genetic properties of *Bunyaviridae* has done much to explain the confusing pattern of serological cross-reactions which has been built up over the years. All *Bunyaviridae* have three major virion proteins, one minor virion protein, and a transcriptase. The major proteins are the two envelope glycoproteins, G1 and G2, and the nucleocapsid protein, N. The minor protein L is a large molecular weight protein associated with the nucleocapsid. The genetic control of these proteins is as follows: the small sized genomic S RNA segment codes for the nucleocapsid protein N, the medium sized M RNA segment codes for the two envelope glycoproteins, G1 and G2, and the large L RNA segment codes for the L protein, and possibly for the transcriptase also. Infection of the same cell with two different *Bunyaviridae* permits genomic reassortment in progeny virions, there being 2^3, or eight different possible ways in which the three genomic segments can occur. This mechanism of recombination almost certainly explains the very large number of antigenically different *Bunyaviridae* encountered in nature, and selective pressures in a wide variety of different vertebrate and invertebrate hosts probably contribute to this diversity. Genetic reassortment has been demonstrated between several different pairs of viruses in the Bunyamwera and California serogroups (Gentsch and Bishop, 1979, Bishop 1979, Iroegbu and Pringle 1981), but there appear to be constraints against recombination between viruses in different genera.

Antigenic relationships

The antigen detected in complement-fixation tests is predominantly the nucleocapsid protein N, whereas the external envelope G1, and to a lesser extent, G2, is the dominant antigen measured in neutralization and haemagglutination-inhibition tests. When the three viruses, Bunyamwera, California and Guaroa were compared, Guaroa virus was clearly placed in the Bunyamwera serogroup on the basis of complement-fixation tests, but in the California group on the basis of neutralization and haemagglutination-inhibition tests (Whitman and Shope 1962). The explanation for this apparent paradox is that the N protein of Guaroa virus is very closely related to the N protein of Bunyamwera virus, whereas the G1 glycoprotein of Guaroa virus is much more closely related to that of California virus than to Bunyamwera G1 protein (Klimas *et al.* 1981).

The genome

While all members of the family have a three segment genome, there is some variation in the molecular weights of each of the three segments. Within a single genus, there is considerable uniformity for each segment, but there are quite substantial differences in RNAs between genera. Similarly, the molecular weights of the virion proteins are conserved within the same genus, but differ between genera; thus the N proteins of Nairoviruses are substantially larger than the N proteins of viruses in other genera (Clerx *et al.* 1981).

Genetic control of virulence

Less is known with certainty of the genetic control of virulence, but there is evidence, at least for viruses in the California serogroup of the Bunyavirus genus, that the ability of a virus to be transmitted orally by mosquitoes is under the control of genes on the medium sized segment M of the RNA (Beaty *et al.* 1981).

For more detailed discussion of different aspects of the *Bunyaviridae*, the following may be consulted: History and relationships to other arboviruses (Theiler and Downs 1973); Catalogue of arboviruses (Berge 1975 and Karabatsos 1978); Biochemistry (Obijeski and Murphy 1977); Structure and genetic properties (Bishop and Shope 1979); Taxonomy, (Bishop *et al.* 1980); Infections of animals (Porterfield and Della Porta 1981): Infections of man (Porterfield 1983*a*); Infections of birds (Porterfield 1983*b*); Transovarial transmission (Tesh and Shroyer 1980).

Table 91.1 shows the relationships between the present taxonomic classification of the *Bunyaviridae*, the serological groups within the family, and the principal disease-producing viruses.

Association with disease

About 40 different *Bunyaviridae* have been associated with diseases in man and a considerably smaller number have been shown to cause disease in animals and birds.

California encephalitis and La Crosse virus

In the Bunyavirus genus there are few important human pathogens. Each year between 30 and 160 cases of *California encephalitis* are diagnosed in the USA, but the great majority of these infections are due to the closely related La Crosse virus, with possibly a few to Jamestown Canyon virus. The prototype California encephalitis virus was isolated in 1943 from *Culex tarsalis* mosquitoes collected in the San Joaquin valley, California, and further isolations were made from *Aedes melanimon* (*dorsalis*), *Aedes vexans*, and other mosquito species in California, New Mexico, Texas and Utah (Hammon *et al.* 1952). Subsequently, sera from children in Kern County, California, who had had illnesses with neurological involvement were found to have antibodies against the California virus (Hammon and Sather 1966). La Crosse virus was isolated by Thompson *et al.* (1965) from brain tissue which had been removed from a 4-year-old girl who had died of encephalitis in La Crosse, Wisconsin. This

Table 91.1 The family *Bunyaviridae*, its genera, serological groups, and viruses associated with disease in man or animals

Genus	Serological group	Viruses producing disease in man or animals
Bunyavirus (over 120)[1]	Anopheles A (5)[2]	Tacaiuma
	Anopheles B (2)	
	Bunyamwera (19)	Bunyamwera, Calovo, Germiston, Guaroa, Ilesha, Maguari, Tensaw, Wyeomyia
	Bwamba (2)	Bwamba
	C group (12)	Apeu, Caraparu, Itaqui, Madrid, Marituba, Murutucu, Oriboca, Ossa, Restan
	California (13)	California encephalitis, Inkoo, La Crosse, Tahyna, Trivittatus
	Capim (8)	—
	Gamboa (3)	—
	Guama (12)	Catu, Guama
	Koongol (2)	—
	Minatitlan (2)	—
	Olifantsvlei (3)	—
	Patois (6)	—
	Simbu (21)	Oropouche, Shuni (Aino, Akabane)[3]
	Tete (5)	—
	Turlock (5)	—
Nairovirus (22)	Bandia (2)	—
	Crimean-Congo (3)	Crimean-Congo HF, Hazara
	Dera Ghazi Khan (5)	—
	Hughes (4)	—
	Nairobi sheep disease (3)	Dugbe, Ganjam, Nairobi sheep disease
	Qalyub (2)	
	Sakhalin (5)	
Phlebovirus (30)	Phlebotomus fever (30)	Candiru, Chagres, Phlebotomus fever, Naples and Sicily, Punta Toro, Rift Valley fever
Uukuvirus (7)	Uukuniemi (7)	Uukuniemi
Other *Bunyaviridae* (30)		Bhanja, Hantaan

1. Approximate number of viruses in genus
2. Approximate number of viruses in serological group
3. Disease in animals

virus, although related to California encephalitis virus, is clearly distinguishable from the prototype strain, and has a different geographic distribution, extending from the east coast of the USA to Utah in the west. Many isolations have been made from a variety of mosquito species, *Aedes*, *Culex*, *Anopheles* and *Psorophora*, and from horseflies; other isolations have been made from sentinel animals exposed in woodlands, and from foxes in nature. La Crosse virus has been shown to be maintained by transovarial transmission in *Aedes triseriatus* mosquitoes over a number of years. Small woodland vertebrates such as chipmunks, tree squirrels, flying squirrels and cotton-tailed rabbits, which have high antibody rates in the Wisconsin area, presumably serve as amplifying hosts (Watts *et al.* 1974, Balfour *et al.* 1976). The states of Wisconsin, Iowa, Indiana, Minnesota and Ohio have the greatest number of recognized human infections with La Crosse virus, but 20 other states have had one case or more, although deaths are exceptional (Gunderson *et al.* 1978).

Most cases of California encephalitis that are diagnosed occur in children, males more than females, and there is nearly always a history of exposure to woodland areas where mosquitoes are common. After an incubation period of from 5 to 10 days there is a gradual onset of fever and headache, which is mild at first, but becomes more severe and usually frontal, occasionally progressing to convulsions and coma. Nausea, neck rigidity and vomiting are common. In spite of meningeal signs, which may be severe, paralysis or permanent brain damage are rare (Balfour *et al.* 1973). Since up to 40 per cent of the population in certain parts of Wisconsin have antibodies against La Crosse virus, many infections are symptomless. The diagnosis of California encephalitis is possible only on the basis of laboratory tests, which may be difficult to interpret because of antigenic differences in viruses in the California encephalitis group (Calisher and Bailey 1981).

Oropouche virus fever

Oropouche virus is a Bunyavirus in the Simbu serogroup. The virus was originally isolated in Trinidad from the blood of a forest worker with a mild febrile illness, but since then several substantial epidemics attributable to Oropouche virus, causing at least 165 000 cases, have occurred in Brazil (Pinheiro *et al.* 1982). Three clinical syndromes have been recognized: a febrile illness, a febrile illness with a rash, and meningitis or meningismus. No deaths have been reported.

The virus is believed to exist in a sylvatic cycle comprising primates, sloths and wild birds as vertebrate hosts, and to spread to an urban cycle involving man through the midge, *Culicoides paraensis* and the mosquito *Culex quinquefasciatus*, the former being the more effective vector.

Akabane and Aino viruses

Two other viruses in the Simbu serogroup, *Akabane* and *Aino* viruses, are important pathogens affecting animals, causing congenital malformations, arthrogryposis, hydranencephaly and other deformities, as well as abortions in cattle, sheep and goats in Japan, Australia, South Africa and Israel (Parsonson *et al.* 1981). It is of interest that both *Culex* mosquitoes and *Culicoides* have been implicated with these viruses also.

Crimean-Congo haemorrhagic fever

A number of viruses in the Nairovirus genus have been associated with disease, of which *Crimean-Congo haemorrhagic fever* is the most important affecting man. Crimean haemorrhagic fever was described by Chumakov (1946) as a severe, acute haemorrhagic fever affecting man in the Crimean region of the USSR, which was transmitted by ticks and carried a mortality of 15 to 30 per cent. Congo virus was first isolated from a 13-year-old African boy in what was then the Belgian Congo, and it was later shown to be responsible for other infections, including at least one fatality, in Uganda (Simpson *et al.* 1967). When a strain of Crimean HF virus became available, Casals (1969) showed that it was indistinguishable from Congo virus, and both names are now used together. The known distribution of these viruses includes the Crimean region, Central and South Africa, Iraq (Burney *et al.* 1980, Al-Tikriti *et al.* 1981) and Pakistan, following the distribution of hyalomma and other ticks (Hoogstraal, 1979). Clinically, an incubation period of about one week is followed by the onset of severe headache, joint and back pains, with fever, nausea, vomiting and photophobia. These may be followed by circulatory collapse, shock, and a variety of haemorrhagic manifestations such as epistaxis, haemoptysis, haematemesis, melaena and skin haemorrhages. Some cases present with acute abdominal pain; and operating theatre staff have been infected and died following surgical intervention in sick cases. In such hospital or nosocomial outbreaks the mortality may be far higher than the usual 15 to 30 per cent. Involvement of the central nervous system indicates a poor prognosis.

Nairobi sheep disease

The type species of the Nairovirus genus, *Nairobi sheep disease*, was first described by Montgomery (1917) and the disease as an acute, haemorrhagic gastro-enteritis affecting sheep and goats in East Africa, and spread by the tick, *Rhipicephalus appendiculatus*. In sheep the mortality may be 70 to 90 per cent, but goats are less seriously affected. Herdsmen tending infected animals may themselves develop a mild, febrile illness, and laboratory infections with the virus have been reported. In India, a very closely related virus was isolated from ticks collected from sheep and goats in Orissa state; this virus goes under the name of Ganjam virus. Another related Nairovirus, Dugbe virus, has been isolated many times from cattle and cattle ticks in Nigeria (Kemp *et al.* 1971) and there have been a few isolations of Dugbe virus from man.

Sandfly fever

The Phlebovirus genus derived its name from the *Phlebotomus, or Sandfly fever viruses*, which have been known since Doerr *et al.* (1909) showed that filtrates of serum taken from patients with pappataci fever would reproduce the disease in human volunteers. The Naples strain of virus was isolated from human serum collected during an outbreak of fever in Naples, and the serologically distinct Sicilian strain was isolated from American troops with Sandfly fever in Palermo, Sicily (Sabin *et al.* 1944). Bartelloni and Tesh (1976) have given a good description of the clinical features of the disease. After an incubation period of from 2 to 6 days, the infection starts with general lassitude, malaise and ill-defined body pains. There is a pyrexia of 39 to 40°, usually within 24 hours of the onset, lasting at least 3 days. Headache may be severe, with pains behind the eyes, which show conjunctivitis, and there is an erythematous rash on the exposed parts of the face, neck and chest, but no true rash such as that seen in dengue fever. Gastro-intestinal disturbances are common, and there is usually a bradycardia. The disease is never fatal and, although convalescence may be slow, there are virtually no complications.

Rift Valley fever

The medical importance of the Sandfly fever viruses is slight as compared with that of *Rift Valley fever* virus, which is now placed within the same Phlebovirus genus. Prior to 1977, Rift Valley fever was known to be endemic in wild game animals in Africa, where it produced occasional severe disease in domestic animals, and very exceptionally, mild disease in man. In 1977 the virus was recognized for the first time in Egypt, where it caused a major epizootic in domestic animals, mainly sheep and goats, but some cattle and camels, and many human infections, resulting in about 600 deaths. The explanation for this increased severity of disease in man is at present unexplained. The principal vector in Egypt appeared to be the mosquito, *Culex pipiens* (Swartz *et al.* 1981). A vaccine for use in man is under development, but protection is probably short-lived (Kark *et al.* 1982).

A number of tick-borne viruses in the Hughes and Sakhalin serogroups of the Nairovirus genus, and some members of the Uukuvirus genus, are associated with bird species in many different habitats, but clear evidence that these viruses are pathogenic to birds is lacking.

Haemorrhagic fever with renal syndrome

Korean haemorrhagic fever was recognized as a distinct disease during the Korean war in the 1950s, but the causative virus, now named Hantaan virus, was not known until 1977 (Lee and Lee 1977). The Korean virus has now been cultivated and appears to be a member of the *Bunyaviridae* on both morphological and biochemical grounds (McCormick *et al.* 1982, White *et al.* 1982, Dalrymple *et al.* 1982).

Clinical and serological studies provide evidence that the disease described by Scandinavian workers as nephropathia epidemica, and that described in the Balkans and European Russia as haemorrhagic fever with renal syndrome, are closely related to the Korean disease, but some serological differences exist. Hantaan virus appears to be an infection of wild rodents in the suprafamily Muridae and the genera *Apodemus*, *Clethrionomys*, *Microtus* and *Rattus* (Anon 1982). An antigenically related virus has recently been detected in the lungs of wild rodents in Maryland, USA, where no overt human disease has been seen. (See also Report 1982, Kitamura *et al.* 1983.)

References

Anon. (1982) *Lancet* **ii,** 1375.
Al-Tikriti, S.K. *et al.* (1981) *Bull. Wld Hlth Org.* **59,** 85.
Balfour, H.H., Edelman, C.K., Bauer, H. and Siem, R.A. (1976) *J. infect Dis.* **133,** 293.
Balfour, H.H., Siem, R.A., Bauer, H. and Quie, P.G. (1973) *Pediatrics* **52,** 680.
Bartelloni, P.J. and Tesh, R.B. (1976) *Amer. J. trop. Med. Hyg.* **25,** 456.
Beaty, B.J., Holterman, M., Tabachnick, W., Shope, R.E., Rozhon, E.J. and Bishop, D.H.L. (1981) *Science*, **211,** 1433.
Berge, T.O. (Ed.) (1975) *International Catalogue of Arboviruses including certain other viruses of vertebrates*, 2nd edn. US Dept. Hlth, Educ. Welfare. Publ. No. (CDC) 75-8301. Atlanta, Georgia, USA.
Bishop, D.H.L. (1979) *Curr. Top. Microbiol. Immunol.* **86,** 1.
Bishop, D.H.L. and Shope, R. E. (1979) *Comprehensive Virol.* **14,** 1.
Bishop, D.H.L. *et al.* (1980) *Intervirol.* **14,** 125.
Burney, M.I. *et al.* (1980) *Amer. J. trop. Med. Hyg.* **29,** 941.
Calisher, C.H. and Bailey, R.E. (1981) *J. clin. Microbiol.* **13,** 344.
Casals, J. (1957) *Trans. N.Y. Acad. Sci.* Ser. 11, **19,** 219; (1969) *Proc. Soc. exp. Biol. Med.* **131,** 233.
Casals, J. and Whitman, L. (1960) *Amer. J. trop. Med. Hyg.* **9,** 73.
Chumakov, M.P. (1946) *Vestn. akad. Nauk. SSSR* **2,** 19 (in Russian).
Clerx, J.P.M., Casals, J. and Bishop, D.H.L. (1981) *J. gen. Virol.* **55,** 165.
Dalrymple, J., Hestys, S., Harrison, S. and Schmaljohn, C. (1982) *Abst. 4th Inst. Conf. Comp. Virol.* Banff, Canada, 117.
Doerr, R., Franz, K. and Taussig, S. (1909) *Das Pappatacifieber*. Deuticke, Leipzig.
Gentsch, J.R. and Bishop, D.H.L. (1979) *J. Virol.* **30,** 767.
Gunderson, C. *et al.* (1978) *CDC Morbid Mort. Wkly Rep.* **27,** 279
Hammon, W.M. and Sather, G. (1966) *Amer. J. trop. Med. Hyg.* **23,** 199.
Hammon, W.M. *et al.* (1952) *J. Immunol.* **69,** 493.
Hoogstraal, H. (1979) *J. med. Entomol.* **15,** 307.
Iroegbu, C.U. and Pringle, C.R. (1981) *J. Virol.* **37,** 383.
Karabatsos, N. (1978) *Amer. J. trop. Med. Hyg.* **27,** 372.
Kark, J.D., Aynor, Y. and Peters, C.J. (1982) *Amer. J. Epidemiol.* **116,** 808.
Kemp, G.E., Causey, O.R. and Causey, C.E. (1971) *Bull. epizoot. Dis. Afr.* **19,** 131.
Kitamura, T. *et al.* (1983) *Jap. J. med. Sci. Biol.* **36,** 17.
Klimas, R.A., Ushijima, H., Clerx-Van Haster, C. and Bishop, D.H.L. (1981) *Amer. J. trop. Med. Hyg.* **30,** 876.
Lee, H.W. and Lee, P.W. (1977) *Korean J. Virol.* **7,** 19.
McCormick, J.B., Palmer, E.L., Sasso, D.R. and Kiley, M.P. (1982) *Lancet* **i,** 765.
Matthews, R.E.F. (1982) *Intervirol.* **17,** 1.
Montgomery, R.E. (1917) *J. comp. Path. Therap.* **30,** 28.
Murphy, F.A., Harrison, A.K. and Whitfield, S.G. (1973) *Intervirol.* **1,** 297.
Obijeski, J.F. and Murphy, F.A. (1977) *J. gen. Virol.* **37,** 1.
Parsonson, I.M., Della-Porta, A.J. and Snowdon, W.A. (1981) *Amer. J. trop. Med. Hyg.* **30,** 660.
Pinheiro, F.O. *et al.* (1982) *Science* **215,** 1251.
Porterfield, J.S. (1983*a*) In: *Oxford Textbook of Medicine*, Vol. 1, pp. 5, 105. Ed. by D.J. Weatherall, J.G.G. Ledingham and D.A. Warrell. Oxford University Press, Oxford; (1983*b*) In: *Viral Infections of Birds*. Ed. by J.B. McFerran. Elsevier, Amsterdam.
Porterfield, J.S. and Della-Porta, A.J. (1981) In: *Comparative Diagnosis of Viral Diseases*, Vol. IV, p. 479. Ed. by E. Kurstak and C. Kurstak. Academic Press, New York.
Porterfield, J.S. *et al.* (1975–6) *Intervirol.* **6,** 13.
Report (1982) *Scand. J. infect. Dis.* Suppl. No. 36, pp. 82–95.
Sabin, A.B., Philip, C.B. and Paul, J.R. (1944) *J. Amer. med. Ass.* **125,** 603, 693.
Simpson, D.I.H. *et al.* (1967) *East Afr. med. J.* **44,** 87.
Smithburn, K.C., Haddow, A.J. and Mahaffy, A.F. (1946) *Amer. J. trop. Med. Hyg.* **26,** 189.
Swartz, T.A., Klingberg, M.A., and Goldblum, N. (Eds.) (1981) In: *Rift Valley Fever*. Karger, Basel.
Tesh, R.B. and Shroyer, D.A. (1980) *Amer. J. trop. Med. Hyg.* **29,** 1394.
Theiler, M. and Downs, W.G. (1973) In: *The Arthropod-borne Viruses of Vertebrates*. Yale University Press, Yale.
Thompson, W.H., Kalfayan, B. and Anslow, R.O. (1965) *Amer. J. Epidemiol.* **81,** 245.
Watts, D.M. *et al.* (1974) *Amer. J. trop. Med. Hyg.* **23,** 694.
White, J.D., Shirey, F.G., French, G.R., Huggins, J.W., Brand, O.M. and Lee, H.W. (1982) *Lancet* **i,** 768.
Whitman, L. and Shope, R.E. (1962) *Amer. J. trop. Med. Hyg.* **11,** 691.
World Health Organization (1967) *Tech. Rep. Ser.* No. 369.

92

Arenaviridae

D. I. H. Simpson

Introductory

The arenaviruses take their name from their sand-sprinkled (arenous) appearance when viewed in the electron microscope. They are a group of enveloped single-stranded RNA viruses. Nearly all of the 12 members so far described have rodents as their natural reservoir hosts. Only four are of medical importance, including the prototype member of the group, lymphocytic choriomeningitis (LCM) virus. This virus has been known since 1934. It occurs all over the world as a common contaminant of laboratory mice, rats and hamsters and occasionally infects man. Eight viruses, including the medically important Junin and Machupo viruses, occur only in the New World and two, including Lassa, are confined to Africa. All the New World members, often referred to as the Tacaribe complex, are more or less antigenically related; LCM and Lassa show only distant relationship to the New World viruses of the group and are not closely related to each other.

Although rodents are divided into over 30 families distributed world-wide, arenaviruses are predominantly associated with two major families, the Muridae (e.g. mice, rats) and Cricetidae (e.g. voles, lemmings, gerbils). The nature of the original animal reservoir for LCM virus remains obscure, but it appears to be primarily associated with species of the Muridae family which evolved in the Old World and subsequently spread to most parts of the globe. The natural reservoir of Lassa virus, Mastomys natalensis, is also a member of the Muridae and, in common with the host of LCM, frequents human dwellings and food stores. In contrast, nearly all arenaviruses isolated from the South American continent have been associated with cricetid rodents whose members frequent open grasslands and forest. For example, Argentinian haemorrhagic fever infections occur predominantly during the major harvesting season of April to June when agricultural workers move temporarily into the endemic area. The exception is Tacaribe virus, which was originally isolated from the fruit-bat, *Artibeus literatus*. Tamiami virus, isolated in Florida, USA, from the cotton-rat, *Sigmodon hispidus*, completes the Tacaribe complex of arenaviruses so far characterized. It is interesting to note that Arata and Gratz (1975) pointed out the close relationship between the cotton-rat and the South American cricetid rodents, in particular *Oryzomys* sp., from which the Pichinde virus was isolated; this virus shares an antigenic relationship with Tamiami virus (Table 92.1).

Table 92.1 *Arenaviridae*: Hosts and geographic distribution

Virus	Vertebrate host	Geographic distribution
Tacaribe	*Artibeus lituratus*	Trinidad
	Artibeus jamaicensis	Trinidad
Junin	*Calomys laucha*	Argentina
	Calomys musculinus	Argentina
	Akadon azarae	Argentina
Machupo	*Calomys callosus*	Bolivia
Amapari	*Oryzomys goeldi*	Brazil
	Neucomys guianae	Brazil
Tamiami	*Sigmodon hispidus*	USA
Pichinde	*Oryzomys albigularis*	Colombia
Parana	*Oryzomys buccinatus*	Paraguay
Latino	*Calomys callosus*	Bolivia
LCM	*Mus musculus*	World-wide
Lassa	*Mastomys natalensis*	West Africa
Mopeia	*Mastomys natalensis*	Mozambique, Zimbabwe
Flexal	*Neocomys* spp.	Brazil

Lymphocytic choriomeningitis virus

Clinical features

The infection in many may present as an inapparent infection; as an influenza-like febrile illness; as aseptic meningitis; or as severe meningoencephalomyelitis. The great majority of LCM infections are benign.

The incubation period has been estimated to be from 6 to 13 days. In the influenza-like illness there is fever, malaise, coryza, muscular pains, and bronchitis. In the meningeal form, which is more common, the same symptoms may remain mild and be of short duration and patients recover within a few days but there can be a more pronounced illness with severe prostration lasting for 2 weeks or more. Chronic sequelae have been reported on occasions. They include headache, paralysis, and personality changes. The few deaths that have occurred followed severe meningoencephalomyelitis and occasionally severe bleeding. In one case there was mild pharyngitis and a diffuse erythematous rash followed by haemorrhages and death.

A leukopenia has been a constant finding early in the course of disease, with a lymphocytosis later. In central nervous system disease, the cerebrospinal fluid (CSF) is at increased pressure with slight increase in protein concentration, normal or slightly reduced sugar concentration, and a moderate number of cells (150–400/mm^3). These changes are not peculiar to LCM infections. Virus can be isolated from blood, CSF, and in fatal cases from brain tissue.

Epidemiology

Man is usually infected through contact with mice. Many infections occur in laboratories, where LCM may be a contaminant in laboratory colonies of mice and hamsters. Hamsters kept as pet animals have also played a role in infections of man. The mechanism of transmission of the virus to man is not fully understood but it may be airborne through dust contamination by urine, or through the alimentary tract from the contamination of food and drink. Transmission may also occur through skin abrasions.

Argentinian haemorrhagic fever (Junin virus)

Clinical features

Argentinian haemorrhagic fever has been known since 1943. Junin virus, the causative agent, was first isolated in 1958. The virus causes annual outbreaks of severe illness, with between 100 and 3500 cases, in an area of intensive agriculture known as the wet pampa in Argentina. Mortality in some individual outbreaks has ranged from 10 to 20 per cent, although the overall mortality is generally 3–15 per cent.

After an incubation period of 7–16 days, the onset of illness is insidious with chills, headache, malaise, myalgia, retro-orbital pain, and nausea; this is followed by fever, conjunctival injection and suffusion and an enanthema, exanthema and oedema of the face, neck and upper thorax. A few petechiae are most pronounced in the axilla. There is hypervascularity and occasional ulceration of the soft palate. Generalized lymphadenopathy is common. After a few days in the more severe cases the patient's condition becomes appreciably worse with the development of hypotension, oliguria, haemorrhages from the nose and gums, haematemesis, haematuria, and melaena. Oliguria may progress to anuria and pronounced neurological manifestations may develop. Laboratory findings have included leukopenia, thrombocytopenia and urinary casts. Patients recover when the fever falls by lysis. This is followed by diuresis and rapid improvement. Death may result from anaemic coma or hypovolaemic shock caused by plasma leakage rather than blood loss. Subclinical infections occur but are extremely rare and man-to-man transmission has not been observed. Treatment with immune plasma is said to be efficacious (Maiztegui *et al.* 1979).

Pathology

Lymphadenopathy and lymphocyte depletion of the spleen have been found at post-mortem together with endothelial swelling in capillaries and arterioles in almost every organ. Focal necrosis in the liver with acidophilic necrosis of hepatocytes and Kupffer-cell hyperplasia have been noted. Other lesions which have been described include tubular necrosis in the kidneys, minimal inflammation of the central nervous system and myocardium, and occasional evidence of intravascular coagulation.

Epidemiology

Argentinian haemorrhagic fever is sharply seasonal, coinciding with the maize harvest between April and July, when rodent populations reach their peak. Agricultural workers, particularly those harvesting maize, are most commonly affected.

The main reservoir hosts of Junin virus are *Calomys* species which live and breed in burrows in the maize fields and in the banks which surround them. Other rodent species may also be affected. *Calomys* species have a persistent viraemia and viruria, and virus is also present in considerable quantities in saliva. The mode of transmission of Junin virus to man has not been conclusively established. The virus may be carried in the air from dust contaminated by rodent excreta or may enter via the alimentary tract in similarly contaminated foodstuffs.

Bolivian haemorrhagic fever (Machupo virus)

Clinical features

Bolivian haemorrhagic fever was first recognized in 1959 in the rural areas of the Beni Region in north-eastern Bolivia. The disease has continued to occur in that region more or less annually in sharply localized epidemics. The mortality in individual outbreaks has varied from 5 to 30 per cent. The most notable outbreak affected 700 people in San Joaquin Township between late 1962 and the middle of 1964. The mortality was 18 per cent.

The disease is similar to Argentinian haemorrhagic fever. The incubation period ranges from 7 to 14 days and the onset is insidious. About one-third of patients show a tendency to bleed with petechiae on the trunk and on the palate and bleeding from the gastro-intestinal tract, nose, gums and uterus. Almost half the patients develop a fine intention tremor of the tongue and hands and several patients may have more pronounced neurological symptoms. The acute disease can last 2–3 weeks and convalescence may be protracted, with generalized weakness being the commonest complaint. Clinically inapparent infection is rare.

Machupo virus, the responsible agent, has been readily isolated from lymph nodes and spleen taken at necropsy in fatal cases. Isolation of the virus from acutely ill patients has been difficult. The best results have been obtained with specimens taken 7–12 days after the onset of illness.

Epidemiology

The rodent reservoir of Machupo virus is *Calomys callosus*; over 50 per cent of this species caught during the San Joaquin epidemic were found to be infected. The distribution of cases in the Township was associated with certain houses and *C. callosus* was trapped in all households where cases occurred. Transmission to man is probably by contamination of food and water or by infection through skin abrasions. Transmission from man-to-man is unusual but a small episode took place in 1971 well outside the endemic zone. The index case, infected in Beni, carried the infection to Cochabamba and, by direct transmission, caused five secondary cases, of which four were fatal.

Lassa virus

History

Lassa fever made a dramatic appearance in Nigeria in 1969 as a lethal, highly transmissible, hitherto unknown disease. The first victim was an American nurse who was infected at a small Mission Station in Lassa Township in north-eastern Nigeria, whence the virus and the disease which it causes derive their names. The mode of the nurse's infection was never determined, although it is thought that the disease was acquired through direct contact with a febrile patient in Lassa. When her condition steadily deteriorated she was flown to the Evangel Hospital in Jos, where she died the following day.

While she was in hospital, she was cared for by two other American nurses, one of whom was sequentially infected by direct contact, probably through a skin abrasion. This nurse became unwell after an 8-day incubation period and died after an illness lasting 11 days. The head nurse of the hospital, who had assisted at the post-mortem of the first patient and had cared for the second patient, fell ill 7 days after the second patient died. She was probably infected while nursing the patient rather than at the post-mortem examination.

The third patient was evacuated to the USA by air in the first-class cabin of a commercial airliner with two attendants and screened from economy class passengers by only a curtain. After a severe illness under intensive care she slowly recovered. A virus, subsequently named Lassa, was isolated from her blood by workers at the Yale Arbovirus Unit. One of these virologists became ill but gradually improved, and survived, after an immune plasma transfusion donated by the third case. Five months after this infection, a laboratory technician in the Yale laboratories, who had not been working with Lassa virus, fell ill and died. The mode of infection was never elucidated.

This tragic trail of events not unnaturally earned for Lassa virus a formidable notoriety. This malevolent reputation was sharply enhanced by two more devastating hospital outbreaks, one in Nigeria, the other in Liberia.

The fourth epidemic was seen in Sierra Leone between October 1970 and October 1972. This outbreak was not confined to hospitals, although hospital staff were at considerable risk and several were infected. Most of the patients acquired their illnesses in the community and there were several intrafamilial episodes.

Lassa fever has continued to occur in West Africa, usually as sporadic cases. Between 1969 and 1978, there have been seventeen reported outbreaks affecting 386 patients and causing 105 deaths, which is an average mortality of 27 per cent. Eleven of the episodes were in hospitals, comprising 57 cases and 25 deaths, a case-fatality rate of 44 per cent; two were laboratory infections, two were individual community-acquired outbreaks, and two were prolonged community outbreaks. Eight patients were flown to Europe or North America. One of them was evacuated with full isolation precautions and the remainder, of whom five were infectious, travelled on scheduled commercial flights as fare-paying passengers. Fortunately no contact cases resulted.

Clinical features

Lassa virus causes a wide spectrum of disease from subclinical to fulminating, fatal infection (see Edmund 1980). The incubation period ranges from 3 to 16 days and the illness usually begins insidiously with non-specific symptoms such as feverishness, chilliness, malaise and muscular pains followed by fever, headache, and sore throat. Between the third and sixth days of illness the symptoms dramatically worsen and there is high fever, severe prostration, chest and abdominal pains, conjunctival injection, diarrhoea, dysphagia and vomiting. Chest pain, located substernally and along the costal margins, is often associated with tenderness on pressure and is exacerbated by coughing and deep inspiration. One important physical finding is a distinct pharyngitis with or without a sore throat, and yellow-white exudative spots may occur on the tonsillar pillars together with small vesicles and ulcers. The patient appears intoxicated, lethargic and dehydrated; the blood pressure is low with a narrow pulse pressure, and there is sometimes a relative bradycardia. There may be cervical lymphadenopathy, a coated tongue, puffiness of the face and neck, and blurred vision. Occasionally a faint maculopapular rash may be seen during the second week of illness on the face, neck, trunk and arms. In severe cases, haemorrhages also occur. Cough is a common symptom and light-headedness, vertigo and tinnitus appear in a few patients. Deafness has been noted in about 20 per cent of patients and may be reversible but is more often permanent.

The fever generally lasts for 7–17 days and is often variable. Convalescence begins in the second to fourth weeks, when temperature returns to normal and the symptoms improve. Most patients complain of extreme fatigue for several weeks. Loss of hair and deafness are often observed and brief bouts of fever can occur.

Patients in whom the disease is fatal not uncommonly have a high sustained fever. Acutely ill patients suddenly deteriorate between days 7 and 14 with a sudden drop in blood pressure, peripheral vasoconstriction, hypovolaemia and anuria. The patient is restless and apprehensive and there may be pleural effusions and ascites. In addition, coma, stupor, tremors and myoclonic twitching can occur. Death is due to shock, anoxia, respiratory insufficiency and cardiac arrest.

Subclinical infection or mild attacks of Lassa fever seem to be quite common, at least in the endemic area of Sierra Leone. There, surveys have shown that 6–13 per cent of the population have Lassa antibodies.

Because of the hazards which might result from contact with blood or other body fluids, few laboratory studies have been carried out. In the early stages a leucopenia has been noted but a leucocytosis can occur later in the illness. The differential count has not been particularly helpful but there is usually a neutropenia and a low platelet count. The prothrombin level may be normal or low while results of other blood-clotting studies are usually normal. Albuminuria and granular casts in the urine have been noted in several instances. Serum bilirubin and protein concentrations are not altered.

The absence of characteristic signs and symptoms in the early stage of infection makes clinical diagnosis difficult unless there is a history of contact with a known case or there is a local outbreak. The possibility of Lassa fever should be considered whenever a patient from an endemic area presents with an unexplained febrile illness of relatively slow onset. A detailed travel history is very relevant because Lassa fever is endemic in West Africa. The differential diagnosis in the early acute stage would include malaria, typhoid fever, influenza, yellow fever, infectious mononucleosis, other virus infections, meningococcaemia and septicaemia.

Pathology

Only a limited number of necropsies have been carried out. Lassa virus is pantropic and causes lesions or malfunction in liver, spleen, kidney, myocardium, lung, brain, skeletal muscle, and pleura. Increased capillary permeability leads to interstitial oedema and haemorrhage, while necrotic foci are found in many organs, particularly in the liver and spleen. Changes in the kidneys are minimal, contrasting strangely with the functional impairment. It is still uncertain whether the cellular damage is caused directly by the virus or by the immunological response to infection. The simultaneous presence of Lassa virus and naturally acquired antibodies have been found in patients' blood during the second week of illness. This suggests that some of the pathological changes may be due to the deposition of antigen-antibody complexes.

Epidemiology

Lassa virus has been repeatedly isolated from the multimammate rat *Mastomys natalensis* in Sierra Leone and Nigeria. This rodent is a common domestic and peridomestic species and large populations are widely distributed in Africa south of the Sahara. During the rainy season it may desert the open fields and seek shelter indoors. Some genetic variation has been shown in *Mastomys* populations which may account for the focal distribution of Lassa fever in West Africa. Not all variants of the species may be equally susceptible to the virus. The animals are infected at birth or during the perinatal period. Like other arenaviruses, Lassa virus produces a persistent tolerated infection in its rodent reservoir host with no ill effects and without any immune response. The animals remain infective during their lifetime, freely excreting Lassa virus in urine and other body fluids.

Primary infection in man is thought to be acquired directly from infected rodent urine or indirectly from foodstuffs or dust contaminated by urine. It has been suggested that a low level of sanitation, storage of food within houses, and the ease with which the rodents can infect mud-and-thatch houses increase contact between rodents and man. However, the way the virus is spread from person to person is still not clear. Secondary spread from person to person may occur in overcrowded houses but spread of this kind is particularly important in rural hospitals. Medical attendants or relatives providing direct personal care are most likely to contract the infection and accidental inoculation with a sharp instrument and contact with blood have accounted for a few cases. Airborne spread as well as mechanical transmission can occur. Although in Sierra Leone there has been no evidence of airborne spread in hospital outbreaks, one of the 1970 outbreaks in Nigeria is believed to have been caused by airborne transmission from a woman with severe pulmonary infection (see Woodruff 1978).

With rapid air trâvel, it is all too easy for patients incubating Lassa infection to develop their illness in distant countries where the disease and its natural reservoir hosts do not normally exist.

Laboratory investigations

The diagnosis of Lassa fever is confirmed by isolating the virus or demonstrating a serological response to Lassa virus in serum.

Lassa virus grows readily in Vero cell culture and isolation of the virus can usually be accomplished within 4 days. Virus has been isolated from serum, throat-washings, pleural fluid, and urine; virus excretion has been shown in the pharynx for up to 14 days after the onset of illness and in urine for up to 67 days after the onset. It may be possible to make an early diagnosis of Lassa infection by detecting Lassa-specific antigens in conjunctival cells by using indirect immunofluorescence, but this test system is still under review. Virus isolation should be attempted only in laboratories specially equipped to provide maximum containment to protect the investigator.

The most sensitive serological test for the detection of Lassa antibodies is the indirect immunofluorescence reaction. Antibodies can be detected by this method in the second week of illness. Complement-fixing antibodies develop more slowly and are rarely detectable before the third week after onset. On occasions complement-fixing antibodies have failed to develop in patients from whom Lassa virus has been isolated. (For the management of patients suspected of having Lassa fever and the surveillance of contacts, see Woodruff, 1978.)

Ultrastructure of arenavirus and infected cells

Negative-staining electron microscopy of extracellular virus shows pleomorphic particles ranging in diameter from 80 to 150 nm. The virus envelope is formed invariably from the plasma membrane of infected cells. A significant thickening of both bilayers of the membrane together with an increase in the width of the electron-translucent intermediate layer is characteristic of arenavirus development. Little is known about the internal structure of the arenavirus particle, although thin sections of mature and budding viruses clearly show the ordered, and often circular, arrangement of host ribosomes typical of this virus group. In some particles, a linear structure approximately 20 nm thick appears to connect these structures (Murphy and Whitfield 1975). It is not clear, however, if these are in any way associated with virus RNA. Palmer *et al.* (1977) showed the release of distinct well-dispersed filaments 5 to 10 nm in diameter from detergent-treated Tacaribe virus. Two predominant size classes were present with average lengths of 640 nm and 1300 nm respectively. Each appeared circular and beaded in appearance. Vezza *et al.* (1977*a*, *b*) observed convoluted filamentous strands up to 15 nm in diameter in preparations of spontaneously disrupted Pichinde virus. However, the internal structure of this or any other arenavirus remains to be clearly defined.

It has been shown that arenavirus replication in experimental animals frequently proceeds in the absence of any gross pathological effect. However, cellular necrosis may accompany virus production, not unlike that seen *in vitro* with virus-infected cell cultures. The variable pathological changes associated with arenavirus infections are further complicated by the infrequent appearance of particles in tissue sections which give strong positive reactions with fluorescein-conjugated antisera (Murphy *et al.* 1973). Granular fluorescence in the perinuclear region of acutely infected Vero cells is frequently seen with convalescent serum. In addition, intracytoplasmic inclusion bodies are a prominent feature in virus-infected cells both *in vivo* and *in vitro*. These usually appear early in the replication cycle, consisting largely of single ribosomes which later become condensed in an electron-dense matrix, sometimes together with fine filaments (Murphy and Whitfield 1975).

Pathology of arenavirus infections

The classical example of virus-induced immunopathological disease is exhibited by LCM virus infection of adult mice (Casals 1975), in which intracerebral inoculation causes severe disease and death. In contrast, if mice are infected before or shortly after birth they develop a non-pathogenic life-long carrier state. The newborn mouse is immunologically immature and the virus does not stimulate an immune response; the virus causes no illness. The immunologically mature adult mouse shows an immune response following LCM virus infection and a fatal choriomeningitis results without evidence of neuronal damage (Lehmann-Grube 1971). Immunosuppressant treatment, including neonatal thymectomy and antilymphocytic serum, protects adult mice against fatal LCM infection; the immune disease appears to be cell mediated.

The mechanisms by which arenaviruses cause disease in man are not fully understood. There is no evidence that either immunopathological or allergenic processes play any part in causing disease; it appears much more likely that direct virus damage to cells is the cause. Although the lesions produced in man by LCM virus are little understood, post-mortem studies on patients who died after Junin virus infection have shown generalized lymphadenopathy with endothelial swelling in capillaries and arterioles in almost every organ, together with lymphocyte depletion in the spleen. There are many similarities in the pathological lesions found in man following Junin, Lassa and Machupo virus infections. Focal non-zonal necrosis in the liver has been described in all three conditions (Child *et al.* 1967, Elsner *et al.* 1973, Winn and Walker 1975) with Kupffer cell hyperplasia, erythrophagocytosis and acidophilic necrosis of hepatocytes. Other lesions which have been described include interstitial pneumonitis, tubular necrosis in the kidney, lymphocytic infiltration of the spleen, minimal inflammation of the central nervous system and myocardium and occasionally evidence of intravascular coagulation. Johnson *et al.* (1973) suggested the following pathogenesis of arenavirus infection in man: the virus enters the body either by the alimentary or upper respiratory tract, is gathered up in local lymphoid tissue or lymph nodes where it first replicates; it then invades the reticuloendothelial system to include those cells concerned in the immune and cellular immune responses, thereby impairing the host's defence mechanisms. Either directly or indirectly the virus capillary damage leading to capillary fragility, haemorrhages and hypovolaemic shock. The malfunction of various organs may be due to capillary damage and oedema of the parenchyma rather than to direct cell damage. As the disease regresses no permanent damage follows, since little cytolysis occurs, although in severe cases direct cytopathic damage to cells may be significant.

Host response to arenavirus infections

Four arenaviruses—Junin, Lassa, lymphocytic choriomeningitis (LCM) and Machupo—are capable of causing significant and often severe human disease. Serological evidence suggests that Pichinde virus may also infect man but no serious human disease has been reported to date. LCM virus infection in man may present in one of four ways: as an inapparent infection, as an influenza-like febrile illness, as aseptic meningitis, or as severe meningoencephalomyelitis. The majority of LCM infections follow a benign course. Very few fatalities have occurred after nervous system involvement or very occasionally after a severe generalized illness accompanied by haemorrhages (Smadel *et al.* 1942). Following infection with either Junin, Lassa or Machupo viruses, the diseases produced in man are very similar. The onset of illness is insidious with chills, malaise, headache, pain behind the eyes and in muscles and nausea followed by fever, conjunctival injection and suffusion, an enanthem, exanthem and oedema of the face, neck and upper thorax. Petechiae and lymphadenopathy are common. After a few days the patient's condition becomes appreciably worse with the development of hypotension, oliguria, haemorrhages from the gums and nose, haematemesis, haematuria and melaena. Oliguria may turn to anuria and pronounced neurological manifestations may develop. Death may result from anaemic coma or hypovolaemic shock caused by plasma leakage. In Lassa infections, pharyngitis with ulcerative lesions on the tonsils is a frequent finding. Case-fatality rates in individual outbreaks have varied from 5 per cent. Junin and Machupo are rarely transmitted from person to person, unlike Lassa which has a formidable reputation resulting from frequent man-to-man transmissions. This may be associated with the low titre of circulating virus in patients with Argentinian and Bolivian haemorrhagic fevers in contrast with the high titre found in cases of Lassa infection (J. I. Maiztegui and K. M. Johnson, personal communications). Inapparent and subclinical infections with Lassa are now believed to be quite common, but inapparent infections with Junin and Machupo viruses are extremely rare.

Almost all the arenaviruses cause a fatal illness in suckling mice after intracerebral inoculation. The exceptions are Latino virus which does not infect mice, Parana virus which causes illness but no deaths, and LCM which has no effect on suckling mice but does produce a fatal infection in young adult mice. Junin, Latino, Machupo, Parana and Pichinde virus will kill new-born hamsters while guinea-pigs are susceptible to Junin and LCM virus infection. Machupo and Lassa viruses cause fatal infections in monkeys; Junin virus has not received as much attention in simians.

Overwhelming evidence of early infection in lymphoreticular cells indicates that the arenaviruses may interfere with host immune defences during acute infection. Murphy *et al.* (1973) showed that huge numbers of virus particles were present in the lymph nodes, thymus and spleen of Machupo-infected *Calomys callosus* rodents. Large numbers of virus particles associated with the plasma membrane of large lymphoblastoid cells were a noticeable feature. Similar studies have also been performed with Tamiami virus infection of the cotton-rat (Murphy *et al.* 1976); in the latter study, megakaryotes were also infected. This finding suggests a possible general relation between platelet function and the pathogenesis of arenavirus haemorrhagic fevers. It has been suggested that interferon induced by LCM infection of mice results in activation of natural killer cells (Welsh 1978). In addition, it has been reported by Gee, Clark and Rawls (1979) that adult hamsters susceptible to Pichinde virus show increased levels of natural killer cell-mediated cytotoxicity for syngeneic and allogeneic tumour cells after infection. Separation of spleen cells revealed that fractions rich in natural killer cell activity also contained the majority of the infectious centres in the infected spleen, suggesting that this cell type might be a target for Pichinde virus replication.

Antigenic relationships

Each member of the arenavirus taxon possesses some antigenic relation to other members of the group, although the degree of cross-reactivity largely depends on the assay system used. The complement-fixation test exhibits the broadest relationships (Casals 1975). All the New World members, often referred to as the Tacaribe complex, are more or less related by complement fixation, with Junin, Machupo, Amapari, Parana, Latino, and Tacaribe viruses being most closely related. Pichinde and Tamiami are not so closely related to each other or to other members of the Tacaribe complex by this serological test. All the available evidence indicates that the complement-fixing antigen is associated with the internal nucleoprotein. By this test LCM and Lassa are only very distantly related to the Tacaribe complex but they do show some distant relationship to each other.

The neutralization test is much more specific. In plaque-reduction tests with patients' antisera no cross-neutralization has been observed between viruses of the Tacaribe complex, which are closely related by complement fixation. For example, Junin and Machupo viruses are quite distinct. A similar specificity has been demonstrated with LCM and Lassa and both viruses are readily distinguishable from one another by this technique. The sensitivity of the neutralization test can be increased by incorporating either complement or anti-γ globulin into the test system (Lehmann-Grube and Ambrassat 1977, Oldstone 1975).

The fluorescent antibody test, particularly indirect immunofluorescence, is less specific than the neutralization test. The test generally stains cytoplasmic antigens; but by this technique LCM and Lassa viruses are clearly related to each other but only distantly to members of the Tacaribe complex. A relation has also been shown by this method between Lassa and an isolate from Mozambique (Wulff *et al.* 1977). However Buchmeier and Oldstone (1978) have shown that by immunofluorescent staining of the surface of infected cells no cross-reactivity was found among members of the Tacaribe complex and between those viruses and LCM.

The assay systems currently being used to determine antigenic relationships among the arenaviruses are thus clearly not of sufficient sensitivity, and new techniques are urgently needed. Those showing promise are radioimmunoassay and enzyme-linked immunosorbent assay (ELISA) by using either human or animal antisera, although these methods have yet to be fully evaluated. The cytotoxicity test might also prove of value using infected histocompatible target cells.

Chemical composition of arenaviruses

Protein

There is now fairly common agreement that arenaviruses contain a major nucleocapsid-associated protein of mol. wt 54–68 000 together with one or two glycoproteins in the outer viral envelope. Polypeptide analyses have been reported for all members of the Tacaribe complex except Amapari, Flexal and Latino viruses. No information is yet available for the protein composition of Lassa or the recently-obtained virus isolated from *Mastomys* sp. caught in the Mopeia region of Mozambique. The serological relatedness of the latter viruses to Lassa make them suitable candidates for a thorough analysis of the African arenaviruses if and when their apparent non-pathogenicity for man is accepted.

Several minor non-glycosylated components have been consistently seen in purified arenavirus preparations: a 72 000 mol. wt polypeptide together with a small 15 000 mol. wt protein are found to be associated with the core component of Pichinde virus (Young and Howard 1983). Other, larger proteins are frequently found which may represent the viral RNA polymerase, but as yet this enzyme has not been purified from either virus or virus-infected cells. Among the other components, there is common agreement over the existence of a 20–25 000 mol. wt protein in LCM (Pedersen 1973), Junin (Segovia and de Mitri 1977) and Pichinde virus (Young and Howard 1983). It has been suggested by Pedersen (1979) that the 25 000 mol. wt protein of LCM virus may represent a matrix protein, although it was noted by Young and Howard (1983) that the 22 000 mol. wt component was labelled readily by several iodination procedures, suggesting that at least a portion of this protein was accessible at or near the virion surface.

Evidence is emerging that the major surface glycoproteins of arenaviruses are not primary gene products; Buchmeier and Oldstone (1979) have shown by peptide analysis and pulse-chase experiments that the two LCM virus glycoproteins arise by proteolytic cleavage of a 75 000 mol. wt glycoprotein present in immunoprecipitates of infected cells. In common with the cell-associated glycoprotein reported by Saleh *et al.* (1979) to be present in Tacaribe infected cells, the LCM virion precursor glycopeptide is comparatively rich in mannose and glucosamine. In contrast, the larger 44 000 mol. wt glycoprotein present in mature LCM virus particles contained glucosamine, fucose and galactose, resembling the branched, A-type glycosylation pattern as described by Johnson and Clamp (1971). These findings, together with the observation that the 54 000 mol. wt glycoprotein of Pichinde virus may be specifically radiolabelled after exposure to galactose oxidase (Young and Skelly, unpublished observations) suggests that shortening of at least some mannose-rich carbohydrate side chains of a precursor may occur followed by the addition of sugars such as fucose or galactose prior to cleavage.

Although the nature of arenavirus surface antigens has yet to be defined, it is likely that the surface glycoproteins play an important role in eliciting a protective antibody response. Indirect evidence of this is the finding that serum containing neutralizing antibodies to LCM virus also immunoprecipitates both polypeptide species (Buchmeier *et al.* 1980). It yet remains to be established, however, whether each protein is present in all of the outer projections or whether these morphologically indistinguishable structures are in fact composed of two or more chemically different types. The role of minor structural components in the formation of antigenically reactive surface components is at present unclear.

Nucleic acid

All evidence obtained so far indicates that purified arenavirus preparations contain at least two discrete pieces of single-stranded RNA with different sizes. Pedersen (1970) showed that two major and one minor RNase-sensitive components were released from LCM virus by treatment with 1 per cent SDS and separated in high-salt sucrose gradients into three populations of 28S, 22S, and 18S respectively. LCM RNA was resolved into four fractions by PAGE, two of which comigrated with 28S and 18S cellular ribosomal RNA (Pedersen 1971). The level of radiolabel incorporated into these two viral RNA species was reduced by the additon of 0.15 μg/ml actinomycin D to the cell cultures at the time of infection; the yield of infectious virus was not significantly reduced under these experimental conditions. The remaining components were deemed to be virus-specific and estimated as having S-values of 31 and 21 respectively. Pedersen and Konigshofer (1976) subsequently reported that both RNA species were associated with a heterogeneous population of ribonucleoprotein of average size 123 to 148S. In addition, a second 83S structure was found which contained only the 23S viral RNA. The multicomponent nature of the arenavirus genome has since been confirmed for other arenaviruses including Pichinde and Junin. The total molecular weight of the virus genome is variously estimated to be in the range of 3.2 to 4.8×10^6. In addition, Farber and Rawls (1975) found a 15S RNA species which was thought to be a third viral RNA segment but this finding has not been confirmed.

Considerable evidence is now available to prove that the electron-dense 20 to 25 nm particles visible within purified virions represent host cell ribosomes and the 28S and 18S RNA species within purified virus particles represent associated host RNA and may account for up to 50 per cent of the total RNA content. Carter, Biswal and Rawls (1973) showed that both 28S and 18S RNA extracted from Pichinde virus possessed a high guanosine plus cytosine (G+C) content and methylated bases characteristic of ribosomal RNA. This study was complemented by the direct demonstration of ribosomal RNA associated with host ribosomes released from virus grown in BHK-21 cells pre-labelled for 48 hours with ^{3}H-uridine. These structures possessed similar physicochemical properties to ribosomes from uninfected BHK-21 cells (Farber and Rawls 1975), and further experiments showed that, like cell ribosomes, 80S structures extracted from Pichinde virus were susceptible to EDTA dissociation and contained both host-derived 28S and 18S RNA. Heterogeneous low molecular weight RNA has frequently been found within arenavirus particles and almost certainly represents further ribosomal RNA material (Carter *et al.* 1973, Pedersen 1973). Oligonucleotide analysis has shown conclusively that the 28S and 18S RNA structures of Pichinde virus are identical with those of non-infected host cell ribosomal RNA (Vezza *et al.* 1978*a*). It should be noted, however, that the total amount of ribosomal RNA, and the relative proportion of the two species, vary considerably (Vezza and Bishop 1977, Leung *et al.* 1979). Only small amounts of 28S RNA and non 18S RNA were reported for Pichinde virus grown from a freshly-cloned stock previously passaged at 37°. The frequent lower ratio of 18S to 28S RNA present in purified virus compared to uninfected cell ribosomes may reflect the higher frequency of 18S RNA degradation during phenol extraction (Vezza *et al.* 1978*a*). The total ribosomal RNA content may in turn be influenced by the varying proportions of infectious to non-infectious particles present in virus stocks.

A variable RNA content in arenavirus preparations raised the question as to the possible function of incorporated host cell ribosomes. The addition of actinomycin D at low concentrations inhibits new ribosomal RNA synthesis, although the production of infectious virus remains unchanged (Carter *et al*, 1973); but prolonged treatment of 48 hours or more may reduce virus yield (Segovia and Grazioli 1969). Virus replication does not therefore require newly-synthesized ribosomal RNA. Leung and Rawls (1977) showed that Pichinde virus-associated ribosomes were not required for virus replication. In these experiments, virus was first grown at 33° in ts-14 cells—a line of hamster embryonic lung cells which do not synthesize proteins at 39° owing to a temperature-sensitive defect in the 60S ribosomal sub-unit—and was subsequently replicated successfully in wild-type hamster lung cell cultures at either 33° or 39°.

Analysis of oligonucleotides obtained after T_1 enzyme digestion of the large and small virus-specific RNA components of Pichinde virus has shown that each species is unique (Vezza *et al.* 1978*a*). This is compatible with the observation of Vezza and Bishop (1977) that 12 temperature-sensitive mutants could be categorized into two non-overlapping groups. Preliminary analysis has shown that extracted viral RNA is approximately 20 per cent resistant to RNase activity, although it is doubtful at present whether this represents hybridization of complementary RNA sequences within the same or different RNA strands (Vezza *et al.* 1978*a*). Although circular nucleocapsid-like structures have been observed in preparations of Tacaribe virus (Palmer *et al.* 1977), linear RNA molecules with some circular forms up to 1.5 μm in length were reported by Vezza *et al.* (1978*a*) for this virus.

The incorporation of ribosomes into Pichinde virus particles suggests the possible presence of virus-specific RNA with positive polarity, but all the evidence obtained so far indicates that arenavirus RNA does not possess the properties of eukaryotic mRNA. Leung, Ghosh and Rawls (1977) found that Pichinde virus RNA did not contain any significant tracts of polyadenylic acid, either before or after nuclease digestion; no evidence of either 5′-capping or methylated bases could be found. The failure of the same material to stimulate the synthesis of identifiable virus-specific polypeptides when added to a wheat-germ *in vitro* protein-synthesizing system provided further evidence for the absence of messenger function. Hitherto no evidence is available to confirm these observations for arenaviruses other than Pichinde. Further evidence as to the nature of the arenavirus genome is the presence of an RNA-dependent RNA polymerase in preparations of Pichinde virus (Carter, Biswal and Rawls 1974, Leung, Leung and Rawls 1979). Carter *et al.* (1974) reported that RNA polymerase activity was optimal in 10 nM mg^{2+} at pH 8.5. The addition of actinomycin D was without effect whereas the inclusion of pancreatic RNase abolished enzyme activity. Rate-zonal centrifugation of the product showed two components, a low molecular weight 6S component which was RNase susceptible and a much larger 26S RNA component which was partially RNase resistant. The nature of Pichinde virus-associated RNA polymerase activity has been classified by Leung *et al* (1979), who found that RNA polymerase associated with viral ribonucleoprotein structures catalysed the incorporation of all four ribonucleotides. In contrast, enzyme activity was present in the slowly sedimenting fraction which contained predominantly ribosomal material and preferentially catalysed the incorporation of AMP and UMP nucleotides. This latter activity represented two discrete enzymes, one preferentially catalysing UMP in the presence of Mg^{2+}, and a second Mn^{2+}-dependent activity which preferentially incorporated AMP. Both activities closely resemble similar RNA polymerase moieties associated with uninfected cell ribosomes. A further finding was the probable presence of a nucleoprotein-associated RNase, as manifested by a decrease of acid-precipitable radiolabel after 10 minutes of RNA polymerase activation. The RNA polymerase associated with the virus-specific ribonucleoprotein appeared to be much more labile than the two host-derived enzymes, and RNA-RNA hybridization showed that only about 23 per cent of the virion RNA was copied by the *in vitro* reaction. When this product was added to the fraction containing the Mn-dependent polyadenylate polymerase activity, a homopolymer of poly (A) was terminally added to the virus cRNA product of the first reaction.

Virus replication and intracellular protein expression

Banerjee, Buchmeier and Rawls (1975/76) have shown that Pichinde virus will not replicate in BHK 21 cells previously enucleated by exposure to cytochalasin B. By means of immunofluorescence it was demonstrated that synthesis of viral antigens did not occur in enucleated cells. Enucleation of cells at various times after infection indicated that the requirement of an intact nucleus extended to at least 8 hours after infection. This finding suggests either that transcription of host cell DNA may be required during the early period of the Pichinde virus replication cycle or that the integrity of the nuclear envelope is required as a primary stie for RNA replication. With respect to the first of these possibilities, it should be noted that the yield of virus is not significantly decreased in the presence of actinomycin D.

The synthesis of Tacaribe viral proteins was analysed at varying times after infection of BHK 21 cells by Saleh, Gard and Compans (1979). Owing to the failure of arenaviruses to inhibit host cell protein synthesis, virus-specific products were removed by immunoprecipitation prior to SDS-PAGE. The nucleoprotein appeared to be synthesized as a major gene product from 24 hours post-infection, reaching a maximum at 48 to 60 hours. In addition, a minor 79 000 mol. wt polypeptide present in the virion was also detected in infected cells during this period. Further studies showed that a 70 000 mol. wt glycosylated protein was present in immune precipitates prepared from infected cell extracts during the same period, but was absent in preparations of whole virus. The carbohydrate moiety of this non-structural component appeared to be particularly rich in mannose in comparison with the major 42 000 mol. wt structural glycoprotein. Because antibody against whole virus was used for these studies, the 70 000 mol. wt non-structural glycoprotein may represent an uncloven precursor of the major structural glycoprotein, a hypothesis further substantiated by pulse-chase experiments which suggested a precursor-product relation between these two structures.

Buchmeier and Oldstone (1979), studying LCM virus, have shown that both glycoproteins (mol. wt 44 000 and 35 000 respectively) have a common cell-associated precursor of mol. wt 75 000. Pulse-chase experiments clearly showed that radiolabel incorporated into the precursor 24 hours after infection was present 6 hours later in both structural glycoproteins. Carbohydrate analysis revealed that the precursor was rich in glucosamine and mannose—a finding similar to that obtained with Tacaribe virus.

Analysis of infected cell extracts showed that the nucleocapsid protein was evident as early as 6 hours in BHK 21 cells infected with LCM virus (Buchmeier, Elder and Oldstone 1978). The appearance of this protein immediately prior to virus maturation suggests that it is a late gene product required for new virus production. Virus antigens, particularly the 44 000 mol. wt structural glycoprotein appeared on the surface of LCM virus-infected cells from 7.5 hours after infection. In addition, several virus-specific proteins have been detected in arenavirus-infected cells by use of appropriate antisera. In an early study, Brown and Kirk (1969) described a heat-sensitive CF antigen in LCM-infected BHK 21 cells, a proportion of which was associated with the virus after centrifugation of cell extracts. Bro-Jørgensen (1971) found two distinct precipitin lines between LCM antiserum and a sucrose-acetone extract of infected cells. The smaller of the two antigens was found to be thermostable, resistant to the action of trypsin, and of approximate mol. wt 48 000. Pichinde viral antigens in infected cell extracts can also be readily detected by complement-fixation tests 2 to 4 days after infection. Buchmeier, Gee and Rawls (1977) detected two virus-specific proteins by agar gel diffusion. The predominant antigen was stable at 56° and resistant to pronase; in contrast, the minor antigen was found to be thermolabile and destroyed by pronase treatment. Buchmeier *et al.* (1978) found that a purified preparation of the major heat-stable antigen contained two polypeptides with mol. wt of 15 000 and 20 000 respectively. Antigenic identity was established between this preparation and a solubilized fraction of Pichinde virus containing the nucleocapsid and large virion glycopeptide; the disparity in size between these structures and the polypeptides which constitute the major intracellular antigen was suggested as having occurred as a result of proteolytic cleavage of nascent structural polypeptides within the infected cell cytoplasm. Several studies have indicated that major antigen detected by complement fixation is identical with the antigen specificities detected in the cytoplasm by immunofluorescence (Buchmeier *et al.* 1977, Rutter and Gschwender 1973).

Welsh and Buchmeier (1979) have analysed the properties of LCM defective interfering (DI) particles purified from cultures of persistently infected BHK 21 and L 929 cells. In sucrose gradients, DI particles were found to be approximately $0.01 \, g \, cm^{-3}$ less dense, although the protein composition of these particles was similar to standard infectious virus. These authors also found that treatment of infected cells with DI virus 4 hours prior to challenge with infectious virus inhibited production of the viral nucleoprotein. In addition, the total rate of viral protein synthesis in persistently-infected cultures was found to be very low, and viral antigens could not be detected at the plasma membranes. (For reports on haemorrhagic fevers characterized by a renal syndrome, see Report 1982, Kitamura *et al.* 1983, and Morita *et al.* 1983.)

References

Arata, A. A. and Gratz, N. G. (1975) *Bull. Wld Hlth Org.* **52,** 621.

Banerjee, S. N., Buchmeier, M. J. and Rawls, W. E. (1975/76) *Intervirology* **6,** 190.

Bro-Jørgensen, K. (1971) *Acta path. microbiol. scand. B.* **79,** 466.

Brown, W. J. and Kirk, B. E. (1969) *Appl. Microbiol.* **18,** 469.

Buchmeier, M. J., Elder, J. H. and Oldstone, M. B. A. (1978) *Virology* **89,** 133.

Buchmeier, M. J., Gee, S. R. and Rawls, W. E. (1977) *J. Virol.* **22,** 175.

Buchmeier, M. J. and Oldstone, M. B. A. (1978) In: *Negative Strand Viruses and the Host Cell*, pp. 91–97. Ed. by B. W. J. Mahy and R. D. Barry. Academic Press, London; (1979) *Virology* **99,** 111.

Buchmeier, M. J., Welsh, R. M., Dutko, F. J. and Oldstone, M. B. A. (1980) *Advanc. Immunol.* **30,** 275.

Carter, M. F., Biswal, N. and Rawls, W. E. (1973) *J. Virol.* **11,** 61; (1974) *Ibid.* **13,** 577.

Carter, M. F., Murphy, F. A., Brunschwig, J. P., Noonan, C. and Rawls, W. E. (1973) *J. Virol.* **12,** 33.

Casals, J. (1975) *Yale J. Biol. Med.* **48,** 115.

Child, P. L., McKenzie, R. B., Valverde, L. R. and Johnson, K. M. (1967) *Arch. Path.* **83,** 434.

Eddy, G. A. and Cole, F. E., Jr (1978) In: *Ebola Virus Haemorrhagic Fever*, pp. 237–42. Ed. by S. R. Pattyn. Elsevier, Amsterdam.

Eddy, G. A., Wagner, F. S., Scott, S. K. and Mahlandt, B. J. (1975) *Bull. Wld Hlth Org.* **52,** 723.

Elsner, B., Schwarz, E., Mando, O. G., Maiztegui, J. and Vilches, A. (1973) *J. trop. Med. Hyg.* **22,** 229.

Edmond, R. T. D. (1980) *Roy. Soc. Hlth J.* **100,** 48.

Farber, F. E. and Rawls, W. E. (1975) *J. gen. Virol.* **26,** 21.

Gee, S. R., Clark, D. A. and Rawls, W. E. (1979) *J. Immunol.* **123,** 2618.

Johnson, I. and Clamp, J. R. (1971) *Biochem. J.* **123,** 739.

Johnson, K. M. (1975) *Bull. Wld Hlth Org.* **52,** 729.

Johnson, K. M., Webb, P. A. and Justines, G. (1973) In: *Lymphocytic Choriomeningitis Virus and Other Arenaviruses*, p. 241–58. Ed. by F. Lehmann-Grube, Springer-Verlag, Vienna.

Kiley, M. P., Lange, J. V. and Johnson, K. M. (1979) *Lancet* **ii,** 738.

Kitamura, T. *et a..* (1983) *Jap. J. med. Sci. Biol.* **36,** 17.

Lehmann-Grube, F. (1971) *Virology Monographs* Vol. 10, Springer-Verlag, Vienna.

Lehmann-Grube, F. and Ambrassat, J. (1977) *J. gen. Virol.* **37,** 85.

Leung, W. C., Ghosh, H. P. and Rawls, W. E. (1977) *J. Virol.* **22,** 235.

Leung, W. C., Leung, M. F. K. L. and Rawls, W. E. (1979) *J. Virol.* **30,** 98.

Leung, W. C. and Rawls, W. E. (1977) *Virology* **81,** 174.

Maiztegui, J. I., Fernandez, N. J. and de Damilano, A. J. (1979) *Lancet* **ii,** 1216.

Morita, C. *et al.* (1983) *Jap. J. med. Sci. Biol.* **36,** 55.

Murphy, F. A. and Whitfield, S. G. (1975) *Bull. Wld Hlth Org.* **52,** 409.

Murphy, F. A., Whitfield, S. G., Webb, P. A. and Johnson, K. M. (1973) In: *Lymphocytic Choriomeningitis Virus and Other Arenaviruses*, p. 273–85. Ed. by F. Lehmann-Grube, Springer-Verlag, Vienna.

Murphy, F. A., Winn, W., Walker, D. H., Flemister, M. R. and Whitfield, S. G. (1976) *Lab. Invest.* **34,** 125.

Oldstone, M. B. A. (1975) *Progr. med. Virol.* **19,** 84.

Palmer, E. L., Obijeski, J. F., Webb, P. A. and Johnson, K. M. (1977) *J. gen. Virol.* **36,** 541.

Pedersen, I. R. (1970) *J. Virol.* **6, 414; (1971)** ***Nature (Lond.) New Biol.*** **234,** 112; (1973) *J. Virol.* **11,** 416; (1979) *Advanc. Virus Res.* **24,** 277.

Pedersen, I. R. and Konigshofer, E. P. (1976) *J. Virol.* **20,** 14.

Rawls, W. E. and Leung, W. C. (1979) In: *Comprehensive Virology*, Vol. 14, pp. 157–92. Ed. by H. Fraenkel-Conrat and R. R. Wagener, Plenum Press, New York.

Report (1982) *Scand. J. Infect. Dis. Suppl.* No. 36, 82–95.

Rutter, G. and Gschwender, H. H. (1973) In: *Lymphocytic Choriomeningitis and Other Arenaviruses*, pp. 51–59. Ed. by F. Lehmann-Grube, Springer-Verlag, Berlin.

Saleh, F., Gard, G. P. and Compans, R. W. (1979) *Virology* **93,** 369.

Segovia, Z. M. and Grazioli, F. (1969) *Acta virol.* **13,** 264.

Segovia, Z. M. de M. and de Mitri, M. I. (1977) *J. Virol.* **21,** 579.

Smadel, J. E., Green, R. H., Paltauf, R. M. and Gonzales, T. A. (1942) *Proc. Soc. exp. Biol. Med. (N.Y.)*, **49,** 683.

Stephen, E. L. and Jahrling, P. B. (1979) *Lancet* **i,** 268.

Vezza, A. C. and Bishop, D. H. L. (1977) *J. Virol.* **24,** 712.

Vezza, A. C., Clewley, J. P., Gard, G. P., Abraham, N. Z., Compans, R. W. and Bishop, D. H. L. (1978*a*) *J. Virol.* **26,** 485.

Vezza, A. C., Gard, G. P., Compans, R. W. and Bishop, D. H. L. (1977) *J. Virol.* **23,** 776.

Vezza, A. C., Gard, G. P., Compans, R. W. and Bishop, D. H. L. (1978*b*) In: *Negative Strand Viruses and the Host Cell*, pp. 73–90. Ed. by B. W. J. Mahy and R. D. Bary, Academic Press, London.

Weissenbacher, M. C., de Guerrero, L. B. and Boxaca, M. C. (1975) *Bull. Wld Hlth Org.* **52,** 507.

Welsh, R. M., Jr (1978) *J. exp. Med.* **148,** 163.

Welsh, R. M. Jr and Buchmeier, M. J. (1979) *Virology* **96,** 503.

Winn, W. C., Jr and Walker, D. H. (1975) *Bull. Wld Hlth Org.* **52,** 535.

Woodruff, A. W. (1978) In: *Modern Topics in Infection*, p. 240. Ed. by J. D. Williams, Wm. Heinemann Medical Books Ltd., London.

Wulff, H., McIntosh, B. M., Hamner, D. B. and Johnson, K. M. (1977) *Bull. Wld Hlth Org.* **55,** 441.

Young, P. R. and Howard, C. R. (1983) *J. gen. Virol.* **64,** 833.

93

Marburg and Ebola viruses

D. I. H. Simpson

Introductory

Marburg and Ebola viruses are both indigenous to Africa causing severe, distinctive haemorrhagic disease in man. They share a unique morphology but show only minimal or no antigenic cross-reactivity. Both viruses are extremely virulent and for these reasons fundamental data necessary for taxonomic purposes have only recently emerged.

Historical

Marburg virus disease, sometimes referred to as vervet monkey or green monkey disease, was first recognized in 1967, when it caused three simultaneous outbreaks in Europe: at Marburg, Frankfurt and Belgrade. There were thirty-one cases, of which twenty-five were primary infections; seven of the primary cases died, but there were no deaths among the six secondary cases. A hitherto unknown virus was isolated. All the primary cases were laboratory personnel who had come into direct contact with blood, organs, or tissue cultures from one particular consignment of vervet monkeys imported from Uganda. Four of the secondary infections were hospital personnel who had come into close contact with patients' blood. The wife of a Yugoslav veterinary surgeon was infected through blood contact with her husband, while the sixth case was the wife of a patient who transmitted the disease during sexual intercourse 83 days after the onset of illness. Marburg virus was detected in his seminal fluid. There were no tertiary cases and no spread of the disease to the community.

The disease next appeared in South Africa in 1975 in a young Australian man who had been hitch-hiking through Central and Southern Africa. He died shortly after his admission to a Johannesburg hospital. His female travelling companion and a nurse who looked after him also contracted the disease. Both women survived. Virological investigations confirmed that the virus isolated from these three cases was morphologically and antigenically identical with Marburg virus.

In 1980, Marburg virus re-appeared, this time in Kenya. A 58-year-old man was admitted to a Nairobi hospital with an 8-day history of progressive fever, myalgia, and backache. On admission he was in a state of peripheral vascular failure and was bleeding profusely from the gastro-intestinal tract. He died within 6 hours of admission. Marburg virus particles were seen by electron microscopy in liver and kidney tissues removed at post-mortem. Nine days later a male doctor who had attended this patient and had attempted resuscitation became ill with a similar disease syndrome. He survived and Marburg infection was confirmed serologically.

Between June and November 1976, outbreaks of severe and often fatal haemorrhagic fever occurred in the equatorial provinces of the Sudan and Zaire. In the Sudan there were 284 known cases with 151 deaths, a case-fatality rate of 53 per cent, whilst in the Zaire outbreaks there were 318 known cases with 280 deaths, a case-fatality rate of 88 per cent. The virus strains isolated from patients in both these outbreaks were found to be morphologically identical with Marburg virus but antigenically distinct. The name **Ebola** was given to the new strain after the river that runs through the epidemic area in Zaire.

A second outbreak of Ebola haemorrhagic fever occurred in the southern Sudan during August and September 1979, in the same area as the original 1976 outbreak. There were 34 reported cases, of which 22 were fatal. The clinical diagnosis was confirmed by virus isolation and serology.

A sporadic case occurred recently in Zaire at a distance of 400 km from the centre of the first outbreak.

Clinical features

The illness caused in man by Marburg and Ebola viruses are virtually indistinguishable. The incubation period was from 3 to 9 days in both the German and South African outbreaks of Marburg disease, but in the two Ebola epidemics a wider range of 4 to 16 days was recorded. Both infections have abrupt onset with severe frontal and temporal headache, followed by high fever and generalized pains, particularly in the back. A relative bradycardia was often one of the early symptoms. The patient rapidly becomes prostrated and some develop severe watery diarrhoea leading to rapid dehydration and weight loss. Diarrhoea, abdominal pain with cramping, nausea, and vomiting often persist for a week. In the Sudanese outbreak, knife-like chest and pleuritic pain was an early symptom and many patients complained of a very dry, rather than sore, throat accompanied by cough. On white skin, a characteristic non-itching maculopapular rash appears between days 5 and 7, lasts 3 to 4 days, and is followed by a fine desquamation. On pigmented skin, the rash, often described as measles-like, is not so obvious and can often only be recognized later with the appearance of skin desquamation.

Conjunctivitis was a regular feature in all outbreaks. An enanthema of the palate was reported in the Marburg outbreak in Germany, but was not seen in the three South African cases. In the Sudanese Ebola outbreak, pharyngitis was commonly noted and the throat was found to be dry and accompanied by fissuring with open sores on the tongue and lips. Genital infection with irritation and inflammation of the scrotum or labia majora was common and orchitis occurred in a few patients. Pancreatitis was noted in several instances.

In the Ebola outbreaks, patients were admitted to hospital as a rule on the fifth day of illness and their general appearance was described as 'ghost-like', with drawn, anxious features, expressionless faces, deep-set eyes, a greyish pallor, and extreme lethargy. Central nervous system infection may be apparent in a number of cases, with signs of meningeal irritiation, paraesthesia, lethargy, confusion, irritability and aggression.

A large number of patients in both the Marburg and Ebola outbreaks developed severe bleeding between days 5 and 7. The gastro-intestinal tract and lungs were most often affected with haematemesis, melaena and sometimes the passage of fresh blood in the stools. There was also bleeding from the nose, gums and vagina; subconjunctival haemorrhages were common, as were petechiae and bleeding from needle puncture sites.

Laboratory studies of clinical cases have necessarily been limited to investigations performed on patients from the first two Marburg outbreaks. In some cases laboratory investigations suggested a disseminated intravascular coagulation with subsequent kidney failure. A leucopenia early in the course of illness has been a constant feature, followed by a leucocytosis and a low erythrocyte sedimentation rate. The acquired Pelger-Huet anomaly of the neutrophils and atypical mononuclear cells were a feature in several patients. A thrombocytopenia was recorded in most patients from about day 3 onwards.

Biochemical investigations showed that all patients had severe liver damage, with both SGOT and SGPT activities considerably raised. Bilirubin values were only slightly increased, if at all. ECG changes were compatible with a myocarditis or other damage to the myocardium.

Death generally occurred between days 7 and 16, usually preceded by severe blood loss and shock. In surviving patients, recovery was slow with debility persisting for many weeks.

Pathology

Marburg and Ebola viruses are pantropic and produce lesions in almost every organ, with the liver and spleen the most conspicuously affected. The pattern of disease is that of stimulation of the reticuloendothelial system, inhibition of the lymphatic system, and vascular changes leading to occlusions of the vessels and the formation of thrombi and haemorrhages. Macroscopic findings are similar in all cases. The stomach and parts of the intestines are usually filled with blood. Petechiae are seen in the mucosa of the stomach and the small intestines. In a number of cases the liver and spleen are enlarged and dark in colour. Severe degeneration of lymphoid tissue and necrosis in spleen and liver result in large accumulations of cellular and nuclear debris. Severe congestion and stasis are obvious in the spleen. In the liver hyalin-necrotic-eosinophil bodies similar to the Councilman bodies of yellow fever are often seen. Accumulation of mononuclear cells occurs in the peripheral spaces. Even at the height of the necrotic process in the liver there is evidence of liver-cell regeneration. By electron microscopy high concentrations of virus are found in the necrotic areas.

Mononuclear transformation of lymphoid tissue as well as necrotic lesions are also found in the pancreas, gonads, adrenals, hypophysis, thyroid, kidney and skin. The lungs show few lesions except for circumscribed haemorrhages and evidence of endoarteritis, especially in the small arterioles. Neuropathological changes are confined mainly to the glial elements scattered throughout the brain.

Epidemiology

There is a strong suspicion that the disease is a zoonosis. Monkeys were originally implicated in the three Marburg outbreaks, but there is no strong evidence to suggest that primates are included in the natural reservoir cycle of the virus. In 1967 the monkeys had probably been incidentally infected. After the South African outbreak in 1975, an extensive search for a reservoir was carried out without success. In 1977 large numbers of small mammals were caught in the epidemic areas of the Sudan and Zaïre and blood and tissues removed for virological investigation in an attempt to throw some light on the natural reservoir for these viruses, again without success.

Virological studies in both these infections have shown no evidence of appreciable virus shedding by any other route than that of haemorrhage. There appears to be very little virus in the throat or urine but the persistence of virus in some body fluids for periods up to 83 days does pose a risk of late transmission. One of the South African Marburg patients developed uveitis two months after recovery and Marburg virus was cultured from fluid aspirated from the anterior chamber.

The mechanism of transmission of infection in the outbreaks was mainly by direct contact with infected blood, by very close and prolonged contact with an infected patient, or by inoculation by accident and through the use of a contaminated syringe and needle. There was no evidence to suggest that there was any respiratory spread of infection in the community. All the major outbreaks have been associated with spread within hospitials and, in the German outbreak of 1967, within the laboratory.

Both in the Sudan and in Zaïre there was serological evidence of small numbers of minor or even subclinical infections. It may be assumed that sporadic cases of Marburg/Ebola virus disease occasionally occur without secondary spread.

In the two 1976 epidemics, the attack rate in infected communities varied from 3.5 per 1000 to 15.3 per 1000 in the Sudan and from 8 per 1000 in the centre of the epidemic in Zaïre to less than 1 per 1000 in neighbouring communities. This indicates that the virus is not as highly transmissible as previously thought.

The secondary attack rate was about 15 per cent in Zaïre. In the Sudan active cases documented showed a secondary spread of 13 per cent, a tertiary spread of 14 per cent, and a quaternary spread of 9 per cent. Transmission stopped spontaneously after four generations, but in exceptional circumstances at least eight generations could be documented. The epidemics were readily brought under control by isolating the patients and instituting strict barrier nursing with gowns, gloves, masks and the effective treatment of patients' excreta with disinfectants such as formaldehyde and hypochlorite.

Laboratory investigations

Specific diagnosis requires isolation and identification of the virus or evidence of antibody development between paired serum samples. The isolation of the virus is best achieved by the inoculation of acute-phase blood intraperitoneally into young guinea-pigs and into various tissue culture cell lines. In the guinea-pigs high concentrations of virus are found in the blood during the febrile period. The liver, spleen and lymph nodes are the organs most often affected. Identification is made by indirect immunofluorescent staining techniques. In early passage in cell culture systems, none produces specific cytopathic effects but a more obvious CPE may appear after several passages. Most workers, however, have preferred to base their evidence of cell infection on the appearance of characteristic intracytoplasmic inclusion bodies demonstrated by immunofluorescence. Attempts to isolate the virus must be carried out in high containment laboratories.

Properties of the virus particle

Nucleic acid The virion contains one molecule of single-stranded RNA with a molecular weight (mol. wt) of approximately 4.2×10^6. Virion RNA is not infectious, does not bind to oligo-dT-cellulose and is therefore thought to be a negative-sense strand.

Protein Purified virions contain at least five polypeptides with the basic pattern being the same for both viruses. The proteins are designated VPO, VP1, VP2, VP3 and VP4 with corresponding Ebola mol. wt of *ca.* 190 000, 125 000, 104 000, 40 000 and 26 000. Marburg virions contain similar proteins with mol. wt of *ca.* 190 000, 140 000, 98 000, 38 000 and 22 000. In both viruses, VP1 is a glycoprotein and is probably the major component of the virion spikes. Both VP2 and VP3 are associated with the 1.32 g/cm^3 viral ribonucleoprotein (RNP) obtained by detergent-high-salt treatment of virions.

Lipid Lipid solvents destroy viral infectivity and release a 1.32 g/cm³ RNP from the virions. The percentage of virion lipid is unknown.

Carbohydrate Sugar is a component of one virion protein and possibly glycolipid.

Physicochemical properties The virion has a mol. wt of approximately (300–600) $\times 10^6$. Larger particles have a very high sedimentation coefficient; that of a uniform bacilliform particle is approximately 1400S. Its density is ca. 1.14 g/cm³ in potassium tartrate. Infectivity is quite stable at room temperature but is destroyed in 30 min at 60°. Virus is inactivated by UV and gamma irradiation and by 1 per cent formalin and β-propiolactone. Both viruses are inactivated by brief exposure to commercial phenolic disinfectants and are very sensitive to lipid solvents.

Morphology

The virions of Ebola and Marburg are very similar in morphology; the following general morphological description is derived from several sources. By electron microscopy, Marburg and Ebola particles are pleomorphic, appearing as long filamentous forms, sometimes with extensive branching, or as U-shaped, '6'-shaped, or circular forms. The particles vary greatly in length (up to 14 000 nm), but have a uniform diameter of approximately 80 nm. There are spikes on the particle surface approximately 70 Å in length and 100 Å from one another.

Beneath the virion envelope lies a complex nucleocapsid structure consisting of a dark central axis of 20 nm in diameter surrounded by a helical tubular capsid of approximately 50 nm bearing cross-striations with a periodicity of about 5 nm. Because its diameter is that of the tubular structures found in intracellular inclusions, the 20-nm dark central axis is presumed to be the virion RNP. This probably corresponds to the 1.32 g/cm³ structure released from virions by detergent treatment and consisting of virion RNA and the VP2 and VP3 virion proteins. Within the nucleocapsid is an axial channel of approximately 10–15 nm.

Though the lengths of Marburg and Ebola particles vary over a wide range, a recent study with rate-zonal, sucrose-gradient-purified particles indicates that the unit length associated with peak infectivity for Marburg is 790 nm and that of Ebola is 970 nm. Ebola virions were approximately 1.2 times as long as Marburg virions.

Antigenic properties

Details of the specific antigenic composition of Marburg and Ebola are not available. Nevertheless, information has been obtained regarding the antigenic properties of these viruses. Casals (1971), using the complement-fixation (CF) test, determined that Marburg virus was not serologically related to a variety of arboviruses and rhabdoviruses.

Immunofluorescent techniques, which have been found to be faster and more sensitive than the CF test, demonstrate that there is little or no antigenic cross-reaction between Marburg and Ebola viruses. The lack of antigenic relatedness is also indicated by the fact that previous immunity to Marburg virus in laboratory animals does not protect against subsequent infection with Ebola virus.

Immunofluorescent and radioimmune precipitation techniques, cross-protection studies, and T_1 oligonucleotide mapping studies also indicate that, though the Sudan and Zaïre strains of Ebola virus share many antigens, there are also significant antigenic differences.

Replication

The morphogenesis of Marburg and Ebola viruses has been examined in several laboratories. Whether investigators studied human liver, guinea-pig organs, monkey organs, or tissue culture, a typical pattern emerged. Virions are constructed from preformed nucleocapsids and envelopes which are added by budding through cellular membranes.

Virus-infected cells contain prominent cytoplasmic inclusion bodies consisting of viral nucleoprotein material. These inclusion bodies are complex and distinct, consisting of a finely fibrillar or granular ground substance which condenses into tubular structures, or nucleocapsids. As infection proceeds, they grow and become highly structured, even at sites remote from the cell membranes. Budding of completed virions takes place at cell membranes into which virion spikes have been inserted. It appears that nucleocapsids, at the time of budding, may orient in any plane from perpendicular to parallel to cell membranes, and may produce the branching seen with these viruses.

The classification of these two viruses is still undecided. A new family, *Filoviridae*, has been suggested for them, but this name has not yet received the sanction of the International Committee on Taxonomy of Viruses.

Further Reading

Bowen, E.T.W., Lloyd, G., Harris, W.J., Platt, G.S., Baskerville, A. and Vella, E.E. (1977) Viral haemorrhagic fever in southern Sudan and northern Zaire. Preliminary studies on the aetiological agent. *Lancet* **i** 571.

Bowen, E.T.W., Platt, G.S., Lloyd, G., Raymond, R.T. and Simpson, D.I.H. (1980). A comparative study of strains of Ebola virus isolated from southern Sudan and northern Zaire in 1976. *J. med. Virol.* **6**, 129.

Casals, J. (1971) Absence of serological relationship between the Marburg virus and some arboviruses. In: *Marburg Virus Disease*, Martini Ed. by G. and Siegert, R. pp.98–104 Springer, New York.

Ebola haemorrhagic fever in Sudan, 1976: Report of a WHO/International Study Team. (1978). *Bull. Wld Hlth Org.* **56,** 247.

Ebola haemorrhagic fever in Zaire, 1976: Report of an International Commission. (1978). *Bull Wld Hlth Org.* **56,** 271.

Ellis, D.S., Simpson, D.I.H., Francis, D.P., Knobloch, J., Bowen, E.T.W., Lolik, P. and Deng, I.M. (1978). Ultrastructure of Ebola virus particles in human liver. *J. clin. Path.* **31,** 301.

Ellis, D.S. *et al.* (1979). Ebola and Marburg viruses. II. Their development within Vero cells and the extracellular formation of branched and torus forms. *J. med. Virol.*, **4,** 213–15.

Gear, J.S.S. *et al.* (1975). Outbreak of Marburg virus disease in Johannesburg. *Brit. med. J.* **iv,** 489.

Johnson, K.M., Lange, J.V., Webb, P.A. and Murphy, F.S. (1977). Isolation and partial characterisation of a new virus causing acute haemorrhagic fever in Zaire. *Lancet* **i,** 569.

Martini, G. and Siegert, R. (Eds.) (1971). *Marburg Virus Disease.* Springer, New York.

Kiley, M.P., Regnery, R.L., and Johnson, K.M. (1980). Ebola virus; identification of virion structural proteins. *J. gen. Virol.* **49,** 333.

Pattyn, S.R. (Ed.) (1978). *Ebola Virus Haemorrhagic Fever.* Elsevier/North Holland Biomedical Press, Amsterdam.

Pattyn, S., Van der Groen, Jacob, W., Piot, P, and Courteille, G. (1977). Isolation of Marburg-like virus from a case of haemorrhagic fever in Zaire. *Lancet* **i,** 573.

Regnery, R.L., Johnston, K.M., Kiley, M.P. (1980). Virion nucleic acid of Ebola virus. *J. Virol.* **36,** 465.

Simpson, D.I.H. (1969) Marburg agent disease in monkeys. *Trans. R. Soc. trop. Med. Hyg.* **63,** 303; (1980) Marburg fever. *Roy. Soc. Hlth. J.* **100,** 52.

94

Rubella

J. E. Banatvala and Jennifer M. Best

Introduction and historical review

Rubella was first described by two German physicians, de Bergen and Orlow in the mid-eighteenth century (Emminghaus 1870). At that time it was frequently known by the German name 'Röteln', and it was due to the early interest of the German physicians and the general acceptance of a German name, that the disease subsequently became known as 'German measles'. However, for many years German measles was frequently confused with measles and scarlet fever, other infectious diseases presenting with rash and at one time was considered to be a cross between them. The clinical differences between these diseases were recognized by Maton (1815) and Veale (1866), and rubella was accepted as a distinct disease by an International Congress of Medicine in London in 1881 (Smith 1881). It was Veale who introduced the name 'rubella'. The disease received comparatively little attention for infection was generally mild and severe complications rare. It received more attention in the 1940s when the association between maternal infection and congenital defects was first recognized.

The first recognition of the association of rubella with congenital defects and further retrospective studies

In 1941, N. McAlister Gregg, an Australian ophthalmologist, published his now famous retrospective study 'Congenital Cataract following German measles in the mother, in which he showed that, if acquired in early pregnancy, rubella could cause congenital malformation. Seventy-eight babies, all with a similar type of congenital cataract were born in New South Wales after an extensive rubella epidemic there in 1940 and all but ten of the mothers gave a history of rubella, usually in the first or second month of pregnancy. Congenital defects of the heart were also recorded in 44 of the 67 cases whose cardiac condition was recorded. These findings were soon confirmed in Australia by Swan *et al* (1943) who, in addition to reporting cataracts and heart defects, noted that many congenitally infected infants were deaf and some also had microcephaly. Deafness occurred in children whose mothers, on average, had had a rash at 2.1 months duration of pregnancy, whereas the mothers of children with cataract reported their illness at a slightly earlier gestational age (1.5 months). Congenital deafness was also reported by Gregg (1944), who in addition observed dental defects and low birth weight.

Despite confirmation of Gregg's original observation, an annotation in the Lancet (Report 1944) suggested that additional studies were required, as it could not be proved that the illness with rash experienced by these mothers was in fact rubella; and that it was unlikely that such an association would have previously gone unnoticed. However, Hope-Simpson (1944) then reported in the Lancet congenital cataract and heart defects in two babies in Dorset after epidemic rubella. In fact, similar defects had been noted before Gregg's original observation but their significance had not been appreciated. Within a short time additional retrospective studies reporting congenital defects induced by rubella in early pregnancy came from Australia as well as other countries (Hanshaw and Dudgeon 1978).

The problems associated with the terminology of rubella and the current knowledge of its pathogenesis were summarized in an excellent review by Wesselhoeft (1947*a*, *b*) in the *New England Journal of Medicine* in 1947.

However, as might be expected, retrospective studies in which the starting point for investigations was an infant with one or more rubella-induced deformities suggested that a very high proportion of mothers who had rubella during pregnancy were delivered of infants with congenital malformations. The outcome of pregnancies in which maternal rubella was followed by the birth of normal infants was not recorded. Thus, in 1940, when rubella was at its peak incidence in New South Wales, Gregg and his colleagues (1945) reported that 111 of 116 (96 per cent) children whose mothers had had rubella in early pregnancy, suffered from congenital defects which were confined to cases in which maternal rubella had occurred before the sixteenth week of gestation.

Prospective studies

In order to obtain a more accurate assessment of the risks of maternal rubella, prospective enquiries were carried out during the 1950s and early 1960s, in which the outcome of pregnancy of women having a history of rubella at different stages of their pregnancy was assessed. These studies showed that the incidence of congenital malformations following maternal rubella was much less than in the previously conducted retrospective enquiries, the risks of major malformation after maternal infection in the 1st trimester varying from 10.4 to 54.2 per cent (see p. 286). However, these studies may have underestimated the incidence of congenital malformation, since they were conducted before laboratory diagnosis was available. Since it is now known that a diagnosis of rubella based on clinical criteria alone is often inaccurate, many women who did not have rubella and who were delivered of normal babies may have been included in these studies.

Early experimental studies

Monkeys and human volunteers were used for the early work on the characterization and transmission of rubella, since no convenient experimental system was available. Hess (1914) was the first to postulate that rubella was caused by a virus, and this concept was more firmly established by experiments conducted by Hiro and Tasaka (1938) and Habel (1942). Hiro and Tasaka obtained nasal washings and blood from patients with rubella at the time of rash and passed them through a Berkefeld N filter. Symptoms of rubella infection were observed in children who had been inoculated with these filtered specimens. Using monkeys, Habel was able to transmit a mild disease, characteristic of rubella, by means of bacteria-free filtrates.

From studies on human volunteers inoculated with nasal washings from cases of rubella, Anderson (1949) showed that the incubation period of rubella varied between 13 to 20 days. Although only five of his 25 volunteers had a previous history of rubella, only 11 developed clinical rubella. He therefore postulated that the other 9 had experienced subclinical infections. However, it is more likely that some of the 9 volunteers were already immune, having experienced subclinical infection at some time in the past. Krugman *et al.*, (1953) also, using human volunteers with no previous history of rubella, produced better evidence for subclinical infection, since secondary cases of rubella with rash occurred in contacts of experimentally inoculated volunteers who did not develop a rash.

Isolation of rubella virus

Attempts to isolate rubella virus in fertile hens' eggs and cell cultures were unsuccessful (Steinmaurer 1938, Habel 1942, Anderson 1954) until 1962, when Weller and Neva (1962) and Parkman *et al.* (1962) published simultaneous reports of the isolation of rubella virus in different cell cultures. Weller and Neva (1962) isolated the virus from specimens of blood and urine taken from four typical cases of rubella. Unique cytoplasmic effects and cytoplasmic inclusions were detected in cell cultures of primary human amnion after one or two passages of the specimens in these cell cultures. The specificity of the cytopathic effect was confirmed by neutralization with convalescent sera from patients suffering from rubella. Parkman *et al.* (1962) detected the presence of rubella virus in cultures of primary vervet monkey kidney (VMK) by means of the interference technique, an indirect method previously employed for the detection of rhinoviruses. Parkman and his colleagues inoculated into VMK cultures throat washings taken at the time of rash from more than 30 military recruits with rubella. Although no cytopathic effects were observed on initial or second passages, when cultures were challenged 7–14 days after inoculation with 10^4TCD_{50} ECHO-11 virus, the usual clear rapidly developing cytopathic effect of this virus was not produced. This suggested the presence of an interfering agent. The interfering agents could be neutralized by rabbit antiserum produced against one of the isolates. When the two groups of workers exchanged the agents isolated in the two different cell culture systems, they found them to be identical, providing further evidence that both agents were indeed rubella virus.

The interference produced by rubella was soon confirmed by other workers. It was also shown that rubella virus would induce interference in continuous lines of monkey kidney and that viruses other than ECHO 11, e.g. Coxsackie A9, could be used as the challenge virus. In 1963, McCarthy *et al.* reported that rubella virus produced a cytopathic effect in cultures of RK13, a continuous line of rabbit kidney (Fig.

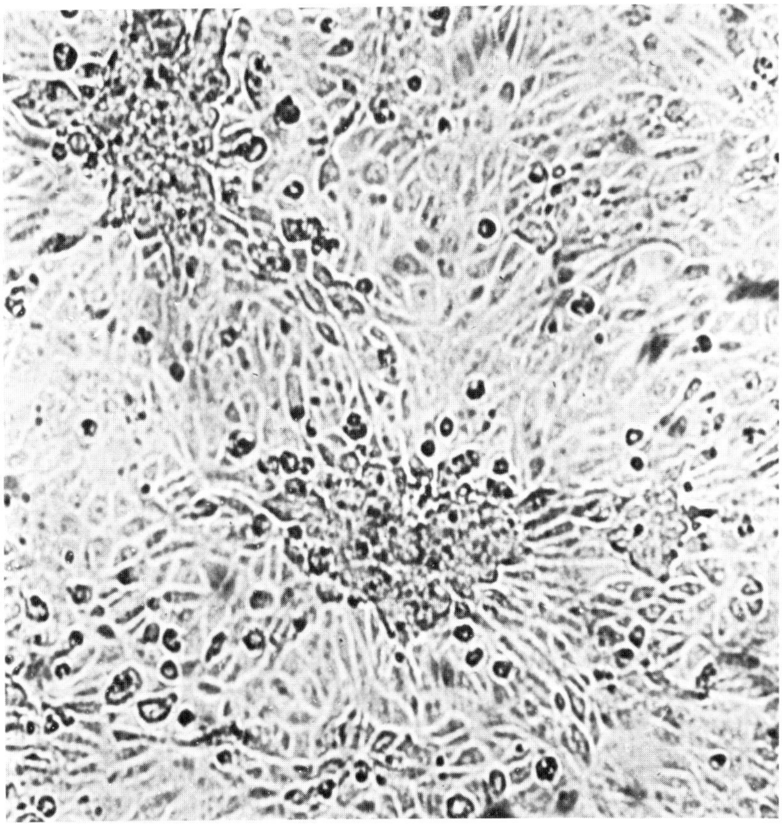

Fig. 94.1 Cytopathic effect produced by rubella virus in RK13 cell cultures. (× 80.)

94.1). Using these cells, they isolated rubella virus from throat washings and blood from typical cases of rubella and passage material from rubella-infected VMK cultures. Since these original observations, many other primary and continuous cultures have been reported to support the growth of rubella virus (Herrmann 1979). Two continuous cell lines, baby hamster kidney (BHK-21) and a continuous line of vervet monkey kidney (Vero) are extensively used to produce the high serum titres of rubella (Vaheri *et al.* 1967, Liebhaber *et al.* 1967) that are necessary to study the properties of the virus and to produce both complement-fixing (CF) and haemagglutinating (HA) antigens.

Early serological studies

The interference technique reported by Parkman and his colleagues (1962 and 1964) was adapted to detect neutralizing antibody responses, and was used for diagnostic purposes as well as for serological surveys. Such studies showed that the acquisition of rubella antibodies was related to age and social class, and that, in general, about 80 per cent of women of child-bearing age living in urban areas in Western countries were immune (Givan *et al.* 1965, Rawls *et al.* 1967). The presence of rubella antibodies was associated with a high order of protection against reinfection.

Rubella epidemic in the USA 1963–4

One of the most extensive epidemics of rubella of recent times occurred in the USA during the winter and spring of 1963/4. Many workers exploited the newly developed laboratory techniques, and their findings led to a greater understanding of the pathogenesis and clinical and virological features of congenitally acquired infection. Inevitably many pregnant women were infected; it was estimated that about 30 000 rubella damaged babies were born during this epidemic (Cooper 1975). It was shown that, when acquired in early pregnancy, maternal rubella induced in the fetus a generalized infection which persisted throughout the remaining gestational period as well as during infancy. Multi-system involvement was common, and the range of abnormalities was much wider than had hitherto been observed (see pp. 286–9). Despite active synthesis of rubella antibodies, infected babies exhibited prolonged excretion of virus and often transmitted infection to susceptible contacts (Report 1965, Plotkin *et al.* 1965).

Development of attentuated vaccines

This epidemic emphasized the importance of developing an effective vaccine to prevent rubella. The first attenuated vaccine (HPV77) was produced at the National Institutes of Health in the USA by 77 serial passages in VMK cell cultures—of a rubella isolate originally derived from a military recruit with rubella (Parkman *et al.* 1966). This attenuated vaccine was shown to induce protective immunity in monkeys. Virus could not be recovered from the pharynx or the blood and was not transmitted to cage controls. Controlled trials in groups of institutionalized children also showed that the virus was attenuated and immunogenic (Meyer *et al.* 1966). HPV77 was further attenuated in duck embryo fibroblasts to give the HPV77-DE5 vaccine, since vaccines produced in avian cells are less likely to contain extraneous agents than those produced in monkey cells (Buynak *et al.* 1968). Within a short time, other attenuated vaccines were developed (see pp. 294 *et seq.*). The accumulated data from a number of earlier trials showed that these vaccines produced an immune response in about 95 per cent of susceptible vaccinees, were generally well tolerated and, although virus was excreted by a variable proportion of vaccinees, it was not transmitted to susceptible contacts. The assessment of vaccine-induced immune responses was greatly facilitated by the development of a haemagglutination inhibition (HAI) test (Stewart *et al.* 1967), which soon replaced the more cumbersome and time-consuming neutralization test. Rubella vaccines were licensed in the USA in late 1969 and in the UK a few months later.

Properties of the virus

1. Morphological characters and classification

Rubella virus is an RNA virus, 40–70 nm in diameter with a lipoprotein envelope (Figs. 94.2 and 94.3). Owing to its structural characteristics it has been classified as a non-arthropod-borne togavirus and has been placed by itself in the genus Rubivirus (Horzinek 1973, Andrewes *et al.* 1978). Using HAI tests, Mettler *et al.* (1968) failed to show any antigenic relationship between rubella and more than 200 alphaviruses and flaviviruses, while Horzinek (1981) reported no cross-reactions with pestiviruses or equine arteritis virus. The nucleocapsid is 30–40 nm in diameter and contains a single molecule of RNA which is infectious when extracted under appropriate conditions (Hovi and Vaheri 1970, Sedwick and Sokol 1970). Owing to the instability of the nucleocapsid, symmetry is difficult to establish. However, icosahedral symmetry (Holmes *et al.* 1969, Horzinek *et al.* 1971) and a central

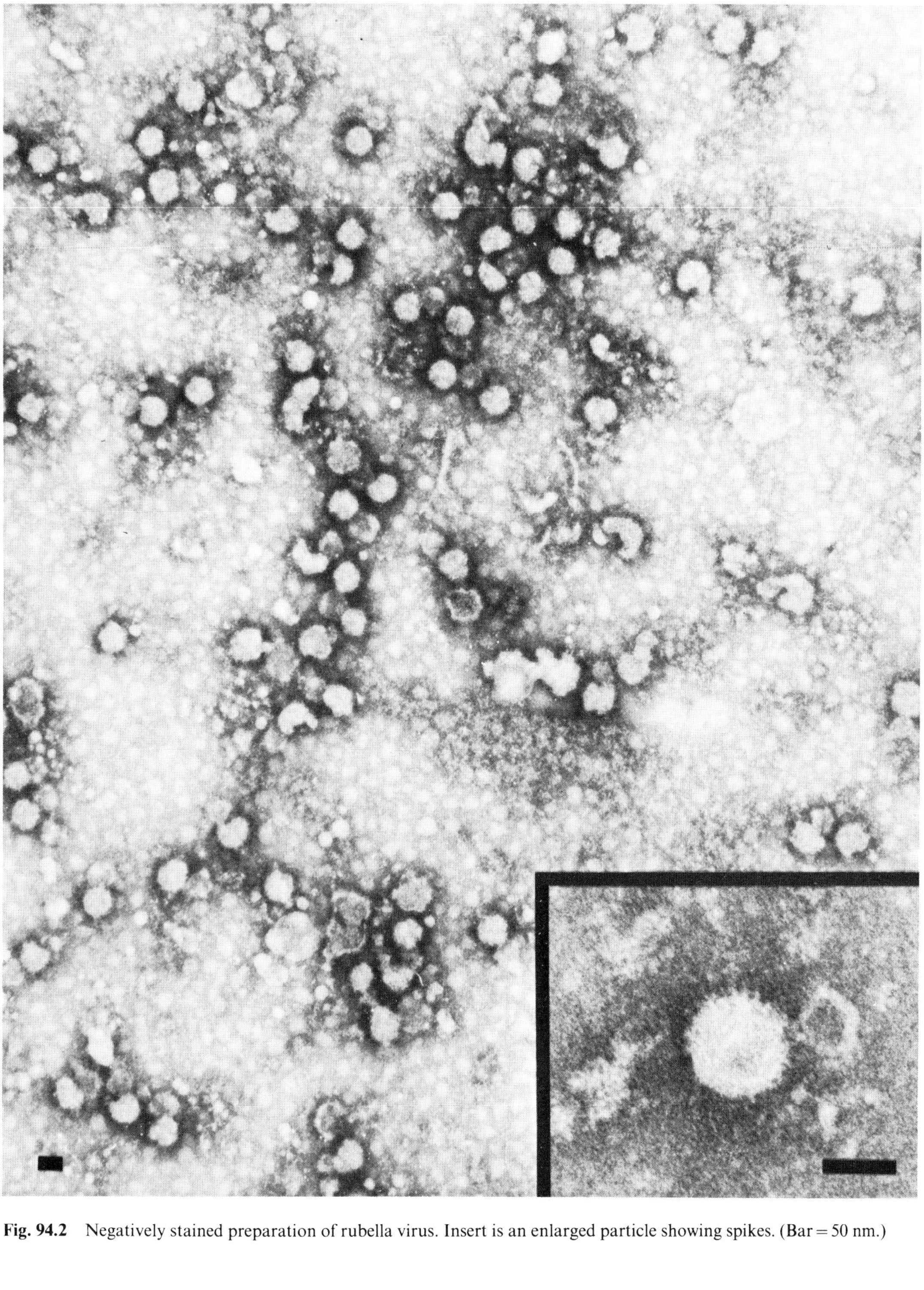

Fig. 94.2 Negatively stained preparation of rubella virus. Insert is an enlarged particle showing spikes. (Bar = 50 nm.)

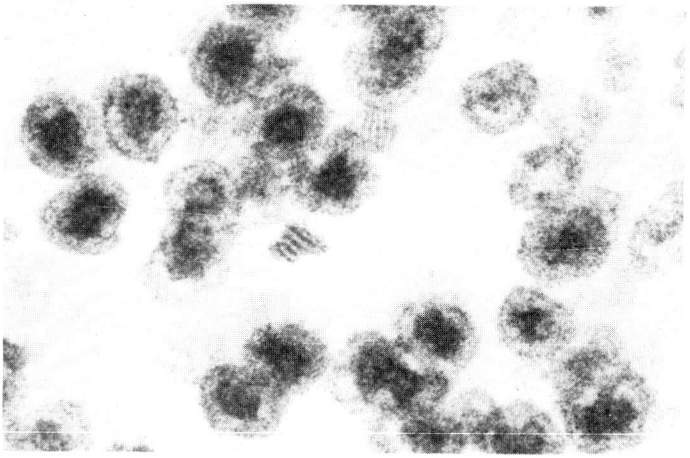

Fig. 94.3 Rubella virus particles in thin section of a monolayer of Vero cells infected with the 14th passage of a homogenate of brain from a rubella-infected fetus.

Table 94.1 Physical and Morphological properties of rubella virus

Virus particle:		
Diameter	40–70 nm	Reviewed by Horzinek (1973)
Buoyant density	in sucrose 1.16–1.19 g/ml	
	in $CsCl_2$ 1.20–1.23 g/ml	Reviewed by Horzinek (1973)
Sedimentation coefficient	240S	Thomssen, Laufs and Müller (1968)
	342S	Russell *et al.* (1967)
	$350 \pm 50S$	Bardeletti, Kessler and Aymard-Henry (1975).
Nucleocapsid:		
Diameter	30–40 nm	Reviewed by Horzinek (1973)
Symmetry	Icosahedral	Reviewed by Horzinek (1973)
Buoyant density		
Sedimentation coefficient	150S	Vaheri and Hovi (1972)
mol. wt	$2.6–4.0 \times 10^6$	Kenney *et al.* (1969)
Nucleic acid:	Single strand of RNA	
Buoyant density	1.634 g/ml	Hovi and Vaheri (1970)
Sedimentation coefficient	38–40S	Sedwick and Sokol (1970)
mol. wt.	$3.2–3.5 \times 10^6$ daltons	
Length of surface projections:	5–6 nm	Holmes *et al* (1969)
		Smith and Hobbins (1969)
Chemical composition:	RNA 2.4% Lipid 18.8%	Voiland and Bardeletti (1980)
	Proteins 74.8% Carbohydrates 4%	
Major polypeptides: (mol. wt in daltons)	Envelope 62.5 47.5 $\times 10^3$	Vaheri and Hovi (1972)
	58 48 16.5 $\times 10^3$	Bardeletti *et al.* (1975)
	63 50 $\times 10^3$	Payment *et al.* (1975)
	55 46 $\times 10^3$	Ho-Terry & Cohen (1980)
	58 47/52 $\times 10^3$	Oker-Blom *et al* (1983)
	Nucleo-capsid 31–35 $\times 10^3$	
Thermal stability:		
4°	Stable for ≥ 7 days	Fabiyi *et al.* (1966)
37°	Inactivated at 0.1–0.4 $\log_{10}$ $TCID_{50}$/ml per hour	Parkman (1965)
56°	Inactivated at 1.5–3.5 $\log_{10}$ $TCID_{50}$/ml per hour	Parkman (1965)
70°	Inactivated at 5.5 $\log_{10}$ $TCID_{50}$/0.1 ml per ½-hour	Kistler and Sapatino (1972)
−70°	Stable	Parkman *et al.*, (1964)
Freeze drying:	Stable	Parkman *et al.*, (1964)
pH sensitivity:	Stable at pH 6.0–8.1 Unstable at more acid and alkaline pH.	Schell and Wong (1966): Norrby (1969)
UV sensitivity:	Inactivated within 40 seconds	Fabiyi *et al.*, (1966)
1350 W/cm²	Inactivated at 7.0 $\log_{10}$ $TCID_{50}$/0.1 ml per hour	Kistler and Sapatino (1972)
Photosensitivity:	Labile $K = 0.07\ min^{-1}$ in PBS	Booth and Stern (1972)
Sonication:	Stable for ≥ 9 minutes	Schell and Wong (1966)

core component, 10–20 nm in diameter have been described (von Bonsdorff and Vaheri 1969, Payment *et al.* 1975). The viral envelope which has been reported to have the triple layered structure of a 'typical unit membrane' is acquired by budding from the host cell (von Bonsdorff and Vaheri 1969, Edwards *et al.* 1969). The pleomorphic character of the virus particle is due to the non-rigid delicate character of the envelope; elliptical and oblong virus particles and particles bearing finger-like protrusions have been described. The envelope bears ill-defined 5–6 nm surface projections probably composed of viral glycoproteins which carry the pH-dependent HA activity (Holmes *et al.* 1969, Bardeletti *et al.* 1975). The protein composition of rubella virus is similar to that of the alphaviruses. At least three major polypeptides have been described, two glycoproteins associated with the envelope, and a non-glycosylated core protein (Table 94.1). A comprehensive study by Oker-Blom *et al.* (1983) described a core protein and three envelope glycoproteins, E1, E2a and E2b, E2a and b being related. Other minor polypeptides have been described (Liebhaber and Gross 1972, Chantler and Tingle 1980, van Alstyne *et al.* 1981), which may represent viral structural proteins, precursor polypeptides, or host proteins which have remained associated with the purified virus. Differences in the molecular weight and number of polypeptides reported by different authors are probably due to variations in technique used to prepare the virus and electrophoresis.

Other physical properties are shown in Table 94.1 and have been reviewed by Horzinek (1973, 1981). The stability of the virus is enhanced by the addition of proteins to the suspending medium. Thermostability is improved in the presence of $MgSO_4$.

2. Sensitivity to chemical treatment

Owing to the lipid content of the viral envelope, rubella virus is inactivated by detergents and organic solvents. The effects of these and other chemicals have been reviewed (Herrmann 1979, Horzinek 1973, 1981, Norrby 1969, Parkman *et al.* 1964, Plotkin 1969).

3. Antigenic composition

I. Antigens

Haemagglutinating, complement-fixing, precipitating and platelet aggregating antigens have been described. Rubella virus also has haemolytic activity.

(a) Haemagglutinin (HA)

This antigen is probably associated with the surface projections on the viral envelope. It can be obtained from the supernatant fluid of infected BHK-21 cell cultures if these are maintained on serum-free medium or if non-specific inhibitors of haemagglutination (serum lipoproteins) are first removed by kaolin-treatment (Stewart *et al.* 1967) from the serum included in the maintenance medium. Alternatively, treatment of the rubella virus harvest with EDTA will separate HA from non-specific inhibitors by removing Ca^{++} (Furukawa *et al.* 1967). Rubella HA can also be prepared by alkaline extraction of infected BHK-21 or Vero cell cultures (Halonen *et al.* 1967, Liebhaber *et al.* 1969). Although treatment of virus with ether destroys the HA activity, treatment with ether and Tween 80 retains this activity and increases the titre of the antigen, owing to the formation of subunits. Ca^{++} are required for the attachment of HA to erythrocytes. Erythrocyte receptors for rubella virus are not destroyed by treatment with neuraminidase. These and other characters of HA prepared by different methods have been described by Furukawa *et al.* (1967) and Laufs and Thomssen (1968). Optimum conditions for haemagglutination and further details of the non-specific inhibitors of HA and methods for their removal from test sera have been reviewed by Herrmann (1979).

(b) Complement-fixing (CF) antigens

High-titred preparations of rubella virus can be used as CF antigens. These are usually prepared by either concentration of infected cell culture fluids or by alkaline extraction of infected cells (Herrmann 1979). BHK-21 cells are most frequently used for the production of suitable high-titred virus. Three CF antigens have been described.

(i) A large particle antigen with a density of 1.19–1.23 g/ml in sucrose gradients and associated with the infectivity and HA activity.
(ii) A small particle ('soluble') antigen (CF-S), density 1.08–1.14 g/ml in sucrose gradients (Schmidt *et al.* 1967). This is probably a subunit of the protein coat of the virus (Schmidt and Styk 1968).
(iii) A 150 S particle, which appears to be associated with the ribonucleoprotein core of the virus (Vesikari 1972).

(c) Precipitating antigens

Two or more precipitating antigens have been described by Le Bouvier (1969), Salmi (1969), and Grandien *et al.* (1976). The two antigens described by Le Bouvier were designated *theta* and *iota*. They appear to be structural components of the virus, but also appear as soluble proteins in infected cell cultures. Antibodies to *theta* appear quickly in parallel with HAI antibodies after naturally acquired infection and after vaccination with Cendehill, HPV77-DE5 and RA27/3 vaccines. Anti-*iota* appear more slowly and are detected only after naturally acquired infection and vaccination with RA27/3 (Le Bouvier and Plotkin

1971, Paul *et al.* 1974). Salmi (1969) described three precipitating antigens (*a*, *b* and *c*). Antigen *a* appeared to have a close relation to the HA and may be the same as *theta*. Antigen *b* is also probably derived from the viral envelope and is probably the same as *iota*. Antigen *c* was not studied in detail owing to its low yield in infected cell cultures.

(d) Platelet-aggregating (PA) antigen

The platelet-aggregation test is based on the aggregation of platelets by the joint action of antibody and small-sized antigens (Penttinen and Myllylä 1968). The PA test for rubella is more sensitive than the CF test. High-titred rubella virus has PA activity, as do the subunits of the viral envelope bearing HA activity, the smaller subunit (soluble antigen CF-S) which sediments at 3–4 S, and the ribonucleoprotein component which sediments at 150 S (Vesikari 1972).

(e) Haemolytic activity

Rubella virus has been shown, by using a two-step procedure, to lyse chick erythrocytes; namely by adsorption of the virus to erythrocytes in the presence of Ca^{++}, and cell lysis by chelating the ion with EDTA (Kobayashi 1978).

Vesikari (1972) has suggested that rubella virus has three main antigenic components. Two are associated with the viral envelope and are firmly attached to each other: these may be the polypeptides mol. wt 62 500 and 47 500 described by Vaheri and Hovi (1972). The larger polypeptide probably has HI, CF, *theta*, *a* and PA activity; the smaller antigen may be responsible for the CF-S, *iota*, *b* and PA-S activities, and also for inducing neutralizing activity (Ho-Terry and Cohen 1980). The third antigen is associated with the ribonucleoprotein core, which has CF and PA activity.

Ho-Terry and Cohen (1979), using a radioimmune-precipitation technique, found IgG and IgA antibodies to the ribonucleoprotein component of rubella virus in early convalescent sera, but not in sera from persons infected some years previously. IgG antibodies to the ribonucleoprotein were detected in only 1 of 16 RA27/3 vaccinees (Ho-Terry and Cohen 1981).

II. Antigenic variation

Cross-HAI tests and kinetic HAI tests have failed to reveal any antigenic variation among rubella virus strains (Banatvala and Best 1969, Best and Banatvala 1970), although differences have been observed by cross-neutralization tests and neutralization kinetics (Fogel and Plotkin 1969, Oxford 1969, Gould and Butler 1980). Although there is a need to distinguish between wild and attenuated strains of rubella, no test has proved to be reliable for this purpose, but differences in plaque morphology (Fogel and Plotkin 1969, Sato *et al.* 1979) and buoyant density (Oxford and Potter 1969) have been reported. Differences in biological activity have also been reported. Thus, unlike two laboratory strains of rubella and the Cendehill and HPV 77 vaccine strains, the RA27/3 attenuated strain does not induce intrinsic interference, with inhibition of superinfection with NDV in human fibroblasts (Kleiman and Carver 1977). In addition, the Cendehill strain has lower infectivity for rabbits (Oxford and Potter 1970, Gill and Furesz 1973), mice, ferrets, gerbils, guinea-pigs and monkeys (Zygraich *et al.* 1971). There is some evidence to suggest that rubella virus strains isolated in Japan have a lower teratogenic capacity for rabbits (Kono *et al.* 1969) and, in contrast to other strains, Japanese strains of rubella induce high concentrations of interferon in human placental cultures (Potter *et al.* 1973). Further information on antigenic variation may come from future studies with monoclonal antibodies and polypeptide analysis.

Rubella virus has been found to hybridize with a latent virus found in BHK-21/WI-2 cells to produce a rubella variant with DNA polymerase activity (Sato *et al.* 1978).

4. Growth in cell cultures

Rubella virus grows in a wide range of cell cultures, both primary and continuous (Herrmann 1979). A cytopathic effect (CPE) is produced only in some, and then only when incubation conditions are carefully controlled. RK13, SIRC (rabbit cornea), Vero (vervet kidney) and primary VMK cells are most frequently used for virus isolation. The virus is identified by its cytopathic effect in RK13 and SIRC cells and by interference in primary VMK. Vero cells are useful for virus isolation as they do not produce interferon and an increase in titre is therefore obtained quickly. Passage into another type of cell culture may be required for identification of rubella virus as it does not produce CPE in all sublines of Vero cells. Haemadsorption can be demonstrated in cell cultures heavily infected with rubella virus and is most easily demonstrated in BHK-21 cells (Schmidt *et al.* 1968).

Rubella virus grown in cells such as VMK will interfere with the replication of other viruses, such as enteroviruses and VSV. This interference is probably mediated through the interferon pathway, although interferon cannot always be detected in infected cell cultures (Desmyter *et al.* 1969, Horzinek 1981). Interference is not induced by rubella virus in Vero cell cultures, which are defective for interferon production (Desmyter *et al.* 1968). Rubella virus also induces intrinsic interference, which is a viral genome-induced resistance to infection by high multiplicities of NDV (Marcus and Carver 1967).

Both Vero cells and BHK-21 cells are used extensively for producing high titres of rubella virus as required for antigens in CF, HA, RIA, and ELISA tests. Vero cells have been found to release a larger amount of infectious virus into the supernatant fluid than BHK-21 cells, since the maturation of the virus occurs solely at the cell membrane in these cells, while the virus buds from both the cell membrane and into vesicles of the Golgi apparatus in the cytoplasm of BHK-21 cells (von Bonsdorff and Vaheri 1969, Bardeletti *et al.* 1979). Virus replication is entirely cytoplasmic in all cells.

Rubella virus can be induced to form plaques in RK13 (Taylor-Robinson *et al.* 1964, Fogel and Plotkin 1969), SIRC (Rhim *et al.*1967), BHK-21 (Vaheri *et al.* 1965) and Vero (Liebhaber *et al.* 1967, Sato *et al.* 1979) cells. However, plaque formation is influenced by such conditions as pH, presence of agar inhibitors and the condition, subline and source of the cells employed. A plaque test employing haemadsorption in BHK-21 cells (Schmidt *et al.* 1969) and haemadsorption-negative plaque tests using intrinsic interference (Marcus and Carver 1965) have also been described.

5. Growth in organ cultures

Rubella virus will multiply in organ cultures of human fetal trachea and nasal mucosa without loss of ciliary activity as well as in fragments of human fetal skin, brain, lung, kidney, heart, liver, spleen, lens and retina (Best *et al.* 1971). It will also multiply in ferret tracheal organ cultures and fragments of hamster and rabbit lung (Oxford and Schild 1967).

6. Pathogenicity for animals

Rubella virus has been shown to infect Vervet (*Macaca mulatta*) and Rhesus (*Cercopithicus aethiops*) monkeys, ferrets, rabbits, hamsters and suckling mice (reviewed by Herrmann 1979). Experimental infection of *Erythrocebus patas* monkeys, marmosets (*Sanguinus* species) chimpanzees and baboons has also been described (Draper and Laurence 1969, Horstmann 1969, Patterson *et al.* 1973). Monkeys usually develop a subclinical infection with virus excretion, viraemia and an immune response which is similar to that described in man (see p. 282). A persistent and chronic infection has also been established in suckling rabbits, ferrets and adult hamsters (Fabiyi *et al.* 1967, Oxford and Potter 1971).

There have been numerous attempts to reproduce the teratogenic effects of rubella virus in an animal model. Although various small laboratory animals have been employed, the monkey is the only animal whose reproductive process is similar to the human. However, results in monkeys have been inconsistent. Parkman *et al.* (1965) isolated rubella virus from 3 of 6 conceptions but found no evidence of malformation. Delahunt and Rieser (1967) reported an increased rate of spontaneous abortion and lenticular changes in two fetuses inoculated during the period of ocular development, while Sever and his colleagues (1966) found no evidence of fetal infection or malformation, although antibody was detected in the cord blood of all four offspring studied. Congenital infection has also been reported in rabbits (Kono *et al.* 1969, Cohen *et al.* 1971), ferrets (Rorke *et al.* 1968), and rats (Cotlier *et al.* 1968, Avila *et al.* 1972). However, most of these reports are unconfirmed, the occurrence of malformation is inconsistent, and the studies were not adequately controlled for such factors as the possible effects of increased handling, nutrition, virus passage history, species-adaptation, route of inoculation and adequacy of control animals (Elizan *et al.* 1969).

Postnatal infection

1. Epidemiology

Rubella has a world-wide distribution. In countries with temperate climates infection occurs most commonly during the spring and summer. Although all age groups may experience infection, rubella is uncommon in pre-school children, but outbreaks among school children and adolescents and young adults in educational and training institutions are common.

Unlike measles, rubella does not exhibit any characteristic epidemic periodicity. However, episodes of somewhat increased incidence occur every 3–4 years, and more extensive outbreaks each decade or so. World-wide pandemics may occur at approximately 30-year intervals (Witte *et al.* 1969). Such outbreaks occurred in the USA, Australia and the UK in the early 1940s, and again between 1963 and 1965; in the USA the latter epidemic resulted in the delivery of about 30 000 rubella-damaged babies (Cooper 1975). During 1978, and to a somewhat lesser extent in 1979, many parts of Europe and the UK experienced a striking increase in rubella infection, the full impact of which may not be evident for many years (Goldwater *et al.* 1978). This was reflected in the increased incidence of cases recorded by the Royal College of General Practitioners, a doubling in the number of therapeutic abortions carried out under Category 4 (which includes maternal rubella) and, compared with the previous two years, an approximately 4-fold increase in the number of cases reported to the Communicable Disease Surveillance Centre (Fig. 94.4). In contrast, the USA apparently did not experience increased

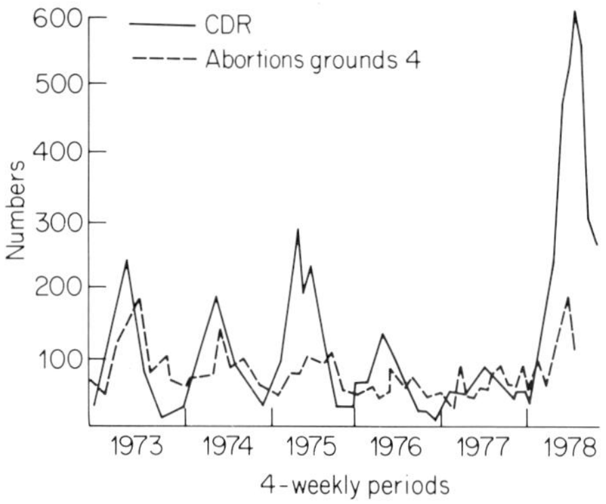

Fig. 94.4 Number of cases of rubella in England and Wales reported to Communicable Disease Surveillance Centre and Category 4 therapeutic abortions, 1973–1978 (unpublished).

rubella during this time; this may reflect the fact that a high proportion of children in the younger age group had been immunized and this had not only protected them, but prevented spread of virus to susceptible contacts. Thus, after the licensing of rubella vaccines in the USA in 1969 there was a reduction in the incidence of both postnatally as well as congenitally acquired rubella (Report 1978).

Since in most countries, including the UK, rubella is not a notifiable disease and, because a diagnosis based on clinical criteria alone is often inaccurate as persons may be infected subclinically, sero-epidemiological studies provide more reliable information on the incidence of rubella in different geographical areas and in different age groups. The accumulated data from a number of surveys in which sera were tested from persons in different age groups from all over the world produced remarkably consistent results despite serological tests being carried out by different techniques and in different laboratories. There was a progressive increase in the proportion of seropositive persons with increasing age; and, in general, among women of childbearing age about 80–85 per cent were immune. However, certain island and rural populations, particularly in the tropics in which rubella is non-endemic, have a higher proportion of seronegative persons than in most mainland populations. Thus in Hawaii, Trinidad, Barbados and Jamaica, among women of mixed ethnic origin, the proportion of immune women aged 20–24 ranged from 28–52 per cent (Dowdle, *et al.* 1970; Evans *et al.* 1974; Halstead *et al.* 1969; Sever *et al.* 1965). In Hong Kong, 36 per cent of Chinese women aged 15–40 were seronegative compared with 20 per cent of Caucasians (Shortridge and Osmund 1979). Although it is not surprising that in rural tropical mainland areas a high proportion of women of childbearing age may be susceptible, this phenomenon may also occur in certain tropical cities. Thus, in such Far Eastern cities as Bangkok and Kuala Lumpur, 47 per cent and 49 per cent of women aged 20–24 were seronegative (Thongcharoen *et al.* 1970; Lam 1972). In Central or South America, 40 per cent of women in this group in Panama City and 33 per cent in urban Peru were susceptible (Dowdle *et al.* 1970). The reasons for the high rubella susceptibility rate on islands is not really understood but such factors as population density, opportunities for reintroduction of virus into the community and climate may be of importance (Harcourt *et al.* 1979). Thus, in Taiwan, some areas of which are densely populated, transmission of rubella ceased during the hot summer months (Gale *et al.* 1969). In Japan there is considerable geographical variation in the prevalence of rubella antibodies among women of childbearing age. In the Northern prefectures 9–14 per cent are susceptible, whereas in the Southern prefectures, which have a hotter climate, the proportion varies between 14–25 per cent (Best *et al.* 1974).

2. Clinical and virological features

After a 14–21 day incubation period the characteristic features of rubella, rash and lymphadenopathy may appear. In young children, the onset of illness is usually abrupt. However, such constitutional symptoms as fever and malaise may be present for a day or two before onset of the rash but they usually subside rapidly after its appearance. Older children and adults may experience more pronounced constitutional symptoms 3–4 days before the rash appears, and during this prodromal phase, an enanthem consisting of erythematous pin-point lesions on the soft palate may be present. The exanthem is usually discrete and is in the form of pin-point maculo-papular lesions. It appears first on the face and then spreads rapidly to the rest of the body; lesions on the body may coalesce. The rash usually persists for about 3 days, occasionally longer, but in many patients it may be fleeting. The mechanism by which rash is induced has not been established. Although immunopathological mechanisms may be responsible, rubella virus has been isolated not only from skin biopsy specimens taken from areas with rash, but also from parts of the skin without rash and from the skin of patients with subclinical infection (Heggie 1978). Furthermore, the development of rash may be prevented by the administration of pooled human immunoglobulin, although this does not prevent viraemia.

Patients may complain of tender lymphadenopathy at or just before the rash appears. Follow-up studies on susceptible persons exposed to rubella have shown that lymphadenopathy may be present 7–10 days before the onset of rash, and sometimes for an even longer period after it has disappeared. Sub-occipital, post-auricular and cervical lymph nodes are most frequently affected. Rubella is rarely associated with severe complications. However, encephalitis may occur in approximately 1 in 10 000 cases but the prognosis is good (Krugman and Ward 1968). Very occasionally, rubella may be associated with thrombocytopenia, which may result in purpuric rash, epistaxis, haematuria and gastro-intestinal bleeding. The commonest complication of postnatally acquired rubella is joint involvement and, although this is rare among children and adult males, it may occur in up to 60 per cent of post-pubertal females. Symptoms generally develop as the rash subsides and vary in their severity from mild stiffness of the small joints of the hands to a frank arthritis, with severe pain, joint swelling and limitation of movement. The finger joints, wrists, knees and ankles are most frequently affected. The duration of these symptoms is usually of the order of about 3 days but occasionally symptoms may persist for up to a month. Rubella-induced arthralgia is not associated with any sequelae.

Arthralgia also occurs commonly in post-pubertal females after administration of rubella vaccine. The mechanism by which naturally acquired and vaccine-induced infection causes arthralgia is probably complex. Thus, joint symptoms may result from direct infection of the synovial membrane by virus, for rubella virus has been isolated from the joint aspirates of vaccinees with vaccine-induced arthritis (Weibel *et al.* 1969). Furthermore, *in-vitro* studies have shown that attenuated virus strains will replicate in human synovial membrane cell cultures (Grayzel and Beck 1971). However, an immune mechanism is probably also concerned for, in addition to virus, joint aspirates have been shown to contain rubella-specific IgG (Ogra and Herd 1971), which suggests that joint symptoms may be induced by immune complexes. It is therefore of interest that the presence of rubella antibody containing immune complexes in the serum has been associated with a high incidence of joint symptoms following rubella vaccination (Coyle *et al.* 1982). However, hormonal factors may also play a role, for, in addition to being common in post-pubertal females, the development of joint symptoms appears to be related to the menstrual cycle; after rubella vaccination, they are most likely to occur within 7 days of the onset of the cycle (Harcourt *et al.* 1979).

Rubella is often difficult to diagnose clinically, since the illness may present atypically with minimal lymphadenopathy and an evanescent rash, and conversely typical rubelliform rashes may be induced by other

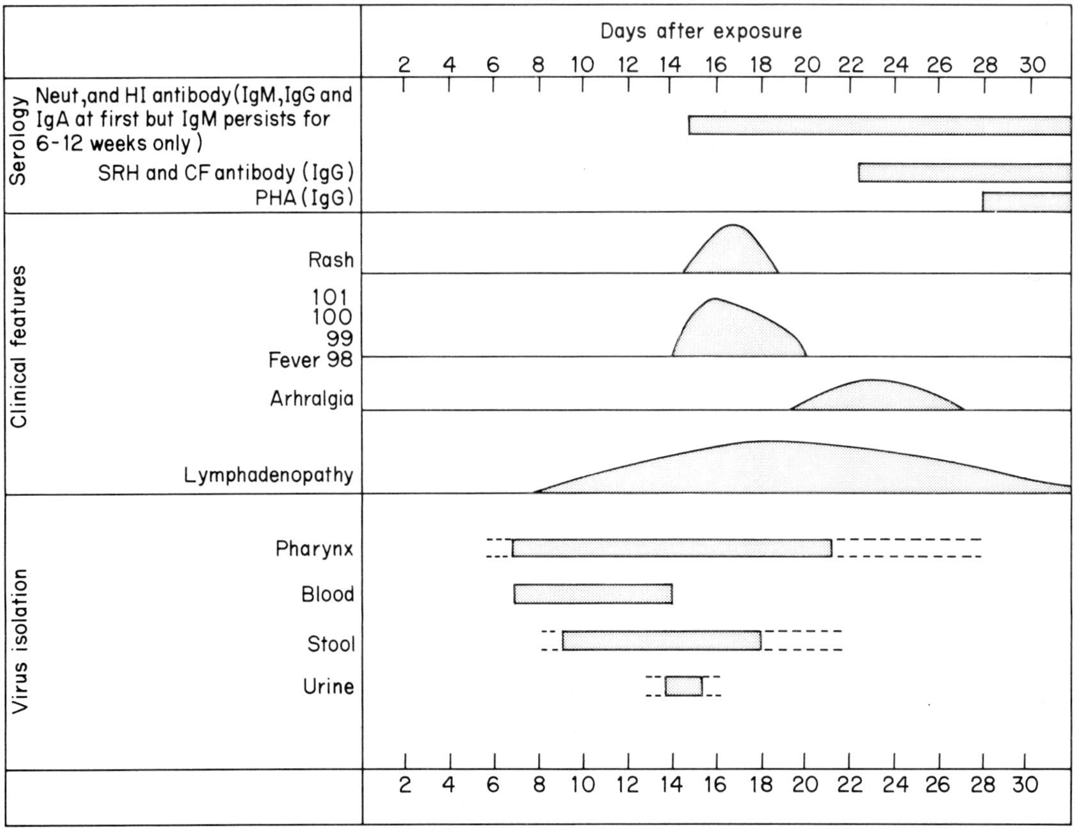

Fig. 94.5 Relation between clinical and virological features of post-natally acquired rubella.

viruses, e.g. enteroviruses. Serological studies have shown that there is a poor correlation between a past history of rubella and immune status, particularly among adults (Brown *et al.* 1969). This is not surprising since not only may rubella present atypically but disease without rash or subclinical infection may occur. Thus Krugman and Ward (1954) showed that of 13 children with experimentally induced rubella, two did not have a rash, although such features as fever and lymphadenopathy occurred. Green *et al.* (1965) followed up 24 children closely exposed to rubella and, although 22 developed neutralizing antibody responses, 8 had no rash. Thus, if the results of these studies are representative, about 25 per cent of children may suffer from subclinical infection or infection without rash.

The relation between clinical and virological features is shown in Fig. 94.5. The patient is potentially infectious for a prolonged period, since pharyngeal virus excretion may occur for up to a week before onset of rash and persist thereafter for 7–10 days. Virus may also be detected occasionally in the stools and urine but excretion occurs over a shorter duration. These are not suitable specimens from which to isolate virus and do not play an important role in virus transmission.

Viraemia is present for about a week before the onset of rash, but as this appears rubella antibodies develop and viraemia terminates. At this stage the antibodies may be detected by haemagglutination inhibition (HAI), neutralization or immunofluorescence (IMF). A week or so later antibodies which can be detected by complement fixation (CF) or single radial haemolysis (SRH) begin to develop. Antibodies may also be detected by passive haemagglutination (PHA) but this response may be delayed for 3-4 weeks after the onset of rash.

3. Assessment of the risk to women who are exposed to or develop rubella-like illness in pregnancy

Since a diagnosis may be made more reliably and rapidly by serological methods than by virus isolation, which may take up to three weeks to confirm, serological techniques are used in the assessment of women who have been exposed to or who have developed rubella-like illnesses in pregnancy.

Precise details on the date of onset of illness, presence and distribution of such clinical features as rash, lymphadenopathy and arthralgia, should be obtained from pregnant women who present with rubella-like clinical features. In pregnant patients who are contacts of possible rubella, results of virological investigations can be interpreted reliably only when precise information on the date, duration and type of contact (e.g. casual contact or more prolonged household contact) are obtained. In addition, enquiry about results of previous screening or history of rubella vaccination should be made.

Blood should be collected from pregnant women with rubella-like features as soon as possible after the onset of symptoms. A significant rise in antibody titre can often be detected by HAI within 5–7 days of the onset of illness; occasionally the response may be more delayed and it may be necessary to test further samples at 3–4 day intervals. Although the presence of HAI antibodies in the acute-phase serum sample collected within the first 48 hours of onset may suggest that antibodies were pre-existing and the patient's illness was not rubella-induced, some patients may occasionally have an appreciable HAI antibody titre early in the course of illness. Nevertheless, a significant rise in antibody titre is usually detectable if further blood samples are collected; furthermore, such patients will have a rubella-specific IgM response (see p. 293). Many patients may present in the post-acute phase of the illness at which time antibody titres may have reached their maximum levels. It is important to emphasize that there is no particular titre of antibody by any test which can be regarded as indicative of recent or current infection. However, antibodies detectable by complement fixation or single radial haemolysis develop somewhat later than HAI antibodies, and antibodies by passive haemagglutination may be delayed for even longer (Fig. 94.5). However, as a virological diagnosis is usually required urgently for obstetric reasons, patients who present in the post-acute phase of the illness should have their sera tested for the presence of rubella-specific IgM, its presence being indicative of recent primary infection by rubella.

Patients who have been exposed to rubella must be carefully followed up both clinically and serologically since retrospective studies have shown that approximately 45 per cent of women delivered of infants with congenitally acquired rubella gave no history of rash during pregnancy (Sheppard *et al.* 1977).

Patients who present within the incubation period and who have antibodies may be reassured, although it is often wise to obtain a second blood sample in, say, 7–10 days' time to make certain that there is no rise in antibody titre, since interpretation of results may depend on the accuracy of the history given by the patient. If there is no change in antibody titre the patient can be reassured with confidence that she was already immune and no further action need be taken.

Patients who present after an interval greater than the incubation period and who are seropositive are more difficult to assess since antibodies may already have reached their maximum titre. The sera of such patients should therefore be tested for the presence of rubella-specific IgM.

However, since rubella-specific IgM tests are expensive and often time-consuming, every attempt should be made to obtain an accurate history so that it is not necessary to resort to this investigation merely because precise details of rubella contact were not obtained.

Since rubella is unlikely to be acquired as a result of casual or brief contact, e.g. while shopping or in public transport, seronegative patients who experience this type of exposure can often be reassured that the risks of acquiring rubella are small, although it is essential that follow-up serum samples are obtained. Seronegative patients are at greater risk of acquiring infection when they have been exposed more closely over a longer period (e.g. family contact). However, they may nevertheless be reassured, particularly during non-epidemic times, by being told that they may not necessarily have been exposed to rubella. In such cases it may even be possible to allay anxiety by carrying out serological studies on the index case to determine whether or not the illness was caused by rubella virus. Patients who have been followed-up but who remain seronegative should be offered rubella vaccination in the immediate post-partum period.

4. Reinfection

Natural infection is followed by a very high order of protection from reinfection. However, antibody titres may be reinforced after natural and experimental exposure to rubella, although the infection is almost invariably asymptomatic (Horstmann *et al.* 1970, Vesikari 1972). Cradock-Watson *et al.* (1981) studied 34 pregnant women who experienced rubella reinfections; these were associated with increases in rubella specific IgG, but no IgM was detected. There was no serological evidence of pre-natal infection in any of the infants delivered to these mothers, which strongly suggests that reinfection is harmless to the fetus if pre-existing maternal immunity is naturally acquired.

Some earlier reports of reinfection based on the presence of HAI antibodies prior to infection must be interpreted with caution, since the apparent presence of antibody may merely have reflected failure to remove non-specific inhibitors adequately (Banatvala and Best 1973, Haukenes *et al.* 1973). However, there have been reports of reinfection accompanied by such features as rash, arthralgia and, among pregnant patients, fetal infection, indicating that reinfection may be associated with viraemia. It must be stressed that such reports are extremely rare; indeed, when such cases occur they are usually recorded as case reports in the medical literature. It is of interest that four of the five reported cases of probable reinfection had low or undetectable levels of neutralizing antibody (Best *et al.* 1981). Since reinfection in which viraemia is likely to have occurred is extremely rare, patients who have been screened for rubella antibodies by reliable tests may be reassured that they have long-lasting immunity. Reinfection following rubella vaccination will be discussed on page 290.

Congenitally acquired infection

1. Pathogenesis

The fetus is at risk during the period of maternal viraemia, since at this time placental infection may occur. The most likely source of virus is from the maternal viraemia. Virus may also be excreted via the cervix for up to 6 days after the onset of rash (Seppälä and Vaheri 1974); and since virus may exist in the genital tract for even longer, placental infection by direct contact or from ascending genital infection cannot be excluded.

After infection in early pregnancy rubella induces a generalized and persistent virus infection in the fetus which may result in multi-system disease. Töndury and Smith (1966) conducted histopathological studies on the products of conception of clinically diagnosed rubella-infected mothers; anomalies were present in 68 per cent of 57 fetuses when maternal rubella was contracted in the first trimester; and when contracted in the first month of pregnancy, 80 per cent were abnormal, sporadic foci of cellular damage being present in the heart, inner ear, lens, skeletal muscle and teeth. They suggest that rubella enters the fetus via the chorion in which it induces necrotic changes in the epithelial cells as well as in the endothelial lining of the blood vessels; the damaged endothelial cells are desquamated into the lumen of the vessel and then transported as virus-infected 'emboli' into the fetal circulation to settle in and infect various fetal organs. Lesions in the chorion were present as early as the tenth day after the onset of maternal rash. Fetal endothelial damage was distributed widely and probably resulted from viral replication rather than from antibody-mediated damage, since the most extensive histopathological changes were present at a gestational period before the fetal immune defence mechanism was sufficiently mature to be activated. Indeed a characteristic feature of rubella embryopathy following maternal rubella in early gestational life is the notable absence of an inflammatory cell response (Töndury and Smith 1966).

At least two mechanisms are involved in producing fetal damage, namely a virus-induced retardation in cell division, and tissue necrosis. *In-vitro* studies on embryonic cell cultures and rubella infected fetuses suggest that rubella virus may induce chromosomal damage and cause cells to divide more slowly than those which are uninfected (Plotkin *et al.* 1965). This may be due to a specific protein which reduces the mitotic rate of infected cells (Plotkin and Vaheri 1967). If retardation of cell division occurs during the critical phase of organogenesis, it is likely to result in the development of congenital malformations. It has also been shown that the organs of rubella-infected infants are smaller and contain fewer cells than those of uninfected infants (Naeye and Blanc 1965). The fetal endothelial damage induced by rubella infection may cause haemorrhages in small blood vessels leading to tissue necrosis and further damage of malformed organs over a longer period. Such organs as the liver, myocardium and organ of Corti may be affected. Studies on the products of conception obtained from virologically confirmed cases of rubella during the first trimester have shown that the fetus is almost invariably infected regardless of the time at which infection has occurred during this period (Rawls 1968, Thompson and Tobin 1970).

However, were all such pregnancies to proceed to term it is unlikely that all infected fetuses would have subsequently exhibited congenital malformations, which suggests that the fetus may be able to limit or even to overcome rubella infection. Whether or not maternal rubella in early pregnancy induces severe fetal damage may depend on such factors as the gestational age at which infection occurs, the degree of maternal viraemia, and possibly genetic factors (see below).

Rubella virus is isolated infrequently from neonates whose mothers developed post-first trimester infection, possibly because by then fetal immune mechanisms can effectively terminate infection. More mature fetal tissues do not have a reduced susceptibility to infection, for *in-vitro* studies have shown that rubella virus will replicate as well in organs derived from fetuses of 12–13 weeks gestational age as in those of younger fetuses (Best *et al.* 1968). Nevertheless, even though severe congenital anomalies are rarely encountered after post-first trimester rubella, serological evidence of fetal infection has been shown to occur in 25–33 per cent of infants whose mothers acquired maternal rubella between the sixteenth and twenty-eighth week of gestation (Vejtorp and Mansa 1980, Cradock-Watson *et al.* 1980).

2. Virus persistence

Although after infection *in utero* in early pregnancy rubella virus persists throughout the gestational period, and in those infants who survive for a limited period in infancy, no consistent defect in immune mechanisms has been found to explain this phenomenon. Possible explanations for virus persistence are that there is a defect in cell-mediated immunity, or lack of interferon production, or that small numbers of infected cells in the fetus give rise to clones of infected cells which persist for a limited period (Rawls 1968, Rawls 1974, Simons 1968). These possibilities were reviewed by Banatvala (1977*a*). Since then, more information has become available about the influence of rubella virus on cell-mediated immune (CMI) responses. *In-vitro* studies with rubella virus have shown that it replicates in T lymphocytes and macrophages and persists without replicating in B lymphocytes, causing inhibition of host-cell protein synthesis (Chantler and Tingle 1980, van der Logt *et al.* 1980). It has been suggested that infection of macrophages may interfere with their interaction with T cells. Postnatally acquired infection causes a transient reduction in lymphocyte responsiveness to phytohaemagglutinin (PHA) (Buimovici-Klein *et al.* 1976, Maller *et al.* 1978, Vesikari 1980) and a decrease in T-cell numbers (Niwa and Kanoh 1979). Thus a more persistent reduction in responsiveness might be anticipated in congenital infection. Indeed, Buimovici-Klein *et al.* (1979) showed that CMI responses to a range of PHA concentrations and to purified rubella virus were significantly lower in 20 congenitally infected 1 to 12-year-old children than in healthy children. Results varied with gestational age of intrauterine infection, the impairment of the CMI response to both PHA and rubella virus stimulation being more severe in children infected in the first two months of pregnancy than at later stages of gestation.

There may be a genetic influence on the ability of the fetus to eliminate rubella virus, since CMI responses and cell sensitivity to interferon may be influenced by genetic factors. Thus, studies on congenitally infected infants in Australia have shown that there is a statistically significant increase in the prevalence of such HLA antigens as HLA-1, HLA-3, HLA-B5 and HLA-B8 in infants with congenitally acquired rubella. The rubella syndrome has been reported in the southernmost archipelago of Japan (Ryukyu Islands) where it was shown that there was a significantly increased frequency of HLA-B15 among the mothers of congenitally infected infants (Kato *et al.* 1980), and that persons with this HLA type responded with high antibody production when given attenuated rubella vaccine. If this is an expression of a higher concentration of circulating virus it may explain the increased frequency of congenitally acquired rubella in infants of mothers with this HLA antigen.

3. Risks to the fetus

Whether maternal rubella induces fetal damage which causes intrauterine death or the birth of a malformed infant may depend on a combination of such factors as the gestational age at which maternal rubella occurs, the degree of maternal viraemia, and perhaps the genetic susceptibility of fetal cells to infection (see above). Maternal rubella may result in spontaneous abortion in up to 20 per cent of cases (Siegel *et al.* 1971); this occurs most commonly when maternal infection is acquired during the first 8 weeks of pregnancy. To this must be added fetal wastage from therapeutic abortion following virologically confirmed rubella. Failure to detect an appreciable increase in the number of cases of congenital malformation after the extensive 1978/1979 rubella epidemic in the UK may be a reflection of this.

Patients and clinicians want to know the risk of abnormalities likely to be met with at different periods of gestation. A precise answer to this question is not possible, because most prospective studies were carried out before laboratory confirmation of the clinical diagnosis was available. These studies, results of which are still widely quoted, may have considerably underestimated the risks of congenital malformation, since not only did they include infants delivered of some women who did not have rubella, but they also varied considerably in the duration and quality of follow-up studies in at-risk babies. Thus it is now known that many infants, although apparently normal at birth, may, when followed-up over a period of many years, be subsequently found to have such defects as perspective deafness or minor CNS anomalies. Hanshaw and Dudgeon (1978), in an analysis of data from 6 prospective studies carried out in different parts of the world, all but one of which were conducted in the pre-laboratory diagnosis era, showed that the risk of major malformation during the first trimester varied from 10.4 to 54.2 per cent, the risks being greatest in the first 8 weeks of pregnancy (Table 94.2). When acquired at this gestational age, infection may result in such features as low birth weight, neonatal purpura, and severe multiple malformations. Neonatal mortality rates of up to 20 per cent may occur in this group of patients (Cooper *et al.* 1965). However, after the first trimester the risks of severe congenital malformations decline markedly, although a small but significant risk extends as far as the twentieth week of pregnancy. Perhaps a more accurate assessment of the risks of maternal rubella was provided by Siegel and colleagues (1971), when they followed the progress of 381 children up to the age of 5 years whose mothers had had rubella between 1957 and 1963, and during the extensive 1964 US outbreak. During the earlier period the incidence of congenital anomalies following first trimester maternal rubella was 33 per cent; during the 1964 epidemic it was 54.2 per cent. However, in both periods a high proportion of major anomalies occurred among those infected during the second month of gestation, being 70 per cent and 80 per cent respectively for the period before and during the 1964 epidemic. Although it has been suggested that the high incidence of congenital malformations following the 1964 epidemic may have been due to an alteration in the biological behaviour of strains circulating at that time, this is difficult to prove since there is no reliable animal model with which to assess rubella teratogenicity. Perhaps the increased incidence of malformations following the 1964 outbreak merely reflected a greater accuracy in the diagnosis of both maternal and congenitally acquired rubella. More recently, Miller *et al.*(1982) followed up infants delivered of mothers who acquired virologically-confirmed rubella between 1976 and 1978, a period which included an extensive epidemic of rubella. Serological evidence of intrauterine infection occurred in more than 80 per cent of infants whose mothers were infected in the first trimester. When infection occurred between 13 and 16 weeks, 54 per cent of infants were infected. Rubella defects, principally heart and hearing defects, occurred in all infants infected before the 11th week of gestation. There was also a high incidence of deafness among infants whose mothers acquired infection between the 13th and 16th week (35 per cent). However, although

Table 94.2 Incidence of congenital rubella defects following maternal rubella

Study Country and year of occurrence	Gestational age when maternal rubella occurred 1st month (%)	2nd month (%)	3rd month (%)	4th month (%)	5th month and later (%)	Risks in first trimester (%)
Lundström (1962) Sweden, 1951	13/113 (11.5)	18/150 (12.0)	15/186 (8.1)	3/204 (1.5)	2/468 (0.4)	46/449 (10.2)
Liggins and Phillips (1963) New Zealand, 1959	2/2 (100)	6/12 (50.0)	2/15 (13.3)	2/16 (12.5)	0/44 (—)	10/29 (34.4)
Manson *et al.* (1960) Great Britain, 1950–52	7/45 (15.6)	12/61 (19.7)	10/77 (13.0)	3/72 (4.2)	5/288 (1.8)	29/183 (15.8)
Pitt (1961) Australia, 1950–55	3/5 (60.0)	5/12 (41.7)	5/33 (15.2)	1/37 (2.7)	0/20 (—)	13/50 (26.0)
Pitt and Keir (1965) Australia 1950–55	3/5 (60.0)	4/12 (33.3)	11/32 (34.4)	2/35 (5.7)	0/19 (—)	18/49 (36.7)
Siegel *et al.* (1971) United States, 1957–63	0/4 (—)	9/11 (81.8)	6/34 (17.6)	7/34 (14.3)	4/31 (12.9)	15/49 (30.6)
United States, 1964	2/7 (28.6)	18/26 (69.2)	25/50 (50.0)	11/62 (17.7)	4/33 (12.1)	45/83 (54.2)

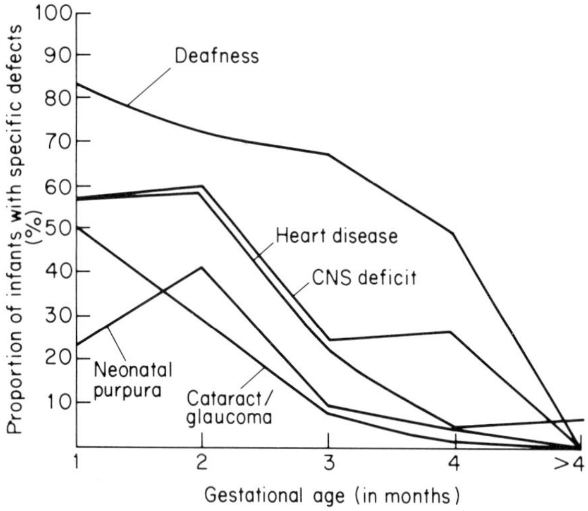

Fig. 94.6 Relation of clinical manifestations of congenital rubella to time of maternal infection, extrapolated from Cooper *et al.* 1969.

these proportions are high, the number of infants followed up in the series was relatively small, owing to the large number of therapeutic and spontaneous abortions.

The incidence of specific malformations is related to the gestational age at which maternal infection occurs. Thus, cardiac defects and such eye defects as cataract and glaucoma occur when maternal rubella is acquired during the first 8 weeks of gestational life, whereas deafness and retinopathy are more evenly distributed throughout the first 16 to 20 weeks of gestation (Cooper *et al.* 1969, Peckham 1972, Ueda *et al.* 1979).

Figure 94.6 relates the gestational age of maternal rubella to the clinical manifestations of congenitally acquired disease among 376 infants infected *in utero* during the 1964 US epidemic. Today, since virologically confirmed rubella in pregnancy usually results in termination, it seems unlikely that additional information which will help to determine the risks of congenitally acquired malformations more accurately after first trimester infection will be forthcoming.

Although the risks of fetal damage following first trimester rubella are appreciable, the risks associated with pre-conceptual and post-first trimester rubella are less clearly defined. Since viraemia is no longer present after the development of the rash, if conception occurs at this stage, there should be little risk of the fetus being infected. However, rubella virus may be detected in cervical excretions for up to six days after the onset of the rash (Blattner 1974), and it is possible that virus persists elsewhere in the genital tract for even longer; indeed, rubella virus has been isolated from the products of conception from a susceptible vaccinee 7 weeks after administration of attenuated rubella vaccine (Fleet *et al.* 1974).

However, there have been only a few reports of congenital rubella following pre-conceptual maternal infection, which suggests that it is a rare occurrence. Furthermore, some of those which have been reported were before virological techniques were available to confirm diagnosis or were poorly documented virologically. However, two infants with congenital malformation consistent with intrauterine infection by rubella were reported when maternal rubella was acquired 10 days and 6 weeks before conception (Hall 1946, Wesselhoeft 1947*b*). Sever *et al.* (1969) recorded details of 7 patients with maternal rubella from 0–28 days before conception. One mother, who had rubella 21 days before conception, gave birth to an infant with peripheral pulmonary stenosis and microcephaly and in three other mothers pregnancy resulted in abortion, stillbirths, or the birth of an infant who experienced multiple infections during the first year of life. The three remaining pregnancies resulted in the delivery of infants who were apparently normal. Hardy (1973) reported two fetal deaths and the birth of an infant with severe deafness following pre-conceptual rubella. Although an accurate assessment of the degree of risk of pre-conceptual rubella would be invaluable, this could only be established by carrying out multi-centre prospective clinical and virological studies on as many infants as possible whose mothers had pre-conceptual rubella and who elected to go to term.

The incidence of congenital defects following post-first trimester rubella is generally considered to be low, the most consistently reported anomalies being deafness and retinopathy. However, serological evidence of intrauterine infection as manifested by the presence of rubella-specific IgM in cord or neonatal blood samples has been recorded in 25–33 per cent of infants whose mothers had rubella between the 16th and 20th week of pregnancy (Vejtorp and Mansa 1980, Cradock-Watson *et al.* 1980). That fetal infection can occur at this stage of pregnancy is not surprising, since experimental *in vitro* studies have shown that rubella virus will replicate as well in organs derived from fetuses of 12–23 weeks gestational life as in those derived from younger fetuses (Best *et al.* 1968). The rarity of severe and multiple malformations following maternal rubella in late pregnancy is probably a reflection of the fact that organogenesis is by then complete and that in the more mature fetuses immune responses may have effectively terminated infection. Thus in contrast to infection acquired in early gestation, rubella virus is seldom isolated from infants whose mothers had had post-first trimester rubella. Despite these findings, Hardy *et al.* (1969) in the USA showed not only that 10 of 23 (45.5 per cent) of infants born alive to mothers who contracted rubella between the 13th and 31st week of pregnancy had serological evidence of intrauterine infection, but by the age of 2–3 years, fifteen (68.2 per cent) had such minor defects as speech disorders, developmental retardation, poor physical growth and hearing defects, although all were apparently normal at birth. This study is difficult to assess since, despite serological evidence of intra-uterine infection, apart from deafness in four children, many of the observed defects were not those commonly associated with congenitally acquired rubella. In contrast, Peckham (1972) in the UK detected no defects in seropositive 1 to 4-year old children whose mothers had had rubella after the 20th week of their pregnancies, although deafness and very occasionally retinopathy were present in 10 of 113 (8.8 per cent) of infants whose mothers contracted rubella between the 13th and 20th week of their pregnancies.

Vejtorp and Mansa (1980) obtained an even lower figure, for they detected only one child with a possible rubella-induced defect (unilateral deafness) among 204 children whose mothers had virologically confirmed post-first trimester rubella. However, in contrast to Peckham's study, the children in this study had been followed up for only a relatively short time (mean age 6 months). These findings emphasize the importance of conducting careful follow-up studies on all infants who have been exposed to rubella *in utero* after the first trimester, particular importance being placed on early recognition of hearing defects.

4. Clinical features

Although the early retrospective enquiries emphasized the frequency and importance of such defects as congenital anomalies of the heart and eyes, and deafness, it was not until the follow-up studies had been carried out on infants whose mothers had had rubella during the extensive 1963/64 US outbreak that it was fully appreciated that congenital rubella frequently caused widespread multi-system disease. However, follow-up studies showed that congenitally acquired rubella was not a static disease and that prolonged careful evaluation of infants at risk was necessary before some or all of the features of congenitally acquired rubella were apparent. The broader range of anomalies described after the US 1963/64 and subsequent outbreaks were probably due not to any change in viral virulence but rather to more careful and prolonged observation, since careful scrutiny of the records of infants with congenitally acquired rubella who were born before these outbreaks revealed that such anomalies as thrombocytopenic purpura and osteitis, although not reported in the literature, occurred fairly frequently.

Cooper (1975) divided clinical features associated with rubella infection into those which were transient, developmental or permanent (Table 94.3). The pathogenesis of transient lesions is not understood but these are usually present only during the first few weeks of life, do not recur, and are not associated with the development of permanent sequelae. Intra-uterine growth retardation resulting in low birth weight, but at a normal gestational age (small for dates babies) is among the commonest of the transient features. Thus, Cooper *et al.* (1965) found that about 60 per cent of infected infants fell below the 10th and 90 per cent below the 50th percentile.

A petechial or purpuric rash is also common, particularly among infants whose mothers had had maternal rubella in early pregnancy (Cooper *et al.* 1965; Horstmann, *et al.* 1965) (Figure 94.7). However, low birth weight and a purpuric rash are seldom the sole manifestations of congenital rubella. These infants may have such other anomalies as congenital heart and eye defects, although these may not always be apparent at birth. Infants with thrombocytopenic purpura generally have a platelet count ranging from 3000 to 100 000/mm^3, this being associated with a decreased number of megakaryocytes, but of normal morphology, in the bone marrow. In general, the platelet count rises spontaneously during the first month of life, although rarely some infants die from such complications as intracranial haemorrhage.

Bone lesions may be detected by x-ray. Irregular areas of translucency are present in the metaphyseal portion of the long bones but without any evidence of periosteal reaction in over 20 per cent of infants with congenital rubella (Cooper *et al.* 1965). These lesions generally resolve within 1–2 months. Cooper *et al.* (1965) detected these characteristic radiological changes in a fetus of 18 weeks gestational age, suggesting that the process inducing such changes begins in early gestational life. Approximately 25 per cent of infants who present at birth with clinical evidence of congenitally acquired rubella have CNS involvement,

Table 94.3 Clinical features associated with congenitally acquired rubella

Transient	Developmental	Permanent
Common	*Common*	*Common*
Low birthweight	Sensorineural deafness	Sensorineural deafness
Thrombocytopenic purpura	Peripheral pulmonary stenosis	Peripheral pulmonary stenosis
Hepatosplenomegaly	Mental retardation	Pulmonary valvular stenosis
Bone lesions	Central language defects	Patent ductus arteriosus
Meningoencephalitis		Ventricular septal defect
		Retinopathy
		Cataract
		Microphthalmia
		Psychomotor retardation
		Microcephaly
		Cryptorchidism
		Inguinal hernia
Uncommon	*Uncommon*	*Uncommon*
Cloudy cornea	Severe myopia	Severe myopia
Hepatitis	Diabetes mellitus	Diabetes mellitus
Generalized lymphadenopathy	Thyroiditis	Thyroid disorders
Haemolytic anaemia	Hypothyroidism	Dermatoglyptic abnormalities
Pneumonitis	Growth hormone deficiency	Glaucoma
	'Late onset disease'	Myocardial abnormalities

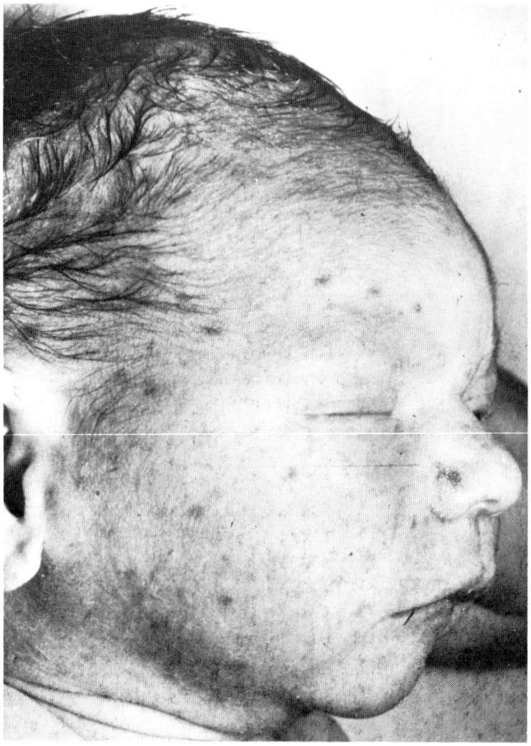

Fig. 94.7 Purpuric rash in newborn infant with congenitally acquired rubella, who was subsequently found to have congenital heart disease and cataract as well.

this usually being in the form of a meningoencephalitis. Such infants are often lethargic at birth but may become irritable and often exhibit evidence of photophobia. They have a full anterior fontanelle, pleocytosis, and an increased amount of protein in the CSF (Desmond *et al.* 1967). The outcome is variable and unpredictable. Although about 25 per cent of infants presenting with meningoencephalitis at birth may by the age of 18 months be severely retarded, suffering from communication problems, ataxia or spastic paresis, some infants appear to progress well neurologically despite poor development in the first six months of life.

Some developmental defects may take many months or years to become apparent, but then persist permanently. Failure to recognize such defects in early infancy may not always be the result of difficulty in their detection. There is evidence which suggests that such defects as perceptive deafness, CNS anomalies, and some ocular defects, may actually develop or become increasingly noticeable some considerable time after birth. Thus Peckham (1972) showed that some two-year-old children with apparently normal hearing had severe perceptive deafness when examined later. Menser and Forrest (1974) showed that it might take up to four years before the first rubella defects were recognized; further defects might continue to be recognized up to the age of 8. The progressive nature of congenitally acquired disease is emphasized by the finding that children with previously stable congenital rubella-induced defects developed a widespread subacute panencephalitis with progressive motor retardation as late as the second decade in life (Townsend *et al.* 1975, Weil *et al.* 1975); rubella virus was isolated from the brain of one such 12-year-old child (Weil *et al.* 1975).

Diabetes mellitus has been the most frequently reported endocrine disturbance. Persons who present with diabetes mellitus exhibit a wide age range (14 months to 36 years). A particularly high frequency of diabetes mellitus has been observed among survivors of the extensive 1940/41 rubella epidemic in Australia. When these patients were reviewed in 1977, 9 of 45 (20 per cent) had diabetes mellitus (Menser *et al.* 1978). Despite these findings, Hanshaw and Dudgeon (1978) failed to encounter diabetes among a series of 287 patients with congenital rubella in the UK. This may perhaps reflect the younger age of the patients in this series. However, islet-cell antibodies have been detected in the sera of 70 per cent of patients who have not yet acquired diabetes (Ginsberg-Fellner *et al.* 1980).

Between the ages of about 3–12 months, some infants may present with such features as a chronic rubelliform rash, persistent diarrhoea, and pneumonitis. Marshall (1973) has referred to this syndrome as 'late onset disease'. Although mortality is high, some infants show a dramatic response to treatment with corticosteroids. This syndrome may reflect an immunopathological phenomenon; circulating immune complexes which appear to contain rubella antigen have recently been demonstrated in infants with late onset disease (Tardieu *et al.* 1980), while Coyle and colleagues (1982) demonstrated rubella antibody containing immune complexes in children with congenital rubella who developed new clinical problems some years after birth.

Of the permanent defects, the commonest is sensorineural deafness. This results from rubella-induced damage to the Organ of Corti. However, central auditory impairment may also occur. Hearing loss, which may be unilateral, bilateral, mild or profound, may sometimes be the only rubella-induced congenital anomaly.

Peckham (1972) followed up 218 children who were apparently normal at birth, but who had been exposed to rubella *in utero*. When assessed for hearing loss at the age 1–4 years, 50 (23 per cent) were deaf; when 85 were re-examined between the ages of 6–8 years, further hearing defects were detected in another 9 children. Ninety per cent of the children with hearing defects were seropositive. Since rubella antibodies are uncommon before the age of 4, it is particularly important to follow up infants with persistent rubella antibody, so that hearing defects can be recognized as early as possible.

Congenital anomalies of the cardiovascular system are responsible for much of the high perinatal mortality associated with congenitally acquired rubella. Numerous studies have shown that the commonest lesions are persistence of a patent ductus arteriosus, proximal (valvular) or peripheral pulmonary artery stenosis, and a ventricular septal defect (Cooper 1975, Hastreiter *et al.* 1967, Sperling and Verska

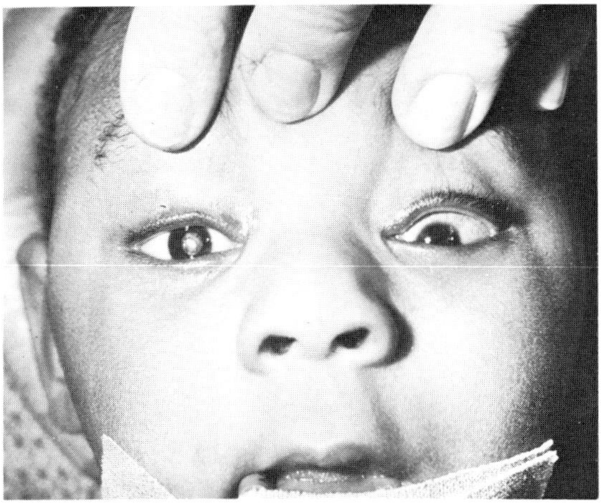

Fig. 94.8 Congenital rubella cataract in a 9 month old infant. Cataract was also present in the left eye but was surgically removed.

1966). Occasionally neonatal myocarditis is found, often associated with other cardiac malformations (Korones *et al.* 1965). Rubella-induced damage to the intima of the arteries may result in obstructive lesions of the renal and the pulmonary arteries (Rorke and Spiro 1967, Phelan and Campbell 1969).

Many of the ocular defects characteristic of congenitally acquired rubella were described by Gregg (1941), who drew particular attention to pigmented retinopathy and cataract. Pigmented retinopathy may be present in up to 50 per cent of infants with congenitally acquired rubella (Menser and Reye, 1974) and may provide a useful aid in clinical diagnosis. The macular area of the retina is generally affected but the lesions rarely impair vision. Cataracts, although usually present at birth, may not be visible until several weeks later (Murphy *et al.* 1967). Lesions may be sub-total consisting of a dense pearly-white central opacity (Fig. 94.8) or alternatively total with a more uniform density throughout the lens. Micropthalmus is often associated with congenital cataract, but glaucoma is rare although important to recognize, since it may rapidly cause blindness.

It will be obvious that as a result of the extent and severity of rubella-induced congenital malformations, children who survive will need continuous specialized management, education, and rehabilitation. However, a study carried out on fifty 25-year-old patients with congenitally acquired rubella born in Australia after the 1940/41 epidemic showed that, although many were deaf or had eye defects, they had developed far better than had been anticipated when assessed in early childhood. Many had married and produced normal children, and all but four were employed, most patients being of average intelligence (Menser *et al.* 1967). Follow-up studies so far reported on children with congenitally acquired rubella following the 1963/4 US rubella epidemic suggest that this may have had a more catastrophic impact on the lives of affected children (Cooper 1975). This may be a reflection on the more modern methods of treatment available to the children born after this more recent epidemic; many might not have survived previously.

5. Virological diagnosis

A diagnosis of congenitally acquired rubella can be established by:

(a) isolation of rubella virus from the infected infant during early infancy.
(b) detection of rubella-specific IgM in serum samples obtained in early infancy, and/or,
(c) detection of persistent rubella antibody in the infant. i.e. the presence of antibodies at a time beyond which maternal antibodies are usually no longer detected (approximately 6 months of age).

Rubella virus can be isolated from most of the organs obtained at autopsy from severely infected infants who die in early infancy, as well as from the stools, urine, tears, CSF and nasopharyngeal secretions of those who survive. It can also be isolated from the nasopharynx of almost all severely infected infants at birth, but by the age of 3 months the proportion declines to about 60 per cent, and by 9–12 months, to approximately 10 per cent (Cooper and Krugman, 1967) (Fig. 94.9). However, rubella virus may persist in other sites

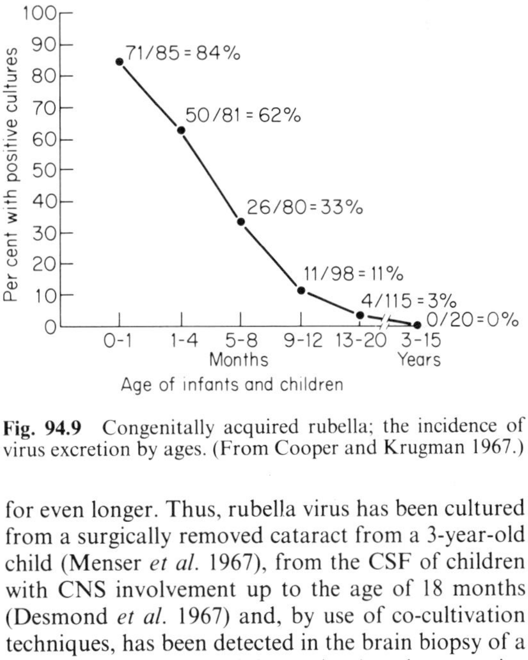

Fig. 94.9 Congenitally acquired rubella; the incidence of virus excretion by ages. (From Cooper and Krugman 1967.)

for even longer. Thus, rubella virus has been cultured from a surgically removed cataract from a 3-year-old child (Menser *et al.* 1967), from the CSF of children with CNS involvement up to the age of 18 months (Desmond *et al.* 1967) and, by use of co-cultivation techniques, has been detected in the brain biopsy of a congenitally infected child who developed progressive degenerative brain disease in the second decade of life (Weil *et al.* 1975).

Babies excreting virus may transmit infection to susceptible contacts and therefore until virus is no longer being excreted, women of childbearing age, some of whom may be in the early stages of pregnancy, should be dissuaded from visiting such babies until serological tests confirm that they are immune.

Figure 94.10 illustrates the pattern of antibody responses in infancy in congenitally acquired rubella. At birth, both maternally derived specific IgG antibodies as well as specific IgM and IgG synthesized *in utero* by the fetus are present.

Rubella-specific IgM can usually be detected in most congenitally infected infants during the first six months of life but after this the number declines.

Employing a sensitive radioimmunoassay technique, Cradock-Watson *et al.* (1979) detected rubella specific IgM in all of 14 congenitally infected infants aged 14 weeks to 6 months, 9 of 17 aged 6½ to 11 months, 6 of 14 aged 1 year to 23 months, and 1 of 12 between the ages of 2 to 3 years. Although rubella specific IgG may persist for many years, studies on 223 children with congenital rubella following the 1963/4 US epidemic showed that the HAI antibodies declined more rapidly among congenitally infected children than among their mothers; by the age of 5, 20 per cent of infants with congenital rubella no longer had HAI antibodies. Nevertheless, seronegative children failed to develop an immune response or excrete virus when challenged with an attenuated rubella vaccine (Cooper *et al.* 1971), which suggests that residual immunity was present despite antibody being undetectable by HAI.

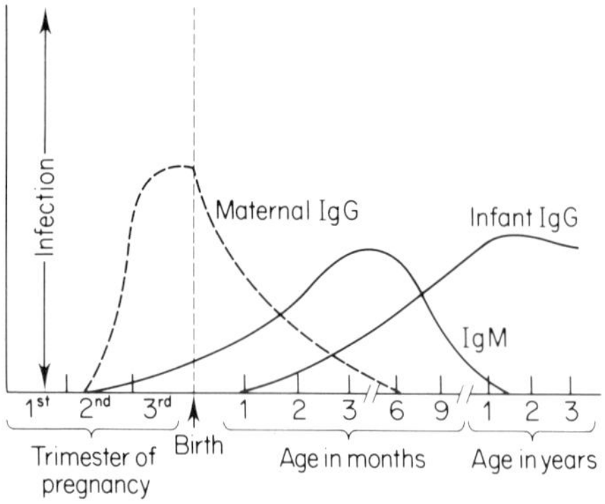

Fig. 94.10 Immune responses in congenitally acquired rubella. (From Herrmann 1979.)

Very occasionally, rubella-specific IgM may not be detectable in the immediate neonatal period, but if further serum samples are tested during the next few weeks, the response may become apparent.

Laboratory techniques

1. Virus isolation and identification

Virus isolation techniques are rarely used for the diagnosis of postnatally acquired rubella, since this may more reliably and rapidly be established by serological methods. However, virus isolation is of importance in order to determine the duration of excretion in congenitally infected infants, since they may transmit infection to susceptible contacts. Attempts may also be made to isolate virus from the products of conception of women whose pregnancies are terminated because of naturally acquired infection in early pregnancy or inadvertent vaccination (Modlin *et al.* 1976). Rubella virus can be cultured from fetal tissues by use of explant techniques (Herrmann 1979) or by mincing and shaking the tissues (Thompson and Tobin 1970). The latter method has produced excellent results and is simpler to use. However, since rubella is a labile virus, there is little chance of isolating virus from material that has been in transit for more than 72 hours.

Rubella virus can be identified by the production of CPE in RK13, SIRC and certain sublines of Vero cells, by interference in primary VMK and other monkey kidney cultures, or by using immunofluorescence (IMF) or immunoperoxidase (IMP) for detection of antigen in such cells as RK13, BHK-21 and Vero. RK13 cell cultures have been used extensively in the UK for the isolation and titration of rubella virus (Best and Banatvala 1967, McCarthy *et al.* 1963, Pattison 1982*a*). The CPE produced by rubella virus after primary inoculation in these cells may be confused with the non-specific effects caused by cellular material present in the specimen. However, after further passage in RK13 cell cultures, a CPE will be observed 3–4 days later, providing a high titre of virus is present (see Fig. 94.1). The use of SIRC cells for rubella virus isolation was initially described by Leerhøy (1965); these cells are now commonly used for rubella virus isolation in Scandinavian countries (Leerhøy 1966). Rubella virus is very fastidious and will produce CPE in RK13 and SIRC cultures only if they are maintained under carefully controlled conditions, using serum which has previously been tested and found to be satisfactory.

Probably the most sensitive method for rubella virus isolation is by inoculation of specimens into Vero cell cultures (see pp. 278–9) followed by passage in RK13 or SIRC cells, virus being identified in these cells by its characteristic CPE. Alternatively, the initial passage in Vero cell cultures can be followed by passage into RK13 or BHK-21 cells grown on cover slips or Lab-Tek eight-chamber slides (Lab-Tek Products Division, Miles Laboratories Inc.), rubella virus being identified by indirect immunofluorescent or immunoperoxidase techniques which show characteristic cytoplasmic and perinuclear staining (Fig. 94.11). The use of IMF for diagnosis of rubella has been reviewed by Gardner and McQuillin (1980) and Pattison (1982*a*).

Since the original description of the isolation of

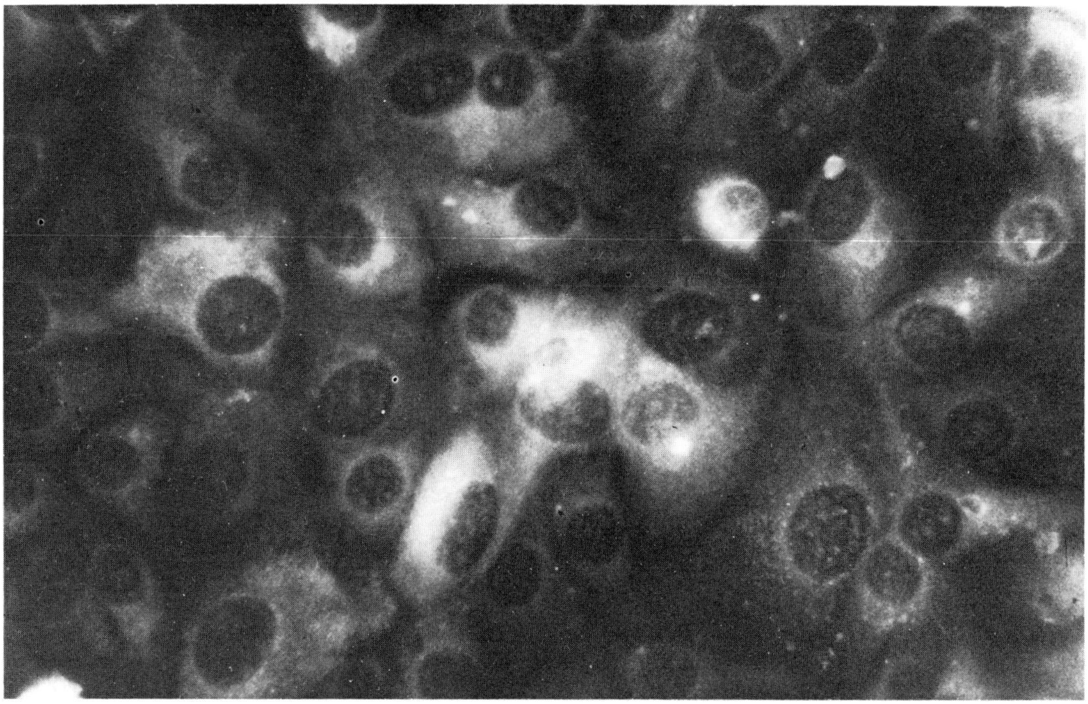

Fig. 94.11 Rubella virus antigen in RK13 cell cultures detected by immunofluorescence. Note cytoplasmic and peri-nuclear fluorescence.

rubella virus by interference in primary VMK (Parkman *et al.* 1962), this technique has been used extensively for rubella virus isolation, particularly in the USA. The use of these cultures and other monkey kidney cell cultures has been reviewed by Herrmann (1979).

The identity of isolates in cell cultures, whether detected by CPE or the interference technique, should be confirmed as rubella by neutralization tests or immunofluorescence with virus-specific immune animal sera (Herrmann 1979, Pattison 1982*a*).

2. Serological techniques for Rubella antibody screening and diagnosis

There is an increasing demand for rubella serology. This is due to (1) the recognition of the importance of conducting investigations on women who have been exposed to or who have developed rubella-like illnesses in pregnancy, (2) to the importance of screening adults prior to rubella vaccination (see p. 295), and (3) to the need for studies to determine the development and persistence of rubella antibodies after vaccination.

(i) Tests for rubella antibody screening

Although the HAI test has been extensively used for screening, it has now been replaced by single radial haemolysis (SRH) (Fig. 94.12) in most laboratories in the UK, since false positive results may occur in the HAI test due to non-specific inhibitors of haemagglutination present in test sera—a problem which does not occur in SRH. Radioimmunoassay (RIA), enzyme-linked immunosorbent assay (ELISA), passive haemagglutination (PHA), latex agglutination and an immunofluorescent test, FIAX (International Diagnostic Technology) are also suitable for screening purposes, but have not yet been extensively used. Some of the characteristics of these tests are shown in Table 94.4. HAI, SRH, RIA and ELISA have been described in detail by Pattison (1982 *a*, *b*) and HAI, PHA and RIA by Herrman (1979). Kits containing all the necessary reagents for HAI, ELISA, PHA and FIAX are available from commercial sources, as are individual reagents for some of these tests. However, it must be stressed that the use of kits in inexperienced hands may lead to erroneous results.

(ii) Serological diagnosis by demonstration of a significant rise in antibody titre

Neutralizing antibodies appear rapidly after onset of illness (see Fig. 94.5), persist indefinitely and are generally believed to confer protection. However, when it was demonstrated that rubella virus contained an HA (Stewart *et al.* 1967), the HAI test generally replaced the neutralization test for screening and diag-

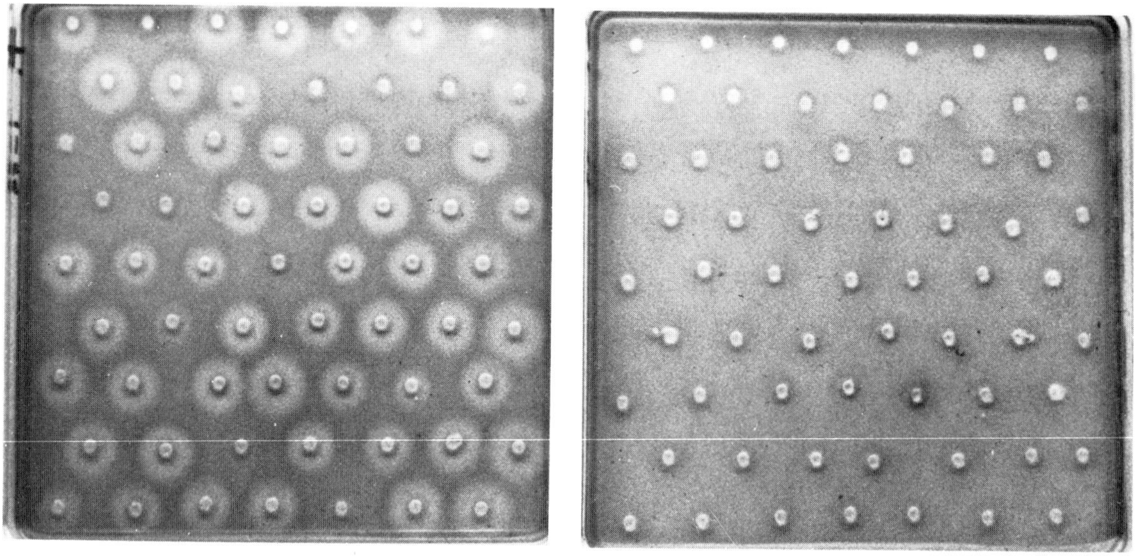

Fig. 94.12 Single radial haemolysis. Test plate on left, control plate containing non-sensitized erythrocytes on the right.

nosis, since this test obviated the need for cell cultures and could be performed more rapidly than neutralization. Although neutralizing and HAI antibodies generally agree well, some sera found negative by conventional HAI tests have a low titre of antibodies when tested by sensitive neutralization tests (Balfour *et al.* 1981) and by SRH, RIA and ELISA (Best *et al.* 1980).

HAI antibodies appear soon after onset of illness and may even be present on the first day of rash. Since they quickly rise to their peak titre, it may be difficult to make a diagnosis if the first serum sample is not obtained within a few days of onset of illness. When the first serum is obtained later, it may be necessary to use HAI in combination with SRH, complement-fixation (CF) or PHA, tests in which the appearance of rubella antibodies may be more delayed (see Fig. 94.5). Alternatively such sera can be tested for rubella-specific IgM (see below).

The major drawback of the HAI technique is that serum lipoproteins act as non-specific inhibitors of haemagglutination. Thus, before they can be tested, sera must be pre-treated to remove both serum lipoproteins and serum agglutinins. There are several different methods available for removing the serum lipoproteins; these methods, together with the different diluents and red cells which can be used in this test, have been described in detail elsewhere (Herrmann 1979, Palmer *et al.* 1977, Pattison 1982*a*).

Although until recently CF tests were used exten-

Table 94.4 Characteristics of tests available for rubella antibody screening

Test	Labour intensive	Speed	Sensitivity	Cost of materials/ test (p)	Capital cost	Comments
HAI	+ + +	Same day	+	15	Minimal	Decreasing use—problem with non-specific inhibitors.
SRH	+	Overnight	+ +	15	Minimal	Method of choice for routine large scale screening
RIA	+ +	Same day or* overnight	+ + +	> 50	γ-counter (£7000–£16000)	With ELISA, most sensitive.†
ELISA	+ +	Same day	+ + +	> 50	Spectrophotometer (£1000–£6000)	With RIA, most sensitive.†
PHA	+ +	2–3 hours	+	50	Minimal	Occasionally difficult to read. Possible value as diagnostic supporting test.
IMF (FIAX)	+ +	2–3 hours	+	60	Fluorimeter (£3000)	Potentially very useful test, but only preliminary results as yet available.

* Time taken depends on type of γ-counter used.

† Useful for determination of immune status when other tests give equivocal results.

sively to support HAI tests for rubella diagnosis, CF antibody responses exhibit considerable variation in their titres and the time at which they appear. This partly reflects the different methods employed to prepare CF antigens, which may vary considerably in their potency. Fewer laboratories are now employing CF tests for diagnosis, placing more reliance on such tests as SRH and on the detection of rubella-specific IgM. The CF test is of no value for screening purposes, since CF antibodies are time-dependent and may persist for only a few years after infection (Sever *et al.* 1967).

Although the main application of SRH is for rubella antibody screening (see p. 291), it can also be used for diagnostic purposes providing the wells are of consistent size and shape, and an accurate volume of serum is added to each well. The IgM class of antibodies is not detected by this technique and sera collected during the acute and early convalescent phase of infection may produce a hazy zone of haemolysis.

More recently, RIA and ELISA techniques have been developed, which are more sensitive methods for measuring rubella antibodies. Since a wider range of titres or readings is experienced, it is often possible to detect a rise in titre when the first serum is obtained too late after onset of illness to detect a rising titre by HAI (Forghani and Schmidt 1979, Vaheri and Salonen 1980). In some laboratories, RIA or ELISA have already replaced HAI for the diagnosis of rubella. Several ELISA tests are available from commercial sources but have not yet been thoroughly assessed for diagnostic use.

(iii) Serologic diagnosis by detection of rubella specific IgM

All laboratories involved in the assessment of the risk to women exposed to rubella-like illnesses during pregnancy will need to carry out tests to detect rubella-specific IgM. Until recently, sucrose density gradient fractionation followed by HAI testing of the fractions, was the test used extensively for this purpose and the standard against which new tests were compared. Despite the development of new techniques, in experienced hands this method is still a very sensitive technique for the detection of rubella specific IgM (Table 94.5). Employing this technique, rubella specific IgM has been detected for as long as a year after naturally acquired and vaccine induced infection (Al Nakib *et al.* 1975). Furthermore, since sera are fractionated, interference from high levels of IgG and anti-globulins such as IgM-rheumatoid factor (RF) are not encountered. However, as this test is laborious and an ultracentrifuge is required, it is unsuitable for testing large numbers of sera. Although the separation of immunoglobulins by gel filtration is a technique which requires less sophisticated and expensive apparatus, its sensitivity is less than that attained with sucrose density gradient fractionation and it is therefore seldom used.

Indirect IMF is a sensitive method for detecting rubella-specific IgM. However, high quality reagents are required and care must be taken to avoid false positive results due to IgM-RF and other anti-globulins and false negative results due to interference by rubella-specific IgG. When used to test sera which have previously been fractionated in a sucrose density gradient, this has provided a very sensitive test for the diagnosis of congenital rubella (Cradock-Watson *et al.* 1976).

It is probable that most laboratories will soon employ RIA or ELISA for detecting rubella-specific IgM. Indirect RIA and ELISA are similar in principle to indirect IMF, with rubella and control antigens attached to a solid phase (Kangro *et al.* 1978, Leinikki and Passilä 1977, Meurman *et al.* 1977). Although these

Table 94.5 Sensitivity of some well established techniques for the detection of rubella-specific IgM

Test	Sensitivity	Rapidity	Problems
RIA (anti-μ) (Mortimer *et al.* 1981)	+ + + + +	⩽24	Sera from patients with infectious mononucleosis may give false-positive results.
RIA (Kangro *et al.* 1978)	+ + + + +	⩽24	False-positives due to RF.
SDGF and IF (Cradock-Watson *et al.* 1976)	+ + + + +	24	Cumbersome and time consuming.
SDGF and HAI (prolonged incubation) (Al-Nakib *et al.* 1975)	+ + + +	48	Care needed to remove non-specific inhibitors. If complete separation of IgM and IgG is not obtained, confirmation with 2ME is necessary.
ELISA (Ziegelmaier, Behrens and Enders 1981)	+ + +	24	False positives due to RF. ?Competition from rubella-specific IgG.
SDGF and HAI (short incubation) (Vesikari and Vaheri 1968)	+ + +	24	If complete separation of IgM from IgA is not obtained, confirmation with 2ME is necessary.
Gel filtration and HAI (prolonged incubation) (Pattison and Mace 1975)	+ +	24	Care needed to remove non-specific inhibitors.
SPIT (Krech and Wilhelm 1979)	+ +	24	Efficacy for detecting low levels of rubella-specific IgM and vaccine induced IgM not established.

assays have many advantages, including sensitivity, speed, objectivity and potential for automation, it cannot be over-emphasized that it is essential to use reagents of high specificity and that the test must be very carefully standardized. Owing to the increased sensitivity of these tests the removal of RF-like anti-immunoglobulins from sera before testing is essential, since very small quantities can produce a false-positive result. RIA has been found to be slightly more sensitive for the detection of low levels of rubella-specific IgM than sucrose density gradient fractionation with IMF and is a quicker and more economical test to perform (Cradock-Watson *et al.* 1979). Alternatively, the IgM 'antibody capture' technique may be used, in which the solid phase is coated with anti-human IgM. The anti-human IgM 'captures' IgM antibodies present in the test sera. Rubella HA is then added and will attach if rubella-specific IgM is present. Bound rubella HA is then detected by the addition of ^{125}I– or enzyme-labelled rubella antiserum. False-positive reactions due to RF do not appear to occur in these tests, since all rubella-specific IgM in the test serum should be washed away Mortimer *et al.* 1981, Chantler *et al.* 1982). However, sera from patients with infectious mononucleosis may give a false-positive result (Morgan-Capner *et al.* 1983).

Table 94.5 compares the sensitivity, rapidity and some of the problems associated with the different techniques. The technical details of most of these assays have recently been published in a PHLS monograph (Pattison 1982*a*).

Commercial ELISA kits are becoming available for the detection of rubella-specific IgM. Again it must be stressed that, even if these kits have been thoroughly evaluated by experienced laboratories, their use by those inexperienced in rubella serology may lead to erroneous results in a variable but significant proportion of specimens. All available tests present problems from time to time and it is essential that a diagnosis of rubella in pregnancy, particularly if termination is contemplated, be made on the basis of considering clinical data as well as the results of all serological tests carried out.

Prevention

1. Rubella vaccination

(i) Vaccines available

Vaccination now makes congenitally acquired rubella a potentially preventible disease. Rubella vaccines were licensed in the USA and UK in 1969 and 1970 respectively. Table 94.6 provides details of the derivation of vaccine strains and their passage history. Although the HPV77-DE5 (Meruvax), Cendehill (Cendevax) and RA27/3 (Almevax) vaccines are licensed in the USA, only the RA27/3 vaccine is currently available for use; in the UK both the Cendehill and RA27/3 vaccines are licensed. Vaccines are administered subcutaneously but RA27/3 will induce an immune response when administered intranasally (Ingalls *et al.* 1970, Plotkin *et al.* 1968); however, it has not been licensed for use by this route.

(ii) Vaccination policies

The US and UK vaccination programmes are directed principally towards different groups. In the USA the vaccine programme has until recently been aimed at

Table 94.6 Attenuated rubella vaccines

Vaccine	Strain derivation	Attenuation	Reference
HPV77	Army recruit with rubella (1961)	Vervet monkey kidney (77)	Parkman *et al.* 1966 Meyer *et al.* 1966
HPV77-DE5	As above	Vervet monkey kidney (77); duck embryo (5)	Buynak *et al.* 1968
Cendehill	Urine from a case of postnatally acquired rubella (1963)	Vervet monkey kidney (3); primary rabbit kidney (51)	Peetermans and Huygelen 1967; Martin du Pan *et al.* 1968
RA27/3	Kidney of rubella-infected fetus (1964)	Human embryonic kidney (4); WI-38 fibroblasts (17–25)	Plotkin *et al.* 1967, 1968
To-336	Pharyngeal secretion of child with postnatally acquired rubella, Toyama, Japan (1967)	Vervet monkey kidney (7); primary guinea-pig kidney (20); primary rabbit kidney (3)	Takeda Chemical Co., Osaka, Japan (Personal Communication).
QEF (MEQ_{11})*	Throat washing from patient in Osaka (1966) = Matsuura strain	VMK (14); Chick amnion (65); Japanese quail embryo fibroblast cells (11)	Minehawa *et al.* 1973

* Two other rubella vaccines have also been licensed in Japan (Shishido and Ohtawara 1976).

pre-pubertal children of both sexes (Report 1972*a*), the programme being designed to substitute vaccine-induced infection for naturally acquired disease, thereby reducing the risk of children transmitting rubella to pregnant women. This programme also reduces the hazard of inadvertent vaccination during pregnancy or vaccination at an age at which reactions are more likely to occur. Since rubella vaccines are often given in infancy in the USA, combined attenuated rubella, measles, and mumps vaccines have been used and shown to induce antibodies to all three viruses which are similar in their level and persistence as those with monovalent vaccines (Buynak *et al.* 1969; Weibel *et al.* 1978). The US programme has resulted in a decline in the incidence of both postnatally and congenitally acquired rubella (Fig. 94.13), although congenitally acquired rubella may be

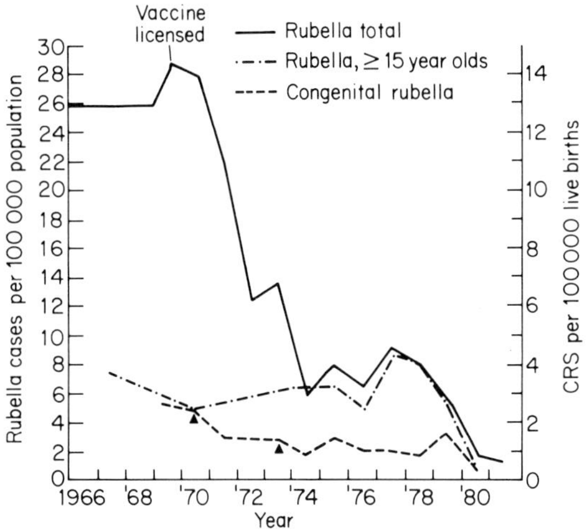

Fig. 94.13 Declining incidence of postnatally acquired rubella in USA, 1966–81.

under-reported. A very high rubella vaccination uptake rate has been achieved. This is a reflection of the fact that US children receive the combined measles, mumps-rubella vaccine and that children cannot enter primary school unless they have proof of measles vaccination. There has until recently been a high rate of susceptibility among adolescents and young adults (Preblud *et al.* 1980*b*; Preblud *et al.* 1980*a*); and outbreaks of rubella among university students and military and hospital personnel have of late been reported (Report 1978, 1979*a*; Polk *et al.* 1980, Strassburg *et al.* 1981). As in the UK, attention has therefore been focused on giving rubella vaccine to susceptible adolescents and adult women who missed vaccination in childhood, and this has resulted in a decline in the incidence of rubella among such persons.

In the UK, the vaccination programme was at first directed towards 11 to 14-year-old schoolgirls (Report 1970*a*), although shortly after the vaccine was licensed it was recommended that women at special risk of contracting rubella, e.g. nurses, doctors, and schoolteachers, should also be offered vaccine provided that prior antibody screening showed that they were seronegative (Report 1972*b*). The programme was later extended to women found to be seronegative while attending family planning and antenatal clinics. Women in the latter, when found seronegative, were offered rubella vaccination in the immediate post-partum period (Report 1976). In contrast with the USA, the UK Congenital Rubella Surveillance Programme shows that there has so far been no reduction in the incidence of congenitally acquired rubella (Report 1981*a*). However, the schoolgirl vaccination programme is unlikely to take effect until the mid-1980s and then only if a much higher proportion of schoolgirls are vaccinated than at present. Recent data suggest that vaccine uptake levels among schoolgirls have increased from about 69 per cent in 1976 to approximately 84 per cent in 1980. Studies among university students, blood donors (Clarke *et al.* 1983) and antenatal patients in Scotland. Ross and McCartney 1979 have shown that the proportion of seronegative females has declined. More emphasis is now being directed towards vaccinating susceptible women of child-bearing age. The extensive 1978–9 rubella epidemic in the UK which resulted in many pregnant women being infected and a large number of terminations of pregnancy due to maternal rubella was a reflection of the vulnerability of women in this age group (Goldwater *et al.* 1978). Although it has been suggested that rubella vaccine should be offered to pre-school chidren in the UK, with perhaps a second dose at 10–14 years, such a policy may have drawbacks. Thus Knox (1980) and Anderson and May (1983) have pointed out that failure to vaccinate a very high proportion of pre-school children might only result in a delay in the age at which rubella is acquired and which in turn might result in infection occurring more frequently among adolescents and young adults. The current rate of acceptance of measles vaccine in the UK (< 50 per cent) does not provide an optimistic forecast for high rubella vaccination uptake rates among pre-school children.

(iii) Immune responses

The HAI test has been used more than any other test to assess immune responses following rubella vaccination. HAI antibodies usually become evident 10–21 days after vaccination; appreciable quantities are present about 8 weeks later. Although levels of HAI antibody may vary, depending on such factors as the vaccine employed and variation in individual responses, vaccine-induced antibody titres are generally some 4–8 fold lower than those which result from recent naturally acquired rubella. In general, detectable antibodies persist although, as with naturally acquired infection, there may be some decline with time. Follow-up studies have shown that antibodies are more likely to decline to unacceptably low levels after administration of HPV77-DE5 and Cendehill vaccines than after naturally acquired infection and vaccination with RA27/3 (Banatvala 1977*b*). More recently Zealley and Edmond (1982) and O'Shea *et al.* (1982) showed that only 5–6.5 per cent vaccinees were seronegative by SRH 6–16 years after vaccination; the majority of the seronegative vaccinees had received Cendehill vaccine. Furthermore, follow-up studies of persons with naturally acquired or vaccine-induced immunity suggest that those who have been naturally infected or given RA27/3 are less likely to be reinfected as shown by a rise in antibody titre (Banatvala 1977*b*, Harcourt *et al.* 1980). Some rather disquieting evidence was produced by Horstmann (1975) and Balfour and Amren (1978), who showed that 29 of 342 (8.5 per cent) and 58 of 159 (36 per cent) children respectively lacked HAI antibodies 3 to 9 years after receiving HPV77-DE5 vaccine. However, when challenged with RA27/3 vaccine it was evident that these seronegative children had residual immunity (Balfour *et al.* 1981). Antibody levels appear to be more durable after vaccination with RA27/3. Rubella vaccines also vary in the qualitative aspects of the immune response they induce; such differences may be of considerable long-term importance. The accumulated data from a number of vaccine trials suggest that the RA27/3 vaccine induces an immune response which more closely resembles that induced by naturally acquired infection than other vaccines, perhaps because this strain has had a less extensive passage history in cell culture (Table 94.6). Thus, serum and nasopharyngeal specific IgA responses following RA27/3 more closely resemble those induced by naturally acquired infection than those induced by other vaccines (Best *et al.* 1979), as do neutralizing, CF (Plotkin *et al.* 1973) and anti-*iota* precipitating antibodies (see p. 277).

(iv) Reinfection

Evidence of reinfection is generally obtained serologically by demonstrating an increase in antibody titre, the incidence of reinfection being highest among those vaccinated with Cendehill and HPV77-derived vaccines and lowest in those with naturally acquired immunity (MacDonald *et al.* 1978, Fogel *et al.* 1978). Reinfection would be hazardous only if viraemia occurred and, as with naturally acquired immunity, this has been rarely documented (Forrest *et al.* 1972). Reinfection is not usually associated with a rubella-specific IgM response in those whose immunity is naturally acquired, but rubella specific IgM has been detected among those whose immunity is vaccine induced. However, compared with primary infection, a relatively small amount of rubella specific IgM is present (Balfour *et al.* 1981, Cradock-Watson *et al.* 1981, Harcourt *et al.* 1980, MacDonald *et al.* 1978). It may seem encouraging that when children whose antibodies had declined to undetectable levels when tested by HAI and SRH were challenged with RA27/3 three to nine years after vaccination with HPV77-DE5, all showed typical secondary immune responses, low or negative specific IgM responses, and no viraemia (Balfour *et al.* 1981). Although this may suggest that revaccination is unnecessary in those with low or undetectable antibody titres after vaccination with HPV-derived vaccines, it must be remembered that this study was carried out less than 10 years after vaccination. O'Shea *et al.* (1983), however, detected viraemia in one of 19 vaccinees with low ($\geqslant 15$ i.u.) levels of rubella antibody following experimental challenge with RA27/3. This emphasizes the importance of carrying out long-term studies on those given different rubella vaccines at different ages.

(v) Virus excretion

Many of the early vaccine trials showed that a variable but significant proportion of vaccinees excreted virus via the nasopharynx, usually one to four weeks after vaccination. However, when particular care is taken to ensure that good specimens are collected, and sensitive methods for virus isolation are used, the proportion of vaccinees excreting virus, regardless of which vaccine is given, may range between 65–100 per cent (Best *et al.* 1974, Mallinson personal communication). The accumulated data obtained from numerous trials conducted in both institutionalized and open communities in which susceptible contacts were closely exposed to vaccinees have shown that virus transmission is so rare as to make the risk of transmitting infection to susceptible persons negligible. Lack of communicability may be related to the small amounts of virus excreted by vaccinees, but a proportion may excrete amounts of virus similar to those detected in naturally acquired infection (Best *et al.* 1974). Perhaps attenuation results in some alteration of the biological properties of the virus whereby it replicates less favourably in the respiratory mucosa of susceptible contacts.

(vi) Vaccine reactions

Rubella vaccines are generally well tolerated. However, such reactions as lymphadenopathy, rash, arthralgia (painful joints), or arthritis (joint swelling or limitation of movement) may occur some two to four weeks after vaccination, but such reactions are usually less severe than those following naturally acquired disease. Although occasionally vaccinees may complain of enlarged and tender lymph nodes, lymphadenopathy is seldom noticed by the vaccinee. Rash is uncommon, but when it does occur it is usually macular, faint, and fleeting. Reactions, particularly joint involvement, are rare in children but are more commonly encountered in post-pubertal females, increasing with age. The knees, finger joints, wrists and ankles are most frequently affected; symptoms are generally mild and do not usually persist for more than 3–4 days. Only rarely do these symptoms result in sickness absence (Best *et al.* 1974). RA27/3 and HPV77-derived vaccines are more likely to induce arthralgia or arthritis than the Cendehill vaccine. The pathogenesis of rubella-induced arthralgia is discussed on page 281.

(vii) Vaccine failures

Rubella is a relatively labile virus and failure to follow the manufacturer's instructions in the storage and reconstitution of the vaccine may result in its inactivation. A small proportion of vaccinees (about 5 per cent) who fail to sero-convert after vaccination usually respond satisfactorily when revaccinated. However, a few may fail to do so, or respond poorly, because they have a pre-existing low level of antibody which is undetectable by conventional tests (Best *et al.* 1980, Buimovici-Klein *et al.* 1980). Occasionally some vaccinees experience a delayed response, antibodies appearing a month or more later than is usual. Passively acquired antibody may interfere with vaccine uptake. Thus, Balfour and Amren (1978) showed that vaccine failure occurred more commonly among infants vaccinated before 14 months of age, at a time when some of these infants may have had residual maternal antibody. Immunization failure may also occur if rubella vaccine is given shortly after blood transfusion (Grillner and Forssman 1974, Watt and McGucken 1980).

(viii) Contraindications

Since rubella vaccine is a live vaccine, it should not be given to patients whose immunological response is deficient, whether as a result of disease (e.g. malignancy) or of treatment with corticosteroids, radiotherapy or cytotoxic drugs. Since rubella vaccine may be transmitted transplacentally, pregnancy is an absolute contraindication and should be avoided for three months after vaccination. As with other live vaccines, it is better to postpone rubella vaccination if the patient is suffering from a febrile illness. When it is necessary to give another live vaccine at the same time, the two should be given simultaneously but at different sites; alternatively the two vaccinations should be separated by an interval of at least three weeks. A three-week interval should also be allowed to elapse between the administration of a live virus vaccine, such as rubella, and BCG. Rubella vaccines contain traces of antibiotics (neomycin and/or polymyxin), and the manufacturer's leaflet should therefore be carefully studied before patients with known hypersensitivity are vaccinated. Vaccination should be deferred for about three months after a blood transfusion, or a dose of normal human immunoglobulin, since passively acquired antibodies might interfere with the vaccine response. However, previous administration of anti-D immune globulin is not a contraindication to post-partum vaccination.

(ix) Risks of vaccination during pregnancy

Examination of the products of conception of rubella-susceptible women vaccinated inadvertently during pregnancy has shown that rubella virus may be recovered from the placenta (Larson *et al.* 1971, Fleet *et al.* 1974), kidney (Vaheri *et al.* 1972), and bone marrow (Ebbin *et al.* 1972). Furthermore, histopathological studies have shown that changes in the placenta, decidua and fetal eye are similar to those occurring in naturally acquired infection. Rubella virus was recovered from the decidua and placenta of one fetus 69 days after vaccination (Larson *et al.* 1971), and from the bone marrow of another 94 days after vaccination (Ebbin *et al.* 1972), which suggests that, as with naturally acquired infection, rubella vaccine strains may induce a persistent infection. However, clinical and serological follow-up studies carried out on women in the USA who elected to go to term after inadvertent rubella vaccination in early pregnancy, have provided encouraging results (Preblud *et al.* 1981 Report, 1982). Thus, none of 143 infants delivered of mothers known to be rubella susceptible and given rubella vaccine within a period ranging from 3 months before to 3 months after conception had major malformations. Many of these infants have been followed up and, though about 10 per cent have serological evidence of intra-uterine infection, all have remained healthy. The risk of rubella-induced major malformations among infants delivered of susceptible mothers was calculated (based on a 95 per cent confidence limit) as 3.3 per cent—a figure similar to the risk of normal pregnancies. The United States Immunization Practices Advisory Committee still recommends that pregnant women should not be given rubella vaccine, but now states that inadvertent vaccination should no longer be a reason to recommend termination of pregnancy routinely (Report 1981*b*). However, most of the data collected in the United States relate to HPV77-DE5 and Cendehill vaccines and as yet we know less about the teratogenic potential of RA27/3, although information is accumulating rapidly. Unlike other vaccine strains it was first isolated from a rubella-infected fetus, and during attentuation had a less extensive passage history in cell culture than other attenuated vaccine strains (perhaps this is why it is a particularly good immunogen). The report from the Centre for Disease Control records only 65 live births to rubella-susceptible women given RA27/3 (Report, 1982). Though all the infants were healthy, only 21 mothers had been vaccinated at the critical period between one week before the date of conception to the end of the first month of pregnancy. These preliminary findings are encouraging, but rather more data are required before the teratogenic potential of this vaccine strain can also be established as negligible. Until then it would still be prudent to adhere to the recommendation that rubella vaccine should be given only to women who have been screened and found to be seronegative, after which they should be advised not to become pregnant for three months (Chief Medical Officer 1979). This precaution may therefore reduce considerably the risks of inadvertent vaccination during pregnancy for which an unacceptably high proportion of terminations are carried out. Although the prior immune status of many of these pregnant women had not been determined, it is now possible to do so retrospectively by testing post-vaccination serum samples for rubella-specific IgM (Banatvala *et al.* 1977, Mortimer, *et al.* 1981).

2. Passive immunization, administration of normal immune globulin

The results of many trials conducted before laboratory techniques were available for the diagnosis of rubella are of limited value, since it was not possible to confirm the diagnosis or detect subclinical infection, and no estimate of the antibody content of different batches of immune globulin could be obtained. However, studies conducted by the Public Health Laboratory Service (Report 1968, 1970*b*) suggest that normal human immune globulin does not confer any clear cut protection; this preparation may reduce the incidence of clinically overt infection, but subclinical infection may nevertheless occur. Normal immune globulin may prolong the incubation period considerably (Aylsworth and Monif 1971, Forrest and Honeyman 1973). Since inapparent infection is accompanied by viraemia, fetal damage is not prevented. However, Peckham (1974) produced evidence to show that infants of mothers given normal human immune globulin who experienced subclinical rubella in early pregnancy were less likely to be infected *in utero* than those infants whose mothers were not given immune globulin. It is therefore possible that the administration of normal human globulin reduces the level of fetal infection, damage, or both.

Dudgeon advocated the use of immune globulin in susceptible pregnant women who were determined to proceed to term after exposure to rubella (Hanshaw and Dudgeon 1978). He suggests that a dose of 1500 mg of immune globulin should be given intramuscularly as soon as possible after exposure. This should be followed by an additional dose of 1500 mg three to four days later (i.e. about 6 days after contact), which might be just prior to the onset of viraemia. However, such patients should be carefully followed up serologically to detect evidence of subclinical infection. Human immune globulin induces only a transient antibody response which will not interfere with the interpretation of serological investigations.

Pepsin-treated normal human immune globulin given intravenously (Martin du Pan *et al.* 1972) and preparations of high-titre rubella immune globulin given intramuscularly (Schiff 1969) have been shown to be effective in preventing rubella when given before exposure. High-titred rubella immune globulin has been used experimentally to determine whether infection induced by rubella vaccine can be prevented (Urquhart *et al.* 1978). Results were encouraging in that 8 out of 20 volunteers (40 per cent) given high-titre immune globulin and rubella vaccine simultaneously failed to sero-convert, and the remaining 12 exhibited delayed antibody responses compared with volunteers given vaccine alone. This preparation, which is in short supply, is available via the Scottish Blood Transfusion Service, but has not been properly evaluated in the field. It may be recommended for the few susceptible pregnant women who come into contact with clinical rubella and for whom therapeutic abortion is unacceptable.

References

Al-Nakib, W., Best, J. M. and Banatvala, J. E. (1975) *Lancet* **1**, 182.

Alstyne D. van, Krystal, G., Kettyls, G. D. and Bohn, E. M. (1981) *Virology* **108**, 491.

Anderson, R. N. and May, R. H. (1983) *J. Hyg. Camb.* **90**, 259.

Anderson, S. G. (1949) *J. Immunol.* **62**, 29; (1954) *Lancet* **ii**, 1107.

Andrewes, C., Pereira, H. G. and Wildy, P. (1978) *Viruses of Vertebrates*, 4th edn., pp. 98–100. Bailliere, London.

Avila, L., Rawls, W. E. and Dent, P. B. (1972) *J. infect. Dis.* **126**, 585.

Aylsworth, A. A. and Monif, G. R. G. (1971) *Obstet. Gynec.* **38**, 752.

Balfour, H. H. and Amren, D. P. (1978) *Amer. J. Dis. Child.* **132**, 573.

Balfour, H. H., Groth, K. E. and Edelman, C. K. (1980) *Amer. J. Dis. Child.* **134**, 350.

Balfour, H. H., Groth, K. E., Edelman, C. K., Amren, D. P., Best, J. M. and Banatvala, J. E. (1981) *Lancet* **i**, 1078.

Banatvala, J. E. (1977*a*) In: *Infections and Pregnancy*, Ed. by C. R. Coid, Academic Press, London; (1977*b*) In: *Recent Advances in Clinical Virology*, pp. 171–190. Ed. by A. P. Waterson, Churchill Livingstone, Edinburgh.

Banatvala, J. E. and Best, J. M. (1969) *Lancet* **i**, 695; (1973) *Lancet* **i**, 1452.

Banatvala, J. E., Druce, A., Best, J. M. and Al-Nakib, W. (1977) *Brit. med. J.* **ii**, 1263.

Bardeletti, G., Kessler, N. and Aymard-Henry, M. (1975) *Arch. Virol.* **49**, 175.

Bardeletti, G., Tektoff, J. and Gautheron, D. (1979) *Intervirology* **11,** 97.

Best, J. M. and Banatvala, J. E. (1967) *J. Hyg. Camb.* **65,** 263; (1970) *J. gen. Virol.* **9,** 215.

Best, J. M., Banatvala, J. E., Bowen, J. M. (1974) *Brit. med. J.* **iii,** 221.

Best, J. M., Banatvala, J. E. and Moore, B. M. (1968) *J. Hyg. Camb.* **66,** 407.

Best, J. M., Banatvala, G. C., Smith, M. E. (1971) *J. Hyg. Camb.* **69,** 223.

Best, J. M., Harcourt, G. C., Banatvala, J. E. and Flewett, T. H. (1981) *Brit. med. J.* **282,** 1235.

Best, J. M., Harcourt, G. C., Druce, A., Palmer, S. J., O'Shea, S. and Banatvala, J. E. (1980) *J. med. Virol.* **5,** 239.

Best, J. M., Harcourt, G. C., O'Shea, S. and Banatvala, J. E. (1979) *Lancet* **ii,** 690.

Blattner, R. J. (1974) *Amer. J. Dis. Child.* **128,** 781.

Bonsdorff G. H. von and Vaheri, A. (1969) *J. gen. Virol.* **5,** 47.

Booth, J. C. and Stern, H. (1972) *J. med. Microbiol.* **5,** 515.

Brown, T., Hambling, M. H. and Ansari, B. M. (1969) *Brit. med. J.* **iv,** 263.

Buimovici-Klein, E., Lang, P. B., Ziring, P. R., and Cooper, L. Z. (1979) *Pediatrics* **64,** 620.

Buimovici-Klein, E., O'Beirne, A. J., Millian, S. J. and Cooper, L. Z. (1980) *Arch. Virol.* **66,** 321.

Buimovici-Klein, E., Vesikari, T., Santangelo, C. F. and Cooper, L. Z. (1976) *Arch. Virol.* **52,** 323.

Buynak, E. B., Hilleman, M. R., Weibel, R. E. and Stokes, J. (1968) *J. Amer. med. Ass.* **204,** 195.

Buynak, E. B., Weibel, R. E., Whitman, J. E., Stokes, J. and Hillemann, M. R. (1969) *J. Amer. med. Ass.* **207,** 2259.

Chantler, S., Evans, C. J, Mortimer, P. P., Cradock-Watson, J. E., and Ridehalgh, M. K. S (1982). *J. virol. Meth.*, **4,** 305.

Chantler, J. K., Tingle, A. J. (1980) *J. gen. Virol.* **50,** 317.

Chief Medical Officer (1979) CMO Letters 79/6.

Clarke, M., Stitt, J., Seagroatt, V., Schild, G. C., Pollock, T. M., Miller, C., Finlay, S. E. and Barbara, J. A. H. (1983) *Lancet* **i,** 667.

Cohen, S. M., Collins, D. N., Ward, G. and Deibel, R. (1971) *Appl. Microbiol.* **21,** 76.

Cooper, L. Z. (1975) *Progr. clin. biol. Res.* **3,** 1.

Cooper, L. Z., Forman, A. L., Ziring, P. R. and Krugman, S. (1971) *Amer. J. Dis. Child.* **122,** 397.

Cooper, L. Z., Green, R. H., Krugman, S., Giles, J. P. and Mirick, G. S. (1965) *Amer. J. Dis. Child.* **110,** 416.

Cooper, L. Z. and Krugman, S. (1967) *Arch. Ophthalmol.* **77,** 434.

Cooper, L. Z., Ziring, P. R., Ockerse, A. B., Fedun, B. A., Kiely, B. and Krugman, S. (1969) *Amer. J. Dis. Child.* **118,** 18.

Cotlier, E., Fox, J., Bohigian, G., Beauty, C. and Du Pree, A. (1968) *Nature* **217,** 38.

Coyle, P. K., Wolinsky, J. S., Buimovici-klein, E., Moucha, R. and Cooper, L.Z. (1982) *Infect. and Immun.*, **36,** 498.

Cradock-Watson, J. E., Ridehalgh, M. K. S., Anderson, M. J. and Pattison, J. R. (1981) *J. Hyg. Camb.* **87,** 147.

Cradock-Watson, J. E., Ridehalgh, M. K. S., Anderson, M. J., Pattison, J. R. and Kangro, H. O. (1980) *J. Hyg. Camb.* **85,** 381.

Cradock-Watson, J. E., Ridehalgh, M. K. S. and Chantler, S. (1976) *J. Hyg. Camb.* **76,** 109.

Cradock-Watson, J. E., Ridehalgh, M. K. S., Pattison, J. R., Anderson, M. J. and Kangro, H. O. (1979) *J. Hyg. Camb.* **83,** 413.

Delahunt, C. S. and Rieser, N. (1967) *Amer. J. Obst. Gynae.* **99,** 580.

Desmond, M. M. *et al.* (1967) *Pediatr.* **44,** 445.

Desmyter, J., De Somer, P., Rawls, W. E. and Melnick, J. L. (1969) *Symp. Series Immunobiol. Stand.* **11,** 139.

Desmyter, J., Melnick, J. L. and Rawls, W. E. (1968) *J. Virol.* **2,** 955.

Dowdle, W. R. *et al.* (1970) *Bull. World Hlth Org.* **42,** 419.

Draper, C. C. and Laurence, G. D. (1969) *J. med. Microbiol.* **2,** 249.

Ebbin, A. J., Wilson, M. G., Wehrle, P. F., Chin, J., Emmons, R. W. and Lennette, E. H. (1972) *Lancet* **ii,** 481.

Edwards, M. R., Cohen, S. M., Bruno, M. and Deibel, R. (1969) *J. Virol.* **3,** 439.

Elizan, T. S., Fabiyi, A. and Sever, J. L. (1969) *J. Mnt. Sinai Hosp.* **36,** 108.

Emminghaus, H. (1870) *Jahrb. f. Kinder* **4,** 47.

Evans, A. *et al.* (1974) *Int. J. Epidemiol.* **3,** 167.

Fabiyi, A., Gitnick, G. L. and Sever, J. L. (1967) *Proc. Soc. exp. Biol. Med.* **125,** 766.

Fabiyi, A., Sever, J. L., Ratner, N. and Caplan, B. (1966) *Proc. Soc. exp. Biol. Med.* **122,** 392.

Fleet, W. F., Benz, E. W., Karzon, D. T., Lefkowitz, L. B. and Herrmann, K. L. (1974) *J. Amer. med. Ass.* **227,** 621.

Fogel, A., Gerichter, Ch. B., Barnea, B., Handsher, R. and Heeger, E. (1978) *J. Pediatr.* **92,** 26.

Fogel, A. and Plotkin, S. (1969) *J. Virol.* **3,** 157.

Forghani, B. and Schmidt, N. J. (1979) *J. clin. Microbiol.* **9,** 657.

Forrest, J. M. and Honeyman, M. C. (1973) *Med. J. Aust.* **i,** 745.

Forrest, J. M., Menser, M. A., Honeyman, M. C., Stout, M. and Murphy, A. M. (1972) *Lancet* **ii,** 399.

Furukawa, T., Plotkin, S. A., Sedwick, W. D. and Profeta, M. L. (1967) *Proc. Soc. exp. Biol. Med.* **126,** 745.

Gale, J. L., Detels, R., Kim, K. S. W., Beasley, R. P. and Grayston, J. T. (1969) *Amer. J. Dis. Child.* **118,** 143.

Gardner, P. S. and McQuillin, J. (1980) In: *Rapid Virus Diagnosis. Application of Immunofluorescence.* 2nd Edn. pp. 211–222. Butterworths, London.

Gill, S. D. and Furesz, J. (1973) *Arch. ges. Virusforsch.* **43,** 135.

Ginsberg-Fellner, F. *et al.* (1980) *Pediatr. Res.* **14,** 572.

Givan, K. F., Rozee, K. R. and Rhodes, A. J. (1965) *Canad. med. Ass. J.* **92,** 126.

Goldwater, P. N., Quiney, J. R. and Banatvala, J. E. (1978) *Lancet* **ii,** 1298.

Gould, J. J. and Butler, M. (1980) *J. gen. Virol.* **49,** 423.

Grandien, M., Espmark, A. and Norrby, E. (1976) *Acta path microbiol. scand.* Sect. C. **84,** 153.

Grayzel, A. I., Beck, C. (1971) *Proc. Soc. exp. Biol. Med.* **136,** 496.

Green, R. H., Balsamo, M. R., Giles, J. P., Krugman, S. and Mirick, G. S. (1965) *Amer. J. Dis. Child.* **110,** 348.

Gregg, N. McA. (1941) *Trans. ophthal. Soc. Aust.* **3,** 35; (1944) *Trans. ophthal. Soc., Austr.* **4,** 119.

Gregg, N. McA., Beavis, W. R., Heseltine, M., Machin, A. E., Vickery, D. and Meyers, E. (1945) *Med. J. Aust.* **ii,** 122.

Grillner, L. and Forssman, L. (1974) *Brit. med. J.* **iv,** 47.

Habel, K. (1942) *Publ. Hlth. Rep. Wash.* **57,** 1126.

Hall, M. B. (1946) *Brit. med. J.* **i,** 737.

Halonen, P. E., Ryan, J. H. and Stewart, J. A. (1967) *Proc. Soc. exp. Biol. Med.* **125,** 162.
Halstead, S. B., Diwan, A. R. and Oda, A. I. (1969) *J. Amer. med. Ass.* **210,** 1881.
Hanshaw, J. B. and Dudgeon, J. A. (1978) *Viral Diseases of the Fetus and Newborn.* W. B. Saunders Co., Philadelphia, London, Toronto.
Harcourt, G. C., Best, J. M. and Banatvala, J. E. (1980) *J. infect. Dis.* **142,** 145.
Harcourt, G. C., Best, J. M., Banatvala, J. E., Kennedy, L. A. (1979) *J. Hyg. Camb.* **83,** 405.
Hardy, J. B. (1973) *Arch. Otolaryngol.* **98,** 230.
Hardy, J. B., McCracken, G. H., Gilkeson, M. R. and Sever, J. L. (1969) *J. Amer. med. Ass.* **207,** 2414.
Hastreiter, A. R., Joorabchi, B., Pujatti, G., Horst, R. van der, Patgesil, G. and Sever, J. L. (1967) *J. Pediat.* **71,** 59.
Haukenes, G., Haram, K. O. and Solberg, C. O. (1973) *New. Engl. J. Med.* **289,** 429.
Heggie, A. D. (1978) *J. infect. Dis.* **137,** 74.
Herrmann, K. L. (1979) *Diagnostic Procedures for Viral Rickettsial and Chlamydial Infections*, 5th edn., pp. 725–766. Amer. Publ. Health Ass., Washington D.C.
Hess, A. F. (1914) *Arch. intern. Med.* **13,** 913.
Hiro, T. and Tasaka, S. (1938) *Mschr. Kinderheilk.* **76,** 328.
Holmes, I. H., Wark, M. C., Warburton, M. V. (1969) *Virology* **37,** 15.
Hope-Simpson, R. E. (1944) *Lancet* **i,** 483.
Horstmann, D. M. (1969) *Ann. N.Y. Acad. Sci.* **162,** 594; (1975) *Ann. intern. Med.* **83,** 412.
Horstmann, D. M., Liebhaber, H., Le Bouvier, G. L., Rosenberg, D. A. and Halstead, S. B. (1970) *New Engl. J. Med.* **283,** 771.
Horstmann, D. M. *et al.* (1965) *Amer. J. Dis. Child.* **110,** 408.
Horzinek, M. C. (1973) *Progr. med. Virol.* **16,** 109; (1981) *Non-arthropod-borne Togaviruses.* Academic Press, London.
Horzinek, M. C., Maess, J. and Laufs, R. (1971) *Arch. ges. Virusforsch.*, **33,** 306.
Ho-Terry, L. and Cohen, A. (1979) *J. med. Microbiol.* **12,** 441; (1980) *Arch. Virol.* **65,** 1; (1981) *J. med. Microbiol.* **14,** 141.
Hovi, T. and Vaheri, A. (1970) *J. gen. Virol.* **6,** 77.
Ingalls, T. H., Plotkin, S. A., Philbrook, F. R. and Thompson, R. F. (1970) *Lancet* **i,** 99.
Kangro, H. O., Pattison, J. R. and Heath, R. B. (1978) *Brit. J. exp. Path.* **59,** 577.
Kato, S., Kimura, M., Takakura, I., Tsuji, K. and Veda, K. (1980) *Tissue Antigens* **15,** 80.
Kenney, M. T., Albright, K. L., Emery, J. B. and Bittle, J. L. (1969) *J. Virol.* **4,** 807.
Kistler, G. S. and Sapatino, V. (1972) *Arch. ges. Virusforsch.* **38,** 11.
Kleiman, M. B. and Carver, D. H. (1977) *J. gen. Virol.* **36,** 335.
Knox, E. G. (1980) *Int. J. Epidemiol.* **9,** 13.
Koboyashi, N. (1978) *Virology* **89,** 610.
Kono, R., Hibi, M., Hayakawa, Y. and Ishii, K. (1969) *Lancet* **i,** 343.
Korones, S. B., Ainger, L. E., Monif, G. R. G., Roane, J., Sever, J. L., Fuste, F. (1965) *Amer. J. Dis. Child.* **110,** 434.
Krech, U. and Wilhelm, J. A. (1979) *J. gen. Virol.* **44,** 281.
Krugman, S. and Ward, R. (1954) *J. Pediat.* **44,** 489; (1968) *Infectious Diseases of Children.* 4th edn., p. 285. C. V. Mosby Co., Saint Louis.
Krugman, S., Ward, R., Jacobs, K. G. and Lazar, M. (1953) *J. Amer. med. Assoc.* **151,** 285.
Lam, S. K. (1972) *Bull. World Hlth Org.* **47,** 127.
Larson, H. E., Parkman, P. D., Davis, W. J., Hopps, H. E. and Meyer, H. M. (1971) *New Engl. J. Med.* **284,** 870.
Laufs, R. and Thomssen, R. (1968) *Arch. ges. Virusforsch.* **24,** 164.
Le Bouvier, G. L. (1969) *Proc. Soc. exp. Biol. Med.* **130,** 51.
Le Bouvier, G. L. and Plotkin, S. (1971) *J. infect. Dis.* **123,** 220.
Leerhøy, J. (1965) *Science* **149,** 633; (1966) *Acta path. microbiol. scand.* **67,** 158.
Leinikki, P. and Pässilä, S. (1977) *J. infect. Dis.* **136,** 294.
Liebhaber, H. and Gross, P. A. (1972) *Virology* **47,** 684.
Liebhaber, H., Pajot, T. and Riordan, J. T. (1969) *Proc. Soc. exp. Biol. Med.* **130,** 12.
Lieberhaber, H., Riordan, J. T. and Horstmann, D. M. (1967) *Proc. Soc. exp. Biol. Med.* **125,** 636.
Liggins, G. C. and Phillips, L. I. (1963) *Brit. med. J.* **i,** 711.
Logt, J. T. M. van der, Loon, A. M. van, Veen, J. van der, (1980) *Infect. and Immun.* **27,** 309.
Lundström, R. (1962) *Acta paediat. Uppsala* **51,** (Suppl.) 133.
McCarthy, K., Taylor-Robinson C. H. and Pillinger, S. E. (1963) *Lancet* **ii,** 593.
MacDonald, H., Tobin, J. O'H., Cradock-Watson, J. E., Lomax, J., and Bourne, M. S. (1978) *J. Hyg. Camb.* **80,** 337.
Maller, R., Frydén, A. and Soren, L. (1978) *Acta path. microbiol. scand.* Sect. C. **86,** 93.
Manson, M. M., Logan, W. P. D. and Loy, R. M. (1960) *Rep. publ. Hlth med. Subj.* London, **101.**
Marcus, P. I. and Carver, D. H. (1965) *Science* **149,** 983; (1967) *J. Virol.* **1,** 334.
Marshall, W. C. (1973) In: *Intrauterine Infections.* Ciba Foundation Symposium 10, pp. 3–12. Ass. Scient. Publs., Amsterdam.
Martin Du Pan, R. M., Huygelen, C., Peetermans, J. and Prinzie, A. (1968) *Amer. J. Dis. Child.* **115,** 658.
Martin Du Pan, R., Koechli, B. and Douath, A. (1972) *J. infect. Dis.* **126,** 341.
Maton, W. G. (1815) *Med. Trans. Coll. Phycns.* London **5,** 149.
Menser, M. A., Dods, L. and Harley, J. D. (1967) *Lancet* **ii,** 1347.
Menser, M. A. and Forrest, J. M. (1974) *Med. J. Aust.* **i,** 123.
Menser, M. A., Forrest, J. M. and Bransby, R. D. (1978) *Lancet* **i,** 57.
Menser, M. A., Harley, J. D., Hertzburg, R., Dorman, D. C. and Murphy, A. M. (1967) *Lancet* **ii,** 387.
Menser, M. A. and Reye, R. D. K. (1974) *Pathology* **6,** 215.
Mettler, N. E., Petrelli, R. L. and Casals, J. (1968) *Virology* **36,** 503.
Meurman, O. H., Vilhanen, M. K. and Granfors, K. (1977) *J. clin. Microbiol.* **5,** 257.
Meyer, H. M., Parkman, P. D. and Panos, R. (1966) *New Engl. J. Med.* **275,** 575.
Miller, E., Cradock-Watson, J. E. and Pollock, T. M. (1982) *Lancet*, **ii,** 781.
Minehawa, Y., Suzuki, N., Osame, J., Yamanishi, K., Takahashi, M. and Okuno, Y. (1973) *Biken J.* **16,** 155.
Modlin, J. F., Herrmann, K., Brandling-Bennett, A. D., Eddins, D. L. and Hayden, G. F. (1976) *New Engl. J. Med.* **294,** 972.

Morgan-Capner, P., Tedder, R. S. and Mace, J. E. (1983) *Lancet* **i,** 589.

Mortimer, P. P., Tedder, R. S., Hambling, M. H., Shafi, M. S., Burkhardt, F. and Schilt, U. (1981) *J. Hyg. Camb.* **86,** 139.

Murphy, A. M. *et al.* (1967) *Amer. J. Ophthalmol.* **64,** 1109.

Naeye, R. L. and Blanc, W. (1965) *J. Amer. med. Ass.* **194,** 1277.

Niwa, Y. and Kanoh, T. (1979) *Clin. exp. Immunol.* **37,** 470.

Norrby, E. (1969) *Virol. Monogr.* **7,** 115.

Ogra, P. L. and Herd, J. L. (1971) *J. Immunol.* **107,** 810.

Oker-Blom, C., Kalkkinen, N., Kääriäinen, L., and Petterson, R. F. (1983) *J. Virol.* **46,** 964.

O'Shea, S., Best, J. M. and Banatvala, J. E. (1983) *J. infect. Dis.* **148,** 639.

O'Shea, S., Best, J. M., Banatvala, J. E., Marshall, W. C. and Dudgeon, J. A. (1982) *Brit. med. J.* **285,** 253.

Oxford, J. S. (1969) *Symp. Series Immunobiol. Stand.* **11,** 181.

Oxford, J. S. and Potter, C. W. (1969) *J. gen. Virol.* **5,** 565; (1970) *J. Immunol.* **105,** 818; (1971) *Arch. ges. Virusforsch.* **34,** 75.

Oxford, J. S. and Schild, G. C. (1967) *Arch. ges. Virusforsch.* **22,** 349.

Palmer, D. F., Cavallaro, J. J. and Herrmann, K. L. (1977) *Immunology, Series No. 2* Revised. Center for Disease Control, Atlanta, Georgia.

Parkman, P. D. (1965) *Arch. ges Virusforsh* **16,** 401

Parkman, P. D., Buescher, E. L. and Artenstein, M. S. (1962) *Proc. Soc. exp. Biol. Med.* **111,** 225.

Parkman, P. D., Buescher, E. L., Artenstein, M. S., McCowan, J. M., Mundon, F. K. and Druzd, A. D. (1964) *J. Immunol.* **93,** 595.

Parkman, P. D., Meyer, H. M., Kirschstein, B. L. and Hopps, H. E. (1966) *New Engl. J. Med.* **275,** 569.

Parkman, P. D., Phillips, P. E. and Meyer, H. M. (1965) *Amer. J. Dis. Child.* **110,** 390.

Patterson, R. L., Koren, A. and Northrop, R. L. (1973) *Lab. Anim. Sci.* **23,** 68.

Pattison, J. R. (1982*a*) *Laboratory Investigation of Rubella.* Publ. Health Lab. Serv., Monograph Series No. 16, H.M.S.O., London; (1982*b*) *J. Hyg. Camb.* **88,** 149.

Pattison, J. R. and Mace, J. E. (1975) *J. clin. Pathol.* **28,** 670.

Paul, N. R., Rhodes, A. J., Campbell, J. B., and Labzoftsky, N. A. (1974) *Arch. ges. Virusforsch.* **45,** 335.

Payment, P., Ajdukovic, D. and Pavilanis, V. (1975) *Canad. J. Microbiol.* **21,** 703.

Payment, P., Roy, L., Gilker, J-C. and Chagnon, A. (1975) *Canad. J. Microbiol.* **21,** 289.

Peckham, C. S. (1972) *Arch. Dis. Childh.* **47,** 571; (1974) *Brit. med. J.* **i,** 259.

Peetermans, J. and Huygelen, C. (1967) *Arch. ges. Virusforsch.* **21,** 133.

Penttinen, K. and Myllylä, G. (1968) *Ann. med. exp. Biol. Fenn.* **46,** 188.

Phelan, D. and Campbell, P. (1969) *J. Pediat.* **75,** 202.

Pitt, D. B. (1961) *Med. J. Aust.* **i,** 881.

Pitt, D. and Keir, E. H. (1965) *Med. J. Aust.* **ii,** 647.

Plotkin, S. A. (1969) *Diagnostic Procedures in Viral and Rickettsial Diseases*, 4th edn., pp. 364–413. *Amer. Publ. Health Ass.*, Washington D.C.

Plotkin, S. A., Farquhar, J., Katz, M. and Ingalls, T. H. (1967) *Amer. J. Epidemiol.* **86,** 468.

Plotkin, S. A., Farquhar, J. D. and Ogra, P. L. (1973) *J. Amer. Med. Ass.* **225,** 585.

Plotkin, S. A., Ingalls, T. H., Farquhar, J. D. and Katz, M. (1968) *Lancet* **ii,** 934.

Plotkin, S. A., Oski, F. A., Harknett, E. M., Hervada, A. R., Friedman, S. and Gowing, J. (1965) *J. Pediatr.* **67,** 182.

Plotkin, S. A. and Vaheri, A. (1967) *Science* **156,** 659.

Polk, B. F., White, J. A., Degirolami, P. C. and Modlin, J. F. (1980) *New. Engl. J. Med.* **303,** 541.

Potter, J. E., Banatvala, J. E. and Best, J. M. (1973) *Brit. med. J.* **i,** 197.

Preblud, S. R., Serdula, M. K., Frank, J. A., Brandling-Bennett, A. D. and Hinman, A. R. (1980*a*) *Epidemiol. Rev.* **2,** 171.

Preblud, S. R., Serdula, M. K., Frank, J. A., and Hinman, A. R. (1980*b*). *J. infect. Dis.* **142,** 776.

Preblud, S. R., Stetler, H. C., Frank, J. A., Greaves, W. L., Hinman, A. R. and Herrmann, K. L. (1981) *J. Amer. med. Ass.* **246,** 1413.

Rawls, W. E. (1968) *Progr. Med. Virol.* **10,** 238; (1974) *Ibid.* **18,** 273.

Rawls, W. E., Desmyter, J. and Melnick, J. L. (1968) *J. Amer. med. Ass.* **203,** 627.

Rawls, W. E. *et al.* (1967) *Bull. World Hlth Org.* **37,** 79.

Report (1944) *Lancet* **i,** 316; (1965) *Amer. J. Dis. Child.* **110,** No. 4; (1968) *Brit. med. J.* **iii,** 206; (1970*a*) *Circular 9/70. London. DHSS*; (1970*b*) *Brit. med. J.* **ii,** 497; (1972*a*) *Morbid. and Mortal.* **21** (Suppl.) 23; (1972*b*) *Circular 17/72. London, DHSS*; (1976) *Circular 76/4. London, DHSS*; (1978) *Morbid. and Mortal.* **27,** No. 49; (1979) *Ibid.* **28,** No. 3; (1981*a*) *Brit. med. J.* **i,** 282, 234; (1981*b*) *Morbid. and Mortal.* **30,** 37–42; (1982) *Morbid. and Mortal.* **31,** 477.

Rhim, J. S., Schell, K. and Huebner, R. J. (1967) *Proc. Soc. exp. Biol. Med.* **125,** 1271.

Rorke, L. B., Fabiyi, A., Elizan, T. S. and Sever, J. L. (1968) *Lancet* **ii,** 153.

Rorke, L. B. and Spiro, A. J. (1967) *J. Pediat.* **70,** 243.

Ross, C. A. and McCartney, A. (1979) *Brit. med. J.* **i,** 1636.

Russell, B., Selzer, G. and Goetz, H. (1967) *J. gen. Virol.* **1,** 305.

Salmi, A. A. (1969) *Acta path. microbiol. scand.* **76,** 271.

Sato, H., Albrecht, P., Krugman, S. and Ennis, F. A. (1979) *J. clin. Microbiol.* **9,** 259.

Sato, M. *et al.* (1978) *Arch. Virol.* **56,** 89.

Schell, K. and Wong, K. T. (1966) *Nature* **212,** 621.

Schiff, G. M. (1969) *Amer. J. Dis. Child.* **118,** 322.

Schmidt, N. J., Dennis, J. and Lennette, E. H. (1968) *Arch. ges. Virusforsch.* **25,** 308.

Schmidt, N. J., Lennette, E. H. and Dennis, J. (1967) *J. Immunol.* **99,** 399; (1969) *Proc. Soc. exp. Biol. Med.* **132,** 128.

Schmidt, N. J. and Styk, B. (1968) *J. Immunol.* **101,** 210.

Sedwick, W. D. and Sokol, F. (1970) *J. Virol.* **5,** 478.

Seppälä, M. and Vaheri, A. (1974) *Lancet* **i,** 46.

Sever, J. L., Fabiyi, A., McCallin, P. F., Chu, P. T., Weiss, W., Gilkeson, M. R. (1965) *Amer. J. Obstet, Gynecol.* **92,** 1006.

Sever, J. L., Hardy, J. B., Nelson, K. B. and Gilkeson, M. R. (1969) *Amer. J. Dis. Child.* **118,** 123.

Sever, J. L., Meier, G. W., Windle, W. F., Schiff, G. M., Monif, G. R. and Fabiyi, A. (1966) *J. infect. Dis.* **116,** 21.

Sever, J. L. *et al.* (1967) *Pediatrics* **40,** 789.

Sheppard, S., Smithells, R. W., Peckham, C. S., Dudgeon, J. A. and Marshall, W. C. (1977) *Hlth Trends* **9,** 38.

Shishido, A. and Ohtawara, M. (1976) *Japan J. med. Sci. Biol.* **29,** 227.

Shortridge, K. F. and Osmund, I. F. (1979) *J. Hyg. Camb.* **83,** 397.
Siegel, M., Fuerst, H. T. and Guinee, V. F. (1971) *Amer. J. Dis. Child.* **121,** 469.
Simons, M. J. (1968) *Lancet* **ii,** 1275.
Smith, J. L. (1881) *Trans. internat. med. Congr.* **4,** 14.
Smith, K. O. and Hobbins, T. E. (1969) *J. Immunol.* **102,** 1016.
Sperling, D. R. and Verska, J. J. (1966) *Calif. Med.* **105,** 340.
Steinmaurer, H. (1938) *Monatschr. f. Kinderheilk.* **75,** 98.
Stewart, G. L., Parkman, P. D., Hopps, H. E., Douglas, R. D., Hamilton, J. P. and Meyer, H. M. (1967) *New. Engl. J. Med.* **276,** 554.
Strassburg, M. A. *et al.* (1981) *Obstet. Gynecol.* **57,** 283.
Swan, C., Tostevin, A. L., Moore, B., Mayo, H. and Black, G. H. B. (1943) *Med. J. Aust.* **ii,** 201.
Tardieu, M., Grospierre, B., Durandy, A. and Griscelli, C. (1980) *J. Pediat.* **97,** 370.
Taylor-Robinson, C. H., McCarthy, K., Grylls, S. G. and O'Ryan, E. M. (1964) *Lancet* **i,** 1364.
Thompson, K. M. and Tobin, J. O'H. (1970) *Brit. med. J.* **ii,** 264.
Thomssen, R., Lauffs, R. and Müller, J. (1968) *Arch. ges.. Virusforsch.* **23,** 332.
Thongcharoen, P. *et al.* (1970) *Far East med. J.* **6,** 285.
Töndury, G. and Smith, D. W. (1966) *J. Pediat.* **68,** 867.
Townsend, J. J. *et al.* (1975) *New. Engl. J. Med.* **292,** 990.
Ueda, K., Nishida, Y., Oshima, K. and Shepard, T. H. (1979) *J. Pediat.* **94,** 763.
Urquhart, G. E. D., Crawford, R. J. and Wallace, J. (1978) *Brit. med. J.* **ii,** 1331.
Vaheri, A. and Hovi, T. (1972) *J. Virol.* **9,** 10.
Vaheri, A. and Salonen, E-M. (1980) *J. med. Virol.* **5,** 171.
Vaheri, A., Sedwick, D. and Plotkin, S. (1967) *Proc. Soc. exp. Biol. Med.* **125,** 1086.
Vaheri, A., Sedwick,W. D., Plotkin, S. A. and Maess, R. (1965) *Virology* **27,** 239.
Vaheri, A. *et al.* (1972) *New Engl. J. Med.* **286,** 1071.
Veale, H. (1866) *Edinb. J. Med.* **12,** 404.
Vejtorp, M. and Mansa, B. (1980) *Scand. J. infect. Dis.* **12,** 1.
Vesikari, T. (1972) *Scand. J. infect. Dis.* **4,** 11, (1980) *Ibid.* **12,** 7.
Vesikari, T. and Vaheri, A. (1968) *Brit. med. J.* **i,** 221.
Voiland, A. and Bardeletti, G. (1980) *Arch. Virol.* **64,** 319.
Watt, R. W. and McGucken, R. B. (1980) *Brit. med. J.* **281,** 977.
Weibel, R. E., Buynak, E. B., McLean, A. A. and Hilleman, M. R. (1978) *Pediatrics* **61,** 5.
Weibel, R. E., Stokes, J. Jr., Buynak, E. B., and Hilleman, M. R. (1969) *Amer. J. Dis. Child.* **118,** 226.
Weil, M. L., Habashim, H. H., Cremer, N. E., Oshiro, L. S., Lennette, E. H. and Carney, L. (1975) *New Engl. J. Med.* **292,** 994.
Weller, T. H. and Neva, F. A. (1962) *Proc. Soc. exp. Biol. Med.* **111,** 215.
Wesselhoeft, C. (1947*a*) *New Engl. J. Med.* **236,** 943; (1947*b*) *Ibid.* **236,** 978.
Witte, J. J. *et al.* (1969) *Amer. J. Dis. Child.* **118,** 107.
Zealley, H. and Edmond, E. (1982) *Brit. med. J.* **i,** 382.
Ziegelmaier, R., Behrens, F. and Enders, G. (1981) *J. biol. Stand.* **9,** 23.
Zygraich, N., Peetermans, J. and Huygelen, C. (1971) *Arch. ges. Virusforsch.* **33,** 225.

95

Orbiviruses

Robert F. Sellers

Introductory

Orbiviruses form a genus of the family *Reoviridae* (Verwoerd *et al*, 1979). Their size is 75 nm to 85 nm, they contain double-stranded RNA and they are sensitive to pH values of 6.0 and below. They are also inactivated to a certain extent by lipid solvents, and contain a diffuse outer envelope. They have been classified into eleven groups by Gorman (1979) (Table 95.1) and include viruses that cause clinical signs in man, in horses, mules and donkeys, in cattle, sheep, goats and wild ruminants as well as inapparent infections in rodents, bats and birds (Table 95.1). Most of them have been shown to be arboviruses with *Culicoides* midges, mosquitoes, phlebotomids and ticks as arthropod vectors. In man, the most important virus is Colorado tick fever, distributed in the north-western United States, which causes fever and encephalitis. Viruses of the Kemerovo group, especially Kemerovo and Lipovnik viruses, have been associated in Siberia and Central Europe with meningitis and meningoencephalitis. A characteristic of the viruses giving rise to Colorado tick fever, bluetongue and African horse sickness is their association with red blood cells and red blood cell precursors.

Morphology of the virus

Borden *et al.* (1971) proposed the name Orbivirus for the genus based on the large, round, doughnut-shaped capsomeres seen on the surface of the virus particles. These authors and Murphy *et al.* (1971) examined the physicochemical and morphological relations of a number of relatively solvent-resistant arboviruses. These viruses included, among others, Colorado tick fever, Wad Medani, Chenuda, Irituia, Palyam, Lebombo, epizootic haemorrhagic disease of deer, and bluetongue. The authors found that the viruses were 65–80 nm in diameter and had an icosahedral symmetry with 32 capsomeres. Before that time there was some doubt about the size of the viruses; for example, sizes of 55–110 had been found. Subsequent work by Verwoerd and Huismans (1972) and Martin and Zweerink (1972) showed that there were two distinct

Table 95.1 Orbivirus serological groups

Group		Types	Hosts	Vector
A	African horse sickness	1–9	Horses, donkeys, zebras, elephant	Culicoides
	Bluetongue	1–20+	Sheep, goats, cattle, wild ruminants	Culicoides
B	Epizootic haemorrhagic disease of deer	8	Deer, ruminants	Culicoides
	Ibaraki	3	Cattle	Culicoides
	Others	3		
C	Colorado tick fever	2	Man, rodents	Ticks
D	Palyam	6	Cattle, sheep, goats	Mosquitoes, Culicoides
E	Changuinola	2	Man, rodents	Phlebotomus
F	Corriparta	3	Birds, man, marsupials, ungulates	Mosquitoes
G	Kemerovo	16	Man, cattle, rodents, birds	Ticks
H	Warrego	2	Marsupials	Culicoides
I	Wallal	2	Marsupials	
J	Equine encephalosis	5	Horses	
Ungrouped	Lebombo	4	Man, rodents, birds, bats	Mosquitoes

types of particle—a larger, 69 nm, with a buoyant density of 19 nm and a smaller, 63 nm, with a buoyant density of 63. The larger consists of a diffuse outer layer within which lies the nucleocapsid (the smaller). Particles larger than 69 nm probably contain varying amounts of host material.

The viruses in the genus are relatively resistant to lipid solvents and sodium deoxycholate, with a fall in titre of $10^{1.5}$ ID_{50} or less. The viruses do not survive exposure to pH 3 for 3 hours. These findings contrast with reoviruses, which are completely resistant to lipid solvents and to exposure at pH 3.

Most of the work on the chemical nature of the genus has been done with bluetongue virus but work on other members of the group has been carried out. The viral genetic material is RNA, double-stranded and segmented (10 segments). The molecular weight of the total genome varies from 11.0 to 12.5×10^6 and the segments from 3.0 to 0.22 (Gorman 1979). The capsid of bluetongue virus contains four major and three minor polypeptides. Polypeptides 2 and 5 form the outer diffuse layer; polypeptides 1, 3, 4 and 6 are located on the outer surface of the nucleocapsid, while polypeptide 7 is on the inner nucleocapsid. African horse sickness virus has an inner capsid particle containing polypeptides 1, 2, 4, 6 and 7 surrounded by an outer protein layer consisting of polypeptides 3 and 5 (Bremer 1976). Epizootic haemorrhagic disease of deer virus has a further minor polypeptide, which is located in the outer layer (Huismans *et al.* 1979). Ibaraki virus is similar. The genome of Colorado tick fever virus is composed of 12 segments of double-stranded RNA (Knudson 1981) (Table 95.2).

Replication

Viruses of the genus multiply in a number of cell types. After adsorption, the virus enters the cell by pinocytosis (Lecatsas 1968). Morphogenesis takes place in the cytoplasm in association with dense granular or reticular matrices. Tubules and filaments have also been found at sites of virus multiplication.

An RNA-dependent RNA polymerase activity has been shown to be present in bluetongue virus after removal of the outer layer. Probably this is responsible for the transcription of the double-stranded RNA into the single-stranded RNA species. These in turn are responsible for synthesis of protein. RNA segments 2

Table 95.2 Properties of orbiviruses

	Diameter (nm)	RNA segments	Polypeptides
African horse sickness	70	10	7
Bluetongue	60–65	10	7
Epizootic haemorrhagic disease of deer		10	8
Colorado tick fever		12	
Palyam	65–70	10	
Changuinola	60–65	10	
Corriparta	63–71	10	
Kemerovo		10	
Warrego	63–71	10	
Wallal	63–71	10	
Equine encephalosis		10	
Lebombo	65–70	10	

and 6 are transcribed and translated into polypeptides 2 and 5; segments 1, 3, 4 and 7 into polypeptides 1, 3, 4 and 6; and segment 9 into polypeptide 7. A further two polypeptides have been identified in cells infected with bluetongue types 10 or 20—5a, associated with the tubular structures, and 6a. No function has yet been found for 6a (Gorman 1979). The method of assembly of virus particles is obscure, but it appears that they pass into cytoplasmic vesicles and are released by extrusion across the plasma membrane. Budding may also occur.

Infection with more than one strain of virus may lead to genetic reassortment (Gorman 1979). Interferon is also produced in some cells.

African horse sickness

African horse sickness is a peracute, acute and subacute disease of horses, mules and donkeys. The orbivirus responsible is transmitted by blood-sucking insects, more especially *Culicoides* midges (Howell 1963*a*).

Characteristics of the virus

In addition to the general orbivirus characters mentioned previously, the virus is acid-sensitive, being inactivated at pH values below 6.0. The virus survives better at $-70°$ and $4°$ than at $-20°$. In infected blood the virus is destroyed after 5 minutes at 70° and after 10 minutes at 55°.

Nine immunological types of virus have been demonstrated by cross-neutralization tests, but the types share a common group antigen, as shown in complement-fixation, agar gel precipitin and fluorescent antibody tests.

Adult and suckling mice are susceptible to African horse sickness (AHS) virus given by the intracerebral route; paralysis develops and the mice may die. Sucking mice are more sensitive. Guinea-pigs develop paralysis after inoculation.

AHS virus can be adapted to grow and cause death in the chick embryo by yolk sac or intravenous inoculation. The virus grows with cytopathic effect in BHK-21 cells, a stable monkey line and other cell lines such as Vero. A mosquito cell line from *Aedes albopictus* supports the growth of virus and cytoplasmic inclusions develop.

Clinical disease

Four forms of clinical disease have been distinguished in the horse—the peracute, acute, subacute and horse-sickness fever.

In the peracute (pulmonary) form the incubation period is 3–5 days. The first signs are fever followed by respiratory distress which becomes worse, with rapid breathing. The animal stands with legs astride, head distended with abdominal breathing, and sweats profusely. Death may occur within a few hours, the animal having drowned in its own serum. The mortality rate is 95 per cent.

The acute form is a mixture of the pulmonary form and the cardiac (subacute) form. Either the animal shows mild pulmonary signs which are followed by oedema of the head and neck, or oedema is seen first, followed by acute dyspnoea. The incubation period is 5–7 days. Death occurs in about 80 per cent of horses 3–6 days after the rise in temperature.

In the subacute form the incubation period is 7–14 days. The first sign is fever, which lasts for 3–6 days. This is followed by oedematous swellings most prominent on the face in the temporal and supraorbital fossae. The rest of the face becomes involved and the oedema may extend down the neck to the chest. The animal may suffer from colic. About 50 per cent of affected animals die 4–8 days after the onset of fever. In the remainder the swellings subside.

The peracute form usually affects fully susceptible horses exposed to virulent strains of virus; the subacute form affects those that have some degree of heterologous immunity or are exposed to less virulent virus. Horse-sickness fever is seen in immune horses or in donkeys and mules.

The main feature after death in the pulmonary and cardiac forms is oedema. In the pulmonary form the oedema is seen in the lungs and lymph nodes. In the cardiac form the oedema is found in the spaces between the skin and muscles of the head and neck and in the pericardial sac. Petechial haemorrhages are seen in the lungs (pulmonary form) and/or in the heart (cardiac form).

African horse sickness has been found to cause disease in dogs fed with infected horsemeat.

Zebras have also been shown to be susceptible, with development of fever and oedema of the supraorbital fossae.

Pathogenesis Infection in the field is caused by the bite of a blood-sucking insect. The virus is then carried to a local lymph node, which forms the initial site of multiplication. From there it is distributed throughout the body and multiplies in the spleen, the lungs, lymph nodes and other blood cell forming organs (Erasmus 1973). It also multiplies in the endothelial cells lining the blood vessels. A viraemia with titres of $10^{6.8}$ LD_{50} per ml is reached and the virus is found associated with the red cells. The virus may persist at low titre in the red blood cells in the presence of disease for long periods after recovery.

The oedema seen during the disease and at post-mortem probably results from the increased permeability of the blood vessels as a result of viral multiplication.

Epidemiology

African horse sickness is endemic in West, Central, East and Southern Africa and has been found in Yemen and along the Red Sea coast. At intervals epidemics occur when the virus is introduced into certain areas; for example, into the Cape Province of South Africa, to Egypt (1928 and 1943) and Palestine (1944), to the Middle East and the Indian subcontinent (1959–60) and to North Africa and Spain (1965–6) (Bourdin 1973).

Nine types have been determined by cross-immunity tests in horses and by neutralization tests in mice with horse and rabbit antisera. Types 1–8 have been found in South Africa and type 9 in Sudan, the Middle East, India and Pakistan, West Africa and Spain. Other untyped strains have been found in Central and West Africa and Egypt.

Hosts Strains of virus vary in their virulence for horses, donkeys and zebras. In addition, in endemic areas the disease is mild or subclinical in local animals but, when horses are introduced from outside, rapid and fatal disease results; for example, introductions of French and British horses into West Africa. If the virus spreads from the endemic area, the disease is severe, as seen in the epidemics in North Africa and the Middle East.

Virus is found in the blood of horses up to 4 days before the end of the incubation period (5–16 days) and titres of $10^{4.0}$ LD_{50} per ml or greater are found for 4–5 days around the time of clinical disease. This is a sufficient amount of virus in the blood for midges to become infected. The virus titre in the blood of zebras is lower, i.e. $10^{2.5}$ LD_{50} per ml over 3–4 days, which is lower than the threshold for midges ($10^{4.0}$ ID_{50} per ml). Viraemia also occurs in dogs.

Vector The main biological vector of African horse sickness is the midge, *Culicoides imicola*, in which multiplication of AHS virus has been demonstrated. It was found possible to infect certain species of mosquito but it is unlikely that they would transmit the virus to any great extent in the field. Mechanical transmission by *Stomoyxs calcitrans* and *Lyperosia minuta* has also been suggested.

Midges lay eggs 2–6 days after a blood meal in damp, muddy areas. The eggs hatch in 2–3 days and larval stages take 12–16 days, with adults emerging 2–3 days after pupation. The next day the adults take a blood meal and this is repeated every 3–4 days provided the ambient temperatures remain between 15° and 35° (Sellers *et al.* 1977).

If the blood meal contains virus, the virus multiplies in the midge, reaching a titre of $10^{6.0}$ LD_{50} per midge or greater by the fifth day. Transmission by biting a horse takes place from the seventh day onwards until the end of the midge's life which may be as long as 70 days. The daily survival rate for midges is 0.8 and it has been calculated that after 18 days 1.8 per cent of a population will have survived. In this time a midge infected on the first day could have transmitted the virus by 4 further bites on the 18th day.

The numbers of midges vary during the year, depending on the presence of suitable temperatures and moisture. In areas where the temperature is warm throughout the year, breeding occurs all the time but peaks are reached at the beginning and end of the rainy season. In other areas, breeding sites are provided by the leaks from irrigation channels and at the edge of waterholes, rivers and streams. No breeding occurs in winter in areas with a cold winter but during summer there may be peaks in numbers at different times. Numbers of midges are important in spread and maintenance of African horse sickness. Given a daily survival rate of 0.8 and a low infection rate, the chances of infecting further hosts will be small without large numbers of midges.

Optimum temperatures for midge activity lie between 13° and 35°. Below and above this temperature, hibernation or aestivation of adults takes place.

Two types of flight behaviour can be seen. There is short flight unaided when the air is still or at wind speeds less than $2\,m\,s^{-1}$; this usually takes place between dusk and dawn. In addition, circumstantial evidence indicates that midges may be carried on the wind for long distances, 40 to 700 km. Such flights take from one hour to twenty hours and are carried out at heights up to 1.5 km with temperatures between 15° and 35° (Sellers *et al.* 1977).

Climate Midges are cold-blooded and, as mentioned before, temperatures between 15° and 35° are required for feeding, oviposition and flying to take place. Moisture is important for their breeding and winds influence their movement from one place to another.

The climates where AHS virus occurs include the tropical, with a wet and dry season, as in West and Central Africa and Sudan. There may be a warm summer with rains but with a colder winter, as in Southern Africa and the Indian subcontinent, or a dry summer with a cooler rainy winter, as in the Mediterranean and the Middle East.

Behaviour of the virus: developmental cycle

The virus undergoes a cycle in the hosts and vectors.

In warm climates the cycle continues throughout the year, although in the dry season the numbers of midges are reduced. During periods of cold or heat the virus survives in the hibernating or aestivating adult midge or in the host. Spread of the virus occurs by movement of horses, mules, donkeys and zebras and by movement of infected midges.

In the endemic areas the cycle continues all the time; infected midges are carried to and fro on local winds and on northerly and southerly movements of the Intertropical Convergence Zone. Occasionally the virus is taken out of this area to epidemic areas either by movement of infected horses or by unusual winds carrying infected midges.

In epidemic areas severe disease is caused and the virus may persist during a number of years until, as a result of antibodies developing in the hosts or through adverse conditions for the midges, such as a long winter, infection falls below a critical level. Later, when climatic conditions improve and availability of virus host and vector coincide, the virus may again enter these areas.

Diagnosis Diagnosis of African horse sickness is based on clinical grounds, especially the presence of oedema. The peracute form has been confused with anthrax, and other forms have been mistaken for equine influenza, infectious anaemia, trypanosomiasis and equine arteritis.

Diagnosis can be confirmed in the laboratory by isolation of virus in suckling mice or BHK-21 cell cultures and testing the isolate by agar gel precipitin and serum neutralization tests. Rise in antibody in complement-fixation and serum-neutralization tests can also be demonstrated.

Control Control is based on measures to interrupt the midge/horse cycle. Those taken to prevent midges biting include housing susceptible animals from dusk to dawn, treating breeding sites with insecticides and animals with insect repellants. In addition, in some countries horses can be kept in areas not reached by midges.

Spread of disease can be prevented by a ban on movement of horses. When they must be moved, virological and serological tests and a quarantine period should be enforced. Little can be done to control the movement of midges on the wind and some insects may recolonize sites that have previously been sprayed.

Vaccines have been developed; inactivated vaccines have not been so successful as attenuated ones. Attenuation was successfully carried out by passage of virus in adult mice and the vaccine made from the mouse brains. Attenuated strains have also been produced as a result of passage in BHK-21 cells and selection of clones (Erasmus 1965).

It is necessary to determine the types of virus present in a country before deciding which types to include in a vaccine. Interference between types has occurred with polyvalent vaccines, resulting in horses not being protected against some types. Recently, the use of vaccines containing two compatible strains given every 3–4 weeks to cover the number of types present has been advocated.

Bluetongue and related viruses

Bluetongue is an infectious disease of ruminants, especially sheep, characterized by congestion, oedema and haemorrhage in the affected animal. Two other members of the group, epizootic haemorrhagic disease of deer and Ibaraki disease, give rise to similar clinical signs in deer and calves, respectively. The orbiviruses responsible are transmitted by *Culicoides* midges (Howell 1963*b*, Bowne 1971, Howell and Verwoerd 1971).

Characteristics of the virus

In addition to the characteristics mentioned earlier, the viruses are more stable at $-70°$ and $4°$ than $-20°$. Infectivity is destroyed by heating at $60°$ for 30 minutes. At pH values of 5.5 and below the virus is rapidly inactivated.

Bluetongue virus multiples in the fertile hen's egg at $33.5°$ after inoculation of 10–11 day old embryos by the intravenous route (Goldsmit and Barzilai 1968) and 8-day-old embryos by the yolk sac route. The embryos die two or three days later and show characteristic haemorrhages. Ibaraki virus also multiplies in fertile eggs, but epizootic haemorrhagic disease of deer virus has to be adapted. The viruses will also multiply in 1–4 day old mice after intracerebral inoculation, causing paralysis and death. BHK-21, Vero and primary lamb kidney cells support the multiplication of bluetongue virus, in many instances after previous adaptation to the chick embryo. Epizootic haemorrhagic disease of deer and Ibaraki viruses multiply in BHK-21 and Vero cell cultures without adaptation (Fig. 95.1).

Antigenic variation

Bluetongue virus has been classified into 20 subtypes. Types 1–12 were originally classifed on the basis of cross-immunity tests in sheep but types are now determined on the basis of cross-neutralization tests with hyperimmune guinea-pig serum. In addition, there are probably other serotypes that have been isolated but not classified.

Epizootic haemorrhagic disease of deer virus contains 8 types. Ibaraki disease virus has been shown to be related to EHD virus by cross-neutralization tests (Campbell *et al.* 1978). Other members of the complex are Eubenangee, Piry and Tilligerry viruses.

Type-specific tests have been carried out by serum neutralization, plaque reduction and plaque inhibition

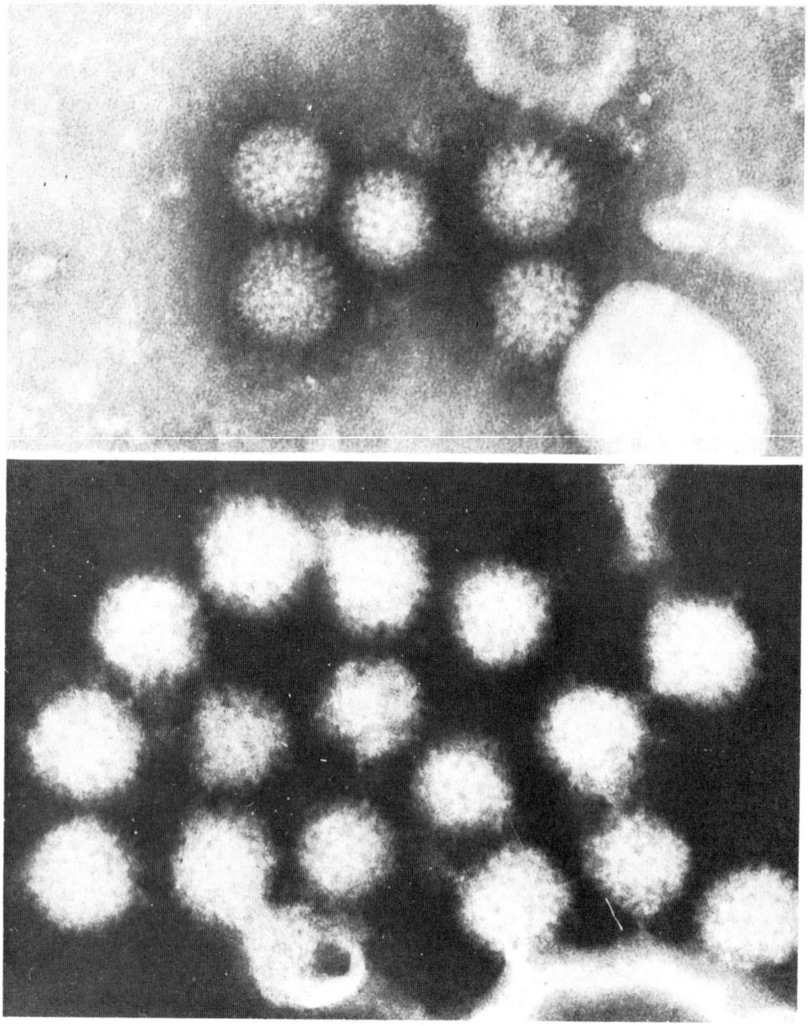

Fig. 95.1 Electronmicrograph of bluetongue virus. *Upper plate*—nucleocapsids; *lower plate*—virus particles. (Photograph by C. J. Smale.)

tests. The type-specific antigen appears to be located in the outer coat. As mentioned before, the outer coat consists of two polypeptides, P2 and P5, in the case of bluetongue and P2, P5 and P3A in that of EHD virus. P2 appears to be the main type-specific antigen in bluetongue and EHD virus and possibly for other members of the bluetongue complex.

All the bluetongue group share a common group antigen located in polypeptide 7. The group antigen was demonstrated by complement fixation, agar gel precipitin and fluorescent antibody tests. A closer relationship has been found between the types of bluetongue virus than between bluetongue and other members of the complex.

Further tests are being developed to analyse antigens and serological relationships, namely, RIA and ELISA tests, but these still have to be evaluated.

Work has also been carried out on the components of the genome by hybridization tests. Extensive variation between segments within the same virus as well as between viruses has been found but as yet this has not been correlated with serological variation (Gorman 1979). However, as genetic reassortment often occurs with dual infection, it should have a bearing on the evolution of new serological types.

Hyperimmune guinea-pig serum has been the choice for the differentiation of types. It tends to be as specific as sheep serum and of greater specificity than rabbit serum or cattle serum. Neutralization tests with the latter two sera tend to cross-react between types. The question of type specificity, classification and preparation of vaccines needs to be investigated in order to clarify the situation from the structural, epidemiological and immunogenic points of view.

Clinical signs

Sheep are the main species affected by bluetongue virus (Erasmus 1975*a*). The incubation period in sheep lasts from 5 to 12 days and sometimes longer, with the majority of sheep showing clinical signs between the sixth and eighth days. The first sign is a rise in temperature and the sheep appears depressed. This is followed by hyperaemia of the oral cavity and swelling of the mucous membranes, leading to oedema of the lips, tongue and intermandibular spaces. There is lachrymation, a serous nasal discharge which becomes mucopurulent, and salivation. At a later stage there is necrosis of the epithelium, with excoriation.

Oedema may extend to the ears and brisket. Leucopenia is found. The feet are often affected with coronitis and laminitis and the animal is lame. Torticollis may occur in some sheep. In pregnant sheep abortions may occur at the time of infection; mummification of the fetus may take place and at term deformed or weak lambs are born.

The extent of clinical signs varies from a subclinical infection, through fever to severe clinical disease. This depends on the strain of virus responsible, the breed of sheep and whether it is in an epidemic or endemic area. In endemic areas the disease is rarely seen except in animals introduced from epidemic or non-bluetongue areas. The local breeds presumably become infected early in life. In South Africa, Israel and India the Merino breed of sheep is especially susceptible.

In cattle the disease is milder and in most animals inapparent infection takes place (Hourrigan and Klingsporn 1975). In clinical cases the incubation period is 6–7 days and the signs are fever, hyperaemia of the oral mucous membranes and other epithelial surfaces, and oedema of the muzzle. Salivation is seen. Ulcers of the mouth and sloughing of the epithelium sometimes occur. Often the animal is lame as a result of coronitis. When the animal is pregnant, infection of the embryo may occur, leading to abortion, congenital malformation or stunting. Calves that are born often develop slowly.

Disease is rarely seen in goats (Erasmus 1975*b*). Mild fever may occur and hyperaemia of the mucous membranes is seen. White-tailed deer after infection with bluetongue virus show similar signs to those caused by EHD virus.

Epizootic haemorrhagic disease of deer virus causes severe clinical disease in white deer (Frank and Willis 1975). The signs are similar to those seen in sheep affected with bluetongue. In the field the deer are often found dead, with bleeding from the orifices and haemorrhages in mucous membranes. Few or no clinical signs are seen in cattle or sheep.

Ibaraki virus affects cattle only. Most cattle are unaffected but haemorrhages and oedema of the oral and nasal mucous membranes are found. The lesion often becomes necrotic. Lameness also develops. Swallowing is difficult in about 20–30 per cent of affected animals (Inaba 1975).

Necropsy examination of animals that have died of bluetongue shows petechial haemorrhages and ecchymoses in the heart and the major vessels. Oedema is found in the muzzle, around the nostrils, in the intermandibular space, the ear and the brisket. The lymph nodes are hyperaemic and the spleen enlarged. Often, however, the animal dies of secondary infection, especially pneumonia. In these cases the lungs will show a broncholobular pneumonia. Post-mortem, lesions in animals dying of EHD and Ibaraki diseases are very similar.

Pathogenesis

Virus is introduced into the animal by the bites of infected *Culicoides* midges. It is carried to the local lymph nodes, where it multiplies, subsequently spreading via the blood stream to other lymph nodes, to the spleen and to the bone marrow, where it multiplies further (Pini 1976). It also multiplies in the endothelial cells lining blood vessels. This leads to damage to the lining and extrusion of fluid, thus causing oedema. The virus also multiplies in monocytes and macrophages (Lawman 1979). Virus to titre 10^6–10^7 is found in the blood mainly associated with the red blood cells. Virus associated with red cells has been isolated from animals for 100 days in cattle and 30 days or longer in sheep and goats.

If the animal which has not previously suffered from an infection is pregnant, the virus crosses the placenta and multiplies in the tissues of the fetus. The extent of damage caused depends on the time of infection in relation to gestation. During early pregnancy the embryo may die; at later stages malformation, including hydranencephaly, may occur (Barnard and Pienaar 1976). In the middle third of pregnancy the fetus may survive and the young be born with virus occurring in the blood (Gibbs *et al.* 1979). Some calves become tolerant to this virus, which may persist in the animal for long periods. If a male animal is infected, virus may be present in the semen and be transmitted to the female.

The pathogenesis of epizootic haemorrhagic disease of deer and of Ibaraki disease in deer and cattle is very similar to that of bluetongue.

Epidemiology

Bluetongue virus is transmitted in a cycle involving, as hosts, cattle, sheep, goats and wild ruminants and, as vectors, *Culicoides* midges. It is also transmitted vertically by passage across the placenta.

Evidence of the virus or disease has been found in North, Central and South America, Africa, the Iberian peninsula, Greece, Italy, Cyprus, Turkey, the Middle East, the Arabian peninsula, the Indian subcontinent, South-East Asia, Australasia and Japan. Seventeen of the 20 types have been found in Southern Africa but information on the types present in other countries is incomplete. Serological surveys have been carried out in many parts of the world but, because of cross-reactions, especially in cattle, the findings are difficult to interpret in the absence of virus isolates or demonstration of a rise in antibody.

Hosts The hosts are cattle, sheep, goats and wild ruminants. As mentioned before, they vary in their response to virus infection. In sheep, virus to a titre of 10^4 per ml or greater is present in the blood from about 4 days before to 4 days after the appearance of disease and similar titres may be found in cattle and goats (Sellers *et al.* 1978). Movement of hosts or their products occurs as a result of migration, transhumance, import and export of animals, export of wild ruminants for zoos, and import and export of semen or embryos.

Vectors The species of *Culicoides* midge responsible for transmission are *C. variipennis* in North America, *C. imicola* in Africa and Western Asia and *C. brevitarsis* in South-eastern Asia and Australia. The midge species responsible in Central and South America may be *C. insignis*. Other species of midge may act as vectors, e.g. *C. milnei*, *C. tororoensis*, *C. schultzei*, *C. cornutus*, *C. puncticollis* and *C. zuluensis* in Africa, *C. obsoletus* in Cyprus and *C. actini* and others in Australia.

The sheep louse, *Melophagus ovinus*, has been shown to harbour and transmit the virus, and mosquitoes, biting flies and ticks might be possible vectors.

The life cycle of the midges has already been described in the section on African horse sickness. After an infected blood meal the virus multiplies to a titre of $10^{5.5}$–$10^{6.0}$ per ml on the sixth day and remains high for the rest of the midge's life. Transmission of virus can occur at the time of feeding every 3–4 days after 7 days.

The two types of flight have been described. In short flights midges are attracted by ruminants to start feeding. For long flights there is circumstantial evidence that midges have been carried on the wind from Morocco to Portugal (200–300 km) and from Turkey and Syria to Cyprus (100–200 km) (Sellers *et al.* 1978, Sellers *et al.* 1979).

Behaviour of the virus: developmental cycle

The virus undergoes a cycle in the host and the vector. Where hosts and vectors are available and where temperatures and other conditions are suitable, the cycle may continue throughout the year. However, where temperatures are greater than 35° or below 13° aestivation or hibernation occurs. During this period the virus 'overwinters' either in surviving adult midges or in the blood of infected ruminants, cattle and wild game. In certain areas it may persist in the fetus and in the newborn animal (Luedke *et al.* 1977).

In addition, the virus may be reintroduced to an area by movement of animals, semen or ova, or inoculation of biological materials. Midges are carried to and fro on the winds; these winds may be local winds or they may be regular movements such as the displacement of the Intertropical Convergence Zone—south-easterlies, south-westerlies, north-easterlies and north-westerlies of the tropic latitudes. Other winds outside the limits of the Intertropical Convergence Zones include the northward or southward movements of warm tropical air as the result of passage of depressions or anticyclones in the middle latitudes (Sellers 1980).

Parts of the world where bluetongue occurs can be divided into endemic, epidemic and free areas. In the endemic areas the climate is (i) warm throughout the year with rain in all months, (ii) warm throughout the year with a rainy and dry season, (iii) warm with a mild winter and rain during the summer or winter. In areas (i) and (ii) bluetongue disease is not usually seen. Animals become infected after the loss of maternal antibody. Peaks of infection are seen at the beginning and end of the rainy season in area (ii). However, sheep imported from temperate climates may suffer disease. In area (iii) disease may or may not be seen, depending on the breeds of sheep present. In these areas the virus overwinters in surviving adult midges or in infected cattle or wild game.

In the epidemic areas the climate is warm during the summer, with a cool winter. The virus is introduced from endemic areas by movement of animals or by carriage of infected midges on the wind. Introduction of virus does not occur each year. The disease may be seen in sheep. In addition, pregnant animals which have not been infected previously may abort as a result of infection or give birth to malformed young or normal animals infected with virus. The virus does not usually overwinter in these areas, as the winter may be too long for sufficient midges to survive or for virus to persist in the blood of cattle.

Free areas are those areas where temperatures are insufficiently high during the summer for transmission to occur and the winters too long for survival of virus in animals or adult midges.

Diagnosis Diagnosis is based on clinical signs in the animal affected correlated with the time of year at which they occur. Virus may be isolated from blood, especially from carriers in the early stages. Sonicated washed red blood cells are inoculated into sheep, fertile eggs, tissue cultures or mice. The virus isolates can then be identified by complement fixation, agar gel precipitin and fluorescent antibody tests for group antigen and neutralization, plaque inhibition and plaque reduction tests for type-specific antigen. Rising antibody may be demonstrated by agar gel precipitin, complement fixation, fluorescent antibody (group-specific) and by neutralization tests (type-specific).

Where deaths or abortions occur, attempts at virus isolation may be made from the spleen and from the lymph nodes and bone marrow. In addition, the serum from the dam may be tested for antibodies.

Control

Control is effected by action against the vector or by protection of the host.

Vectors can be controlled by spraying insecticides or applying larvicides to breeding sites. However, this may have only a temporary effect, since midges may recolonize the site later, but it has the effect of breaking a cycle and reducing the numbers of midges. More permanent effects can be achieved by suitable control of breeding sites, e.g. proper drainage, control of leaks in irrigation lines, but often this is not feasible.

The host may be protected for a short period by the use of insect repellents. In addition, movement to high ground or housing at night may reduce the chances of being bitten. Since in some areas the midges responsible are more attracted to cattle than to sheep, running of cattle with sheep has been advocated. However, a cycle is still maintained in the cattle.

Humoral and cell immunity are concerned in protection against bluetongue. Antibodies (IgM and IgG) develop as a result of infection. Complement-fixation antibodies are found which persist for a shorter time than fluorescent or neutralizing antibodies. Lymphocytes, macrophages and monocytes are also involved.

Live attenuated vaccines have been developed by passage of virus in fertile hens' eggs (Howell and Verwoerd 1971). Subsequently, the strains have been grown in lamb kidney culture for vaccine production. Plaque selection in BHK-21 cells has also been carried out. These strains multiply in the sheep, causing a viraemia but to a lower titre than the virulent strain, thus avoiding the possibility of transmission by midge bites. Sheep infected during pregnancy may, however, abort or give rise to malformed fetuses or lambs (Osburn *et al.* 1971). In countries such as South Africa, where 17 serotypes occur, care has to be given to the types in a vaccine. This problem has been solved to a certain extent by giving three vaccines with an interval of 3 weeks between vaccinations. Each vaccine contains five serotypes. Antibodies usually develop well to such vaccines.

Inactivated vaccines have also been developed by inactivation of the virus with betapropiolactone or binary ethyleneimine (Parker *et al.* 1975, Stott *et al.* 1979). In some animals such as those in Cyprus, there is a primary response with neutralizing antibodies developing. In others, the animals are sensitized and antibodies are apparent only on challenge or as a result of a second dose of vaccine.

With both types of vaccine it is not yet clear whether the exposure to virulent virus will result in persistence of the virus, since it has been shown to persist in red cells for at least 100 days in cattle.

No vaccines have been developed for epizootic haemorrhagic disease of deer or Ibaraki viruses.

Colorado tick fever

Colorado tick fever virus is an orbivirus which affects people resident in or visiting the north-western United States and the neighbouring part of Canada. It is transmitted by a tick, *Dermacentor andersoni*, and causes an acute febrile illness in man, especially in the spring or early summer when the ticks are active.

Characteristics of the virus

The virus genome contains 12 segments of double-stranded RNA (Knudson 1981) but otherwise it shares many of the physico-chemical characteristics of other members of the orbiviruses. It survives for long periods in serum at 4° and also at −70°.

The virus causes death in 1 to 4-day-old mice when given by the intraperitoneal or intracerebral route and has been adapted to 3 to 4-week-old mice by the intracerebral route. The virus can also be adapted to grow in the fertile hen's egg by yolk sac passage; most virus is found in the nervous tissue of the embryo.

Cytopathic effects are produced as a result of virus growth in human KB cells, baby hamster kidney cells (BHK-21) and Vero cells. Plaques have been produced in a line of hamster kidney cells. Tick cultures of *D. andersoni* have been used for growth of virus, as well as a mosquito cell line of *Aedes albopictus* (Yunker and Cory 1969).

Viraemia has been found after inoculation of hamsters, ground squirrels (Columbian and golden), chipmunks, deer mice, rhesus monkeys, porcupines and 1-day-old chicks. Highest titres were present in chipmunks and porcupines.

Clinical symptoms and signs

The incubation period is 3–6 days after a tick bite. The

onset of disease is sudden with chilliness, generalized aching, headache with retro-orbital pains. Pyrexia, myalgia, lumbar backache, nausea and vomiting are also found. After two days the temperature falls for 2–7 days. Then there is a recrudescence with pyrexia for a further 2–3 days. Occasionally a macular papular rash develops. In some cases symptoms of encephalitis such as drowsiness and neck stiffness develop. Aseptic meningitis and gastrointestinal symptoms have been reported in some patients (Goodpasture *et al.* 1978). Recovery occurs in 7–14 days but convalescence may be prolonged for as long as three weeks to several months.

Some cases of Colorado tick fever can be severe or fatal. Disseminated intravascular coagulation was found in one fatal case. Myocardial infarction, CNS disease, spontaneous abortion and diminished visual acuity have been reported (see Hoogstraal 1980).

No disease has been seen in wild rodents.

In laboratory animals killed after experimental infection, cytoplasmic inclusions have been found. Changes in spleen cells of hamsters have also been noted. In mice there is necrosis of the brain and of cardiac muscle fibres.

Pathogenesis It is believed that this may be similar to that of bluetongue or African horse sickness. After the tick bite the virus multiplies in the haemopoietic organs, such as lymph nodes, spleen, liver and bone marrow. Persistence of virus is found associated with the red cells in man and in mice for up to 120 days after infection or onset of symptoms (Emmons *et al.* 1972, Hughes *et al.* 1974, Oshiro *et al.* 1978). Antibodies are present from 14–21 days onwards.

Colorado tick fever virus has been shown to pass the placenta of pregnant mice and give rise to teratogenic effects in the embryo (Harris *et al.* 1975).

Epidemiology

The virus is involved in a cycle between wild rodents and ticks (Burgdorfer 1977). Man is infected occasionally through being bitten by ticks. Man may also be infected by transfusion of infected donor blood.

The disease is seen in persons in the areas where the tick, *Dermacentor andersoni*, is distributed, namely in the north-western United States, British Columbia and Alberta. At 900–1200 metres the infection coincides with peak activity of the ticks in April and May. At higher altitudes infection occurs later, during May and June. Some infections have occurred between March and October.

The preferred hosts in nature appear to be the gold-mantled squirrel, ground squirrels, chipmunks, pine squirrels and deer mice, from all of which the virus has been isolated (Burgdorfer 1977). Virus circulates in the first three at titres higher than the amount required to infect ticks (10^3 LD_{50} per ml) and for a sufficient period.

Dermacentor andersoni ticks are the main vectors but virus has also been isolated from *D. occidentalis*, *D. parumapertus*, *D. albipictus* and *Otobius lagophilus* (Eklund *et al.* 1955). *D. andersoni* readily bites man and small rodents.

The time of infection coincides with the tick rise in the spring, but the virus is maintained by persistence in the tick, in the red cells of wild rodents and by regular cycles between the *Dermacentor* ticks and the wild rodents.

Diagnosis Diagnosis is based on a history of exposure to ticks, febrile response with general aching, the biphasic temperature response and leucopenia.

Virus isolation may be made from blood in 1 to 4-day-old mice and the brain examined in neutralization tests in KB cells. Rise in serum antibody may also be demonstrated by complement fixation and neutralization tests. Early serological diagnosis is based on an indirect fluorescent antibody test with infected BHK-21 or Vero cells (see Hoogstraal 1980).

Control One attack leads to lasting immunity. There is no satisfactory vaccine, and only measures against ticks are satisfactory. The public must be educated to the dangers. Care should be taken not to use blood for transfusion from persons who have had the disease or been exposed to ticks, owing to the persistence of virus associated with blood cells.

Kemerovo group

Viruses of the Kemerovo group were isolated in 1962 and 1963 from *Ixodes persulcatus* and *Ixodes ricinus* ticks in Siberia and Slovakia respectively and also from patients with meningitis and meningo-encephalitis who had previously been bitten by ticks (Libikova *et al.* 1978). The viruses in the group include: Kemerovo (Asia, Africa and Europe), Tribec (Europe), Lipovnik (Europe), Chenuda (Africa), Mono Lake (North America), Huacho (South America), Wad Medani (Africa, Asia and North America), Bauline (North America), Yaquina Head (North America) and others. They share *the properties* of other orbiviruses as regards size, acid sensitivity, sensitivity to lipid solvents and development in cells.

The Kemerovo viruses grow in baby mice 1–4 days old by the intracerebral route, causing fatal encephalitis. New-born rats are also sensitive. In fertile hens' eggs, virus inoculated into the yolk sac or allantoic cavity causes death of the embryo in 2–3 days. The viruses grow in chick embryo cultures, human and hamster primary cells, and in cell lines, L cells HeLa as well as BHK-21 and Vero cells. The virus also grows in tick tissue cultures and in *Aedes albopictus* cell line.

The members of the group are related to a lesser or greater extent by cross-complement fixation tests. Neutralization tests have been used to differentiate strains within the group.

The viruses were originally isolated during work on ticks in areas where tick-borne encephalitis occurred (Chumakov *et al.* 1963). Subsequent isolates of virus have been made from ticks in many parts of the world (Lvov 1980). Many persons in such areas have antibodies to members of the Kemerovo group, and it has been suggested that in some areas simultaneous infection with tick-borne encephalitis virus and a member of the Kemerovo group may lead to encephalitis (Libikova *et al.* 1978). Most isolations of virus and disease in Europe and Siberia occur during spring and early summer, at the time of maximum tick rise.

Viruses have been isolated from ticks of the *Ixodes* genus in association with birds, small mammals, cattle and horses. The species of tick are *Ixodes persulcatus*, *I. ricinus* and *I. uriae*. The first named is found mainly in Siberia, the second in Europe and the third in subarctic and subantarctic regions (see also Berge 1975).

Equine encephalosis and related orbiviruses

A number of viruses of the orbivirus group were isolated from horses between 1967 and 1976 and shown by complement-fixation tests to belong to the equine encephalosis group of orbiviruses (Erasmus *et al.* 1970, Erasmus *et al.* 1976). The first viruses were isolated from liver, spleen, brain and blood samples from horses that had died as a result of a peracute illness; subsequent virus isolations have been made from the blood of horses showing signs varying from fever and inappetence to cardiac failure (Bryanston virus) and from fetuses aborted after 6–6½ months' gestation. No insect vector has been demonstrated.

The virus has similar physico-chemical and morphological characters to other orbiviruses. It grows in BHK-21 cells, with production of cytopathic effect, and can be adapted to grow in suckling mice inoculated intracerebrally.

The main post-mortem lesions are general venous congestion, fatty liver degeneration, brain oedema and catarrhal enteritis. With the Bryanston virus, local fibrosis was found in the myocardium.

So far, the viruses have been isolated only in South Africa and nothing is known of how they are maintained. Antibodies have been found in horses and not in other animals.

Diagnosis is based on clinical signs together with isolation of the virus in BHK-21 cells and serological tests. No vaccine is available.

Other groups

Of those viruses associated with man, Changuinola virus was isolated in suckling mice from a mosquito catcher in Panama who had a febrile illness. A strain of virus related to Lebombo virus was isolated from human plasma in Nigeria.

References

Barnard, B. J. H. and Pienaar, J. G. (1976) *Onderstepoort J. vet. Res.* **43,** 153.

Berge, T. O. (1975) *International Catalogue of Arboviruses*, 2nd edn. Communicable Diseases Centre. Dept. Hlth, Educ. and Welfare, Washington D.C.

Borden, E. C., Shope, R. E. and Murphy, F. A. (1971) *J. gen. Virol.* **13,** 261.

Bourdin, P. (1973) *Proc. 3rd int. Conf. Equine Infectious Diseases*, p. 12. Karger, Basel.

Bowne, J. G. (1971) *Advanc. vet. Sci.* **15,** 1.

Bremer, C. W. (1976) *Onderstepoort J. vet. Res.* **43,** 193.

Burgdorfer, W. (1977) *Acta trop.* **34,** 103.

Campbell, C. H., Barber, T. L. and Jochim, M. M. (1978) *Vet. Microbiol.* **3,** 15.

Chumakov, M. P. *et al.* (1963) *Vop. Virus* **8,** 440.

Eklund, C. M., Kohls, G. M. and Brennan, J. M. (1955) *J. Amer. med. Ass.* **157,** 335.

Emmons, R. W., Oshiro, L. S., Johnson, H. N. and Lennette, E. H. (1972) *J. gen. Virol.* **17,** 185.

Erasmus, B. J. (1965) *Bull. Off. int. Epiz.* **64,** 697; (1973) *Proc. 3rd int. Conf. Equine Inf. Dis.*, p. 1. Karger, Basel; (1975*a*) *Aust. vet. J.* **51,** 165; (1975*b*) *Ibid.* **51,** 209.

Erasmus, B. J., Adelaar, T. F., Smit, J. D., Lecatsas, G. and Toms, T. (1970) *Bull. Off. int. Epiz.* **74,** 781.

Erasmus, B. J., Boshoff, S. T. and Pietrese, L. M. (1976) *Proc. 4th Conf. Equine Infect. Dis., Lyon* p. 447.

Frank, J. F. and Willis, N. G. (1975) *Aust. vet. J.* **51,** 174.

Gibbs, E. P. J., Lawman, M. J. P. and Herniman, K. A. J. (1979) *Res. vet. Sci.* **27,** 118.

Goldsmit, L. and Barzilai, E. (1968) *J. comp. Path.* **78,** 477.

Goodpasture, H. C., Poland, J. D., Francy, D. B., Bowen, G. S. and Horn, K. A. (1978) *Ann. intern. Med.* **88,** 303.

Gorman, B. M. (1979) *J. gen. Virol.* **44,** 1.

Harris, R. E., Morahan, P. and Coleman, P. (1975) *J. inf. Dis.* **131,** 397.

Hoogstraal, H. (1980) *Zbl. Bakt. Suppl.* **9,** 49.

Hourrigan, J. L. and Klingsporn, A. L. (1975) *Aust. vet. J.* **51,** 170.

Howell, P. G. (1963*a*) *FAO Agricultural Studies*, No. 61, p. 71. Food and Agriculture Organization, Rome; (1963*b*) *FAO Agricultural Studies*, No. 61, p. 111. Food and Agriculture Organization, Rome.

Howell, P. G. and Verwoerd, D. W. (1971) *Virology Monographs* **9,** 35. Springer Verlag, Vienna.
Hughes, L. E., Casper, E. A. and Clifford, C. M. (1974) *Amer. J. trop. Med. Hyg.* **23,** 530.
Huismans, H., Bremer, C. W. and Barber, T. L. (1979) *Onderstepoort J. vet. Res.* **46,** 95.
Inaba, Y. (1975) *Aust. vet. J.* **51,** 178.
Knudson, D. L. (1981) *Virology* **112,** 361.
Lawman, M. J. P. (1979) *Ph.D. Thesis.* Univ. of Surrey, Guildford, UK.
Lecatsas, G. (1968) *Onderstepoort J. vet. Res.* **35,** 139.
Libikova, H., Heinz, F., Ujhazyova, D. and Stuenzner, D. (1978) *Med. Microbiol. Immunol.* **166,** 253.
Luedke, A. J., Jones, R. H. and Walton, T. E. (1977) *Amer. J. trop. Med. Hyg.* **26,** 313.
Lvov, D. K. (1980) *Zbl. Bakt. Suppl.* **9,** 35.
Martin, S. A. and Zweerink, H. J. (1972) *Virology* **50,** 495.
Murphy, F. A., Borden, E. C., Shope, R. E. and Harrison, A. (1971) *J. gen. Virol.* **13,** 273.
Osburn, B. I., Johnson, R. T., Silverstein, A. M., Prendergast, R. A., Jochim, M. M. and Levy, S. E. (1971) *Lab. Invest.* **25,** 206.
Oshiro, L. S., Dondero, D. V., Emmons, R. W. and Lennette, E. H. (1978) *J. gen. Virol.* **39,** 73.
Parker, J., Herniman, K. A. J., Gibbs, E. P. J. and Sellers, R. F. (1975) *Vet. Rec.* **96,** 284.
Pini, A. (1976) *Onderstepoort J. vet. Res.* **43,** 159.
Sellers, R. F. (1980) *J. Hyg. Camb.* **85,** 65.
Sellers, R. F., Gibbs, E. P. J., Herniman, K. A. J., Pedgley, D. E. and Tucker, M. R. (1979) *J. Hyg. Camb.* **83,** 547.
Sellers, R. F., Pedgley, D. E. and Tucker, M. R. (1977) *J. Hyg. Camb.* **79,** 279; (1978) *J. Hyg. Camb.* **81,** 189.
Stott, J. L., Osburn, B. I. and Barber, T. L. (1979) *Proc. 83rd Ann. Mtg US Anim. Hlth Assoc.* p. 55.
Verwoerd, D. W. and Huismans, H. (1972) *Onderstepoort J. vet. Res.* **39,** 185.
Verwoerd, D. W., Huismans, H. and Erasmus, B. J. (1979) In: *Comprehensive Virology*, Vol. 14, p. 285. Plenum Press, New York and London.
Yunker, C. E. and Cory, J. (1969) *J. Virol.* **3,** 631.

96

Influenza

Geoffrey C. Schild

Introductory

The causative viruses of influenza are classified as *Orthomyxoviridae* and belong to a group of agents known as myxovirus, i.e. viruses with affinity for mucins. The orthomyxoviruses are pleomorphic in shape, existing as spherical or filamentous particles, and contain RNA in a helical nucleocapsid surrounded by a lipid-containing envelope. Other myxoviruses, including those which cause measles and mumps, respiratory syncytial virus and the parainfluenza viruses, share certain characteristics in common with the influenza viruses but differ in other respects and have been placed in a separate family, *Paramyxoviridae*. An important taxonomic feature of the orthomyxoviruses is that the diameter of the nucleocapsid is 9 nm while for paramyxoviruses the diameter of the nucleocapsid is 18 nm. The orthomyxoviruses have a further distinct feature; their genomes consist of eight segments of single negative stranded RNA with a total molecular weight of approximately 5×10^6. Orthomyxoviruses are classified into three types, influenza type A, type B and type C viruses on the basis of the structure of their nucleoprotein antigen. The natural hosts of influenza viruses include man, lower terrestrial and aquatic mammals and birds. In man influenza viruses cause pandemics and epidemics of respiratory disease associated with considerable morbidity and mortality. Characteristic features of influenza virus, associated with their unique epidemiological behaviour, are their high degree of antigenic and genetic variability and their propensity for genetic reassortment which readily occurs between strains of the same type in the laboratory and apparently in nature.

The influenza A and B viruses are major causes of human respiratory disease producing a considerable epidemiological impact in all areas of the world where surveillance has been conducted (Assaad *et al.* 1973, Pereira 1979, 1980, McGregor *et al.* 1979). During the present century, world-wide outbreaks, pandemics, caused by influenza A viruses have occurred with an irregular periodicity of some 10 to 39 years and have been associated with high levels of morbidity and mortality. Historic accounts (Creighton 1894) suggest that epidemics of influenza-like illness may have occurred in previous centuries and descriptions of epidemics from mediaeval times are strongly suggestive of influenza. Between pandemics less widespread outbreaks of influenza A infection occur. Such inter-pandemic epidemics probably produce a total impact on morbidity and mortality which cumulatively, over several years, may be greater than that which occurs during pandemics (Dowdle 1976). The influenza B virus produces epidemics which are less widespread and less frequent than those due to influenza A and which are occasionally associated with raised mortality (Assaad *et al.* 1973, Dowdle, Coleman and Gregg 1974). Influenza A epidemics typically affect all age groups of the population but deaths occur largely in elderly persons and those who are at high risk owing to pre-exisiting chronic disease of the respiratory or circulatory system. Secondary bacterial pneumonia is a major factor in fatal infections (Stuart-Harris and Schild 1976). Acute deaths due to primary viral pneumonia are relatively infrequent; however, their occurrence in previously healthy young persons is noted at times when large numbers of influenza cases occur, as for example in the Asian influenza pandemic of 1957 (Stuart-Harris and Schild 1976). Influenza C virus, although apparently a common infectious agent in man, does not appear to be an important cause of respiratory disease.

Influenza has justifiably been called 'the last great plague' (Beveridge 1977), since it is the only remaining truly pandemic disease of man at a time when many previously important infectious diseases, e.g. smallpox, have been eliminated or reduced to insignificant proportions as a result of improved standards of public health, by vaccination or by chemoprophylaxis.

Associated with the success of the influenza viruses as human pathogens and their capricious epidemiological behaviour is their high degree of antigenic and genetic adaptability. That exhibited by influenza A and B viruses is unique amongst viruses of man. The study of antigenic variation is an important aspect in our attempts to understand the epidemiological behaviour of the disease. This feature of the virus also presents serious problems for the control of the disease by immunoprophylaxis.

The control of influenza by vaccination or chemoprophylaxis has not yet been achieved, but prospects for this are ultimately dependent upon full understanding of the biology, chemistry and epidemiology of the virus and the nature of immunity towards it. The technology now exists for the preparation of inactivated vaccines of high degree of purity and antigenicity which are used for the protection of 'high risk' groups of the population, and important advances are being made in the development of live attenuated vaccines. Effective world-wide surveillance of influenza by the World Health Organization, particularly aimed at the early detection of new antigenic variants, is an essential aspect of attempts to understand the epidemiology of the disease and to institute effective vaccination programmes.

In addition to man, influenza A viruses occur in certain species of lower mammals and in domestic and wild birds; these viruses are discussed in the present chapter. Influenza A viruses from non-human sources show close antigenic similarities to human influenza A viruses. Influenza A viruses from non-human hosts may provide useful experimental model systems relevant to the study of human influenza but their main significance is as possible progenitors for new pandemic influenza viruses of man. This possibility has stimulated much interest in animal and avian influenza A viruses and there is now extensive information, which is discussed below, on the natural history and ecology of these viruses and on comparisons of their genetic and antigenic characters with those of human influenza A viruses.

The influenza virus: isolation

Although the earliest isolations of influenza from man were not made until 1932, viruses had been isolated earlier from non-human sources which only subsequently were shown to be influenza A strains. The first clinical descriptions of fowl plague were reported in Italy in 1878 and the disease was shown in 1901 by

Centanni and Avonuzzi (see Stubbs 1965) by experimental transmission studies to be due to a 'filterable virus'. Fowl plague remained of little interest to influenza research workers for over 50 years until Schäffer (1955) showed that the agent possessed the antigenic and morphological attributes of influenza A viruses isolated from man—and was indeed a true influenza virus. Similarly swine influenza virus first transmitted from pig to pig with bacteria-free filtrates (Shope 1931) was shown (Andrewes *et al.* 1935) to have antigenic similarities with human influenza A viruses. Since these early observations numerous isolations of type A influenza viruses from non-human sources have been recorded and much has been learned of the antigenic and biological spectrum of influenza A viruses among lower mammals and birds and their relations to human influenza A viruses.

Conclusive evidence that influenza B or C viruses exist in non-human hosts is lacking. Although on a few occasions evidence of influenza B infections in animals has been reported, these findings have not generally been confirmed.

Sir Macfarlane Burnet recalls (Burnet 1979) that the first successful attempts to isolate influenza virus were 'first foreshadowed at a tea-table discussion at University College Hospital Medical School during a pathology meeting some time in winter 1932–3. I was at a table with some of my seniors, of whom I can remember clearly only W W C Topley and P P Laidlaw. There had already been significant influenza in London, and I found myself listening with fascination as Topley laid down the law to Laidlaw as to how the influenza virus should be sought. By 1932 everyone felt certain that there was a virus responsible for influenza—and another for the common cold. It is also well to remember that in those days tissue culture was in its infancy and viruses were isolated by the injection of bacteria-free filtrates into experimental animals. Topley's advice was to make filtrates from throat washings of influenza patients and inoculate them into the widest possible range of animals. Laidlaw had just completed a valuable series of studies on dog distemper using ferrets—whether as a result of that talk with Topley or not, Laidlaw started work that winter with Wilson Smith and Andrewes as collaborators, using ferrets as experimental animals, with immediate success.'

Thus the viral aetiology of influenza in man was firmly established at the National Institute for Medical Research when Wilson Smith, Andrewes and Laidlaw attempted to transmit the disease to various laboratory animals using garglings from patients with influenza (Smith *et al.* 1933). Ferrets inoculated intranasally developed a febrile respiratory disease which was contagious to their cage-mates. Other animals used in the studies including rabbits and guinea-pigs appeared refractory to the infection. It is anecdotal that influenza was transmitted from an infected ferret to one of the senior scientists from whom a virus was re-isolated to provide the classical WS strain of influenza A virus. Later it was found that infection could be artificially transmitted from ferrets to mice producing a sometimes fatal pneumonia (Andrewes *et al.* 1934). These findings led to rapid progress in our knowledge of the immunological and pathological properties of the influenza virus. In 1940, a second virus was isolated which was antigenically distinct from the 1933 strains (Francis 1940). This virus was later designated influenza B to distinguish it from that of Smith *et al.* (1933) which became known as influenza A. A third type, influenza C, was isolated in 1949 (Taylor 1949). Alternatives to experimental animals for the isolation and cultivation of influenza viruses became available later when Burnet (1941) discovered that the virus would grow in the amniotic and allantoic cavities of the developing chick embryo. This method also led to the discovery of the haemagglutinating properties of the virus (Hirst 1941), and to the use of haemagglutination-inhibition techniques for its serological study. These two techniques have contributed greatly to the knowledge of the influenza virus. Later biochemical work (Gottschalk 1957) provided the observation that influenza viruses contained the enzyme neuraminidase which, with the haemagglutinin, is a major antigenic component of the virus particle.

Morphology, polypeptide composition and genome structure

Detailed studies of the molecular structure of the influenza viruses, and their comparison with those of other viruses, may provide the basis for the understanding of the unique epidemiological behaviour of influenza and the extreme antigenic and genetic variability exhibited by the causative viruses. They may also provide scientific information on the understanding of basic biological systems such as the structure and function of viral genes and of biological membranes and membrane proteins.

The influenza viruses of types A, B and C form a group of agents united by their generally similar morphological, chemical and biological properties and have been accorded the family status, *Orthomyxoviridae* (Melnick 1973). Influenza A and B viruses are classified as separate genera within the family. Influenza C virus differs from influenza A and B in certain respects, including envelope structure and the absence of neuraminidase activity, and has been accorded provisional genus status (Dowdle *et al.* 1975).

Orthomyxoviruses are enveloped viruses which possess the property of haemagglutination. They possess a virion-associated, RNA-dependent RNA polymerase enzyme (RNA transcriptase), and their genetic material is single-stranded RNA, which is present in

Table 96.1 Comparative properties of ortho- and para-myxoviruses

	Orthomyxoviruses	Paramyxoviruses
Morphology	Lipid-containing, enveloped viruses with surface projections, 'spikes'; particles roughly spherical; filamentous forms also exist	
Particle diameter	80–110 nm	150–200 nm
Biological properties:		
haemagglutination activity	+	+
neuraminidase activity	+	+
cell fusion and lytic activity	–	+
Polypeptide compositon of virion	Eight species of polypeptide. Range of 25 000–100 000 mol.wt Haemagglutinin and neuraminidase activities are associated with distinct glycoprotein subunits	Five species of polypeptide. Range of 38 000–69 000 mol.wt Haemagglutinin and neuraminidase activities are associated with the same glycoprotein subunits
Possession of polypeptide 'F' associated with cell fusion and lysis	–	+
Genome:		
RNA composition	Single-stranded, segmented RNA genome total 5×10^6 mol.wt, genome segmented into 8 RNA fragments mol.wt range 2×10^5 to 9×10^5 mol.wt	Single stranded, unsegmented RNA or 7–8×10^6 mol.wt
RNA function	Negative strand RNA in virion complementary to mRNA of infected cells 'Naked' RNA isolated from virus particles not infective	
Replication:		
maturation	Particles mature by budding at the cell surface	
role of cell nucleus in replication	+	–
inhibition by actinomycin D	+	–

the virion in several discrete segments. A transcript of this RNA constitutes the messenger RNA of the virus. A further group of viruses, the paramyxoviruses, resemble orthomyxoviruses in some respects. They, too, are enveloped viruses which contain haemagglutinin and neuraminidase proteins and an RNA polymerase enzyme. However, they differ from orthomyxoviruses in that their RNA genome is not segmented. In contrast to the influenza viruses, paramyxoviruses are antigenically relatively stable. There are also fundamental differences in the mode of replication of orthomyxoviruses and paramyxoviruses; whereas the replication of influenza virus is inhibited by low concentrations of actinomycin D, indicating the probable dependence on cell nuclear DNA at some stage of the replication cycle, paramyxoviruses are not so inhibited. In addition synthesis of influenza virus-specific polypeptides does not occur at all in enucleated cells, whereas such cells are susceptible to infection with paramyxoviruses. The differences between the characteristics of orthomyxoviruses and paramyxoviruses (reviewed by Choppin and Compans 1975) are summarized in Table 96.1.

In the present chapter limitation of space does not allow a full description of the structure and composition of the influenza viruses. For the reader requiring more detailed information several comprehensive review articles are available (Choppin and Compans 1975, Schulze 1975, Stuart-Harris and Schild 1976, Wrigley 1979, Palese *et al.* 1980, Scholtissek 1980, Laver *et al.* 1980, Skehel *et al.* 1980, Hay *et al.* 1980, Skehel and McCauley (1980), Mahy *et al.* 1980, Ghendon and Markushin 1980, Rott 1980, Laver and Air 1980) which provide accounts of the morphology, protein and nucleic acid structure and replication of the influenza viruses.

Morphology

Influenza A and B viruses share common morphological characteristics; influenza C virus differs in certain respects from the other two types. Figure 96.1 illustrates a model of an influenza virus particle showing the arrangement of the major structural components. Negatively stained preparations of influenza virus examined by transmission electron microscopy (Fig. 96.2**a**) reveal approximately spherical particles having an external diameter of 80–110 nm and an inner electron dense core of approximately 70 nm. In some preparations, particularly those of freshly isolated influenza A viruses from both human and non-human sources, filamentous particles 1000 nm or more in length may be abundant (Fig. 96.2**b**). For influenza A and B viruses the particles are covered with densely arranged radial projections—'spikes', of two distinct morphologies (Laver and Valentine 1969, Wrigley 1979). These are the two biologically active glycoprotein components of the virus, the haemagglutinin and

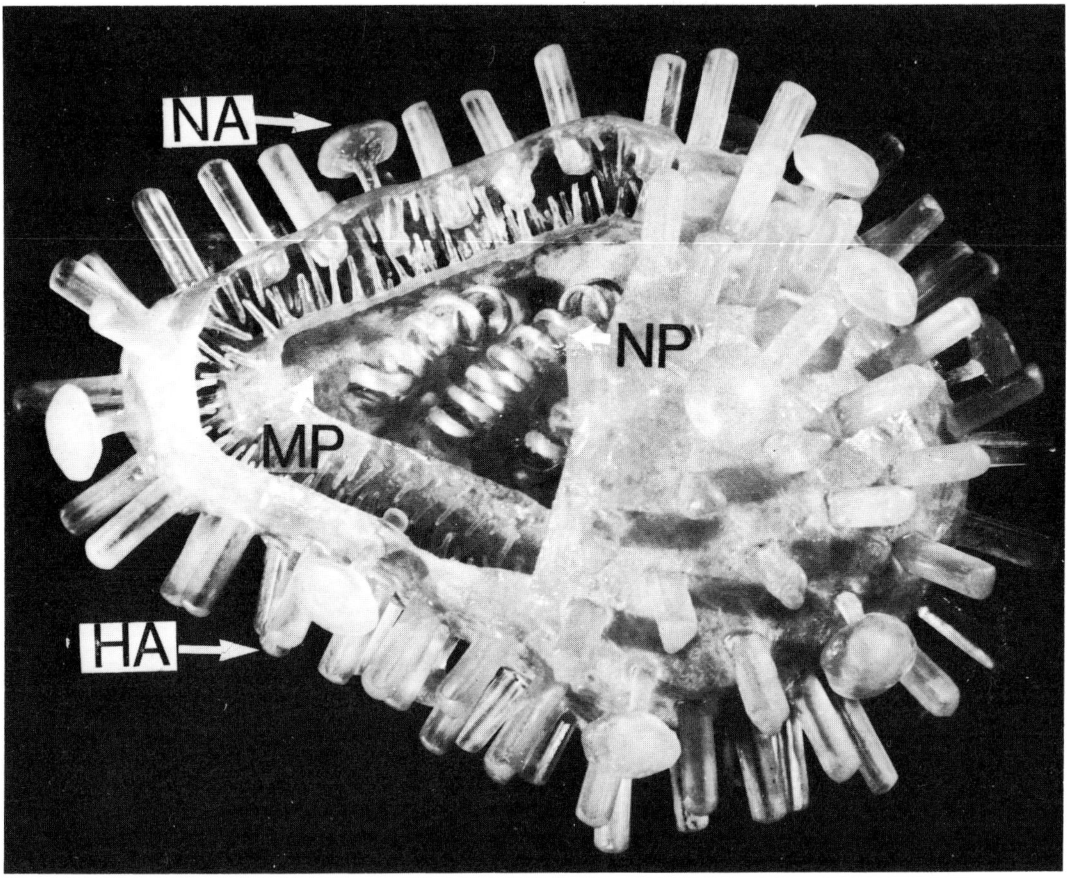

Fig. 96.1 Model of influenza virus particle showing major structural features. The haemagglutinin (HA) and neuraminidase (NA) are inserted into the bilayer lipid membrane by their distal ends. Internally the nucleoprotein (NP) is associated with the virus RNA to form the helical nucleocapsid. The matrix protein (MP) underlies the lipid membrane.

the neuraminidase. The haemagglutinin protein exists as tapered projections or 'spikes' with their broadest end outermost and their narrow end inserted in the lipid membrane. The neuraminidase consists of mushroom-shaped projections inserted in the lipid membrane by a narrow fibre representing the mushroom 'stalk'. The chemical and antigenic properties of these two important surface antigens of the virus will be described in detail below. By means of freeze-drying and freeze-etching techniques, which may give a more realistic picture of virus morphology than that obtained by negative staining, Nermut and Frank (1971) obtained evidence that the influenza virus particle might exhibit regular 'plastic' icosahedral morphology.

Virus polypeptides

Electrophoretic analysis of purified influenza virions reveals eight distinct virus-coded polypeptides (Tables 96.2 and 96.3) of molecular weights from 25 000 to 100 000 daltons (Skehel and Schild 1971, McGeoch, Fellner and Newton 1976). The three largest polypeptides designated P1, P2 and P3 (Kilbourne *et al.* 1972*b*) of molecular weight approximately 80–100 000 daltons are present in small quantities in the virus core and are concerned with the RNA polymerase activity of the virus and replication of the RNA. A major polypeptide of molecular weight approximately 50 000 daltons corresponds to the nucleoprotein antigen. The virion polypeptide of lowest molecular weight, 25–27 000 daltons, corresponds to the matrix protein.

The three remaining components are glycopolypeptides present on the surface of the virus. Of these, a polypeptide of molecular weight 45–65 000 is the neuraminidase protein. The other surface protein, haemagglutinin, is synthesized as a single polypeptide of molecular weight 70–80 000 which, during virus maturation, is cloven by trypsin-like enzymes in infectious virus particles grown in eggs, i.e. is present as two polypeptides, HA1 and HA2, of molecular weights 55 000 and 28 000 respectively. The carbo-

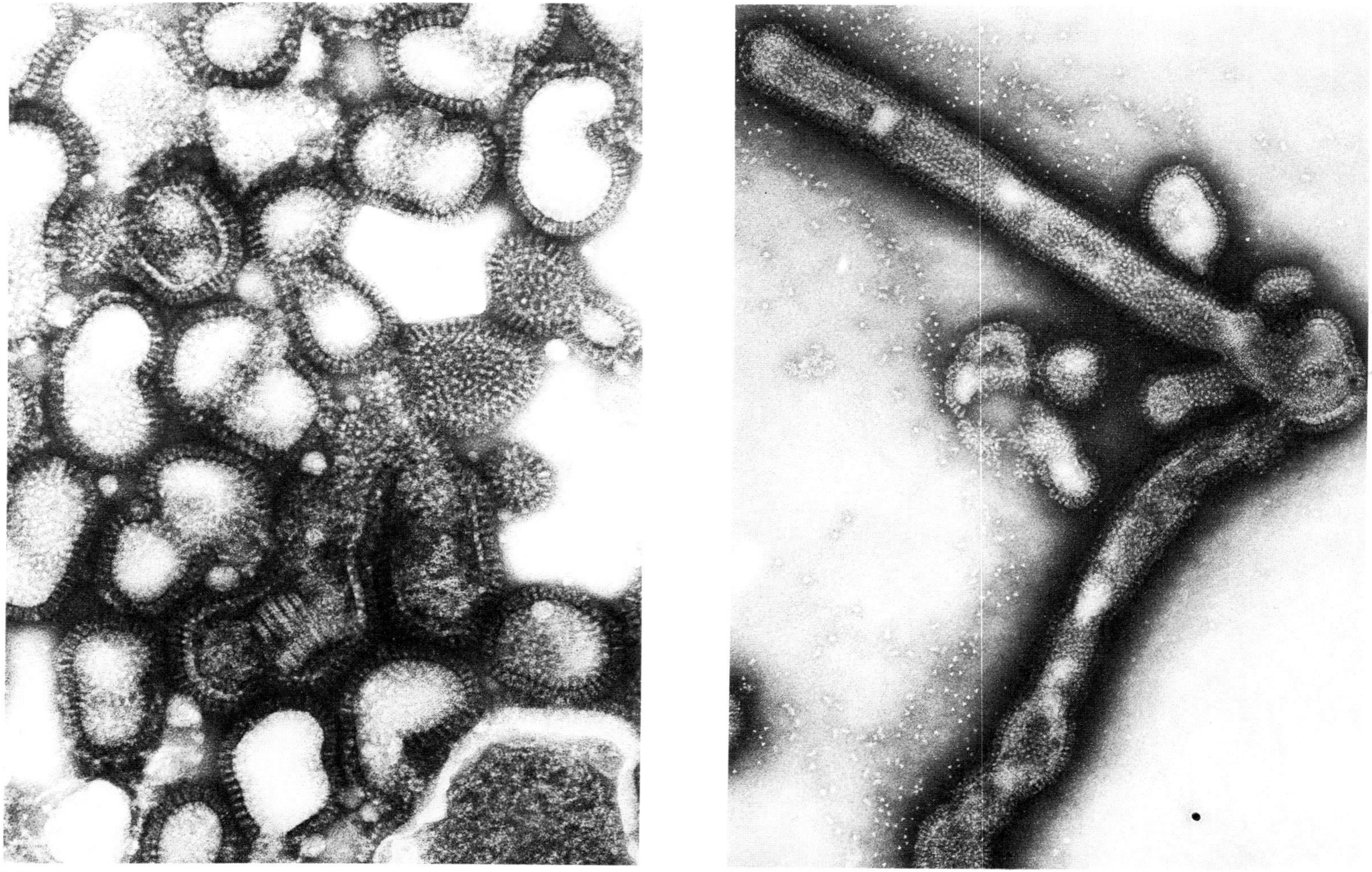

(a) **(b)**

Fig. 96.2 Electronmicrographs of negatively stained influenza virus particles. (**a**) Particles of A/Hong Kong/68 (H3N2) virus showing exposed helical nucleocapsid (× 150 000). (**b**) Early passage avian influenza A virus particles showing filamentous forms (× 98 000).

Table 96.2 RNA gene segments of influenza A virus and their assignment to virus specific polypeptides

Genome segment in order of molecular weight	Molecular weight ($\times 10^{-5}$)	Gene product			
		Designation	Molecular weight ($\times 10^{-3}$)	Approximate no. of molecules per virus particle	Assignment to structural and functional and antigen components of virus
I	8.9	P3 (PB2)*	80	50	Minor internal non-glycosylated proteins of unknown antigenic specificity. Associated with RNA polymerase activity and genome transcription. With the NP form 'polymerase complex'
II	8.4	P1 (PB1)*	97		
III	8.6	P2 (PA)*	86		
IV	6.6	HA	72	1000	The major surface glycoprotein. Morphologically tapered rod with triangular section. Biologically active subunit of mol.wt 210 000 contains 3 HA1 and 3 HA2 polypeptide chains linked by multiple disulphide bridges. Responsible for haemagglutination activity of virus and virus attachment to cell surface. Antigenically variable 'shift' and 'drift'. Important to immunogenicity
V	5.6	NP	55	1000	Major internal, non-glycosylated protein associated with RNA and probably P proteins to form helical nucleocapsid with RNA. Antigenically type specific for influenza A, B and C but shows some antigenic variability
VI	4.8	NA	45	100–200	Surface glycoprotein, neuraminidase enzyme activity. Morphologically knob and fibre. Active subunit mol.wt 200 000 contains 4 polypeptide chains. Antigenically variable 'shift' and 'drift'. Contributes to immunogenicity
VII	2.8 (coding capacity for two polypeptides)	MP	27	3000	Major internal non-glycosylated protein. Associated with inner surface of lipid membrane. Antigenically type specific for influenza A and B (and C)
VIII	2.1 (coding capacity for two polypeptides)	NS1 NS2	23 11	not present not present	Non-structural virus coded protein synthesized in cytoplasm of infected cells migrates to nucleus. Antigenic specificity unknown
Total	48				

* Alternative designations for P1, P2 and P3 have been proposed by Horisberger, M. A. (1980) *Virology*, **107**, 302–5.

Table 96.3 Gene segments of influenza B virus and their assignment to virus specific polypeptides

Genome segment in order of molecular weight	Molecular weight ($\times 10^{-5}$)	Gene product	
		Designation	Molecular weight ($\times 10^{-3}$)
I, II, III	8.8–9.1	P1,P2,P3	8.0–10.2
IV	7.0	HA	8.4
V	6.8	NP	6.6
VI	5.5	NA	6.6
VII	3.8	MP	2.5
VIII	3.3	NS	4.0
Total	5.3		

hydrate side chains of these three glycoproteins are specified primarily by the host cell and vary in composition according to the host. The influenza virus particle does not contain proteins derived directly from host cell components.

Viral infection of cells also induces the formation of two polypeptides which are not included in virions, NS1 of molecular weight approximately 23 000 and NS2 of molecular weight 11 000 daltons (Skehel 1972, Lamb and Choppin 1980). Their function is not yet understood.

There may be small amounts of other virus specific polypeptides which yet await identification.

Strains of influenza A and B virus give generally rather similar polypeptide patterns, but different subtypes of viruses often show differences in the migration rates of specific polypeptides (Oxford, Schild and Alexandrova 1980). Such differences can be very valuable in identifying strains.

The genome

In influenza viruses the genetic information is carried by a segmented single-stranded RNA genome of total molecular weight approximately 5×10^6 daltons (Table 96.2). The genome consists of eight pieces of RNA ranging in molecular weight from approximately 2×10^5 to 9×10^5 daltons (Palese *et al.* 1980, Racaniello and Palese 1979, Desselberger and Palese 1978, Inglis *et al.* 1979, Allen *et al.* 1980). Table 96.2 and 96.3 based on the published work of these authors summarizes information on the molecular weight of the RNA segments of influenza A and B viruses and proteins for which they code.

The assignment of coding functions for the eight species of RNA has been established (Table 96.2).

There is now much information on the base sequence of the haemagglutinin genes of several influenza strains (reviewed in Laver and Air 1980, Webster *et al.* (1982). The haemagglutinin genes are 1742–55 nucleotides in length and code for 563–5 amino acids.

Information is also available on the complete or partial base sequences of the genes which code for the matrix and NS proteins of several influenza A virus strains (Porter, Smith and Emtage 1980, Lamb *et al.* 1980, Air 1980, Air and Hackett 1980, Hall and Air 1981, Both and Air 1979). In addition, the complete sequence of the gene coding for the nucleoprotein of influenza A/PR8/32 (H1N1) virus has also been determined (Winter and Fields, 1981). This is 1565 nucleotides long and codes for a protein of 498 amino acids in length.

Emtage *et al.* (1980) reported synthesis of a protein with specific antigenic reactivity with antisera to fowl plague haemagglutinin following insertion of a gene sequence for fowl plague virus haemagglutinin into an *Escherichia coli* plasmid. Other workers have observed full expression of DNA copies of influenza haemagglutinin genes in *Esch. coli* cells after insertion into bacterial plasmids (Davis *et al.* 1981) or in eukaryotic cells after insertion in a vector SV40 virus gene (Gething and Sambrook 1981). In the latter case biologically functional haemagglutinin molecules were produced. These important findings foreshadow the possibility of producing influenza vaccines by genetic engineering techniques.

The virus antigens: chemical structure, serological and immunological properties

Structural features

The haemagglutinin

The haemagglutinin is the major envelope glycoprotein and an important antigenic component of the virus. It constitutes 29–32 per cent of the total protein of the virus particles (Oxford, Corcoran and Hugentobler 1981). In each virus particle there are some 1000 haemagglutinin 'spikes' (Wrigley 1979) which are 14 nm in length and 4–6 nm in diameter and appear to be of triangular cross-section. Figure 96.3 shows a representation of the three-dimensional structure of the haemagglutinin as proposed by Wilson, Skehel and Wiley 1981). Each haemagglutinin 'spike' has a hydrophobic tip by which it is inserted in the virus membrane and by means of which the spikes aggregate together when isolated in pure form (Laver and Valentine 1969). The binding sites for host cell receptors, responsible for the biological function of haemagglutination, and the variable antigenic sites are situated on the distal tip of the spike.

The chemical structure of influenza haemagglutinin has been reviewed by Schulze (1973), Wrigley (1979), Waterfield *et al.* (1979), Wiley, Wilson and Skehel (1981), Wilson *et al.* (1981), and Caton *et al.* (1982). (See Fig. 96.4)

The crystallization of haemagglutinin (Brand and Skehel 1972) has enabled detailed studies to be made of its physical structure. The application of high resolution x-ray diffraction technique to such crystals (Wilson *et al.* 1981, Wiley *et al.* 1981) has provided precise information on the three-dimensional structure of haemagglutinin. The trimeric structure contained two distinct regions—a triple-stranded coiled

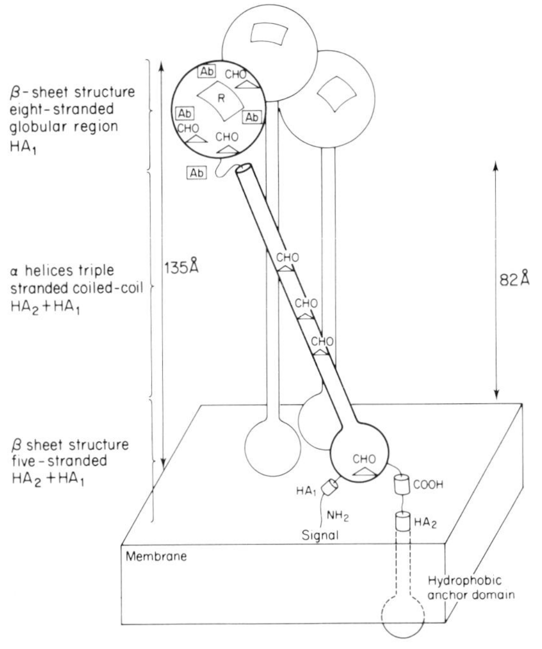

Fig. 96.3 Schematic diagram of trimeric structure of 'mature' haemagglutinin on virus surface. The cell surface receptor binding site (R), antibody binding sites (Ab) and carbohydrate side chains (CHO) are indicated. The anchoring regions of the molecule in the membrane, the NH_2 terminal region of HA and the C-terminal region of HA_2 are indicated. (Modified from Wilson, Skehel and Wiley, 1981.)

coil of alpha-helices extending 76Å from the virus membrane supporting a globular region of anti-parallel beta-sheet which contained the binding site for receptors on cells or erythrocytes and antigenic determinants (Fig. 96.3).

Several laboratories have studied the amino-acid sequences of haemagglutinins of numerous influenza virus strains. For a review see Laver and Air (1980).

The general conclusions from these studies is that certain regions of the haemagglutinin molecules of antigenically different strains show considerable sequence homologies which have been attributed to the need for conservation of function and structural requirements, e.g. Table 96.4. An overall conservation of amino-acid sequences between haemagglutinins of

Table 96.4 Conservation of amino acid and nucleotide sequences in strains of different H subtypes

	% amino acid conservation			
	H1	H2	H3	H7
H1		58(79)	35(53)	33(51)
H2	61(72)		36(50)	35(53)
H3	45(58)	45(57)		36(65)
H7	44(58)	46(59)	45(66)	

% nucleotide conservation

The conservation in the HA1 subunit is given first, with the conservation in the HA2 subunit shown in parentheses. Amino acid conservation is listed above the diagonal and nucleotide conservation below.
Data from Winter *et al.* 1981.

different subtypes of 33–36 per cent has been reported (Winter *et al.* 1981) (Table 96.5, p. 328). Homologies between the amino-acid sequences and the nucleic acid sequences of haemagglutinins of different subtypes are shown in Table 96.4. In contrast, areas in the region of the distal tip associated with antigenic determinants show considerable variability (Wiley *et al.* 1981).

During major antigenic changes in the haemagglutinin molecule (antigenic shift), the peptide composition of both polypeptides HA1 and HA2 change radically, while in minor antigenic changes (drift) only small modification of the peptide composition of these components is detected (Webster and Laver 1975). Substitution of even single amino acids in the molecule may be accompanied by changes in antigenic specificity. The three-dimensional positions of the variable antigenic sites on the distal tip of the haemagglutinin molecule and their structural identification with amino-acid sequences has been described (Wiley *et al.* 1981). Studies by Caton *et al.* (1982) have led to the identification of five antigenic regions in the haemag-

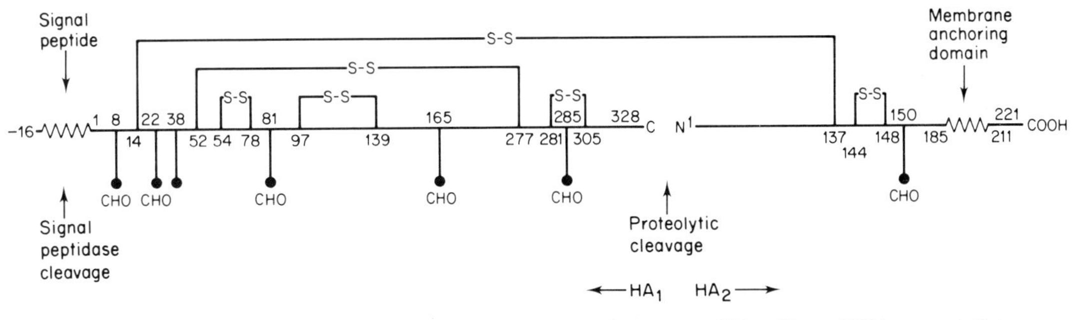

Fig. 96.4 Haemagglutinin precursor sequence. Schematic diagram of structure of Hong Kong (H3) haemagglutinin precursor protein showing positions of disulphide bridges (S-S), carbohydrate side chains (CHO) membrane anchoring domain, signal peptide and HA1–HA2 cleavage site. The numbers represent amino-acid residues of HA1 and HA2 from N-terminal ends of HA1 and HA2 respectively. (Modified from Wilson, Skehel and Wiley, 1981.)

glutinin molecule which have been precisely located within the three-dimensional configuration of the molecule.

The neuraminidase

Less detail is known of the structure and antigenic characters of neuraminidase than of haemagglutinin. The enzymic action of neuraminidase is the hydrolysis of N-acetyl neuraminic acid (sialic acid) residues from specific glycoprotein substrates. Such glycoproteins form specific receptor sites for haemagglutinin on the surface of host cells or erythrocytes. Though there is evidence that neuraminidase may have a function in mediating the release of virus from the host cell surface receptors after replication (Gottschalk 1966), its precise role in virus replication is unknown. It has also been suggested that viral neuraminidase may play a role in maturation of the virus particle (Klenk, Compans and Choppin 1970). The neuraminidase protein exists in the form of mushroom-shaped projections on the virus surface, these being fewer than the haemagglutinin spikes.

The chemical structure of neuraminidase has been reviewed by Bucher and Palese (1975). Purified neuraminidase can be isolated from certain strains of influenza virus by the electrophoretic separation of viral components after disruption of the virus by sodium dodecyl sulphate detergent (Laver and Webster 1976), or by treating virus with proteolytic enzymes such as trypsin (Noll *et al.* 1962, Wrigley *et al.* 1973). In the presence of detergent these are 8 nm long and 4 nm wide with a centrally attached fibre some 10 nm in length. On the removal of detergent the neuraminidase subunits become aggregated, joining together by the tips of their fibres to form characteristic 'rosette' structures (Laver and Valentine 1969, Wrigley 1979). The neuraminidase molecule, like haemagglutinin, appears to possess a hydrophobic end, the tip of the fibre, by which it is anchored to the lipid membrane of the virus and a hydrophilic end, the 'head' of the subunit, which contains the enzymically active site and antigenic determinants. The neuraminidase molecule is orientated with its N-terminus anchored in the viral membrane. The three-dimensional structure of influenza neuraminidase has been established and the positions on the molecule of the catalytic and antigenic sites identified (Varghese *et al.* 1983, Coleman *et al.* 1983)

The neuraminidase is antigenic and like the haemagglutinin is capable of undergoing antigenic variation leading to complete changes in antigenic structure, antigenic 'shift' and antigenic 'drift', but independently of changes in the haemagglutinin. The antigenic character of the neuraminidase, like the haemagglutinin, is taken into account in the classification of influenza A viruses as described below (Table 96.6).

The nucleoprotein and matrix protein

The nucleoprotein in the influenza virus particle is closely associated with the RNA genome of the virus and forms a RNA-protein complex, the nucleocapsid (Pons 1975). Preparations of this complex contain approximately 10 per cent RNA and 90 per cent protein (Pons *et al.* 1969). Electron microscopy of these complexes (reviewed by Schulze 1973) shows them as co-polymers of RNA, and protein subunits arranged in a complex manner, the molecules being turned back on themselves and twisted. The structures show terminal loops at one or both ends and a helical double-stranded mid-section. They probably represent single RNA molecules with nucleoprotein subunits attached along their lengths. In thin sections of virus particles, coiled structures of the same dimensions as those of isolated nucleoprotein may be observed (Nermut 1972).

Skehel and Schild (1971) have estimated that there are approximately 1000 molecules of nucleoprotein per influenza virus particle.

The 'core membrane' (Nermut 1972) consists of the major internal protein component of the influenza virus, the matrix protein. The probable association of this protein with the inner surface of the lipid bilayer and its abundance in the particle has led to its designation as 'M', indicating major, membrane or matrix protein (Kilbourne *et al.* 1972*a*). The matrix protein is the most abundant species of protein in the virus particle—about 33 per cent of the total protein of the virus. There are approximately 3000 molecules per virus particle (Skehel and Schild 1971). The matrix protein supports the outer membrane of the virus and determines the shape of the virion while also enclosing the nucleoprotein complex.

Little is known of the antigenic determinants of matrix protein, but its type specific nature for influenza A and B viruses has been established (Schild 1972).

Antigenic properties of virus components

The antigenic composition and properties of the influenza A virus are probably the most fully understood of all animal viruses. The virus particle contains seven virus-coded components of which four have been identified as antigens. These are the two antigenically variable envelope proteins, haemagglutinin and neuraminidase, and the two major internal proteins, the nucleoprotein and the matrix protein, which is the most abundant antigen of the virus particle. A complete picture of the host response to infection or immunization with influenza antigens should include studies of antibody directed against each of these antigens. However, since antibody to the two surface antigens is associated with immunity and since it is these antigens which undergo extensive antigenic variation, they have been more fully studied than the internal antigens.

Antibody to haemagglutinin

Antibody to haemagglutinin is an important part of the immune response to infection or immunization. Such antibodies provide a useful index of immunity to infection in man (Schild and Dowdle 1975, Potter and Oxford 1979, Tyrrell and Smith 1979) and also play a major functional role in immunity as indicated by studies on the induction of immunity by the passive administration of anti-haemagglutinin antibody in experimental animals (Virelizier 1975).

Antibody to haemagglutinin is effective in neutralizing virus infectivity. Virus-neutralization tests are readily performed by pre-incubating antisera and known challenge doses of influenza virus and subsequently inoculating the residual virus into susceptible cultures. This test is highly strain-specific and largely dependent upon antibody to haemagglutinin, particularly if excess antibody is removed from cultures shortly after inoculation (Schild and Dowdle 1975).

A further property of specific antibody to haemagglutinin is haemagglutination-inhibition. Inhibition of the agglutination of erythrocytes is brought about by attachment of antibody at or near the biologically active site of the haemagglutinin molecule, presumably in the region of the tip of the 'spike' (Wrigley 1979). The standard haemagglutination-inhibition assay test currently in routine use is based on that described by Salk (1944). The test, although largely specific to antibody to haemagglutinin, is subject to a number of variable factors (Schild and Dowdle 1975) including the presence in sera of several species of non-specific inhibitors of haemagglutination, interference by anti-neuraminidase antibody, and variability amongst the test strains of virus in their avidity for erythrocytes and antibody. Amongst non-specific inhibitors are the Francis (or alpha) inhibitor, the Chu (or beta) inhibitor and the gamma inhibitor. Although alpha and beta inhibitors are found in human sera, gamma inhibitors are restricted to animal sera (horse or guinea-pig). These non-specific inhibitors all appear to be glycoproteins in the alpha 2-macroglobulin fraction. The treatment of sera to remove non-specific inhibitors is discussed in detail elsewhere (Schild and Dowdle 1975). Anti-neuraminidase antibody may inhibit haemagglutination, presumably owing to steric interference with the biological activity of haemagglutinin at the virus surface (Schulman and Kilbourne 1969). This effect may be eliminated by the use of appropriate recombinant viruses possessing the haemagglutinin antigen against which it is desired to assay antibody, but an 'irrelevant' neuraminidase antigen against which there is no antibody in the serum samples under test. In addition, antibody to the so-called 'host' antigen of influenza viruses (Haukenes, Harboe and Martenssen-Egnund 1965) which may be present in the sera of animals hyperimmunized with egg-grown virus, is effective in inhibiting haemagglutination. Such antibodies, however, are not present in post-infection sera.

In more recent years further tests for antibody to haemagglutinin have been employed (Schild and Dowdle 1975). The immuno-double-diffusion test provides a useful method of carrying out qualitative comparisons of the specificities of haemagglutinin antigens and their corresponding antibodies (Schild *et al.* 1980). In addition, single-radial-diffusion tests (Schild, Aymard-Henry and Pereira 1972*a*) may be used for the specific assay of anti-haemagglutinin antibody and for detailed antigenic comparisons of the haemagglutinins of virus isolates. A further method is based on the haemolysis of influenza virus-treated erythrocytes mediated by specific anti-haemagglutinin antibody and complement (Schild, Pereira and Chakraverty 1975). This assay system, single-radial-haemolysis, provides a sensitive, rapid and reliable method for the assay of antibody to haemagglutinin which is not affected by non-specific inhibitors and does not require dilution of the test sera.

The haemagglutinin molecule is antigenically complex and bears at least two sets of antigenic determinants (reviewed by Schild 1979). One set of determinants is antigenically 'common' within a single subtype of haemagglutinin and the antibody corresponding to these common determinants reacts broadly with haemagglutinin antigens of all members of the subtype irrespective of antigenic 'drift'. This population of anti-haemagglutinin molecules has been termed 'cross-reactive' (CR). In addition, a further population of anti-haemagglutinin antibody is relatively strain-specific (SS), reacting only with the haemagglutinin antigens of the homologous or very closely related virus strains but not with those of antigenic variants showing considerable 'antigenic drift' from the homologous virus. Antibody to both determinants is known to attach to haemagglutinin in a region just behind the proximal tip (Wrigley, Laver and Downie 1977).

In mouse protection studies Virelizier (1975) found that CR antibody molecules were only weakly protective against influenza infection when passively administered to mice, while SS antibodies conferred complete protection. Other studies (Haaheim and Schild 1979) have shown that CR and SS antibodies differ considerably in their capacity to neutralize virus infectivity. In these investigations, IgG molecules of CR specificity were found to be ten- to a thousand-fold less effective in virus neutralizing activity than was strain-specific IgG. It was shown by Virelizier *et al.* (1974) that antibody formation in mice to both CR and SS determinants was thymus-dependent requiring the helper effect of T-lymphocytes. However, the thymus dependence of SS antibody production was more pronounced than that of the CR propulation, since production of CR, but not of SS antibody, could be obtained after repeated immunization of thymectomized, x-irradiated, bone-marrow reconstituted (TXBM) mice. These studies indicated an indirect role of T-lymphocytes in antibody response to influenza virus haemagglutinin. In contrast to the thymus dependence of antibody formation, the development of antigenic memory for haemagglutinin antigens in mice was shown to be thymus-independent (Virelizier *et al.* 1974, Virelizier, Allison and Schild 1974). In mice, T-lymphocytes appear to be essential in providing a 'triggering' mechanism to enable the transformation of primed haemagglutinin antigen-primed B-lymphocytes into antibody secreting cells.

The term 'original antigenic sin' has been used (Davenport, Hennessy and Francis 1953) to describe the phenomenon which occurs when sequential infection of a host occurs with two different but antigenically related strains of influenza A virus. The anti-haemagglutinin antibodies produced as a result of the second ('reinforcing') infection react more readily with the virus encountered in the first ('priming') infection than with the 'reinforcing' virus which stimulated the response. Analysis of this system in mice (Virelizier, Allison and Schild 1974) showed that when mice were 'primed' by immunization with the haemagglutinin antigen of one virus and given a reinforcing dose of antigenically distantly related haemagglutinin of the same subtype (H1), the secondary antibody response led to an increase of CR antibodies reacting with the haemagglutinin antigens of both virus strains. In addition, SS antibodies directed against strain-specific haemagglutinin determinants of the first haemagglutinin were stimulated by the reinforcing injection. This was a paradoxical immune response, since SS antigenic determinants of the priming virus were not present in the 'boosting' antigen. Further studies indicated that the secondary response was dependent upon B-lymphocytes.

Since CR and SS antibodies differ in their protective properties as described above, this paradoxical 'recall' phenomenon may have important implications for immunity to influenza in man. In the interpandemic periods persons are sequentially infected with successive variants of influenza A virus which arise as a result of antigenic 'drift'. The 'priming' effect of infection with the first encountered member of a series of antigenic variants may 'condition' the immunological system so that when later variants are encountered, the antibody response includes production of anti-haemagglutinin antibody of CR specificity and, in addition, SS antibody directed against the SS determinants of the haemagglutinin of the first strain. This latter antibody would be redundant in its protective effect. Evidence that this actually occurs in man after infection or immunization with inactivated influenza vaccines has been obtained (Schild *et al.* 1977, Oxford, Schild and Jennings 1979). After immunization with inactivated A/Port Chalmers/73 (H3N2) vaccines in 1974 the anti-haemagglutinin antibody elicited led to the production of CR antibodies in most recipients. Many persons developed SS antibody but this was more frequently directed against the SS determinant of A/Hong Kong/68 haemagglutinin than against the haemagglutinin of the vaccine virus A/Port Chalmers/73. Thus natural infection and immunization may often induce antibody responses which are 'outdated' in terms of immunity to the virus strain inducing them; this may, in part, explain the epidemiological success of influenza viruses and also the incomplete immunity (Stuart-Harris and Schild 1975) induced by inactivated vaccines.

Antibody to neuraminidase

Antibody to neuraminidase is frequently stimulated by natural infection or immunization with influenza virus in man and experimental animals and there is clear evidence that this antibody plays a significant role in protection (Schulman, Khakpour and Kilbourne 1968, Schild and Dowdle 1975, Potter and Oxford 1979, Tyrrell and Smith 1979). Unlike anti-haemagglutinin, antibody to neuraminidase does not directly neutralize virus infectivity unless present in very high concentrations. However, it does inhibit neuraminidase enzyme activity under certain conditions and is routinely assayed by neuraminidase-inhibition tests using fetuin as substrate (Aymard-Henry *et al.* 1973, Schild and Dowdle 1975). Although antibody to neuraminidase completely inhibits the activity of the enzyme when high molecular weight substrates (40 000 or greater molecular weight) are used, e.g. fetuin, there is little inhibition of enzyme activity for small substrates, such as sialo-lactose, with molecular weights of a few hundred daltons. Thus it is concluded that neuraminidase inhibition is due to 'steric inhibition' of enzyme activity by antibody, that the antibody does not combine directly with the enzymic site, and that the antigenic site of the molecule is not identical with the enzymic site.

A further property of antibody to neuraminidase is its ability to inhibit the spread of virus from cell to cell giving rise to apparent neutralization. This has been clearly demonstrated by the fact that anti-neuraminidase antibody reduced the size of influenza virus plaques in cell cultures (Jahiel and Kilbourne 1966) but not the number of plaques. Dowdle, Downie and Laver (1974) have shown that antibody to neuraminidase is capable of inhibiting the virus replication in egg-bit cultures provided it is present during the whole period of virus growth. However, such inhibition occurred even though the enzymic effect of the viral neuraminidase on substrates at the surface of the host cell was not inhibited. This finding suggests that the apparent inhibitory effects of anti-neuraminidase antibody on virus growth do not depend upon inhibition of enzyme activity. The mechanism of antibody in restricting virus replication is probably based on its ability to form immunological complexes resulting in the binding of neuraminidase sub-units, inserted in the plasma membranes of infected cells, to neuraminidase on budding virus particles, thus preventing the release of newly formed virus particles into the medium.

In the experimentally infected mouse (Schulman *et al.* 1968) there is evidence that antibody to neuraminidase restricts the dissemination of virus and correspondingly the severity of virus pneumonia. The degree of shedding and transmission of virus from the upper respiratory tract is also lowered. Similarly in human volunteers immunized with inactivated, neuraminidase specific antigens (Couch *et al.* 1974, Kilbourne 1976) susceptibility to infection by challenge was not affected but shedding of virus from the respiratory tract was reduced. Kilbourne *et al.* (1972*a*) have suggested the use of neuraminidase specific vaccines to control influenza in man. Such vaccines might have the advantage of reducing the severity of illness and restricting transmission of virus, but not blocking infection and consequently the development of natural immunity.

Antibody to nucleoprotein and matrix protein

The nucleoprotein is antigenically relatively stable and provides the basis for the classification of influenza viruses into types A, B and C. Antibody to influenza nucleoprotein may be detected by complement-fixation tests (Lief and Henle 1959) or by single-radial-immunodiffusion tests with disrupted virus particles (Schild, Aymard-Henry and Pereira 1972*a*), and is of valuable epidemiological and diagnostic significance in providing evidence of past exposure to infection. Antibody to nucleoprotein does not neutralize virus infectivity or inhibit virus release from infected cells, and there is no evidence that it is associated with immunity. Experimental animals with high levels of antibody to nucleoprotein induced by immunization with highly purified nucleoprotein antigen are fully susceptible to influenza infection (Oxford and Schild 1976).

Minor antigenic differences have been detected among the nucleoprotein antigens of influenza A viruses by means of immuno-double-diffusion tests with potent antisera prepared against purified antigens (Schild, Oxford and Newman 1979). The antigens of early human influenza A strains isolated before 1940 were distinguishable from those of strains isolated from 1943 onwards. In addition, avian influenza A viruses contained nucleoprotein antigens which were homogeneous amongst all avian viruses isolated between the early 1900s and 1980 but clearly distinguishable from those of all human influenza A viruses. Antigenic differences in nucleoprotein antigens of different influenza A virus strains have also been detected by using monoclonal antibodies (van Wyke *et al.* 1980).

The matrix protein is also a type specific antigen of the influenza virus (Schild 1972). The antigen is common to all strains of influenza A virus but differs from that of influenza B viruses. The matrix protein is thus a second type-specific internal antigen of the virus. Potent antibody preparations specific for influenza matrix protein do not neutralize virus infectivity or other biological properties of the virus and the antibody is probably not associated with immunity. Experimental animals possessing high titres of antibody to matrix protein antigen are readily infected with influenza viruses (Oxford and Schild 1976). Although antibody to matrix protein is infrequently detected in convalescent human serum after influenza A infection, the individuals in whom such antibody is detected are generally those who suffered severe clinical illness (Mostow *et al.* 1975).

Cellular immunity

The role of cell-mediated immune factors in protection against influenza is uncertain. Most information derives from studies in experimental mice. Immunization of mice with influenza virus results in the development of cytotoxic T lymphocytes which are able to destroy 'target' cells infected with influenza virus. The studies of Doherty, Effros and Bennink (1977) indicated that the cytotoxic T lymphocytes were of wide specificity reacting with target cells infected with any influenza A virus, but not with influenza B and were restricted functionally by lymphocyte cell surface antigens (HLA). In contrast, Ennis, Martin and Verbonitz (1977) found evidence of cytotoxic T lymphocytes which were haemagglutinin subtype specific. The studies of Ennis *et al.* (1981) suggest that such T-cell responses contribute to the recovery of mice from pneumonia. The differences in specificity detected in different studies may reflect differences in the nature of the influenza antigens exposed on the surfaces of the various target cells used in the studies. Cytotoxic T-cell responses also occur in man. Ennis *et al.* (1981) have detected HLA restricted influenza virus specific cytotoxic T-cell responses in a high proportion of human volunteers immunized with whole virus or purified surface antigen vaccines or infected with attenuated H1N1 influenza viruses.

Whether such T-cell responses in man contribute to the protective effects of vaccination by live or inactivated vaccines is not known. However, there are a number of instances where protection in man or experimental animals cannot be readily attributed to circulating antibody to the haemagglutinin or neuraminidase. Firstly, mice previously infected with one subtype of influenza A virus show partial immunity to challenge with another influenza A subtype bearing entirely different surface antigens (Schulman and Kilbourne 1965, Werner 1966). Secondly, in some studies, live influenza vaccines have appeared to confer significant protection against homologous or closely related influenza A viruses even when levels of circulating antibody to haemagglutinin or neuraminidase were low or undetectable. Thirdly, the H1N1 virus when it reappeared in 1977 to cause epidemics (see page 336) did not infect persons over the age of 23 years or so. Older individuals appear to have been protected because of infection with H1N1 viruses some 20–30 years previously, but in many such older individuals antibody to the surface antigens of the 1979 H1N1 are absent or at levels which are undetectable by conventional techniques. Thus it seems possible that mechanisms, other than the presence of circulating antibody to haemagglutinin or neuraminidase, may contribute to immunity to influenza infections. Whether immunological memory phenomena, such as that described for antibody production in mice by Virelizier *et al.* (1974), or cellular immune responses contribute to such unexplained protection remains to be established.

Classification and nomenclature of influenza viruses

The influenza viruses are classified into types A, B and C on the basis of the antigenic character of their nucleoprotein antigens. Influenza type A viruses are further divided into subtypes based on the antigenic character of their haemagglutinin and neuraminidase antigens. In the system of nomenclature used from 1971 to 1980 (WHO 1971) the haemagglutinin antigen subtypes of human influenza A viruses were designated H0, H1, H2 and H3, representative of the human influenza A viruses prevalent from 1932–46 (H0N1), 1947–1957 (H1N1), 1957–68 (H2N2, 'Asian') and 1968 to the present time (1984) (H3N2, 'Hong Kong'). One haemagglutinin subtype of swine influenza virus (Hsw1), two subtypes of equine virus (Heq1 and Heq2) and eight haemagglutinin subtypes of avian influenza virus (Hav1 to Hav8) were designated (Table 96.8). Since 1971, however, it has become apparent that antigenic relations exist between certain viruses which were classified into different haemagglutinin antigen subtypes (Schild 1970, Baker *et al.* 1973, Schild *et al.* 1980). These have been demonstrated by a variety of methods but particularly by immunodouble-diffusion tests in gels with potent specific antisera to isolated haemagglutinin antigens (Schild and Dowdle 1975). Two additional haemagglutinin antigen subtypes of avian influenza A viruses are now recognized (Webster *et al.* 1976, Hinshaw and Webster 1979). (See the revised nomenclatural system (WHO 1979, 1980) which is followed in Tables 96.5–96.11.) The neuraminidase antigens were likewise divided into antigenic subtypes employing neuraminidase-inhibition tests (Aymard-Henry *et al.* 1973) and immunodouble-diffusion tests with neuraminidase specific sera. Among human influenza A viruses there were two neuraminidase antigen subtypes, N1 and N2, represented by the viruses prevalent from 1932–56 H1N1 and from 1957 to the present (1984) H2N2 and H3N2 (Table 96.5). Among swine influenza viruses there was one subtype (N1), closely related to the human N1 subtype (Table 96.6). Among equine influenza viruses two distinct N antigen subtypes, designated Neq1 and Neq2, were described (Table 96.7). For avian influenza A strains there were eight subtypes of NA antigen; two of these (N1 and N2) were shared with human influenza A viruses, two (Neq1, Neq2)

Table 96.5 Reference strains for the subtypes of haemagglutinin and neuraminidase antigens of influenza A viruses isolated from man. Revised nomenclature WHO 1980)

H and N subtypes	Reference strains
H1N1	A/PR/8/34 (H1N1)
	A/Weiss/43 (H1N1)
	A/FM1/47 (H1N1)
	A/England/1/51 (H1N1)
	A/Denver/1/57 (H1N1)
	A/New Jersey/8/76 (H1N1)
	A/USSR/90/77 (H1N1)
H2N2	A/Singapore/1/57 (H2N2)
	A/Japan/305/57 (H2N2)
	A/England/12/64 (H2N2)
	A/Tokyo/3/67 (H2N2)
H3N2	A/Hong Kong/1/68 (H3N2)
	A/England/42/72 (H3N2)
	A/Port Chalmers/1/73 (H3N2)
	A/Victoria/3/75 (H3N2)
	A/Texas/1/77 (H3N2)

Table 96.6 Reference strains for the subtypes of haemagglutinin and neuraminidase antigens of influenza A viruses isolated from swine (WHO 1980)

H and N subtypes	Reference strains
H1N1	A/Swine/Iowa/15/30 (H1N1)
	A/Swine/Wisconsin/67 (H1N1)
H3N2	A/Swine/Taiwan/1/70 (H3N2)

Table 96.7 Reference strains for subtypes of haemagglutinin and neuraminidase antigens of influenza A viruses isolated from horses (WHO 1980)

H and N subtypes	Reference strains
H7N7	A/equine/Prague/1/56 (H7N7)
H3N8	A/equine/Miami/1/63 (H3N8)

with equine viruses and four subtypes (Table 96.9) (Nav1 to Nav4) were confined to viruses of avian origin. Since 1971 evidence for two additional neuraminidase subtypes among avian subtypes has become available (Downie and Laver 1973, Webster *et al.* 1976). (Table 96.9.)

A comprehensive antigenic analysis of a large collection of prototype strains of influenza A virus of human, swine, equine and avian origin has recently been completed (Schild *et al.* 1980) employing precipitin tests with haemagglutinin and neuraminidase specific antisera. This and other recent information has permitted re-evaluation of the nomenclature of the influenza A viruses (WHO 1979). In a revised system adopted for use in 1980 (WHO 1980) several of the subtypes described in the 1971 system have been merged, resulting in 12 subtypes of haemagglutinin antigen and 9 subtypes of neuraminidase antigen.

Other considerations for taxonomy

Antigenic similarities between the surface antigens of influenza A viruses of different subtypes have been observed even in the absence of demonstrable serological cross-reactions. These include evidence of relations, based on cross-protection, antigenic memory or cell-mediated immunity (Doherty *et al.* 1977). How-

Table 96.8 Reference strains for subtypes of haemagglutinin antigens of influenza A viruses isolated from avian species (WHO 1980)

New subtype designation	Previous subtype designation (WHO 1971)	Reference strains	Other strains and related antigens
H1	Hsw1	A/duck/Alberta/35/76 (H1N1)	A/duck/Alberta/97/77 (H1N8)
H2	H2	A/duck/Germany/1215/73 (H2N3)	A/duck/Germany/1/72 (H2N9)
H3	Hav7	A/duck/Ukraine/1/63 (H3N8)	A/duck/England/62 (H3N8) A/turkey/England/69 (H3N2)
H4	Hav4	A/duck/Czechoslovakia/56 (H4N6)	A/duck/Alberta/300/77 (H4N3)
H5	Hav5	A/tern/South Africa/61 (H5N3)	A/turkey/Ontario/7732/66 (H5N9) A/chick/Scotland/59 (H5N1)
H6	Hav6	A/turkey/Massachusetts/3740/65 (H6N2)	A/turkey/Canada/63 (H6N8) A/shearwater/Australia/72 (H6N5) A/duck/Germany/1868/68 (H6N1)
H7	Hav1	A/fowl plague virus/Dutch/27 (H7N7)	A/chick/Brescia/1902 (H7N1) A/turkey/England/63 (H7N3) A/fowl plague virus/Rostock/34 (H7N1)
H8	Hav8	A/turkey/Ontario/6118/68 (H8N4)	—
H9	Hav9	A/turkey/Wisconsin/1/66 (H9N2)	A/duck/Hong Kong/147/77 (H9N6)
H10	Hav2	A/chick/Germany/N/49 (H10N7)	A/quail/Italy/1117/65 (H10N8)
H11	Hav3	A/duck/England/56 (H11N6)	A/duck/Memphis/546/74 (H11N9)
H12	Hav10	A/duck/Alberta/60/76 (H12N5)	—

Table 96.9 Reference strains for subtypes of neuraminidase antigens of influenza A viruses isolated from avian species (WHO 1980)

New subtype designation	Previous grouping (WHO 1971)	Reference strains	Other strains with related N antigens
N1	N1	A/chick/Scotland/59 (H5N1)	A/duck/Alberta/35/76 (H1N1) A/duck/Germany/1868/68 (H6N1)
N2	N2	A/turkey/Massachusetts/3740/65 (H6N2)	A/turkey/Wisconsin/66 (H9N2) A/turkey/England/69 (H3N2)
N3	Nav2 Nav3	A/tern/South Africa/61 (H5N3) A/turkey/England/63 (H7N3)	A/duck/Germany/1215/73 (H2N3)
N4	Nav4	A/turkey/Ontario/6118/68 (H8N4)	A/duck/Wisconsin/6/74 (H6N4)
N5	Nav5	A/shearwater/Australia/1/72 (H6N5)	A/duck/Alberta/60/76 (H12N5)
N6	Nav1	A/duck/Czechoslovakia/56 (H4N6) A/duck/England/56 (H11N6)	—
N7	Neq1	A/fowl plague virus/Dutch/27 (H7N7)	A/chick/Germany/N/49 (H10N7)
N8	Neq2	A/quail/Italy/1117/65 (H10N8)	A/turkey/Canada/63 (H6N8) A/duck/England/62 (H3N8)
N9	Nav6	A/duck/Memphis/546/74 (H11N9)	A/turkey/Ontario/7732/66 (H5N9)

ever these relations, where they are not supported by directly demonstrated serological cross-reactions between the surface antigens such as those detected in immunoprecipitation tests have not been considered relevant to the subtyping of virus strains.

The RNAs of several influenza virus strains and their recombinants have been characterized by various methods including RNA-RNA hybridization (Scholtissek 1978) and RNA-DNA hybridization and oligonucleotide analysis (Young and Palese 1979). Results obtained using RNA-RNA hybridization techniques have indicated that the genes coding for the haemagglutinin antigens of viruses previously designated (WHO 1971) Hsw1, H0, and H1 subtypes were closely related, thus supporting the inclusion of these antigens in a single subtype H1 in the revised (WHO 1980) nomenclatural system. The genes coding for the HA antigens of viruses previously designated H3, Hav7 and Heq2 subtypes in the 1971 system also exhibited a high base-sequence homology, as did those coding for Heq1 and Hav1. These antigens have likewise been allocated to common subtypes, H3 and H7 respectively. The results of the analyses of genes coding for the Nav2 and Nav3 proteins supported the inclusion of these antigens in a single group. Based on similar analyses Nav6 appeared to constitute a separate subtype (Scholtissek 1978).

Knowledge of the comparative primary structure of influenza virus proteins, obtained by tryptic peptide, nucleic acid and amino-acid sequence analyses of influenza viruses, is at present mainly restricted to the haemagglutinin molecule. The results of comparative tryptic peptide analyses, particularly of the HA2 components of the haemagglutinins so far obtained, are consistent with the revised (WHO 1980) nomenclatural system. The nucleotide sequences that have been determined indicate that the haemagglutinins within a subtype are much more closely related to each other than to those of other subtypes (Porter *et al.* 1979, Min Jou *et al.* 1980, Gething *et al.* 1980, Sleigh *et al.* 1980). Similar data supporting the 1980 sub-classification system for neuraminidase antigens have been provided by partial sequences of the genes coding for neuraminidase antigens of influenza viruses from human and non-human sources (Blok and Air 1980). However, an anomalous finding has been the detection of a higher degree of homology between the amino-acid sequences and nucleotide sequences of the haemagglutinins of subtypes H1 and H2 than between other apparently unrelated haemmagglutinin subtypes (Winter *et al.* 1981, Hiti *et al.* 1981). The homologies determined between a number of haemagglutinin subtypes are shown in Table 96.4. Comparisons of the sequences of the genes coding for the matrix and NS proteins of human influenza A viruses isolated from 1940 to 1972 indicate that these genes are well conserved (Hall and Air 1981).

The 1980 classification system

A scientifically meaningful system for the classification and nomenclature of influenza viruses is important for the description and understanding of epidemiological and immunological phenomena (WHO 1979). A full description of the revised system of nomenclature is therefore not out of place in this chapter.

In the revised system of nomenclature recommended by WHO for use from 1980, the strain designation for influenza viruses contains the following information:

1 A description of the antigenic type of the virus based on the antigenic specificity of the NP antigen (type A, B, or C). Since 1971, a further type-specific internal antigen of the influenza A and B viruses, the matrix (M) protein, has been described (Schild 1972). Typing of influenza A and B viruses based on the M protein is consistent with the results obtained with NP antigen.

2 The host of origin is not indicated for strains isolated from human sources, but is indicated for all strains isolated from non-human hosts, e.g. swine, horse (equine), chicken, turkey. For viruses from non-human species, both the Latin binomial nomenclature and the common name of the host of origin should be recorded in the original publication describing the virus isolate, e.g. *Anas acuta* (pintail duck). Thereafter, the common name of the species should be used for the strain, e.g. A/duck/USSR/695/76 (H2N3). When viruses are isolated from non-living material the nature of the material should be specified, e.g. A/lake water/Wisconsin/1/79.

3 Geographical origin

4 Strain number

5 Year of isolation

For influenza A viruses the antigenic description includes:

(a) An index describing the antigenic character of the haemagglutinin, i.e. H1, H2, H3, H4 etc. The numbering of subtypes is a simple sequential system which applies uniformly to influenza viruses from all sources (Table 96.10).

Table 96.10 Subtypes of haemagglutinin antigens of influenza A viruses

Proposed subtypes	Previous subtypes (1971 system)
H1	H0, H1, Hsw1
H2	H2
H3	H3, Heq2, Hav7
H4	Hav4
H5	Hav5
H6	Hav6
H7	Heq1, Hav1
H8	Hav8
H9	Hav9
H10	Hav2
H11	Hav3
H12	Hav10

(b) An index describing the antigenic character of the neuraminidase, i.e., N1, N2, N3, N4, etc. applied uniformly to all influenza A viruses (Table 96.11).

Table 96.11 Subtypes of neuraminidase antigens of influenza A viruses

Proposed subtypes	Previous subtypes (1971 system)
N1	N1
N2	N2
N3	Nav2, Nav3
N4	Nav4
N5	Nav5
N6	Nav1
N7	Neq1
N8	Neq2
N9	Nav6

It is implicit that a given HA or NA subtype designation will include strains exhibiting some antigenic variation within the subtype (antigenic 'drift'). The exact antigenic character of an influenza virus variant may be defined by indicating similarities to designated reference strains.

The 1980 nomenclature system does not provide for the description of distinct subtypes of influenza B and C viruses because of the lack of clear-cut antigenic distinctiveness between members of these types. The description of these viruses is therefore limited to strain designation, e.g. B/England/5/66, C/Paris/1/67.

Examples of reference strains of influenza A viruses of human, swine, equine and avian influenza viruses and their new subtype designations are given in Tables 96.5–9.

The 1980 nomenclature system was not designed to provide information on the host range or virulence of influenza viruses. The isolation from different hosts of antigenically similar influenza A viruses is well established and examples in which one subtype of the surface antigens has been found in influenza viruses from different species are numerous (Hinshaw *et al.* 1979). Representatives of each of the neuraminidase antigenic subtypes of influenza viruses from man, pigs and horses (N1, N2, N7 and N8) have also been isolated from birds. Similarly representatives of each of the haemagglutinin subtypes of man, pigs and horses (H1, H2, H3 and H7) have been isolated from avian species.

Clinical manifestations of influenza in man

The influenza virus is constantly changing in its antigenic characters but the clinical disease it produces may have remained relatively constant over several centuries according to some of the earliest descriptions of the disease. Thus, influenza has been described as 'an unchanging disease due to a changing virus' (Kilbourne 1975, 1980).

The clinical aspects of influenza have been reviewed by Douglas (1975) and Stuart-Harris and Schild (1976). In uncomplicated influenza the commencement of symptoms occurs fairly suddenly some 2–3 days after infection. The first symptoms are headache, shivering and a dry cough accompanied by a sudden onset of fever. Malaise and aching of the limb muscles and back may occur especially in the adult patient. In some patients, the symptoms may subside rapidly after the first 24 hours of illness. In others, the disease takes a more prolonged course, the temperature remaining high for 2–5 days, and the patient having residual weakness and a cough for some days. In the absence of complications, the patient is usually sufficiently recovered to return to work within 7–10 days from the onset of illness. Fever, 38–40° (100–104°F), and sudden onset are the clinical features which most characteristically distinguish the disease from the common cold (see Chapter 97).

There are few physical signs in the patient enabling the clinician to make a firm diagnosis of influenza; the disease has no unique features distinguishing it from infections with other respiratory viruses. The diagnosis is made largely on the basis of the patient's symptoms in association with knowledge of epidemiological features of the outbreak in the community. Unequivocal identification of the disease is entirely dependent upon virus isolation or the detection of a diagnostic rise in specific influenza antibodies. In general, childhood attack rates are high. In the very young, under 4 years of age, convulsions, croup and vomiting and neck stiffness may occur accompanied sometimes by a prolonged high fever; influenza is undoubtedly an important cause of severe clinical illness in these age groups. Death is uncommon in the healthy child. In older children, the prevalent symptoms may be coughing and the onset of a croup-like illness. Muscular aching is a more prevalent feature of the disease in older children and adults. In major outbreaks, such as occurred in 1957 due to the Asian (H2N2) virus, a number of deaths have occurred in previously healthy pregnant women, particularly those in the last trimester, but there is no clear evidence that the virus has teratogenic effects resulting in the development of congenital defects in the babies of mothers infected during pregnancy.

The major complications of influenza affect the lower respiratory tract and the circulatory and nervous systems. Complications of the lower respiratory tract may be due directly to the virus, as in primary viral pneumonia, or to secondary bacterial infections. Bronchitis, bronchiolitis or pneumonia may occur in some 10 per cent of all adult influenza patients; the frequency of these complications is also higher in patients with a previous history of chronic pulmonary or cardiac disorder. Complications in the elderly contribute the greater part of the considerable mortality attributed to the disease in major epidemics. Most bacterial complications of influenza are due to the pneumococcus, though staphylococcal influenzal pneumonia does occur during epidemics and may, if untreated, result in extremely rapid death. Bacterial complications may frequently be amenable to treatment with antibiotics, but there is currently no routinely used specific antiviral drug to prevent or treat the primary infection. The lack of bacterial antibiotics in the times of earlier pandemics, e.g. 1918, may have contributed to their much higher mortality rates in contrast to those epidemics which occurred in the present era of routine use of antibiotics. Chronic heart disease itself is a predisposing factor in the development of severe complications of influenza, but influenza itself may produce various cardiac complications ranging from temporary changes in electrocardiogram patterns to inflammatory changes in the heart muscles (myocarditis). The pathogenesis of these phenomena is unknown. Reye's syndrome is an encephalopathy associated with liver degeneration which is thought to be associated with influenza B infection in children. This disease, though fortunately rare, is usually fatal. Diabetics who suffer from influenza usually exhibit a worsening of their metabolic state, and influenza deaths among diabetic patients are not uncommon. Although detailed comparisons of the clinical differences between the features of the disease in different countries have not been made, there is evidence in some areas that exceptionally high mortality rates are associated with influenza. For example during August 1975, an outbreak in Papua and New Guinea caused by A/Victoria/3/75 (H3N2) virus resulted in 400 deaths.

(For a fuller account of the clinical picture of influenza, see pp. 2501–3 of the 6th edition and Report 1979.)

International surveillance of influenza by the World Health Organization

In 1947, worldwide surveillance of influenza was initiated by WHO as part of the organization's influenza programme. At first, surveillance was on a modest scale with one international influenza centre in London and a few regional centres in other countries. The system has now grown into an extensive network of laboratories. The main purpose of the programme is the early detection and isolation of new antigenic variants of influenza virus and the collection of epidemiological data leading to a fuller understanding of the epidemiology and natural history of the disease. The early and accurate information available to the WHO on the behaviour of the virus provides the basis for recommendations to be given to national authorities enabling the preparation to be undertaken of appropriate public health and prophylactic measures for combating the disease (Schild and Dowdle 1975). The activities of the programme are coordinated by the WHO in Geneva. There are two WHO international influenza centres, in London and in Atlanta, Georgia, together with over 100 national influenza laboratories in 73 different countries.

When new influenza virus isolates showing significant antigenic change are detected, they are made available by the international centres to laboratories involved in research, epidemiological surveillance and vaccine production.

Epidemiology of influenza in man

Antigenic variation in epidemic strains—'drift' and 'shift' (*See also* Chapter 85)

The surface antigens of the influenza A virus prevalent in man undergo two types of antigenic change. Major antigenic changes, i.e. changes in antigen subtype termed antigenic 'shift', occur at infrequent intervals independently in the haemagglutinin and neuraminidase antigens. The emergence of the 'Asian' (H2N2) influenza virus in 1957 and its rapid replacement of the previously prevalent H1N1 virus is an example of simultaneous antigenic 'shift' in the haemagglutinin and neuraminidase antigens. In contrast, in 1968 the appearance of the 'Hong Kong' (H3N2) virus which quickly replaced the H2N2 virus is an example of antigenic shift occurring in the haemagglutinin alone. The subtypes of influenza A virus prevalent since 1918 are shown in Table 96.12. It is seen that the H1N1 subtype which probably emerged as a human pandemic virus in 1918 (see page 339) was prevalent for 39 years. In contrast the H2N2 virus was prevalent for only 10 years, 1957–68. The H3N2 virus which, at the time of writing this chapter is still circulating in man, has been prevalent for at least 13 years. Viruses containing N2 neuraminidase have been prevalent from 1957–81, a period of 25 years.

Antigenic 'shift' appears to be a requirement for the appearance of pandemic influenza, at least as far as the limited epidemiological experience of the present century indicates. Pandemic disease, however, is not a necessary result of the introduction of a new antigenic subtype of influenza A virus into man. The outbreaks associated with the appearance of swine (H1N1) influenza A virus in man in 1976 (see page 334) were limited in extent and time. An entirely unexpected event in the epidemiology of influenza was the reappearance in 1977, after an absence of some 20 years, of H1N1 influenza viruses antigenically and genetically close to the H1N1 strains which circulated in 1950. The H1N1 strain between 1977 and 1980

Table 96.12 Prevalence of antigenic subtypes of influenzia A virus in man, 1918–81

Antigenic designation: haemagglutinin (H) and neuraminidase (N) antigen subtype	Period of prevalence	Type of antigenic variation	Probable mechanism of origin of new virus
	1918	Shift	Unknown, but swine (H1N1) virus was probably epidemic in pigs in the USA as early as 1918
H1N1	1918–1957	progressive antigenic drift	
	1957	Shift	Genetic recombination between former human H1N1 virus and animal influenza A virus from animal or avian sources with surface antigens H2N2. Genes for H2N2 transferred to new pandemic virus
H2N2	1957–1968	progressive antigenic drift	
H3N2	1968	Shift	Genetic recombination between former human H2N2 virus and an influenza A virus from animal or avian sources with surface antigens H3N? Gene for H3 transferred to new pandemic virus
	1968–?	progressive antigenic drift	
H1N1	1977–?	reappearance of 1950 H1N1 virus	Derived from 1950 H1N1 virus, preserved in nature by unknown mechanism

caused widespread outbreaks of influenza in children and young adults in many countries (Pereira 1979) but did not produce illness in persons over the age of approximately 23 years who would have been expected to have been naturally infected with H1N1 virus during the earlier period of prevalence of this subtype up to 1956. The epidemiology of influenza in man since 1977 has been unusually complex with the prevalence in the world of two distinct subtypes of influenza A virus H1N1 and H3N2 together with influenza B virus.

At more frequent intervals minor changes, termed antigenic 'drift', take place in one or both surface antigens. Although each episode of 'drift' is in itself minor, the effects are additive and over a period of several years result in a virus showing a considerable degree of antigenic difference from the original pandemic virus. In antigenic 'shift' both 'common' and 'strain-specific' determinants (see page 323) of the haemagglutinin antigen undergo change, while in antigenic 'drift' only strain-specific determinants change, the 'common' determinants are conserved and enable the haemagglutinin to be identified as belonging to a specific subtype, e.g. H1, H2 or H3.

The viruses of subtypes H1N1, H2N2 and H3N2 underwent progressive antigenic 'drift' in their haemagglutinin and neuraminidase antigens during their periods of epidemic activity in man from 1918–56, 1957–67 and 1968–81 respectively (see Pereira 1979, 1980). Table 96.13 shows the results of representative haemagglutinin-inhibition tests on successive variants of influenza H3N2 virus isolated from 1968–79 which clearly show evidence of antigenic 'drift' in the haemagglutinin antigen. Changes somewhat less frequent than these occurred in the neuraminidase antigen in the same period.

Table 96.13 Antigenic drift of influenza A (H3N2) viruses between 1968 and 1979 (Haemagglutination-inhibition tests with post-infection ferret sera.)

Virus Strain	A/Hong Kong/1/68	A/England/42/72	A/Port Chalmers/1/73	A/Scotland/840/74	A/Victoria/3/75	A/Texas/1/77	A/Bangkok/1/79
A/Hong Kong/1/68	2560	2560	320	80	<20	<20	<20
A/England/42/72	80	1280	320	160	40	<20	<20
A/Port Chalmers/1/73	20	640	640	160	40	40	<20
A/Scotland/840/74	<20	40	40	640	<20	320	40
A/Victoria/3/75	<20	<20	<20	20	1280	320	160
A/Texas/1/77	<20	<20	<20	<20	80	1280	640
A/Bangkok/1/79	<20	<20	<20	<20	20	320	2560

Antigenic 'drift' occurs as a result of the natural selection of mutants under the selective pressure of increasing levels of immunity in the population. The mechanism of 'shift' remains unknown. It seems unlikely, however, that 'shift' is due to genetic mutation of the previously prevalent strain. There is circumstantial evidence (Laver and Webster 1979) that new subtypes of influenza A virus appearing in man may arise by genetic 'recombination' between different influenza A viruses of human and/or non-human origin resulting in a re-assortment of their genes (RNA segment) and the production of new viruses differing from the parental strains in their genetic, antigenic and biological properties and potential for epidemic spread. This subject is discussed in further detail below (see page 339).

In contrast to influenza A, influenza B viruses exist as a single subtype and the progressive antigenic changes which the prevalent strains have undergone since their first isolation in 1940 are attributed entirely to antigenic 'drift'.

The epidemiological impact of influenza A and B viruses

Since 1933 when the laboratory study of influenza virus was first established, recurrent epidemics of influenza A or influenza B infection have been identified in all countries where surveillance has been carried out.

In countries for which epidemiological information is available, years in which no influenza outbreaks are recorded are rare (Assaad *et al.* 1973). Most epidemics and all pandemics are due to the influenza A virus. Influenza A, perhaps most characteristically an infection of children and young adults (Stuart-Harris and Schild 1976), does affect all age groups of the population. However, the main impact on mortality is in the older age groups of the population, 70 years and over, and in the medically debilitated. In contrast, influenza B outbreaks are four to six times less frequent than those caused by influenza A virus, and are relatively localized, producing their main impact in children (Dowdle, Coleman and Gregg 1974, Stuart-Harris and Schild 1976). Raised mortality frequently accompanies influenza A outbreaks but is not a constant feature of influenza B epidemics. Figure 96.5 on page 335 shows data from the USA on the relative frequencies of influenza A and B epidemics between 1934 and 1975 and the increases in mortality rates due to influenza and pneumonia during epidemic periods. The greater frequency of outbreaks due to influenza A than influenza B epidemics is seen. Excess mortality associated with influenza A epidemics was of a considerably higher order than for influenza B, with the exception of 1935–6, when high mortality rates were apparently associated with influenza B epidemics.

The history of influenza epidemics and pandemics

From as early as the twelfth century onwards (Hirsch 1883, Creighton 1894) descriptions of extensive outbreaks of a disease resembling influenza appear in the literature. Severe outbreaks were reported in Europe in 1510, 1562, 1693 and 1729. Later outbreaks spreading around Europe were in 1781–2, 1803, 1833, 1837, 1847–8, 1889–90, 1898–1900 and 1918–19. The outbreaks of 1847–8 and 1889–90 may have been pandemic like those of 1918–19, 1957–8 and 1968–70. The outbreaks of 1781–2, 1847–8 and 1889–90 are known to have spread from Asia to Europe.

The unpredictability of the disease in mediaeval times, as now, is emphasized by the fact that its name is derived from the Italian 'influenze', suggesting that the disease appeared under the influence of extraterrestrial forces. During outbreaks of disease in the Court of Mary Queen of Scots in 1562, the disease earned the title of the 'Newe Acquayntance', with a similar significance.

Although the influenza virus was not isolated until 1933, there is circumstantial evidence, discussed below, that the major outbreaks of 1889–90, 1898–1900 and 1918–19 were due to this virus.

The 1918–19 pandemic

The years 1918–19 saw the most severe pandemic of influenza yet recorded, the so-called 'Spanish influenza' pandemic. The infection probably spread from Asia to Europe. The first wave in the spring of 1918 was relatively mild, the attack rate being 20–40 per cent in those aged up to 50 years but lower in the aged. The second, in the autumn of 1918, was exceptionally severe, producing enormous mortality in the 20–40-year age groups. A less severe wave occurred in early 1919. The total mortality for the pandemic is estimated at 15–50 million, probably about 20 million, deaths. Evidence (see Stuart-Harris and Schild 1976) suggests that this pandemic was caused by the classical swine (H1N1) influenza virus or a virus antigenically close to it. First, the initial appearance of swine influenza in pigs in the USA coincided with the onset of human pandemics. Second, antibody to the swine (H1N1) virus is present in a high proportion of persons who were alive in 1918.

A limited outbreak caused by a similar strain occurred in military recruits in New Jersey, USA, in January 1976 (Topp and Russell 1977). There was concern, in the event unjustified, that this outbreak heralded the onset of a new pandemic era; so in the winter 1976–7, 40 million doses of vaccine were administered to persons of all age groups in the USA (Neustadt and Fineberg 1978).

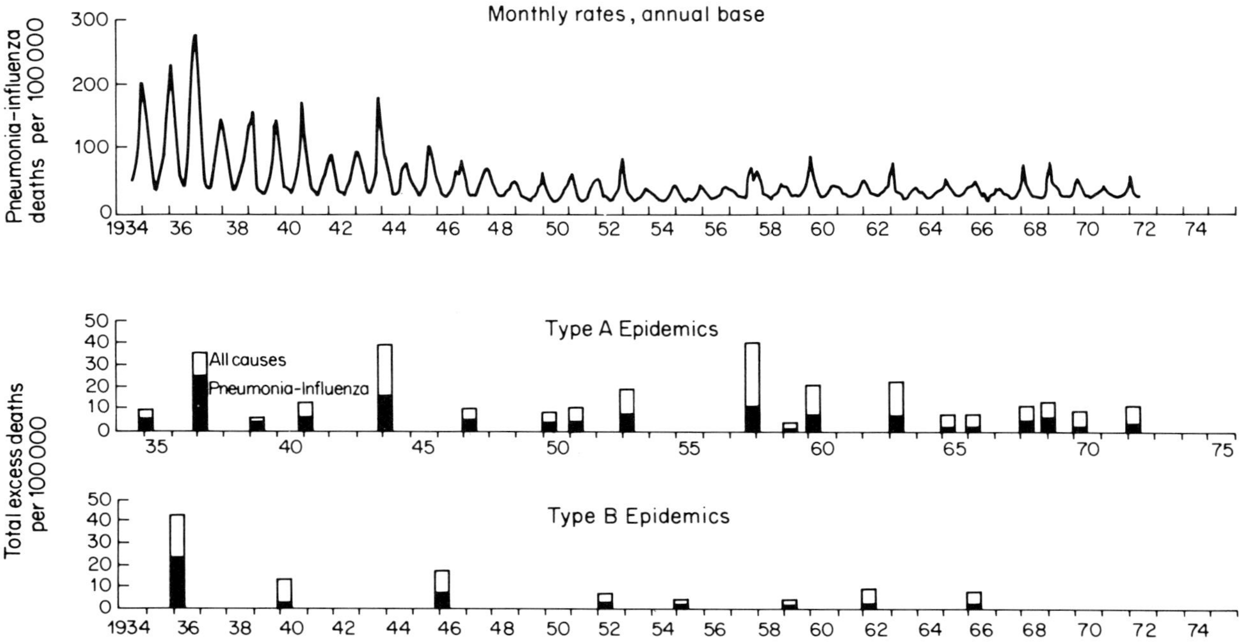

Fig. 96.5 Pneumonia and influenza death rates and excess mortality during influenza A and influenza B epidemic periods in the USA, 1934–70 (Dowdle *et al.*, 1974).

The 'Asian' influenza pandemic and the period 1957–68

The Asian (H2N2) influenza virus first appeared in Central China in February 1957 (Jensen 1957) and possessed haemagglutinin and neuraminidase antigens showing complete antigenic differences (antigenic 'shift') from the formerly prevalent H1N1 strain and probably representing the first antigenic shift in the influenza A virus since the pandemic of 1918 (see Table 96.12). The new virus spread rapidly and virtually every country had been infected by the end of November 1957. The virus had thus behaved in a truly pandemic manner. Some countries, notably Japan and the USSR, in early September 1957 and the United Kingdom and USSR in early 1958, experienced second waves of infection.

The 'Asian' virus produced high attack rates during the 1957–8 pandemic period which varied from 20–80 per cent in different countries. Highest rates of attack were in the younger age groups of the populations and in areas of high population density (reviewed by Stuart-Harris and Schild 1976). In the United Kingdom and USA approximately 50 per cent attack rates were recorded in the 5–15 age group.

In 1957, before the circulation of the 'Asian' virus, some 30 per cent of serum samples from persons aged 75–85 years contained antibody to the 'Asian' strain (Mulder and Masurel 1958). This has been taken to indicate that the 'Asian' virus of 1957 had circulated some 68–70 years previously. As already mentioned, pandemics occurred in the period 1889–90 which thus might have been associated with an earlier prevalence of the H2N2 virus.

Between 1958 and 1968 many epidemics of H2N2 infection occurred. In general, mortality and morbidity were low. By 1968 progressive antigenic 'drift' had occurred in the haemagglutinin and neuraminidase antigens of the prevalent strains. These were typified by the A/England/12/64 (H2N2) strain in 1965 and the A/Tokyo/3/67 (H2N2) strain in 1967–8.

The 'Hong Kong' pandemic and the period 1968–76

The 'Hong Kong' (H3N2) virus was first isolated during influenza outbreaks in Hong Kong in July 1968 (Chang 1969). The virus spread to India, Japan and Australia by the early autumn (Cockburn *et al.* 1969). However, only the USA suffered major outbreaks with increased mortality rates during the winter of 1968–9; in other countries epidemics were mild. The full effect of the Hong Kong pandemic was not felt until the next winter 1969–70 in Europe, Asia and most other countries; but mortality and morbidity were less severe than in the Asian pandemic in 1957. The antigenic changes in the virus may provide a clue to its epidemiological behaviour. Though the haemagglutinin antigen (H3) was completely different from that of the Asian virus (H2), the neuraminidase was of the same antigenic subtype (N2). Thus the Hong Kong virus showed only partial antigenic 'shift'. The common neuraminidase antigen of the Asian and Hong Kong virus may thus have contributed to a cross-immunity between these strains which modified the epidemiological impact of the Hong Kong virus.

In most of the years 1971 to 1981 annual outbreaks of influenza occurred in many countries (Pereira and Chakraverty 1977) associated with several successive antigenic variants of influenza H3N2 virus (see Table 96.13 p. 333).

Reappearance of H1N1 influenza virus in 1977

In May 1977 influenza A viruses of the H1N1 subtype were isolated in the Northern region of the People's Republic of China (Pereira 1979, 1980). The outbreak appeared to affect only school children and others under the age of 23 years. In these groups it had a high attack rate but the disease it produced was generally mild.

The infection spread rapidly around the world but nowhere were significant numbers of clinical infections reported to occur in persons older than 25 years, and no country reported increased mortality rates associated with this virus. There is evidence based on serological investigations of frequent subclinical infections in older persons with H1N1 virus between 1977 and 1981.

Conventional antigenic analyses demonstrated a close antigenic resemblance between the haemagglutinin and neuraminidase antigens of the 'new' H1N1 strains and those of a virus isolated in the USA in 1950, A/Fort Warren/50 (H1N1). An analysis of the genetic composition of the virus by Nakajima, Desselberger and Palese (1978), using oligonucleotide mapping techniques of virus RNA, showed that all 8 gene segments of the 1977 H1N1 virus had a high degree of sequence homology with those of the A/Fort Warren/50 strain. The ways in which the genome of the H1N1 virus may have been highly conserved between 1950 and 1977 are discussed below (p. 339). (For further discussion of the epidemiology of influenza, see Hope-Simpson 1979, 1981, 1983.)

Environmental factors in influenza

Influenza epidemics occur notably but not exclusively in the colder half of the year in both the northern and southern hemispheres. One exception was the 1957 summer and early autumn outbreak due to the Asian virus. In tropical areas, influenza may coincide with the wet season (Schild *et al.* 1977).

It is not clear how climate and other factors may affect the epidemiological behaviour of the virus (Kilbourne 1975). Some factors may influence the susceptibility of the host to become infected, to develop disease, and to shed virus, while others may influence the virus itself. Davey and Reid (1972) found a positive correlation in successive winters between the incidence of acute respiratory disease including influenza and the coldness of the winter. Other studies have suggested that rapid fluctuation in temperature values may be more important than actual values. Similarly, the effect of humidity is not well established. The influenza virus itself survives best in aerosols at low relative humidities. In experimental mice (Schulman and Kilbourne 1963), which were maintained under even conditions of temperature and humidity in winter and summer, transmission of influenza was significantly higher in winter possibly suggesting that other unidentified factors were responsible. The spread of the virus in the winter of temperate countries may be helped by several factors such as low exterior temperature, low interior humidity and high local population density. There is evidence that spread of infection is generally more effective in densely populated urban areas than in rural areas. Other factors may be the distribution of non-influenzal microbial factors such as organisms which produce secondary bacterial pneumonia (pneumococcus, staphylococcus, haemophilus) which may have a season-dependent distribution in the population. For a more detailed review of non-specific factors in influenza epidemics, see Kilbourne (1975) and Hope-Simpson (1980, 1981); for a general review of influenza including vaccination against the disease, see Report (1979), and for vaccination in particular, see Chapters 85 and 86.

Influenza in non-human hosts

Swine

In 1918, the sudden onset of epizootics of respiratory disease among swine herds in Iowa was noted (Koen 1919). They appeared to coincide with the onset among human beings of the great pandemic of influenza in Mid-West USA. The clinical similarities between the diseases of swine and man led to the description 'swine influenza'. Koen emphasized the fact that the swine influenza in 1919 was apparently a new clinical entity. Although the primary infectious agent was undoubtedly a virus (Shope 1931), a bacterial commensal, *Haemophilus influenzaesuis*, appeared to be important in the pathogenesis of the disease. Retrospective serological studies in man (Davenport, Hennessey and Francis 1953), and subsequent investigations (reviewed by Stuart-Harris and Schild 1976), provided strong circumstantial evidence that the Shope strain was antigenically close to the virus responsible for the 1918–19 human pandemic.

Since 1918–19 outbreaks of influenza associated with classical swine (H1N1) influenza A virus have occurred in pigs almost annually in the USA, and recent intensive surveillance in that country has indicated that the virus is widespread throughout the states and may circulate at all times of the year (Hinshaw *et al.* 1978). International surveys carried out since 1976 suggest that pigs in the USA may be the major reservoir in the world of swine (H1N1) viruses. Investigations of pigs in other countries have indicated a low prevalence of infection in some areas, including Japan, Italy, Israel and Central Europe, but these countries may have acquired the viruses from the USA. Evidence of infection in the UK with viruses resembling the classical swine influenza A virus has not been apparent since 1939 and pigs in several other European countries appear to be free from the disease. Swine influenza virus is probably endemic in pigs from Hong Kong, Singapore, and the People's Republic of China (Shortridge and Webster 1979). Detailed serological studies (Schild *et al.* 1972*b*, Kendal, Noble and Dowdle 1977) have shown that a moderate degree of antigenic 'drift' has occurred in the prevalent swine (H1N1) influenza viruses between 1930 and 1977. Genetic studies of swine influenza viruses isolated from pigs have revealed considerable heterogeneity in the RNAs of the virus (Hinshaw *et al.* 1978). The swine (H1N1) viruses from different farms were usually distinctive suggesting that transmission of virus between farms was restricted.

There is widespread evidence of infection of pigs with human Hong Kong (H3N2) viruses. A/Hong Kong/68 (H3N2)-like viruses were first isolated in 1970 from pigs in Taiwan (Kundin 1970), and subsequently several of the later antigenic variants of human H3N2 virus have been detected in pigs in almost every country where studies have been made. Although the studies (Schild *et al.* 1972*b*, Harkness *et al.* 1972) have indicated high frequency of infection, it appears that viruses of the H3N2 subtype cause no significant disease in pigs. It is thought that H3N2 virus infection is commonly transmitted from man to pigs. However, viruses closely resembling the 1968 prototype strain of A/Hong Kong/68 (H3N2) virus were isolated from pigs in Hong Kong in 1976 (Shortridge *et al.* 1977*b*) several years after their disappearance from man. Thus, it is possible that pigs may serve as a repository of viruses of past human pandemics as well as a possible source of genetic information for the production of recombinant viruses between human and porcine strains of influenza A virus (Kaplan and Webster 1977).

Serological studies (D.H. Roberts unpublished observation) of pigs in the United Kingdom in 1980 have provided evidence of frequent infection with H1N1 viruses resembling A/USSR/77 (H1N1) strains which have circulated in man since 1977. Such infections do not appear to be associated with overt disease.

Horses

Although epizootics of influenza-like illness in horses were reported in the seventeenth to nineteenth centuries, the specific viral aetiology of equine influenza was not established as an influenza A virus until 1956 (Heller, Espmark and Vriden 1956, Sovinova *et al.* 1958). In the following years virus isolations and serological evidence indicated that infection by the same agent occurred among horses in other central European countries followed by the USA and Western Europe. The 1956 isolate designated A/equine/Prague/1/56 (H7N7) was found, on its subsequent antigenic characterization, to possess haemagglutinin and neuraminidase antigens unrelated to those of human influenza A viruses. H7N7 viruses continue to produce outbreaks in horses in many parts of the world. The most recently documented outbreak was in England (Powell *et al.* 1974) and was shown to be caused by a virus exhibiting a minor degree of antigenic 'drift' over the past 18 years. It is now clearly established (Tumova and Pereira 1968, Tumova and Schild 1972) that there are close antigenic relations between the neuraminidases of equine H7N7 virus and those of several avian influenza viruses and, in the 1980 system, they are classified in a single neuraminidase subtype (N7). Similarly Schild *et al.* (1980) showed that the haemagglutinin antigen of the 1956 equine virus was related to that of fowl plague virus and thus both were classified as H7N7.

In 1963 a second subtype of equine influenza was isolated (Waddell, Teigland and Siegel 1963) from extensive outbreaks of influenza in horses in the USA. Later this virus was detected in several South American countries and in Europe. Serological evidence of infection has been detected among the large populations of horses in Mongolia. The new isolate designated A/equine/Miami/1/63 (H3N8) possesses haemagglutinin and neuraminidase antigens unrelated to those of the equine (H7N7). The disease in horses caused by the equine (H3N8) virus is essentially an acute upper respiratory tract infection, but epidemics tend to be more severe and widespread than those caused by H7N7 virus.

Up to 1969 isolates of equine (H3N8) virus were antigenically close to the prototype 1963 isolate. In 1969 H3N8 isolates from horses in Brazil (Pereira *et al.* 1972) were found to have a significant degree of antigenic difference from the 1963 strain. Later, severe epidemics among racehorses in Tokyo yielded strains showing even greater differences (Kono *et al.* 1972).

Several workers have shown antigenic relations between the haemagglutinin antigen of the 1963 equine virus and that of human A/Hong Kong/1/68 virus providing the basis for their inclusion in a single subtype, H3. It has also been clearly demonstrated. (Tumova and Schild 1972, Schild *et al*, 1980) that the neuraminidase antigen of the equine (H3N8) virus is of the same subtype as the neuraminidases of several avian influenza strains.

Although all breeds and ages of horses as well as donkeys and mules are susceptible to infection with equine influenza viruses, racehorses appear to be particularly affected. Routine vaccination of horses with inactivated equine influenza virus vaccines is practised in several countries, particularly among racing animals and breeding stock.

Birds

Avian species are the most abundant source of influenza A viruses from non-human sources both numerically and in terms of agents widely different in biological and antigenic characters. Reference strains of avian influenza viruses are listed in Tables 96.8 and 96.9. Some of these agents, like fowl plague (H7N7) virus, are responsible for rapidly fatal infections associated with viraemia and pantropic dissemination of virus. In other cases infection is apparently symptomless. The pathology of influenza infections in birds has been reviewed extensively by Easterday and Tumova (1972, 1975). Fowl plague was shown long ago to be due to a virus (Stubbs 1965) which was subsequently identified as an influenza A virus (Schäffer 1955). Later other agents were isolated from ducks with acute sinusitis in Czechoslovakia (Koppel *et al.* 1956), and in England (Andrewes and Worthington 1959, Roberts 1964) were also identified as being type A influenza.

Since these early findings influenza viruses have been isolated in many countries from a very wide range of birds, including chickens, ducks, turkeys, quail, pheasant, sea birds and passerines (reviewed by Pereira 1969, Stuart-Harris and Schild 1976, Easterday 1975, Laver and Webster 1979, Easterday 1980; Hinshaw, Webster and Rodriguez 1979). More recently isolations have been made in the USA and Western Europe from several species of imported exotic aviary birds including myna birds, parrots, parakeets, cockatoo and weaver birds (Slemons *et al.* 1974).

Becker (1963) isolated influenza virus, A/tern/South Africa/61 (H5N3), from an outbreak of disease with a high fatality rate which occurred in common tern (*Sterna hirundo*) off Cape Town. In contrast, an influenza virus (A/Shearwater/E. Australia/72 (H6N5) isolated off the Australian Barrier Reef, from shearwater (*Puffinus pacificus*), a seabird pelagic in the Pacific Ocean (Downie and Laver 1973) was apparently avirulent in its natural host.

Although most earlier investigations of avian influenza were made on domestic or cage birds, it is apparent from recent ecological investigations that wild bird populations harbour a wide range of antigenic varieties of type A influenza virus. During a study of migratory wild duck populations in California, Slemons *et al.* (1974) isolated 43 strains of avian influenza A virus from the respiratory tract in several thousand wild duck. A wide range of different influenza A viruses have also been isolated from faecal specimens of ducks in Hong Kong and on the Mississippi (Shortridge *et al.* 1977*a*, Hinshaw, Webster and Rodriguez 1979). It has been shown (Webster *et al.* 1978) that avian strains of influenza virus replicate in the lungs and also in the cells lining the intestinal tracts of ducks. The viruses reach the lower intestinal tract despite the low pH of the gizzard and are shed in high concentrations in the faeces. Human influenza A viruses replicate in ducks, but do so only in the upper respiratory tract, and not in the intestinal tract.

In some studies on wild ducks in North America a high proportion of healthy birds have yielded influenza A viruses from cloacal samples. Among the influenza viruses isolated from ducks are included strains possessing many of the possible combinations of haemagglutinin and neuraminidase antigen subtypes. Influenza A viruses have been isolated from water samples collected from lakes frequented by ducks in Canada at the time when the birds were assembling before migration; viruses were also isolated from faecal samples collected on the shores of these lakes (Hinshaw, Webster and Rodriguez 1979). The high concentration of duck influenza viruses in faecal material, the isolation of influenza viruses from faecal material, and from lake water offer a possible mechanism for transmission of avian influenza viruses from feral ducks to domestic avian and mammalian species.

As indicated above in the section on nomenclature, the influenza A viruses from avian species so far characterized are divided into twelve haemagglutinin antigen subtypes (H1 to H12) and nine neuraminidase antigen subtypes (N1 to N9). Of the HA subtypes four (H1, H2, H3, H7) also have counterparts among influenza A viruses of human, swine or equine origin, the remaining eight are so far unique to avian species. For the neuraminidase subtypes four (N1, N2, N7 and N8) are represented among human, swine or equine viruses, while five are unique to viruses from avian species.

An important finding in respect of avian influenza viruses, and quite unlike the general situation with human, swine or equine viruses, is that a given haemagglutinin antigen is not always associated with the same neuraminidase antigen. Thus among strains possessing a common neuraminidase antigen may be found several subtypes of haemagglutinin, and strains which possess common haemagglutinin antigens may have different neuraminidase antigens. There is no correlation between antigenic character and host of origin; antigenically identical strains have been isolated from several different species of bird (Pereira 1969, Hinshaw, Webster and Rodriguez 1979).

Influenza in other species

Influenza infections in non-human species, other than swine, horses and birds, have been infrequently confirmed. However one recent observation is worthy of note concerning epizootic pneumonia in seals.

Between November 1979 and June 1980 (Webster *et al.* 1981) approximately 500 common seals were found to have died on the New England coast of the USA. Apparently up to 20 per cent mortality occurred in the seals and the fatalities peaked in January 1980. Influenza A virus designated as A/seal/Massachusetts/1/80 (H7N7), antigenically close to the Dutch strain of fowl plague virus, was isolated from several individuals and was present in high titres from the lungs. Isolates were also made from brain tissue. Pathological examination of the lungs of dead animals showed evidence of primary viral pneumonia. The virus was found to be avirulent for domestic avian species, including chickens, ducks and turkeys. The outbreak represents the first evidence of an apparently avian influenza A virus causing fatal infections in a mammalian species.

Evidence that influenza A viruses from non-human sources may be progenitors of human influenza

This fascinating area of influenza virology has been reviewed in detail elsewhere (see Stuart-Harris and Schild 1976, Laver and Webster 1979). Since 1918 three pandemic viruses have circulated in man—H1N1 virus probably from 1918–56, Asian (H2N2) virus from 1957–68 and Hong Kong (H3N2) virus from 1968. In 1977 H1N1 viruses closely resembling antigenically and genetically human H1N1 strains which previously circulated around 1950 (Nakijima *et al.* 1978, Young and Palese 1979) re-emerged as epidemic viruses of children and young adults and circulated for the next several years (Pereira 1979). The source of the H1N1 virus in 1977 is uncertain but several possible explanations for the preservation of its genome in nature have been proposed. These include its escape from laboratory sources, its persistence in infectious form in animals or birds or possible latency in a human or non-human host, or even as a parasite of man.

Circumstantial evidence that the causative virus of the 1918–19 pandemic was antigenically closely related to swine (H1N1) virus is described above (p. 336). In 1976 confirmation of the potential of H1N1 swine (H1N1) virus from pigs to infect man was obtained. Viruses antigenically close to swine (H1N1) virus were isolated from a limited outbreak in young military recruits at Fort Dix, New Jersey (Topp and Russell 1977), and circumstantial evidence indicated that the source of infection was pigs. Unequivocal evidence for the infection of human beings by swine (H1N1) viruses from porcine sources has been obtained by the isolation of genetically identical 'swine' viruses from pigs and persons on a farm in Wisconsin (Hinshaw *et al.* 1978).

Concerning the Asian (H2N2) pandemic virus there is evidence (reviewed by Laver and Webster 1979, Scholtissek 1978) that it possessed four genes probably derived from the former human H1N1 virus, i.e. genes I, V, VII and VIII coding for polypeptides P1, NP, MP and the NS proteins (see Table 96.12). The remaining gene segments II, III, IV, VI coding for P2, P3, haemagglutinin and neuraminidase were probably derived from an unidentified influenza A virus possessing H2N2 surface antigens. Isolates of avian influenza A virus with antigens of the H2 and N2 subtype have been described (Schild *et al.* 1980), and thus circumstantial evidence exists that the pandemic Asian virus may have arisen as a recombinant between the former human H1N1 virus and a virus possessing H2N2 genes possibly of avian origin.

In 1968 the Hong Kong (H3N2) virus was found to possess an antigenically distinct haemagglutinin antigen (H3) but its neuraminidase antigen was close to that of the human H2N2 viruses of 1967. Serological evidence and biochemical analyses (see Laver and Webster 1979) have shown that H3 haemagglutinin is closely related to the HA of certain avian influenza viruses, A/duck/Ukraine/1/63 (H3N7) and also to that of A/equine/Miami/1/63 (H3N8) virus. Other studies suggest that the human Hong Kong (H3N2) virus contains seven gene sequences closely similar to the corresponding genes of the human H2N2 virus, only gene IV, coding for haemagglutinin not being homologous for the two viruses. It is thus possible that the H3N2 virus arose as a result of recombination between the human H2N2 virus which donated 7 of the 8 genes and a virus from another source which donated the gene for H3. These findings provide probably the strongest circumstantial evidence yet available that lower mammals or birds may play a role in the origin of human influenza pandemics. However, it is clear that the mechanism of genetic recombination cannot be invoked as a universal explanation for the emergence of new human viruses, since the re-emerged human H1N1 virus of 1977 probably did not originate in this manner. (For a review of influenza viruses in animals, and their reaction to human influenza, see Alexander 1982.)

Vaccination against influenza

Killed influenza virus vaccines, as currently used, are prepared by growing the chosen strain(s) in the allantoic cavities of fertile hens' eggs, and inactivating virus infectivity with formalin or beta-propiolactone. The virus is subsequently concentrated and purified by ultracentrifugation (Reimer *et al.* 1967). While some vaccines contain intact virus particles, in others the virus particles are chemically disrupted into their component protein subunits by treatment with ether and/or detergents. In some cases further purification results in a vaccine containing essentially haemagglutinin and neuraminidase molecules. The virus antigens may be adsorbed into $Al(OH)_3$ in the final vaccine.

Recent work has investigated the use of live attenuated virus strains as vaccines. Vaccine strains under investigation include temperature-sensitive viruses, cold-adapted viruses selected by prolonged cultivation at low temperatures ($\simeq 25°$) and genetic hybrid viruses attenuated by recombination between a virulent epidemic strain and an attenuated laboratory strain. Although their general safety for the individual and the community, and their efficacy have yet to be clearly demonstated, these products may be of great value in future prophylaxis (Tyrrell and Smith 1979).

This chapter has discussed the innate variability of influenza viruses and the likelihood that each year will see the appearance of a new strain or strains which, antigenically, have shifted or drifted from the former strains. Populations or individuals immunized previously by either natural infection or vaccination will be more susceptible to the new strains than to the old. New batches of killed vaccine are made each year, in preparation for the expected advent of an influenza epidemic. The strains chosen for incorporation in the vaccines are decided by the manufacturers, after consideration of advice from expert virologists and the World Health Organization. The data gathered by the WHO World Influenza Centres and their many collaborating laboratories are crucial in formulating this advice to manufacturers and to National Health Authorities.

The objective of this policy is to match the antigenic patterns of the haemagglutinin and neuraminidase of the vaccine viruses as nearly as possible with those of the newly emerging epidemic strains of influenza virus, on the assumption that this is likely to provide the best protection against infection. But contrary arguments have been advanced for the value of using older established strains in a vaccine, which take account of the dictum or hypothesis of 'original antigenic sin' described elsewhere in this chapter (see Tyrrell and Smith 1979). Consideration is also given to technical problems, such as the immunogenicity in man of the antigen on offer and the amount of antigen produced by the different candidate vaccine strains. Genetic hybrid (reassortant) viruses produced from a rapidly growing and high-yielding laboratory strain A/PR8/34 virus and a newly isolated strain possessing antigenic characters relevant to current immuno-prophylaxis have been widely used for vaccine production (Kilbourne and Murphy 1960). Such high-yielding vaccine virus strains enable many thousands of vaccine doses to be produced rapidly in the face of an impending epidemic, as was dramatically illustrated during the Swine (H1N1) influenza campaign in the USA in 1976 (Parkman *et al.* 1977, Neustadt and Fineberg 1978).

Purified surface antigen vaccines represent an important step towards refinement of the vaccine, as they consist mainly of haemagglutinin or neuraminidase antigens, the essential immunogens of the virus and may be less reactogenic and pyrogenic than vaccines containing whole virus, particularly in young children (Bachmeyer 1975). Their efficacy as vaccines in unprimed individuals has been questioned however, and the intact virions of whole virus vaccines may play some part in determining the immune response of the vaccinee particularly if he should be unprimed (Parkman *et al.* 1977).

There are biological and logistic difficulties in vaccination against influenza. Firstly, there is the evidence discussed above that host response to a new antigen may be determined more by the pattern of antigen the host originally experienced in childhood than by the pattern of the antigen now presented in the vaccine, at least in the formation of strain-specific antibody to haemagglutinin (Schild *et al.* 1977). Vaccines are now routinely standardized for potency by means of single-radial-diffusion tests to measure directly the antigen content in micrograms of HA per vaccine dose (Schild *et al.* 1975). The immunizing potency of a vaccine is often estimated by either the protective effect in experimental animals such as mice, or by the anti-haemagglutinin serum antibody response in a group of human volunteers. Though the latter is generally taken to be a measure of immunity (Freestone *et al.* 1972), a definite estimate of vaccine efficacy should be based on protection of a group of volunteers against natural infection during an epidemic. Such a test is most difficult to organize, particularly when time is pressing. Secondly, since influenza epidemics spread widely and rapidly, it is often impossible to produce sufficient vaccine soon enough to immunize all those who are considered highly susceptible—for example the aged, debilitated and perhaps the very young. In this regard it must be remembered that any vaccine, influenza or other, must be rigorously tested for potency and safety before it is issued for general use, and these procedures are often very time consuming.

Estimates of the effectiveness of influenza vaccines vary considerably, and some strains and some batches are better immunogens than others, and some test populations react better than others. Estimates of potency between 40 per cent and 80 per cent are common. The reappearance of the H1N1 epidemic virus in 1977, after its disappearance in 1957, indicated that natural immunity to influenza virus is often long lasting. Persons over 23 years of age in 1977—those who had experienced the 1957 epidemic—were resistant to infection with H1N1 in 1977; those under 23 years of age were largely susceptible (Editorial 1978). By contrast it is generally believed that vaccination immunity is much more transient (Eickhoff 1971, Smith *et al.* 1975).

Any vaccination procedure will induce unwanted side effects, either local or general, in a proportion of vaccinees and an estimate of the value of a vaccine must set the benefits against the disadvantages of these side effects. A particularly unpleasant and sometimes fatal side effect became apparent during the large vaccination programme with swine (H1N1) influenza vaccine in the USA in 1976—the occurrence of the Guillain-Barré syndrome, a neurological disease of some consequence (Longmuir 1979). One estimate was that 1 in 120 000 vaccinees might develop this serious illness as a complication (Tyrrell and Smith 1979). It is yet uncertain whether this was due entirely to the particular virus antigen in the vaccine or whether it was attributable to influenza vaccination in general.

The last few years have seen great advances in understanding both the mechanism of the immune response and the structure and antigenic formation of the components of the influenza virus, including its genome. There have also been major technical advances including notably the use of cell fusion techniques and DNA replication techniques. These promise a much greater understanding of the natural variation of the influenza virus and of the host's available defences. It can be confidently hoped that the near future will see great advances in the safety and efficacy of influenza vaccines. (For a description of the para-influenza viruses see Chapter 98.)

References

Air, G. M. (1979) *Virology* **97,** 468; (1980) *Develop. cell Biol.* **5,** 135.

Air, G. M. and Hackett, J. A. (1980) *Virology* **103,** 291.

Alexander, D. J. (1982) *J. roy. Soc. Med.* 75, 799.

Allen, H., McCauley, J., Waterfield, M. and Gething, M. J. (1980) *Virology* **107,** 548

Anderson, N. G. (1967) *J. Virol.* **1,** 1207.

Andrewes, C. H., Laidlaw, P. P. and Smith, W. (1934) *Lancet* **ii,** 859; (1935) *Brit. J. exp. Path.* **16,** 566.

Andrewes, C. H. and Worthington, G. (1959) *Bull. Wld Hlth Org.* **20,** 435.

Assaad, F., Cockburn, C. W. and Sundareson, T. K. (1973) *Bull. Wld Hlth Org.* **49,** 219.

Assaad, F. and Reid, D. (1971) *Bull. Wld Hlth Org.* **45,** 113.

Aymard-Henry, M. *et al.* (1973) *Bull. Wld Hlth Org.* **48,** 199.

Bachmeyer, H. (1975) *Intervirology* **5,** 260.

Baker, N., Stone, H. O. and Webster, R. G. (1973) *J. Virol.* **11,** 137.

Becker, W. B. (1963) *Virology* **20,** 318.

Beveridge, W. I. B. (1977) In: *Influenza, the Last Great Plague*, Heinemann, London.

Blok, J. and Air, G. M. (1980) *Virology* **107,** 50.

Both, G. W. and Air, G. M. (1979) *Europ. J. Biochem.* **96,** 363.

Brand, C. M. and Skehel, J. J. (1972) *Nature (Lond.) New Biology* **238,** 145.

Bucher, D. and Palese, P. (1975) In: *The Influenza Viruses and Influenza*, p.84, Ed. by E. D. Kilbourne, Academic Press, New York.

Burnet, F. M. (1941) *Aust. J. exp. Biol. med. Sci.* **19,** 291; (1979) *Intervirology* **11,** 201.

Burnet, F. M. and Clarke, E. (1942) *Influenza.* Monogr. No. 4. Walter and Eliza Hall Institute, Melbourne, Australia.

Caton, A. J., Brownlee, G. G., Yewdell, J. W. and Gerhard, W. (1982) *Cell* **31,** 417.

Chang, W. K. (1969) *Bull. Wld Hlth Org.* **41,** 349.

Choppin, P. W. and Compans, R. W. (1975) *Comprehens. Virol.* **4,** 95.

Cockburn, W. C., Delan, P. J. and Fereira, W. (1969) *Bull. Wld Hlth Org.* **41,** 345.

Coleman, P. M., Varghese, J. N. and Laver, W. G. (1983) *Nature (Lond.)* **303,** 41.

Couch, R. B. *et al.* (1974) *J. infect. Dis.* **129,** 411.

Creighton, C. (1894) In: *A History of Epidemics in Britain*, 2nd edn. 1965, Cass, London.

Davenport, F. M., Hennessy, A. V. and Francis, T., Jr (1953) *J. exp. Med.* **98,** 641.

Davey, M. L. and Reid, D. (1972) *Brit. J. prev. soc. Med.* **26,** 28.

Davis, A. R. *et al.* (1981) *Proc. nat. Acad. Sci. (Wash.)* **78,** 5376.

Desselberger, J. and Palese, P. (1978) *Virology* **88,** 394.

Doherty, P. C., Effros, R. B. and Bennink, J. (1977) *Proc. nat. Acad. Sci. (Wash.)* **74,** 1209.

Dopheide, T. A. and Ward, C. W. (1980) *Develop. cell Biol.* **5,** 21.

Douglas, R. G., Jr (1975) In: *The Influenza Viruses and Influenza*, p. 395. Ed. by E. D. Kilbourne, Academic Press, New York.

Dowdle, W. R. (1976) In: *Influenza Virus Vaccines and Strategy*. Ed. by P. Selby, Academic Press, London.

Dowdle, W. R., Coleman, M. T. and Gregg, M. B. (1974) *Progr. med. Virol.* **17,** 91.

Dowdle, W. R., Downie, J. and Laver, W. G. (1974) *J. Virol.* **13,** 269.

Dowdle, W. R. *et al.* (1975) *Intervirology* **5,** 245.

Downie, J. C. and Laver, W. G. (1973) *Virology* **51,** 259.

Easterday, B. C. (1975) In: *The Influenza Viruses and Influenza*, pp. 449–77. Ed. by E. D. Kilbourne, Academic Press, New York; (1980) In: *Influenza*, pp. 145–51. The Royal Society, London.

Easterday, B. C. and Tumova, B. (1972) In: *Diseases of Poultry*, 6th edn. Ed. by M. S. Hofstrad, Bailliere, London; (1975) In: *Ibid*, 7th edn. Ed. by M. S. Hofstrad, Bailliere, London.

Editorial (1978) *Brit. med. J.* **ii,** 230.

Eickhoff, T. C. (1971) *J. infect. Dis.* **123,** 446.

Emtage, J. S. *et al.* (1980) *Nature (Lond.)* **283,** 171.

Ennis, F. A., Martin, W. J. and Verbonitz, M. W. (1977) *J. exp. Med.* **146,** 893.

Ennis, F. A. *et al.* (1981) *Lancet* **ii,** 887.

Fields, S., Winter, G. and Brownlee, G. G (1981) *Nature (Lond.)* **290,** 213.

Flewett, T. H. and Apostolov, K. (1967) *J. gen. Virol.* **1,** 297.

Francis, T. Jr (1940) *Science* **92,** 405.

Freestone, D. S., Haerton-Smith, S., Schild, G. C., Bucheard, R., Chinn, S. and Tyrrell, D. A. J. (1972) *J. Hyg. (Camb.)* **70,** 531.

Gerhard, W. (1982) *Cell* **31,** 417.

Gerhard, W. Yewdell, J., Frankel, M. E. and Webster, R. G. (1981) *Nature (Lond.)* **290,** 713.

Gething, M. J. and Sambrook, J. (1981) *Nature (Lond.)* **293,** 620.

Gething, M. J. *et al.* (1980) *Develop. Cell Biol.* **5,** 1.

Ghendon, Y. Z. and Markushin, S. G. (1980) *Phil. Trans. R. Soc. B* **288,** 95.

Gottschalk, A. (1957) *Physiol. Rev.* **37,** 66: (1966) In: *The Glyproteins. Their Composition, Structure and Function.* Elsevier, Amsterdam.

Haaheim, L. and Schild, G. C. (1979) *Acta path. microbiol. scand. B.* **87,** 291.

Hall, R. M. and Air, G. M. (1981) *J. Virol.* **38,** 1.

Harkness, J. W., Schild, G. C., Lamont, P. B. and Brand, C. M. (1972) *Bull. Wld Hlth Org.* **46,** 709.

Haukenes, G., Harboe, A. and Mortenssen-Egnund, K. (1965) *Acta path. microbiol. scand.* **64,** 534.

Hay, A. J., Skehel, J. J. and McCawley, J. (1980) *Phil. Trans. R. Soc. B* **288,** 53.

Heller, L., Espmark, A. and Vriden, P. (1956) *Arch. ges. Virusforsch* **7,** 120.

Hinshaw, V. S., Bean, W. J. Jr, Webster, R. G. and Easterday, B. C. (1978) *Virology* **84,** 51.

Hinshaw, V. S. and Webster, R. G. (1979) *J. gen. Virol.* **45,** 751.

Hinshaw, V. S., Webster, R. G. and Rodriguez, R. J. (1979) *Arch. Virol.* **62,** 281.

Hirsch, A. (1883) *Handbook of Geographical and Historical Pathology*, Vol. 1 p. 7. New Sydenham Society, London.

Hirst, G. K. (1941) *Science* **94,** 22.

Hiti, A. L., Davis, A. R. and Nayak, D. P. (1981) *Virology* **111,** 103.

Hope-Simpson, R. E. (1979) *J. Hyg. (Camb.)* **83,** 11; (1980) *Biometeorological Survey* 1973–1978. Part A **1,** 170; (1981) *J. Hyg. (Camb.)* **86,** 35; (1983) *Ibid.* **91,** 293.

Inglis, S. C., Barrett, T., Brown, C. M. and Almond, J. W. (1979) *Proc. nat. Acad, Sci. (Wash.)* **76,** 3790.

Jahiel, R. I., and Kilbourne, E. D. (1966) *J. Bact.* **92,** 1521.

Jackson, D. C. *et al.* (1979) *Virology* **93,** 458.

Jensen, N. E. (1957) *J. Amer. med. Ass.* **164,** 2025.

Kaplan, M. M. and Webster, R. G. (1977) *Scientific American* **237,** 88.

Kendal, A. P. (1975) *Virology* **65,** 87.

Kendal, A. P., Noble, G. R. and Dowdle, W. R. (1977) *Virology* **82,** 111.

Kilbourne, E. D. (1975) In: *The Influenza Viruses and Influenza*, p. 1. Ed. by E. D. Kilbourne, Academic Press, New York; (1976) *J. infect. Dis.* **132,** 384; (1980) *Phil. Trans. R. Soc. B.* **288,** 3.

Kilbourne E. D. and Murphy, J. S. (1960) *J. exp. Med.* **111,** 387.

Kilbourne, E. D. *et al.* (1972*a*) *J. infect. Dis.* **125,** 447, (1972*b*) In: *International Virology* **2,** 118. Karger, Basel.

Klenk, H. D., Compans, R. W. and Choppin, P. W. (1970) *Virology* **42,** 1158.

Klenk, H. D. *et al.* (1975) *Virology* 68, 426.

Koen, J. S. (1919) *J. vet. Med.* 14, 468.

Kono, Y., Ishikawa, K., Fukunaya, Y. and Fujino, M. (1972) *Nat. Inst. Animal Hlth Quart.* **12,** 183.

Koppel, Z. J., Vriak, V. M. and Spiesz, S. (1956) *Veterinarstvi* **6,** 267.

Kundin, W. D. (1970) *Nature (Lond.)* **228,** 857.

Lamb, R. A. and Choppin, P. W. (1980) *Phil. Trans. R. Soc. B* **288,** 39.

Lamb, R. A. *et al.* (1980) *Develop. cell. Biol.* **5,** 91.
Langmuir, A. D. (1979) *J. roy. Soc. Med.* **72,** 660.
Laver, W. G. and Air, G. (1980) *Develop Cell. Biol.* **5.**
Laver, W. G. and Valentine, R. C. (1969) *Virology* **38,** 105.
Laver, W. G. and Webster, R. G (1976) *Virology*, **69,** 511.
Laver, W. G. and Webster, R. G. (1979) *Brit. med. Bull.* **35,** 29.
Laver, W. G. *et al.* (1979) *Proc. nat. Acad. Sci.* (*Wash.*) **76,** 1425, (1980) *Develop. cell Biol.* **5,** 295.
Lazarowitz, S. G. and Choppin, P. W. (1975) *Virology* **68,** 440.
Lief, F. S. and Henle, W. (1959) *Bull. Wld. Hlth Org.* **20,** 411.
McCauley, J. *et al.* (1979) *FEBS Letters* **108,** 422.
McGeoch, D., Fellner, P. and Newton, C. (1976) *Proc. nat. Acad. Sci.* (*Wash.*) **73,** 3045.
McGregor, I. *et al.* (1979) *Brit. med. Bull.* **35,** 15.
Mahy, B. W. J. *et al.* (1980) *Phil. Trans R. Soc. B* **288,** 61.
Melnick, J. (1973) *Progr. med. Virol.* **15,** 380.
Min Jou, W. *et al.* (1980) *Develop. Cell. Biol.* **5,** 63.
Mostow, S. R. *et al.* (1975) *J. clin. Microbiol.* **2,** 531
Mulder, J. and Masurel, N. (1958) *Lancet*, **i,** 810.
Nakajima, K., Desselberger, C. and Palese, P. (1978) *Nature* (*Lond.*) **274,** 334.
Nermut, M. V. (1972) *J. gen. Virol.* **17,** 317.
Nermut, M. V. and Frank, H. (1971) *J. gen. Virol.* **10,** 37.
Neustadt, R. E. and Fineberg, H. V. (1978) In: *The Swine Flu Affair – Decision-making on a Slippery Disease.* U.S. Dept. of Health, Education and Welfare.
Noll, H., Aoyagi, T. and Orlando, J. (1962) *Virology*, **18,** 154.
Oxford, J. S., Corcoran, T. and Hugentobler, A. (1981) *J. biol. Stand.* **9,** 483.
Oxford, J. S. and Schild, G. C. (1976) *Virology*, **74,** 394.
Oxford, J. S., Schild, G. C. and Alexandrova, G. (1980) *Arch. Virol.* **65,** 277.
Oxford, J. S., Schild, G. C., Potter, C. W. and Jennings, R. (1979) *J. Hyg. Camb.* **82,** 51.
Palese, P. *et al.* (1980) *Phil. Trans. R. Soc. B* **288,** 11.
Parkman, P. D., Hopps, H., Rastogi, S. C. and Meyer, H. M. Jr (1977) *J. infect. Dis.* **136,** Suppl. S722.
Pereira, H. G. (1969) *Progr. med. Virol.* **11,** 46.
Pereira, H. G., Takimoto, S., Piegas, N. S. and Ribeiro Do Valle (1972) *Bull Wld Hlth Org.* **47,** 465.
Pereira, M. S. (1979) *Brit. med. Bull.* **35,** 9; (1980) *Phil. Trans. R. Soc. B* **288,** 135.
Pereira, M. S. and Chakraverty, P. (1977) *J. Hyg. Camb.* **79,** 77.
Pons, M. W. (1975) In: *The Influenza Viruses and Influenza*, p. 145. Ed. by E. D. Kilbourne, Academic Press, New York.
Pons, M. W. *et al.* (1969) *Virology* **39,** 250.
Porter, A. G., Smith, J. C. and Emtage, J. S. (1980) *Proc. nat. Acad. Sci.* (*Wash.*) **77,** 5074.
Porter, A. G. *et al.*, (1979) *Nature* (*Lond.*) **282,** 471.
Potter, C. W. and Oxford, J. S. (1979) *Brit. med. Bull.* **35,** 69.
Powell, D. G., Thompson, G. R., Plowright, W., Burrows, R. and Schild, G. C. (1974) *Vet. Rec.* **94,** 282.
Racaniello, V. R. and Palese, P. (1979) *J. Virol.* **29,** 361.
Reimer, C. B., Barker, R. S., Frank, R. M. van, Newlin, T. E., Clive, G. B. and Anderson, N. G. (1967) *J. Virol.* **1,** 1207.
Report (1979) *Brit. med. Bull.* **35,** 1–91.
Roberts, D. H. (1964) *Vet. Rec.* 76, 470.
Robertson, J. S., Schubert, M. and Lazzarini, R. A. (1981) *J. Virol.* **38,** 157.
Rott, R. (1980) *Phil. Trans R. Soc. B* **288,** 105.
Salk, J. E. (1944) *J. Immunol.* **49,** 87.
Schäffer, W. (1955) *Z. Naturforsch* **10,** 81.
Schild, G. C. (1970) *J. gen. Virol.* **9,** 197; (1972) *J. gen. Virol.* **15,** 99; (1979) *Postgrad. med. J.* **55,** 87.
Schild, G. C., Aymard-Henry, M. and Pereira, H. G. (1972*a*) *J. gen. Virol.* **16,** 231.
Schild, G. C., Brand, C. M., Harkness, J. W. and Lamont, P. B. (1972*b*) *Bull. Wld Hlth Org.* **46,** 720.
Schild, G. C. and Dowdle, W. R. (1975) In: *The Influenza Viruses and Influenza*, p. 316. Ed. by E. D. Kilbourne. Academic Press, New York.
Schild, G. C., Oxford, J. S. and Newman, R. W. (1979) *Virology* 93, 569.
Schild, G. C., Pereira, M. S. and Chakraverty, P. (1975) *Bull. Wld Hlth Org.* **52,** 43.
Schild, G. C., Smith, J. W. G., Cretescu, L. and Newman, R. W. and Wood, J. M. (1977) *Develop. biol. Stand.* **39,** 273.
Schild, G. C. Wood, J. M. and Newman, R. W. (1975) *Bull. Wld Hlth Org.* **52,** 223.
Schild, G. C. *et al.* (1977) *Bull. Wld Hlth Org.* **55,** 1, (1980) *Arch. Virol.* **63,** 184.
Scholtissek, C. (1978) *Curr. Top. Microbiol. Immunol.* **80,** 139; (1980) *Phil. Trans. R. Soc. B* **288,** 19.
Schulman, J. L., Khakpour, M. and Kilbourne, E. D. (1968) *J. Virol.* **2,** 778.
Schulman, J. L. and Kilbourne, E. D. (1963) *J. exp. Med.* **118,** 267; (1965) *J. Bact.* **89,** 170; (1969) *Proc. nat. Acad. Sci.* (*Wash.*) **63,** 326.
Schulze, I. T. (1973) *Advanc. Virus Res.* **18,** 1; (1975) In: *The influenza Viruses and Influenza*, p. 53. Ed. by E. D. Kilbourne, Academic Press, New York.
Schulze, I. T. *et al.* (1970) In: *The Biology of Large RNA Viruses*, p. 324. Ed. by R. D. Barry and B. W. Mahy, Academic Press, New York.
Shope, R. E. (1931) *J. exp. Med.* **54,** 349.
Shortridge, K. F. Butterfield, W. K., Webster, R. G. and Campbell, C. H. (1977*a*) *Bull. Wld Hlth Org.* **55,** 15.
Shortridge, K. F. and Webster, R. G. (1979) *Intervirology* **11,** 9.
Shortridge, K. F., Webster, R. G., Butterfield, W. K. and Campbell, C. H. (1977*b*) *Science* (*New York*) **196,** 1454.
Skehel, J. J. (1971) *J. gen. Virol.* **11,** 103; (1972) *Virology* **49,** 23.
Skehel, J. J. and Hay, A. J. (1978) *J. gen. Virol.* **39,** 128.
Skehel, J. J. and Schild, G. C. (1971) *Virology* **44,** 396.
Skehel, J. J. *et al*,, (1980) *Phil. Trans. R. Soc. B* **288,** 47.
Sleigh, M. J. *et al.* (1980) *Develop. Cell Biol.* **5,** 11.
Slemons, R. D., Johnson, D. C., Osborn, J. S. and Hayes, F. (1974) *Avian Dis.* **18,** 119.
Smith, J. W. G., Fletcher, W. B., Pekers, M., Westwood, M. and Perkins, F. T. (1975) *J. Hyg. Camb.* **74,** 251.
Smith, W., Andrewes, C. H. and Laidlaw, P. P. (1933) *Lancet* **ii,** 66.
Sovinova, O., Tumova, B., Pouska, F. and Nemec, J. (1958) *Acta Virol. Prague* **2,** 52.
Stuart-Harris, C. H. and Schild, G. C. (1976) In: *Influenza, the Viruses and the Disease.* Edward Arnold, London.
Stubbs, E. L. (1965) In: *Diseases of Poultry*, 5th edn. Ed. by H. E. Biester and D. D. Schwarte, Iowa State University Press.
Taylor, R. M. (1949) *Amer. J. publ. Hlth* **39,** 171.
Topp, F. H. Jr and Russell, P. K. (1977) *J. infect. Dis.* **136,** 376.
Tumova, B. and Pereira, H. G. (1968) *Virology* **27,** 253.

Tumova, B. and Schild, G. C. (1972) *Bull. Wld Hlth Org.* **47,** 453.
Tyrrell, D. A. J. and Smith J. W. G. (1979) *Brit. med. Bull.* **35,** 77.
Tyrrell, D. A. J. *et al.* (1981) *Bull. Wld Hlth Org.* **59,** 165.
van Wyke, K. L., Hinshaw, V. S., Bean, W. J. and Webster, R. G. (1980) *J. Virol.* **35,** 24.
Varghese, J. N., Laver, W. G. and Coleman, P. M. (1983) *Nature (Lond.)* **303,** 35.
Verhoeyen, M. *et al.* (1980) *Nature (Lond.)* **286,** 771.
Virelizier, J. L. (1975) *J. Immunol.* **115,** 434.
Virelizier, J. L., Allison, A. C. and Schild, G. C. (1974) *J. exp. Med.* **140,** 1571.
Virelizier, J. L. *et al.* (1974) *J. exp. Med.* **140,** 1559.
Waddell, G. H., Teigland, M. B. and Siegel, M. M. (1963) *J. Amer. vet. Assn.* **143,** 587.
Ward, C. W. and Dopheid, T. A. (1980) *Develop. cell Biol.* **5,** 27.
Waterfield, M. D., Espelic, K., Elder, K. and Skehel, J. J. (1979) *Brit. med. Bull.* **35,** 57.
Waterfield, M. D., Gething, M. J., Scrance, G. and Skehel, J. J. (1980) *Develop. cell Biol.* **5,** 27.
Webster, R. G., Hinshaw, V. S., Bean, W. J., Wyke, K.L., Geraci, J. R., St Aubin, D. J. and Petersson, G. (1981) *Virology*, **113,** 712.
Webster, R. G. and Laver, W. G. (1967) *J. Immunol.* **99,** 49; (1975) In: *The Influenza Viruses and Influenza*, p. 270. Ed. by E. D. Kilbourne, Academic Press, New York; (1980) *Develop. cell Biol.* **5,** 283.
Webster, R. G., Laver, W. G., Air, G. M. and Schild, G. C. (1982) *Nature (Lond.)* **296,** 115.
Webster, R. G., Yakhno, M., Hinshaw, V.S., Bean, W. J. and Murti, K. G. (1978) *Virology* **84,** 268.
Webster, R. G. *et al.* (1976) *J. gen. Virol.* **32,** 217: (1981) *Virology* **113,** 712.
Werner, G. H. (1966) *C. R. Acad. Sci. (Paris)* (D) **263,** 1913.
Wiley, D. C., Skehel, J. J. and Waterfield, D. M. (1977) *Virology* **79,** 446.
Wiley, D. C., Wilson, I. A. and Skehel, J. J. (1981) *Nature (Lond.)* **289,** 373.
Wilson, I. A., Skehel, J. J. and Wiley, D. C. (1981) *Nature (Lond.)* **289,** 366.
Winter, G. and Fields, S. (1981) *Virology* **114,** 423.
Winter, G., Fields, S. and Brownlee, G. G. (1981) *Nature (Lond.)* **292,** 72.
World Health Organization (1971) *Bull. Wld Hlth Org.* **45,** 119; (1979) *Ibid.* **57,** 227; (1980) *Ibid.* **58,** 585.
Wrigley, N. G. (1979) *Brit. med. Bull.* **35,** 35.
Wrigley, N. G., Laver, W. G. and Downie, J. C. (1977) *J. molec. Biol.* **109,** 405.
Wrigley, N. G., Skehel, J. J., Charlwood, P. A. and Brand, C. M. (1973) *Virology* **51,** 525.
Young, J. F. and Palese, P. (1979) *Proc. nat. Acad. Sci. (Wash.)* **76,** 6547.

97

Respiratory disease: rhinoviruses, adenoviruses and coronaviruses

E. J. Stott and David J. Garwes

Introductory

Respiratory disease is a major cause of human mortality and morbidity, particularly in the developed countries of the world. The intractable nature of the problem has recently been emphasized by Cockburn (1979) in a review of the causes of mortality recorded in the World Health Organization Data Bank between 1955 and 1975. While most infectious diseases decreased dramatically, by between 70 and 99 per cent during the 20-year period, acute respiratory infections showed little reduction and are now responsible for more than 99 per cent of deaths from infectious diseases in the developed world.

Table 97.1 Acute respiratory disease syndromes

Upper respiratory tract	Lower respiratory tract
Influenza	Croup (laryngo-tracheo-bronchitis)
Sore throat	Acute bronchiolitis of infants
Febrile cold	Lobar pneumonia
Common cold	Bronchopneumonia
	Pneumonitis

Accurate statistics on the morbidity due to respiratory disease are not available but most studies indicate that adults experience between 2 and 6 colds each year. Although generally mild, these illnesses cause a reduction in efficiency at work, if not actual absence from work, and therefore have considerable economic importance. Furthermore, among children under 5 years respiratory disease is the commonest cause of admission to hospital in developed countries.

In the agricultural industry the modern trend towards intensive rearing of livestock has led to large numbers of animals being collected into relatively small areas. Consequently, respiratory disease has become a major problem in poultry, pigs and cattle, resulting in substantial economic losses.

The classification of diseases of the respiratory tract is difficult. A number of different clinical illnesses may all be produced by one virus and, conversely, several quite distinct viruses may cause the same clinical picture. Several disease syndromes (Table 97.1) were defined by a Medical Research Council study group (Report 1965). These are not distinct and one syndrome may merge into another during the course of an illness. The clinical findings in patients infected with three viruses are shown in Table 97.2. There is considerable similarity in the diseases produced although the viruses are widely different taxonomically. Thus, it is quite impossible in a single case to diagnose a specific infection on clinical criteria. Furthermore, viruses are not the only cause of respiratory disease. Bacteria, mycoplasmas and chlamydias alone, or in association with viruses, frequently play a part.

In this chapter, three families of viruses which play a major role in mild upper respiratory disease will be described. The physicochemical and biological properties of the agents will be outlined, and their role in disease and the prospects for its prevention will then be considered.

Table 97.2 Clinical features of illnesses produced by rhinovirus, adenovirus and coronavirus

Feature	Rhinovirus	Adenovirus	Coronavirus
Mean incubation period (days)	2.1	NK	3.2
Mean duration (days)	10	7	6
Average number of handkerchiefs used daily	18	NK	21
Malaise (%)	25	18	47
Headache (%)	56	40	53
Chill (%)	15	20	18
Pyrexia (%)	18	80	21
Nasal discharge (%)	80	40	62
Sore throat (%)	73	46	79
Cough (%)	56	46	44
Nausea and vomiting (%)	1	8	0
Abdominal pain (%)	1	20	0
Diarrhoea (%)	1	10	0

NK = not known.
Data from Tyrrell, 1965; Bradburne *et al.* 1967.
Rhinovirus and Coronavirus data from volunteers; adenovirus data from natural infections.

Rhinoviruses

History

Kruse (1914) first demonstrated that a cold could be transmitted by bacteria-free filtrates and his observations were subsequently confirmed by Foster (1916) and Dochez and colleagues (1930). This work intensified in 1946 when the Common Cold Research Unit at Salisbury was established. The *in vitro* cultivation of rhinoviruses resulted from prolonged investigation of transmission of the common cold to human volunteers. The DC strain was passaged serially in explants of human embryonic lung in 1953 (Andrewes *et al.* 1953) and was subsequently shown to be a strain of rhinovirus type 9 (Conant *et al.* 1968). Meanwhile, in the United States, two cytopathogenic agents, JH and 2060, were described and shown to be serologically related (Price *et al.* 1959). They were originally classified as ECHOvirus 28 because of properties they shared with enteroviruses but subsequently became rhinovirus 1A (Kapikian *et al.* 1967). In 1960, Tyrrell and Parsons found that other viruses which caused common colds could be propagated in cultures of human embryonic kidney cells provided they were rolled at 33° in medium with low pH. This observation and the introduction of semicontinuous strains of diploid human embryonic lung fibroblasts (Hayflick and Moorhead 1961) led to a rapid increase in the number of viruses isolated. A variety of names was suggested including rhinoviruses (Andrewes *et al.* 1961), coryzaviruses (Hamparian *et al.* 1961), muriviruses and respiroviruses (Mogabgab 1962), ERC group (Ketler *et al.* 1962) and enteroviruses (Johnson and Rosen 1963). Rhinovirus became the accepted name (Virus Subcommittee of the International Committee on Nomenclature of Viruses, 1963), deriving from the special adaptation of these viruses to growth in the nose. During this period rhinoviruses were also isolated from horses (Plummer 1963) and cattle (Bögel and Böhm 1962). In 1963 rhinoviruses were defined as etherstable, RNA-containing viruses which were smaller than 30 nm in diameter and labile below pH 5 (Tyrrell and Chanock 1963).

It soon became clear that many different serotypes of human rhinoviruses could be distinguished by neutralization with specific antisera. The World Health Organization, therefore, initiated a Rhinovirus Collaborative Programme which established a numbering system based on reciprocal cross-neutralization studies (Kapikian *et al.* 1967, 1971).

Taxonomy

Rhinoviruses constitute one of the four genera within the family *Picornaviridae* (Cooper *et al.* 1978). Like other picornaviruses, the rhinovirus particle has a naked, ether-resistant, icosahedral capsid, 22–30 nm in diameter. The capsid consists of 60 structural units which are probably composed of one each of the four major structural polypeptides. The virus particle contains one piece of single-stranded RNA whose molecular weight is about 2.5×10^6. This RNA is infectious, is the message for protein translation, and carries a polyadenylated tract at the 3-prime end which is added during transcription. Virus multiplies in the cytoplasm producing by translation a large precursor polypeptide which is subsequently processed and cloven to yield functional proteins.

Rhinoviruses are distinguished from other genera of the *Picornaviridae*, primarily by their instability below pH 5.6 (though the foot-and-mouth virus is unstable below pH 7), their buoyant density in caesium chloride (1.38 – 1.41 g/cm^3), and their optimal temperature for replication (33°).

Although human and bovine rhinoviruses belong to the genus rhinovirus, the two serotypes of equine rhinoviruses remain unclassified because they differ in sedimentation coefficient, caesium chloride density and RNA properties from each other and from human rhinoviruses (Newman *et al.* 1977). They also differ in being more stable in acid, and replicating optimally at 37°.

Structure and properties of the virion

The virus particle consists of a protein capsid enclosing a single piece of RNA.

The apparent molecular weights of the four polypeptides that compose the capsid of rhinovirus 1A are 35 000; 30 000; 25 000 and 8000; similar values have been obtained for serotypes 2, 4 and 5 (Butterworth *et al.* 1976). In rhinovirus 1A up to eleven copies of polypeptides VP2 and VP4 may be replaced by their uncloven precursor VP0 but this does not occur in rhinovirus 2 (Lonberg-Holm and Butterworth 1976; McGregor and Rueckert, 1977). The organization of polypeptides in the rhinovirus capsid is probably similar to that of enteroviruses and cardioviruses. One molecule of each of the four polypeptides combine to form a protomer. Five protomers are bonded together in a pentameric subunit and twelve of these combine to form the sixty subunit capsid. In the intact virion VP1 is probably located on the surface and VP4 is an internal protein.

The molecular weight of rhinovirus 2 RNA is between 2.2 and 2.8×10^6. Polyadenylic acid sequences are found at the 3-prime end of the RNA which has an unusually high adenine content (Brown *et al.* 1970) but the polycytidylic acid tract found in the RNA of cardioviruses and aphthoviruses is not present in rhinoviruses.

Although the role of the polyadenylic acid sequence is not clear, it is important for RNA infectivity. At the 5-prime terminus of rhinovirus RNA the 'capping' group usually found on most eukaryotic messenger RNA is replaced by a small covalently-linked protein, VPg. This protein may act as a primer in RNA synthesis and also have a role in virus morphogenesis (Sangar 1979). Although in the intact virion RNA is protected from RNAse, it is partly exposed since it binds caesium ions and this accounts for the greater density of rhinoviruses in caesium chloride.

Weak bonding in the capsid may also account for the acid lability of rhinoviruses. Acidification of rhinoviruses produces A-particles which sediment at 135S and lack VP4, and B particles which sediment at 80S and lack both VP4 and RNA. Although A-particles contain RNA and are morphologically similar to native virions, they are not infectious because they cannot attach to receptors on the host cells (Butterworth *et al.* 1976).

Inactivation of rhinoviruses by heat occurs by two different mechanisms depending on the temperature. Prolonged incubation below 39° fragments virion RNA *in situ* but does not change antigenicity (Gauntt and Griffith 1974). At 50° A-particles are rapidly produced, containing infectious RNA but antigenically different from native virions. At 56° B-particles are produced.

Rhinoviruses exist in two antigenic forms: the native N- or D-antigen which is infectious and the H- or C-antigen which may be produced by heating mature virions and is composed of A- or B-particles (Butterworth *et al.* 1976). The antigens are detected by neutralization, haemagglutination-inhibition, immunodiffusion and complement-fixation reactions.

The neutralization of rhinovirus infectivity by antisera is largely type-specific and is the basis of the numbering system by which rhinoviruses are classified. Hence it is a reaction involving the N- or D-antigen. Minor serological variations may exist within a serotype. Within type 1 there are two subtypes A and B; the low-level cross-reaction between these subtypes may be reciprocal or one-way depending on the strain of rhinovirus 1A used (Monto and Johnson 1966). 'Prime' strains of type 22 which are antigenically broader than the prototype strain have been isolated (Schieble *et al.* 1970). Strains of type 51 isolated in 1962 were antigenically distinct from those isolated in 1965–1966, suggesting either that there was a progressive antigenic drift or that two variants circulated successively (Stott and Walker 1969).

Cross-reactions between different serotypes were first reported by Fenters and colleagues (1966) but subsequent work, though confirming that heterologous reactions occur, has not been consistent in defining which serotypes cross-react (Schieble *et al.* 1974; Cooney *et al.* 1975). Fox (1976) has suggested that there may be families of antigenically related rhinoviruses. However, cross-neutralization by hyperimmune sera does not prove that common D-antigens exist on different serotypes (Butterworth *et al.* 1976). Native virions carrying D- determinants may aggregate with A- or B- particles carrying C- determinants which will cross-react, giving apparent neutralization. Although most antigen-antibody aggregates are homogeneous, mixed aggregates of 'full' and 'empty' particles have been observed by immune electron microscopy (Kapikian *et al.* 1972).

Haemagglutination of sheep erythrocytes by certain rhinovirus serotypes is a property of the native virion and therefore haemagglutination-inhibition is a type-specific reaction like neutralization (Stott and Killington 1972*a*).

Both the D- and C- antigenic forms of rhinoviruses react in immunodiffusion and complement-fixation tests. Since the C- antigens of several rhinoviruses and other picornaviruses share determinants, their reactions are not type-specific (Butterworth *et al.* 1976; Smith 1978). The C- antigenic form occurs naturally as a less dense particle separated from virions by density gradient centrifugation or may be produced from native virions by gentle treatment with heat, acid or urea. The two antigenic forms may also be separated by isoelectric focusing. The change from D to C antigenicity probably results from a change in conformation of the virion polypeptides (Butterworth *et al.* 1976).

The infectivity of most rhinoviruses for cells is rigidly species specific (Stott and Killington 1972*b*). Apart from a limited number of human rhinoviruses which will replicate in monkey cells, human and bovine rhinoviruses infect cells only of the homologous species. Equine rhinoviruses are exceptional and grow readily in human, simian or rodent, as well as in equine, cells. The intrinsic infectivity of human rhinoviruses is between 1000 and 6000 physical particles per plaque-forming unit.

Replication

The growth cycle of several rhinovirus serotypes has been investigated in suspension or monolayer cultures of human continuous cell lines or semi-continuous cell strains (Butterworth *et al.* 1976). These studies show that the replication of rhinoviruses is similar to that of other picornaviruses with the important difference that human and bovine rhinoviruses multiply optimally at 33° rather than 37°.

The adsorption of rhinoviruses to trypsin-sensitive receptors on the cell surface is temperature-dependent and is enhanced by a high concentration of magnesium chloride. Different serotypes attach at different rates although there are the same number of receptors (approximately 10^4 per cell). There are at least two different receptors for human rhinoviruses. Serotypes 1A and 2 belong to one receptor family and types 3, 5, 14, 15, 39, 41 and 51 to another family which also includes the enterovirus Coxsackievirus A21.

Infectious rhinovirus attached to cells rapidly becomes non-infectious or 'eclipsed'. Fifty per cent of rhinovirus 2 is eclipsed in 5 min at 34.5°. Eclipse probably involves conversion of 150S native virion to 135S A particle lacking VP4 and subsequently to 80S B particle from which RNA has been released into the cytoplasm.

The infecting virion RNA acts as messenger for the synthesis of proteins required to produce both minus strand template RNA and new plus strands. Thus, the infecting RNA must act as both message for translation and template for transcription. As with other picornaviruses, rhinovirus RNA replicates on smooth endoplasmic reticulum probably via replicative intermediate and double-stranded RNA or replicative form, although Macnaughton and colleagues (1976) have suggested that the latter is an artefact induced by actinomycin D. Supra-optimal temperatures appear to inhibit rhinovirus replication by accelerating the degradation of newly formed RNA (Killington, Stott, and Lee 1977).

Most picornaviruses, including rhinoviruses, inhibit cell protein synthesis. This action is probably related to the difference between the 5-prime end of cell RNA which is methylated and that of picornavirus RNA which carries the protein VPg. Thus, in cells co-infected with poliovirus and rhinovirus 1A, translation of cell RNA is inhibited but translation of rhinovirus RNA is not. The rhinovirus RNA is translated as a single message yielding a precursor polypeptide of approximately 100 000 mol. wt which is cloven to give three main families of proteins. The capsid precursor is translated first and cloven to give the four capsid polypeptides. In rhinovirus 1A, a stable protein of 38 000 mol. wt is translated next and may be a protease involved in subsequent cleavages. Finally, a polypeptide of 84 000 mol. wt is produced and further cloven in a number of alternative ways (McLean, Matthews, and Rueckert 1976). The function of these non-structural proteins is not known but they are likely to be part of the viral RNA polymerase. The proteases responsible for cleavage may be host-derived, virus-induced or both. Their action on rhinovirus polypeptides can be prevented by zinc ions which probably stabilize regions of the precursor polypeptide (Butterworth and Korant 1974).

The assembly of protein and RNA into infectious rhinovirus particles is probably similar to that of enteroviruses (McGregor and Rueckert 1977). Twelve 14S subunits containing VP0, VP1 and VP3 combine to form an 80S empty capsid which is converted to a mature virion by the addition of RNA and cleavage of VP0 to VP2 and VP4.

Diagnosis

Rhinovirus infections are normally diagnosed by isolation of virus from respiratory secretions. Serological techniques have only limited value because of the large number of serotypes and absence of a specific rhinovirus group antigen.

Rhinoviruses are most frequently isolated from nasal washings but throat or nasal swabs and sputa are suitable alternative samples. Virus is rarely isolated from blood. Specimens should be transported to the laboratory in a buffered medium containing about 1 mg per ml of protein, such as bovine plasma albumen, and inoculated into cell cultures as soon as possible. Most rhinoviruses are rigidly species specific. Therefore, apart from the few M strains which grow in monkey kidney cells, human rhinoviruses will grow only in cells of human origin. Semi-continuous cell strains from human embryonic lung are most widely used for virus isolation, although certain virus strains are more readily isolated in diploid cells from human embryonic kidney or tonsil. Semi-continuous cell strains from different human embryos vary considerably in their sensitivity to rhinoviruses and these differences are not always revealed by titration of laboratory passaged viruses (Brown and Tyrrell 1964, Stott and Walker 1967, Fox *et al.* 1975). A continuous line of HeLa cells, sensitive to rhinoviruses, may also be used for primary isolation (Strizova *et al.* 1974, Cooney and Kenny 1977). Bovine rhinoviruses are usually isolated in secondary calf kidney cells. All cell cultures inoculated with rhinoviruses should be rolled at 33° in medium with pH below 7.6. Virus-infected cultures develop small areas of refractile cells, usually within 7 days. Such cytopathic agents are tentatively identified as rhinoviruses by demonstrating that they resist chloroform or ether, pass through a 50 nm filter, grow in the presence of bromodeoxyuridine (a DNA inhibitor) and are destroyed at pH 5. Passage of human rhinoviruses in monkey kidney cells to identify M strains is a useful preliminary classification. Final identification of a rhinovirus serotype requires neutralization by specific antiserum, but, with 113 human serotypes, this can be simplified by combining antisera into a scheme of pools (Kenny *et al.* 1970).

Organ cultures of human embryonic trachea or nasal epithelium detect some rhinoviruses not isolated in cell cultures (Higgins *et al.* 1969). Although most of these viruses can subsequently be passaged in cell culture, some, such as the HS strain apparently replicate only in organ cultures (Hoorn and Tyrrell 1966).

Serological diagnosis of rhinovirus infection by demonstrating rising antibody titres is practicable only when the serotype of the infecting virus is known. The microneutralization test has been widely used for the detection of rhinovirus antibodies (Gwaltney 1966, Monto and Bryan 1974). The haemagglutination-inhibition test is particularly useful for rapidly screening volunteers for antibodies before inoculation (Reed and Hall 1973) but can, of course, be used only for serotypes which haemagglutinate. This restriction does not apply to the passive haemagglutination test in which virus antigen is bound to erythrocytes by chromic chloride; these sensitized cells are then agglutinated by antibodies. However, extensive cross-reactions, presumably due to C antigens, make interpretation of the specificity of this test difficult (Faulk *et al.* 1971). The same limitation applies to the complement-fixation test.

Association with disease

Before their true identity was known rhinoviruses were established as a cause of the common cold by experiments in human volunteers (Andrewes 1965). Subsequent epidemiological studies established that this group of viruses was the major cause of mild upper respiratory disease in adults. They may also be responsible for some lower respiratory illness in young children and in adults with chronic respiratory disorders.

When human volunteers are inoculated intranasally with nasal secretions or tissue culture fluid containing rhinoviruses about 30 per cent develop colds whereas less than 5 per cent of volunteers given uninfected material have symptoms (Tyrrell 1968). Many volunteers fail to develop colds because of protection acquired by previous experience with the infecting serotype. When virus is given to adults without antibody as many as 100 per cent may develop colds (D'Allessio *et al.* 1976). At least 15 different serotypes have been given to volunteers with essentially the same results.

Epidemiological surveys indicate that rhinoviruses are responsible for most mild upper respiratory disease in adults. In studies of large numbers of university students (Gwaltney and Jordan 1966, Hamre *et al.* 1966, Phillips *et al.* 1968), industrial workers (Gwaltney *et al.* 1966, Hamparian *et al.* 1964), military personnel (Bloom *et al.* 1963) and civilian families (Elveback *et al.* 1966, Hendley *et al.* 1969, Higgins *et al.* 1966, Monto and Cavallaro 1972, Fox *et al.* 1975) rhinoviruses were isolated from 7–40 per cent of persons with upper respiratory illness but from less than 2 per cent of healthy persons. These results probably underestimate the importance of rhinoviruses because conventional techniques of virus isolation may detect only half the infections which occur in a population (Ketler *et al.* 1969, Fox *et al.* 1975). Thus, rhinovirus infection is significantly associated with upper respiratory disease. However, these same epidemiological studies also indicate that between 10 per cent and 40 per cent of rhinovirus infections are subclinical (Johnson *et al.* 1965, Hamre *et al.* 1966, Ketler *et al.* 1969, Fox *et al.* 1975), and that there may be substantial differences in pathogenicity between serotypes (Fox *et al.* 1975).

Rhinovirus infections are more common in young children than in adults. Virus may be isolated from 4 to 10 per cent of children with upper respiratory disease and from 2 to 5 per cent of children without symptoms (Bloom *et al.* 1963, Hamparian *et al.* 1964, Hendley *et al.* 1969, Higgins *et al.* 1966). Rhinoviruses are also associated with lower respiratory disease in children. However, in controlled studies of children in hospital rhinoviruses are found in 3 to 8 per cent of patients with lower respiratory illness and from similar proportions of children without respiratory symptoms (Chanock and Parrott 1965, Holzel *et al.* 1965, Mufson *et al.* 1970, Stott *et al.* 1967). Hence, the precise role of rhinoviruses in acute respiratory disease of children is unclear (Cherry 1973). In children with a history of wheezy bronchitis or asthma, rhinoviruses are associated with up to 33 per cent of acute attacks, and virus is found more frequently in sputum than in the nose or throat, suggesting that virus multiplies in the lower respiratory tract (Horn *et al.* 1979, Minor *et al.* 1974, Mitchell *et al.* 1978).

The ability of rhinoviruses to cause lower respiratory tract illness in adults is apparent when volunteers are infected with virus in a fine particle aerosol (Cate *et al.* 1965). Transient abnormal pulmonary function has also been found in a proportion of healthy volunteers after intranasal instillation of rhinovirus and in a minority of natural infections (Blair *et al.* 1976, Cate *et al.* 1973). Although rhinoviruses have been isolated from the lungs of adults who died of pneumonia, large surveys do not implicate rhinoviruses as a significant cause of adult pneumonia (Craighead *et al.* 1969, Person and Herrmann 1970, Mufson *et al.* 1967). However, in persons with chronic bronchitis, rhinovirus infections have been associated with 12 to 43 per cent of exacerbations (Eadie *et al.* 1966, Stenhouse 1968, McNamara *et al.* 1969) and virus frequently appears to invade the lower respiratory tract (Stott *et al.* 1968). Although Buscho and colleagues (1978) found that rhinoviruses could infect patients without exacerbating their chronic bronchitis, such patients are more susceptible to rhinovirus infection than normal persons (Monto and Bryan 1978). Furthermore, Horn and Gregg (1973) found lower respiratory tract signs in 85 to 90 per cent of rhinovirus infections in adults with a previous history of asthma or bronchitis but in only 5 per cent of such infections in normal adults.

Rhinoviruses are responsible for most mild acute upper respiratory disease in man and most infections are confined to the nasopharynx. However, in a minority of cases, particularly the very young or those with a history of chronic respiratory disease, rhinoviruses may invade the lower respiratory tract.

Bovine rhinovirus type 1 has been isolated from cattle with respiratory disease but there is little epidemiological or experimental evidence to show that it is an important cause of such disease (Stott 1975, Mohanty 1978). There is only one reported isolation of bovine type 2 virus (Reed *et al* 1971). Similarly, although infection with equine rhinoviruses is widespread in horses, their precise contribution to the problem of respiratory disease is not yet clear (Powell *et al.* 1978).

Pathogenesis

The mechanism by which rhinoviruses induce disease has largely been deduced from *in vitro* work in organ and cell culture. Rhinoviruses replicate in and destroy the ciliated epithelium of both nasal and tracheal mucosae but grow to highest titre in nasal epithelium (Hoorn and Tyrrell 1965). Ciliated cells are extruded from the epithelium leaving a generally smooth surface (Reed and Boyde 1972). Such virus-damaged tissue is probably more susceptible to secondary invasion by bacteria and mycoplasmas. It has been shown that mycoplasma replication is enhanced in rhinovirus-damaged epithelium (Reed 1972). The excessive mucus characteristically produced during a common cold is probably a response to the loss of cilia. In cases where rhinoviruses invade the lower respiratory tract, the target cell is not known but type II pneumocytes are infected and destroyed by rhinoviruses *in vitro* (Tyrrell *et al* 1979). Rhinoviruses are rarely found outside the respiratory tract probably because they have a tropism for ciliated epithelium and their growth is restricted by temperatures above 35°.

Transmission

The wartime slogan 'coughs and sneezes spread diseases' was an official endorsement of the widely held belief that colds arc primarily spread by airborne droplets. For rhinoviruses it now seems more likely that hand contact is the prime mode of transmission (Gwaltney and Hendley 1978, Gwaltney *et al.* 1978).

Although volunteers can be infected by rhinoviruses in an aerosol, attempts to recover virus from air in rooms occupied by infected persons have failed. Likewise, it has proved impossible to transmit rhinovirus colds between volunteers separated by a double wire mesh which allowed free flow of air but no direct contact, although under the same conditions Coxsackievirus A21 was readily transmitted. One reason for the absence of rhinovirus from the air may be that aerosols produced by talking, coughing and sneezing derive primarily from the saliva which has a low titre of virus and not from the nasal secretions in which rhinovirus titres are highest (Buckland and Tyrrell 1964).

In contrast, rhinovirus can readily be isolated from the hands of persons during the acute stage of a cold (Reed 1975, Gwaltney *et al.* 1978). Furthermore, contact between the hand and eye or nose, both proven portals of entry for the virus, occurs, on average, once every three hours (Hendley *et al.* 1973). Thus, in a comparison of routes of rhinovirus transmission, 11 of 15 hand-contact exposures initiated infection but only 1 of 22 aerosol exposures was successful (Gwaltney *et al.* 1978).

Although these conclusions are based mainly on experimental work, they are consistent with epidemiological findings which indicate that colds spread poorly in offices and other workplaces where many people share the same airspace but direct contact is minimal (Gwaltney *et al.* 1968). Transmission occurs more readily within familes (Hendley *et al.* 1969) where direct contact, particularly with children, is frequent.

Epidemiology

Rhinovirus infections occur throughout the year but are most prevalent in the spring and autumn when sudden changes in temperature most commonly occur. It has been suggested that rapid temperature changes are closely associated with the incidence of colds (Hope-Simpson 1958).

Geographically rhinoviruses are distributed throughout the world and antibodies to them have been detected even in the remote communities of Alaskan Eskimos, Pacific Micronesians and Kalahari Hottentots (Brown and Taylor-Robinson 1966).

Although early reports suggested that rhinoviruses were more frequently isolated from adults than children, the extensive surveys of Fox and colleagues (1975) indicate that rhinovirus infections are both more frequent and more severe in infants and children than in adults.

The multiplicity of rhinovirus serotypes is a major factor in their epidemiology. Several different serotypes circulate simultaneously within a community and two different serotypes may infect even the same individual at the same time (Cooney and Kenny 1977). Some serotypes persist continuously in a population for several years while others may appear for two months and then not be detected again for four years. Several surveys have demonstrated a change with time in the distribution of serotypes isolated. Before 1967 most isolates belonged to the first 55 serotypes; in 1968 and 1969 less than half the isolates were serotypes 1 to 55 and an increasing number were types 56–89. Since 1970, 50 per cent of rhinoviruses isolated do not belong to the first 89 types (Fox 1976). These results suggest that rhinoviruses are continuously changing antigenically, giving rise to new serotypes. This possibility is supported by the antigenic variation detected in serotypes 22 and 51, and antigenic alteration induced in type 2 by laboratory manipulation in the presence of antibody (Acornley *et al.* 1968). Some serotypes such as 1B, 12, 15 and 38 are more commonly isolated, possibly because they spread more effectively. Certainly, the more frequently isolated serotypes do have a higher secondary attack rate in family surveys (Fox *et al.* 1975). However, it must be remembered that frequency of isolation may reflect merely the sensitivity of the cell cultures used rather than the virus distribution in a population.

Prevention and treatment

The complete control of rhinovirus infections will be difficult. Two approaches are being pursued with some success: prevention by vaccination and treatment with antiviral drugs.

The immunology of rhinovirus infections has received little attention but it is established that infection stimulates an antibody response in 48 to 77 per cent of cases (Fox *et al.* 1975). The response is greater in adults than in children. Severe lower respiratory infection induces higher antibody titres than mild upper respiratory infection. Furthermore, certain serotypes (particularly M strains) are more effective antigens than others. Specific antibody in the circulation is related to protection against re-infection. This was shown experimentally when volunteers with pre-existing antibody were challenged with live virus and failed to become infected. The association has also been found in epidemiological surveys. Few pre-infection sera from persons shedding virus contain homotypic antibody. Furthermore, in studies of families where contact with known serotypes had taken place possession of antibody was 52 to 69 per cent effective in preventing infection. Although most correlations have been made between protection and serum antibody it is clear that rhinovirus infections also stimulate locally produced IgA. Whether secretory IgA in the mucus of the upper respiratory tract or circulating antibody in the serum is more important in protection remains contentious (Perkins *et al.* 1969*b*, Douglas and Couch 1972) but the protective value of antibody is clear.

The use of formalin-inactivated rhinovirus 1A vaccine to induce antibody was first described in 1963 (Doggett *et al.* 1963, Mufson *et al.* 1963). Subsequently, similar inactivated rhinoviruses 2 and 13 vaccines have been administered by subcutaneous, intramuscular or intranasal routes and have protected volunteers against live virus challenge (Scientific Committee 1965, Perkins *et al.* 1969*a*, Douglas & Couch 1972). Thus, the principle of vaccination against rhinoviruses was established over 15 years ago, but no large scale trials have been conducted because the problem of numerous serotypes with little or no cross-protection seemed insuperable. However, Fox (1976) believes renewed efforts to develop a rhinovirus vaccine are justified by recent epidemiological and immunological data. The finding that certain serotypes are more frequently encountered and tend to persist indicates that such serotypes account for a disproportionate share of disease and are prime targets for vaccines. Furthermore, the demonstration of many cross-reactions between serotypes suggests that some protection may be stimulated against serotypes not included but related to those in a vaccine. It is thus of interest that a decavalent vaccine has been developed which, although deficient in antigen mass, induced expected heterotypic responses in four of ten volunteers immunized (Hamory *et al.* 1975).

Live attenuated rhinoviruses offer another approach to vaccination but, so far, only type 15 has been attenuated by passage in tissue culture. Although this strain stimulated both nasal and serum antibody when given intranasally, the volunteers were not challenged and thus their protection was not assessed (Douglas *et al.* 1974).

An alternative approach to the control of rhinovirus infection is the use of interferon, interferon-inducers or other antiviral drugs. Both fibroblast and leucocyte interferon inhibit rhinoviruses *in vitro*. When assayed in HeLa cells, the interferon sensitivity of different rhinoviruses may vary by as much as one hundredfold (Came *et al.* 1976). In human volunteers, leucocyte interferon given as a nasal spray 9–12 times daily protected against challenge with rhinovirus type 4. A total dose of 14×10^6 units was given over 4 days (Merigan *et al.* 1973). In contrast, fibroblast interferon given as nasal drops three times daily failed to protect volunteers against rhinovirus type 4 infection, possibly because of its instability in nasal secretions (Reed 1980).

Interferon inducers active in animal models have proved less active in man. High molecular weight double-stranded RNAs such as polyriboinosinic-polyribocytidylic acid (poly I:C) and a fungal virus RNA (BRL 5907) have been used in volunteers challenged with rhinoviruses but the beneficial effects were small. The low molecular weight synthetic interferon inducer, CP 20961, reduced the severity of symptoms in volunteers challenged with rhinoviruses 13, 14, or 21 but subsequent reports on this compound and its successor were not favourable (Reed 1980). The use of live enterovirus vaccine to stimulate interferon has been advocated but when vaccinated volunteers were challenged with rhinovirus 4 no protection was observed (Matthews *et al.* 1974). The main problem in using interferon or its inducers against rhinoviruses is the difficulty of maintaining high concentrations in the upper respiratory epithelium for long periods. Until interferon is readily available and cheap it is unlikely to play a significant role in controlling rhinovirus infection.

Synthetic compounds which specifically inhibit rhinoviruses *in vitro* are often disappointing when tested *in vivo*. Guanidine and the benzimidazoles have been known as powerful inhibitors of picornaviruses for many years. Both compounds appear to inhibit the initiation of viral RNA synthesis without affecting cellular RNA synthesis. However, treatment of enterovirus infections in animals with these compounds has usually been ineffective, probably because of the rapid excretion of guanidine. A number of compounds related to guanidine or the benzimidazoles have been tested against rhinoviruses 4, 30 and 44 in human volunteers or in chimpanzees but in most trials the drugs either failed to inhibit virus growth or caused unacceptable side reactions (Reed 1980). Other anti-rhinovirus compounds which have been tested in man or apes with some success are the imidazo-thiazole RP 19326 and the triazino-indole group of compounds. RP 19326 has variable activity against different serotypes. Nevertheless, virus shedding and symptoms were reduced by the drug in volunteers challenged with the sensitive type 9 virus. The triazine-indole SKF 40491 is active *in vitro* against a wide range of serotypes but although active in gibbons infected with rhinovirus 2 or 1A it was not effective in volunteers challenged with rhinovirus 3 (Pinto *et al.* 1972, Reed *et al.* 1976).

Although vitamin C (ascorbic acid) is not an antiviral compound, it has been repeatedly recommended as a prophylactic or therapeutic agent for the common cold. Over 20 trials designed to assess the efficacy of vitamin C have been described but many of them failed to include appropriate controls or 'double-blind' assessment of the results. The danger of suggestion affecting the beneficial results reported has been convincingly described (Lewis *et al.* 1975). It is generally concluded that the effect of vitamin C supplementation on colds is, at best, small (Chalmers 1975, Dykes and Meier 1975). Rhinovirus infection does, however, significantly reduce the excretion of ascorbic acid in man (Davies *et al.* 1979).

Two newer compounds, an acetylated hydantoin CP 196J (Ajdukovic *et al.* 1978) and the benzimidazole-related LY 122771–72 (DeLong *et al.* 1978) have much greater antirhinoviral activity in HeLa or human fibroblast cells than any compounds previously tested. However, when tested against rhinovirus 31 in organ cultures of human embryo nasal epithelium, LY 122771–72 was highly inhibitory whereas CP 196J was relatively ineffective. This observation, that a drug which is active in cell culture may be ineffective in organ culture, may in part explain some of the disappointing results from clinical trials of earlier compounds. A further problem with many earlier compounds was their low solubility and poor absorption and distribution in man. These properties necessitated intranasal administration of the drug which was then rapidly removed by highly efficient mucociliary clearance. Thus, if the more recent compounds with greater activity *in vitro* possess even moderately favourable pharmacokinetic properties, there is real hope that they may offer effective therapy for rhinovirus infections (Reed 1980).

Adenoviruses

History

The first adenoviruses were isolated in 1953 (Rowe *et al.* 1953) from human adenoids which underwent spontaneous degeneration when maintained in culture for several weeks. The following year, Hilleman and Werner (1954) reported the isolation of three adenoviruses from the throats of American soldiers suffering from acute respiratory illness and suggested they might be responsible for some outbreaks of respiratory disease. In the subsequent 25 years, 35 serotypes of human adenoviruses have been described (Stadler *et al.* 1977) and they have been shown to infect man frequently and often persistently. This wide prevalence has hampered efforts to prove their aetiological role. However, it is now clear that adenoviruses types 1–7 cause about 5 per cent of respiratory disease in the general population and a much greater proportion in military recruits. Types 8 and 19 are responsible for epidemic keratoconjunctivitis (Jawetz *et al.* 1955, Sprague *et al.* 1973) and types 11 and 21 are associated with haemorrhagic cystitis in children (Mufson and Belshe 1976). Most of the remaining human adenovirus serotypes have no proven aetiological role, and, although diarrhoea, intussusception, pericarditis and congenital abnormalities have all been associated with adenoviruses, their role is far from clear.

In 1962 Huebner and colleagues reported the oncogenic effects in hamsters of adenovirus types 12 and 18. This finding stimulated intense structural, biochemical and genetic studies of certain members of the group. As a result the detailed structure of the virus is known, a genetic map has been constructed and functions have been ascribed to most regions of the DNA.

Taxonomy

Adenoviruses are members of the family *Adenoviridae* (Norrby *et al.* 1976). The virion has a naked icosohedral capsid 70–90 mm in diameter composed of 252 capsomeres with projections at the vertices. The particle contains double-stranded linear DNA with molecular weight of $20–30 \times 10^6$ and matures in the nucleus. Some members of the family are oncogenic in certain hosts. The family is subdivided into two genera which have distinct group-specific antigens; mastadenovirus comprising mammalian adenoviruses and aviadenovirus comprising the avian adenoviruses. Further subdivision of the mammalian adenoviruses may be necessary if it is confirmed that certain bovine adenoviruses do not contain the group-specific antigen found in all other mastadenoviruses.

In addition to the 35 adenovirus serotypes of human origin, there are 24 simian types, 1 equine, 8 bovine, 3 ovine, 2 canine, 2 murine and 4 porcine. The aviadenoviruses consist of 9 serotypes of fowl origin, 3 goose types and 2 serotypes from turkeys.

Adenoviruses from man have been further divided into four subgroups according to their haemagglutination of monkey and rat erythrocytes (Rosen 1958). This subdivision also loosely relates to length of fibres on the virions, guanine and cytosine content of the DNA, percentage of DNA-DNA homology between types, and oncogenic potential (Table 97.3).

Structure and properties of the virion

The virion consists of a protein capsid surrounding a dense core. The core is 60–65 mm in diameter and contains all the DNA and about 20 per cent of the total protein of the virion.

The 252 capsomeres which make up the capsid are arranged into an icosahedron having 20 triangular faces and 12 vertices (Horne *et al.* 1959). The 12 capsomeres at the vertices have five neighbours and are called pentons. Each penton carries a fibre which consists of a rod-like projection with a knob attached at the distal end. The 240 non-vertex capsomeres have six neighbouring capsomeres and are called

Table 97.3 Classification of adenoviruses of human origin

		Haemagglutination		Fibre length (nm)	GC in DNA (%)	DNA-DNA homology between types (%)	Oncogenicity	
Subgroup	Serotypes	Pattern	Cells				Animals	Cell culture
I	3, 7, 11, 14, 16, 21	Complete	Monkey	9–11	50–52	70–95	Moderate	Moderate
II	8–10, 13, 15,17, 19, 20, 22–28	Complete	Rat	12–13	57–61	NK	Low or none	Moderate
III	4	Partial	Rat	17–18	57–59	85–95	Low or none	Low
	1, 2, 5, 6			23–31				
IV	12, 18, 31	Little or none		28–31	48–49	80–85	High	Moderate

hexons (Ginsberg *et al.* 1966). Each capsomere is 7–9 nm in diameter with a hollow centre of 2.5 nm.

Proteins comprise 87 per cent by weight of mastadenoviruses and 83 per cent of aviadenoviruses. There are 8 to 10 polypeptides in the virion. Two are core proteins of 48 500 mol. wt and 18 500 mol. wt respectively; the latter is rich in arginine. The hexon consists of 3 identical polypeptides of 120 000 mol. wt. Five polypeptides of 85 000 mol. wt make up the penton base and three 62 000 mol. wt polypeptides compose the fibre. There are also three hexon-associated polypeptides.

Double-stranded DNA makes up 13 per cent of the adenovirus virion weight and has a molecular weight of $20–25 \times 10^6$ in mastadenoviruses, with distinct differences between types, and 30×10^6 in aviadenoviruses. The sedimentation coefficient of DNA varies between 29S and 35S. The base ratios of human adenovirus serotypes vary and correlate with their oncogenic potential (Pina and Green 1965). Types 12 and 18 with the highest oncogenic capacity have the lowest content of guanine and cytosine (48–49 per cent). Moderately oncogenic types 3 and 7 have a GC content of 49–52 per cent and weakly or non-oncogenic types 1, 2, 8 and 9 have 57–59 per cent GC (Table 97.3). Human adenoviruses within subgroups show more than 70 per cent DNA-DNA homology by hybridization techniques (Table 97.3), whereas between DNA of members of different subgroups, or between human and simian adenovirus DNA, only 10–25 per cent homology is found. Isolated adenovirus DNA is infectious and specific fragments representing only 6 per cent of the total genome will transform cells *in vitro*. Physiochemical properties of the virion include a density in caesium chloride of 1.33 g/cm^3 and a sedimentation constant of 795 S. Virus infectivity is stable between pH 6.0 and 9.5 and is undiminished after 70 days at 4°. Heating purified adenovirus type 5 at 56° for 10 min released the pentons and their five neighbouring capsomeres from each of the 12 vertices of the virion, making the viral DNA accessible to DNAse (Russell *et al.* 1967). Adenoviruses are resistant to chloroform, ether and fluorocarbon, but are disrupted by acetone, yielding nucleoprotein cores. Chlorine at 1 part per ten million in water inactivates adenoviruses.

Biological properties of adenoviruses include haemagglutination, antigenicity and infectivity. Most human and some animal adenoviruses agglutinate erythrocytes and the reactions between Rhesus monkey or rat red cells and human adenoviruses form the basis of a classification (Table 97.3). Virus attaches to specific receptors on the red cell membrane through the type-specific antigen on the fibre and haemagglutination is therefore inhibited by type-specific antibody. Haemagglutination may, however, be enhanced by the addition of heterotypic antibody which cross-links through group-specific antigens.

The antigenic composition of adenoviruses is complex. The hexon carries four antigenic specificities: a group-specific activity demonstrable by complement fixation and shared by all mastadenoviruses, a type-specific activity shown in neutralization and, probably, intrasubgroup and intersubgroup activities (Norrby and Wadell 1969). The penton base carries toxic activity and weak group-specific antigenicity as well as inter- and intrasubgroup activity as demonstrated by haemagglutination enhancement. The fibre has type-specific antigenicity carried in the knob region and demonstrated by virion haemagglutination-inhibition. Fibres longer than 12 nm (Table 97.3) also have intrasubgroup activity in their proximal region (Wadell and Norrby 1969). The major core protein contains 23 per cent arginine and like other basic proteins is only weakly antigenic.

The infectivity of adenoviruses is not species specific. Strains of human origin will infect monkey, rabbit, mouse and calf cells, although they are most readily propagated in continuous cell lines derived from human epithelial cells. Primary human embryonic kidney cells are also particularly sensitive. Adenovirus infection of cells is usually lytic and results in the release of up to 10^6 progeny particles per cell with an intrinsic infectivity of between 10 and 2000 particles per infectious unit. Persistent and abortive infections also occur yielding little or no infectious virus. Simultaneous infection of monkey cells with human adenoviruses type 7 and SV40 produces defective hybrid viruses containing parts of the adenovirus and SV40 genomes within an adenovirus capsid. Finally, adenovirus infection may transform cells and, in this case, infectious virus is only rarely recovered but virus-specific DNA sequences are found integrated into the host genome. Only the productive infection will be considered in this chapter: details of other types of infection may be found in the review of Philipson and colleagues (1975).

Replication

Replication of adenoviruses has been studied most intensively in suspension cultures of HeLa or KB cells. At high multiplicity of infection the growth cycle may be divided into two phases. In the first period, lasting 6–8 hours, early antigens are produced but only about 40 per cent of the viral genome is expressed. During the second phase viral DNA is produced and most of the genome is expressed resulting in a change in the production of viral mRNA (Philipson *et al.* 1975).

The first event is attachment of virus particles to specific receptors of which there are about 10^4 per KB cell. Within the next few minutes uncoating of the virion begins, probably by removal of penton capsomeres, and virus is taken into the cell. Virions are then transported to the nuclear pores, probably by microtubules, where the adenovirus core is exposed by removal of hexon capsomeres. Inside the nucleus, intact virus DNA is freed of all remaining virion proteins. The complete uncoating process takes 1–2 hours.

Complementary viral RNA production in the nucleus begins as soon as the DNA is fully uncoated. Several species of RNA are produced ranging in molecular weights from 3×10^5 to 3×10^6, and all of them are polyadenylated in the nucleus. By 5 hours after infection virus-specific RNA accounts for 10–15 per cent of mRNA associated with polyribosomes. Because most of the proteins synthesized early after infection are host coded, it is difficult to detect virus-specific protein synthesis. However, by means of antisera from hamsters bearing adenovirus-induced tumours, an early T- or tumour antigen has been detected. There is also a P-antigen.

About 8 hours after infection, viral DNA synthesis begins in the nucleus and within five hours accounts for 90 per cent of the total DNA synthesis in the cell. There is also an increase and qualitative change in viral mRNA synthesis. Large mRNA molecules are synthesized and cloven in the nucleus to give smaller molecules in the cytoplasm. These are of two main classes, sedimenting at 22 S and 26 S respectively. Coincident with the production of viral DNA and changes in viral mRNA synthesis, capsid proteins are first produced on polyribosomes in the cytoplasm. There is probably some minor cleavage of precursor polypeptides immediately after synthesis or during assembly. Late in the cycle large amounts of structural proteins accumulate in the nucleus where about 10 per cent of the viral DNA and protein are assembled into virions. Assembly is dependent on an adequate arginine concentration. At this stage virus material may be observed as paracrystalline arrays in the nucleus. Virus is released by disruption of the nuclear membrane.

Diagnosis

Adenovirus infections are diagnosed either by virus isolation, detection of viral antigen or by serological tests.

Virus may be isolated from respiratory secretions or tissue, ocular secretions, urine or faeces. However, in the diagnosis of respiratory infection, which is the subject of this chapter, a nasal or throat swab is preferable and should be broken into transport medium immediately after collection. A sample of the transport medium containing respiratory secretion and cells is then inoculated on to young monolayer cultures of HeLa, HEp-2 or KB cells and when available human embryonic kidney cells. Since adenovirus replication is dependent on arginine, it is important to use medium containing adequate concentration of this amino acid and to prevent contamination of cells with mycoplasmas which metabolize arginine. Adenovirus cytopathic effects may appear in one to four weeks and are often characterized by rounding and clumping of cells giving the appearance of 'bunches of grapes', before they slough from the glass. The medium also turns yellow owing to excess acid production. Such cytopathic agents may be identified as adenoviruses by complement fixation of the tissue culture fluid with standard antiserum. Isolates may be further characterized into subgroups by their haemagglutination with rat and monkey erythrocytes, and into serotypes by haemagglutination-inhibition or neutralization with specific antisera. Adenoviruses may also be isolated from excised lymphoid tissue, such as tonsil, but this frequently requires cultivation of the tissue for several weeks.

Infection may be diagnosed more rapidly but with less precision by detecting viral antigen in cells collected from the respiratory tract by means of fluorescein-labelled antiserum.

Serological diagnosis depends on demonstrating a fourfold or greater rise in antibody titre between paired serum samples. Complement fixation is most commonly used because the antibodies measured are group specific. If the serotype of the infecting adenovirus is known, antibodies may also be titrated by haemagglutination-inhibition or neutralization.

Association with disease

Although adenoviruses have been associated with diarrhoea, intussusception, meningitis, nephritis and cystitis, their most frequent and strongest aetiological association is with respiratory disease.

Infection of volunteers by swabbing the conjunctiva with adenovirus types 1–6 showed that these viruses could cause catarrhal conjunctivitis often associated with febrile or afebrile pharyngitis. When virus was administered by intranasal drops conjunctivitis was not seen. Adenoviruses 26 and 27 also caused conjunctivitis in volunteers but, although these viruses colonized the gut, no other clinical signs were recorded (Tyrrell 1965; Knight *et al.* 1963). When volunteers were infected by adenovirus type 4 in a small particle aerosol, febrile upper and lower respiratory disease was produced (Couch *et al.* 1966).

Epidemiologically, it is difficult to prove that natural adenovirus infections cause respiratory symptoms, because adenoviruses are found in apparently normal tonsils and adenoids and may be shed intermittently for long periods after infection, particularly by children, especially types 1, 2 and 5. A causal relation between illness and infection by types 1, 3 and 5 was first proved by a painstaking survey of children in an American orphanage (Bell *et al* 1962). The same survey also showed that a child was very likely to become ill when first infected, but upon reinfection (or reactivation of virus shedding) was much less likely to have an illness. Subsequent surveys of families indicate that 45–49 per cent of adenovirus infections are associated with illness but that most of these occur in children and account for only a small proportion of the total respiratory disease problem. Thus, only 5 per cent of illnesses in the first year of life were associated with adenoviruses and this figure declined to 1 per cent in 5 to 9-year-old children (Fox *et al.* 1977). However, in a small number of cases adenovirus infection may result in severe and sometimes fatal pneumonia, particularly in children (Brandt *et al.* 1969). Disseminated fatal adenovirus infections have also been observed in patients on immunosuppressive therapy. Adenoviruses have their most striking impact on military recruits. Between 30 and 80 per cent of this population become infected, usually with types 4 or 7, and 20 per cent may require hospitalization (van der Veen 1963). The reasons for such explosive outbreaks in military training camps are not clear. It seems likely that a high concentration of susceptible persons in crowded conditions are important factors. The same reason may account for an explosive outbreak of fever and pharyngitis caused by adenovirus type 7B in a boarding school in which 75 out of 294 boys had to be admitted to the sanatorium (Report 1981).

The main clinical syndromes associated with adenovirus infections in man and the predominant serotypes involved are listed in Table 97.4. Simian, bovine, ovine, canine and equine adenoviruses are all associated with respiratory disease in their respective hosts. Certain foals of the Arabian breed, which have an inherited combined immunodeficiency, possess virtually no immunoglobulin and drastically reduced numbers of T and B lymphocytes. In these animals generalized adenovirus infection is the major cause of death (Perryman *et al.* 1978). Canine adenovirus causes a fatal hepatitis in puppies and acute fatal encephalitis in foxes. Among avian adenoviruses, chicken embryo lethal orphan (CELO) virus causes quail bronchitis and GRL virus is responsible for liver necrosis in chickens. (For winter vomiting disease, See 6th edn., p. 2291.)

Table 97.4 Association between clinical syndromes and predominant adenovirus serotypes

Syndrome	Predominant serotypes
Respiratory disease in children	1, 2, 5, 6
Sore throat, febrile cold Pneumonia	3, 4, 7, 14, 21
Epidemic keratoconjunctivitis (shipyard eye)	8
Respiratory epidemics in military recruits	4, 7, 21

Pathogenesis

The pathogenesis of mild adenovirus infections has not been studied in any detail. In organ cultures of human embryonic nasal mucosa and trachea, adenoviruses infect and destroy the ciliated epithelium and it is likely that the same process occurs *in vivo*. Many infections are abortive with no illness and little virus excretion; other infections are associated with illness and prolonged excretion of virus in faeces.

In fatal pneumonia of children or adults, bronchial and tracheal epithelia may be denuded and deeply basophilic nuclear inclusions may be seen in degenerating cells. Hyaline membranes often line the alveolar air spaces, replacing the alveolar lining cells.

Although organs outside the respiratory tract may be affected during severe adenovirus infection, virus replication rarely occurs unless the patient is immunodeficient. Thus, it has recently been suggested that the damage to brain, liver, skeletal muscle and mycocardium in fatal adenovirus type 7 infection of children may be due to the penton antigen found in their sera and known to be cytotoxic *in vitro* (Ladisch *et al.* 1979).

Transmission

Although transmission of adenoviruses has received less attention than that of rhinoviruses, certain facts are known. Adenoviruses can infect man via the conjunctiva or the nasal mucosa. However, when virus is administered in an aerosol of particles 0.3–2.5 μm in size, infection is more effective and the human infectious dose is equivalent to one 50 per cent tissue culture infecting dose. The relative roles of contact and airborne transmission in adenovirus epidemiology have not been explored. Nevertheless, adenovirus infections that are followed by persistent faecal excretion spread to a higher proportion of susceptible contacts than purely respiratory infections. This implies that faecal-oral spread occurs, particularly among children.

Epidemiology

Seasonal variation of adenovirus epidemics is well recognized. Most outbreaks of pharyngoconjunctival fever in school-age children occur in the summer. In contrast, adenovirus epidemics in military recruits occur almost exclusively in the winter, as does adenovirus pneumonia of children. Endemic adenovirus infections associated with upper respiratory illness, or no disease at all, occur throughout the year but most frequently in winter and spring. Geographically adenoviruses occur throughout the world.

Adenovirus infections occur early in life and by 5 years of age practically all children have been infected by at least one serotype and over half of them by four types. In a civilian population about 7 per cent of children and 2 per cent of adults have evidence of adenovirus infection in a year. Age also affects the type of illness associated with adenovirus infection. Lower respiratory disease is more common in children than adults. Pharyngoconjunctival fever is characteristically found in children of school age. In adults adenovirus infections causing upper respiratory illness occur sporadically.

The epidemiology of individual serotypes varies. Of over 12 000 infections diagnosed worldwide 95 per cent were serotypes 2, 1, 7, 3 and 5, in that order of importance. Infection with types 1 and 2 tends to occur early in life and later with types 3 and 5. All these serotypes are endemic in the population. However, types 4, 14 and 21 tend to occur in explosive epidemics in semi-closed communities such as military recruitment camps.

Prevention and treatment

Vaccination has been used, with some success, to control epidemics of adenovirus infection in military recruits. However, although a number of compounds inhibit adenovirus growth *in vitro*, no effective chemotherapy has yet been developed.

The protective effect of serum antibody against adenovirus infection was first demonstrated in volunteer experiments (Bell *et al.* 1956). Subsequent epidemiological surveys have confirmed and extended this observation. Homotypic antibody is 85 per cent effective in preventing infection and probably reduces the risk of illness when infection does occur. There is no evidence that heterotypic antibody affects either the occurrence of infection or the subsequent development of disease (Fox *et al.* 1977). Furthermore, it is clear that serum antibody is protective in the absence of any local antibody produced in the upper respiratory tract. These aspects of immunity to adenovirus infections have influenced vaccine strategy.

Inactivated and live adenovirus vaccines have been used in man. The first effective vaccine produced within two years of the discovery of adenoviruses was formalin-inactivated types 4 and 7 virus given intramuscularly (Hilleman *et al.* 1956). When tested in over 300 military recruits this vaccine reduced hospitalization from 24 per cent in the placebo group to 5 per cent in the vaccinated group and virus shedding was reduced from 75 to 33 per cent. This preparation was subsequently abandoned because of poor potency standards, contamination with simian viruses, and potential oncogenicity of adenoviruses particularly when hybridized with simian virus 40 frequently found in monkey kidney cell cultures.

Alternative inactivated vaccines consisting of purified capsid proteins of adenovirus type 5, freed from DNA, have been tested in volunteers. A hexon vaccine gave partial protection from illness on subsequent challenge and a fibre vaccine gave complete protection in 8 volunteers (Couch *et al.* 1973).

Live adenovirus vaccines administered by an atypical route have proved to be the most successful. Initial development concentrated on type 4 because it lacks oncogenic potential and causes widespread epidemics of acute respiratory disease. Unattenuated virus was fed to volunteers in enteric-coated capsules thus bypassing the respiratory tract in which symptoms are most frequently detected and establishing an asymptomatic infection in the lower alimentary tract. Moderate amounts of neutralizing antibody were produced. The infection did not spread from volunteers or military recruits to susceptible contacts despite prolonged exposure. Subsequent studies indicate that vaccine virus may be transmitted between marital partners but that secondary infection does not give rise to respiratory tract disease. The type 4 vaccine, grown in human diploid cells, has been given to several hundred thousand military recruits with no untoward side effects. It has significantly reduced the incidence of febrile respiratory disease and hospital admissions while the incidence of a febrile respiratory disease was unchanged. Addition of type 7 virus to the type 4 vaccine increased its effectiveness in controlling disease; interference between the two viruses does not occur when they are administered simultaneously. A triple vaccine incorporating types 4, 7 and 21 is now being assessed (Dudding *et al.* 1973, Jackson and Muldoon 1973, Chanock 1974, Takafuji *et al* 1979).

Live adenovirus vaccines have effectively reduced respiratory disease in military recruits. It is, however, still questionable whether the same success can be expected in the civilian population where a greater number of serotypes are endemic and where adenovirus associated illness constitutes a much smaller proportion of the total respiratory disease problem.

Coronaviruses

The coronaviruses comprise a monogeneric family, the *Coronaviridae*, with similar morphology and biochemical structure. They infect a wide range of animals causing diseases of the gastrointestinal and respiratory tracts, liver, kidneys and nervous system. Though many coronaviruses are able to grow in the lung and upper respiratory tract, only three are considered to be the cause of respiratory diseases: human respiratory coronaviruses (HCV), avian infectious bronchitis virus (IBV), and rat pneumonia virus (RCV). We shall be concerned here principally with the human viruses but will include a section on the avian and rat viruses for comparison.

History and taxonomy

The first report of successful cultivation of a virus, subsequently classified as a coronavirus, from a patient with a common cold came from Tyrrell and Bynoe in 1965. They had been studying a virus isolated from a boy in 1960 and passed in human volunteers at the MRC Common Cold Research Unit. The virus, designated B814, could not be cultivated in a wide range of tissue culture systems but was shown to grow in human embryonic tracheal organ culture (HETOC). From the inability of the virus to replicate in cell cultures and the absence of antibodies to known human respiratory viruses in inoculated volunteers, it was concluded that B814 was a new type of common cold virus which might be related to the myxoviruses.

During the same period of time, Hamre and Procknow (1966) in the USA were examining a virus, 229E, that had been isolated from a student in 1962. The virus was readily adapted to grow in human embryo kidney cells where it produced an obvious cytopathic effect. The authors showed that the virus was ether-labile, approximately 89 nm in diameter and probably contained RNA. Serological tests suggested that the virus was not related to any known myxoviruses or paramyxoviruses. Electron microscopical examination of these two isolates in 1967 revealed negatively stained preparations with virus particles that resembled avian infectious bronchitis virus, suggesting the existence of a previously unrecognized group of viruses.

Continued use of HETOC for cultivation of human respiratory viruses resulted in the isolation of several more viruses morphologically resembling B814 and 229E. These included OC38 and OC43 from McIntosh and his colleagues (1967) and the British isolates LP and EVS from the MRC Common Cold Research Unit (Tyrrell *et al.* 1968*b*). The isolation of these viruses and the demonstration that they shared a common morphology not only with avian infectious bronchitis virus but also with mouse hepatitis virus prompted a group (Tyrrell *et al.* 1968*a*) to suggest that all these viruses should be classified in a new 'coronavirus' group, deriving its name from the appearance of a halo (L. = corona) surrounding the virion. The virus group acquired the status of a family, '*Coronaviridae*', in 1975 (Tyrrell *et al.* 1975); and more recent work on the structure and replication strategy of the coronaviruses established them as a unique taxonomic group characterized by a polyadenylated genomic RNA with messenger function, complexed with a basic protein to form a helical nucleocapsid and surrounded by a lipid envelope (Tyrrell *et al.* 1978, Garwes 1979).

Structure and properties of the virion

Morphology

The initial observations of HCV strains 229E, B814 and OC16 by electron microscopy of negatively stained preparations established the morphology that has been found to be characteristic of coronaviruses. The particles are circular, rather pleomorphic with a diameter in the range 80–120 nm (Almeida and Tyrrell 1967) or 120–200 nm (McIntosh *et al.* 1967). The surface of the particle is covered with bulbous projections that are approximately 20 nm long and 10 nm wide at their distal end (Fig. 97.1).

In a more recent study, Davies and Macnaughton (1979) showed that the size of 229E could be affected by the choice of negative stain used to observe them. Table 97.5 shows these differences, confirming at the same time the dimensions previously published.

Strain 229E was examined in thin sections of infected WI38 cells by Becker and his colleagues (1967). Particles having a mean diameter of 82 nm (66–112 nm range) were seen in vacuoles; no surface projection could be detected by this method but the virus particle appeared to have a complex internal structure. The lipid envelope, 7–8 nm thick, surrounded an electron-translucent zone that was 4–8 nm across. Within this was an inner shell of 9–17 nm width enclosing a central, amorphous area of variable density. The 229E particles were similar in appearance to those of IBV except that the avian virus lacked the electron-translucent zone.

It could be postulated that the inner shell within the 229E virion might comprise a nucleocapsid containing the viral genome. Recent studies from several laboratories have examined internal structures from human coronaviruses and have identified the nucleocapsid. Kennedy and Johnson-Lussenburg (1975/76) disrupted 229E with the non-ionic detergent NP40 and purified a viral component by equilibrium centrifugation in sucrose. RNA-containing material banding at a density of 1.27 $g.cm^{-3}$ (compared with 1.19 $g.cm^{-3}$ for intact virus) was examined by electron microscopy after negative staining. Loosely twisted, helical, continuous strands were seen, having an average width of 9 nm. Similar structures were reported by Caul and his associates (1979) from untreated and NP40-lysed purified 229E. The diameters of the

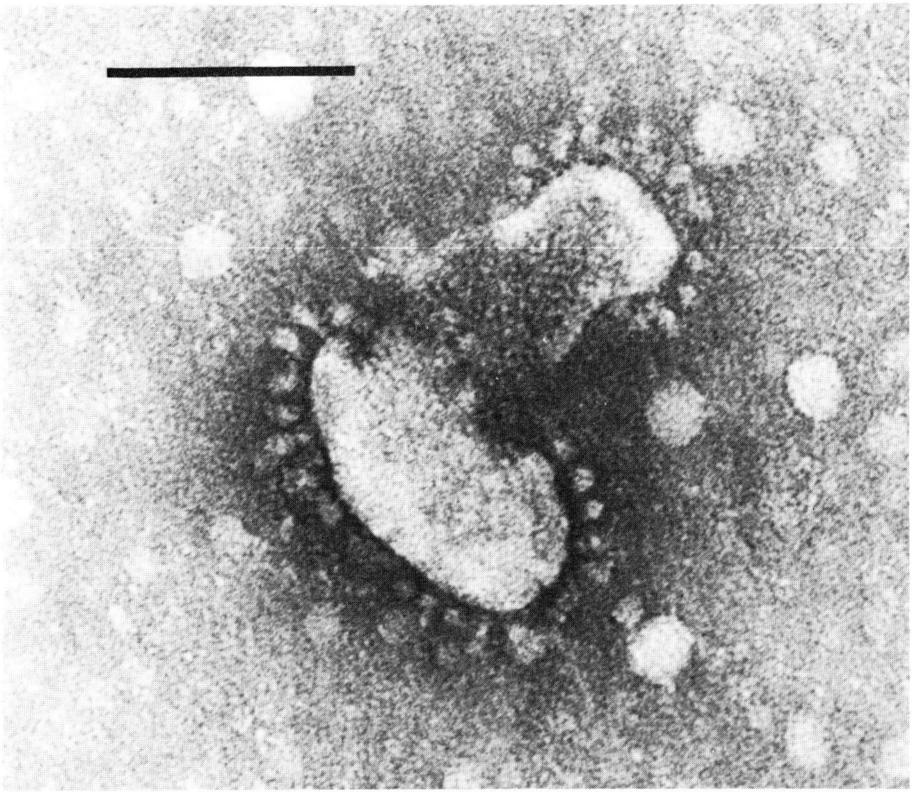

Fig. 97.1 Electron micrograph of a negatively stained particle of HCV 229E. The bar represents 100 nm. (Courtesy of E. O. Caul)

Table 97.5 Dimensions of HCV-229E negatively stained with potassium phosphotungstate or uranyl acetate

	Potassium phosphotungstate 2%, pH 6.5				Uranyl acetate 0.5%, pH 4.4			
	No. of observations	Mean	Standard error	Range	No. of observations	Mean	Standard error	Range
Envelope diameter	49	89.6*	1.5	67–123	61	108.9	2.3	75–152
Surface projection length	89	20.3	0.19	15.6–23.8	88	10.9	0.29	7.2–19.5
Surface projection width	90	8.7	0.09	5.8—10.5	66	9.8	0.17	7.7–13.0

* All dimensions in mm.
Adapted from Davies & Macnaughton (1979).

strands were 9–11 nm after negative staining with potassium phosphotungstate and 11–13 nm with uranyl acetate. In addition, a central canal of 3–4 nm diameter was seen running through the centre of the strand (Fig. 97.2).

These clearly represent the coronaviral nucleocapsid but differ from structures reported by Macnaughton, Davies and Nermut (1978). Incubation of purified 229E at 23° for several hours liberated an internal component that had a diameter of 14–16 nm. The strands appeared to be composed of globular subunits with their long axes, 5–7 nm, normal to the axis of the helix. The relation between these structures of 14–16 nm diameter and those reported to be 9–11 nm is not clear but it is thought that they represent the same structure in two different configurations.

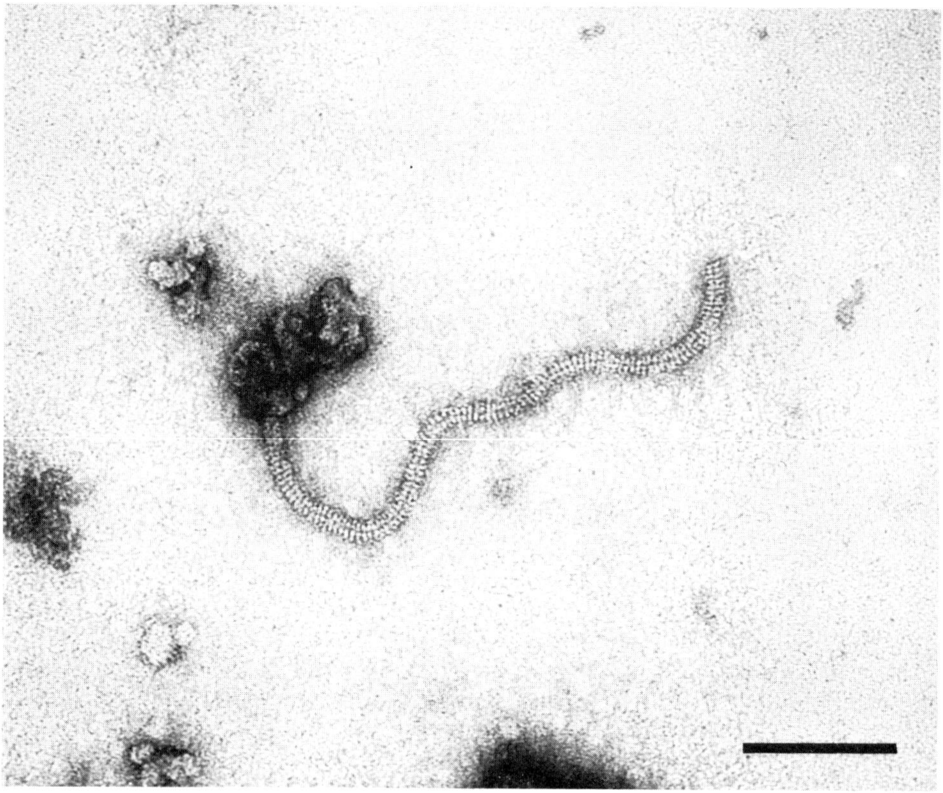

Fig. 97.2 Electron micrograph of isolated nucleoprotein of HCV 229E. The bar represents 100 nm. (Courtesy of E. O. Caul)

Composition

Nucleic acid

It was shown that the cell-culturable human coronaviruses were resistant to the action of the DNA inhibitors 5-iodo-deoxyuridine, 5-bromodeoxyuridine and 5-fluorodeoxy-uridine at the time of their isolation (Hamre and Procknow 1966, Kapikian *et al.* 1969, Bradburne 1969), suggesting that their genomes contained RNA. Similar work on other members of the coronavirus group confirmed this but it was several years later before the characteristics of the virion RNA were studied.

Tannock and Hierholzer (1977) extracted the RNA from purified HCV-OC43 that had been grown in mouse brain in the presence of ^{32}P. The material had a sedimentation coefficient of 70S, corresponding to a molecular weight of approximately 6.1×10^6. The genome appeared to be in the form of a heat-dissociable complex, however, breaking down into smaller molecules in the range 70–4S on heating at 60°.

The following year however, Macnaughton and Madge (1978) showed that the RNA from HCV-229E contained a single molecule with molecular weight of 5.8×10^6. The molecule contained tracts of polyadenylic acid [poly (A)] at or near the 3′ terminus and the presence of poly (A) in the genome of HCV-OC43 was confirmed in the same year by Tannock and Hierholzer (1978). Additionally, these authors could detect no RNA polymerase activity in the HCV-OC43 virion suggesting, together with the finding of genome poly (A), that the RNA constitutes a messenger strand that can code for polypeptides directly after entry into the host cell. This feature has been confirmed for several other coronaviruses and clearly distinguishes the group from the myxoviruses, paramyxoviruses, and rhabdoviruses. The only other taxonomic group having a lipid-enveloped, helical nucleocapsid with poly (A)-RNA is that of the retroviruses but this group replicates via a DNA intermediate synthesized by virion reverse transcriptase. This enzyme has not been detected in any coronavirus and recent studies on members of the group suggest that coronaviruses replicate via a double-stranded RNA intermediate.

Protein

The first description of the polypeptide composition of a coronavirus was that of HCV-OC43 by Hierholzer and his associates in 1972. After polyacrylamide gel electrophoresis of sodium dodecyl sulphate-dissociated purified virus, a total of 6 polypeptides could be reliably detected (Fig. 97.3). Of these, 4 reacted in a test for carbohydrate, suggesting that they consisted of glycopolypeptides and that the largest polypeptide also contained lipid. Treatment of the virus with the proteolytic enzyme bromelain removed the surface pro-

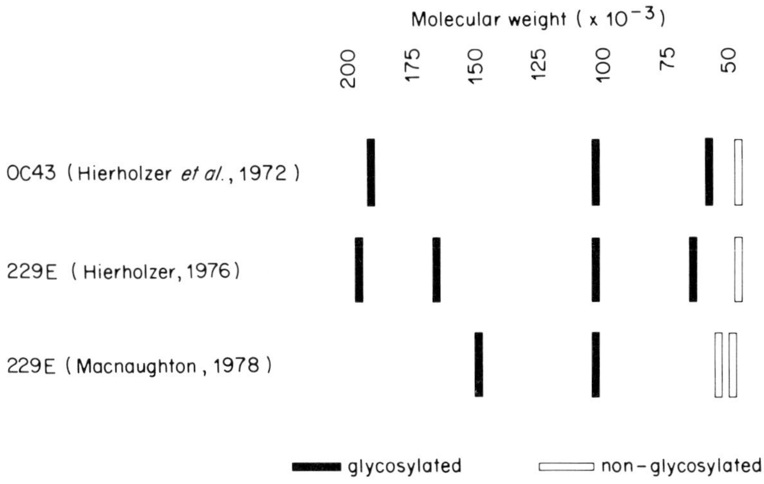

Fig. 97.3 Polypeptide composition of human coronaviruses.

jections and glycopolypeptides 2 and 6, indicating that these were associated with the surface projections.

Hierholzer followed this in 1976 with a similar study of HCV-229E. He found a similar pattern of polypeptides to that of HCV-OC43 (Fig. 97.3) with small differences in molecular weight and in the glycosylation pattern.

A comparative study of HCV-229E with avian infectious bronchitis virus and mouse hepatitis virus (Macnaughton 1978) produced a simpler polypeptide pattern for the human coronavirus than that demonstrated by Hierholzer (Fig. 97.3). Macnaughton found 5 structural polypeptides of which 3 were glycosylated. These differences observed between the two laboratories may, however, only reflect differences in purification and analytical procedures; the predominance of glycoproteins in coronaviruses tends to give more variable results in gel electrophoresis than would be obtained with viruses which lacked glycopolypeptides. The HCV patterns fit with the general scheme of protein composition established with a number of different coronaviruses: a nucleocapsid, containing a non-glycosylated protein of molecular weight 50–60 000, surrounded by a lipid envelope in which are located glycopolypeptides of < 50 000 molecular weight and surface projections composed of glycopolypeptides of > 90 000 molecular weight (Garwes 1979).

Haemagglutination

The organ culture isolates OC38 and OC43 agglutinated erythrocytes after the viruses had been passaged several times through mouse brain (Kaye and Dowdle 1969) but HCV-229E did not. Both OC38 and OC43 reacted similarly in their ability to agglutinate erythrocytes from different species at different temperatures (Table 97.6). Unlike haemagglutination with myxoviruses and paramyxovirus, however, erythrocyte neuraminic acid did not seem to be necessary for the coronavirus activity. Treatment of the viruses with trypsin or ether totally destroyed haemagglutination activity and infectivity; heating at 56° for 30 minutes considerably reduced it. Antibodies prepared against the virus inhibited the haemagglutination activity, suggesting that the haemagglutinin formed an integral part of the virion. Further evidence for this was provided by Hierholzer and his colleagues (1972) who showed that removal of the surface projections from the virion of OC43 with bromelain also totally destroyed haemagglutinin activity.

As seen in Table 97.6, the human coronaviruses will not haemagglutinate human 'O' cells at room temperature or 37°. This feature was used by Hierholzer and his associates (1972) to purify the particles, since virus bound to human erythrocytes at 4° would elute at 37°. Attempts by these workers to identify a viral neuraminidase were unsuccessful, however; and, together with the lack of involvement of erythrocyte neuraminic acid, they suggest that the phenomenon of elution of OC38/43 is different from that observed with the myxoviruses and paramyxoviruses.

Table 97.6 Agglutination titres of various erythrocyte species by strains of human coronaviruses at different temperatures

Erythrocyte species[a]	Strain and temperature: OC38 4°	OC38 RT[b]	OC38 37°	OC43 4°	OC43 RT	OC43 37°
Rhesus	< 10	< 10	< 10	< 10	< 10	< 10
Guinea pig	< 10	< 10	< 10	< 10	< 10	< 10
Human O	80	< 10	< 10	320	< 10	< 10
Vervet	80	< 10	< 10	320	< 10	< 10
Chicken	80	80	80	640	320	320
Rat	160	320	320	640	1280	1280
Mouse	160	320	320	640	1280	1280

[a] All erythrocytes made up as 0.4% suspensions except chicken, which was 0.5%.
[b] RT = Room temperature.

Haemadsorption was demonstrated in OC43-infected cell culture (Kapikian *et al.* 1972). This presented an interesting problem since it had been clearly shown that coronaviruses did not mature at the plasma membrane, so that viral haemagglutinin would not be expected to be located at the cell surface. An explanation came from Bucknall, Kalica and Channock (1972) who suggested that the haemadsorption phenomenon resulted from progeny virus being liberated from the host cell and then rebinding on to the surface, giving rise to a process they named 'pseudo-haemadsorption'.

Replication

The process by which a virus replicates in its host cell is most readily investigated in cell culture systems. For this reason most of the original studies on the human coronaviruses concentrated on strain 229E. After the successful adaptation of strain OC43 to growth in BSC-1 cell monolayers by Bruckova *et al.* 1970, however, the comparative study of virus morphogenesis was conducted.

Bradburne (1972) examined the multiplication of strains 229E and LP in suspension cultures of L132 cells, a continuous line derived from human lung. He found that it was difficult to infect all the cells in a culture, owing in part to the low multiplicity of infection and also to an inefficient adsorption process by the virus. The cultures behaved essentially as a single cycle of virus replication, however, and enabled him to deduce the temporal aspects of the growth cycle. Adsorption of the virus to the cell surface was followed by an eclipse period of about 4.5 hr, in which newly formed virus could not be detected. Virus production could then be demonstrated, initially as cell associated infectivity, but shortly followed by cell-free virus. Virus infectivity titres increased to reach a maximum by 12 hr after infection and maintained their levels although there were uninfected cells which could have become infected and further increased the virus yield.

The effect of growth temperature was carefully examined by Bradburne. He found that maximum titres were reached when the infected cells were held at 33° while the titre was greatly reduced at 30° and 39°. He showed that the temperature-sensitive step occurred towards the end of the eclipse phase, but the difficulty experienced in infecting all the cells prevented an accurate determination of the time point.

The scheme of virus multiplication described above is supported by the studies on virus morphogenesis with electron microscopy. Two reports of HCV-229E development in WI-38 cells (Hamre, Kindig and Mann 1967; Becker *et al.* 1967) and a later report of HCV-OC43 replication in WI-38 cells (Bucknall *et al.* 1972) are in essential agreement with each other and with reports of other coronaviruses.

No morphological changes are visible in the infected cells for 4–6 hr after infection, corresponding to the eclipse phase described by Bradburne (1972). After 6 hr, however, 229E-infected cells show some virus particles located within cytoplasmic vacuoles and these increase in number up to 12 hr after infection. Virus starts to be seen on the outside of cells by 12 hr and the process of release continues by the virus-filled vacuoles moving to the plasma membrane and emptying their contents. Similar stages were seen by Bucknall and his associates (1972) with HCV-OC43 but the time scale was extended as compared with that of HCV-229E.

The mechanism by which virus is formed in the vesicles is not completely clear but several authors have suggested that it occurs by a budding maturation. Areas of 'thickening' or electron density appear in the lipid bilayer surrounding the vesicles, these areas curving to form crescents and the rounding-up continuing to form the core of the virus. The membrane constricts until the virion is attached to the vesicle by a narrow stalk which finally pinches off to liberate the virus into the vesicle. This process has been seen with HCV-229E but Bucknall and co-workers could find no sections of cells infected with HCV-OC43 that showed budding and concluded, therefore, that the budding event probably occurred rapidly, thereby minimizing the chance of fixing a cell during that stage.

Diagnosis

Diagnosis of human coronavirus infections relies on two approaches; isolation and subsequent identification of the virus, and serological methods to detect specific antibodies and antigens.

Isolation

The most successful system for isolation of HCV strains has proved to be organ culture derived from human embryonic trachea (HETOC). As discussed above, the first report of a human coronavirus isolation, B814, was by HETOC, as was the series of OC isolates reported by McIntosh and his associates in 1967. That HETOC has the ability to support the growth of all the known serological groups of HCV is demonstrated by the isolation of strain LP, related to 229E, by Tyrrell, Bynoe and Hoorn (1968*b*) and the strain 692 by Kapikian's group (1973), reported to be serologically distinct from strains B814, OC43 and 229E.

When the initial isolations of 'IBV-like' viruses were being made, there was a clear responsibility to show that the virus being cultivated could induce clinical signs of a common cold in volunteers. Once the characteristics of the human coronaviruses were established it became generally adequate to demonstrate that the isolate resembled a coronavirus by electron microscopy and reacted with antisera prepared against one of the known serotypes.

No suitable animal model is available for the isolation of human coronaviruses or for testing for pathogenicity except, of course, for human volunteers. Adaptation of the isolates OC38 and OC43 to suckling mouse brain (McIntosh *et al.* 1967) provided a convenient way to cultivate the viruses but was probably not related to human pathogenicity. It did, however, allow for serological testing of sera for antibodies to the OC38/43 serotype of HCV.

Cell cultures of several types have been reported for primary isolation of human coronaviruses but it would appear that only viruses of the 229E serotype can be detected by this method (see Kapikian 1975). After their isolation in HETOC, strains OC43 and LP were adapted to grow in primary and continuous cultures of human cells.

Serology

Antibodies to HCV have been detected and quantified in serum by a variety of procedures including virus neutralization, complement fixation, indirect haemagglutination and a modification of it called immune-adherence haemagglutination, by radial haemolysis, and by the enzyme-linked immunosorbent assay (ELISA). Rising titres to at least one HCV serotype following clinical signs of respiratory tract infection is usually accepted as evidence of HCV involvement in disease but it is generally acknowledged that this method lacks specificity, as it cannot distinguish between coronaviruses within a serologic group. Thus there may be strains of HCV that are different from 229E in virulence and minor antigens but are still detected as '229E-related'.

Coronaviral antigens in nasopharyngeal secretions have been detected by immune electron microscopy (Kapikian *et al.* 1973) although identification of coronavirus particles by direct electron microscopy has had limited application owing to the frequent difficulty experienced in distinguishing coronaviruses from fragments of cellular debris.

Immunofluorescence of epithelial cells in nasopharyngeal secretions of symptomatic volunteers promised a method that would simplify clinical diagnosis of HCV infection, although a survey of 106 children hospitalized with respiratory disease failed to detect any that were positive by this test (McIntosh *et al.* 1978).

Association with disease

The evidence that coronaviruses are the causal agents of at least some of the common cold infections seen in the population comes from two lines of study. Most of the HCV strains, originally isolated from clinical cases, have been inoculated into human volunteers where they all produce symptoms of respiratory disease (Bradburne, Bynoe and Tyrrell 1967, Bradburne and Somerset 1972). Only rarely did every volunteer become infected but approximately 50 per cent of the experimental inoculations resulted in clinical signs. Additionally, although less convincingly, infection of tracheal organ cultures with the HCV strains interferes with ciliary movement of the epithelial cells, suggesting that a similar cytopathology may occur *in vivo* and result in illness.

The other aspect of viral aetiology concerns the relation between serological evidence of infection with HCV and respiratory illness. A large proportion of the adult population has antibodies to HCV strains but a 4-fold rise in antibody level is considered to be evidence of current infection. Several large surveys have been conducted and many of these have been reviewed by Kapikian (1975). In those studies that sampled patients suffering from upper respiratory tract infections, rather than the population at large, there was statistical evidence of association of serum antibody rises and illness. During an outbreak of 229E infection in Techumseh, Michigan, between 1966–1967, Monto's group showed that illness was significantly more common in those with serological evidence of infection than among matched individuals without infection. Similarly, Dorothy Hamre and her associates investigated 229E infection among medical students in Chicago between 1961 and 1967 and showed that there was a statistical association with illness.

The proportion of colds that can be associated with coronaviruses is in the range 2–10 per cent (as shown in Table 97.7). As mentioned above, however, these figures may not reflect the true incidence, since there may be human coronaviruses that are not related to 229E and OC43 and that are not readily isolated by cell or organ cultures.

The role of human coronaviruses in lower respiratory tract disease is less well defined. McIntosh and his associates (1974) identified HCV infection in 34 out of 417 children under 18 months hospitalized with lower respiratory tract disease. This incidence of 8.2 per cent was similar to that of 7.7 per cent in 13 controls. The authors admitted that the figures did not provide evidence of a causal relation but believed that when the considerations of increased lower respiratory tract disease during a 229E outbreak and the isolation of a 229E-like strain from two cases of pneumonia in children were added to the epidemiological evidence there was a suggestion of association that deserved further study.

Such a study was presented by a group from Utah (Smith *et al.* 1980) who examined 150 subjects over an 8-year period. By comparing the evidence of infection with several respiratory viruses and *Mycoplasma pneumoniae* during 1030 periods of illness and 1398 illness-free periods, they were able to show that coronavirus infections were significantly associated with illness. Of the 17 illness-associated infections with

Table 97.7 Proportion of colds associated with coronaviruses

Study group	Location	Duration of study	No. of patients	Association with 229E	OC43
Hospital employees	Chicago, Illinois, USA	1967–68	88	7(8%)	NT
Insurance company employees	Charlottesville, Virginia, USA	1963–70	592	10 (1.7%)	
			620		15 (2.4%)
5 to 19-yr-old Church home residents	Atlanta, Georgia, USA	1960–67	1328		44 (3.3%)
Military group	Gloucestershire, England	1966–67	91	0	0
Outpatient infants and children	Sao Paulo, Brazil	1969	36	4 (11%)	NT
NIH employees	Bethesda, Maryland, USA	1962–64	256	5 (2%)	4 (1.6%)
		1965–67	317	24 (7%)	23 (7%)
		1965–70	541	32 (6%)	31 (6%)
College students	College Park, Maryland, USA	1960–61	110	12 (11%)	7 (6%)

Adapted from Kapikian, 1975.
NT = not tested.

229E and/or OC43-related viruses, 8 (47 per cent) were identified with lower respiratory tract disease.

In summary, the evidence suggests that coronaviruses can and do cause acute respiratory illness in children and adults. The most frequent symptoms are those of the common cold but there is a significant association with illness of the lower respiratory tract.

Pathogenesis

A comparison of the clinical features presented by volunteers infected with HCV strains 229E and B814 and rhinovirus type 2 and OC was made by Bradburne, Bynoe and Tyrrell (1967) at the Common Cold Research Unit and is presented in Table 97.2.

In general, the symptoms caused by coronaviruses are very similar to those produced by rhinoviruses. The incubation period was significantly longer while the duration was shorter. There was a tendency for more paper handkerchiefs to be used however and malaise was more pronounced. The differences between the colds induced by the two coronavirus strains were minimal but when they did occur, as in the higher incidence of headache and lower incidence of mucopurulent nasal discharge for 229E, they could be explained by the low number of volunteers available in the study.

Bradburne also reported that colds induced by strains LP, EVs and the OC isolates in volunteers were all fairly similar (Bradburne and Somerset 1972). Differences in the number of colds and the severity of illness produced by the various strains were thought by the authors to reflect variation in the dose of virus given and the immune status of the volunteer rather than differences of virulence.

Transmission

It is generally assumed that the human respiratory coronaviruses are transmitted by the respiratory route. Intranasal inoculation of volunteers has been used in the studies of experimental infection, although this procedure does not fully rule out involvement of the alimentary tract as the primary site of infection. There are no animal reservoirs of infection known and the epidemiological studies of Monto (1976) in Tecumseh suggested to him that transmission was by aerosol and large droplets. This conclusion was based on the observation that the speed at which the disease spread was similar to that seen with influenza virus whereas rhinovirus infection spread more slowly via large droplets and direct contact.

Epidemiology

Most of the sero-epidemiological surveys have been reviewed by Kapikian (1975) while Monto (1976) has summarized his findings from his Tecumseh study; the reader is referred to these two publications for detailed analysis of coronavirus epidemiology. Here, we shall outline the areas in which most workers agree and discuss those aspects in which they differ.

It is an unfortunate fact that epidemiological studies need to sample large populations for the representative facts to emerge. Small populations are affected by changes in local conditions and the numbers do not lend themselves to adequate statistical analysis. This means that the epidemiologist needs to select a diagnostic method that can be applied to thousands of samples over a long period of time and, for viruses as fastidious as the coronaviruses, this tends to rule out virus isolation in every case. Consequently, most of the epidemiological surveys that have been published rely heavily on serology for evidence of infection. This is complicated by the fact that, although four or more distinct coronavirus serotypes are thought to exist, the antigenic tests are available for only two of them, 229E and OC43. Inevitably, published reports probably lack the data for coronavirus infections of different serotypes as well as being unable to distinguish between, for example, OC38 and OC38-like agents. It is as well to bear these points in mind when considering the sero-epidemiologic studies and, in fairness, it is true to say that many of the workers in this area have recognized the limitations in the surveys.

Coronavirus antibody is present in the majority of adults in N. America and the UK and reports of coronavirus infections have been published from S. America, the Soviet Union, China and many other parts of the world in which the viruses appear to be endemic.

There has been some disagreement between workers on the incidence of infection within different age groups. From the Tecumseh study, Monto concluded that OC43 infection was equally common in children and adults whereas 229E was less frequent in children than in the 15 to 29-year age group. Whether this represented a true difference between the two agents remains to be determined; however, since a recent study of OC43 infections in Hamburg, GDR Sarateanu and Ehrengut (1980) found the highest incidence in the 15 to 24-year age group. This appears to be the general pattern for human respiratory coronaviruses with little infection up to the age of 4 years and then a steady increase up to the late teens followed by a general decline in incidence up to 60 plus. Several authors have suggested, however, that the age incidence figures may be caused by a poorer immune response in young children rather than by fewer infections.

It is clear that these data do not represent a single infection with each agent. Evidence of increasing antibody titres in paired sera, generally accepted as evidence of infection, has frequently been associated with individuals with pre-existing serum antibody, suggesting that a previous immune response does not fully protect a person against reinfection with the same, or a related, virus. This would explain why the outbreaks of disease may appear in communities in which antibodies were present from previous exposure, as illustrated by the cyclic behaviour of these viruses (see Fig. 97.4). This figure also shows that outbreaks occur every 2–4 years and may, as in the case of 229E, appear in several localities during the same period. Local variation does occur, however, as with OC43 infections during 1964–1969.

As with many other respiratory diseases, the highest incidence of coronavirus outbreaks occurs during the colder months. The prevalence during October to May has been well documented (see Kapikian 1975); as an example, Kaye and Dowdle (1975) presented data of seroresponse to 229E in children between 1960–1968. Of the 168 positive seroresponses, 40 (23.8 per cent) occurred in the fall (September–November), 75 (44.6 per cent) during the winter months (December–Feb-

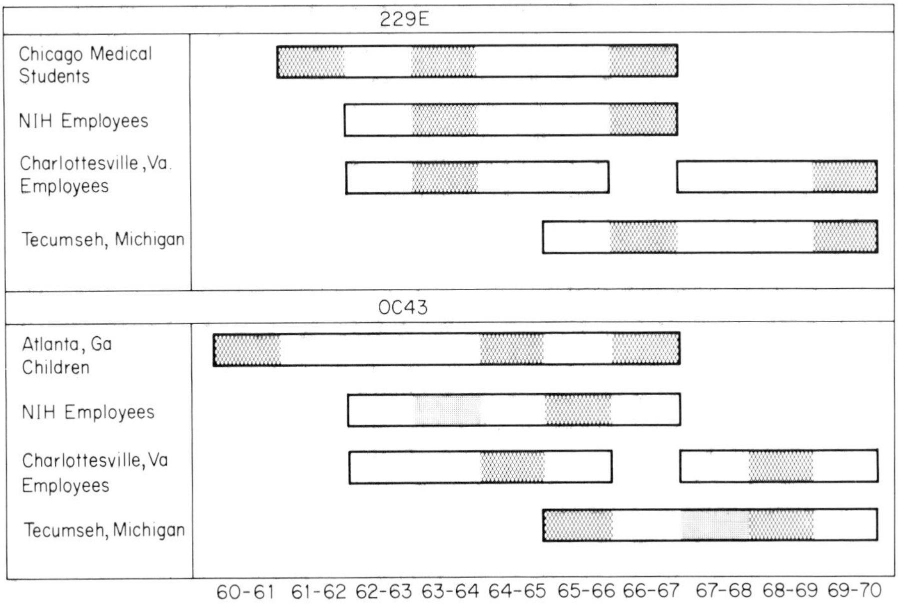

Fig. 97.4 Incidence of infections with human coronaviruses.

ruary), 43 (25.6 per cent) in spring (March–May) and only 10 cases (6.0 per cent) through the summer (June–August). The reasons for this seasonal variation are probably several and complex, including enhanced survival of the agent in the air, owing to lower temperatures and intensity of light, and the tendency of people to congregate indoors, reducing the distance that the virus needs to travel to reach a new host.

Prevention and treatment

Although the role of respiratory coronaviruses in causing a proportion of common colds is established beyond doubt, this knowledge does not help at present to prevent or treat the disease.

The development of a vaccine is complicated by two major difficulties: antigenic variation between isolates and the apparent possibility that serum antibodies do not protect against reinfection. As stated earlier in this chapter, at least four distinct serotypes are known to exist and it is believed by several researchers that even isolates related by such tests as complement fixation may vary in other antigens. Until this complex serology is clearly defined it would be fruitless to attempt to formulate a vaccine that would be expected to protect against every respiratory coronavirus challenge.

The second difficulty, that of the possibility of infection in the presence of circulating antibody, may be valid or may reflect the confusion in serotypes that exists. The sero-epidemiological evidence shows that antibodies which react with 229E do not prevent a 229E-like agent from establishing infection but the two viruses may be recognized as quite distinct by the human immune system. Clearly, vaccines may be developed but further work on coronavirus serology is required before that can occur.

The use of antiviral chemotherapy is equally unproductive at present. No compounds have been reported that can be used successfully against either 229E or OC43, even though a great deal of effort is being expended by the major drug companies. Whether or not interferon might be an effective treatment remains to be established.

Avian infectious bronchitis virus

Avian infectious bronchitis virus (IBV) has received more research attention over a longer period than any other coronavirus, owing to its economic importance in the poultry industry. Our intention here is to summarize the characters of the virus and the disease it causes and to compare these features with those of the human respiratory coronaviruses described above. The reader is referred to the excellent reviews by Hofstadt (1965) and Cunningham (1970) for a more detailed treatment of the subject.

History

Infectious bronchitis was first recognized as a distinct disease of poultry in 1930 in North Dakota, USA, by Schalk and Hawn (1931) who described a respiratory disease of baby chicks. The aetiologic agent which was established as a virus in 1936 was first cultivated *in ovo* in the following year and an immunization programme started in 1941 by use of egg-passaged attenuated virus. The morphology of the virus was described by means of negative staining and electron microscopy during the 1960s and it became the reference type for the IBV-like viruses isolated from the human respiratory tract. After the recognition of the coronavirus taxonomic group in 1969, IBV became the prototype of the group. During the 1970s, when the structural characters of many coronaviruses were being established, there was some concern that IBV was substantially different from the mammalian coronaviruses and should be separated into a subgroup, but recent reports have tended to show the similarity between the avian and mammalian coronaviruses. At the time of writing, IBV retains its position as the prototype strain of the *Coronaviridae*.

Structure and properties of the virion

Morphology

Electron microscopy of shadow cast preparations in 1957 showed rounded particles of 60–100 nm diameter but no surface features could be distinguished. Use of negative staining in electron microscopy, however, revealed large surface projections, 9–11 nm long, radiating from pleomorphic particles 80–120 nm in diameter (Berry *et al.* 1964). It can be seen from Fig. 97.5 that IBV particles are not substantially different from those of human respiratory coronavirus. A difference is revealed, however, when preparations of IBV are treated to liberate nucleocapsid material and examined by electron microscopy. Whereas human, murine and porcine coronaviruses show nucleocapsid structures, no similar structures are seen in IBV. It seems unlikely that IBV lacks a nucleocapsid similar to those found in other members of the virus family, particularly when a helically arranged thread-like structure was seen in thin sections of purified virus (Apostolov *et al.* 1970). It is probable that the nucleocapsid is fragile and is breaking down to subunits and fragments.

There have been several studies of IBV in thin sections of infected cells, reviewed by McIntosh (1974), showing condensation of viral precursors at the endoplasmic reticulum and maturation of the virus by budding through the membrane. These studies present evidence of virions having an outer

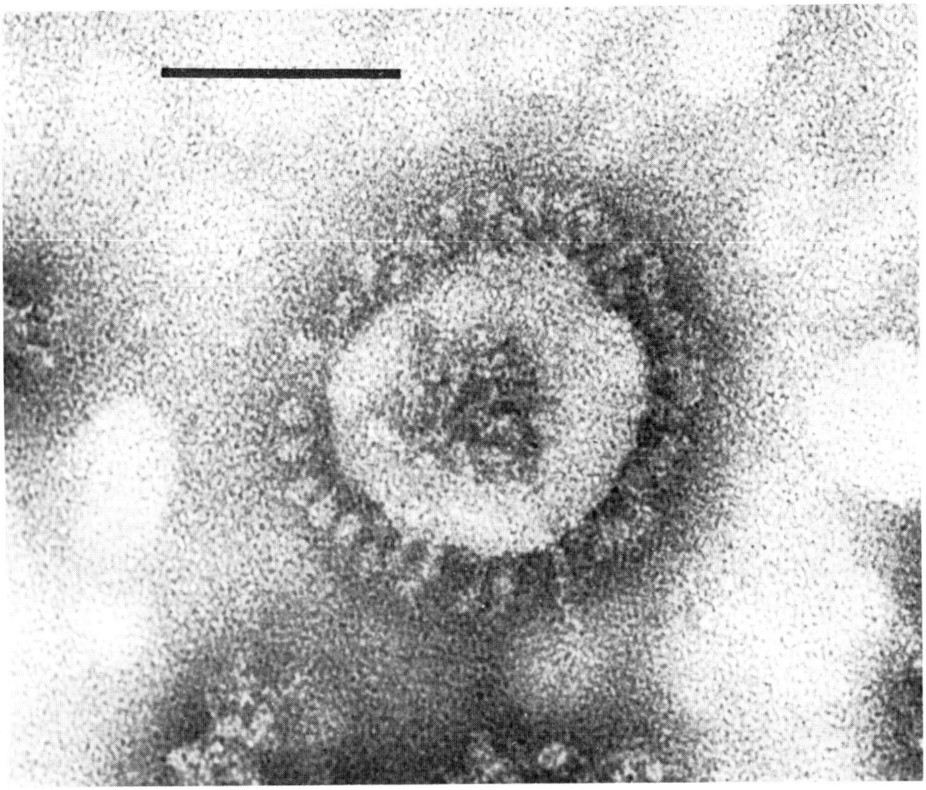

Fig. 97.5 Electron micrograph of a negatively stained particle of IBV. The bar represents 100 nm. (Courtesy of E. O. Caul)

envelope surrounding the inner shell with an electron-lucent centre, resembling the human coronaviruses described above. There was some variation between reports concerning the size of the space between the envelope and the inner shell but this might represent an artefact brought about by drying rather than an important structural feature of the virus. More recently, Chasey and Alexander (1976) published photographs of IBV in thin section which clearly showed the surface projections radiating from the virus, thus fully confirming the observations made with negative staining.

Chemical composition

Nucleic acid

IBV was shown to contain an RNA genome in 1963 by fluorescence with acridine orange, its susceptibility to ribonuclease and its ability to replicate in the presence of various inhibitors of DNA synthesis (see Cunningham 1970).

Little more was reported of the IBV genome until 1973 when Tannock examined the size and configuration of the nucleic acid and found it to be single stranded and fragmented. Over the following four years there were a number of reports (see Garwes 1979) which concluded that IBV RNA was an infectious single-stranded, linear molecule of approximately 8×10^6 molecular weight, having a tract of polyadenylic acid at the 3′-end.

Recently Stern and Kennedy (1980) have described 5 subgenomic molecules of IBV RNA in infected cells. They range in mol. wt, from 0.8×10^6 to 2.6×10^6, are all polyadenylated and are thought to be the functional messenger molecules responsible for viral translation.

Polypeptides

As with the IBV genome, the structural polypeptides of the virus have presented a confusing picture for several years. The earliest reports suggested that IBV had at least 16 structural proteins, whereas some later studies found as few as four polypeptides (summarized by Robb and Bond (1979); cf. also Lanser and Howard (1980)) (Fig. 97.6). It was considered that the large number of polypeptides might have arisen as artefacts during degradation of the virus prior to electrophoresis but a study of the degradation process has tended to discount this hypothesis. An alternative explanation is that egg-grown virus contains a large number of host components but this, again, seems unlikely as Lanser and Howard (1980) found only four polypeptides in highly purified egg-grown virus. At present it appears that IBV probably has only 4–7 structural polypeptides and that there are 10 or more associated polypeptides derived either from the host,

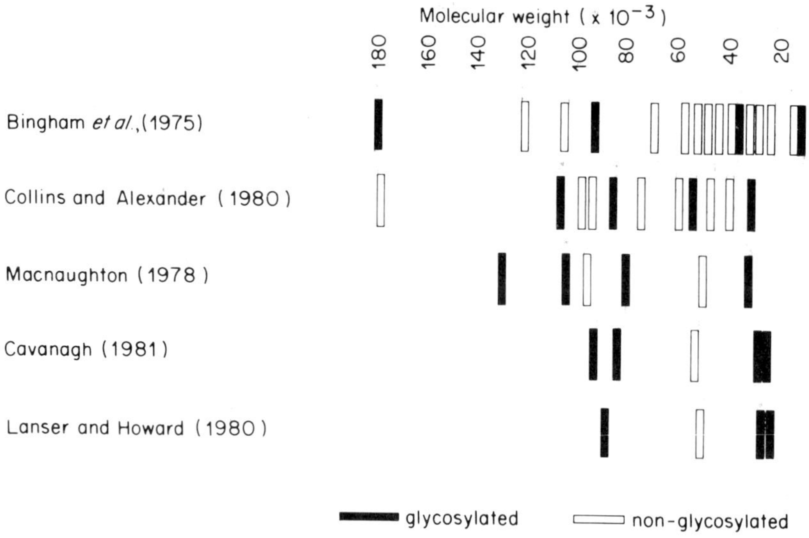

Fig. 97.6 Polypeptide composition of infectious bronchitis virus.

as demonstrated by Cavanagh (1981), or from degraded or precursor forms of the major structural proteins. This would be consistent with the general finding for coronaviruses of a peplomer glycopolypeptide of molecular weight 180–200 000 or its 2 cleavage products around 90–120 000 mol. wt, a nucleocapsid phosphopolypeptide of 50–60 000 mol. wt and one or two envelope glycopolypeptides of 20–30 000 mol. wt size range.

Haemagglutination

In common with human coronavirus OC38/43, IBV will agglutinate erythrocytes from a range of animal species but this property has presented some difficulties. The first description of haemagglutination came in 1959 from Corbo and Cunningham, who showed that IBV would agglutinate chicken red cells only after the virus had been treated with trypsin. Whether proteolysis was required to activate a structural haemagglutinin or to remove an inhibitor was not clear but the interpretation of the finding was difficult since the haemagglutination could not be inhibited by immune serum.

It was perhaps surprising, then, that a report in 1975 by Bingham, Madge and Tyrrell described haemagglutination by an untreated Connecticut strain of IBV. The virus was purified and concentrated before use, however, and this may have removed inhibitors. The Massachusetts strain would also haemagglutinate but only after treatment with phospholipase C, which may have altered the lipid envelope to allow mobility of the proteins or may have contained a contaminating enzyme responsible for the activity. The haemagglutination could be inhibited by antisera to IBV, suggesting that the activity was more similar to that of OC38/43 than that described by Corbo and Cunningham. The virus-erythrocyte complexes formed with the Connecticut strain of IBV appeared to be stable at temperatures between 4° and 37° over 2 hours whereas the Massachusetts strain gave haemagglutination patterns that tended to break down at 21° and 37° over 2 hours. Whether this represents active elution or poor initial affinity is not established but neuraminidase has not been found in purified IBV preparations.

Replication

The processes involved in the replication of IBV are essentially identical with those described above for the human coronaviruses. Early work on IBV was complicated by the necessity to use eggs for cultivation but more recently the use of primary chick kidney cells enabled Chasey and Alexander (1976) to study the morphogenesis of the virus. They described the entry of the virus into the host cell by viropexis and subsequent maturation by budding into either the cisternae of the endoplasmic reticulum or cytoplasmic vacuoles. They suggested that the surface projections were attached to the viral envelope during the budding process.

The events that occur between viropexis and maturation cannot be visualized by electron microscopy but Stern and co-workers (1981) have suggested a scheme that incorporates what is at present known of the molecular biology of coronavirus replication. This is diagrammatically represented in Fig. 97.7. It is assumed that the genome RNA is translated to produce a polymerase, as no enzymes have been found in

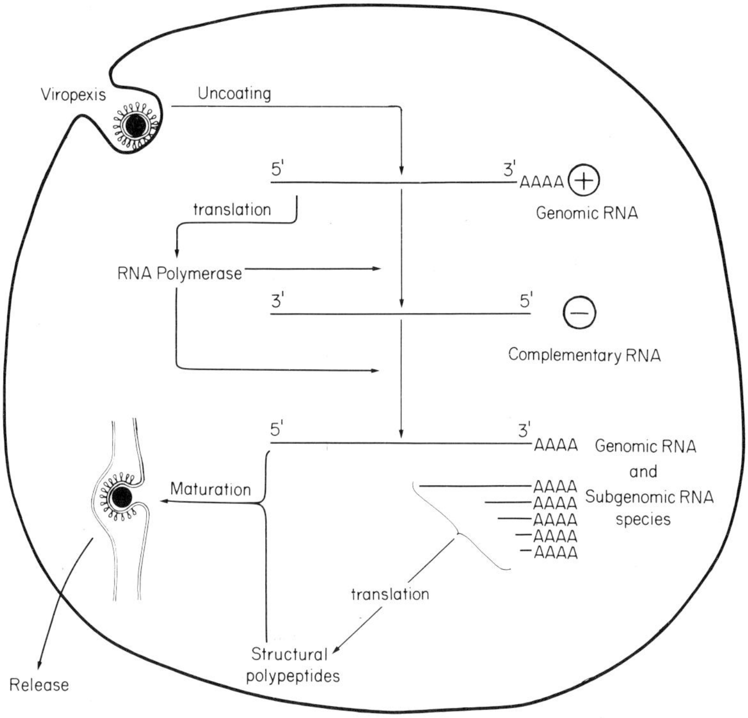

Fig. 97.7 Schematic representation of coronavirus replication.

association with virions. The polymerase produces more polyadenylated virion RNA molecules, presumably via an intermediate complementary strand, and 5 subgenomic molecules which are thought to act as the messengers for the structural polypeptides. The components assemble at the endoplasmic reticulum and maturation continues by budding followed by release from the cell.

Diagnosis

Since the clinical features caused by IBV infection may be confused with those of Newcastle disease and laryngotracheitis, particularly in the early stages, diagnosis of disease relies mainly on serological tests, and, to a lesser extent, isolation of the virus.

Virus can be isolated by inoculation of infected respiratory tissue into uninfected chickens, fertile eggs or, more recently, tracheal organ cultures. Isolation of the virus is not of itself sufficient to diagnose infection unless the isolated virus can be characterized as IBV, so that serological tests are frequently used in conjunction, such as for specific antibodies in the serum of inoculated chickens or by use of IBV-neutralizing antibodies to detect protection of organ cultures.

The effects of IBV on fertile chicken eggs have been described (see Cunningham 1970) and were frequently used for virus diagnosis but this has been generally superseded by cheaper, faster and more reliable serological methods. Most of the standard serological procedures have been employed, including neutralization, immunodiffusion, immunofluorescence and complement fixation (reviewed by Cunningham 1970) but haemagglutination inhibition is now widely used. The test uses virus that has been treated with phospholipase C to expose the haemagglutinin, as described above, and comparison with other diagnostic tests has shown that it is sensitive, reliable and readily applied to large numbers of samples. Tests of this sort, which can be semi-automated, will be preferred to older methods, and in this context the enzyme immunoassay has good potential as a diagnostic procedure. Results from several research centres have shown this assay to be very sensitive and be easily scaled up to accommodate large numbers of serum samples.

Diagnosis of IBV not only necessitates differentiation of the disease from other agents that produce similar clinical signs but also identification of the serotype of IBV causing the infection, since vaccine against one serotype may be of limited efficacy against another. Serum neutralization tests can be used to distinguish strains; Darbyshire and his associates (1979) at the Houghton Poultry Research Station have used this in organ cultures to show the relations between 24 strains of the virus, illustrated in Fig. 97.8. It is clear that large numbers of relatively minor anti-

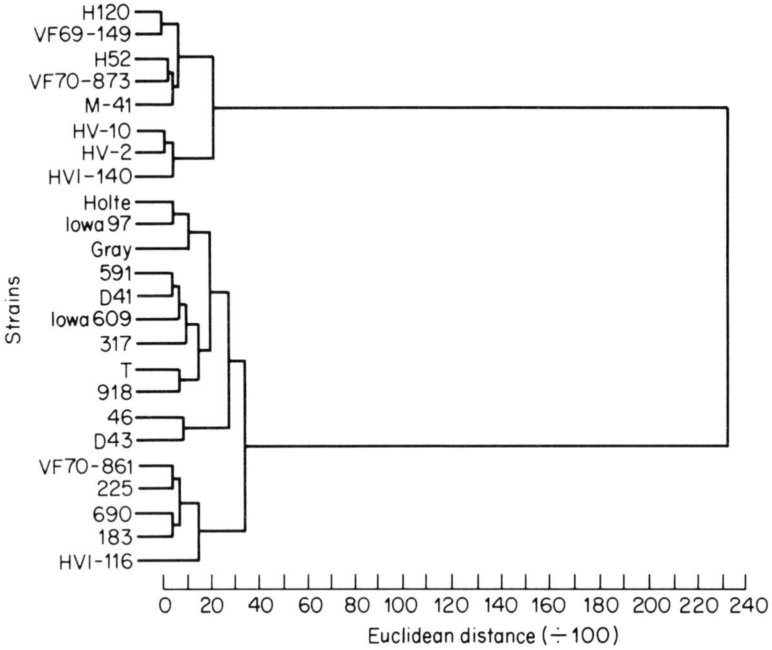

Fig. 97.8 Dendrogram showing relationships between IBV strains.

genic changes occur in IBV, giving rise to the wide diversity of serotypes that have been, and are continuing to be, isolated from the field.

Association with disease

Whereas the human coronaviruses presented some difficulties in proving an association with disease, requiring extensive epidemiological studies and volunteer experiments in which previous exposure to the virus was likely to affect the clinical response, IBV can be unequivocally linked to infectious bronchitis. Chickens that have not been exposed to the virus are available in large numbers, and age-related studies can be conducted.

Pathogenesis

Typical clinical signs of IBV infection in chicks aged less than 6 weeks are respiratory distress, tracheal râles and nasal discharge, with coughing and sneezing. Weight gain and food consumption are both reduced and the birds become weak and depressed as the disease progresses. Morbidity is high and mortality is frequently greater than 25 per cent.

In chickens over 6 weeks, the clinical signs are similar to those in the chick but are less pronounced. Morbidity may be fairly high but death is usually infrequent. Food consumption drops, resulting in reduced weight gain, but the most obvious characteristic is the drastic drop in egg production which occurs as the disease progresses. The eggs produced during this period may be abnormal in shape, surface texture and thickness of shell. The extent and duration of the decrease in egg production depends on the age and general condition of the birds; occasionally egg production returns to normal but usually the preinfection rate is never reached.

Serous or catarrhal exudates are usually found throughout the respiratory tract with yellow caseous plugs in the lower trachea and bronchi of chicks that die. The tracheal mucosa has been reported to undergo cyclic changes of acute phase (epithelial hypertrophy and oedema), reparative phase (epithelial hyperplasia and cellularity of the propria) and immune phase (epithelial restoration with lymphoid infiltration of the propria).

Transmission and epidemiology

IBV resembles HCV in that aerosols appear to be the mechanism of transmission, although the influence of droplet size has not been reported. It is likely that virus-contaminated feed, water, equipment and personnel may also play a part in the spread of disease within and between flocks. Other avian species are not known to be naturally infected but the role of recovered chickens as carriers is not clear. It has been established that virus can be re-isolated from chickens over 4 months after infection (Cook 1968) and these may be instrumental in transmission on occasions where birds are moved between flocks.

The epidemiology of IBV shows several features in common with that of the human respiratory diseases. Chickens of both sexes and all ages appear to be susceptible and within any given flock infection spreads rapidly to many individuals. Reinfection can occur in the presence of circulating antibodies. Whether specific antibody in serum is not protective for infection of the respiratory mucosa or whether there are serological variants, known to be very common for IBV, which can establish infection in the presence of group-specific antibodies, are questions already raised for HCV.

Prevention and treatment

Prevention of IBV infection is more feasible than its human counterpart. Modern poultry farming techniques use specially designed buildings to house the chickens and procedures can be taken to keep these specific-pathogen free. Careful management can certainly reduce the chance of introduction of infection into a disease-free flock but the cost of maintaining a fully effective barrier would not be economically sound.

Vaccination is widely used throughout the world and is generally effective. Inactivated virus has the major advantage of being safe but has proven unsuitable in most cases. Whether this is due to inadequacy in the resulting antibody level or immunoglobulin class has not been established.

Live viruses are the most commonly administered form of vaccine available. Several strains of IBV have been selected for low virulence or passaged in chick embryos for at least 25 passes to attenuate them and these are given to the birds in their drinking water. It was found that extensive passage in chick embryos lowered the virulence to a point where the virus would barely establish an infection in the chicken sufficient to provide any immunity. The point at which a strain is sufficiently attenuated to prevent pathogenesis but will generate useful amounts of antibodies appears to vary between isolates. A commonly used procedure is to infect the birds with a lowly virulent, poorly immunogenic strain and follow this with a strain of higher virulence that is highly immunogenic, relying on the immune response against the first strain to protect the bird against the second.

The choice of strains is clearly central to any vaccination programme. The variety of field strains that exist and the apparent ease with which new strains can arise present problems that have been discussed earlier in the human disease section. The possibility of polyvalent vaccines, incorporating several serotypes, has been examined but does not appear particularly promising owing to factors such as interference (reviewed by Winterfield and Fadley 1975). At present the most useful form of vaccine would seem to be that using a strain that cross-reacts most widely between isolates; research is continuing to recognize such strains by means of analyses such as that shown in Fig. 97.8. There is evidence, however, that strains which are not particularly close by serological tests may still cross-protect on challenge in the chicken; more needs to be known of the immune defence mechanisms against this disease for fully effective vaccines to be developed.

Rat pneumonia coronavirus

This virus was isolated by Parker and his associates (1970) in a programme intended to identify the agent stimulating the production of antibodies to mouse hepatitis virus (MHV) in rats (Hartley *et al* 1964). The rat virus resembled a coronavirus in its morphology and chemical composition, and was related to MHV by complement fixation and, to some extent, by neutralization tests.

The virus has a widespread distribution. Rats from 11 conventional colonies were seropositive, as were 3 of 5 specific-pathogen free colonies. Wild rats from N. American dumps also had antibodies, although no attempts were made to demonstrate that the same virus was the cause in each case. Identification is complicated by an antigenically related coronavirus isolated from rats with sialodacryoadenitis (Bhatt *et al.* 1972) although the two agents can be distinguished by their pathogenesis.

RCV was identified by inoculation of lung extracts from asymptomatic rats into weanling rats by a combined intranasal-intraperitoneal route followed by complement-fixation testing of their sera for antibody to MHV 28 days later. Initial isolation was then accomplished by harvesting lungs from weanling rats 7 days after intranasal inoculation with a suitable extract.

The lung isolate was passaged in rat kidney cell cultures in which it produced cytopathogenesis; other primary and continuous cultures were tried without success. Virus from cell culture was inoculated into newborn rats, resulting in respiratory distress by 5 days and subsequent death 1–7 days later. Rats less than 2 days old were most susceptible, animals aged 1–2 weeks developing respiratory disease but recovering and rats older than 3 weeks showing no clinical signs. Differences between rat strains were noted;

Fischer rats showed almost 100 per cent mortality by the 12th day whereas Wistar rats never exceeded 25 per cent mortality, usually after the 12th day.

Disease caused by RCV was characterized by widespread interstitial pneumonitis and focal atelectasis. Viral antigen was demonstrated in the mucosal epithelium of the upper respiratory tract and in pulmonary alveolar septa of 9 to 10-week-old rats (Bhatt and Jacoby 1977) providing evidence that the agent was a primary pathogen of the rat respiratory tract. Similar studies with sialodacryoadenitis virus (SDAV) by Bhatt and Jacoby (1977) suggested that the two coronaviruses were closely related but showed some differences in their tissue tropisms and pathology (see Table 97.8).

Table 97.8 Comparison of the major features of experimental infection with SDAV and RCA in adult axenic CD rats.

Feature	SDAV	RCV
Clinical signs		
Photophobia	Yes	No
Sneezing	Yes	No
Cervical swelling	Yes	No
Virus replication		
Respiratory system	Yes	Yes
Salivary glands	Yes	Trace
Lacrimal glands	Yes	Trace
Lesions		
Acute rhinotracheitis	Yes	Yes
Focal interstitial pneumonia	No	Yes
Sialoadenitis	Yes	Trace
Dacryoadenitis	Yes	No
Antibody response		
Complement fixing	Yes	No
Neutralizing	Yes	Yes

From Bhatt and Jacoby (1977).

There have been no studies reported on the transmission and epidemiology of RCV although the widespread distribution within colonies of caged rats makes it appear likely that an aerial route is responsible. There are no prescribed methods for prevention or treatment.

References

Acornley, J. E., Chapple, P. J., Stott, E. J. and Tyrrell, D. A. J. (1968) *Arch. ges. Virusforsch.* **23,** 284.
Ajdukovic, D. *et al.* (1978) 28th Annual Meeting of Canadian Society for Microbiology, Abs. V-17
Almeida, J. D. and Tyrrell, D. A. J. (1967) *J. gen. Virol.* **1,** 175.
Andrewes, C. H. (1965) *The Common Cold.* Weidenfeld and Nicolson, London.
Andrewes, C. H., Chaproniere, D. M., Gompels, A. E. H., Pereira, H. G. and Roden, A. T. (1953) *Lancet* **ii,** 546.
Andrewes, C. H. *et al.* (1961) *Virology* **15,** 52.
Apostolov, K., Flewett, T. H. and Kendal, A. P. (1970) *The Biology of Large RNA Viruses.* Academic Press, London.
Becker, W. B., McIntosh, K., Dees, J. H. and Chanock, R. M. (1967) *J. Virol.* **1,** 1019.
Bell, J. A., Rowe, W. P. and Rosen, L. (1962) *Amer. J. publ. Hlth* **52,** 902.
Bell, J. A., Ward, T. G., Huebner, R. J., Rowe, W. P., Suskind, R. G. and Paffenbarger, R. S. (1956) *Amer. J. publ. Hlth* **46,** 1130.
Berry, D. M., Cruickshank, J. G., Chu, H. P. and Wells, R. J. H. (1964) *Virology* **23,** 403.
Bhatt, P. N. and Jacoby, R. O. (1977) *Arch. Virol.* **54,** 345.
Bhatt, P. N., Percy, D. H. and Jonas, A. M. (1972) *J. infect. Dis.* **126,** 123.
Bingham, R. W., Madge, M. H. and Tyrrell, D. A. J. (1975) *J. gen. Virol.* **28,** 381.
Blair, H. T., Greenberg, S. B., Stevens, P. M., Bilunos, P. A. and Couch, R. B. (1976) *Amer. Rev. resp. Dis.* **114,** 95.
Bloom, H. H., Forsyth, B. R., Johnson, K. M. and Chanock, R. M. (1963) *J. Amer. med. Ass.* **186,** 38.
Bögel, K. and Böhm, H. (1962) *Zbl. Bakt.* Orig. Abt. 1. **187,** 2.
Bradburne, A. F. (1969) *Nature, Lond.* **221,** 85; (1972) *Arch. ges. Virusforsch.* **37,** 397.
Bradburne, A. F., Bynoe, M. L. and Tyrrell, D. A. J. (1967) *Brit. med. J.* **iii,** 767.
Bradburne, A. F. and Somerset, B. A. (1972) *J. Hyg., Camb.* **70,** 235.
Brandt, C. D. *et al.* (1969) *Amer. J. Epidem.* **90,** 484.
Brown, F., Newman, J. F. E. and Stott, E. J. (1970) *J. gen. Virol.* **8,** 145.
Brown, P. K. and Taylor-Robinson, D. (1966) *Bull. World Hlth Org.* **34,** 895.
Brown, P. K. and Tyrrell, D. A. J. (1964) *Brit. J. exp. Path.* **45,** 571.
Bruckova, M., McIntosh, K., Kapikian, A. Z. and Chanock, R. M. (1970) *Proc. Soc. exp. Biol., N.Y.* **135,** 431.
Buckland, F. E. and Tyrrell, D. A. J. (1964) *J. Hyg., Camb.* **62,** 365.
Bucknall, R. A., Kalica, A. R. and Chanock, R. M. (1972) *Proc. Soc. exp. Biol., N.Y.* **139,** 811.
Buscho, R. O., Saxton, D., Shultz, P. S., Finch, E. and Mufson, M. A. (1978) *J. infect. Dis.* **137,** 377.
Butterworth, B. E., Grunert, R. R., Korant, B. D., Lonberg-Holm, K. and Yin, F. H. (1976) *Arch. Virol.* **51,** 169.
Butterworth, B. E. and Korant, B. D. (1974) *J. Virol.* **14,** 282.
Came, P. E., Schafer, T. W. and Silver, G. H. (1976) *J. infect. Dis.* **133,** Suppl. A136.
Cate, T. R., Couch, R. B., Fleet, W. F., Griffith, W. R., Geron, P. J. and Knight, V. (1965) *Amer. J. Epidem.* **81,** 95.
Cate, T. R., Roberts, J. S., Russ, M. A. and Pierce, J. A. (1973) *Amer. Rev. resp. Dis.* **108,** 858.
Caul, E. O., Ashley, C. R., Ferguson, M. and Egglestone, S. I. (1979) *F.E.M.S. Microbiol. Letters* **5,** 101.
Cavanagh, D. (1981) *J. gen. Virol.* **53,** 93.
Chalmers, T. C. (1975) *Amer. J. Med.* **58,** 532.
Chanock, R. M. (1974) *Preventive Med.* **3,** 466.
Chanock, R. M. and Parrott, R. H. (1965) *Pediatrics* **36,** 21.
Chasey, D. and Alexander, D. J. (1976) *Arch. Virol.* **52,** 101.
Cherry, J. D. (1973) *Advanc. Pediatr.* **20,** 225.
Cockburn, W. C. (1979) *J. infect.* **1,** Suppl. **2,** 3.

Collins, M. S. and Alexander, D. J. (1980) *Arch. Virol.* **63,** 239.

Conant, R. M., Hamparian, V. V., Stott, E. J. and Tyrrell, D. A. J. (1968) *Nature, Lond.* **217,** 1264.

Cook, J. K. A. (1968) *Res. vet. Sci.* **9,** 506.

Cooney, M. K. and Kenny, G. E. (1977) *J. clin. Microbiol.* **5,** 202.

Cooney, M. K., Wise, J. A., Kenny, G. E. and Fox, J. P. (1975) *J. Immunol.* **114,** 635.

Cooper, P. D. *et al.* (1978) *Intervirology* **10,** 165.

Corbo, L. J. and Cunningham, C. H. (1959) *Amer. J. vet. Res.* **20,** 876.

Couch, R. B., Cate, T. R., Douglas, G., Gerone, P. J. and Knight, V. (1966) *Bact. Rev.* **30,** 517.

Couch, R. B., Kasel, J. A., Pereira, H. G., Haase, A. T. and Knight, V. (1973) *Proc. Soc. exp. Biol., N.Y.* **143,** 905.

Craighead, J. E., Meier, M. and Cooley, M. H. (1969) *New. Eng. J. Med.* **281,** 1403.

Cunningham, C. H. (1970) *Advanc. vet. Sci.* **14,** 105.

D'Allessio, D. J., Peterson, J. A., Dick, C. R. and Dick, E. C. (1976) *J. infect. Dis.* **133,** 28.

Darbyshire, J. H., Rowell, J. G., Cook, J. K. A. and Peters, R. W. (1979) *Arch. Virol.* **61,** 227.

Davies, H. A. and Macnaughton, M. R. (1979) *Arch. Virol.* **59,** 25.

Davies, J. E. W., Hughes, R. E., Jones, E., Reed, S. E., Craig, J. W. and Tyrrell, D. A. J. (1979) *Biochem. Med.* **21,** 78.

DeLong, D. C. *et al.* (1978) *Abstracts Annual Meeting, American Society for Microbiology*, p. 234.

Dochez, A. R., Shibley, G. S. and Mills, K. C. (1930) *J. exp. Med.* **52,** 701.

Doggett, J. E., Bynoe, M. L. and Tyrrell, D. A. J. (1963) *Brit. med. J.* **i,** 34.

Douglas, R. G. and Couch, R. B. (1972) *Proc. Soc. exp. Biol., N.Y.* **139,** 899.

Douglas, R. G., Couch, R. B., Baxter, B. D. and Gough, M. (1974) *Infect. Immunol.* **9,** 519.

Dudding, B. A., Top, F. H., Winter, P. E., Buescher, E. L., Lamson, T. H. and Leibovitz, A. (1973) *Amer. J. Epidem.* **97,** 187.

Dykes, M. H. M. and Meier, P. (1975) *J. Amer. med. Ass.* **231,** 1073.

Eadie, M. B., Stott, E. J. and Grist, N. R. (1966) *Brit. med. J.* **ii,** 671.

Elveback, L. R., Fox, J. P., Ketler, A., Brandt, C. D., Wasserman, F. E. and Hall, C. E. (1966) *Amer. J. Epidem.* **83,** 436.

Faulk, W. P., Vyas, G. N., Phillips, C. A., Fudenberg, H. H. and Chism, K. (1971) *Nature, Lond.* **231,** 161.

Fenters, J. D., Gillum, S. S., Holper, J. C. and Marquis, G. C. (1966) *Amer. J. Epidem.* **84,** 10.

Foster, G. B. (1916) *J. Amer. med. Ass.* **66,** 1180.

Fox, J. P. (1976) *Amer. J. Epidem.* **103,** 345.

Fox, J. P., Cooney, M. K. and Hall, C. E. (1975) *Amer. J. Epidem.* **101,** 122.

Fox, J. P., Hall, C. E. and Cooney, M. K. (1977) *Amer. J. Epidem.* **105,** 362.

Garwes, D. J. (1979) *Viral Enteritis*, Les colloques de l'I.N.S.E.R.M. **90,** 141.

Gauntt, C. J. and Griffith, M. M. (1974) *J. Virol.* **13,** 762.

Ginsberg, H. S., Pereira, H. G., Valentine, R. C. and Wilcox, W. C. (1966) *Virology* **28,** 782.

Gwaltney, J. M. (1966) *Proc. Soc. exp. Biol., N.Y.* **122,** 1137.

Gwaltney, J. M. and Hendley, J. O. (1978) *Amer. J. Epidem.* **107,** 357.

Gwaltney, J. M., Hendley, J. O., Simon, G. and Jordan, W. S. (1966) *New Engl. J. Med.* **275,** 1261; (1968) *Amer. J. Epidem.* **87,** 158.

Gwaltney, J. M. and Jordan, W. S. (1966) *Amer. Rev. resp. Dis.* **93,** 362.

Gwaltney, J. M., Moskalski, P. B. and Hendley, J. O. (1978) *Ann. intern. Med.* **88,** 463.

Hamory, B. H., Hamparian, V. V., Conant, R. M. and Gwaltney, J. M. (1975) *J. infect. Dis.* **132,** 623.

Hamparian, V. V., Kelter, A. and Hilleman, M. R. (1961) *Proc. Soc. exp. Biol., N.Y.* **108,** 444.

Hamparian, V. V., Leagus, M. B., Hilleman, M. R. and Stokes, J. (1964) *Proc. Soc. exp. Biol., N.Y.* **117,** 469.

Hamre, D., Connelly, A. P. and Procknow, J. J. (1966) *Amer. J. Epidem.* **83,** 238.

Hamre, D., Kindig, D. A. and Mann, J. (1967) *J. Virol.* **1,** 810.

Hamre, D. and Procknow, J. J. (1966) *Proc. Soc. exp. Biol., N.Y.* **121,** 190.

Hartley, J. W., Rowe, W. P., Bloom, H. H. and Turner, H. C. (1964) *Proc. Soc. exp. Biol., N.Y.* **115,** 414.

Hayflick, L. and Moorhead, P. S. (1961) *Exp.* cell. *Res.* **25,** 585.

Hendley, J. O., Gwaltney, J. M. and Jordan, W. S. (1969) *Amer. J. Epidem.* **89,** 184.

Hendley, J. O., Wenzel, R. P. and Gwaltney, J. M. (1973) *New Engl. J. Med.* **288,** 1361.

Hierholzer, J. C. (1976) *Virology* **75,** 155.

Hierholzer, J. C., Palmer, E. L., Whitfield, S. G., Kaye, H. S. and Dowdle, W. R. (1972) *Virology* **48,** 516.

Higgins, P. G., Ellis, E. M. and Boston, D. G. (1966) *Mthl. Bull. Minist. Hlth Lab. Serv.* **25,** 5.

Higgins, P. G., Ellis, E. M. and Woolley, D. A. (1969) *J. med. Microbiol.* **2,** 109.

Hilleman, M. R., Stallones, R. A., Gauld, R. L., Warfield, M. S. and Andersen, S. A. (1956) *Proc. Soc. exp. Biol., N.Y.* **92,** 377.

Hilleman, M. R. and Werner, J. H. (1954) *Proc. Soc. exp. Biol., N.Y.* **85,** 183.

Hofstad, M. S. (1965) *Diseases of Poultry*, 5th edn. Iowa State University Press, Ames.

Holzel, A. *et. al.* (1965) *Brit. med. J.* **i,** 614.

Hoorn, B. and Tyrrell, D. A. J. (1965) *Brit. J. exp. Path.* **46,** 109.

Hoorn, B. and Tyrrell, D. A. J. (1966) *Arch. ges. Virusforsch.* **18,** 210.

Hope-Simpson, R. E. (1958) *Proc. Roy. Soc. Med.* **58,** 267.

Horn, M. E. C. and Gregg, I. (1973) *Chest*, **63,** Suppl. 44.

Horn, M. E. C., Reed, S. E. and Taylor, P. (1979) *Arch Dis. Child.* 54, 587.

Horne, R. W., Brenner, S., Waterson, A. P. and Wildy, P. (1959) *J. molec. Biol.* **1,** 84.

Huebner, R. J., Rowe, W. P. and Lane, W. T. (1962) *Proc. nat. Acad. Sci., Wash.* **48,** 2051.

Jackson, G. G. and Muldoon, R. L. (1973) *J. infect. Dis.* **128,** 814.

Jawetz, E., Kimura, S., Nicholas, A. N., Thygeson, P. and Hanna, L. (1955) *Science* **122,** 1190.

Johnson, K. M., Bloom, H. H. and Forsyth, B. R. (1965) *Amer. J. Epidem.* **81,** 131.

Johnson, K. M. and Rosen, L. (1963) *Amer. J. Hyg.* **77,** 15.

Kapikian, A. Z. (1975) *Develop. biol. Stand.* **28,** 42.

Kapikian, A. Z., Almeida, J. D. and Stott, E. J. (1972) *J. Virol.* **10,** 142.

Kapikian, A. Z., James, H. D., Kelly, S. J., King, L. M. and Chanock, R. M. (1972) *Proc. Soc. exp. Biol., N.Y.* **139,** 179.

Kapikian, A. Z., James, H. D., Kelly, S. J. and Vaughn, A. L. (1973) *Infect. Immun.* **7,** 111.

Kapikian, A. Z. *et al.* (1967) *Nature, Lond.* **213,** 761; (1969) *J. infect. Dis.* **119,** 282; (1971) *Virology* **43,** 524.

Kaye, H. S. and Dowdle, W. R. (1969) *J. infect. Dis.* **120,** 576; (1975) *Amer. J. Epidem.* **101,** 238.

Kennedy, D. A. and Johnson-Lussenburg, C. M. (1975/76) *Intervirol.* **6,** 197.

Kenny, G. E., Cooney, M. K. and Thompson, D. J. (1970) *Amer. J. Epidem.* **91,** 439.

Ketler, A., Hall, C. E., Fox, J. P., Elveback, L. and Cooney, M. K. (1969) *Amer. J. Epidem.* **90,** 244.

Ketler, A., Hamparian, V. V. and Hilleman, M. R. (1962) *Proc. Soc. exp. Biol., N.Y.* **110,** 821.

Killington, R. A., Stott, E. J. and Lee, D. (1977) *J. gen. Virol.* **36,** 403.

Knight, V. *et al.* (1963) *Amer. Rev. resp. Dis.* **88,** Suppl. 135.

Kruse, W. V. (1914) *Münch. med. Wschr.* **61,** 1547.

Ladisch, S. *et al.* (1979) *J. Pediat.* **95,** 348.

Lanser, J. A. and Howard, C. H. (1980) *J. gen. Virol.* **46,** 349.

Lewis, T. L. *et al.* (1975) *Ann N.Y. Acad. Sci.* **258,** 505.

Lonberg-Holm, K. and Butterworth, B. E. (1976) *Virology* **71,** 207.

McGregor, S. and Rueckert, R. R. (1977) *J. Virol.* **21,** 548.

McIntosh, K. (1974) *Curr. Top. Microbiol. Immunol.* **63,** 86.

McIntosh, K., Chao, R. K., Krause, H. E., Wasil, R., Mocega, H. E. and Mufson, M. A. (1974) *J. infect. Dis.* **130,** 502.

McIntosh, K., Dees, J. H., Becker, W. B., Kapikian, A. Z. and Chanock, R. M. (1967) *Proc. nat. Acad. Sci., Wash.* **57,** 933.

McIntosh, K., McQuillin, J., Reed, S. E. and Gardner, P. S. (1978) *J. med. Virol.* **2,** 341.

McLean, C., Matthews, T. J. and Rueckert, R. R. (1976) *J. Virol.* **19,** 903.

McNamara, M. J., Phillips, I. A. and Williams, O. B. (1969) *Amer. Rev. resp. Dis.* **100,** 19.

Macnaughton, M. R. (1978) *Abstracts, 4th Int. Congr. Virol.* The Hague.

Macnaughton, M. R., Cooper, J. A. and Dimmock, N. J. (1976) *J. Virol.* **18,** 926.

Macnaughton, M. R., Davies, H. A. and Nermut, M. V. (1978) *J. gen. Virol.* **39,** 545.

Macnaughton, M. R. and Madge, M. H. (1978) *J. gen. Virol.* **39,** 497.

Matthews, T. H. J., Reed, S. E. and Tyrrell, D. A. J. (1974) *Arch. ges. Virusforsch.* **45,** 106.

Merigan, T. C., Reed, S. E., Hall, T. S. and Tyrrell, D. A. J. (1973) *Lancet* **i,** 563.

Minor, T. E., Dick, E. C., De Meo, A. N., Ouellette, J. J., Cohen, M. and Reed, C. E. (1974) *J. Amer. med. Ass.* **277,** 292.

Mitchell, I., Inglis, J. M. and Simpson, H. (1978) *Arch. Dis. Child.* **53,** 106.

Mogabgab, W. J. (1962) *Amer. J. Hyg.* **76,** 15.

Mohanty, S. B. (1978) *Advanc. vet. Sci. comp. Med.* **22,** 83.

Monto, A. S. (1976) *Virus Infections of Humans.* Plenum Press, New York.

Monto, A. S. and Bryan, E. R. (1974) *Proc. Soc. exp. Biol., N.Y.* **145,** 690; (1978) *Amer. Rev. resp. Dis.* **118,** 1101.

Monto, A. S. and Cavallaro, J. J. (1972) *Amer. J. Epidem.* **96,** 352.

Monto, A. S. and Johnson, K. M. (1966) *Proc. Soc. exp. Biol., N.Y.* **121,** 615.

Mufson, M. A. and Belshe, R. B. (1976) *J. Urol.* **115,** 191.

Mufson, M. A., Chang, V., Gill, V., Woods, S. C., Romansky, M. J. and Chanock, R. M. (1967) *Amer. J. Epidem.* **86,** 526.

Mufson, M. A., Krause, H. E., Mocega, H. E. and Dawson, F. W. (1970) *Amer. J. Epidem.* **86,** 526.

Mufson, M. A. *et al.* (1963) *J. Amer. med. Ass.* **186,** 578.

Newman, J. F. E. *et al.* (1977) *Intervirology* **8,** 145.

Norrby, E. and Wadell, G. (1969) *J. Virol.* **4,** 663.

Norrby, E. *et al.* (1976) *Intervirol.* **7,** 117.

Parker, J. C., Cross, S. S. and Rowe, W. P. (1970) *Arch. ges. Virusforsch.* **31,** 293.

Perkins, J. C., Tucker, D. N., Knopf, H. L. S., Wenzel, R. P., Kapikian, A. Z. and Chanock, R. M. (1969*b*) *Amer. J. Epidem.* **90,** 519.

Perkins, J. C. *et al.* (1969*a*) *Amer. J. Epidem.* **90,** 319.

Perryman, L. E., McGuire, T. C. and Crawford, T. B. (1978) *Amer. J. vet. Res.* **39,** 1043.

Person, D. A. and Herrmann, E. C. (1970) *Mayo Clinic. Proc.* **45,** 517.

Philipson, L., Pettersson, U. and Lindberg, U. (1975) *Virol. Monogr.* **14,** 1.

Phillips, C. A., Melnick, J. L. and Grim, C. A. (1968) *Amer. J. Epidem.* **87,** 447.

Pina, M. and Green, M. (1965) *Proc. nat. Acad. Sci., Wash.* **54,** 547.

Pinto, C. A., Bahnsen, H. P., Ravin, L. J., Haff, R. F. and Pagano, J. F. (1972) *Proc. Soc. exp. Biol., N.Y.* **141,** 467.

Plummer, G. (1963) *Arch. ges. Virusforsch.* **12,** 694.

Powell, D. G. *et al.* (1978) *Equine Infectious Diseases IV.* Ed. by J.T. Bryans and H. Gerber. Veterinary Publications, Princeton.

Price, W. H., Emerson, H., Ibler, I., Lachaine, R. and Terrell, A. (1959) *Amer. J. Hyg.* **69,** 224.

Reed, S. E. (1972) *J. comp. Path.* **82,** 267; (1975) *J. Hyg., Camb.* **75,** 249; (1980) In: *Developments in Antiviral Therapy,* p. 157. Ed. by L. H. Collier and J. Oxford, Academic Press, London.

Reed, S. E. and Boyde, A. (1972) *Infect. Immun.* **6,** 68.

Reed, S. E., Craig, J. W. and Tyrrell, D. A. J. (1976) *J. infect. Dis.* **133,** Suppl. A128.

Reed, S. E. and Hall, T. S. (1973) *Infect. Immun.* **8,** 1.

Reed, S. E., Tyrrell, D. A. J., Betts, A. O. and Watt, R. G. (1971) *J. comp. Path.* **81,** 33.

Report (1965) *Brit. med. J.* **ii,** 319; (1981) *CDR, Colindale* 81/48.

Robb, J. A. and Bond, C. W. (1979) *Comprehensive Virology,* **14,** Plenum Press, New York.

Rosen, L. (1958) *Virology* **5,** 574.

Rowe, W. P., Huebner, R. J., Gilmore, L. K., Parrott, R. H. and Ward, T. G. (1953) *Proc. Soc. exp. Biol., N.Y.* **84,** 570.

Russell, W. C., Valentine, R. C. and Pereira, H. G. (1967) *J. gen. Virol.* **1,** 509.

Sangar, D. V. (1979) *J. gen. Virol.* **45,** 1.

Sarateanu, D. E. and Ehrengut, W. (1980) *Infection* **8,** 70.

Schalk, A. F. and Hawn, M. C. (1931) *J. Amer. vet. med. Ass.* **78,** 413.

Schieble, J. H., Fox, V. L., Lester, F. and Lennette, E. H. (1974) *Proc. Soc. exp. Biol., N.Y.* **147,** 541.

Schieble, J. H., Lennette, E. H. and Fox, V. L. (1970) *Proc. Soc. exp. Biol., N.Y.* **133,** 329.

Scientific Committee on Common Cold Vaccines (1965) *Brit. med. J.* **i,** 1344.

Smith, C. B. (1978) *Arch. Virol.* **57,** 231.

Smith, C. B., Golden, C. A., Kanner, R. E. and Renzetti, A. D. (1980) *Amer. Rev. resp. Dis.* **121,** 225.

Sprague, J. B., Hierholzer, J. C., Currier, R. W., Hattwick, M. A. W. and Smith, M. D. (1973) *New Engl. J. Med.* **289,** 1341.

Stadler, H., Hierholzer, J. C. and Oxman, M. N. (1977) *J. clin. Microbiol.* **6,** 257.

Stenhouse, A. C. (1968) *Brit. med. J.* **iii,** 287.

Stern, D. F., Burgess, L., Linesch, S. and Kennedy, S. I. T. (1981) *Biochemistry and Biology of Coronaviruses.* Plenum Press, New York.

Stern, D. F. and Kennedy, S. I. T. (1980) *J. Virol.* **34,** 665.

Stott, E. J. (1975) *Develop. biol. Standard* **28,** 65.

Stott, E. J., Bell, E. J., Eadie, M. B., Ross, C. A. C. and Grist, N. R. (1967) *J. Hyg., Camb.* **65,** 9.

Stott, E. J., Grist, N. R. and Eadie, M. B. (1968) *J. med. Microbiol.* **1,** 109.

Stott, E. J. and Killington, R. A. (1972*a*) *Lancet* **i,** 1369; (1972*b*) *Annu. Rev. Microbiol.* **26,** 503.

Stott, E. J. and Walker, M. (1967) *Brit. J. exp. Path.* **48,** 544; (1969) *Nature, Lond.* **224,** 1311.

Strizova, V., Brown, P. K., Head, B. and Reed, S. E. (1974) *J. med. Microbiol.* **7,** 433.

Takafuji, E. T., Gaydos, J. C., Allen, R. G. and Top, F. H. (1979) *J. infect. Dis.* **140,** 48.

Tannock, G. A. (1973) *Arch. ges. Virusforsch.* **43,** 249.

Tannock, G. A. and Hierholzer, J. C. (1977) *Virology* **78,** 500; (1978) *J. gen. Virol.* **39,** 29.

Tyrrell, D. A. J. (1965) *Common Colds and Related Diseases.* Edward Arnold, London; (1968) *Monogr. in Virology* **2,** 67.

Tyrrell, D. A. J. and Bynoe, M. L. (1965) *Brit. med. J.* **i,** 1467.

Tyrrell, D. A. J., Bynoe, M. L. and Hoorn, B. (1968*b*) *Brit. med. J.* **i,** 606.

Tyrrell, D. A. J. and Chanock, R. M. (1963) *Science* **141,** 152.

Tyrrell, D. A. J., Mika-Johnson, M., Phillips, G., Douglas, H. J. and Chapple, P. J. (1979) *Infect. Immun.* **26,** 621.

Tyrrell, D. A. J. and Parsons, R. (1960) *Lancet* **i,** 239.

Tyrrell, D. A. J. *et al.* (1968*a*) *Nature, Lond.* **220,** 650; (1975) *Intervirology* **5,** 76; (1978) *Intervirology* **10,** 321.

van der Veen, J. (1963) *Amer. Rev. resp. Dis.* **88,** Suppl. 167.

Virus Subcommittee of the International Nomenclature Committee (1963) *Virology* **19,** 114.

Wadell, G. and Norrby, E. (1969) *J. Virol.* **4,** 671.

Winterfield, R. W. and Fadley, A. M. (1975) *Amer. J. vet. Res.* **36,** 524.

98

The *Paramyxoviridae*

Robert F. Sellers

Introductory

The family *Paramyxoviridae* has been divided into three genera: Paramyxovirus, which contains mumps, parainfluenza 1–4, Newcastle disease and other paramyxoviruses; Morbillivirus, to which measles, distemper, rinderpest and peste des petits ruminants viruses belong; and Pneumovirus, which contains human and bovine respiratory syncytial viruses and pneumonia virus of mice (Kingsbury *et al.* 1978). All three genera contain RNA that is single-stranded, non-infective and possibly negative in polarity, 6–7 polypeptides as protein and lipid. However, Paramyxovirus is the only genus that contains neuraminidase (Table 98.1 and Fig. 98.1).

The viruses multiply intracellularly in the cytoplasm but so may they also in the nucleus. Maturation is by budding at the cell surface membrane. The formation

Table 98.1 Some characteristics of the genera of the family *Paramyxoviridae*

Character	Paramyxovirus	Morbillivirus	Pneumovirus
RNA single stranded	+	+	+
Protein (6–7 polypeptides)	+	+	+
Neuraminidase	+	−	−
Haemagglutinin	+	+ Measles ? Others	− RSV + PVM
Haemolysin	+	+	?
Cell fusion	+	+	?
Lipid	+	+	+
Carbohydrate	+	+	+
pH stability (4.5)	Stable	Labile	Labile
Site of multiplication	Cytoplasm	Cytoplasm	Cytoplasm
Persistent infection of culture	−	+	?
Disease	Respiratory Generalized	Generalized Respiratory Alimentary Neurological	Respiratory

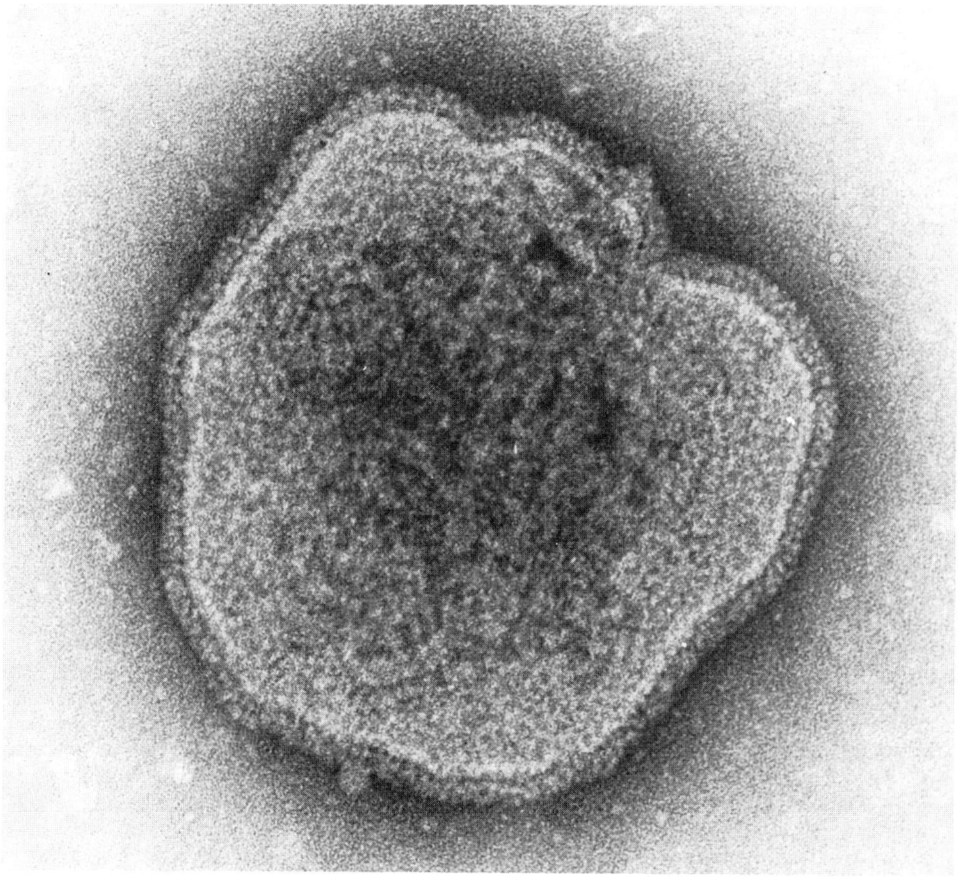

Fig. 98.1 Electronmicrograph of paramyxovirus type 3. (Photograph by C. J. Smale.)

of syncytia in tissue culture, haemolytic activity and haemagglutination are found in members of each genus. Persistent infection of tissue cultures has been established.

The nature of the disease caused by members of the family varies from generalized infection to restriction to the respiratory tract. Involvement of the nervous system is common and this may lead to neurological symptoms later in life. Although other routes are used, the main route of infection with many of the family is by the respiratory tract.

The host range covers mammals, including man, and birds (Paramyxovirus). Immunity depends on both humoral and cell-mediated systems and these systems influence the nature of disease and the development of vaccines. Although successful live attenuated vaccines have been developed for distemper, rinderpest and other diseases in the group, difficulties have been experienced with inactivated vaccines for measles and respiratory syncytial virus and exacerbation of disease has resulted. The successful protection of the respiratory tract continues to present problems and may depend on the elicitation of certain protective mechanisms without giving rise to toxic or immunological reactions harmful to the host (McClelland 1980; Lennette 1981).

Paramyxoviruses

The genus Paramyxovirus includes mumps, parainfluenza 1–4 and avian paramyxovirus 1–6, Newcastle disease being avian paramyxovirus 1.

Mumps

Mumps is an acute infectious disease of man characterized by enlargement of the parotid and submaxillary glands, which may be accompanied by complications. It is caused by a virus (Johnson and Goodpasture 1935) whose size ranges between 150 and 1250 nm in diameter. It consists of a nucleocapsid containing negatively charged single-stranded RNA which is surrounded by a lipid and glycoprotein envelope with projections on the outside (Horne *et al.* 1960). It has proteins similar to Newcastle disease virus, namely, haemagglutinin, neuraminidase (HN), cell fusion haemolysis (F) as glycoproteins, together with RNA polymerase (P), nucleocapsid (NP), large (L) and membrane (M) proteins.

The virus has been grown in the fertile egg after amniotic or yolk sac inoculation and incubation at 33–36° (Habel 1945). However, tissue cultures are more sensitive and cultures of chick embryo, HeLa and monkey kidney cells have been used. The growth of virus is detected by lysis of cells, by syncytia formation, by haemadsorption and by the development of intracytoplasmic eosinophilic inclusion bodies. The extent to which this occurs depends on the degree of adaptation to the tissue culture and whether the virus has been passaged in eggs. Syncytia formation may also occur in cells without multiplication when large amounts of virus are inoculated (Henle *et al.* 1954). Plaques can be produced under agar after 7 to 10 days.

Haemagglutination of red blood cells of chickens, man or guinea-pigs is found, but little or no elution takes place. The haemagglutinin is inhibited by specific antiserum and also, to a lesser extent, by antisera of Newcastle disease and parainfluenza viruses. Haemagglutination does not readily occur with newly isolated strains; these are detected more readily in tissue cultures by haemadsorption with chick or guinea-pig erythrocytes. Virus in high concentrations induces lysis of red cells.

Two complement-fixing antigens corresponding to the viral nucleocapsid (S antigen—presumably containing the NP protein) and the envelope (V antigen—containing the whole virus particle and the projections and envelope) can be identified. Antibodies to the S antigen develop in man more rapidly than antibodies to the V antigen but disappear sooner.

Other tests used for detection include plaque neutralization and agar gel precipitin tests. Mumps exists as a single antigenic type and no antigenic variation has been described.

Clinical signs

The virus causes swelling of parotid and submaxillary salivary glands and may cause meningitis or meningo-encephalitis (Wilfert 1969). Gastro-enteritis, oophoritis and orchitis may also occur; and occasionally deafness, mastitis, pancreatitis, haemolytic anaemia, subacute thyroiditis, diabetes, nephritis and arthritis have been found.

The disease is usually benign and lasts only a short time (Modlin *et al.* 1975). Sometimes inapparent infection occurs. Deaths are rare and second attacks uncommon. Mumps during pregnancy may cause abortion but fetal abnormalities are not seen.

Similar clinical signs have been produced experimentally in Rhesus monkeys and other primates, with fever and swelling of the parotid gland after inoculation into Stenson's ducts and orchitis after intratesticular inoculation.

The virus enters mouth and nose by droplet infection and multiplies first in the respiratory tract. A viraemia then occurs and the virus spreads to different parts of the body. The incubation period ranges from 12 to 26 days, with 18 days the most usual. The patient is infectious from 4 to 5 days before the appearance of symptoms to 7 to 9 days afterwards. During this period saliva carries virus (Ennis and Jackson 1968). Virus may also be found in the cerebrospinal fluid, urine (14 days from onset of disease), blood and milk (Kilham 1948, Kilham 1951, Utz *et al.* 1964) and the salivary glands of healthy contacts.

Experimentally, the virus has been adapted to multiply in hamsters, mice and rats inoculated by the intracerebral route, as well as in monkeys. No signs of illness are seen in these animals but antibodies develop. The virus was adapted to suckling mice by the intracerebral route and ferrets and hamsters were inoculated by the intranasal route.

Mumps is world-wide. It is commonest in children 4 to 15 years of age but rarely attacks infants under one year old. About 80 per cent become infected at the time of adolescence. Mumps has been seen also in young adults in fighting services or those in isolated communities. It spreads directly through infected saliva either as aerosol or possibly on surfaces. Most cases occur in the winter and spring; reported cases are few in the summer and autumn (Modlin *et al.* 1975).

Diagnosis

Diagnosis is carried out by isolation of the virus from the saliva, throat washings, swabs from the parotid ducts or cerebrospinal fluid as soon as possible after the clinical symptoms have appeared. The material is inoculated into the amniotic sac of eggs or into monkey kidney or HeLa tissue cultures. Presence of virus is detected in eggs by the production of haemagglutinin and in tissue cultures by haemadsorption.

Complement-fixation and haemagglutination-inhibition tests on paired sera are carried out and neutralization tests may also be performed.

Control

Two kinds of vaccine have been designed for control. Killed vaccines incorporating virus inactivated with formalin or ultraviolet light have been used (Henle *et al.* 1959). Attenuated strains have been developed but vary in the extent to which they multiply when given to man. With some, multiplication may be limited and the attenuated vaccine may act in the same way as an inactivated one. The extent of use of live vaccine in the United States has been reported by Modlin and others (1975). Studies on the persistence of antibodies after administration of combined live attenuated measles, mumps and rubella vaccines demonstrated that antibodies to mumps were produced after inoculation and remained for at least ten years and afforded protection against the disease (Weibel *et al.* 1980*a*; Weibel *et al.* 1980*b*).

Parainfluenza viruses

Properties

Parainfluenza viruses were originally called haemadsorption viruses. They are now classified into four antigenic types, parainfluenza types 1, 2, 3 and 4 (Kingsbury *et al.* 1978) (Table 98.2).

Like other members of the genus, the viruses are enveloped and pleomorphic, with diameters ranging from 120 to 300 nm. The nucleocapsid contains a single-stranded, negatively charged RNA and the envelope has projections. Among the polypeptides present are two glycoproteins on the surface projections associated with the haemagglutinin-neuraminidase activity (HN) and the fusion-haemolytic activity (F). The viruses replicate in the cytoplasm and mature by budding through the surface membrane.

Parainfluenza type 1 contains two virus strains—HA-2 isolated from man (Chanock *et al.* 1958) and

Table 98.2 Replication of parainfluenza viruses 1, 2, 3 and 4 in tissue culture, eggs and laboratory animals together with haemagglutination

	Virus	Growth in cells	CPE	Syncytia	Inclusions	Haemadsorption	HA red blood cells	Eggs	Laboratory animals
1	HA2	Mammalian	(+)	+	+	+	Human, Guinea-pig, Chick	(+) Allantois	Mice
	Sendai	Chick						+	Guinea-pigs
2	CA, SV5	Mammalian	(+), +	+	+	+	Chick	(+), +	
3	Human, Bovine	Mammalian	+	+	+	+	Human, Chick, Guinea-pig	+	Hamster
4	A, B	Mammalian cells and Chick	(+)	(+)	+	+	Guinea-pig	—	—

Sendai, which is a murine strain. Both grow in a number of mammalian and avian primary tissue culture cells. On first isolation the cytopathic effect is difficult to elicit but after passage both strains give rise to a cytopathic effect, formation of syncytia and intracytoplasmic inclusions. Presence of virus can be demonstrated by haemadsorption. The strains will grow in eggs after amniotic inoculation, and haemagglutinins can be demonstrated with human O and guinea-pig erythrocytes for human strains and with erythrocytes of many origins for murine strains. Haemolysis of chicken erythrocytes often occurs. The Sendai virus will kill mice when inoculated by the intracerebral route but passage of virus is not successful. Sendai virus also causes fetal death in pregnant guinea-pigs.

Parainfluenza type 2 includes the human croup-associated virus (Chanock 1956) and the simian strains SV5 and DA, as well as others. A strain isolated from dogs is also related to SV5 and DA. Strains grow in a number of mammalian primary tissue cultures; monkey kidney monolayers are used for isolation of human and monkey strains, and dog kidneys for dog strains. The cytopathic effect is difficult to obtain at first. Syncytia and cytoplasmic inclusions are found. Presence of virus is demonstrated by haemadsorption. Human strains can be adapted to eggs by the amniotic route but simian strains multiply after amniotic or allantoic inoculation. Haemagglutination with chicken and human red blood cells has been demonstrated. Cell cultures derived from kidneys of apparently healthy monkeys may contain simian viruses.

Parainfluenza type 3 includes human and bovine strains (Chanock *et al.* 1958, Reisinger *et al.* 1959). Human strains grow in monkey kidney cultures and can be adapted to grow in other mammalian cells, in cell lines and in chick embryo tissue cultures. Bovine strains grow in bovine tissue cultures and in cells derived from other bovid species. Cytopathic effect, syncytia formation and development of intracytoplasmic inclusions are seen. In addition, bovine strains produce intranuclear inclusions. Haemadsorption can be demonstrated in cultures. The strains grow in fertile eggs when inoculated by the amniotic route. Haemagglutinins can be demonstrated with human, guinea-pig and chick erythrocytes (human strains) and with a range of mammalian and avian cells (bovine strains). Haemolysis of red cells is also found.

Parainfluenza type 4 (Johnson *et al.* 1960) grows in monkey kidney cells, with the production of cytopathic effect and inclusions. Haemadsorption is also found and haemagglutination with guinea-pig erythrocytes has been demonstrated.

Antigenic differences are based on differences in the envelope and nucleocapsid antigens.

With parainfluenza type 1 a closer relation between Sendai and HA-2 strains is shown in complement-fixation tests than in cross-haemagglutinin inhibition and neutralization tests. With parainfluenza type 2 common antigens are found in human and animal strains. Strains of human and bovine origin in parainfluenza virus type 3 can be distinguished by complement fixation, haemagglutination inhibition, neutralization and precipitin tests. In parainfluenza type 4 two antigenic subtypes (A and B) have been shown and a cross-reaction with mumps virus has been demonstrated.

Clinical disease

In man all four types can give rise to disease of varying severity; probably the least severe is that due to parainfluenza type 4. All types probably cause upper respiratory symptoms similar to the common cold; influenza-like symptoms may also occur. Most severe cases are usually found in children (Chanock and Parrott 1965, Clarke 1973, Glezen and Denny 1973).

Types 1 and 2 have been isolated from children with acute laryngotracheitis (croup). Type 3 has been associated with bronchiolitis, laryngitis and pneumonia in children. Type 1 has also been isolated from adults with pneumonia. Antibodies to all types can be demonstrated in older children and adults.

Epidemics of types 1, 2 and 3 occur from time to time (Hope-Simpson and Higgins 1969, Monto 1973, Martin *et al.* 1978, Hope-Simpson 1981). Type 3 can be isolated all the year round but peak times in the northern hemisphere are in the summer (June and July) and autumn (October and November). Epidemics of types 1 and 2 are less frequent than type 3 and viruses are isolated in summer and winter. Apparently the viruses are maintained in nature by passage between children, especially during epidemics. However, persistent infection of adults also occurs, as is shown by the demonstration of the persistence of parainfluenza type 3 in the Antarctic (Parkinson *et al.* 1980).

Parainfluenza type 3 causes respiratory disease in cattle, sheep and monkeys (Fig. 98.1). In cattle it is associated with 'shipping fever' that occurs as a result of the collection of animals followed by transport over some distance. Respiratory disease also occurs in calves and parainfluenza virus type 3 is one of the viruses responsible, the others being infectious bovine rhinotracheitis, bovine virus diarrhoea and respiratory syncytial virus. Infection by more than one of these viruses occurs and the signs of disease may be exacerbated by subsequent infection with other micro-organisms, especially bacteria. Respiratory disease from parainfluenza 3 has also been seen in sheep (Hore *et al.* 1968). An outbreak of pneumonia in patas monkeys was associated with isolation of parainfluenza 3 virus. Parainfluenza type 2 has been associated with respiratory illness in dogs.

Parainfluenza viruses are distributed world-wide, although their incidence in different countries varies. Antibodies can be demonstrated in man and animal species even though no clinical disease may be apparent.

Diagnosis Diagnosis is made by the isolation of virus from nasal or throat swabs, from bronchial secretions or, after death, from extracts of affected lungs. The material is inoculated in tissue cultures or fertile eggs. Presence of virus is shown by haemadsorption or by haemagglutination and the type identified with specific antisera. Rise in antibodies in serological tests is also used.

Control Inactivated vaccines have been developed for man but their use is not extensive. Inactivated PI3 vaccines have been used in cattle but their effectiveness is controversial.

Newcastle disease

Newcastle disease was first seen in Java, Indonesia, in 1926 but it derives its name from an outbreak in chickens on a farm near Newcastle-on-Tyne, England (Doyle 1927). Since that time the disease has been recognized in most parts of the world where poultry, cage and game birds are reared or where wild birds exist. In addition, many different clinical signs have been observed which have subsequently been classified into four distinguishable forms.

Physico-chemical properties of the virus

The virus is termed avian paramyxovirus 1 (Alexander 1980). Newcastle disease virus is pleomorphic. Most particles are spherical; the diameter varies from 100 to 200 nm, although larger particles have been seen (up to 600 nm in diameter). Filamentous forms are also found.

As with other members of the family *Paramyxoviridae*, the virus consists of an internal nucleocapsid surrounded by a lipoprotein envelope which is covered with projections or spikes 8 nm long. The nucleocapsid shows helical symmetry and has a diameter of 17–18 nm with a central hole of 4–5 nm (Horne and Waterson 1960, Waterson and Cruickshank 1963). The RNA is a single continuous strand negatively charged, with a molecular weight of 4.8×10^6 to 7.5×10^6. It forms 1 per cent by weight of the virus particle. Protein forms 67 per cent, lipid 27 per cent and carbohydrates 7 per cent by weight. Seven to eight polypeptides have been described, viz. (mol. wt $\times 10^3$) 180, 75, 55, 55, 53, 51, 49 and 42 (Moore and Burke 1974, Hightower *et al.* 1975, Alexander 1980). The function of the 75 000 polypeptide is that of haemagglutination and neuraminidase (HN); it is a surface glycoprotein associated with spikes. One of the 55 000 mol. wt polypeptides is a surface glycoprotein whose function is cell fusion and haemolysis (F). The other 55 000 mol. wt polypeptide is the nucleocapsid subunit (NP) and the 42 000 mol. wt polypeptide forms the membrane protein (M). The functions of the other polypeptides have not been determined, although one may be associated with an RNA polymerase.

Biological properties of the virus

Newcastle disease virus is readily grown in fertile chicken eggs after inoculation on to the chorioallantoic membrane or into the allantois. Growth is detected by the production of haemagglutinin and, depending on the strain, death of the embryo with production of haemorrhages. High titres of virus are produced. Quail eggs are also susceptible. Mice and hamsters can be infected by intranasal or intracerebral inoculation and Rhesus monkeys by intracerebral inoculation.

Newcastle disease virus grows in a wide range of cells including primary chick embryo cells, BHK-21 and Hep-2 cell lines as well as in other avian, mammalian and reptilian tissue cultures. The amount of virus produced and the extent of cytopathogenicity vary with the cell and the strain of Newcastle disease virus used. Cell fusion and haemadsorption are also seen in tissue cultures. Plaques are formed under overlays and vary in size and colour on staining. Persistent infection of tissue cultures may occur.

Haemagglutinin and neuraminidase activity are associated with the 75 000 mol. wt glycoprotein (HN). Haemagglutination occurs with chicken, mouse, human and guinea-pig erythrocytes; it varies between strains and has been used to distinguish between viral isolates. Differences between virulent and avirulent strains are based on heat resistance and ability to agglutinate horse erythrocytes. Neuraminidase activity is generally less variable between strains.

Variation in Newcastle disease virus

Variation in viruses is generally of two kinds, antigenic variation and variation in virulence. With Newcastle disease virus most variation occurs in virulence.

Studies on morphology and buoyant density show that strains of Newcastle disease virus are similar and there are no significant differences in the RNA. The polypeptides of the strains may migrate at varying rates but this has not yet been correlated with antigenic variation or with virulence. Serological studies with haemagglutinin inhibition and neuraminidase inhibition tests indicate that there may be significant differences (Alexander 1980). Differences between strains have also been demonstrated in neutralization tests, but not to the extent that they affect vaccination (Allan *et al.* 1978). However, differences occur especially when kinetic neutralization tests are used (Pennington 1978).

Variation in virulence is found, as can be seen from the classification into velogenic-viscerotropic, velogenic-neurotropic, mesogenic and lentogenic strains (see later). Three tests have been devised for assessing strains (Allan *et al.* 1978):

1. Mean death time in eggs (MDT). This is the average time in hours in which embryos inoculated with 1 minimum lethal dose die. Velogenic strains take 40–60 hours, neurogenic 60–90 hours and lentogenic 90 hours or longer; some strains do not kill the embryo.

2. Intracerebral pathogenicity in chicks (ICPI). This is the time taken for one-day-old chicks to die or to show signs of disease after intracerebral inoculation.

3. Intravenous pathogenicity index (IVPI). This is similar to ICPI except that six-week-old chicks are inoculated intravenously.

These tests do not appear to be entirely satisfactory. The route of infection is not that normally encountered in the field, e.g. inhalation or ingestion. Such tests, however, are difficult to standardize under laboratory conditions. In addition, there may be other factors in the field which may add to or subtract from the virulence of a particular strain.

Virulence has also been correlated with physicochemical and biological tests. For example, with some avirulent strains the haemagglutinin is more heat-stable and agglutinates horse erythrocytes more readily. Growth of virulent virus in certain tissue cultures, development of haemadsorption and of cell fusion, is more rapid and extensive. In addition, there may be an association between plaque size and virulence (Hanson 1974).

If one attempts to correlate structure and function, it would appear that variation in virulence can be correlated with the surface glycoproteins responsible for haemagglutination and neuraminidase activity (HN—mol. wt 75 000) and for cell fusion and haemolysis (F—mol. wt 55 000).

Clinical signs

Four forms of clinical disease have been described—the viscerotropic-velogenic, the neurotropic-velogenic, the mesogenic and the lentogenic (Hanson 1978).

The viscerotropic-velogenic (Doyle's form—Doyle 1927) is the type of disease that was originally seen in Indonesia and in Newcastle. It is characterized by being very lethal to chickens and giving rise to severe haemorrhage of the intestine. Mortality is 90 per cent or higher. Oedema of the eyes, nasal and ocular discharges, difficulties in breathing, torticollis and diarrhoea are found, but often the birds are discovered dead.

The neurotropic-velogenic form was originally described by Beach (1944) and called avian pneumoencephalitis. The disease is characterized by respiratory signs including difficulty in breathing. The nervous system may also be attacked, leading to paralysis of the legs or wings and to torticollis. Ten to 15 per cent of the flock die. Lesions of the intestine are not seen.

Mesogenic strains (Beaudette's form—Beaudette and Black 1946) cause a milder form of the neurotropic-velogenic disease. Respiratory signs are seen and nervous signs may develop. Young chicks are most commonly affected and may die, but mortality in adults is not seen.

Lentogenic strains cause only a very mild or inapparent form of disease (Hitchner and Johnson 1948), although egg production may be affected.

Essentially, the virus attacks the epithelium of the respiratory tract, causing destruction of the ciliated cells. Endothelial cells in the interstitial tissues of the lung are also affected. Sinusitis and inflammation of the air sacs also occur. In the alimentary tract the virus gives rise to proventricular haemorrhages, splenic necrosis and ulceration of the intestine, especially around the duodenum and ileocaecal valves. In the brain, encephalomyelitis characterized by perivascular cuffing, endothelial proliferation and vasculitis are seen. The extent to which these lesions occur depends on the form of virus affecting the bird.

Other birds such as turkeys, ducks and geese are also affected by Newcastle disease virus but the signs of disease are milder and usually only young birds are affected. Pheasants, partridges, quail, pigeons and doves as well as wild birds and pet birds such as parrots, parakeets, canaries and others are affected to varying degrees by the different forms of disease. Infection of mink has also been recorded; in them Newcastle disease virus gave rise to meningo-encephalitis and death.

Pathogenesis

The chicken takes in the virus by inhalation of aerosols either in the air or caused by droplets from drinking water; infection by ingestion also occurs and the virus may enter through the conjunctiva. Sometimes the virus may be introduced parenterally or as an aerosol through contamination of vaccines with Newcastle disease virus (Hanson 1978).

During an incubation period ranging from 2 to 11 days the virus spreads and multiplies within the chicken. At the time of disease virus is found in the tissues and is present in the expired air from the birds. Excretion in the faeces occurs for two weeks or longer but the chicken is not usually infectious after one month. However, virus may be isolated from tracheal explants at a later date.

Virus is also found in the ovary and oviduct and may infect the egg.

Newcastle disease in man

A number of cases of Newcastle disease in man have

been recorded (Brandly 1964, Lancaster 1966, Hanson 1975). These have been among workers in chicken houses containing infected chickens, administering aerosol vaccines, working in laboratories where the virus is investigated or vaccines prepared, in plants handling dead infected poultry carcases or in kitchens (Brandly 1951; Trott and Pilsworth 1965; Nelson *et al.* 1952). Infection occurs mainly by two routes, the conjunctiva or by inhalation. Virus enters the conjunctiva either by splash of infective material or through the person rubbing the eye with contaminated hands or arms. Infection by inhalation occurs as a result of breathing in air in poultry houses containing virus as a result of aerosol vaccine administration or of disease. With both routes the incubation period is 1 to 4 days.

Conjunctivitis can be acute and is usually in one eye. It is accompanied by pain and lachrymation. The local preauricular lymph gland is often swollen. Systemic symptoms are rare and the patient recovers in a week.

Infection by inhalation can lead to a more generalized type of disease. Pharyngitis may occur, accompanied by influenza-type symptoms with headaches, malaise and disturbance of the upper respiratory tract. A slight fever may be found. In more severe cases pneumonitis may occur and mild signs of encephalitis have been described.

The disease has resulted from exposure to all forms of the virus—velogenic, mesogenic and lentogenic. Antibody studies on human sera indicate that subclinical infections often occur. However, spread from infected persons to in-contacts is not usual, although transfer by workers to children has been recorded (Trott and Pilsworth 1965).

Diagnosis is based on the isolation of virus from conjunctival fluids, saliva and nasopharyngeal washings. Virus is sometimes found in the blood and urine. Diagnosis can also be based on the demonstration of a rise in antibody by haemagglutinin-inhibition and complement-fixation tests. The antibodies in some patients also react with mumps virus. However, not all patients develop antibodies as a result of the disease and in some there may be only a transient rise (Rasmussen 1963). Reinfection of some persons has been recorded 3 to 4 months after recovery from infection.

There is no specific treatment. The use of respirators during administration of vaccine and the handling of virus under suitable containment will reduce the chances of infection. Control of disease in the field helps in reducing the risk in poultry slaughter-houses.

Epidemiology

Geographical distribution Newcastle disease is world-wide. There is practically no country where poultry are kept that has not experienced losses from the disease. An account of the distribution and spread throughout the world has been given by Lancaster (1966, 1977) and by Lancaster and Alexander (1975).

From 1926 until the early 1940s the velogenic-viscerotropic form described by Doyle (1927) was predominant and this occurred in Asia, Africa and Europe. However, in the United States in the 1940s the dominant form was the velogenic-neurotropic, characterized by neurological and respiratory signs of disease. During the 1940s, 1950s and 1960s this latter form and also the mesogenic and lentogenic strains of virus appeared. In the 1960s subclinical Newcastle disease was seen in Northern Ireland and mild and subclinical forms in Australia.

However, in 1962 the velogenic-viscerotropic form appeared in Indonesia and subsequently this form was seen in Iran in 1966, countries of the Near and Middle East and south-eastern Europe in the late 1960s and in Central and Western Europe, including the United Kingdom, from 1970 onwards. At the end of the 1960s and in the early 1970s it appeared in South America, and from there spread to the United States and Canada through the importation of parrots (Hanson 1974, Lancaster and Alexander 1975).

Means of spread The virus spreads in a number of ways. It may be spread by direct contact between infected and healthy chickens, by the movement of infected chicks and infected or contaminated eggs, by contact with contaminated poultry material or equipment, or carriage in infected transport or on people. In addition, contact with infected poultry residues may lead to intake of virus (Lancaster and Alexander 1975). Infected chickens excrete virus into the air and airborne spread also occurs between poultry farms (Hugh-Jones *et al.* 1973).

Vaccinated birds may become infected without clinical signs and then excrete the virus. In addition, Newcastle disease vaccines, and vaccines against other poultry diseases, may be contaminated with live Newcastle disease virus.

Turkeys and other poultry may spread the virus in similar ways. Introduction of infected game birds and cage birds from different parts of the world or the same country may also spread Newcastle disease, as may migration of wild birds (Hanson 1974).

The relative importance of these methods of spread varies with each country, according to the poultry industry, the game and cage bird industry and the species of wild birds present (Hanson 1974, Lancaster and Alexander 1975).

Undoubtedly, the growth of intensive husbandry of chickens as a source of protein has led to large numbers of birds being housed together, and spread occurs through the trade in eggs and birds. Where production units are in tropical countries, there may be many chances of disease passing from wild psittacines to the chickens and vice versa. In other countries cage birds are reared and aviaries may be situated near intensive chicken farms. Game birds reared for shooting may also be close to poultry houses. Thus, the opportunities for spread within and between the various types of industry and wildlife have increased in recent years.

Other avian paramyxoviruses

Newcastle disease virus is PMV-1 serotype of the avian paramyxovirus. Other serotypes have been isolated from wild, captive, caged and domestic birds during outbreaks of disease or during surveillance for avian influenza infection. PMV-2 viruses are worldwide and occur in wild or captive birds and especially in turkeys. PMV-3 virus occurs in turkeys and cage birds. PMV-4 occurs mainly in ducks, PMV-5 in budgerigars and PMV-6 in ducks. There are also a number of unclassified viruses. The viruses are not generally pathogenic, although in some instances they have been isolated from sick birds. Disease, however, may be the result of a synergistic action with other avian viruses or micro-organisms. A review of avian paramyxoviruses has been written by Alexander (1980).

Diagnosis Diagnosis of Newcastle disease is based on clinical signs in a flock, post-mortem examination and isolation of virus from tissues, secretions and excretions in fertile eggs or in tissue cultures. The growth of virus is demonstrated by the presence of haemagglutinins or cytopathic effect in tissue cultures. Evidence of infection is obtained from blood samples by showing a rise in neutralizing or haemagglutinin-inhibiting antibody (Hanson 1978).

Control The chief measures for control are the maintenance of a high standard of hygiene, control of importation of possibly infected birds and vaccination (Hanson 1978). Although killed vaccines have been used, the need for widespread coverage and the cost of administration has led to the extensive use of live vaccines (Allan *et al.* 1978).

Strains used for live vaccines are strains that have reduced pathogenicity for chickens and include lentogenic and mesogenic strains. Among lentogenic strains the least pathogenic is F, followed by B_1 and La Sota; the mesogenic strains in order of increasing pathogenicity are Roakin, Komarov, Hertfordshire and Mukteswar. The lentogenic strains are generally administered to birds of all ages in drinking water, intranasal or intraocular instillation or by dusting or spraying as an aerosol. The mesogenic strains are given by ingestion or by wing-web stab to older birds.

The live vaccines may be grown in fertile eggs or in mammalian cell culture systems to which they have been adapted. In some instances an adjuvant such as aluminium hydroxide gel is added.

Inactivated vaccines are prepared from virus grown in eggs that is then inactivated by formalin or betapropriolactone. Adjuvants in the form of an aluminium hydroxide gel or oil emulsion are incorporated in the vaccines.

The choice of vaccine depends on a number of factors. Of great importance is the prevalent form of disease and its extent in an area. For example, with the possibility of greater challenge in the field, an attenuated strain which retains some of its virulence is used in order to obtain the required level of immunity. Where the challenge is likely to be mild, a lentogenic attenuated strain would be suitable. In many instances more than one dose of vaccine is given; the first dose may be a lentogenic strain, which is then followed by a mesogenic strain or an inactivated vaccine. An extensive discussion of Newcastle disease vaccines, their production, use and application, together with a discussion of the various regimens possible, has been published by Allan *et al.* (1978).

Morbilliviruses

The genus Morbillivirus includes measles, distemper, rinderpest and peste des petits ruminants. Measles has been recognized for a long time as an infectious disease of children and in recent years subacute sclerosing panencephalitis (SSPE) has been shown to be one of its complications in later life. Distemper is also one of the oldest diseases and affects dogs and members of the canine and mustelid families. Rinderpest is primarily a disease of cattle known in Asia, Europe and Africa. Peste des petits ruminants (PPR), a rinderpest-like disease of sheep and goats, has only recently been proposed as a separate member of the genus (Gibbs *et al.* 1979). There may be additional members of the family (Hall *et al.* 1980).

Physico-chemical properties

Morbilliviruses are pleomorphic, with diameters 100 to 700 nm (Waterson *et al.* 1961, Plowright *et al.* 1962, Norrby *et al.* 1963, Gibbs *et al.* 1979). Virus consists of a nucleocapsid surrounded by an outer layer. The nucleocapsid is a single strand of RNA (50S) to which are attached about 2000 molecules of nucleocapsid protein. The outer layer or envelope consists of lipid or glycoprotein 5-8 nm in width with protruding spikes 9-15 nm in length. The glycoproteins are termed the H protein (mol.wt 76 000-80 000), associated, in measles, with haemagglutination and the F_1 and F_2 proteins associated with haemolysis and cell fusion (F_1 mol.wt 40 000 and F_2 mol.wt 16-23 000). Between the outer layer and the nucleocapsid is the membrane protein (M—mol.wt 34–39 000). There are two nucleocapsid phosphorylated proteins, P (mol.wt 66–73 000) and NP (mol.wt 58–62 000). A further protein is the L protein (mol.wt 100–200 000) but its role has not been determined. There may be an RNA-dependent RNA polymerase present. Most of the investigations have been carried out on measles and canine distemper viruses (Graves *et al.* 1978, Hall *et al.* 1980, Tyrrell and Norrby 1978, Rima *et al.* 1979, Örvell 1980, Underwood and Brown 1974).

The virion attaches to the cell by the glycoproteins and fusion may occur (Choppin *et al.* 1981). After entry into the cell, the negative strand RNA is transcribed into m-RNA, which by translation forms the various proteins. At the same time, positive RNA strands are found which replicate to form the viral RNA. The viral RNA undergoes encapsidation into the nucleocapsid, which on maturation receives the membrane protein and glycoproteins as it passes through the cell membrane (Choppin and Compans 1975).

Biological properties of the virus

Measles virus has been adapted to monkeys but there are no definite clinical signs. Ferrets are used as laboratory animals for distemper virus; this virus can also be adapted to suckling mice and hamsters. After adaptation, rinderpest virus will multiply in rabbits. Measles, distemper and rinderpest viruses can all be adapted to grow in fertile eggs.

Measles virus grows in human and monkey kidney cells and in established cell lines such as Vero, HeLa and Hep-2 (Fraser and Martin 1978). Distemper virus grows in dog and ferret macrophages and can be adapted to dog kidney cells and canine tumour cell lines and others (Appel and Gillespie 1972, Appel 1978). Rinderpest virus multiplies in calf, sheep and goat kidney and testis cells, in calf thyroids and lung macrophages. It has also been grown in MDCK, HeLa and MG cell lines (Plowright 1968). PPR virus is best grown in lamb or goat kidney cells but also multiplies in Vero cells (Hamdy *et al.* 1976, Taylor and Abegunde 1979). Culture of tissues from infected animals also leads to multiplication of virus. In the case of SSPE virus, co-cultivation is necessary (ter Meulen *et al.* 1972).

Virus grown in tissue cultures is recognized by the production of intracytoplasmic and intranuclear eosinophilic inclusions and the production of multinucleated giant cells. In some instances persistent infection may be established. In some cultures little or no cytopathic effect or production of infective virus is found; virus antigen may be present and the cells of the culture continue to multiply at rates similar to uninfected cells. Presence of antigen is detected by immunofluorescence on cells (Fraser and Martin 1978, Rustigian 1966, Rima and Martin 1977).

Serological tests are available for these viruses such as neutralization, complement fixation and haemagglutination (measles) and haemolysis inhibition (measles). ELISA and RIA tests are being devised and monoclonal antibodies are being produced (Giraudon and Wild 1981, Togashi *et al.* 1981, ter Meulen *et al.* 1981).

Tests to detect cell-mediated immunity are being used for measles and distemper virus.

Antigenic relationships have been demonstrated between measles, distemper, rinderpest and peste des petits ruminants viruses (Imagawa 1968, Plowright 1968, Appel and Gillespie 1972, Örvell and Norrby 1974, Gibbs *et al.* 1979, Stephenson and ter Meulen 1979, Hall *et al.* 1980). Relationships between measles and distemper have been demonstrated between the M, NP and F polypeptides but less with the H polypeptides (Örvell and Norrby 1980).

On the practical side, inactivated or live preparations of heterologous virus have been used for protection against disease, e.g. killed measles vaccine against distemper and attenuated rinderpest vaccine against PPR. In cross-protection studies the contribution of the humoral and cell-mediated arms in immunity has still to be elucidated.

There have been attempts to determine the relation of measles and canine distemper to multiple sclerosis (Field *et al.* 1972). The antibody titre to measles is higher in sera from patients with multiple sclerosis than that from patients without it. These antibodies appear to be specific for measles and not for canine distemper (Stephenson *et al.* 1980). The relationship is far from clear.

Measles

Clinical symptoms and signs

After infection, usually through the respiratory route but possibly via the conjunctiva, the incubation period may last 10–14 days. Occasionally in adults this may be prolonged to 3 weeks. The first symptoms are termed the prodromal or catarrhal phase and are characterized by a slight fever, cough, nasal and ocular discharge, together with conjunctivitis and photophobia. Koplik's spots also appear as greyish white dots in the mouth usually around the lower molar teeth; they disappear in 24 hours. Shortly afterwards the later stage of the disease appears with the development of a papular macular rash, first seen on the neck and face and subsequently found all over the body. At the same time there is a rise in temperature and the lymph nodes become enlarged.

In some cases there are no further symptoms, and with the fading of the rash the patient recovers. However, complications occur with the development of otitis media and bronchopneumonia (a giant-cell pneumonia). In about one patient in 2000 encephalitis may also be seen—on average on the sixth day after the appearance of the rash but the interval may vary from the first to the fifteenth day.

Morbidity and mortality are variable but the disease is worst in tropical countries, especially in West Africa, where mortality is high in very young children (Morley 1969).

Normally, sequelae after recovery are unlikely; nevertheless, in about one in a million among children and young adults a subacute sclerosing panencephalitis (SSPE) develops, characterized initially by unusual behaviour and a deterioration in learning ability, which leads to paralysis and dementia (Agnarsdottir 1977, 1980).

Pathogenesis

After entry through the respiratory route or the conjunctiva, the virus is spread through the body via the lymphatics and blood and multiplies most probably in the lymphoid and mucosal epithelial tissues. During the incubation period there is a decrease in the numbers of circulating lymphocytes. With the onset of the prodromal phase virus is found in the bodily secretions, tears, throat swabs, urine and in the blood. Nucleocapsids have been identified in the tissues of the Koplik spots (Suringa *et al.* 1970).

At the time of the appearance of the rash humoral antibodies are found. IgM is found first and reaches a peak at 10 days and then declines. IgG develops more slowly but persists. IgA is also found in the secretions. Cell-mediated immunity is also involved and is probably more important than humoral immunity, since agammaglobulinaemic children recover, whereas those with impaired cellular immune reactions do not (Gatti and Good 1970). At first, however, the lymphatic system is attacked by the virus. The blastogenic response of lymphocytes is depressed. Normal killer lymphocytes are present as well as sensitized T lymphocytes and in addition, after the appearance of the rash, blood mononuclear cells show cellular cytotoxicity, both antibody-dependent and antibody-independent (Whittle and Werblinska 1980). The various contributions of the immune system to recovery from infection still have to be completely elucidated. It is not clear whether virus enters the brain to cause encephalitis. If it did, it could be by mononuclear cells infiltrating nervous tissue, as described by Summers *et al.* (1978) rather than by entry across the capillaries.

In SSPE, virus very similar if not identical with strains of measles virus has been isolated from the brain by co-cultivation of brain cells with established cell lines (ter Meulen *et al.* 1972, Agnarsdottir 1977, Fraser and Martin 1978). Patients with SSPE have high levels of antibody to measles, including IgM in the blood and cerebrospinal fluid. However, no antibody to the M protein of the virus has been found. Three suggestions for the lack of M protein have been put forward:

(1) lack of synthesis of M protein in the cells of the brain,

(2) degradation of protein before it stimulates the immune system, and

(3) a specific immunological defect in which the protein was not recognized by the immune system (Choppin *et al.* 1981). The first possibility seems the most likely, since in direct studies of measles protein in the brain all other proteins except the M protein were found. Thus, M protein was not being synthesized and an abortive infection had occurred. Most patients with SSPE have a history of measles before the age of two. Possibly at that age measles virus entering the brain might multiply in the cells. At a later age the brain cells develop a resistance and an abortive infection occurs.

Epidemiology

Measles occurs world-wide and has been known to exist since the first century BC. Much of its epidemiology has been learnt from outbreaks in isolated communities such as islands. Measles is spread through contact between people. Spread by contaminated fomites is unlikely to occur since the virus is labile. No animals are concerned in transmission. In communities, epidemics usually occur every two to five years and last for 3–4 months. Babies obtain protection from their mothers through maternal antibody, which persists for 6 months. After that time they are susceptible and so the numbers of susceptibles in a population are regularly renewed. By the age of 18 most persons in the population have antibodies. In the northern hemisphere most cases occur in April and May. There is a critical size of population for measles virus to maintain itself. Below 300 000 measles disappears from time to time in populations (Bartlett 1960, Black 1965). Epidemics come at shorter intervals with increasing population size. Population density also affects the duration, it being longer where persons are more dispersed through the country. Subacute sclerosing panencephalitis occurs between the ages of 2 and 21, with most cases between 4 and 12, with a mean of 7. There is no opportunity for fully infectious virus to be formed and hence to spread.

Diagnosis Clinical diagnosis is based on the presence of Koplik spots in the prodromal stage, on catarrhal symptoms and the appearance of the rash.

Laboratory diagnosis can be obtained by immunofluorescent staining of giant cells (see 6th edn, p. 2535) obtained from nasal or pharyngeal collections (Fulton and Middleton 1975). Virus isolation is possible in the prodromal phase but difficult in the later stages, owing to the development of antibody. It is accomplished in primary tissue cultures of human or monkey kidney. Rises in antibody are demonstrable in haemagglutination-inhibition tests. RIA and ELISA tests are also being developed.

Laboratory diagnosis of SSPE is based on the high antibody level with lack of antibodies to the M protein. Electron microscopy of autopsy material from brain has also been performed.

Control: vaccination

Since measles is widespread, little can be done to prevent its spread other than by immunization.

Two types of vaccine have been developed, an inactivated and a live attenuated. The inactivated vaccine induced the production of antibodies and gave a short-lived protection. However, hypersensitive reactions or atypical measles often occurred, especially in response to exposure to natural infection or live vaccine. It is now thought that the formalin or Tween-ether inactivated vaccines induced haemagglutinating-inhibiting (HN) antibodies but not haemolysin-inhibition or fusion (F) antibodies (Norrby and Pettinen 1978). The latter are important in preventing spread by fusion between cells. Hence, after challenge by natural infection or live vaccine, multiplication of virus and spread by fusion between cells occurred; thus antigens were being produced at the same time as a secondary immune response, so giving rise to the symptoms. It is suggested that this could be overcome by incorporating F protein in the vaccine (Choppin *et al.* 1981, Merz *et al.* 1980, McClelland 1980). Live virus vaccine was developed by attenuating the virus by passage in the chick embryo (Katz *et al.* 1960, Krugman 1971). Some of the earlier trials showed reactions in children after the vaccine but these were overcome by the simultaneous administration of human immunoglobulin. Later, further attenuation occurred and the present vaccine is given without the immunoglobulin. Occasionally in some children a rise in temperature, a faint rash and a feeling of malaise occur. Limited multiplication of the virus may occur and antibodies are induced which persist for 10 years (Weibel *et al.* 1980*a*). It is also given as a combined vaccine with mumps and rubella attenuated strains. (See Christensen *et al.* 1983)

Distemper

Clinical signs

Infection of dogs is usually by the respiratory route and after an incubation period of 4–5 days the first signs of distemper are seen. There is a rise in temperature, followed by a fall and a subsequent rise. Nasal and ocular discharges are accompanied by conjunctivitis. Vomiting, diarrhoea, skin eruptions and bronchopneumonia are commonly seen. Nervous signs with fits, muscle jerks, incoordination and changes in behaviour may occur. Keratitis of the feet—hard pads—is also seen occasionally. Mortality is variable; the disease is sometimes fatal to puppies. Some may recover but with permanent damage to one or more of the body systems; others may lose weight and become dehydrated but recover without complications. Slight or inapparent infections may occur.

Old dogs may develop encephalitis, which is ultimately fatal after deterioration in the dog's condition and behaviour. Older dogs, however, may be infected late in life with distemper virus.

The disease also affects other members of the dog family, e.g. fox, wolf, jackal and coyote, and also members of the Mustelid family i.e. ferrets, weasels, mink, skunk, etc. In ferrets the disease is usually fatal.

Pathogenesis

After exposure to infection by the respiratory route, the virus multiplies in the macrophages and lymphocytes found in the respiratory tract and lymph nodes. It then spreads via the lymphatics to the spleen, thymus, bone marrow and other lymph nodes and subsequently to other organs through the blood stream and to the brain, possibly through infiltration of nervous tissue by infected mononuclear cells (Summers *et al.* 1978).

The initial replication in lymphocytes leads to a lymphopenia and also to a general impairment of the immune response, so that levels of humoral antibodies are reduced. However, if the impairment is not extensive, there is an immune response after the spread to the lymphatic organs, and spread to other organs is prevented, with subsequent early recovery of the affected animal (Appel and Gillespie 1972).

Among the humoral antibodies, IgM develops earlier and persists for only 3 months; IgG, however, persists longer, often for many years. IgA is also found. The presence of humoral antibodies is not necessary for protection, since dogs without antibody survive challenge (Gillespie 1965). Cell-mediated immunity found includes antibody-dependent cell-mediated cytotoxicity (Ho and Babiuk 1979) sensitized T lymphocytes (Krakowka and Wallace 1979) and probably other mechanisms.

In the brain, infection may lead to demyelination of the nervous tissues and persistence of the virus. In such animals, antibodies may be found in the cerebrospinal fluid and higher concentrations of IgG and IgM in the blood are found (Cutler and Averill 1969).

Epidemiology

Distemper virus is shed from affected dogs through secretions and excretions and is present in urine and faeces. The virus spreads by aerosols over a short distance from one animal to another (see Gorham 1966). It does not appear to survive in the environment.

Transplacental infection may occur, although infection early in pregnancy in non-immune bitches leads to abortion (Krakowka *et al.* 1974, 1977). However, bitches that have recovered from the disease or have been vaccinated supply maternal immunity to the pups which may last for an average of 9 weeks (vaccinated animals) or longer. After that, the pups are susceptible and hence the population of susceptibles is continually being renewed unless immunization is carried out. Persistence of virus occurs in carrier dogs (Gorham 1966).

Wild living members of the dog and mustelid families maintain the virus among themselves and this may spill over into domestic animals through contact.

There appears to be no connection between the occurrence of distemper in dogs and the occurrence of multiple sclerosis in man (Stephenson *et al.* 1980).

Diagnosis

Diagnosis is based on the clinical signs together with post-mortem and laboratory examination. At post-mortem bronchopneumonia may be found and intracytoplasmic inclusions may be seen in cells from the epithelial cells of the respiratory and alimentary tracts. Both intranuclear and intracytoplasmic inclusions may be found in brain cells.

Immunofluorescent staining of leucocytes, cells of the cerebrospinal fluid and biopsy or post-mortem tissues may be carried out (Appel and Gillespie 1972).

Virus isolation may be attempted by inoculation of ferrets or by cultivation of lung macrophages or post-mortem material.

The detection of IgM antibody as an indication of recent infection is suggested and the ELISA test has been used for this (Noon *et al.* 1980). The ELISA test has also been used to detect IgG.

Control

As with measles, the best method of control is prevention of the disease by immunization.

Various types of vaccination were originally developed (Laidlaw and Dunkin 1928). However, the most successful are the live vaccines attenuated in fertile eggs by Haig (1948) and Cabasso and Cox (1949) or in dog kidney cells by Rockborn (1960). The chick embryo adapted strains have subsequently been adapted to grow in chick embryo cultures (Cabasso *et al.* 1962, Prydie 1968).

Such vaccines protect dogs against disease and against infection with distemper virus. The vaccines are given simultaneously with live or inactivated canine hepatitis virus vaccine and canine leptospirosis vaccines.

However, maternal antibody may interfere with the development of immunity up to 12 weeks after birth and vaccination is postponed until then. To cover the period from 6 to 9 weeks in situations where exposure to virulent virus may occur, inactivated measles vaccines have been given. These will protect against disease in 6–12-week-old puppies but will not prevent infection (Schultz *et al.* 1977).

Rinderpest

Clinical signs

The clinical signs in cattle affected with rinderpest are variable, depending on the breed, whether they are in an endemic or epidemic area and the nature of the strain of rinderpest virus circulating (Plowright 1965).

The route of infection is respiratory and the incubation period may last between 3 and 15 days, with 8–11 days being the most common. The incubation period is followed by the prodromal phase, during which there is a pyrexia followed by the development of clinical signs. These include, as well as fever, depression, anorexia, congestion of the mucosal lining of the mouth and other orifices, a serous discharge from the nose and eyes and a drop in milk yield. The mucosal phase which follows is characterized by the development of small areas of necrotic epithelium, especially on the under surface of the tongue but also in other parts of the mouth, on the lips and gums. This leads to salivation. The nasal discharge becomes mucous, often tinged with streaks of blood. Diarrhoea develops; the faeces are watery and contain mucus. Skin lesions are found on the mucosa surrounding the genitalia. In severe cases the animal may die; in others recovery takes place and during the convalescent phase the necrotic areas in the mouth are shed and the epithelium regenerates; however, permanent damage may be done to the buccal papillae. The diarrhoea stops and the faeces return to normal consistency. The animal starts to regain weight.

In mild cases of disease in cattle, all that may be apparent are slight lesions in the mouth and traces of diarrhoea. Water buffalo may also be affected.

Sheep and goats may likewise be affected with rinderpest (Scott 1955, Rao *et al.* 1974). The disease may be mild or severe. In severe cases animals may be depressed, listless and anorexic. Discharges from the nose and eye are apparent and necrosis of the epithelium occurs in the nasal cavity and mouth and on the tongue. Diarrhoea also occurs and a cough and bronchopneumonia may develop. In severe cases death results. In mild cases there is a transient rise in temperature together with mild lesions in the mouth and slight diarrhoea.

Pigs are also affected. In pigs of European origin there is a transient infection with a rise in temperature, depression and anorexia (Scott *et al.* 1962). Pigs in Asiatic countries are more severely affected (Bansal *et al.* 1974).

Pathogenesis

The virus enters by the respiratory route and probably through the epithelium and the tonsil; from there it spreads by the lymphatics to the local lymph nodes (Taylor *et al.* 1965). Multiplication of virus occurs in the lymph nodes and spread occurs by the blood stream to all lymph nodes, spleen, lung, bone marrow and epithelial mucosae (Liess and Plowright 1964, Plowright 1964). Before the onset of clinical signs virus is present in secretions such as nasal secretions and saliva. Later, virus is present in the urine and faeces and also in the mouth.

With the onset of clinical signs humoral antibodies are found and the amounts of virus present in secretions are reduced. The role of cell-mediated immunity in rinderpest has still to be determined. The virus induces a suppression of humoral and cell-mediated immunity initially, and it may be that the position is similar to that with measles and distemper (Penhale and Pow 1970, Yamanouchi *et al.* 1974).

Epidemiology

Rinderpest is essentially an Asiatic disease which has over the centuries spread into Europe or into Africa (Scott 1964, Plowright 1968). In Africa it has maintained itself in the countries north of the equator and in Asia it is mainly in the Indian subcontinent, but in recent years it has been found in the Middle East and in the Arabian peninsula.

During and after the waning of maternal immunity young animals become susceptible (Provost 1972) and thus the disease maintains itself by spread among them. Spread is by contact with infected animals and occasionally by contact with contaminated faeces. Pigs may be infected by ingestion of meat. Outside the body the virus is rapidly inactivated.

Movements of animals affect the distribution of the disease. The trade in cattle has brought about spread from Somalia to Southern Arabia and from Afghanistan to Iran and other parts of the Middle East. Nomadic movements influence spread, as in West Africa and Sudan, and game animals may also have an influence although they do not appear to act as a reservoir.

Diagnosis Diagnosis is based on the clinical signs and post-mortem examination. However, laboratory confirmation is required. The most rapid is the agar gel precipitin test using lymph nodes specially collected from the affected animal and hyperimmune rinderpest antiserum (Brown and Scott 1960).

Virus isolation may be attempted from blood, lymph nodes, spleen or mucosal erosions by inoculation of tissue cultures of bovine kidney (Plowright 1962). Cultivation of leucocytes collected from infected animals may also be attempted.

Serological tests can be used to demonstrate a rise in neutralizing antibodies.

Control Eradication of rinderpest in many European countries was effected by control of movements and slaughter of affected animals. Reintroduction of infection is prevented by restriction on the import of animals and animal products.

Where eradication is not feasible, vaccination of cattle has been practised. The vaccines used are attenuated and include attenuated goat-adapted virus, attenuated lapinized virus and tissue culture adapted virus (Edwards 1928, Nakamura and Miyamoto 1953, Plowright and Ferris 1959). The virus adapted by passage in primary calf kidney cells (Plowright and Ferris 1962) has been widely used in Africa and contributed much to the control of the disease. It is also used in the Middle East and other parts of Asia, but goat-adapted virus is still used in the Indian subcontinent and the lapinized vaccine in South-East Asia.

The tissue culture adapted vaccine gives rise to a long-lasting immunity. However, if the virus in the vaccine is insufficient or deteriorates, or if the animal's immune system is depressed, failure to produce immunity may result (Provost *et al.* 1969, Plowright quoted by Parker 1976). The vaccine virus multiplies in lymphoreticular tissues. On exposure to virulent virus, local multiplication of virulent virus may occur (Plowright quoted by Parker 1976); hence the suggestion to administer it by the intranasal route (Provost and Borredon 1972).

Peste des petits ruminants

Clinical signs

The route of infection is respiratory. This is followed by an incubation period of 4 to 5 days, although variation may occur. A rise in temperature is the first sign and the sheep or goat appears depressed, does not eat and may lie away from the rest of the flock or herd. There is nasal discharge, at first serous and then mucoid; crusts of discharge are seen around the nose and necrosis of the epithelium of the mouth and tongue is apparent. The conjunctiva and the mucous epithelium are congested. Diarrhoea is often a feature, and there is frequently a cough accompanied by bronchopneumonia. Death occurs in severe cases. Often, however, the disease may be mild. In some areas goats may be affected more than sheep. Abortion may be caused in pregnant animals.

No disease is seen in cattle but a rise in temperature and oral erosions have been seen in calves. Pigs do not suffer from the disease (Nawathe and Taylor 1979).

Pathogenesis Probably the route of infection and the multiplication of virus within the animal are similar to that of rinderpest. However, little work has been done. Virus is found in the mucosa of the small intestine, in the lymph nodes and in the lung and lung macrophages. Virus is also present in the saliva, nasal secretions and faeces (Mornet *et al.* 1956, Whitney *et al.* 1967, Johnson and Ritchie 1968, Hamdy *et al.* 1976, Taylor and Abegunde 1979).

Epidemiology Peste des petits ruminants is present in West Africa. In parts of Nigeria it is known as kata, although this term may also include goats and sheep infected with contagious pustular dermatitis, other micro-organisms and parasites. It has recently been identified in Sudan. Serological evidence of infection has also been found in Southern Arabia.

The virus is present in the secretions and excretions of the affected animal and spreads by contact between animals or by indirect carriers. Spread is effected by the movement of animals either to markets or to the slaughterhouse. These movements often result in epidemics.

Diagnosis Diagnosis is based on clinical signs and on post-mortem examination, where engorgement of the blood capillaries in the large colon produces the so-called 'zebra striping'.

Laboratory diagnosis is based on the agar gel precipitin test with the use of lymph nodes or scrapings from the mucosa of the large intestine as antigen and hyperimmune rinderpest serum as antiserum.

Isolation of virus is by inoculation of primary lamb or calf kidney cells, with the use of material from the mucosa of the large intestine. A neutralization test on the isolate has to be carried out to distinguish between PPR and rinderpest viruses (Hamdy *et al.* 1976, Taylor and Abegunde 1979).

Serological tests can also be carried out. Sheep and goats previously infected with PPR virus have higher antibody titres to the virus compared with rinderpest virus (Taylor 1979*a*).

Control Control in epidemic areas by the use of the tissue culture attenuated rinderpest virus vaccine has been carried out (Bourdin 1973, Taylor 1979*b*, Bonniwell 1980). This protects against the disease but does not prevent multiplication and appearance of PPR virus in nasal secretions (Gibbs *et al.* 1979). Extensive trials are being carried out in parts of West Africa to assess its efficacy. However, it is necessary to carry out differential diagnosis of breakdowns to establish the cause.

Pneumoviruses

The genus Pneumovirus includes respiratory syncytial virus (human and bovine strains) and the pneumonia virus of mice (Kingsbury *et al.* 1978).

Respiratory syncytial virus

The respiratory syncytial virus causes respiratory illnesses in man, especially in early childhood (human strains), and in cattle (bovine strains).

The virus is pleomorphic, roughly spherical or filamentous, with a diameter of 80–500 nm. It consists of a nucleocapsid containing RNA which is negatively charged and has helical symmetry; the nucleocapsid differs from other members of the family in that it is 12–14 nm in diameter with a pitch of 6.5–7 nm. On the surface of the envelope which surrounds the nucleocapsid are projections 12 nm long. The 6–7 polypeptides found differ from those of other paramyxoviruses; neuraminidase and haemagglutinin activity have not been demonstrated (Berthiaume *et al.* 1974, Levine 1977).

The human strains of virus grow readily in Hep-2 and HeLa cells as well as in other cell lines but not so well in primary monkey kidney cells. Syncytia are formed together with cell degeneration, and cytoplasmic inclusions are present that contain virus antigen. Bovine strains grow in cells derived from bovine kidney, testis and lung; growth in other mammalian cells has been described.

Complement fixation, serum neutralization and fluorescent antibody tests have been used to distinguish the strains of virus. Human strains are related to each other. Bovine strains are related among themselves and also to human strains. No antigenic relation has been found with pneumonia virus of mice.

Clinical signs

The human virus was first isolated from a chimpanzee with a cold and was known as the chimpanzee coryza agent (CCA) (Morris *et al.* 1956).

In man the disease is essentially one of childhood, with the main symptoms being bronchiolitis. Infants 1–6 months of age are most commonly affected, although children up to 2 years of age also suffer severe disease. Milder forms similar to a common cold may occur in older children and in adults.

After an incubation period of 4–5 days a running nose is seen; this is followed by a rise in temperature and a cough, together with rapid breathing, wheezing and dyspnoea. The bronchiolitis leads to narrowing of the nasal airways as a result of oedema of the mucosa, mucosal cell debris and mucus. Pneumonia may also be present. In infection of babies under 3 weeks of age the clinical picture is different: the babies have upper respiratory symptoms with lethargy and irritability. Pneumonia is uncommon and bronchiolitis very rare. In infants death may occur. Recovery occurs in many children but in some there is a chance of developing recurrent wheezing in later life. Chronic lung disease in adults may also result.

Although serum antibodies develop as a result of

exposure, these disappear over some months and children are subject to reinfection. By the age of five, 90 per cent of all children have been infected or reinfected.

In contrast to human respiratory syncytial virus, the bovine strain affects cattle of all ages (Woods 1974). The main clinical signs are fever, nasal discharge and cough, together with bronchopneumonia (Inaba *et al.* 1970, Paccaud and Jacquier 1970). Anorexia and depression may also be present and the respiratory involvement may lead to pneumonia and emphysema. It has been suggested that the disease may be more severe in calves which have maternal antibody but this is not always the case (Pirie *et al.* 1981).

Experimental infection of calves has been described (Elazhary *et al.* 1980); high titres of interferon were found initially as well as fluorescent and neutralizing antibodies.

Pathogenesis

The pathogenesis of the disease is obscure. It has been suggested that an immunopathologic mechanism is involved. Some support is given to this possibility by the finding that, after administration of inactivated vaccines, a more severe type of disease was observed in the vaccinated children. It has been suggested that there is a reaction of serum antibody with virus in the respiratory tract; however, no correlation with circulating antibody can be made. Sensitization by prior infection has been put forward, but prior sensitization may not be necessary in every case to produce the symptoms. Others argue that cell-mediated immunity is concerned; in support of this a proliferation of lymphocytic responses has been found. Whatever the cause, it can be pointed out that the most severe forms of the disease occur at a time when developments in the immune system of the child are occurring (McIntosh and Fishaut 1980, Hall 1980, Cranage *et al.* 1981).

Epidemiology

Respiratory syncytial viruses have been isolated world-wide. There is an annual epidemic with most severe morbidity among infants under six months of age. In temperate climates the epidemics occur during the winter months (Hope-Simpson and Higgins 1969, Glezen and Denny 1973, Martin *et al.* 1978, McIntosh and Fishaut 1980), and there is a negative correlation with winter temperature and hours of winter sunlight. The epidemics usually reach a peak after 2 months but may last 3–5 months. Virus isolations are made during the winter months and also during part of the summer but in the USA isolation of virus is rare in August and September. The virus multiplies in the respiratory tract of older children and adults and, since the neutralizing antibodies are short-lived and reinfection occurs, presumably there are ample opportunities for it to be maintained in the community. Spread is by aerosols, although infection through fomites and from contamination of hands is also postulated (Hall *et al.* 1980).

Disease due to bovine syncytial virus is world-wide. When it is introduced into a country for the first time or where it has not been present for some time it may cause severe disease, as seen in Switzerland and Japan (Inaba *et al.* 1970, Paccaud and Jacquier 1970). In areas where it is endemic it may be maintained in a population, causing severe disease only in young animals. Often during outbreaks of respiratory disease in herds, both parainfluenza virus type 3 and respiratory syncytial virus may be present (Bryson *et al.* 1979).

Serological surveys have shown antibody in sheep but no disease has been seen either in the field or experimentally. In cattle it appears that antibodies are detectable for a longer period than they are in man.

Diagnosis Diagnosis is carried out by isolation of virus in tissue cultures. Growth of the virus is detected by the production of syncytia in the cultures. Nasopharyngeal washings can be stained by fluorescent antibody techniques. Neutralizing and complement-fixing antibodies are also used for diagnosis but they decline after some months.

Neutralizing, complement-fixing and precipitating antibodies have been demonstrated in cattle infected with respiratory syncytial virus.

Control A killed vaccine has been used, with unsuccessful results. A live attenuated vaccine has been tested but progress is slow. It is not clear what marker can be used to measure resistance to infection, since in infants and children there appears to be no correlation between protection and the presence of IgA or interferon in secretions (Hall *et al.* 1978, McIntosh 1978, McIntosh *et al.* 1978), although cytotoxic antibody can be related to IgG (Cranage *et al.* 1981). Again, even after natural disease reinfection of children can occur. The duration of immunity is not known.

However, experimental inactivated and live vaccines are being developed for cattle. Among them one containing persistently infected bovine nasal mucosa cells inactivated with glutaraldehyde stimulated antibody formation and inhibited multiplication of respiratory syncytial virus on challenge (Stott *et al.* 1981).

Pneumonia virus of mice

Pneumonia virus of mice, which also belongs to the Pneumovirus genus, differs in that haemagglutination can be demonstrated with mouse or hamster erythrocytes. The virus grows in BHK-21 cells. It is maintained in mouse colonies as an inapparent infection and gives rise to lung lesions after passage in mice of lung suspensions.

References

Agnarsdottir, G. (1977) *Rec. Adv. clin. Virol.* **1,** 21.
Alexander, D. J. (1980) *Vet. Bull.* **50,** 737.
Allan, W. H., Lancaster, J. E. and Toth, B. (1978) *Newcastle Disease Vaccines: Their Production and Use.* FAO, Rome.
Appel, M. (1978) *J. gen. Virol.* **41,** 385.
Appel, M. and Gillespie, J. H. (1972) *Virol. Monog.* **11,** 1.
Bansal, R. P., Joshi, R. C. and Kumar, S. (1974) *Bull. Off. int. Epiz.* **81,** 305.
Bartlett, M. S. (1960) *J. R. Stat. Soc. Ser. A.* **123,** 37.
Beach, J. R. (1944) *Science* **100,** 361.
Beaudette, F. R. and Black, J. J. (1946) *Proc. U.S. Livestock Sanit. Ass.* **49,** 49.
Berthiaume, L., Joncas, J. and Pavilanis, V. (1974) *Arch. ges. Virusforsch.* **45,** 39.
Black, F. L. (1965) *Theor. Biol.* **11,** 207.
Bonniwell, M. A. (1980) *Bull. Off. int. Epiz.* **92,** 1233.
Bourdin, P. (1973) *Revue Elev. Méd. vét. Pays trop.* **26,** 71*a*.
Brandly, C. A. (1951) *Public Health Rep.* **66,** 668; (1964) *Lab. Amin. Care* **14,** 433.
Brown, R. D. and Scott, G. R. (1960) *Vet Rec.* **72,** 1055.
Bryson, D. G., McFerran, J. B., Ball, H. J. and Neill, S. D. (1979) *Vet. Rec.* **104,** 45.
Cabasso, V. J. and Cox, H. R. (1949) *Proc. Soc. exp. Biol. Med.* **71,** 246.
Cabasso, V. J., Kiser, K., Stebbins, M. R. and Cooper, H. K. (1962) *Amer. J. vet. Res.* **23,** 394.
Chanock, R. M. (1956) *J. exp. Med.* **105,** 555.
Chanock, R. M. and Parrott, R. H. (1965) *Pediatrics* **36,** 21.
Chanock, R. M., Parrott, R. H., Bell, J. A., Rowe, W. P. and Huebner, R. J. (1958) *Publ. Hlth Rep. Wash.* **73,** 193.
Choppin, P. W. and Compans, R. W. (1975) In: *Comprehensive Virology*, Vol. 4, pp. 95–178. Plenum, New York.
Choppin, P. W., Richardson, C. D., Merz, D. C., Hall, W. W. and Scheid, A. (1981) *J. infect. Dis.* **143,** 352.
Christensen, B., Böttger, M. and Heller, L. (1983) *Brit. med. J*, **ii,** 389.
Clarke, S. K. (1973) *Postgrad. Med. J.* **49,** 792.
Cranage, M. P., Gardner, P. S. and McIntosh, K. (1981) *Clin. exp. Immunol.* **43,** 28.
Cutler, R. W. P. and Averill, D. R. (1969) *Neurology* **19,** 1111.
Doyle, T. M. (1927) *J. comp. Path. Ther.* **40,** 144.
Edwards, J. T. (1928) *Agric. Jour. India* **23,** 185.
Elazhary, M. A. S. Y., Galina, M., Roy, R. S., Fontaine, M. and Lamothe, P. (1980) *Canad. J. comp. Med.* **44,** 390.
Ennis, F. A. and Jackson, D. (1968) *J. Pediat.* **72,** 536.
Field, E. J., Cowshall, S., Narang, H. K. and Bell, T. M. (1972) *Lancet* **ii,** 280.
Fraser, K. B. and Martin, S. J. (1978) *Measles Virus and its Biology*. Academic Press, London and New York.
Fulton, R. E. and Middleton, P. J. (1975) *J. Pediatr.* **86,** 17.
Gatti, J. M. and Good, R. A. (1970) *Med. Clin. North Am.* **54,** 281.
Gibbs, E. P. J., Taylor, W. P., Lawman, M. J. P. and Bryant, J. (1979) *Intervirology* **11,** 268.
Gillespie, J. H. (1965) *Cornell Vet.* **55,** 3.
Giraudon, P. and Wild, T. F. (1981) *J. gen. Virol.* **57,** 179.
Glezen, W. P. and Denny, F. W. (1973) *New Engl. J. Med.* **288,** 498.
Gorham, J. R. (1966) *J. Amer. vet. med. Ass.* **149,** 610.
Graves, M. C., Silver, S. M. and Choppin, P. (1978) *Virology* **86,** 254.
Habel, K. (1945) *Pub. Health Rep.* **60,** 201.
Haig, D. A. (1948) *Onderstepoort J. vet. Sci.* **23,** 149.
Hall, C. B. (1980) *Rev. infect. Dis.* **2,** 384.
Hall, C. B., Douglas, G. and Geiman, J. M. (1980) *J. infect. Dis.* **141,** 98.
Hall, C. B., Douglas, R. G., Jr., Simons, R. L. and Geiman, J. M. (1978) *J. Pediatr.* **93,** 28.
Hall, W. W., Lamb, R. A. and Choppin, P. W. (1980) *Virology* **100,** 433.
Hamdy, F. M., Dardiri, A. H., Nduaka, O. Breese, S. S. and Ihemelandu, E. C. (1976) *Canad. J. comp. Med.* **40,** 276.
Hanson, R. P. (1974) *Advanc. vet. Sci. comp. Med.* **18,** 213; (1975) In: *Diseases Transmitted from Animals to Man*, p. 851. Ed. by W. T. Hubbert, W. F. McCulloch and P. R. Schnurrenberger, Charles C Thomas, Ill.; (1978) In: *Diseases of Poultry*, 7th edn., p. 513. Iowa State University Press, Ames, Iowa, USA.
Henle, G., Deinhardt, F. and Girardi, A. (1954) *Proc. Soc. exp. Biol. Med.* **87,** 386.
Henle, W., Crawford, M. N., Henle, G., Tabi, H. F., Deinhardt, F., Charav, A. G. and Olshin, I. J. (1959) *J. Immunol.* **83,** 17.
Hightower, L. E., Morrison, J. G. and Bratt, M. A. (1975) *J. Virol.* **16,** 1599.
Hitchner, S. B. and Johnson, E. P. (1948) *Vet. Med.* **43,** 525.
Ho, C. K. and Babiuk, L. A. (1979) *Immunology* **37,** 231.
Hope-Simpson, R. E. (1981) *J. Hyg. Camb.* **87,** 393.
Hope-Simpson, R. E. and Higgins, P. G. (1969) *Progr. med. Virol.* 11, 354.
Hore, D. E., Stevenson, R. G., Gilmour, N. J. L., Vantsis. J. T. and Thompson, D. A. (1968) *J. comp. Path.* **78,** 259.
Horne, R. W. and Waterson, A. P. (1960) *J. molec. Biol.* **2,** 75.
Horne, R. W., Waterson, A. P., Wildy, P. and Farnham, A. E. (1960) *Virology* **11,** 79.
Hugh-Jones, M., Allan, W. H., Dark, F. A. and Harper, G. J. (1973) *J. Hyg. Camb.* **71,** 3251.
Imagawa, D. T. (1968) *Prog. med. Virol.* **10,** 160.
Inaba, Y., Tanaka, Y., Sato, K., Ito, H., Omori, T. and Matumoto, M. (1970) *Jap. J. Microbiol.* **14,** 246.
Johnson, C. D. and Goodpasture, E. W. (1935) *Amer. J. Hyg.* **21,** 46.
Johnson, K. M., Chanock, R. M., Cook, M. K. and Huebner, R. J. (1960) *Amer. J. Hyg.* **71,** 81.
Johnson, R. H. and Ritchie, J. S. D. (1968) *Bull. epizoot. Dis. Afr.* **16,** 411.
Katz, S. L., Kempe, C. H., Black, F. L., Lepow, M. L., Krugman, S., Haggerty, R. J. and Enders, J. F. (1960) *New Engl. J. Med.* **273,** 180.
Kilham, L. (1948) *Proc. Soc. exp. Biol. Med.* **69,** 99; (1951) *J. Amer. med. Assn.* **146,** 1231.
Kingsbury, D. W. *et al.* (1978) *Intervirology* **10,** 137.
Krakowka, S., Confer, A. and Koestner, A. (1974) *Amer. J. vet. Res.* **35,** 1251.
Krakowka, S., Hoover, E. A., Koestner, A. and Ketring, K. (1977) *Amer. J. vet. Res.* **38,** 919.
Krakowka, S. and Wallace, A. L. (1979) *Amer. J. vet. Res.* **40,** 669.
Krugman, S. (1971) *J. Pediatr.* **78,** 1.
Laidlaw, P. P. and Dunkin, G. W. (1928) *J. comp. Path.* **41,** 209.
Lancaster, J. E. (1966) *Monogr. No. 3.* Canad. Dep. Agric., Ottawa; (1977) *Wld. poult. Sci. J.* **33,** 155.

Lancaster, J. E. and Alexander, D. J. (1975) *Monogr. No. 11.* Can. Dep. Agric., Ottawa.

Lennette, E. H. (1981) *Bull. World Hlth Org.* **59,** 305.

Levine, S. (1977) *J. Virol.* **21,** 427.

Liess, B. and Plowright, W. (1964) *J. Hyg. Camb.* **62,** 81.

McClelland, A. J. (1980) *Nature, Lond.* **284,** 404.

McIntosh, K. (1978) *J. Pediatr.* **93,** 33.

McIntosh, K. and Fishaut, J. M. (1980) *Progr. med. Virol.* **26,** 94.

McIntosh, K., Masters, H. B., Orr, I., Chao, R. K. and Barkin, R. M. (1978) *J. infect. Dis.* **138,** 24.

Martin, A. J., Gardner, P. S. and McQuillin, J. (1978) *Lancet* **ii,** 1035.

Merz, D. C., Scheid, A. and Choppin, P. W. (1980) *J. exp. Med.* **151,** 275.

Modlin, J. F., Orenstein, W. A. and Brandling-Bennett, A. D. (1975) *J. infect. Dis.* **132,** 106.

Monto, A. (1973) *Amer. J. Epidemiol.* **97,** 338.

Moore, N. F. and Burke, D. C. (1974) *J. gen. Virol.* **28,** 275.

Morley, D. (1969) *Proc. R. Soc. Med.* **57,** 846.

Mornet, P., Orue, J., Gilbert, Y., Thiery, G. and Mamadou, S. (1956) *Revue Elev. Méd. vét. Pays trop.* **9,** 313.

Morris, J. A., Blount, R. E., Jr., and Savage, R. E. (1956) *Proc. Soc. exp. Biol. Med.* **92,** 544.

Nakamuru, J. and Miyamoto, T. (1953) *Amer. J. vet. Res.* **14,** 307.

Nawathe, D. R. and Taylor, W. P. (1979) *Trop. Anim. Hlth Prod.* **11,** 120.

Nelson, C. B., Pomeroy, B. S., Schrall, K., Parks, W. E. and Lindeman, R. J. (1952) *Amer. J. Public. Health* **42,** 672.

Noon, K. F., Rogul, M., Binn, L. N., Keefe, T. J., Marchwicki, R. H. and Appel, M. J. (1980) *Amer. J. vet. Res.* **41,** 605.

Norrby, E., Friding, B., Rockborn, G. and Gard, S. (1963) *Arch. ges. Virusforsch.* **13,** 335.

Norrby, E. and Penttinen, K. (1978) *J. infect. Dis.* **138,** 672.

Örvell, C. (1980) *Arch. Virol.* **66,** 193.

Örvell, C. and Norrby, E. (1974) *J. Immunol.* **113,** 1850; (1980) *J. gen. Virol.* **50,** 231.

Paccaud, M. and Jacquier, C. (1970) *Arch. ges. Virusforsch.* **30,** 327.

Parker, J. (1976) Commission of the European Communities. *Information on Agriculture* No. 16, p. 92. CEC, Brussels.

Parkinson, A. J., Muchmore, H. G., McConnell, T. A., Scott, L. V. and Miles, J. A. R. (1980) *Amer. J. Epidem.* **112,** 334.

Penhale, W. J. and Pow, I. A. (1970) *Clin. exp. Immunol.* **6,** 627.

Pennington, T. H. (1978) *Arch. Virol.* **56,** 345.

Pirie, H. M., Petrie, L., Pringle, C. R., Allan, E. M. and Kennedy, G. J. (1981) *Vet. Rec.* **108,** 411.

Plowright, W. (1962) *Bull. Off. int. Epiz.* **57,** 1; (1964) *J. Hyg. Camb.* **63,** 497; (1965) *Vet. Rec.* **77,** 1431; (1968) *Virol. Monogr.* **3,** 25.

Plowright, W., Cruickshank, J. G. and Waterson, A. P. (1962) *Virology* **17,** 118.

Plowright, W. and Ferris, R. D. (1959) *J. comp. Path.* **69,** 173; (1962) *Res. vet. Sci.* **3,** 172.

Provost, A. (1972) *Revue Elev. Méd. vet. Pays trop.* **25,** 155.

Provost, A. and Borredon, C. (1972) *Revue Elev. Méd. vet. Pays trop.* **25,** 141.

Provost, A., Maurice, Y. and Borredon, C. (1969) *Revue Elev. Méd. vet. Pays trop.* **22,** 453.

Prydie, J. (1968) *Res. vet. Sci.* **9,** 443.

Rao, M., Devi, T. I., Ramachandran, S. and Scott, G. R. (1974) *Indian vet. J.* **51,** 439.

Rasmussen, A. J. Jr. (1963) quoted by Brandly, C. (1964).

Reisinger, R. C., Heddleston, K. L. and Manthei, C. A. (1959) *J. Amer. vet. med. Ass.* **135,** 147.

Rima, B. K. and Martin, S. J. (1977) *Med. Microbiol. Immunol.* **162,** 89.

Rima, B. K., Martin, S. J. and Gould, E. A. (1979) *J. gen. Virol.* **43,** 603.

Rockborn, G. (1960) *J. small Anim. Pract.* **1,** 63.

Rustigian, R. (1966) *J. Bacteriol.* **92,** 1805.

Schultz, R. D., Appel, M., Carmichael, M. E. and Farrow, B. (1977) In: *Current Therapy: Small Animal Practice*, 6th edn., p. 1271. Saunders, Philadelphia, Pennsylvania, USA.

Scott, G. R. (1955) *Bull. epizoot. Dis. Afr.* **3,** 117; (1964) *Advanc. vet. Sci.* **9,** 113.

Scott, G. R., DeTray, D. E. and White, G. (1962) *Amer. J. vet. Res.* **23,** 452.

Stephenson, J. R. and Ter Meulen, V. (1979) *Proc. Nat. Acad. Sci., Wash.* **76,** 6601.

Stephenson, J. R., Ter Meulen, V. and Kiessling, W. (1980) *Lancet* **ii,** 772.

Stott, E. J., Thomas, L. H. and Taylor, G. (1981) *Abstract P44/23: 5th Int. Congr. Virol.* p. 407.

Summers, B. A., Greisen, H. A. and Appel, M. J. G. (1978) *Lancet* **ii,** 187.

Suringa, D. W. R., Bank, L. J. and Ackerman, A. B. (1970) *New Engl. J. Med.* **283,** 1139.

Taylor, W. P. (1979*a*) *Res. vet. Sci.* **26,** 236; (1979*b*) *Res. vet. Sci.* **27,** 321.

Taylor, W. P. and Abegunde, A. (1979) *Res. vet. Sci.* **26,** 94.

Taylor, W. P., Plowright, W., Pillinger, R., Rampton, C. S. and Staple, R. F. (1965) *J. Hyg. Camb.* **63,** 497.

Ter Meulen, V., Löffler, S., Carter, M. J. and Stephenson, J. R. (1981) *J. gen. Virol.* **57,** 357.

Ter Meulen, V., Müller, D., Käckell, M. Y., Katz, M. and Meyermann, R. (1972) *Lancet* **ii,** 1172.

Togashi, T., Örvell, C., Vartdal, F. and Norrby, E. (1981) *Arch. Virol.* **67,** 149.

Trott, D. G. and Pilsworth, R. (1965) *Brit. med. J.* **ii,** 1514.

Tyrrell, D. L. J. and Norrby, E. (1978) *J. gen. Virol.* **39,** 219.

Underwood, B. and Brown, F. (1974) *Med. Microbiol. Immunol.* **160,** 125.

Utz, J. P., Houf, V. N. and Alling, D. W. (1964) *New Engl. J. Med.* **270,** 1283.

Waterson, A. P. and Cruickshank, J. G. (1963) *Z. Naturf.* **18B,** 114.

Waterson, A. P., Cruickshank, J. G., Laurence, G. D. and Kanarek, A. D. (1961) *Virology* **15,** 379.

Weibel, R. E., Buynak, E. B., McLean, A. A., Roehm, R. R. and Hilleman, M. R. (1980*a*) *Proc. Soc. exp. Biol. Med.* **165,** 260.

Weibel, R. E., Carlson, A. J., Jr., Villarejos, V. M., Buynak, E. B., McLean, A. A. and Hilleman, M. R. (1980*b*) *Proc. Soc. exp. Biol. Med.* **165,** 323.

Whitney, J. C., Scott, G. R. and Hill, D. H. (1967) *Bull. epizoot. Dis. Afr.* **15,** 31.

Whittle, H. C. and Werblinska, J. (1980) *Clin. Exp. Immunol.* **42,** 136.

Wilfert, C. M. (1969) *New Engl. J. Med.* **280,** 855.

Woods, G. T. (1974) *Advanc. vet. Sci. comp. Med.* **18,** 274.

Yamanouchi, K., Fukuda, A., Kobune, F., Yoshikawa, Y. and Shino, F. (1974) *Infection and Immunity* **9,** 206.

99

Enteroviruses: polio-, ECHO-, and Coxsackie viruses

D. R. Gamble

Introductory

Polioviruses, Coxsackieviruses and ECHOviruses (Enteric Cytopathogenic human orphan) were together designated enteroviruses in 1957 on the basis of their shared physical, chemical and biological characteristics. The viruses are small (20–30 nm), icosahedral, RNA-containing particles, which are stable at pH 3–5 and resistant to ether; they are of human origin, infecting and multiplying in the alimentary tract. Infection occurs most commonly in summer and autumn in temperate climates; it is generally inapparent or associated with only minor illness, but occasionally causes serious or even fatal illness, usually through involvement of the central nervous system.

Poliomyelitis was probably known in antiquity but its emergence as a serious epidemic disease in western Europe and North America in the late nineteenth century initiated a period of intensive investigation. At first, this was limited to epidemiological studies, but experimental infection became possible after the successful transmission of infection to monkeys in 1908 (Landsteiner and Popper 1909), and later to rodents (Armstrong 1939). The pathogenesis of poliomyelitis was elucidated in animal experiments, and three antigenically different types of virus were distinguished by serological studies.

In 1948, during the investigation of a poliomyelitis outbreak in the town of Coxsackie in New York State, viruses isolated from the faeces of two children were found to induce paralysis in newborn mice and hamsters (Dalldorf and Sickles 1948). The isolation of further viruses with similar properties was soon reported (Dalldorf 1949, Melnick *et al.* 1949). Pathogenicity for newborn mice was used to define this new group of viruses, which were named Coxsackieviruses. The group was later subdivided into groups A and B; 24 serotypes were assigned to group A, and 6 to group B.

The discovery that poliovirus could be cultivated *in vitro* in non-neural primate cell cultures (Enders *et al.* 1949) led to the isolation of many new viruses from faecal material. It became clear that some of these viruses were neither polioviruses nor Coxsackieviruses; they differed both serologically and in their non-pathogenicity for newborn mice. Because at first they lacked an association with any specific disease they were described as orphan viruses and, in 1955, were given the acronym ECHOviruses, derived from the description Enteric, Cytopathic, Human, Orphan (Report 1955*a*). Of the 33 serotypes originally included in the ECHOvirus group, type 10 was reclassified as reovirus type 1 and type 28 was found to be a rhinovirus. Experience has shown that the accepted taxonomic criteria do not always permit a clear cut distinction between Coxsackieviruses and ECHOviruses. Thus, some strains of ECHOviruses have proved pathogenic for newborn mice and, conversely, some Coxsackieviruses, when first isolated in tissue culture, have failed to multiply in newborn mice. For example Coxsackievirus A23, was found to be serologically identical with ECHOvirus 9 and some strains of ECHOvirus 9 have proved pathogenic for newborn mice. These difficulties led to the recommendation that new enterovirus serotypes should be given numbers only and should not be assigned to a particular subgroup (Rosen *et al.* 1970). To date, four new enterovirus serotypes have been accepted and designated enteroviruses 68 to 71.

In the current virus classification, enteroviruses are accorded generic status within the *Picornaviridae* family, in which rhinoviruses and a number of animal picornaviruses are also included.

Polioviruses

Epidemiology

Prevalence Paralytic poliomyelitis occurs both sporadically and epidemically, but until the end of the nineteenth century the disease was almost entirely confined to sporadic cases in infants or young children. The first recorded outbreak was in St. Helena about 1830. A few other small outbreaks were reported in Europe in the middle of the nineteenth century, and increasingly towards the end of the century in Sweden, Norway and France (see Paul 1955). A larger outbreak in Sweden in 1887 (Medin 1891) was followed over the next 60 years by the dramatic emergence of epidemic poliomyelitis in northern Europe and other developed countries. Sweden, Norway and the United States were among the first to suffer and many other European countries were affected in the 1930s, but major epidemics did not reach Germany and the UK until 1947, and still later France and Belgium. The first outbreaks were usually small but major epidemics eventually appeared and continued with increasing frequency until the introduction of vaccination in the 1950s. In the United States, for example, a small outbreak was recorded in Vermont in 1894 and a low incidence persisted until 1907 when larger outbreaks became regular annual occurrences. Major epidemics appeared in 1916, 1931, 1943, 1946 and 1948 (Fig. 99.1). Cases generally occurred in infants in the earlier outbreaks but there was a progressive rise in the age incidence in subsequent outbreaks. Thus, in the 1916 epidemic in New York, 80 per cent of cases were in children under 5 years of age, whereas in subsequent epidemics the peak age incidence was in 5 to 9-year-olds and about one-third of all cases were in patients over the age of 15 years (Lavinder *et al.* 1918, Sabin 1949).

Several explanations for the emergence of epidemic poliomyelitis have been advanced but opinion has wavered as to which, if any, is the most important. Firstly, there has been an increase in the proportion of susceptible subjects in many of the developed countries. The earlier sporadic pattern of disease was undoubtedly due to the widespread dissemination of poliovirus among infants in overcrowded communities with poor sanitation and hygiene. Infection at this early age, when resistance to the disease appears to be greatest, leads to only sporadic paralytic cases and, by symptomless infection, to immunity in most older members of the population. In developing countries where poliomyelitis is still sporadic, serological studies show that most children acquire antibodies in infancy (Paul 1955). The numerous outbreaks that occur among Europeans and American visitors to these countries testify to the presence of virulent virus which failed to cause outbreaks in the indigenous population. In western countries improvements in sanitation

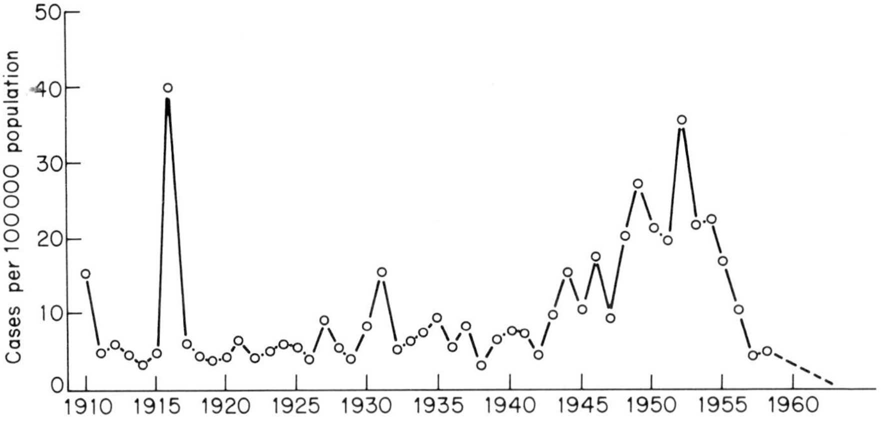

Fig. 99.1 Annual incidence of poliomyelitis in the United States 1910–1958. (After Langmuir.)

and hygiene are thought to have reduced dissemination of infection among infants, so that susceptibility in children and adults has increased to the point where epidemics have become possible.

Secondly, the severity of illness caused by poliovirus infection increases with age; so also does the risk of paralysis. It has been suggested that the increase in the age at which infection occurred was a major factor in the emergence of epidemics. However, Melnick and Ledinko (1953) showed that the ratio of paralytic cases to infections increased only about three-fold between infancy and 14 years, and that a much greater increase would be necessary for the appearance of epidemics (Nathanson and Martin 1979).

Thirdly, epidemics may have been due to the appearance of more virulent virus strains. Different strains of poliovirus vary in their neurovirulence for monkeys (Shelekov *et al.* 1959), but direct comparison of recent strains with those of past centuries is not possible. There is, however, circumstantial evidence of a recent change in poliovirus virulence. In some developing countries paralytic poliomyelitis increased at about the time of World War II. Gear (1948) observed that this took place on the overland route from South Africa to the Middle East and suggested that more virulent polioviruses had been introduced from other areas by increased population movements during and after the war. Further circumstantial evidence is provided by the paucity of 'virgin soil' outbreaks in isolated communities before the twentieth century. Poliovirus tends to die out in such communities and the reintroduction of virus after a long period of freedom from infection may produce very high attack rates. A number of such outbreaks have been reported since 1900. For example, an outbreak in 1949 among the Eskimos of Chesterfield Inlet caused paralytic illness in 57 per cent of the 275 members of the community (Peart 1949). Isolated communities were more common in past centuries and they should have provided opportunities for 'virgin soil' outbreaks. It is surprising that none was recorded before 1900, assuming that the polioviruses endemic in the towns were fully virulent.

The incidence of poliomyelitis declined in most western countries after the introduction of poliovirus vaccination in 1955. In the United States paralytic cases fell from 31 582 in 1955 to a few hundred a year in the early 1960s and to an almost negligible number by 1970 (Fig. 99.2). The pattern was similar in most European countries; Out of 22 countries 10 had no cases in 1970 and 5 others recorded three cases or less (Report 1971). By contrast, in some developing countries in which poliomyelitis had previously been endemic, epidemics have appeared (Freyche and Nielsen

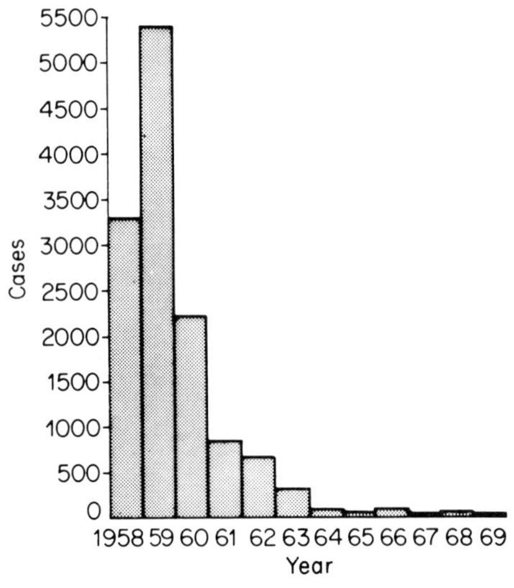

Fig. 99.2 Best available paralytic poliomyelitis case count by year, United States of America, 1958–1969. (From US CDC Annual Poliomyelitis Summary 1969.)

1955, Gear 1955, Seal and Gharpure 1961, Melnick 1958). This has usually but not always been associated with rising standards of living and declining infant mortality rates (Gear 1955, Sabin 1963).

Seasonal variation Although the poliomyelitis season corresponds roughly with seasonal climatic changes, a close correlation with either temperature or humidity has not been found (Armstrong 1950, Spicer 1959, Bradley and Richmond 1953). The seasonal increase in temperature and humidity in fact precedes the rise in incidence of poliomyelitis and reaches an earlier peak. However, Yorke and his collaborators (1979) have pointed out that changes in transmission rates of infection must precede changes in incidence of poliomyelitis by an interval which they calculate to be about two months; the rise in temperature and humidity in spring would therefore coincide with the increase in transmissibility of infection that would be necessary to generate an outbreak in summer and autumn. Moreover, Serfling and Sherman (1953) have shown that the seasonality of poliomyelitis in different climatic regions of the United States is strikingly correlated with seasonal variations in temperature and relative humidity, both in the degree of fluctuation and the duration of the cycle. The reason for the climatic influence in poliovirus transmission is still uncertain, but the increased survival of the virus in relative humidities of over 40 per cent (Hemmes *et al.* 1962) suggests that humidity may be the determining factor.

Mode of transmission Infection is spread by human association. Poor hygienic conditions and overcrowding increase dissemination. As living conditions are improved, infant mortality falls but the incidence of poliomyelitis increases. It has been shown that when the number of deaths per 1000 livebirths falls below 75, poliomyelitis often changes from an endemic to an epidemic disease (Paul 1958*a*, *b*). Close association is necessary for spread of infection to be transmitted by patients and healthy carriers (Wickman 1907). Children are the most frequent disseminators. The infection rate among contacts within households approaches 100 per cent in children and only slightly less in adults, although multiple paralytic cases in families are uncommon (Aycock 1941). The incidence of paralysis varies inversely with the degree of overcrowding in households (Report 1950). The higher incidence in lower social classes and in urban rather than rural areas is presumably a reflection of population density. Before its viral aetiology was established, Wickman (1907) investigated the Swedish outbreak of 1905 and found that the disease spread along lines of communication, railways and roads. Poliovirus has been isolated from flies and cockroaches and from food contaminated by flies (Ward *et al.* 1945, Gear 1952); but it is unlikely that insect vectors are an important source of infection, since this occurs readily in Arctic regions in the absence of flies. Poliovirus is found in sewage reflecting the prevalence of infection in the community (Melnick 1947). In view of the increasing use of recycled waste water for drinking purposes, concern has been expressed that 'low level transmission' of virus by drinking water may be responsible for seeding populations with isolated cases of infection which become foci of outbreaks (Berg 1967). Drinking water that has been adequately treated to conventional bacteriological standards contains few enteroviruses and there is no evidence that it has ever been responsible for human infection; even if it were it would probably be an insignificant source of infection in comparison with person-to-person spread (see Gamble 1979).

It is uncertain whether transmission of the virus occurs by droplets directly from the pharynx, or by the faecal-oral route, or both. Virus is present in the oropharynx from about 5 days before to 5 days after the onset of symptoms; and epidemiological evidence shows that this is the period of greatest infectivity. Faecal excretion commences a few days after onset of symptoms; it continues for several weeks and occasionally for months. Transmission before onset of symptoms is therefore likely to be due to droplet spread from the pharynx; transmission occurring more than a week after onset is most probably due to faecal virus. Both mechanisms probably operate. The association of endemic poliomyelitis with poor sanitation suggests that faeco-oral transmission between young children is chiefly responsible; and that oral droplets may be more important among older people in communities with higher standards of hygiene.

Only a proportion of infected persons suffer from clinical illness and few develop paralysis. In an outbreak in North Carolina, Melnick and Ledinko (1953) estimated case-infection ratios as ranging from 1:150 in infants aged less than one year to 1:42 in children aged 5–9 years, Table 99.1. Other estimates have varied from as much as 1:5 in a 'virgin soil' outbreak (Peart 1949) to less than 1:1000 in England and Wales in a non-epidemic year (Spicer 1961). The ratio varies with the age of the patients and possibly with the virulence of the virus. Tonsillectomy performed during an inapparent poliovirus infection may lead to severe paralytic illness (Aycock 1942). There may also be a greater risk or paralysis in persons who have had their tonsils removed in the past (Anderson and Rondeau 1954). The removal of IgA-producing tissue at a site of primary implantation of the virus may be the reason. This is supported by the observation that poliovirus-specific IgA antibody titres in nasopharyngeal washings have fallen after tonsillectomy (Ogra 1971). Pregnancy also seems to increase susceptibility (Siegel *et al.* 1957), possibly because of hormonal changes or the increased exposure of pregnant women to young children. The incidence of paralysis is increased in infections occurring within a few weeks of certain immunizing injections; paralysis tends to occur more frequently in the injected limb (McCloskey 1950,

Table 99.1 The case:infection ratio as estimated from a prospective study during an outbreak of poliomyelitis in North Carolina, 1948* (*From* Nathanson, N and Martin, R. (1979)) *Amer. J. Epidem.* **110,** 672 (based on the data of Melnick, J. L. and Ledinko, N. (1953) *Amer. J. Hyg.* **58,** 207.)

Age group (years)	Population	Paralytic cases	Sero-converters	Estimated total converters	Cases/100 converters
<1	1800	3	5/20	450	0.66
1–2	3900	10	10/39	1000	1.00
3–4	3600	12	7/34	741	1.62
5–9	7300	25	8/56	1042	2.40
10–14	6300	13	5/44	716	1.82

* Seroconverters: a grab sample was taken of the population and these subjects were bled before and after the outbreak. Converters were the proportion of all subjects (negative and positive) who converted from negative to positive.

Hill and Knowelden 1950). Strenuous physical exertion during the preparalytic stage also increases the likelihood of severe paralysis (Russell 1949, Horstmann 1950).

Transmission to animals

Poliomyelitis was first transmitted to laboratory animals by the intraperitoneal inoculation of Rhesus monkeys with spinal cord from two human cases (Landsteiner and Popper 1909). Flexner and Lewis (1910) infected monkeys by using Berkefeld filtrates of infected tissue and pharyngeal washings, and transmitted the disease by serial passage. Monkeys and chimpanzees are most readily infected by injections directly into the brain or spinal cord, but infection may be induced by most parenteral routes. Cynomolgus monkeys (Kling *et al.* 1934) and chimpanzees (Howe and Bodian 1941) can be infected orally. The disease in monkeys is similar to that in human beings. The incubation period is 6–10 days after intracerebral inoculation and 1–2 weeks after oral infection. In orally infected chimpanzees, virus can be demonstrated in the presymptomatic stage in the pharyngeal lymphoid tissue and deep cervical lymph nodes, and in the Peyer's patches of the ileum and mesenteric nodes. A viraemia occurs, terminated by the appearance of circulating antibodies. In paralysed animals the lesions are chiefly in the grey matter of the cord, with degeneration of the ganglion cells and inflammatory cell infiltration. The infiltrate is largely lymphocytic with some polymorphs but few plasma cells; it is diffuse in the grey matter with perivascular accumulations (see Bodian and Horstmann 1965).

Cotton-rats, mice and hamsters may be infected by intracerebral or intraspinal inoculation with adapted virus strains, but other animals are insusceptible. Armstrong (1939) transmitted the monkey-adapted type 2 Lansing strain to cotton-rats; Jungeblut and Sanders (1940) then transferred it to white mice. After inoculation, the incubation period ranged from 2–10 days or longer, and was followed by paralysis and death. Adaptation of type 2 virus to newborn mice and to fertile hens' eggs, and types 1 and 3 virus to adult mice has also been accomplished.

Characters of the virus

Poliovirus is an icosahedral, non-enveloped particle of about 25 nm diameter, which is stable at pH3 and has a buoyant density in caesium chloride of 1:34. The nucleic acid is single-stranded RNA and constitutes about 30 per cent of the particle mass. The capsid has 60 capsomeres and incorporates 4 polypeptides which are designated VP1, VP2, VP3, and VP4 (see Rueckert 1971). Phenol-extracted poliovirus RNA when brought into contact with previously uninoculated cells can induce the formation of new whole infectious particles, even in cells ordinarily insusceptible (Mountain and Alexander 1959).

The *heat-lability* of the virus varies according to circumstances. At 37° its titre declines about ten-fold in 24 hours. The virus however survives for days at room temperature, for weeks at +4° and indefinitely when frozen at −30° or lower. In aqueous suspension, it is inactivated within 30 minutes at 50°. Proteins are protective, so that when mixed with milk, cream or ice-cream, it may survive temperatures of 60° or higher (Kaplan and Melnick 1952). In the presence of M magnesium chloride, survival is prolonged at all temperatures. The virus is resistant to ether (Andrewes and Horstmann 1949) and to sodium deoxycholate (Theiler 1957). It is resistant to repeated freezing and thawing, sonic vibration and fluorocarbon treatment. It is slowly inactivated by alcohol, phenol and formalin. Antigenicity is retained after formalin inactivation—a fact exploited in vaccine manufacture. The virus is rapidly inactivated by halogens, ozone, ultraviolet light and drying; moisture favours its survival.

The existence of more than one immunological type of poliovirus was first recognized by Burnet and Macnamara (1931) and of three types by Bodian *et al.* (1949). These were numbered types 1, 2 and 3 on the basis of their frequency in the first hundred strains examined (Report 1951). An antigenically distinct virus isolated from paralysed children in Russia (Chumakov *et al.* 1956) was initially proposed as a fourth type, but was subsequently identified as Coxsackievirus A7 (Habel and Loomis 1957). Comparison of the genomes of poliovirus by RNA hybridization has shown that about 30 per cent of the genome is shared by the three types, whereas there is only 5 per cent homology among enteroviruses as a whole (Young 1973).

Heating changes the physical and biological properties of the virus; part of the capsid protein is split off and there is a loss of infectivity. This may be accompanied by the loss of the nucleic acid core, leaving an empty shell (McGregor and Mayor 1971, Breindl 1971). Infectious virus preparations contain two type-specific antigens detectable by complement-fixation and precipitin tests, and separable by density gradient centrifugation. They are designated C and D (or N and H) antigens. The more dense D or N (native) antigen is infectious and is associated with intact virus; the lighter C or H (heated) antigen contains little or no RNA and is not infectious.

Poliovirus multiplies only in cultured primate cells having specific receptor sites (see Holland 1964). Cells not bearing these receptors may be infected by naked RNA extracted from poliovirus. After attachment the virus is taken into the cell; uncoating, which may commence before engulfment (Joklik and Darnell 1961), releases RNA into the cytoplasm (Dales 1965, Dunnebacke *et al.* 1969). The single-stranded RNA acting as its own messenger RNA is translated into protein, which is broken up, yielding viral capsid proteins and the RNA polymerase (see Baltimore 1971, Baltimore *et al.* 1971). Viral polymerase initiates the formation of complementary RNA from viral RNA and of new viral RNA from the complementary strand (see Weissmann *et al.* 1968, Spiegelmann *et al.* 1968). Assembly of viral RNA and capsid proteins takes place in the cytoplasm of the cell and mature virions are released by cell lysis. New virus first appears within 3–4 hours of infection and intracellular virus is maximal in 6–8 hours, but its release in quantity from a cell monolayer and the appearance of cytopathic changes occur later (Ackermann *et al.* 1958).

Natural infection and immunity

The minimal infective dose of poliovirus for man is small. In experimental studies with attenuated poliovirus, infection has been produced with doses as low as 2 PFU administered to adults in gelatin capsules (Koprowski 1956), or one TCD 50 given to newborn premature infants by gastric tube (Katz and Plotkin 1967). Naturally occurring infections may require larger doses, but a dose of virus sufficient to infect a tissue culture, perhaps even a single infective particle, is probably sufficient to initiate infection if it happens to encounter a receptor on a susceptible cell.

The portal of entry is the alimentary tract and primary multiplication takes place in the pharynx and ileum. Virus can be isolated from these sites both in chimpanzees in the preparalytic stage (Bodian and Horstmann 1965) and in man after the onset of paralysis (Wenner and Rabe 1951). It is uncertain whether primary multiplication occurs in epithelial cells or lymphoid tissue. In favour of lymphoid tissue is the finding of virus particularly in areas rich in such tissue: but this would probably be concentrated beneath those areas of mucosa that are most susceptible to microbial invasion. In favour of the superficial cells is the continued formation and excretion of virus from the alimentary tract long after the appearance of antibody, and the inability of some strains to multiply in the lymphatic system within the body when inoculated parenterally, even though the same strains multiply readily when given by mouth. Furthermore, in cynomolgous monkeys infected orally with type 1 poliovirus, fluorescent antibody staining reveals specific viral fluorescence, first in the pharyngeal epithelium, and later in the submucosa and local lymph nodes of the pharynx and gut (Kanamitzu *et al.* 1967). The virus passes from these sites of multiplication to the regional lymph nodes and into the blood stream.

With strains of low virulence there is little if any further spread, but virulent strains, aided in some cases by deficiencies in host defences, may spread to the central nervous system. Whether virus reaches the brain or spinal cord via the blood stream or by neural pathways from peripheral nerve ganglia is uncertain, but blood-borne infection seems most likely. Although it has been shown experimentally that virus can travel along neural pathways from the periphery to the central nervous system (Bodian and Howe 1940), it is not known whether this occurs in man. An earlier view that virus reached the brain by the olfactory nerve route has been abandoned; but bulbar poliomyelitis following tonsillectomy might be the result of spread from the pharynx via the exposed nerve endings. Neuronal damage seems to be the primary change in the central nervous system, both in infected monkeys and in human cases of poliomyelitis (Bodian 1948, 1949). Chromatolysis and dissolution of the Nissl granules are followed by necrosis and neuronophagia. Early degenerative changes are sometimes reversible with recovery of some neurones, but how cells recover from poliovirus infection has not been explained. The surrounding tissue is usually infiltrated with inflammatory cells, often intensely, and a perivascular distribution is common. The cellular response appears to be secondary to neuronal damage rather than responsible for it. Neuronal damage and virus multiplication sometimes occur in the absence of inflammatory infiltration, and virus introduced into regions of the monkey brain depleted of nerve cells by retrograde degeneration does not induce a cellular inflammatory response (Bodian and Howe 1941). Clinically, the illness is often biphasic. Initially, there is a minor illness associated with the viraemia, then a remission, followed in a minority of patients by a recurrence of fever and the development of paralysis (Horstmann 1953). Orally infected cynomolgus monkeys and chimpanzees have a similar illness in which viraemia occurs 4–6 days before the onset of paralysis.

After natural poliomyelitis infection, type-specific neutralizing antibody appears in the blood. It is considered to provide lifelong protection. The antibody is specific, and its presence is an indication of past infection, clinical or inapparent. It reaches its maximum titre in the blood soon after infection and declines gradually over several years without further exposure to the virus. Lennette and Schmidt (1957) found that type I poliovirus antibody titres fell to 25 per cent of the highest titre in the first year, and to a further 7 per cent in the second year after infection. The fall, however, varies with individuals. The decline in antibody titre after immunization with inactivated vaccine follows a similar pattern (Salk 1958). Even a small quantity of serum antibody is considered to be protective against viraemia and paralysis, but it does not necessarily interfere with infection or reinfection of the alimentary tract, except perhaps in very young infants. After natural infection the alimentary tract becomes almost completely resistant for a time to reinfection with the same virus type. Live poliovirus vaccine induces a similar but less pronounced effect; inactivated vaccine induces only a very slight degree of resistance but this varies with the potency of the vaccines used. Ghendon and Sanakoyeva (1961) found that, of infants given type I oral vaccine, those who had previously suffered from paralytic poliomyelitis were completely resistant to reinfection of the gut; those previously given live vaccine were fairly resistant, excreting virus in low titre for 4–6 days; those who had received inactivated vaccine were generally susceptible, excreting virus heavily for an average of 12 days; and unvaccinated children who had no antibody were most susceptible, excreting the virus in still higher titres for an average of 20 days. Since the serum antibody titre may be much the same in the first three categories, it presumably plays only a minor role in resistance of the gut, which is more probably associated with IgA secretory antibody (Tomasi and Bienenstock 1968).

Diagnosis

Poliovirus can be isolated from the pharynx and faeces during the acute stage of the illness by the inoculation of susceptible cell cultures. Isolation from the cerebrospinal fluid is unusual, but virus can be isolated from brain and spinal cord tissue in fatal cases. Although viraemia occurs, virus isolation from the blood is not usually attempted. Many primate tissues provide susceptible cells, in primary and secondary culture; but numerous cell lines derived from human or monkey tissues are susceptible and Hela cells have been widely used. The standard neutralization test is generally employed for virus typing, but complement-fixation and precipitation tests may also be used. Identification of different strains within the same type is sometimes possible; its main value is in helping to distinguish between vaccine and wild strains of virus. In these studies sensitive neutralization tests based on plaque reduction or neutralization kinetics have been employed (Nakano and Gelfand 1962). The ability of wild poliovirus strains but not vaccine strains, to grow in tissue culture at temperatures of 39.8° or more, has also proved a useful marker (Lwoff 1959; Cossart 1967, 1977).

Retrospective serological diagnosis based on a significant rise of specific neutralizing antibody is often possible, but heterologous reactions between the different antigenic types may make interpretation difficult. Such reactions have been observed, particularly in types I and 3 infections, when type 2 antibody may also temporarily appear. Equally, possession of type 2 antibody may afford some protection against other types. Neutralization tests are based on cytopathic inhibition or metabolic inhibition methods. Complement-fixing antibody appears later and disappears sooner than neutralizing antibody but is sometimes useful in diagnosis (see Melnick *et al.* 1979). Group reactions may be troublesome. Specificity is dependent on the use of intact virus particles or D-antigens.

General prophylaxis

In outbreaks in poorly immunized communities, quarantine of contacts is ineffective, owing to the frequency of latent infections and the long duration of virus excretion. Social activities which congregate people should be restricted; and travel to or from infected areas is best avoided. Residential schools should not be closed, partly because dispersing the children may spread infection to other areas and partly because the chances of the disease spreading in a school, camp or institution are not high. During outbreaks excessive fatigue should be avoided, particularly by children with fever; operations on the nose and throat, and prophylactic injections against diphtheria and whooping cough, should be postponed.

In well vaccinated communities, only sporadic cases or small outbreaks are likely to occur and, apart from vaccination of contacts, few prophylactic measures are necessary in the community. In hospital, patients should be isolated for about 7 days and pharyngeal discharges and faeces should be treated as infectious.

Seroprophylaxis The presence of neutralizing antibody in the blood before infection reaches the viraemic stage is usually sufficient to prevent the development of paralysis. Controlled trials of the prophylactic use of immunoglobulin in the United States (Hammon *et al.* 1954) showed it to have little effect when given after the first few days of the incubation period. Any protection conferred had disappeared after four weeks. Its use is therefore an ineffective method of controlling outbreaks.

Vaccination

The widespread use of vaccines in the last 20 years has led to the eradication of epidemic poliomyelitis in most developed countries. However, in many parts of the world the disease remains endemic and, even in countries where it is now uncommon, problems still arise.

The inactivated vaccine

Introduced by Salk *et al.* (1955) this vaccine was prepared from the three types of poliovirus grown in monkey cell cultures. Virus suspensions were exposed to 1:4000 formalin at 37° for a number of days to destroy infectivity, but not antigenicity. The efficacy of the vaccine was tested in controlled trials in the United States (Report 1955*b*) and Great Britain (Report 1957). A number of vaccine-associated outbreaks of poliomyelitis followed its first use on a large scale (Syverton *et al.* 1956, Nathanson and Langmuir 1963, Wilson 1967) but further precautions in the manufacturing process (Report 1956) rendered it safe, with little risk to recipients and their contacts, even when they were immunodeficient. Satisfactory immunity takes some time to develop fully; it is dependent on an adequate antigenic stimulus that requires three or four injections at proper intervals with a vaccine of high potency.

Vaccination can be incorporated in the routine paediatric immunization schedule with the usual bacterial antigens. Further doses may be necessary to maintain immunity. There have been many serological assessments of its effectiveness and of antibody persistence. The vaccine clearly provides a high level of protection. Although circulating antibodies produced by the vaccine prevent viraemia and spread of infection to the central nervous system, they do not prevent infection of the alimentary tract and excretion of virus. There may be diminished excretion particularly from the pharynx (Marine *et al.* 1962) but naturally occurring polioviruses can multiply in the gut and spread through partly vaccinated communities causing disease in unvaccinated subjects. Most western countries now use live vaccines but Sweden, Finland and the Netherlands have relied exclusively on inactivated vaccine. Sweden and Finland, with immunization rates of close to 100 per cent, have enjoyed almost total freedom from poliomyelitis since 1960. In the Netherlands, however, about 20 cases a year have continued to occur, mainly among unimmunized religious groups who have refused any vaccination. In 1978, an outbreak of 110 cases occurred in these communities (Report 1979). The inactivated vaccine is costly to produce and administer; but it will afford complete protection so long as virtually the whole population is vaccinated and kept immune.

Attenuated live vaccine

The attenuated live vaccine is given by mouth and is now in common use throughout most of the world. Oral poliomyelitis immunization was first employed by Koprowski and his colleagues (Koprowski *et al.* 1952, Koprowski 1960) using rodent-adapted virus strains. Sabin and his associates (Sabin *et al.* 1954, Sabin 1959) further developed the method by investigating the characters of viruses grown in monkey kidney cell cultures. Investigations of their neurotropism in monkeys and chimpanzees, and eventually in human volunteers enabled constituent strains of the oral vaccine to be selected on the basis of their limited neurovirulence. Frequent unacceptable presence of simian viruses in monkey kidney cell cultures led to their replacement by human embryonic lung diploid cell cultures (Hayflick and Moorhead 1961) for vaccine production.

The attenuated vaccine virus infects and multiplies in the gut and is excreted in the faeces. It stimulates antibody production in a similar manner to natural infection and the antibody so produced is probably as long lasting, so that reinforcing doses should not be required. Antibody first appears 7 to 10 days after feeding; immunity is by this means rapidly acquired. After infection the alimentary tract becomes temporarily resistant to reinfection with the same type of virus. Secretory IgA is believed to be responsible for this gut immunity. There is, moreover, some evidence of heterotypic resistance. Oral vaccination tends to stop the circulation of wild viruses in the community. On the other hand, vaccine virus excreted by vaccines may spread to contacts. In some circumstances this may be an advantage by increasing the immunization rate; but reversion of the vaccine to virulence during growth in vaccinees may cause paralytic poliomyelitis in the recipients or their contacts. The simultaneous vaccination of susceptible contacts is therefore advisable. Fortunately, vaccine-associated cases of poliomyelitis are uncommon. In a five-year study by the WHO (Assaad *et al.* 1976), attack rates were estimated as 0 to 2 recipients and 0.1 to 0.6 contacts per million doses distributed. Most of the recipients affected were children infected with type 3 virus. The affected contacts were often type 2 infections in parents of vaccinated infants. About 10 per cent of vaccine-associated cases have been found to be associated with hypogammaglobulinaemia. Inactivated vaccine is for this reason preferable for immuno-deficient or immuno-suppressed patients and their household contacts.

Though oral vaccine has given excellent results among western peoples, children in the tropics have often responded poorly (see John and Jayabal 1972, Domok 1974). Interference by other enteroviruses present in the gut at the time vaccine is fed may be the explanation, since these children have a very high incidence of infection. It may also be due to antibody-like substances in infants' saliva. These may be neutralized by the simultaneous administration of equine antibody to human gammaglobulin and the vaccine. In babies, immaturity of the immune system and maternal antibodies may interfere with the response to vaccine. Primary vaccination is, therefore, usually deferred till the age of 2 months in the United States and 6 months in the UK. Four or five doses are usually given to complete the immunization, but schedules vary in different countries. The rapid development of immunity in response to oral vaccine, and the resistance to reinfection of the gut by other viruses while it is colonized with vaccine virus, are of value when poliomyelitis occurs. In outbreaks, mass vaccination of communities may reduce or terminate their progress (Hale *et al.* 1959). When isolated cases occur in vaccinated communities, vaccine should be given to all known contacts, particularly to children.

Live and inactivated poliovirus vaccines each have their advantages and disadvantages. Inactivated vaccine is claimed to have the merit of freedom from cases of vaccine-associated poliomyelitis and, in countries which can afford its use and can maintain a high level of vaccination, there is no good reason to abandon it. On the other hand, the cost of inactivated vaccine would be prohibitive in many countries and the high immunization rate necessary for its effective use would be unattainable. In these countries live vaccine is preferable. In countries which have already reduced poliomyelitis to almost negligible levels by the use of live vaccine, there is at present no reason to change.

Coxsackie and ECHO viruses

Epidemiology

Prevalence Coxsackie and ECHOviruses have a worldwide distribution (see Gelfand 1961). They have been found in every country in which they have been sought for; and antibodies to the majority of common types can be found in all but the most isolated of communities. In temperate regions infections occur mainly in outbreaks during summer and autumn. Serotypes vary in their secular distribution patterns. These appear to be determined by the interaction of many factors, including the immune state of the population, the infectivity of the virus, season, climate, geographical area, hygiene, socio-economic conditions, and interference by other enterovirus types that are circulating simultaneously in the population.

Much of our knowledge of the distribution of these viruses is based on isolations from patients during the past 30 years; but there are many sources of sampling bias in the selection of patients for laboratory investigation, and widespread infection sometimes occurs in the absence of detectable illness (Clemmer *et al.* 1966, Nelson *et al.* 1979). The frequency of virus excretion by healthy subjects, the identification of enteroviruses in sewage and surface waters, and surveys of enterovirus antibody distribution have provided more representative information. Enteroviruses that often cause inapparent infection, particularly the ECHOviruses, are isolated less frequently from patients than in other types of survey. Apart from this, most surveys have shown broadly similar distributions of serotypes.

The picture that has emerged is of a continually changing pattern of infection (see Gelfand 1961, Melnick 1976). Some serotypes are endemic, infecting very young children, but the majority show some degree of epidemicity. Outbreaks recur periodically, usually at intervals of two to five years, presumably reflecting the time required for the recruitment of enough susceptibles to permit the development of a new outbreak. Several epidemic serotypes predominate in each year; but the prevailing types vary from year to year and often from area to area in the same year (Froeschle *et al.* 1966).

Although communities are the natural ecological units for enterovirus circulation, the growth of communications has facilitated dissemination, and global spread of novel virus strains may occur. In 1957, an outbreak of '*hand-foot-and-mouth disease*' due to Coxsackieviruses A16 was reported in Toronto by Robinson and his colleagues (1958). Further outbreaks soon followed in the United States, Europe, South Africa, Australia, New Zealand and Japan (see Cherry and Nelson 1966, Fujamiya *et al.* 1974). Before 1957 this serotype had seldom been isolated and a serological study in Wisconsin showed that it had not been active in that area before the outbreak in 1966 (Cherry and Jahn 1966). Events suggest that a new variant emerged about 1957, followed by a world-wide spread. Similarly, in 1969 an outbreak of *acute haemorrhagic conjunctivitis* broke out in Ghana and quickly spread along the west coast of Africa through eight different countries. Further outbreaks followed in the next three years in Africa, Asia and Europe (see Grist *et al.* 1978). A previously unknown enterovirus isolated in these outbreaks has now been classified as enterovirus 70 (Mirkovic *et al.* 1973, Hierholzer *et al.* 1975).

Antibody surveys have generally revealed evidence of previous exposure to almost all serotypes investigated (see Gelfand 1959, 1961, Sato *et al.* 1972), except in isolated communities. The proportion of seropositive subjects usually rises steeply with increasing age, often to 50 per cent or more during the first five years of life, but sometimes not till later childhood in communities that have not experienced a recent outbreak. This antibody pattern is found for the most frequently isolated serotypes, which are evidently widely disseminated in early childhood by epidemics every few years. Antibodies to less common types have a lower prevalence, increasing slowly with age to reach only 20 or 30 per cent in adults.

Seasonal variation In northern temperate climates enterovirus infections generally have their highest incidence in late summer and autumn, but they continue at a lower rate for the rest of the year. In the southern hemisphere isolations are most frequent from November to February (Assaad and Cockburn 1972). Gelfand (1959), in a study of virus excretion by healthy children, found that the seasonal cycle of ECHOviruses occurred about four weeks later than that of Coxsackieviruses. Virus isolations from patients in the United Kingdom reported to the Public Health Laboratory Service show the same phenomenon (Fig. 99.3). The reason for this difference is uncertain but it may be due to the lower infectivity of ECHOviruses (Kogon *et al.* 1969). In tropical regions outbreaks occur irregularly in any part of the year (Vanderpitte 1960). Less seasonal fluctuation is found in the warmer southern states of the United States than in the northern states (Gelfand *et al.* 1957, Froeschle *et al.* 1966).

Temperature or a related climatic variable is clearly an important factor in the development of enterovirus epidemics; and humidity, by favouring virus survival, seems the factor most likely to increase transmissibility (Yorke *et al.* 1979). Epidemics are not, however confined to the warmest months, nor to the hottest countries. In temperate countries the peak incidence is usually in August or September, but it may be as late as December or January. When epidemics persist for two or more years, the incidence in the intervening winters may remain well above the normal winter level. In Arctic communities outbreaks occur in all seasons, including winter when living conditions favour the spread of infection (Banker and Melnick 1951, Faulkner *et al.* 1957). Similarly in nurseries and childrens' institutions, where infections are readily disseminated, outbreaks may occur in any season, whenever new infections are introduced (Bell *et al.* 1961). These examples serve to illustrate that the infectivity of the virus, the level of immunity in the commnunity, and the social and hygienic conditions which determine the opportunities for spread of infection are the most important determining factors of epidemics; but

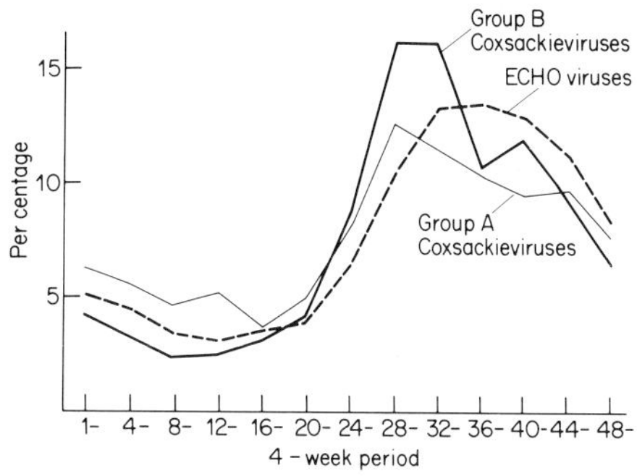

Fig. 99.3 Seasonal distribution of enteroviral infections in England and Wales reported to the Public Health Laboratory Service in the years 1967–1979. Data include 15,980 Echoviruses 3707 group A Coxsackievirueses and 9282 group B Coxsackieviruses. (By permission of the Communicable Disease Surveillance Centre London.)

climatic factors, by affecting the transmissibility of infection, probably play a role in triggering the onset of many epidemics in temperate regions and in reducing their impact in the colder months.

Viruses in faeces and sewage Coxsackie and ECHOviruses are usually detectable in faeces of infected patients for two to four weeks but less regularly and for shorter periods in oro-pharyngeal secretions. Coxsackieviruses are shed for longer than ECHOviruses; this may explain their higher infectivity. In faeces, Coxsackieviruses are often present for a month and occasionally for two months or more. ECHOviruses are usually found in the faeces in the first week of infection, but excretion diminishes in the second week and rarely persists for more than a month (Fox 1964, Kogon *et al.* 1969). The presence of virus in the pharynx is variable. It occurs more frequently in respiratory illnesses than in subclinical infections and in Coxsackievirus infections more than in ECHOvirus infections (Cole *et al.* 1951, Kogon *et al.* 1969, Wenner and Behbehani 1968). Virus may be detected in respiratory secretions a few days before onset of illness and for one or two weeks after onset. Unusual modes of virus shedding are associated with certain serotypes. Enterovirus 70 is found in conjunctival secretions of patients with acute haemorrhagic conjunctivitis (Yin-Murphy and Lim 1972) and Coxsackievirus A21, which causes common cold-like infections, is isolated from respiratory secretions more often than from faeces (Parsons *et al.* 1960). The group A Coxsackieviruses associated with vesicular skin rashes may be isolated from vesicular fluid and occasionally from skin ulcers (Mitchell and Dempster 1955, Evans and Waddington 1967). Enteroviruses may be present in small numbers in the urine in the acute stage of infection (Utz and Shelekov 1958); but faecal or oropharyngeal virus is the most usual source of transmission.

Virus concentrations in faeces are highest early in

the infection. For several ECHOviruses estimates range from 10^3 to $10^{5.5}$ TCD_{50} per g of faeces (see Wenner and Behbehani 1968). Virus has been detected on the skin of infected children (Gelfand 1961), and faecal contamination appears to be the most important mode of transmission. The upper respiratory tract is generally a less profuse source of virus, particularly for ECHOviruses; it is probably of minor importance in the transmission of infection with a few possible exceptions. ECHOvirus 9 has been found in throat swabs in amounts of up to 10^5 TCD_{50}, compared with less than $10^2 TCD_{50}$ per g in faeces (Wigand and Sabin 1962), and oral transmission seems likely. Coxsackievirus A21 and enterovirus 70 are probably disseminated by respiratory secretions; and other Coxsackieviruses may be, particularly those that localize in the mouth and pharynx.

Coxsackie and ECHOviruses are regularly present in sewage, in which their prevalence, seasonal variation and serotype distribution reflect the experience of the community (Melnick *et al.* 1954, Kelly *et al.* 1955, Bloom *et al.* 1959). Conventional sewage treatment procedures decrease its virus content, but enteroviruses can be regularly isolated from sewage-contaminated surface water. Contamination of recreational waters has been reported (McLean *et al.*, 1965, Hawley *et al.*, 1973) but there is no evidence that infection has been acquired from this source. Standard purification processes for drinking water remove or inactivate most of the viruses present in the water intake. The available evidence suggests that, when treated to conventional bacteriological standards, drinking water contains few, if any, detectable viruses in 10 litre samples (see Gamble 1979). Enteroviruses have been isolated from flies of several species and from their excreta (Melnick and Penner 1952), as well as from mosquitoes, cockroaches, pigs, cats, dogs and a fox. Antibodies that neutralize human enteroviruses have been found in a variety of animals (see Gelfand 1961, Christie 1974). However endemic infections in man constitute a huge reservoir of enteroviruses and direct person-to-person transmission clearly accounts for most infections. There is virtually no evidence at present of significant transmission by other routes, but indirect transmission by sewage, flies or contaminated food may sometimes supplement direct spread, particularly in communities with poor hygienic standards.

Animal reservoirs Although there appears to be no important animal reservoir of human enteroviruses, many vertebrate species including cats, pigs, cattle and other domestic animals each harbour a different set of enteroviruses. In spite of the ability of many of these viruses to infect more than one host species the extent to which they do so in nature is unknown. Thus, Coxsackieviruses infect chimpanzees, monkeys, and a number of rodents in the laboratory, but naturally occurring infections have not been reported. However, encephalomyocarditis virus which is a natural parasite of the mouse, has on rare occasions caused infection in man; and, in a recent outbreak of swine vesicular disease (SVD) caused by a porcine enterovirus, several human infections occurred among workers in close contact with the animals (Brown *et al.* 1976). This porcine virus proved to be closely related to Coxsackievirus B5. Pigs exposed to Coxsackie B5 virus developed neutralizing antibodies to both Coxsackievirus B5 and SVD virus, although it is uncertain whether virus replication occurred (Graves 1973, Garland and Mann 1974). Antigenic studies by neutralization tests, gel diffusion and RNA hybridization experiments showed that substantial antigenic variation had occurred and that currently circulating Coxsackie B5 viruses differed significantly from the Faulkner prototype B5 strain isolated in 1952 (Brown and Wild 1974, Brown *et al.* 1973, 1976). The relatively recent appearance of SVD in pigs and the antigenic relation of the virus to Coxsackievirus B5 has led to the suggestion that the transmission of a human virus to pigs may have occurred. The possibility that animal enteroviruses may sometimes be transmitted to man is of equal interest. It seems improbable that enteroviruses affected primitive human beings who lived in communities too small to sustain endemic or epidemic infection. It is more likely that enteroviruses first appeared in animals. Man may therefore have acquired enteroviruses from animals in agrarian or urban societies. Transmission between animals and man may still be responsible for the appearance of novel enteroviruses.

Infection rates Infection spreads from person to person and primarily affects young children (Cole *et al.* 1951, Gelfand 1961). The highest excretion rates have been in 1-year-old children, with gradually decreasing rates at ages 2 to 5 years (Froeschle *et al.* 1966). Infection in the first year of life is less frequent, because of protection afforded by maternal antibody and the sheltered existence at this age. Antibody studies show that 50 per cent of school children normally have evidence of past infection with many of the common serotypes (Gelfand 1959, Sato *et al.* 1972). The diminishing frequency of infection from age 1 to 5 years is probably due to acquired immunity in older children. Indeed, the infection rate in non-immune children probably increases through early childhood with increasing association during play and at school. In some studies virus excretion has been detected slightly more often in boys than in girls. The cause may be a higher infection rate among boys, because of greater physical contact during play, or boys may shed virus more profusely or persistently than girls. If there is a sex difference it is small and probably of little significance.

Although children are the primary hosts of enteroviruses, infection introduced into the household spreads rapidly to other members of the family. In New York, Kogon and his colleagues (1969) found that Coxsackieviruses spread to 76 per cent of susceptible family contacts and ECHOviruses to 43 per cent. Infection rates within families greatly exceed those in the community as a whole. An outbreak of Coxsackievirus B3 in Louisiana affected 11 households in which infection spread to 80 per cent of children, while in the same period 11 other households remained unaffected (Clemmer *et al.* 1966). Transmission of infection by children between households therefore occurs far less readily than spread within households, illustrating the importance of close association in virus transmission. Spread in communities tends to be localized. Cole and his associates (1951) identified clusters of infected households in a suburban community in which contact between children could usually be demonstrated.

Social class, climate, and hygiene are important factors in virus dissemination. In two American cities, Melnick (1957) found that enterovirus excretion was 3 to 6 times more frequent in lower socio-economic districts than in middle and upper class districts; this difference was greater for ECHOviruses than Coxsackieviruses. Similar results were obtained in other American cities (Gelfand *et al.* 1957, 1963, Froeschle *et al.* 1966). Virus isolation rates in the warmer southern cities were about four times as high as those in the more northern cities and a greater variety of circulating serotypes was associated with both low socio-economic class and the warmer southern climate. In countries where a warm climate, poor hygiene and low socio-economic class are combined, infection is endemic and very high prevalence rates of enterovirus have been reported in children. In Karachi 80 per cent of infants were found to be excreting enteroviruses. Two different enteroviruses were found in the stools of 4 per cent of these children, three in 14 per cent and in two infants four different viruses were present (Parks *et al.* 1967).

Interference between different enteroviruses is probably one of a number of interacting ecological factors that determine secular changes in their distribution but evidence is largely circumstantial. Dalldorf and Albrecht (1955) observed that outbreaks of poliomyelitis occurred later in the year than usual when preceded by an outbreak of pleurodynia; and that the number of reports of paralytic poliomyelitis was inversely related to the frequency of Coxsackievirus B infections in the same year. Furthermore, when live poliovirus vaccine was introduced it was found that excessive wild enterovirus infection in a community could interfere with the effectiveness of vaccination (Ramos-Alvarez *et al.* 1959). On the other hand, the vaccine strain of poliovirus sometimes interfered with the circulation of wild enteroviruses (Gelfand *et al.* 1963, Fox *et al.* 1961). Although interference between different Coxsackieviruses and ECHOviruses has not been demonstrated epidemiologically, it probably occurs; indeed the earlier seasonal rise of Coxsackievirus infection may, through prior seeding of the community, interfere with the establishment of ECHOvirus infection. This may be one reason for the greater epidemicity of the Coxsackieviruses compared with most ECHOviruses.

Clinical manifestations Most ECHOvirus infections are clinically inapparent or produce only minor illness, but there is a wide variation from outbreak to outbreak. Inapparent infection appears to be more common with ECHOviruses than with Coxsackieviruses; in New York families, Kogon and his colleagues (1969) found that between 9 and 24 per cent of Coxsackievirus infections and from 9 to 18 per cent of ECHOvirus infections were accompanied by minor illness. Other estimates range from two illnesses among 51 Coxsackievirus B3 infections (Clemmer *et al.* 1966) to much higher figures in 'virgin soil' outbreaks in which the higher proportion of infections in adults results in a greater incidence of illness (Dippe *et al.* 1975). Serious illness is uncommon. In an epidemic of ECHOvirus 9 infection in Milwaukee, Sabin and his colleagues (1958) estimated that only about 0–4 per cent of over 28 000 patients were admitted to hospital suffering from *aseptic meningitis*; 85 per cent of all infections were thought to result in clinically recognized illness, so only about 1: 250 infections gave rise to serious illness affecting the central nervous system.

The *clinical manifestations* of infection are related to the age of the patients and, to a lesser extent, their sex; in general, more serious illness is commoner in children and adults than in infants, and in males than in females (Table 99.2). Exceptions to this generalization are severe infections in the first year of life when the distribution of virus receptors and immunological immaturity may be responsible. Such illnesses usually occur in the neonatal period in the form of meningoencephalitis, myocarditis, or disseminated illness. Mild respiratory illnesses and skin eruptions have been most commonly reported in children, particularly in those under 5 years of age (Assaad and Cockburn 1972, Karzon *et al.* 1961, Landsman and Bell 1962, Sabin *et al.* 1958). Aseptic meningitis on the other hand is more frequent in older children and adults than in infants. Males outnumber females by about 2: 1 in children, but the sexes are more equally distributed in adult cases, with perhaps a small excess of females (Johnsson 1954, Assaad and Cockburn 1972, Lennette *et al.* 1962, Froeschle *et al.* 1966, McNair Scott 1961). The reason for this age distribution is not known but, as with poliovirus infection, susceptibility of the central nervous system clearly increases with age. Winkelstein and his colleagues (1957) estimated that the attack rate of aseptic meningitis during an ECHOvirus 6 epidemic was 3 to 4 times greater among older children than in those below 4 years of age, thus

Table 99.2 Age and sex distribution of coxsackievirus B infections by clinical diagnosis (Based on notifications of virologically diagnosed illnesses in England and Wales in 1962–1973 reported to the Public Health Laboratory Service) (By permission of the Epidemiological Research Laboratory of the PHLS)

Diagnosis		Age in years						
		1	1–14	5–9	10–14	15–40	40	All ages
Minor illness	Male	395	655	348	121	307	44	1870
	Female	273	588	279	69	248	20	1477
CNS Disease	Male	84	147	233	144	317	7	932
	Female	67	79	90	75	324	17	652
Myalgia	Male	2	26	53	43	196	20	340
	Female	6	13	34	33	117	10	213
Cardiac Disease	Male	13	6	10	8	110	17	164
	Female	9	6	8	4	32	10	69
Totals		849	1520	1055	497	1651	145	5717

resembling the 3-fold increase from infancy to age 14 in the attack rate of paralytic disease in poliovirus infection (Melnick and Ledinko 1953). A disproportionately large number of reported cases of aseptic meningitis occur in adults, compared with children. It would be expected that most adults would have been infected in childhood so that adult infections should be rare. There may be an increased susceptibility to neurological involvement in adults, or many adult cases may be the result of reinfection. At present this question cannot be answered but, as primary infection generally occurs in childhood, antigenic drift in enteroviruses over a period of 20 years might well modify some viruses sufficiently to permit reinfection in adult life. The striking excess of male cases in childhood suggests that hormonal changes are mainly responsible but the absence of a male excess in adult cases implies that other factors must be involved. The increase in the number of cases among young adults following the decline in cases after childhood may reflect acquisition of infection by parents from children, and the small excess sometimes reported in adult females may be due to their greater exposure to children (Froeschle *et al.* 1966). Other illnesses associated with enterovirus infection that have a high frequency among adults and a male preponderance are Bornholm disease and myocarditis (Sylvest 1934, Johnsson 1954); but no explanation for this distribution has yet been offered.

Transmission to animals

Group A Coxsackieviruses

The Coxsackieviruses are highly infectious for the newborn of many species including mice, hamsters, cotton-rats, rabbits, merions, squirrels, ferrets and bats. Newborn mice are generally used for virus isolation and the study of the experimental disease. Group A viruses induce a generalized *myositis* of striated muscle, sparing the heart and tongue. Affected mice suffer from weakness followed by flaccid paralysis and death within a few days. Susceptibility is maximal during the first 3 or 4 days of life, after which it progressively declines (Lerner *et al.* 1962). Mice may be infected orally or by any parenteral route; but larger doses are required for oral infection (Kaplan and Melnick 1952).

Post mortem, the muscles often have whitish streaks, and histologically show hyaline degeneration and fragmentation, with inflammatory cell infiltration and mineral deposition in the later stages (Gifford and Dalldorf 1951, Godman *et al.* 1952). Variants of group A viruses that affect adult mice have been described but they are uncommon (Dalldorf 1957, Lerner 1965).

Infection of chimpanzees and monkeys with Coxsackie A viruses usually results in virus excretion without signs of illness, but infections of Cynomolgous monkeys with types 7, 14 and 16 have been associated with lesions in the central nervous system similar to those produced by poliovirus. The most severe lesions accompanied by paralytic illness have been associated with types 7 and 16 (Dalldorf 1957, Habel and Loomis 1957, Lou and Wenner 1962).

Meningo-encephalitis in rabbits, guinea-pigs and mice

Although this is not a virus disease, attention is drawn to it here because of the confusion it may cause during virus investigation. The disease, usually symptomless, may occur in epizoötics. It is spread by cage infection and also as a result of experimental inoculation with brain suspension or sedimented urine. It is cause by a microsporidium—*Nosema cuniculi.* Histologically there are in the brain infiltrations with lymphocytes and plasma cells around the vessels, in the meninges, and in the cerebral substance. In some animals there are focal areas of necrosis in which the parasites may be observed within rounded or oval cyst-like bodies. These areas are surrounded by epithelioid and mononuclear cells (see Oliver 1922, McCartney 1924). Five human cases are on record (Report 1974).

Coxsackievirus group B

Group B viruses typically cause an encephalitis in newborn mice, leading to spastic paralysis and death. The most regular feature at necropsy is *necrosis of the brown fat*, particularly in the cervical, axillary and interscapular regions. There may also be focal necrosis of striated muscle, pancreatitis, myocarditis, hepatitis and parotitis. Opaque lesions in the interscapular fat pad are often visible to the naked eye, and histologically there is degeneration of the brown fat commencing at the periphery of the lobules, followed by inflammatory infiltration, necrosis and calcium deposition. In the brain, there may be focal necrosis with cyst formation in mice that have survived the first week of infection. In the pancreas, a massive necrosis of the acinar tissue sometimes occurs, but the pancreatic ducts and islets of Langerhans are usually spared. Lesions of the heart, liver and muscle generally consist of scattered foci of degeneration or necrosis. The morphological changes in mouse tissues have been described in detail elsewhere (Gifford and Dalldorf 1951, Godman *et al.* 1952, Pappenheimer *et al.* 1951). Susceptibility to group B viruses is greatest within 24 hours of birth and declines sharply thereafter. Illness may not be apparent for a week or more after inoculation. As with group A viruses, mice are susceptible to oral infection, but large doses of virus may be required. When viruses are introduced directly into the stomach by a gastric tube, the infectious dose is similar to that required for parenteral infection. In feeding experiments, animals presumably fail to swallow the whole dose (Loria *et al.* 1974).

The most constant features of Coxsackievirus B infection in mice are the characteristic lesions in the brain and brown fat of newborn mice. Older mice are normally resistant but severe infections of the heart, liver or pancreas sometimes occur. This variability in the outcome of infection is influenced by viral, host and environmental factors. Susceptibility is primarily dependent on the availability of virus receptors. Kunin (1962) detected an association between the relative abundance of receptors on mouse tissues *in vitro* and susceptibility to infection *in vivo*. Thus, receptor activity of brain tissue from newborn mice for Coxsackievirus B *in vitro* declined sharply with age reflecting the age-dependent susceptibility to Coxsackievirus encephalitis *in vivo*. On the other hand, liver, heart and kidney tissues from newborn mice also absorb Coxsackievirus more actively than those from adults. The greater susceptibility of adult mice to Coxsackievirus-induced myocarditis and hepatitis would thus appear to imply factors other than receptor distribution.

Different strains of Coxsackievirus B vary in their tissue tropism in infections of adult mice. Certain strains produce severe myocardial necrosis (Grodums and Dempster 1959, Kilbourne *et al.* 1956); others induce hepatitis (Minkowitz and Berkovich 1970), pancreatitis (Dalldorf and Gifford 1952), Pappenheimer *et al.* 1951) or diabetes (Notkins 1977). Susceptibility to these diseases also varies in different strains of mice (Grodums and Dempster 1959; Ross and Notkins 1974); and for some diseases it is greater in male than in female mice. For example, the strain of Coxsackievirus B1 studied by Minkowitz and Berkovich (1970) induced a severe hepatitis in CD1 mice, affecting 73 per cent of males but only 9 per cent of females. The abolition of the sex difference by castration suggests that it is mediated by hormones rather than by a difference in the distribution of virus receptors. Several other hormones influence susceptibility; corticosteroids (Kilbourne *et al.* 1956) increase it, whereas thyroxine (Boring and Ungar 1962) and growth hormone (Behbehani *et al.* 1962) may decrease it. Extrinsic factors also affect susceptibility. A low environmental temperature (Boring *et al.* 1956), malnutrition (Woodruff 1970 *a*, *b*) and strenuous exercise (Federici *et al.* 1963) have increased the severity of some experimental Coxsackievirus infections.

Although Coxsackievirus infections are cytolytic, there is growing evidence that myocardial damage in adult mice is partly caused by the inflammatory response. Virus growth in the tissues reaches its maximum in the first week of infection, but chronic inflammatory changes in the heart may persist for six months or more (Wilson *et al.* 1969). Woodruff and his colleagues found that thymus-derived lymphocyte (T cell) depletion protected mice against the severe myocardial necrosis caused by a strain of Coxsackievirus B3 (see Woodruff 1980). They also found that T cells from Coxsackievirus-infected mice lysed syngeneic fibroblasts; and cardiac muscle fibres that had been infected with Coxsackievirus. These and other studies suggest that, in some Coxsackievirus infections, an immune response to persisting viral antigen or to altered tissue antigens may contribute to the resultant disease (see Woodruff 1980).

Cynomolgus monkeys and chimpanzees have been infected with group B Coxsackieviruses (Melnick and Ledinko 1950, Wenner *et al.* 1960). Infection is usually subclinical, but virus is present in the pharynx and faeces for a week or two after oral infection; a viraemia occurs and type-specific antibodies are formed. Carditis has been reported by several investigators in monkeys infected with Coxsackie virus B4 (see Lou *et al.* 1960) Sun *et al.* 1967). Lesions in the central nervous system of infected monkeys have sometimes been reported but the difficulty of distinguishing viral damage from traumatic damage due to cisternal punctures makes it difficult to assess neuropathogenicity. (For benign myalgic encephalomyelitis see 6th edn, p. 2449)

ECHOvirus

ECHOviruses are by definition non-pathogenic for mice and none of the prototype strains multiplies in newborn mice. However, naturally occurring variants that produce paralytic infection in newborn mice have been reported, particularly in outbreaks of ECHOvirus 9 infection (Archetti *et al.* 1956, Eggers and Sabin 1961). These strains differ from the prototype 'Hill' strain in their growth characters in tissue culture. Antigenically they are prime strains with a broader constitution than the prototype. There has been one report of a mouse-pathogenic strain of ECHOvirus 6 which produced a spastic paralysis in both suckling and adult mice with lesions similar to those produced by group B Coxsackieviruses (Vasilenko and Atsev 1965). Clinically inapparent infections have been produced in chimpanzees and monkeys with a number of serotypes (Itoh and Melnick 1957, Lou and Wenner 1962). Focal lesions with neuronal destruction have sometimes been found in the central nervous system of monkeys, particularly after inoculaton into the brain or cord, but occasionally after intramuscular injection.

Properties of enteroviruses

Morphology, antigenic structure and replication

Coxsackie and ECHOvirus share the physical characters of enteroviruses. They are 20–30 nm icosahedral particles, with a buoyant density of 1.32–.35 g/cm^3. They are resistant to ether and to pH 3–5. The RNA core is a single-stranded molecule of 2.6 million daltons, and is indistinguishable in size and base constitution from that of other enteroviruses (see Rueckert 1971). The extent of homology of nucleotide sequences in the genomes of enteroviruses has been investigated in hybridization experiments between viral RNA and complementary strands obtained by denaturing double-stranded replicative form RNA from infected cells (Young 1973). Though all enteroviruses showed at least 5 per cent homology in their genomes, there was greater homology between different virus types of the same group than between viruses of different groups. Thus, there was 36–52 per cent homology between different types of poliovirus, but only 4–8 per cent between poliovirus and either group A or B Coxsackieviruses or ECHOviruses. In general, 30–50 per cent of the genome was shared by members of the same group compared with 4–20 per cent among members of different groups. These results broadly support the original biological classification of enterovirus into poliovirus, ECHOvirus, and Coxsackievirus A and B groups, but it is of interest that Coxsackievirus B4 showed greater homology with ECHOviruses than with group A Coxsackieviruses. This anomaly is clearly the result of the arbitrary attribution in the biological classification of greater significace to virus multiplication in newborn mice than in tissue culture. Hybridization experiments have also demonstrated significant differences between virus strains of the same serotype. In two strains of poliovirus type 1, Young (1973) found only 74 per cent homology of nucleotide sequences, and in two strains of Coxsackievirus B5, Brown and Wild (1974) found 100 per cent homology but both strains showed only 50 per cent homology with the prototype (Faulkner) B5 strain—a difference as great as that usually found between virus types.

The viral capsid of Coxsackievirus B1 was shown to contain four polypeptides designated VP1, VP2, VP3, and VP4 of similar molecular weights to those of other picornaviruses (Rueckert 1971). All picornaviruses probably have a similar capsid structure with 60 protein capsomeres, made up of the polypeptide chains VP1, VP2, and VP3. Serological tests show cross-reactivity reflecting shared surface antigens. In neutralization tests cross-reactions have regularly been found between Coxsackieviruses A3 and A8, A11 and A15, A13, and A18, and ECHOviruses 1 and 8, 12 and 29, and 6 and 30. Antigenic variation has led to the appearance of a number of prime strains of both Coxsackieviruses and ECHOviruses (Wigand and Sabin 1962, Melnick 1957). These variants have a broader antigenic constitution than the prototype strains and, though they are poorly neutralized by antiserum to the prototype strain, they induce antibody which neutralizes both prototype and prime strain equally. Purified suspensions of Coxsackieviruses contain two distinct antigens detectable by precipitin or complement-fixation tests which, like the similar antigens found in poliovirus suspensions, have been designated C and D antigens (Schmidt *et al.* 1963, Frommhagen 1965). These antigens are separable by gel diffusion or density-gradient centrifugation. The C antigen has group-specific precipitating and complement-fixing activity with both monkey and human sera. This antigen appears to be responsible for the non-specific reactivity of many human sera in complement-fixation tests. The D antigen is associated with infectious virus particles and with type-specific complement-fixing activity with homotypic monkey serum but not with human sera. After heating at 56° for 30 minutes the D antigen fraction behaves like the C antigen fraction, with group-specific complement-fixing activity with human sera.

Some strains of several types of Coxsackievirus and ECHOvirus agglutinate human red cells (Goldfield *et al.* 1957); Coxsackievirus A7 agglutinates vaccinia-agglutinable fowl cells (Grist 1960). Not all strains of these virus types have this property; it is more common in some types than others and may be lost when the virus is passaged in tissue culture. Agglutination is specifically inhibited by antibody, and absorption of virus suspensions with red cells removes both the haemagglutinating ability and infectivity. Haemagglutination is therefore a property of the intact virus except in Coxsackievirus A7 where it is associated with a 'soluble' haemagglutinin. Attachment of the virus to the red cell occurs in two stages: the first is temperature—indepedent and reversible; the second is temperature-dependent and irreversible suggesting an enzymic process.

Studies of the attachment of representative picornaviruses to HeLa cells suggest that they carry different receptors for the different subgroups of enteroviruses. The number of Coxsackievirus B receptors is approximately 10^5 per cell. Saturation of the receptors with a group B Coxsackievirus prevents attachment of other group B Coxsackieviruses but not of group A Coxsackieviruses, polioviruses or ECHOviruses (see Crowell and Siak 1978). Only representative virus strains have so far been studied and the full sensitivity spectra of receptors have yet to be defined. The finding that Coxsackievirus A21, which produces a common cold-like illness, shares receptors with a number of rhinoviruses, and that Coxsackievirus B2 and adenovirus 2 likewise share receptors, indicates the occurrence of interesting exceptions to the broad correspondence between receptor sensitivity and the main biological groupings (Lonberg-Holm *et al.* 1976).

Uncoating of the virus probably commences after attachment while it is still outside the cell. Coxsackievirus B viruses eluted from cells were shown to have lost VP4 capsid polypeptide and were unable to reattach to cells (Crowell and Philipson 1971). It is thought that the capsid may be weakened sufficiently by the loss of VP4 to allow the release of RNA within the cell cytoplasm. Uncoating may be paralleled by the seqence of events following the disruption of Coxsackievirus at alkaline pH, which leads first to the release of VP4 and the loss of ability to attach to receptors and then to loss of VP2 with the release of RNA leaving an empty capsid (Crowell and Siak 1978). The presence of virus receptors is not the only pre-requisite of cell susceptibility. Certain cell lines derived from mouse tissues attach Coxsackievirus B but fail to support their replication. In experiments with hybrid cell lines derived from susceptible human cells and resistant mouse cells, Medrano and Green (1973) showed that group B Coxsackieviruses multiplied in and killed human-mouse hybrid cells, whereas the virus attached but failed to replicate in pure mouse cells. Human gene products were therefore necessary for the virus multiplication irrespective of the availability of virus receptors. The mode of replication of the non-polio enteroviruses, although less extensively studied, appears to be essentially similar to that of polioviruses which has already been outlined.

Resistance

Enteroviruses remain viable for long periods at $-20°$ and $-70°$, but lose infectivity slowly at room temperature and with increasing rapidity as the temperature is raised. A high relative humidity favours their survival. They are rapidly inactivated by drying. Thermal inactivation is decreased by the presence of glycerol, milk, ice-cream, or molar magnesium chloride in the medium. Alcoholic and phenolic disinfectants are ineffective; but formaldehyde, ozone, chlorine, bromine and elemental iodine are viricidal. The activity of chlorine is diminished by a high pH, low temperature and the presence of organic substances. Liu and his colleagues (1971) found a wide range of resistance of 14 enteroviruses to 0.5 mg/litre of residual free chlorine in river water at 2°. Poliovirus was the most resistant, but 99.99 per cent perished in 40 minutes; ECHOviruses were the least resistant and Coxsackieviruses fell in between the two.

Enteroviruses are resistant to antibiotics and chemotherapeutic agents, but 2 (α-hydroxybenzyl)—benzimidazole (HBB) inhibits their replication with the exception of Coxsackieviruses A7, 11, 13 16, and 18, and ECHOviruses 22 and 23. HBB depresses viral RNA synthesis at about one-tenth of the concentration required to affect cellular RNA (Bucknall 1967), but the precise mechanism of its action is uncertain. Guanidine hydrochloride also depresses enteroviral multiplication in cultured cells (see Baltimore 1968); it appears to act synergically with HBB (Eggers and Tamm 1963). Guanidine interferes with RNA synthesis and with cleavage of the capsid precursor polypeptide VP0, so that empty capsids accumulate in the cell. Virus passaged in the presence of guanidine rapidly produces mutants that are first resistant and eventually dependent on the compound; this is probably the explanation for its lack of activity *in vivo*.

Natural infection and immunity

It is generally accepted that the portal of entry is the alimentary tract, but there are probably a few exceptions. Coxsackievirus A24 and enterovirus 70, the usual causes of haemorrhagic conjunctivitis, may gain entry directly through the conjunctiva. Similarly, in those enteroviral infections that predominantly affect the upper respiratory tract, the virus may be implanted directly in the mucous membranes affected. Human volunteers have been infected experimentally with Coxsackievirus A21 instilled into the nose, but colonization of the human intestine by large doses of virus administered in enteric-coated capsules has proved difficult (Parsons *et al.* 1960, Johnson *et al.* 1962). Infection produces a common cold-like illness in which virus is regularly recovered from the throat, but rarely from the faeces, indicating a respiratory rather than an enteric infection. The sequence of events in most enteroviral infections seems to be similar to that of poliovirus infection. Primary colonization probably occurs in the epithelium or lymphoid tissue of the alimentary tract, followed by further virus multiplication in regional lymph nodes and viraemia, by which infection may be transferred to other sites. Although this sequence has not been established in man, it is suggested by the initial shedding of oral and faecal virus, by the frequent occurrence of a prodromal illness and viraemia, and by the subsequent appearance of infection in other organs remote from the alimentary tract.

Type-specific antibodies are formed in response to infection and, as in many other virus infections, antibody appears sequentially in IgM, IgG, and IgA immunoglobulin classes. Ogra (1970) found that in patients with ECHOvirus 6 infection IgM antibody was detectable in serum within a few days of onset, reaching its peak titre in one to two weeks, and declining to an undetectable amount within six weeks of onset. IgG antibody appeared about a week after onset and increased in titre over the following six weeks. Only two of 16 patients formed IgA antibody, which was present in low titre by two weeks after onset. IgA antibody was present in the nasopharynx and rectum earlier, and in higher titre than in serum, and was therefore presumed to be locally produced secretory antibody not derived from serum immunoglobulin. Antibody in the spinal fluid was of the IgG class and, since it was present in patients both with and without meningeal involvement and was accompanied by a similar increase in poliovirus-specific IgG antibody in the serum, its presence presumably reflected transudation of immunoglobulin from the plasma. It has been shown however, that in poliovirus infection antibody is produced locally by plasma cells in the brain (Morgan *et al.* 1947, Esiri 1980). Both mechanisms probably operate in enterovirus infection.

While specific antibody gives protection from enterovirus infection, its formation would appear to occur too late to prevent virus from multiplying and spreading within the body. Indeed, viraemia precedes the rise in serum-neutralizing antibody and is terminated by it. However, patients with hypogammaglobulinaemia, but with intact cell-mediated immunity, are prone to experience severe and fatal ECHOvirus infections (Wilfert and Buckley 1977); antibody clearly modifies the course and severity of infection. Adult mice treated with cyclophosphamide form little antibody after Coxsackievirus B3 infection, and die from severe infection of the heart and pancreas; but specific virus antiserum given after infection protects them against lethal infection (Rager-Zisman and Allison 1973*a*). On the other hand, suppression of cell-mediated immunity by neonatal thymectomy does not increase susceptibility and in some circumstances may reduce it (Woodruff and Woodruff 1974). Although antibody does not prevent the spread of virus from the site of initial multiplication to other organs, it probably restricts it and prevents more extensive tissue damage. Low titres of antibody, usually of the IgM class, are detectable early in infection in mice; but this early antibody neutralizes virus poorly and other host defences are probably of more importance at this stage. Peritoneal exudate cells transferred from uninfected adult mice protect newborn mice against the lethal effects of Coxsackievirus B3 infection; the effect is enhanced by specific antibody (Rager-Zisman and Allison 1973*b*). Furthermore, intraperitoneal injection of silica particles, which are selectively taken up by macrophages and impair their function, increases susceptibility of adult mice to infection. Macrophages may therefore constitute an important host defence in the early stages of infection, and their activity may partly depend on the presence of antibody.

Thus it appears that immune defences first limit the spread of infection and, within two or three weeks, eliminate it. However, infection does not always follow this typical course. Relapses have often been noted, particularly in Bornholm disease (Sylvest 1934). Chronic cases of hand-foot-and-mouth disease (Evans and Waddington 1967), myositis (Freudenberg *et al.* 1952, Tang *et al.* 1975) and myocardiopathy (Lerner *et al.* 1975) have been described in which virus or viral antigen apparently persisted for a year or more. Why some infections follow a chronic course is not understood. In agammaglobulinaemic patients ECHOvirus infection is sometimes persistent and lethal (Wilfert and Buckley 1977), and immune deficiency may explain some other atypical illnesses.

The diversity of the clinical response to enteroviral infection in man resembles that seen in experimental infections in animals; and some factors that affect the outcome of infection in animals also do so in man. Different strains of virus vary in their effects. Thus, a rash was often a feature of ECHOvirus 9 infection in outbreaks in 1956–7, but rarely in previously reported outbreaks (see Wenner and Behbehani 1968). As in animals, susceptibility to infection may be partly determined by the distribution of virus receptors on the tissues; Holland (1961) found that this was so in human fetal tissues. Epidemiological studies show that spread of infection to the heart and central nervous system varies with age and sex of the patients. Whether some diseases are partly due to the immune response is uncertain. The late appearance of the rash in some ECHOvirus infections (Neva 1956, Kibrick 1964), and of cardiac manifestations in Coxsackievirus infections suggest that it may be.

Clinical syndromes associated with infection

The many publications on the association of Coxsackieviruses and ECHOviruses with human disease have been comprehensively reviewed by Kibrick in 1964 and by Grist and his colleagues 1978, and the reader is referred to these reviews for full details of sources. A brief summary here will suffice.

Minor illness Most infections are symptomless or produce only minor illness. Typically there is a short febrile illness which may be asymptomatic or accompanied by a rash, minor upper respiratory symptoms, or lymphadenitis. Group A Coxsackieviruses, particularly types 1,2,3,4,5,6,8,10 and 22, characteristically produce outbreaks of *herpangina*, but other Coxsackieviruses of groups A and B and some ECHO-viruses have been associated with sporadic cases. The throat is sore and papules and small ulcers with red areolae appear on the fauces, uvula and adjacent palate; these lesions are scanty rarely numbering more than about ten, and they clear without treatment within a few days. Lesions around the vaginal orifice have been described (Mitchell and Dempster 1955). Lymphonodular pharyngitis is a rare manifestation associated with Coxsackievirus A10 (Steigman *et al.* 1962); small nodules appear briefly on the fauces and uvula but do not ulcerate.

Rashes The frequent association of various rashes with infections due to Coxsackieviruses A9, A16 and B5, and ECHOviruses 4, 9, and 16 has established a causal relation with these viruses. Less frequent associations have been reported with many other enteroviruses but, as both rashes and enterovirus infections are common in childhood, some associations are likely to be fortuitous. Rashes are commonest in young children. They are generally macular, or maculo-papular, but those due to Coxsackievirus A16 are vesicular. Petechial rashes have occasionally been reported and when accompanied by meningeal illness confusion with meningococcal infection has occurred. The rash caused by Coxsackievirus A16 has been called '*hand-foot-and-mouth disease*', describing its typical distribution, but lesions sometimes appear elsewhere, particularly on the buttocks in infants. Although most outbreaks have been due to Coxsackievirus A16, they have also been associated with Coxsackieviruses A5, A9, A10, B2, B5, and enterovirus 71. The skin lesions, at first papular, soon become vesicular. In the mouth, red macules, white vesicles and small ulcers are found on the fauces, tongue, palate and inside the cheeks (Robinson *et al.* 1957).

The ability to cause a rash probably varies for different strains of virus. Little is known of the pathogenesis of the skin lesions, which may be due either to direct viral cytolysis or to immunopathological damage. In ECHOvirus 9 infections, the rash appears early in the illness, but in ECHOvirus 16 infections it follows the illness, appearing several days after the temperature has subsided (Neva *et al.* 1954, Neva 1956). A similar sequence has been noted in ECHOvirus 11 infection (Kibrick 1964). Again, in hand-foot-and mouth disease antibodies to Coxsackievirus A16 may be present when the rash appears (see Seddon and Duff 1971).

Bornholm disease

This disease, which is also known as epidemic myalgia or pleurodynia and by several colloquialisms including the descriptive 'devil's grippe', was described by Sylvest (1934), who studied an outbreak in 1930 on the Danish island of Bornholm. It is characterized by fever and acute paroxysms of stitch-like pain, usually in the chest or abdomen, which may recur over a period of from one or two days to three weeks. Sylvest's classic monograph (published in translation—Sylvest 1934) includes descriptions of many earlier outbreaks in Europe and America in which there was often a remarkably high incidence of the illness. In Norway in 1872 for example, 346 of the 1950 inhabitants of the village of Drangedal were affected and 135 of 400 of those in Krogen. Although virological diagnosis was not available to Sylvest, there is little doubt that the disease he described was the same as that which still bears the same name. Recent outbreaks no longer seem to have the high incidence of those described by Sylvest, suggesting that immunity must now be more widespread. Most outbreaks are due to group B Coxsackieviruses, but muscle pain is sometimes a symptom in infections with some Coxsackie A viruses and ECHOviruses (see Grist *et al.* 1978).

Diseases of the central nervous system

Aseptic meningitis has been associated sporadically with almost all enterovirus serotypes; outbreaks have been caused by Coxsackievirus A7 and 9, by all of the Coxsackievirus B types, and by many of the ECHO-viruses particularly types 4,6,9,11,14,16 and 30. (Table 99.3) Meningitis is sometimes accompanied by other characteristic enteroviral syndromes like myalgia or myocarditis, but otherwise indistinguishable from meningitis caused by other viruses. There are often prodomal symptoms and the illness itself may be distinctly biphasic. Rarely there may be muscle weakness or paralysis but patients almost always recover completely. It is often difficult to be sure that paralysis in sporadic cases is due to a virus concurrently isolated from faeces, so that some of the reported associations are of uncertain aetiological significance; Coxsackievirus A7 is the only enterovirus other than poliovirus to have caused outbreaks of paralytic disease, and these have been small and infrequent (Voroshilova

Table 99.3† Neurological disease associations of Coxsackievirues, ECHOviruses and 'new' enteroviruses (After Grist, N. R., Bell, E. J. and Assaad, F. 1978 (included in ref. list)

Syndrome or clinical feature	Virus types[1]
Meningitis	Coxsackieviruses A_1_, _2_, _3_, _4_, _5_, _6_, _7_*, 8, _9_*, _10_, 11, _14_, _16_, 17, 18, _22_, _24_ Coxsackieviruses B_1_*, _2_*, _3_*, _4_*, _5_*, 6* ECHOviruses 1, _2_*, _3_*, _4_*, _5_, _6_, _7_*, _9_*, _11_*, 12, 13, _14_*, _15_, _16_*, _17_*, _18_*, _19_*, 20, 21, 22, _23_, _25_*, 27, _30_*, _31_, _33_* enterovirus _71_*
Paralytic disease	Coxsackieviruses _A4_, 6, 7, 9, 11, 14, 21 Coxsackieviruses B1, _2_, _3_, 4, 5, 6 ECHOviruses 1, _2_, 3, _4_, _6_, _7_, _9_, _11_, 14, _16_, 18, _19_, 30 enterovirus 70, _71_
Encephalitis	Coxsackieviruses A2, 5, 6, 7, 9 Coxsackieviruses B1, 2, _3_, _5_, _6_ ECHOviruses _2_, 3, 4, _6_, 7, _9_, 11, 14, _17_, 18, _19_, 25 enterovirus _71_
Ataxia	Coxsackieviruses A4, 7, 9 ECHOvirus _9_

[1] Italicized figures indicate virus isolation from CSF or other parenteral source.
* = Outbreaks reported.
† = Compiled from published information from all countries

and Chumakov 1959, Grist and Bell 1970). Encephalitis is fortunately a rare complication which occurs most commonly in infants, often as part of generalized neonatal infection. The illness may be severe leading to coma and death, particularly in infants and agammaglobulinaemic children (Wilfert and Buckley 1977) and occasionally adults (Heathfield *et al.* 1967). In children and adults cerebellar ataxia has often been noted.

Some investigations have revealed little evidence of residual brain damage in patients who have had enteroviral meningitis, but Sells and his colleagues (1975) found significant neurological sequelae and a reduced head circumference in children who had had meningitis before the age of one year. Neurological sequelae were also reported in three children with meningitis in an outbreak caused by Coxsackievirus B5 (Heathfield *et al.* 1967). Several cases of the Guillain-Barré syndrome have been associated with Coxsackievirus A and ECHOvirus infections. In one case virus was isolated from the brain but the cause of this disease remains uncertain (see Kibrick 1964).

Cardiac disease

Severe myocarditis is usually a prominent feature of disseminated enterovirus infections of neonates, but a less severe myocarditis or pericarditis may follow infection of older children and adults. Proof of aetiological association is difficult unless virus or virus antigens can be found in heart tissue or pericardial fluid. On these criteria we can say that carditis is caused by Coxsackieviruses A4, A16, B1–5, and ECHOviruses 9 and 16. An association with other serotypes is suspected but not proven. Some of these induce cardiac lesions in monkeys similar to those found in man (Abelmann 1973, Lerner and Wilson 1973, Grist 1977).

The relation of enteroviral infection to chronic cardiac diseases is uncertain. There is no doubt that enteroviruses cause some cases of acute myocarditis or pericarditis and that these diseases sometimes progress to chronic congestive cardiomyopathy or constrictive pericarditis; but whether the chronic cases are those which start as acute enteroviral infections is difficult to prove. There are many reports of chronic cardiac disease following presumptive enteroviral myocarditis or pericarditis in which diagnosis was based on virus excretion or serological evidence (Strachan 1963, Sainani *et al.* 1968, Howard and Maier 1968, Matthews *et al.* 1970, Smith 1970, Levi *et al.* 1977), but none in which the diagnosis was unequivocally established by isolation of virus from pericardial fluid or cardiac biopsy. Retrospective diagnosis after chronic disease is established has been attempted, but it may then be too late to identify the cause. There is however one report of the demonstration of Coxsackievirus B4 antigen in the myocardium by immunoelectronmicroscopy in a man who died a year after an acute pancarditis. A 64-fold rise in Coxsackievirus B4 serum antibody occurred during the course of the illness (Lerner *et al.* 1975). Persistent Coxsackievirus B antigens have been detected in the heart by immunofluorescence by Burch and his colleagues (1967), but their findings are unconfirmed. Higher titres of Coxsackievirus B neutralizing antibodies have been found in patients with chronic cardiomyopathy than controls (Kawai 1971, Cambridge *et al.* 1979); but some investigators have found no difference (Sanders 1963, Fletcher *et al.* 1968). Higher antibody titres have been associated with patients giving a shorter clinical history and the negative findings of earlier studies were based on patients with longer histories (Cambridge *et al.* 1979).

Neonatal infection Infections of the newborn are acquired before, during or after birth. Illness usually appears within two or three weeks (see Kibrick and Benirschke 1958, Kibrick 1961, 1964, Horstmann 1969, Grist *et al.* 1978). Infection may be symptomless or accompanied by illness of varying severity; often there is just a mild pyrexial or respiratory illness, or a gastro-intestinal disturbance. Severe generalized fatal infections are most frequently caused by group B Coxsackieviruses or ECHOvirues 9, 11 or 19. The illness is often predominantly a myocarditis, but sometimes a meningo-encephalitis. At autopsy there is usually severe inflammation of the myocardium; there may also be pericarditis and meningo-encephalitis and necrotizing or haemorrhagic lesions in the liver, bone marrow, pancreas and adrenals. Outbreaks of mild or severe illness have occurred in nurseries and special-care baby units in which infection was probably introduced by a mother or a member of the staff who may themselves have had a minor illness or subclinical infection. Spread of infection in these institutions can be rapid, extensive and disastrous.

Conjunctivitis Enterovirus infections of the upper respiratory tract may include mild conjunctivitis, but enteroviral outbreaks in which conjunctivitis is the predominant manifestation of infection were not reported before 1969. In 1969 to 1971, epidemics of *acute haemorrhagic conjunctivitis* affected large numbers of persons, mainly in Africa, South-East Asia, Japan and India. Infections were caused by Coxsackievirus A24 in some outbreaks (Lim and Yin-Murphy 1971), and by a new serotype designated enterovirus 70 (Mirkovic *et al.* 1973) in others. The conjunctivitis due to Coxsackievirus A24 varies in severity. Subconjunctival haemorrhages occur in only a minority of cases. That due to enterovirus 70 has a more sudden onset, with an incubation period of about 24 hours; subconjunctival haemorrhage is its main feature. In enterovirus 70 infections, transient corneal affection and lumbar radiculomyelopathy have been rare complications. The virus is isolated from conjunctival swabs or scrapings, but not from faeces. The very rapid spread of infection among adults suggests that it is transmitted by tears or upper respiratory secretions. In 1977, ECHOvirus 7 was isolated from five patients in an epidemic of conjunctivitis in Sweden (Sandelin *et al.* 1977).

Congenital abnormalities The relation of enterovirus infection during pregnancy to the development of congenital abnormality has been studied in several investigations but results have been inconclusive. There is no doubt that intra-uterine infection sometimes occurs; documented cases have been in the perinatal period when effects are similar to those of neonatal infection (Horstmann 1969). When intra-uterine infection occurs in the first trimester, the cytolytic character of enterovirus infection might be expected to lead to fetal destruction and abortion rather than to developmental abnormality. However, Brown and his colleagues (Brown and Evans 1967, Brown and Karunas 1972), using serological methods, found more evidence of Coxackievirus B infection in neonates with developmental defects and in their mothers than in matched controls; but the differences were small, and other studies have shown no evidence of association with either Coxsackievirus B or ECHOvirus 9 infcction.

Pancreatitis Although group B Coxsackievirses regularly produce pancreatitis in adult mice, it has only rarely been observed in infections in man. The pancreas is often affected as part of a generalized infection in neonates (Kibrick and Benirschke 1958). A few cases of acute pancreatitis have been described in adults during infection with Coxsackieviruses B4 (Murphy and Simmul 1964) and B5 (Ursung 1973). Subclinical infection of the pancreas may however be much more common. Nakao (1971) found raised serum amylase concentrations in 31 per cent of Coxsackievirus B5 and 23 per cent of Coxsackievirus A infections, but not in ECHOvirus 4 or 6 infections; and evidence of elevated serum immunoreactive trypsin concentrations was found in patients with recent Coxsackievirus B infection (Gamble *et al.* 1979).

Insulin-dependent diabetes The possibility that virus infection is a cause of insulin-dependent diabetes has recently attracted much interest (see Craighead 1975, Notkins 1977, Gamble 1980). The ability of several picornaviruses to induce diabetes in animals has focused attention on these viruses; and several serological investigations in man have revealed a small excess of antibodies to Coxsackievirus B4 in some newly diagnosed cases, but this has not been confirmed in other studies (see Gamble 1980). However, in 1979, Notkins and his colleagues (see Yoon *et al.* 1979) identified Coxsackievirus B4 antigen by immunofluorescence in the islets of Langerhans of a child who died within a few days of the development of acute diabetes. Moreover the virus isolated from this patient induced diabetes in mice. Another fatal case investigated by Gladisch and associates (1976) was also probably associated with Coxsackievirus B4 infection, although the evidence was less complete. There seems little doubt therefore that this virus can invade the islet cells and cause diabetes, but how often it does so is unknown. The lack of any marked epidemicity of insulin-dependent diabetes excludes a frequent association with recent picornavirus infections of any one type, but association with a number of different viruses is a possibility. Other evidence favours a complex aetiology. Susceptibility is inherited and probably associated with an immunological abnormality (see Cudworth 1978). However, frequent discordance of diabetes in identical twins implies an environmental cause (Tattersall and Pyke 1972). The ability of picornaviruses to invade the islets of Langerhans and the evidence linking them with diabetes in animals make them plausible candidates.

Other diseases A number of other diseases have been reported in association with enterovirus infection in rare or isolated instances. Orchitis has been noted as an infrequent complication of Coxsackievirus B infection, most commonly in association with Bornholm disease; in one case Coxsackievirus B5 was isolated from testicular biopsy tissue (Craighead *et al.* 1962). Parotitis has been reported in several cases of Coxsackievirus A infection (Howlett *et al.* 1957) and in one case of Coxsackievirus B3 infection (Bertaggia *et al.* 1976). Several reports of the presence of picornavirus-like particles in muscle fibres in cases of chronic myositis and the isolation of Coxsackievirus A9 from muscle tissue in two such cases (Freudenberg *et al.* 1952, Tang *et al.* 1975) has suggested that enteroviruses may play some part in the aetiology of this disease.

Many investigations have attempted to establish a connection between enterovirus infection and diarrhoea (see Grist *et al.* 1978), usually with confused or negative results. Though it seems likely that diarrhoea is sometimes a symptom of Coxsackievirus and ECHOvirus infection, it is an inconstant symptom and other symptoms are usually more prominent. There is no evidence at present that any of the known enteroviruses causes outbreaks of diarrhoea or gastroenteritis in which diarrhoea or vomiting is the predominant symptom, but some of the small round viruses recently isolated from such outbreaks may eventually be classified as enteroviruses.

Diagnosis

Interpretation of the results of laboratory investigations of enterovirus infections is often difficult. The isolation of a virus from the throat or faeces is indicative of infection, but not necessarily of a causal association with the patient's illness. Simultaneous outbreaks of infection with several serotypes are usual, and patients may be infected with two or more viruses at the same time. Moreover, infection is often clinically inapparent (Nelson *et al.* 1967).

The isolation of a virus from affected patients may therefore have no significance, unless it can be shown that unaffected members of the community of the same age and social background are not similarly affected. The detection of virus in body fluids or tissues is stronger evidence of aetiological significance, particularly when involvement of the organs concerned is supported by clinical or histological evidence. Serological findings may be misleading. All the available tests on human sera are subject to some degree of non-specificity and, even with the highly specific neutralizing antibody test, anamnestic antibody responses may lead to ambiguity. Thus, a Coxsackievirus infection may provoke an antibody response not only to the infecting serotype, but also to any previously encountered Coxsackievirus or ECHOvirus. Though serological tests are valuable in epidemiological surveys, they are of less value in the investigation of individual patients, but may provide useful supporting evidence to virus isolation tests. They sometimes establish the time when infection occurred with greater precision than virus isolation. A rising antibody titre in paired serum samples is evidence that infection was systemic rather than a superficial colonization of the gut. In the remainder of this section the common diagnostic laboratory procedures are outlined, but comprehensive details are available elsewhere (Melnick *et al.* 1979).

For the isolation of viruses, specimens of faeces and throat swabs are usually examined. In appropriate cases, cerebrospinal fluid, pericardial fluid, vesicle fluid, nose and conjunctival swabs and autopsy specimens are required. After suitable preparation, specimens are inoculated into newborn mice and a range of tissue cultures. Mice should be less than 24–48 hours old and inspected daily for 14 days for evidence of disease. Group A viruses produce a flaccid paralysis, often confined to the hind limbs, but sometimes affecting the whole body. A characteristic foot-drop is often seen in the fore limbs. Group B viruses produce weakness and tremors followed by generalized spastic paralysis with spasm of the limbs, dyspnoea, cyanosis and death. The typical disease sometimes develops only after one or more passages of the virus. For cultivation in tissue culture, the use of two or more cell types extends the range of viruses isolated. Primary monkey kidney and human embryonic kidney have the broadest range of sensitivity, but human amnion and thyroid, and human cell lines such as W1-38, are often used. Apart from Coxsackieviruses A7, 9 and 21, most group A viruses grow poorly or not at all in tissue culture. RD cells, a cell line derived from a human rhabdomyosarcoma (Schmidt *et al.* 1975), has extended the range of group A viruses growing in tissue culture, but their sensitivity is inferior to that of newborn mice. The characteristic cytopathic effect of enteroviruses is the same in all susceptible cell types, namely shrinkage and rounding up of the cells, with increased refractility and nucear pyknosis. In ECHOvirus 22 and 23 infection, changes have been described in the nucleus in which balloon-like vesicles appear (Shaver *et al.* 1961), but other enteroviruses cannot be distinguished by their cytopathology in cell monolayers. Under an agar overlay, plaques are formed in sensitive cell monolayers, and plaque morphology has been used as an as an aid to identification (see Melnick 1965).

Virus identification is most reliably established by neutralization with type-specific antisera. For economy of labour and reagents, these may be incorporated in a number of serum pools in combinations so designed that the identity of the virus can be inferred from the pattern of neutralization in tests against the pools (Lim and Benyesh-Melnick 1960, Schmidt *et al.* 1961). By this method any one of 42 enteroviruses can be identified in tests against only eight serum pools. Mixtures of two or more viruses are not neutralized by single antisera and inconclusive patterns are produced in tests using serum pools; in such cases purification by terminal dilution or plaque selection is necessary. Some strains of virus are poorly neutralized by antisera prepared against prototype viruses. These may be 'prime strains', which are variants with a broader antigenic constitution than the prototype strain. Their identification requires cross-neutralization tests with antisera prepared against both wild and prototype strains. Virus aggregation is another cause of poor neutralization. It has been found, for example, that conventionally prepared suspensions of the ECHOvirus 4 prototype strain (Pesascek) contain viral aggregates and are poorly neutralized by homologous antiserum. After removal of aggregates by filtration through millipore membranes, the 'monodispersed' virus suspensions are neutralized normally. Other strains of ECHOvirus 4, such as the 'du Toit' strain, do not form aggregates to an appreciable extent, and this strain is preferred for use in neutralizing-antibody tests (Wallis and Melnick 1967). Haemagglutination-inhibition, complement-fixation, immunofluorescence and gel diffusion tests have all been used for virus identification. Although their use may occasionally save time, the greater specificity of the neutralization test makes it the most reliable procedure.

The profusion of enterovirus serotypes precludes the use of serological tests for routine diagnosis in individual cases, unless the virus responsible for the infection has been identified. A few diseases—notably pleurodynia, hand-foot-and-mouth disease, and acute haemorrhagic conjunctivitis—are so regularly associated with particular serotypes viz. Coxsackievirus B1-6, Coxsackievirus A16 and enterovirus 70, respectively, that their serological diagnosis is often feasible. In other cases investigation is best restricted to virus isolation. The neutralizing-antibody test against a constant virus dose, with the cytopathic endpoint, plaque reduction, pH change, or mouse pathogenicity as an index, is widely preferred. Cross-reactivity of the viruses is limited and well documented, and a rising titre is evidence of recent infection, but the occurrence of heterotypic anamnestic antibody responses allows only a presumptive diagnosis of the type of the infecting virus, unless it has been isolated and identified. Neutralizing antibodies generally appear within two weeks of infection but they may be delayed, particularly in infections with some group A Coxsackieviruses in which an antibody rise may not be detected for four or more weeks after infection. Antibody generally persists for at least several years, sometimes at high titre. A high titre therefore in a single specimen of serum cannot be regarded as evidence of recent infection, however high it may be. Many enteroviruses agglutinate erythrocytes, and a haemagglutination-inhibition test has been used in serological studies. Coxsackieviruses of types A20, A21, A24, B1, B3, B5, and B6, and 14 of the 30 ECHOvirus types agglutinate human group 0 cells whereas Coxsackievirus A7 agglutinates vaccinia-agglutinable fowl cells. Antibodies to these viruses may be titrated in paired serum specimens by a conventional haemagglutination-inhibition test. Human serum often contains non-specific inhibitors which must be removed by preliminary absorption. Haemagglutination-inhibiting antibodies usually appear within two or three weeks of infection and persist for at least six months. Although heterotypic antibody responses are common, the test provides similar information to the neutralizing-antibody test.

A **passive haemagglutination test** for the assay of antibody to a 'group-specific' enterovirus antigen has also been described. Antibodies are detected by their ability to agglutinate guinea-pig red cells coated with an enterovirus group antigen, which is usually but not necessarily derived from ECHOvirus 11. The test lacks type specificity and rising titres are found in paired serum specimens from patients with enterovirus infection irrespective of the type of the virus. This appears to be a simple and useful test, but requires further evaluation. The complement-fixation test lacks specificity but may be useful as another simple group-specific test. If diseases like myocardiopathy are late manifestations of enterovirus infection, laboratory investigation may be possible only when antibodies have already developed and when virus excretion is diminishing. At this stage the demonstration of virus-specific antibodies in the IgM immunoglobulin fraction may offer the best prospect of establishing a diagnosis. This approach has not been much exploited but Schmidt and her colleagues (1973) were able to detect IgM antibody to group B Coxsackieviruses in 27 per cent of 259 patients with pericarditis, myocarditis, or pleurodynia. These workers relied mainly on a gel diffusion test to detect IgM antibody, but several alternative techniques have been described.

Prophylaxis

The success of the poliovirus vaccination campaign shows that prophylaxis for other enteroviruses is certainly feasible, but has not so far been contemplated because of the profusion of serotypes and their low morbidity. The most serious illnesses generally occur in neonates and very young infants, and outbreaks in nurseries and special care baby units can be devastating. In these circumstances, closure of the unit to new admissions until the outbreak subsides is an important preventive measure. The demonstrable effects of hygienic conditions in the epidemiology of enteroviral infection suggests that hygienic precautions in handling infants may usefully contribute to their protection from infection. The administration of attenuated poliovirus vaccine as an interfering agent has been tried by Farmer and Patten (1968); although its efficacy was not proven, it may be a useful measure, as may the administration of human pooled immunoglobulin, to prevent viraemia.

References

Abelmann, W. H. (1973) *Annu. Rev. Med* **24,** 145.

Ackermann, W. W., Payne, F. E. and Kurtz, H. (1958) *J. Immunol.* **81,** 1.

Anderson, G. W. and Rondeau, J. L. (1954) *J. Amer. med. Ass.* **155,** 1123.

Andrewes, C. H. and Horstmann, D. M. (1949) *J. gen. Microbiol.* **3,** 290.

Archetti, I. *et al.* (1956) *Sci. Med. Ital.* **5,** 321.

Armstrong, C. (1939) *Publ. Hlth Rep., Wash.* **54,** 1719; (1950) *Amer. J. publ. Hlth* **40,** 1296.

Assaad, F. and Cockburn, W. C. (1972) *Bull. World Hlth Org.* **46,** 329.

Assaad, F. A, Cockburn, W. C. and Perkins, F. T. (1976) *Bull. World Hlth Org.* **53,** 319.

Aycock, W. L. (1941) *Viral and Rickettsial Diseases* p. 555. Harvard University Press, Cambridge, Mass.; (1942) *Medicine* **21,** 65.

Baltimore, D. (1968) In: *Medical and Applied Virology.* p. 340. St. Louis, Missouri; (1971) *Perspect. Virol.* **7,** 1.

Baltimore, D., Huang, A., Manly, K. F., Rekosh, D. and Stampfel, M. (1971) *Strategy of the Viral Genome.* p. 101. Churchill Livingstone, Edinburgh.

Banker, D. O. and Melnick, J. L. (1951) *Amer. J. Hyg.* **54,** 383.

Behbehani, A. M, Sulkin, S. E. and Wallis, C. (1962) *J. infect Dis.* **110,** 147.

Bell, J. A. *et al.* (1961) *Amer. J. Hyg.* **74,** 267.

Berg, G. (1967) *Transmission of Viruses by the Water Route.* Wiley, New York.

Bertaggia, A., Meneghetti, F. and Carretta, M. (1976) *G. Mal. infett.* **28,** 188.

Bloom, H. H.,Mack, W. N. and Krueger, B. J. and Mallmann, W. L. (1959) *J. infect. Dis.* **105,** 61.

Bodian, D. (1948) *Bull. Johns. Hopk. Hosp.* **83,** 1; (1949) *Amer. J. Med.* **6,** 563.

Bodian, D. and Horstmann, D. M. (1965) *Viral and Rickettsial Infections of Man* 4th edn., p. 430. Pitman Medical, London.

Bodian, D. and Howe, H. A. (1940) *Brain* **63,** 135; (1941) *Bull. Johns. Hopk. Hosp.* **68,** 58.

Bodian, D., Morgan, I. M. and Howe, H. A. (1949) *Amer. J. Hyg.* **49,** 234.

Boring, W. D. and Ungar, R. F. (1962) *J. Bact.* **83,** 694.

Boring, W. D., Zurhein, G. M. and Walker, D. L. (1956) *Proc. Soc. exp. Biol. N.Y.* **93,** 273.

Bradley, W. H. and Richmond, A. E. (1953) *Mon. Bull. Minist. Hlth Lab. Serv.* **12,** 2.

Breindl, M. (1971) *J. gen. Virol.* **11,** 147.

Brown, F., Goodridge, D. and Burrows, R. (1976) *J. comp. Path.* **86,** 409.

Brown, F., Talbot, P. and Burrows, R. (1973) *Nature, Lond.* **245,** 315.

Brown, F. and Wild, T. F. (1974) *Intervirology* **3,** 125.

Brown, G. C. and Evans, T. N. (1967) *J. Amer. med. Ass.* **199,** 183.

Brown, G. C. and Karunas, R. S. (1972) *Amer. J. Epidem.* **95,** 207.

Bucknall, R. A. (1967) *J. gen. Virol.* **1,** 89.

Burch, G. E., Sun, S. C., Colcolough, H. L., Sohal, R. S. and De Pasquale, N. P. (1967) *Amer. Heart J.* **74,** 13.

Burnet, F. M. and Macnamara, J. (1931) *Brit. J. exp. Path.* **12,** 57.

Cambridge, G. MacArthur, G. G. C., Waterson, A. P., Goodwin, J. F. and Oakley, C. M. (1979) *Brit. Heart J.* **41,** 692.

Cherry, F. J. and Nelson, D. B. (1966) *Clinical Pediatrics* **5,** 659.

Cherry, J. D. and Jahn, C. L. (1966) *Pediatrics, Springfield* **37,** 637.

Christie, A. B. (1974) *Infectious Diseases: Epidemiology and Clinical Practice,* 2nd edn., p. 535. Churchill Livingstone, Edinburgh.

Chumakov, M. P., Voroshilova, M. K., Zhevandrova, V. I., Mironova, L. L., Itzelis, F. I. and Robinzon, I. A. (1956) *Probl. Virol.* **1,** 16.

Clemmer, D. I., Li, F., Le Blanc, D. R., and Fox, J. P. (1966) *Amer. J. Epidem.* **83,** 123.

Cole, R. M., Bell, J. A., Beeman, E. A. and Huebner, R. J. (1951) *Amer. J. publ. Hlth* **41,** 1342.

Cossart, Y. E. (1967) *J. Hyg. Camb.* **6,** 67; (1977) *Brit. med. J.* **ii,** 1621.

Craighead, J. E. (1975) *Progr. med. Virol.* **19,** 161.

Craighead, J. E., Mahoney, E. M., Carver, D. H., Naficyk, K. and Fremont-Smith, P. (1962) *New Engl. J. Med.* **267,** 498.

Crowell, R. L. and Philipson, L. (1971) *J. Virol.* **8,** 509.

Crowell, R. L. and Siak, J. S. (1978) *Perspect. Virol.* **10,** 39.

Cudworth, A. G. (1978) *Diabetologia* **14,** 281.

Dales, S. (1965) *Progr. med. Virol.* **7,** 1.

Dalldorf, G. (1949) *N.Y. St. J. Med.* **49,** 1330; (1957) *J. exp. Med.* **106,** 69.

Dalldorf, G. and Albrecht, R. (1955) *Proc. nat Acad. Sci., Wash.* **41,** 978.

Dalldorf, G. and Gifford, R. (1952) *J. exp. Med.* **96,** 491.

Dalldorf, G. and Sickles, G. M. (1948) *Science,* **108,** 61.

Debré, R. *et al.* (1964) *Bull World Hlth Org.* **30,** 663.

Dippe, S. E., Bennett, P. H., Miller, M., Maynard, J. E. and Berquist, K. R. (1975) *Lancet* **i,** 1314.

Domok, I. (1974) *Bull. World. Hlth Org.* **51,** 333.

Dunnebacke, T. H., Levinthal, J. O. and Williams, R. C. (1969) *J. Virol.* **4,** 505.

Eggers, H. J. and Sabin, A. B. (1961) *J. exp. Med.* **110,** 951.

Eggers, H. J. and Tamm, I. (1963) *Nature, Lond.* **199,** 513.

Enders, J. F., Weller, T. H., and Robbins, F. C. (1949) *Science,* **109,** 85.

Esiri, M. M. (1980) *Clin. exp. Immunol.* **40,** 42.

Evans, A. D. and Waddington, E. (1967) *Brit. J. Dermat.* **79,** 309.

Farmer, K. and Patten, P. T. (1968) *N.Z. med. J.* **68,** 86.

Faulkner, R. S., McLeod, A. J. and van Rooyen, C. E. (1957) *Canad. med. Ass. J.* **77,** 439.

Federici, E. E., Lerner, A. M. and Abelmann, W. H. (1963) *Proc. Soc. exp. Biol., N.Y.* **112,** 672.

Fletcher, G. F., Coleman, M. T., Feorino, P. M., Marine, W. M. and Wenger, N. K. (1968) *Amer. J. Cardiol.* **21,** 6.

Flexner, S. and Lewis, P. A. (1910) *J. exp. Med.* **12,** 227.

Fox, J. P. (1964) *Amer. J. publ. Hlth* **54,** 1134.
Fox, J. P., Gelfand, H. M., Leblanc, D. R., Potash, L., Clemmer, D. I. and Lapenta, D. (1961) Papers and Dicussions *Int. Polio Congr.* **5** p. 368. J. B. Lippincott Co., Philadelphia.
Freudenberg, V. E., Roulet, F. and Nicole, R. (1952) *Ann. Paediatr.* **178,** 150.
Freyche, M. J. and Nielsen, J. (1955) *WHO Monograph No. 26*, pp. 59–106. WHO, Geneva.
Froeschle, J. E., Feorino, P. M. and Gelfand, M. (1966) *Amer. J. Epidem.* **83,** 455.
Frommhagen, L. (1965) *J. Immunol.* **95,** 818.
Fujamiya, Y., Oyama, S., Numazaki, Y. and Ishida, N. (1974) *Jap. J. Microbiol.* **18,** 379.
Gamble, D. R. (1979) *Lancet* **i,** 425; (1980) *Epidem. Rev.* **2,** 49.
Gamble, D. R., Moffatt, A. and Marks, V. (1979) *J. clin. Path.* **32,** 897.
Garland, A. J. M. and Mann, J. A. (1974) *J. Hyg. Camb.* **73,** 85.
Gear, J. H. S. (1948) *Proc. 4th int. Congr. trop. Med. Malar., Wash. D.C.* **1,** 555;(1952) *Second Int. Polio. Conf. Philadelphia*, J. B. Lippinott Co., Philadelphia; (1955) *WHO Monograph No. 26*, pp. 31–58. WHO, Geneva.
Gelfand, H. M. (1959) *Sth. med. J.* **52,** 819; (1961) *Progr. med. Virol.* **3,** 193.
Gelfand, H. M., Fox, J. P., and Leblanc, D. R. (1957) *Amer. J. trop. Med.* **6,** 521.
Gelfand, H. M., Holguin, A. H., Marchetti, G E. and Feorino, P. M. (1963) *Amer. J. Hyg.* **78,** 358.
Ghendon, Y. Z. and Sanakoyeva, I. I. (1961) *Acta virol.* **5,** 265.
Gifford, R. and Dalldorf, G. (1951) *Amer. J. Psth.* **27,** 1047.
Gladisch, R., Hoffman, W. and Waldherr, R. (1976) *Z. Kardiol.* **65,** 837.
Godman, G. C., Bunting, H. and Melnick, J. L. (1952) *Amer. J. Path.* **28,** 223.
Goldfield, M., Srihongse, S. and Fox, J. P. (1957) *Proc. Soc. exp. Biol., N.Y.* **96,** 788.
Graves, J. H. (1973) *Nature, Lond.* **245,** 314.
Grist, N. R. (1960) *Lancet* **i,** 1054; (1977) *Recent Advances in Clinical Virology* No. 1, p. 141. Churchill Livingstone, Edinburgh.
Grist, N. R. and Bell, E. J. (1970) *Arch. environm. Hlth* **21,** 382.
Grist, N. R., Bell, E. J., Assaad, F. (1978) *Progr. med. Virol.* **24,** 114.
Grodums, E. I. and Dempster, G. (1959) *Canad. J. Microbiol.* **5,** 605.
Habel, K., and Loomis, L. N. (1957) *Proc. Soc. exp. Biol., N.Y.* **95,** 597.
Hale, J. H., Doraisingham, M., Kanagaratnam, K., Leong, K. W. and Montiero, E. S. (1959) *Brit. med. J.* **i,** 1541.
Hammon, W. M. *et al.* (1954) *J. Amer. med. Ass.* **156,** 21.
Hawley, H. B., Morin, D. P., Gerachty, M. E., Tomkow, J., and Phillips, C. A., (1973) *J. Amer. med. Ass.* **226,** 33.
Hayflick, L. and Moorhead, P. S. (1961) Exp. Cell Res. **25,** 585.
Heathfield, K. W. G., Pilsworth, R., Wall, B J. and Corsellis, J. A. N., (1967) *Quart J. Med.* **36,** 579.
Hemmes, J. H., Winkler, K. C. and Kool, S. M. (1962) *Ant. v. Leeuwenhoek.* **28,** 221.
Hierholzer, J. C., Hilliard, K. A. and Esposito, J. J. (1975) *Amer. J. Epidem.* **102,** 533.
Hill, A. B. and Knowelden, J. (1950) *Brit. med. J.* **ii,** 1.
Holland, J. J. (1961) *Virology* **15,** 312; (1964) *Bact. Rev.* **28,** 3.
Horstmann, D. M. (1950) *J. Amer. med. Ass.* **142,** 236; (1953) *Bull. NY. Acad. Med.* **29,** 910; (1969) *Yale, J. Biol. Med.* **42,** 99.
Howard, E. J. and Maier, H. C. (1968) *Amer. Heart. J.* **75,** 247.
Howe, H. A. and Bodian, D. (1941) *Bull. Johns Hopk. Hosp.* **69,** 149.
Howlett, J. G., Somlo, F. and Kalz, F. (1957) *Canad. med. Ass. J.* **77,** 5.
Itoh, H., and Melnick, J. L., (1957 *J. exp. Med.* **106,** 677.
John, T. J. and Jayabal, P. (1972) *Amer. J. Epidem.* **96** 263.
Johnson, K. M., Bloom, H. H., Chanock, R. M., Mufson, M. A. and Knight, V. (1962) *Amer. J. publ. Hlth* **52,** 93.
Johnsson, T. (1954) *Arch. ges. Virusforsch.* **5,** 384.
Joklik, W. K. and Darnell, J. E. (1961) *Virology* **13,** 439.
Jungeblut, C. W. and Sanders, M. (1940) *J. exp. Med.* **72,** 407.
Kamitzu, M , Kasamaki, A., Ogawa, M., Kasahara, S. and Imamura, M. (1967) *Jap. J. med. Sci. Biol.* **2,** 175.
Kaplan, A. S. and Melnick, J. L. (1952) *Amer. J. publ. Hlth* **42,** 525.
Karzon, D. T., Eckert, G. L., Barron, A. L., Hayner N. S. and Winkelstein, W. (1961) *Amer. J. Dis. Child.* **101,** 601.
Katz, M. and Plotkin, S. A. (1967) *Amer. J. publ. Hlth* **57,** 1837.
Kawai, C. (1971) *Jap. Circulation Journal* **35,** 765.
Kelly, S., Clark, M. E and Coleman, M. B. (1955) *Amer. J. publ. Hlth* **45,** 1348.
Kibrick, S. (1961) *Perspect. Virl.* **2,** 140; (1964) *Progr. med. Virol.* **6,** 27.
Kibrick, S. and Benirschke, K. (1958) *Paediatrics, Springfield* **22,** 857.
Kilbourne, E. D., Wilson, C. B. and Perrier, D. (1956) *J. clin. Invest.* **35,** 362.
Kling, C., Levaditi, C. and Hornus, G. (1934) *Bull. Acad. nat. med.* **111,** 709.
Kogon, A. *et al.* 1969 *Amer. J. Epidem.* **89,** 51.
Koprowski, H. (1956) *Amer. J. trop. Med.* (Hyg) **5,** 440; (1960) *Trans. Stud. Coll. Phycns. Philad.* **27,** 95.
Koprowski, H., Jervis, G. A. and Norton, T. W. (1952) *Amer. J. Hyg.* **55,** 108.
Kunin, C. M. (1962) *J. Immunol.* **88,** 556.
Landsman, J. B. and Bell, E. J. (1962) *Brit med. J.* **i,** 12.
Landsteiner, K. and Popper, E. (1909) *Z. Immunforsch.* **2,** 377.
Lavinder, C. H., Freeman, S. W. and Frost, W. H. (1918) *USPHS Hyg. Lab. Bull.* No. 91.
Lehner, T. and Barnes, C. G. (1979) *Behçet's Syndrome.* Academic Press, London.
Lennette, E. H., Magoffin, R. L. and Knouf, E. G. (1962) *J. Amer. med. Ass.* **179,** 687.
Lennette, E. H. and Schmidt, N. J. (1957) *Amer. J. Hyg.* **65,** 210.
Lerner, A. M. (1965) *Progr. med. Virol.* **7,** 97.
Lerner, A. M., Levin, H. S. and Finland, M. (1962) *J. exp. Med.* **115,** 745.
Lerner, A. M. and Wilson, F. M. (1973) *Progr. med. Virol.* **15,** 63.
Lerner, A. M., Wilson, F. M. and Reyes, M. P. (1975) *Mod.-Conc.-Cardiov. Dis.* **44,** 11.
Levi, G. F., Proto, C., Quadri, A. and Ratti, S. (1977) *Amer. Heart J.* **93,** 419.

Lim, K. A. and Benyesh-Melnick, M. (1960) *J. Immunol.* **84,** 309.
Lim, K. A. and Yin-Murphy, M. (1971) *Singapore med. J.* **12,** 247.
Liu, O. C., Seraichekas, H. F., Akin, E. W. Brashear, D. A., Katz, E. L. and Hill, J. R. (1971) *Proc. 13th Water Quality Conf.* p. 171. Urbana-Champaign: Univ. Illinois.
Lonberg-Holm, K., Crowell, R. L. and Philipson, L. (1976) *Nature. Lond.* **259,** 679.
Loria, R. M., Kibrick, S. and Broitman, S. A. (1974) *J. infect. Dis.* **130,** 225.
Lou, T. Y., and Wenner, H. A. (1962) *Arch. ges. Virusforsch.* **12,** 233, 303.
Lou, T. Y, Wenner, H. A. and Kamitsuka, P. S. (1960) *Arch. ges. Virusforsch.* **10,** 451.
Lwoff, A. (1959) *Bact. Rev.* **23,** 109.
McCartney, J.E. (1924) *J. exp. Med.* **39,** 51.
McCloskey, B. P. (1950) *Lancet* **i,** 659.
McGregor, S. and Mayor, H. D. (1971) *J. gen. Virol.* **10,** 203.
McLean, D. M., Larke, R. P. B., McNauthton, G. A., Best, J. M. and Smith, P. (1965) *Canad. med. Ass. J.* **92,** 658.
McNair Scott (1961) *Advanc. Virus Res.* **8,** 165.
Marine, W. M., Chin, T. D. Y. and Gravelle, C. R. (1962) *Amer. J. Hyg.* **76,** 173.
Matthews, J. D., Cameron, S. J. and George, M. (1970) *Thorax* **25,** 624.
Medin, O. (1891) *Int. med. Kongr. 1890, Berlin* 2, Abt. **6,** p. 37.
Medrano, L. and Green, H. (1973) *Virology* **54,** 515.
Melnick, J. L. (1947) *Amer. J. Hyg.* **45,** 240; (1957) *Cellular Biology, Nucleic Acids and Viruses* Vol. 5. p. 365. Special Publ. New York Acad. Sci; (1958) *Amer. J. publ. Hlth* **48,** 1170; (1965) *Viral and Rickettsial Infections of Man*, 4th edn., p. 513. Pitman Medical, London; (1976) *Viral Infections of Humans. Epidemiology and Control*, p. 163. Plenum Publ. Corp., New York.
Melnick, J. L., Emmons, J., Coffy, J. H. and Schoof, H. (1954) *Amer. J. Hyg.* **59,**164.
Melnick, J. L. and Ledinko, N. (1950) *J. Immunol.* **64,** 101; (1953) *Amer. J. Hyg.* **58,** 207.
Melnick, J. L. and Penner, L. R. (1952) *J. exp. Med.* **96,** 255.
Melnick, J. L., Shaw, E. W. and Curnen, E. C. (1949) *Proc. Soc. exp. Biol., N.Y.* **71,** 344.
Melnick, J. L., Wenner, H. A. and Phillips, C. A. (1979) *Diagnostic Procedures for Viral, Rickettsial and Chlamydial Infections.* 5th edn., p. 471. Amer. publ. Hlth Ass. Inc., Washington D.C.
Minkowitz, S. and Berkovich, S. (1970) *Arch. Path.* **89,** 427.
Mirkovic, R. R., Kono, R., Yin-Murphy, M., Sohier, R., Schmidt, N. J. and Melnick, J. L. (1973) *Bull. Wld Hlth. Org.* **49,** 341.
Mitchell, S. C. and Dempster, G. (1955) *Canad. med. Ass. J.* **72,** 117.
Morgan, I., Howe, H. A. and Bodian, D. (1947) *Amer. J. Hyg.* **45,** 390.
Mountain, I. M. and Alexander, H. E. (1959) *Proc. Soc. exp. Biol., N.Y.* **101,** 527.
Murphy, A. M. and Simmul, R. (1964) *Med. J. Aust.* **ii,** 443.
Nakano, J. H. and Gelfand, H. M. (1962), *Amer. J. Hyg.* **75,** 363.
Nakao, T. (1971) *Lancet.* **ii,** 1423.
Nathanson, N and Langmuir, A. D. (1963) *Amer. J. Hyg.* **78,** 16.
Nathanson, N. and Martin, J. R. (1979) *Amer. J. Epidem.* **110,** 672.
Nelson, D., Hiemstra, H., Minor, T. and D'Alessio, D. (1979) *Amer. J. Epidem.* **109,** 352.
Nelson, D. B., Circo, R. and Evans, A. S. (1967) *Amer. J. Epidem.* **86,** 641.
Neva, F. A. (1956) *New Engl. J. Med.* **254,** 383.
Neva, F. A., Feemster, R. F. and Gorbach, I. J. (1954) *J. Amer. med. Ass.* **155,** 544.
Notkins, A. L. (1977) *Arch. Virol.* **54,** 1.
Ogra, P. L. (1970) *Infection and Immunity* **2,** 150; (1971) *New Engl. J. Med.* **284,** 59.
Oliver, J. (1922) *J. Infect. Dis.*, **30,** 91.
Pappenheimer, A. M., Kunz, L. J. and Richardson, S. (1951) *J. exp. Med.* **94,** 45.
Parks, W. P., Quieroga, L. T. and Melnick, J. L. (1967) *Amer. J. Epidem.* **85,** 469.
Parsons, R., Bynoe, M. L. and Pereira, M. S. (1960) *Brit. med. J.* **i,** 1776.
Paul, J. R. (1955) *WHO Monograph No. 26*, pp. 9–30 WHO, Geneva; (1958*a*) *Bull. World Hlth Org.* **19,** 747; (1958*b*) *Science* **127,** 1062.
Peart, A. F. W. (1949) *Canad. J. publ. Hlth* **40,** 405.
Rager-Zisman, B. and Allison, A. C. (1973*a*) *J. gen. Virol.* **19,** 329; (1973*b*) *J. gen. Virol.* **19,** 339.
Ramos-Alverez, M., Gomez Santos, F., Rangel Rivera, L. and Mayes, O. (1959) *First Int. Conf. Live Poliovirus Vaccines, Pan American Sanitary Bureaux, Washington, D.C.* Scientific Publication No. 44 p. 483.
Report (1974) *J. Amer. med. Ass.*, 228, 555.
Report (1950) *Poliomyelitis. A survey of the outbreak in Scotland in 1947.* HMSO Edinburgh; (1951) *Amer. J. Hyg.* **54,** 191; (1955*a*) *Science* **122,** 1187.
(1955*b*) *Amer. J. publ. Hlth* **45,** No. 5, Pt. 2; (1956) Regulations for Biological Products. Additional Standards. Poliomyelitis Vaccine 4922. *U.S. Public Health Service*; (1957) *Brit. med. J.* **i,** 1271; (1971) *Wkly. epidem. Rec.* **46,** 337; (1979). *Ibid.* **54,** 363.
Robinson, C. R., Doane, F. W. and Rhodes, A. J. (1958) *Canad. med. Ass. J.* **79,** 615.
Rosen, L., Melnick, J. L., Schmidt, N. J. and Wenner, H. A. (1970) *Arch. ges. Virusforsch.* **30,** 89.
Ross, M. E. and Notkins, A. L. (1974) *Brit. med. J.* **ii,** 226.
Rueckert, R. R. (1971) *Comparative Virology*, p. 256. Academic Press, New York.
Russell, W. R. (1949) *Brit. med. J.* **i,** 465.
Sabin, A. B. (1949) In: *Poliomyelitis. Papers and Discussions presented at the First International Poliomyelitis Conference. Philadelphia*, pp. 3–33. Lippincott; (1959) *Brit. med. J.* **i,** 663; (1963) *Trop. geogr. Med.* **15,** 38.
Sabin, A. B., Hennessen, W. A. and Winsser, J. (1954) *J. exp. Med.* **99,** 551.
Sabin, A. B., Krumbiegel, E. R. and Wigand, R. (1958) *Amer. J. Dis. Child.* **96,** 197.
Sainani, G. S., Krompotic, E. and Slodki, S. J. (1968) *Medicine, Baltimore* **47,** 133.
Salk, J. E. (1958) *J. Amer. med. Ass.* **167,** 1.
Salk, J. E., Lewis, L. J., Bennett, B. L., Ward, E. N., Krech, U., Youngner, J. S. and Bazeley, P. L. (1955) *Amer. J. publ. Hlth* **45,** 151.
Sandelin, K. Tuomioja, M. and Erkkila, H. (1977) *Scand. J. infect. Dis.* **9,** 71.
Sanders, V. (1963) *Arch. intern. Med.* **112,** 661.

Sato, S., Sato, H., Kawana, R. and Matumoto, M. (1972) *Japan, J. med. Sci. Biol.* **25,** 355.
Schmidt, N. J., Dennis, J., Frommhagen, L. and Lennette, E. H. (1963) *J. Immunol.* **90,** 654.
Schmidt, N. J., Guenther, R. W. and Lennette, E. H. (1961) *J. Immunol.* **87,** 623.
Schmidt, N. J., HO, H. H. and Lennette, E. H. (1975) *J. clin. Microbiol.* **2,** 183.
Schmidt, N. J., Magofiin, R. L., Lennette, E. H. (1973) *Infect. Immun.* **8,** 341.
Seal, S. C. and Gharpure, P. V. (1961) *Bull. Wld Hlth Org.* **24,** 123.
Seddon, J. H. and Duff, M. F. (1971) *N.Z. med. J.* **74,** 368.
Sells, C. J. Carpenter, R. L. and Ray, C. G. (1975) *New Engl. J. Med.* **293,** 1.
Serfling, R. E. and Sherman, I. L. (1953) *Publ. Hlth Rep., Wash.* **68,** 453.
Shaver, D. N., Barron, A. L. and Karzon, D. T. (1961) *Proc. Soc. exp. Biol. N.Y.* **106,** 648.
Shelekov, A., Habel, K. and Mackinstry, D. H. (1959) *Proc. N.Y. Acad. Sci.* **61,** 998.
Siegel, M., Greenberg, M. and Bodian, J. (1957) *New Engl. J. Med.* **257,** 998.
Smith, W. G. (1970) *Amer. Heart. J.* **80,** 34.
Spicer, C. C. (1959) *Brit. J. prev. soc. Med.* **13,** 139; (1961) *J. Hyg. Camb.* **59,** 143.
Spiegelmann, S. *et al.* (1968) *Cold Spring Harbor Symp. quant. Biol.* **33,** 101.
Steigman, A. J., Kokko, U. P. and Braspennickx, H. (1962) *J. Pediat.* **61,** 331.
Strachan, R. W. (1963) *Scot. med. J.* **8,** 402.
Sun, S. C., Sohal, R. S., Burch, G. E., Chu, K. C. and Colcolough, H. L. (1967) *Brit. J. exp. Path.* **48,** 655.
Sylvest, E. (1934) *Epidemic Myalgia: Bornholm Disease.* Levin and Munksgaard, Copenhagen.
Syverton, J. T., Brunner, K. T., Tobin, J. O'H. and Cohen, M. M. (1956) *Amer. J. Hyg.* **64,** 74.
Tang, T. T., Sedmar, G. V., Siegesmund, K. A. and McCreadie, S. R. (1975) *New Engl. J. Med.* **292,** 608.
Tattersall, R. B. and Pyke, D. A. (1972) *Lancet* **ii,** 1120.
Theiler, M. (1957) *Proc. Soc. exp. Biol. N.Y.* **96,** 380.
Tomasi, T. B. and Bienenstock, J. (1968) *Advanc. Immunol.* **9,** 1.
Ursung, B. (1973) *Brit. med. J.* **iii,** 524.
Utz, J. P. and Shelekov, A. I. (1958) *J. Amer. med. Ass.* **168,** 264.
Vandeputte, M. (1960) *Congo Bull. WHO/OMS.* **22,** 313.
Vasilenko, S. and Atsev, S. (1965) *Acta Virol.* **9,** 541.
Voroshilova, M. L. and Chumakov, M. P. (1959) *Progr. med. Virol.* **2,** 106.
Wallis, C. and Melnick, J. L. (1967) *J. Virol.* **1,** 478.
Ward, R., Melnick, J. L., and Horstmann, D. M. (1945) *Science* **101,** 493.
Weissmann, C., Feix, G. and Slor, H. (1968) *Cold Spring Harbour Symp. quant. Biol.* **33,** 83.
Wenner, H. A. and Behbehani, A. M. (1968) *Virology Monographs I,* p. 52. Springer-Verlag, Wien and New York.
Wenner, H. A., Lou, Te-Y and Kamitsuka, B. A. (1960) *Arch. ges. Virusforsch* **10,** 428.
Wenner, H. A. aand Rabe, E. F. (1951) *Amer. J. med. Sci.,* **222,** 292
Wickman, O. I. (1907) *Beiträge zur Kenntniss der Heine-Medinschen Krankheit.* Berlin.
Wigand, R. and Sabin, A. B. (1962) *Archiv. ges. Virusforsch* **12,** 29.
Wilfert, C. M. and Buckley, R. H. (1977) *New Engl. J. Med.* **296,** 1485.
Wilson, F. M., Miranda, Q. R., Chason, J. L. and Lerner, A. M. (1969) *Amer. J. Path.* **55,** 253.
Wilson, G. S. (1967) *The Hazards of Immunization,* p. 44. Athlone Press, London.
Winkelstein, W., Karzon, D. J., Barrow, A. L. and Haynor, N. S. (1957) *Amer. J. publ. Hlth* **47,** 741.
Woodruff, J. F. (1970*a*) *J. infect. Dis.* **121,** 137; (1970*b*) *Ibid.* **121,** 164; (1980) *Amer. J. Path.* **101,** 427.
Woodruff, J. F. and Woodruff, J. J. (1974) *J. Immunol.* **113,** 1726.
Yin-Murphy, M. and Lim, K. H. (1972) *Lancet* **ii,** 857.
Yoon, J. W., Austin, M., Onodera, T. and Notkins, A. L. (1979) *New Engl. J. Med.* **300,** 1173.
Young, N. A. (1973) *J. Virol.* **11,** 832.
Yorke, J. A. *et al.* (1979) *Amer. J. Epidem.* **109,** 103.

100

Other enteric viruses

C. R. Madeley

Introductory

All the viruses to be discussed in this chapter are found in stools but they are by no means the only ones to be found there. Leaving aside reoviruses for the moment, they form a miscellany with only three things in common. 1. They have all been discovered in the past 12 years. 2. They were all discovered as a result of a search for viruses causing diarrhoea. 3. They do not grow in the cell cultures routinely used for isolating viruses from diagnostic specimens. Nevertheless it should not be forgotten that other viruses may be isolated from stools. These include the enteroviruses (see Chapter 99), adenoviruses (Chapter 97) and hepatitis A (Chapter 101) all of which, as far as is known, replicate in the cells lining the gut. With the exception of some serotypes of adenoviruses, they have not yet been seriously implicated as causes of diarrhoea, although occasional common-source outbreaks have been reported. The very rarity of such outbreaks, however, highlights the lack of evidence linking them to diarrhoea. The fact that other viruses replicate in the same organ without apparently causing disease, although they can and do cause disease elsewhere in the body, means that the evidence linking the newer viruses with disease has to be evaluated carefully. Establishing valid criteria for proving a causative role is not easy and this problem is discussed below.

The existence of the viruses discussed in this chapter was predicted long before any of them was identified. Diarrhoea is a common affliction of all ages and can be due to a variety of causes with only a proportion due to infection with micro-organisms. If the severe bacterial causes of enteritis (cholera, typhoid and dysentery) and chronic disease (ulcerative colitis, Crohn's disease, etc.) are excluded, diarrhoea in adults is not usually a major problem. In infants and young children this is not so, and life-threatening diarrhoea is much more common, particularly in the tropics. In this chapter, any diarrhoea mentioned can be taken to be in children under 5 years old unless otherwise indicated.

Of those cases where an infective cause is likely (acute onset, absence of other non-infective causes and other related cases) only about one-third have been shown to have a bacterial cause. In temperate zones parasites account for a small additional percentage, but the remainder was thought on clinical evidence to be due to 'viruses' but without positive virological findings to support this conclusion. The discovery of new viruses by direct electron microscopy (EM) of stool specimens came therefore as the fulfilment of a prophecy, and because they were found where expected they have been readily accepted as pathogens. As evidence has accumulated a straightforward cause and effect relationship is too simple to fit all the facts and now needs to be modified.

Among these newly discovered viruses is one that is not new. This is the reovirus genus which has been known for over 20 years and which also differs from the other viruses of this chapter in growing in cell culture. The reoviruses may be isolated from both the respiratory and alimentary tracts but have still to be linked unequivocally with disease. As viruses to be found in the gut and with an uncertain role in disease, they fit more appropriately into this chapter than might be thought at first sight. Although they are related to the orbiviruses, they are also very similar to rotaviruses and it is necessary to distinguish them in a stool specimen. Orbiviruses have yet to be recorded in stools and cause disease in man only through contact with ticks or sandflies.

Since the discovery of these 'new' viruses is both recent and has had a considerable influence on how they have been regarded, it is necessary to start by recounting their history briefly.

Historical background

With the development of cell cultures in the 1950s for isolating viruses it became possible to isolate viruses from a variety of sites on and inside the body. The viruses that were recovered from stool specimens included polioviruses, other enteroviruses (Coxsackie A and B, ECHOviruses) and adenoviruses. Initially, isolations were mostly made from cases of neurological disease but, as the need for more epidemiological information became apparent, surveys showed excretion of viruses with no evidence of illness. This was particularly true in children especially those in day nurseries, and it became evident that extension of an infection from the gut to the central nervous system was relatively uncommon. The concept of disease-free carriage of gut viruses was recognized although no one was able to ascribe a positive benefit to the host from such carriage, except for the possibility that the presence of one apparently harmless virus might prevent superinfection with a more virulent one by occupying the appropriate cells or their surface receptors.

By comparison with most other viruses this failure to cause disease at their main site of replication in the cells of the gut was surprising, all the more so because the epidemiological evidence had suggested strongly that viruses did damage the gut with resultant diarrhoea. Admittedly the evidence was negative, owing to failure to demonstrate any other infective cause for acute non-bacterial 'gastro-enteritis'.

From time to time adenoviruses and enteroviruses were isolated from common-source outbreaks of diarrhoea. Examples are ECHOvirus 18 (Eichenwald *et al.* 1958), adenovirus 7 (Gardner *et al.* 1961) and ECHOvirus 19 (Cramblett *et al.* 1962), although the evidence that these particular viruses were truly the cause of the outbreak was not overwhelming. It was equally possible that the virus concerned had spread in a susceptible community and that this was made easy by diarrhoea in children too young to understand the hygiene necessary to discourage transmission.

Apart from these outbreaks, surveys of diarrhoea, endemic and epidemic, yielded a number of virus isolates, particularly in children, but the frequency was not substantially greater than in controls of a similar age; moreover no single virus type predominated. As a result most virus diagnostic laboratories ceased investigating diarrhoea and discouraged clinicians from sending specimens. No one was satisfied with this situation but further investigation by standard techniques seemed a frustrating and fruitless exercise.

This deadlock was broken by the use of electron microscopy. Immune electron microscopy (IEM), the use of immune serum to coat or aggregate virus particles, had been the means of identifying rhinoviruses and was used to investigate an outbreak of diarrhoea and vomiting in a school in Norwalk, Ohio, during 1969. Convalescent serum from patients and, later, volunteers was shown to aggregate small numbers of ill-defined particles 27 nm in diameter. Acute sera produced much less aggregation but the first 'new' stool virus had been recognized. Possibly because the virus was scanty in the stool extracts examined, the virus was not clearly identifiable and the technique necessary to find it elaborate, the report of this work (Kapikian *et al.* 1972) attracted little attention, although two further similar viruses were later described in Hawaii and Montgomery county, Alabama. Much more interest was aroused a year later when Ruth Bishop and her colleagues in Melbourne (1973) took duodenal biopsies from nine children with diarrhoea aged between 4 and 31 months. In six of these biopsies virus particles about 67 nm in diameter were seen by thin section electron microscopy in the cytoplasm of the cells lining the gut.

This report was soon followed by another (Flewett *et al.* 1973) reporting similar viruses in stool extracts examined by negative contrast electron microscopy

(EM) after careful extraction and purification. In these stool extracts virus was frequently present in vast numbers and it became evident that elaborate methods to extract and concentrate the virus were not necessary. This second virus was subsequently found readily in the stools of children, but much less commonly in those of adults, in every country where it was sought. As a widely endemic virus found where expected in the stools of patients with the disease it was soon being hailed as *the* virus of infantile diarrhoea and there was disagreement only over its name. In appearance it resembled the reoviruses but as experience grew it was seen that the new virus, when complete, had a smooth outer rim never found on reoviruses and, with no agreed name at first, a variety were used: reovirus-like (RVL), orbivirus-like, infantile gastro-enteritis virus (IGV). A more specific name was necessary, particularly as other viruses were being found by EM, and Flewett *et al.* (1974) proposed *rotavirus* from the virus's resemblance to a wheel with a wide hub, numerous short spokes and a narrow rim (Latin, rota = wheel). The Melbourne group countered with *duovirus* (Davidson *et al.* 1975) but, after some debate, rotavirus was adopted officially.

As often happens, the veterinary virologists had learnt a great deal about rotaviruses before the medical virologists discovered the human virus. The virus of epidemic diarrhoea of infant mice (EDIM) had been recognized as being orbivirus-like in 1963 and Mebus and his colleagues had published micrographs in 1969 of a calf rotavirus, Nebraska calf diarrhoea virus (NCDV). Subsequently morphologically indistinguishable viruses, or serological evidence for them, were found in most animal species in which they were sought, and the list now includes at least two avian species.

The veterinarians, in their investigations into diarrhoea, had one advantage over their medical counterparts in that they could deliberately infect newborn animals, particularly gnotobiotic ones, and show that their strains of rotavirus were capable of inducing disease. This could not be done ethically with babies and there is now enough evidence to doubt whether the result would have been the same if the experiments had been possible. A number of investigations have now reported that newborn babies may excrete rotaviruses from a few hours after birth without evidence of gastro-intestinal upset. This has not been the case with newborn animals which are more likely to develop disease at this age than they are later.

The discovery of rotaviruses prompted many electron microscopists to search for them. They were found frequently but the search soon showed that other viruses or virus-like particles could also be present in faeces. Sometimes they were found in the same stool as rotaviruses but more often alone. The list of newly discovered viruses has gradually been extended and now includes previously unknown adenoviruses, coronaviruses and a variety of small round virus-like particles of different sizes and surface morphologies. All have been discovered by electron microscopy and none of them grows in the cell cultures used routinely by diagnostic laboratories. Some evidence of partial growth in animal cells has been obtained with adenoviruses, coronaviruses and astroviruses as well as rotaviruses but not for the remainder.

Even where some growth takes place either no new infectious virus is released or the virus fails to adapt to cells and dies out after a few passages. It became clear that some factor was missing from cultured cells but so far its nature has remained elusive despite much hard work.

This failure to grow in cell culture is typified by the adenoviruses. Recognized serotypes had been isolated regularly from faeces previously but with no clear association with gastro-enteropathy and it was surprising to find that adenoviruses could be present in very large numbers in stools from which no virus was isolated. Occasionally other viruses, such as enteroviruses, could be isolated from such stools showing that there was no inhibitory factor in the stool which prevented growth. This was confirmed later by direct experiment (Kidd 1980).

Adenoviruses, with an unmistakable appearance in the electron microscope, posed no problems of identification and no one has ever doubted their identity, regardless of whether or not they grew in cell culture. Other particles have posed more problems and partly accounts for their being less frequently reported. Coronaviruses have an enveloped structure with the surface of the envelope studded with pin-like peplomers. These were originally reported from India (Mathan *et al.* 1975) and Bristol (Caul, Paver and Clarke, 1975) and subsequently from other places. They have not, however, been found universally but whether this reflects an uneven distribution or the difficulty in identifying them is not clear.

Astroviruses were first described in 1975 by Madeley and Cosgrove and posed a different problem. They were often present in faeces in very large numbers and this made them more credible as viruses, but their star-shaped surface of configuration from which they took their name had not been seen before and has yet to be understood fully. Morphologically similar viruses have since been found in the faeces of lambs and calves with diarrhoea, while reports of the human virus have been made in several parts of the UK and abroad.

Human caliciviruses, on the other hand, were similar to viruses which had been found in cats, pigs and sea-lions. In none of these animal species did caliciviruses cause diarrhoea, although a second pig virus was found after the description of a human virus, and this has been associated with diarrhoea. The first report of human caliciviruses (Madeley and Cosgrove 1976) noted that they were less common than rotaviruses and astroviruses and the numbers in a stool were usually smaller. This report also showed a less strong association with gastro-enteropathy, but subsequent ones have indicated that it is a cause of outbreaks of 'winter vomiting' in primary schools (Cubitt, McSwiggan and Moore 1979).

The viruses which have been discussed so far have all been associated with endemic rather than epidemic diarrhoea with the exception of Norwalk virus and, possibly, calicivirus. By no means all outbreaks of diarrhoea and vomiting have been investigated virologically but examination of stool specimens from some have shown that virus-like objects can be present. Two basic morphological types exist and may be seen in stools in sufficient numbers and consistency in size to leave little doubt in the microscopist's mind that they are viruses and not spherical debris. The two types may be either plain featureless spheres, very similar in size and appearance to enteroviruses (25 nm in diameter and now referred to as small round viruses or SRVs) or slightly larger with an ill-defined periphery and surface structure (about 33 nm in diameter and now referred to as small round structured viruses or SRSVs). Both morphological types of particle have been associated with outbreaks, some of which are food-borne usually through shell-fish. Within each outbreak the particles have all been of one type but the size of the particles quoted by the authors has varied between one episode and another. Whether these differences are real or due to observer error is not known but this is discussed later in the chapter.

To provide a convenient working name each of these viruses has been called after the place where it was first recorded or the presumed vehicle of transmission. Hence Ditchling and cockle agents are SRVs and Parramatta agent is an SRSV. Similar particles have been seen in faeces from sporadic cases of diarrhoea, generally in smaller numbers, and may prove to be the source of outbreaks.

The current list of viruses in faeces is completed by tailed bacteriophages. These have a typical head and tail lollipop structure but between stools differ in details of structure and size. In an environment full of bacteria, their absence would perhaps be more surprising than their presence but with the potential of some 'phages for altering the pathogenic potential of their host bacteria they should not be ignored as irrelevant to the causation of diarrhoea. They are usually present in only small numbers but occasionally have been found in very large numbers. What the significance of this may be is unknown so far but any episodes are worth recording.

Starting from one newly discovered virus in 1969 a substantial catalogue of hitherto unknown viruses has since been built up. The list has gradually extended but the most frequently observed virus remains the rotavirus. Frequency of sighting should not be equated with causation or even pathogenic potential but rotavirus probably accounts for well over half the total numbers. With the expense of electron microscopes and the difficulties of maintaining them in the tropics, there have been efforts to find alternative methods of detection based on the use of antibody to recognize the virus. However, this has led to the emphasis being laid more on investigating the virus than the disease, a specialization that may have been a little premature.

The viruses

The various viruses to be found in specimens of stool have been introduced but, before discussing their role in disease, it is necessary to describe them in more detail. As will become apparent, by far the most is known about the members of the family Reoviridae, the rotaviruses and the reoviruses (which are discussed separately at the end of the chapter). This reflects the intense interest in the first genus and the long interval since the second was originally described. Under each virus will be listed, as far as they are known: EM appearance, size, nucleic acid, polypeptides, physicochemical properties, related viruses and serological relationships. These are summarized in Table 100.1.

1 Rotavirus

i **EM appearance**—by negative contrast, the virions appear spherical, although where the support membrane curls away from a tear it is evident that there has been considerable flattening of the particles. The virus is seen in one of 3 forms:

(a) Complete, smooth. These have a smooth outer layer which is said to have icosahedral symmetry, although no substructure is visible, and appears as a featureless shell (see Fig. 100.1a, arrowed).

(b) Incomplete, rough. These are slightly smaller with a toothed outer margin. The teeth resemble short wheel spokes attached to a wide hub and the complete virion therefore resembles a short-spoked wheel with a narrow rim; hence the name rotavirus (Latin, rota = wheel). The surface of the rough particle (Fig. 100.1c), which can also be seen inside the outer layer of the complete particle, has a structure which can be seen clearly on the majority of particles, except in stool extracts containing a large amount of protein and debris. On the surface a number of ring-like structures may be seen. These are about 12 nm in diameter and overlap. The number visible on any single particle is

Table 100.1 Properties of the viruses to be found in faeces by electron microscopy

Virus	Nucleic acid	Size (nm)	No. of serotypes	Buoyant density g/cm³ CsCl	Symmetry	Group antigen
Adenovirus	ds DNA	74	§G 38 F 3+? NG ?	1.34	cubic	Yes
Astrovirus	RNA?	28	2?	1.39	—	—
Calicivirus	RNA*	33	3?	—	cubic†	—
Coronavirus	RNA*	Pleomorphic 150+	—	—	helical†	—
Norwalk	RNA?	27**	3	1.38	cubic†	No
Reovirus	ds RNA (segmented)	75	3	—	cubic	No (but some cross-reactivity
Rotavirus	ds RNA (segmented)	Smooth 68–70 Rough 55 Cores 28	2+	Smooth 1.36 Rough 1.38 Empties 1.29–1.30	cubic	Yes
SRVs	—	25–35	—	—		—
W agent	—	22**	—	1.34–1.42 }		—
Ditchling	—	24–25**	—	1.39–1.41 }	cubic†	—
cockle virus	—	24–25**	—	1.39–1.41 }		—
AAV	DNA*	22	—		cubic*	—
Parramatta	—	23–26**	—	—	cubic†	—
SRSVs	—	29–35	—	1.36–1.41	cubic†	—
Atlanta	—	29	—	1.35†	cubic†	—
Taunton	—	30–35		1.36–1.41	cubic†	—
Harlow	—	30–35	—	—	cubic†	—
Dumfries	—	30	—	—	cubic†	—

— No information.
* By analogy with similar viruses.
† Probable.
‡ In glycerol-pot. tartrate gradient.
§G: Growable in routine cell culture.
NG: Not growable, so far.
** Estimates by other authors.

variable but rarely exceeds four or five. They form part of a surface 'net' which is thought to be made up from 180 structural units. These are not resolved on intact particles but may be visible on broken ones. Detailed analyses, with possible models, have been published by Martin *et al.* (1975), Palmer *et al.* (1977) and Stannard and Schoub (1977).

(c) Stain-penetrated particles. As with most spherical viruses a proportion are penetrated by the stain and appear empty. Such penetration with rotaviruses reveals an inner membrane (arrowed, Fig. 100.1b). This membrane and its core can be extracted from intact particles by treatment with a trypsin/versene mixture. Consequently the complete virion consists of at least three layers. Extracted cores are fragile but appear to have an icosahedral structure.

The three forms may be separated on centrifuged density gradients (see v. below) and 'infectivity' is then associated mostly with the complete particles. The rough and, in particular, the 'empty' particles show much lower infectivity but this may be due to a lower resistance to the methods used in separation.

In addition to the three spherical forms of the virus, tubular forms are also found in some stools and, with some animal species, in infected cells. These have a similar diameter to that of the rough particles (Fig. 100.1c). The length is variable and the diameter also varies more than the spherical particles. These tubular forms appear to be formed from surplus coat protein which is probably self-assembled. Similar structures have been described with some orbiviruses and self-assembly of purified coat protein around pieces of nucleic acid has been described with some plant viruses. It is not known whether the tubular forms contain RNA but since they are irregular in width they may lack the template function which could be provided by the nucleic acid. The tubular forms do not have the smooth outer layer.

Thin-section electron microscopy of cells infected with the animal strains shows that large numbers of particles are produced in the cytoplasm. They consist usually of a darkly stained core surrounded by a less dense zone and an outer darkly stained shell. These correspond roughly in size to the negatively stained core and the outer rough layer. A larger, stained 'envelope' is also seen on a proportion of particles which have apparently been released by budding into cisternae in the endoplasmic reticulum. These enveloped particles are appreciably larger than the smooth complete particles seen by negative contrast and a structure corresponding in size to this smooth outer layer has not been seen in thin sections. In negatively stained stool extracts occasional particles are seen in a kind of membranous bag but whether this is the same as the envelopes seen in thin section is unknown. These anomalies between the appearance in thin section and negative contrast have still to be resolved.

ii **Size**—owing to differences in preparative technique,

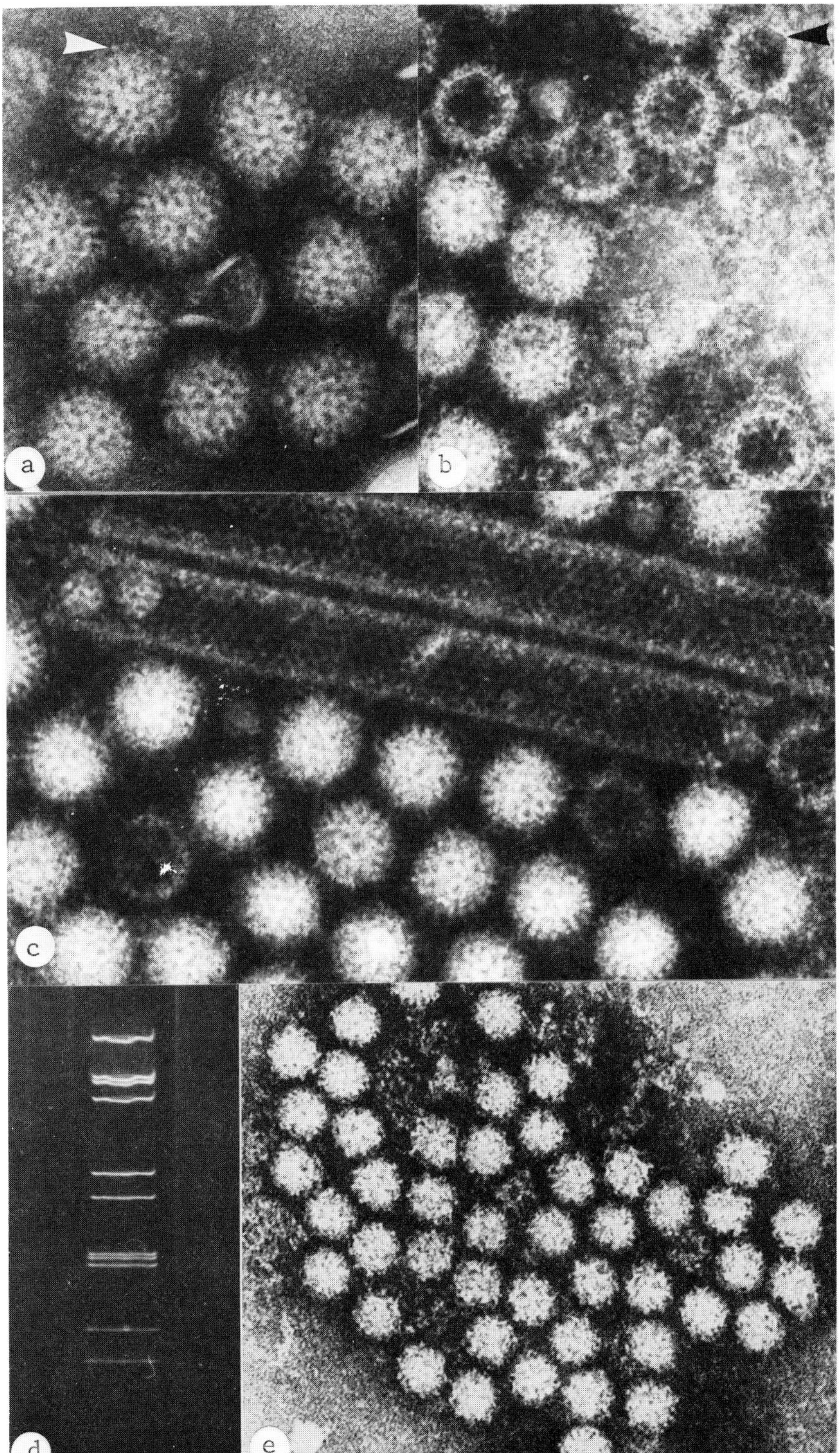

Fig. 100.1 EM-detectable viruses in faeces (1). a. Rotaviruses, mostly complete with outer membrane (arrowed). b. Rotaviruses with some empty particles. The stain-penetrated centre is bounded by a well-defined membrane (arrowed). c. Rotaviruses without the outer membrane. Note the tubular forms, probably formed by self-assembled subunits. d. Polyacrylamide gel electrophoresis of rotavirus RNA segments. Note eleven bands in four groups—I: 1, 2, 3, 4; II: 5, 6; III: 7, 8, 9; IV: 10, 11. 7.5 per cent acrylamide stained with ethidium bromide (courtesy of Dr R. B. Moosai) e. Norwalk virus. 'Immune' clump of virus using acute phase serum and scored as having minimal quantity of antibody present. (Reproduced from Kapikian *et al.* 1972, by permission of the Editor, *Journal of Virology*). All micrographs are negative contrast, PTA pH 7, with a magnification of × 200 000.

and variations in the calibration of microscopes, in measurement and in the amount of distortion of individual particles, it is not possible to quote absolute figures for the size of virus particles. In one microscope under similar conditions, it can be said that reoviruses are slightly larger than complete rotavirus particles which are larger than rough incomplete rotavirus particles. There are no reports of significant variation between rotaviruses of different species and, despite the sources of variability listed above, there is considerable agreement between different authors. The complete particles are variously estimated to be between 65 nm and 75 nm in diameter with most estimates being 68–70 nm. The rough particles are 50–60 nm in diameter with the cores being about 38 nm in diameter. The thickness of the outer smooth membrane is difficult to measure accurately but appears to be 1–2 nm.

The figures quoted in Table 100.1 are those obtained in the author's laboratory on a calibrated microscope. They are the mean values of at least 100 particles and are included to give a comparison between the different viruses under similar circumstances.

iii **Nucleic acid**—this is double-stranded RNA in eleven segments, ranging in size from a molecular weight of 2.04×10^6 to 230 000 and totalling $10–11 \times 10^6$. By polyacrylamide gel electrophoresis of phenol extracts of stool extracts these segments may be separated readily to give a characteristic pattern (Fig. 100.1d). A careful comparison of the patterns seen in different recognizates shows that the mobility of the segments is not constant. Individual patterns are reproducible and these variations are likely to form the basis for at least one classification of rotaviruses. These differences in the RNA segments will be reflected in variations in the structure or antigens of the particles themselves but little direct evidence has yet been obtained. Recent 'fingerprint' analyses by two-dimensional chromatography of RNA fragments from endonuclease digestion have shown that the differences between strains are even more complex. Differences have been found even with strains giving identical patterns in one-dimensional electrophoresis.

One consequence of a multipartite genome is that genetic reassortment between strains may occur; evidence has been obtained that this can happen between animal strains, producing hybrids. There appears to be no reason why reassortment between human and animal strains should not also take place. As it is, animal strains can cross species barriers and the potential patterns of transmission and infection are complex and will take much patient research to unravel.

iv **Polypeptide**—owing to the difficulties of separating virus from contaminating protein in stool extracts the results of polypeptide analysis are not so clear as those of the virus RNA.

The genome of rotaviruses is in 11 segments and it is therefore reasonable to look for an equivalent number of virus-specified polypeptides as gene products in infected cells. Despite the fact that most rotaviruses do not produce infectious virus in cell cultures, eleven such products have been recognized in rotaviruses from several species (Thouless 1979). By general consent, these are four inner capsid polypeptides (I_1–I_4), four outer ones (O_1–O_4) and three non-structural ones (NS_1–NS_3). The appearance of these polypeptides as bands on polyacrylamide gels is not entirely constant and sometimes may appear as more than one band suggesting that they may be cloven after synthesis. Alternatively, these duplications may be due to the recognizate being a mixture of two or more different strains. The polypeptides can be expected to vary in size if the RNA segments which code for them do so. For a more detailed discussion of this topic the reader is referred to Thouless (1979).

It now seems probable that one or more of the polypeptides of the outer shell (O_3?) is glycosylated. This may be the reason why the complete smooth particles with the outer shell have a lower buoyant density than the incomplete rough particles (see below, section v).

The non-structural polypeptides are not found in digests of purified virus but only in those from infected cells and are probably virus-specified enzymes promoting virus replication.

v **Physico-chemical properties**

(a) Density—the three spherical forms of the virus can be separated by density gradient ultracentrifugation. At equilibrium on a caesium chloride gradient they form three separate bands. The rough particles are the most dense at 1.38 g/cm³, the complete are intermediate at 1.36 g/cm³ and the empties are the lightest at 1.29–1.30 g/cm³. These results are compatible with the complete particles having a greater protein to RNA ratio (reducing the overall density) and the empties lacking RNA.

(b) **Stability**—owing to the difficulties of assaying the infectivity of a virus which does not grow in cell culture, the evidence so far is patchy and has not been assessed for more than a few rotavirus strains from different species. The threshold of heat lability of rotavirus from different species appears to be 50–55°. Thermal stability of the simian virus, SA11, is improved by magnesium sulphate but not by magnesium chloride; information about other species is lacking. At room temperature or below, crude stool extracts are usually stable, in particular the calf virus.

Stability to acid pH appears to extend down to about pH3, which is similar to reoviruses, better than orbiviruses but poorer than enteroviruses.

Infectivity is generally resistant to lipid solvents such as ether, chloroform or fluorocarbons. Rotaviruses also resist a variety of proteases under normal physiological conditions. However, it has been shown that the outer layer of the human virus can be removed by α-chymotrypsin in the presence of caesium ions.

This does not happen with the calf and SA11 viruses and for this reason the results of infectivity studies on virus purified on caesium chloride gradients should be interpreted cautiously.

(c) **Haemagglutinin**—a haemagglutinin is associated with the calf and simian (SA11) viruses grown in cell culture. It agglutinates red blood cells from a variety of sources including human group O cells, and is inhibited by convalescent sera. It appears to be associated only with the complete smooth virions. No mention is made in the published reports whether stool extracts also showed an HA but initial searches for an HA in human strains were negative. A later paper from Japan reported an HA in some human stools active against chicken red cells but this has to be confirmed, as has the exact relation of the HA to the virion. There is evidence that the HA and complement-fixing antigens are different.

vi **Related viruses**—Rotaviruses have been detected in the faeces, or serum antibody to rotavirus has been found, in all species of mammal in which they have been seriously sought. The list now includes calves, piglets, lambs, mice, foals, rabbits, kids, guinea-pigs, chimpanzees, dogs and monkeys as well as man. There seems to be no reason why the list should not extend eventually to all mammals. In addition to these species, rotaviruses have now been found in several avian species including chickens and turkeys. Consequently rotaviruses join enteroviruses, adenoviruses and herpesviruses as being widespread in the environment.

vii **Serological relations**—morphologically, strains of rotavirus from whatever source are indistinguishable and most are related antigenically. The rough particles carry a group specificity which is similar in most recognizates regardless of the species of origin, whereas the outer layer appears to carry type specificity. Recently, however, recognizates from pigs and chickens have been found that do not possess this group antigen and there is now no known antigen found in all rotaviruses. At first it was thought that each animal species had its own type-specific rotavirus but three observations have undermined this straightforward concept: 1. there are probably several human serotypes; 2. some strains will cross species barriers (this applies particularly to human, bovine and porcine rotaviruses); 3. genetic re-assortment can occur. Whether later evidence will complicate the relationship still further remains to be seen but there is already a difficulty over nomenclature of new strains. As far as information extends at present, neutralization is type specific whereas complement fixation is group-specific. (For cultivation, see Report 1982).

2 Norwalk agent

i **EM appearance**—by negative contrast the virus is difficult to identify because it is normally present in only small numbers and its appearance is not particularly characteristic. It has not been observed by thin section electron microscopy at all.

The virus was first observed by immune electron microscopy using negative contrast and, in the micrographs published by Kapikian *et al.* (1972), it was always mixed with human serum, either acute or convalescent. Since the virus appeared to form clumps even with acute-phase sera it is difficult to know what the virus would look like without some antibody on the surface. It is clear that the amount of antibody was often small and Fig. 100.1e, taken from Kapikian *et al.* (1972), was scored by the authors as having minimal visible antibody. Consequently these particles are probably naked virions with a roughly circular outline having an ill-defined edge. There is a hint of surface structure although it is not possible to distinguish any regular arrangement. It would be extremely difficult to identify single particles with certainty as being identical with Norwalk or even as virus particles at all. It is not surprising therefore that other workers have failed to identify it. Similar viruses have been reported in the UK, and Norwalk particles have some resemblances to the caliciviruses. This point is discussed in more detail under vi below.

ii **Size**—this is estimated at 27 nm but with a proportion appearing elliptical (Kapikian *et al.* 1972). If ellipticity is not an artefact, this will be unique for viruses of this size.

iii **Nucleic acid**—for some time after its original description, Norwalk was referred to as parvovirus-like. This was from its estimated size and buoyant density. If truly a parvovirus the nucleic acid would have been DNA. However it is too large in size and does not look like a genuine parvovirus. (See descriptions of adeno-associated virus below and elsewhere). The density is similar but that is its sole claim to resemblance. In general properties it is much closer to the small RNA viruses than to those containing DNA but there is no direct evidence as to which nucleic acid it contains.

iv **Polypeptides**—recent work (Greenberg *et al.* 1981) has shown that Norwalk virions contain a single major capsid polypeptide with a molecular weight of about 59 000. This is similar in size to that found in caliciviruses. However its size, as with that of the virions themselves at 27 nm, is just sufficiently smaller than the equivalent dimensions of the calicivirus (33 nm diameter, polypeptide of mol. wt 66 000) for it not to be possible to assume that they are the same virus (or even of the same family) without considerable further work.

v **Physico-chemical properties**—the buoyant density of members of the group in a caesium chloride gradient is about 1.38 g/cm^3 which is similar to the caliciviruses and some DNA viruses. The small amount of virus present in stools and the absence of an assay for infectivity other than in adult volunteers has made it difficult to assess its resistance to inactivating agents. Norwalk is resistant to ether, low pH (2.7 for 3 hr at room temperature) and shows good resistance to heat, with some infectivity surviving 60° for 30 minutes. Details of resistance to other chemicals and enzymes have not been published.

vi **Related viruses**—Norwalk virus was first identified in a school outbreak. Similar agents were then recognized in two other outbreaks, in Hawaii and in Montgomery County, Maryland. Evidence was obtained that the Montgomery County agent was antigenically related to Norwalk but Hawaii was dissimilar. Norwalk or a closely similar agent has been implicated in other outbreaks investigated by Kapikian's group but it is not known how many other viruses may also be related either by having a similar structure or serologically. No animal strains have been discovered.

The Norwalk group of agents may be related to other small round structured viruses (SRSVs) in general and caliciviruses in particular, although there is no direct evidence at present. With some cell-culture grown strains of feline caliciviruses, the surface hollows on the virions characteristic of the genus (see Fig. 100.2e) are seen only rarely, with the majority being generally similar to the Norwalk virus in appearance. Further evidence may show that the Norwalk virus and caliciviruses both belong to the same genus but such evidence is lacking at present.

3. Astrovirus

i **EM appearance**—by negative contrast on stool extracts the astrovirus appears as a small round virus with a smooth outer margin. Often present in very large numbers, the viruses can be found in ordered, semi-crystalline arrays with regular spacing (Fig. 100.2a). At first sight they resemble enteroviruses but closer inspection of the particles in the microscope or on a high quality micrograph shows that most particles have some surface substructure which in a minority is resolved as an unstained five or six-pointed star (Fig. 100.2b) p. 429. The star extends right across the presenting surface of the particle. This description is of astroviruses stained with potassium phosphotungstate, pH 7.0. Other stains at other pH values give slightly different appearances. In particular, the outer margin may appear 'lumpy' but the virus is still distinguishable from caliciviruses (Madeley 1979).

Occasional empty particles are seen but these are rare. No internal substructure has been observed but close examination of some crystalline assays has revealed structures between the particles that could be surface projections or even antibody bridges. They are close to the limits of resolution and a definite attachment to the surface of the virions cannot be seen (Fig. 100.2c). The number of these structures per particle and their significance, if any, is unknown.

In thin sections of infected cells, astroviruses appear in regular arrays of dense dark-staining objects in the cytoplasm with a centre-to-centre spacing of 28 nm. As new infectious virus is not produced, the relation between these objects and extracellular virus particles is uncertain.

ii **Size**—28 nm, with little variation.

iii **Nucleic acid**—not known, but probably RNA.

iv **Polypeptides**—there are no published data on the human virus but a similar virus recovered from lamb faeces contains two polypeptides with a molecular weight of 33 000.

v **Physico-chemical properties**—again there are no published data on the human virus but the lamb one has a buoyant density of 1.39 g/cm^3, which is in agreement with several personal communications.

vi **Related viruses**—there are probably at least two human serotypes. A morphologically indistinguishable virus has been isolated from an outbreak of diarrhoea in lambs and induces diarrhoea in newborn gnotobiotic lambs. Viruses of similar size and appearance have been seen in stool extracts from calves and pigs and are also probably astroviruses.

Studies have shown no cross-reactivity between antisera to the human, lamb and calf viruses in immunofluorescence tests on sections of lamb gut infected with the lamb virus.

4. Calicivirus

i **EM appearance**—these viruses are about 10 per cent larger than astroviruses, have a much more indistinct outline by negative contrast and in practice *look* quite dissimilar. Descriptions of the two viruses *sound* very similar, however, but there is no substitute for a careful comparison of known examples of the two viruses in order to fix the differences in the observer's mind. The main point of confusion is that with each virus, some of the virions exhibit a six-pointed star on their surface. On the calicivirus it resembles a 'Star of David' made from two interlocking equilateral triangles (Fig. 100.2d). The dark stain-filled hollow in the centre of the calicivirus star is never found on an astrovirus. A proportion of the calicivirus particles show several of these stain-filled surface hollows which appear round or oval, whereas those of astroviruses appear triangular. A more detailed comparison of the features of these two viruses has been made (Madeley 1979).

Caliciviruses may be observed in clumps in stool extracts but they are never seen in ordered crystalline arrays as are often found with astroviruses; the clump is quite irregular, and the numbers are usually much smaller (Fig. 100.2e).

Although they are indistinguishable from animal strains in the electron microscope, more information is needed before the ICTV can accept them officially as members of the family.

ii **Size**—about 30–35 nm in diameter with uncertainty

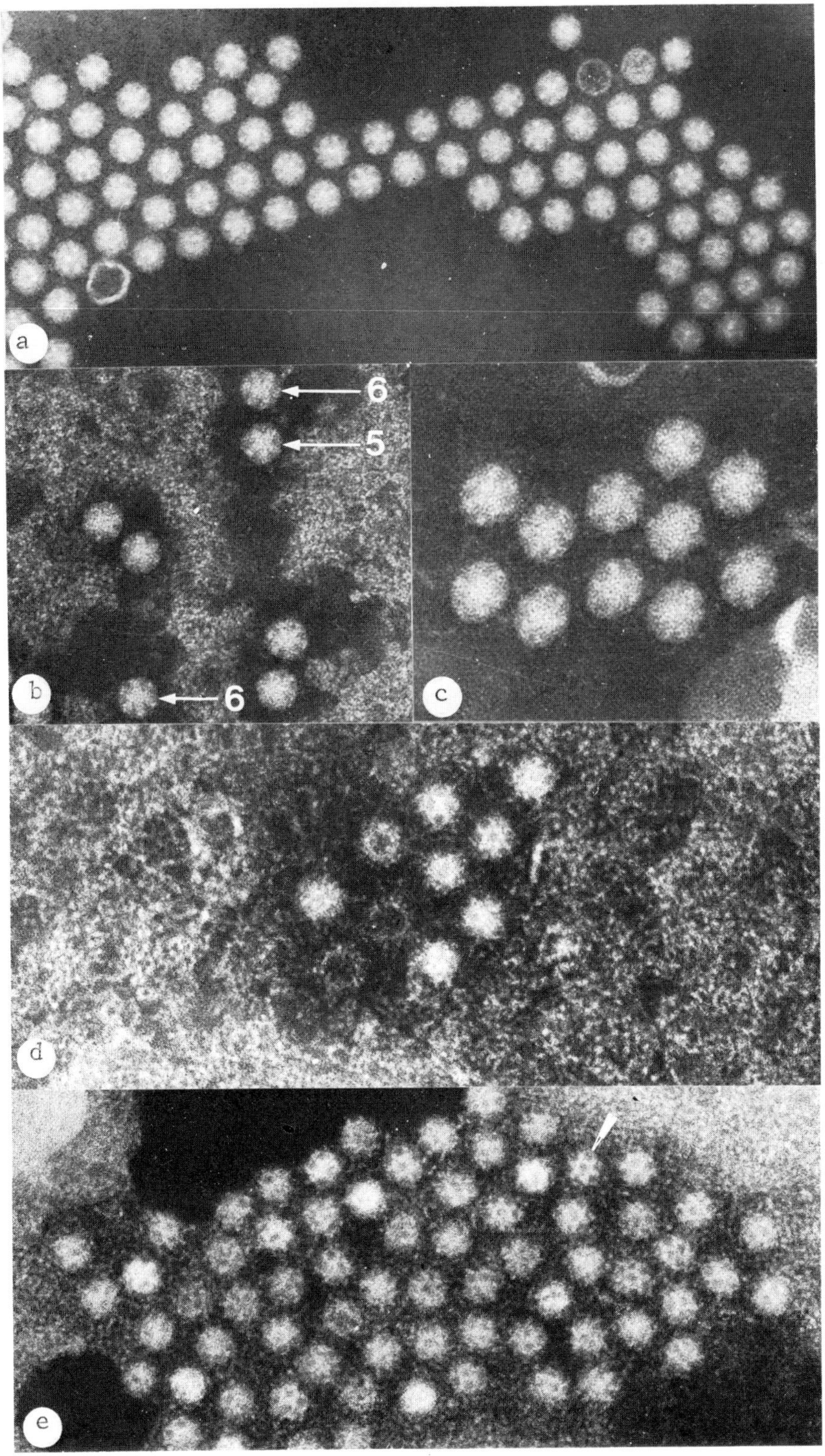

Fig. 100.2 EM-detectable viruses in faeces (2). a. Crystalline array of astroviruses in a stool extract. Note (i) the orderly arrangement of the particles, (ii) solitary 'empty' particles (they are uncommon with astroviruses), (iii) the star-shaped surface configuration on a proportion of the particles. b. Astroviruses showing 5- and 6-pointed surface stars. Note the absence of a central hollow to either form. c. Astroviruses at higher magnification (× 300 000) to show bridge-like structures between the particles. They are not clearly attached to the virus particles. d. Caliciviruses. Note the diagnostic six-pointed star of David with a central hole (arrowed). e. Caliciviruses. An irregular clump but no antibody has been added. Compare with aggregates of astrovirus. All micrographs are negative contrast, PTA pH7, with a magnification (except for c) of × 200 000. (d. Reproduced by permission of the Editor, *The Lancet*)

about the exact figure owing to the ill-defined edge of the particle.

iii **Nucleic acid**—this has not been assessed for the human virus, but animal strains all contain RNA. To be acceptable as caliciviruses human strains must also contain RNA.

iv **Polypeptides**—all officially accepted caliciviruses have one major capsid polypeptide with a mol. wt of about 66 000. There is also one minor polypeptide but no detailed analysis has so far been published on a human strain.

v **Physico-chemical properties**—no assessments on the human virus have been made but it should be similar to animal strains (buoyant density of 1.36–1.39 g/cm^3 in caesium chloride, sensitive to low pH(3), resistant to lipid solvents and can be readily inactivated at 50° in the presence of Ca^{++} or Mg^{++}).

vi **Related viruses**—there are probably at least three human serotypes. Accepted caliciviruses have been isolated from kittens (in which they cause a respiratory infection with glossitis), pigs (causing a vesicular exanthem), and possibly sea-lions and northern fur-seals. The sea-lion and swine viruses each have several distinct serotypes; while the cat virus has many subtypes which show considerable cross-reactivity and it is not clear whether these are variants of the same virus or distinct strains. The human virus will not grow in fetal kitten cells which readily support the growth of one of the strains of cat virus.

5. Small Round Viruses (SRVs)

Some stool extracts contain virus-like particles 25–35 nm in diameter; some examples are shown in Fig. 100.3. Appearance, size and numbers vary between one stool and another but in any one stool extract the appearance and size are consistent. Some resemble enteroviruses closely but no virus can be isolated in the cells in which enteroviruses grow routinely. Where such extracts have been inoculated into newborn mice only the very occasional Coxsackie A virus has been isolated. A proportion of these SRVs are probably small cubic bacteriophages whose bacterial host species is unknown, but the rest are probably human viruses and may be responsible for some cases of diarrhoea. It is necessary to record their presence, as part of the mosaic of infection, with whatever characteristics they exhibit. Those that are important will be recognized as such in time.

i **EM appearance**—these virus-like objects fall into two groups according to whether any surface structure can be seen:

(a) **Plain**—these are round, without distinguishable surface features and resemble enteroviruses (Fig. 100.3a and b). Name is abbreviated to SRV to distinguish them from structured varieties (below). Occasionally they may appear smaller, and isolated particles may have a hexagonal outline. A careful search will often then reveal an occasional adenovirus and these SRVs are probably true parvoviruses, adeno-associated viruses (AAVs). See, for example, Fig. 100.3c.

Good quality micrographs taken with a calibrated microscope are essential for comparison of both appearance and size with other similar recognizates.

(b) **Structured**—these show some surface structure which may either be an intrinsic part of the virus or antibody attached to the surface (Fig. 100.3d and e). If it is very dense, antibody is unmistakable and will obscure surface detail with a 'hedge' up to 15 nm wide, equivalent to the long dimension of IgG molecules. If the antibody coating is less dense the particles will appear irregular in size and the surface features will lack sharpness.

Such particles have been given several names by different authors: minireovirus, minirotaviruses, even 'fuzzy-wuzzies'. For record purposes, the UK Public Health Laboratory Service has decided to refer to them as SRSVs (small round structured viruses).

Although the observer can be fairly certain that there is some surface structure on SRSVs, it is difficult to be sure what it is and how it is organized. Possible relationships with other viruses, particularly Norwalk and caliciviruses, are discussed under vi below.

ii **Size—**Plain SRVs are usually between 20 nm and 25 nm. Featureless spheroids over 30 nm are unlikely to be viruses, as all the known cubic viruses larger than this show easily recognized surface subunits which can be easily distinguished from the background grain of the preparation. SRSVs are normally larger than SRVs whose size is 25–30 nm, with the majority being over 30 nm. Although they may vary between one stool and another, the size of the particles, of either type, show little variation in any one stool. If the observed variation is greater than 10 per cent of the mean diameter, the objects are unlikely to be cubic viruses. The surface detail of SRSVs is too fine and irregular for measurement or description to be worthwhile.

iii **Nucleic acid**
iv **Polypeptides**
v **Physico-chemical properties**
vi **Related viruses**
} No information available

(a) SRVs—both enteroviruses and AAVs have been mentioned as possible identities. Small size (22 nm) and a hint of a hexagonal outline will suggest an AAV or other parvovirus. As already mentioned, some may be bacteriophages. Further exploration of their relationships will be hampered by lack of material until the trick of growing them can be found.

(b) SRSVs—it has been found that some cell culture grown strains of feline calicivirus do not show the characteristic structure in the EM and look like SRSVs. This raises the possibility that caliciviruses and SRSVs from stool are variants of the same virus.

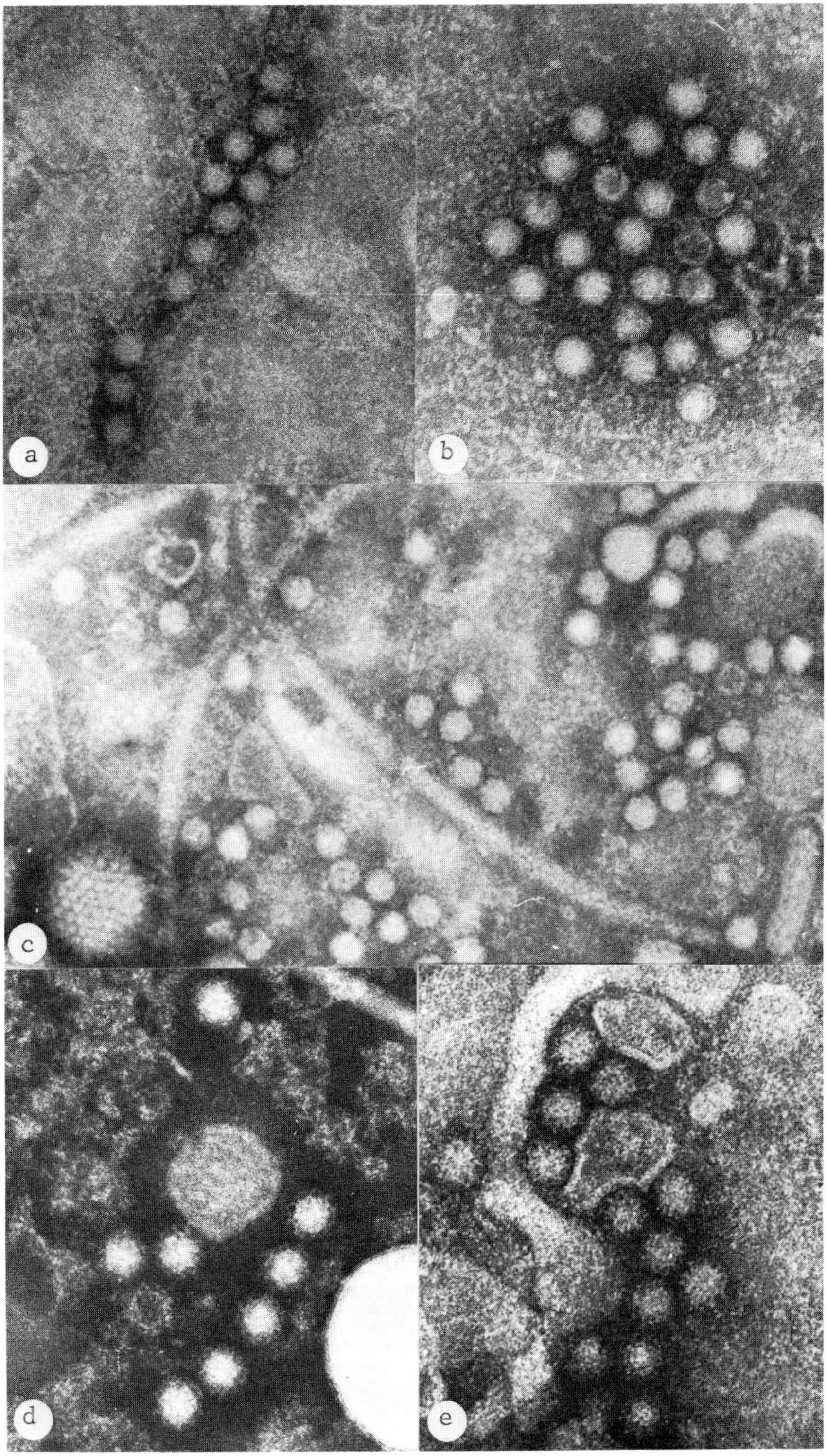

Fig. 100.3 EM-detectable viruses in faeces (3). a. Typical small round viruses (SRVs), with a smooth entire outline and no clearly discernible surface structure. b. More SRVs. Some exhibit some surface structure which is poorly defined. They *could* be astroviruses but, in the absence of typical particles, must be classified as SRVs. c. Adenovirus-associated (satellite) virus. Note: (i) adenovirus, (ii) the smaller size compared to a. and b. and (iii) the hexagonal outline of some of the particles. d. Small round structured viruses (SRSVs). Note that they are consistent in size, larger than the SRVs but do not resemble caliciviruses closely (Fig. 100.2d and e). Compare also with Norwalk virus (Fig. 100.1e). e. More SRSVs. All micrographs are negative contrast, PTA pH7, with a magnification of × 200 000.

If so, SRSVs will be found to have other properties characteristic of caliciviruses but there has been insufficient material in positive stools for these to be assessed.

The results of polypeptide or nucleic acid analysis will help to separate the small round viruses into groups with similar characteristics. Such classification is necessary because, as will be seen below, the SRVs are the viruses most frequently associated with epidemics and they must come from somewhere, most probably from endemic cases. Complete records of these endemic cases and their associated viruses are therefore essential.

6. Coronavirus

i **EM appearance**—Coronavirus-like objects found in stool specimens differ from previously described respiratory strains in that the projections which make up the surface fringe resemble pins rather than clubs, being longer and with narrower stalks (Fig. 100.4a and b). These stalks are sometimes not visible throughout their length and it is only the constant distance of the terminal knob from the surface membrane that shows the two to be connected. These projections are 20–28 nm long and sometimes appear as a two-tier structure.

It is necessary to distinguish the virus particles from other membranous debris and this can be very difficult. Careful point-by-point consideration of the image will usually resolve the problem.

ii **Size**—the virions are very pleomorphic without a constant size but in the region of 150–300 nm.

iii **Nucleic acid**—if similar to other coronaviruses then the stool virus should contain RNA, but no direct assessment has yet been made.

iv **Polypeptides**
v **Physico-chemical properties** } No information yet

vi **Related viruses**—in man coronaviruses are recognized as pathogens of the respiratory tract but the number of serotypes and the pathogenic potential of each one still have to be established. These respiratory strains have the classic club-shaped projections of the family and do not resemble those from stools. Coronaviruses with club-shaped projections are known to cause diarrhoea in animals. For example, although it is not absolutely identical in morphology, the transmissible gastro-enteritis of swine virus is a well established gut pathogen.

There is no evidence that any of the human gut coronavirus(es) are antigenically related to any of these.

7. Adenovirus

It has long been possible to isolate adenoviruses from stool specimens in cell culture, but such adenoviruses have only rarely been implicated as causes of diarrhoea. Most adenovirus serotypes, of which 35 are known, have been isolated from faeces at one time or another although types 1–7 are the most frequently recorded.

Direct electron microscopy of stool extracts soon showed that adenoviruses were often present and also that only rarely was an adenovirus isolated from these virus-positive stools. Paradoxically, the more virus there was in a stool, the less likely was it to grow in cell culture. What was seen was not grown and what was grown was not seen. More recent work, however, has shown that an adenovirus can be cultivated with some difficulty from some EM-positive stools in less commonly used cell types, but the growth is sluggish without the usual increase in adaptation to the cells with each passage. Frequently such growth as there is dies out completely and there are a considerable number which do not grow at all. The established growing (G) serotypes (1–35) are dealt with in detail elsewhere (Chapter 97). In this chapter they will only be referred to while the non-growing (NG) and fastidious (F) strains will be described in more detail.

i **EM appearance**—these are typical adenoviruses (Fig. 100.4c) with a sharply defined and quite characteristic morphology that is not easy to miss or to confuse with another virus.

The virions have the typical icosahedral appearance with triangular facets on the surface with clearly visible capsomers. Except for the absence of visible apical fibres, a feature rarely seen on adenoviruses from any source, these F and NG adenoviruses are indistinguishable from the established G strains. It is inconceivable that these viruses are not some form of adenovirus, for they have been observed by a large number of experienced electron microscopists working in many different countries without any hint of disagreement over their identity. It has been found, however, that a proportion of these stool adenoviruses may appear to be round rather than hexagonal in outline (Fig. 100.4d). Close inspection, reveals the typical capsomeric surface structure but whether this rounding of the outline is due to contaminating debris in the extract blurring it or whether it is due to some other factor is unknown at present.

ii **Size**—there is no significant difference from the approximate 75 nm quoted for the G adenoviruses. If there is a difference it will be shown only by detailed analysis of a large number of particles and is unlikely to be detectable in the virions in any single stool specimen.

iii **Nucleic acid**—analysis of F virions concentrated from stools has shown that the genome, which is unequivocally DNA, can be digested by restriction endonucleases to give a pattern of DNA pieces on polyacrylamide gel electrophoresis which differs from those found in G serotypes. With a mol. wt of 20×10^6, the genome is similar in size to that of G strains.

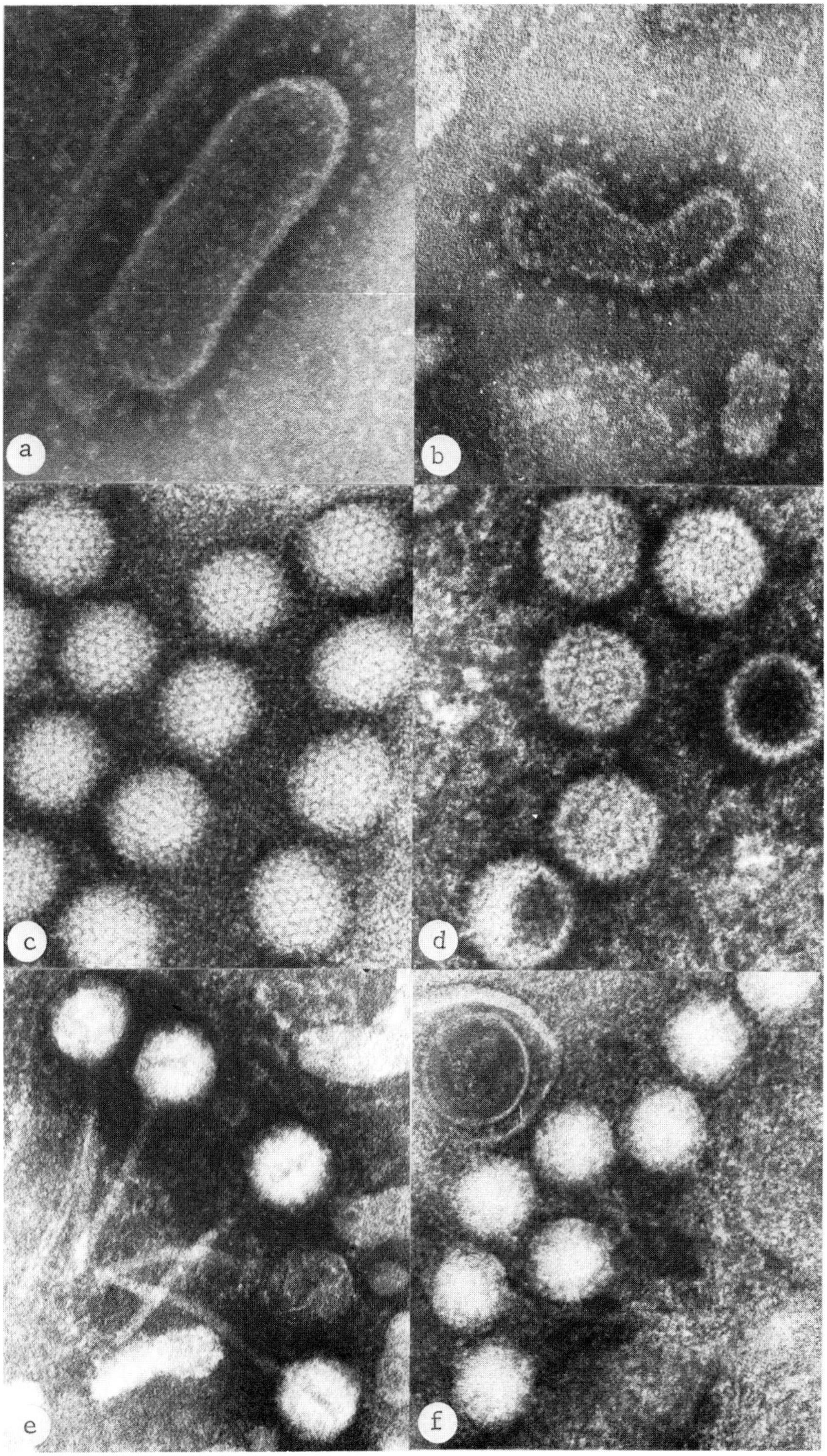

Fig. 100.4 EM-detectable viruses in faeces (4). a. and b. Faecal coronaviruses. Note pin-like projections on pleomorphic particles. The projections are constant in length and regularly spaced. c. and d. Fastidious adenoviruses. Note that the appearance of the particles in c. is typically that of an adenovirus although the debris obscures the hexagonal outline. In d. the particles appear more round and the surface capsomers are less well defined. e. and f. Tailed bacteriophages with typical 'lollipop' appearance. All micrographs are negative contrast, PTA pH7, with a magnification of × 200 000. (b. Reproduced by permission of the Editor, *Journal of Clinical Pathology*.)

iv **Polypeptides**—at least 10 polypeptides have been found in the adenoviruses of growable strains but no studies have been reported on fastidious or non-growable ones. Five of these polypeptides correspond to hexon, penton base, fibre and two core proteins. Given a similar basic structure, these polypeptides are likely to be found in the fastidious and non-growable strains as well but the presence/location of the others is more conjectural.

v **Physico-chemical properties**—these have not been fully investigated. There is no evidence to date that they differ substantially from those of the G strains.

vi **Related viruses**—human G adenoviruses all possess a common group antigen located on the hexons. Present evidence suggests that F and NG adenoviruses also possess it. There is now evidence that the F strains include two previously unknown serotypes, 36 and 38. These have not yet been accepted officially and the position of other F and NG strains is unknown.

8. Tailed bacteriophages

Stool extracts frequently contain EM-detectable amounts of typical tailed bacteriophages (Fig. 100.4e and f). They vary in size and detailed morphology but there is no reason to doubt their identity as bacterial viruses since no animal viruses with this morphology have been described. The only reason that they are discussed here is that occasionally the numbers present are very large and in excess of 10^9/g of faeces. In such stools all have the same morphology and are likely to be the same strain. Consequently it is worth considering the effect of such a substantial infection on the bacterial flora of the gut. This might include an increase in toxic products within the gut lumen following massive bacterial lysis, a shift in the balance of particular bacterial species, or even the transduction of normally non-pathogenic bacteria into those of greater virulence. There is at present no evidence that any of these mechanisms operates but each might precipitate diarrhoea if it did. Consequently, in investigating gastro-enteropathy a record of the clinical details associated with episodes of high levels of phage excretion should be kept until more is known about their effects on gut bacteria. With an enormous choice of potential host bacteria, it is unlikely that the natural host of more than a very few of them will ever be identified. Nevertheless any opportunity to discover more about their ecology should be seized.

The detailed structure of most of these phages has not been examined but it is probably similar to other phages with the same morphology. They are likely to contain large molecular weight DNA and have a complex structure and life-cycle.

Relation to disease

Viruses are not recognized as part of the normal commensal flora of the gut or elsewhere in the body. They have not been known to confer any direct benefit on their host (except possibly that a variegated tulip may be prized more highly than a plain one). Where clearly associated with disease, they are usually found in the body only during the active phase of the illness. For these reasons there will be an assumption, to put it no higher, that the finding of a virus has some significance and that, if signs of illness are present in the same organ as that from which the virus came, the two are probably causally related.

However, the experience of virologists who culture children's stools for enteroviruses, etc., is that a number of viruses, capable of causing illness in others, may be isolated from the faeces of healthy children, particularly those in day nurseries and other crowded groups. To these virologists, the mere presence of a virus does not prove that it is causing harm and the relation of virus to disease must be considered in more detail.

All the 'newer' stool viruses have also been seen in the faeces of healthy subjects at one time or another and it is clear that disease is not invariably associated with their presence. Virus may be associated with diarrhoea only because the increased activity of the gut flushes out virus already present and being excreted in numbers too small to be detected. Alternatively there could be considerable variation in susceptibility to the virus between one child and another. There may be babies who are constitutionally more liable to gastro-intestinal illness than others although no one has yet described such a phenomenon except as a result of congenital defect. There is no evidence that viruses are involved in any of the manifestations of lactose intolerance, etc.

There is some support for the 'flushing out' concept from the absence of a constant relation in time between virus excretion and diarrhoea. When examined closely the virus may be found in faeces at any time before, during and after the period of loose stools. There is an appreciable transit time between the sites of virus replication in the upper gut and the rectum during which the diarrhoea and the virus could become separated. It is not easy to see how this could occur and leaves open the possibility that virus and illness are related only casually. If doubts can exist over whether a virus has caused illness in an individual, this may be extended to the community as a whole, although the more frequently the same virus is found in association with diarrhoea the stronger the evidence for causation becomes.

A similar argument could be applied to the amount of virus in a stool extract. The more virus present the greater the likelihood of disease being caused, but here, too, it is difficult to decide about individuals.

Owing to the impossibility of making a standard preparation of a standard extract of a standard stool, the amount of virus seen on the microscope grid does not bear a constant relation to the concentration of virus in the faeces in the descending colon or rectum. Failure to reabsorb as much water in the large gut or the addition of urine will dilute the concentration in the specimen of stool, while absorption of much of the fluid into a nappy will leave a more concentrated residue of solid matter. A formed stool may not be homogeneous and the concentration of virus will vary depending on which part is extracted; a soft or watery stool can be extracted more easily than a hard one. These factors make it difficult to compare one stool with another quantitatively even when care is taken to make the preparations as similar as possible. Nevertheless in general it is reasonable to suppose that the differences in extraction and preparation between two stools are small compared with differences in numbers which can be seen. The amount of virus actually observed in preparations of two different stools may differ by several orders of magnitude and differences as great as these are probably real although the number of virus particles released from infected cells does not have to be proportional to the amount of damage caused to the gut.

Other possible criteria of causation which have been proposed include seroconversion at the right time and the evidence from finding a similar virus in animals where the direct experiment of infecting them deliberately can be done.

The production of antibody means only that an antigenic stimulus has been registered at a level high enough to induce a response. With very large numbers of virus particles in the gut it would be much more surprising if no stimulation took place, especially as a good antibody response can follow oral poliovirus immunization in which the concentration of the vaccine virus in the stools does not reach EM-detectable levels. Similarly it is possible to demonstrate antibody production to some gut bacteriophages which do not invade the gut cells at all. The value of such seroconversion is that it pinpoints the time when the first stimulus to antibody production has been registered, but cannot prove that the stimulus was associated with overt disease.

A number of animal rotaviruses have been shown to cause illness in the natural host or in other species after administration of the virus to gnotobiotic newborns. Under these laboratory conditions, diarrhoea can be induced regularly by rotaviruses from several species including man, and the lesions of the gut wall are similar to those seen following natural infection. However it is less easy to induce a comparable diarrhoea in newborns reared normally even when they have been deprived of colostrum.

It is very tempting to extrapolate from animals to man but the parallels are not exact. In human babies the finding of excretion of rotavirus by healthy newborns is common, and disease is more likely in babies infected several weeks or months later. In contrast, animal newborns are liable to diarrhoea from birth onwards and disease-free excretion is rare. They become refractory to infection at the time that infants become susceptible.

From these arguments it becomes evident that proof of causation is difficult to obtain if the subject is man. The direct challenge experiments cannot be done and it is not certain what the outcome would be were it possible. There seems now to be little doubt that viruses may cause diarrhoea if the circumstances are right but we do not yet know what they are. The consequence of a virus assault on the body lies on a continuum between inapparent infection at one end and death at the other. The exact position on this continuum for an individual episode will depend on factors we cannot yet measure, but more complete epidemiological figures would allow an estimate of probability (percentage associated with illness) to be made.

Methods

The methods to be described in this section can be divided into those for the detection of virus or viral antigen and those for the detection of antibody. Almost all the published methods were first developed to detect rotavirus with some additions for the Norwalk group of viruses.

Detection of virus or virus antigen

All the agents discussed in this chapter except the reoviruses have been discovered by electron microscopy. Since none of them grows in routine cell cultures and electron microscopy is expensive, unsuitable for screening large numbers and insensitive, there has been a considerable need for alternative techniques to detect virus or antigen in faeces. Nevertheless electron microscopy is, and is likely to remain, an essential part of investigating acute non-bacterial diarrhoea for some time.

1. Electron microscopy (EM)

The advantages and disadvantages of EM are listed in Table 100.2. Although the disadvantages are formidable and, for poorer countries, overwhelming in terms of cost, the advantages are also considerable. As will be seen, all the other methods so far evolved depend on the use of specific antibody which can be made only if the virus has already been identified. Consequently EM is the sole method which can detect whatever virus is present in sufficient quantity. As double, triple and multiple infections are not uncommon in young children it is important to record all of them to get a

Table 100.2 Advantages and disadvantages of EM in detecting faecal viruses

Advantages
1. Will detect any type of virus present.
2. Will detect multiple infections.

Disadvantages
1. Expensive: a. High initial cost.
   b. High maintenance costs.
   c. Skilled operator needed.
2. Insensitive (at least 10^6 particles/g of faeces needed).
3. Will not distinguish different serotypes (but see text).
4. Unsuitable for screening large numbers.
5. Will detect only intact particles, antigen is not detected.
6. Some viruses are difficult to detect with certainty (e.g. some SRVs and coronaviruses).

balanced view of the patterns of infection. Only EM can provide this.

Electron microscopes are expensive to buy and maintain. Recognition and identification of virus particles depends on the fine surface detail being visible and this means that small EMs with limits of resolution above 1 nm (point to point) are not suitable. Although it can be reasonably argued that structures as small as 1 nm cannot be distinguished on the surface of the virus even with high-resolution machines, in practice other penalties of poorer illumination and less sharpness of the image come with lower resolution and make these machines more trouble than they are worth. If the amount of virus is small it is important to give the operator the best possible chance of finding it. This specifies a machine with high resolution (better than 0.5 nm).

The insensitivity of the EM is a major problem, and comes from the small amount of specimen that finally remains on the specimen grid. This is about 1 μl of an aqueous suspension and this cannot be easily increased. If the extract is purified and concentrated the relative amount of the original stool can be increased, but this involves increasing the amount of preparative work on each stool to a degree which makes it too labour-intensive for screening large numbers. In addition, a second virus may be lost in the purification procedure and thus negate one of the principal advantages of the EM.

Alternatively antibody may be used either to coat the grid or to clump the virus (immune electron microscopy, IEM), but these will bias the method in favour of that virus and distort the results. Immune electron microscopy has a number of other limitations which reduce its usefulness in recognizing or typing virus: the presence of group antigens (and antibody), the difficulties in deciding whether the virus found in a clump is aggregated by antibody or by some other factor, and the possibility that the antibody will produce a few large virus aggregates which fail to reach or remain on the grid. These make this technique not nearly as useful as some of its advocates have claimed. Immune electron microscopy works best with purified virus and least well with crude extracts. Stools may also contain copro-antibody which can interfere with any added antibody. Usually the virus particles appear free of antibody at any stage of the disease but they can have a surface fuzziness that *might* be due to attached antibody. Presence of an antibody coating is not, however, a likely explanation for the failure of these viruses to grow in cell culture.

Stool extracts may contain very large amounts of virus and these will be scored as positive after only a few seconds scanning in the microscope. Others contain much less virus or none at all; these will require much longer scrutiny. Even strongly positive extracts should be searched carefully for the possible presence of a second virus. Consequently each specimen will require at least a quarter of an hour's search. This plus operator fatigue will limit the number of specimens that can be examined in any one day to 12–15 and makes EM unsuitable for screening large numbers of stools.

The EM can be used only to detect virtually complete particles. The only known exception to this is surplus rotavirus coat protein which may assemble itself into tubular forms (Fig. 100.1c), although where they are present there will also be large numbers of characteristic virions as well. Other antigen(s) which do not form recognizable polymers might be present in large amounts but the microscopist would be unable to recognize them.

If antigen is unrecognizable, then some 'viruses' are not much easier to define. A single particle of a rotavirus, reovirus, adenovirus, calicivirus and sometimes an astrovirus can be recognized with certainty but a single SRV (plain or structured) or a coronavirus can leave even experienced microscopists undecided. Examination of a second preparation of the extract may sometimes resolve the dilemma. This is a problem to admit but not to get out of proportion. The threshold between detection and non-detection is uncertain and variable. It depends on factors, such as the texture of the stool, which are beyond the control of the investigator. It is, as has been said already, more important to record photographically any particles that *might* be virus. Their significance will emerge later.

Preparation of specimens for EM

This section is not intended to provide detailed recipes for the reader, who should consult the scientific literature for these. However, some general principles can be discussed.

Some stools contain vast numbers of virus particles. For these the method of preparation is not critical and even crude extracts will be readily found to be positive. The others present more of a problem, and some concentration of the extract will be necessary. Whatever method is used should attempt to remove large debris including bacteria, salt and large molecular weight protein, all of which will help to obscure virus particles on the microscope grid. This can be done by initial clarification in a bench centrifuge followed by high speed centrifugation, salt precipitation or the use of a hygroscopic compound (polyacrylamide or polyethylene glycol). Generally, viruses survive all the methods well but coronaviruses may lose their surface projections in centrifugation and some smaller viruses may be adsorbed on to the precipitating salt or into the hygroscopic compound.

These methods will result in a virus suspension still heavily contaminated with debris, pieces of bacterial flagella and occasional bacteria. Sometimes the centrifuge pellet may be so large that after resuspending in a salt-free diluent it is necessary to dilute it considerably to obtain a thin enough preparation. Although it is called a concentration step, centrifugation or the abstraction of water is more important for the removal of salt and soluble protein than for increasing the concentration of virus. This accounts for some reports that centrifugation, etc. is not necessary, although large amounts of salt or protein can obscure most of the specimen grid.

In addition to differential centrifugation isopyknic density gradient centrifugation has been used both to concentrate virus and to assess its buoyant density. This is too elaborate to do on all stool extracts and its main use will be in common-source outbreaks involving SRVs. Initially it will necessitate making several grids until the place of the virus in the gradient is established (and it will not always be concentrated as a visible band). It must also be remembered that the virus morphology may be altered by caesium and other salts and what infectivity there is will be reduced.

Concentrated specimen and negative stain are then either combined before addition to the grid or added to it sequentially. Different microscopists usually have their preferred stain but 2–3 per cent potassium phosphotungstate at a neutral pH is probably the best all-round routine stain.

Opinions differ over the need to disinfect the grid before examining it. There are good electron microscopic reasons for not touching the grid before or after it has been in the microscope and the author is unaware of anyone having become infected by virus from a grid. Nevertheless infectious virus has been recovered from a grid even after it has been in the microscope and most authorities recommend treatment with a disinfectant or high intensity ultraviolet irradiation. The disinfectants, including glutaraldehyde, which have been tried are all unsuitable because they damage the morphology of the virus enough to lower significantly the chances of making a positive identification. Ultraviolet light, too, has its problems, as it is necessary to use a high intensity source placed very close to the grid (i.e. about 5 cm away) to be reasonably certain of inactivating all the virus. It is also necessary to monitor the source regularly as the emission in the ultraviolet part of the spectrum falls off more quickly than the visible light.

The microscope must be calibrated if estimates of size of any particles seen are to be made. The most suitable method is to use glutaraldehyde-fixed beef liver catalase which is adequately stable and gives an assessable standard at the magnification used to look for and photograph virus. How often the microscope should be standardized will depend on its electronic stability, which should be assessed frequently at first until its performance is known.

2. Other methods

All other methods for detecting virus or viral antigen except one (polyacrylamide gel electrophoresis of nucleic acid) use antibody to fix or otherwise identify it. This is a fundamental difference and has several effects:

i Only one virus can be sought at a time. The test must be repeated to look for each additional virus.

ii The virus used to prepare the antibody must have been obtained previously and purified sufficiently to make the chances of significant contamination with another virus too small to be considered. Such separation is unlikely to be successful if both viruses are very similar in size and density.

iii None of these methods can detect a 'new' virus, unless it has some cross-reactivity with a previously known one.

The tests can therefore be complementary. The first generation test, EM, will have a place in research for some time as the only one able to detect any virus regardless of size, structure, nucleic acid or antigenic make-up, provided that enough virus is present in the stool. The second and later generation antibody capture-based tests can follow up the EM results and investigate the individual viruses. Only those research workers using EM can reasonably claim to be investigating the illness, while others can only be looking at the role of individual viruses in it.

Table 100.3, p. 438, lists (by groups) the tests available. Each group contains variations on a similar theme. Each test has been given a rating ranging from + to + + + + according to the usefulness of the technique in detecting rotavirus. Most of them have not been tried with the other viruses, but where they have been is indicated below. Since several factors have been taken into account and others may make different assessments, some amplification of these ratings is necessary.

Group I tests are poorly rated for several reasons. Immune electron microscopy is more tedious to do than straightforward EM and is not very much more sensitive, despite claims to the contrary. The best results are obtained with 'clean' virus and a concentration of antibody optimum for that amount of virus.

Table 100.3 Antibody-based tests for detecting virus in stools

Group	Type	Test	Usefulness in detecting virus*
I	Direct	Immune EM	+ +
		FA test on stool extracts	+
		FA test on fixed smears	+
II	Direct in gels	Immunodiffusion	+
		Immunoelectro-osmophoresis (IEOP, CIE, etc.)	+ +
III	Complement fixation	CFT	+ +
IV	Cell cultures	Organ culture + FA	+
		Cell culture + FA	+ + +
		a. ± Trypsin	+ + +
		b. ± Centrifugation	
V	Capture antibody	Radioimmunoassay (RIA)	+ + + +
		Enzyme-linked immunosorbent-assay (ELISA)	+ + + +
		Solid phase aggregation of coated erythrocytes (SPACE)	+ + +
		Immune adherence haemagglutination (IAHA)	+ + +

* On a scale from + to + + + +, in terms of ease of performance and suitability for screening large numbers of stools.

With crude extracts containing variable amounts of virus it is difficult to judge the right conditions and several workers, including myself, have found the technique disappointing. It was used successfully, however, to identify the Norwalk agent and its true place may be in identifying agents whose morphology is not highly characteristic. It is acceptable to spend time getting the conditions right for identifying a new virus where it would not be for screening routine specimens. Fluorescent antibody tests, direct or indirect (sandwich), can only be read reliably when intact cells are present in the preparation. Dead cells are liable to give auto-fluorescence and therefore the interpretation is difficult for the inexperienced. Immune electron microscopy is unsuitable for routine use but the other Group I tests could be used on large numbers of preparations once the practical problems have been satisfactorily solved.

The Group II tests use a gel phase in which virus and antibody form a visible precipitate. Immuno-diffusion (ID) is too insensitive for practical use and the use of IEOP does not increase it very substantially. The specificity of any line in a gel can be tested by excising and examining it in the EM, but this cannot be done on every specimen.

In Group III tests complement fixation (CF) has a traditional role in diagnostic virology laboratories. The rotavirus antigen involved appears to be the group-specific one and therefore it is not possible to distinguish different serotypes by this method or even to identify the species from which the virus originated. However, where extracts of human stools and cell-culture grown calf virus have been compared in complement fixation studies on human sera, a higher percentage reacted with the human virus, suggesting that some type-specificity is concerned in the reaction. The use of CF is also limited by a proportion of stool extracts being anti-complementary. This may be overcome by purifying the 'virus' and repeating the test. To purify a substantial number of stools is not practicable and the method is not significantly more sensitive than EM.

The Group IV tests all depend on the virus going through partial replication *in vitro*. Rotaviruses, astroviruses, and F adenoviruses have all been reported to induce new antigens detectable by immunofluorescence (IF) in cells in which no new virus is produced. The F adenoviruses vary from isolate to isolate in the amount of growth shown in various cell types. In general, though, this does not appear to lead to a progressively productive infection capable of providing as much virus as the researcher can need. With the rotaviruses, the number of IF-detectable cells is proportional to the dilution of the stool and is therefore, presumably, proportional to the 'infectious' virus present. This method can be used to quantify the virus although a number of factors limit the accuracy of the technique:

i The cultures must be fixed before staining so that a progressive increase in the effect of the virus cannot be observed directly.

ii All preparations have to be read on a microscope which is (a) tedious and (b) liable to error (has that cell been counted already?).

iii The number of focus-forming units (FFU) in a stool varies with the cells used and the conditions under which they are brought into contact with virus (see below).

Although the production of IF-detectable cells by astrovirus and F adenoviruses has been demonstrated, these phenomena have not been investigated as thoroughly as they have with rotaviruses.

The number of foci produced in cell monolayers by rotaviruses can be increased in two ways, neither fully understood:

i Centrifugation. If the cultures, to which virus-containing extract has been newly added, are centrifuged lightly (about 500 g) for 60 minutes, the number of FFU is increased 500 per cent. The amount of centrifugation is insufficient to sediment the virus and the effect is presumed to be on the cells, possibly by helping the virus to enter them.

ii The count of FFU can also be raised by the addition of 1 μg/ml trypsin and/or DEAE-dextran to the inoculum and maintenance is obscure. With trypsin, the enzyme appears to act on either the virus or the virus cell combination but the mechanism involved is not understood (R.B. Moosai, unpublished).

It is apparent, from these and other observations, that it has not so far been possible to reproduce in cell cultures *in vitro* the conditions found in the gut and its cells *in vivo*. The missing factor(s) may be enzymic (other proteolytic enzymes?), some other constituent of normal gut contents or absence of a hormonal or other influence on the cells. Failure may be due to not being able to reproduce the metabolic rate or the systems found in the gut cells. These are normally very active and, from four or five months of age onwards, will be adapting to new components of the diet.

Cell cultures are not organized tissues and the nearest approach to this *in vitro* is fetal gut organ culture. Such material is increasingly difficult to obtain and even then it has not provided a complete culture medium where it has been tried. There may be several reasons for this: for example, organ culture cells are deprived of a normal blood supply and hormonal influences, lack gut contents in contact with them and the normal villous architecture rapidly reverts to undifferentiated fibroblasts. Nevertheless evidence has been obtained that faecal coronaviruses can replicate, with the production of complete virus, in fetal gut organ culture. Some F adenoviruses can also be maintained over several passages requiring enough dilution of the culture fluid for mere carry-over to be ruled out.

The group V tests are third generation tests and differ from the earlier ones in being more sensitive and more suitable for screening larger numbers of faecal extracts. Like several of the preceding tests they depend on antibody to detect the virus and this means that they can be used only to detect viruses whose existence is already known well enough for a specific antiserum to be prepared.

RIA, ELISA and SPACE are very similar in principle differing only in the method used to tag the marker antibody (Fig. 100.5). RIA has a radioactive ^{131}I label, ELISA has an enzyme (horseradish peroxidase or alkaline phosphatase) and SPACE uses trypsinized goat erythrocytes treated with chromic chloride to attach the antibody. There are technical problems with each that mean that their specificity must be carefully evaluated before routine use. This is particularly true of ELISA where variation between batches of microtitre plates has hindered its introduction more widely into diagnostic virology in other contexts than the investigation of stool viruses. Nevertheless it is being used by a number of laboratories and seems likely to be the test of choice for rotavirus in the future.

Owing to the cost and the additional laboratory hazard introduced by radioactivity, the use of RIA has been confined to a few laboratories. This failure to adopt it more widely is probably also due in part to

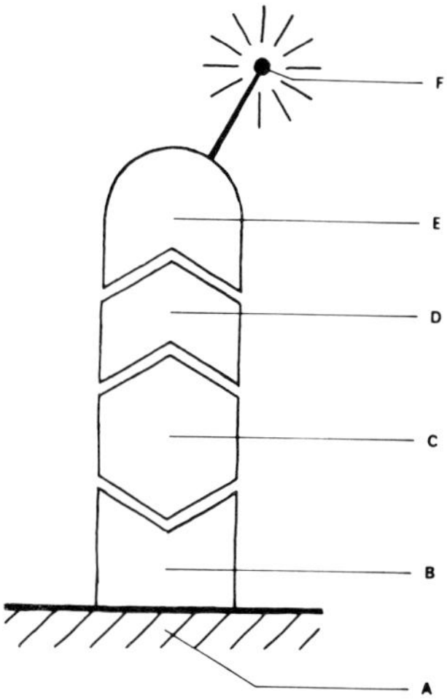

	RIA	ELISA	FAT
A	Solid phase (polystyrene plate or sphere)		Microscope slide
B	Capture antibody		Fixed cell
C	Antigen (stool extract)		Surface antigen on cell
D		Detecting antibody (made in different species to B)	
E		Anti-species antibody (made against antibody D)	
F Label	^{131}Iodine	Enzyme (Peroxidase or alkaline phosphatase)	Fluorochrome (FITC or RB200)

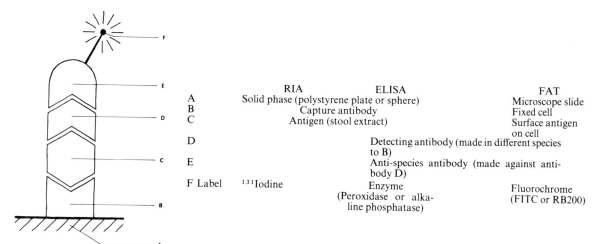

Fig. 100.5 Diagrammatic representation of components of radioimmunoassay (RIA), enzyme-linked immunosorbent assay (ELISA) or immuno-fluorescent antibody test (FAT).

the need for a highly sensitive test being less urgent than, for example, hepatitis B surface antigen for which purpose it is now used in many centres.

Immune adherence haemagglutination (IAHA) was originally developed as a test for detecting hepatitis A antigen in faeces and was adapted by Kapikian *et al.* (1978) to detect Norwalk agent in faecal extracts. In principle it is a mixture of complement fixation and the other group V tests, making use of complement bound to antigen-antibody complexes to adsorb human group O red cells after treatment of the complement/antigen/antibody mixtures with dithiothreitol. Because there are comparatively small amounts of Norwalk and related agents in faeces there have been considerable problems in finding a satisfactory test for virus or antigen. Consequently recent research has concentrated more on diagnosis by seroconversion. IAHA has been given a + + + rating although its sensitivity may be similar to RIA and ELISA. This lower rating (and a similar one for SPACE) is given because neither test has been widely adopted which suggests that they may have more technical problems than were realized at first.

Polyacrylamide gel electrophoresis (PAGE) of the eleven segments of RNA extracted from rotavirus shows a pattern of bands which is unique to this genus (Fig. 100.1d, p. 425). The patterns have been found to vary between different strains and those from different species although individual patterns are reproducible. When the gels had to be stained with ethidium bromide and viewed by high-intensity ultraviolet light this technique offered little promise in diagnosis. More recent silver-staining methods (Herring *et al.* 1982) have increased the sensitivity to the level where it is comparable to electron microscopy, IF and ELISA (R.B. Moosai, J.S.M. Peiris and C.R. Madeley, unpublished observations). Although the technique requires specialized apparatus, it is not particularly difficult and offers another method for recognizing rotavirus; it also shows some promise as a means of classifying recognizates directly. Methods for exploiting this possibility are under investigation in several laboratories and will be the subject of numerous reports in the next few years.

So far there is no evidence that this method is applicable to other viruses. Only reovirus (see p. 447) is known to have a multipartite double-stranded genome and the conditions for visualizing single-stranded genomes with larger molecular weights have yet to be worked out, and it is by no means certain that any patterns detected will be useful in diagnosing infection with particular viruses or strains of virus. However it is known that the double-stranded DNA of F adenoviruses gives characteristic patterns after digestion with endonucleases (Wadell *et al.* 1978). This is a difficult expensive technique that, so far, seems unlikely to be useful in diagnosis.

Future lines of development are likely to include means of typing individual recognizates of rota- and other viruses, and simple, reliable methods to detect viruses. The former are needed for epidemiological studies and the latter for investigations in the poorer (and often tropical) parts of the world. To be useful they should require no expensive equipment to prepare the test or to read the result, and the reagents should be stable in a tropical climate, preferably without refrigeration. Methods to improve detection by EM are most needed for SRVs and SRSVs, because existing methods will detect adequately the other more easily identifiable viruses. Finding ways of separating SRVs from other debris of similar size and density that are also suitable for use on large numbers of specimens presents a considerable challenge.

Detection of antibody

Except for electron microscopy, PAGE, all the methods described for the detection of virus or viral antigen use a known positive antiserum as part of the test. In theory, therefore, all could be used to detect antibody by substituting a known antigen for the stool extract and dilutions of the patient's serum for the positive serum. In practice, however, only the Groups III–V tests are suitable, particularly for antibody surveys. However such tests can be used only where there is an adequate supply of the appropriate virus or antigen; and it should be remembered that none of these viruses can be grown reliably to high titre and some not at all. This problem can be overcome to some extent with rotaviruses by using animal strains, particularly of the calf virus, that do grow in cell culture to provide a source of antigen. Since they are not identical with the human strains, the positivity rates obtained by using them may be lower than the true levels. The alternative approach of growing large amounts of human rotavirus in gnotobiotic piglets is not available to more than a very few laboratories.

Although it may seem simple, to substitute the patient's serum for a positive one in converting a test for antigen to one for antibody, to do so will often mean that the test has to be completely re-worked. As a result the alternative approach of doing a preliminary neutralization or blocking of a standard amount of antigen and then testing for residual antigen in the standard test makes practical sense; both methods have been used, particularly with immunofluorescence and radioimmunoassay. Complement fixation with both human and calf strains, indirect immunofluorescence, ELISA and immune adherence haemagglutination have been used in surveys of rotavirus antibodies. Immune adherence haemagglutination and immune electron microscopy have been used to detect anti-Norwalk antibodies, and indirect immunofluorescence has been used in a small survey of astrovirus antibody. Surveys of antibodies to other viruses have not yet been reported.

Because they are sensitive, require comparatively small quantities of dilute antigen and can be semi-automated, the Group V tests are likely to be the most used in future although they will not all detect the same antibodies. For example the antibodies to Norwalk virus detected by IAHA and IEM are not identical, and some sera show a response in only one of the tests (Kapikian *et al.* 1978).

Association with disease: epidemiology

The problems of proving causation discussed above mean that the evidence linking the viruses with disease is no more than circumstantial. Nevertheless there is a considerable amount of it which suggests strongly that viruses play a part in diarrhoea. Still to be established are the circumstances in which an infection becomes severe enough to induce disease; and how the existence of different serotypes, with possibly different degrees of virulence, can account for the apparent discrepancies in the evidence so far collected.

As the evidence about each virus is different, few generalizations are possible and the role of each will be discussed separately. (See also Chapter 71, Volume 3)

1. Rotavirus

Discovered in Australia, the virus has been found in every country in which it has been sought. It is so frequently found in the stools of young children, particularly those with diarrhoea, that it must be one of the most widely distributed viruses in nature and, given the necessary equipment, one of the easiest to find. There is little hard evidence about the exact number of serotypes of the human virus that exist but it is at least clear that there is more than one. All the evidence to be quoted in this sub-section should be viewed in this light. As the virus has a characteristic appearance it is probable that the overwhelming majority of recognitions have been correct. Occasionally they will have been reoviruses (see below) but this is sufficiently rare to be mentioned only for the sake of completeness and the figures will not be seriously altered.

Rotaviruses are found more frequently in the faeces of children with diarrhoea than in the faeces of normal children or adults. Most, but by no means all, of these patients are between 6 months and 2 years of age. The majority of children whose stools have been examined have been hospital in-patients, with the stool taken usually a day or two after onset of the disease. Typical of such evidence is the summary of the first two years evaluation published by Davidson and his colleagues in 1975. The percentage of positive patients of about 50 per cent is average for such studies, with some findings as low as 20 per cent and others as high as 80 per cent. These variations will be due to several factors: difficulty in defining diarrhoea, or its severity, the prevalence of the virus in the community, and the possible variations in virulence between one antigenic type and another. Evidence from the USA is that their human type 2 is more virulent than their type 1, but the questions of how many serotypes of human rotavirus there are and their relation to each other have yet to be resolved. Until this has been done and the results generally accepted, any epidemiological information must be accepted with caution.

For some years after their discovery, rotaviruses were thought to cause infections exclusively in children under the age of 5 years, with the occasions when they were observed in the stools of older patients being too rare to be of any significance. This was confirmed by the difficulties of causing disease in adult volunteers and in serological surveys which showed by several techniques (complement fixation, ELISA and immunofluorescence) that the majority of the populations of the UK, Australia and the USA had antibody from early childhood (Fig. 100.6). Subsequently outbreaks

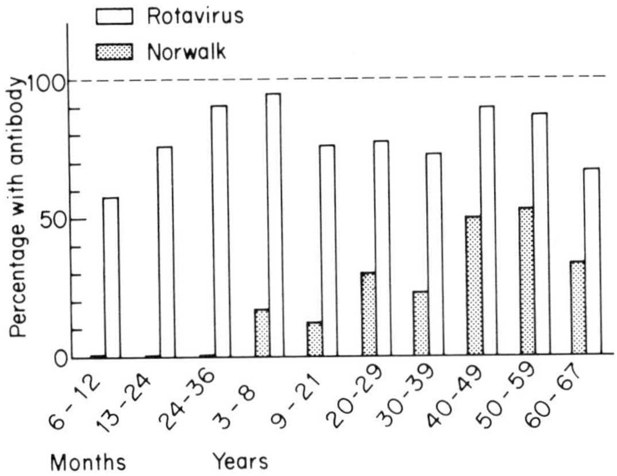

Fig. 100.6 Acquisition of antibody to rotavirus and Norwalk virus with age in Washington DC. (Adapted, with permission, from Greenberg and Kapikian, 1978.)

of diarrhoea and vomiting in adults have been reported in which the only potential pathogen to be recorded has been a rotavirus. These include a large outbreak, probably water-borne, in Sweden (Lycke *et al.* 1978) and minor ones in hospitals and old people's homes. The tests used in the serological surveys all detected group-specific antibodies, although type-specific antibodies may have been reacting as well. These surveys cannot therefore be good predictors of immunity to individual serotypes and it is possible that these outbreaks have involved a serotype not prevalent in the area previously. Another intriguing possibility is that a 'new' serotype arose by reassortment with an animal strain as has been shown to occur with influenza A, another virus with a segmented RNA genome and a substantial reservoir of animal strains. If this can happen, then one must also postulate that such novel strains are also unstable and do not spread beyond the initial outbreak. Neither explanation is entirely satisfactory and other factors must be concerned. There is probably considerable subclinical spread of virus amongst healthy (and immune?) older children and adults whose gut may possess enough immunity to other serotypes of rotavirus than the original infecting strain to prevent a second infection developing into clinical disease. However, this immunity may not be enough to prevent a superficial infection of the respiratory tract and release of amounts of virus too small to detect but enough to infect susceptible persons in the vicinity. Initiation of overt disease will be dose-related and an outbreak may follow only when in a susceptible person the virus has multiplied to a level capable of overwhelming any immunity present in contacts.

Apart from outbreaks, however, evidence of sporadic disease in adults is rare. If infection takes place, the effects are trivial. At present one can speculate about adults but direct evidence of infection is lacking. However, in very young children evidence that infection without disease can occur has been obtained in several places. In England, Scotland and Australia three separate groups found that neonates could excrete EM-detectable quantities of rotavirus without gastrointestinal upset. This immunity to disease does not apparently depend on breast-feeding although it has been observed by many that breast-fed babies are less likely to suffer from diarrhoea. Studies in London by Chrystie *et al.* (1978) showed that both breast and bottle-fed babies might excrete rotaviruses (although the former were less likely to than the latter) but such excretions were only rarely associated with diarrhoea. This early refractoriness to rotaviruses in man is in contrast to newborn animals which are usually most susceptible in the first few days of life and become rapidly and permanently resistant. There is a further difference in that there is often a significant mortality in newborn animals, even in temperate climates, that is not found in human babies except when fluid loss leads rapidly to dehydration. How this early resistance in babies is acquired is not known or how it is mediated. Pooled *human* gammaglobulin given orally as a supplement to bottle feeds was capable of preventing disease in lambs although they excreted virus and seroconverted (Snodgrass *et al.* 1977). This has been found to work in man as well. Barnes and his colleagues (1982) gave pooled human gammaglobulin orally to chidren in a premature ward with a high incidence of natural infection and showed there was a reduction in the amount and severity of illness although most of the babies still excreted virus. We know little about how maternal immunity is transferred to the child in any virus context and rotavirus is no exception. There is no doubt, from both serological and electron microscopic findings, that the rotavirus is a widespread and a very common virus in most parts of the world, so much so that failure to find virus-positive stools in any study area would cause more astonishment than finding them. The fact that the virus is perhaps the commonest virus of all does not prove that it is responsible for causing clinical disease and any evidence linking the virus with diarrhoea should be examined critically. In particular the evidence from studies in animals, even using strains of virus from human faeces, should not be accepted unconditionally. Studies in gnotobiotic animals demonstrate quite clearly that rotaviruses are capable of damaging the gut wall with loss of the normal villous architecture and reduction of the normal ratio between villus height and crypt depth. Further, virus may be readily demonstrated in the cells of the villi. Nevertheless, the outcome in normal babies may not be, and often is not, the same.

Evidence from volunteer studies is not very helpful. Few such studies have been done and had to be done in adults. It proved difficult to induce disease except by the direct introduction of virus into the duodenum and even then the resultant illness was mild. The failure probably reflected prior exposure although it is a common finding that adults of any species are not generally susceptible and the mechanisms involved may not be solely immunological.

2. Norwalk virus

Detection of Norwalk and its related viruses, Montgomery County and Hawaii, has been confined almost entirely to outbreaks rather than to endemic cases. The difficulties of recognizing the virus without using the far from satisfactory IEM technique have led the workers at the National Institute of Health, Bethesda, Md. to concentrate more on the detection of antibody responses by use of an RIA blocking test and IAHA. By these tests they have found evidence of seroconversion in some of those involved in 11 out of 25 outbreaks investigated, some retrospectively, since 1966 (Greenberg *et al.* 1979). The pathogenic potential suggested by these results has been reinforced by volunteer studies. Unlike rotavirus, the Norwalk agent can induce illness in adults and 10/16 of a series of volunteers suffered from overt illness (Dolin *et al.* 1971). The virus caused illness over two passages in the volunteers and also after three passages in organ culture.

Despite the number of outbreaks associated with this group of viruses in the USA, they have rarely been implicated elsewhere although antibody surveys showed that antibody could be found in sera obtained in Europe (Yugoslavia), Asia (Bangladesh) and S. America (Ecuador). (See also Baron *et. al.* 1982). This may reflect the difficulties in finding particles in faeces without the use of IEM or other tests with antisera which have not been widely available. It may also reflect the problems of nomenclature and the difficulties of comparing micrographs from different sources; this is discussed further under Conclusions (below).

The volunteer studies uncovered an interesting paradox. A group of 12 subjects were challenged and six became ill. They were rechallenged 27–41 months later when the six who had been ill on the first occasion became ill again, while the other six remained well (Parrino *et al.* 1977). A third challenge of four of those who had already been ill twice after a further interval of seven months showed three of them to be resistant but the fourth was again ill. The same strain of virus was used throughout and this was not apparently contaminated with another virus or another strain. With only insensitive methods of detection it is difficult to guarantee no contamination but it would be difficult to see how the second virus passed through the volunteers during the first administration without invoking some form of immune response. Interestingly, neither resistance nor susceptibility correlated with titres of antibody in the serum or jejunal fluid. It is not possible to predict which babies will develop diarrhoea after exposure to rotavirus and this could be interpreted as indicating that some are inherently less likely to develop diarrhoea. If it is indeed so, it remains to be seen whether this is a consequence of genetic or environmental (feeding?) factors. The results of challenge experiments with Norwalk virus may indicate that such susceptibility may be more important than immunity. If a general phenomenon, in deciding whether an individual becomes ill, it will make it very difficult to develop an effective vaccine.

The pattern of acquisition of antibody to Norwalk virus is also different from that to rotavirus. Antibody is acquired more slowly and, in Washington DC at least, the percentage of the population with detectable levels did not reach 50 per cent until the fifth decade of life (Fig. 100.6, p. 441).

3. Adenovirus

The problem of discussing adenovirus and diarrhoea is, except in outbreaks, of knowing which virus is the subject. It is by no means rare to isolate more than one serotype of adenovirus from a single stool. The type of adenovirus isolated in cell culture may not be the one seen by EM; and prolonged excretion is common, particularly in children, making it difficult to relate virus to illness (if any) in endemic cases. Where an outbreak occurs, a number of related cases each excreting adenoviruses raises the index of suspicion considerably but fails to explain why such outbreaks are so uncommon. Adenoviruses are probably the commonest human respiratory viruses and they may even surpass rotaviruses as common stool viruses. With 35 known cultivable serotypes plus at least two fastidious ones and possibly other non-growing serotypes, the choice of potential pathogens is large. Adenoviruses have been reported to be implicated in several UK episodes; a long stay children's ward, twins (one fatal) in London and in the married quarters of an armed forces unit near Bristol. For a common virus, this is hardly a very impressive record but is not very different from its record in the respiratory tract where it can be difficult to relate isolation to clinical disease. Investigations in Newcastle-upon-Tyne have suggested that detecting virus in desquamated cells from the respiratory tract is of greater significance than isolating virus only without detecting it in the cells. A parallel situation may occur in the gut but obtaining viable infected cells from the target organ is not ethically possible in every case. Alternatively it would help to be able to type reliably the adenoviruses detected by EM or to find a way of growing the F and NG adenoviruses to high enough titre to make typing by conventional methods feasible. Neither is possible at the moment and consequently the patterns of infection are far from clear.

Some criteria of causation and their limitations have been discussed above. With viruses which may be excreted over a period of several months it may be very difficult to pinpoint the start of the infection and in this situation recording the time at which seroconversion occurs may be valuable, although it is likely to show that most gut infections with adenoviruses are trivial and not related to episodes of diarrhoea. As with other viruses, outbreaks may be associated with particular serotypes which have yet to be identified. Figures in several studies on childhood diarrhoea have noted that between 5 per cent and 10 per cent of unselected stools will contain EM detectable adenoviruses. In Glasgow approximately the same figure was found whichever way stool specimens were separated into study groups. There was nothing in these figures to suggest that adenoviruses caused a substantial part of endemic diarrhoea although from time to time they were found in diarrhoeal stools.

To date, no volunteer studies have been attempted with the strains detected by EM and, because adult excretion is uncommon, such experiments may not help in establishing the pathogenicity of particular serotypes.

4. Astrovirus

Compared with rotaviruses and adenoviruses, the astrovirus has been found less frequently and by fewer laboratories. It has been occasionally reported in association with small outbreaks but much more frequently in the stools of endemic cases of diarrhoea as well as from those of well babies. Its superficial resem-

Table 100.4 Association of identifiable faecal viruses with gastro-intestinal disease. Glasgow figures, 5 years from June 1974 to June 1979. (Madeley, Cosgrove and Miller, unpublished data)

Virus	Associated with diarrhoea and/or vomiting (%)	?Associated* (%)	Not associated† (%)	Total (%)
Rotavirus	190‡(76)	36 (14.4)	24 (9.6)	250 (100)
Astrovirus	74 (80.4)	7 (7.6)	11 (12)	92 (100)
Calicivirus	25 (64)	6 (15.4)	8 (20.6)	39 (100)
Adenovirus		5–8% of all stools		

* Association with disease doubtful because, either
1 Not the only virus present, or
2 Not the first virus to be found, or
3 Virus not present in first stool examined after onset.
† Virus found in faeces from healthy child or child admitted to hospital for an illness not affecting the gut and in whom no gastrointestinal upset was recorded.
‡ No. of patients.

blance to an SRV may have meant that some isolates have been mis-reported. However, infections by astroviruses have one thing in common with rotaviruses and adenoviruses. All three may result in the excretion of very large numbers of virus particles such that it is difficult to believe that an electron microscopist could miss them. Nevertheless there is a significantly smaller chance of observing viruses less than 50 nm in diameter, particularly if the observer's eye is *expecting* to find larger particles. With any virus by far the most difficult particle to recognize is the first. After that it is progressively easier.

There are few published figures comparing the frequency of recognition of astroviruses and rotaviruses. Table 100.4 gives figures from examination of stools over a five-year period in Glasgow. The figures are crude and the diagnosis of illness is taken from the request cards accompanying the specimens. 'Diarrhoea' will include every degree of illness from a semi-formed stool to frequent foul-smelling or watery stools. Ill-defined though some of them must have been, they are probably no more vague than the clinical categories on which many published stool virus studies are based and there is no reason to suspect bias in favour of one virus more than another. With these provisos, the figures suggest that the astrovirus is found with about a quarter of the frequency of rotaviruses but the evidence linking both to infantile diarrhoea is similar in quality. About 80 per cent of recognizates were made from stools labelled *diarrhoeal*; in about another 10 per cent the association was less strong (another virus present in the same stool, the virus not present in the stools examined on the first or second day of illness, or not the first virus to be found), leaving about 10 per cent in which there was no suggestion of diarrhoea or vomiting within a reasonable interval (2–3 days) before or after observing the virus. On this basis there is a *prima facie* case that, if less frequent, the astrovirus is as likely to be associated with gastro-intestinal illness as the rotavirus.

There has been only one study of the prevalence of antibody to rotavirus. This was made by Kurtz and Lee (1978) in Oxford. Although the numbers were small (87 children, between birth and ten years old) the distribution suggested that a large proportion (70–75 per cent) had antibody from 3 years of age upwards, although this should be confirmed by a larger study.

One volunteer study has been published which showed that it was possible to transmit the virus to adults, and from one adult to another, although few of those receiving virus showed more than very transient symptoms (Kurtz *et al.* 1979). These results are reminiscent of those obtained with rotavirus to which adults were also far from being universally susceptible.

The main problem in collecting more complete information on the astrovirus as a pathogen is the difficulty found in recognizing and recording it. There have been reports of calicivirus labelled as an astrovirus and the rarity of reports from most parts of the world is unexpected when compared to the UK. As will be seen below, a similar patchy distribution has been found with human faecal coronaviruses.

Animal strains of astro-like viruses have been found in lambs, calves and possibly piglets. Those from lambs and calves are indistinguishable from the human strains and the lamb virus has been shown to cause diarrhoea in gnotobiotic lambs, although the disease was mild. Viruses from man, lamb and calf are antigenically distinct and there is evidence that there is more than one serotype of human virus, although this possibility has not been fully explored.

5. Calicivirus

When a new catch-all technique such as electron microscopy is being applied to a particular system in search of new viruses, the order in which new types are reported ought to reflect their increasing rarity. This is probably true of caliciviruses which can be conveniently used to illustrate some useful points about stool viruses and their epidemiology. Because animal viruses indistinguishable in morphology have been familiar for over ten years, their identity as viruses of animal origin, as opposed to phages of bacterial origin, has not been seriously questioned although they have yet to be officially welcomed into the calicivirus fold.

They were recognized in faeces almost simultaneously in Scotland and England and did not present, in Scotland at least, as strong candidates for causing diarrhoea (Table 100.4). They were less common than astroviruses and only a little over 50 per cent of them were found in diarrhoeal faeces.

Subsequently they have been the only likely pathogens implicated in several outbreaks of vomiting and diarrhoea in England and Japan. These have mostly been in school children but occasionally adults have also been affected. Caliciviruses have therefore been implicated in the complete epidemiological spectrum from epidemics to single cases to disease-free excretion. If any micro-organism is to cause outbreaks of disease then it must be disseminated first. Usually the first recognition of its presence is from an epidemic, the calicivirus story is unusual in that its endemic presence was recognized first.

Unlike rotaviruses, adenoviruses or astroviruses, caliciviruses have not yet been shown to undergo even partial replication in cell cultures. Consequently the only way of comparing different recognizates antigenically has been by the unsatisfactory technique of immune electron miscroscopy. Few comparisons have been made but they suggest that recognizates from different outbreaks differ. The animal strains can be divided into two groups, the swine and sea-lion viruses in which different serotypes show little cross-reactivity, and the cat viruses where the many isolates show considerable cross-reactivity but without a clearly demonstrable group antigen. Which pattern of relationship will be found with the human strains remains to be seen.

The evidence so far is that caliciviruses are more important as causes of small epidemics of diarrhoea than as endemic causes. Where particles with characteristic morphology are present, identification by EM presents no problems but if they are absent then an alternative method of recognition is needed and is not yet available.

It is noteworthy that, with one exception, the animal caliciviruses do not cause gut infections in their natural hosts. The exception is that a second calicivirus has been found in pig faeces which is clearly not vesicular exanthem virus. It has been associated with diarrhoea but it still remains to be established whether it can actually cause it.

6. Coronavirus

Among a collection of isometric viruses of one sort or another, the coronaviruses are unique in being enveloped. As most envelopes contain lipid, which is usually soluble in bile salts, their survival in the gut environment is worth comment although it is by no means unique, since well characterized coronaviruses are also found as causes of diarrhoea in swine and other animals. Human faecal coronaviruses are found more often in adult than in childhood faeces but prolonged excretion is usual, often extending over several months. This makes establishing a causal role as difficult as it is with adenoviruses.

As with several other viruses, coronaviruses have not been recorded in every laboratory investigating diarrhoea. The virus particles are enveloped with surface projections which can be lost if too roughly handled. Consequently the preparative method must be fairly gentle if the virus is not to be damaged. Whether inappropriate handling is the sole reason why some laboratories never record them may be doubted but no other explanation is available at present.

Coronaviruses have been reported in India, the United Kingdom and Australia. Although adults are affected, no volunteer studies have been done so far, possibly because of the difficulty of preparing adequate quantities free from other viruses. Although growth of virus in organ cell culture has been reported the amount of virus released appears to be small and the problems of proving a rule in disease are just as formidable as with the other viruses. They were first recognized as associated with small outbreaks but have never been implicated in large epidemics.

A considerable obstacle to proving a role for coronaviruses in diarrhoea is that excretion, particularly in adults, is often prolonged over several months. This makes it difficult to be sure when the first positive stool was passed.

7. Small round viruses

These are most readily recognized when implicated in common-source outbreaks. Examples are the W agent in a school epidemic, Ditchling agent, cockle agent, Parramatta agent and various other reported and unreported episodes. Within each outbreak the agent concerned is consistent in size, appearance and, when measured, buoyant density. As far as can be ascertained each agent has been different but occasionally they might be different serotypes of the same virus. For example, the Ditchling and cockle agents have similar physical characteristics when assessed in the same laboratory (Dr Hazel Appleton, personal communication) and these are sufficiently different from those found for W agent and others for them to be kept separate in the records.

By electron microscopy the SRVs are put in two groups, smooth and structured (Caul and Appleton, 1982). The smooth ones can probably be further subdivided into two groups, smaller and larger, without a clear dividing line between them at present (Table

Table 100.5 Small round viruses and disease

Outline	Division	Name (location)	Outbreak	Reference
Smooth	Smaller (approx. 22nm)	W (UK)	Vomiting, severe and mild diarrhoea. 142/830 pupils and 2 staff in a boys' boarding school.	Clarke *et al.* 1972
		Adeno-associated virus (widespread)	Casual finding in faeces from time to time often, but not invariably, found with an adenovirus in the same stool. No outbreaks reported.	
	Larger (approx. 24–25 nm)	Ditchling (UK)	'Winter vomiting disease.' 33/129 pupils and staff of a primary school affected. Food poisoning suspected but no vehicle identified. Virus particles seen in 7/8 specimens examined 1–5 days after onset. IEM suggested some cross-reactivity with W agent.	Appleton *et al.* 1977
		Cockle (various places in UK)	3 outbreaks associated with eating lightly cooked cockles. 8/9, 4/5 and 5/6 faecal specimens positive in the three outbreaks respectively. Cross-reaction with W agent not demonstrated.	Appleton & Pereira, 1977
		Parramatta (Australia)	Outbreak in Australia, involving 207/381 children and 9/18 staff of a primary school. 14/23 faecal specimens were positive for an SRV. 4 patients still excreting virus six weeks later. No cross-reactions with other agents demonstrated.	Christopher *et al.* 1978
Structured	29 nm	Atlanta (Ga. USA)	Virus in 6/6 faecal specimens from babies less than 2 years hospitalized with 'gastro-enteritis'.	Palmer *et al.* 1979

Note: Other outbreaks have been reported but insufficient data were published for inclusion in this table.
For an account of winter vomiting disease, see 6th edition, p. 2291.

100.5). These smooth-outlined viruses are different from the structured ones; this is unlikely to be due to antibody on the surface of smooth particles. The elements of the surface structure, although not clearly visualized, are more substantial than the more thread-like antibody molecules. Comparison of a group of structured small round viruses with an immune complex shows the two to be distinguishable.

The SRVs, plain or structured, are the viruses most likely to be involved in outbreaks. While there seems to be no reason why rotaviruses, astroviruses and adenoviruses should not cause epidemics they rarely seem to do so and the need therefore to know more about the SRVs means that they should be recorded whenever seen with as much detail about them as can be obtained. There is little doubt that particles resembling SRVs are also found in a proportion of endemic stools, normal and diarrhoeal, but often only in small numbers. This makes finding out more about them difficult but the technique of using caesium chloride density gradients to concentrate them (Appleton *et al.* 1977) also has a bonus of giving an indication of their density at the same time.

At present it is not possible to present a coherent picture of the epidemiology of the SRVs although some predictions are discussed under Conclusions below. By the time the next edition is written much that is now confused will be clearer.

Conclusions

A great deal has been discovered about the viruses discussed in this chapter in the past twelve years. Acute non-bacterial diarrhoea is a world-wide disease carrying a substantial mortality in the poorer and hotter parts of the globe. The lack of human or veterinary therapeutic agents to treat more than a minority of viral infections has turned research towards prevention with vaccines. The variety of viruses already identified as potential causative agents means that the task of reducing the morbidity substantially will not be easy.

The biggest handicap to further research is the inability to grow these viruses in cell culture. The reason for this is unknown and is particularly puzzling as several viruses undergo partial replication in some cells and some adenoviruses which were previously thought not to grow at all will grow, albeit not well and with little evidence of increasing adaptation, in some uncommonly used cell lines. Despite these successes, the general picture is of refractory viruses failing to grow in cells which must therefore lack some essential factor.

A further problem is that it is common for more than one virus to be present in a single stool. As many as five can be identified at a time, some by culture, some by EM. There is, however, little evidence that viruses act in synergy in this situation and multiple infection may not be accompanied by any evidence of gastro-intestinal upset.

EM identification of any virus which is present is not a problem provided some characteristic particles are seen. This condition is not always fulfilled and the operator may be left with only a probability. Additional methods to identify difficult viruses would be valuable but are not available, particularly with small round viruses.

The SRVs are of two types, smooth and structured. If one subtracts the adenovirus-associated viruses, small ($\sim$22 nm) with a slightly hexagonal outline and a fairly high proportion of empty particles, the descriptions of the remaining smooth SRVs are not incompatibly different. Their published sizes vary but the difficulties of making accurate estimates are such that these differences may prove not to be real. Where they have been measured their buoyant densities (higher than most viruses at $\sim$1.40 g/cm^3) are in the same range. The nearest virus in size and density to this group which has a definite identity is the astrovirus. To suggest, even for the sake of argument, that all smooth SRVs are really astroviruses is to ignore some of the facts about both kinds of virus. Astroviruses are frequently present in very large numbers, often in crystalline arrays, and this does not occur with any of the other SRVs. Also astroviruses are probably larger than some other SRVs and empty particles are rare. Nonetheless it is as well to keep an open mind.

There is little doubt that the SRSVs form a different group. Although they *might* be SRSVs coated with antibody this is not likely for several reasons. During the course of an infection only SRVs are found, even in the early stages; the structure frequently has a hint of order to it; its components appear more substantial than thread-like antibody, and clumps of particles do not appear to be joined together. They are often adjacent but joining threads are seen only rarely and do not appear to be part of an antibody 'hedge'.

The size of most SRSVs is greater than 30 nm but not usually more than 35 nm. This is compatible with their being caliciviruses which show no characteristic particles. The finding that Norwalk virus has a single major structural polypeptide would suggest that it, too, could be a calicivirus. Unfortunately the size of its polypeptide (at a mol.wt of 59 000) and the particle itself (at 27 nm) are both just too small. The differences are too large to be regarded as due to experimental variation and it would not be possible to accept Norwalk virus as a convincing calicivirus without a revised estimate of its size.

It would be tidier to be able to put all the smooth SRVs into one group and all the SRSVs into another. This may prove to be the final classification but there are too many incompatibles at present for it to seem very likely.

It will be necessary to know eventually whether the apparently greater prevalence of particular viruses in the catchment area of the laboratories interested in that virus is due to a greater awareness or to real variations between one area and another. Interest, after all, may have been aroused by a greater preponderance. At present it seems clear that some laboratories are more adept at finding particular viruses than others and it is not difficult to establish a tradition of finding the favourite virus of the laboratory. Nevertheless knowledge of the true distribution of each virus is important but it may be necessary to exchange particular workers and/or stools between laboratories to verify differences.

Finally there is the problem of the adenoviruses. Do they cause diarrhoea and, if so, which serotypes are responsible and are they new ones not recognized hitherto? There is little doubt that most infections due to adenoviruses produce no significant disease but the occasional one probably does. Which strains are more virulent cannot be identified at present; they can be recognized only as unequivocal adenoviruses. That there may be at least two new serotypes is intriguing. Culture of both respiratory and gut specimens has shown that several serotypes can be found in individual specimens from each source. It seems unlikely that all the strains which do not grow in cell culture are all of one novel serotype.

There is still a considerable amount of work to be done before the role of viruses in infantile diarrhoea is completely understood and, where necessary, modified by the use of effective vaccines. At the present rate of accumulation of new knowledge, there will be much to include in the next Edition. It will be interesting to see how many of today's problems will be solved by then.

Reovirus

Reoviruses are the original members of the family *Reoviridae*, which contains three mammalian genera and some insect and plant viruses. The mammalian ones are the orbiviruses (see Chapter 95), rotaviruses and the orthoreoviruses. Known insect viruses include several cytoplasmic polyhedrosis viruses, and the plant viruses include wound tumour virus, maize rough dwarf virus, rice dwarf virus and others.

The features that this apparently disparate group have in common are a roughly spherical morphology 60–80 nm in diameter and a segmented double stranded RNA genome. It is this latter feature that accounts for the enormous interest shown by virologists in the orthoreoviruses, those to be found in man, which far outweighs their importance as pathogens. The orbiviruses are important causes of disease in animals and some of them can injure man if he strays into their orbit. In general however the only genus that may be a frequent pathogen in man is the rotavirus and the reader will by now be familiar with the evidence implicating it.

The name reovirus is an acronym from *R*espiratory *E*nteric *O*rphan virus. They have been known for over 20 years and during the past ten years their structure, mode of replication and genetics have been very thoroughly investigated. A great deal is now known about them as laboratory material but in contrast they have obstinately remained viruses in search of a disease.

Originally they were thought to be ECHOviruses and reovirus type 1 was first designated ECHOvirus type 10. However, its different size and affinity to human 'O' red cells obliged a reconsideration and reoviruses were then put in a family of their own. Their name reflects recovery from the respiratory and alimentary tracts of man and the 'orphan' to the lack of disease association. The name was proposed in 1959 and in the years that followed they were isolated by many diagnostic virology laboratories with varying frequency and success but are still no more linked with disease than when they were first recognized.

1. Association with disease

There are three serotypes of mammalian reoviruses which may be found in a wide variety of domestic and wild animals as well as in man. In addition there are five avian serotypes which have been associated with respiratory or enteric disease in poultry. In man, reoviruses have been found in mild fevers and respiratory illness in children; diarrhoea has also been recorded. Since similar illnesses are common in children *without* excreting reoviruses, these associations have never been very convincing. Given that reoviruses are widely distributed and that skill in recovering them is variable, the evidence for reoviruses as pathogens is so unconvincing that the isolation of a reovirus is, in most laboratories, taken as evidence that the systems are working well but otherwise occasions little interest and no excitement.

Experimental infection of human volunteers has not been reported but inoculation of mice, particularly newborn ones, leads to steatorrhoea, jaundice and 'oily hair'. The outcome may be fatal and suggests that the virus is capable of causing damage in some hosts when presented in an appropriate way. It is therefore difficult to predict how human volunteers would react and, in the light of present evidence, there seems little point in finding out. In newborn mice the lesions are widespread in brain, spinal cord and liver and may be dose-related. The difference in man may be a result of a poorer production of virus in the gut or respiratory tract, although the concentration of virus in the faeces may occasionally reach EM-detectable levels ($>10^6$ particles/g of faeces).

2. Structure and properties of the virus

The virus particles are 75 nm in diameter (Fig. 100.7*a*). They are circular with a surface structure that may

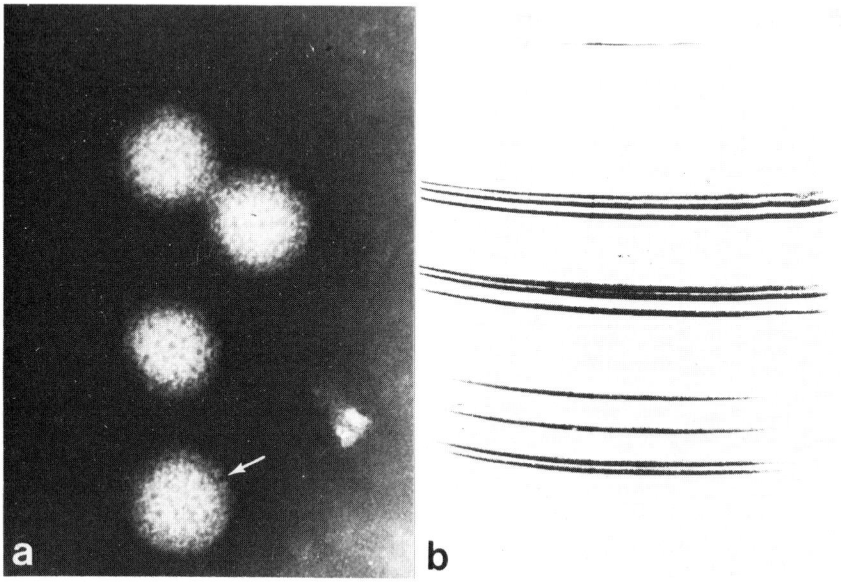

Fig. 100.7 a. Reovirus type 3 from cell culture. Note square peripheral castellations, apparently not connected to the virion (arrowed), and the clear five-fold symmetry on one of the particles. Negative contrast, PTA pH7, with a magnification of × 200 000 (reproduced from Virus Morphology by permission of Churchill Livingstone). b. Polyacrylamide gel electrophoresis of reovirus RNA extracted from cell cultures. Note 10 segments which give a very different pattern from that given by rotavirus (Fig. 100.1d, p 425). (Reproduced from McCrae and Joklik, 1978 *Virology*, **89**, 578–93.)

occasionally show clear cut order but usually does not. The periphery of the virions shows squarish capsomeres that often appear not to be directly attached to the rest of the particles. It is particularly this feature which, together with a slightly larger size and a less well-defined surface structure, distinguishes reoviruses from rotaviruses. From time to time all members of the *Reoviridae* may show ring-like surface structures but these are more rare on reoviruses than the other two mammalian genera.

Inside the outer capsid shell, there is an inner one about 50 nm in diameter which forms the outer surface of the core; the core encloses a potential cavity 38 nm in diameter, within which are the 10 segments of double stranded RNA. After entry into susceptible cells the outer shell is removed, probably by lysosomal enzymes and this releases a virion RNA transcriptase, although the core remains intact. Reoviruses are unique in that the RNA is not uncoated completely and is replicated within the core.

The ten RNA segments have a total mol.wt of about 15×10^6 and can be separated into three size groups by polyacrylamide gel electrophoresis (PAGE). They have been designated large (L1–L3, mol.wt 2.5–2.6 $\times 10^6$), medium (M1–M3, mol.wt 1.2–1.4 $\times 10^6$) and small (S1–S4, mol.wt 0.6–0.8 $\times 10^6$) (McCrae and Joklik 1978). The patterns observed (Fig. 100.7b) are quite distinct from those of rotaviruses (Fig. 100.1d p. 425) and could be used to distinguish the two in a situation where electron microscopy was equivocal.

The virions have a buoyant density in caesium chloride of 1.36–1.38 g/cm^3, and are relatively pH stable (down to pH 2.2). As expected they are ether stable and relatively resistant to chemical disinfectants. Interestingly, virus infectivity is enhanced by treatment with chymotrypsin—probably by removing the outer shell and facilitating RNA replication—in contrast to rotaviruses whose 'infectivity' for some cell lines is enhanced by trypsin but not chymotrypsin.

All three mammalian types agglutinate human O red cells but the avian types do not. The haemagglutinin is mainly type-specific with some cross-reactivity between the types. Neutralization is also type specific but the relative ease of haemagglutination inhibition makes this the method of choice for identifying isolates.

3. Growth in cell cultures

Unlike the other viruses discussed in this chapter, reoviruses will grow in cell cultures without too much difficulty although some laboratories isolate more than others. The reason for this is partly that the cytopathic effect (CPE) can be mistaken for a toxic effect. This toxicity is not uncommon with stool extracts and may require a sharp eye to distinguish a specific CPE from non-specific toxicities. Also, unlike the other viruses, a reovirus reaching EM-detectable levels in stool extracts is just as likely to grow in cell cultures.

A wide variety of both primary and continuous cells will support the growth of reoviruses. For routine isolation probably the best are Vero (continuous African green monkey kidney) cells.

4. Consequences of a segmented genome

The ten RNA segments each code for one polypeptide whose sizes do not exactly correspond to the sizes of the segments. Of these, eight are incorporated into the virus particles, one being cloven into smaller portions beforehand. The other two are non-structural proteins, presumably enzymes. Because the RNA segments are separate, infection of a cell with two viruses of different types might result in progeny virus of both original types plus some hybrids resulting from reassortment of the segments. The hybrids will have some of the properties of both parents but are likely to have predominantly the antigenic identity of one. A similar mechanism has been shown to occur with influenza A, another virus with a segmented genome. Experimental marker rescue has been demonstrated with reoviruses by means of one irradiated parent strain and a temperature-sensitive (*ts*) mutant strain as the other. Progeny virus without the *ts* modification but with some of the properties of the irradiated parent were recovered.

Reassortment of this kind may be a way of overcoming some of the difficulties experienced in growing rotaviruses. Some strains of the calf virus will grow in cell cultures and, when the correct conditions can be found, some of the characteristics, particularly antigenic identity, of human strains could be 'rescued' by the reassortment process. This might lead, with perseverance, to tailor-made viruses with desirable characteristics and particularly the production of strains of varying virulence for man. Production of hybrids and mixed infections has already been demonstrated with some bovine strains and some attempts have been made with viruses from human faeces.

5. Conclusions

Orthoreoviruses are a good example of a successful host-parasite relationship. The parasite is widely distributed in a variety of susceptible hosts, none of which it appears to damage. Interestingly, this does not hold true with the insect and plant members of the family.

The other viruses of this chapter seem likely, but not invariably, to cause malfunction of the gut as a consequence of infection, resulting in diarrhoea and/ or vomiting. Reoviruses would make an interesting comparison and, by locating the cells of the gut infected by the various viruses, it might be possible to get useful information about how viruses induce diarrhoea.

(For an account of winter vomiting disease and other adenovirus-associated diseases, see 6th edition, pages 2291 and 2523).

References

Appleton, H., Buckley, M., Thom, B.T., Cotton, J.L. and Henderson, S. (1977) *Lancet* **i**, 409.

Appleton, H. and Pereira, M. S. (1977) *Lancet* **i**, 780.

Barnes G. L. *et al.* (1982) *Lancet* **i**, 1371

Baron, R. C. *et al.* (1982) *Amer. J. Epidem.*, **115**, 163

Bishop, R. F., Davidson, G. P., Holmes, I. H. and Ruck, B. J. (1973) *Lancet* **ii**, 1281.

Caul, E. O. and Appleton, H. (1982) *J. med. Virol.* **9**, 257.

Caul, E. O., Paver, W. K. and Clarke, S. K. R. (1975) *Lancet* **i**, 1192.

Christopher, P. J., Grohman, G. S., Millsom, R. H. and Murphy, A. M. (1978) *Med. J. Aust.* **1**, 121.

Chrystie, I. L., Totterdell, B. M. and Banatvala, J. E. (1978) *Lancet* **i**, 1176.

Clarke, S. K. R. *et al.* (1972) *Brit. med. J.* **iii**, 86.

Cramblett, H. G. *et al.* (1962) *Arch. intern. Med.* **110**, 574.

Cubitt, W. D., McSwiggan, D. A. and Moore, W. (1979) *J. clin. Pathol.* **32**, 786.

Davidson, G. P., Bishop, R. F., Townley, R. R. W., Holmes, I. H. and Ruck, B. J. (1975) *Lancet* **i**, 242.

Dolin, R. *et al.* (1971) *J. infect. Dis.* **123**, 307.

Eichenwald, H. F., Ababio, A., Arky, A. M. and Hartman, A. P. (1958) *J. Amer. med. Ass.* **166**, 1563.

Flewett, T. H., Bryden, A. S. and Davies, H. A. (1973) *Lancet* **ii**, 1497.

Flewett, T. H. *et al.* (1974) *Lancet* **ii**, 61; Flewett, T. H. (1978) In: *Modern Topics in Infection.* Ed. by J. D. Williams. Wm Heinemann Medical Books Ltd., London.

Gardner, P. S., Wright, A. E. and Hale, J. H. (1961) *Brit. med. J.* **ii**, 424.

Greenberg, H. B. and Kapikian, A. Z. (1978) *J. Amer. vet. med. Assoc.* **173**, 620 (1979) *J. infect. Dis.* **139**, 564; (1981) *Perspect. Virol.* **11**, 163.

Herring, A. J., Inglis, N. F., Ojeh, C. K., Snodgrass, D. R. and Menzies, J. R. (1982) *J. clin. Microbiol.* **16**, 473.

Kapikian, A. Z., Wyatt, R. G., Dolin, R., Thornhill, T. S., Kalica, A. R. and Chanock, R. M. (1972) *J. Virol.* **10**, 1075.

Kapikian, A. Z. *et al.* (1978) *J. med. Virol.* **2**, 281.

Kidd, A. H. (1980) *PhD Thesis*, University of Glasgow.

Kurtz, J. and Lee, T. W. (1978) *Med. Microbiol. Immunol.* **166**, 227.

Kurtz, J. B., Lee, T. W., Craig, J. W. and Reed, Sylvia E. (1979) *J. med. Virol.* **3**, 221.

Lycke, E., Blomberg, J., Berg, G., Erikson, A. and Madsen, L. (1978) *Lancet* **ii**, 1056.

McCrae, M. A. and Joklik, W. K. (1978) *Virology* **89**, 578.

Madeley, C. R. (1979) *J. infect. Dis.* **139**, 519.

Madeley, C. R. and Cosgrove, B. P. (1975) *Lancet* **ii**, 124; (1976) *Ibid.* **i**, 199.

Martin, M. L., Palmer, E. L. and Middleton, P. J. (1975) *Virology* **68**, 146.

Mathan, M., Mathan, V. I., Swaminathan, S. P., Yesudoss, S. and Baker, S. J. (1975) *Lancet* **i**, 1068.

Mebus, C. A., Underdahl, N. R., Rhodes, M. B. and Twiehaus, M. J. (1969) *Univ. Nebraska agric. exp. Station Res. Bull.* No. 233.

Palmer, E. L., Martin, M. L., Hatch, M. H. and Gary, G. W. Jr. (1979) *J. gen. Virol.* **44**, 833.

Palmer, E. L., Martin, M. L. and Murphy, F. A. (1977) *J. gen. Virol.* **35**, 403.

Parrino, J. A., Schreiber, D. S., Trier, J. S., Kapikian, A. Z. and Blacklow, N. R. (1977) *New Engl. J. Med.* **297**, 86.

Report (1982) WHO/CDD/VID/82.3.

Snodgrass, D. R., Madeley, C. R., Wells, P. W. and Angus, K. W. (1977) *Infect. and Immun.* **16**, 268.

Stannard, L. M. and Schoub, B. D. (1977) *J. gen. Virol.* **37**, 435.

Thouless, M. E. (1979) *J. gen. Virol.* **44**, 187.

Wadell, G., Varsányi, T., Hammarskjöld, M-L. and Winberg, G. (1978) *Proc. IVth. Int. Cong. Virol.* p. 258. The Hague, The Netherlands.

101

Viral hepatitis

Colin R. Howard

1 Introductory

Many viruses infect the liver of animals and man and may produce severe disease. Viruses that are important in animals include infectious canine hepatitis, duck and turkey hepatitis, Rift Valley fever virus, ectromelia and the group of mouse hepatitis viruses. In man, hepatitis A and B viruses and yellow fever virus are the most important causes of acute inflammation and necrosis of the liver. Progress in the specific diagnosis of these agents has now proceeded to such an extent that it has revealed the existence of additional agents unrelated to hepatitis A and B. The 'non-A, non-B' hepatitis viruses now constitute the most common type of post-transfusion hepatitis in Europe and the USA and sporadic infection by these agents has occurred. Hepatitis is also frequently associated with other common viral infections such as cytomegalovirus (Chapter 88) and EB virus (Chapter 88). In addition there are a number of viruses that do not normally cause liver damage but nevertheless occasionally display increased hepatotropism, resulting sometimes in jaundice and a clinical picture which may be primarily that of the systemic infection or of hepatitis. This group of viruses includes the adenoviruses, rubella virus, the paramyxovirus, a number of enteroviruses, particularly Coxsackie A and B virus, and herpes simplex virus.

This chapter outlines the properties and virology of hepatitis virus A and B, the detection of these viruses and their antibodies and the involvement of 'non-A, non-B' agents in human viral hepatitis. For further details and comprehensive reviews of these agents, the reader is referred to Zuckerman and Howard (1979), Robinson (1979) and Bianchi *et al.* (1980).

2 Nomenclature of hepatitis viruses and their antigens

Although the common forms of viral hepatitis were referred to as hepatitis A and B as long ago as 1947, it is only within the last decade that these terms, together with a uniform system for the nomenclature of virus-specific antigens and antibodies, have become widely accepted. Hepatitis A replaces synonyms such as infectious hepatitis, epidemic jaundice and catarrhal jaundice, while hepatitis B is the accepted term for serum hepatitis, homologous serum jaundice and syringe jaundice.

The aetiological agent of hepatitis A is the hepatitis A virus (HAV). At present, infection is diagnosed by the recognition of the complete 27 nm particle in stools or the presence of circulating antibody (anti-HAV). There is no evidence as yet of multiple serotypes of this virus, there being a wide cross-reactivity between virus and antibody specimens collected from different parts of the globe.

Hepatitis B antigen (HBAg) is a general term to describe antigenic material produced during the expression of the genome of the hepatitis B virus (HBV). Previously used terms for HBAg include Australia antigen, SH antigen, Au/SH and hepatitis-associated antigen (HAA).

During the past few years, several studies have shown the hepatitis B surface antigen (HBsAg) to be antigenically complex. A 'group' specificity, *a*, is thought to be shared by all samples of HBsAg. In addition, the particles generally carry two sub-specific determinants which belong to two sets of generally mutually exclusive determinants, *d*/*y* and *w*/*r*. Thus at least four phenotypic combinations of 'subtypes' are possible—HBsAg/*adw*, HBsAg/*adr*, HBsAg/*ayw* and HBsAg/*ayr*. In addition, phenotypic variations or subgroups within the *w* determinant have been described. These subtypes provide valuable epidemiological markers for distinguishing the source of any particular infection. The different subtypes, however, are not associated with particular forms of liver disease. A second, unrelated antigenic system has been extensively described, the determinants of which are not generally exposed in fresh serum. This has been designated the hepatitis B core antigen (HBcAg) by virtue of its enclosure within the virus by an outer coat of HBsAg reactive material. Should phenotypic variations of HBcAg become identified, these could be indicated in a similar way to the HBsAg phenotypes.

An additional antigen, designated '*e*', is found in some HBsAg-positive sera and is specific for HBV infection. This determinant is referred to as HBeAg, and may also contain complex heterogeneous determinants designated as HBeAg/1, HBeAg/2 and HBeAg/3. It should be noted, however, that the nomenclature of *e* antigen determinants may vary between laboratories (see Zuckerman and Howard 1979, p 87). Antibodies to these various determinants are designated anti-HBs, anti-HBs/*adw*, anti-HBc, anti-HBe, etc. As yet (March 1984) no clearly identified virus or viral antigen systems have been associated with non-A, non-B hepatitis.

3 Clinical features and pathology

The clinical pattern of viral hepatitis varies in presentation from inapparent or subclinical infection, slight malaise, mild gastro-intestinal symptoms to acute icteric illness, severe prolonged jaundice, chronic liver disease, and acute fulminant hepatitis. Many cases of acute viral hepatitis remain unrecognized, the ratio of anicteric cases being as high as ten to one, particularly among children.

Acute anicteric hepatitis is accompanied by anorexia, nausea and mild gastro-intestinal disturbances. The liver is enlarged and tender, bile is present in the urine and biochemical tests reveal a rise in the content of serum transaminase. A transient rise in conjugated serum bilirubin may also be detected.

Acute icteric hepatitis is preceded by the appearance of dark urine and clay-coloured stools. Anorexia and nausea are particularly prominent symptoms, and may be accompanied by headache, generalized myalgia, arthralgia, skin rashes and fever. The jaundice usually persists from one to two weeks, with recovery following within a few months. In a proportion of patients, sequelae may occur such as subacute hepatitis, chronic hepatitis or post-hepatic cirrhosis. Although hepatitis A and B infections are impossible to distinguish on clinical or histological grounds, it is generally agreed that hepatitis B infections tend to be more severe and occasionally progress to chronicity. There is no evidence of chronic liver disease following acute hepatitis A.

A number of factors make it difficult to describe the morphological changes in the liver in a disease which varies in its severity from an asymptomatic or anicteric infection to massive necrosis. Biopsy of the liver is not required during the incubation period of the infection; in anicteric patients or in patients with the typical mild illness: moreover, biopsy should not be carried out in cases of severe liver damage or on patients with disturbed blood clotting. Some generalizations are therefore inevitable in describing the structural changes in the liver during the course of viral infection; little is known of the events in the early stages of infection. Two features are constant in acute viral hepatitis: parenchymal cell necrosis and histiocytic periportal inflammation. In general the reticulin framework of the liver is well preserved except in some cases of massive and submassive necrosis. The liver cells show various forms of necrotic changes, the necrotic areas are usually multifocal, but necrosis frequently tends to be zonal with the most severe changes in the centrilobular areas. Individual hepatocytes are commonly swollen and may show ballooning, but they can shrink. Swollen cells typically show a granular 'ground glass' appearance in the cytoplasm. Shrunken cells give rise to acidophilic bodies, and dead or dying rounded liver cells are extruded into the perisinusoidal space. The variations in the size of the nuclei and in their staining quality are useful in diagnosis. Fatty changes in the liver are conspicuous by their absence.

A mononuclear cellular infiltration, which is particularly marked in the portal zones, is the characteristic mesenchymal reaction. This is also accompanied by some proliferation of the bile ducts. In some cases, during the early phases of the illness, polymorphonuclear leukocytes and eosinophils may be prominent. The mononuclear inflammatory changes are scattered throughout the sinusoids and in parts of the lobules involved in focal necrosis. The lost hepatocytes are replaced mainly by mononuclear cells.

The Kupffer cells and endothelial cells proliferate and become enlarged, often containing excess lipofuscin pigment. In the icteric phase of the average case of hepatitis, the wall of the tributaries of the hepatic vein may be thickened and frequently infiltrated; the proliferations of the lining cells in the terminal hepatitis veins would justify the term 'endophlebitis'.

Cholestasis may occur in the early stages of viral hepatitis and plugs of bile thrombi may be seen in the bile canaliculi. Occasionally cholestatic features dominate the picture but spotty necrosis is almost invariable.

Spotty or focal necrosis with the associated mesenchymal reaction may also be found in anicteric hepatitis, although on the whole the lesions tend to be less severe. At the other end of the scale, in fulminant hepatitis, there is massive necrosis of the liver cells.

At a later stage of the evolution of the lesion of acute hepatitis there is often a variable degree of collapse and condensation of reticulin fibres, with an accumulation of ceroid pigment and available iron in large phagocytic cells, first within the lobules and later also in the portal tracts.

Repair of the liver lobule occurs by regeneration of the hepatocytes. Frequent mitoses, polyploidy, atypical cells and binucleated cells are found. There is gradual disappearance of the mononuclear cells from the portal tracts, but elongated histiocytes and fibroblasts may remain. Some scarring may follow.

The outcome of acute viral hepatitis may be morphologically complete resolution, recurrence, fatal massive necrosis, chronic persistent or aggressive hepatitis, or resolution with scarring and cirrhosis. Cirrhosis may result from extensive confluent necrosis leading to passive septum formation, architectural distortion and nodular hyperplasia, or it may follow chronic aggressive hepatitis, or both processes may occur together. Chronic liver disease following acute viral hepatitis may thus be the result of either necrosis, collapse of the reticulin framework, or the formation of scars and nodular hyperplasia. Other factors may include complex immunological processes and/or the persistence of the agents of hepatitis in the liver.

The electron microscope has been applied extensively to the examination of the fine structure of the liver biopsy material obtained from patients with viral hepatitis and in the search for virus particles in hepatocytes. Most of the changes in fine structure appear to be non-specific; indeed the changes are not dramatic and tend to be variable, ranging from minor alterations to total necrosis.

The first response of the hepatocyte to any injury is an alteration in the appearance of the rough (granular) endoplasmic reticulum, which becomes dilated and disrupted; the ribosomes, which usually line the membranes, become detached. The degree to which this dilation occurs varies from cell to cell and the swelling may be associated with formation of small vesicles as a result of disruption. When the swelling is extreme it is responsible for the appearance of the 'balloon cell' which is readily recognized in the light microscope. The changes in the rough endoplasmic reticulum are a non-specific reaction to injury. The smooth (agranular) endoplasmic reticulum is also affected and hypertrophy occurs. In some liver cells, aggregates of smooth membranes occupy large areas of the cytoplasm; these may take the form of whorls, which are seen under the light microscope as eosinophilic bodies; these may be a form of focal cytoplasmic degradation of the cell organelles. The number and size of lysosomes is increased in hepatitis. In some cells the cytosomes are large and represent autophagic vacuoles or cytoplasmic degradation. Primary mitochondrial change, particularly swelling, is usual in any type of advanced cell damage and is particularly conspicuous in the 'balloon cells'. Loss of the outer membranes of mitochondria may also be observed in the early stages of the disease. The cytoplasm of infected cells may occasionally be packed with mitochondria.

In contrast to the cytoplasmic changes, nuclear abnormalities are not usually pronounced, although pyknosis is sometimes seen. Nuclear inclusions resulting from cytoplasmic invaginations are also found.

The microvilli on the sinusoidal surface of the liver cells are decreased in number and those present are frequently oedematous. The changes in the secretory apparatus of bile are related to the degree of cholestasis. Aggregates of bile may be found in normal hepatocytes as well as in ballooned cells. Bile thrombi are found in the canaliculi, and the cell membranes in contact with the thrombi are usually damaged. The changes in the fine structure of the bile canaliculi are not specific to viral hepatitis, and thus non-specific changes are frequently found in the biliary epithelium.

The Kupffer cells are increased in size and number. The hypertrophy is due to swelling of the cytoplasm with extensive vacuole formation. Some of the vacuoles contain fragments of hepatocytes or acidophilic bodies, whereas others contain glycogen or lipofuscin. Other cells implicated in the inflammatory response are found. The endothelial layer is often multi-layered in areas of hepatocellular injury.

4 Hepatitis A virus

4.1 Epidemiology of hepatitis A

Although all age groups are susceptible to infection, hepatitis A is most common in infancy and childhood. There is a noticeable shift in age incidence, however, in countries where there has been a significant improvement in socio-economic conditions. In temperate zones there is a characteristic seasonal trend with a rise in incidence during the late summer and autumn months immediately after the holiday season, although this seasonal trend has disappeared in the USA. Hepatitis A is typically a mild disease with an estimated ten subclinical cases for every patient with clinical jaundice, and it is doubtful whether more than 50 per cent of the latter are reported to public health authorities.

Hepatitis A virus is spread by the intestinal-oral route, most commonly by close contact. Food or waterborne transmission may occur in areas with underdeveloped sanitation and inadequate treatment of sewage. In addition, ingestion of shellfish harvested from polluted water is associated with a high risk of infection. Hepatitis A is not transmitted by the use of blood or blood products, although parenteral transmission has been demonstrated experimentally in human volunteers. There is no evidence to indicate the chronic persistence of hepatitis A virus in persons exposed to this virus.

The incubation period of hepatitis A is generally between three and five weeks, although that of the well-documented MS-1 strain was found to be somewhat longer, namely five to seven weeks, when experimentally transmitted to volunteers. The observation by Deinhardt and colleagues (Deinhardt *et al.* 1967) that marmosets are susceptible to HAV provided an experimental model for hepatitis A. Although histological and biochemical changes in experimentally infected marmosets resemble those seen in man, the infection is essentially mild and infectivity is neutralized by human convalescent serum. Repeated passage of HAV in marmosets results in adaptation and shortening of the incubation period to only seven days. High titres of HAV are consistently found in the liver tissue of infected animals. Mild hepatitis similarly follows experimental infection of chimpanzees. Susceptibility of these animals strongly suggests that HAV infection was responsible for previously reported outbreaks of hepatitis A-like illness among primate handlers.

4.2 Laboratory tests of hepatitis A infection

The finding by Feinstone *et al.* (1973) of virus-like particles measuring 27 nm in diameter in faecal extracts of hepatitis A patients immediately before or during the acute illness has led to the characterization of HAV as a member of the *Picornaviridae*. The specificity of these particles may be confirmed by the formation of immune complexes after reaction with convalescent hepatitis A serum; no aggregates are found after the addition of control serum collected from experimentally infected volunteers or chimpanzees before the onset of illness. An important observation is that there is no serological cross-reaction between the antigen on the virus particles and any of the documented hepatitis B antigens. The use of immune electron microscopy was a major step in the development of hepatitis A serodiagnosis. The development of antibodies to the 27-nm particles was demonstrated in patients experimentally infected with the MS-1 strain of HAV and in patients from several different outbreaks of hepatitis A. Owing to the relatively low sensitivity of this procedure (10^4–10^5 particles/ml), the requirement for expensive electron microscopy apparatus, and the obvious limitation of the number of samples that may reasonably be accommodated, immune electron microscopy has been rapidly replaced by more convenient and more sensitive immunoassays.

Owing to the limited quantities of virus in the stools of infected patients, the successful transmission of HAV to non-human primates has been of importance in providing sufficient reagents for the development of specific serological tests. Studies in infected chimpanzees and marmosets have shown that virus is shed in large amounts only during the prodromal period of the infection; this explains the frequent failure to find virus in the faeces of patients after the onset of jaundice. Serodiagnosis of HAV infection is therefore based on the detection of serum anti-HAV known to be already present during the acute phase.

The development of complement-fixing antibody is closely associated with the appearance of neutralizing antibody. By means of an antigen extract from infected marmoset liver, it has been shown that the highest titre by complement-fixation methods is seen within the first months after the onset of the acute illness; this antibody may persist for several years. Anti-complementary activity may be detected in early serum samples, probably owing to circulating antigen-antibody complexes during or immediately after a brief viraemia. Immune adherence haemagglutination may also be used with success, being more specific, more sensitive and easier to perform than the complement-fixation test. For optimal immune adherence, the human erythrocytes should be selected with care, as red cells vary in their suitability for this type of assay.

Radioimmunoassay (RIA) and related enzyme-linked immunoassay (ELISA) procedures are now becoming widely adopted for the detection of anti-HAV. In particular, these tests may be modified for the detection of anti-HAV IgM, enabling accurate diagnosis to be established on a single serum specimen for up to several months after onset of illness. Absence of anti-HAV IgM in a sample positive for anti-HAV IgG indicates immunity due to a previous infection. Hepatitis A is only rarely diagnosed in Europeans over the age of 40, because of the high percentage of immune persons, although as hygienic standards improve an increasing number of young adults become susceptible.

The limited availability of viral antigen from human or primate sources inevitably restricts the use of these techniques. However, two groups have independently reported the isolation of HAV in tissue culture. Provost and Hilleman (1979) successfully infected fetal Rhesus monkey kidney cells with the marmoset-adapted CR 326 strain of HAV; and Frösner *et al.* (1979) detected by immunofluorescence virus-specific antigen in the cytoplasm of the PLC/PRF/5 hepatoma cell line exposed to a virus-containing human stool extract. It has recently been reported that hepatitis A virus gives rise to an abortive infection in Vero cells; although all four structural proteins may be identified within the infected cell cytoplasm, no cell-free infectious virus is produced (Locarnini *et al.* 1981).

4.3 Properties of the hepatitis A virus (HAV)

Particles containing hepatitis A antigen are found in immune complexes formed by the reaction of convalescent hepatitis A antiserum with faecal extracts and suspensions. These virus-like structures are spherical, being 27 nm in diameter and of uniform morphology (Fig. 101.1). A proportion appear empty as a result of negative stain penetration and it has been suggested that this proportion is greater during the early period of clinical illness. Complete particles possess a buoyant density of 1.34 g/ml and a sedimentation value of 160S, properties closely resembling those of poliovirus and other members of the *Picornaviridae*. Similar full and empty particles have been seen in the liver, bile and serum of experimentally infected marmosets.

Biochemical analyses of purified virus have confirmed the similar chemical composition of the agent of hepatitis A and other enteroviruses. Four polypeptides are present with approximate molecular weight of 34 000, 25 000, 23 000 and 14 000, being of similar

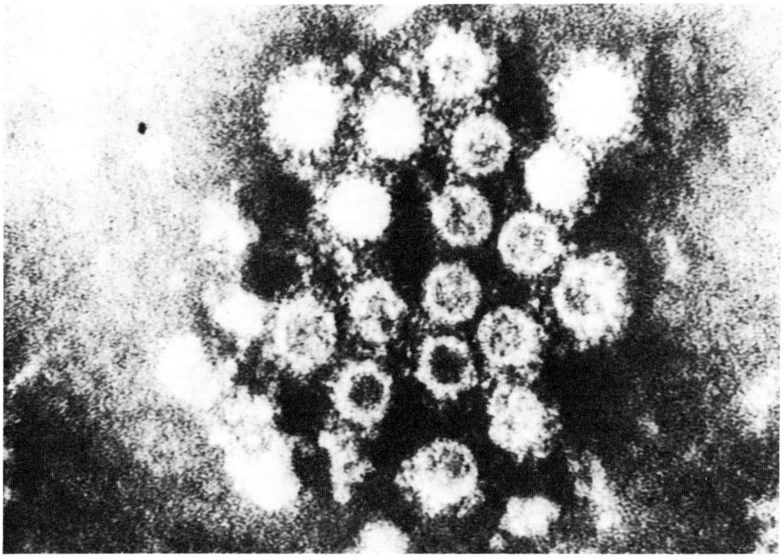

Fig. 101.1 Negative-staining electron microscopy of hepatitis A virus-convalescent-antibody immune complexes. Virus was obtained from the faeces of a chimpanzee during the acute phase of illness. Note the presence of both full and empty 27-nm diameter virus particles. (Micrograph by courtesy of Mrs A. Thornton and Professor A. J. Zuckerman.)

size to the VP1, VP2, VP3 and VP4 proteins of picornaviruses (Coulepis *et al.* 1980; Feinstone *et al.* 1978). In addition, a 59 000-mol. wt protein is present in small quantities which may resemble the smaller VPO component of poliovirus. Direct electron microscopy of solubilized virus has shown that the hepatitis A viral genome consists of a single RNA strand with a length approximately that of poliovirus RNA. Although exhaustive comparisons have yet to be performed, there seems to be little antigenic variation between different isolates. The universal protection of gamma globulin would appear also to suggest that HAV is antigenically similar throughout the world.

Little is known about the thermostability of HAV. Early transmission studies showed that infectivity was destroyed by boiling for 20 minutes but heating at 60° for one hour had little effect. The virus may also be inactivated by water chlorination, ultraviolet light and formaldehyde, but resists exposure at pH 3 for three hours.

5 Hepatitis B virus

5.1 Epidemiology

Perhaps the most remarkable epidemiological feature of hepatitis B infection is the incubation period, which may extend from two to six months before the development of clinical disease. Shorter incubation periods have been reported, however, and there may be some overlap within the range associated with hepatitis A. The enormous reservoir of HBV in persistently infected persons poses a major public health problem. The prevalence of HBsAg in adults ranges widely from 0.01–0.1 in northern Europe, North America and Australia to approximately 5 per cent in countries bordering the Mediterranean, parts of Eastern Europe, the Middle and Far East, and to 15 per cent or more in tropical areas. It has been conservatively estimated that there are some 176 million carriers of the virus in the world (Szmuness 1978). With the introduction of sensitive tests for the markers of HBV, it has become increasingly apparent that modes of transmission must exist other than direct inoculation of blood and blood products. HBsAg has been detected in saliva, semen, breast milk, amniotic fluid and in various tissue and body fluids contaminated with blood. Occasionally, the antigen has also been found in bile and urine. Close personal contact may therefore be of primary importance in the spread of hepatitis B. The practice of tattooing, ritual scarification, ear piercing and the use of inadequately sterilized instruments for other procedures such as acupuncture are other contributing factors. Syringe-transmitted infection is particularly common among those addicted to narcotics and where parenteral drug abuse is a serious problem in large metropolitan areas.

Hepatitis B is uncommon in children in western communities with the exception of children with Down's syndrome in institutions, multiply-transfused children and children with immunological defects. In contrast, in tropical areas and developing countries a significant proportion of children are asymptomatic carriers of the virus or have serological evidence of previous infection. In remaining populations, the prevalence of HBsAg increases notably in males with a general increase in the incidence of type B hepatitis among young adults.

Additional factors may be of importance in the transmission of hepatitis B in tropical areas. In particular, haemophagous insect vectors may play a role by passive transfer of virus from persistently infected carriers to susceptible persons. There has been no evidence, however, that HBV replicates in an arthropod vector. Laboratory studies of a number of mosquito species, including *Aedes aegypti*, have shown that ingested HBsAg disappears in parallel with digestion of the blood meal. It has been estimated from field studies that the minimum field infection rate for mosquitoes is of the order of 0.5 per cent. In similar studies, HBsAg has been found in bed bugs over four weeks; juvenile bugs may retain the antigen after moulting, at a time when the insect begins to search for a host and re-feed. Approximately 10 to 30 per cent of bed bugs of the species *Cimex hemipterus*, the predominant species in West Africa, have been found on capture to possess demonstrable markers of hepatitis. The particularly close association of these bugs with bed occupants who are carriers may be of prime epidemiological importance in the transmission of HBV in endemic areas. Poor socio-economic conditions generally may also encourage transmission by hitherto unidentified means, there being a close association between the prevalence of infection and socio-economic status. Vertical transmission almost certainly plays a major role in certain areas of the world, in particular the Far East where infection of neonates born to infected mothers is very common. Vertical transmission does not appear to be so extensive in African communities, although it still is an important epidemiological feature of hepatitis B in the African continent.

5.2 Morphology of hepatitis B antigens

Examination by negative staining and electron microscopy of HB antigen-containing sera reveals at least three discrete virus-like forms (Fig. 101.2), all of which are agglutinated by anti-HBs. By far the most common

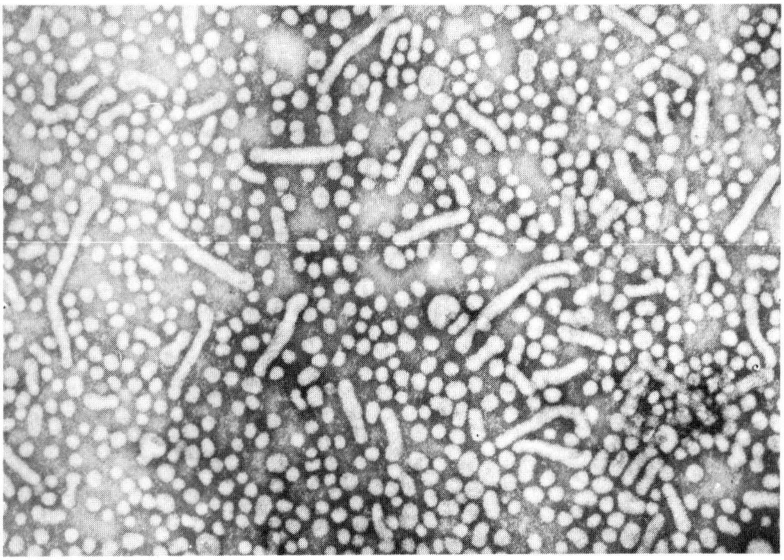

Fig. 101.2 Negative staining electron microscopy of hepatitis B antigens. Small 22-nm HBsAg spherical particles and filaments are present in the sera of persistently infected individuals. The hepatitis B virus is occasionally seen as a larger 42-nm particle with a distinct 27-nm core.

is a roughly spherical particle of variable diameter in the range 16 to 25 nm. Examination of immune aggregates shows these small particles to possess antigenic determinants in common with long filamentous forms which are also a characteristic feature of HB antigen-containing sera. Although possessing a diameter close to 20 nm, the length of these filaments may vary widely from less than 50 nm to over 200 nm. The presence of long filaments can make for difficult recovery from rate zonal gradients of a larger double-shelled particle described by Dane and colleagues (1970), seen in far fewer numbers compared to HBsAg filaments or the abundant smaller spherical particles. The detection of these particles is enhanced by the use of immune electron microscopy techniques, which reveal immune aggregates containing all three forms. Occasionally, aggregates consisting entirely of double-shelled particles are seen in serum, and it has been suggested that an additional antibody-antigen system may be present on the surface of these particles. This 42 nm double-shelled particle, a proportion of which contains the HBV genome, represents the virus of hepatitis B.

5.3 Laboratory tests for hepatitis B

There are now a large number of laboratory methods for the detection of present or past hepatitis B virus (HBV) replication. The detection of the surface antigen HBsAg (previously known as Australia or SH-antigen) remains the primary serological method for both the serodiagnosis of acute hepatitis B and the detection of persistently-infected carriers. Since the discovery of Australia antigen by the Ouchterlony double diffusion method, immunoassays of greater sensitivity have become available. Techniques such as complement fixation, immune adherence, immune electron microscopy, immunoelectrophoresis, latex agglutination and passive haemagglutination all offer reproducible and sensitive screening assays for large numbers of clinical and donor specimens. However, these 'second generation' tests for HBsAg have now been superseded by the introduction of solid-phase immunoassays, currently the most sensitive techniques for the screening of samples for hepatitis B antigens and antibodies. Radioimmunoassay techniques require expensive handling and counting apparatus; the replacement of the radioactive ligand with an enzyme-antibody conjugate, however, have made solid-phase immunoassays more widely available. These enzyme-linked immunoassay systems (ELISA) have long stability at 4° and require only the more basic laboratory equipment.

Although less sensitive than radioimmunoassay or ELISA methods, counter-immunoelectrophoresis remains a useful and simple rapid method for the detection of HBsAg and anti-HBs. Results can be obtained within two hours, as the migration of antigen and antibody into an agarose gel is enhanced in an electrophoretic field. False 'negative' results may occur as a result of either antigen or antibody excess, so that it is important to determine the optimal size of wells and/or dilution of each reagent. The use of two or more dilutions of the test sample is also advised in order to optimize the sensitivity of this technique. Although counter-immunoelectrophoresis may be up to ten times more sensitive than immunodiffusion, the results may vary according to the quality of the test reagents and applied technical skill.

Among the 'second generation' tests, reverse passive haemagglutination is perhaps the most useful and widely used procedure. Human or animal erythrocytes coated with purified anti-HBs are readily available from commercial sources for HBsAg screening. The presence of antigen in test samples is detected as a positive haemagglutination reaction after addition of sera to the coated erythrocytes. Non-specific false-positive results may occur owing to species-specific red cell agglutinins, and confirmatory tests are therefore required. This method has been recommended by the WHO Expert Committee on Viral Hepatitis for use as a screening test for HBsAg (WHO 1977). Erythrocytes coated with HBsAg may similarly be used for detecting the presence of anti-HBs. This test is very sensitive, being as much as 10 000 times more sensitive than the Ouchterlony gel diffusion test, although the relatively short shelf life of the reagent is a notable disadvantage.

Solid-phase radioimmunoassay is performed in a two-step procedure whereby the test sample is incubated with a polystyrene bead, tube or plate previously coated with anti-HBs. After extensive rinsing, the solid phase is incubated a second time with specific anti-HBs IgG radiolabelled with ^{125}I. The amount of residual radioactivity present is measured after a second wash and is directly proportional to the quantity of antigen-antibody complexes remaining on the polystyrene solid phase. This 'sandwich-type' technique has also been used for the subtyping of HBsAg samples by means of monospecific labelled anti-HBs, and may be reversed for the detection of antibody in serum samples.

This latter technique, however, does not produce the equivalent degree of sensitivity for anti-HBs as is obtained for the antigen screening. ELISA techniques are essentially similar in design, although the use of microtitre plates as the solid-phase is generally favoured. The presence of bound HBsAg is detected after the addition of a colourless enzyme substrate; enzyme hydrolysis results in a coloured product which can be assessed visually or measured by spectrophotometry.

The detection of anti-HBc has become increasingly recognized as a valuable adjunct to hepatitis B serodiagnosis. Of several approaches available, the most suitable is a solid-phase radioimmunoassay or ELISA technique similar in design to those developed for HBsAg detection. Core antigen is prepared either from infected liver tissue or by the disruption of circulating HBV with the aid of non-ionic detergents and subsequent purification of the released HBcAg material. The amount of antibody to HBcAg in a test sample is determined by competition with labelled anti-HBc, generally of human origin, for the adsorbed HBcAg on the solid phase. After rinsing, the amount of residual label present is in indirect proportion to the amount of competing unlabelled antibody present in the test sample. The detection of anti-HBc of the IgM sub-class is becoming of increasing value for the serodiagnosis of acute infection in those rare cases of hepatitis B where HBsAg is not detected by established methods. This antibody is also particularly prominent in the sera of HBsAg-positive donors which support a high level of virus activity as indicated by the dual presence of virus-associated DNA polymerase and HBeAg.

For a detailed description of the use of radioimmunoassay and ELISA techniques, the reader is referred to Howard (1982).

5.4 Association of hepatitis B antigens with hepatitis B infection

Experimental studies carried out by Krugman and colleagues at the Willowbrook State School in the 1960s (see Krugman *et al.* 1967) clearly demonstrated the existence of at least two forms of viral hepatitis. Each one clinically and epidemiologically resembled those described previously in adult human volunteer experiments. Two serum pools (MS-1 and MS-2) obtained from one individual in the Willowbrook study on two separate occasions produced short- and long-incubation period hepatitis respectively in human transmission studies. Further experiments showed that there was no cross-immunity between hepatitis induced by MS-1 (infectious hepatitis) and the disease induced by MS-2 (serum hepatitis). Although parenteral inoculation is the major route of serum hepatitis transmission, these studies clearly showed that MS-2 could be infectious when administered orally, confirming the long-standing clinical observation of secondary infection in the absence of apparent parenteral inoculation in close contacts of cases of serum hepatitis.

The identification of an antigen closely associated with the causative agent of serum hepatitis represented a considerable advance in the understanding of type B viral hepatitis. This antigen (Australia antigen) was found incidentally by Blumberg (1964) during an investigation of lipoprotein allotype precipitins. The new precipitin was identified in the serum of a haemophiliac as a result of its affinity for an antigen present in the serum of an Australian aborigine. Although initially regarded as a recessive trait, a relation was soon recognized between Australia antigen and leukaemia, Down's syndrome, lepromatous leprosy and hepatitis. The electron microscopy of MS-2 serum has shown this material to contain both the small 16–25-nm diameter spherical and the tubular forms of HBsAg in addition to the complex 42-nm double-shelled particle. These forms morphologically and serologically resembled two of the structures described originally in sera containing Australia antigen. The antigenic determinants associated with these virus-like structures are now referred to as HBsAg and HBcAg (see p. 452).

HBcAg is first detected in the serum of an infected person on average four weeks prior to clinical or laboratory evidence of liver damage and may persist in most cases until the onset of symptoms and liver

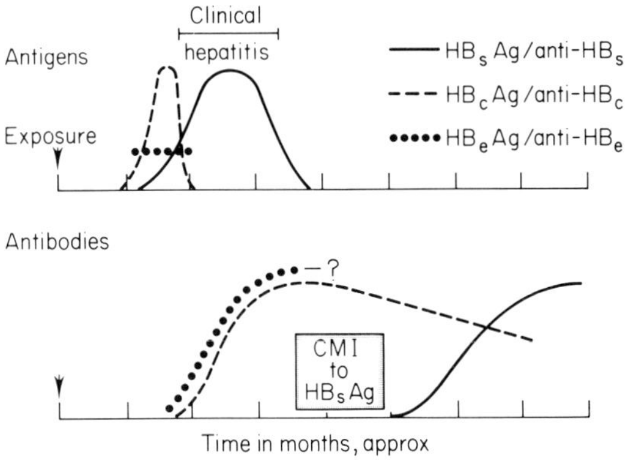

Fig. 101.3 Diagrammatic representation of the appearance of antigens (upper panel) and antibodies (lower panel) during acute hepatitis B.

dysfunction (Fig. 101.3). In the majority of naturally occurring cases of serum hepatitis, HBsAg is most likely to be detected during the first week of the acute phase of illness and may persist from a few days to several weeks. Persistence of antigenaemia is known to occur in a small percentage of cases; HBsAg has been continually detected in the serum of some persons over periods as long as 20 years.

Anti-HBs generally develops some weeks or months after recovery, albeit at a low titre detectable only by sensitive assay methods such as radioimmunoassay or passive haemagglutination. During antigenaemia, anti-HBs can occasionally be detected by electron microscopy as circulating antigen-antibody complexes. Analysis of these complexes has shown precipitating antibody to HBsAg in acute hepatitis to contain IgG, IgM and IgA. No significant change in the content of immunoglobulin is found at the onset of the acute phase, but subsequent serial determinations may reveal a fall in IgG. This decrease may reflect an immunosuppression process related to the evolution of chronic antigenaemia as observed in some patients with HBsAg-positive hepatitis. A high anti-HBs titre may occur when there is a history of repeated exposure to the antigen, often in the absence of clinical disease.

The use of a sensitive radioimmunoassay technique has shown that over 80 per cent of haemophiliacs in the USA possess circulating anti-HBs as opposed to 15 per cent or less in the general blood donor population.

In parallel with the development of a humoral response to HBsAg, specific cell-mediated immunity has been demonstrated. Delayed hypersensitivity to HBsAg, as measured by the leucocyte migration test, has been shown in patients recovering from type B viral hepatitis. The positive cell-mediated response appears to be transient, beginning two to three months after the onset of disease and is more or less simultaneous with the clearing of HBsAg from the circulation. There is no evidence as yet, however, that cytotoxic T cells play any role in the immunopathology of acute hepatitis B.

The presence of antibody in convalescent sera to the inner HBcAg component of the complex 42-nm double-shelled particle may be clearly demonstrated. HBcAg immune complexes closely resemble those that may be obtained from homogenates of infected liver. It has been suggested that recovery from HBV infection is accompanied by a relatively short lived anti-HBs response, while a normal immune response of greater longevity is produced against HBcAg. This hypothesis is supported by the observation of core-like particles confined to the nuclei of infected hepatocytes. The availability of infected chimpanzee liver containing HBcAg has led to the development of a complement-fixation test to this antigen. Using this procedure it has been demonstrated that anti-HBc appears during or immediately after HBsAg antigenaemia. A strong association with the persistence of HBsAg suggests that anti-HBc is produced in response to the active replication of this virus. The relatively high frequency of anti-HBc at low titre in the population indicates that HBV infection may frequently be inapparent and probably transient.

However, a mild form of acute hepatitis may provide an unusually good background for the development of severe HBsAg-associated chronic hepatitis. In addition, extra-hepatic lesions may be associated with the presence of HBsAg, e.g. polyarteritis nodosa, hepatocellular carcinoma and some cases of glomerulonephritis in children.

Chronic hepatitis may develop in 10 per cent of patients admitted to hospital with acute icteric type B hepatitis. In one study, a third of these patients suffered from chronic active hepatitis, displaying a spectrum of hepatic lesions and sporadic episodes of jaundice. The remaining two-thirds showed signs of persistence of HBsAg associated with a continuing elevation of transaminase levels, but were otherwise in good health. In the latter group, resolution of chronic persistent hepatitis may occur over one to three years, although serum HBsAg persists. Comparison of HBsAg titres between the two groups has shown a significantly higher titre of circulating HBsAg in cases of chronic persistent hepatitis.

In general terms, the pathology of HBsAg-associated chronic aggressive hepatitis closely resembles the clinical syndrome of active chronic hepatitis in which 15–20 per cent of patients possess circulating HBsAg. However, of the remaining HBsAg-negative active chronic hepatitis cases, over 60 per cent are found to possess a significant delayed hypersensitivity response to HBsAg, suggesting that a past exposure to HBV may have represented an important event in the development of chronic liver disease. This is further indicated by clinical observations of patients progressing from HBsAg-positive acute to HBsAg-negative active chronic hepatitis. Successful immunosuppressive therapy for the treatment of active chronic hepatitis has implicated autoimmunity as an important factor in the pathogenesis of this condition. Eddleston and Williams (1974) suggested that the stimulation of sensitized T-cell lymphocytes to the surface of normal uninfected hepatocytes is one result of viral-induced changes at the plasma membrane of infected cells. Immunofluorescence has frequently been used to study the distribution of HBV gene products *in vivo*. HBcAg reactivity appears by this technique to be restricted to the nucleus or perinuclear region, whereas HBsAg is confined to the cytoplasm. This duality of reactivity has additionally been observed in experimentally infected chimpanzees. During the early stages of acute hepatitis, both reactivities are present in liver tissue. HBsAg is typically distributed in a diffuse fashion in the cytoplasm of hepatocytes throughout the liver. The restriction of HBcAg to the nucleus is accompanied by enlargement of the nucleoli and extensive proliferation of the chromatin. A spectrum of antigen expression has been reported in patients with type B acute and chronic hepatitis as assessed by immunofluorescence, ranging from extensive HBsAg reactivity in cases of chronic persistent hepatitis accompanied by few HBcAg-positive nuclei to focal areas of limited HBsAg and HBcAg in every nucleus seen in heavily immunosuppressed transplant patients. The pattern of fluorescence in cases of chronic aggressive hepatitis may be seen as intermediate with equal expression of each antigen in focal areas. These findings indicate that the immune response is of paramount importance in determining the course of the disease. The presence of the HBeAg antigenic moiety distinct from the HBs/HBc systems appears to predispose the patient to the development of the chronic disease. HBeAg is often present in the sera of persistent carriers of HBsAg found in haemodialysis units, but only infrequently among chronic carriers in the donor populations. A close relation has been shown between HBeAg and the presence of HBcAg. In contrast, carriers of HBsAg in the donor population show no histological or biochemical signs of liver disease. Hence the presence or absence of HBeAg may be an indication of HBV infectivity and the subsequent course of the disease after infection.

Studies of the HBsAg subtypes associated with type B viral hepatitis showed that the *ay* subtype predominated in outbreaks associated with haemodialysis units, whereas the predominant subtype in asymptomatic carriers was *ad*. It has been suggested, therefore, that an HBs/*ad* antigenaemia promotes the formation of detectable quantities of precipitating antibodies against other specificites such as HBcAg. However, in geographical areas where *ayw* is the predominant subtype, its occurrence is associated with all categories of acute and chronic hepatitis as well as with asymptomatic carriage. The opposite appears to be true for zones where *adw* predominates. However, in mixed-zone populations, *adw* is found at a high frequency among volunteer blood donors as well as patients with chronic aggressive hepatitis, whereas *ayw* is more frequently encountered in haemodialysis units in addition to drug abusers and their contacts. Both subtypes *adw* and *ayw* may be associated with all forms of acute and chronic type B viral hepatitis, although the geographical origin of the HBsAg-positive individual and possibly the route of infection may influence the subtype findings in any one area.

5.5 The properties of hepatitis B antigen

5.5.1 Hepatitis B surface antigen (HBsAg)

The overwhelming mass of HBsAg-reactive material in the sera of hepatitis B infected persons is present as small 22-nm lipoprotein particles which are morphologically distinct from the 42-nm hepatitis B virus particle. The relation of the filamentous HBsAg structures to either whole virus or 22-nm particles remains unclear. Purified 22-nm particles are readily obtained by ultracentrifugation, the relatively slowly sedimenting small particles having a mean sedimentation coefficient ($S_{20,W}$) in the range of 33 to 45S. The molecular weight of these structures is estimated at 2.5×10^6, of which up to 30 per cent may be accounted for by the lipid moiety. The physicochemical properties of these particles are given by Howard and Burrell (1976).

Early studies involving the treatment of purified HBsAg with organic solvents and dissociating reagents showed that HBsAg immunoreactivity was remarkably stable in the presence of compounds promoting denaturation, in particular diethyl ether, urea, sodium dodecyl sulphate and resistant to various proteolytic enzymes. There was no loss of HBsAg reactivity after treatment with 50 per cent chloroform or diethyl ether. However, there was a complete loss of reactivity after exposure to ethanol. A similar loss has been reported after treatment with butanol. Several studies have also shown HBsAg to be stable for many hours at an acidic pH.

The reduction of disulphide bonds results in the substantial loss of HBsAg reactivity, although considerable antigenic activity may be regained by the alkylation of free sulphydryl groups with iodoacetamide. After alkylation, intact particles with a sedimentation constant of 31S are re-formed. Reduction-sensitive and reduction-resistant components of HBsAg have been defined. The group determinant *a* is destroyed by exposure to dithiothreitol at concentrations below 10 mmol. At higher concentrations, resistance to reduction is serologically detected in all HBsAg preparations, regardless of the subtype determinants present.

The reactivity of HBsAg is remarkably heat-stable,

there being no loss of reactivity after heating purified antigen for ten hours at 60°, but heating for five minutes at 100° completely abolishes affinity for anti-HBs. The *a* group specific determinant is stable at 60° for periods of up to 21 hours, whereas the *d* and *y* subtype reactivities are reduced after only three hours of incubation at the same temperature. The stability of HBsAg at high temperatures, together with resistance to protease digestion, strongly suggests the presence of carbohydrate, a 90 per cent reduction in the serological activity of purified HBsAg particles occurring after treatment with 0.01 molar sodium periodate for four hours at 37°.

Widely differing results have been reported on the number and size of the polypeptides found in HBsAg 22-nm particles purified from donor serum and plasma. Early studies suggested the presence of two or possibly three major polypeptides in the molecular weight range of 22 000 to 40 000. Although more exhaustive analyses have shown that up to seven polypeptides may be present in HBsAg of different subtypes, only two with molecular weights of 23 000 and 28 000 represent major virus-specific proteins. The most notable of the remaining components is serum albumin which may be present in considerable quantities; purified HBsAg 22-nm particles may therefore represent a mixture of host and virus-specific material. The major 23 000 and 28 000 mol. wt polypeptides have a similar amino-acid composition, the presence of carbohydrate side-chains accounting for the larger size of the 28 000 mol. wt component. Nucleotide sequence data are now available for the HBV genome and the gene coding for the 23 000 mol. wt protein. The full amino-acid sequence predicted from these studies shows that the HBsAg viral polypeptides contain highly hydrophilic sequences compatible with intrinsic membrane proteins (Fig. 101.4). From the evidence obtained so far, it would appear that HBsAg particles contain at least two virus-specific polypeptides resulting from transcription of the same gene, one of which is glycosylated, and together form a sub-unit structure via disulphide bonds. Other components, e.g. albumin, presumably become highly associated with HBsAg particles during maturation within the infected hepatocyte. Delipidation is insufficient to release any protein components, indicating the importance of protein-protein interactions in maintaining both the structural and serological integrity of HBsAg.

Little attention has been paid so far to either the

	1								10										
H_2N.	met	glu	asn	ile	thr	ser	gly	phe	leu	gly	pro	leu	leu	val	leu	gln	ala	gly	phe
20										30									
phe	leu	leu	thr	arg	ile	leu	thr	ile	pro	gln	ser	leu	asp	ser	trp	trp	thr	ser	leu
40										50									
asn	phe	leu	gly	gly	ser	pro	val	cys	leu	gly	gln	asn	ser	gln	ser	pro	thr	ser	asn
60										70									
his	ser	pro	thr	ser	cys	pro	pro	ile	cys	pro	gly	tyr	arg	trp	met	cys	leu	arg	arg
								thr											
80										90									
phe	ile	ile	phe	leu	phe	ile	leu	leu	leu	cys	leu	ile	phe	leu	leu	val	leu	leu	asp
100										110									
tyr	gln	gly	met	leu	pro	val	cys	pro	leu	ile	pro	gly	ser	thr	thr	thr	ser	thr	gly
														ser					
120										130									
pro	cys	lys	thr	cys	thr	thr	pro	ala	gln	gly	asn	ser	met	phe	pro	ser	cys	cys	cys
		arg			*met*		*thr*				*thr*			*tyr*					
140										150									
thr	lys	pro	thr	asp	gly	asn	cys	thr	cys	ile	pro	ile	pro	ser	ser	trp	ala	phe	ala
																			gly
160										170									
lys	tyr	leu	trp	glu	trp	ala	ser	val	arg	phe	ser	trp	leu	ser	leu	leu	val	pro	phe
	phe		*ser*					*ala*											
180										190									
val	gln	trp	phe	val	gly	leu	ser	pro	thr	val	trp	leu	ser	ala	ile	trp	met	met	trp
														val					
200										210									
tyr	trp	gly	pro	ser	leu	tyr	ser	ile	val	ser	pro	phe	ile	pro	leu	leu	pro	ile	phe
									leu				*leu*						
220						226													
phe	cys	leu	trp	val	tyr	ile.	**COOH**												

Fig. 101.4 Predicted amino acid sequence of the HBsAg 23 000–25 000 mol. wt polypeptide, subtype *adw* as reported by Valenzuela *et al.* (1979). The corresponding sequence for the *ayw* subtype gene suggests a total of 14 amino acid changes (in italics). Carbohydrate side chains may be attached via asparagine (*asn*) residues at positions 3, 59 and 146. There is a long hydrophobic sequence of 19 amino acids between positions 80 and 98 inclusive.

biochemical or immunochemical nature of the HBsAg material found in the outer coat of the hepatitis B virus particle. Immune electron microscopy suggests the presence of at least some common antigenic determinants between the complete virus particles and HBsAg 22-nm particles and serological studies have confirmed the presence of either *d* or *y* HBsAg subdeterminants. Alberti *et al.* (1978) have reported the existence of a determinant unique to the surface of the hepatitis B virion. Confirmation of this finding will be of paramount importance for the future design of experimental hepatitis B vaccines. As yet, there is little biochemical evidence to suggest the existence of correspondingly unique polypeptides in the outer surface coat.

5.5.2 The inner core component

The removal of the outer surface coat of HBV with non-ionic detergents results in the release of serologically reactive core antigen. Two density populations of core particles may be released in this manner from circulating virus particles; the heavier particles banded at 1.35–1.36 g/cm^3 in caesium chloride correspond to the 110S DNA polymerase-positive core component described by Kaplan *et al.* (1973). A more heterogeneous population is also frequently recovered in the lighter density range of 1.28–1.32 g/cm^3; this represents free cores without DNA polymerase in addition to cores containing DNA polymerase complexed with core antibody. It thus seems that hepatitis B virus of higher buoyant density gives rise after detergent treatment to DNA containing core particles with DNA polymerase activity, whereas the lighter band of virus particles contains predominantly low density core components with little or no DNA content and may represent incomplete or defective virus particles. This is consistent with the enhanced degree of penetration by negative stain observed with particles of lower density.

HBcAg prepared from serum containing *e* antigen produces two peaks by equilibrium centrifugation in contrast to only one peak from serum without detectable *e* antigen. This suggests that core antigen may exist in different conformations independent of whether or not DNA is present.

Although much attention has been devoted to the structure and nature of surface antigen polypeptides, fewer attempts have been made to characterize in detail the protein composition of the virus core. In the past this has been due partly to the scarcity of suitable assays but also to the considerable difficulty of either separating sufficient quantity of the antigen from circulating virus particles or preventing the formation of immune complexes during the process of extracting core antigen from infected livers.

The staining of separated components with Coomassie Blue shows the presence of a major 17–19 000 mol. wt component, together with minor components in the range of 35 000, 70 000 and 80 000 mol. wt respectively. The association of DNA with the core antigen does not appear to require a gross change in polypeptide composition. Periodate-Schiff staining of separated components did not reveal the presence of glycoproteins at any stage.

A protein kinase activity has recently been identified in association with hepatitis B core particles obtained either from infected liver or complete virus particles (Albin and Robinson 1980). A proportion of the major core polypeptide was identified as the primary phosphorylated product migrating with a slightly increased molecular weight of 20 600 as estimated by gel electrophoresis. During the course of these studies, it was noted that this phosphorylated component underwent cleavage to give polypeptides with molecular weights of 14 700 and 6000. This process was accelerated by the specific addition of anti-HBc, suggesting that immune complex formation activated a protease activity, possibly located within the core component. It remains to be established what role this enzyme plays in hepatitis B virus replication, although it should be noted that many enveloped viruses possess protein kinase activities for modification of virus proteins closely associated with viral nucleic acids.

The immunoreactivity of HBcAg is affected by solvents such as 50 per cent methanol, ethanol and butanol. In addition, the reactivity of core antigen with specific antibody appears to be dependent on the presence of intact sulphydryl groups and possibly internal hydrophobic bonding. The morphology and immunoreactivity of HBcAg particles is destroyed by exposure to pH below 3.0 or by heating at 100°.

Recent evidence concerning the nature of hepatitis B *e* antigen has led to the suggestion that this antigenic specificity may be closely associated with HBcAg. Although the two antigens do not give lines of identity by immunodiffusion, several recent studies have clearly shown that solubilization of either core components from infected liver or intact hepatitis B virus particles results in the release of material which reacts with *e* antibody.

The core component polypeptides with molecular weights of 45 000 and 19 000 respectively have been found to react with anti-HBe, and solubilized virus particles elicit an anti-HBe response in immunized rabbits (Takahashi *et al.* 1980). HBcAg activity has been detected in *Esch. coli* cells after transfection with plasmids or phage vectors containing the whole, or fragments of, HBV DNA (Burrell *et al.* 1979; Pasek *et al.* 1979). Although detailed amino acid analyses for purified HBcAg polypeptides are not available, Pasek and colleagues have suggested that HBcAg activity resides in a 183 amino acid long polypeptide. Arginine-rich regions are predicted towards the carboxyl-terminus of the 21 000 mol. wt polypeptide, indicating that the carboxyl-terminal region binds with the DNA genome within the virus particle while the remaining regions fulfil a structural role. In this context, it is interesting to note that *e* antigen released from disrupted virions has been found to be maximum for fractions containing DNA polymerase and the DNA genome, suggesting that polypeptides bearing this antigen may function during virus assembly as a maturation protein.

Other proteins have been described, in particular a 45 000 mol. wt protein which is reactive for HBeAg and a 68 000 mol. wt protein as a minor component of intact cores. This closely resembles in size the newly described hepatitis B 'δ' antigen. This new antigen is found by immunofluorescence in the hepatocyte nucleus, where it gives rise to a staining pattern similar, but unrelated, to HBcAg. High titres of antibody to delta antigen occur simultaneously in the circulation of those chronically infected patients who exhibit hepatic antigen (Rizetto *et al.* 1979). Extraction of infected liver has been successful in providing a partly purified delta antigen preparation with a molecular weight of 68 000. Prolonged exposure to chaotropic salts or low pH led to a rapid loss of antigenicity, however; it is possible that delta antigen therefore represents an intermediate component of hepatitis B virus assembly. Other alternatives yet to be excluded are suggestions that delta antigen is a component of a hepatitis B variant or represents an agent which is defective and interferes with hepatitis B virus replication.

Other intermediate components of core assembly may exist which express both HBeAg and HBcAg determinants, with the latter being essentially a conformationally-dependent antigen. Although structures morphologically resembling cores are never seen free in HBsAg-positive sera, proteins bearing these determinants may exist as immune complexes. Indeed, the particularly high titres of anti-HBs in persistently infected persons may be explained by a continual release of excess core components from infected cells. The heterogeneity of the HBeAg system as found by the immunodiffusion analysis of positive sera may also be explained in terms of free and bound antigen.

5.5.3 The hepatitis B virus genome

The discovery of an endogenously-primed DNA polymerase activity within hepatitis B virions provided the initial step in characterizing the genome and ultimately led to a description of its complete nucleotide sequence. Kaplan *et al.* (1973) demonstrated that incorporation of deoxynucleotides into an acid insoluble product followed the removal of the outer surface antigen coat by treatment with a non-ionic detergent. The reaction proceeds in the absence of an exogenous template and is stimulated by the presence of magnesium and ammonium ions. Although it is by no means clear that hepatitis B DNA polymerase is specified by the viral genome, a comparison of the ionic requirements of the enzyme and its sensitivity to *N*-ethylmaleimide suggest that this polymerase is unique to hepatitis B infection (see Zuckerman and Howard 1979, Chapter 11).

Despite the somewhat specialized nature of the assay, the detection of DNA polymerase is a useful marker of virus replication and in some infections may be the only marker of hepatitis B, especially during the prodromal period of acute illness. The precise conditions required for the assay of the enzyme for diagnosis of hepatitis B have been reviewed by Howard (1978). It has also been noted that there is frequently a close association between DNA polymerase activity and the titre of HBsAg in serum.

The finding of an apparently virus-specific nucleic acid polymerase with unique properties stimulated laboratory and clinical investigation of a number of antiviral compounds which might be of value in the treatment of chronic hepatitis B infection. These include leucocyte and fibroblast interferon, adenine arabinoside and others. Another advantage of complementing the current serological methods is that the assay of DNA polymerase activity may discriminate between complete and defective virus particles.

It has been suggested that the production of defective particles is necessary for the establishment and maintenance of virus persistence. The frequent fluctuation in the enzyme levels often observed in serial samples from persistently infected patients and the finding of apparently 'empty' virus particles is consistent with this hypothesis. It is interesting that consistently high levels of DNA polymerase activity have often been noted in the serum of surface antigen-positive patients treated by maintenance haemodialysis. Possible variations occurring in the natural course of infection should therefore be considered during the monitoring of patients in clinical trials of antiviral agents.

The product of the endogenously primed DNA polymerase reaction has been characterized as DNA with an approximate sedimentation coefficient of 15S. The reaction is not stimulated by the addition of a number of synthetic or natural DNA structures, indicating that the reaction template is sequestered within the core of the 42-nm virus particle. Circular molecules with a mean length of 0.78 μm may be extracted from core particles. The size of the circular forms is compatible with an estimated molecular weight of 1.6–2.0×10^6, which is considerably smaller than the genome of any other known DNA animal virus. It also appears to be smaller than the total amount of nucleic acid required to code directly for the many virus-specific antigenic determinants found in the sera of acutely and persistently infected individuals. It has been estimated that full expression of this structure would permit the synthesis of unique polypeptides with an approximate total molecular weight of 125 000 (Robinson 1977).

Undoubtedly the biggest advance in our understanding of the hepatitis B virus genome has resulted from the application of gene cloning technology. At least four groups have recently reported the successful introduction of hepatitis B DNA fragments into *Esch. coli* by the use of plasmid or phage vectors. Although there are some similarities and differences between DNA cloned from sources containing different HBsAg subtypes, differences between preparations expressing HBsAg of the same subtype have been found. Important findings of these studies include (a) confirmation that the circular nature of the genome is due to base pairing of the 5′ ends of each strand, (b) the location of the single-stranded region in one strand and a nick in the other larger DNA strand, (c) the location of the gene coding for HBsAg determinants is on the fully replicated strand and extends into the single-strand region, and (d) *Esch. coli* transfected with plasmids containing HBV DNA sequences synthesize HBcAg. Although this approach promises to provide a potentially excellent source of HBsAg for the development of a hepatitis B vaccine, attempts to demonstrate the synthesis of this HBV gene product in transfected *Esch. coli* are still in an early stage.

Despite the experimental difficulties, several attempts have been made to determine whether the DNA of hepatitis B virus contains either unique virus-specific sequences or heterogeneous DNA of host origin which is incorporated at random during some stage of virus replication. The specific hybridization of viral DNA to cellular DNA extracted from the liver of patients infected with hepatitis B virus has been examined. The DNA probe re-anneals with a Cot 1/2 value compatible with observations that 25 to 50 per cent of the circular molecule is used as a template during the DNA polymerase reaction, thereby providing further evidence of the uniformity of the template molecules. In addition, unlabelled DNA from patients with hepatitis B infection has been found to reduce the rate of reassociation by a factor of 2 or more when compared with the control reactions; thus DNA from infected livers appears to contain virus specific base sequences.

The high frequency of hepatitis B surface antigen and other markers of hepatitis B virus infection in patients with primary hepatocellular carcinoma suggests that the viral genome may play some role in the oncogenic transformation of hepatocytes. Examination of DNA extracted from tumour, cirrhotic and metastatic tissues from patients with primary hepatocellular carcinoma has suggested that integration of the genome of hepatitis B virus into every cell is not required for the maintenance of the neoplastic transformation of hepatocytes.

The presence of hepatitis B viral DNA in the PLC/PRF/5 human hepatoma cell line has been demonstrated. DNA extracted from these cells showed an enhancement in the reassociation of the ^{32}P-HBV DNA probe prepared by nick translation of virion DNA. It has been established that up to four copies of sequences representative of the entire HBV genome are present in high molecular weight DNA extracts. Furthermore, virus-specific RNA transcripts from all parts of the virus genome are present, indicating that there is full genome expression in these transformed cells.

These studies suggest that the mode of replication of this virus may be unique in contrast to the more familiar strategies of viral genomes. The finding that the genome of *Φ*X174 virus may code for different polypeptide species by selective initiation at different points along the virus genome may help to overcome the objection that the coding capacity of the virus genome does not allow for the coding of the multiple antigen determinants expressed during infection. An inspection of the nucleotide sequence data published so far indicates that at least eight open reading frames may exist on the viral genome with frequent overlapping of possible transcription sequences (Galibert *et al.* 1979). In addition, post-transcriptional splicing of viral RNA may occur prior to synthesis of virus-specific polypeptides. Although the hepatitis B virus may therefore possess the smallest genome among human pathogens, much remains to be learnt about the strategy of the viral nucleic acid within the infected cell.

The repeated failure to isolate hepatitis B virus in generally available cell lines may reflect the incapacity of certain cells to support virus replication at any stage from genome penetration to virus maturation and release. A particularly exciting finding, however, was the successful production of HBsAg in mouse cells by introducing a plasmid containing two randomly arranged virus genomes (Dubois *et al.* 1980). By using the technique of co-transformation with the cloned herpes simplex virus thymidine kinase gene, several copies of the plasmid were integrated into high molecular weight cellular DNA. Detectable amounts of HBsAg were found in the cell culture medium 20 days after transformation, and electron microscopy showed the presence of typical 22 nm HBsAg particles. This study demonstrates that at least the mouse L cell line supports transcription and translation of hepatitis B-specific gene products. A similar approach has also been reported by Hirschman *et al.* (1980), who found expression of both HBsAg and HBcAg in cells of non-hepatic origin receiving cloned hepatitis B virus DNA of the *ayw* genotype.

Hepatitis B research until recently has been confined to a further understanding of the virus immunopathology and sequelae. However, the application of molecular biology techniques to the hepatitis B virus genome has added a new and important dimension to the investigation of the processes accompanying replication of human hepatitis viruses. It is anticipated that the availability of monoclonal antibodies to hepatitis B antigens, for example, will complement this newly acquired information, allowing a clearer picture of hepatitis B virus replication to emerge whereby accurate predictions from the genome sequence will be combined with the identification of virus specific products by means of sera of single specificity. It has often been stressed that the virus of hepatitis B is unique; this concept has now been revised and hepatitis B virus should be regarded as one of a family of variously related viruses with similar morphology, proteins and genome structure. These newly described agents are pathogens of widely differing animal species including American woodchucks, prairie dogs, ground squirrels and Pekin ducks. It has been proposed that these viruses should be grouped in the new family *Hepadnaviridae*.

5.6 Hepatitis B virus and primary liver cancer

Primary liver cancer is one of the most widespread of carcinomas and occurs with a high incidence in areas of Africa and Southern Asia where hepatoma may represent the most common form of cancer among native populations, especially in young men (for review, see Szmuness 1978). Hepatitis B virus is also endemic in the same geographical areas with a high percentage of primary liver cancer patients positive for serological markers of hepatitis B. Histologically, macronodular, post-hepatic cirrhosis is also a common feature associated with hepatoma.

The causal relation between HBV and primary liver cancer may be explained by either the direct oncogenic potential of the virus or the induction of hepatocellular changes which predispose the liver to transformation by a second unrelated agent. Although the second of these hypotheses cannot yet be discarded, recent evidence now points to the direct oncogenic properties of the HBV genome. It is envisaged that acute hepatitis B may under certain circumstances progress to chronic hepatitis with antigenaemia. This carrier state may progress further to hepatoma via an intermediate state of liver cirrhosis, although the development of chronicity and subsequent cirrhosis may not necessarily always precede the tumorigenic state. The factors responsible for the initial development of hepatitis B persistence in newborns delivered from carrier mothers are not clear; temporary or long-lasting immunosuppression arising from either immune defects or parasitic infection, or genetic factors may be involved singly or together. Age of initial infection may be particularly important with children born to HB *e* antigen positive carrier mothers being at high risk of subsequently developing chronic hepatitis. It is estimated that there are over 176 million carriers of HBV, the majority of which reside in those areas where primary liver cancer incidence is high.

HBsAg is found in 30 to 90 per cent of patients with primary liver cancer, although the titre may be low. There appears to be no significant correlation with HBsAg subtype and these patients are rarely positive for HB *e* antigen or antibody. These findings suggest that virus replication and formation of new infectious virus proceeds at a low rate in these patients. Several studies have shown that HBsAg prevalence may be related to age, with fewer patients 50 years of age or older having circulating antigen. In the absence of HBsAg, the presence of anti-HBc may indicate HBV involvement. In a recent study conducted in East Africa, a significant percentage of HBsAg negative patients with primary liver cancer were positive for anti-HBc. In general terms, it appears that these patients possess serological markers of HBV infection with a 5-fold increase in frequency compared to control populations.

Although the direct oncogenic potential of HBV has not been demonstrated *in vitro*, a number of HBsAg-secreting cell lines have been established from hepatoma tissue. The most notable of these is the PLC/PRF/5 line developed by Alexander (Macnab *et al.* 1976). Detailed study of these cells has indicated that 4–6 copies of the complete HBV genome are integrated into the host DNA of these cells. However, HBsAg is the sole virus-specific product produced in culture, optimal titres being found in culture fluids when essential metabolites have been exhausted. Qualitative studies indicate therefore that HBsAg production is terminally related, with maximum titres of antigen occurring in cultures with a decreasing viable cell count.

6 Non-A, non-B hepatitis

6.1 Epidemiology

The introduction of sensitive immunoassays for the diagnosis of both hepatitis A and hepatitis B has revealed the presence of one or more previously unrecognized agents of hepatitis in man. Until these agents are positively identified, the term 'non-A, non-B' by general usage refers to hepatitis excluded by serological criteria from being associated with either HAV or HBV infection. This is now the most common form of post-transfusion hepatitis in many areas of the world: a particularly high incidence of infection has been documented after the administration of blood products such as Factor VIII and Factor IX preparations (Zuckerman 1980). Non-A, non-B hepatitis occurs in renal dialysis and other specialized units and may also occur as sporadic outbreaks in the general population. There is now considerable evidence to indicate that this infection, like hepatitis B, may progress to chronic liver disease.

After the widespread introduction of radioimmunoassay for HBsAg in the USA, it has been estimated that more than 80 per cent of post-transfusion hepatitis is now directly attributable to non-A, non-B hepatitis agents. The incubation periods range from 2 to 15 weeks between transfusion and the onset of hepatitis. As many as half of these patients exhibit biochemical evidence of liver damage six months after infection and liver biopsy invariably reveals histological evidence of chronic liver disease. Several workers have shown that prophylaxis with normal immunoglobulin may be effective in preventing non-A, non-B hepatitis—an observation which suggests that non-A, non-B viruses are common in the general population.

Epidemiological studies of sporadic hepatitis and outbreaks in haemodialysis units have shown that a significant number of patients suffer repeated attacks of acute hepatitis. In particular, multiple attacks have been frequently observed in drug addicts, and these are accompanied by the clinical features of acute infection without evidence of the recurrence of underlying chronic disease.

Transmission studies on susceptible chimpanzees have confirmed the existence of at least two agents of non-A, non-B hepatitis, as distinguished by the length of the incubation periods. Several sera from acutely or chronically infected patients, in addition to 'high-risk' products such as pooled plasma, fibrinogen and antihaemophilic factors, have been documented as inducing histologically proven hepatitis within 2 to 6 weeks of infection. Animals convalescent from the short incubation form of non-A, non-B hepatitis are not immune to a second non-A, non-B infection which is characterized by a much longer incubation period of 8 to 12 weeks. Conversely, animals convalescent from the long incubation form are still susceptible to infection by the heterologous agent while having demonstrable protection against the homologous virus. It is conceivable that either or both agents may be present in a contaminated product owing to the large number of donations required to produce sufficient volumes for plasma fractionation. As with hepatitis B, commercial plasma pools are obtained by plasmaphoresis of paid donors and therefore carry a high risk of causing transfusion hepatitis.

6.2 Laboratory markers of non-A, non-B hepatitis

There is an urgent need for specific laboratory tests for markers of non-A, non-B hepatitis, both for specific diagnosis and for the large-scale screening of blood donations. Although several serological tests have been proposed as providing specific markers of non-A, non-B hepatitis, there is no widely available serological test for either of the hitherto identified non-A, non-B infections. Methods employing human or chimpanzee convalescent sera have been published from several laboratories, but the specificity of the reagents has yet to be widely reproduced or confirmed. These procedures commonly use either immunodiffusion or radioimmunoassay methods. Use of biopsy material as a positive substrate in an immunofluorescence test has shown the most promise to date, as the presence of circulating virus-specific antigens in either acute or chronic non-A, non-B hepatitis has not as yet been demonstrated. This may in part be due to the masking of viral antigens within circulating immune complexes which have been detected in non-A, non-B patients, both during acute infection and the prodromal period. The immunology of non-A, non-B hepatitis would thus appear similar to hepatitis A and hepatitis B infections in so far as antibody against viral antigens may be present at the onset of clinical illness; indeed, animal transmission experiments have shown the presence of infectious virus in the serum of infected chimpanzees 12 days before clinical illness.

Examination by electron microscopy of thin sections of liver infected with non-A, non-B hepatitis reveals a number of cytological changes both in the nuclei and cytoplasm of infected cells (reviewed by Zuckerman 1980). During the period of peak serum alanine aminotransferase levels the endoplasmic reticulum becomes altered in configuration to produce tubular forms bonded either by a single or a double membrane. However, virus particles are not seen in association with these structures, so that they may represent non-specific changes frequently found in a large variety of tissues and cultured cell lines. Nuclear changes appear to occur earlier, consisting of an abnormal shape and extensive condensation of chromatin. Intranuclear aggregates of particles measuring 20–27-nm in diameter are frequently seen, although at least a proportion of these particles lacks the morphological uniformity in shape of virus particles. Attempts to develop a specific immunofluorescence test for non-A, non-B hepatitis have shown that convalescent sera may react specifically with an antigen within the nucleus of these cells. It is not known whether this reactivity is a soluble antigen and/or associated with maturing virus particles. However, one study of infected liver obtained from a chimpanzee infected with the short incubation form of non-A, non-B hepatitis has shown that homogenates containing particles 25–30 nm in diameter with an average buoyant density of 1.31 g/cm^3 may be purified and subsequently shown to induce non-A, non-B hepatitis in susceptible animals after an incubation period of 15 days (Bradley *et al.* 1979).

7 Immunization against viral hepatitis

7.1 Passive immunization

Passive immunization against hepatitis A has been successfully demonstrated by the administration of normal immunoglobulin, provided this occurs prior to exposure to hepatitis A virus or early during the incubation period. This approach is particularly useful in the control of epidemic hepatitis in large institutions and other semi-closed communities and is often recommended for travellers briefly visiting endemic areas, and for health care workers who may be exposed to the virus. However, the repeated use of immunoglobulin may not be warranted in certain circumstances, as there is a risk of anicteric or subclinical cases not being recognized, thus leading to extensive spread of the virus in the community. This concern applies particularly to the repeated administration of immunoglobulin to healthy children (World Health Organization 1977).

The exclusion of contaminated blood and blood products by screening for markers of hepatitis B has led to a dramatic decrease in the incidence of post-transfusion hepatitis wherever such measures have been implemented. Attempts to identify infected patients, particularly those requiring long-term hospitalization, have proved more difficult. The high cost of such programmes is often prohibitive, and thus staff remain at risk of infection as a result of contact with infectious blood and body fluids. This risk of hepatitis B transmission is particularly great in haemodialysis units which either do not pursue a policy of segregating known infected patients and/or adopt barrier techniques. The passive immunization of susceptible persons by administration of immunoglobulin has therefore been of considerable interest in hepatitis B immunoprophylaxis. There has been controversy surrounding the respective merits of normal immunoglobulin preparations and those prepared from donations selected because they contain a high titre of anti-HBs. Various studies have attempted to define the efficacy of these preparations in persons who have been exposed to small amounts of infected material, e.g. following accidental skin puncture with contaminated syringe needles. These studies have largely proved conflicting, either owing to the absence of adequate randomized control groups or because the efficacy of high titre preparations has been contrasted with 'normal' preparations of immunoglobulin that often contain only small quantities of anti-HBs. This uncertainty is enhanced by the finding that more recent preparations of normal immunoglobulin may frequently contain a higher titre of anti-HBs than preparations designated as 'hepatitis B' immunoglobulin.

Early studies with hepatitis B immunoglobulin were encouraging, its use in post-exposure prophylaxis has recently been re-assessed. Clinical symptoms and extent of hepatic injury may occur less frequently in groups receiving hepatitis B immunoglobulin, but the incidence of acute hepatitis B appears to be similar in those persons receiving either immunoglobulin with a high anti-HBs titre or normal immunoglobulin. Indeed, several investigators have expressed the opinion that passive administration of viral antibodies in high titre may actually suppress immunity to the hepatitis B virus. In addition, several studies have indicated that chronic hepatitis B is often preceded by a mild or asymptomatic acute infection. Nevertheless, the administration of hepatitis B immunoglobulin within 48 hours of exposure remains the recommended procedure for the treatment of persons accidentally exposed to small amounts of virus, at least until the general introduction of a safe and effective vaccine becomes possible.

Passive immunization of infants born to persistently infected mothers or mothers acutely infected in the third trimester of pregnancy has also been advocated. At present, the optimal dosage and duration of treatment remains to be clarified. A thorough study in Taiwan, where vertical transmission is particularly important, has shown that infants who received hepatitis B immunoglobulin within the first 48 hours after birth suffered from significantly fewer persistent infections during the first year of life.

The most significant observation, however, was the delay in the onset of antigenaemia in this group. The timing of hepatitis B immunoglobulin administration appears to be important, since neonates receiving treatment later than 48 hours after birth have a much greater likelihood of becoming persistently infected. This is in accord with the concept of the risk of infection from the mother being greatest during labour and delivery.

The combined use of specific immunoglobulin and inactivated HBsAg vaccine material is being considered as a basis of inducing 'passive-active' immunity to hepatitis B. No data are yet available to evaluate the relative efficacy of this approach in cases of accidental exposure to the virus, although there is recent evidence to suggest that HBsAg present in certain immunoglobulin preparations has induced a primary immunization response in recipients. Immune complexes of HBsAg and antibody may therefore have a role in stimulating an effective protecting antibody response to the hepatitis B virus.

Immunotherapy with hepatitis B immunoglobulin has proved of little value in the treatment of HBV positive chronic liver disease. Although intravenous infusion may result in HBsAg clearance, the effect is transient and appears to be inadequate in eliminating the virus.

7.2 Active immunization

At the time of writing, there is no candidate vaccine for hepatitis A, although recent advances suggest that this position may change rapidly in the next few years. The replication, though limited, of hepatitis A virus in conventional tissue culture cell lines (Deinhardt *et al.* 1981), together with the molecular cloning of the viral genome (von der Helm *et al.* 1981), are important advances in the development of a vaccine for hepatitis A.

A hepatitis B vaccine is urgently required for the protection of groups at high risk of acquiring the infection. These groups include persons requiring repeated transfusions of blood or blood products, prolonged in-patient care, patients who require frequent tissue penetration or who have a natural or acquired immune deficiency, or who are suffering from malignant diseases. Viral hepatitis is also a major occupational hazard among staff of haemodialysis units and institutions for the mentally retarded and for dental surgeons. In addition, high rates of infection occur in drug addicts, homosexuals and prostitutes. In certain areas of the world, the high carriage rate of HBV is most probably maintained by the maternal transmission of virus to neonates; the availability of a hepatitis B vaccine is therefore also desirable for the effective reduction of the numbers persistently infected with this virus.

Early attempts to vaccinate against hepatitis B with infectious serum heated at 98° for one minute prevented or modified hepatitis B in 69 per cent of 29 volunteers subsequently challenged with the original infectious serum four to six months later. The results were essentially the same after one, two or three inoculations and established the fact that serum from persons infected with hepatitis B virus could be used as a source of viral antigens for inducing a protective immune response, although a recent re-evaluation of this study now indicates that HBV was partly inactivated before administration (Krugman *et al.* 1979). In another study with HBsAg, serum from a persistently infected donor was heated at 60° for ten hours but the virus was not effectively inactivated, as shown by the development of HBsAg accompanied by raised alanine aminotransferase levels in some of the recipients. Electron microscopy has shown that such sera contain both complete HBV particles and numerous 22-nm HBsAg particles. In the absence of conventional tissue culture systems for the *in vitro* growth of HBV, plasma collected from large numbers of infected blood donors is the major source of virus-containing material for vaccine development. The difficulties associated with the use of positive serum pools and standardization of candidate vaccines have been summarized by Gerety *et al.* (1979).

Human HBV has been successfully transmitted to chimpanzees and, although the infection is generally mild the biochemical, histological and serological responses in these primates are very similar to those in man. In addition, the antigenic markers of virus replication are antigenically indistinguishable from the homologous antigens and antibodies found in infected human beings. Chimpanzees therefore offer an ideal model for the evaluation of experimental hepatitis B vaccines. Vaccines prepared from the 22-nm HBsAg particles after purification from plasma and inactivation by formaldehyde have been shown to be both safe and protective in susceptible primates, leading to the production of antibody to HBsAg. The use of a vaccine derived from infected persons represents a major departure from conventional approaches to preventive medicine. The World Health Organization Expert Committee on Viral Hepatitis has therefore proposed guidelines and criteria for the safe development of hepatitis B vaccines. The more important of these include the recommendation that HBsAg positive blood should be drawn from donors devoid of circulating HBV as assessed by the absence of virus-associated DNA polymerase activity and HBeAg. The absence of these markers is associated with a lower risk of infectivity, although the association is not absolute. After removal of any residual HBV particles, the guidelines propose that the final vaccine preparations should be treated by accepted inactivation procedures, e.g. by exposure to formaldehyde, and the absence of infectious HBV confirmed by appropriate safety tests in non-immune chimpanzees.

Other considerations include minimizing the risk of agents of non-A, non-B hepatitis being present by the repeated use of a small number of donors and careful design of clinical trials for the evaluation of hepatitis B vaccines in man. The Expert Committee has recommended that the safety of any vaccine preparation must be demonstrated in a small group of healthy adult volunteers before administration to a larger group. The long term follow-up of recipients is regarded as an essential criterion in the development of hepatitis B immunoprophylaxis, with careful monitoring of specific anti-HBs levels and frequent examination for non-specific immune responses to contaminating agents, development of hepatitis and the presence of auto-immune markers.

The first human trial with a partly pure 22-nm HBsAg preparation was reported by Maupas and colleagues (1978), who administered vaccine to the patients and staff of several haemodialysis units in France. Although this study did not include a control group, there was subsequently a significant difference in the incidence of hepatitis B infection between the immunized group and a group of patients and staff who had not received the vaccine. Hilleman and colleagues (1978) have also conducted a trial in a small number of volunteers using a purified 22-nm preparation treated with formaldehyde which had been previously shown to protect chimpanzees against challenge with HBV. In this early study, the subjects received the vaccine in either aqueous form or with an alum adjuvant; in both groups the preparation elicited antibody to HBsAg in individuals with and without pre-existing homologous antibody. Carefully controlled field trials with similar preparations have now been completed in the USA and include various groups at high risk such as patients and staff of haemodialysis and young male homosexuals. Szmuness *et al.* (1980) have published preliminary figures on the use of a purified 22-nm vaccine preparation in the latter group. Over 96 per cent of immunized male homosexuals developed a high anti-HBsAg titre and showed a notably reduced attack rate of hepatitis B infection 18 months later as compared with the control group.

Biochemical analyses have repeatedly shown that the 22-nm particle is a complex structure consisting of lipid and several proteins of either host or viral origin. In particular, a close association between human serum albumin and the 22-nm particle has been established. Hepatitis B-specific antigenic determinants are associated with a non-glycosylated polypeptide having a molecular weight in the range of 22–24 000 and a glycosylated polypeptide with a molecular weight in the range 26–29 000. Both polypeptides may exist as a 49 000 mol. wt complex under non-reducing conditions, and have identical amino-acid sequences at amino- and carboxyl-terminals, indicating that the larger polypeptide represents a glycosylated form of the smaller non-glycosylated polypeptide. Although the 22-nm HBsAg particles share a common antigen with the outer coat of the HBV particle, there exists a possibility that the presence of host proteins, in particular serum albumin, might either depress the level and intensity of the anti-HBs response or induce undesirable immunological side reactions.

In anticipation of a successful outcome of clinical trials with intact 22-nm particles, several approaches are being considered for the formulation of hepatitis B vaccines which exclusively contain virus-specific material. It is considered that vaccines prepared from the constituent polypeptides of HBsAg would have an added margin of safety, since the preparations would be chemically well defined and even less likely to contain contaminating virus or host components.

Solubilization of the integral proteins of the 22-nm particles may be readily accomplished by disruption of the lipid bilayer with non-ionic detergents such as Triton X-100 (Skelly *et al.* 1979). In general, the biological and antigenic properties of viral proteins are preserved in contrast to the use of strongly ionic detergents. Both virus-specific polypeptides can be extracted as a 3.9S complex from intact HBsAg 22-nm particles by Triton X-100 solubilization and affinity chromatography on Concanavalin A-Sepharose with retention of antigenic activity. After rate-zonal centrifugation in detergent-free sucrose gradients, HBsAg polypeptides treated by this procedure form micellar structures with an average diameter of 120 nm. The immunogenicity of the HBsAg micelles has been compared to that of intact HBsAg 22-nm particles by inoculation of the SWR/J strain mice in the presence of an alum adjuvant. It has been suggested that the serological response in mice may be a useful indicator of the immunogenic potential of candidate hepatitis B vaccines. The titre of anti-HBs induced by the micelles greatly exceeded that induced by intact 22-nm particles. Furthermore, the slopes of the dilution curves obtained by radioimmunoassay indicated an enhanced antibody affinity in the sera of mice receiving HBsAg micelles. This physical form may indeed also be suitable for vaccines prepared by using HBsAg polypeptides produced by the expression of cloned hepatitis B virus in *Esch. coli* or other hosts.

8 Antiviral therapy of chronic hepatitis B

8.1 Interferon

Several studies have indicated that the administration of human interferon, both in man and chimpanzees, has an inhibitory effect on continuing hepatitis B virus replication (Merigan and Robinson 1978). Transient changes in markers of virus activity have been described when leucocyte interferon is given for less than two weeks, although more prolonged treatment of patients with chronic active hepatitis has resulted in a drop in the number of circulating virus particles which persisted for up to 15 weeks after stopping treatment. A more extensive study of the use of leucocyte interferon has been reported where eight patients with chronic hepatitis B were treated for periods of five to eight weeks and, in one case, for five months (Scullard *et al.* 1979). In one patient, there was a fall in the number of virus particles in the blood, which coincided with the disappearance of HBeAg and a reduction in aspartate aminotransferase levels. In three other patients, a similar effect was seen but proved to be transient, whereas in a fourth, hepatitis B-associated DNA polymerase activity continued to rise despite treatment. Although there was no significant change in HBsAg titres in these patients, a decrease in cytotoxic T-cell activity towards HBsAg-coated target cells was found. This important observation suggests that the antiviral effect of leucocyte interferon may be offset by a depression of cell-associated responses to infected hepatocytes, resulting in impaired clearance of the virus, despite a decrease in the extent of hepatitis B virus replication. Persistently infected female patients appear to respond better than males to interferon treatment.

Similar results with fibroblast interferon have also been reported, although interferon produced by fibroblasts appears to be somewhat less potent; in one study, a two-week course of treatment of four patients with interferon from human fibroblasts failed to produce any significant change in markers of hepatitis B virus replication.

Leucocyte interferon has been shown to be considerably more immunosuppressive than fibroblast interferon, augmenting natural killer cell activity. At present, there is a notable lack of information on the relative level of circulating interferon in patients with active chronic hepatitis. The natural killer cell system may be of paramount importance in recovery from acute disease with persistent virus arising as a result of a deficiency in either interferon production or natural killer cell activity.

8.2 Ara-A (adenosine arabinoside)

Ara-A is an analogue of the adenine deoxyribonucleoside and, in contrast to other pyrimidine analogues, is virtually non-toxic, being rapidly de-aminated *in vivo* and converted to hypoxanthine arabinoside. The antiviral activity of ara-A in the treatment of herpes infections is well documented and has encouraged the use of this drug for the treatment of chronic active hepatitis B (Pollard *et al.* 1978). Patients have received two courses of treatment for up to two weeks each. One patient responded with a rapid decrease in the level of hepatitis B-associated DNA polymerase activity during both courses of treatment, although a return to pre-treatment levels was seen during the seven-week period between courses and after cessation of the second course. In the second patient, the reduction in DNA polymerase activity was more prolonged, the second course resulting in a reduction in DNA polymerase activity to an undetectable level which persisted 12 months beyond the treatment period. The effects of ara-A have been described as transient in patients with chronic liver disease treated with the drug for a single shorter period. The response to ara-A treatment is often variable; although circulating virus is often reduced during treatment these changes do not persist in all patients treated with ara-A. Similar temporary effects in persistently infected chimpanzees treated with ara-A have also been found. It remains to be established whether continued treatment with this drug alone, or its phosphorylated derivative, or in combination with interferon, will have a more lasting effect on hepatitis B virus replication.

References

Alberti, A., Diana, S., Scullard, G. H., Eddleston, A. L. W. F. and Williams, R. (1978) *Brit. med. J.* **ii,** 1056.

Albin, C. and Robinson, W. S. (1980) *J. Virol.* **34,** 297.

Bianchi, L., Gerok, W., Sickinger, K. and Stalder, G. A. (Eds) (1980) *Virus and the Liver*. MTP Press, Lancaster.

Blumberg, B. S. (1964) *Bull. N.Y. Acad. Med.* **40,** 377.

Bradley, D. W. *et al.* (1979) *J. med. Virol.* **3,** 253.

Burrell, C. J., Mackay, P., Greenaway, P. J., Hofschneider, P. H. and Murray, K. (1979) *Nature* **279,** 43.

Coulepis, A. G., Locarnini, S. A. and Gust, I. D. (1980) *J. Virol.* **35,** 572.

Dane, D. S., Cameron, C. H. and Briggs, M. (1970) *Lancet* **i,** 695.

Deinhardt, F., Holmes, A. W., Capps, R. B. and Popper, H. (1967) *J. exp. Med.* **125,** 673.

Deinhardt, F., Scheid, R., Gauss-Müller, V., Frösner, G. G. and Siegl, G. (1981) *Prog. med. Virol.* **27,** 109.

Dubois, M-F., Pourcel, C., Rousset, S., Chany, C. and Tiollais, P. (1980) *Proc. nat. Acad. Sci. Wash.* **77,** 4549.

Eddleston, A. L. W. F. and Williams, R. (1974) *Lancet* **ii,** 1543.

Feinstone, S. M., Kapikian, A. Z. and Purcell, R. H. (1973) *Science* **182,** 1026.

Feinstone, S. M., Moritsugu, Y., Shih, J. W-K., Gerin, J. L. and Purcell, R. H. (1978) In: *Viral Hepatitis* pp. 41–48. Ed. by G. N. Vyas, S. N. Cohen and R. Schmid. Franklin Institute Press, Philadelphia.

Frösner, G. G. *et al.* (1979) *Infection* **7,** 303.

Galibert, F., Mandart, E., Fitoussi, F., Tiollais, P. and Charnay, P. (1979) *Nature* **281,** 646.

Gerety, R. J., Tabor, E., Purcell, R. H. and Tyeryer, F. J. (1979) *J. infect. Dis.* **140,** 642.

von der Helm, K., Winnacker, E. L., Deinhardt, F., Frösner, G., Gauss-Müller, V., Bayerl, B., Scheid, R. and Siegl, G. (1981) *J. virol. Methods* **3,** 37.

Hilleman, M. R. *et al.* (1978) In: *Viral Hepatitis*, pp. 525–537. Ed. by G.N. Vyas, S. N. Cohen and R. Schmid. Franklin Institute Press, Philadelphia.

Hirschman, S. Z., Price, P., Garfinkel, E., Christman, J. and Acs, G. (1980) *Proc. nat. Acad. Sci. Wash.* **77,** 5507.

Howard, C. R. (1978) *J. med. Virol.* **3,** 81–86; (Ed.) (1982) *Newer Development in Practical Virology*. Alan R. Liss Inc., New York.

Howard, C. R. and Burrell, C. J. (1976) *Prog. med. Virol.* **22,** 36.

Kaplan, P. M., Greenman, R. L., Gerin, J. L., Purcell, R. H. and Robinson, W. S. (1973) *J. Virol.* **12,** 96.

Krugman, S., Giles, J. P. and Hammond, J. (1967) *J. Amer. med. Ass.* **200,** 365.

Krugman, S. *et al.* (1979) *New Engl. J. Med.* **300,** 101.

Locarnini, S. A., Coulepsis, A. G., Westaway, E. D. and Gust, I.D. (1981) *J. Virol.* **37,** 216.

Macnab, G. M., Alexander, J. J., Lecatsas, G., Bey, E. M. and Urbanowicz, J. M. (1976) *Brit. J. Cancer* **34,** 509.

Maupas, P., Goudeau, A., Coursaget, P., Drucker, J. and Bagros, P. (1978) *Intervirology* **10,** 196.

Merigan, T. C. and Robinson, W. S. (1978) In: *Viral Hepatitis*, pp. 575–579. Ed. by G. N. Vyas, S. N. Cohen and R. Schmid. Franklin Institute Press, Philadelphia.

Pasek, M. *et al.* (1979) *Nature* **282,** 575.

Pollard, R. B. *et al.* (1978) *J. Amer. med. Ass.* **239,** 1648.

Provost, P. J. and Hilleman, M.R. (1979) *Proc. Soc. exp. Biol. Med.* **160,** 213.

Rizetto, M. *et al.* (1979) *Lancet* **ii,** 986.

Robinson, W. S. (1977) *Ann. Rev. Microbiol.* **31,** 375; (1979) In: *Comprehensive Virology*, Vol. 14, pp. 471–526. Ed. by H. Fraenkel-Conrat and R. W. Wagner. Plenum Press, New York.

Scullard, G. H., Alberti, A., Wansborough-Jones, M. H., Howard, C. R., Eddleston, A. L. W. F., Zuckerman, A. J., Cantell, K. and Williams, R. (1979) *J. clin. Lab. Immunol.* **1,** 277.

Skelly, J., Howard, C. R. and Zuckerman, A. J. (1979) *J. gen. Virol.* **44,** 679.

Szmuness, W. (1978) *Prog. med. Virol.* **24,** 40.

Szmuness, W. *et al.* (1980) *New Engl. J. Med.* **303,** 833.

Takahashi, K. *et al.* (1980) *J. gen. Virol.* **50,** 49.

Valenzuela, P., Gray, P., Quiroga, M., Zaldivar, J., Goodman, H. M. and Rutter, W. J. (1979) *Nature* **280,** 815.

World Health Organization (1977) *Tech. Rep. Series* No. 602, Geneva.

Zuckerman, A. J. (1980) In: *Recent Advances in Clinical Virology*, Vol 2, pp. 41–51. Ed. by A. P. Waterson. Churchill Livingstone, Edinburgh.

Zuckerman, A. J. and Howard, C. R. (1979) *Hepatitis Viruses of Man*. Academic Press, London.

102

Rabies

G. S. Turner

Introductory

The history of rabies extends far into antiquity, reference to human deaths from the bites of mad dogs occurring in the Babylonian legal codes established in 2300 BC (Goetze 1948). Many observations of human and animal rabies, of the venomous nature of the saliva of mad dogs, and of the value of cauterizing bite wounds were recorded between 500 BC and AD 100. More comprehensive descriptions of the disease and of bizarre attempts to cure it accompany the progress of medicine into the Middle Ages where Fracastoro's prescient theory of contagion is strikingly illustrated by his observations on rabies (Wilkinson 1977). These early accounts may suggest that the disease was common, but the incidence and historical impact of human rabies appears insignificantly low when compared with that of bubonic plague or smallpox. Past observers provided much accurate information but many were no doubt influenced by the awesome clinical manifestations of human rabies and frequently contributed more fable than fact (Smithcors 1958).

Few substantial advances in knowledge of the disease occurred until the nineteenth century when experimental transmission of rabies by infectious saliva from dogs and man substantiated past conjecture (Zinke 1804, Magendie 1821, Galtier 1879). The classic experiments of Pasteur and his co-workers (1881, 1882) dispelled much of the rabies mystique and forecast the ultra-microscopic nature of the infectious agent; its propagation and attenuation by passage in rabbit brain led to the successful preparation and use of rabies vaccine. Improvements in vaccines and in the laboratory diagnosis and control of rabies continued during the first half of this century, but almost 80 years elapsed before further major advances occurred.

The successful adaptation of rabies virus to growth in cultures of non-neural cells by Kissling (1958) was rapidly extended by others (Fenje 1960, Abelseth 1964, Wiktor *et al.* 1964). The use of cell culture methods for the large-scale propagation of rabies virus permitted a degree of purification hitherto impossible with virus produced in either animal brain or avian embryos. During the last two decades, these methods have been vigorously exploited for studies of the virus itself and for the preparation of safe and much more efficacious vaccines.

Morphology and properties of rabies virus

Most of the physico-chemical and biological properties of rabies virus so far established are those of fixed strains purified and concentrated from large scale cell cultures. Typically the virus is bullet-shaped with average external dimensions 75 × 180 nm (Fig. 102.1). From the surface inwards the virus displays an array of knobbed projections 6 to 8 nm long protruding from a bilaminar membrane which covers the whole particle except at the flat end, which is frequently invaginated. The membrane envelops a nucleocapsid which contains single-stranded RNA. The latter is associated with protein N as a ribonucleoprotein which is coiled in a right-handed helix with a periodicity of approximately 7.5 nm; the dimensions of the intact core are 50 × 165 nm (Sokol *et al.* 1969, Murphy 1975, Schneider and Diringer 1976).

Atypical elongated filaments and V- or Y-shaped forms may be seen in preparations harvested from cell cultures late after infection. Short or truncated (T) particles also appear in cell cultures infected with some strains of rabies virus. They contain the normal structural proteins but only part of the genome. In undiluted serial passage these short forms specifically interfere with the normal replication of the virus and have been termed defective interfering (DI) particles. They appear to be analogous to the short, non-infectious DI particles of vescular stomatitis virus. Little is known of their role in natural infections (Crick and Brown 1974, Kawai *et al.* 1975, Wiktor *et al.* 1977*a*).

The virus has a sedimentation coefficient ($S_{20}W$) of 550 to 650 and a buoyant density in CsCl of approximately 1.2 cm^3 (Neurath *et al.* 1966). It contains approximately 67 per cent protein. 26 per cent lipid, 4.0 per cent RNA and 3 per cent carbohydrate; detailed analyses of these individual constituents have been reported (Sokol *et al.* 1969, 1971, Diringer *et al.* 1973, Schlumberger *et al.* 1973). Investigation of components isolated after selective disruption of the virus with detergents has made it possible to ascribe functions to particular structures (Matsumoto 1970, Kuwert 1970, Schneider and Diringer 1976, Sokol 1973, Atanasiu and Gamet 1978).

Of five polypeptides isolated from rabies virus, the glycoprotein (G) forms the basic constituent of the surface projections. G-protein is strongly antigenic, inducing and reacting with virus-neutralizing antibody, and also immunizing animals against challenge infection with the virus. Immunofluorescent antibody to G-protein stains only the cell membranes of rabies infected cells at sites where virus matures. In the presence of complement, G-antibody 'lyses' infected cells in culture (Wiktor *et al.* 1968, Schneider and Diringer 1976). Pure G-protein has been isolated; its potential as a virion-free vaccine is under investigation (Dietzschold 1977, Schneider and Diringer 1976). The virulence of rabies virus appears to be associated with the structure of its glycoprotein. Chemical analysis of G-protein from pathogenic parent virus and non-pathogenic variants showed them to differ in amino acid composition (Dietzschold *et al.* 1983). No antigenic

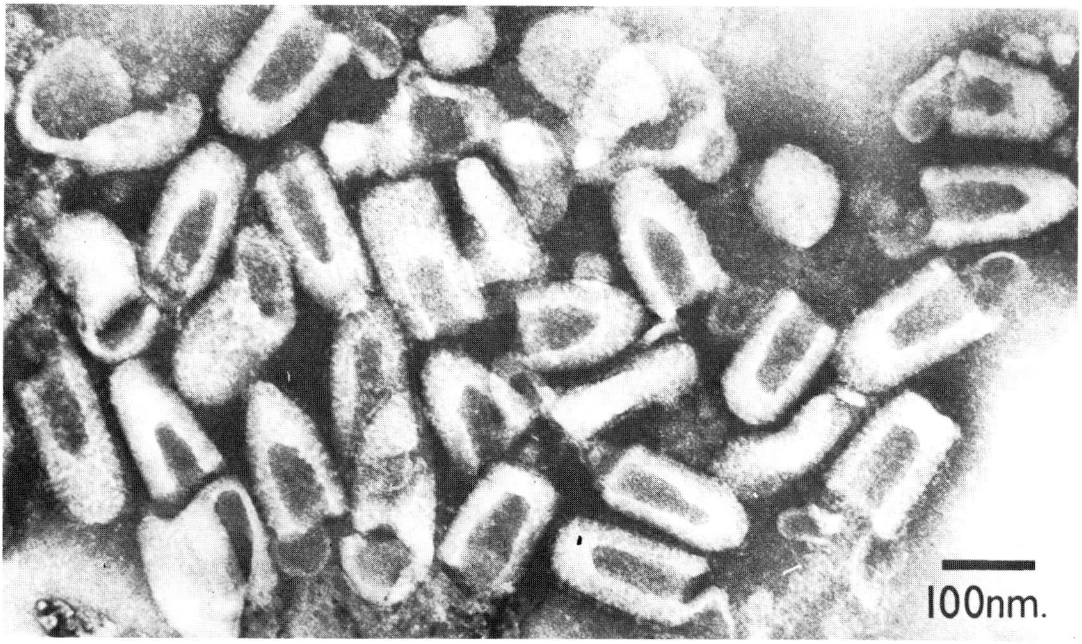

Fig. 102.1 Electronmicrograph of rabies virus. (By courtesy of C. J. Smale.)

functions are yet attributed to the remaining proteins except that of a nucleoprotein (N), associated with the nucleocapsid. In contrast to the antibodies against G-protein, antisera to pure N protein have no virus neutralizing activity. In immunofluorescent conjugates they specifically stain intracytoplasmic inclusions (*Negri bodies*), and in complement-fixation tests they cross-react with the N proteins from 'rabies-related' viruses. It is currently believed that the N protein is a group-specific antigen, and that G-protein is type-specific (Schneider *et al.* 1973).

In common with other negative strand RNA viruses, the isolated nucleic acid of rabies is non-infectious. The specific virus-associated RNA polymerase required for transcribing the parental RNA is probably associated with proteins designated L & NS (Kawai 1977). The base composition of rabies virus RNA is similar to that of vesicular stomatitis virus, the type species of the rhabdovirus group. However, apart from the rabies subgroup, genetic relatedness between rabies and other rhabdoviruses appears unlikely (Aaslestad and Urbano 1972, Repik *et al.* 1974).

Most laboratory strains of rabies virus, grown in a variety of cell types in the absence of serum, have haemagglutinating properties (HA), optimally demonstrable with goose erythrocytes between 0–4° at pH 6.8 (Halonen *et al.* 1968). Minimal infectivity titres of 10^6 are necessary before HA is detectable, but above this level infectivity and HA are directly related. HA is not readily demonstrated with virus recovered from animal brain (Turner and Kaplan 1967, Kuwert *et al.* 1968, Ardoin *et al.* 1977). Haemagglutinating properties may be extracted from purified virus with saponin; extracts possess all the immunogenic properties of whole virions and are specifically inhibited by homologous antibody (Schneider *et al.* 1971). The precise nature of rabies HA is unclear, but it is probably a lipoprotein. The necessity for high-titred virus and the inhibition of HA by non-specific inhibitors in serum has prevented its wide use in antibody measurement.

Rabies virus is thermolabile, with a half-life of approximately 4 hr at 40° and 35 seconds at 60°. Thermal inactivation is less in the presence of serum proteins and chelating agents (Michalski *et al.* 1976). The virus remains stable for several days at 0–4° indefinitely at −70° and when freeze-dried. In moist saliva in temperate climates it will survive for about 24 hr. The virus is inactivated below pH 4.0 and above pH 10, by oxidizing agents, most organic solvents and surface-active agents—notably, quaternary ammonium compounds, soaps and detergents. It is rapidly inactivated by proteolytic enzymes and by ultraviolet and x-irradiation (Kaplan *et al.* 1966).

Classification of rabies virus

Several early reports suggested that rabies virus had a close affinity with the myxoviruses. However, the characteristic features of the myxoviruses are their adsorption and inhibition by mucoprotein receptors and the possession of neuraminidase—properties not shared by rabies viruses (Almeida and Waterson 1966). Rabies virus is now classified with the *Rhabdoviridae*, a family whose 75 or so members all have bullet-shaped or bacilliform morphology as well as similar protein and RNA structures; their hosts include a wide range of vertebrates, invertebrates and plants (Brown *et al.* 1979, Matthews 1979).

Table 102.1 Lyssavirus genus of *Rhabdoviridae*. (From 3rd report of the International Committee on Taxonomy of Viruses Matthews 1979).

Lyssavirus	Distribution	Host
Rabies	World-wide	All warm-blooded animals
Lagos bat	Nigeria	Fruit bats
Mokola	Nigeria	Man, shrews
Duvenhage	S. Africa	Man, bat?
Obodhiang?	Nigeria	Culicoides
Kotonkan?	Sudan	Mosquitoes

The isolation of several 'rabies-like' viruses from insects, bats, shrews or man in Africa has destroyed the concept of rabies virus as being antigenically unique. A genus 'lyssavirus' within the *Rhabdoviridae* now includes rabies virus, Lagos bat, Mokola, Duvenhage and possibly Obodhiang and Kotonkan viruses (Table 102.1). These viruses do not necessarily produce 'rabies-like' disease, but they do share nucleocapsid antigens with classical rabies virus that cross-react in complement-fixation and immunofluorescent tests and in radio-immunoassays. Their differences are expressed in neutralization tests by surface antigens, notably glycoprotein (Shope 1975). The use of monoclonal antibodies has revealed further antigenic differences among strains of rabies virus isolated in different areas of the world (Wiktor and Koprowski 1978, Sureau and Rollin 1982), the immunological significance of which is at present obscure.

Growth and morphogenesis of rabies virus

Virus isolated from natural infections was called 'street virus' (*virus des rues*) by Pasteur. After a variable number of intracranial (ic) passages in animals the properties of most street viruses are modified. The incubation period shortens and becomes 'fixed' at 5–8 days. 'Fixed' virus remains neurotropic and its virulence for the central nervous system may be enhanced; it produces paralytic rather than furious symptoms. Its infectivity by peripheral inoculation is diminished, as is its ability to induce inclusion bodies in the central nervous system.

Prolonged passage is necessary to adapt virus to growth in cell cultures but the broad growth characteristics of several strains of fixed rabies virus in various cell systems are now well established (Wiktor

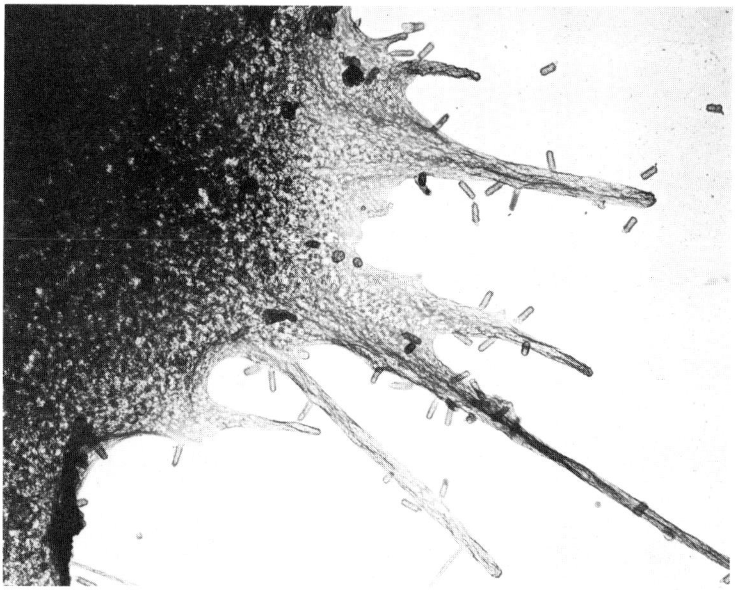

Fig. 102.2 Electronmicrograph of rabies virus (strain HEP) budding from the microvilli of a mouse neuroblastoma cell. (By courtesy of Drs Y. Iwasaki and H F. Clark, (1977). *Laboratory Investigation* **36,** 578.)

1973, Wiktor and Clark 1975). Either productive or carrier type cycles of virus growth are described. Cytopathic effects and plaque formation occur in some cells infected with certain virus strains (Wiktor 1973), but in most cell types cytological changes are absent. Adsorption to and penetration of cells is rapid and is enhanced by polycations like diethylamino-ethyldextran. Maximum yields of virus are produced by cells infected in suspension and incubated at 31–33° for 12–72 hr.

Virus release from infected cells is preceded by the formation of an unbounded matrix of unorganized viral nucleocapsid protein. This is formed in excess of requirement and is subsequently coiled and enveloped as virus particles mature at host cell membranes. In cell cultures infected with fixed strains, virus may bud either intracellularly from the membranes of the endoplasmic reticulum into adjacent cysternae or into extracellular spaces from the plasma membrane (Fig. 102.2). The latter process has also been observed in cultures of mouse ganglia and in neuroblastoma cell lines. Since these observations were made with fixed strains in particular cell lines, their relevance to the effects of streeet virus on cells *in vivo* is questionable.

The neurones of animals infected with street virus form characteristic *inclusion bodies* which have long been considered pathognomic of rabies (Negri 1903). Such cells show little cytological damage. Virus synthesis appears to be restricted to the endoplasmic reticulum where virus accumulation may be seen associated with the matrix of the inclusion (Fig. 102.3). Maturation, budding and extracellular release of large quantities of virus have been observed in the mucogenic acinar cells of the salivary gland of foxes and skunks infected with street virus (Fig. 102.4), providing abundant material for bite transmission (Dierks *et al.* 1969, Murphy 1975).

In animals infected with fixed virus, damage to nerve cells is more extensive, matrix formation is under-developed and typical Negri bodies are not visible by light microscopy. Nevertheless, the Negri bodies seen in the neurones of animals infected with street virus are similar in composition to those found in cell cultures infected with fixed virus *in vitro* (Matsumoto 1970). The morphogenesis of both forms of rabies virus is now considered to be basically similar (Murphy 1975, Schneider and Diringer 1976).

Epidemiology in animals

Rabies is present in all continents except Australia and Antarctica. The virus can infect all warm-blooded animals including birds, but the latter are very rarely sources of infection. Although some 'rabies-like' viruses have been isolated from mosquitoes in Nigeria and Sudan (see Table 102.1), extensive surveys have provided little evidence that arthropods are either reservoirs or transmitters of rabies (Aubert and Andral 1975).

Among the mammals, the medium sized and smaller carnivores, particularly *Canidae*, are of world-wide distribution and most frequently affected. Two major epidemiological cycles are recognized: (1) urban dog rabies, which is now largely confined to the less developed countries of the world; and (2) a sylvatic or wildlife cycle, which predominates in North America and throughout most of Europe. The species involved in sylvatic rabies vary throughout the world.

In the United States and Canada, skunks, foxes and racoons are the wildlife species most often found rabid

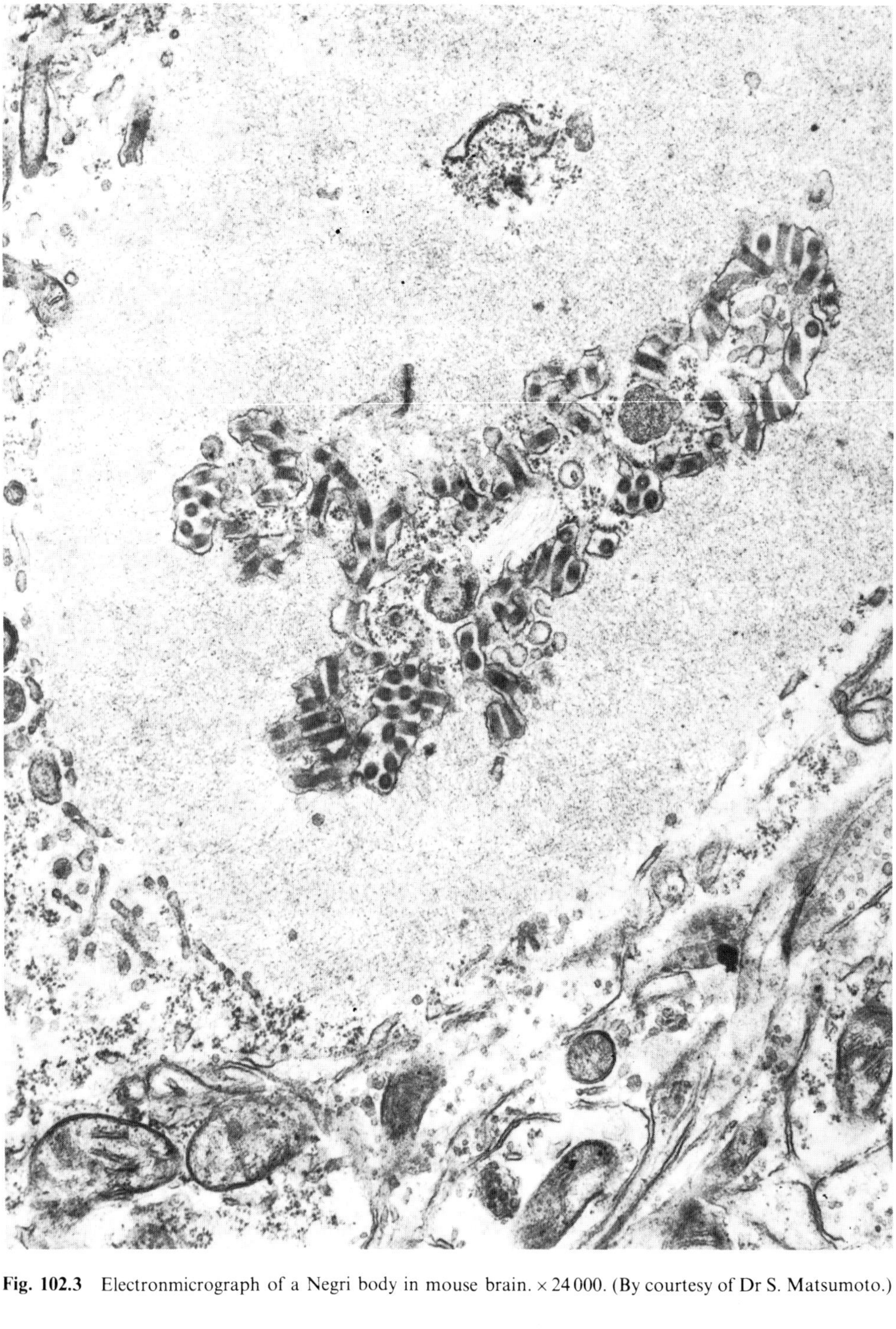

Fig. 102.3 Electronmicrograph of a Negri body in mouse brain. ×24 000. (By courtesy of Dr S. Matsumoto.)

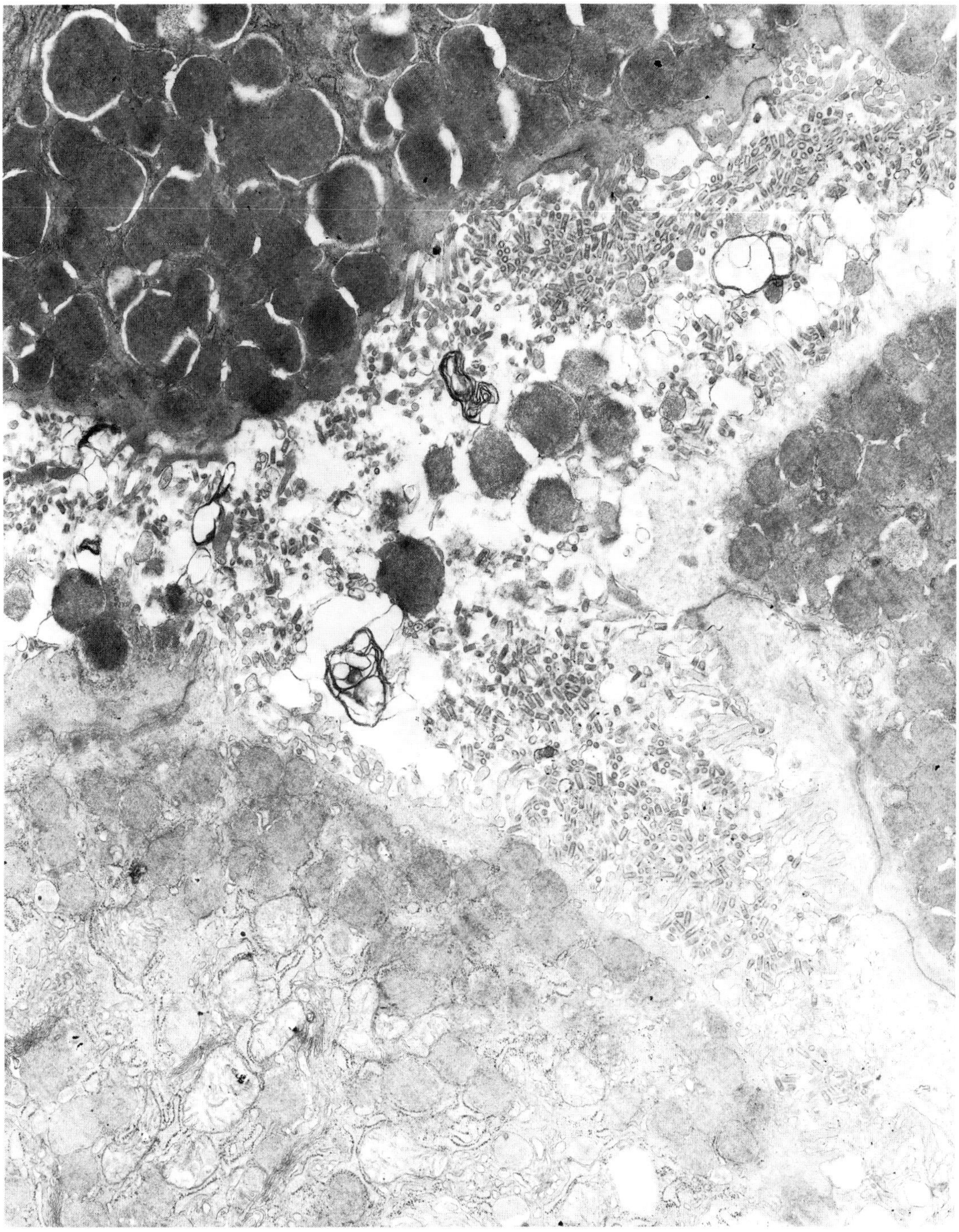

Fig. 102.4 Electronmicrograph of rabies virus in fox salivary gland. × 15 000 (By courtesy of Dierks, R. E., Murphy, F. A. and Harrison, Alyne K. (1969) and *The American Journal of Pathology* **54,** 25.)

Table 102.2 Species responsible for human exposures to rabies (per cent of total). (From *Rabies Bulletin Europe* (1980*b*) and World Survey of Rabies XX (1980–81) WHO/Rabies/**82,** 193.)

	Dogs	Cats	Cattle	Foxes	Others	Total
Europe (1980)	65	12	4	6	12	12,888
World wide (1980 to 1981)	82	10	1	2	5	83,701

(Report 1978); in the Arctic region, Arctic foxes; in southern Africa, the yellow mongoose and jackals; in eastern Europe, the red fox and racoon dog; and in central and western Europe foxes account for 60–80 per cent of all cases of rabies reported in animals (Report 1980*a*).

There is little evidence that latent or abortive rabies infection occurs in foxes in Europe or mustelids in North America. The highly neurotropic virus rapidly kills these species, and its maintenance in them is related to their habit, population density, and annual reproductive capacity. Other wild species may be infected, but they are apparently unable to sustain an enzoötic in the absence of rabies in the main host (Report 1973, Steck and Wandeler 1980).

The incidence of rabies in wildlife is directly related to that in domestic animals. Cattle are particularly vulnerable victims for rabid carnivores. However, herbivores in general have little significance in rabies epidemiology, and human rabies transmitted by rabid bovines is rare. Nevertheless, exposure frequently occurs through attempts to examine or treat sick animals.

In Latin America virus transmitted by vampire bats is responsible for the annual deaths of hundreds of thousands of cattle (Baer 1975*b*). Numerous human deaths are also attributed to vampire bites. Early reports suggested that a symptomless salivary carrier state existed in vampires. Most evidence, however, indicates that vampires usually die of the disease (Report 1973). Rabies has also been transmitted to man by insectivorous bats. The disease has been reported in 30 of 39 bat species resident in North America (Constantine 1979). Bat rabies is largely confined to the new world. Extensive surveys of bats in Asia and Africa have revealed no rabies; and the very rare isolations from European bats suggest that they make little contribution to the epidemiology of the disease.

Rodents constitute the largest mammalian order, but their significance as rabies vectors varies throughout the world. In south eastern Europe, the Middle and Far East and parts of tropical America, rabies in common rats and mice is apparently a frequent source of human exposure (Report 1979). In the United States and western Europe very few positive identifications of rabies have been made in extensive surveys of a wide variety of rodents (Winkler 1972, Forster *et al.* 1977). Some atypical strains of rabies virus have been identified in shrews and voles in Czechoslovakia, but their epidemiological significance remains undetermined (Sodja *et al.* 1971).

Non-human primates appear to play a negligible role in the epidemiology of wildlife rabies. World surveys record bites by monkeys in various parts of the world, in which some of the animals were shown to be rabid and some were not (Report 1979). The exportation of primates for zoo or research purposes makes vigilance necessary in their handling and maintenance.

Rabies in the *Canidae* and dogs in particular has been frequently described (Bedford 1976, Minor 1977). Changes in behaviour are early signs of rabies in dogs and many other species, friendly animals becoming aggressive and wild animals losing their fear of man. Notable features of canine rabies are the absence of hydrophobia and the preponderance of the dumb paralytic rather than the classical furious form, which occurs in only 25 per cent of cases. Excessive salivation occurs in both forms and virus may be present in the saliva for 3 or 4 days before clinical signs appear.

Non-fatal rabies in dogs in which asymptomatic virus shedding occurs is fortunately very rare but has nevertheless been reported (Fekadu 1975, Veeraraghavan *et al.* 1969). The difficulty it would present in control of the disease is too obvious to need stressing. Oulou fato, the mild paralytic canine disease of West Africa, is frequently cited as an example of non-fatal rabies infection in dogs (Bouffard 1912). Although the virus was originally identified as rabies, it apparently has closer affinity with the 'rabies-like' viruses than with true rabies (Shope 1975).

Domestic cats are usually infected by a 'spillover' from other domestic animals, or from those in the wild. In some parts of Europe the incidence of human exposures from cats has exceeded that from any other species.

Human rabies is occasionally acquired by direct contact with rabid wild animals in the western world, but in these areas it is now a rare disease. Accurate estimates of its incidence are difficult to obtain (Turner 1976*a*). Nevertheless, world figures suggest that man's domestic pets, dogs, and cats still account for more than 90 per cent of all human exposures (Table 102.2).

Human rabies

Human rabies is a prime example of a zoonosis. Exposure to the virus invariably occurs by the bites of rabid animals in which infected saliva is deposited on to recently abraded skin or the mucous membranes of the mouth, conjunctiva, genitalia or anus (Hattwick and Gregg 1975). The inhalation of massive virus aerosols generated in bat caves or in laboratory accidents has also been reported (Constantine 1962, Winkler *et al.* 1973, Report 1977). Oral infection has been experimentally induced in some animals, but no evidence so far suggests that it is common (Afshar 1979).

Human exposure to rabies virus does not necessarily result in infection. Contrary to popular belief, man is not highly susceptible to the virus. The incidence of human rabies after bites by known rabid dogs averages about 15 per cent. It varies, however, from 0.1 per cent after the contamination of minor wounds to more than 60 per cent after severe bites on the face. The probability of rabies developing after exposure varies with the location of the bite. Clinical rabies follows bites on the head and neck more frequently than after those on hands or arms, and still less frequently after those on feet or legs. Other variables include the severity of the injury, whether it occurs through clothing, and whether the animal is secreting virus at the time of the bite. The salivary concentration of virus reputedly varies in different rabid species and is, for example, higher in skunks and foxes than in dogs.

The highest human mortality is reported after wolf bites in which deep lacerated wounds, liberally contaminated with saliva, are inflicted on the head and neck of victims (Hattwick and Gregg 1975). The disease in man usually follows an incubation period of 1 to 2 months. It may be as short as 9 days, and in rare cases be as long as a year or more. It is shorter in children than in adults, and shorter after bites on the head than after those on the extremities.

In general, early symptoms are non-specific, but abnormal sensation at the site of injury is recorded in a substantial proportion of the cases and is regarded as an early specific sign. After a prodromal period of one to ten days, disturbance of the central nervous system becomes apparent. The disease may develop as 'furious' rabies characterized by hydrophobia and the classical violent episodes of arousal and lucidity followed by coma and death (Fig. 102.5). More rarely the disease presents itself as 'dumb' rabies characterized by ascending paralysis, absence of hydrophobia and a more prolonged illness (Warrell 1976). Paralytic rabies occurred in all cases of vampire transmitted rabies in the Trinidad epidemic described by Hurst and Pawan (1968).

Man-to-man transmission of rabies is rarely recorded. Virus may be secreted in the saliva of human rabies victims just as in that of animals, and before death may be found in urine and other secretions. Corneal transplants from donors with undiagnosed rabies have provided two recent bizarre examples of human transmission (Houff *et al.* 1979, Report 1980*b*). In treatment of the disease, intensive supportive care has been widely used since its apparently successful application in the rare cases in which the patient has recovered (Hattwick *et al.* 1972, Porras *et al.* 1976). The initial optimism engendered by these successes has had little subsequent justification, however; indeed the procedures of intensive therapy may well increase the risks to the personnel responsible for their application (Cundy 1980). For practical purposes the disease in man may be regarded as uniformly fatal.

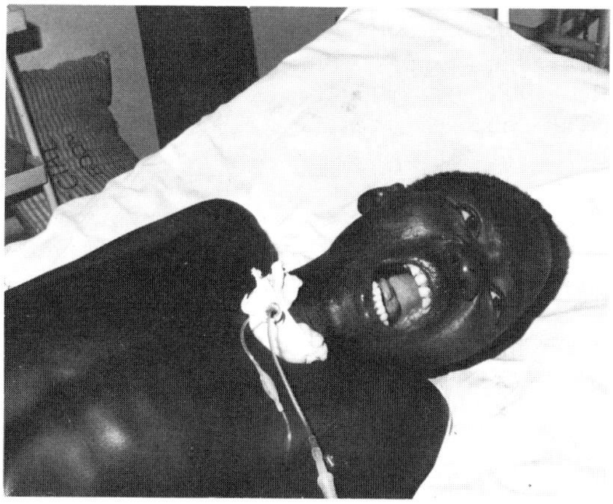

Fig. 102.5 Furious rabies in man. Eighteen-year-old Nigerian male bitten on the temple by a dog. Thirteen days later he developed fever, headache, vomiting and dysphagia. Later he had hydrophobia. On the third day of illness he began to have attacks of agitation and delirium (illustrated) with shouting and struggling. (By courtesy of Dr D. A. Warrell (1976) and the *Trans. roy. Soc. trop. Med. Hyg.* **70,** (3) 190.)

Rabies pathogenesis

The inoculation of many animal species has firmly established the neurotropism of rabies virus; whatever route of inoculation is used, the ultimate objective of the virus is the central nervous system (Baer 1975*a*).

Movement of virus from peripheral sites to the central nervous system is effected by passive transport within the axoplasm of nerves. There is little evidence that haematogenous or other modes of centripetal spread are concerned in rabies infection (Baer 1975*a*, Murphy 1977). Footpad infection in laboratory rodents, followed by sequential amputation or section of the nerve supply of the infected limb, has allowed estimates of the rate of this centripetal movement.

With some virus strains this approximates to 3 mm/hr; with others it is longer and more variable (Baer 1975*a*).

Experiments designed to mimic bite exposure suggest that muscle cells or epithelial cells in the lower layers of the epidermis provide sites for an initial cycle of viral replication before neural invasion begins. It is also considered that in infections with long incubation periods these are likely places for the sequestration of virus (Murphy 1977). After transport within peripheral nerves virus may undergo further replication or amplification in the neurones of dorsal root ganglia before its ascent to the brain from the spinal cord.

Extensive replication occurs within the brain and, terminally, neurones in all parts may be infected (Schneider 1975*a*). However, the appearance of specific symptoms at various stages of the disease is probably related to progressive viral-induced dysfunction of particular areas of the brain. The furious and aggressive behaviour necessary for the natural perpetuation of the virus is associated, for example, with greater virus localization in the limbic system (Johnson 1971).

In street virus infections, the brain frequently shows no obvious cell damage; neurones remain intact and, apart from inclusions, cells exhibit few changes in particular organelles. There is nevertheless great variability in the microscopic changes reported in the brain. These vary from minor perivascular infiltrates, common to many viral encephalitides, to neuronal degeneration and neuronophagia (Perl 1975).

After proliferation in the brain the virus moves centrifugally by the same axoplasmic routes as those used for centripetal transit. Organs close to the CNS are rapidly infected; thus, nerve end-organs in oral and nasal cavities, the retina, cornea, and cutaneous branches to the sebaceous glands and hair follicles of the head, are frequently infected early in the disease. Demonstration of viral antigen in corneal scrapings and skin biopsies has provided a means of ante-mortem diagnosis of rabies (Schneider 1975*b*).

The rich innervation of the salivary glands provides a direct viral route to the secretory cells. Virus titres in salivary glands frequently exceed those in brain tissue, suggesting that viral replication occurs within them, and that salivary secretion of virus in some animals may precede clinical signs of the disease (Dierks 1975).

The isolation of virus from many extraneural tissues and organs at death, is widely reported. Those from which virus might be excreted include the respiratory system and the urinary tract. Virus excretion in the urine is reputedly high in foxes and the 'nosing' of urine used by foxes to mark territory may be a 'non-bite' transmission mechanism in sylvatic rabies. Rabies virus has also been isolated from the urine of human cases of rabies (Jonnescu 1927). Virus in the alimentary tract appears to be limited to intermittent infection of the autonomic nerve plexuses (Schneider 1975*b*).

Some response of the host to infection with rabies virus may be concerned in the pathogenesis of the disease. Studies *in vitro* have shown that immune thymus-derived (T) lymphocytes are strongly cytotoxic for rabies-infected susceptible cells (Wiktor *et al.* 1977*b*). Similarly infected cells are lysed by rabies antibody in the presence of complement (Koprowski 1974). Antibody may also be implicated *in vivo* in the phenomenon described as 'early death', in which immunized animals with low antibody titres die earlier after experimental challenge than animals with no antibody (Sikes *et al.* 1971). Whether these host responses are pathogenic or protective is largely speculative.

Rabies immunology

Though rabies virus is capable of stimulating most host defence mechanisms, it is apparently insulated from the immune system throughout the long incubation period of street virus infection. Few, if any, antigenic stimuli are presented until massive amounts of virus are released late in infection, when the brain is invaded and clinical illness begins (Murphy 1977). In man, for example, the mean period from the onset of illness to death is 8 to 10 days. The latter time frequently coincides with the first appearance of neutralizing antibody. In those cases where survival is prolonged, either naturally or by intensive therapy, antibody in blood, brain and CSF reaches very high titres and is related to the disappearance of infectious virus (Hattwick and Gregg 1975). High antibody titres in the CSF of animals are indicative of abortive or non-fatal infections, which occur rarely in nature but can be readily induced experimentally in animals (Bell *et al.* 1966).

The application of vaccine after exposure circumvents the long period of immunological inactivity. Post-exposure vaccination has now been practised for almost a century as a means of preventing rabies infection from becoming a fatal disease. The doubt associated with this procedure in the past has been dispelled by evidence from recent animal models. Mice, hamsters, and monkeys can be protected after infection by single doses of potent cell culture vaccine (Sikes *et al.* 1971), Baer and Cleary 1972, Wiktor *et al.* 1972). Nevertheless, the mechanism by which protection is induced remains obscure.

Humoral responses. Antibody is protective when present before exposure; and prompt vaccination after exposure induces resistance associated with antibody production. Nevertheless, the role of antibody remains uncertain. The high antibody titres recorded in some human victims are only rarely associated with survival. In animals treated solely with antibody after infection, the incubation period may be prolonged but mortality is not significantly reduced. By contrast, experiments in immunosuppressed animals suggest that the presence of humoral antibody is essential for resisting rabies infection (Miller *et al.* 1978, Turner 1979). Furthermore, both in animals and man survival rates are increased when vaccine treatment is supplemented with antiserum (Cabasso 1975).

Cell-mediated responses. Whether cell-mediated immunity is a critical component of the host defence in rabies infections is also unclear. Specific blastogenesis has been demonstrated in the splenic or circulatory lymphocytes of animals and man immunized with a variety of rabies antigens (Wiktor *et al.* 1974, Nozaki and Atanasiu 1976, Nicholson *et al.* 1979). Delayed-type hypersensitivity to rabies virus may occur (Lagrange *et al.* 1978); and rabies antigens are recognized as T-cell dependent (Kaplan *et al.* 1975; Turner 1976*b*). Furthermore, the cytotoxic responses already mentioned are similarly attributed to immune T-lymphocyte activity. While these observations all point to the participation of T-lymphocytes in the host response, many were demonstrated under special experimental conditions and require further examination.

Interferon. The growth of rabies virus is inhibited by interferon *in vitro*; and both fixed and street strains of the virus induce interferon *in vivo*. In experimentally infected animals interferon production is most abundant in brain tissue just before death, and is detectable in blood and other tissues. Interferon was absent throughout the entire clinical course of one fatal human case, but systematic examination for interferon in human rabies infection is rare (Bhatt *et al.* 1974). Interferon induction, like the antibody response, apparently occurs too late in the natural disease to be protective.

However, evidence which has accumulated over more than 20 years shows that the early induction of endogenous interferon, or the administration of the exogenous product, will protect infected animals. Interferon is most effective when it is applied at the site of infection within a short time of exposure. Interferon might well complement the delayed but more durable effects of immunization in the post-exposure treatment of man. It is significant that protection induced by vaccines in experimental models occurred only with products that stimulated the production of interferon. However, vaccines that are considered suitable for use in man are inconsistent in their possession of this property (for reviews see Sulkin and Allen 1975, Turner 1977*a*, Baer 1978).

Laboratory diagnosis

Three techniques are generally used: histological examination and demonstration of *Negri bodies* in sections of brains (Fig. 102.6); virus isolation by animal inoculation; and the immunochemical detection of rabies antigen. The last test competently performed with rabies-specific fluorescent antibody is regarded as the best rapid method available. Peroxidase-labelled antibody, similar in principle, has also been used. (For assessments and detailed accounts of these tests see Kaplan and Koprowski 1973, Atanasiu 1975, and Kissling 1975). Intracytoplasmic Negri bodies are mainly localized in the pyramidal cells of Ammon's horn, in the Purkinje cells of the cerebellum, and in the medulla and various ganglia. Not all workers are convinced that Negri bodies are an infallible diagnostic sign (Derakshan *et al.* 1978).

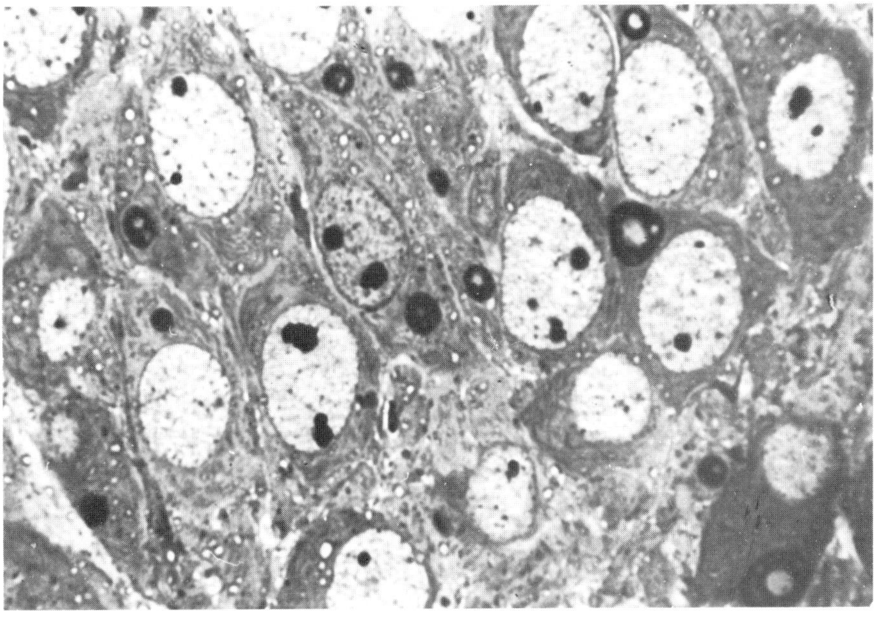

Fig. 102.6 Negri bodies in Ammon's Horn; street virus; Toluidine Blue stain × 1600. (By courtesy of Professor P. Atanasiu, Pasteur Institute, Paris.)

Although other animals may be used, sucking mice, inoculated intracerebrally, are the most sensitive for virus isolation. Inoculated mice are observed for 21 days, but this period may be shortened by fluorescent antibody staining of the brains of mice killed 4 to 7 days after inoculation. These techniques are applied, when possible, to the brains and salivary glands of suspect animals that have bitten human beings. The results assist in determining whether post-exposure antirabies treatment of those bitten should be continued, modified or terminated—a consideration of great importance with some vaccines still in common use.

The detection of rabies infection before death in animals and man can be accomplished by the inoculation of mice with specimens of saliva, material from throat and nasal swabs, and occasionally cerebrospinal fluid. Rabies antigen has also been demonstrated by fluorescent antibody staining of corneal scrapings and sections of skin obtained by biopsy. However, none of these sources provides consistently positive results; negative tests do not eliminate the possibility of infection (Report 1973, Reis *et al.* 1976).

The value of cell culture in the diagnostic isolation of rabies virus remains experimental. Promising results have been obtained with cultures of BSC-1, CER and BHK cells inoculated with street virus. Although no cytological changes were seen, rabies antigen could be detected by fluorescent antibody staining 24 to 48 hr after inoculation (Smith *et al.* 1977, Rudd *et al.* 1980). The titration of antibody normally has little value as an index of infection, because death usually occurs before significant antibody increases are detectable.

Rabies vaccines and vaccination

The immunoprophylaxis of human rabies instituted by Pasteur almost a century ago is based upon the belief that an active immune response can be stimulated within the incubation period. The latter is frequently prolonged and the time of the animal bite can usually be determined with accuracy.

Pasteur's vaccine was prepared from suspensions of the spinal cord of rabbits dying from experimental rabies. When dried over KOH the cords gradually lost their virulence. He was able to protect dogs against natural and experimental rabies with a series of inoculations prepared from cords that had been dried for varying periods. Inoculations began with the least virulent and ended with the most virulent. This technique was first applied to man in July 1885; Joseph Meister, a boy aged 9 years had received multiple bite wounds from a rabid dog; he was given a course of 13 subcutaneous injections spread over a period of 11 days. The course began 60 hr after exposure with cords dried for 14 days; those used subsequently were dried for periods decreasing to 1 day. The boy survived, and this historic event focused attention upon the practical and theoretical possibilities of immunization against infectious disease. It established an enduring model for the preparation of many other viral vaccines.

The 'Pasteur' technique continued to be used in Paris until the 1950s but many modifications were made to it in the intervening period (for accounts of these early vaccines see van Rooyen and Rhodes 1948). Among those that may be mentioned are the methods of Högyes, Babès, Alivisato, Hempt, Fermi and Semple. Most of these modifications were designed to reduce or remove the virulence or infectivity of nervous tissue suspensions either by dilution or treatment with heat, ether, formalin or phenol. However, the balance between safety and the maintenance of antigenic potency was difficult to achieve. Rabies infection attributable to live virus in vaccines was a contentious subject in Pasteur's time and occasional accidents with this type of vaccine have been reported since then (Para 1965). Rabies vaccines containing live virus are no longer recommended for human use (Report 1973). Few of the early modifications of Pasteur's vaccine are still prepared except for Semple vaccine. The latter is a 10 to 20 per cent suspension of rabies infected rabbit, sheep or goat brain in which the virus is inactivated with phenol or β-propiolactone; it is still prepared and used in many parts of Asia and Africa. The dose schedules considered necessary for this type of vaccine require repeated and painful injections and their efficacy has long been controversial. Furthermore, these crude preparations have been notorious for the fearful neuroparalytic accidents associated with their administration. Reports of the incidence of these neurological side effects vary but can be as high as 1:500 vaccinations. (For further information on the complications of rabies vaccination see Wilson 1967).

The substance responsible for these neuro-allergic reactions is associated with the myelin component of brain tissue. Attempts to remove these dangerous impurities appear to have been neither rewarding nor economic, and have seldom been successful without losing some antigen. Most attention has been directed at alternative substrates for producing rabies vaccines. The brain tissues of newborn animals contain little if any of the factors responsible for inducing paralysis, and they also provide high yields of virus. Suckling mouse brain vaccine of high potency is much used in Latin America as well as in some parts of France (Fuenzalida and Palacios 1955). Neurological accidents associated with the use of this vaccine have been reported, although the incidence is at least fivefold lower than that associated with vaccine prepared from adult animal brain.

Vaccine produced from virus adapted to growth in avian embryos has had wide use in man and animals. Live vaccines prepared from the 'Flury' strain of rabies virus adapted to chick embryos were successfully used in the mass immunization of dogs in several rabies eradication campaigns. Low egg passage (LEP) vaccine prepared from virus passaged 40 to 50 times in chick embryos is safe and antigenic in adult dogs, but retains some pathogenicity for cattle, cats and puppies. Flury HEP vaccine, prepared from virus at the high (180th) egg passage level is more attenuated and is recommended for the prophylactic inoculation of cattle and other animals in areas where rabies is a serious problem. Flury HEP vaccine was tested extensively in man and, although apparently safe, it was poorly antigenic unless large amounts were injected; it is not now recommended for human use.

Vaccine prepared from virus propagated in duck embryos and inactivated with β-propiolactone was claimed to be as immunogenic as Semple vaccine and relatively free from neuroallergens. The latter claim was substantially true. In a study of approximately 420 000 subjects the incidence of neurological reactions was estimated to be only 3.1 per 100 000 (Rubin *et al.* 1973). The vaccine was used for almost 20 years in Europe and the United States. However, the responses to the 14 to 21 doses recommended were frequently poor, and its content of duck protein produced a high incidence of allergic reactions (For reviews see Turner 1969, Clark *et al.* 1975, Turner 1977*b* Plotkin and Wiktor 1978).

The use of vaccines prepared from virus grown in cell cultures appears to have solved the difficulties of both safety and antigenicity associated with earlier rabies vaccines. The vaccine first produced by Wiktor and co-workers (1964) in a human diploid cell strain (HDCS) has had extensive human trials (Report 1980*c*), and its efficacy and safety in human pre- and post-exposure immunization is firmly established (Aoki *et al.* 1975, Kuwert *et al.* 1976, Bahmanyar *et al.* 1976, Plotkin and Wiktor 1978). Because of its superior immunogenicity, the quantity administered is much less than with former vaccines. Only six doses given intramuscularly in 1.0 ml volumes on days 0, 3, 7, 14, 30 and 90 after exposure are required (Report 1980*c*). The immune responses to vaccine administered intradermally in 0.1 ml doses are similar to those produced by 1.0 ml i.m. doses (Turner *et al.* 1976); Nicholson *et al.* 1978). These findings suggest that even fewer doses might suffice.

Pre-exposure immunization, widely recommended for those at high risk, was either unsafe or uncertain with earlier vaccines. No such problems have been experienced with the cell culture vaccine now available. Two doses given a month apart are adequate, with annual or less frequent reinforcing doses dependent on the degree of risk.

Sero-vaccine therapy

At the end of the last century, it was shown that a combination of active and passive immunization could protect animals against swine erysipelas or anthrax. Contemporaneously, Babès and Lepp (1889) showed that protection against rabies could be transferred to dogs and rabbits with blood from dogs vaccinated against the disease. Babès and Cerchez (1891) also reported that 12 persons severely bitten by a rabid wolf survived after receiving 6 injections of immune blood from man or dogs combined with a course of Pasteur's vaccine! Numerous attempts to protect man and experimental animals by the optimal use of serum and vaccine were reported subsequently (For review see Cabasso 1975).

The provision of early antibody coverage, in the vulnerable period before active response to vaccine occurs, is now recommended practice in all severe human exposures. The superiority of combining antiserum and vaccine treatment was amply demonstrated in the remarkable human trial recorded in Iran (Baltazard and Bahmanyar 1955). Hyperimmune antirabies serum was formerly of equine origin; its use was complicated by a high incidence of serum sickness. Human rabies immune globulin (HRIG) is now available, so that serum sickness can be avoided. Passively administered antibody inhibits the normal response to vaccine and therefore careful adjustment of the balance between antiserum and vaccine is required. A dose of 20 iu/kg HRIG is currently recommended, half of which is infiltrated around the wound and the remainder given intramuscularly (see Report 1973). Dose requirements for HRIG were largely determined with vaccines less potent than HDCS, and might bear re-examination with the newer vaccine (see Cabasso *et al.* 1971, Hattwick *et al.* 1974).

First-aid treatment to eliminate virus from the site of infection should precede the administration of serum and vaccine. Vigorous washing of wounds is regarded as an urgent and integral part of all post-exposure anti-rabies treatment. Copious flushing with soap and water or ethyl alcohol (42 per cent or higher) should be instituted immediately, and repeated later with 20 per cent soap solution or 1–2 per cent solutions of quaternary ammonium compounds (Report 1973).

The dilemma imposed either by the risk of rabies or the dangers of rabies vaccination should be eliminated when the new cell culture products are used. Nevertheless, unnecessary vaccination is to be avoided and the decision to vaccinate must be assessed from all the information available for each incident. The major factors to be considered are:

1 Whether or not exposure to rabies virus actually occurred and, if so, whether it was severe. Immediate treatment is recommended after multiple bites to the head, face and neck, the upper extremities and after licks on mucous membranes. Treatment should begin, even though the animal appears healthy and behaves normally at the time, and is being kept under observation.
2 The species of animal responsible is also important. Most wild carnivores should be regarded as rabid and their bites require immediate treatment. This

should be discontinued only if the animal has been captured, killed and laboratory tests are negative. Of other animals small rodents and rabbits are considered unlikely to be rabid. Most bites are inflicted by domestic dogs and cats and the need for immediate treatment is decided after finding the owner, tracing the animal and placing it under observation. If it remains healthy for 10 days, it will not have had virus in its saliva at the time of the bite.

3 The risk to man is greater in countries where canine rabies persists; but specific antirabies treatment is considered necessary wherever rabies is enzootic and the biting animals are stray or untraceable. In countries free from rabies and in England in particular, antirabies treatment is seldom necessary except for subjects bitten abroad. In these, decisions may be difficult because the required information is frequently unreliable and the tracing and surveillance of the biting animal beyond control.
However, whenever circumstances indicate that specific treatment is necessary it should not be withheld, despite a delay of weeks or even months after the bite (For further information see Gardner 1979).

Rabies control

Effective control of the disease in domestic animals, particularly dogs, has been described as the 'corner stone' in the prevention of human rabies. Wherever concerted and sustained programmes are instituted a reduction in canine rabies is followed by a decline in human cases. Rabies was common in England and Wales until the present century. Attempts were made to control the 'spread of canine madness' as early as 1831, but these and subsequent measures for the detention and destruction of dogs were feebly enforced. However, by the firm and general application of the Muzzling Orders of 1885 and 1889 imposed by the Government at the instance of Sir Victor Horsley, rabies was successfully controlled; together with the imposition of quarantine, the disease was eradicated by 1903. The occasional incidents that have occurred subsequently have brought the regulations under close scrutiny and usually resulted in stricter controls to prevent re-entry of the disease (Report 1971, Hill 1971, Wilson 1979). From 1922 to 1980 27 cases of rabies have occurred in animals held in quarantine; current figures show that approximately 4500 dogs and 1800 cats enter quarantine annually (Report 1980*d*). From 1955 to 1980, 16 patients died of rabies in England; in every case the disease occurred in persons who had sustained animal bites outside the United Kingdom.

Canine vaccination

In countries with adjacent land frontiers, where rabies exists in dogs or indigenous wild species, quarantine is either insufficient or impracticable, and must be supplemented by the vaccination of animals likely to provide a link between man and the wildlife reservoirs. Immunization of at least 70 per cent of the dogs in an area in the shortest possible period is recommended, together with the elimination of stray animals. Canine vaccination has been successfully instituted in many parts of the world (for review see Tierkel 1975). Numerous vaccines, both live and inactivated, are available for the purpose. The choice depends upon the age and species of animal to be immunized. Modified live virus vaccines often provide more durable immunity, but are pathogenic for some species; they are not recommended for use in exotic pets or zoo animals (see Report 1973).

Wildlife control

The control of rabies in wildlife presents problems that have so far not been solved. Population reduction by gassing, poisoning, trapping and shooting are commonly used. Where the population of foxes, for example, can be reduced to 1 per 5 km^2, transmission is broken and rabies disappears (Report 1973). However, these measures have had limited success, even in areas where the major host species is well defined and topographical conditions are favourable (Müller 1974).

The incorporation of either antifertility agents or vaccine into suitable baits provides a feasible method for controlling rabies in some wild species. Oral vaccination is under investigation in foxes (Baer *et al.* 1971). However, live virus vaccine is required for successful immunization, and its possible pathogenicity for other than the species to be vaccinated requires caution before its use becomes widespread. The high population turnover of foxes suggests that annual revaccination would be necessary for a long period.

Conventional large-scale vaccination of cattle in Latin America is still the major weapon used to combat the depredations of vampire bat rabies. However, recent control methods using the anticoagulant diphenadione have taken advantage of the communal grooming which is part of the social behaviour of vampires. Bats have been netted, smeared with the anticoagulant in vaseline and released to return to their roosts. Some 38 dead vampires have been found for each one treated and complete colonies have been destroyed. This method may well prove less costly than vaccinating cattle. The contiguous land borders of many countries make the control of rabies in wildlife an international problem, dependent upon the concerted and sustained efforts of wildlife experts and animal ecologists, as well as veterinary and public health authorities.

References

Aaslestad, H. G. and Urbano, C. (1971) *J. Virol.* **8,** 922.
Abelseth, M. K. (1964) *Canad. vet. J.* **5,** 84.

Afshar, A. (1979) *Brit. vet. J.* **135,** 142.

Almeida, J. and Waterson, A. P. (1966) Int. symp. rabies Talloires 1965. *Symp. Ser. immunobiol Stand.* **1,** 27. S. Karger, Basel.

Aoki, F. Y., Tyrrell, D. A. J., Hill, L. E. and Turner, G. S. (1975) *Lancet* **i,** 660.

Ardoin, P. de Lalun, E. and Gamet, A. (1977) *Ann. Microbiol.* (*Inst. Pasteur*) **128b,** 553.

Atanasiu, P. (1975) In: *The Natural History of Rabies*, Vol. 1, p. 373. Ed. by G. M. Baer. Academic Press, New York.

Atanasiu, P. and Gamet, A. (1978) In: *Handbook of Clinical Neurology*, 34, (11) 235. Ed. by P. J. Vinken and G. W. Bruyn. North Holland Pub. Co., Amsterdam.

Aubert, M. F. A. and Andral, L. (1975) *Symbiosis* **7,** 57.

Babès, V. and Cerchez, T. (1891) *Ann. Inst. Pasteur, Paris* **10,** 625.

Babès, D. V. and Lepp, (1889) *Ann. Inst. Pasteur, Paris* **3,** 384.

Baer, G. M. (Ed) (1975*a*) *The Natural History of Rabies*, Vol. 1, p. 181. Academic Press, New York; (1975*b*) *Ibid.* Vol. 2, p. 155; (1978) *Amer. J. clin. Path.* **70** (Suppl.), 185.

Baer, G. M., Abelseth, M. K. and Debbie, J. G. (1971) *Amer. J. Epidem.* **93,** 487.

Baer, G. M. and Cleary, W. F. (1972) *J. infect. Dis.* **125,** 520.

Bahmanyar, M., Fayaz, A., Nour-Salehi, S., Mohammadi, M. and Koprowski, H. (1976) *J. Amer. med. Ass.* **236,** 2751.

Baltazard, M. and Bahmanyar, M. (1955) *Bull. World Hlth Org.* **13,** 347.

Bedford, P. G. C. (1976) *Vet. Rec.* **99,** 160.

Bell, J. F., Lodmell, D. L., Moore, J. G. and Raymond, G. H. (1966) *J. Immunol.* **97,** 747.

Bhatt, D. R., Hattwick, M. A. W., Gerdsen, R., Emmons, R. W. and Johnson, H. N. (1974) *Amer. J. Dis. Child.* **127,** 862.

Bouffard, G. (1912) *Ann. Inst. Pasteur.* **26,** 727.

Brown, F. *et al.* (1979) *Intervirol.* **12,** 1.

Cabasso, V. J. (1975) In: *The Natural History of Rabies*, Vol. 2, p. 319. Ed. by G. M. Baer. Academic Press, New York.

Cabasso, V. J., Loofburouw, J. C., Roby, R. E. and Anuskiewicz, W. (1971) *Bull. World Hlth Org.* **45,** 303.

Clark, H. F., Wiktor, T. J. and Koprowski, H. (1975) In: *The Natural History of Rabies*, Vol. 2, p. 341. Ed. by G. M. Baer. Academic Press, New York.

Constantine, D. G. (1962) *Publ. Hlth Rep., Wash.* **77,** 287; (1979) *J. Wild. Dis.* **15,** 347.

Crick, J. and Brown, F. (1974) *J. gen. Virol.* **22,** 147.

Cundy, J. M. (1980) *Anaesthesia* **35,** 35.

Derakshan, I., Bahmanyar, M., Nour-Salehi, S., Fayaz, A. and Mohammadi, M. (1978) *Lancet* **i,** 302.

Dierks, R. E. (1975) In: *The Natural History of Rabies*, Vol. 1, p. 303. Ed. by G. M. Baer. Academic Press, New York.

Dierks, R. E., Murphy, F. A. and Harrison, A. K. (1969) *Amer. J. Path.* **54,** 251.

Dietzschold, B. (1977) *J. Virol.* **23,** 286.

Dietzschold, B. *et al.* (1983) *Proc. nat. Acad. Sci., Wash.*, **80,** 70.

Diringer, H., Kulas, H. P., Schneider, L. G. and Schlumberger, H. D. (1973) *Z. Naturf.* **23***c*, 90.

Fekadu, M. (1975) Lancet **i,** 569.

Fenje, P. (1960) *Canad. J. Microbiol.* **6,** 479.

Forster, von U., Wachendörfer, G. and Krokel, H. (1977) *Berl. Münch. tierärztl. Wschr.* **90,** 335.

Fuenzalida, E. and Palacios, R. (1955) *Boln. Inst. bact. Chile.* **8,** 3.

Galtier, P. V. (1879) *C. R. Acad. Sci., Paris* **89,** 444.

Gardner, S. J. (1979) In: *Virus Diseases*, pp. 87–96. Ed. by R. B. Heath. Pitman, London.

Goetze, A. (1948) 'Sumer' *J. Archaeol. Iraq.* **4,** 63.

Halonen, P. E., Murphy, F. A., Fields, B. N. and Reese, D. R. (1968) *Proc. Soc. exp. Biol. Med.* **127,** 1037.

Hattwick, M. A. W. and Gregg, M. B. (1975) In: *The Natural History of Rabies*, Vol. 2, p. 281. Ed. by G. M. Baer. Academic Press, New York.

Hattwick, M. A. W., Rubin, R. H., Music, S., Sikes, R. K., Smith, J. S. and Gregg, M. B. (1974) *J. Amer. med. Ass.* **227,** 407.

Hattwick, M. A. W., Weiss, T. T., Stechschulte, C. J., Baer, G. M. and Gregg, M. B. (1972) *Ann. intern Med.* **76,** 931.

Hill, F. J. (1971) *Proc. roy. Soc. Med.* **64,** 231.

Houff, S. A. *et al.* (1979) *New Engl. J. Med.* **300,** 603.

Hurst, E. W. and Pawan, J. L. (1968) *Caribb. med. J.* **30,** 17.

Johnson, R. T. (1971) In: *Proceedings of working conference rabies, 1970*, p. 59. Ed. by Y. Nagano and F. M. Davenport. University of Tokyo press, Tokyo.

Jonnescu, D. (1927) *C. r. Soc. Biol.* **97,** 1731.

Kaplan, M. M. and Koprowski, H. (1973) *Laboratory Techniques in Rabies*, 3rd edn. World Hlth Org., Geneva.

Kaplan, M. M., Wiktor, T. J. and Koprowski, H. (1966) *Bull. World Hlth Org.* **34,** 293; (1975) *J. Immunol.* **114,** 1761.

Kawai, A. (1977) *J. Virol.* **24,** 826.

Kawai, A., Matsumoto, S. and Tanabe, K. (1975) *Virology* **67,** 520.

Kissling, R. E. (1958) *Proc. Soc. exp. Biol. Med.* **98,** 223; (1975) In: *The Natural History of Rabies*, Vol. 1, p. 401. Ed. by G. M. Baer. Academic Press, New York.

Koprowski, H. (1974) *Symp. ser. immunobiol. Stand.* 21, 89, S. Karger, Basel.

Kuwert, E. K. (1970) *Zbl. Bakt. I Abt. Orig. Suppl.* **3,** 1.

Kuwert, E. K., Marcus, I. and Höher, P. G. (1976) *J. biol. Stand.* **4,** 249.

Kuwert, E. K., Wiktor, T. J., Sokol, F. and Koprowski, H. (1968) *J. Virol.* **2,** 1381.

Lagrange, P. H., Tsiang, H., Hurtrell, B. and Ravisse, P. (1978) *Infect. Immun.* **21,** 931.

Magendie, F. (1821) *J. Physiol. exp. Path.* **1,** 40.

Matsumoto, S. (1970) *Advanc. Virus Res.* **16,** 257.

Matthews, R. E. F. (1979) *Classification and Nomenclature of Viruses*, pp. 222–6. S. Karger, Basel.

Michalski, F., Parks, N. F., Sokol, F. and Clark, H. F. (1976) *Infect. Immun.* **14,** 135.

Miller, A., Morse, H. C., Winkelstein, J. and Nathanson, N. (1978) *J. Immunol.* **121,** 321.

Minor, R. (1977) *Vet. Rec.* **101,** 516.

Müller, J. (1974) Int. Sym. Rabies, Lyons 1972, *Symp. ser. immunobiol. Stand.* **21,** 18. S. Karger, Basel.

Murphy, F. A. (1975) In: *The Natural History of Rabies*, Vol. 1, p. 33, Ed. by G. M. Baer. Academic Press, New York; (1977) *Arch. Virol.* **54,** 279.

Negri, A. (1903) *Z. Hyg. InfektKr.* **43,** 507.

Neurath, A. R., Wiktor, T. J. and Koprowski, H. (1966) *J. Bact.* **92,** 102.

Nicholson, K. G., Cole, P. J., Turner, G. S. and Harrison, P. (1979) *J. Infect. Dis.* **140,** 176.

Nicholson, K. G., Turner, G. S. and Aoki, F. Y. (1978) *J. infect. Dis.* **137,** 783.

Nozaki, J. and Atanasiu, P. (1976) *Ann. Microbiol.* (*Inst. Pasteur*) **127A,** 429.

Para, M. (1965) *Bull. World Hlth Org.* **33,** 177.

Pasteur, L., Chamberland, C., Roux, E. and Thuillier, L. (1881) *C. R. Acad. Sci., Paris* **92,** 1259; (1882) *Ibid.* **95,** 1187.

Perl, D. P. (1975) In: *The Natural History of Rabies*, Vol. 1, p. 235. Ed. by G. M. Baer. Academic Press, New York.

Plotkin, S. A. and Wiktor, T. J. (1978) *Annu. Rev. Med.* **29,** 583.

Porras, C., Barboza, J. J., Fuenzalida, E., Lopez-Adaros, H., Diaz, A. M. O. and Furst, J. (1976) *Ann. intern. Med.* **76,** 931.

Reis, R. *et al.* (1976) *Rev. Inst. Med. trop. S. Paulo* **18,** 393.

Repik, P., Flamand, A., Clark, H. F., Obijeski, J. F., Roy, P. and Bishop, D. H. L. (1974) *J. Virol.* **13,** 250.

Report (1971) *Report of the committee of enquiry on Rabies. Comnd. No. 4696.* H.M.S.O., London; (1973) 6th Rep. Expert committee on rabies. Tech. Rep. Series No. 5. World Hlth Org., Geneva; (1977) *Morb. Mort. Wkly. Rep.* **26,** (22), 183; (1978) Rabies Surveillance Annu. Summary, 1977. Centre for Disease Control, Atlanta, USA; (1980*a*) *Rabies Bulletin Europe* (Eds. L. G. Schneider and H. Jackson) 3, 5, 1979, World Hlth Org. Tübingen; (1980*b*) *Morb. Mort. Wkly. Rep.* **29,** (3), 25; (1980*c*) Consultation on cell culture rabies vaccines and their protective effect on man. Essen Mar. 1980, unpublished WHO document, WHO/Rab. Res./80, 7, Geneva 1980; (1980*d*) *Rabies Prevention and Control. Guidance Notes.* Min. Agric. Fish, Food. Tolworth, England; (1982) World Survey of Rabies XX (for years 1980/81) unpublished WHO document, WHO/rabies/82, 193, Geneva.

Rubin, R. H., Hattwick, M. A. W., Jones, S., Gregg, M. B. and Schwartz, V. D. (1973) *Ann. intern. Med.* **78,** 643.

Rudd, R. J., Trimarchi, C. V. and Abelseth, M. K. (1980) *J. clin. Microbiol.* **12,** 590.

Schlumberger, H. D., Schneider, L. G., Kulas, H. P. and Diringer, H. (1973) *Z. Naturf.* **28***c*, 103.

Schneider, L. G. (1975*a*) In: *The Natural History of Rabies*, Vol. 1, p. 199. Ed. by G. M. Baer, Academic Press, New York; (1975*b*) *Ibid.* Vol. 1, p. 273.

Schneider, L. G., Dietzschold, B., Dierks, R. E., Matthaeus, W., Enzmann, P. J. and Ströhamer, K. (1973) *J. Virol.* **11,** 748.

Schneider, L. G. and Diringer, H. (1976) *Curr. Top. Microbiol. Immun.* **75,** 153.

Schneider, L. G., Horzinek, M. and Novicky R. (1971) *Arch. ges. Virusforsch* **34,** 360.

Shope, R. (1975) In: *The Natural History of Rabies*, Vol. 1, p. 141. Ed. by G. M. Baer. Academic Press, New York.

Sikes, R. K., Cleary, W. F., Koprowski, H., Wiktor, T. J. and Kaplan, M. M. (1971) *Bull. World Hlth Org.* **45,** 1.

Smith, A. L., Tignor, G. H., Mifune, K. and Motohashi, T. (1977) *Intervirol.* **8,** 92.

Smithcors, J. F. (1958) *Vet. Med.* **53,** 149, 267, 435.

Sodja, I., Lim, D. and Matouch, O. (1971) *J. Hyg. Epidem., Microbiol., Immunol., Praha* **15,** 231.

Sokol, F. (1973) In: *Laboratory Techniques in Rabies*, 3rd edn., p. 165. Ed. by M. M. Kaplan and H. Koprowski. World Hlth Org., Geneva.

Sokol, F., Schlumberger, H. D., Wiktor, T. J., Koprowski, H. and Hummeler, K. (1969) *Virology* **38,** 651.

Sokol, F., Stancek, D. and Koprowski, H. (1971) *J. Virol.* **7,** 241.

Steck, F. and Wandeler, A. (1980) *Epidem. Rev.* **2,** 71.

Sulkin, S. E. and Allen, R. (1975) In: *The Natural History of Rabies*, Vol. 1, p. 536. Ed. by G. M. Baer. Academic Press, New York.

Sureau, P. and Rollin, P. E. (1982) *Comp. Immunol. Microbiol. infect. Dis.* **5,** 109.

Tierkel (1975) In: *The Natural History of Rabies*, Vol. 2, p. 189. Ed. by G. M. Baer. Academic Press, New York.

Turner, G. S. (1969) *Brit. med. Bull.* **25,** 136; (1976*a*) *Trans. roy. Soc. trop. Med. Hyg.* **70,** 175; (1976*b*) *J. gen. Virol.* **33,** 335; (1977*a*) *Immunität u Infekt.* **5,** 208; (1977*b*) *Rec. Advanc clin. Virol.* **1,** 79; (1979) *Arch. Virol.* **61,** 321.

Turner, G. S., Aoki, F. Y., Nicholson, K. G., Tyrell, D. A. J. and Hill, L. E. (1976) *Lancet* **i,** 1379.

Turner, G. S. and Kaplan, C. (1967) *J. gen. Virol.* **1,** 537.

van Rooyen, C. E. and Rhodes, A. J. (1948) *Virus Diseases of Man*, 2nd edn., pp. 854–906. Nelson, New York.

Veeraraghavan *et al.* (1969) *Sci. Rep. Pasteur Inst. Southern India.* Coonoor, p. 66.

Warrell, D. A. (1976) *Trans. roy. Soc. trop. Med. Hyg.* **70,** 188.

Wiktor, T. J. (1973) In: *Laboratory Techniques in Rabies*, 3rd edn., p. 101. Ed. by M. M. Kaplan and H. Koprowski. World Hlth Org., Geneva.

Wiktor, T. J. and Clark, H. F. (1975) In: *The Natural History of Rabies*, Vol. 1, p. 155. Ed. by G. M. Baer. Academic Press, New York.

Wiktor, T. J., Dietzschold, B., Leamnson, R. N. and Koprowski, H. (1977*a*) *J. Virol.* **21,** 626.

Wiktor, T. J., Doherty, P. C. and Koprowski, H. (1977*b*) *Proc. natl. Acad. Sci., Wash.* **74,** 334.

Wiktor, T. J., Fernandez, M. V. and Koprowski, H. (1964) *J. Immunol.* **93,** 353.

Wiktor, T. J., Kamo, I. and Koprowski, H. (1974) *J. Immunol.* **112,** 2013.

Wiktor, T. J. and Koprowski, H. (1978) *Proc. natl. Acad. Sci., USA* **75,** 3938.

Wiktor, T. J., Kuwert, E. and Koprowski, H. (1968) *J. Immunol.* **101,** 1271.

Wiktor, T. J., Postic, B., Ho, M. and Koprowski, H. (1972) *J. infect. Dis.* **126,** 408.

Wilkinson, L. (1977) *Med. Hist.* **21,** 15.

Wilson, G. S. (1967) *The Hazards of Immunization*, pp. 60 and 179. Athlone Press, London; (1979) *J. Hyg. Camb.* **82,** 508.

Winkler, W. G. (1972) *J. infect. Dis.* **126,** 565.

Winkler, W. G., Fashinell, T. R., Leffingwell, L., Howard, P. and Conomy, J. P. (1973) *J. Amer. med. Ass.* **226,** 1219.

Zinke, G. G. (1804) In: *Neue Ansichten der Hundswuth: ihrer Ursachen und Folgen nebst eines sichern Behandlungsart der von tollen Tieren gebissenen Menschen.* 184–194 C. E. Gabler, Jena.

103

Slow viruses: conventional and unconventional

Richard H. Kimberlin

Introductory

Much of this volume is organized on the basis of the major taxonomic groups of viruses or else in terms of the target organs that become affected, e.g. the respiratory or enteric viruses. The slow viruses form neither a distinct morphological group nor do they cause disease in just a single tissue system. They are in fact an extremely heterogeneous collection of viruses associated with a variety of target organs and different types of pathogenic mechanism. They are linked mainly by the very long incubation period and the chronic progressive nature of some of the diseases they cause. Hence, the word 'slow' in the title of this chapter is a descriptive term which only loosely classifies viruses in the manner that the terms acute, chronic, latent and persistent are used to describe other infections.

These terms are not strictly applicable to the viruses themselves because slowness, persistence, or latency are features of the interaction between virus and host, and for a given virus, this interaction may vary with circumstances. A good example is measles, which causes an acute disease in the classical infection, but in which, occasionally, when the virus is not completely cleared by the immune system, persistence can occur leading to post-infectious encephalomyelitis; or alternatively, a third type of disease may develop several years after primary infection, namely subacute sclerosing panencephalitis. The latter illness is one that is commonly regarded as a slow virus disease.

Other slow diseases are caused by viruses in the parvo, papova, and oncorna groups, all well known conventional viruses (see later). However there is another, less well characterized group of so-called unconventional viruses which are also associated with slow diseases. These include the causal agents of scrapie in sheep, transmissible mink encephalopathy, and kuru and Creutzfeldt-Jakob disease in man. Little is known of their morphology or chemical structure but they appear to differ from conventional viruses in several respects, notably, in having a high degree of physicochemical stability and no detectable antigenicity. Because of these differences, many workers (the author included) prefer the term 'agent' to 'virus' on the grounds that 'agent' does not prejudge the physicochemical structure. Others have suggested that scrapie and related agents may be animal equivalents of the plant viroids (see Diener 1979), but the evidence for this is unconvincing (page 501). The issue will not be resolved until we know the structure of the scrapie-like agents. Meanwhile it is convenient to regard them as a separate group of transmissible agents, particularly as their investigation requires the use of some 'unconventional' methods (page 499).

The concept of slow virus diseases is best understood in its historical context. The story is a remarkable one which started in 1933, when 20 apparently healthy Karakul sheep were brought by the Icelandic Government from Halle, Germany. The sheep were kept in quarantine for two months and then sent to 14 farms in different parts of the country (see Pálsson 1976). Two of the rams carried infections that gave rise to epizootics starting from foci in the north and in the south-west. One ram in the latter area carried the viruses causing two different lung diseases, *jaagsiekte* (infectious adenomatosis) and *maedi*. Initially, jaagsiekte caused extremely heavy losses but maedi was largely unnoticed. This was because the incubation period of jaagsiekte, although long, was shorter than that of maedi which is rarely seen in sheep less than 4 years of age. About six years after the original importation the incidence of jaagsiekte declined rapidly and maedi came into prominence in the south-west, and also in the north where the second ram had been sent. At the peak of the maedi outbreak in the early 1940s about 60 per cent of the sheep-rearing districts in Iceland were affected, and the total number of winter fed sheep in Iceland had declined by about 40 per cent.

A third disease, visna, appeared in the south-west in the early 1940s but only in flocks where maedi had already caused losses for some years. Visna is a disease of the central nervous system and is clinically and pathologically different from maedi, although subsequent studies have shown that both diseases are caused by the same viral infection (see Pálsson 1976).

Sigurdsson and his colleagues made extensive field and laboratory studies of these diseases and also of a fourth sheep disease called 'rida' which was unrelated to the 1933 importation; *rida* is the Icelandic equivalent of scrapie and had been known for many decades. All four diseases were characterized by extremely long incubation periods. However, in contrast to the unpredictable clinical course of many chronic diseases which show a pattern of appearance of signs followed by remission and recurrence, Sigurdsson observed that the early clinical signs of maedi, jaagsiekte, visna and rida usually progressed in a predictable way and invariably ended in death. He coined the phrase 'slow infections' to describe them, and suggested that they could be distinguished by the following criteria (Sigurdsson 1954):

1. very long initial period of latency lasting from several months to several years.
2. A rather regular protracted course after clinical signs have appeared, usually ending in serious disease or death.
3. Limitation of the infection to a single host species, and anatomical lesions in only a single organ or tissue system.

As already explained, the taxonomic value of Sigurdsson's criteria is limited, but historically his concept was of the greatest importance. By drawing attention to a group of animal virus diseases with extremely long incubation periods, Sigurdsson emphasized the possibility that many chronic progressive diseases of man might also have a viral aetiology. The search for these has been going on ever since. So far, five have been identified: subacute sclerosing panencephalitis, progressive multifocal leucoencephalopathy, progressive rubella panencephalitis, kuru, and Creutzfeldt-Jakob disease. Some believe that multiple sclerosis should be added to the list, but the evidence for a viral aetiology for this disease is still far from convincing.

Among animal diseases, the following seven are commonly listed as belonging to the slow virus group: scrapie, jaagsiekte, visna, maedi (progressive pneumonia), Aleutian disease of mink, transmissible mink encephalopathy and Borna disease. All except jaagsiekte, maedi and Aleutian disease are diseases of the central nervous system. This may be significant for the following reasons (see Brooks *et al.* 1979). First, although there is a barrier to viral invasion, the c.n.s. is isolated from the systemic immune system, so that once a virus has gained entry it is reasonably safe from immune clearance. Secondly, the c.n.s. contains several distinct populations of cells with different susceptibilities to viral infection and damage. Thirdly, cells of the c.n.s. have a high rate of metabolic activity but little cell turnover, which favours persistence of viruses; and limited regeneration which makes it more likely that virus-induced damage will progress and eventually lead to neurological disease.

In what follows only the most interesting and best known diseases will be discussed. The next section gives a brief account of four slow diseases caused by conventional viruses; and here the emphasis is placed on pathogenesis. This is followed by a description of each of the diseases caused by unconventional agents. Of these, scrapie is by far the best understood. The rest of the chapter is devoted to a detailed discussion of the problems and successes of scrapie research.

The interested reader who wants further information on slow viruses should consult the books on the subject edited by: Hotchin (1974), Zeman and Lennette (1974), Kimberlin (1976*a*), ter Meulen and Katz (1977), Prusiner and Hadlow (1979) and Tyrrell (1979).

Progressive multifocal leucoencephalopathy (PML)

Progressive multifocal leucoencephalopathy is a rare, chronic demyelinating disease of man, with a progressive course leading to death, usually within a year. It was first described as a separate neuropathological syndrome by Åström and co-workers (1958). The demyelination is multifocal and is most prominent in subcortical white matter. There is loss of oligodendroglia and myelin with little damage to the axons. Astrocytes within the demyelinated areas are enlarged with pleomorphic, hyperchromatic nuclei. There is little or no inflammation. Cavanagh and his colleagues (1959) suggested a viral aetiology, mainly because intranuclear inclusions were present in enlarged oligodendroglia, and partly because PML was usually associated with diseases causing impaired immune responsiveness. This suggestion received strong support from the ultrastructural studies of ZuRhein and Chou (1965), and Silverman and Rubinstein (1965), which showed the presence of large numbers of papova-like virus particles in the nuclei of abnormal oligodendrocytes. This observation has been repeatedly confirmed and is now regarded as a characteristic feature of PML (Richardson 1974).

Many attempts were then made to isolate viruses from the brains of PML patients by inoculation into animals or cultured cells (see ter Meulen and Hall 1978). These were unsuccessful until 1971 when Padgett and her colleagues (1971) isolated a new papovavirus (*JC virus*) by inoculating brain homogenates into explant cultures of human fetal brain. Subsequently, JC virus has been isolated or identified in the brain tissue of over 20 cases of PML (Padgett and Walker 1976). Immunofluorescent staining of frozen sections of PML brain has repeatedly shown the presence of JC virus antigen in the nuclei of oligodendroglial cells associated with lesions. Hence there is persuasive evidence that JC virus causes PML (see Padgett and Walker 1976).

However it is interesting to note that *SV40 virus* has been isolated from the brains of two patients with PML (Weiner *et al.* 1972). Also, there have been two naturally occurring cases of PML in Rhesus monkeys due to SV40 virus (Holmberg *et al.* 1977). Although there is an antigenic relation between JC virus and SV40, they are quite distinct viruses in many of their antigenic (see Padgett and Walker 1976) and molecular properties (Osborn *et al.* 1974). This suggests that two different papovaviruses may cause PML. A third papovavirus, called BK, distinct from JC and SV40 viruses, was isolated from man by Gardner and her colleagues (1971) and was later shown to cause a common, human infection (Gardner 1973), unassociated with PML (Gardner 1977).

There are two well documented findings which suggest an underlying mechanism for the pathogenesis of PML. First, sero-epidemiological studies have shown that infection with JC virus is very common in man but not in other species (Padgett and Walker 1976). In one study, antibodies to JC virus were found in 65 per cent of persons by the age of 14 years, and in later life, in 70–80 per cent (Padgett and Walker 1973). This suggests that infection is most frequently acquired in childhood. Secondly, over 100 cases of PML have been reported. Most have occurred either as a complication of disorders affecting the reticuloendothelial system, e.g. leukemia, Hodgkin's disease or sarcoidosis, or in patients therapeutically immunosuppressed for other diseases or for renal transplantation. Occasionally PML occurs in patients without any other accompanying disease, but immunological studies have often shown a defect in cell-mediated immunity (see Johnson *et al.* 1977, Brooks *et al.* 1979.)

Taken together, these observations suggest that PML is caused by a widespread, opportunistic papovavirus (JC), which in an immunologically defective host causes a lytic infection of oligodendroglia cells that leads to demyelination (Walker *et al.* 1974). However, nothing is known of the earlier stages of pathogenesis; several possibilities can be considered. After childhood infection, JC virus may persist extraneurally with immunosuppression leading to invasion of the c.n.s. and rapid multiplication in oligodendrocytes. Alternatively, persistence may occur in the c.n.s. A third possibility is that the patient who develops PML is one of the minority who escapes childhood infection and becomes exposed to the virus when immunologically defective (see Johnson *et al.* 1977). If this last suggestion is correct, then PML may not have a long incubation period.

It will be difficult to answer many questions of the pathogenesis of PML without experimental models of the disease. Intracerebral inoculation of JC virus, as also SV40 and BK viruses, into hamsters can cause tumours (see Padgett and Walker 1976), but no disease resembling PML has yet been induced experimentally in animals. However reports of spontaneous PML occurring in macaques and in Rhesus monkeys (Gribble *et al.* 1975, Holmberg *et al.* 1977) raise hopes for methods of inducing PML in animals.

Subacute sclerosing panencephalitis (SSPE)

Subacute sclerosing panencephalitis is a rare progressive disease of the central nervous system that occurs in children and young adults. It was originally described under different names by different workers, but principally by Dawson (1933) and van Bogaert

(1945). The various clinicopathological descriptions were unified under the term SSPE by Greenfield (1950). Pathological changes occur only in the c.n.s. and consist of perivascular cuffing and a diffuse mononuclear cell infiltration of the grey and white matter. Proliferation of microglial cells and hypertrophy of astrocytes are common features; demyelination also occurs (see Agnarsdóttir 1977). However, the most important diagnostic feature is the presence of intranuclear, Cowdry type A and B inclusion bodies in neurons, astrocytes and, most frequently, in oligodendroglial cells.

The occurrence of inflammation and inclusion bodies suggested a viral aetiology. The first clear evidence of this was obtained by Bouteille and co-workers (1965), who found that the inclusion bodies contained tubules that resembled nucleocapsids of paramyxovirus ultrastructurally. This finding was rapidly confirmed by Tellez-Nagel and Harter (1966) and by Perier and Vanderhagen (1966). Soon afterwards, Connolly and his colleagues (1967) found raised antibody titres to measles virus in serum and cerebrospinal fluid (c.s.f.) and also demonstrated measles antigen in brain by direct immunofluorescence. Further studies by immunofluorescence methods confirmed the presence of measles virus antigen in neurons and glia (see ter Meulen and Hall 1978).

Very high antibody titres to measles virus are found in serum and c.s.f. of SSPE cases. Indeed, 10 to 20 per cent of total serum IgG and about 75 per cent of c.s.f. IgG is measles specific antibody (Mehta *et al.* 1977). There is a reduced serum: c.s.f. ratio of measles antibodies, and oligoclonal IgG specific for measles is present in c.s.f. These findings indicate a local production of measles virus antibodies within the c.s.f. (Vandvik and Norrby 1973).

Despite the presence of measles antigen in brain, early attempts to isolate a virus were unsuccessful. Katz and co-workers (1968) produced an encephalomyelitis in ferrets by intracerebral inoculation of SSPE brain suspension. However, the important development came when measles-like virus (SSPE virus) was isolated by co-cultivating or fusing brain cells from SSPE cases with continuous cell lines of non-neural origin (Horta-Barbosa *et al.* 1969, Payne *et al.* 1969, Barbanti-Brodano *et al.* 1970). Although not successful every time, SSPE virus has been repeatedly isolated from brain and also from lymph nodes of patients (see Agnarsdóttir 1977).

Subacute sclerosing panencephalitis therefore presents an interesting example of a very rare disease—about 1 case per million per annum—clearly associated with a very common infection. The mechanisms by which measles virus, or a variant, causes SSPE are still the subject of investigation. Three main hypotheses have been advanced (see Brooks *et al.* 1979).

First, SSPE may be the result of an abnormal host immune response to measles. Aside from the hyperimmune reaction against all demonstrable measles antigens, there appear to be no generalized alterations in humoral or cell-mediated responses in SSPE patients (see Agnarsdóttir 1977, ter Meulen and Hall 1978; Johnson and ter Meulen 1978; K. B. Fraser 1979). However, there are indications that host immune factors may be important in establishing the persistent infection of measles virus. For example, several attempts have been made to produce animal models of SSPE (see Agnarsdóttir 1977). In one of the more successful of these, Byington and Johnson (1972) inoculated weanling golden hamsters with SSPE virus and produced a chronic encephalitis with persistent cell-associated virus. The disappearance of cell-free virus coincided with the development of serum antibody to measles. No viral persistence occurred when mature hamsters were inoculated, and, in newborn animals, an acute encephalitis occurred. Chronic SSPE-like disease has also been produced in ferrets (Thormar *et al.* 1977) and Rhesus monkeys (Albrecht *et al.* 1977) by inoculation of SSPE virus into animals that had been immunized with measles virus. Most SSPE patients have a prior history of measles infection, and over 50 per cent have had measles before two years of age (see Brooks *et al.* 1979). Together, these findings suggest that persistence of measles virus in man may be due to an age-dependent susceptibility of certain cells and to the presence of maternal antibody, or both.

The second hypothesis supposes that SSPE is caused by an atypical measles virus, in which case one might expect to see geographical clusters of SSPE cases. This has not been reported. Therefore if a variant of measles virus is responsible, it probably arises after infection. Several comparative studies have been made using biological, cell culture, and immunological methods, but no clear cut differences between measles and SSPE virus have emerged (see Agnarsdóttir 1977; ter Meulen and Hall 1978). However, the need for co-cultivation or cell-fusion techniques to isolate SSPE virus in cell culture implies that the virus is defective in some way. Electron microscopic studies of brain biopsies and cell cultures from SSPE patients reveal nucleocapsids but no enveloped virions (Dubois-Dalcq *et al.* 1974*a*, *b*, 1976). Virion assembly requires an interaction between nucleocapsids and the cell membrane, and this interaction is mediated by the virus specified matrix (M) protein. Several recent studies have indicated that M proteins of SSPE viruses are larger than those found in measles strains (Schluederberg *et al.* 1974; Wechsler and Fields 1978, Hall *et al.* 1978). These findings suggest that SSPE strains may arise from measles virus after a mutation that leads to an abnormal M protein.

According to the third hypothesis, there may be an interaction between measles virus and a second agent. This is consistent with the greater preponderance of SSPE cases among rural males (Brody and Detels 1970). One group has reported papovavirus-like particles in brain cell cultures of SSPE patients (Koprowski *et al.* 1970), but no supporting evidence has been forthcoming.

In *summary*, there is strong evidence that SSPE is caused by measles virus. It seems likely that host-immune responses at the time of infection contribute to persistence of virus, and that neutralizing antibodies provide a selective pressure favouring the appearance of cell-associated defective strains that may arise by mutation or even recombination with a second, unknown virus. However detailed studies of suitable animal models of SSPE are needed to establish this hypothesis and to discover how the disease develops.

Visna

Visna was first described in Iceland after the importation of Karakul sheep in 1933 (page 488). Initially its occurrence was largely unnoticed because of the greater prevalence of maedi (Pálsson 1976) and because, for a time, it was confused with rida, the Icelandic form of scrapie (Sigurdsson *et al.* 1957). Visna has since been observed in sheep and sometimes goats, in several countries (Pálsson 1976, Pétursson *et al.* 1979). It occurs only in flocks that have maedi, but even then it is a far less common disease than maedi. The introduction of the maedi eradication scheme led to the disappearance of visna from Iceland in 1951 (Pálsson 1976).

Clinical cases of visna show a progressive paresis of the hind limbs leading to complete immobility and death. Histological lesions are confined mostly to the c.n.s. and consist of meningitis, perivascular infiltration and inflammatory lesions of grey and white matter. Areas of intense inflammation are often associated with necrosis of axons and demyelination (Sigurdsson *et al.* 1962, Georgsson *et al.* 1976).

Between 1949 and 1957 it was shown that visna could be serially transmitted by intracerebral inoculation of brain and spinal cord from natural cases, but the incubation periods were very long (Sigurdsson *et al.* 1957). Virus isolations were first made in cell cultures of normal sheep brain by incubating with infected brain homogenate. Later the virus was grown in explant cultures of choroid plexus from visna-affected sheep (Sigurdsson *et al.* 1960). In both types of culture, cytopathic changes appeared after about 3 weeks, characterized by the formation of multinucleated giant cells. Inoculation of sheep with virus passaged in cell culture caused the clinical signs and characteristic lesions of visna (see Thormar *et al.* 1974).

The relative ease with which visna virus can be grown and assayed in cells of sheep choroid plexus has made possible its biochemical characterization. Visna belongs to the family of retroviruses. In size, morphology and physicochemical properties it is similar to the oncornaviruses (Haase 1975), but recent studies by molecular hybridization and serological techniques have shown that visna virus belongs to a distinct group of retroviruses, the lentiviruses (Harter *et al.* 1973; Weiss *et al.* 1977; Stowring *et al.* 1979). In addition to visna, the lentiviruses include maedi and progressive pneumonia viruses. However, it now seems that visna and maedi viruses are probably identical, thus explaining the natural and experimental association of these two diseases.

Sigurdsson and his colleagues (1957 and 1962) noted the similarities between visna and human demyelinating diseases such as multiple sclerosis and post-infectious encephalomyelitis. These observations coupled with the transmissibility of visna to sheep encouraged detailed studies of the pathogenesis of the disease. The first study was initiated by Gudnadóttir and Pálsson in 1960. Because of the long and variable incubation period of visna—from a few months to greater than lifespan—their observations were continued for almost 11 years (Gudnadóttir and Pálsson 1966; Gudnadóttir 1974). Surprisingly, these and other studies established the fact that virus multiplication, antibody production and the appearance of lesions all started within a few weeks of infection. Hence most investigations have focused on the early events in pathogenesis (Narayan *et al.* 1974; Pétursson *et al.* 1976; Griffin *et al.* 1978a).

Visna virus can be recovered at virtually all times after intracerebral infection from 1–2 weeks onwards. It can be found in blood and c.s.f., but is most consistently recovered from brain, especially choroid plexus, and lymphoid organs. However virus is nearly always cell-associated, and isolation often requires the use of tissue explants and one or more blind passages. It is difficult to find visna virus particles in infected tissues by electron microscopy or immunofluorescence. This failure contrasts with the rapid growth of visna virus in sheep cells cultured *in vitro* (Thormar 1976). Haase and co-workers (1977) applied *in situ* hybridization techniques and found visna proviral DNA in up to 18 per cent of choroid plexus cells *in vivo*, but in only 0.025 per cent of the same cells stained for the p30 antigen of visna virus by immunofluorescence. However, 14 per cent of these cells produced virus when grown *in vitro*. These studies show that many cells contain the virus genome but production of virus is severely restricted *in vivo*, and the cells become highly permissive only when grown in culture.

Complement-fixing antibodies can be detected in serum a few weeks after infection, but specific neutralizing antibodies do not develop until 2–3 months. Titres may remain fairly high for years. Neutralizing antibodies can also be found in the c.s.f. of some infected sheep; there is evidence that they are produced locally (Griffin *et al.* 1978*b*; Nathanson *et al.* 1979). Also, oligoclonal bands of IgG have been demonstrated in c.s.f. These findings are consistent with an increased protein content of c.s.f. and with a pleocytosis that starts 1–2 weeks after infection, reaching a maximum after about 1 month. There is, in addition, some evidence for a specific cell-mediated response to visna virus infection but this is short-lived and, as with the humoral response, it fails to eliminate infection from the host (Griffin *et al.* 1978a).

Pathological changes can be seen in brains as early as 2–4 weeks after intracerebral injection of visna virus into sheep. In a serial study carried out over 13 months, there was little evidence for the progression of lesions beyond the first month. There was, however, a strong correlation between the severity of histological lesions and the frequency of virus recovery, suggesting a causal relation between virus multiplication and the appearance of lesions (Pétursson *et al.* 1976, Nathanson *et al.* 1979). The inflammatory nature of the lesions suggested that they were immunologically mediated. Immunosuppression by antithymocyte serum and cyclophosphamide had no effect on the frequency of virus isolation from the c.n.s., but the development of lesions was strikingly suppressed (Nathanson *et al.* 1976). Similarly, immunopotentiation of visna-infected sheep by injecting high doses of virus in Freund's adjuvant increased the severity of lesions (Pétursson *et al.* 1979). The frequency of virus isolation from the c.n.s. and the severity of lesions was also increased by infecting sheep with very large doses of virus (Pétursson *et al.* 1979). These results suggest that the antigens for the immunological attack are viral proteins, at least during the development of early lesions. Auto-immune reactions to myelin or other host antigens may develop at the later stages of disease (Panitch *et al.* 1976).

Until recently, it was hard to reconcile these fairly rapid events with the fact that visna is a slow disease. However, Gudnadótťir (1974) observed that some virus isolates, made several years after infecting sheep, were poorly neutralized by antiserum collected early after infection. She suggested that antigenic drift in the presence of an immune response might enable virus to remain active for a long time in sheep. This observation was confirmed by Narayan and co-workers (1977*b*). Other studies have indicated that many mutants can arise months after infection, and that there is a sequential development of antibody to parental virus and later to each mutant strain (Narayan *et al.* 1978). This slow process of mutation and selection can be accelerated *in vitro* by growing virus in cultures in the presence of antibody. A stable mutant can be produced by this method within a few weeks (Narayan *et al.* 1977*a*).

Taken together, these findings suggest a reasonably coherent picture of the pathogenesis of visna. Intracerebral injection of virus leads to early infection of the c.n.s. and the development of immunologically mediated lesions. Virus persists as an integrated DNA provirus producing only a few virus particles; and spread of infection is probably limited by the continuing presence of neutralizing antibody. However, infectious antigenic variants arise which initially escape neutralization, thus giving new waves of infection and causing c.n.s. damage. Presumably it is the slow accumulation of irreversible c.n.s. lesions which eventually passes a critical level and leads to clinical disease.

Aleutian disease of mink

In 1941, a mutation arose in mink bred in Oregon which produced a blue-grey coat colour similar to that of the Aleutian blue fox. The trait in mink is due to an autosomal recessive gene. This variant became so valuable commercially that mink of the Aleutian genotype were rapidly distributed throughout the mink-raising areas of the world (see Gorham *et al.* 1976). During the early 1940s several breeders reported losses due to a disease which seemed to be confined to Aleutian mink. In some cases the losses were devastating (Hartsough and Gorham 1956). For a time it was believed that Aleutian disease was of genetic origin until a number of serious outbreaks occurred several months after the use of 'homemade' distemper vaccines prepared by grinding infected spleens with dilute formalin (see Gorham *et al.* 1965). These epizootics showed that Aleutian disease was caused by a transmissible agent (Hartsough and Gorham 1956, Helmboldt and Jungherr 1958). Subsequent studies with cell-free extracts indicated that a virus was probably responsible (Henson *et al.* 1962; Karstad and Pridham 1962, Trautwein and Helmboldt 1962).

Although Aleutian disease was first recognized in mink of the eponymous genotype, Hartsough and Gorham (1956) observed that it could occur in non-Aleutian mink. It was also reported in ferrets (Kenyon *et al.* 1967, Daoust and Hunter 1978). The disease occurs in mink of either sex and of all ages beyond about 4 months. The clinical signs include severe weight loss, polydipsia and anaemia, and is nearly always fatal. Pathologically the most striking feature is a systemic plasmacytosis affecting the bone marrow, spleen, lymph nodes, liver and kidney (Helmboldt and Jungherr 1958, Obel 1959, Leader *et al*, 1963; Henson *et al.* 1966*a*). This is paralleled by a profound hypergammaglobulinaemia (Henson *et al.* 1961), which in one study averaged five times the normal levels (Porter and Larsen 1964). Microscopic lesions occur in many tissues. There is a widespread arteritis, but glomerulonephritis is the dominant histological abnormality. Several studies using light, fluorescent, and electron microscopy (see Henson *et al.* 1976) have shown that virus antibody complexes are deposited in the kidney causing alterations in the basement membranes. There is stimulation of the mesangial cells and mesangial matrix, probably in an effort to catabolize the immune complexes. The size of the mesangium progressively increases, eventually leading to compression of capillaries and renal failure. This is the usual cause of death in Aleutian disease.

Detailed characterization of Aleutian disease virus (ADV) has proved difficult. Original attempts to isolate ADV in cell culture gave conflicting results; and in no case were viral titres obtained *in vitro* as high as those found in tissues of infected mink (Yoon *et al.* 1973, 1975, Hahn *et al.* 1977 *a*, *b*, Porter *et al.* 1977). Therefore, most studies of ADV have been based on *in vivo* sources of virus, assayed by inoculation of mink, followed by immunological or histological tests for infection or clinical diagnosis of the disease (see Cho 1976). Virus has been partly purified by various methods and shown to have an icosahedral structure of 23–25 nm diameter. However, this structure and various *in vitro* properties led one group of workers to suggest that ADV was a picornavirus (Kenyon *et al.* 1973, Yoon *et al.* 1973, 1975, Notani *et al.* 1976), whereas two other groups considered it to be a parvovirus (Cho and Ingram 1973, Chesebro *et al.* 1975).

The most recent studies have been carried out with highly purified virus which was shown to contain a single stranded DNA of mol. wt 1.2×10^6. Four polypeptides were also identified with mol. wt of 30 000, 27 000, 20 500 and 14 000; these were present in the ratios of 10:3:10:1, respectively (Cho 1977, Shahrabadi *et al.* 1977, Cho and Porter 1978). There is now sound evidence that ADV can be isolated in feline renal cells maintained at reduced temperatures. After several *in vitro* passages the virus can be grown at 37°. Cultured virus will reproduce Aleutian disease in mink and virus can then be re-isolated *in vitro* (Porter *et al.* 1977). The properties of the cultured virus and of ADV purified from mink strongly indicate that Aleutian disease is caused by an autonomous parvovirus, although the polypeptide pattern differs from that of previously studied parvoviruses (Shahrabadi *et al.* 1977).

The transmissibility of Aleutian disease was firmly established at an early stage and many detailed studies were made of its pathogenesis (see Porter and Larsen 1974; Henson *et al.* 1976). Animals with clinical signs of disease always show the same pattern of lesions but the proportion of clinical cases and the incubation period varies with genotype. Most Aleutian mink injected intraperitoneally with high doses of virus die 50 to 300 days later, with a mean death time of about 100 days (Padgett *et al.* 1968, Eklund *et al.* 1968). The mean death time in non-Aleutian mink can be much longer (200–300 days), but many animals never reach the clinical stage. Clinical disease rarely occurs in ferrets infected experimentally (Kenyon *et al.* 1966).

ADV replicates rapidly in all genotypes of mink after experimental infection (Eklund *et al.* 1968; Porter *et al.* 1969). Virus titres reach a maximum 1–2 weeks later and can be as high as 10^8–10^9 ID_{50} units per gram of spleen, liver and lymph nodes, with about 10^5–10^7 units in blood. Titres steadily fall to about 10^5 ID_{50} units in spleen and 10^4 in blood within 2 months. Most mink remain persistently infected for the rest of their life, which may be several years in some non-Aleutian animals. However about 20–25 per cent of mink of this genotype can apparently clear the virus, by a mechanism that is not genetically determined (Larsen and Porter 1975, Henson *et al.* 1976).

Hypergammaglobulinaemia is detectable after 3–4 weeks in infected Aleutian mink and continues to increase till death, sometimes reaching values as high as 10 g/100 ml (Eklund *et al.* 1968, Henson *et al.* 1976). The half-life of serum IgG is decreased by one-third in infected mink (Porter *et al.* 1965); hence the hypergammaglobulinaemia is due to a continuous overproduction of IgG. A similar pattern occurs in many non-Aleutian mink except that the progression is slower in keeping with extended death times in these animals (Eklund *et al.* 1968; Lodmell *et al.* 1971, 1973). In non-persistently infected mink, the hypergammaglobulinaemia is transient (Larsen and Porter, 1975).

These changes are mirrored by correspondingly high concentration of specific ADV antibody, which can be detected in the second week after infection. Geometric mean antibody titres can reach 100 000 by immunofluorescence (Porter *et al.* 1969) and 260000 by complement fixation (McGuire *et al.* 1971). Despite this phenomenal immune response, ADV is not neutralized by IgG, but circulates in the form of infectious immune complexes (Henson *et al.* 1966*b*, Porter *et al.* 1969). Porter and Larsen (1967) removed these from serum by treating with an excess of antiserum to mink IgG. Infectivity was decreased about a hundredfold, showing that much of the antibody was specific for ADV, but precisely how much was not discovered.

The enormous increase in IgG may be caused by a breakdown of the control mechanisms of antibody synthesis or by a chronic stimulation arising from persistently high titres of ADV (Henson *et al.* 1976). About 10 per cent of mink still surviving a year after infection develop monoclonal antibody with virus-specific titres of 10^5–10^6 (Porter *et al.* 1965, Tabel and Ingram 1970), but again the proportion of IgG that is virus-specific is not known. There is antigenic competition with ADV antigens—and the antibody response to unrelated antigens is depressed in infected mink (Porter *et al.* 1965, Kenyon 1966, Tabel *et al.* 1970, Lodmell *et al.* 1970, 1971, Trautwein *et al.* 1974). There is also limited evidence that the cellular immune response may be depressed in Aleutian disease (Lodmell *et al.* 1971, Perryman *et al.* 1975).

High titres of ADV *per se* cause few if any ill effects. Lesions develop because of the immune response, which leads to the production and accumulation of virus-antibody complexes. This view is supported by considerable evidence. There is a strong association between antibody titres, or IgG content, the presence of immunoglobulin-containing complexes in tissue sections, and the severity of lesions in infected mink. The same is broadly true for infected ferrets, except that the severity of these processes is much less and few animals die (see Porter and Larsen 1974; Henson *et al.* 1976). Cyclophosphamide treatment of infected mink for several weeks had no effect on virus titres, but antibody production was suppressed and the development of lesions prevented. Cessation of treatment was followed by the formation of lesions (Cheema *et al.* 1972). ADV antibody given passively to mink at the peak of virus replication enhanced the production of lesions (Porter *et al.* 1972). An even greater effect was obtained by injecting mink with inactivated ADV vaccine before infection with live virus; both antibody titres and lesion scores were increased by immunization.

In conclusion, it is clear that, at the tissue level, Aleutian disease develops quite rapidly and is slow only in the development of clinical signs. Animals die mainly because of accumulated damage in the kidneys leading to renal failure. The speed of this process varies with genotype and species of host and is presumably a reflection of three factors. First, the persistence or clearance of virus; secondly, the rate and extent of antibody production; and thirdly, the efficiency of removal of immune complexes. Aleutian mink are at one end of the spectrum and develop a persistent infection and progressively increased amounts of antibody. These animals may have an impaired ability to catabolize immune complexes, thus hastening the clinical disease (Gorham *et al.* 1976). Some non-Aleutian mink and ferrets represent the other end of the spectrum, because they can either clear virus or limit the accumulation of immune complexes so that clinical disease never develops.

Slow diseases caused by unconventional agents

Scrapie

Scrapie is a fatal disease of the c.n.s. that affects most breeds of sheep and also goats (Dickinson 1976). It has been a continuing problem in parts of Western Europe for over 200 years but has been of worldwide occurrence, often following the importation of pedigree sheep (Hourrigan *et al.* 1979). The disease occurs in sheep of either sex, but is most commonly seen in breeding ewes between the ages of $2\frac{1}{2}$ and $4\frac{1}{2}$ years, with about 20 per cent of cases outside this range. Clinically affected animals show either hind-limb incoordination or signs of intense pruritus leading to loss of wool and to skin lesions. Commonly, both types of abnormality occur. The disease follows an ingravescent course over weeks or months and is invariably fatal (Stamp 1962). The incidence of scrapie in experimental flocks can be as high as 60–70 per cent per annum (Kimberlin 1979a), but is usually much lower than this under commercial conditions.

Besnoit and Morel made the first histopathological studies of scrapie in 1898. Lesions occur only in the c.n.s.; diagnostically, the most important are vacuolated neurons in the medulla and brain stem (Holman and Pattison 1943, Zlotnik 1958). Interstitial spongy degeneration usually accompanies vacuolation of neuronal perikarya, and there may also be some neuronal loss. Hypertrophy of astrocytes occurs as an additional but non-specific lesion (see Fraser 1976). Demyelination is either slight or absent. There are no inflammatory lesions to indicate the presence of an infectious agent.

For many years the strong familial pattern in scrapie encouraged the belief that scrapie was a genetic disease. However in 1936, Cuillé and Chelle showed that the disease could be transmitted to sheep by injecting c.n.s. material from an affected animal. This observation was inadvertently confirmed on a spectacular scale when it was discovered that the injection of 18 000 sheep with a crude louping-ill vaccine, prepared from formalin-treated sheep brain, was responsible for several hundred cases of scrapie. In retrospect it was shown that some of the vaccine was prepared from sheep incubating scrapie (Gordon 1946).

Pioneering work on scrapie was carried out by Wilson and his colleagues (1950: *see* Dickinson 1976), who in the next 20 years established four key features of the disease (Greig 1950, Gordon 1957, Stamp *et al.* 1959). First, scrapie is transmissible by inoculation, with an incubation period of about 1 to more than 5 years, though in genetically selected lines of sheep it may be as short as 5 months (Dickinson *et al.* 1968*b*, Nussbaum *et al.* 1975). Secondly, the transmissible agent replicates *in vivo* and can be experimentally passaged in sheep indefinitely. Thirdly, the agent can be recovered in a 410 nm gradocol membrane filtrate but not after filtration through a 27 nm membrane. Finally, the agent is not totally inactivated by exposure to 0.35 per cent formalin for at least 3 months or by boiling for 8 hours. Thus, scrapie is caused by a virus-like agent, but one that has unusual physicochemical stability.

Other studies during this period indicated the absence of specific antibodies to scrapie in sheep and goats (Chandler 1959, Gardiner 1965, Clarke and Haig 1966). Unsuccessful attempts were made to isolate scrapie agent in cell culture systems. A turning point came when the disease was experimentally transmitted from goats to mice (Chandler 1961, 1963), and finally from sheep to mice (Zlotnik and Rennie 1962, Dickinson and Smith 1966). This led to a variety of different mouse models of scrapie (see Dickinson and Fraser 1977), and to a very short incubation model in hamsters (Kimberlin and Walker 1977). Quantitative bioassays of scrapie infectivity became possible in laboratory animals (page 499); most of our detailed understanding of the disease dates from this time. Two major findings relevant to the natural disease are (a) that many biologically different strains of scrapie agent exist (page 499) and (b) that at least some of these can undergo mutation (page 500).

Scrapie is an infectious disease. Studies have shown

that between 14 and 39 per cent of exposed sheep may suffer from scrapie after contact with diseased animals (Brotherston *et al.* 1968, Dickinson *et al.* 1974; Hourrigan *et al.* 1979, Pálsson 1979). This occurs, for example, when sheep are close-penned at lambing or during prolonged winters; but the precise routes of infection are not firmly established. Little if any agent is found in body fluids or secretions, but many tissues outside the c.n.s. contain the agent and may be sources of infection (Stamp *et al.* 1959, Hadlow *et al.* 1979). In particular, fetal membranes contain quite a high content of the agent (Pattison *et al.* 1974, Dickinson and Fraser 1979). Animals can be experimentally infected by oral dosing (Gordon 1966, Pattison *et al.* 1974), scarification (Stamp *et al.*, 1959) and via the conjunctiva (Haralambiev *et al.* 1973). These methods probably indicate the natural routes of infection. The extreme physicochemical stability of the scrapie agent (page 501) supports the circumstantial evidence that sheep may become infected from contaminated buildings (Pálsson 1979) or pastures (Greig 1950).

Maternal transmission plays a major role in the natural spread of the disease (Dickinson *et al.* 1965, 1974; Hourrigan *et al.* 1979). In part this is due to infection from the dam occurring after parturition. There is some evidence that the incidence of scrapie in progeny increases with the length of exposure (Hourrigan *et al.* 1979). However, sheep removed from their dams at birth can also develop scrapie, implying infection before or during parturition. Lambs born to ewes experimentally infected at conception develop scrapie exceptionally early indicating pre-natal infection (Gordon 1966, Dickinson *et al.* 1966). This may occur transplacentally, since fetal membranes contain appreciable amounts of agent at parturition (Pattison *et al.* 1974).

A high proportion of sheep with scrapie become infected before weaning, so that the age-incidence pattern of the natural disease reflects the incubation period. In support of this are reports of brain lesions (Dickinson *et al.* 1965) or of scrapie agent in lambs 4 to 11 months of age, well in advance of clinical signs which rarely appear in sheep less than 18 months old (Renwick and Zlotnik 1965, Hourrigan *et al.* 1979). A serial study of clinically normal lambs born in a high-incidence scrapie flock showed that agent was often detectable in lymphoid tissues and in the intestines, but the frequency of isolation was much higher from lambs born to ewes that later developed scrapie than from those whose mothers remained clinically normal (Hadlow *et al.* 1979).

Although scrapie is caused by an infectious agent there is strong evidence that host-genetic factors control the disease (Dickinson *et al.* 1965 and 1974). However, maternal transmission, coupled with an inability to measure the degree of exposure to scrapie agent, have made a detailed genetic analysis of natural scrapie impossible. These problems do not arise when sheep are infected experimentally. Two major studies have shown that Cheviot and Herdwick sheep can be selectively bred for increased (positive line) or decreased (negative line) incidence of scrapie in response to subcutaneous injection of a standard source of scrapie agent, known as SSBP/1. Line-crossing experiments have shown that the response to SSBP/1 is mainly controlled by a single gene with the dominant allele conferring susceptibility (Dickinson *et al.* 1968*b*, Nussbaum *et al.* 1975, Dickinson 1976, Dickinson and Fraser 1979).

At an early stage in this work it appeared that the selection of negative-line animals might be the basis of breeding methods for controlling the natural disease (Hoare *et al.* 1977). Two additional findings, however, indicate that such a plan would probably not work and could even be disastrous (Kimberlin 1979*a*). First, sheep selected for a given response to infection with the SSBP/1 source of agent may respond quite differently when injected with a different source such as CH 1641 (Dickinson and Fraser 1979). It is well established that the genetic control of scrapie exercised by *Sinc* gene in mice varies with the strain of agent (page 500). Secondly, *Sinc* gene controls length of incubation period rather than susceptibility or resistance; this control is exerted on the process of agent replication (page 500). In some situations mice can appear to be resistant to scrapie but, in fact, agent replication is so prolonged that animals die of old age before the clinical disease becomes apparent (Dickinson *et al.* 1975*b*). There is evidence that the same type of genetic control probably operates in sheep (see Dickinson and Fraser 1979). Hence it is possible that apparently resistant sheep could be infected carriers and, as such, a source of infection to others.

Although not proven, the concept of a carrier state has far-reaching implications for the epidemiology of scrapie. For example, the sudden appearance of scrapie in a flock may not necessarily be due to the introduction of contaminated stock. It could equally well be caused by mutation of agent in a pre-existing carrier state to give a more pathogenic variant. Alternatively, a stable carrier state could be broken by the widespread use of a ram which introduces a more favourable genetic background for agent replication.

The possibility of a carrier state and the persistence of agent in the environment are two possible reasons why attempts to eradicate endemic scrapie from the USA and Iceland have failed (Pálsson 1979). However, enough is known of the natural history of the disease to attempt control. The least disruptive method is by selective culling in the female line to limit the effects of maternal transmission (Dickinson 1976). Careful husbandry at lambing time (e.g. prompt removal of afterbirths) also helps by reducing the opportunities for contagious spread of infection (see Kimberlin 1981).

More detailed studies of the natural disease are hampered by two main problems. First, there is no diagnostic test such as is needed to establish the existence of a carrier state, to study genetic control, and to make assays of infectivity easier. Secondly, infectious agent can be assayed only by injecting sheep or mice and waiting for clinical signs. Genetic variation and restricted numbers confound bioassays in sheep (Dickinson 1976); and the species-barrier effect makes assays of sheep agent in mice insensitive (Kimberlin 1979*b*). Hence detailed studies of the agent and the disease have been restricted to experimental models of scrapie in mice and other rodents.

Transmissible mink encephalopathy (TME)

Transmissible mink encephalopathy was first observed in a Wisconsin mink herd in 1947 (Hartsough and Burger 1965). Fourteen years later, the disease appeared on five mink ranches which shared a common source of feed. Further outbreaks occurred in 1963 in Wisconsin, Idaho, and Ontario (Hartsough and Burger 1965, Hadlow and Karstad 1968). A total of 14 occurrences of TME have been recorded in the world (see Marsh and Hanson 1979). Although a rare disease, some outbreaks have affected most adult animals on a ranch.

The classic studies of Hartsough and Burger (Hartsough and Burger, 1965, Burger and Hartsough 1965) showed that TME occurred in breeding mink, usually of more than 1 year of age. The onset is insidious with subtle behavioural changes progressing to locomotor incoordination, somnolence, and invariably death. Histological changes are found only in the c.n.s. and consist of widespread vacuolation similar to that found in natural scrapie, except that neuronal vacuolation is less prominent. Like scrapie, there is hypertrophy of astrocytes but no inflammatory response. Transmissible mink encephalopathy can be experimentally transmitted to mink by inoculation of brain from affected animals, and serially passaged by different routes of administration including intracerebral, subcutaneous and oral dosing (Burger and Hartsough 1965, Marsh *et al.* 1969*b*). The minimum incubation period of the experimental disease is five months and the lesions closely resemble those found in natural TME (Eckroade *et al.* 1979). The disease has also been experimentally transmitted to other species, including golden hamsters and squirrel monkeys (Marsh *et al.* 1969*a*, Eckroade *et al.* 1970, 1973, Hanson and Marsh 1974).

Several other studies have revealed close similarities between scrapie and TME. For example, the TME agent is filtrable and remarkably resistant to boiling and to treatment with formalin, ether and u.v. light (Burger and Hartsough 1965, Marsh and Hanson 1969). There is also evidence for different strains of TME agent (Marsh and Hanson 1979). Agent is present in many tissues outside the c.n.s. in infected mink (Marsh *et al.* 1969*b*), but not in circulating lymphocytes (Marsh *et al.* 1973); and there is no detectable antibody to TME (Marsh *et al.* 1970). A detailed study of the pathology and pathogenesis of TME and scrapie in a common host—the golden hamster—confirmed the close similarity of the two diseases (Kimberlin and Marsh 1975; Marsh and Kimberlin 1975).

Epidemiological studies have shown that TME does not maintain itself in a mink population. There is no vertical or horizontal transmission of agent, apart from rare instances when young mink cannibalize their mothers who have died of the disease (Hanson and Marsh 1974). In some outbreaks, there is good circumstantial evidence that infectious agent was introduced through the feed (Hartsough and Burger 1965). Although there is no proof that TME arises from feeding mink with scrapie-contaminated tissue, this seems the most likely explanation. Mink can be infected orally, and intracerebral injection of a brain suspension from scrapie-affected sheep can produce a disease in mink indistinguishable from TME (Hanson *et al.* 1971).

In two outbreaks of TME it was possible to calculate a maximum incubation of one year, based on the transfer dates of mink between ranches (Marsh 1976). This is shorter than the incubation period found when passaging TME in mink by oral dosing. However, intramuscular or intradermal infection can produce TME in as little as 24–40 weeks. This finding led Marsh and Hanson (1979) to suggest that fighting between littermates at feeding time results in the 'injection' of contaminated feed via bite wounds.

In summary, TME is a rare disease which almost certainly represents scrapie in mink. Therefore it is perhaps surprising that no one has succeeded in transmitting TME to mice. However, not all natural isolates of scrapie are transmissible to mice (Dickinson 1976), and the same holds for the hamster passaged agent 263K (Kimberlin and Walker 1978*a*). Hence the rare occurrence of TME may be related to a number of factors such as adequate exposure to certain strains of scrapie, perhaps coupled with mutation to give a variant that is pathogenic for mink.

Kuru

Kuru occurs in the eastern highlands of Papua New Guinea, affecting the Fore speaking people and those neighbouring linguistic groups with whom the Fore intermarry, a total population of about 30000. The disease was discovered in 1955 by Zigas while on a medical field trip (Zigas and Gajdusek 1957, Gajdusek and Zigas 1957). In the late 1950s and early 1960s kuru was by far the commonest cause of death in the Fore people, the annual mortality rate being about 1 per cent. The disease occurred in all age groups beyond 4 years, affecting children of either sex equally, but with a striking predominance of female cases in adults (Alpers and Gajdusek 1965).

Clinical disease is heralded by unsteady balance and clumsy movements. These signs progress to disorders of cerebellar function and culminate in recumbency and death. In adults the afebrile clinical course usually occupies a few months but can last for 18 months. In children, the disease progresses more quickly, and usually produces bulbar and brain stem disturbances as well. There is a remarkable uniformity of clinical signs, which are quite different from those of Creutzfeldt-Jakob disease (see Hornabrook 1979).

The pathology of kuru is also remarkably uniform, with degenerative changes occurring only in the c.n.s., particularly in the cerebellum. The lesions consist of neuronal vacuolation and degeneration, status spongiosus, and reactive changes in astrocytes and microglia. Characteristic amyloid plaques occur in most cases. There are no inflammatory lesions, and demyelination is secondary and minimal (Klatzo *et al.* 1959, Beck and Daniel 1979).

In 1959, Hadlow realized that the neuropathology of kuru was strikingly similar to scrapie, and suggested that attempts should be made to transmit kuru to primates (Hadlow 1959). In 1966, a disease clinically and pathologically similar to human kuru was reported in chimpanzees, 18 to 21 months after intracerebral inoculation of affected brain (Gajdusek *et al.* 1966, Beck *et al.* 1966). Since then kuru agent has been isolated from many patients and passaged in chimpanzees and other species of primate, including the marmoset (Peterson *et al.* 1973, Gibbs *et al.* 1979). Incubation periods vary, but are usually a year or more. Kuru has not been transmitted to mice or other convenient laboratory animals (Gibbs and Gajdusek 1973). Studies in primates have shown that the kuru agent is similar to that of scrapie in being filtrable and unusually resistant to moist heat and ionizing radiation (Gajdusek and Gibbs 1973, Gibbs *et al.* 1978). There is also preliminary evidence for the existence of different strains of kuru agent (Gibbs *et al.* 1979).

Kuru is important in being the first chronic degenerative c.n.s. disease of man which has been shown to be caused by a transmissible agent. It also has an interesting epidemiology which, because of its occurrence in a small population, has been studied in detail (Alpers 1979).

Between 1956 and 1978, 2514 cases of kuru occurred in the New Guinea population at risk. In the late 1950s, 90 per cent of adult deaths in South Fore women were due to kuru, indicating a high susceptibility to the disease. However, since 1957, over 300 orphaned children born to kuru mothers have been observed and none has developed the disease (Masters *et al.* 1979). Hence there appears to be no vertical transmission of infection and in this respect kuru resembles TME but contrasts with scrapie.

Before 1956 endocannibalism of dead relatives was a rite of mourning among the Fore and their neighbours (Gajdusek 1977). The ceremony was performed by the women accompanied by young children, but not by males over 6 years old. In retrospect it is clear that this practice was the main, if not the sole, source of kuru infection. It accounts for the distinctive age and sex pattern of kuru that was originally observed. Endocannibalism largely ceased in 1956, and the number of cases has steadily declined since then. There has also been a change in the pattern of the disease. The juvenile form of kuru has disappeared; in fact, no cases under the age of 20 have occurred since 1973. Among adults, only 7 patients were born since 1956; and the average age of onset continues to rise indicating that the incubation period can be more than 25 years. The female: male in adults ratio has also declined dramatically and now resembles the 1:1 ratio originally seen in young children. The disease is expected to disappear completely in another 10 years or so (Alpers 1979).

Although transmission of kuru by endocannibalism satisfactorily explains the epidemiological data, direct evidence is limited. The titres of kuru agent are low in peripheral tissues and body fluids of patients but high in brain, which is presumed to be the source of infection (Gajdusek 1977). However feeding large amounts of infected brain to chimpanzees apparently does not produce experimental kuru. Hence it is suggested that infection occurs via damaged skin and mucous membranes (Gajdusek 1979).

In summary, kuru represents an interesting case where the social behaviour of a community provides the means of transmitting an infection. Because the disease is slow to develop, the connection between endocannibalism and kuru went unsuspected by the patients. However, the origin of kuru remains unknown. It has been suggested that kuru arose from a spontaneous case of Creutzfeldt-Jakob disease (Gajdusek 1977). Alternatively the kuru epidemic may have been caused by a pathogenic variant of an agent that is not normally associated with disease and that still may be present in the population.

Creutzfeldt-Jakob disease (CJD)

Creutzfeldt-Jakob disease was first described by Creutzfeldt (1920) and Jakob (1921) as a rare degenerative disease of man. In 1968, Kirschbaum published his detailed review of the disease based on the 150 cases that were known at the time. Ten years later Masters *et al.* (1979) presented their analysis of over 1400 cases. The difference reflects the greatly increased awareness of CJD, which started when Klatzo and co-workers (1959) noted that the pathology of kuru resembled that of CJD.

In 1968, the transmission of CJD to a chimpanzee was reported, following the intracerebral inoculation of material from a brain biopsy (Gibbs *et al.* 1968). Since then 112 isolates of CJD have been made in various species of primate (Zlotnik *et al.* 1974, Gibbs *et al.* 1979). The CJD agent has been repeatedly passaged *in vivo*, usually with an incubation period in excess of 1 year. It remembles the scrapie agent in being filtrable and highly resistant to ionizing radiation (Gajdusek and Gibbs 1973; Gibbs *et al.* 1978). There is evidence that different strains of CJD agent exist (Gibbs *et al.* 1979).

In 1975, Manuelidis reported the transmission of CJD from man to guinea-pigs and this has now been confirmed (Matthews *et al.* 1979). There are also experimental models of CJD in cats (see Gibbs *et al.* 1979) hamsters, rats and mice (see Manuelidis and Manuelidis 1979, Gibbs *et al.* 1979, Tateishi *et al.* 1979). Hence detailed quantitative studies of experimental CJD will now be possible.

The clinical signs and brain lesions of human CJD are more variable than those of kuru. CJD has been described under various names with the suggestion that it constitutes a number of disorders. However it is now generally regarded as a single nosological syndrome (Beck *et al.* 1969). The early clinical symptoms of CJD are vague, consisting of visual complaints, confusion and depression. Within weeks or months, a progressive dementia appears with clinical signs of cerebellar, basal ganglia, pyramidal and lower motor neuron involvement. Myoclonus is common later on, and most patients show characteristic e.e.g. changes (see Malmgren *et al.* 1979). The clinical course usually lasts less than a year and ends in death. The pathological picture closely fits the pattern that is characteristic of the group of scrapie-like diseases (Beck *et al.* 1969, Beck and Daniel 1979). Interestingly a small proportion of CJD cases show 'kuru type' amyloid plaques in the brain.

Creutzfeldt-Jakob disease occurs equally in men and women of different races throughout the world. Ninety per cent of all cases occur between the ages of 35 and 65, most of them in the sixth decade (Masters *et al.* 1979, Brown *et al.* 1979). There is a fairly uniform, world-wide incidence of about 1–2 cases per million per year. The disease occurs sporadically with only occasional spatial or temporal clustering. With one exception, the significance of the few clusters reported is hard to assess; they may well have occurred by chance (Malmgren *et al.* 1979). The exception is an annual incidence of over 30 cases of CJD per million of Libyan Jews living in Israel (Kahana *et al.* 1974; Alter and Kahana 1976). Detailed studies of this ethnic group suggest that there may be an enhanced genetic susceptibility to CJD (Neugut *et al.* 1979). Alternatively, a higher than normal exposure to CJD infection has been proposed, but the putative sources of infection are completely unknown. About 15 per cent of CJD cases throughout the world show a familial pattern which has also been interpreted in genetic terms or as examples of common exposure among relatives. However the rarity of CJD makes this type of analysis extremely difficult.

Iatrogenic transmission of CJD has been reported on two occasions. The first followed corneal transplantation to a recipient who died of CJD 2 years later. The donor was diagnosed as having CJD at autopsy (Duffy *et al.* 1974). Subsequent studies of guinea-pig CJD and hamster TME have shown that quite high titres of infectious agent are present in the corneas of affected animals (Marsh and Hanson 1974, Manuelidis *et al.* 1977). On the second occasion, two patients contracted CJD after the use of implanted e.e.g. electrodes which had previously been used on a case of CJD. Given the physico-chemical stability of the unconventional agents, it is clear that the electrodes had been inadequately sterilized (Bernoulli *et al.* 1977).

The medical implications of these accidents are obvious; it has been suggested that iatrogenic transmission may account for other cases of CJD. This is possible in some instances, but there is no clear evidence that CJD is generally associated with patients who have undergone surgical operations on the brain or other organs, or who were in the so-called health professions (Masters *et al.* 1979, Brown *et al.* 1979).

An important question about CJD concerns the natural reservoir of infection. Because of the rarity of CJD and its close similarities to scrapie, some investigators have compared it with TME, suggesting that CJD represents scrapie in man. There is now considerable evidence against this.

Firstly, the world-wide occurrence of CJD differs from the distribution of scrapie and the consumption of sheep products. A few examples of this are (a) a very low occurrence of CJD in Iceland, which has scrapie and a high consumption of sheep meat (Pálsson 1979); (b) an exceptionally high occurrence of CJD in Libyan Jews in Israel where scrapie has never been reported (Masters *et al.* 1979); (c) the occurrence of CJD in Australia, which has an enormous sheep population but no scrapie; and in Japan, which essentially has neither (Tateishi *et al.* 1979).

Secondly, two major epidemiological surveys have analysed the incidence of CJD in relation to the eating habits (e.g. brain) and environment (e.g. urban or rural) of various occupations, such as those of shepherds, butchers and veterinarians. These and other studies have failed to establish a link between scrapie and CJD (see Masters *et al.* 1979, Brown *et al.* 1979). Unless some other external source of infection can be identified, it seems more likely that man is the reservoir of CJD agent. If so, then CJD may resemble PML or SSPE in being a rare disease associated with a common infection that is readily maintained in the human population.

Since some strains of scrapie agent undergo mutation (Bruce and Dickinson 1979), the sporadic nature of CJD could be explained on the basis of random mutations of a fairly common, non-pathogenic agent leading to pathogenic variants. If the selection of variants is influenced by host genetic factors, then the incidence of disease might be higher than average in certain populations. Unfortunately it will be difficult to test this hypothesis without practicable methods to diagnose infection with CJD agent.

Experimental scrapie in laboratory animals

Approaches to scrapie research

General problems

There are obvious difficulties in studying any disease which has a long incubation period but, compared with visna and Aleutian disease, the problems with scrapie are more serious. There are no electronmicroscopic or immunological methods for detecting the presence of the scrapie agent, and no cell-culture methods of assay are available. Amounts of the agent can be measured by titrating serial tenfold dilutions of inoculum in animals and calculating an LD_{50} end point; this requires large numbers of animals. In some circumstances, a more economical assay can be performed by injecting a single dilution and carefully measuring the incubation period, which is proportional to the dose of agent if the infectivity of the agent remains constant. In both cases, the assays depend on the development of clinical disease, and even with combinations of agent strain and host genotype giving short incubation periods this can take 100–200 days. With slower models the assay time can be in excess of 500 days and a logical sequence of 3 or 4 experiments may take several years to perform. Hence scrapie research requires careful planning and the efficient use of high quality, large-scale animal facilities.

Cell cultures. A persistently infected cell line (SMB) was originated by Clarke and Haig (1970*a*, *b*) from cultures of scrapie mouse brain. Agent replication appears to be synchronous with cell division, so that the number of infectious units per 100 cells is constant at about 1 mouse intracerebral LD_{50} unit. This restriction is consistent with *in vivo* evidence (page 503) that agent replication is limited by a finite number of replication sites. It may not be possible to obtain high titres of the agent *in vitro* if this restriction is generally applicable.

Attempts have been made to infect established cell cultures with scrapie. Some success was obtained with L cells, but the infectivity titres were very low (Clarke and Millson 1976*a*). Recently, cell lines were obtained from the brains of mice fed on a diet containing cuprizone and then exposed to diluted homogenates of scrapie brain. Three cultures have been developed, and in two of these the infectivity titres were similar to those found in SMB cells (Clarke 1979). However, none of these cell cultures show cytopathic effects, and the presence of the agent can be detected only by inoculating mice.

Pathology. Many histological, ultrastructural and biochemical studies have been made in attempts to identify the fundamental brain lesions in scrapie and related diseases (see Fraser 1976, Kimberlin 1976*a*). These have been unsuccessful, partly because of the difficulty of distinguishing primary events associated with scrapie from non-specific secondary changes. The extreme variation in the severity and distribution of vacuolar lesions in scrapie has raised strong doubts about the significance of vacuolation *per se* in the development of clinical disease (Kimberlin 1976*c*, Fraser 1979*a*). These doubts have been increased by comparative studies of TME in mink of different genotypes. Vacuolation was virtually absent in old mink of the Chediak-Higashi genotype, even though clinical signs, incubation periods, and agent titres in brain were the same as in other genotypes of mink which had severe vacuolation (Marsh *et al.* 1976).

Although vacuolation may not be relevant to clinical scrapie, Fraser has shown that the severity and distribution of lesions can be measured in defined grey and white matter areas of brain to give a 'lesion profile', which is characteristic of a given experimental model of scrapie when all the relevant variables are controlled (Fraser and Dickinson 1968, 1973, Fraser 1976, 1979*a*, *b*). The variables include the strain of donor mouse and the tissue source of inoculum, the strain of agent, the route of inoculation, and the sex and strain of the recipient mouse. However, the most important application of the lesion 'profile' system has been to identify different strains of the agent.

Agents such as 22A, 22C and ME7 are easily distinguishable by their grey matter 'profiles' in C57BL and VM mice, which differ in their *Sinc* genotype. These strains produce little or no vacuolation in white matter in contrast to 79A and 139A which produce a lot. There are other qualitative differences between strains of the agent which help in their identification. For example, 22L in most strains of mice produces an extraordinarily severe spongiform degeneration throughout the cerebellar cortex. Strains 79A and 22C are never associated with cerebral amyloidosis, whereas 87A and 87V produce many amyloid plaques and show some asymmetric vacuolation in certain areas of brain. The combined use of these quantitative and qualitative methods provides a sensitive and reliable means of agent-strain typing.

Biological studies

Classical experiments by Dickinson have shown that the incubation period of scrapie in mice is under the control of a gene, *Sinc* which has two alleles, s7 and p7 (Dickinson *et al.* 1968*a*; Dickinson and Meikle

1971, Dickinson and Fraser 1977, 1979). One group of agent strains behaves like ME7 and has a fairly short incubation period in mice of the *Sinc* genotype s7s7, and a much longer incubation period in mice homozygous for *Sinc*^{p7}. Another group, exemplified by 22A, shows the reverse properties. However, many members of both the ME7 and 22A groups can be distinguished when the incubation periods in the heterozygotes are compared with those in the parental homozygous genotypes. For example, within the ME7 group of agents, ME7 itself shows no dominance, 22C shows partial dominance of the s7 allele, 79A partial dominance of p7 and 139A, overdominance of p7. This diversity of interactions between strains of the agent and the alleles of *Sinc* gene provides a second way of agent-typing which is independent of the 'lesion profile' system (page 499). About 20 strains of the scrapie agent have been identified by these methods, which have also been of crucial importance to studies of the biological stability of agents (page 499) and of competition between agent strains *in vivo*. A third method relies on comparing the relative pathogenicity of strains in different hosts. This method is essential to studies of those strains which have not been transmitted to mice. It has been used to investigate strain selection on passage in a new host (Kimberlin and Walker 1978*a*, Kimberlin 1979*c*), and to demonstrate strain variation in TME (Marsh and Hanson 1979), kuru, and Creutzfeldt-Jakob disease (Gibbs *et al.* 1979).

Studies of scrapie pathogenesis fall into two categories. The direct approach, pioneered by Eklund and co-workers (1967), demands measurement of the dynamics of agent replication in various tissues. The indirect approach, described in detail by Outram (1976), requires the attempted modification of incubation period, or other disease characters, by a variety of means. A simple example of this approach has been the use of splenectomy or genetic asplenia to show that the spleen is an important site of agent replication when mice are infected by a peripheral route (see page 502). Other studies have shown that the pathogenesis of scrapie in mice infected immediately after birth is more variable; incubation periods are generally longer than in mice infected after weaning. These results indicate the incomplete maturation of some, as yet unknown, populations of cells in newborn mice which are needed to establish scrapie infection (Outram *et al.* 1973; Hotchin and Buckley 1977, Kimberlin and Walker 1978*b*).

Various drug treatments have been used to study the pathogenesis of scrapie. A short course of injections of anti-inflammatory compounds, such as prednisone acetate or arachis oil, have a suppressive effect on peripherally injected scrapie in mice (Outram *et al.* 1974, 1975). An antiviral agent, ammonium 5-tungsto-2-antimoniate has also been found to suppress scrapie infection (Kimberlin and Walker 1979*c*). In contrast, single intraperitoneal injections of phytohaemagglutinin (Dickinson *et al.* 1978) or of the residue left after methanol extraction of BCG vaccine (Kimberlin and Cunnington 1978) shorten the incubation period by the intraperitoneal route. These compounds have a stimulatory effect on the lymphoreticular system. There is also a report of reduced intracerebral incubation periods in mice injected intracerebrally with interferon inducers (Allen and Cochran 1977). However,it is debatable whether interferon induction is the important effect, since several studies have shown that interferon is not concerned in the pathogenesis of scrapie (Gresser and Pattison 1968, Katz and Koprowski 1968, Field *et al.* 1969*a*, Worthington 1972).

Some important negative findings emerge from this type of study. The failure of prolonged immunosuppression with cyclophosphamide (Worthington and Clark 1971) to alter the incubation period argues against the direct involvement of B cells in scrapie pathogenesis, as does the failure to detect specific antibody to scrapie (Gardiner 1965; Gibbs *et al.* 1965, Clarke and Haig 1966; Gibbs 1967; Porter *et al.* 1973, Cunnington *et al.* 1976) and the normal response of infected mice when challenged with other antigens (Clarke 1968, Gardiner and Marucci 1969). Similarly there is strong evidence from studies of the effects of neonatal and adult thymectomy that T cells are not important in scrapie pathogenesis (McFarlin *et al.* 1971, Fraser and Dickinson 1978). It is reasonable to conclude that, although the scrapie agent replicates in lymphoid organs, cells other than T and B are probably concerned.

Properties of the scrapie agent

Biological properties

Careful studies carried out for over 15 years have shown that some strains of the scrapie agent, like ME7, are completely stable after 10 or 11 serial passages in either *Sinc* ^{s7} or *Sinc* ^{p7} mice. ME7 has been passaged 5 times in *Sinc* ^{s7} mice, once in Cheviot sheep, and then re-isolated in mice with no alteration of biological properties. This is described as class I stability by Bruce and Dickinson (1979). Class II agents such as 22A are stable when passaged in the *Sinc* genotype in which they were isolated, but gradually change their properties on later passage in mice of the other homozygous genotype. This pattern is indicative of accumulated point mutations over several passages. There is also a group of class III agents which includes 87A and four other isolates associated with a high incidence of amyloid plaques and with asymmetrical vacuolation in brain. These agents are unstable even on serial passage in the mouse genotype in which they were isolated. They show an unpredictable and discontinuous change of properties but in all cases so far studied, the same new or modified strain is obtained, namely 7D. It is interesting that 7D has properties similar to, though not identical with, ME7, and that ME7 is the agent most frequently isolated from sheep (Dickinson 1976). This suggests that ME7, and perhaps others in class I, represents a minimal form of scrapie agent. However it should not be regarded as a laboratory artefact, since ME7, and several other mouse passaged agents, can still produce scrapie when injected into sheep or goats (Dickinson and Fraser 1979).

Physico-chemical properties

Studies on sub-cellular fractionation of brains and spleen from scrapie mice have demonstrated that most infectivity is associated with microsomal fractions (Millson *et al.* 1971, Semancik *et al.* 1976, Prusiner *et al.* 1977). Studies on the SMB cell line (page 499) showed that the highest scrapie titres were in the plasma membrane, although titres were also high in the endoplasmic reticulum (Clarke and Millson 1976*b*). Many attempts have been made to purify infectious agent from membranes. The most recent have employed detergents (Millson and Manning 1979), density gradient centrifugation (Siakotos *et al.* 1979, Brown *et al.* 1978), a combination of detergents and differential centrifugation (Prusiner *et al.* 1978), or electrophoresis in agarose gels (Prusiner *et al.* 1980). None of these has succeeded beyond a ten- to fifty-fold purification with respect to protein. A much greater degree of purification may not be possible unless the scrapie agent exists as a discrete virus-like nucleoprotein structure (Kimberlin 1976*c*; Millson and Manning 1979). At the moment there is no ultrastructural or other direct evidence for this.

Physico-chemical studies of scrapie infectivity have of necessity been performed with crude tissue extracts. This raises questions about 'protection' of agent (Kimberlin 1979*c*), and also about the effects of various treatments on tissue components which in some cases alter the relation between infectivity titre and amount of agent (Kimberlin 1977). Hence, some of the published data are difficult to interpret, particularly when infectivity titres differ by only 1 to 2 $\log_{10}$ units.

However, in confirmation of the early sheep work (page 494) there is no doubt that the infectious scrapie agent is highly stable to wet and dry heat, ionizing radiation and exposure to alkylating agents, organic solvents, concentrated salt solutions and many detergents (see Millson *et al.* 1976, Hunter and Millson 1977). Substantial infectivity in 10 per cent brain homogenates containing 22A persists after autoclaving at 110° for 30 minutes (Dickinson and Taylor 1978). This illustrates the need for exceptional standards of disinfection in scrapie work. The physicochemical stability of scrapie agent may perhaps be due to its apparent location in membranes (Hunter 1972, Kimberlin 1976b, Prusiner *et al.* 1979). Treatments that destroy most scrapie infectivity, e.g. 80 per cent 2-chloroethanol, 90 per cent phenol, 5 per cent sodium dodecyl sulphate, 6 M guanidinium hydrogen bromide (see Millson *et al.* 1976) also disrupt the hydrophobic interactions that contribute to the intrinsic stability of membrane structures.

Centrifugation studies have shown that the infectious agent is extremely heterogeneous in size, presumably reflecting the different sizes of membrane fragments with which it is associated. Attempts to estimate the size of the minimum infectious unit by filtration, sedimentation and electrophoretic methods (Kimberlin *et al.* 1971, Prusiner *et al.* 1977, 1980) suggest a minimum particle size of less than 50 nm.

The high resistance of scrapie, and related agents, to ionizing radiation has been thoroughly documented (Alper *et al.* 1966; Alper and Haig 1968; Field *et al.* 1969*b*, Gibbs *et al.* 1978; Latarjet 1979). On the assumption that the radiation target is nucleic acid and that there is no repair of damaged molecules *in vivo*, these results indicate that the putative nucleic acid of scrapie may have a mol. wt of ca. 10^5. This figure is broadly consistent with the high levels of u.v. radiation needed to inactivate the agent (Alper *et al.* 1966). The u.v. resistance of scrapie is also similar to that of potato spindle tuber viroid (PSTV) irradiated in crude suspensions (Diener 1973), and only ten times higher than that of purified PSTV RNA (Diener *et al.* 1974). The molecular weight of infectious PSTV is 127 000 (Sanger *et al.* 1976). This raises the possibility that the nucleic acid of scrapie may resemble viroids at least in size; but it must be remembered that if repair takes place, the size of the scrapie nucleic acid might be ten times larger (Latarjet 1979).

Again, on the assumption that the agent has a small nucleic acid, there is plenty of evidence to indicate that the scrapie agent is not simply an animal viroid, i.e. a small infectious nucleic acid. For example, nucleic acid extracts of scrapie tissues do not produce disease when tested in animals (Marsh *et al.* 1974, Ward *et al.* 1974, Hunter *et al.* 1976). The operational size of the scrapie infectious unit appears to be larger than 10^5 (Kimberlin *et al.* 1971). The greater sensitivity of the scrapie agent at wavelengths of 210 or 230 nm than at 250 or 280 nm suggests that a nucleic acid chromophore must be firmly associated with other molecules forming a complex (Latarjet *et al.* 1970, Latarjet 1979). This is supported by the enhanced inactivation of scrapie suspensions when exposed to ionizing radiation in the presence of oxygen (Alper *et al.* 1978). Also, the evidence cited earlier (page 499) concerning the association of infectivity with membranes and the loss of titre when membranes are disaggregated should not be forgotten.

In summary, the nature of the scrapie agent is still unknown. It may be surmised that it consists of a small specific nucleic acid, possibly of the same size as viroid RNA, functionally associated with hydrophobic proteins to give an infectious complex which *in vivo* is located in cell membranes. A small nucleic acid would have a limited coding potential and may not make a protein at all, as with viroids (see Diener 1979), and would presumably use host components to form an infectious unit. This would account for the heterogeneity of the scrapie agent and the difficulties encountered in its purification. It would also explain the absence of a specific immune response to infection. The fact that different strains of agent exist and can undergo mutation is a strong *a priori* reason for believing that a scrapie-specific nucleic acid exists. However, apart from an unconfirmed report that it may be DNA (Marsh *et al.* 1978), there is still no direct evidence for this. (For a further discussion on the nature of the scrapie agent, see Kimberlin 1982.)

Pathogenesis of scrapie

When all the variables relating to infection are carefully controlled, the incubation periods of even the slow models of mouse scrapie are extremely predictable, with a standard error of only about 1–2 per cent of the mean. This clockwork-like precision is consistent with the absence of specific immune responses to infection (page 500). However such predictability implies some kind of host control; this is most clearly seen in the profound effect *Sinc* gene has on the incubation period. Therefore, studies of scrapie pathogenesis have been focused on agent replication in various tissues and on the nature of the control exerted on this process by the host.

A classic study by Eklund and co-workers (1967) first showed that the pathogenesis of the short incubation models of mouse scrapie, by using a peripheral route of infection, could be divided into two stages. In stage 1, the agent replicates initially in spleen and later in lymph nodes, thymus and other lymphoid organs. The early replication in the spleen has been confirmed by subsequent studies (Dickinson *et al*, 1968*a*, 1975*b*, Clarke and Haig 1971, Hunter *et al.* 1972). Its importance in pathogenesis is shown by the longer incubation period that occurs when the spleen is removed before or soon after infection (Clarke and Haig 1971, Dickinson and Fraser 1972, Fraser and Dickinson 1970, 1978). Stage 2 follows many weeks later and consists of a long period of agent replication in the c.n.s., which continues until clinical signs appear.

Several detailed studies have been carried out with either the 139A or the ME7 strain injected peripherally into *Sinc* ^{s7} mice, so as to give incubation periods in the range of 150 to 250 days. A more complete account of pathogenesis is now possible.

Short incubation models

Comparisons of the titres of a standard scrapie homogenate with different routes of inoculation give a measure of the degree of infectivity. The intracerebral (i.c.) route is always the most infective, followed by intravenous (i.v.) intraperitoneal (i.p.) and subcutaneous (s.c.); though with the last route the infectivity varies with the site. The i.v. route is only ten times less effective than the i.c. route, but it can be over 1000 times more so than s.c. injection (Kimberlin and Walker 1978*b*). A prior i.p. injection of BCG or the methanol extraction residue of BCG will increase the infectivity of the i.p. route by seven- to twenty-fold (Kimberlin 1979*b*). The nature of the agent-cell interactions concerned in establishing infection are unknown, but the net result is to determine the effective dose of the agent. This is shown by the similar dose-response curves obtained with i.v. i.p. and s.c. routes of infection when correction is made for the degree of infectivity (Kimberlin and Walker 1978*b*). The same is true for the agent replication curves in spleen and brain (Kimberlin and Walker 1979*b*, 1980).

Infectivity can be detected in blood within minutes of injecting large doses of the agent, but the amounts decrease rapidly over the next few hours (Field *et al.* 1968, Millson *et al.* 1979). Within half an hour, some infectivity appears in tissues remote from the site of injection, even with the i.c. route. Hence infection of tissues occurs rapidly, probably by haematogenous spread of the agent from the site of injection. There are also indications that only a small proportion of the injected dose is needed to establish infection (Kimberlin and Cunnington 1978, Millson *et al.* 1979).

With large doses of the agent, replication in the spleen starts almost immediately, although some of this may be due to accumulation of the original inoculum. A week or two later, the concentration of the agent exceeds the dose injected. When low doses are used, there is a 'zero-phase' of two weeks before replication in the spleen can be detected. In other words, the time of onset is dose-dependent (Clarke and Haig 1971, Kimberlin and Walker 1979*b*). Once established, replication proceeds rapidly in the spleen, equivalent to a doubling time of a few days (Dickinson *et al.* 1968*a*, 1975*a*, Clarke and Haig 1971, Hunter *et al.* 1972, Kimberlin and Walker 1979*b*). After about 4 to 6 weeks, the concentration in the spleen reaches a plateau level; this persists, or perhaps declines a little, during the remainder of the incubation period.

There is some evidence that the agent can enter the c.n.s. directly from the blood stream when high doses are injected i.v., and the brain is simultaneously damaged by 'jagging' with a needle (Fraser 1979*b*). However, it is not certain that the agent normally enters the c.n.s. from the vascular system. Apart from the first few hours after injection, it is either undetectable in the blood (Eklund *et al.* 1967); or the amounts are so low that they could be due to contamination from infected tissues damaged during the collection of blood samples (Clarke and Haig 1967; Dickinson *et al.* 1969). Studies of 139A agent have shown that replication starts in the brain and lumbar spinal cord about halfway through the incubation period, irrespective of the dose of agent and the peripheral route of inoculation used. However, this is always preceded by replication in the thoracic cord, which occurs after about a third of the way through the incubation period (Kimberlin and Walker 1979*b*, 1980). This pattern suggests that the agent may enter the c.n.s. by neural spread along sympathetic fibres running from the spleen and other visceral sites of replication to the thoracic cord via the splanchnic nerves. However, direct evidence for this pathway has yet to be obtained; and haematogenous spread of infection to the c.n.s. may occur in some cases as indicated by studies of CJD in guinea-pigs (Manuelidis *et al.* 1978).

Agent replication in the brain occupies the second half of the incubation period in mice infected peripherally. When the i.c. route is used, the same sequence of events occurs in the spleen and spinal cord, but these are irrelevant to pathogenesis because after i.c. infection replication in brain occurs much sooner (Kimberlin and Walker 1979*b*). With all routes of infection, titres of the agent in the brain increase progressively to reach concentrations at least ten-fold higher than in the spleen and other peripheral organs. For 139A, the concentration in the brain when clinical signs develop is always the same, irrespective of the dose, the route of inoculation, or the age of the mouse at infection (Kimberlin and Walker 1978*b*). Similar results have been obtained with other strains of mouse-passaged scrapie (A.G. Dickinson, unpublished data), and with 263K in hamsters (Kimberlin and Walker 1977). Hence, the sharp onset of clinical signs appears to be related to the attainment of a threshold concentration of agent in the brain or particular parts of it. This concept implies that functional changes occur in the brain at earlier stages, before the accumulated damage reaches the threshold—a stage at which compensatory systems break down and gross clinical abnormality develops (see Kimberlin 1976*a*; Outram 1976). This is evident from studies of behaviour (Savage and Field 1965, Suckling *et al.* 1976, MacFarland and Hotchin 1980), electrophysiology (Court *et al.* 1979), and, particularly, of drinking and feeding responses (Outram 1972). It has been suggested that the gross histological and biochemical lesions of scrapie may not be directly related to the development of clinical disease (page 499). Therefore, scrapie may best be regarded as a disease of cell dysfunction rather than cell degeneration, possibly manifested by a failure in neurotransmission (McDermott *et al.* 1978). The fundamental pathological process may be reversible for most of the incubation period, provided that dysfunction does not lead to neuronal death (Fraser 1979*c*).

Long incubation models

There have been fewer studies of the long incubation models of scrapie. Present evidence suggests that the extraneural and neural stages of agent replication occur much as described for the quicker models except that there is a long 'zero phase' before replication becomes detectable (see Dickinson and Outram 1979, Fraser 1979*c*, Kimberlin 1979*b*). In part, the zero phase may reflect an insensitivity of the infectivity assay. This could account for the increased length of zero phase with decreased doses of agent in short incubation models of scrapie (Kimberlin and Walker 1979*a*). However in these models, the zero phase lasts only a few weeks. In others, a year may elapse between infection and the onset of replication in the spleen: in other words, the zero phase can occupy half the lifespan of the mouse (Dickinson *et al.* 1975*a*). The zero phase differs from viral latency in that a secondary event is not necessary to terminate it and initiate replication. In fact the duration of zero phase in the long incubation models of scrapie is controlled by *Sinc* gene, which is exerted on the onset of replication in both brain and spleen (Dickinson and Fraser 1969, Dickinson *et al.* 1969).

A prolonged zero phase can occur in other situations. For example, a zero phase of about one year was found when mice were infected i.p. on the day of birth by using a combination of agent strain and mouse genotype giving a short incubation period in weanling mice (Hotchin and Buckley 1977). Studies have also been made of the species barrier effect in scrapie (Kimberlin *et al.* 1975, Kimberlin and Walker 1978*a*). For instance, it was shown that the extended incubation period at the first passage in the new species of host, compared to the second, was entirely due to a prolonged zero phase lasting 175 days (Kimberlin and Walker 1979*a*). The cellular events associated with the zero phase remain completely unknown, but understanding them is a major goal of scrapie research.

The control of replication of scrapie agent

Early observations of the slow and unpredictable onset of natural scrapie in sheep created the impression of a disease of complex pathogenesis. This was largely based on ignorance of the genetic factors in both agent and host that influence the appearance of clinical signs. Later studies of experimental scrapie in sheep selectively bred to give a uniform response to infection showed that it was not necessary to equate slowness with variability (Dickinson *et al.* 1968*b*, Nussbaum *et al.* 1975). The same conclusion can be drawn from studies of scrapie in mice, in which both the slowness and predictability are due to an overriding control exerted by *Sinc* gene on the dynamics of the extraneural and neural stages of agent replication. Studies of the interaction between strains of the agent and the alleles of *Sinc* have shown the occurrence of overdominance (page 500). This means that the alleles of *Sinc* do not act independently. Dickinson has suggested that each allele contributes a monomer to a multimeric structure concerned in agent replication (Dickinson and Meikle 1971). These suggestions have been developed into the 'Scrapie Replication Site Hypothesis', which states that the slow development of clinical scrapie in mice is due to a restriction in the number of agent replication sites, which are multimeric structures specified by the alleles of *Sinc* gene (see Dickinson and Outram 1979). The concept of a limited number of these sites is consistent with the restriction of agent replication to dividing cells in the SMB line (page 499). There is some evidence that the replication process *in vivo* can become saturated, because increasing the dose of i.c. injected agent beyond a certain point does not cause further shortening of the incubation period (Mould *et al.* 1970, Kimberlin and Walker 1978*b*). The dominant role of the spleen in the pathogenesis of peripherally injected scrapie is probably due to the high proportion of extraneural replication sites contained in this tissue. Splenectomy removes these sites, which are not replaced (Fraser and Dickinson 1978), and therefore incubation is prolonged. However, the hypothesis predicts that, when the dose of injected agent is less than the number of replication sites in the spleen, splenectomy will have a reduced effect so that the titres of a standard scrapie homogenate will be similar in splenectomized and intact mice. So far, both these predictions have been fulfilled (Dickinson and Outram 1979).

Further evidence comes from experiments showing competition between strains of agent (see Dickinson and Outram 1979). The basic experiment is to inject mice with an agent strain—blocking agent—that is operationally slow in the genotype used and, after an interval, to inject an agent that is quicker—killing agent. Competition is seen as an increased incubation period of the second agent which can be identified by its 'lesion profile' (page 499). The original experiments (Dickinson *et al.* 1972, 1975*a*) were carried out with agents from the ME7 and 22A groups; by changing the genotype of mouse, the 'blocking' and 'killing' roles of the agents could be reversed. However, blocking has now been demonstrated between agents within the ME7 group itself (Dickinson and Outram 1979). Blocking can be demonstrated either by the i.c. or the i.p. route, and is independent of the mouse genotype in which the agents were passaged.

The simplest and most convincing explanation of these results is that the first agent occupies a proportion of a small number of agent replication sites, so that fewer are available when the second agent is injected. As a consequence the effective dose of the second agent is reduced and incubation is prolonged. In some cases blocking may be total. The extent of blocking can be precisely controlled by varying the doses of the two agents. For example, blocking is greater when the dose of the first agent is increased and that of the second reduced. Blocking is also greater when the interval between injections is increased, because this allows more time for the slower blocking agent to replicate and occupy sites that would otherwise be used by the killing agent.

Blocking occurs only with an agent that, under the conditions of the experiment, can replicate. For example, a dose that is too low to produce infection will not block. No blocking has been demonstrated with the related TME agent, which is not transmissible to any of the 14 strains of mice tested (Dickinson and Outram 1979). Neither does blocking occur with strain 22A that has been inactivated by chemical or physical treatment (R.H. Kimberlin and A.G. Dickinson, unpublished). Hence there is no immediate hope of a scrapie 'vaccine' to prevent agent replication, but the discovery of a chemical blocking agent might have clinical applications. It could also lead to the identification of the scrapie replication site and the normal function of *Sinc* gene.

Postscript

It is clear that there is nothing particularly slow about many of the slow viruses. Visna and Aleutian disease viruses multiply rapidly soon after infection. Even with natural scrapie in sheep, agent replication occurs in lymphoid organs at an early stage of the incubation period (Hadlow *et al.* 1979).

However the responses of the immune system to infection vary widely in slow virus diseases. At one extreme there is a phenomenal production of antibody, for example, to ADV, whereas the response to scrapie seems to be non-existent—although a recent report of increased content of IgG in the serum in some scrapie sheep may prove to be an interesting exception (Collis *et al.* 1979). The immune system appears to play a protective role in PML, and clinical disease is associated with immunosuppression. The immune system also has a limiting effect on visna virus and perhaps SSPE, but this is eventually overcome by the appearance of mutant strains that are less effectively neutralized. In Aleutian disease, the immune response has no protective effect at all; it is in fact the direct cause of disease. The pathogenesis of visna is also immunologically mediated. In both diseases the development of lesions is enhanced by vaccination, making this an inappropriate prophylactic measure. The only method of controlling these diseases is by selective culling of infected animals. This is most effective with Aleutian disease and maedi because of the availability of sensitive tests for infection (Crawford *et al.* 1977, Gorham *et al.* 1976, de Boer *et al.* 1979, de Boer and Houwers 1979). However, prevention or cure of the human slow virus diseases is impossible at the moment.

The studies of mouse scrapie have shown that there can be host control of infection without apparently the need for specific immune responses. There could be a superficial resemblance between the control by *Sinc* gene of replication of the scrapie agent in mice and the low permissiveness of sheep cells to visna virus multiplication *in vivo*. However, it is difficult to pursue these comparisons until more is known of the molecular biology of the scrapie agent and its mode of replication. The demonstration of competition between different strains of the scrapie agent may also have its parallels in conventional virology, particularly in the interference between defective and non-defective virus particles. But there is nothing defective about the 'blocking' agents in scrapie competition experiments and, again, there is not enough detailed information in scrapie for useful comparisons to be made. Despite the considerable progress of scrapie research in the last 20 years, most of the subject is still at that early stage when the most useful definitions and descriptions are made in purely operational terms. These can always be redefined as knowledge increases.

This brings us back to the point, made at the beginning of this Chapter, about 'scrapie virus' or 'scrapie agent'. There are sound reasons for suspecting that scrapie and related agents may be different from conventional viruses. Certainly many of the standard virological approaches to scrapie research are either inapplicable or inappropriate (page 499). The scrapie agent might fit into an ecological niche lying between viruses and viroids, i.e. an agent with a nucleic acid genome but no agent-specified coat proteins, so that the nucleic acid uses whatever host proteins are suitable to form an infectious agent. This might be dubbed the 'hermit-crab hypothesis' for scrapie. If it should happen to be correct, then by analogy with 'neutrinos', the group of scrapie-like agents could be referred to as 'virinos', since they appear to be 'small immunologically neutral particles with high penetration properties but needing special criteria to detect their presence' (Dickinson and Outram 1979). The idea that there is 'no immunology' in scrapie because there are 'no protein antigens' has an aesthetically appealing simplicity. However, it must be emphasized that these are merely speculations with little foundation in reality.

Scrapie, TME, kuru and CJD are closely related, even by the limited criteria available for making such comparisons. Hence, scrapie is the prototype for the group of unconventional agents. This relation has also raised the possibility that CJD in man may represent rare infections with scrapie agent, but the evidence is against it (page 498). Nevertheless, since iatrogenic transmission of CJD has been reported on two occasions (page 498), it is prudent to take some precautions when working with CJD in hospitals and in laboratories. The risks involved and the precautions needed have been carefully analysed in two excellent papers on the subject (Corsellis 1979, Chatigny and Prusiner 1979).

Scrapie can also provide experimental models relevant to a much larger group of diseases, namely the senile and presenile dementias of man. Alzheimer's disease is one of the commonest of these; a dominant feature is the occurrence of cerebral amyloid (see Terry and Davies 1980). Amyloid plaques are often found in the brains of kuru patients (page 497), and sometimes in cases of CJD (page 498). However, some combinations of scrapie strain and mouse genotype produce large amounts of cerebral amyloid covering the whole range of morphological types, including amorphous deposits, 'kuru-type' stellate plaques, and neuritic plaques (Bruce and Fraser 1973, 1975, Wiśniewski *et al.* 1975, Bruce *et al.* 1976). At the moment, scrapie provides the only experimental models for studying the pathogenesis of cerebral amyloid in dementia. Recently a type of degeneration has been seen in scrapie mouse brains which is strikingly similar to the loss of granule cells and microgliosis which occurs in the hippocampus of many Alzheimer patients (Fraser 1979*c*). These similarities do not necessarily imply that dementia is caused by a scrapie-like agent, but the possibility is worth investigating (see Dickinson *et al.* 1979).

References

Agnarsdóttir, G. (1977) *Recent Advances in Clinical Virology*, p. 21. Ed. by A.P. Waterson. Churchill Livingstone, Edinburgh.

Albrecht, P., Burnstein, T., Klutch, M.J., Hicks, J.T. and Ennis, F.A. (1977) *Science, N.Y.* **195,** 64.

Allen, L.B. and Cochran, K.W. (1977) *Ann. N.Y. Acad. Sci.* **284,** 676.

Alper, T. and Haig, D.A. (1968) *J. gen. Virol.* **3,** 157.

Alper, T., Haig, D.A. and Clarke, M.C. (1966) *Biochem. biophys. Res. Commun.* **22,** 278; (1978) *J. gen. Virol.* **41,** 503.

Alpers, M.P. (1979) In: *Slow Transmissible Diseases of the Nervous System*, Vol. 1, p. 67. Ed. by S.B. Prusiner and W.J. Hadlow, Academic Press, New York.

Alpers, M. and Gajdusek, D.C. (1965) *Amer. J. trop. Med. Hyg.* **14,** 852.

Alter, M. and Kahana, E. (1976) *Science, N.Y.* **192,** 428.

Åström, K.E., Mancall, E.L. and Richardson, E.P. Jr. (1958) *Brain* **81,** 93.

Barbanti-Brodano, G., Oyanagi, S., Katz, M. and Koprowski, H. (1970) *Proc. Soc. exp. Biol.* **134,** 230.

Beck, E. and Daniel, P.M. (1979) In: *Slow Transmissible Diseases of the Nervous System*, Vol. 1, p. 253. Ed. by S.B. Prusiner and W.J. Hadlow, Academic Press, New York.

Beck, E., Daniel, P.M., Alpers, M., Gajdusek, D.C. and Gibbs, C.J. Jr. (1966) *Lancet* **ii,** 1056.

Beck, E. *et al.* (1969) *Brain* **92,** 699.

Bernoulli, C. *et al.* (1977) *Lancet* **i,** 478.

Besnoit, C. and Morel, C. (1898) *Rev. Vet., Toulouse* **23,** 397.

Bogaert, L. van (1945) *J. Neurol. Neurosurg. Psychiat.* **8,** 101.

Bouteille, M., Fontaine, C., Vedrenne, C. and Delarue, J. (1965) *Rev. Neurol.* **113,** 454.

Brody, J.A. and Detels, R. (1970) *Lancet* **ii,** 500.

Brooks, B.R., Jubelt, B., Swarz, J.R. and Johnson, R.T. (1979) *Annu. Rev. Neurosci.* **2,** 309.

Brotherston, J.G., Renwick, C.C., Stamp, J.T., Zlotnik, I. and Pattison, I.H. (1968) *J. comp. Pathol.* **78,** 9.

Brown, P., Cathala, F. and Gajdusek, D.C. (1979) *Ann. Neurol.* **6,** 438.

Brown, P., Green, E.M. and Gajdusek, D.C. (1978) *Proc. Soc. exp. Biol. Med.* **158,** 513.

Bruce, M.E. and Dickinson, A.G. (1979) In: *Slow Transmissible Diseases of the Nervous System*, Vol. 2, p. 71. Ed. by S.B. Prusiner and W.J. Hadlow, Academic Press, New York.

Bruce, M.E., Dickinson, A.G. and Fraser, H. (1976) *Neuropathol. appl. Neurobiol.* **2,** 471.

Bruce, M.E. and Fraser, H. (1973) *Lancet* **i,** 617; (1975) *Neuropathol. appl. Neurobiol.* **1,** 189.

Burger, D. and Hartsough, G.R. (1965) *J. infect. Dis.* **115,** 393.

Byington, D.P. and Johnson, K.P. (1972) *J. infect. Dis.* **126,** 18.

Cavanagh, J.B., Greenbaum, D., Marshall, A.H.E. and Rubinstein, L.J. (1959) *Lancet* **ii,** 524.

Chandler, R.L. (1959) *Vet. Rec.* **71,** 58; (1961) *Lancet* **i,** 1378; (1963) *Res. vet. Sci.* **4,** 276.

Chatigny, M.A. and Prusiner, S.B. (1979) In: *Slow Transmissible Diseases of the Nervous System*, Vol. 2, p. 491. Ed. by S.B. Prusiner and W.J. Hadlow, Academic Press, New York.

Cheema, A., Henson, J.B. and Gorham, J.R. (1972) *Amer. J. Pathol.* **66,** 543.

Chesebro, B., Bloom, M., Hadlow, W. and Race, R. (1975) *Nature Lond.* **254,** 456.

Cho, H.J. (1976) In: *Slow Virus Diseases of Animals and Man*, p. 159. Ed. by R.H. Kimberlin, North Holland, Amsterdam; (1977) *Canad. J. comp. Med.* **41,** 215.

Cho, H.J. and Ingram, D.G. (1973) *Nature, New Biol.* **243,** 174.

Cho, H.J. and Porter, D.D. (1978) *Persistent Viruses*, Vol. 11, Part 2, p. 711. Ed. by J.G. Stevens, G.J. Todaro and C.F. Fox, Academic Press, New York.

Clarke, M.C. (1968) *Res. vet. Sci.* **9,** 595; (1979) In: *Slow Transmissible Diseases of the Nervous System*, Vol. 2, p. 225. Ed. by S.B. Prusiner and W.J. Hadlow, Academic Press, New York.

Clarke, M.C. and Haig, D.A. (1966) *Vet. Rec.* **78,** 647; (1967) *Ibid.* **80,** 504; (1970*a*) *Nature, Lond.* **225,** 100; (1970*b*) *Res. vet. Sci.* **11,** 500; (1971) *Ibid.* **12,** 195.

Clarke, M.C. and Millson, G.C. (1976*a*) *Nature, Lond.* **261,** 144; (1976*b*) *J. gen. Virol.* **31,** 441.

Collis, S.C., Kimberlin, R.H. and Millson, G.C. (1979) *J. comp. Pathol.* **89,** 389.

Connolly, J.H., Allen, I.V., Hurwitz, L.J. and Millar, J.H.D. (1967) *Lancet* **i,** 542.

Corsellis, J.A.N. (1979) *Brit. J. Psychiat.* **134,** 553.

Court, L., Cathala, F., Bouchard, N., Breton, P. and Gourmelon, P. (1979) In: *Slow Transmissible Diseases of the Nervous System*, Vol. 1, p. 305. Ed. by S.B. Prusiner and W.J. Hadlow, Academic Press, New York.

Crawford, T.B., McGuire, T.C., Porter, D.D. and Cho, H.J. (1977) *J. Immunol.* **118,** 1249.

Creutzfeldt, H.G. (1920) *Z. ges. Neurol. Psychiat.* **57,** 1.

Cuillé, J. and Chelle, P.L. (1936) *C.R. Acad. Sci., Paris* **203,** 1552.

Cunnington, P.G., Kimberlin, R.H., Hunter, G.D. and Newsome, P.M. (1976) *IRCS med. Sci.* **4,** 250.

Daoust, P.Y. and Hunter, D.B. (1978) *Canad. vet. J.* **19,** 133.

Dawson, J.R. (1933) *Amer. J. Path.* **9,** 7.

De Boer, G.F., and Houwers, D.J. (1979) In: *Aspects of Slow and Persistent Virus Infections*, p. 198. Ed. by D.A.J. Tyrell, Martinus Nijhoff, The Hague.

De Boer, G.F., Terpstra, C., Houwers, D.J. and Hendriks, J. (1979) Res. vet. Sci. 26, 202.

Dickinson, A.G. (1976) In: *Slow Virus Diseases of Animals and Man*, p. 209. Ed. by R.H. Kimberlin, North Holland, Amsterdam.

Dickinson, A.G. and Fraser, H. (1969) *J. comp. Pathol.* **79,** 363; (1972) *Heredity, London* **29,** 91; (1977) *Slow Virus Infections of the Central Nervous System*, p. 3. Ed. by V. ter Meulen and M. Katz, Springer Verlag, New York; (1979) In: *Slow Transmissible Diseases of the Nervous System*, Vol. 1, p. 367. Ed. by S.B. Prusiner and W.J. Hadlow, Academic Press, New York.

Dickinson, A.G., Fraser, H. and Bruce, M.E. (1979) In: *Alzheimer's Disease; Early Recognition of Potentially Reversible Defects*, p. 42. Ed. by A.I.M. Glen and L.J. Whalley, Churchill Livingstone, Edinburgh.

Dickinson, A.G., Fraser, H., McConnell, I. and Outram, G.W. (1978) *Nature, Lond.* **272,** 54.

Dickinson, A.G., Fraser, H., McConnell, I., Outram, G.W., Sales, D.I. and Taylor, D.M. (1975*a*) *Nature, Lond.* **253,** 556.

Dickinson, A.G., Fraser, H., Meikle, V.M.H. and Outram, G.W. (1972) *Nature, New Biol., London* **237,** 244.

Dickinson, A.G., Fraser, H. and Outram, G.W. (1975*b*) *Nature, Lond.* **256,** 732.

Dickinson, A.G. and Meikle, V.M.H. (1971) *Molec. gen. Genet.* **112,** 73.

Dickinson, A.G., Meikle, V.M.H. and Fraser, H. (1968*a*) *J. comp. Pathol.* **78,** 293; (1969) *Ibid.* **79,** 15.

Dickinson, A.G. and Outram, G.W. (1979) In: *Slow Transmissible Diseases of the Nervous System*, Vol. 2, p. 13. Ed. by S.B. Prusiner and W.J. Hadlow, Academic Press, New York.

Dickinson, A.G. and Smith, W. (1966) *Report of Scrapie Seminar* p. 251, ARS 91–53, US Dept. Agric., Wash., DC.

Dickinson, A.G., Stamp, J.T. and Renwick, C.C. (1974) *J. comp. Pathol.* **84,** 19.

Dickinson, A.G., Stamp, J.T., Renwick, C.C. and Rennie, J.C. (1968*b*) *J. comp. Pathol.* **78,** 313.

Dickinson, A.G. and Taylor, D.M. (1978) *New Engl. J. Med.* **299,** 1413.

Dickinson, A.G., Young, G.B. and Renwick, C.C. (1966) *Report of Scrapie Seminar*, p. 244. ARS 91–53, US Dept. Agric., Wash., DC.

Dickinson, A.G., Young, G.B., Stamp, J.T. and Renwick, C.C. (1965) *Heredity* **20,** 485.

Diener, T.O. (1973) *Ann. clin. Res.* **5,** 268; (1979) *Viroids and Viroid Diseases*. John Wiley, New York.

Diener, T.O., Schneider, I.R. and Smith, D.R. (1974) *Virology* **57,** 577.

Dubois-Dalcq, M., Barbosa, L.H., Hamilton, R., Sever, J.L. (1974*a*) *Lab. Invest.* **30,** 241.

Dubois-Dalcq, M., Coblentz, J.M., Fleet, A. (1974*b*) *Arch. Neurol. Chicago* **31,** 355.

Dubois-Dalcq, M., Reese, T.S., Murphy, M., Fuccillo, D. (1976) *J. Virol.* **19,** 579.

Duffy, P., Wolf, J., Collins, G., Devoe, A.G., Streeten, B. and Cowen, D. (1974) *New Engl. J. Med.* **290,** 692.

Eckroade, R.J., ZuRhein, G.M. and Hanson, R.P. (1973) *J. Wildl. Dis.* **9,** 229; (1979) In: *Slow Transmissible Diseases of the Nervous System*, Vol. 1, p. 409. Ed. by S.B. Prusiner and W.J. Hadlow, Academic Press, New York.

Eckroade, R.J., ZuRhein, G.M., Marsh, R.F. and Hanson, R.P. (1970) *Science, N.Y.* **169,** 1088.

Eklund, C.M., Hadlow, W.J., Kennedy, R.C., Boyle, C.C. and Jackson, T.A. (1968) *J. infect. Dis.* **118,** 510.

Eklund, C.M., Kennedy, R.C. and Hadlow, W.J. (1967) *J. infect. Dis.* **117,** 15.

Field, E.J., Caspary, E.A. and Joyce, G. (1968) *Vet. Rec.* **83,** 109.

Field, E.J., Farmer, F., Caspary, E.A. and Joyce, G. (1969*b*) *Nature, Lond.* **222,** 90.

Field, E.J., Joyce, G. and Keith, A. (1969*a*) *J. gen. Virol.* **5,** 149.

Fraser, H. (1976) In: *Slow Virus Diseases of Animals and Man*, p. 267. Ed. by R.H. Kimberlin, North Holland, Amsterdam; (1979*a*) In: *Slow Transmissible Diseases of the Nervous System*, Vol. 1, p. 387. Ed. by S.B. Prusiner and W.J. Hadlow, Academic Press, New York; (1979*b*) *Progress in Neurological Research*, p. 194. Ed. by P.O. Behan and F. Clifford Rose, Pitman Medical, Bath; (1979*c*) In: *Aspects of Slow and Persistent Virus Infections*, p. 30. Ed. by D.A.J. Tyrrell, Martinus Nijhoff, The Hague.

Fraser, H. and Dickinson, A.G. (1968) *J. comp. Pathol.* **78,** 301; (1970) *Nature, Lond.* **226,** 462; (1973) *J. comp. Pathol.* **83,** 29; (1978) *Ibid.* **88,** 563.

Fraser, K.B. (1979) In: *Aspects of Slow and Persistent Virus Infections*, p. 76. Ed. by D.A.J. Tyrrell, Martinus Nijhoff, The Hague.

Gajdusek, D.C. (1977) *Science, N.Y.* **197,** 943; (1979) In: *Slow Transmissible Diseases of the Nervous System*, Vol. 1, p. 7. Ed. by S.B. Prusiner and W.J. Hadlow, Academic Press, New York.

Gajdusek, D.C. and Gibbs, C.J. Jr. (1973) *Perspect. Virol.* **8,** 279.

Gajdusek, D.C., Gibbs, C.J. Jr. and Alpers, M. (1966) *Nature, Lond.* **209,** 794.

Gajdusek, D.C. and Zigas, V. (1957) *New Engl. J. Med.* **257,** 974.

Gardiner, A.G. (1965) *Res. vet. Sci.* **7,** 190.

Gardiner, A.G. and Marucci, A.A. (1969) *J. comp. Pathol.* **79,** 233.

Gardner, S.D. (1973) *Brit. med. J.* **1,** 77; (1977) In: *Recent Advances in Clinical Virology*, p. 93. Ed. by A.P. Waterson, Churchill Livingstone, Edinburgh.

Gardner, S.D., Field, A.M., Coleman, D.V. and Hulme, B. (1971) *Lancet* **i** 1253.

Georgsson, G., Nathanson, N., Pálsson, P.A. and Pétursson, G. (1976) In: *Slow Virus Diseases of Animals and Man*, p. 61. Ed. by R.H. Kimberlin, North Holland, Amsterdam.

Gibbs, C.J. Jr. (1967) *Curr. Top. Microbiol.* **40,** 44.

Gibbs, C.J. Jr. and Gajdusek, D.C. (1973) *Science, N.Y.* **182,** 67.

Gibbs, C.J. Jr., Gajdusek, D.C. and Amyx, H. (1979) In: *Slow Transmissible Diseases of the Nervous System*, Vol. 2, p. 87. Ed. by S.B. Prusiner and W.J. Hadlow, Academic Press, New York.

Gibbs, C.J. Jr., Gajdusek, D.C. and Latarjet, R. (1978) *Proc. natn. Acad. Sci., Wash.* **75,** 6268.

Gibbs, C.J. Jr., Gajdusek, D.C. and Morris, J.A. (1965) *Slow Latent and Temperate Virus Infections*, p. 195. Ed. by D.C. Gajdusek, C.J. Gibbs Jr. and M. Alpers, US Gov. Print. Off. Wash., DC.

Gibbs, C.J. Jr. *et al.* (1968) *Science, N.Y.* **161,** 388.

Gordon, W.S. (1946) *Vet. Rec.* **58,** 516; (1957) *Ibid.* **69,** 1324; (1966) *Report of Scrapie Seminar*, p. 8 and 19. ARS 91-53, US Dept. Agric., Wash., DC.

Gorham, J.R., Henson, J.B., Crawford, T.B. and Padgett, G.A. (1976) In: *Slow Virus Diseases of Animals and Man*, p. 135. Ed. by R.H. Kimberlin, North Holland, Amsterdam.

Gorham, J.R., Leader, R.A., Padgett, G.A., Burger, D. and Henson, J.B. (1965) *Slow, Latent and Temperate Virus Infections*, p. 279. Ed. by D.C. Gajdusek, C.J. Gibbs, Jr. and M. Alpers, US Gov. Print. Off., Wash., DC.

Greenfield, J.G. (1950) *Brain* **73,** 141.

Greig, J.R. (1950) *J. comp. Pathol.* **60,** 263.

Gresser, I. and Pattison, I.H. (1968) *J. gen. Virol.* **3,** 295.

Gribble, D.H., Haden, C.C., Schwartz, L.W. and Henrickson, R.V. (1975) *Nature, Lond.* **254,** 602.

Griffin, D.E., Narayan, O. and Adams, R.J. (1978*a*) *J. infect. Dis.* **138,** 340.

Griffin, D.E., Narayan, O., Bukowski, J.F., Adams, R.J. and Cohen, S.R. (1978*b*) *Ann. Neurol.* **4,** 212.

Gudnadóttir, M. (1974) *Progr. med. Virol.* **18,** 336.

Gudnadóttir, M. and Pálsson, P.A. (1966) *J. Immunol.* **95,** 1116.

Haase, A.T. (1975) *Curr. Top. Microbiol. Immunol.* **72,** 101.

Haase, A.T., Stowring, L., Narayan, O., Griffin, D. and Price, D. (1977) *Science, N.Y.* **195,** 175.

Hadlow, W.J. (1959) *Lancet* **ii,** 289.

Hadlow, W.J. and Karstad, L. (1968) *Canad. vet. J.* **9,** 193.

Hadlow, W.J., Race, R.E., Kennedy, R.C. and Eklund, C.M. (1979) In: *Slow Transmissible Diseases of the Nervous System*, Vol. 2, p. 3. Ed. by S.B. Prusiner and W.J. Hadlow, Academic Press, New York.

Hahn, E.C., Ramos, L. and Kenyon, A.J. (1977*a*) *Infect. & Immun.* **15,** 204; (1977*b*) *Arch. Virol.* **55,** 315.

Hall, W.W., Kiessling, W. and ter Meulen, V. (1978) *Nature, London* **272,** 460.

Hanson, R.P., Eckroade, R.J., Marsh, R.F., ZuRhein, G.M., Kanitz, C.L. and Gustafson, D.P. (1971) *Science, N.Y.* **172,** 859.

Hanson, R.P. and Marsh, R.F. (1974) In: *Slow Virus Diseases*, p. 10. Ed. by W. Zeman and E.H. Lennette, Williams and Wilkins, Baltimore.

Haralambiev, H., Ivanov, I., Vesselinova, A. and Mersmerski, K. (1973) *Zbl. VetMed., B.* **20,** 701.

Harter, D.H., Axel, R., Burny, A., Gulati, S., Schlom, J. and Spiegelman, S. (1973) *Virology* **52,** 287.

Hartsough, G.R. and Burger, D. (1965) *J. infect. Dis.* **115,** 387.

Hartsough, G.R. and Gorham, J.R. (1956) *Natl. Fur News* **28,** 10.

Helmboldt, C.F. and Jungherr, E.L. (1958) *Amer. J. vet. Res.* **19,** 212.

Henson, J.B., Gorham, J.R., Leader, R.W. and Wagner, B.M. (1962) *J. exp. Med.* **116,** 357.

Henson, J.B., Gorham, J.R., McGuire, T.C. and Crawford, T.B. (1976) *In. Slow Virus Diseases of Animals and Man*, p. 175. Ed. by R.H. Kimberlin, North Holland, Amsterdam.

Henson, J.B., Leader, R.W. and Gorham, J.R. (1961) *Proc. Soc. exp. Biol. Med.* **107,** 919.

Henson, J.B., Leader, R.W., Gorham, J.R. and Padgett, G.A. (1966*a*) *Path. Vet.* **3,** 289.

Henson, J.B., Williams, R.C. and Gorham, J.R. (1966*b*) *J. Immunol.* **94,** 344.

Hoare, M., Davies, D.C. and Pattison, I.H. (1977) *Vet. Rec.* **101,** 482.

Holman, H.H. and Pattison, I.H. (1943) *J. comp. Pathol.* **53,** 231.

Holmberg, C.A., Gribble, D.H., Takemoto, K.K., Howley, P.M., Espana, C. and Osburn, B.I. (1977) *J. infect. Dis.* **136,** 593.

Hornabrook, R.W. (1979) In: *Slow Transmissible Diseases of the Nervous System*, Vol. 1, p. 37. Ed. by S.B. Prusiner and W.J. Hadlow, Academic Press, New York.

Horta-Barbosa, L., Fuccillo, D.A., Sever, J.L. and Zeman, W. (1969) *Nature, Lond.* **221,** 974.

Hotchin, J. (Ed.) (1974) *Progress in Medical Virology: Slow Virus Diseases.* Karger, Basel.

Hotchin, J. and Buckley, R. (1977) *Science, N.Y.* **196,** 668.

Hourrigan, J., Klingsporn, A., Clark, W.W. and de Camp, M. (1979) In: *Slow Transmissible Diseases of the Nervous System*, Vol. 1, p. 331. Ed. by S.B. Prusiner and W.J. Hadlow, Academic Press, New York.

Hunter, G.D. (1972) *J. infect. Dis.* **125,** 427.

Hunter, G.D., Collis, S.C., Millson, G.C. and Kimberlin, R.H. (1976) *J. gen. Virol.* **32,** 157.

Hunter, G.D., Kimberlin, R.H. and Millson, G.C. (1972) *Nature, New Biol., London* **235,** 31.

Hunter, G.D. and Millson, G.C. (1977) In: *Recent Advances in Clinical Virology*, Vol. 1, p. 61. Ed. by A.P. Waterson. Churchill Livingstone, Edinburgh.

Jakob, A. (1921) *Z. ges. Neurol. Psychiat.* **64,** 147.

Johnson, R.T. and ter Meulen, V. (1978) *Advanc. int. Med.* **23,** 353.

Johnson, R.T., Narayan, O., Weiner, L.P. and Greenlee, J.E. (1977) In: *Slow Virus Infections of the Central Nervous System*, p. 91. Ed. by V. ter Meulen and M. Katz. Springer Verlag, New York.

Kahana, E., Alter, M., Braham, J. and Sofer, D. (1974) *Science, N.Y.* **183,** 90.

Karstad, L. and Pridham, T.J. (1962) *Canad. J. comp. Med.* **26,** 97.

Katz, M. and Koprowski, H. (1968) *Nature, Lond.* **219,** 639.

Katz, M., Rorke, L.B., Masland, W.S., Koprowski, H. and Tucker, S.H. (1968) *New Engl. J. Med.* **279,** 793.

Kenyon, A.J. (1966) *Amer. J. vet. Res.* **27,** 1780.

Kenyon, A.J., Gander, J.E., Lopez, C. and Good, R.A. (1973) *Science, N.Y.* **179,** 187.

Kenyon, A.J., Howard, E. and Buko, L. (1967) *Amer. J. vet. Res.* **28,** 1167.

Kenyon, A.J., Magnano, T., Helmboldt, C.F. and Buko, L. (1966) *J. Amer. vet. med. Ass.* **149,** 920.

Kimberlin, R.H. (Ed.) (1976*a*) *Slow Virus Diseases of Animals and Man.* North Holland, Amsterdam; (1976*b*) *Ibid.* p. 307; (1976*c*) *Scrapie in the Mouse.* Meadowfield Press Ltd., Shildon, Co. Durham, England; (1976*d*) *Science Progress, Oxford* **63,** 461; (1977) *Trends Biochem. Sci.* **2,** 220; (1979*a*) *Livestk. Prod. Sci.* **6,** 233; (1979*b*) In: *Slow Transmissible Diseases of the Nervous System,* Vol. 2, p. 33. Ed. by S.B. Prusiner and W.J. Hadlow, Academic Press, New York; (1979*c*) In: *Aspects of Slow and Persistent Virus Infections,* p. 4. Ed. by D.A.J. Tyrrell, Martinus Nijhoff, The Hague; (1981) *Brit. vet. J.* **137,** 105; (1982) *Nature* **297,** 107.

Kimberlin, R.H. and Cunnington, P.G. (1978) *FEMS Microbiol. Letts.* **3,** 169.

Kimberlin, R.H. and Marsh, R.F. (1975) *J. infect. Dis.* **131,** 97.

Kimberlin, R.H., Millson, G.C. and Hunter, G.D. (1971) *J. comp. Path.* **81,** 383.

Kimberlin, R.H. and Walker, C.A. (1977) *J. gen. Virol.* **34,** 295; (1978*a*) *Ibid.* **39,** 487; (1978*b*) *J. comp. Path.* **88,** 39; (1979*a*) *J. gen. Virol.* **42,** 107; (1979*b*) *J. comp. Pathol.* **89,** 551; (1979*c*) *Lancet* **ii,** 591; (1980) *J. gen. Virol.* **51,** 183.

Kimberlin, R.H., Walker, C.A. and Millson, G.C. (1975) *Lancet* **ii,** 1309.

Kirschbaum, W.R. (1968) *Jakob-Creutzfeldt Disease.* Elsevier, New York.

Klatzo, I., Gajdusek, D.C. and Zigas, V. (1959) *Lab. Invest.* **8,** 799.

Koprowski, H., Barbanti-Brodano, G. and Katz, M. (1970) *Nature Lond.* **225,** 1045.

Larsen, A.E. and Porter, D.D. (1975) *Infect. & Immun.* **11,** 92.

Latarjet, R. (1979) In: *Slow Transmissible Diseases of the Nervous System,* Vol. 2, p. 387. Ed. by S.B. Prusiner and W.J. Hadlow, Academic Press, New York.

Latarjet, R., Muel, B., Haig, D.A., Clarke, M.C. and Alper, T. (1970) *Nature, Lond.* **227,** 1341.

Leader, R.W., Wagner, B.M., Henson, J.B. and Gorham, J.R. (1963) *Amer. J. Path.* **43,** 33.

Lodmell, D.L., Bergman, R.K., Hadlow, W.J. and Munoz, J.J. (1971) *Infect. & Immun.* **3,** 221; (1973) *Ibid.* **8,** 769.

Lodmell, D.L., Hadlow, W.J., Munoz, J.J. and Whitford, H.W. (1970) *J. Immunol.* **104,** 878.

McDermott, J.R., Fraser, H. and Dickinson, A.G. (1978) *Lancet* **ii,** 318.

McFarland, D. and Hotchin, J. (1980) *Biol. Psychiat.* **15,** 37.

McFarlin, D.E., Raff, M.C., Simpson, E. and Nehlsen, S.H. (1971) *Nature, Lond.* **233,** 336.

McGuire, T.C., Crawford, T.B., Henson, J.B. and Gorham, J.R. (1971) *J. Immunol.* **107,** 1481.

Malmgren, R., Kurland, L., Mokri, B., Kurtzke, J. (1979) In: *Slow Transmissible Diseases of the Nervous System,* Vol. 1, p. 93. Ed. by S.B. Prusiner and W.J. Hadlow, Academic Press, New York.

Manuelidis, E.E. (1975) *Science, N.Y.* **190,** 571.

Manuelidis, E.E., Angelo, J.N., Gorgacz, E.J., Kim J.H. and Manuelidis, L. (1977) *New Engl. J. Med.* **296,** 1334.

Manuelidis, E.E., Gorgacz, E.J. and Manuelidis, L. (1978) *Science, N.Y.* **200,** 1069.

Manuelidis, E.E. and Manuelidis, L. (1979) In: *Slow Transmissible Diseases of the Nervous System,* Vol. 2, p. 147. Ed. by S.B. Prusiner and W.J. Hadlow, Academic Press, New York.

Marsh, R.F. (1976) In: *Slow Virus Diseases of Animals and Man,* p. 359. Ed. by R.H. Kimberlin, North Holland, Amsterdam.

Marsh, R.F., Burger, D., Eckroade, R., ZuRhein, G.M. and Hanson, R.P. (1969*a*) *J. infect. Dis.* **120,** 713.

Marsh, R.F., Burger, D. and Hanson, R.P. (1969*b*) *Amer. J. vet. Res.* 30, 1637.

Marsh, R.F. and Hanson, R.P. (1969) *J. Virol.* **3,** 176; (1974) *Science, N.Y.* **187,** 656; (1979) In: *Slow Transmissible Diseases of the Nervous System* Vol. 1, p. 451. Ed. by S.B. Prusiner and W.J. Hadlow, Academic Press, New York.

Marsh, R.F. and Kimberlin, R.H. (1975) *J. infect. Dis.* **131,** 104.

Marsh, R.F., Malone, T.G., Semancik, J.S., Lancaster, W.D. and Hanson, R.P. (1978) *Nature, Lond.* **275,** 5676.

Marsh, R.F., Miller, J.M. and Hanson, R.P. (1973) *Infect. & Immun.* **7,** 352.

Marsh, R.F., Pan, I.C. and Hanson, R.P. (1970) *Infect. & Immun.* **2,** 727.

Marsh, R.F., Semancik, J.S., Medappa, K.C., Hanson, R.P. and Rueckert, R.R. (1974) *J. Virol.* **13,** 993.

Marsh, R.F., Sipe, J.C., Morse, S.S. and Hanson, R.P. (1976) *Lab. Invest.* **34,** 381.

Masters, C.L., Harris, J.O., Gajdusek, D.C., Gibbs, C.J. Jr., Bernoulli, C. and Asher, D.M. (1979) *Ann. Neurol.* **5,** 177.

Matthews, W.B., Tomlinson, A.H. and Hughes, J.T. (1979) *Lancet* **ii,** 752.

Mehta, P.D., Kane, A. and Thormar, H. (1977) *J. Immunol.* **118,** 2254.

Meulen, V. ter and Hall, W.W. (1978) *J. gen. Virol.* **41,** 1.

Meulen, V. ter and Katz, M. (Eds.) (1977) *Slow Virus Infections of the Central Nervous System.* Springer Verlag, New York.

Millson, G.C., Hunter, G.D. and Kimberlin, R.H. (1971) *J. comp. Path.* **81,** 255; (1976) In: *Slow Virus Diseases of Animals and Man,* p. 243. Ed. by R.H. Kimberlin, North Holland, Amsterdam.

Millson, G.C., Kimberlin, R.H., Manning, E.J. and Collis, S.C. (1979) *Vet. Microbiol.* **4,** 89.

Millson, G.C. and Manning, E.J. (1979) In: *Slow Transmissible Diseases of the Nervous System,* Vol. 2, p. 409. Ed. by S.B. Prusiner and W.J. Hadlow, Academic Press, New York.

Mould, D.L., Dawson, A.McL. and Rennie, J. (1970) *Nature, Lond.* **228,** 779.

Narayan, O., Griffin, D.E. and Chase, J. (1977*a*) *Science, N.Y.* **197,** 376.

Narayan, O., Griffin, D.E. and Clements, J.E. (1978) *J. gen. Virol.* **41,** 343.

Narayan, O., Griffin, D.E. and Silverstein, A.M. (1977*b*) *J. infect. Dis.* **135,** 800.

Narayan, O., Silverstein, A.M., Price, D. and Johnson, R.T. (1974) *Science, N.Y.* **183,** 1202.

Nathanson, N., Panitch, H., Pálsson, P.A., Pétursson, G. and Georgsson, G. (1976) *Lab. Invest.* **35,** 444.

Nathanson, N., Pétursson, G., Georgsson, G., Pálsson, P.A., Martin, J.R. and Miller, A. (1979) *J. Neuropathol. exp. Neurol.* **38,** 197.

Neugut, R.H., Neugut, A.I., Kahana, E., Stein, Z. and Alter, M. (1979) *Neurology* **29,** 225.

Notani, G.W., Hahn, E.C., Sarkar, N.H. and Kenyon, A.J. (1976) *Nature, Lond.* **261,** 56.

Nussbaum, R.E., Henderson, W.M., Pattison, I.H., Elcock, N.V. and Davies, D.C. (1975) *Res. vet. Sci.* **18,** 49.

Obel, A.L. (1959) *Amer. J. vet. Res.* **20,** 384.

Osborn, J.E., Robertson, S.M., Padgett, B.L., ZuRhein, G.M., Walker, D.L. and Weisblum, B. (1974) *J. Virol.* **13,** 614.

Outram, G.W. (1972) *J. comp. Path.* **82,** 415; (1976) In: *Slow Virus Diseases of Animals and Man*, p. 325. Ed. by R.H. Kimberlin, North Holland, Amsterdam.

Outram, G.W., Dickinson, A.G. and Fraser, H. (1973) *Nature, Lond.* **241,** 536; (1974) *Ibid.* **249,** 855; (1975) *Lancet* **i,** 198.

Padgett, B.L. and Walker, D.L. (1973) *J. infect. Dis.* **127,** 467; (1976) *Progr. med. Virol.* **22,** 1.

Padgett, B.L., Walker, D.L., ZuRhein, G.M., Eckroade, R.J. and Dessel, B.H. (1971) *Lancet* **i,** 1257.

Padgett, G.A., Reiquam, C.W., Henson, J.B. and Gorham, J.R. (1968) *J. Path. Bact.* **95,** 509.

Pálsson, P.A. (1976) In: *Slow Virus Diseases of Animals and Man*, p. 17. Ed. by R.H. Kimberlin, North Holland, Amsterdam; (1979) In: *Slow Transmissible Diseases of the Nervous System*, Vol. 1, p. 357. Ed. by S.B. Prusiner and W.J. Hadlow, Academic Press, New York.

Panitch, H., Pétursson, G., Georgsson, G., Palsson, P.A. and Nathanson, N. (1976) *Lab. Invest.* **35,** 452.

Pattison, I.H., Hoare, M.N., Jebbett, J.N. and Watson, W.A. (1974) *Brit. vet. J.* **130,** 65.

Payne, F.E., Baublis, J.V. and Itabashi, H.H. (1969) *New Engl. J. Med.* **281,** 585.

Perier, O. and Vanderhagen, J.J. (1966) *Rev. Neurol.* **115,** 250.

Perryman, L.E., Banks, K.L. and McGuire, T.C. (1975) *Fed. Proc.* **34,** 870.

Peterson, D.A., Wolfe, L.G., Deinhardt, F., Gajdusek, D.C. and Gibbs, C.J. Jr. (1973) *Intervirol.* **2,** 14.

Pétursson, G., Martin, J.R., Georgsson, G., Nathanson, N. and Pálsson, P.A. (1979) In: *Aspects of Slow and Persistent Virus Infections*, p. 165. Ed. by D.A.J. Tyrrell, Martinus Nijhoff, The Hague.

Pétursson, G., Nathanson, N., Georgsson, G., Panitch, H. and Pálsson, P.A. (1976) *Lab. Invest.* **35,** 402.

Porter, D.D., Dixon, F.J. and Larsen, A.E. (1965) *Blood.* **25,** 736.

Porter, D.D. and Larsen, A.E. (1964) *Amer. J. vet. Res.* **25,** 1226; (1967) *Proc. Soc. exp. Biol. Med.* **126,** 680; (1974) *Progr. med. Virol.* **18,** 32.

Porter, D.D., Larsen, A.E., Cox, N.A., Porter, H.G. and Suffin, S.C. (1977) *Intervirology* **8,** 129.

Porter, D.D., Larsen, A.E. and Porter, H.G. (1969) *J. exp. Med.* **130,** 575; (1972) *J. Immunol.* **109,** 1.

Porter, D.D., Porter, H.G. and Cox, N.A. (1973) *J. Immunol.* **111,** 1407.

Prusiner, S.B., Garfin, D.E., Baringer, J.R. and Cochran, S.P. (1979) In: *Slow Transmissible Diseases of the Nervous System*, Vol. 2, p. 425. Ed. by S.B. Prusiner and W.J. Hadlow, Academic Press, New York.

Prusiner, S.B., Groth, D., Bildstein, C., Masiarz, F.R., McKinley, M.P. and Cochran, S.P. (1980) *Proc. natn. Acad. Sci., Wash.* **77,** 2984.

Prusiner, S.B. and Hadlow, W.J. (Eds.) (1979) *Slow Transmissible Diseases of the Nervous System*, Vols. 1 and 2. Academic Press, New York.

Prusiner, S.B., Hadlow, W.J., Eklund, C.M. and Race, R.E. (1977) *Proc. nat. Acad. Sci., Wash.* **74,** 4656.

Prusiner, S.B. *et al.* (1978) *Biochemistry* **17,** 4993.

Renwick, C.C. and Zlotnik, I. (1965) *Vet. Rec.* **77,** 984.

Richardson, E.P. Jr. (1974) *Ann. N.Y. Acad. Sci.* **230,** 358.

Sanger, H.L., Klotz, G., Riesner, D., Gross, H.J. and Kleinschmidt, A.K. (1976) *Proc. nat. Acad. Sci., Wash.* **73,** 3852.

Savage, R.D. and Field, E.J. (1965) *Anim. Behav.* **13,** 443.

Schluederberg, A., Chavanich, S., Lipman, N.B. and Carter, C. (1974) *Biochem. biophys. Res. Commun.* **58,** 647.

Semancik, J.S., Marsh, R.F., Geelen, J.L.M.C. and Hanson, R.P. (1976) *J. Virol.* **18,** 693.

Shahrabadi, M.S., Cho, H.J. and Marusyk, R.G. (1977) *J. Virol.* **23,** 353.

Siakotos, A.N., Raveed, D., Traub, R.D. and Longa, G. (1979) *J. gen. Virol.* **43,** 417.

Sigurdsson, B. (1954) *Brit. vet. J.* **110,** 255, 307, 341.

Sigurdsson, B., Pálsson, P.A. and Grímsson, H. (1957) *J. Neuropath. exp. Neurol.* **16,** 389.

Sigurdsson, B., Pálsson, P.A. and Van Bogaert, L. (1962) *Acta Neuropathol.* **1,** 343.

Sigurdsson, B., Thormar, H. and Pálsson, P.A. (1960) *Arch. ges. Virusforsch.* **10,** 368.

Silverman, L. and Rubinstein, L.J. (1965) *Acta Neuropath.* **5,** 215.

Stamp, J.T. (1962) *Vet. Rec.* **74,** 357.

Stamp, J.T., Brotherston, J.G., Zlotnik, I., Mackay, J.M.K. and Smith, W. (1959) *J. comp. Path.* **69,** 268.

Stowring, L., Haase, A.T. and Charman, H.P. (1979) *J. Virol.* **29,** 523.

Suckling, A.J., Bateman, S., Waldron, C.B., Webb, H.E. and Kimberlin, R.H. (1976) *Brit. J. exp. Path.* **57,** 742.

Tabel, H. and Ingram, D.G. (1970) *Canad. J. comp. Med.* **34,** 329.

Tabel, H., Ingram, D.G. and Fletch, S.M. (1970) *Canad. J. comp. Med.* **34,** 320.

Tateishi, J., Sato, Y., Koga, M., Ohta, M. and Kuroiwa, Y. (1979) In: *Slow Transmissible Diseases of the Nervous System*, Vol. 2, p. 175. Ed. by S.B. Prusiner and W.J. Hadlow, Academic Press, New York.

Tellez-Nagel, I. and Harter, D.H. (1966) *Science, N.Y.* **154,** 899.

Terry, R.D. and Davies, P. (1980) *Annu. Rev. Neurosci.* **3,** 77.

Thormar, H. (1976) In: *Slow Virus Diseases of Animals and Man*, p. 97. Ed. by R.H. Kimberlin, North Holland, Amsterdam.

Thormar, H., Arnesen, K. and Mehta, P.D. (1977) *J. infect. Dis.* **136,** 229.

Thormar, H., Lin, F.H. and Trowbridge, R.S. (1974) *Progr. med. Virol.* **18,** 323.

Trautwein, G.W. and Helmboldt, C.F. (1962) *Amer. J. vet. Res.* **23,** 1280.

Trautwein, G., Schneider, P. and Ernst, E. (1974) *Zbl. VetMed.* **21,** 467.

Tyrrell, D.A.J. (Ed.) (1979) *Aspects of Slow and Persistent Virus Infections.* Martinus Nijhoff, The Hague.

Vandvik, B. and Norrby, E. (1973) *Proc. Nat. Acad. Sci., Wash.* **70,** 1060.

Walker, D.L., Padgett, B.L., ZuRhein, G.M., Albert, A.E. and Marsh, R.F. (1974) *Slow Virus Diseases*, p. 49. Ed. by W. Zeman and E.H. Lennette. Williams and Wilkins, Baltimore.

Ward, R.L., Porter, D. D. and Stevens, J.G. (1974) *J. Virol.* **14,** 1099.

Wechsler, S.L. and Fields, B.N. (1978) *Nature, Lond.* **272,** 458.

Weiner, L.P. *et al.* (1972) *New Engl. J. Med.* **286,** 385.

Weiss, M.J., Zeelon, E.P., Sweet, R.W., Harter, D.H. and Spiegelman, S. (1977) *Virology* **76,** 851.

Wilson, D.R., Anderson, R.D. and Smith, W. (1950) *J. comp. Path.* **60,** 267.

Wiśniewski, H.M., Bruce, M.E. and Fraser, H. (1975) *Science, N.Y.* **190,** 1108.

Worthington, M. (1972) *Infect. & Immun.* **6,** 643.

Worthington, M. and Clark, R. (1971) *J. gen. Virol.* **13,** 349.

Yoon, J.-W., Dunker, A.K. and Kenyon, A.J. (1975) *Virology* **64,** 575.

Yoon, J.-W., Kenyon, A.J. and Good, R.A. (1973) *Nature, New Biol.* **245,** 205.

Zeman, W. and Lennette, E.H. (Eds.) (1974) *Slow Virus Diseases.* Williams and Wilkins, Baltimore.

Zigas, V. and Gajdusek, D.C. (1957) *Med. J. Aust.* **ii,** 745.

Zlotnik, I. (1958) *J. comp. Path.* **68,** 148.

Zlotnik, I., Grant, D.P., Dayan, A.D. and Earl, C.J. (1974) *Lancet* **ii,** 435.

Zlotnik, I. and Rennie, J.C. (1962) *J. comp. Path.* **72,** 360.

ZuRhein, G.M. and Chou, S.M. (1965) *Science, N.Y.* **148,** 1477.

104

Oncogenic viruses

John A. Wyke

Introductory

Viruses that produce tumours in animals have been known for over 70 years, but it is only in the last 20 years that they have excited great interest. The reasons for this are twofold. Firstly, evidence has accumulated implicating viruses as a cause of cancer in a wide range of vertebrate species and it was hoped that they might be of similar importance in the aetiology of human neoplasia. This concept was attractive, not least because it suggested that, if the causative organisms could be identified and characterized, then cancer incidence might be reduced by the prophylactic measures so successful against other infectious agents. Secondly, viruses appeared promising tools for laboratory investigation into the basic mechanisms underlying neoplastic change. This promise rested on several earlier developments in laboratory research. Inbreeding of laboratory animals produced some strains which showed high incidences of various virus-associated neoplasms. In addition, inbred strains permitted transplantation studies, which demonstrated that a single donor cell could produce a tumour in a recipient animal. This focused attention on changes which occurred at a sub-cellular level and provided the rationale for undertaking many studies on oncogenesis *in vitro* rather than in the whole animal. This switch in emphasis was facilitated by new tissue culture techniques and, at the same time, it was found that tissue culture was an ideal way of studying and measuring the cytopathic effects of viruses.

From these beginnings has burgeoned an enormous amount of work, a comprehensive survey of which is beyond the scope of this chapter, but which is summarized in standard texts on the subject (Gross 1970; Essex, Todaro and zur Hausen 1980, for clinical studies; and Tooze 1980, Klein 1980 and Weiss *et al.* 1982 for basic laboratory investigations). This discussion will consider the two broad divisions of tumour virology: 1) clinical and laboratory investigations into the viral aetiology of cancer, and 2) basic laboratory studies on the molecular mechanisms of virus-induced neoplasia. The two fields overlap considerably; clinical studies have used methods originally developed for basic laboratory work; and cell and molecular biologists are continually extending their investigations to new clinical phenomena.

Part 1 Virus-induced cancer in animals, including man

A Diversity of relations between oncogenic viruses and host animals

Virus-induced neoplasms have been described in members of all vertebrate classes. A diverse collection of viruses has been implicated and the neoplasms show a wide range of histological types (Tables 104.1 and 104.2). There is no obvious relation between the taxonomic status of the causative viruses and the types of tumour it induces. Members of all the maor subdivisions of DNA-containing viruses have been implicated in oncogenesis (Table 104.2, p. 515), but only one family of RNA-containing viruses—the retroviruses. (Table 104.1). The reasons for this lie in similarities between retroviruses and the oncogenic DNA viruses at certain stages of their life cycles. These must be understood before the epidemiology of viral oncognesis can be appreciated.

When DNA viruses infect susceptible cells virus multiplication and cell death are the usual results (Table 104.3, p. 516). However, in some instances, particularly of herpesvirus infection, a permissive cell may harbour a virus in an unexpressed form. These latent viruses are occasionally activated and complete their lytic life cycle but the mechanisms that determine the initiation, maintenance and breakdown of latency are unknown (Darby 1980; Marsden 1980). Another outcome of DNA virus infection is the induction of neoplasia in the cell and its descendants. This occurs only when virus infection can be initiated but not completed, either because the virus is defective or because the cell does not allow the later stages of virus replication. In many cases the tumour cells produced by virus infection carry all or part of the viral DNA inserted into their chromosomal DNA. The tumour cells do not produce virus but the viral genetic material is passed on at cell division in the same way as normal cell genes. This phenomenon resembles the insertion of prophage DNA into bacterial chromosomes, and tumour virologists have adopted the bacterial geneticist's name for it, integration.

After retroviruses penetrate the cell the RNA genome of the virion is converted into a double-stranded DNA copy, the provirus, by a virion enzyme, RNA-dependent DNA polymerase (reverse transcriptase) (Temin and Baltimore 1972). The DNA provirus is then inserted permanently into the host chromosome. This obligatory integration of viral into cellular DNA is the single crucial way in which retroviruses resemble many oncogenic DNA viruses and differ from all other RNA-containing viruses. Some proviruses remain latent in the cell but usually progeny viral RNA is transcribed from the integrated provirus and new virions are released by budding through the cell membrane. This process is seldom cytopathic; so retrovirus-induced tumour cells can shed virus as well as transferring it to their progeny in proviral form. However, virus is not always produced; as with the DNA viruses, the tumours may be induced by defective viruses or the tumour cells may be of a type that do not allow the virus to complete its life cycle (see Table 104.3, p. 516).

B The epidemiology of oncogenic viruses in animals: general considerations

The pattern of disease induction by different tumour viruses in different animal and bird species varies greatly. Before we detail examples in a systematic account of the various tumour viruses (see below) we should consider the basic principles governing the incidence of virus-induced tumours in host populations.

The process of viral oncogenesis can be divided arbitrarily into three successive stages, each of which must be completed before clinical neoplasia is detected in the host. Stage one covers the events leading to the stable presence of viral DNA in the host cell, often in integrated form. This DNA is a prerequisite for the initiation, though not necessarily for the maintenance, of neoplasia (see above). The second stage comprises the factors that determine whether the presence of viral DNA leads to the conversion of a normal to a neoplastic cell. During stage 3 the neoplastic cell and its descendants must survive and multiply to form a tumour.

The major factors in stage one are those that also

Table 104.1 Some oncogenic RNA-containing viruses (retrovirus family)

Genus	Virus	Host of origin	Associated neoplasms
Type C oncovirus	Rous sarcoma virus	Fowl	Fibrosarcoma, spindle-cell sarcoma, glioma.
	Fujinami virus	Fowl	Myxosarcoma
	Lymphoid leucosis viruses	Fowl	Lymphoid leukaemia (B cells), erythroblastosis, osteopetrosis, nephroblastoma, sarcoma
	Myeloblastosis viruses	Fowl	Myeloblastosis, erythroblastosis
	Myelocytomatosis viruses (MC29, CM11, MH2, OK10)	Fowl	Myelocytomas, liver and kidney carcinomas, endothelioma, sarcomas of mesenchymal and haemopoietic tissues
	Erythroblastosis viruses	Fowl	Erythroblastosis, sarcomas
	Reticulo-endotheliosis virus complex	Fowl	Reticulo-endotheliosis, lymphoid leukaemia (B cells)
	Gross leukaemia viruses	Mouse	Thymic lymphosarcomas, lymphoid and myeloid leukaemias
	Graffi leukaemia virus	Mouse	Myeloid leukaemia
	Rauscher, Kirsten, Friend erythro-leukaemia viruses	Mouse	Erythroblastosis, polycythaemia, lymphoid leukaemia
	Moloney leukaemia virus	Mouse	Lymphoid leukaemia (T cells)
	Abelson leukaemia virus	Mouse	pre-B-cell lymphosarcoma
	Harvey, Kirsten sarcoma viruses	Mouse/Rat	Pleomorphic sarcomas
	Moloney sarcoma virus	Mouse	Rhabdomyosarcoma
	Myeloproliferative virus	Mouse	Erythroleukaemia, myeloid leukaemia
	FBJ sarcoma virus	Mouse	Osteosarcoma
	Feline leukaemia virus	Cat	Lymphosarcoma (mainly T cell)
	Feline sarcoma virus	Cat	Fibrosarcoma
	Bovine lymphosarcoma virus	Cattle	Lymphosarcoma, leukaemia (B cell)
	Simian sarcoma virus	Woolly Monkey	Fibrosarcoma
	Gibbon ape leukaemia viruses	Gibbons	Myeloid (granulocytic) leukaemia, lymphosarcoma
	Human T-cell leukaemia virus	Man	Adult T-cell leukaemia-lymphoma
Type B oncovirus	Mammary tumour virus	Mouse	Mammary adenocarcinoma

govern the spread of other infectious diseases; the amount of infectious virus in the hosts' environment and the susceptibility to infection of individual animals. These two factors are related, the density of the susceptible population helping to determine the virus levels in the environment. Individual susceptibility depends firstly on the genetic constitution of the host, which dictates whether or not the early stages of infection can occur. Secondly, susceptibility depends on whether the animal shows active or passive immunity to infection. In some retrovirus infections and in Marek's disease of fowls, where tumour-bearing hosts are often the source of infectious virus, the relation between virus spread and tumour incidence is obvious. With certain DNA virus infections this relationship may be absent or obscure because a virus that is able to replicate might not be tumorigenic and the animals with tumours do not serve as a source of infection, either because they do not permit virus replication or because the virus that infects them is defective.

The importance to tumour formation of stages two and three makes the epidemiology of oncogenic viruses more complex than that of many cytopathic agents. The first sources of complexity are the variable mechanisms by which viruses might induce tumours (summarized in Table 104.4, p. 528 and considered in detail in Part 2B). Thus the viruses with a gene encoding a protein that itself promotes tumour formation are more rapid and potent tumour inducers than those viruses whose oncogenic effect is by disruption of normal cell gene expression or whose action is indirect, either by suppressing host immunity or by stimulating abnormal cell proliferation. Not only do these latter agents induce tumours inefficiently, or with long latency, but their effect may also depend on other agents acting as co-carcinogens. Another factor that is important at stage two is the extent to which the host cell can regulate expression of integrated viral DNA. Examples will be mentioned below in which this regulation is affected by the genetic constitution, age and hormonal status of the host and by exposure to chemicals or x-irradiation.

There are probably many factors that influence stage 3, the growth of a neoplastic cell into a tumour, but they are poorly understood. Immune responses, both humoral and cellular, to novel antigens on tumour cells are probably important, for many viruses can only cause tumours experimentally in newborn or immunodeficient animals. However, the role of the immune system in affecting tumour growth is complex and in some instances may enhance rather than suppress the development of neoplasia. Concurrent disease of other kinds, nutritional state, age and hormonal factors may also play a role in the development of some tumours, both directly and by means of their effect on the immune system.

The interplay of different factors at each of these three stages of viral oncogenesis explains the variations in epidemiology of different virus-induced neoplasms. Where the incidence of disease is high enough to permit study three basic mechanisms of virus spread have been identified: 1 horizontal transmission by contact with infected animals or a contaminated environment; 2 vertical transmission from mother to offspring by congenital infection *in utero* or *in ovo*; 3 vertical transmission, from either parent to offspring, of viral DNA integrated in germ line cell DNA. The first two modes of transmission are not peculiar to tumour viruses; the third, in which a virus genome is inherited in the same way as normal cell genes, is probably unique to retroviruses.

C The epidemiology of cancer in man and the problem of identifying human tumour viruses

The epidemiology of viral neoplasia in animals has been studied mainly in domestic species (including fowl, cattle and cats) and laboratory animals (chiefly rodents and primates). With these species intensive husbandry and/or inbreeding have resulted in populations in which virus-induced tumours are common. Many of these tumours are caused by retroviruses. Haemopoietic and connective tissue are common targets for oncogenesis, and neoplasia frequently occurs in young adults as well as aged hosts. The pattern of tumour incidence in man is markedly different (Cairns 1978). The bulk of human tumours are carcinomas, whereas neoplasms of haemopoietic and connective tissues are rare. Moreover, tumours are very rare in young adults and their incidence increases logarithmically with age. This pronounced non-linear age dependence of tumour incidence suggests that multiple events are required. The early stages of neoplasia may occur many years before a tumour is detected, and hence may be very hard to identify. Furthermore, there is wide geographical variation in the incidence of most human tumours, and environmental rather than genetic differences seem to play the major role in this variation. The conclusions drawn from these and other epidemiological characteristics of human neoplasia are that physical or chemical carcinogens, whose levels vary in different human societies, are important causative factors; viruses probably play only a secondary role, if any, in the genesis of many human tumours.

It may, in fact, be very difficult to detect specific viruses in human tumours even if they are concerned in tumour aetiology. For example, a wide range of DNA-containing viruses has been shown to induce tumours in animals (Table 104.2); and related viruses, particularly in the *Adenoviridae*, *Herpesviridae* and *Papovaviridae* families, commonly infect a high proportion of human beings. Since most of these infections are cytocidal, neoplasia might result from a rare infection with a defective representative of these common human pathogens (Table 104.3, p. 516). If other events are necessary for tumour formation, this infection might have occurred long before a tumour was detected making it extremely difficult to identify the infection or assess its significance. In the few human tumours where there is good evidence for viral involvement, the virus implicated is indeed a common human pathogen that is neoplastic only in special circumstances or in combination with other contributory factors.

Even when viruses are detected in human tumours there are still difficulties in showing that these agents are causally related to neoplasia and are not contaminants or non-pathogenic passenger viruses whose growth is encouraged in tumour cells. The virus and the host cell must exist in one of the specific non-cytocidal relations outlined above (Table 104.3). If the agent detected is a DNA virus, then it will either be defective or the host will usually not permit its replication, so it will be stranded in the tumour, incapable of transmission. The same situation might apply to many retroviruses, but even if the virus can be isolated from the tumour it may be difficult to demonstrate its oncogenicity in test animals for two reasons. Firstly, a virus tumorigenic in one species may be quite innocuous in others. Secondly, if the virus is only one of several factors in oncogenesis, then it might produce

Table 104.2 Some oncogenic DNA-containing viruses

Family	Virus	Host of origin	Associated neoplasms
Papovavirus	Papilloma viruses	Man	Cutaneous, genital and laryngeal warts; some may progress to carcinomas
		Cattle (horse)	Cutaneous, genital and alimentary papillomas and fibropapillomas, carcinoma, haemangiomas, equine sarcoid (?)
		Other mammals	Papillomas, may progress to carcinomas (Shope papilloma)
	Polyoma virus	Mouse	Adenomas, adenocarcinomas, haemangiomas, fibromas, fibrosarcomas
	SV40	Old-world monkeys	Fibrosarcomas, ependymomas (in hamsters)
	BK virus, JC virus	Man	Tumours (mainly neurectodermal) in rodents and primates
Adenovirus	Various strains of human, simian, bovine and avian origin		Sarcomas (in hamsters)
Herpesvirus	Frog herpesvirus	Leopard frog	Adenomas, adenocarcinomas, osteochondroma (?)
	Marek's disease	Fowl	Neurolymphomatosis
	Guinea-pig herpesvirus	Guinea-pig	Lymphocytic leukaemia
	Herpesvirus ateles	Spider monkey	Lymphoma, lymphoblastic leukaemia
	Herpesvirus saimiri	Squirrel monkey	Lymphoma, lymphoblastic leukaemia
	Herpes simplex 1	Man	Squamous cell carcinoma (??)
	Herpes simplex 2	Man	Cervical carcinoma (?)
	Epstein – Barr virus	Man	Burkitt's lymphoma, nasopharyngeal carcinoma
	Cytomegalovirus	Man	Kaposi's sarcoma (?)
Hepatitis virus	Hepatitis B group	Man, apes, rodents, ducks	Hepatocarcinoma
Poxvirus	Shope fibroma	Rabbit	Fibroma*
	Yaba virus	Rhesus monkey	Fibroma-like nodular hyperplasia*
	Molluscum contagiosum	Man	Nodular epidermal hyperplasia*

The evidence that these viruses are causally associated with the tumours listed is, in most cases, strong if not complete. However (?) indicates reasonable doubt and (??) considerable doubt that the virus plays an aetiological role in tumour production. Cell responses that are hyperplastic rather than neoplastic are denoted by.*

a tumour only in concert with other contributory factors, and these may not be known.

These considerations make it likely that a human tumour virus may be unable to fulfil Koch's postulates. Henle (1971) suggested a modification of Jacob Henle's (1840) original postulates. These are:

1 the virus, virus-determined antigen(s), or virus nucleic acid must be present in all tumour cells:
2 antibodies to virus-determined antigens must occur at higher frequency and/or titres in subjects with malignancy than in control individuals:
3 the virus must be able to transform normal cells *in vitro*:
4 the virus must induce tumours in subhuman primates.

To these one might add a requirement that the incidence of the tumour is that expected of an infectious disease.

Any virus fulfilling these criteria would be strongly, though nor certainly, implicated in the causation of the tumour. Nevertheless a virus concerned in a human neoplasm might yet fail to fulfil these postulates, particularly criteria 3 and 4. These drawbacks limit the usefulness of Henle's postulates and this should be borne in mind as we now survey those agents proven or claimed to be oncogenic in various species.

D A brief catalogue of clinically important oncogenic viruses

i Avian retroviruses Retroviruses have been isolated from fish, reptiles, birds and mammals, but their pathogenic potential has been convincingly demonstrated only in the latter two classes. The *Retroviridae* family is defined by the possession of RNA-dependent polymerase which converts the RNA genome into a DNA provirus. In this DNA form retroviruses can

Table 104.3 Interactions between tumour viruses and hosts

DNA viruses (Papova, Adeno, Herpes)	1) Competent virus, permissive cell. a) **Replication and cell lysis** (e.g. polyoma virus in mouse) b) **Latent infection** (e.g. Herpes simplex virus, Epstein–Barr virus) 2) Competent virus, non-permissive cell **No replication, rare tumour formation** (e.g. polyoma in hamster) 3) Defective virus, permissive or non-permissive cell **No replication, rare tumour formation** (e.g. polyoma mutants, inactivated herpesviruses)
RNA viruses (Retro)	1) Competent virus, permissive cell a) **Replication only** (non-tumorigenic virus or tumorigenic virus in non-'target' cell e.g. avian erythroblastosis virus + helper in macrophages) b) **Replication and tumour formation** (e.g. Rous sarcoma virus in chickens) c) **Latent infection** (e.g. some endogenous viruses) 2) Competent virus, non-permissive cell **No replication, rare tumour formation** (e.g. Rous sarcoma virus in mammals) 3) Defective virus, permissive or non-permissive cell **No replication, tumour formation** (e.g. most transforming viruses such as avian erythroblastosis virus, murine sarcoma virus, in the absence of helper).

enter into a chronic non-cytocidal association with the host cell that is probably crucial to their oncogenicity (see above). The oncogenic avian retroviruses were the first tumour viruses to be discovered, and in many aspects of tumour virus research they have maintained their pre-eminence.

The first generally accepted induction of neoplasia by inoculation of a cell-free tumour filtrate was accomplished by Rous (1911), studying a transplantable spindle cell sarcoma of a chicken. Fujinami and Inamoto (1914) and others, as well as Rous, then demonstrated that filtrable agents were responsible for other sarcomas in fowl (reviewed in Weiss *et al.* 1982). The virus isolated from Rous's tumour 1 has now given rise to a number of distinct laboratory strains; it has been shown to be a retrovirus and is known as Rous sarcoma virus (RSV). Similar studies defined a group of retroviruses, related to RSV and to one another, which cause lymphoid leucosis (visceral lymphomatosis) in chickens. This is a disease in which affected birds are found to have an enlarged liver and spleen, and nodules or diffuse infiltrations of primitive B lymphoid cells in the viscera, with or without leukaemia. Erythroblastosis is also seen, and some strains of 'lymphoid leucosis' viruses seem to elicit mainly osteopetrosis or nephroblastomas. With the exception of some osteopetrosis tumours, these diseases have a long and variable latency of several months, distinguishing the lymphoid leucosis viruses from a serologically related group which are sometimes called acute leukaemia viruses, because they produce haemopoietic neoplasms in a few weeks.

The acute leukaemia viruses are of three types (Graf and Beug 1978, Graf and Stehelin 1982). (a) One type causes diffuse myeloblastosis, often associated with leukaemia and anaemia. (b) A second type is a complex of closely related viruses, MC29, CM11, MH2 and OK10 that cause myelocytomas, endotheliomas, hepatocarcinomas, adenocarcinomas and soft tissue sarcomas. (c) The third type of acute leukaemia virus causes erythroblastosis, with an accompanying erythroblastic leukaemia. The viral aetiology of avian erythroblastosis was described three years before the discovery of RSV (Ellermann and Bang 1908), but it received little attention and the disease was not thought to be neoplastic. Intravenous inoculation of any of the acute leukaemia viruses results in the rapid appearance of the appropriate neoplasm. However, when injected subcutaneously these viruses, with the exception of the myeloblastosis virus, will also induce sarcomas at the site of inoculation. These agents are sometimes called defective leukaemia viruses, because they all lack some of the functions needed for their replication, and can cause spreading infections only when they are accompanied by a 'helper' virus, usually a lymphoid leucosis virus, which supplies the missing functions.

The natural occurrence of avian sarcomas and acute leukaemias is rare and sporadic. Lymphoid leucosis, however, was once a common and commercially important disease. Horizontal spread of virus occurs between non-immune fowls, but the adult chickens normally produce an immune response and do not develop tumours. However, virus may persist in ovarian follicles and oviducts, infecting the embryos, which are then tolerant to the virus and are viraemic on hatching (Rubin *et al.* 1962). These birds, and those chicks contacting them early in life, develop leucosis about 6 months later, so the disease spreads mainly by congenital infection and horizontal transmission soon after hatching.

Pathogenic avian sarcoma and leucosis viruses are closely related to non-pathogenic or very slightly pathogenic agents, which are derived from proviruses that are present in latent form in all domestic chickens. These endogenous viruses are vertically transmitted from parent to offspring in the germ line DNA. Their gene functions are expressed to varying degrees in different strains of chickens. In some strains they are inactive; in others they express the whole viral genome, and generate progeny virus which produces a harmless viraemia in the host (Vogt and Friis 1971). Endogenous proviruses seem to have been acquired by chance horizontal infection from an unknown source during evolution of the domestic fowl. Closely related birds in the genus *Gallus* may lack such endogenous viral genomes, whereas more distantly related birds have acquired them (Frisby *et al.* 1979). Since these viruses are ubiquitous in chickens it is not surprising that they are harmless. Natural selection would operate against any birds carrying pathogenic proviruses unless the expression of the pathogen was strongly suppressed or unless the virus conferred some benefit on the host. However, it is possible to breed healthy chickens that have lost their endogenous viral genomes, so they appear to be of no importance to the individual host and are simply exquisitely well adapted passengers. They may none the less play a role in the evolution of the exogenously infecting pathogenic avian retroviruses.

Members of a more distantly related group of retroviruses have been implicated in reticuloendotheliosis of turkeys, leucosis of chickens and infectious anaemia and spleen necrosis in ducks (Weiss *et al.* 1982). Biochemical tests show resemblances between reticuloendotheliosis complex viruses and mammalian retroviruses, so reticuloendotheliosis viruses may have spread to birds from mammalian hosts or *vice versa.*

ii Murine leukaemia and sarcoma viruses Studies on inbred mouse lines have revealed variation in the incidence of leukaemia. Strains such as C58 and AKR show a very high incidence; in strains such as C3H/He the disease is rare. Early attempts to transmit leukaemia in cell-free filtrates from high to low incidence strains were unsuccessful until Gross (1951), using newborn C3H mice as recipients for filtrates from AKR leukaemia cells, showed that thymic lymphosarcomas developed several months after inoculation (in thymectomized recipients myeloid leukaemias may develop instead). Similar murine leukaemia viruses (MLV) were isolated from the leukaemias which arose rarely in low leukaemia incidence strains, such as C3H, either late in life or after exposure to x-irradiation. Unlike the pathogenic avian retroviruses, which spread horizontally or congenitally, these MLV are transmitted vertically from parent to offspring as endogenous DNA proviruses integrated in the host germ line DNA. Thus the first of the three stages of viral oncogenesis outlined above has already occurred in mice.

Several factors explain the persistence of these endogenous viruses despite their association with disease. A number of genes, both viral and host, determine the transcription of proviral DNA and the suceptibility of host cells to spreading virus infection (reviewed by Lilly and Mayer 1980, Pincus 1980). As a result viral oncogenesis is blocked at stage 2, and these viruses remain latent in most mice unless activated by agents such as chemicals or x-irradiation. Even inbred strains that have virus activation early in life and high leukaemia incidence do not show disease until late in life (Rowe and Pincus 1972, Staal, Hartley and Rowe 1977), so their ability to reproduce is not greatly impaired and natural selection is less effective in eliminating populations carrying the provirus. Moreover, most vertically transmitted endogenous proviruses, like those of birds, are not themselves pathogenic. However, a number of different proviruses are present in the host genome and when virus expression occurs in high leukaemia strains these proviruses appear to recombine genetically with one another. Some of these recombinant viruses are tumorigenic and their appearance immediately precedes the development of neoplasia (Hartley *et al.* 1977). It is these agents that cause leukaemia after experimental transmission to newborn mice.

Genetic recombination may likewise have played a role in the appearance of other murine retroviruses which, unlike Gross MLV, were obtained not from naturally occurring leukaemias but from transplantable tumours such as sarcomas and Ehrlich ascites cells. It was thought that these agents, the Graffi, Moloney, Rauscher, Friend and Kirsten viruses, might induce neoplasms similar to those from which they were isolated. However, they were found to induce haemopoietic tumours of various sorts (see Table 104.1, p. 513), and so probably played no causal role in their tumour of origin. Another isolate, Abelson virus, was derived from B-cell lymphosarcomas produced by infecting chemically thymectomized mice with Moloney MLV; lymphosarcoma is the disease it produces on transfer. All these new MLV differ from Gross MLV. For example, the Friend and Abelson viruses produce tumours with short latency and are unable to reproduce themselves fully, requiring a lymphoid leukaemia 'helper' virus. In these respects they resemble the defective avian acute leukaemia viruses.

Other defective viruses which are derived from MLV induce sarcomas in recipient animals. The Harvey and Kirsten strains of murine sarcoma virus were obtained after passaging a Moloney MLV and Kirsten erythroleukaemia virus respectively through rats. They appear to have acquired the ability to produce sarcomas by genetic recombination between MLV, rat genes and material in the rat genome which resembles an endogenous retrovirus (Scolnick *et al.* 1973, Scolnick and Parks 1974). Moloney sarcoma virus was likewise obtained from a rhabdomyosarcoma induced in BALB/c mice by injecting high doses of Moloney MLV. However, type FBJ osteosarcoma virus differs in that it was isolated from a naturally occurring murine osteosarcoma.

iii Oncogenic retroviruses of other species Leukaemia viruses similar to MLV have been isolated from rats, hamsters and guinea-pigs. Of more interest are the viruses which appear to be causally related to feline lymphosarcoma and fibrosarcoma and bovine lymphosarcoma. In both cats and cattle lymphosarcomas are the commonest malignant tumours. They are of social and economic importance; and they both show a clustered incidence, suggesting that a transmissible agent is responsible. In cats this agent has been identified as a retrovirus that can be spread horizontally producing disease in non-immune contacts (Hardy *et al.* 1973). The virus can be isolated from many, though not all, clinical cases (Essex *et al.* 1975). Infection is common, particularly among cats living in colonies or under semi-feral conditions; and it stimulates three types of response. The majority, particularly of adult animals, produce an immune response to viral structural antigens and to a non-viral antigen, FOCMA, found on tumour cell membranes. These animals are resistant to virus and remain disease free. Some cats develop immunity to viral antigens but not to FOCMA. This population eliminates the virus but is at risk of developing virus-negative lymphosarcoma after long latency. A small proportion of cats do not produce an effective immune response. They remain viraemic and among this group are the animals that develop the commoner, virus positive, lymphosarcoma and other virus associated diseases (various authors in Essex, Todaro and zur Hausen 1980). Since cats live in close proximity to man and since two subgroups of feline leukaemia virus can grow in human cells, these agents may yet have some bearing on human disease, though evidence for this has been sought and not found.

In cattle, transmission of leukaemia may occur both congenitally and horizontally. Spread occurs by contact and also by veterinary manipulations with instruments such as needles and cannulae and transmission of the virus seems to require contact between infected and uninfected cells. Retrovirus-like particles have been observed in leukaemic animals, but their causal relation to the disease is not yet fully understood. Virus infection is necessary, though not sufficient, for the causation of enzootic adult bovine leukosis; but the role of the virus is less certain in a similar disease that occurs sporadically in cattle of all ages (Burny *et al.* 1980). Retrovirus-like particles have also been observed in leukaemia of sheep and in lymphosarcomas of dogs and horses, but in the latter two species there is less evidence that they are concerned in the aetiology of the tumour.

Another important disease of ruminants is sheep pulmonary adenomatosis, a bronchoalveolar cell carcinoma that is horizontally transmitted. A retrovirus has been identified in these tumours and it is stated that virus-containing fractions from tumours can induce disease experimentally (reviewed by Perk and Hod 1982). If confirmed, this would extend the oncogenic effect of retroviruses to epithelia of endodermal origin, tissue types that are common sites for neoplasia in man.

Retroviruses isolated from sub-human primates form another interesting group, not least because closely related viruses have been isolated from human leukaemia, though there is little reason to associate these viruses causally with any human disease. Endogenous retroviruses have been obtained from both Old World and New World monkeys and appear to be non-pathogenic. An exogenous retrovirus-simian sarcoma virus, isolated from a pet woolly monkey with multiple fibrosarcomas, causes sarcoma when transmitted to squirrel monkeys and some marmosets (Theilen *et al.* 1973). Other exogenous agents, called gibbon ape leukaemia viruses, have been isolated on several occasions from captive gibbons (reviewed by Deinhardt 1980). The animals yielding virus have sometimes been clinically normal, but more often they have suffered from granulocytic leukaemia or lymphosarcoma. Moreover, most of the animals had been subject to experimental manipulation, either exposure to low doses of x-irradiation, injection of human malaria or leukaemic blood, or injection of brain extract from cases of human Kuru (see Chapter 103). The significance of these observations awaits elucidation.

iv Human T-cell leukaemia virus Over the last decade there have been many reports of retroviruses isolated from human tumours. Whilst some isolates are related to the primate viruses mentioned above, others have simply been laboratory contaminants. Recently, however, novel retroviruses, named human T-cell leukaemia viruses (HTLV), have been detected in tumour cells cultured from American patients with forms of cutaneous T-cell lymphoma (mycosis fungoides) and leukaemia (Sezary syndrome) (Poiesz *et al.* 1980, 1981). These aggressive tumours occur in adults; they comprise relatively mature T-lymphoid cells and they are rare in most other western countries. However, a related but distinguishable syndrome, adult T-cell leukaemia (ATL), occurs in Japan, with its incidence concentrated in the south western island of Kyushu and in people originating from that region. It was soon shown that tumour cells from ATL patients could also yield a retrovirus (Miyoshi *et al.* 1981) and that sera from these patients reacted with HTLV (Robert-Guroff *et al.* 1982). Furthermore, another focus of HTLV-associated T-cell lymphoma-leukaemia was identified in patients of Caribbean origin (Catovsky *et al.* 1982). These findings have spurred investigations into the connection between HTLV infection and T-cell tumours (for reviews see Essex 1982 and Weiss *et al.* 1982).

The HTLV viruses isolated from Japanese ATL and from T-cell tumours in other parts of the world appear for the most part very similar. They are not endogenous human viruses and they are distinct from other known retroviruses (their closest relatives, on the basis of genome and virion structure, seem to be bovine leucosis virus and the avian retroviruses). When normal T-cells are co-cultivated with leukaemic T-cells the former become infected with HTLV and their growth in culture then becomes independent of T-cell growth factor, a medium constituent essential for the culture of normal T-cells and many malignant derivatives (Miyoshi *et al.* 1981). HTLV is thus a transforming virus and, like bovine leukaemia virus, its infectivity is probably low and reliant upon cell to cell contact (it has been pointed out that HTLV may thus pose a risk in blood transfusion).

Epidemiology provides additional indications for the role of HTLV in neoplasia: in cases of both Japanese and Caribbean origin evidence for virus infection is almost universal. However, in these patients virus is found in T-cells but not in B-lymphocytes (Gallo *et al.* 1982), suggesting that natural infection spreads horizontally and that the virus shows a preference for infecting T-rather than B-lymphocytes. Moreover, about a quarter of the normal population from HTLV endemic areas also shows serological evidence for virus infection and exhibits viral antigens in cultured T-lymphocytes (Gotoh, Sugamura and Hinuma 1982), whereas HTLV-infection outside these areas seems rare. This finding helps to explain the clustering of adult T-cell lymphoma-leukaemia cases and also suggests that man may be an adequate reservoir for HTLV infection, although natural infection of Japanese monkeys has also been observed.

In summary HTLV is a locally common horizontally transmitted virus that is implicated, although not yet conclusively, in the causation of a rare tumour. As with other human tumours associated with virus infection (see below) we do not yet understand why only a very small proportion of infected individuals develop neoplasia, but additional factors such as immune deficiency and others mentioned in Section 1B are likely to be important. We are also ignorant of the prevalence of HTLV, and its role in neoplasia, outside the endemic areas that have been identified so far. Preliminary evidence suggests that HTLV-like viruses may be more widespread and may be associated with a wider range of neoplasms than is so far indicated (see Report 1982*a*), but only further work will permit a full assessment of the importance of these viruses.

v Mouse mammary tumour virus The agents described above all belong to the type C genus of the oncovirus sub-family of retroviruses. Another genus of the sub-family *Oncovirinae* includes type B viruses such as mouse mammary tumour virus (MMTV), whose mature virions are distinguished from type C virions by possession of an eccentric rather than a centrally located nucleoprotein core.

The incidence of mammary adenocarcinoma, like leukaemia, shows great variation between different inbred mouse strains, Bittner (1936) showed that newborn mice of high incidence strains showed a low incidence when nursed on low incidence foster-mothers, and that low incidence strains showed a high incidence when fostered on high incidence mice. This implicated a milk-borne factor in the disease, subsequently shown to be a virus. This virus, MMTV-S, is thus transmitted horizontally to newborns, a situation analogous to oviduct infection by avian leucosis viruses. Other strains of MMTV are now known which are transmitted vertically as endogenous proviruses in the germ line (like murine leukaemia viruses) or both vertically and horizontally. The different MMTV strains show variable oncogenicity and some of the induced tumours are interesting in showing a hormone dependence e.g. tumours produced by the MMTV-P strain arise only during pregnancy (Hageman, Calafat and Daams 1972). As with the murine leukaemia viruses, host genetic factors are also important in determining susceptibility to MMTV and age of onset of the disease.

Retrovirus-like agents have been observed in both canine and simian mammary carcinomas, and in some human milks. There is no good evidence to show that they are responsible for the disease, but their presence may yet be of diagnostic or prognostic value in mammary carcinoma (Spiegelman *et al.* 1980).

vi Retroviruses and non-neoplastic disease Retroviruses are widespread and common among vertebrates, and have been implicated in disorders other than neoplasia. Thus type C viruses are occasionally associated with autoimmune disease. Immune complexes containing type C retroviral glycoproteins have been observed in autoimmune glomerulonephritis of mice. Complexes of antibody and retroviral proteins have also been reported in human systemic lupus erythematosus (Panem and Reynolds 1979), and in a similar syndrome in dogs. The feline leukaemia/sarcoma viruses that infect most cats cause neoplasia in only a minority. Other infected animals may show anaemia, glomerulonephritis, reproductive failure, immunosuppression, and secondary infections resulting from immune deficiency (various authors in Essex, Todaro and zur Hausen 1980).

The retrovirus family contains two sub-families in addition to the oncoviruses: the *Spumavirinae* (Foamy viruses) and the *Lentivirinae* (slow viruses). The former are apparently harmless, but cause syncytium formation and cell degeneration in tissue culture. The latter include the viruses causing maedi (interstitial pneumonitis) and visna (diffuse encephalomyelitis and demyelination) in sheep, and similar syndromes, together with arthritis, in goats (see Chapter 103).

vii Papovaviruses: Papillomavirus genus The name of the Papovaviridae family derives from the first two letters of the names of three of its members, *pa*pilloma virus, *po*lyoma virus and *va*cuolating agent, an early name for SV40. The family is unusual among DNA viruses in that most of its members are oncogenic, or potentially so. The papilloma viruses cause infectious warts in many mammals (see Table 104.2, p. 515) including man, and some birds. Papilloma virus infection induces cellular proliferation in the Malpighian layer of the skin. Very little virus is produced in these multiplying cells, but as the cells in the outer zone of the stratum Malpighi lose the ability to divide and start to keratinize, virus production often, though not always, increases. Many viruses also induce a dermal proliferative response which leads to the production of fibropapillomas. Papilloma viruses are unusual in their persistence in tumour cells in free DNA form, not integrated in the cell chromosome like most other tumour viruses.

Papilloma viruses show strong species and even tissue specificity. For example, Shope rabbit papilloma virus produces papillomas only in rabbit epidermis—and experimentally in hares—but not when inoculated elsewhere, whereas a separate virus species causes papillomas of rabbit oral mucosa. Twelve different types of human papilloma virus (HPV) have been identified so far (Tooze 1980; Campo 1982) and each induces a distinguishable spectrum of warts. For example, the viruses that cause common warts (HPV-2, HPV-4), deep plantar warts (HPV-1, HPV-4) and flat warts (HPV-3) are all distinct from HPV-6, the cause of both venereally-transmitted genital condylomata acuminata and laryngeal warts. The bovine papilloma viruses (BPV) show a similar complexity with fibropapillomas of skin and gut and papillomas of gut, bladder, penis and teats being caused by at least six different viruses (Jarrett 1981). Papilloma-like viruses have also been observed in the hyperplastic lesions that precede bovine ocular squamous cell carcinomas, but their causative role in this important cattle tumour has not yet been demonstrated (Ford *et al.* 1982). Bovine papilloma viruses have been identified in the common invasive dermal fibrosarcoma of horses, sarcoid (Lancaster, Olson and Meinke 1977), possibly providing a rare example of a single papilloma virus causing tumours naturally in different host species. However, cells from sarcoid have also been shown to contain a transforming retrovirus (Fatemi-Nainie, Anderson and Cheevers 1982) and the significance of these two agents remains in doubt.

Papilloma virus infection spreads horizontally, usually by contact with an infected animal or contaminated environment, though sometimes by an arthropod vector. The tumours are usually benign but may cause inconvenience. Papillomas on the teats of cows or the penis of bulls can, for instance interfere with the functions for which these animals were domesticated. Occasionally the tumours may progress to malignancy. Thus, it has been suggested that HPV-6 may play a role in some malignant neoplasms of the external genitalia (zur Hausen 1976); a concept supported by recent discoveries of HPV DNA in several anogenital cancers (Green *et al.* 1982; Zachow *et al.* 1982). Moreover, it has recently become apparent that many mild and moderate cervical dysplasias are associated with HPV infection in the absence of typical acuminate genital warts (reviewed in Campo 1982; Report 1982*b*). It is not yet known whether these HPV-induced lesions, resembling as they do cervical intraepithelial neoplasia, are among those that progress to malignant carcinoma.

The type of virus, the genetic constitution and immune status of the host and environmental factors may all be important in the development of malignancies associated with papillomavirus infections. This requirement for multiple factors in the oncogenic progression is well illustrated by examples from rabbits, cattle and man.

Shope papilloma virus is harder to isolate from warts in domestic rabbits than from wild cottontail rabbits. In the former the warts commonly develop into carcinomas. In the latter an immune response often leads to tumour regression and malignancy.

Many cattle in upland areas of Britain, particularly in western Scotland, have multiple upper alimentary tract papillomas. In 2–5 per cent of beasts these progress to carcinomas, sometimes with concomitant bladder malignancies (Jarrett 1980, 1981). In contrast, although 20 per cent of lowland cattle have alimentary papillomas, these are rarely multiple and do not become malignant. Alimentary carcinomas occur in areas infested with bracken fern, a plant that contains an unidentified radiomimetic toxin and a flavinoid, quercetin, that is tumorigenic in laboratory animals. Bracken ingestion is known to cause bladder tumours and it may thus provide the co-carcinogens for BPV-associated malignancies. Experiments to test this possibility suggest that an immunosuppresant drug, azathioprine, increases the number of papillomas in animals, and the bracken radiomimetic toxin may do the same, but the effects of quercetin and bracken itself, with or without BPV, have yet to be evaluated. Since bracken is an item of human diet in parts of the Far East, this work is of more than veterinary significance.

Like these bovine cancers, the rare human disease epidermodysplasia verruciformis, suggests roles for both immune suppression and co-carcinogens in papillomavirus-induced malignancy (Orth *et al.* 1980). Epidermodysplasia verruciformis is characterized by disseminated papillomas in childhood, typically of two types: flat warts caused by HPV-3 and reddish pityriasis-like plaques caused by HPV-5. The latter lesions, but not the former, may convert in early adult life to carcinomas, particularly in areas of the skin exposed to sunlight. Thus a physical co-carcinogen may be important, but only with the appropriate HPV strain. This concept is supported by the observation that the pityriasiform lesions in African cases of epidermodysplasia verruciformis seem to be due to a different virus, HPV-8 (Pfister *et al.* 1981) and rarely become malignant. Cases of epidermodysplasia verruciformis show some familial clustering, suggesting a genetic component, and they also usually show impaired cell-mediated immunity. This latter defect is probably important and it is interesting that patients under immune suppressive therapy (such as kidney transplant recipients) may develop widespread papillomavirus infections as well as infections with human polyomaviruses, cytomegalovirus and others (see below). These immune suppressed individuals also show a higher incidence of certain tumours. As we mentioned above for HTLV and as we shall stress below, virus-associated malignancy in man is frequently, if not invariably, linked with a greater or lesser degree of immune deficiency and often accompanies exposure to environmental carcinogens.

viii Papovaviruses: Polyomavirus genus Members of the genus Polyomavirus are smaller than, but otherwise similar to papilloma viruses and include polyoma, K virus, which causes pneumonia and liver lesions in baby mice, SV40, and some related human isolates. Polyoma was discovered by Gross (1953) while studying murine leukaemia, as an agent inducing salivary gland carcinomas. It was soon shown to produce a variety of tumours in several species of experimental animals (see Table 104.2, p. 515); hence its name (Stewart, Eddy and Borgese 1958). In its natural host, the mouse, infection is widespread and usually inapparent, the population being protected firstly by maternal antibodies and later in life by an effective immune response. However, newborn or immunosuppressed animals may show not only tumours after infection but also runting, anaemia, paralysis, nephritis and conjunctivitis.

Like polyoma, SV40 virus appears to cause inapparent infections in its natural hosts, Rhesus, Cynomolgus and cercopithecus monkeys, but it induces sarcomas in experimental animals, notably hamsters. Both polyoma and SV40 will transform a wide range of cells in tissue culture (see below), including human cells, suggesting that such events may occur more readily in the absence of immune responses. SV40 was at one time a common contaminant of commercially used simian tissue culture cells, and many persons who received inactivated or attenuated polio vaccines have antibodies to it. There has so far been no evidence of adverse effects among the enormous number of those who received these vaccines, but it is perhaps still too early to conclude that the virus is entirely innocuous to man.

Two viruses of human origin, BK and JC (reviewed by Howley 1980), are closely related to SV40. BK virus has been isolated only from immunodeficient persons, either cases of Wiskott-Aldrich syndrome or allograft recipients undergoing immunosuppressive therapy. JC has been isolated from the brains of patients with progressive multifocal leucoencephalopathy. Both these viruses are oncogenic in newborn hamsters and can transform certain cells *in vitro*, though there is as yet no evidence that they are pathogenic in man. Exposure to both viruses is common, seroconversion often occurring in childhood, so that up to 80 per cent of adults are said to have antibodies. However, as the case of Epstein-Barr virus illustrates (see below), the fact that an agent is nearly ubiquitous does not mean that it is always harmless.

ix Adenoviruses Since adenoviruses are common causes of inapparent or minor respiratory tract and conjunctival infections in man (see Chapter 97), the demonstration that some of these agents induced tumours in newborn hamsters was of great interest (Trentin, Yabe and Taylor 1962). Human adenoviruses can be divided into three groups on the basis of their oncogenicity; highly oncogenic (types 12, 18 and 31); weakly oncogenic (types 3, 7, 14, 16, 21), and non-oncogenic (types 1, 2, 5, 6, 11). The non-oncogenic types, though incapable of inducing tumours, can none the less transform cells in tissue culture. Adenoviruses of other species can also induce tumours in laboratory rodents. There is no evidence that any of these adenoviruses are involved in human cancer, nor is there evidence for enhanced adenovirus expression in human tumours (Green and Mackey 1977).

Vaccines against human adenoviruses have been prepared in monkey tissue culture cells, and, as with polio vaccines, have been contaminated with SV40. The SV40 appears to be an obligatory contaminant, 'helping' the growth of human adenovirus in the simian cells, which are otherwise non-permissive. Recombinant adeno-SV40 hybrid viruses have arisen in this way. These hybrid viruses have been a useful research tool, but so far seem to have no pathological significance.

x Lymphotropic herpesviruses (Gamma-herpesviruses) (see Chapter 88) This group contains two viruses of great importance to man: Marek's disease virus, the causative agent of a costly disease in commercial chicken flocks, which at one time caused losses of 200 million dollars per annum in the United States alone; and Epstein-Barr, virus, a significant human pathogen.

Marek's disease (fowl paralysis or neurolymphomatosis; reviewed by Nazerian 1980), like other herpesvirus infections, spreads horizontally, infecting birds by 4–6 weeks of age. The virus mainly replicates in, and causes degeneration of, feather follicle epithelium, and by this means is transmitted in dust or litter. A few days after exposure the birds develop a cytopathic infection of lymphocytes followed by splenomegaly and atrophy of the thymus and bursa of Fabricius. Though infected birds initially produce a humoral and cell-mediated immune response, the pathological changes lead to permanent immune suppression in susceptible hosts. Circulating lymphocytes contain the virus in latent form and about a week after infection T-lymphoblasts, bearing tumour-associated surface antigens, begin to proliferate and infiltrate viscera and nerves. The victims clinically show paralysis, often with ocular infiltration by white cells, and die a few weeks later. Birds infected with virus of low virulence may survive longer, and in some cases a high incidence of virus-associated atherosclerosis has been observed 3 to 7 months after infection (Minick *et al.* 1979).

The immune suppression early in disease is probably important to tumour development, for experimentally induced immune deficiency enhances virus tumorigenicity. Moreover, some inbred chicken strains are resistant to disease; these birds produce a competent immune response without permanent immunosuppression, even though they continue to harbour latent virus.

Epstein-Barr virus (EBV; reviewed in Epstein and Achong 1979) was first identified in biopsies from *Burkitt's lymphoma* (*BL*), a malignant B-cell lymphoma that is the commonest childhood tumour in Central Africa and New Guinea but is rare and sporadic elsewhere. Burkitt's lymphoma occurs almost entirely in children from the age of 2 years onwards. Epstein-Barr virus is extremely common in all parts of the world, many persons acquiring antibodies to it early in life with no sign of clinical disease. Once infected, the host carries the virus in a latent form in B-lymphocytes. Occasional spontaneous activation occurs, resulting in the release of virus into the saliva and its spread by contact to non-immune hosts. In countries with good hygienic practices infection is often delayed until adolescence or early adulthood. In these cases the virus causes an overt disease, infectious mononucleosis (glandular fever), typified by fever, sore throat, cervical lymphadenopathy, transient lymphoid proliferation, sometimes with spleen and liver involvement and shedding of virus. Patients produce antibodies to viral proteins, notably the viral capsid antigen (VCA) and the diffuse nuclear and cytoplasmic component (*D*) of the early antigen (EA), as well as heterophile antibodies which form the basis of the diagnostic Paul-Bunnell reaction. Later in infection antibodies appear to the Epstein-Barr nuclear antigen (EBNA); these remain high throughout life. Many features of the pathogenesis of infectious mononucleosis are unknown, particularly those during the incubation period. For example, the primary target cells for the virus have not yet been identified. None the less, the causal relation of EBV to this disease is unequivocal.

In contrast, the evidence that EBV plays a role in BL, though strong, remains circumstantial. The postulates of Henle, outlined above, are fulfilled in African BL as follows.

1 Multiple copies of EBV DNA are found in 98 per cent of tumours: whether or not any of this DNA is integrated in the host genome is still undecided. Viral antigens are not seen in tumours, but appear upon culturing cells, presumably as a result of virus activation in a small number of cells.

2. Antibody titres against EBV VCA and the restricted cytoplasmic component (*R*) of EA are higher in African BL cases than in controls. Retrospective studies have shown that these titres are of prognostic significance. When titres to *R* antigen decline after treatment the prognosis is far better than in cases where titres remain high. In these latter cases remission is usually succeeded by eventually fatal tumour recurrence. Prospective studies indicate that anti-VCA titres are often high months or years before the onset of disease, whereas anti-*R* titres rise with tumour appearance. Children whose anti-VCA titres are four times or more higher than the control population have a thirty-fold higher risk of developing BL.

3 Cultures of B-lymphocytes are 'immortalized' to permanently dividing lymphoblasts. EBV transformation thus differs from the transformation of morphology and behaviour which many other tumour viruses induce in already dividing tissue culture cells (described later).

4 EBV injected into marmosets or some other New World primates may induce inapparent infections or transient lymphocytosis, but frequently it induces lymphomatous tumours which resemble BL in many respects.

Despite this satisfactory fulfilment of Henle's postulates, several features of this association of EBV with BL counsel caution before assigning an aetiological role to the virus. Firstly, 2 per cent of African BL cases and most of the sporadic BL cases seen elsewhere in the world show no evidence of the virus in the tumours. This might be explained by postulating two forms of the disease: one that occurs commonly in Africa, and probably New Guinea, but rarely elsewhere, in which EBV is causally involved; the other occurring sporadically everywhere including Africa, and not associated with EBV. Secondly, though most of the world's population becomes infected with EBV, BL is restricted mainly to Central Africa and New Guinea. This lack of correspondence between the epidemiology of EBV and that of BL suggests that co-factors are concerned in tumour formation. The most likely co-factor is holoendemic malaria, which occurs only in Africa and New Guinea over a climatically determined range matching that of BL incidence. Moreover, since malaria is immunosuppressive and stimulates proliferation of B-cells, the targets for EBV infection, one can envisage the two agents as having synergistic effects on these cells. However, though nearly all African children in BL areas are exposed early in life to EBV and most are affected by malaria, only one case of BL occurs per 10000 children per year. Moreover, the tumours, when they occur, are apparently monoclonal, suggesting that a successful neoplastic change is very unusual. The few BL victims are not unusual cases of late primary exposure to EBV, because the prospective survey, mentioned above, shows that they have VCA titres higher than the normal population long before tumour onset. One suspects that other co-factors are responsible, such as genetic predisposition. It is worth mentioning here that most BL tumour cells show a cytogenetic marker, a translocation from chromosome No. 8 to the long arm of chromosome No. 14. The same translocation is seen in some other haematopoietic neoplasms, but not in EBV lymphoblastoid lines transformed *in vitro*.

Thus it is not pathognomic of either BL or EBV infection, but may have some relation to lymphoid malignancy in general, and may be the crucial step in this process (Klein 1979; see Part 2C). If so, then the cooperative effect of EBV and malaria in stimulating B cell proliferation may serve mainly to increase the number of targets for this cytogenetic alteration.

Since malaria and possibly other co-factors, reflected perhaps in the cytogenetic abnormalities, appear to be associated with BL, and since EBV is ubiquitous in the population at risk, can one argue that the virus is constantly associated with the tumour but not concerned in its aetiology? Such reasoning is unconvincing. Since the tumours are monoclonal and all tumour cells carry the virus, EBV is either present at tumour inception or the cells all become infected during the growth of the tumour clone. However, transformed B lymphocytes form only a small minority of the lymphocyte population in infected persons, and only a minute fraction of these transformed lymphocytes are activated to release virus. It is unlikely that this virus is sufficient to infect all the tumour cells, even if they are exceedingly susceptible. Other good evidence for a causal role for EBV comes from serology, namely the effect of high antiviral titres on both the risk of BL development and its prognosis. These, however, may simply be a consequence of factors which independently also affect the appearance and progression of BL. Clarification can come only from epidemiological evidence, such as the demonstration that BL can be eradicated by a vaccine against EBV. An effective vaccine might well be of use against infectious mononucleosis in developed countries, but it is harder to predict its effect on BL incidence in subjects whose immune responses are already distorted by malaria and whose exposure to the virus may occur very early in life. Indeed, it has been pointed out that BL would best be controlled by malaria eradication. Were this successful, it would then be hard to verify the true role of EBV in the disease.

Observations similar to those made with BL have associated EBV with another tumour, *nasopharyngeal carcinoma* (*NPC*), which, though particularly common in Southern China and North and East Africa, occurs sporadically elsewhere. Various histological types of NPC have been described. Epstein-Barr virus DNA has invariably been found in epithelial tumour cells of those classified as undifferentiated squamous cell carcinomas, whether of common or sporadic incidence, but not in other types. Moreover, these cases of NPC showed raised titres of EBV antibodies, particularly against VCA and the *D* component of EA. As with BL, co-factors have been invoked to explain variations in incidence of this tumour. Environmental influences such as dietary nitrosamines may be important but genetic predisposition also seems a major factor. The remarkably high incidence of the disease in Cantonese Chinese is maintained in first-generation immigrants to other parts of the world, but children of mixed marriages show an intermediate incidence.

An association of NPC with an increased frequency of certain histocompatibility antigens has been proposed for the Chinese but this association does not apply to North Africans in whom the disease is also common.

As with BL, a number of intriguing problems remain to be resolved before EBV can be accepted unequivocally as a causal agent in NPC. The reason why epithelial cells should be the target in this neoplasm, whereas the virus is found in B-lymphocytes in infectious mononucleosis and BL, is unknown. This resembles the situation in Marek's disease, in which the virus transforms T-lymphocytes but replicates in feather follicle epithelium. Moreover, though high-risk populations are exposed to EBV early in life, the tumour occurs mainly in middle-aged adults, more commonly men, so the time between first exposure and neoplasia is very long.

A prospective survey of NPC is needed to answer these questions. Nasopharyngeal carcinoma is the major neoplasm in men of Cantonese origin, the incidence in middle-aged males being 100 per 100000. It is also common in women and occurs in some other parts of the world. Since there are many millions of Cantonese extraction, NPC is a major health problem. In the absence of clearly defined environmental cofactors, there is a great incentive to prevent the disease by acting against EBV, the only controllable precipitating factor.

Epstein-Barr virus infection has also been implicated in the polyclonal lymphoproliferative disorders seen in immune-deficient persons and patients undergoing immunosuppression. Evidence that EBV plays a part in causing these 'lymphomas' is discussed by Klein and Purtilo (1981).

We have dwelt at length with the problems of implicating EBV in neoplasia. In part this is to illustrate the difficulties in assigning a pathogenic role to any candidate human tumour virus except the wart-forming papilloma virus, and in part because EBV is so far the best authenticated viral risk factor in any human malignancy.

Two other lymphotropic herpesviruses—herpesvirus saimiri (HSV) from squirrel monkeys and herpesvirus ateles (HVA) from spider monkeys—are of interest to experimentalists (Deinhardt and Deinhardt 1979, Sugden, Kintner and Mark 1979). These viruses cause non-pathogenic infections in their natural hosts early in life, but on transfer to several New World primates they cause lymphomas and lymphoblastic leukaemias which, unlike BL, are polyclonal. HVS and HVA are unrelated to EBV and also to several EBV-like agents reported in Old World monkeys and apes. They also differ from EBV in that their target cells are T- and not B-lymphocytes.

xi Other herpesviruses (see Chapter 88) Among the alpha herpesviruses, herpes simplex virus of man (HSV) types 1 and 2, equine herpesvirus 1, and infectious bovine rhinotracheitis virus have all been shown to transform cells *in vitro* (see part 2 of this chapter), and sero-epidemiological evidence has been used to link HSV with two human cancers. The evidence associating HSV type 1 with squamous cell carcinoma is extremely tenuous; the connection between HSV type 2 and carcinoma of the uterine cervix is stronger but still controversial (Goldberg and Gravell 1976; Thomas and Rawls 1978). Firstly, both cervical carcinoma and HSV type 2 infection are more prevalent in promiscuous women of lower social groups, suggesting that predisposition to both the tumour and the virus might be venereally transmitted. Secondly, an increased incidence of cervical anaplasia has been noted in women with detectable HSV-2 genital infections, and carcinoma cases tend to have higher titres of HSV-2 antibodies than controls. Thirdly, a large proportion of cervical carcinomas and/or pre-cancerous lesions have been reported to contain RNA homologous to part of the HSV-2 genome, or HSV-2 associated antigens. None the less, the association between virus infection and cervical carcinoma is by no means consistent, and workers in this field have reached conflicting conclusions. Even were the association constant this would not necessarily implicate HSV-2 causally with the tumour. However, the evidence so far is sufficiently persuasive to warrant major prospective studies and the consideration of vaccine prophylaxis, particularly since cervical carcinoma is one of the commonest neoplasms in women.

The beta herpesvirus *cytomegalovirus* (CMV), like EBV, usually causes latent infections in man, the majority in most populations showing evidence of this. However it may cause more serious disease, particularly in the fetus and newborn, and in persons with immune deficiency, where the virus may either establish new infections or be re-activated. Cytomegalovirus not only shows a predilection for the immunologically-incompetent but it can exacerbate this state, increasing the susceptibility to other infections. This virus has been implicated in the genesis of Kaposi's sarcoma and, although there is no firm evidence that it plays a causative role, its frequent detection in tumour cells (Boldogh *et al.* 1981) and its ability to transform fibroblasts in tissue culture (see Part 2A), suggest that it is not a chance infection and is not acting solely as an immune suppressant.

Kaposi's sarcoma is a variable tumour that is thought to arise from endothelial cells. In equatorial Africa, where it is epidemic and can comprise 10 per cent of all malignancies in certain areas, it occurs in patients of all ages, its prognosis depending on which of several pathological forms it presents (Taylor *et al.* 1971). One form, the lymphadenopathic variety, resembles Burkitt's lymphoma, occurring mainly in the young and pursuing an aggressive course. Outside Africa the tumour is rare: until recently it occurred typically as an indolent neoplasm in elderly men of Jewish or Mediterranean origin and it was then noted that immunosuppressed patients, such as renal allograft recipients, also had an unusually high incidence of the disease. Kaposi's sarcoma has now gained prominence as one component of a rapidly growing disease problem known as the acquired immune deficiency syndrome (AIDS), occurring mainly among male homosexuals, notably in the United States (reviewed by Ziegler 1982; Groopman and Gottlieb 1982).

Patients with this disease show an impaired cellular immunity. The reasons for this are unknown, but the association of the syndrome with sexual promiscuity and drug-taking and the frequent history of recurrent infections by viruses such as hepatitis B, HSV and CMV, often introduced venereally or intravenously, suggest that chronic immune stimulation by infectious agents may play a role. These immunodeficient patients often die from opportunistic infections by protozoa, fungi, atypical mycobacteria or various viruses. They also frequently develop Kaposi's sarcoma which is unlike the typical non-African form of the disease in affecting young men and in pursuing an aggressive course. Evidence of CMV infection is almost universal among homosexuals and since the virus may be a cause of AIDS it is not surprising that the Kaposi's sarcoma cases tested showed signs of CMV infection. However, the detection of viral genomes in tumour but not in normal tissue from the same patient (Drew *et al.* 1982) supports the findings on the classic form of the tumour that indicate a causal role for CMV.

Three other types of tumour that have been associated with virus infection with greater or lesser justification (see above) have also been reported at increased frequency in male homosexuals: Burkitt's lymphoma (EBV), oral squamous cell carcinoma (HSV-1) and rectal squamous cell carcinoma (HSV-2 and papillomaviruses) (see Ziegler *et al.* 1982). Thus, again, we see a link between immune deficiency and virus-associated neoplasms, and the homosexual with AIDS shares these hazards with the allograft patient receiving immunosuppressive drugs and to a lesser extent, the cancer patient undergoing chemotherapy. However, the homosexual population poses a greater problem in the numbers at risk, in the lack of medical surveillance and in the fact that they are common donors of blood products in the United States, perhaps explaining cases of AIDS in male haemophiliacs (Groopman and Gottlieb 1982). Moreover, there are still many puzzles about the syndrome. For instance, it is not known why AIDS has only recently been observed, but perhaps a variant of a virus such as CMV or a novel virus, similar to HTLV (Section 1D iv), is the explanation.

Two herpesviruses are associated with tumours in lower vertebrates (Deinhardt and Deinhardt 1979), namely carp pox virus, which is found in benign epitheliomas of *Cyprinus carpio*, and Lucke frog herpesvirus, which probably causes adenocarcinomas and possibly other tumours in leopard frogs (*Rana pi-*

piens). The latter virus is interesting, because it is associated with an epithelial tumour, and the tumours tend to appear during the summer. At low temperatures the virus maintains a lytic life cycle, but when the temperature is raised virus replication is arrested and tumorigenesis *in vivo* or cell transformation *in vitro* ensues. This is reminiscent of the finding that cell transformation by HSV and CMV can be brought about by using temperature-sensitive viruses at supraoptimal temperatures (Rapp and Reed 1976).

xii Hepatitis B virus (see Chapter 101) Epidemiological evidence suggests a link between persistent hepatitis B virus (HBV) infection and primary hepatocellular carcinoma (PHC) (reviewed by Zuckerman and Howard 1979, Blumberg and London 1980, Report 1981). This tumour is very common in parts of Africa and south east Asia—areas where HBV infection is prevalent—and it is often, but not always, associated with cirrhosis (the role of cirrhosis in PHC is not understood). The tumour is frequent in young adults as well as older patients and has been estimated to cause each year 500 000 to 1 000 000 deaths worldwide, making it, perhaps, the commonest fatal neoplasm of man. Persons particularly at risk are chronic carriers of the virus who were infected very early in life, perhaps even congenitally. In patients with PHC serological markers of HBV infection are more frequent than in controls, and viral DNA, RNA, surface (HBs) antigen and core (HBc) antigen have been demonstrated in affected livers. Moreover, a major prospective survey of more than 20000 male Taiwanese showed that the incidence of PHC was 1158 per 100 000 in HBs antigen positive men but only 5 per 100 000 in those that were negative (Beasley *et al.* 1981).

Early studies showed that viral antigens in PHC were located mainly in liver cells around the tumour and were difficult to find in the tumour cells themselves. Blumberg and London (1980) have thus postulated that the regenerative response to the chronically diseased liver provides a high probability of environmental carcinogens or 'random' mutations inducing a neoplastic change. These cells develop into tumours because they fail to respond to HBV infection, unlike surrounding normal cells that are subject to both viral damage and host responses to viral antigens. This ingenious suggestion differs from most other modes of viral oncogenesis, since the virus is not affecting the tumour cells themselves but is acting indirectly by its deleterious effect on normal liver. There are problems in accepting this concept; one is the finding of HBV DNA covalently integrated into the cellular DNA of some tumour cells (Brechot *et al.* 1980, Chakraborty *et al.* 1980, Edman *et al.* 1980). Such integration is not invariable in PHC tumour cells and it can also occur in normal liver cells of PHC patients and in some HBs antigen carriers without PHC, whilst other carriers possess HBV DNA in a free form (Brechot *et al.* 1981). However, these findings are consistent with HBV integration being necessary but not sufficient for PHC induction, and integration of the viral genome into that of the cell is significant because it is a common property of other tumour viruses. However, whatever the role of HBV in primary liver cancer, it is probable that environmental, genetic or other co-factors are also important in determining the incidence of this tumour. Toxins of various kinds may be significant, notably fungal aflatoxin, which often contaminates food in areas where primary liver carcinoma is common (see, for example, Peers, Gilman and Linsell 1976), and cigarette smoking seems an independent risk factor (Trichopoulos *et al.* 1980, Lam *et al.* 1982). It is also not known why hepatocellular carcinoma and, to a lesser extent, chronic HBV carrier states, are commoner in men than women.

Answers to some of these problems may come from convenient animal model systems that have recently been described. One is a virus, closely related to HBV, that infects 10–20 per cent of wild populations of American woodchucks (*Marmota monax* (Snyder and Summers 1980). In this rodent, virus infection is associated with a high incidence of chronic hepatitis and primary liver cancer—a situation analogous to that in man. Other HBV-related agents have been found in ground squirrels, prairie dogs and ducks, and investigations on their molecular biology may provide insights into the role of HBV in carcinogenesis (reviewed by Varmus 1982*a*).

The HBV-like viruses have a life cycle reminiscent of retroviruses, for at one stage single stranded RNA is transcribed into double stranded DNA. With retroviruses this occurs straight after infection, but with HBV it is a late stage of the intracellular cycle and the virion contains the DNA product of reverse transcription. In both cases this DNA can integrate, but for HBV this may be a non-essential and imprecise event. Nonetheless, since we know a great deal about mechanisms of retrovirus oncogenesis (Part 2) it will be interesting to see if similar concepts apply to tumour induction by HBV.

xiii Poxviruses The poxviruses are large DNA viruses that replicate in the cytoplasm of infected cells (see Chapter 87), and encode their own enzymes for nucleic acid synthesis without necessarily shutting off equivalent cellular synthesis. Indeed, a degree of cellular proliferation usually precedes the cytolytic stages of poxvirus infections. With some poxviruses this proliferative response can last weeks or months. Such virus-induced cell proliferation should perhaps be regarded as hyperplasia, rather than neoplasia, for the growths apparently never become malignant and all eventually regress. Moreover, unlike most tumour-bearing viruses the poxvirus genomes are probably not integrated into the host cell chromosomes. None the less, it would be interesting to know whether poxviruses and tumour viruses show common features in the mechanisms by which they induce cell proliferation.

Part 2 The molecular mechanisms of virus-induced neoplasia

A The transformed cell

The first fifty years of tumour virus research were occupied by studies of tumour production in whole experimental animals. These observations provided much information on host responses to viral tumorigenesis, but gave no clues to the basic mechanisms by which the viruses induced tumours. It was the development of appropriate tissue culture systems that permitted a molecular analysis of virus-induced neoplasia.

Cells removed from an animal have a limited life span in tissue culture. However, certain cells from some species undergo a spontaneous change (transformation) *in vitro* to become cell lines. These resemble normal cells but they have become immortal in tissue culture (Fig. 104.1). In the right conditions many tumour viruses and some chemicals can induce changes in the shape and behaviour of both normal tissue culture cells and cell lines (Fig. 104.2). These changes are spoken of as morphological transformation and, since morphologically transformed cells, but not their normal progenitors, can often grow to form tumours in appropriate test animals, this transformation is regarded as the *in vitro* analogue of *in vivo* tumour formation (Fig. 104.1).

Some of the smaller tumour viruses, such as RSV and polyoma, are the most effective in inducing rapid and reproducible cell transformation, and it is infections by agents like these that have been studied most intensively *in vitro*. Virus-transformed cells and their normal counterparts are popular tools for cell biologists attempting to unravel the basic mechanisms of neoplasia. Many changes occur upon morphological transformation, though their relations to one another, and to the cell's ability to form a tumour, are not yet fully understood. Another major use of virus-transformed cells is in trying to decipher the role played in transformation and tumour formation by the virus. With some small tumour viruses this task is progressing well; they have only three or four genes and we already know a great deal about them. With viruses like those of the herpes group, which are 30 to 50 times as big as RSV or polyoma, the problem is commensurately larger and its solution is not yet in sight.

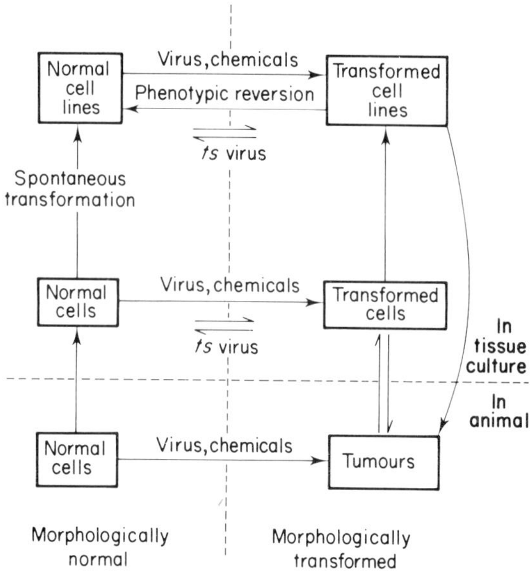

Fig. 104.1 Schematic representation of cellular conversions which occur *in vitro* and *in vivo*. Events above the horizontal dashed line are those that occur in tissue culture, those below the line in suitable test animals. Cells to the right of the vertical dashed line show some or all of the behavioural and biochemical alterations associated with cell transformation, those to its left show none of these changes (except for 'normal' cell lines which show immortality and perhaps some other changes to a minor degree). The effect of temperature-sensitive (*ts*) transforming virus is mentioned in the text, Part 2B. Phenotypic reversion of the cell from transformed to 'normal' is seen particularly in virally transformed cells and is reviewed by Wyke *et al.* (1980) and Tooze (1980).

Note that, although transformation is the *in vitro* analogue of tumour formation, the two phenomena are not necessarily the same. Even hosts that provide the most suitable environment for the growth of transformed cells apply restraints that do not exist in tissue culture. Thus, although transformed cells are frequently tumorigenic many may not be, or they may produce tumours at a very low efficiency. These observations are consistent with the concept that tumorigenesis requires cell alterations in addition to those needed to confer a transformed phenotype on a cell. Transformation may be essential for tumorigenesis but it is often not sufficient (see, for example, Klein 1979).

B The mechanism of virus-induced transformation and tumour formation

Table 104.4 summarizes the ways in which a tumour virus might convert a normal to a neoplastic cell. In the indirect mechanisms the virus need not infect the tumour cell lineage but tumours appear as a consequence of the host response to virus infection. We have already stressed the higher incidence of certain tumours in immune deficient hosts, and some viruses, such as the avian reticuloendotheliosis complex viruses, feline leukaemia virus, Marek's disease virus and cytomegalovirus are known to depress host immunity. This effect is probably important in neoplasia but the viruses also seem to act in one of the direct fashions outlined below. Another host response to virus infection is cell proliferation, and these actively proliferating populations may then be targets for further neoplastic change. The hypothesis that a regenerative response to hepatitis B virus infection is

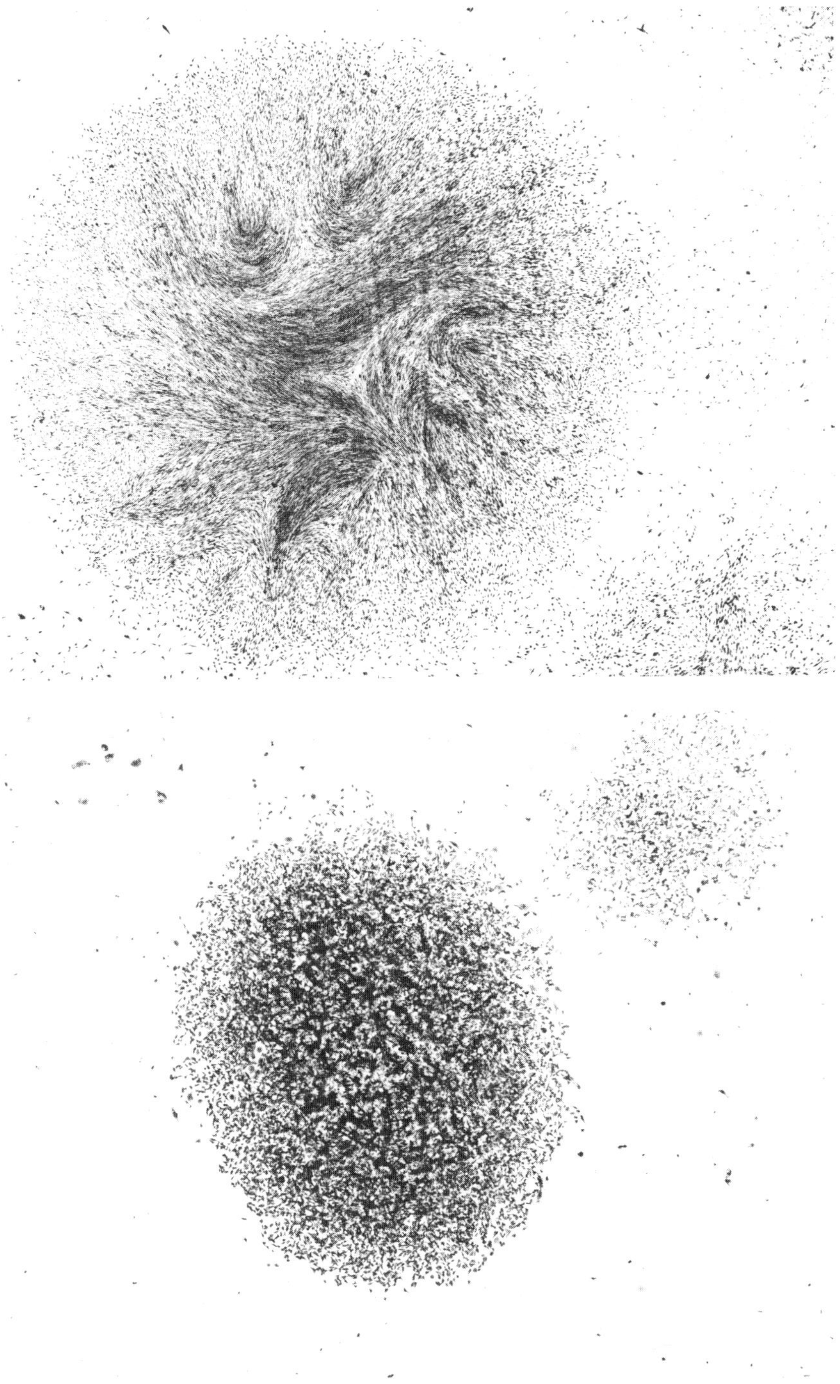

Fig. 104.2 The morphological appearance of colonies of normal and transformed cells in tissue culture (magnification × 20). (*a*) A colony of BHK21/C13 a cell line derived from Syrian hamster kidney. Note that the colony is flat and the cells are lined in parallel array. (*b*) The same cell line after transformation by polyoma virus. The cells are piled upon one another and lack orientation.

important in liver carcinogenesis has been mentioned above. Another example is murine leukaemia, in which it has been suggested that the viral proteins provide a chronic immune stimulation leading to proliferation of a subset of T-lymphocytes (McGrath and Weissman 1979) among which neoplastic changes occur (Lee and Ihle 1981).

In the direct mechanisms viruses infect the tumour cell ancestors, but in some instances these viruses may not persist, implying that viral genetic material or viral proteins induce a heritable alteration in the infected cell that is passed on to all its descendants in the absence of the virus. There are, however, only a few instances in which conversion to neoplasia might involve only a transient association of the virus with the cell, and these are not proven. Some carcinomas derived from virus-induced papillomas lack virus, but the virus appears to have been lost in the progression from a benign to a malignant tumour; its presence seems to have been needed for the development of the benign growth. About 30 per cent of feline leukaemias lack evidence of virus, though they show an epidemiological association with virus infection and contain an antigen, FOCMA, associated with viral leukaemia. In this case, as in some avian leucoses and in the HBV associated hepatomas mentioned above, lack of patent virus in the tumours may reflect selection of cell populations that have lost, or are resistant to virus, and hence escape immune surveillance directed against virion antigens.

In most virus-induced neoplasms all or part of the viral genome persists in the cell, and could be functioning in one of two main fashions (Table 104.4). (a) The virus DNA might act as a mutagen, inserting into the cell DNA in a position where it disrupts the expression of genes regulating normal cell growth and behaviour, either decreasing or increasing their activity (Varmus 1982*b*). One means by which cell gene activity can be increased is by the insertion of regions of viral DNA that control the expression of viral genes, either specific 'promoter' sequences or non-specific 'enhancers'. Cell genes may come under the control of these promoters or enhancers and hence be abnormally expressed. (b) The virus genome might itself encode a protein that directly initiates cell transformation or tumour formation (Vogt 1977).

Several experimental approaches can distinguish these mechanisms. First the virus genome can be manipulated to produce mutants with altered transforming ability. For example, mutants of RSV can be isolated that no longer transform cells (Hanafusa 1977, Vogt 1977). In some instances this defect is temperature (heat)-sensitive, so that infected cells, normal at high incubation temperatures, transform completely in a few hours when shifted to lower temperatures (Wyke 1975). These mutants provide strong evidence that RSV contains a gene whose function maintains the cell in a transformed state; this temperature-sensitive transformation is a useful technique for the cell biologist. Another approach is to determine whether in transformed cells it is obligatory to retain some parts of the viral genome but not others; this again provides evidence for regions of the virus DNA, either genes, enhancers or promoters, essential for the maintenance of transformation. A third method consists in studying the site in cell DNA at which the virus genome is inserted. When this shows any specificity it suggests that the functioning of nearby host genes may be important to the transformed phenotype.

Table 104.4 Possible mechanisms of tumour induction and cell transformation by viruses

A Indirect	Neither the tumour cells nor their ancestors need to have been infected by virus. 1) *Immune suppression.* Virus infection impairs the host immune system, and tumour cells that would otherwise be eliminated develop to clinical neoplasia. 2) *Stimulation of cell proliferation.* Tumours arise among cell populations that are expanding because of a) Regeneration of damaged tissue or b) Mitogenesis of immune competent cells.
B Direct	The tumour cells or their ancestors have been infected by virus. 1) *Virus absent from tumour cells.* Transient infection causes a stable heritable neoplastic change in the tumour cell lineage. 2) *Virus present in tumour cells.* All or part of the viral genome persists. a) Integrated viral DNA disrupts normal regulation of gene expression (insertional mutagenesis) b) Integrated or free viral DNA carries a gene whose product initiates and/or maintains neoplasia (viral oncogenes).

Studies of this kind have revealed a great deal about the mechanism of viral tumour formation (various authors see Cold Spring Harbor Symposium in Quantitative Biology, 1980). Many viruses, including some adenoviruses, polyoma, SV40, and numerous retroviruses, contain regions in their genome whose presence is required to initiate, and perhaps maintain, transformation. In the sarcoma-inducing retroviruses, such as RSV and the acute leukaemia viruses, this region of the genome encodes a protein whose function is either known, or suspected, to be essential for cell transformation. These retroviral transforming genes are different in different viruses; this is not surprising since the viruses induce different types of tumour (reviewed by Hayman 1981). However they have one thing in common. They are related to genes that are present in normal host cells of all vertebrate species tested and it seems that during evolution copies of these various cell genes were incorporated into the genomes of different viruses. In this way their expression became subject to viral rather than cellular controls and the genes themselves may have been altered. When reintroduced into a host cell by virus infection, the abnormal expression of these genes leads to neoplasia (Bishop *et al.* 1980, Bishop 1982).

C Oncogenes of viruses and cells

The cellular genes that have been acquired by viruses and that render them tumorigenic are called oncogenes (Huebner and Todaro 1969, Bentvelzen 1972). More than 15 have been identified and there are probably more, but they may fall into a limited number of classes. For example, several oncogenes whose expression leads to sarcoma formation are functionally related to one another (Langan 1980) and show structural similarities (Bishop 1981). Table 104.5 lists some representative oncogenes, each of which has a three letter designation preceded by *v-* when referring to the oncogene carried by a retrovirus and *c-* when referring to its cell ancestor.

The supposed evolution of *v-onc* genes from *c-onc* ancestors led to the concept that inappropriate *c-onc* expression, or expression of an altered *c-onc*, might be important in many forms of neoplasia, whether or not a retrovirus was concerned in the aetiology. Oncogene sequences from human and animal tumours have been sought in many laboratories and from such work are emerging major advances in our understanding of the molecular basis of all forms of cancer (see Logan and Cairns 1982). In particular, five pieces of evidence suggest that untoward *c-onc* expression can indeed initiate neoplasia (reviewed in Weiss *et al.* 1982, Enrietto and Wyke 1983).

1 That DNA from some human and animal tumours, and from cell lines derived from tumours, contain sequences that morphologically transform recipient mouse cells (called NIH 3T3 cells) and render them tumorigenic. These thus conform to the definition of oncogenes and, since DNA from normal human and animal tissues transforms NIH 3T3 cells far less efficiently, these tumour transforming genes are qualitatively different from their normal cell counterparts, either in their regulation or their function. A number of these transforming genes from tumours have been cloned in bacteria by genetic manipulation and then characterized. Most are unrelated to any known *v-onc* genes, and presumably represent oncogenes that have never been acquired by retroviruses, but about 20 per cent are the cell homologues of *v-onc* sequences. Thus the cell ancestor of the *ras* gene of Kirsten murine sarcoma virus (Table 104.5) is implicated in human lung, colon, pancreas, gall bladder and urinary bladder carcinomas and a rhabdomyosarcoma (Der, Krontiris and Cooper 1982, Pulciani *et al.* 1982), whilst the human homologue of the *ras* gene of Harvey murine sarcoma virus is the transforming sequence in a bladder carcinoma cell line (Parada *et al.* 1982). Recent studies have shown that this bladder carcinoma transforming gene differs from its counterpart in normal cells at a single nucleotide within the coding region of the *c-ras*^{H} product, an important first

Table 104.5 Some representative oncogenes and their associated tumours

Viral oncogene (*v-onc*)				Cell homologue (*c-onc*)		
Gene	Virus	Type of tumour induced	Species in which tumours occur	Gene	Tumours associated with *c-onc* activity	Species in which tumours occur
v-src	Rous sarcoma virus	Fibrosarcoma	Chicken	*c-src*	None known yet	—
v-myc	Myelocytomasis virus MC29	Myelocytoma, carcinomas, sarcomas	Chicken	*c-myc*	B-cell lymphoma, possibly other leukaemias	Chicken, possibly man and mouse
v-erb	Erythroblastosis virus	Erythroblastosis, sarcomas	Chicken	*c-erb*	Erythroblastosis	Chicken
v-ras^{H}	Harvey sarcoma virus	Pleomorphic sarcomas	Mouse	*c-ras*^{H}	Urinary bladder carcinoma	Man
v-ras^{K}	Kirsten sarcoma virus	Pleomorphic sarcomas	Mouse	*c-ras*^{K}	Various carcinomas, rhabdomyosarcoma	Man
v-mos	Moloney sarcoma virus	Rhabdomyosarcoma	Mouse	*c-mos*	None known yet	—
v-abl	Abelson leukaemia virus	Pre-B cell lymphosarcoma	Mouse	*c-abl*	Possibly pre-B cell chronic myeloid and other leukaemias	Man

There are several cell genes in the *ras* group and their nomenclature varies in original publications. This complexity has been ignored here. Harvey and Kirsten sarcoma viruses acquired their ability to induce mouse tumours after passage through rats and their *ras* genes are of rat origin.

demonstration of a neoplastic change resulting from mutation in a defined cell gene (Reddy *et al.* 1982, Tabin *et al.* 1982).

2 Mutation in the coding region may not be the only way in which *c-onc* genes are rendered neoplastic: an increase in their activity, or onset of activity in cells in which they are usually inactive, may suffice. Thus, DNA of the normal *c-onc* genes, *c-mos* and *c-ras* (Table 104.5) can transform NIH 3T3 cells and make them tumorigenic, but only when the genes are joined, by genetic manipulation, to a retroviral promoter that increases their expression (Blair *et al.* 1981, Chang *et al.* 1982).

3 A similar enhancement in *c-onc* expression may occur in some naturally occurring tumours. Avian leukosis, a B-cell lymphoma, is induced by a retrovirus that lacks a *v-onc* gene. However, in more than 90 per cent of lymphomas analysed the provirus has integrated in the vicinity of a *c-onc* gene, *c-myc*, whose expression is concomitantly enhanced. It is assumed that proviral promoter and/or enhancer sequences augment expression of normal *c-myc* and that this event is the initial step in the progression to neoplasia (Hayward, Neel and Astrin 1981, Payne, Bishop and Varmus 1982). Other retroviruses that lack their own *v-onc* sequences are also probably insertional mutagens that enhance expression of unidentified *c-onc* genes. Important examples are murine mammary tumour virus (Nusse and Varmus 1982) and, possibly, human T-cell leukaemia virus and the feline and bovine leukaemia viruses.

4 Some virally induced lymphoid tumours and some cell lines from tumours with no known viral cause show an amplification in the numbers of *c-myc* genes. This clearly is another way in which *c-onc* expression might be increased, and a further possibility for altering expression is to re-locate the *c-onc* gene in a portion of the genome where it is no longer properly regulated (reviewed by Forman and Rowley 1982).

It has recently been pointed out that the causation of many human cancers is not convincingly explained as a result of simple mutation but might require the transposition of large amounts of genetic information (Cairns 1981). For instance, many cancers in addition to BL are characterized by cytogenetic abnormalities. Though transpositions can occur in vertebrate cells, specific elements, transposons, capable of mediating such information shifts, are reliably known only in lower organisms (Calos and Miller 1980). The integrated proviruses of retroviruses are, however, structurally very similar to transposons (Shimotohno, Mizutani and Temin 1980). If they can relocate themselves and associated cell DNA in the same way as transposable elements, this too may initiate neoplastic change. Retrovirus-mediated transposition has, however, not yet been observed, although *c-onc* activation as a result of insertion of a sequence resembling a transposable element has been reported (Rechavi, Givol and Canaani 1982). On the other hand it now appears that some of the chromosome rearrangements typical of certain cancers might lead to the altered activity of *c-onc* genes. The best evidence for this comes from studies on Burkitt's lymphoma lines (Dalla-Favera *et al.* 1982, Taub *et al.* 1982). The translocation from chromosome 8 to chromosome 14 typical of BL and seen in other lymphoid tumours places the *c-myc* gene (located on chromosome 8) in the vicinity of the immunoglobulin heavy chain genes (on chromosome 14) that are very active in B-lymphoid cells. Other translocations seen in BL link *c-myc* to other immunoglobulin genes and in mouse plasmacytomas homologous translocations also join *c-myc* to immunoglobulin genes (Crews *et al.* 1982). In view of the well documented viral enhancement of *c-myc* in chicken B cell lymphomas it is reasonable to suppose that altered *c-myc* regulation as a result of translocation to an active area of chromatin is also important in human B-cell neoplasia. Another oncogene, *c-abl*, is located on chromosome 9 and is translocated to chromosome 22 (the Philadelphia chromosome) in the 9:22 translocation typical of chronic myeloid leukaemia and some acute leukaemias (de Klein *et al.* 1982). The outcome of this translocation is so far less obvious than those involving *c-myc*, for there is only one report of aberrant *c-abl* expression in a human leukaemic cell (Ozanne *et al.* 1982).

5 The last piece of evidence that *c-onc* genes are important in neoplasia comes from experiments that recapitulate, in the laboratory, the presumed evolution of *v-onc* genes from *c-onc* ancestors. Rapp and Todaro (1980) passaged murine leukaemia virus (which has no *v-onc* gene and cannot transform cells) in chemically transformed cells. Among the progeny viruses were some that could transform, presumably because they had acquired unidentified active *onc* genes from their transformed hosts. More direct demonstrations of genetic recombination between retroviruses and *c-onc* genes used mutants of transforming viruses that had lost large portions of their *v-onc* genes and hence showed a reduced tumorigenicity in animals. Such deletion mutants do, however, induce a few tumours and some of these contain viruses in which the *v-onc* genes have been reconstituted, apparently by recombination with their *c-onc* homologues (Karess and Hanafusa 1981, Ramsay *et al.* 1982).

We are only just beginning to reveal the role of *c-onc* genes in neoplasia and many details have yet to be determined. For example, epidemiological and clinical observations suggest that much neoplasia is a multiple stage process (Cairns 1978). A gene that is activated or mutated at any stage can be considered an oncogene, yet its activity in isolation may be insufficient to induce a tumour.

Avian lymphoid leukosis provides a good example. Leucosis virus infection causes hyperplastic lymphoid nodules in the bursa of Fabricius within 6 to 8 weeks, but clonal metastasizing lymphomas in the viscera do not occur until 6 months or more after infection (Neiman *et al.* 1980). Activation of *c-myc* is seen in hyperplastic nodules as well as in lymphomas, so although the *c-myc* function seems necessary for leucosis to develop, additional neoplastic events are presumably required. This conclusion is supported by studies on *in vitro* transformation of NIH 3T3 mouse cells by lymphoma DNA: the lymphoma gene that transforms NIH 3T3 is not *c-myc* but a gene that is presumably active at a later stage of tumour development (Cooper and Neiman 1981). These observations have important implications. For instance, NIH 3T3 respond to only one of the oncogenes activated in the step-by-step progression of avian leucosis and this limited range of response may explain why many tumour DNAs have no effect on these cells. It is also likely that the genesis of some other virally induced tumours is even more complex. Thus, it is postulated that EBV-induced cell proliferation leads to rare chromosomal translocations that are required for the development of Burkitt's lymphoma (Part 1D (x)). If, as now appears, these translocations lead to *c-myc* activation that is itself an early stage in lymphomagenesis, then it is easier to understand the rarity with which even such a common virus as EBV induces tumours.

To unravel these complexities future studies must elucidate the cell processes in which *c-onc* genes participate and where and when in the animal these genes normally function: presumably they do not exist just to induce cancer, but play roles in development or differentiation. Moreover, the ways in which *c-onc* function and regulation are altered in neoplasia must be determined if rational approaches to cancer prophylaxis, diagnosis and therapy are to be designed. We must also explain how the DNA viruses cause tumours. Some, such as polyoma, have transforming genes that stimulate cell growth but are not related to cell genes, being perhaps functional analogues of *c-onc* genes. Others may act indirectly or as insertional mutagens (see Table 104.5, p. 529). Whatever the cause of neoplasia, the functions specified by *c-onc* genes may be common elements in the process.

Part 3 Prophylaxis and therapy of virus-associated tumours

A knowledge of the epidemiology of tumours in which viruses are a risk factor (Part 1) is essential to successful prophylaxis. An understanding of the mechanisms of virus infection and tumour formation (Part 2) expands the prophylactic options and also the prospects for therapy. We shall consider prophylactic or therapeutic measures available at each of the three arbitrary stages of viral oncogenesis defined in Part 1. Thus we shall deal with: A) preventing stable association of the virus with the host cell; B) preventing conversion of the infected cell to a tumour cell; and C) preventing growth of the neoplastic cell to a clinical tumour. Two points should be remembered throughout. Firstly, a wider range of prophylactic measures can be applied in veterinary than in human medicine. With farm and laboratory stock and, to a lesser extent, with pet animals, care for the health of the population can override the survival of the individual. Secondly, for all tumours in which a virus is implicated, the detection of virus infection at any stage of the disease might be of diagnostic or prognostic use. This is clearly so in those few diseases where infection frequently leads to tumour formation. However, even when neoplasia is rare, detailed knowledge of the role played by the virus might identify features of virus infection that are characteristic of hosts at high risk.

A Prevention of the stable association of tumour viruses with host cells

At this stage steps can be taken to: (i) decrease the concentration of virus in the hosts' environment; (ii) increase the natural (genetic) resistance of host populations; and (iii) increase the artificial resistance of the hosts by active or passive immunity.

i The source of virus This is determined by the density of symptomatic or asymptomatic carriers in the population, the ability of the virus to survive outside its host, and whether or not other animals act as natural reservoirs for the virus. Many tumour viruses are species-specific and in practice only the first two factors have been subject to prophylactic measures.

The identification of carriers often requires specialized techniques to detect virus infection. It is thus costly and is effective only when the carrier can be killed, cured, or isolated from the susceptible population. For these reasons, carriers of potential human tumour viruses have so far been identified only for investigational and not for preventive purposes. The one possible exception is the identification of HBV carriers (Chapter 101), but this is usually performed on a limited scale and in parts of the world where hepatocellular carcinoma is rare.

Identification of carriers can be useful in controlling avian, feline, and bovine lymphoid leukaemias. A small proportion of feline leukaemia virus-infected cats do not develop immunity and remain viraemic. Such animals can be removed from colonies, or identified before introduction into virus-free groups. Eradication of enzootic bovine leucosis is greatly assisted by tests for virus carriers. In the past, herd testing was based on haematology, which detected only diseased animals. An immunodiffusion test for antiviral antibodies now being used in West Germany (Mussgay *et al.* 1980) detects animals within 1 to 3 months of infection. It is not certain that this test detects all the carriers, nor that a sero-positive beast is necessarily a virus shedder, but elimination of seropositive animals is reducing the incidence of the disease.

Hygienic practices aimed at reducing the virus source are probably of little significance within susceptible host populations, though, in combination with appropriate husbandry, they can reduce the dissemination of disease between local concentrations of animals, for example, the spread of avian leukaemia from one poultry farm to another. Many tumour viruses are very susceptible to chemical disinfectants, drying, and other inactivating agents, but local horizontal and congenital spread demands such intimate contact between hosts that disinfection is impracticable. The exceptions are papillomas and enzootic bovine leucosis, the viruses of which can be spread on the instruments of stockmen and veterinarians. In man, the hygienic practices of many western countries certainly delay some EBV infections until puberty or later, though this increases the incidence of infectious mononucleosis. A delay in EBV infection in Africa and Asia might reduce the incidence of Burkitt's lymphoma or nasopharyngeal carcinoma, but this has yet to be demonstrated. However, one virus-associated human neoplasm where a change in habits should reduce incidence is carcinoma of the uterine cervix. Like other venereally transmitted infections, the spread of papilloma viruses and HSV type 2 has been favoured by promiscuity and changes in contraceptive practices. Promiscuity and intravenous drug administration might also favour the development of Kaposi's sarcoma in male homosexuals.

ii The genetic susceptibility of host animals The selective breeding of genetically resistant stock is applicable only to animals, particularly those with a short generation time. It has been useful in reducing the incidence of avian lymphoid leucosis, which exists in the field as two major virus subtypes, characterized by the presence of particular glycoproteins on the viral envelope. These glycoproteins interact with specific cell surface receptors and hence the viruses penetrate host cells. Strains of domestic chicken have been bred that lack the surface receptors for the field strains of avian leucosis virus and these birds are resistant to infection (Crittenden and Motta 1969). Some strains partly resistant to lymphoid leucosis apparently permit infection but not tumour development. A similar genetic resistance is seen in production of Marek's disease lymphomas (see above).

Most mice are resistant to spreading infection by their endogenous retroviruses. In this case the resistance operates at a stage after penetration but before provirus integration (Jolicoeur and Baltimore 1976). In wild mouse populations this resistance may limit the spread of a potentially oncogenic retrovirus should one be activated.

iii Immunity to virus infection Some mice whose endogenous virus genomes have been artificially manipulated do not show the genetic resistance to spreading infection mentioned above. Leukaemia in these animals is accompanied by virus spread and an increase in the number of virus genomes in leukaemic cells (Jaenisch 1979). Both spread of virus and the incidence of leukaemia can be prevented by passive immunotherapy with anti-viral serum. With most other tumour viruses where immunoprophylaxis has been attempted, however, vaccines have been used to elicit active immunity.

The first **vaccines** against viral tumours were prepared from wart tissue. Subcutaneous injection of formalin-inactivated autogenous wart vaccine protects against bovine papillomas (Blood and Henderson 1963). However, the first successful vaccines against a neoplasm of great economic importance were those for Marek's disease. Avirulent strains of Marek's disease virus, either natural or attenuated, have been used, as has a related herpesvirus of turkeys that is avirulent in chickens. The vaccine viruses establish a latent infection and induce tumour associated surface antigens in lymphoid cells, with sometimes slight cytopathic effects and nerve inflammation. There is, however, no immune suppression and the vaccinated birds resist challenge with oncogenic virus (Calnek *et al.* 1979), apparently by an immunity directed against both viral and tumour antigens.

Feline lymphosarcoma and associated non-neoplastic diseases are horizontally transmitted disorders for which vaccination is being investigated. Antibodies against viral proteins and the tumour antigen, FOCMA, are both elicited in natural infections. However, some tumours are virus-negative but always bear FOCMA, and animals with tumours have a low anti-FOCMA activity. It seems then, that for prevention of neoplasia anti-FOCMA activity is important and FOCMA antiserum can protect experimentally against tumour formation. Antiviral antibodies may, however, be significant in protection against the non-neoplastic components of this disease complex and, if they could be induced prophylactically, they should prevent infection and hence tumour development.

Birds vaccinated against Marek's disease may be destined for human consumption, and some feline leukaemia viruses infect human cells. Care must be taken to ensure that the manipulations involved in making vaccines do not produce altered viruses of danger to man. With these provisos, however, the decisions to make animal vaccines are straightforward. Any proposal to vaccinate against a virus suspected of being a risk factor in human neoplasia poses additional problems. Where cancer is thought to result from multiple factors, each necessary but not sufficient for tumour formation, prevention of virus infection might be the simplest way to attack the aetiological complex. Indeed, final proof that a virus is causally connected with neoplasia often rests on epidemiological evidence and, in particular, the prophylactic effect of vaccination. However, if the consequences of vaccination are the proof of causation, then one faces the logical dilemma of whether to embark on such prophylactic measures before they can be shown to be necessary. In cases where the virus concerned causes other disease this decision is not so difficult, but where a possibility of neoplasia is the only untoward result of infection, then the expense and risk of vaccination must be weighed against the chance (perhaps remote) that the naturally infected person might develop a tumour. Moreover, since cell transformation often results from infection with defective viruses, some vaccines may have more oncogenic potential than natural infection. This has, in fact, been a problem in accepting proposals for vaccination against agents such as cytomegalovirus (Waterson 1979), which can cause serious non-oncogenic disease, but which are known to transform cells when defective.

Improvements in safety may come from the use of vaccines that contain only those proteins of the virion that are important for immunity. The purity of subunit vaccines made from virus preparations may be in doubt, but the techniques of genetic manipulation should permit the viral genes encoding immunogenic proteins to be cloned in bacteria. Appropriate clones could produce high yields of these proteins, free from potentially pathogenic contaminants. If the amino acid sequence of such a protein is known it is also possible to synthesize an artificial peptide for use as a vaccine. These procedures, however, require precise knowledge of the basis of immunity to any given disease, and extensive testing will be needed to determine the efficacy of such vaccines.

Three human viruses associated with cancer are now likely candidates for vaccine prophylaxis: EBV, HSV type 2, and HBV. We may add to this list HTLV, papillomaviruses and CMV when we understand more about their pathogenesis. The case for vaccination against HBV in areas where primary hepatocellular carcinoma is common is strong, for hepatitis is endemic in these regions. Vaccine trials are under way in West Africa (Maupas *et al.* 1980), and also in the United States, though here liver carcinoma is not a problem (Szmuness *et al.* 1980). HBV has not been grown in tissue culture, and the vaccines consist of HBs antigen from the serum of symptomless donors. Though the vaccines seem safe and efficacious, a genetically manipulated vaccine would reduce any doubts about safety.

For HSV type 2 infections the case for vaccination is less obvious. However, if an indisputably safe and effective vaccine can be produced, it should be useful in populations particularly at risk. Vaccination against EBV is also contentious. The risk of developing Burkitt's lymphoma is low and probably does not justify at present the expense of vaccination, particularly since we do not know what type of vaccine to use and whether it will be effective. The eradication of malaria, a co-factor and a far greater health problem, seems a better alternative, but this too is very expensive. In nasopharyngeal carcinoma arguments of expense for small returns also apply against vaccination, with the added factor that, were vaccination effective, there would be a lag of 40 to 50 years for this effect to be seen. On the other hand no clear-cut avoidable co-factors have yet been identified for this disease.

B Prevention of the conversion of an infected to a tumour cell

With any tumour whose induction depends on multiple factors, each essential but not sufficient for the disease, it is necessary to remove only one of the factors. When a virus is implicated, it is often the focus of attention for two reasons: 1) there are well tried principles for avoiding infection; 2) detecting and measuring virus involvement is far easier than monitoring and assessing other carcinogens. For these reasons, and aside from experimental models such as radiation-induced leukaemia in mice, there are few virus-associated neoplasms in which manipulation of non-viral risk factors has formed the basis for prevention. However, such an approach might be a sensible prophylactic measure in tumours where avoidable co-carcinogens have recently been identified.

In man, control of Burkitt's lymphoma by controlling malaria, mentioned above, is a good example. Other workers have suggested that nitrosamine-contaminated salted fish is a dietary co-factor for nasopharyngeal carcinoma. The association between primary liver carcinoma and food contaminants such as *Aspergillus flavus* aflatoxin has also been stressed, although HBV associated PHC can be common in areas where food toxins are not a problem. Unfortunately, dietary changes may be difficult in man for many reasons and not much easier in animals. Thus, it is sensible to prevent cattle in west Scotland from ingesting bracken, not only because it is associated with the progression of alimentary papillomas to carcinomas but because its radiomimetic toxin also causes more acute illness. The economics of raising beef cattle on hill farms, however, make it hard to implement such measures.

C Prevention of clinical neoplasia

Once a virus-infected cell has become neoplastic the prevention of tumour development is in the realm of therapy rather than prophylaxis. Tests for virus expression or host responses to virus at this stage might aid early diagnosis and also prognosis. Thus, screening populations at risk for high titres to EBV capsid antigen has enabled early detection of nasopharyngeal carcinoma (de-The 1980). It has also been suggested that a human breast cancer antigen that cross-reacts with a mouse mammary tumour virus protein might provide a useful diagnostic or prognostic aid (Spiegelman *et al.* 1980). Many chemical and biological agents used to treat tumours are most effective against low numbers of tumour cells, so that early diagnosis is particularly important to the success of such therapies. Treatment of virus-associated neoplasms, like that for other cancers, can vary from simple ligation or excision, as for infectious warts, to complex combinations of surgery with chemotherapy or radiotherapy. Some therapeutic approaches are of particular significance to virus-induced tumours. First, because virus transformation is rare, many tumours are monoclonal. However, some tumours caused by rapidly transforming retroviruses that carry an oncogene are polyclonal, and virus spread may contribute to their growth. Inhibitors of viral reverse transcriptase have been tested on such tumours but have had little therapeutic effect.

Second, some antimetabolites, including halogenated pyrimidines, may be used in chemotherapy or for treating acute infections with complex viruses such as members of the herpes group. These agents are also known to activate endogenous retroviruses. Such activation probably has no clinical implication, but until we fully understand endogenous viruses this phenomenon should be regarded with caution.

Thirdly, interferon (Chapter 86), an antiviral agent produced by the host, can inhibit both early and late stages of replication of many oncogenic viruses. *In vitro* and *in vivo* studies on the effect of interferon on virus-induced tumour cells suggested that interferon affected not only the virus life cycle but also the cells themselves. It was then shown that interferon reduced the growth *in vitro* of tumour cells that were not virally induced, and normal cell growth could also be inhibited (reviewed by Gresser and Tovey 1978). In test animals non-toxic doses of interferon inhibited tumour growth, particularly when the number of tumour cells was small or other forms of therapy were used as adjuvants. In trials on human patients interferon was beneficial in the treatment of some tumours even when they were large. The therapeutic effect of interferon is not yet fully evaluated, nor is it certain how interferon acts to reduce tumour growth. However, the mechanism of action at present seems complex. Interferon can prevent virus replication and spread in those tumours where a virus is implicated; it can directly inhibit the growth of tumour cells; and it can probably inhibit tumour development in an indirect way by stimulating the complex host responses to neoplasia that delay or decrease tumour growth.

Fourthly, tumours that have a known or suspected association with wart viruses have been treated with tumour extracts. Both infectious warts in animals (Blood and Henderson 1963) and bovine ocular squamous cell carcinoma (Hoffmann, Jennings and Spradbrow 1981) show partial or complete remission after administering extracts of the appropriate allogeneic tumours. The mechanism of this effect is unknown but is presumed to be immunological.

Finally, as pointed out in Section 1, many virus-associated neoplasms show a higher incidence in immune deficient persons. Such tumours may be more likely to appear after treatment for a pre-existing tumour, since cancer therapies can be both mutagenic and immunosuppressant and carry a risk of inducing a second neoplasm.

Conclusions and prospects

I have surveyed both clinical and laboratory studies on tumour viruses. Inevitably, some viruses have been discussed in more detail than others and this extra attention has usually been warranted because they illustrate particularly well a general point or because they have attracted the attention of a larger number of investigators. However, agents that have received only cursory mention, or that have even been omitted, may yet prove to be very significant. Unfortunately, in attempting to foresee future directions of tumour virus research we are limited by the information accrued in past studies, though even within this framework it is clear that the field is capable of rapid advances.

On the clinical side it is obvious that viruses can be potent tumour inducers in both laboratory and domestic animals. However, tumour formation often depends on special circumstances which probably seldom exist in human populations. Thus, although it is likely that potential human tumour viruses will continue to be reported, it is also probable that more emphasis will now be placed on viruses as one factor among several in the causation of tumours, and their interplay with host immune responses and with peculiar environmental and genetic elements will be stressed. Vaccination or other methods aimed at preventing virus infection may be developed for some tumours, particularly in animals, but the elimination of other predisposing factors should also receive attention.

The design of safe vaccines needs a thorough knowledge of the molecular basis of virus-induced neoplasia and one can predict a rapid increase in laboratory studies, particularly on the clinically important, yet complex, oncogenic agents such as EBV. Modern techniques of molecular biology and biochemistry, particularly gene cloning and amplification, now make such studies feasible. The same techniques are also aiding further studies on the far simpler retroviruses and papovaviruses and in the next few years we should: 1, learn a great deal about the functions of their transforming genes and 2, learn more details of the ways in which the viral and cellular genomes interact and affect each other's expression.

Such studies have already stimulated research beyond the strict confines of tumour virology. For instance, the concept that viruses promote inappropriate expression of *c-onc* genes has led to a search for such oncogene expression in various human tumours. Many researchers hope that these genes are implicated in a common path of disordered growth that might be stimulated by a variety of chemical, physical and biological agents. By helping us to locate and, in some cases, by capturing such oncogenes, tumour viruses make them amenable to detailed functional studies which may reveal aspects of development of the organism and differentiation of individual cell lineages that have an importance beyond cancer research. An understanding of *c-onc* genes may also soon have practical implications for the prevention, diagnosis and even treatment of cancer. Prevention may be improved because knowledge of the molecular bases of neoplasia will help to predict which environmental factors are potential risks to different populations and, more importantly, may help in devising rapid and rational tests for carcinogens. Probes for the activity of oncogenes, or the cell molecules with which they interact, may help in diagnosis. Finally, two approaches to therapy can be envisaged. Antibodies to tumour-specific cell surface molecules (either *onc* gene products or cell proteins with which they interact) may be linked to cytotoxic drugs. This will direct the drugs specifically to tumour cells; but to use this technique safely we must understand the pattern of *c-onc* expression in normal cells. It may also prove possible to interfere with the metabolic activity of oncogenes, attenuating their effect and thus reducing tumour growth.

References

Beasley, R. P., Hwang, L.-Y., Lin, C.-C. and Chien, C.-S. (1981) *Lancet* **ii,** 1129.

Bentvelzen, P. (1972) In: *RNA Viruses and the Host Genome in Oncogenesis*, p. 309. Ed. by P. Emmelot and P. Bentvelzen. North-Holland, Amsterdam.

Bishop, J. M. (1981) *Cell* **23,** 5; (1982) *Scientific American* **246,** 69.

Bishop, J. M. *et al.* (1980) *Cold Spring Harbor Symp. quant. Biol.* **44,** 919.

Bittner, J. J. (1936) *Science* **84,** 162.

Blair, D. G., Oskarsson, M., Wood, T. G., McClements, W. D., Fischinger, P. J. and Vande Woude, G. F. (1981) *Science* **212,** 941.

Blood, D. C. and Henderson, J. A. (1963) *Veterinary Medicine* 2nd edn. Bailliere, Tindall and Cox, London.

Blumberg, B. S. and London, W. T. (1980) *Cold Spring Harbor Conf. Cell Proliferation* **7,** 401.

Boldogh, I., Beth, E., Huang, E.-S., Kyalwazi, S. K. and Giraldo, G. (1981) *Int. J. Cancer* **28,** 469.

Brechot, C., Pourcel, C., Louise, A., Rain, R. and Tiollais, P. (1980) *Nature* **286,** 533.

Brechot, C. *et al.* (1981) *Lancet* **ii,** 765.

Burny, A. *et al.* (1980) In: *Viral Oncology*, p. 231. Ed. by G. Klein. Raven Press, New York.

Cairns, J. (1978) *Cancer, Science and Society*. W.H. Freeman, San Francisco; (1981) *Nature* **289,** 353.

Calnek, B. W., Carlisle, J. C., Fabricant, J., Murthy, K.K. and Schat, K. A. (1979) *Amer. J. vet. Res.* **40,** 541.

Calos, M. P. and Miller, J. H. (1980) *Cell* **20,** 579.

Campo, S. (1982) *Nature* **298,** 605.

Catovsky, D. *et al.* (1982) *Lancet* **i,** 639.

Chakraborty, P. R., Ruiz-Opazo, N., Shouval, D. and Shafritz, D. A. (1980) *Nature* **286,** 531.

Chang, E. H., Furth, M. E., Scolnick, E. M. and Lowy, D. R. (1982) *Nature* **297,** 475.

Cooper, G. M. and Neiman, P. E. (1981) *Nature* **292,** 857.

Crews, S., Barth, R., Hood, L., Prehn, J. and Calame, K. (1982) *Science* **218,** 1319.

Crittenden, L. B. and Motta, J. V. (1969) *Poultry Sci.* **48,** 1751.

Dalla Favera, R., Bregni, M., Erikson, J., Patterson, D., Gallo, R. C. and Croce, C. M. (1982) *Proc. nat. Acad. Sci., USA* **79,** 7824.

Darby, G. (1980) *Nature* **285,** 13.

Deinhardt, F. (1980) In: *Viral Oncology*, p. 357. Ed. by G. Klein. Raven Press, New York.

Deinhardt, F. and Deinhardt, J. (1979) In: *The Epstein–Barr Virus*, p. 373. Ed. by M. A. Epstein and B. G. Achong. Springer-Verlag, Berlin, Heidelberg, and New York.

Der, C. J., Krontiris, T. G. and Cooper, G. M. (1982) *Proc. nat. Acad. Sci., USA* **79,** 3637.

Drew, W. L. *et al.* (1982) *Lancet* **ii,** 125.

Edman, J. C., Gray, P., Valenzuela, P., Rall, L. B. and Rutter, W. J. (1980) *Nature* **286,** 535.

Ellermann, V. and Bang, O. (1908) *Zbl. Bakt., Abt. 1* (Orig.) **46,** 595.

Enrietto, P. J. and Wyke, J. A. (1983) *Advanc. Cancer Res.* **39,** 269.

Epstein, M.A. and Achong, B. G. (Eds) (1979) *The Epstein–Barr Virus*. Springer-Verlag, Berlin, Heidelberg, and New York.

Essex, M. (1982) *J. nat. Cancer Inst.* **69,** 981.

Essex, M., Todaro, G. and Zur Hausen, H. (Eds) (1980) *Cold Spring Harbor Conf. Cell Proliferation* **7.**

Essex, M. *et al.* (1975) *J. nat. Cancer Inst.* **55,** 463.

Fatemi-Nainie, S., Anderson, L. W. and Cheevers, W. P. (1982) *Virology* **120,** 490.

Ford, J. N., Jennings, P. A., Spradbrow, P. B. and Francis, J. (1982) *Res. vet. Sci.* **32,** 257.

Forman, D. and Rowley, J. (1982) *Nature* **300,** 403.

Frisby, D. P., Weiss, R. A., Roussel, M. and Stehelin, D. (1979) *Cell* **17,** 623.

Fujinami, A. and Inamoto, K. (1914) *Z. Krebsforsch.* **14,** 94.

Gallo, R. C. *et al.* (1982) *Proc. nat. Acad. Sci., USA* **79,** 5680.

Goldberg, R. J. and Gravell, M. (1976) *Cancer Res.* **36,** 795.

Gotoh, Y.-I., Sugamura, K. and Hinuma, Y. (1982) *Proc. nat. Acad. Sci., USA* **79,** 4780.

Graf, T. and Beug, H. (1978) *Biochim. biophys. Acta.* **516,** 269.

Graf, T. and Stehelin, D. (1982) *Biochim. biophys. Acta.* **651,** 245.

Green, M. and Mackey, J. K. (1977) *Cold Spring Harbor Conf. Cell Proliferation* **4,** 1027.

Green, M. *et al.* (1982) *Proc. nat. Acad. Sci., USA* **79,** 4437.

Gresser, I. and Tovey, M. G. (1978) *Biochim. biophys. Acta.* **516,** 231.

Groopman, J. E. and Gottlieb, M. S. (1982) *Nature* **299,** 103.

Gross, L. (1951) *Proc. Soc. exp. Biol. Med.* **76,** 27; (1953) *Proc. Soc. exp. Biol. Med.* **83,** 414; (1970) *Oncogenic Viruses*. Pergamon Press, New York and London.

Hageman, P. C., Calafat, J. and Daams, J. H. (1972) In: *RNA Viruses and the Host Genome in Oncogenesis*, p. 283. Ed. by P. Emmelot and P. Bentvelzen. North-Holland, Amsterdam.

Hanafusa, H. (1977) In: *Comprehensive Virology*, Vol. 10, p. 410. Ed. by H. Fraenkel-Conrat and R. Wagner. Plenum Press, New York.

Hardy, W. D. Jr., Old, L. J., Hess, P. W., Essex, M. and Cotter, S. M. (1973) *Nature* **244,** 266.

Hartley, J. W., Wolford, N. K., Old, L. J. and Rowe, W. P. (1977) *Proc. nat. Acad. Sci., USA.* **74,** 789.

Hayman, M. J. (1981) *J. gen. Virol.* **52,** 1.

Hayward, W. S., Neel, B. G. and Astrin, S. M. (1981) *Nature* **290,** 475.

Henle, J. (1840) *Pathologische Untersuchungen*, p. 19. Berlin.

Henle, W. (1971) In: *Proc. 1st. int. Symp. Princess Takamatsu Cancer Res. Fund: Recent Advances in Human Tumor Virology and Immunology*, p. 361. Ed. by W. Nakahara, K. Mishioka, T. Hirayama and Y. Ito. University of Tokyo Press, Tokyo.

Hoffmann, D., Jennings, P. A. and Spradbrow, P. B. (1981) *Austr. vet. J.* **57,** 159.

Howley, P. M. (1980) *Nature* **284,** 124.

Huebner, R. J. and Todaro, G. J. (1969) *Proc. nat. Acad. Sci., USA* **64,** 1087.

Jaenisch, R. (1979) *Virology* **93,** 80.

Jarrett, W. F. H. (1980) *Brit. med. Bull.* **36,** 79; (1981) *Recent Advances in Histopathology* **11,** 35.

Jolicoeur, P. and Baltimore, D. (1976) *Proc. nat. Acad. Sci., USA.* **73,** 2236.

Karess, R. E. and Hanafusa, H. (1981) *Cell* **24,** 155.

de Klein, A. *et al.* (1982) *Nature* **300,** 765.

Klein, G. (1979) *Proc. nat. Acad. Sci., USA* **76,** 2442; (Ed.) (1980) *Viral Oncology*. Raven Press, New York.

Klein, G. and Purtilo, D. (1981) *Cancer Res.* **41,** 4302.

Lam, K. C., Yu, M. C., Leung, J. W. C. and Henderson, B. E. (1982) *Cancer Res.* **42,** 5246.

Lancaster, W. D., Olson, C. and Meinke, W. (1977) *Proc. nat. Acad. Sci., USA* **74,** 524.

Langan, T. (1980) *Nature* **286,** 329.

Lee, J. C. and Ihle, J. N. (1981) *Nature* **289,** 407.

Lilly, F. and Mayer, A. (1980) In: *Viral Oncology*, p. 89. Ed. by G. Klein. Raven Press, New York.

Logan, J. and Cairns, J. (1982) *Nature* **300,** 104.

Marsden, H. (1980) *Nature* **288,** 212.

McGrath, M. S. and Weissman, I. L. (1979) *Cell* **17,** 65.

Maupas, P. *et al.* (1980) *Cold Spring Harbor Conf. Cell Proliferation* **7,** 481.

Minick, C. R., Fabricant, C. G., Fabricant, J. and Litrenta, M. M. (1979) *Amer. J. Path.* **96,** 673.

Miyoshi, I. *et al.* (1981) *Nature* **294,** 770.

Mussgay, M. *et al.* (1980) *Cold Spring Harbor Conf. Cell. Proliferation* **7,** 911.

Nazerian, K. (1980) In: *Viral Oncology*, p. 665. Ed. by G. Klein. Raven Press, New York.

Neiman, P. E., Jordan, L., Weiss, R. A. and Payne, L. N. (1980) *Cold Spring Harbor Conf. Cell Proliferation* **7,** 519.

Nusse, R. and Varmus, H. E. (1982) *Cell* **31,** 99.

Orth, G. *et al.* (1980) *Cold Spring Harbor Conf. Cell Proliferation* **77,** 259.

Ozanne, B., Wheeler, T., Zack, J., Smith, G. and Dale, B. (1982) *Nature* **299,** 744.

Panem, S. and Reynolds, J. T. (1979) *Fed. Proc.* **38,** 2674.

Parada, L. F., Tabin, C. J., Shih, C. and Weinberg, R. A. (1982) *Nature* **297,** 474.

Payne, G. S., Bishop, J. M. and Varmus, H. E. (1982) *Nature* **295,** 209.

Peers, F. G., Gilman, G. A. and Linsell, C. A. (1976) *Int. J. Cancer* **17,** 167.

Perk, K. and Hod, I. (1982) *J. nat. Cancer Inst.* **69,** 747.

Pfister, H., Nürnberger, F., Gissmann, L. and Zur Hausen, H. (1981) *Int. J. Cancer* **27,** 645.

Pincus, T. (1980) In: *Molecular Biology of RNA Tumor Viruses*, p. 77. Ed. by J. R. Stevenson. Academic Press, New York.

Poiesz, B. J., Ruscetti, F. W., Gazdar, A. F., Bunn, P. A., Minna, J. D. and Gallo, R. C. (1980) *Proc. nat. Acad. Sci, USA*, **77,** 7415.

Poiesz, B. J., Ruscetti, F. W., Reitz, M. S., Kalyanaraman, V. S. and Gallo, R. C. (1981) *Nature* **294,** 268.

Pulciani, S., Santos, E., Lauver, A. V., Long, L. K., Aaronson, S. A. and Barbacid, M. (1982) *Nature* **300,** 539.

Ramsay, G. M., Enrietto, P. J., Graf, T. and Hayman, M. J. (1982) *Proc. nat. Acad. Sci, USA.* **79,** 6885.

Rapp, F. and Reed, C. (1976) *Cancer Res.* **36,** 800.

Rapp, U. and Todaro, G. J. (1980) *Proc. nat. Acad. Sci., USA*, **77,** 624.

Rechavi, G., Givol, D. and Canaani, E. (1982) *Nature* **300,** 607.

Reddy, E. P., Reynolds, R. K., Santos, E. and Barbacid, M. (1982) *Nature* **300,** 149.

Report (1981) *Lancet* **ii,** 1394; (1982*a*) *Ibid.* **ii,** 1083; (1982*b*) *Ibid.* **ii,** 365.

Robert-Guroff, M., Nakao, Y., Notake, K., Ito, Y., Sliski, A. and Gallo, R. C. (1982) *Science* **295,** 975.

Rous, P. (1911) *J. exp. Med.* **13,** 397.

Rowe, W. P. and Pincus, T. (1972) *J. exp. Med.* **135,** 429.

Rubin, H. L., Fanshier, L., Cornelius, A. and Hughes, W. F. (1962) *Virology* **17,** 143.

Scolnick, E. M. and Parks, W.P. (1974) *J. Virol.* **13,** 1211.

Scolnick, E. M., Rands, E., Williams, D. and Parks, W. P. (1973) *J. Virol.* **12,** 458.

Shimotohno, K., Mizutani, S. and Temin, H. M. (1980) *Nature* **285,** 550.

Snyder, R. L. and Summers, J. (1980) *Cold Spring Harbor Conf. Cell Proliferation* **7,** 447.

Spiegelman, S. *et al.* (1980) *Cold Spring Harbor Conf. Cell Proliferation* **7,** 1149.

Staal, S. P., Hartley, J. W. and Rowe, W. P. (1977) *Proc. nat. Acad. Sci., USA* **74,** 3065.

Stewart, S. E., Eddy, B. E. and Borgese, N. (1958) *J. nat. Cancer Inst.* **20,** 1223.

Sugden, B., Kintner, C. R. and Mark, W. (1979) *Advanc. Cancer Res.* **30,** 239.

Szmuness, W. *et al.* (1980) *New Engl. J. Med.* **303,** 833.

Tabin, C. J. *et al.* (1982) *Nature* **300,** 143.

Taub, R. *et al.* (1982) *Proc. nat. Acad. Sci., USA* **79,** 7837.

Taylor, J. F., Templeton, A. C., Vogel, C. L., Ziegler, J. L. and Kyalwazi, S. K. (1971) *Int. J. Cancer* **8,** 122.

Temin, H. M. and Baltimore, D. (1972) *Advanc. Virus Res.* **17,** 129.

de-The, G. (1980) *Cold Spring Harbor Conf. Cell Proliferation* **7,** 11.

Theilen, G. H. *et al.* (1973) *Bibl. haematol.* **39,** 251.

Thomas, D. B. and Rawls, W. E. (1978) *Cancer* **42,** 2716.

Tooze, J. (Ed.) (1980) *Molecular Biology of Tumor Viruses*, 2nd edn., Part 2: DNA Tumor Viruses. Cold Spring Harbor Laboratory, New York.

Trentin, J. J., Yabe, Y. and Taylor, G. (1962) *Science* **137,** 835.

Trichopoulos, D., Macmahon, B., Sparros, L. and Merikas, G. (1980) *J. nat. Cancer Inst.* **65,** 111.

Various Authors (1980) *Cold Spring Harbor Symp. quant. Biol.* **44.**

Varmus, H. E. (1982*a*) *Nature* **299,** 204; (1982*b*) *Science* **216,** 812.

Vogt, P. K. (1977) In: *Comprehensive Virology*, Vol. 9, p. 341. Ed. by H. Fraenkel-Conrat and R. Wagner. Plenum Press, New York.

Vogt, P. K. and Friis, R. R. (1971) *Virology* **43,** 223.

Waterson, A. P. (1979) *Brit. med. J.* **ii,** 564.

Weiss, R. A., Teich, N., Varmus, H. E. and Coffin, J. M. (Eds) (1982) *Molecular Biology of Tumor Viruses*, 2nd edn., Part 3: RNA Tumor Viruses. Cold Spring Harbor Laboratory, New York.

Wyke, J. (1975) *Biochim. biophys. Acta.* **417,** 91.

Wyke, J. A., Beamand, J. A. and Varmus, H. E. (1980) *Cold Spring Harbor Symp. quant. Biol.* **44,** 1065.

Zachow, K. R. *et al.* (1982) *Nature* **300,** 771.

Ziegler, J. L. (1982) *J. nat. Cancer Inst.* **68,** 337.

Ziegler, J. L. *et al.* (1982) *Lancet* **ii,** 631.

Zuckerman, A.J. and Howard, C.R. (1979) *Hepatitis Viruses of Man*. Academic Press, London.

Zur Hausen, H. (1976) *Cancer Res.* **36,** 794.

105

African swine fever

Walter Plowright

Introductory

African swine fever (ASF) is typically a peracute, generalized disease of domestic swine characterized by haemorrhagic lesions in many tissues and an extremely high mortality. It was originally called 'East African swine fever' by R. Eustace Montgomery who, working in Kenya, recognized its clinicopathological resemblances to European 'swine fever' (N. American 'hog cholera') but also showed clearly that the filtrable agents were not immunologically related. Between 1910 and 1915, 15 outbreaks were reported in Kenya, involving 1366 pigs of which 98.9 per cent died. The E. African outbreaks were not related to movements of sick pigs or fomites but to the presence on pig farms of the warthog, *Phacochoerus aethiopicus*, and to systems of management which allowed the two species to range over the same areas (Montgomery 1921).

It was later recognized by Steyn (1928) and De Kock and others (1940) that a similar, if not identical disease occurred in S. Africa, where the causal agent was demonstrated in the blood and tissues of apparently normal warthogs and bushpigs (*Potamochoerus* sp.). Reports of ASF were also received later from many other countries in Africa, including Angola and Mozambique, so that the specifically 'East' African connotation gradually became irrelevant. It was also recognized that some of the very few pigs surviving outbreaks of ASF continued to carry and periodically to excrete the virus, although they were resistant to challenge, at least with the homologous virus (see Scott 1965*a*).

For decades ASF was entirely an African disease of somewhat infrequent occurrence (Table 105.1), except perhaps in Angola and Mozambique where it became an important problem for the pig industry; it had proved impossible to develop attenuated or inactivated vaccines or antisera which were effective prophylactically. In Angola also, endemicity in swine was apparently first associated with increasing numbers of more chronic cases (Velho 1956). The potential threat of ASF to the more intensive swine industries of the developed world became increasingly recognized and in 1957 De Tray concluded, prophetically as it turned out, 'We believe that ASF is potentially one of, if not the most, serious of the diseases of pigs. This potential threat holds for many parts of Africa and other swine-raising areas of the world . . .'.

Catastrophe struck sooner than he had anticipated. In 1957 the disease was reported in Portugal, probably

Table 105.1 New outbreaks of African swine fever in African countries, 1960–1978

Country	No. of outbreaks of ASF* in 1960/61	1962/63	1964/65	1966/67	1968/69	1970/71	1972/73	1974/75	1976/77	1978	Total 1960/78
Angola	27	10	60	60	60	29	8	11**	16**	29	310
Congo (L)	—	—	—	—	—	—	—	—	1	—	1
Dahomey	—	—	3	1	1	4	3	—	—	—	12
Kenya	—	—	1	—	—	—	—	—	—	—	1
Lesotho	—	—	—	—	—	—	1	—	—	—	1
Malawi	7	6	3	6	14	2	2	5	14	1	60
Mozambique	52	39	1	7	11	2	3	2**	—	—	117
Rhodesia	—	7	—	1	—	2	—	—	—	1	11
Senegal	—	1	—	—	—	—	—	—	—	—	1
Sudan	—	—	—	—	—	—	—	—	—	2	2
S. Africa	3	2	2	—	2	2	10	8†	3	4	36
Tanzania	—	—	—	—	—	1	—	—	—	—	1
Zaire	—	—	3	—	2	1	2	—	5	2	15
Zambia	—	—	2	—	—	—	—	—	—	—	2

* As reported by the Office International des Épizooties.
** Records incomplete.
† Including S.W. Africa.

introduced from Angola by pig-meat products contained in aircraft garbage and fed to swine in the vicinity of Lisbon airport (Manso Ribeiro *et al.* 1958). It was thought to have been eliminated by a 'stamping-out' policy which involved slaughter of 10 637 pigs after 6352 had already died of the disease. However, it reappeared near Lisbon in 1960 and spread to south-west Spain in May of that year; it has been continuously present in the Iberian peninsula since that time, in enzootic or epizootic form, creating an ever-present threat to the swine industry of Europe and elsewhere, as well as causing enormous losses in Spain, estimated by 1977 to total 17 billion pesetas (about £95m).

The number of outbreaks in Spain and Portugal has tended to move in cycles, with peaks at intervals of four to seven years, e.g. 1963, 1967, 1971 and 1978 in Spain (Ordas *et al.* 1983). Although the pathogenicity of the virus undoubtedly declined in the years following its introduction to the peninsula, losses continued to be heavy for a long time. In 1971 nearly 100 000 pigs were slaughtered in Spain bringing losses over the 12 years following introduction of ASF to 737 000; the number of disease foci officially reported from 1957 to 1978 in Portugal was over 22 000 but there has been a rapid decline in the recent past, as also in Spain (Table 105.2).

Table 105.2 New outbreaks of African swine fever in Spain and Portugal

Year	No. of outbreaks of ASF in Spain*	Portugal†
1977	1894	2184
1978	1428	1264
1979	1044	104
1980	447	110
1981	223 (6 months)	25 (7 months)

Source: * Ordas *et al.* 1983
† Vigario *et al.* 1983

The threat of wider dissemination has frequently materialized, as shown in Table 105.3 which summarizes the outbreaks outside Africa which have probably been derived, in most if not all cases, from the Iberian peninsula. France has been invaded by and rapidly cleared of the disease on three occasions, the Italian mainland twice; the islands of Madeira, Sardinia and Malta have all suffered serious outbreaks. In the western hemisphere the disease appeared in Cuba in 1971 and in Brazil and the Dominican Republic in 1978; it spread to Haiti in 1979 and probably from there, with refugees, to Cuba again in 1980 (see Hess 1981).

Morphology and development of the virion of ASF virus

The virus is unique among mammalian pathogens in that the virions are large icosahedral structures, produced in the cytoplasm of infected cells, and possessing a genome consisting of double-stranded DNA. In these respects it resembles, however, the 'iridescent' viruses of insects, lymphocystis virus of fish, and frog virus 3, and it has been suggested that it should be classified with the '*Iridoviridae*' (Matthews 1979) but others consider such a grouping premature (Kelly and Robertson 1973).

The complete ASF virion is 175–215 nm in diameter and, in thin sections, has a hexagonal or pentagonal shell about 12 nm thick, comprising at least two layers and surrounding a dense central nucleoid about 100 nm in diameter (Breese and DeBoer 1966, Haag *et*

Table 105.3 New outbreaks of African swine fever in non-African countries, 1964–1980 (other than Spain and Portugal)

Country	Years with outbreaks	No. of outbreaks	Remarks	References
Brazil	1978/79	226	66 902 pigs slaughtered	Lyra, 1983
Cuba	1971	36	Of 32 524 in affected herds 12 173 pigs died; 536 000 slaughtered. Disease eliminated in 10 weeks.	Oropesa 1971
	1980	56	173 287 pigs slaughtered in 3 provinces. Eradicated in 8 weeks.	Negrin 1983
Dominican Republic	1978/80	374	Island was depopulated and restocked; 190 000 pigs killed.	Rivera 1983 Reichard 1978
France	1964	5	Outbreaks in Pyrenean region and Finisterre.	Larenaudie *et al.* 1964
	1967	1	Outbreak in Pyrenean region.	
	1974		,,	Costes *et al.* 1974
Haiti	1979–	?	Eradication and restocking programme in progress (1983).	
Italy	1967/69 1980	12	These outbreaks were eradicated by destruction of more than 100 000 pigs.	Anon 1967 Hess 1981
Madeira	1965; 1974 1976	?	1965 isolate serologically distinct from Portuguese strains.	Vigario *et al.* 1970
Malta	1978–81	304	Total pig population of ca 80 000 was eventually killed, at a cost of £5m.	Wilkinson *et al.* 1980
Sardinia	1978–81	92	42 500 pigs slaughtered; disease persists.	Contini *et al.* 1983

al. 1966). The nucleoid contains DNA (Breese and DeBoer, 1967, 1969) and some authors describe a thin envelope in the electron-translucent zone between the nucleoid and shell; filaments have been reported to extend from the nucleoid to the inner envelope. Virions released from the cell acquire as they pass through the plasma membrane or through microvilli a lipid-containing envelope which is absent from particles released by cell rupture (Moura Nunes *et al.* 1975, Breese and Pan 1978). Some particles released from intact cells showed, when negatively stained, that the shell appeared to be composed of triangular subunits (trisymmetrons) similar to those described by Wrigley (1970) for *Tipula* iridescent virus (Almeida *et al.* 1967). Thin sections also sometimes show subunits of the shell whose diameter corresponds to those of the trisymmetrons mentioned above (Moura Nunes *et al.* 1977).

The virions develop in areas of the cytoplasm which become 'virus factories'. Electron-dense material accumulates into circular or polygonal segments, the latter finally becoming closed and then acquiring a nucleoid; there is apparently a large excess of circular-capsid segments which are not incorporated into virions. Crystalline arrays of complete particles occur but are uncommon (Moura Nunes *et al.* 1975). The virus 'factories' are equivalent to the cytoplasmic inclusion bodies in infected cells, being DNA positive in preparations stained by the Feulgen or acridine-orange techniques. (Figs. 105.1, 105.2, 105.3, 105.4)

Although there is no evidence for virion development in the nucleus of infected cells, early changes consisting of a clumping of chromatin and condensation on the nuclear membrane occur at about the same time as or shortly before the cytoplasmic alterations. The nucleoli also become vacuolated, fragmented or reduced in size (Haag *et al.* 1966, Moulton and Coggins 1968*b*). The nucleus is essential for virus replication, probably for DNA synthesis (Ortin and Vinuela 1977); ASF-DNA is produced in the nucleus of infected cells and later transported to the cytoplasm. This process is not affected by mitomycin C, which depresses host-cell DNA synthesis (Tabares and Sanchez Botija 1979). It is inhibited by halogenated pyrimidine analogues and by phosphono-acetic acid, which presumably depresses virus-induced DNA polymerase, an enzyme that is increased in infected cells (Moreno *et al.* 1978, Polatnick and Hess 1972).

Released ASFV is rendered rapidly and completely non-infectious by treatment with lipid solvents, such as diethylether, but this cannot be due to their effect on the external lipid envelope, since intracellular virus harvested before envelopment is still infectious (Moura Nunes *et al.* 1975) and rapidly inactivated by ether. ASFV is apparently protected by its lipid coat from the action of proteolytic enzymes, such as trypsin

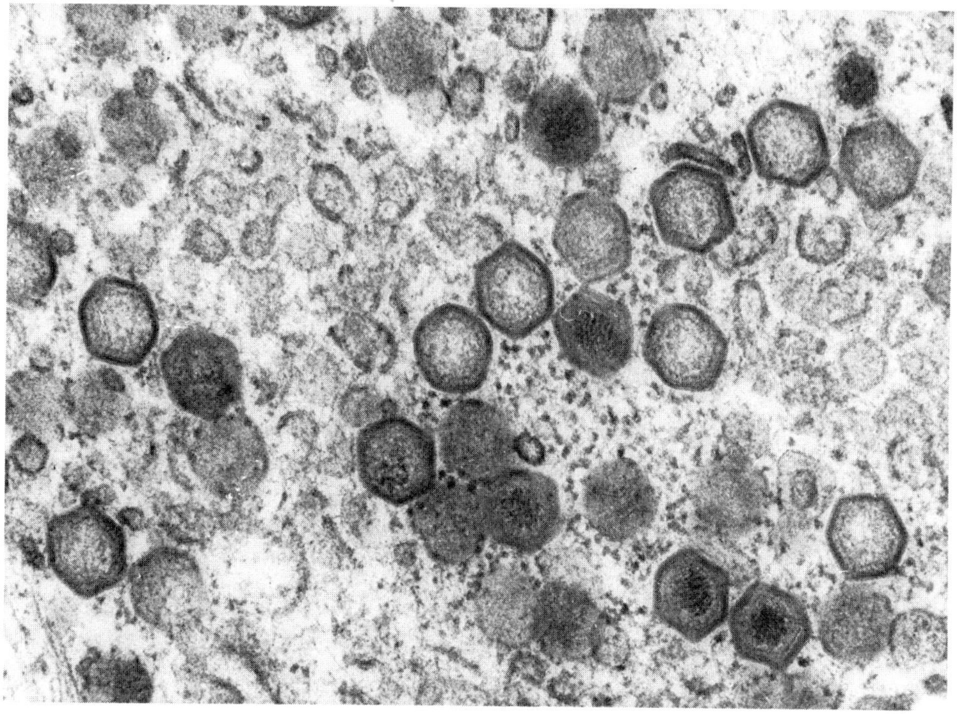

Figure 105.1 CV-1 cells infected 14 hours previously with ASFV and fixed with osmium tetroxide-uranyl acetate. Cytoplasmic 'factory' with prominent polygonal and less dense circular contours. (× 65 000)

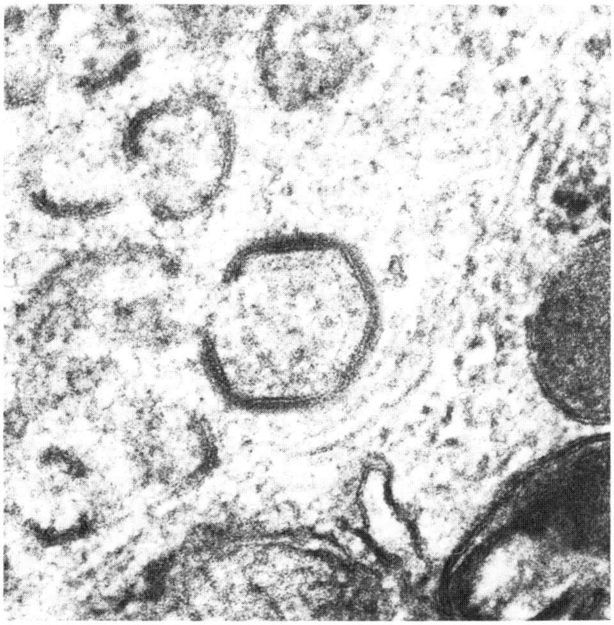

Figure 105.2 Pig bone marrow (PBM) cells, 14 hours post-infection, fixed with glutaraldehyde, post-fixed with osmium tetroxide and uranyl acetate. Inside the dense trilaminar outer shell there is a suggestion of a thinner internal 'envelope'. (× 120 000)

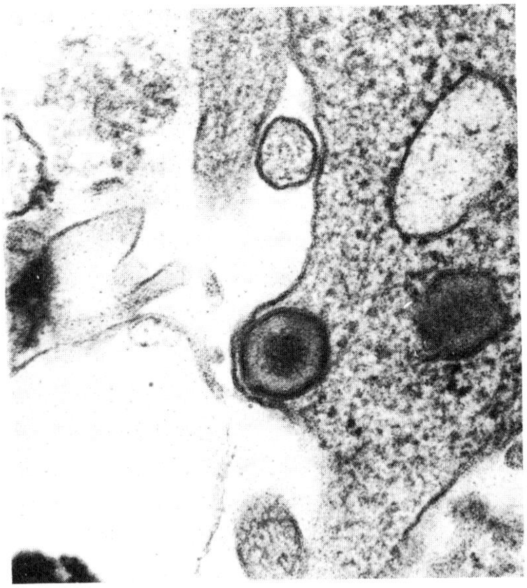

Figure 105.3 Plasma membrane of a PBM cell protruding over a nucleated ASF virion. (× 55 500)

(All figures were kindly supplied by Dr. J. F. Moura Nunes. Figures 105.2 and 105.3 are used with acknowledgements to *Archives of Virology*.)

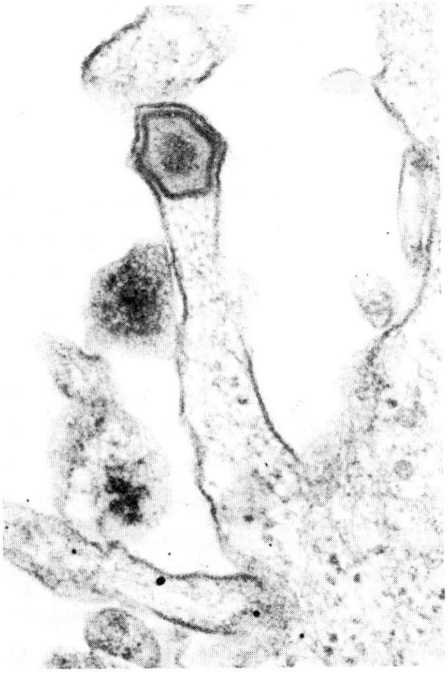

Figure 105.4 A nucleated ASF virion migrating into a microvillus of a CV-1 cell. (× 65 000)

and pepsin, or of nucleases, but is highly sensitive to pancreatic lipase which presumably attacks the envelope (Hess 1971). It is also rapidly inactivated even in the presence of serum by sodium dodecyl sulphate, a surface-active agent, and by oxidizing agents such as sodium hypochlorite.

Infectious double-stranded nucleic acid was first extracted from virions by Adldinger and co-workers (1966). It has a sedimentation coefficient of 60S, buoyant density of 1.700 g/ml, contour length of 58 ± 3 μm and a mol. wt of about 100×10^6. There are cross-links between the strands, towards the end of the molecule (Ortin *et al.* 1979).

Though Black and Brown (1976) reported that ASF virions contained 5 major structural proteins (mol. wt 125, 76, 50, 44 and 39×10^3 respectively), a figure also agreed to by Dalsgaard and colleagues (1976), more recent studies by polyacrylamide gel electrophoresis of intracellular proteins induced by ASFV have shown that they number 34 to 37 of mol. wt 10 to 250×10^3. Of these, 12 are 'early' polypeptides, not dependent on viral DNA synthesis, among which VP12, a highly antigenic structural protein, predominates. At least 9 of the 30+ proteins are antigenic *in vivo*. Intracellular virions contain at least 28 polypeptides of mol. wt 11.5 to 243×10^3. Treatment with NP-40 and 2-mercaptoethanol releases VP15 and VP11.5 to yield subviral particles (CORE I) that have lost some proteins; CORE I particles lose a further major polypeptide, VP73, on NaCl treatment to give a smaller particles (CORE II) with at least 14 proteins (see Hess, 1981; Wardley *et al.* 1983, for a review). Intracellular virus is reported to contain at least 7 glycoproteins; Vigario *et al.* (1977) reported that at least 2 glycoproteins might be associated with the shell.

In addition to the DNA polymerase mentioned above a DNA-dependent RNA polymerase, resembling that in poxviruses, is present in ASF virions; *in vitro* it synthesizes at least 4 classes of capped and polyadenylated RNAs (6 to 14S) which hybridize preferentially with the ends of the DNA (see Hess 1981, and Wardley *et al.* 1983, for reviews).

Treatment of ASFV-DNA with the restriction endonuclease, Eco RI, produces 28 fragments varying from 0.3 to 21.9 Kbp of which 26 (ca 90 per cent of the genome) have been cloned in the vector lambda WES, lambda B and propagated in *Esch. coli*, EK2 strain. According to Talavera *et al.* (1983) the Sal-1 restriction patterns of three isolates from Brazil and the Dominican Republic, grown in swine macrophages, were identical, as were patterns for recent Spanish isolates; there were also similarities between the two groups. Furthermore, genetic variants have been shown to be present in parent virus populations by plaque purification in VERO cells and by comparing their DNA by Sma-1 and EcoR 1 restriction patterns, particularly the latter (Wesley and Pan 1983). Hence, with the proviso that adaptation to 'unusual' cells such as VERO may itself result in variants, DNA restriction patterns do offer a promising means of following the geographical epidemiology and the development of ASFV infections.

The resistance of ASF virus to physico-chemical factors

The virus has long been known to have a considerable resistance to inactivation in the environment (see Scott 1965*a*), serum for example retaining infectivity at room temperature for 18 months (Montgomery 1921); in chilled meat the virus is still infectious after 15 weeks, in processed hams after 5 or 6 months. In the laboratory a fraction of cultured virus remained infectious after heating at 56° for more than an hour in medium with 25 per cent serum, but 60° for 30 minutes killed it completely; stability at these temperatures was reduced by previous storage at 4°. Storage of virus-containing tissues at −20° results in considerable loss of infectivity, whereas it is stable at −70°. Stability at 37° or 4° depends on the presence of protein but, at 4°, is not dependent on Ca or Mg ions. African swine fever virus has a remarkable resistance to extremes of pH; some infectivity was retained after 7 days at pH 13.4 in the presence of 25 per cent serum and one isolate was still infectious after 4 hours at pH 2.7 (Plowright and Parker, 1967, Hess 1971). These data are useful in formulating measures for disinfection or rendering meat products safe for commerce.

In spite of the high resistance of ASFV to heat and pH inactivation, contaminated pig sties were not infectious after as little as 5.5 days in the tropics, whereas they still were at 3 days. This observation may, however, be a reflection of the high dosage required to infect pigs by natural routes (*vide infra*).

The host range of ASF virus

The original vertebrate hosts of ASFV are wild *Suidae* in Africa; of these the most important is the warthog, whose geographical distribution was for so long approximately the same as that of the recorded disease (Anon 1962). The bushpig is less frequently infected (Thomas and Kolbe 1942) and there is one record only of virus isolation from the third African wild pig, the giant forest hog (*Hylochoerus meinertzhagerni*). The wild boar is highly susceptible both naturally (see Scott 1965*a*) and experimentally (Carnero *et al.* 1974); hence it is unlikely to act as a reservoir of infection. Feral pigs (*Sus scrofa*) in Florida (USA) are at least as susceptible as domesticated swine to natural and experimental exposure to ASFV (McVicar *et al.* 1981); and so-called 'wild boars' in Sardinia have also died from naturally acquired disease, as well as being found to be serologically positive (Contini *et al.* 1983).

Wild species shown not to be susceptible when exposed naturally, or to be insusceptible experimentally, include the hippopotamus, hyena, porcupine and white-collared peccary. Cattle, sheep, horses, dogs, cats, mice, hamsters and guinea-pigs are not susceptible experimentally but rabbits and goat kids have been used to establish passage series, as have also fertile eggs (Scott 1965*a*). The role of argasid ticks in the maintenance and transmission of ASFV will be described later.

Infection of African Wild Swine with ASF virus

Most, but not all, warthog populations are infected with ASFV, but those in Zululand, Natal, and Nigeria are an exception (see DeTray 1963, Scott 1965*a*, Pini and Hurter 1975, Taylor *et al.* 1977, Thomson *et al.* 1983). The earlier figures for infection rates in warthogs are unreliable, since often blood and spleen tissue only were examined, whereas more recent observations on the isolation of virus from tissues such as lymph nodes, and serological methods such as agar-gel diffusion precipitation (AGDP) tests, show that virtually all warthogs undergo infection in most enzootic areas. Nevertheless isolation rates fall after the age of one year from peaks which range from 65 to 100 per cent in different areas, down to 8 to 20 per cent in older animals (Plowright 1981). In some regions, e.g. the Serengeti area, young animals are infected very early in life whereas in others, e.g. W. Uganda, for reasons unknown they remain free until after 3 months of age. Viraemia is uncommon—7 per cent only, in one series—and when present, is of low degree—$\lesssim 10^{2.5}$ HAD_{50} per ml except in some neonates (*vide infra*). Virus in naturally infected 'carrier' warthogs occurs predominantly in the lymph nodes, especially the parotid and mandibular nodes, where maximal titres are of the order 10^5–10^6 HAD_{50}/g; liver, kidney and lung are rarely infected and virus is infrequent in the spleen compared with that in some lymph nodes—6–18 per cent v 50–61 per cent (Heuschele and Coggins 1969, Plowright 1981). Figures for other reservoir hosts which are less frequently infected—bushpigs and giant forest hogs—are not available.

No good evidence exists to show that warthogs exhibit any pathological effects as a result of infection, whether naturally acquired or experimentally, though Montgomery (1921) did observe a low-grade pyrexia in the latter. Thomson and colleagues (1980) studied experimental infection in seronegative warthogs 4 months old, but observed no signs of ill health. Viraemia was present between days 3 and 18, reaching $10^{3.7}$ HAD_{50}/ml of blood, and lymphoid tissues attained titres of up to $10^{6.4}$/g. Further experimental infections of seronegative animals are required to investigate warthog-to-warthog transmission and the ability of viraemic warthogs to infect ticks or other vectors.

The method of transmission and maintenance of the virus in wild swine populations for long remained a mystery, since efforts to demonstrate warthog-to-warthog or pig-to-warthog infection failed, perhaps as a result of previously undetected exposure of the recipients. Furthermore, it was not possible to demonstrate transmission from infected warthogs to pigs, a finding which probably reflected the internal localization of virus in the lymphopoietic tissues of the reservoir host, low-level viraemia and insignificant excretion. It was suggested that vertical, i.e. *in utero* or milk transmission, might take place but an examination of 52 fetuses of different gestational ages from 17 warthog sows and of mammary tissues from 5 lactating sows failed to support this hypothesis (Plowright 1981), as did the absence of ASF antibody in fetal sera and amniotic fluids derived from 3 late-term sows (Simpson and Drager 1979). Difficulties in transmission led to suggestions by Montgomery (1921) and others that vector transmission—by lice, biting flies, fleas and ticks, etc.—might be responsible, but no evidence to support this was available until recently.

The first real evidence of tick transmission was eventually obtained outside the old enzootic areas of Africa, namely in Spain. In 1963, Botija reported that the argasid tick, *Ornithodoros erraticus*, which was frequently present in rustic pig shelters, could account for the persistence of virus in a property. The virus was demonstrable in ground-up ticks; and infected tampans (argasid ticks) could transmit the disease in feeding on pigs as well as becoming infected by imbibing viraemic pig blood and retaining the virus for many months. Another argasid tick, *Ornithodoros moubata*, the vector of relapsing fever, was known to be very common in the burrows which are used by warthogs as refuges and homes in many parts of their range. Early attempts to isolate ASFV from *O. moubata* failed, but later it was shown that in many areas of E. and S. Africa these ticks were infected, often to high titre; furthermore, they could readily transmit ASF in feeding on domestic pigs (Plowright *et al.* 1969*a*, *b*) and hence presumably could transfer the virus to susceptible warthogs. Details of the role of ticks in the maintenance and transmission of ASFV will be given in a following section.

Transmission of ASF virus

Montgomery (1921) showed that ASFV was readily transmitted by any parenteral route. He demonstrated by muzzling experiments that close contact between reacting and healthy pigs did not suffice for transmission of the virus; it was necessary for the recipient to nuzzle or ingest infected excreta. He also showed that the route of infection was probably through the upper alimentary or respiratory tracts, since pigs fed with the virus enclosed in a bait did not contract the disease. It is indeed easy to transmit ASF regularly by spraying infected materials into the mouth or dropping it into the nostrils, but the dose of virus required differs with the two routes; the minimal quantity of virus necessary to infect in a single dose by the oral route was 10^5 to $>10^6$ HAD_{50} and the 50 per cent infectious dose was $10^{5.4}$ HAD_{50} (Greig 1972*a*); the requirements by the intranasal route were lower—10^3 to 10^4 HAD_{50}.

A number of experimental studies have shown that the route of penetration of the virus may vary, perhaps with the age of pig, strain of virus and form of presentation. After intranasal infection, the pharyngeal tonsil frequently appears to be the tissue first affected, usually together with the mandibular lymph nodes; the nasal mucosa is not a primary site of virus proliferation. Viraemia leads to generalization of the infection. Similar findings have been reported in pigs infected by the oral route or by natural contact, but Plowright and colleagues (1968) found a regular early localization of virus in the dorsal retropharyngeal mucosa and associated lymph nodes in animals inoculated intranasally.

Though Montgomery (1921) failed to demonstrate, to his satisfaction, airborne infection over a distance as short as 6 inches (15 cm), more recent work has shown that aerosol transmission from heavily infected swine is possible over a distance of 2-3 metres and after 48 hours exposure; the route of infection is then probably through the lower respiratory tract, possibly without affecting a mucosa, to the bronchial lymph nodes (Wilkinson and Donaldson 1977). The lower respiratory route also appears to be implicated in a few cases after natural or intranasal infection; and occasionally the alimentary tract is involved, as shown by early infection of the gastro-hepatic or mesenteric nodes.

Although there is no difficulty in the natural transmission of ASF between reacting and susceptible pigs, it was found by Montgomery and subsequently confirmed by others that in the incubation period or during the first 24 hours of fever pigs do not disseminate the infection to animals in contact, possibly because the virus present in mucosae at this stage is primarily in the blood and rupture of surface vessels is necessary to raise excretion to the level necessary to infect by natural routes. Figures for excretion by various routes are given below.

Systematic studies of excretion of two strains of virus, after natural or experimental (intranasal) infection, showed that the main route of dissemination was from the upper respiratory tract where virus was present in pharyngeal secretions 1–3 days before the onset of pyrexia, reaching 10^2 to 10^3 per swab at the beginning of fever and 10^4 to 10^5 HAD_{50} on the second and third days of reaction. The virus also appeared in decreasing quantities and at later times in nasal, conjunctival, vaginal and oral secretions; urine contained moderate amounts, up to 10^5 HAD_{50}/ml, later in the disease. Faecal excretion was demonstrable from the onset of fever, increasing greatly especially in those animals which survived longer and contracted dysentery.

'Carrier' animals have been investigated less thoroughly but excretion is probably intermittent, often corresponding with periods of recurrent fever and never attaining the degree seen in acute cases. Greig (1980) found that 'recovered' or chronically infected pigs had moderate amounts of virus in lymph nodes, similar to those found in naturally infected warthogs. He also observed that some animals retained low concentrations of virus in the palatal tonsil, and many excreted virus detectable periodically in pharyngeal swabs but not from other sites. Although the level of pharyngeal excretion was low, up to about 10^3 HAD_{50} per swab, this was sufficient to infect susceptible swine placed in contact with the donor for long periods. Pigs inoculated 110 or 135 days previously with Dominican and Brazilian isolates failed to transmit ASF by contact to susceptible controls, although they still retained virus in their tissues (Mebus and Dardiri. 1980). Partly attenuated virus from Malta was no longer isolable from recovered pigs six months after infection (Wilkinson *et al.* 1981).

As already noted, the methods whereby ASFV spreads from wild swine to domestic pigs are still inadequately understood. The disease in Africa usually occurred when warthogs and pigs foraged over the same ground; but there was sometimes circumstantial evidence suggesting that virus transfer took place when parts of warthog carcasses were brought into piggeries or fed to pigs (De Tray 1963, Heuschele *et al.* 1965). Thomson and co-workers (1979) were successful in infecting pigs by feeding minced tissues from experimentally infected warthogs, mixed with pig meal; a dose of as little as 10^5 HAD_{50} was found to be infectious. Others, however, had been unsuccessful with up to $10^{6.1}$ HAD_{50} of warthog virus (Plowright *et al.* 1969*a*).

Argasid ticks and the transmission of ASF virus

The importance of argasid ticks in the epizootiology of ASFV varies considerably in different geographical locations. Thus some areas in E. Africa, e.g. the Serengeti and neighbouring regions, have infected warthog populations and many burrows contain *Ornithodoros moubata*, which is frequently infected. In other areas, e.g. W. Uganda, the warthogs are also infected but the numerous ticks are virtually free of virus. In still other areas, central Kenya for example, warthogs all become infected early in life, but there are no argasid ticks or other proven vectors (Plowright 1977). A clear-cut association is seen only in southern Africa where the occurrence of ASFV in warthogs is determined by the distribution of *O. moubata* in their burrows (Thomson *et al.* 1983). Finally, outside Africa, it is only in S.W. Spain that tick transmission has been proved to be of importance. The disease spreads readily among pigs, without any need for a vector.

In many areas of E. Africa, 30–95 per cent of all warthog burrows are infested with argasid ticks (*Ornithodoros moubata porcinus*), sometimes in enormous numbers ($\geq 10\,000$); approximately 30–60 per cent of infested burrows in the Serengeti and adjacent areas contain ticks infected with ASFV and the rate of tick infection is about 0.45 per cent. Burrow infection rates in the N. Transvaal (S. Africa) are about 40–60 per cent, when samples of $\geq$ 100 ticks are examined (Pini 1977). Virus is found in all stages of the tick life cycle, from unfed first-stage nymphae to adults; infection rates rise, from 0.15 to 2.24 per cent, abruptly between nymphal and adult stages. A maximum of about 5 per cent positives is found in adults from some burrows; in S. Africa similar figures are recorded, 1.62 per cent of nymphae and 3.45 per cent of adults being infected (Pini 1977). In southern Africa infection rates in adult females were about 4-fold greater than in adult males (Thomson *et al.* 1983). Single ticks contain large quantities of virus, usually ranging from $10^{4.5}$ to $10^{7.5}$ HAD_{50}, with a mean of $10^{5.6}$ HAD_{50}. Naturally infected tampans readily transmit virus to experimental pigs, even when feeding singly, and the incubation periods are often short, thus indicating a large dose of ASFV inoculated (Plowright *et al.* 1969*a*, Plowright 1977).

The mechanisms of tick transmission are manifold. Ticks frequently excrete virus (up to $10^{4.7}$ HAD_{50}) in the coxal fluid, which is exuded from fluid-regulatory glands during and after engorgement—a process which takes at most 30–45 minutes. Virus is also sometimes present in considerable quantities in the Malpighian excrement, often voided during feeding, and in the genital secretion of adult female tampans. All of these materials might contaminate the skin of vertebrate hosts and give rise to percutaneous infection, either through pre-existing discontinuities or through wounds inflicted by ticks. Salivary excretion of ASFV can be demonstrated after pilocarpine administration; it is probable that this is the main route of transmission. Finally, pigs ingesting crushed ticks, containing 10^6 to 10^7 HAD_{50} of virus, might also be infected orally. The rapid feeding habits of *O. moubata* ensure that it is not usually carried outside burrows on vertebrate hosts, but this happens occasionally and thus it could reach unfenced ground used for domestic swine.

The mechanism of transfer of virus from wild swine to 'clean' ticks was at first unclear, since the minimal dose of virus necessary to give persistent infection of tampans (10^2 to 10^4 HAD_{50} for different strains, Plowright *et al.* 1970*b*) could be acquired only from blood with a titre of at least 10^3 to 10^4 HAD_{50} per ml. Such levels had not been found anywhere in warthogs of any age until Thomson *et al.* (1980) reported a transient viraemia greater than 10^3 HAD_{50} per ml in experimental animals about 4 months old. It was also found that 6 of 20 neonatal warthogs in S.W. Africa were viraemic and five of these had blood titres which reached or exceeded $10^{3.6}$ HAD_{50} per ml, i.e. they were probably adequate for infection of some ticks.

The degree of proliferation of virus in single ticks can be of the order of 10^5 to 10^6. Primary multiplication occurs in the gut diverticuli, from which it probably spreads to the haemocoele, and then to the salivary gland, coxal sacs, Malpighian tubules and reproductive tract (Greig 1972*b*). The natural occurrence of virus in unfed first-stage nymphae, as well as experimental observations (Plowright *et al.* 1970*a*), show that transovarial infection occurs and infection rates in the offspring can be very high—up to 80 per cent. Virus antigens have been shown to be present in the ova of infected females. Infected male ticks transmit virus to a high proportion of clean females with which they copulate, whereas transfers from female to male are rare. Sexually infected females suffer from a generalized infection and transmit ASFV during feeding, but transovarial infection has not yet been demonstrated in them. Virus is probably excreted in the seminal fluid of the male and transferred in the spermatophore to the uterus of the female (Plowright *et al.* 1974, Greig 1972*b*).

Single infected ticks can retain ASFV for at least three years and remain capable of transmitting the virus all this time. Trans-stadial transmission occurs regularly, although there is evidence that a proportion of experimentally infected ticks cleanse themselves of virus over long periods. There is no doubt that ASFV, at least in many parts of Africa, is well 'adapted' to maintenance in *O. moubata*; and, as it has been reported that this species can live as long as ten years without a blood meal, its potential as a reservoir is very considerable. In Spain, it was considered that 50–55 per cent of new outbreaks in the early 1960s were attributable to ticks; ten years later this had declined to about 5 per cent. In the USA the soft tick *O. coriaceus* was recently proved to be capable of maintaining ASFV and transmitting it to pigs (Groocock *et al.* 1980).

In areas of Africa where ASFV is maintained in wild swine without the assistance of argasid ticks, as also in countries elsewhere with enzootic ASF, attempts have been made to implicate other vectors such as lice but the results have been either inconclusive or negative. Efforts have also failed to demonstrate natural or experimental infection of fleas, ixodid ticks (especially *Rhipicephalus* and *Amblyomma*, spp.) or biting flies. Nevertheless, it is difficult to believe that a disease associated with such a high viraemia is not sometimes transmitted mechanically by blood-sucking Diptera.

Clinical signs of African swine fever

In the natural reservoir hosts, i.e. wild swine and ticks, no significant pathological findings have been reliably reported, but the acute disease in domestic pigs is one of the most devastating known. The incubation period in domestic swine, i.e. the time to first pyrexia after natural 'contact' exposure, is usually 5–15 days. The time to death thereafter is usually 4–5 days, although it is clear that strains freshly isolated from reservoir hosts often require longer to kill (7–8 days). After parenteral inoculation incubation periods are reduced to 2–5 days.

The case-fatality rate is extremely high for the majority of African isolates, often exceeding 99 per cent. Subclinical infections are nevertheless noted occasionally, even with the most virulent strains. As the disease progresses the temperature often reaches 42° or higher and then, after a sustained plateau, falls rapidly. There are increasing areas of cyanosis of the skin of the abdomen and the extremities, particularly ears, snout and limbs. Discrete haemorrhages may also occur in the skin. Vomiting is common and diarrhoea, with or without dysentery, can be a feature with some strains. Pregnant sows usually abort. The breathing becomes rapid and shallow, the pulse rapid; when lung oedema is prominent a blood-stained watery fluid or froth flows from the nares of recumbent animals, which also exhibit dyspnoea.

Acute outbreaks in domestic swine may begin with sudden deaths, without premonitory signs, followed by a lull and second wave within 7–10 days. Anorexia or other easily detected signs are often not displayed during the first 24–48 hours of pyrexia, but this prodromal phase is followed by listlessness, mucopurulent ocular and nasal discharges, excessive thirst, huddling together and disinclination to stand. Some animals show incoordination of the hindquarters, convulsions or muscular tremors; finally they often become comatose or show paresis.

In addition to the acute or hyperacute cases, such as described above, animals are increasingly encountered, even in Africa, which exhibit a subacute or chronic form (De Kock *et al.* 1940, DeTray 1957). Such cases were seen especially in Angola and later in Spain, Portugal and those countries, such as France and Italy, which probably acquired infection from the peninsula. They may begin with the typical acute form or show moderated signs from the outset; irregular and remittent fever is accompanied by stunting, emaciation, oedematous swellings of joints, tendon sheaths and adjacent subcutaneous tissues. Some pigs have blackened areas of skin necrosis over bony protuberances, followed by deep ulcerations, especially when they are recumbent for long periods.

The development of oedema of the intermandibular region is usually a sign of chronic fibrinous pericarditis; many of the animals affected die suddenly, particularly when forced to move quickly or when struggling during handling. Dyspnoea with abdominal breathing is evidence of pulmonary localization.

Whilst the subacute and chronic signs were for long regarded as common to infections caused by both African and Iberian strains, the latter subsequently produced many more inapparent and atypical cases, disease signs and mortality being most frequent in suckling pigs (Sanchez Botija *et al.* 1969). Recent isolates from Brazil and the Dominican Republic produced indefinite fever and lethargy but mortality was still appreciable – 4 out of 9 in one experiment – although the isolates were misleadingly referred to as 'of low virulence' (Mebus *et al.* 1978). In fact, it was characterisitic of many outbreaks in newly infected countries that they were first noticed as foci of acute lethal disease but rapidly changed, following elimination of these foci, to a form dominated by subacute, chronic or inapparent infections (Hess 1981). A considerable variability in the severity of clinicopathological manifestations often persists, however, and great diversity has been observed in experimental pigs infected with recent European or American isolates (Wilkinson *et al.* 1981, Hess 1981).

The reasons for the increase in less acute or inapparent cases in countries with enzootic infection of swine are debatable. First, in Portugal, efforts to develop attenuated live-virus vaccines were followed by their widespread use in the early 1960s. The results were calamitous, as severe delayed reactions occurred frequently when vaccinated animals were stressed and some died of acute disease. The modified virus may well have persisted and reverted to virulence. Second, it is often postulated that ASFV is becoming 'adapted' to domestic swine, a process which was not possible in those African countries which adopted effective control measures. Pini (1977), however, reported natural outbreaks of reduced virulence in the N. Transvaal enzootic area.

Finally, Hess (1981) has put forward the hypothesis that isolates of ASFV, as derived from swine, are mixed populations which become selected for reduced or greater pathogenicity by immunological pressures in the host.

The pathology of ASF

The lesions in animals dying of acute ASF (see Maurer *et al.* 1958, Moulton and Coggins 1968*a*) closely resemble those of acute hog cholera (European swine fever), which is caused by a completely different RNA virus, belonging to the Pestivirus group. The outstanding characteristic of ASF is enlargement and haemorrhagic transformation of lymphoid tissues, including superficial as well as visceral nodes. Those of the head and neck, and particularly the gastro-hepatic and renal groups, may appear as blackened haematoma-like structures; partly affected nodes show predominantly cortical haemorrhage. In subacute or chronic cases of ASF a fleshy or rubbery enlargement of the lymph nodes is frequently reported, sometimes with patchy congestion. Some strains of virus cause enlargement of the spleen, with the pulp soft and intensely congested, but haemorrhagic infarcts along the edges, such as are frequent in acute hog cholera, are rare.

Serous cavities are usually affected in both acute and chronic cases. In the former they generally contain a moderate to large excess of watery fibrin-containing fluid, often with blood clots; the abdominal viscera or lungs frequently exhibit petechiae or larger haemorrhages on the serosae. The heart shows epicardial petechiae or larger blood splashes and the endocardium massive extravasations. In chronic cases the epicardium may be greatly enlarged; and thick continuous deposits of fibrin, often in process of organization or forming adhesions, are found all over the heart surface. Similarly, extensive fibrinous pleurisy, with adhesion formation and massive fluid exudation, is not unusual in subacute and chronic forms.

The nasal cavities often contain exudates, mucopurulent or blood-stained, and the tonsils are frequently haemorrhagic in acute cases. The tracheo-bronchial tree is filled with watery froth and the mucosae are intensely congested, with haemorrhages. The lungs are usually oedematous, with prominent gelatinous interlobular septa and haemorrhages. In chronic cases more or less extensive consolidation is seen, sometimes leading to necrosis or fibrotic changes.

Lesions in the alimentary tract include congestion or haemorrhage of the gastric mucosa with necrosis and erosion, clearly seen in the oesophageal area. The large intestine often shows haemorrhagic or blackened streaking of the mucosa. In subacute or chronic cases raised concentric ('button') ulcers, such as those frequent in hog cholera, are rarely seen. The liver is often swollen and, in acute cases, a gelatinous and haemorrhagic thickening of the wall of the gall-bladder is characteristic.

The kidneys of acute and subacute cases show many cortical or medullary haemorrhages and petechiae are present in the mucosa of the urinary bladder; renal infarcts are rare (cf hog cholera). In chronic cases a form of glomerulonephritis, possibly immune-complex in origin, has been reported.

Meningeal congestion and small haemorrhages may be found macroscopically in the c.n.s.

The pathogenesis of ASF at the cellular level

The distribution of lesions in acute ASF indicates that the virus has a predilection for those cells in the lymphoreticular tissues which are associated with phagocytosis of antigens, i.e. the dendritic reticular cells of the perifollicular, paracortical and stromal areas of lymph nodes, the arteriolar sheaths and perifollicular red pulp of the spleen, and Kupffer cells in the liver sinusoids (Moulton and Coggins 1968*a*, Konno *et al.* 1971, 1972, Mebus *et al.* 1983). In the enlarged hyperplastic nodes of persistently infected survivors a considerable proliferation of reticulum cells has been described. In areas of pneumonic consolidation there is a mononuclear and macrophage infiltration of alveolar walls and spaces, followed sometimes by necrosis and fibrosis. Accumulations of myeloid cells have been reported in the spleen and liver of subacute and chronic cases (Konno *et al.* 1971, 1972).

It has been shown by immunofluorescent staining that viral antigens appear first in reticulum cells and macrophages, whereas the lymphocytes, which often undergo necrosis and karyorrhexis, stain positively later. Megakaryocytes and myeloblasts, blood monocytes, vascular endothelium and the tunica media of vessels also reveal viral antigens (Colgrove *et al.* 1969). An angiitis is, in fact, widely distributed in diseased tissues; it is accompanied by hyaline or fibrinoid medial change, with swelling or karyorrhexis of the endothelium and is sometimes succeeded by thrombosis. The vascular lesions occur not only in the lymphoid tissues but also in the skin, kidneys, alimentary and respiratory tracts. In the CNS perivascular 'cuffing' is constantly present, with neuronal degeneration. It is also significant that cultures of blood monocytes, macrophages, lymphocytes and endothelial cells all support the growth of ASFV *in vitro* (Malmquist and Hay 1960, Wardley *et al.* 1977, Wilkinson and Wardley 1978).

Viral antigens also occur in the leucocytes of the circulating blood (Colgrove *et al.* 1969) and cytoplasmic inclusion bodies can be seen in monocytes in Giemsa-stained blood smears. The viraemia of ASF is, however, largely due to virus associated with the erythrocytes (Plowright *et al.* 1968), more than 90 per cent of circulating virus being present in this fraction of the blood (Wardley and Wilkinson 1977); some virus particles are demonstrable in indentations of the membrane of or even inside erythrocytes (Rodriguez *et al.* 1983).

Haematological changes in ASF include a leucopenia developing with the pyrexia, accompanied by a relative lymphopenia and neutrophil leucocytosis, with a spectacular increase in juvenile forms (De Tray and Scott 1957, Wardley and Wilkinson 1977). According to the latter workers the cytolytic effect of ASFV is predominantly on the B lymphocytes. A defect in blood clotting which occurs in acute ASF may be attributable to the destruction of megakaryocytes (Colgrove *et al.* 1969).

Cultivation of ASF virus and the haemadsorption phenomenon

A major advance in the *in vitro* detection, assay and cultivation of ASFV was reported by Malmquist and Hay (1960), who found that large granular cells in cultures prepared from swine bone marrow and blood (buffy-coat) showed adsorption of erythrocytes, followed by cytolysis; fibroblasts were unaffected. The haemadsorption was inhibited by the serum of survivor pigs, the cytolytic effect was not; however, no haemagglutination occurred with virus-infected fluids. These observations have been confirmed repeatedly, and are still the basis of many diagnostic and assay techniques, especially since prior 'adaptation' of virus is not required. The cells concerned are mononuclear phagocytes, some of which differentiate in culture to become macrophages; the monocytes do not multiply in culture, but degenerate progressively over 10–14 days. For use in assays of ASFV cultures are most sensitive at 2–4 days after seeding. In addition to haemadsorption, infected cells show nucleolar degeneration and DNA-positive cytoplasmic inclusions, which fluoresce with FITC-labelled antibody. Haemadsorption occurs as early as 8–10 hours after infection and several layers of erythrocytes usually adhere to give a large cluster; cytolysis is delayed to 20 hours and affected cells detach progressively over the ensuing 48 hours (Moulton and Coggins 1968*b*). Serially cultivated and differentiated macrophages from the blood, lungs or peritoneal cavity of pigs are more sensitive to attenuated than to virulent virus, but both types can support persistent infection (Wardley *et al.* 1977). Complete destruction of monolayers required 100 times more virulent virus than the attenuated Uganda strain.

Macrophages from man, rabbit, guinea-pigs, hamsters or rats are not susceptible (Enjuanes *et al.* 1977) but cells from goats are (Dardiri 1966). Haemadsorption assays using mononuclear phagocytes from pigs in the wells of microtitre plates have been described (Greig 1975, Enjuanes *et al.* 1977). For the assay of ASFV direct from the tissues of pigs, wild swine or ticks, leucocyte cultures are normally about as sensitive as pig inoculation. Possible exceptions are strains of virus which are non-haemadsorbing (*vide infra*) and, though still cytolytic, may cause difficulties in end-point determination. Apart from their sensitivity, haemadsorption assays in swine phagocyte cultures are ASFV-specific.

The mechanism of haemadsorption is not known. It is not related to the presence of virions near the plasma membrane or in the microvilli of the infected cell (Breese and Hess 1966); it is possibly due to a virus-associated antigen located in the cell membrane whose activity does not survive disruption of the cell. Haemadsorption-inhibiting (HAdI) antibodies, which are 'strain' or 'type' specific, appear in the serum of survivor pigs; and HAdI tests have been used in 'typing' strains (Malmquist 1962, Hess 1971, Vigario *et al.* 1974). It was shown, for example, that the 1957 outbreak of ASF in Portugal was associated with an isolate different from that of the 1960 epizootic, but the latter had subsequently remained the usual type isolated outside Africa. Isolates from Africa differ among themselves and do not usually appear to be consistent in a single geographical area.

Non-haemadsorbing (HAd−) variants of ASFV have been reported in laboratory strains (Hess 1971), and have been isolated from outbreaks of disease in Portugal (Vigario *et al.* 1974) and South Africa, where such isolates were also demonstrated in ticks (Pini

1976). Variation in the haemadsorbing property therefore occurs in viruses maintained in reservoir hosts, being independent of culture or pig passage (Pini 1977). The HAd– character is maintained on serial passage in cell cultures but sometimes reverts on back-passage in pigs. It was suggested that HAd– strains were of lowered virulence, being associated, for instance, with chronic pneumonic forms. However, more recent work has shown that some stable HAd– strains in S. Africa have high virulence; furthermore, HAd– isolates of low virulence can also protect against virulent HAd– isolates. The antigen responsible for haemadsorption is not, therefore, necessary for generating immunity.

Malmquist (1962) demonstrated that, with or without initial propagation in porcine leucocytes, strains of ASFV could be adapted to growth with cytopathogenicity in lines of porcine or even bovine and ovine kidney cells. Virus yields, however, were low in these systems and cytopathic effects at end-point dilution were difficult to detect, although haemadsorption was demonstrable. Subsequently, primary pig kidney or chick embryo cell cultures were shown to be susceptible, though 'adaptation' was necessary to improve virus yields and cytopathic effects.

A number of non-porcine cell lines, including BHK-21 (baby hamster kidney), AGMK, VERO, MS and LLC-MK2 (monkey kidney cells), have also been found useful, VERO and LLC-MK2 requiring much shorter periods for 'adaptation' (Hess 1971, Pini 1977). Prolonged serial passage in these cell lines, as also in primary porcine cultures, may lead to attenuation of the virus (see Coggins 1974). Cytopathic viruses can also be assayed in porcine and other lines by conventional plaque techniques under agarose (Parker and Plowright 1968, Enjuanes *et al.* 1976). Pini (1977) found that unadapted viruses produced cytolytic plaques in LLC-MK2 cells under agarose; others have used immunofluorescence or immunoperoxidase techniques to detect plaques (Tessler *et al.* 1973, Pan *et al.* 1978). The size of plaques varies from <1 mm to 3 mm but these variations have not been related to pathogenicity or haemadsorption.

The growth of ASFV has been studied in primary cultures of mononuclear phagocytes and in a variety of cell lines to which it has been adapted. In the former, DNA synthesis begins 6–7 hours after inoculation and progeny virus first appears intracellularly at 10–12 hours. Release of virus commences soon thereafter and reaches a maximum of ca 10^7 HAD_{50}/ml at 36–72 hours. Cytopathic effects are visible at 20 hr, reaching a maximum at 40 hr; free virus eventually exceeds cell-associated infectivity in this system.

In lines of pig kidney or other susceptible cells well adapted virus shows an eclipse phase of 12–24 hr, followed by a logarithmic increase to 30–48 hr, when a plateau is attained. During this period the titre of cell-associated virus remains higher, by 0.3 to 1.0 $\log_{10}$ units, than that of spontaneously released virus. Cytopathic effects are usually first detected on days 3 to 4 and are extensive or complete by 4–6 days (see Hess 1971 and Coggins 1974).

The antigens of ASF virus

By immunodiffusion tests with the sera of recovered carrier pigs or of warthogs, and antigens prepared from infected cell lines, up to seven lines of precipitation can be obtained which are virus-specific but not isolate-specific. One or two ASFV-specific antigens are also demonstrable in the tissues of infected pigs, especially the spleen, liver, kidney and lymph nodes; the precise number in each case depends on the serum and source of antigen. Similar or identical common antigens are thought to be reactive in CF, immunofluorescence and ferritin-conjugated antibody tests. The antigens reacting in CF-tests are not denatured by treatment with methanol or acetone-ether mixtures (Boulanger *et al.* 1967).

The relation of these antigens to viral or virus-associated proteins is not established, but Stone and Hess (1965) found by isoelectric precipitation that the common antigens were precipitated at pH 5, together with another antigen which was strain-specific and separated from the mixture by methanol. Dalsgaard and co-workers (1976) used crossed immuno-electrophoresis to separate five antigens from ASFV-infected cells, diffused against 'hyperimmune' pig serum; one of these antigens was a glycoprotein.

Immunity to ASF virus

The majority of pigs recovered from infection with one ASFV isolate are clinically resistant to challenge with homologous and, sometimes, related viruses (Malmquist 1963, Manso Ribiero *et al.* 1963, Hess *et al.* 1965). The immunizing virus can be a highly virulent strain or an isolate with naturally or experimentally reduced virulence; repeated passage in cell cultures, usually porcine bone marrow or kidney cells, or line cells such as Vero, for example, has been used to reduce, often notably, the pathogenicity of isolates of ASFV. The protection afforded may, broadly speaking, be restricted to the haemadsorption group to which the viruses belong; thus, attenuated Lisbon/60 virus protects not only against itself but also, clinically at least, against isolates from the Dominican Republic and Brazil (Mebus and Dardiri 1980). It is significant that animals recovered from infection with a Dominican isolate nevertheless showed viraemia after challenge with virulent Lisbon/60 virus; a similar but better defined superinfection was recorded by Thomson *et al.* (1979). These observations indicate the potential dangers of employing protective vaccines in an infected environment, at least when eradication of the disease is the objective. If the superinfecting virus is excreted in adequate quantities, with or without 'activation', then transmission to susceptible pigs might occur.

Warthogs and bushpigs show no clinical or pathological effects of infection, although they acquire ASFV early in life and retain it to maturity. The basis of this resistance is unknown and is in complete contrast to the extremely high and usually rapid mortality induced in domestic swine or the wild boar. Both the African *Suidae* and recovered domestic swine form antibodies that react, for example, in complement-fixation, precipitation (AGDP) and indirect immunofluorescence tests.

These antibodies are 'group-specific', i.e. they react with all isolates of ASFV and do not distinguish between them, as do HAdI antibodies (see Cultivation of ASFV, p. 548). They circulate in quantities sufficient to constitute a hypergammaglobulinaemia in some chronically infected swine and to give positive results in iodine agglutination tests (Pan *et al.* 1970, 1974*a*). However, in spite of the massive amounts of antibody produced and the resistance of such recovered carrier pigs to challenge with homologous virus, no neutralizing activity is consistently demonstrable. Reduced virus doses and the use of long exposure periods do not help; and the addition of complement or antiglobulin to virus-serum mixtures fails to affect the outcome significantly (Parker and Plowright 1968, De Boer *et al.* 1969).

Furthermore, multiple inoculations of inactivated viral antigens with adjuvants, into pigs or non-susceptible heterologous hosts such as rabbits, guinea-pigs and chickens, have usually failed to induce antibodies neutralizing virus infectivity (De Boer 1967, Coggins 1974). That antibodies are directed against constituents of the virion can, nevertheless, be inferred from studies with ferritin-labelled antibody in thin sections of infected cells (Breese *et al.* 1967).

Circulating antibody becomes detectable about seven days after infection with attenuated viruses and reaches a peak at 5–6 weeks; there is an initial and variable production of 19S macroglobulin, followed later by a 7S (IgG) response (see Coggins 1974).

Antibody binding to pig kidney cells (IBRS-2) which have been infected at least 11–12 hr previously is present in pigs recovering from infection with or hyperimmunized by African and Iberian isolates. In an homologous system, with porcine complement, antibody which causes the lysis of ^{51}Cr-labelled infected cells is demonstrable 14–15 days after infection, whereas immunoglobulin attaching to infected cells in RIA tests is detectable as early as 5 to 7 days. The former may have a degree of isolate specificity, resembling that of HAdI antibody, and might be useful epidemiologically. It may help to eliminate from the tissues infected cells, such as macrophages and lymphocytes, which express viral antigens on their surface (Norley and Wardley 1982).

Some of the lesions of chronic ASF, especially those in the lungs, perhaps also in the kidney (Pan *et al.* 1975), together with the hypergammaglobulinaemia which develops in more chronic infections (Pan *et al.* 1970) are suggestive of an immunopathological process, possibly a type of auto-immune reaction. Slauson and Sanchez-Vizcaino *et al.* (1981) found deposits in the glomeruli of pigs surviving more than two weeks after infection. The deposits consisted of ASFV antigens, antibody and C3; they were associated with an IgE-like antibody on the surface of basophil leucocytes which, when complexed with specific antigens, caused degranulation and release of vasoactive amines. There was also a reduction in circulating haemolytic complement but no lesions of an immune-complex glomerulonephritis.

Interferon does not inhibit ASFV replication in cell cultures, nor do ASFV-infected porcine cells produce interferon (Wilkinson 1977, Wardley *et al.* 1979).

The absence of neutralizing antibody in infections by ASFV has focused attention on cell-mediated immune reactions, particularly in pigs given partly attenuated virus and then challenged with the homologous virulent strain. Delayed hypersensitivity is not demonstrable by intradermal inoculations of ASFV 'soluble' antigens in infected pigs; but the migration of blood polymorphs and lymphocytes is inhibited for up to 6 months by these preparations, and 'immune' lymphocytes produce migration-inhibition factor in their presence. At 4–10 weeks after infection there is often a significant lymphocytosis attributable mainly to an increase in 'null' lymphocytes (Hess and Pan 1977, Shimizu *et al.* 1977). Wardley and Wilkinson (1980*b*) found that acute lethal infections caused a lymphocytopenia which affected predominantly B cells; infection with attenuated virus produced within 10 days short-term sensitization of both B and T lymphocytes to ASFV antigens *in vitro*. Cell-mediated responses to other antigens are, like heterologous antibody responses, not inhibited in persistently infected pigs.

The effect of live ASFV on mitogen-driven assays of peripheral blood leucocytes *in vitro* was studied by Wardley (1982). Live attenuated isolates caused a dose- and time-dependent suppression of both B and T lymphocytes, whereas virulent virus caused an increased uptake of ^{3}H-thymidine. Attenuated strains also suppressed IgG production in cultures treated with the B-cell stimulant, poke-weed mitogen (PWM), whereas virulent strains increased secretion of IgG (Wardley 1982)—an effect mirrored in the more rapid appearance of antibody in pigs which died most rapidly from infection with these strains (Wardley and Wilkinson 1980*a*).

In contrast to these findings Sanchez-Vizcaino *et al.* (1981, 1983*b*) observed that T-cell responses to B- and T-cell mitogens decreased in pigs infected with highly virulent strains, whereas they increased in infections by strains of low virulence. Small increases in lymphocyte blastogenesis were also noted in response to crude viral antigens, these reaching a peak at 4 weeks after infection. Norley and Wardley (1983) found that the NK (natural killer) cell activity in peripheral blood mononuclear cells was diminished in early acute ASF but this did not appear to be important in recovery from ASF, and cultured cells maintained their NK activity in the presence of ASFV.

The diagnosis of ASF

Laboratory confirmation of a diagnosis of ASF is particularly important when suspected cases occur in countries hitherto free from the disease. It has become increasingly difficult by means of clinico-pathological criteria, because the virulence of the strains current in pigs has declined and the resemblances to hog cholera or production of indefinite clinical syndromes have increased. The methods used for laboratory confirmation are diverse; they have been summarized by Hess (1981) and include the following:

1 Techniques for ASF virus antigen detection in tissues

These include direct immunofluorescent (IF) staining of smears and cryostat sections with conjugated serum from survivor-carrier pigs. The method is simple, effective and can give positive results in 1 or 2 hours; it is however unsuccessful when free antibody is present, as in late acute or more chronic infections. Thus, the method was very effective in Spain, revealing 76 per cent of positive samples in 1970 but only 33 per cent by 1974 (Botija 1977). The tissues of choice are liver, lymph node and kidney, but smears of blood leucocytes are also used. Direct IF tests are employed to confirm virus infection in cell cultures, particularly with non-haemadsorbing isolates.

Antigens may also be detected by complement-fixation (CF) tests, with or without normal calf serum as an enhancer of fixation (Cowan 1963) or by precipitation in agar-gel AGDP (Coggins and Heuschele 1966); more recently enzyme-linked immunosorbent (ELISA) assays (Hamdy and Dardiri 1979) and radio-immunoassay techniques (Crowther *et al.* 1979) have been developed which have been applied to large-scale antigen or antibody detection tests. They have the disadvantage, however, compared with virus isolation, that they are unable to detect small amounts of infectious virus early in the disease (Wardley *et al.* 1979, Wardley and Wilkinson 1980*a*).

2 Techniques for the detection of infectious ASF virus in tissues

The inoculation of infected blood or lymphoid tissue suspensions into cultures of peripheral blood monocytes or bone marrow cells is a highly sensitive method of isolating haemadsorbing strains of ASFV. Even non-haemadsorbing isolates produce cytopathic effects, the origin of which can be confirmed by direct immunofluorescence on cellular sediments. Sometimes one to three passages are required before an isolate exhibits haemadsorption, but usually positive results are obtained in 1-10 days.

The ultimate method of confirmation is to inoculate pigs, a proportion of which are immunized against hog cholera (HCV); all will succumb if ASFV is present, but concurrent infection with HCV and ASFV is by no means infrequent.

The application of HAdI tests for investigating the serological grouping and origin of isolates may have value epidemiologically.

3 Techniques for the detection of antibodies

The detection of antibodies as a diagnostic procedure has become increasingly important in recent years with the emergence of strains of reduced virulence. Antibodies can be detected in serum or tissue—spleen, lung, lymph node, liver—extracts by a variety of methods, including IIF, CF, AGDP, immunoelectro-osmophoresis (IEOP), reverse radial immunodiffusion (RRID), ELISA (Hamdy and Dardiri 1979), and radioimmunoassay (RIA) techniques.

The application of diagnostic tests for ASF in 'open' laboratories, situated in countries hitherto free of the disease, is facilitated by knowledge that potent antigens for use in CF, AGDP or IEOP tests can be imported safely after prior treatment with substances such as β-propiolactone, acetylethyleneimine glycidaldehyde (Stone and Hess 1967) or binary ethyleneimine (Schloer 1980) which destroy infectivity without seriously impairing antigenicity. Of these, the most widely used methods have been the IIF and IEOP tests, the two together having a combined effectiveness of 98 per cent in Spain; positive results can be obtained within three hours (Botija 1977). The RRID test, which uses culture-propagated antigens in an agar base, with wells cut to take a drop of serum, might be applicable in the field (see Pan *et al.* 1974*b*). The RIA technique is about 100 times as sensitive as the CF test and can detect the equivalent of $10^{1.7}$ to $10^{2.7}$ HAD_{50}/ml of ASFV, as well as being 1000 times more effective than the IEOP for antibody assay (Crowther *et al.* 1979). ELISA tests are of similar sensitivity (Wardley *et al.* 1979). Recent joint studies in Europe of the ELISA technique for detection of antibodies to ASFV have shown that purified protein A (from *Staphylococcus aureus*) provides a better conjugate than labelled anti-pig globulins for the ELISA test; its use, as well as the employment of VP 73 separated from crude virus antigens, reduced the number of false-positive reactions in Spain (Sanchez-Vizcaino *et al.* 1983*a*). False positives have also been noted in the Dominican Republic when using poor quality sera (Hamdy *et al.* 1981).

Control and vaccination

In countries previously free from the disease rapid and reliable confirmation of the presence of ASFV is paramount. As soon as ASF is suspected, local or more general controls of all movements of swine and pigmeat products should be imposed in order to prevent further spread by live animals and fomites, including transport vehicles, clothing of personnel, utensils and foodstuffs. There are increasing dangers that mild or inapparent infections may allow contaminated meat products to enter the 'food chain' undetected and recirculate to swine through swill (food refuse) that has not been adequately heat-treated. It was suspected that the primary outbreaks in Lisbon in 1957 and 1960 and the later introductions to Italy and Brazil were attributable to contaminated food waste from airports. Regulations requiring the boiling of swill for at least 30 minutes are already in force in some countries, including Britain. In African countries, with extensive reservoirs of infection in wild pigs and ticks, contact between these sources and domestic swine can be effectively prevented by double fencing and strict control of the disposal of any warthogs which are killed.

When the presence of ASFV has been confirmed, all animals on infected premises and any exposed to them should be slaughtered and burned or otherwise rendered harmless; experience in some countries suggests that even more ruthless measures, such as depopulation of an area within a radius of say 5–10 km of the infected focus are needed for success in eradicating ASF. Cleansing and disinfection of contaminated premises with an effective virucidal agent such as 2 per cent caustic soda are then mandatory (see also Stone and Hess 1973). Argasid ticks, if present, must be eliminated from the vicinity of pig farms. Restocking is usually forbidden for periods of at least three months, especially when potential vectors are known to have been present. Small test groups are then introduced.

The wide dissemination of subacute or inapparent ASF in countries such as Spain, Portugal and possibly now in Brazil makes slaughter and eradication difficult, both on logistic and socio-economic grounds. Extensive serological testing would be a necessary prerequisite and compensation for slaughter would be imperative. The problems are compounded when hog cholera is also present. Such considerations have prompted a reconsideration of vaccines in spite of the dangers which have already materialized with live attenuated products. Early attempts in Portugal to use as vaccines in the field viruses which had been passaged in cell cultures and were safe and effective immunogens in the laboratory were catastrophic, because of reversion to virulence and reacquisition of the ability to spread when the recipients were subjected to various stresses. Later laboratory studies were no more promising (see Coggins 1974).

Inactivated vaccines have failed to provide protection, although they induce plentiful antibodies of non-neutralizing activity (Stone and Hess 1967). Any future vaccine, whether live attenuated or inactivated, that is effective in preventing naturally acquired disease but is incapable of preventing superinfection with virulent virus will create a persistent threat. This unfortunately would be the probable outcome, judging by the evidence available.

References

Adldinger, H. K., Stone, S. S., Hess, W. R. and Bachrach, H. L. (1966) *Virology* **30,** 750.

Almeida, J. D., Waterson, A. P. and Plowright, W. (1967) *Arch. gesamt. Virusforsch.* **20,** 164.

Anon (1962) *Bull. epiz. Dis. Afr.* **10,** 91; (1967) *Bull. Off. int. Épiz.* **67,** 999.

Black, D. N. and Brown, F. (1976) *J. gen. Virol.* **32,** 509.

Botija, C. Sanchez (1963) *Bull. Off. int. Épiz.* **60,** 895 (1977) In: Hog Cholera/Classical Swine Fever and African Swine Fever. *Eur. 5904 EN. p. 602.* Commission of the European Communities, Brussels.

Botija, C. Sanchez, Ordas, A. and Gonzales, G. (1969). *Bull. Off. int. Epiz.* **72,** 841.

Boulanger, P., Bannister, G. L., Gray, D. P., Ruchetbauer, G. M. and Willis, NiQ. (1967) *Canad. J. comp. Med.* **31,** 7.

Breese, S. S. and DeBoer, C. J. (1966) *Virology* **28,** 420; (1967) *Arch. ges. Virusforsch.* **20,** 164; (1969) *Amer. J. Path.* **55,** 69.

Breese, S. S. and Hess, W. R. (1966) *J. Bact.* **92,** 272.

Breese, S. S. and Pan, I. C. (1978) *J. gen. Virol.* **40,** 499.

Breese, S. S., Stone, S. S., DeBoer, C. J. and Hess, W. R. (1967) *Virology* **31,** 508.

Carnero, R., Gayot, G., Costes, C., Delelos, G. and Plateau, F. (1974) *Bull. Soc. vét. Méd. comp.* **76,** 349.

Coggins, L. (1974) *Progr. med. Virol.* **18,** 48.

Coggins, L. and Heuschele, W. P. (1966) *Amer. J. vet. Res.* **27,** 485.

Colgrove, G. S., Haelterman, E. O. and Coggins, L. (1969) *Amer. J. vet. Res.* **30,** 1343.

Contini, A., Cossu, P., Rutili, D. and Firinu, A. (1983) In: *African Swine Fever.* Eur. 8466, en p.l. Commission of the European Communities, Brussels.

Costes, C., Carnero, R. and Gayot, G. (1974) *Rev. Méd. vét.* **125,** 1119.

Cowan, K. M. (1963) *Amer. J. vet. Res.* **24,** 756.

Crowther, J. C., Wardley, R. C. and Wilkinson, P. J. (1979) *J. Hyg. Camb.* **83,** 353.

Dalsgaard, K., Overby, E. and Sanchez Botija, C. (1976) *J. gen. Virol.* **36,** 303.

Dardiri, A. H. (1966) *Fed. Proc.* **25,** 421.

DeBoer, C. J. (1967) *Arch. ges. Virusforsch.* **20,** 164.

DeBoer, C. J., Hess, W. R. and Dardiri, A. H. (1969) *Arch. ges. Virusforsch.* **27,** 44.

De Kock, G., Robinson, E. M. and Keppel, J. J. G. (1940) *Onderstepoort J.* **14,** 31.

de Paula Lyra, T. M. (1983) In: *African swine fever.* Eur. 8466 en. p. 25. Commission of the European Communities, Brussels.

DeTray, D. E. (1957) *Amer. J. vet. Res.* **18,** 811; (1963) *Advanc. vet. Sci.* **8,** 299.

DeTray, D. E. and Scott, G. R. (1957) *Amer. J. vet. Res.* **18,** 484.

Enjuanes, L., Carrascosa, A. L. and Vinuela, E. (1976) *J. gen. Virol.* **32,** 479.
Enjuanes, L., Cubero, I. and Vinuela, E. (1977) *J. gen. Virol.* **34,** 455.
Greig, A. (1972*a*) *J. comp. Path.* **82,** 73; (1972*b*) *Arch. ges. Virusforsch.* **39,** 240; (1975) *Ibid.* **47,** 287; (1980) Thesis for RCVS Fellowship.
Groocock, C. M., Hess, W. R. and Gladney, W. J. (1980) *Amer. J. vet. Res.* **41,** 591.
Haag, J., Lucas, A., Larenaudie, B., Gonzalvo, F. R. and Carnero, R. (1966) *Rec. Méd. vét.* **142,** 801.
Hamdy, F. M., Colgrove, G. S., de Rodriguez, E. V. and Snyder, M. L. (1981) *Amer. J. vet. Res.* **42,** 1441.
Hamdy, F. M. and Dardiri, A. H. (1979) *Vet. Rec.* **105,** 445.
Hess, W. R. (1971) *Virology Monographs* **9,** 1–33; (1981) *Adv. vet. Sci. comp. Med.* **25,** 39.
Hess, W. R., Cox, B. F. and Heuschele, W. P. (1965) *Amer. J. vet. Res.* **26,** 1441.
Hess, W. R. and Pan, I. C. (1977) In: Hog Cholera/Classical Swine Fever and African Swine Fever. *Eur. 5904 En. p. 602.* Commission of the European Communities, Brussels.
Heuschele, W. P. and Coggins, L. (1969) *Bull. épiz. Dis. Afr.* **17,** 179.
Heuschele, W. P., Stone, S. S. and Coggins, L. (1965) *Bull. épiz. Dis. Afr.* **13,** 157.
Kelly, D. C. and Robertson, R. S. (1973) *J. gen. Virol.* **20,** 17.
Konno, S., Taylor, W. D. and Dardiri, A. H. (1971) *Cornell Vet.* **61,** 71.
Konno, S., Taylor, W. D., Hess, W. R. and Heuschele, W. P. (1972) *Cornell Vet.* **62,** 486.
Larenaudie, B., Haag, J. and Lacaze, B. (1964) *Bull. Acad. vét. France* **37,** 257.
Malmquist, W. A. (1962) *Amer. J. vet. Res.* **18,** 484; (1963) *Ibid.* **24,** 450.
Malmquist, W. A. and Hay, D. (1960) *Amer. J. vet. Res.* **21,** 104.
McVicar, J. W., Mebus, C. A., Becker, H. N., Belden, R. C. and Gibbs, E. P. J. (1981) *J. Amer. vet. Med. Assoc.* **179,** 441.
Manso Ribiero, J., Nunes Petisca, J. L., Lopez Frazao, F. and Sobral, M. (1963) *Bull. Off. int. Epiz.* **60,** 921.
Manso Ribiero, J. *et al.* (1958) *Bull. Off. int.* Épiz. **50,** 516.
Matthews, R. E. F. (1979) *Intervirol.* **12,** 150–296.
Maurer, F. D., Griesemer, R. A. and Jones, T. C. (1958) *Amer. J. vet. Res.* **19,** 517.
Mebus, C. A. and Dardiri, A. H. (1980) *Amer. J. vet. Res.* **41,** 1867.
Mebus, C. A., Dardiri, A. H., Hamdy, F. M., Ferris, D. H., Hess, W. R. and Callis, J. J. (1978) *Proc. 82nd annu. Meeting U.S. Anim. Hlth. Ass.* p. 232.
Mebus, C. A., McVicar, J. W. and Dardiri, A. H. (1983) In: *African Swine Fever.* Eur. 8466, en. p. 183. Commission of the European Communities, Brussels.
Montgomery, R. E. (1921) *J. comp. Path.* **34,** 159, 243.
Moreno, M. A., Carrascosa, A. L., Ortin, J. and Vinuela, E. (1978) *J. gen. Virol.* **93,** 253.
Moulton, J. and Coggins, L. (1968*a*) *Cornell Vet.* **58,** 364; (1968*b*) *Amer. J. vet. Res.* **29,** 219.
Moura Nunes, J. F., Vigario, J. D., Castro Portugal, F. L., Ferreira, C. and Alves de Matos, A. P. (1977) In: Hog Cholera/Classical Swine Fever and African Swine Fever. *Eur. 5904* pp. 543–554. Commission of the European Communities, Brussels.
Moura Nunes, J. F., Vigario, J. D. and Terrinha, A. M. (1975) *Arch. Virol.* **49,** 59.
Negrin, Rosa, E. S. (1983) In: *African Swine Fever.* Eur. 8466 en. p. 36. Commission of the European Communities, Brussels.
Norley, S. G. and Wardley, R. C. (1982) *Immunology* **46,** 75; (1983) *Ibid.* **49,** 593.
Ordas, A., Sanchez Botija, C., Bruyel, V. and Olias, J. (1983) In: *African Swine Fever.* Eur. 8466, en. p. 7. Commission of the European Communities, Brussels.
Oropesa, P. R. (1971) *Bull. Off. int. Épiz.* **75,** 415.
Ortin, J., Enjuanes, L. and Vinuela, E. (1979) *Virology* **31,** 579.
Ortin, J. and Vinuela, E. (1977) *J. Virol.* **21,** 902.
Pan, I. C., DeBoer, C. J. and Heuschele, W. P. (1970) *Proc. Soc. exp. Biol., N.Y.* **134,** 367.
Pan, I. C., Moulton, J. E. and Hess, W. R. (1975) *Amer. J. vet. Res.* **36,** 379.
Pan, I. C., Trautman, R., DeBoer, C. J. and Hess, W. R. (1974*a*) *Amer. J. vet. Res.* **35,** 629.
Pan, I. C. *et al.* (1974*b*) *Amer. J. vet. Res.* **35,** 787.
Pan, I. C., Shimizu, M. and Hess, W. R. (1978) *Amer. J. vet. Res.* **39,** 491.
Parker, J. and Plowright, W. (1968) *Nature* (*Lond.*) **219,** 524.
Pini, A. (1976) *Vet. Rec.* **99,** 479; (1977) *D.V.Sc. Thesis,* University of Pretoria, S. Africa.
Pini, A. and Hurter, L. R. (1975) *Bull. epiz. Dis. Afr.* **17,** 179.
Plowright, W. (1977) In: *Hog cholera/classical swine fever and ASF.* Eur. 5904, en. p. 575. Commission of the European Communities, Brussels; (1981) In: *Infectious Diseases of Wild Mammals.* 2nd Ed. Ed by J. W. Davis, L. H. Karstad and P. O. Trainer. Iowa University Press.
Plowright, W. and Parker, J. (1967) *Arch. ges. Virusforsch.* **21,** 383.
Plowright, W., Parker, J. and Peirce, M. A. (1969*a*) *Vet Rec.* **85.** 668; (1969*b*) *Nature* (*Lond.*) **221,** 1071.
Plowright, W., Parker, J. and Staple, R. F. (1968) *J. Hyg. Camb.* **66,** 117.
Plowright, W., Perry, C. T. and Greig, A. (1974) *Res. vet. Sci.* **17,** 106.
Plowright, W., Perry, C. T. and Peirce, M. A. (1970*a*) *Res. vet. Sci.* **11,** 582.
Plowright, W. Perry, C. T., Peirce, M. A. and Parker, J. (1970*b*) *Arch. ges. Virusforsch.* **31,** 33.
Polatnik, J. and Hess, W. R. (1972) *Arch. ges. Virusforsch.* **38,** 383.
Reichard, R. E. (1978) *Proc. 82nd annu. Meeting, U.S. anim. Hlth. Ass.* p. 226.
Rivera, Eva, M. (1983) In: *African Swine Fever.* Eur. 8466 en. p.17. Commission of the European Communities, Brussels.
Rodriguez, L. C., Andrade, de M., de Silva, G. and Baptista, M. de F. D. (1983) In: *African Swine Fever.* Eur. 8466 en. p. 161. Commission of the European Communities, Brussels.
Sanchez-Vizcaino, J. M., Crowther, J. R. and Wardley, R. C. (1983*a*) In: *African Swine Fever.* Eur. 8466 en p. 297. Commission of the European Communities, Brussels.
Sanchez-Vizcaino, J. M., Mebus, C. A., McVicar, J. W. and Valero, F. (1983*b*) In: *African Swine Fever.* Eur. 8466 en. p. 195. Commission of the European Communities, Brussels.

Sanchez-Vizcaino, J. M., Slauson, D. O., Ruiz-Gonzalvo, F. and Valero, F. (1981) *Amer. J. vet. Res.* **42,** 1335.

Schloer, G. M. (1980) 23rd annu. Proc. Amer. Ass. vet. lab. Diagnosticians. 351.

Scott, G. R. (1965*a*) *Bull. Off. int. Épiz.* **63,** 645; (1965*b*) *Ibid.* **63,** 751.

Shimizu, M., Pan, I. C. and Hess, W. R. (1977) *Amer. J. vet. Res.* **38,** 27.

Simpson, V. R. and Drager, N. (1979) *Vet. Rec.* **105,** 61.

Slauson, D. O. and Sanchez-Vizcaino, J. M. (1981) *Vet. Pathol.* **18,** 813.

Steyn, D. G. (1928) *13–14th Reps. Dir. vet. Educ. Res. S. Africa,* 415.

Stone, S. S. and Hess, W. R. (1965) *Virology,* **26,** 622; (1967) *Amer. J. vet. Res.* **28,** 475; (1973) *Appl. Microbiol.* **25,** 115.

Tabares, E. and Sanchez Botija, C. (1979) *Arch. Virol.* **61,** 49.

Talavera, A., Almendral, J. M., Ley, V. and Vinuela, E. (1983) In: *African Swine Fever.* Eur. 8466, en. p. 254. Commission of the European Communities, Brussels.

Taylor, W. P., Best, J. R. and Couquhoun, I. R. (1977) *Bull. Anim. Hlth. Prod. Africa.* **25,** 196.

Tessler, J., Hess, W. R., Pan, I. C. and Trautman, R. (1973) *Canad. J. comp. Med.* **38,** 443.

Thomas, A. D. and Kolbe, F. F. (1942) *J. S. Afr. vet. Med. Ass.* **8,** 1.

Thomson, G. R., Gainaru, M. D. and van Dellen, A. F. (1979) *Onderstepoort J. vet. Res.* **46,** 149; (1980) *Ibid.* **47,** 19.

Thomson, G. R. and others (1983) In: *African swine fever.* Eur. 8466, en. p. 85. Commission of the European Communities, Brussels.

Velho, E. L. (1956) *Bull. Off. int. Épiz.* **46,** 335.

Vigario, J. D., Castro Portugal, P. L., Festas, M. B. and Vasco, S. G. (1983) In: *African Swine Fever.* Eur. 8466, en. p. 12. Commission of the European Communities, Brussels.

Vigario, J. D., Terrinha, A. M., Bastos, A. L., Moura-Nunes, J. F., Dante Marques and Silva, J. F. (1970) *Arch. ges. Virusforsch.* **31,** 387.

Vigario, J. D., Terrinha, A. M. and Nunes, J. F. (1974) *Arch. ges. Virusforsch.* **45,** 272.

Vigario, J. D., Castro Portugal, F. L., Ferreira, C. A. and Festas, C. B. (1977) In: Hog Cholera/Classical Swine Fever and African Swine Fever. *Eur. 5904, EN.* p. 469. Commission of the European Communities, Brussels.

Wardley, R. C. (1982) *Immunology* **46,** 215.

Wardley, R. C., Abu Elzein, E. M. E., Crowther, J. R. and Wilkinson, P. J. (1979) *J. Hyg. Camb.* **83,** 363.

Wardley, R. C. and Wilkinson, P. J. (1977) *Arch. Virol.* **55,** 327; (1980*a*) *Vet. Microbiol.* **5,** 169; (1980*b*) *Res. vet. Sci.* **28,** 185.

Wardley, R. C., Wilkinson, P. J. and Hamilton, F. (1977) *J. gen. Virol.* **37,** 425.

Wardley, R. C., Hamilton, F. and Wilkinson, P. J. (1979) *Arch. Virol.* **61,** 217.

Wardley, R. C. *et al.* (1983) *Arch. Virol.,* **76,** 73.

Wesley, R. D. and Pan, I. C. (1983) In: *African Swine Fever.* Eur. 8466, en. p. 240. Commission of the European Communities, Brussels.

Wilkinson, P. J. (1977) In: Hog Cholera/Classical Swine Fever and African Swine Fever. *Eur. 5904 EN.* p. 628. Commission of the European Communities, Brussels.

Wilkinson, P. J. and Donaldson, A. I. (1977) *J. comp. Path.* **87,** 497.

Wilkinson, P. J., Johnstone, R. S. and Lawman, M. J. P. (1980) *Vet. Rec.* **106,** 94.

Wilkinson, P. J. and Wardley, R. C. (1978) *Brit. vet. J.* **134,** 280.

Wilkinson, P. J., Wardley, R. C. and Williams, S. M. (1981) *J. comp. Path.* **91,** 277.

Wrigley, N. G. (1970) *J. gen. Virol.* **6,** 169.

Indexes

Most subjects that would normally be entered in the index are included in the contents list at the beginning of each chapter. The present index serves to indicate where subjects not obviously related to any special chapter will be found. To assist the reader who is looking for information on a particular organism we preface the general index by a list of genera and species mentioned in this volume. The main genera figure in the contents lists, but some of the lesser known ones appear only in the text. These, however, are usually referred to in the general index.

Genera of bacteria

Species of bacteria

Cumulative general index for Volumes 1 to 4

To be read in conjunction with the chapter contents lists.
Bold numbers preceding each page number denote the volume.
Volume 1 contains chapters 1–19; volume 2, 20–47; volume 3, 48–78 and volume 4, 79–105.